ge Library

D1131114

Organic Mechanisms
Reactions, Stereochemistry and Synthesis

Reinhard Bruckner

Organic Mechanisms

Reactions, Stereochemistry and Synthesis

Edited by Michael Harmata

With a foreword by Paul A. Wender

Prof. Dr. Reinhard Bruckner
Albert-Ludwigs-Universität Freiburg
Institut für Organische Chemie und Biochemie
Albertstr. 21
79104 Freiburg
reinhard.brueckner@organik.chemie.uni-freiburg.de

Prof. Dr. Michael Harmata
Norman Rabjohn Distinguished Professor of Chemistry
Department of Chemistry
University of Missouri-Columbia
601 S. College Avenue
Columbia, Missouri 65211
harmatam@missouri.edu

Translation: Karin Beifuss

ISBN: 978-3-642-03650-7 e-ISBN: 978-3-642-03651-4
DOI: 10.1007/978-3-642-03651-4
Library of Congress Control Number: 2009938642

Translation of Brückner, R *Reaktionsmechanismen*, 3rd edition, published by Spektrum Akademischer Verlag, © 2007 Spektrum Akademischer Verlag, ISBN 987-3-8274-1579-0

Cover design: KuenkelLopka GmbH

Printed on acid-free paper

987654321

springer.com

Biographies

Reinhard Bruckner (born 1955) studied chemistry at the Ludwig-Maximilians-Universität München, acquiring his doctoral degree under the supervision of Rolf Huisgen. After postdoctoral studies with Paul A. Wender (Stanford University), he completed his habilitation in collaboration with Reinhard Hoffmann (Philipps-Universität Marburg). He was appointed associate professor at the Julius-Maximilians-Universität Würzburg and full professor at the Georg-August-Universität Göttingen before he moved to his current position in 1998 (Albert-Ludwigs-Universität Freiburg). Professor Bruckner´s research interests are the total synthesis of natural products and the development of synthetic methodology. Besides being the author of 150 publications he has written 4 textbooks, for one of which he was awarded the Literature Prize of the Foundation of the German Chemical Industry. He has been a Visiting Professor in the US, Spain, and Japan, and served as an elected peer reviewer of the German Research Foundation and as the Vice-President of the Division of Organic Chemistry of the German Chemical Society.

Michael Harmata was born in Chicago on September 22, 1959. He obtained his A.B. in chemistry from the University of Illinois-Chicago in 1980. He received a Ph.D. from the University of Illinois-Champaign/Urbana working with Scott E. Denmark on the carbanion-accelerated Claisen rearrangement. He was an NIH postdoctoral fellow in the labs of Paul A. Wender at Stanford University, where he focused on synthetic work involving the neocarzinostatin chromophore. He joined the faculty at the University of Missouri-Columbia in 1986 and is now the Norman Rabjohn Distinguished Professor of Chemistry at that institution. Professor Harmata's research interests span a large range of chemistry and include molecular tweezers, [4+3]-cycloadditions, pericyclic reactions of cyclopentadienones and benzothiazine chemistry. He enjoys cooking, reading, stamp collecting, and recently earned his black belt in Taekwondo.

I dedicate this book to my family,
who serve to support me in my pursuit of science
and provide the love that so enriches my life.

Judy L. Snyder
Gail Harmata
Diana Harmata
Alexander Harmata

Foreword

"*Much of life can be understood in rational terms if expressed in the language of chemistry. It is an international language, a language without dialects, a language for all time, a language that explains where we came from, what we are, and where the physical world will allow us to go. Chemical Language has great esthetic beauty and links the physical sciences to the biological sciences.*" from *The Two Cultures: Chemistry and Biology* by Arthur Kornberg (Nobel Prize in Physiology and Medicine, 1959)

Over the past two centuries, chemistry has evolved from a relatively pure disciplinary pursuit to a position of central importance in the physical and life sciences. More generally, it has provided the language and methodology that has unified, integrated and, indeed, molecularized the sciences, shaping our understanding of the molecular world and in so doing the direction, development and destiny of scientific research. The "language of chemistry" referred to by my former Stanford colleague is made up of atoms and bonds and their interactions. It is a system of knowledge that allows us to understand structure and events at a molecular level and increasingly to use that understanding to create new knowledge and beneficial change. The words on this page, for example, are detected by the eye in a series of events, now generally understood at the molecular level. This knowledge of molecular mechanism (photons in, electrons out) in turn enables us to design and synthesize functional mimetics, providing for the development of remarkable retinal prosthetics for those with impaired vision and, without a great leap in imagination, solar energy conversion devices. Similarly, the arrangement of atoms in natural antibiotics provides the basis for understanding how they function, which in turn has enabled the design and synthesis of new antibiotics that have saved the lives of countless individuals. We are even starting to learn about the chemistry of cognition, knowledge that defines not only "what we are" but how we think. We have entered the age of molecularization, a time of grand opportunities as we try to understand the molecular basis of all science from medicine to computers, from our ancient past (molecular paleontology) to our molecular future. From our environment and climate to new energy sources and nanotechnology, chemistry is the key to future understanding and innovation.

This book is a continuation of a highly significant educational endeavor started by Reinhard Bruckner and joined by Michael Harmata. It is directed at understanding the "language of chemistry": more specifically, the structures of organic compounds; how structure influences function, reactivity and change; and how this knowledge can be used to design and synthesize new structures. The book provides a cornerstone for understanding basic reactions in chemistry and by extension the chemical basis for structure, function and change in the whole of science. It is a gateway to the future of the field and all fields dependent on a molecular view for innovative advancement. In an age of instant access to information, Bruckner and Harmata provide special value in their scholarly treatment by "connecting the dots" in a way that converts a vast body of chemical information into understanding and understanding into knowledge. The logical and rigorous exposition of many of the core reactions and concepts of chemistry and the addition of new ones, integration of theory with experiment, the infusion of "thought" experiments, the in-depth attention to mechanism, and the emphasis on fundamental principles rather than collections of facts are some of the many highlights that elevate this new text. As one who has been associated with the education of both the author and the editor, I find this book to be an impressively broad, deep and clear treatment of a subject of great importance. Students who seek to understand organic chemistry and to use that understanding to create transformative change will be well served in reading, studying and assimilating the conceptual content of this book. It truly offers passage to an exciting career and expertise of critical importance to our global future. Whether one seeks to understand Nature or to create new medicines and materials, Bruckner and Harmata provide a wonderfully rich and exciting analysis that students at all levels will find beneficial. Congratulations to them on this achievement and to those embarking upon this journey through the molecular world!

October, 2009

Paul A. Wender
Stanford University

Preface to the English Edition

This book is an attempt to amalgamate physical, mechanistic and synthetic organic chemistry. It is written by a synthetic organic chemist who happens to also think deeply about mechanism and understands the importance of knowing structure and reactivity to synthetic organic chemistry. I helped get the 1st German edition of this book translated into English, for two reasons. First, Reinhard Bruckner has been a friend of mine for over twenty years, ever since we were postdocs in the Wender group in the mid-80s. He was a study in Teutonic determination and efficiency, and I, and a few other Americans, and one Frenchman in particular, have been trying to cure him of that, with some success, I might add, though he remains an extremely dedicated and hard-working educator and scientist. That's a good thing. Second, I especially liked the project because I liked the book, and I thought Reinhard's way of dealing with synthesis and mechanism together was an approach sufficiently different that it might be the "whack on the side of the head" that could be useful in generating new thought patterns in students of organic chemistry.

Well, I was actually a bit surprised to be invited to work on the English translation of the 3rd German edition of the book. I was even more surprised when the publisher gave me editorial license, meaning I could actually remove and add things to the work. This potentially gives the English edition a life of its own. So besides removing as many "alreadys" (schon, in German) as humanly possible and shortening sentences to two lines from the typical German length of ten or so, I was able to add things, including, among others, a word of caution about the reactivity/selectivity principle. Speaking of long sentences…

Will the English-speaking world find the book useful? Time will tell. I see this book as being most appropriate as an organic capstone course text, preparing those who want to go to graduate school or are just starting graduate school, as it makes use not only of strictly organic chemistry knowledge, but of physical and inorganic chemistry as well. I could dream of this becoming the Sykes of the 21st century, but to make that a reality will require a great deal of work. To that end, constructive criticism is necessary. As you read this book, can you tell me what should be added or omitted, mindful of the fact that it should not get any longer and will likely present concepts with the same general format? Most importantly, is it easy and interesting to read? I did not do all I could have done to "spice up" the text, but I was very tempted. I could easily do more. In any case, if you have suggestions, please send them to me at harmatam@missouri.edu; and put the phrase *Bruckner Book* in the subject line. I can't say I will answer, but feedback given in the spirit of the best that our community has to offer will do nothing but good.

One omission that might be considered flagrant is the lack of problems. Time precluded our constructing a problem set with answers. (However, if you are inclined to do one, contact the publisher!) In the meantime, the web is bulging with organic chemistry problems, and it may be redundant to construct a book when so much is out there waiting to be harvested. One website in particular is noteworthy with regard to the variety and quality of advanced organic chemistry problems and that is the one by Dave Evans at Harvard. With the help of students and colleagues, Dave put together a site called *Challenging Problems in Chemistry and Chem-*

ical Biology (http://www2.lsdiv.harvard.edu/labs/evans/problems/index.cgi) and it is a good place to start practicing advanced organic chemistry.

Students! There are a number of things I want to say to you. Don't just read this book, study it. Read novels, study chemistry. This book is typeset with fairly wide margins. Use those margins! Draw structures there. Write down questions. Write down answers, theories, conjectures. We did not supply you with problem sets. Create them. Ask your instructors for help. Or go off on your own. Hone your skills by using resources to search out answers to questions. Searching the literature is not any easier than it used to be, in spite of the space age databases that exist. Developing the skills to find answers to chemical questions can save time and money, always a good thing, especially to those whose money you are spending. You will learn this soon enough if you haven't already done so.

Although this book is being published by Springer, it was initially taken on by Spektrum. I want to thank Ms. Bettina Saglio and Ms. Merlet Behncke-Braunbeck of Spektrum for all of their efforts. I was able to visit with them in Heidelberg and found working with these two lovely people to be a real joy. They gave me a very long leash and I appreciate it! My experience with Springer has just begun. May it be as pleasant and productive.

My work on this book began in earnest in Germany in the spring and summer of 2008. The Alexander von Humboldt Foundation saw fit to "reinvite" me back to Germany for a three month stay. I am grateful for the opportunity and would like to thank Ms. Caecilia Nauderer, who was my liaison at the Humboldt Foundation, for her assistance. It is an honor to serve as a part of the "Atlantik-Brücke", helping, if in only a small way, to build and maintain strong and positive relations between the United States and Germany. I was hosted by my friend and colleague Peter R. Schreiner at the University of Giessen. Thank you, Peter, for your hospitality. But beware: I will return!

Of course, my family must tolerate or endure, as the case may be, my "projects"! Thank you Judy, Gail, Diana and Alexander for your support!

Finally, I must note that ventures of this type are very time consuming. They represent "synergistic activities" and "broader impacts" that would not be possible without my having some funding for a research program of my own. The Petroleum Research Fund and the National Institutes of Health deserve some recognition in this context, but it is by far the National Science Foundation that has allowed me the greatest opportunity to build a research program of which I can be proud. To them and the anonymous reviewers who have supported me, I offer my most sincere thanks.

Learning and creating organic chemistry are joys that only a few are privileged to experience. May your travels into this delightful world be blessed with the thrills of discovery and creativity.

August, 2009

Michael Harmata
University of Missouri–Columbia

Preface to the 1st German Edition

To really understand organic chemistry requires three stages. First, one must familiarize oneself with the physical and chemical properties of organic chemical compounds. Then one needs to understand their reactivities and their options for reactions. Finally, one must develop the ability to design syntheses. A typical curriculum for chemistry students incorporates these three components. Introductory courses focus on compounds, a course on reaction mechanisms follows, and a course on advanced organic chemistry provides more specialized knowledge and an introduction to retrosynthesis.

Experience shows that the *second* stage, the presentation of the material organized according to reaction mechanisms, is of central significance to students of organic chemistry. This systematic presentation reassures students not only that they can master the subject but also that they might enjoy studying organic chemistry.

I taught the reaction mechanisms course at the University of Göttingen in the winter semester of 1994, and by the end of the semester the students had acquired a competence in organic chemistry that was gratifying to all concerned. Later, I taught the same course again—I still liked its outline—and I began to wonder whether I should write a textbook based on this course. A text *of this kind* was not then available, so I presented the idea to Björn Gondesen, the editor of *Spektrum.* Björn Gondesen enthusiastically welcomed the book proposal and asked me to write the "little booklet" as soon as possible. I gave up my private life and wrote for just about two years. I am grateful to my wife that we are still married; thank you, Jutta!

To this day, it remains unclear whether Björn Gondesen used the term "little booklet" in earnest or merely to indicate that he expected *one* book rather than a series of volumes. In any case, I am grateful to him for having endured patiently the mutations of the "little booklet" first to a "book" and then to a "mature textbook." In fact, the editor demonstrated an indestructible enthusiasm, and he remained supportive when I repeatedly presented him increases in the manuscript of yet another 50 pages. Moreover, the reader must thank Björn Gondesen for the two-color production of this book. All "curved arrows" that indicate electron shifts are shown in red so that the student can easily grasp the reaction. Definitions and important statements also are graphically highlighted.

In comparison to the preceding generation, students of today study chemistry with a big handicap: an explosive growth of knowledge in all the sciences has been accompanied in particular for students of organic chemistry by the need to learn a greater number of reactions than was required previously. The omission of older knowledge is possible only if that knowledge has become less relevant and, for this reason, the following reactions were omitted: Darzens glycidic ester synthesis, Cope elimination, S_Ni reaction, iodoform reaction, Reimer-Tiemann reaction, Stobbe condensation, Perkin synthesis, benzoin condensation, Favorskii rearrangement, benzil-benzilic acid rearrangement, Hofmann and Lossen degradation, Meerwein-Ponndorf reduction and Cannizarro reaction.

A few other reactions were omitted because they did not fit into the current presentation (nitrile and alkyne chemistry, cyanohydrin formation, reductive amination, Mannich reaction, enol and enamine reactions).

This book is a highly modern text. All the mechanisms described concern reactions that are used today. The mechanisms are not just *l'art pour l'art.* Rather, they present a conceptual tool to facilitate the learning of reactions that one needs to know in any case. Among the modern reactions included in the present text are the following: Barton-McCombie reaction, Mitsunobu reaction, Mukaiyama redox condensations, asymmetric hydroboration, halolactonizations, Sharpless epoxidation, Julia-Lythgoe and Peterson olefination, *ortho*-lithiation, *in situ* activation of carboxylic acids, preparations and reactions of Gilman, Normant, and Knochel cuprates, alkylation of chiral enolates (with the methods by Evans, Helmchen, and Enders), diastereoselective aldol additions (Heathcock method, Zimmerman-Traxler model), Claisen-Ireland rearrangements, transition metal-mediated C,C-coupling reactions, Swern and Dess-Martin oxidations, reductive lithiations, enantioselective carbonyl reductions (Noyori, Brown, and Corey-Itsuno methods), and asymmetric olefin hydrogenations.

The presentations of many reactions integrate discussions of stereochemical aspects. Syntheses of mixtures of stereoisomers of the target molecule no longer are viewed as valuable—indeed such mixtures are considered to be worthless—and the control of the stereoselectivity of organic chemical reactions is of paramount significance. Hence, suitable examples were chosen to present aspects of modern stereochemistry, and these include the following: control of stereoselectivity by the substrate, the reagent, or an ancillary reagent; double stereodifferentiation; induced and simple diastereoselectivity; Cram, Cram chelate, and Felkin-Anh selectivity; asymmetric synthesis; kinetic resolution; and mutual kinetic resolution.

You might ask how then, for heaven's sake, is one to remember all of this extensive material? Well, the present text contains only about 70% of the knowledge that I would expect from a *really well-trained* undergraduate student; the remaining 30% presents material for graduate students. I have worked most diligently to show the reactions in reaction diagrams that include every intermediate—and in which the flow of the valence electrons is highlighted in color—and, whenever necessary, to further discuss the reactions in the text. It has been my aim to describe all reactions so well, that in hindsight—because the course of every reaction will seem so plausible—the readers feel that they might even have *predicted* their outcome. I tried especially hard to realize this aim in the presentation of the chemistry of carbonyl compounds. These mechanisms are presented in four chapters (Chapters 7–11), while other authors usually cover all these reactions in a single chapter. I hope this pedagogical approach will render organic chemistry readily comprehensible to the reader.

Finally, it is my pleasure to thank—besides my untiring editor—everybody who contributed to the preparation of this book. I thank my wife, Jutta, for typing "version 1.0" of most of the chapters, a task that was difficult because she is not a chemist and that at times became downright "hair raising" because of the inadequacy of my dictation.

I thank my co-workers Matthias Eckhardt (University of Göttingen, Dr. Eckhardt by now) and Kathrin Brüschke (chemistry student at the University of Leipzig) for their careful reviews of the much later "version .10" of the chapters. Their comments and corrections resulted in "version .11" of the manuscript, which was then edited professionally by Dr. Barbara Elvers (Oslo). In particular, Dr. Elvers polished the language of sections that had remained unclear, and I am very grateful for her editing. Dr. Wolfgang Zettlmeier (Laaber-Waldetzenberg) prepared the drawings for the resulting "version .12," demonstrating great sensitivity to my aesthetic wishes. The typsesetting was accomplished essentially error-free by Konrad Triltsch (Würzburg), and my final review of the galley pages led to the publication of "version .13" in book form. The production department was turned upside-down by taking care of all the "last minute" changes—thank you very much, Mrs. Nothacker! Readers who note any errors, awkward formulations, or inconsistencies are heartily encouraged to contact me. One of these days, there will be a "version .14."

It is my hope that your reading of this text will be enjoyable and useful, and that it might convince many of you to specialize in organic chemistry.

August, 1996

Reinhard Bruckner
University of Göttingen

Preface to the 2nd German Edition

Working on the second edition of a textbook is similar to renovating a house: on the one hand, we would like to preserve the existing, but we also know its flaws, and the fact that it isn't any longer "fresh as the morning dew" is perceived as more and more irritating. In both cases, it is unacceptable to simply add new things, since—hoping for enhanced attractiveness—the continued homogeneity of the complete work is a *sine qua non*. Only sensitive remodeling of the existing structure will allow for parallel expansion of the original design in such a way that the final result seems to be cast from the same mold. The tightrope walk this requires makes this endeavor a challenge for an architect or author.

Put in a nutshell, it is certainly worthwhile to buy this book, even if you already own the first edition, since the second edition offers much more! You can tell this by five changes:

1. All misprints, errors in figures, language problems and the few irregularities in the content of the first edition have been eliminated. This would not have been possible, however, without the detailed feedback of many dozens of watchful readers whose comments ranged from a single detail up to the complete inventory of 57 objections (at this point I began to think this list could have been compiled by my Ph.D. supervisor, since the tone reminded me of him, until I learned that Erik Debler, a student in his fifth semester at the Freie Universität Berlin was behind it). All of these comments have been truly appreciated and I am cordially thankful to all these individuals, since they have not only assisted with all this trouble-shooting, but through their feedback have crucially contributed to motivate work on the second edition. Apart from the aforementioned people, these include Joachim Anders, Daniel Bauer, Dr. Hans-Dieter Beckhaus, Privat-Dozent Dr. Johannes Belzner, Bernd Berchthold, Prof. Dr. Manfred Christl (whose question finally led me to have the respective issue experimentally checked by Stefan Müller, one of my co-workers), Marion Emmert, Timm Graening, Dr. Jürgen Hain, Prof. Dr. Mike Harmata, Sören Hölsken, Dr. Richard Krieger, Prof. Dr. Maximilian Knollmöller, Privat-Dozent Dr. Dietmar Kuck, Eva Kühn, Prof. Dr. Manfred Lehnig, Ralf Mayr-Stein, Elisabeth Rank, Prof. Dr. Christian Reichardt (whose criticism regarding the use of the term "transition state" for what should have read "activated complex" was as appropriate, as was the uneasy feeling he had towards analyzing reactions of single molecules instead of macroscopic systems by plotting ΔG as a function of the reaction coordinate … all the same, it did not lead to a more precise conception in the new edition—a concession to the *customary* and more casual handling of these terms), Daniel Sälinger, Dr. Klaus Schaper, Prof. Dr. Reinhard Schwesinger, Konrad Siegel and Dr. Jean Suffert!
2. The majority of the many professors who submitted their comments on the first edition to Spektrum Akademischer Verlag complained about the lack of references. The new edition eliminates this shortcoming—by providing a clearly structured list of review articles for each chapter.
3. One of the key features of the first edition has remained in the new edition: "… this new edition provides the purchaser with a state-of-the-art textbook," which is assured by (1) new mechanistic details on cyclopropanations with heavy-metal carbenoids, (2) detailed

discussions of asymmetric Sharpless epoxidations, the asymmetric Sharpless dihydroxylation and the asymmetric Noyori hydrogenation, which were honored with the Nobel Prize in 2001, (3) the iodine/magnesium exchange reaction with aromatic compounds, (4) the discussion of the structures of organolithium compounds/Grignard reagents/cuprates, (5) the carbocupration of alkynes, (6) instructive findings regarding Grignard reactions via radical intermediates, (7) Myers' 'universal' alkane synthesis, (8) the Kocienski modification of the Julia olefination, (9) proline-catalyzed enantioselective Robinson annulations, (10) enzyme-catalyzed polycyclization/Wagner–Meerwein rearrangement routes to steroid skeletons, (11) the Mukaiyama aldol addition, (12) functionalizations of aromatic compounds of the Ullmann type with carbon- and heteroatom nucleophiles, (13) the Stille and the Sonogashira–Hagihara couplings, (14) the Fürstner indole synthesis and many more. Research findings that have been published after the completion of the first edition have been incorporated in this new edition as changes occured due to scientific progress; these concern modifications in the mechanisms for the osmylation of C=C double bonds, for asymmetric carbonyl group reductions with Alpine-Borane® or Brown's chloroborane, 1,4-additions of cuprates, Heck reactions, the reductive step of the Julia–Lythgoe olefination, the McMurry reaction as well as the S_Ni reaction with thionyl chloride (which was missing in the last edition since it certainly is a standard method for the preparation of primary chlorides—irrespective of its very seldom used stereochemical potential). As in the first edition, great care has been devoted in all figures to give cross-references to the origin of a given substrate and to the further processing of the final product. This is a valuable aid to acquiring knowledge of the interrelated aspects of any chemical reaction.

4. In the preface to the first edition you can find the following 'disclaimer': "We have only refrained from presenting several other reactions (nitrile and alkyne chemistry, formation of cyanohydrin, reductive amination, Mannich reaction, and enol and enamine reactions) to avoid disruption of the coherent structure of the current presentation." Omitting these reactions, however, often led to an undesired effect: frequently students would be left without *any* knowledge in the cited subject areas. Even if one thinks that "I only need *one* book per chemical subject" the claim that "for organic chemistry I only need the 'Bruckner'"—which in my opinion is a forgivable variant—the latter will *not cause any more comparable collective damage* in the future: detailed information is offered in Chapter 7 ("Carboxylic Compounds, Nitriles and Their Interconversion") on the chemistry of nitriles, in the new Section 9.1.3 on the formation of cyanohydrines and aminonitriles and in the new Chapter 12 ("The Chemistry of Enols and Enamines") on enol chemistry (including the Mannich reaction) and enamine reactions.
5. Due to my deepened teaching experience the following areas of the second edition are pedagogically more sophisticated than in the first edition:
   - The former chapter "Additions of Heteroatom Nucleophiles to Heterocumulenes, Additions of Heteroatom Nucleophiles to Carbonyl Compounds and Follow-up Reactions" has been split into two separate chapters: into Chapter 8 "Carbonic Acid Derivatives and Heterocumulenes and Their Interconversion", whose systematic organization should represent a particularly valuable learning aid, and into Chapter 9 "Additions of Heteroatom Nucleophiles to Carbonyl Compounds and Follow-up Reactions—Condensations of Heteroatom Nucleophiles with Carbonyl Compounds."

- The former Chapter "Reaction of Ylides with Saturated or α,β-Unsaturated Carbonyl Compounds" got rid of its three-membered ring formations and the rest strictly remodelled to furnish the new Chapter 11 "Reaction of Phosphorus- or Sulfur-stabilized *C* Nucleophiles with Carbonyl Compounds: Addition-induced Condensations".
- Chapter 1 ("Radical Substitution Reactions at the Saturated C Atom") was also subjected to a novel systematization, making it much more easy for students to also perceive reactions like sulfochlorinations or sulfoxidations as "easily digestible stuff."

In summary, all these modifications also imply that compared to the first edition, the size of the second has increased by 50%, just like its price. This aspect gave *me* the collywobbles because for you this might mean that the price of this book has increased by the equivalent of two visits to an Italian restaurant. But even if this was exactly your plan to crossfinance: it shouldn't give *you* any collywobbles whatsoever, but at best a short-term sense of emptiness in the pit of your stomach.

One may say that the increase in information in the second edition—naturally!—involves a greater part for graduate students (30% rise in volume) rather than that relevant to undergraduates (20% rise in volume). The text contains 60% of the knowledge that I would expect an ideal undergraduate student to acquire; graduate students are addressed by the remaining 40%. In the first edition, this ratio amounted to 70:30. The overall change in emphasis is fully intentional: the broad feedback for the first edition and its translations (*Mécanismes Réactionnels en Chimie Organique*, DeBoeck Université, 1999; *Advanced Organic Chemistry*, Harcourt/Academic Press, 2001) unambiguously revealed that this textbook is not only extensively used in lectures accompanying advanced undergraduate organic chemistry, but also in advanced-level graduate courses. *This* in itself warranted the enlargement of the advanced-level part of the textbook.

Finally, it is my pleasure to thank everybody without whose comprehensive contributions this new edition would not have been possible: Björn Gondesen, with whom I have already completed two books and who for a third time has stayed with me as—in this case freelance—copy editor and let me benefit from his critical review of the entire manuscript; Merlet Behncke–Braunbeck, who has been in charge of this project for Spektrum Akademischer Verlag for quite a long time and applied so much "fur grooming" to the author that he decided to accept her suggestion for this new edition; Dr. Wolfgang Zettlmeier, who corrected the mistakes in the old figures and also prepared the numerous new drawings very carefully and thoroughly (and, by the way, made the author get accustomed to the idea that you do not necessarily have to stick to the drawing standards of the first edition); Bettina Saglio, who is also with Spektrum Akademischer Verlag and who took care of the book from manuscript to final page proofs and managed to meet the increasingly tight deadlines at the very final stage of production, even during the holiday season; and finally my secretary Katharina Cocar-Schneider, who with great perseverance and even greater accuracy assisted me with renumbering all those figures, chapters and page references of the first edition and the subject index.

August, 2002

Reinhard Bruckner
University of Freiburg

Preface to the 3rd German Edition

The second edition of the present textbook appealed to so many readers that it sold out quickly and a reprint became necessary earlier than expected. That it became a *new* edition is owed primarily to those readers who did not only go error-hunting, but kept me informed about their prey. These hunter-gatherers included chemistry students Daniel Sälinger (who submitted a list with suggestions for improvement, the length of which the author prefers to keep private), Philipp Zacharias (who submitted a long list of deficiencies while preparing for his final exams), Birgit Krewer (who also compiled a whole catalog of irregularities) and Georgios Markopoulos (who, too, had conducted a critical error analysis on Chapter 17) as well as my colleague Professor C. Lambert (who found the only incorrect reaction product to date that had sneaked into the book). To all these people I am truly grateful for their assistance with optimizing the contents of this book!

It is due to the commitment of Mrs. Merlet Behncke-Braunbeck of Spektrum Akademischer Verlag that all these issues raised were actually addressed and *resolved*. In connection with these corrections the lists of review articles and web addresses on "name reactions" were updated as well. The total number of corrections reached the four-digit level; they were performed by the successful mixed-double team consisting of Bettina Saglio (Spektrum Akademischer Verlag) and Dr. Wolfgang Zettlmeier (Graphik + Text Studio, Barbing). This is the second time they worked together for the benefit of both the book and its author. The cooperation with all these persons has been very much appreciated.

June, 2004

Reinhard Bruckner
University of Freiburg

Postscriptum:

A suggestion for the enthusiastic Internet users among the readers: before surfing the Internet haphazardly, you can brush up your knowledge of name reactions by visiting the following websites:

- http://www.pmf.ukim.edu.mk/PMF/Chemistry/reactions/Rindex.htm
  (840 name reactions)
  or
- http://themerckindex.cambridgesoft.com/TheMerckIndex/NameReactions/TOC.asp
  (707 name reactions)
  or
- http://en.wikipedia.org/wiki/List_of_reactions
  (630 name reactions)
  or

- http://www. geocities.com/chempen_software/reactions.htm
  (501 name reactions)
  or
- http://www.organic-chemistry.org/namedreactions/
  (230 name reactions with background information and up-to-date references)
  or
- or
- http://www.chemiestudent.de/namen/namensreaktionen.php
  (166 name reactions)
  or
- http://www.monomerchem.com/display4.html
  (145 name reactions)
  or
- http://orgchem.chem.uconn.edu/namereact/named.html
  (95 name reactions)

Contents

Radical Substitution Reactions at the Saturated C Atom

1

In a substitution reaction, a substituent X of a molecule R—X is replaced by a group Y (Figure 1.1). The subject of this chapter is substitution reactions in which a substituent X that is bound to an sp^3-hybridized C atom is replaced by a group Y via radical intermediates. Radicals are usually short-lived atoms or molecules. They contain one or more unpaired electrons. You should already be familiar with at least two radicals, which by the way are quite stable: NO and O_2. NO contains one lone electron; it is therefore a monoradical or simply a "radical." O_2 contains two lone electrons and is therefore a biradical.

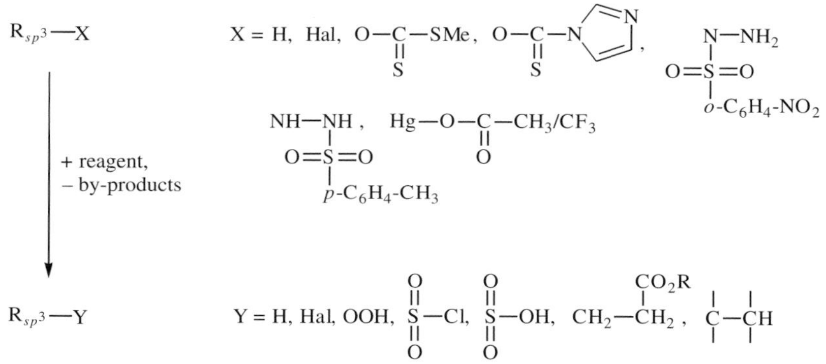

R_{sp^3}—X

X = H, Hal, O—C(=S)—SMe, O—C(=S)—N(imidazolyl), N(—NH_2)—SO_2—o-C_6H_4-NO_2, NH—NH—SO_2—p-C_6H_4-CH_3, Hg—O—C(=O)—CH_3/CF_3

+ reagent,
– by-products

R_{sp^3}—Y

Y = H, Hal, OOH, SO_2—Cl, SO_2—OH, CH_2—CH_2(CO_2R), C—CH

Fig. 1.1. Some substrates and products of radical substitution reactions.

Radical substitution reactions may occur with a variety of different substituents X and Y (Figure 1.1): X may be a hydrogen atom, a halogen atom or a polyatomic substituent, the latter of which may be bound to the carbon atom (where the substitution takes place) through an O, N or Hg atom. The substituents Y, which may be introduced by radical substitutions, include hydrogen, chlorine and bromine atoms, the functional groups –OOH, $–SO_2Cl$ and $–SO_3H$ as well as certain organic substituents with two or more carbon atoms ($C \geq 2$ units).

A radical substitution reaction may proceed through a complex reaction mechanism (Figure 1.2). This can be the case, for example, when the substituent X does not leave the substrate as an intact unit, but in pieces, in which case a fragmentation occurs at some point in the process. An intermediate step of overall radical substitutions may also involve a radical addition, occurring with reaction partners such as elemental oxygen (a biradical), sulfur dioxide (which can become hypervalent, i. e, possess more than eight valence electrons) and alkenes, including α, β-unsaturated esters (by cleavage of a pi bond). As might be expected, there are overall substitutions involving addition(s) *and* fragmentation(s). You will learn more about these options in the following sections.

Bruckner R (author), Harmata M (editor) In: *Organic Mechanisms – Reactions, Stereochemistry and Synthesis*
Chapter DOI: 10.1007/978-3-642-03651-4_1, © Springer-Verlag Berlin Heidelberg 2010

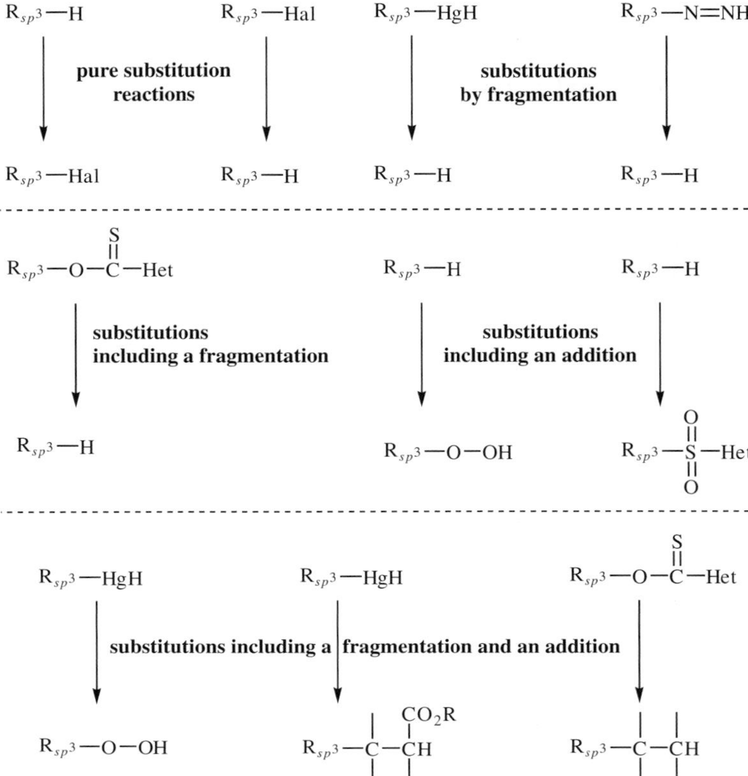

Fig. 1.2. Classification of substitution reactions via radical intermediates, according to the type of reaction(s) involved.

1.1 Bonding and Preferred Geometries in Carbon Radicals, Carbenium Ions and Carbanions

A carbon radical has seven valence electrons, one shy of the octet of a valence-saturated carbon atom. Typical carbon-centered radicals have three substituents (see below). In terms of electron count, they occupy an intermediate position between the carbenium ions, which have one electron less (a sextet and a positive charge), and the carbanions, which have one electron more (an octet and a negative charge). Since both C radicals and carbenium ions are electron deficient, they are more closely related to each other than to carbanions. Because of this, carbon radicals and carbenium ions are also stabilized or destabilized by the same substituents.

Nitrogen-centered radicals ($R_2N\bullet$) or oxygen-centered radicals ($RO\bullet$) are less stable than carbon-centered radicals $R_3C\bullet$. They are higher in energy because of the higher electronegativity of these elements relative to carbon. Such radicals are consequently less common than analogous carbon radicals, but are by no means unheard of.

What are the geometries of carbon radicals, and how do they differ from those of carbenium ions or carbanions? And what types of bonding are found at the carbon atoms of these three species? First we will discuss geometry (Section 1.1.1). and then use molecular orbital (MO) theory to provide a description of the bonding (Section 1.1.2).

We will discuss the preferred geometries and the MO descriptions of carbon radicals and the corresponding carbenium ions or carbanions in two parts. In the first part, we will examine carbon radicals, carbenium ions, and carbanions with three substituents on the carbon atom. The second part treats the analogous species with a divalent central C atom. Things like alkynyl radicals and cations are not really important players in organic chemistry and won't be discussed. Alkynyl anions, however, are extremely important, but will be covered later.

1.1.1 Preferred Geometries

The preferred geometries of carbenium ions and carbanions are correctly predicted by the **valence shell electron pair repulsion (VSEPR) theory**. The theory is general and can be applied to organic and inorganic compounds, regardless of charge.

VSEPR theory can be used to predict the geometry of compounds in the environment of a particular atom. This geometry depends on (a) the number n of atoms or groups ("ligands") attached to this central atom. If the atom under consideration is a C atom, then $n + m \leq 4$. In this case, the VSEPR theory says that the structure in which the repulsion between the n bonding partners and the m nonbonding valence electron pairs on the C atom is as small as possible will be preferred. This is the case when the orbitals that accommodate the bonding and the nonbonding electron pairs are as far apart from each other as possible.

For **carbenium ions**, this means that the n substituents of the cationic carbon atom should be at the greatest possible distance from each other:

- In alkyl cations $R_3C^{\oplus}$, $n = 3$ and $m = 0$. The substituents of the trivalent central atom lie in the same plane as the central atom and form bond angles of 120° with each other (trigonal planar arrangement). This arrangement was confirmed experimentally by means of a crystal structural analysis of the *tert*-butyl cation.
- In alkenyl cations $=C^{\oplus}—R$, $n = 2$ and $m = 0$. The substituents of the divalent central atom lie on a common axis with the central atom and form a bond angle of 180°. Alkenyl cations have not been isolated yet because of their low stability (Section 1.2). However, calculations support the preference for the linear structure.

According to the VSEPR theory, in **carbanions** the n substituents at the carbanionic C atom *and the nonbonding electron pair* must move as far away from each other as possible:

- In alkyl anions $R_3C^{\ominus}$, $n = 3$ and $m = 1$. The substituents lie in one plane, and the central atom lies above it. The carbanion center has a trigonal pyramidal geometry. The bond angles are similar to the tetrahedral angle (109° 28'). The geometry can be considered to be tetrahedral if the lone pair is considered to be a substituent.
- In alkenyl anions $=C^{\ominus}—R$, $n = 2$ and $m = 1$. The substituents and the divalent central atom prefer a bent structure. The bond angle in alkenyl anions is approximately 120°. Basically, one can look at the lone pair as a substituent and get a reasonable idea of the structure of the carbanion.

The most stable structures of alkyl and alkenyl anions predicted with the VSEPR theory are supported by reliable calculations. There are no known experimental structural data, due to the fact that counterions occur with formally carbanionic species and they generally experience some type of bonding with the carbanionic carbon. However, you can often approximate both structure and reactivity by assuming that such spieces (e. g., organolithiums) are carbanions.

Since the VSEPR theory is based on the mutual repulsion of valence electron *pairs*, it formally can't be used to make statements about the preferred geometries of **C radicals**. One might expect that C radicals are structurally somewhere between their carbenium ion and carbanion analogs. In agreement with this, alkyl radicals are either planar (methyl radical) or slightly pyramidal, but able to pass rapidly through the planar form (inversion) to another near-planar structure (*tert*-butyl radical). In addition, some carbon-centered radicals are considerably pyramidalized (e. g., those whose carbon center is substituted with several heteroatoms). Alkenyl radicals are bent, but they can undergo *cis/trans*-isomerization through the linear form very rapidly. Because they are constrained in a ring, aryl radicals are necessarily bent.

Something that comes as a surprise at first glance is that the *tert*-butyl radical is not planar, while the methyl radical is. Deviation from planarity implies a narrowing of the bond angles and thus a mutual convergence of the substituents at the radical center. Nevertheless, the *tert*-butyl radical with its 40% pyramidalization of an ideal tetrahedral center is 1.2 kcal/mol more stable than a planar *tert*-butyl radical.

1.1.2 Bonding

The type of bonding at the C atom of carbenium ions, carbanions, and C-centered radicals follows from the geometries described in Section 1.1.1. From the bond angles at the central C atom, it is possible to derive its hybridization. Bond angles of 109° 28' correspond to sp^3, bond angles of 120° correspond to sp^2, and bond angles of 180° correspond to *sp*-hybridization. From this hybridization it follows which atomic orbitals (AOs) of the C atom are used to form the molecular orbitals (MOs). The latter can be used as bonding MOs, in which case each possesses an electron pair and represents the bond to a substituent of the central atom. On the other hand, one AO of the central atom could be a nonbonding MO, which is empty in the carbenium ion, contains an electron in the radical, and contains the nonbonding electron pair in the carbanion. How the valence electrons are distributed in the molecular orbitals follows from the *Aufbau* principle: they are placed, one after the other, in the MOs, in the order of increasing energy. The Pauli principle is also observed: any MO can have only two electrons and only on the condition that they have opposite spins.

The bonding at the atom of carbenium ions $R_3C^{\oplus}$ is therefore described by the MO diagram in Figure 1.3 (left), and the bonding of the valence-unsaturated C atom of carbenium ions of type $=C^{\oplus}—R$ is described by the MO diagram in Figure 1.4 (left). The MO description of $R_3C^{\ominus}$ carbanions is shown in Figure 1.3 (right), and the MO description of carbanions of type $=C^{\ominus}—R$ is shown in Figure 1.4 (right). The MO description of the radicals R• or =CR• employs the MO picture for the analogous carbenium ions or carbanions, depending on which of these species the geometry of the radical is similar to. In each case, only seven instead of six or eight valence electrons must be accommodated.

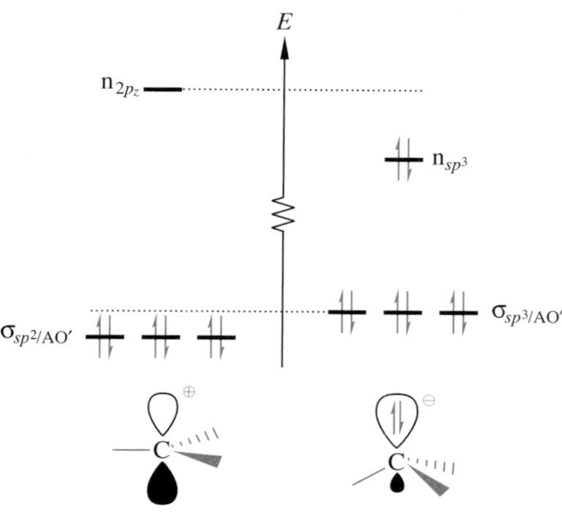

Fig. 1.3. Energy levels and occupancies (red) of the MOs at the trivalent C atom of planar carbenium ions $R_3C^{\oplus}$ (left) and pyramidal carbanions $R_3C^{\ominus}$ (right). The indices of each of the four MOs refer to the AOs from the central C atom.

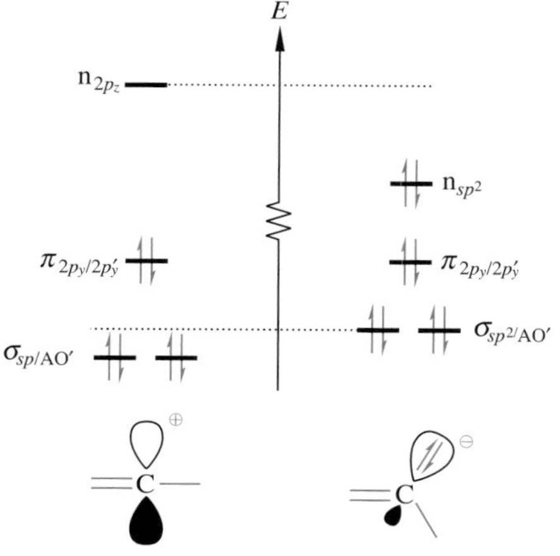

Fig. 1.4. Energy levels and occupancies (red) of the MOs at the divalent C atom of linear carbenium ions $=C^{\oplus}$—R (left) and bent carbanions $=C^{\ominus}$—R (right). The indices of each of the four MOs refer to the AOs from the central C atom.

1.2 Stability of Radicals

Stability in chemistry is not an absolute, but a relative concept. Let us consider the standard heats of reaction ΔH^0 of the homolytic dissociation reaction R—H → R• + H•. It reflects, on the one hand, the strength of this C—H bond and, on the other hand, the stability of the radical R• produced. So the dissociation enthalpy of the R—H bond depends in many ways on the structure of R. But it is not possible to tell clearly whether this is due to an effect on the bond energy of the broken R—H bond and/or an effect on the stability of the radical R• that is formed.

How do we explain, for example, the fact that the dissociation enthalpy of a C_{sp^n}—H bond essentially depends on n alone and increases in the order $n = 3$, 2, and 1, that homolytic cleavage of a C-H bond of an sp^3-hybridized carbon requires a lot less energy than that of an sp-hybridized carbon?

$\frac{DE}{\text{kcal/mol}}$	$HC{\equiv}C_{sp}{-}H$	$C_6H_5{-}C_{sp^2}{-}H$	$H_2C{=}\overset{H}{C}_{sp^2}{-}H$	$H_3C{-}\overset{H_2}{C}_{sp^3}{-}H$
	131	113	111	101

To help answer this question, it is worthwhile considering the following: the dissociation enthalpies of bonds such as C_{sp^n}—C, C_{sp^n}—O, C_{sp^n}—Cl, and C_{sp^n}—Br also depend heavily on n and increase in the same order, $n = 3$, 2, and 1. The extent of the n-dependence of the dissociation energies, though, depends on the *element that is cleaved off*, which implies some important things: (1) The n-dependence of dissociation enthalpies of the C_{sp^n}-element bonds cannot only be due to the decreasing stability of the radical in the $Csp^3\bullet > C\bullet\, sp^2 > Csp\bullet$ series. (2) So the n-dependence, or at least part of it, reflects an n-dependence of the respective C_{sp^n}-element bond, something that should be remembered if one looks at bond dissociation energies in handbooks. For example, a carbon-iodine bond results from the overlap of an sp^n hybrid (2p-like) orbital of the carbon and a 5p or ("5p-like") orbital of the iodine. The overlap is inherently poorer than that which would be found in the overlap of orbitals of similar size and shape.

Overall, however, the homolytic bond dissociation energy of every C_{sp^n}-element bond increases in the order $n = 3$, 2, and 1. This is due to the fact that C_{sp^n}-element bonds become shorter in this order, i.e. $n = 3$, 2, and 1, which, in turn, is due to the fact that the s character of the C_{sp^n}-element bond increases in the same order. Other things being equal, the shorter the bond, the stronger the bond.

An immediate consequence of the different ease with which C_{sp^n}-element bonds dissociate is that in radical substitution reactions, alkyl radicals are more easily formed. Vinyl and aryl radicals are less common, but can be generated productively. Alkynyl radicals do not appear at all in radical substitution reactions. In the following, we therefore limit ourselves to a discussion of substitution reactions that take place via radicals of the general structure $R^1R^2R^3C\bullet$.

1.2.1 Reactive Radicals

If radicals $R^1R^2R^3C\bullet$ are produced by breaking the C—H bond in molecules of the type $R^1R^2R^3C$—H, one finds that the dissociation enthalpies of such C—H bonds differ with molecular structure. These differences can be explained completely by the effects of the substituents R^1, R^2, and R^3 on the stability of the radicals $R^1R^2R^3C\bullet$ formed.

Table 1.1 shows one substituent effect that influences the stability of radicals. The dissociation enthalpies of reactions that lead to R—$CH_2\bullet$ radicals are listed. The substituent R varies from C_2H_5 through H_2C=CH—(vinyl substituent, vin) to C_6H_5— (phenyl substituent, Ph). The dissociation enthalpy is greatest for R = H. It can also be seen that a radical center is stabilized by 12 ± 1 kcal/mol by the neighboring C=C double bond of an alkenyl or aryl substituent.

In the valence-bond (VB) model, this effect results from the fact that radicals of this type can be stabilized by resonance (Table 1.1, right). In the MO model, the stabilization of radical centers of this type is due to the overlap of the π system of the unsaturated substituent with the $2p_z$ AO at the radical center (Figure 1.5). This overlap is called conjugation.

Table 1.1. Stabilization of Radicals by Unsaturated Substituents

	$\frac{DE}{\text{kcal/mol}}$	VB-formulation of the radical R •
H	101	
H	88	
H	90	

Alkynyl substituents stabilize a radical center by the same 12 kcal/mol that on average is achieved by alkenyl and aryl substituents. From the point of view of the VB model this is due to the fact that propargyl radicals exhibit the same type of resonance stabilization as formulated for allyl and benzyl radicals in the right column of Table 1.1. In the MO model, the stability of propargyl radicals rests on the overlap between the one correctly oriented π system of the C≡C triple bond and the $2p_z$ AO of the radical center, just as outlined for allyl and benzyl radicals in Figure 1.5 (the *other* π system of the C≡C triple bond is orthogonal to the $2p_z$ AO of the radical center, thus excluding an overlap that is associated with stabilization).

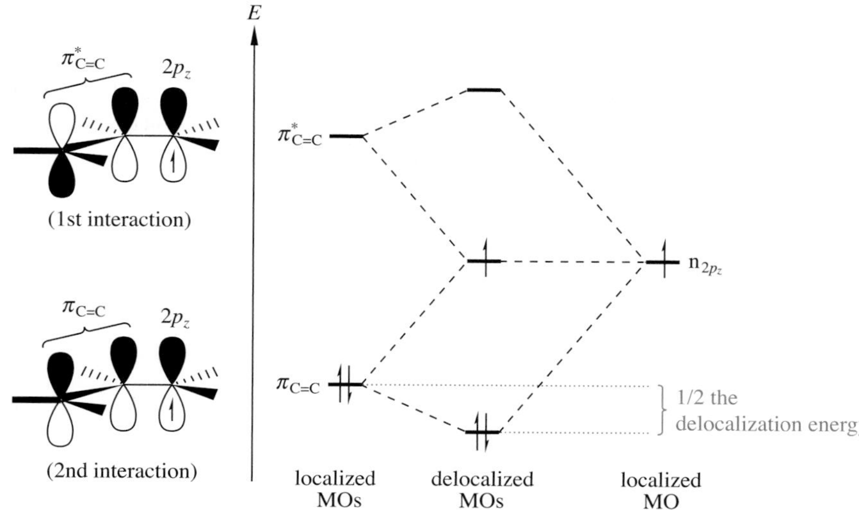

Fig. 1.5. Stabilization by overlap of a singly occupied $2p_z$ AO with adjacent parallel $\pi_{C=C}$ or $\pi^*_{C=C}$ MOs.

Table 1.2 illustrates an additional substituent effect on radical stability. Here the dissociation enthalpies of reactions that lead to (poly)alkylated radicals $(alk)_{3-n}H_nC\bullet$ are listed ("alk" stands for alkyl group). From these dissociation enthalpies it can be seen that alkyl substituents stabilize radicals. A primary radical is by 4 kcal/mol more stable, a secondary radical is by 7 kcal/mol more stable, and a tertiary radical is by 9 kcal/mol more stable than the methyl radical.

Table 1.2. Stabilization of Radicals by Alkyl Substituents

	$\frac{DE}{\text{kcal/mol}}$	VB-Formulation of the radical R •
$H_3C{-}H$	105	
$H_3C{-}H_2C{-}H$	101	i.e., no-bond resonance form per H_β
$(H_3C)_2HC{-}H$	98	⟷ 6 no-bond resonance forms
$(H_3C)_3C{-}H$	96	⟷ 9 no-bond resonance forms

In the VB model, this effect is explained by the fact that radicals of this type, too, can be described by the superpositioning of several resonance forms. These are the somewhat exotic "no-bond" resonance forms (Table 1.2, right). From the point of view of the MO model, radical centers with alkyl substituents have the opportunity to interact with these substituents. This interaction involves the C—H bonds that are in the position α to the radical center and lie in the same plane as the $2p_z$ AO at the radical center. Specifically, $\sigma_{\text{C—H}}$ MOs of these C—H bonds are able to overlap with the radical $2p_z$ orbital (Figure 1.6). This overlap represents a case of lateral overlap between a σ bond, a bond that is 75% p in character based on hybridization, and a p orbital. It is referred to as **hyperconjugation** to distinguish it from lateral overlap between π bonds and p orbitals, which is referred to as conjugation. When the $\sigma_{\text{C—H}}$ bond and the $2p_z$ AO have a dihedral angle χ that is different from that required for optimum overlap (0°), the stabilization of the radical center by hyperconjugation decreases. (In fact, it decreases by the square of the cosine of the dihedral angle χ.)

Table 1.3 illustrates a third radical stabilizing effect of a substituent. Homolyses producing radicals with a heteroatom with free electron pairs adjacent to the radical center are less endothermic than the comparable reaction giving $H_3C{-}H_2C\bullet$. Qualitatively, the more "available" such electrons are, the greater the stabilization. Accordingly, an amino group stabilizes a radical center better than a hydroxyl group because nitrogen is less electronegative than oxygen. However, carbon radicals bearing oxygen are common. They are the key intermediates in the autoxidation of ethers (see Figure 1.38). As might be expected, an even larger stabilization effect on a radical center is achieved through an $O^{\ominus}$ substituent, a better donor than a neutral oxygen atom. In the alkoxide case, the donation of electrons to the radical center leads to the

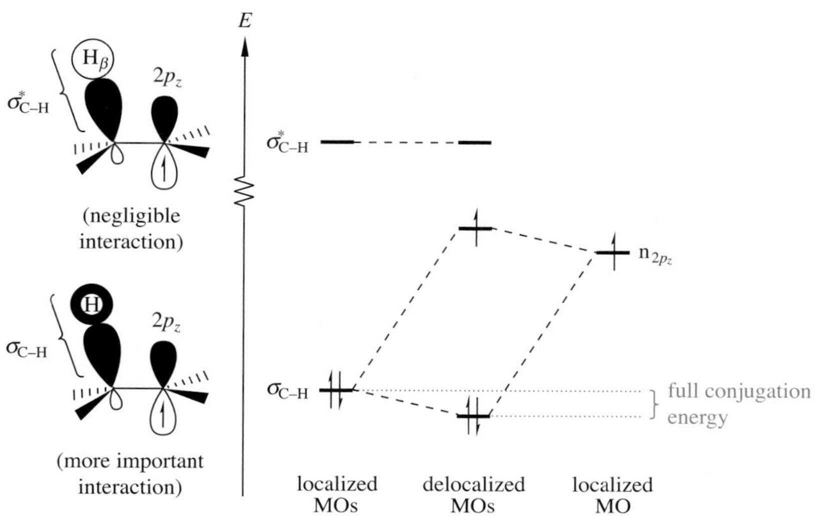

Fig. 1.6. Stabilization by overlap of a singly occupied $2p_z$ AO with vicinal nonorthogonal σ_{C-H} MOs.

delocalization of the excess negative charge and thus reduces the Coulomb repulsion force. You may find $O^{\ominus}$ substituted C radicals rather exotic. Nevertheless, they are well-known as intermediates (so-called ketyl radicals) of the one-electron reduction of carbonyl compounds (Section 17.4.2).

Table 1.3. Stabilization of Radicals by Substituents with Free Electron Pairs

	$\frac{DE}{\text{kcal/mol}}$	VB-Formulation of the radical R •
$H_3C{-}H_2C{-}H$	101	$\left\{ (H_\beta)_3C{-}\dot{C}H_2 \longleftrightarrow \right.$ 3 no-bond resonance forms (see Table 1.2) $\left. \right\}$
$H\ddot{\underset{..}{O}}{-}H_2C{-}H$	98	$\left\{ H_\beta{-}\ddot{\underset{..}{O}}{-}\dot{C}H_2 \longleftrightarrow \right.$ 1 no-bond resonance form $\left. \longleftrightarrow H_\beta{-}\overset{\oplus}{\dot{\underset{..}{O}}}{-}\ddot{C}H_2^{\ominus} \right\}$
$H_2\ddot{N}{-}H_2C{-}H$	93	$\left\{ H_\beta{-}\ddot{N}(H_\beta){-}\dot{C}H_2 \longleftrightarrow \right.$ 2 no-bond resonance forms $\left. \longleftrightarrow H_\beta{-}\overset{\oplus}{\dot{N}}(H_\beta){-}\ddot{C}H_2^{\ominus} \right\}$
$^{\ominus}\!:\ddot{\underset{..}{O}}{-}H_2C{-}H$	unknown	$\left\{ ^{\ominus}\!:\ddot{\underset{..}{O}}{-}\dot{C}H_2 \longleftrightarrow :\dot{\underset{..}{O}}{-}\ddot{C}H_2^{\ominus} \right\}$

In the VB model, the ability of heteroatoms with a free electron pair to stabilize an adjacent radical center is based on the fact that such radicals may be described by *several* resonance forms (Table 1.3, right). In addition to the "C radical" resonance form and the one or two no-bond resonance forms, a zwitterionic resonance form occurs with neutral radicals, and a carbanion/O radical resonance form with the negatively charged ketyl radical. In the MO model, the stabilization of heteroatom-substituted radical centers depends on the overlap between an orbital containing a lone pair of electrons and the half-occupied $2p_z$ orbital of the radical center (Figure 1.7). The result is a small energy decrease, corresponding to the stabilization that the half-occupied $2p_z$ orbital of a radical center experiences as in Figure 1.6 by the overlap with a suitably oriented doubly occupied σ_{C-H} MO.

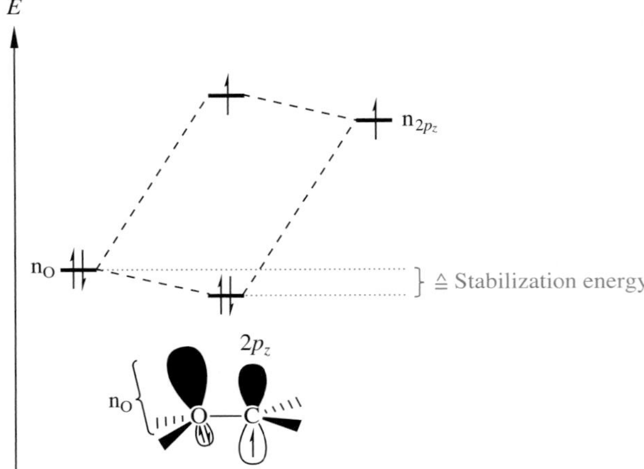

Fig. 1.7. Stabilization by overlap of a singly occupied $2p_z$ AO with the vicinal nonorthogonal free electron pair of an oxygen atom.

1.2.2 Unreactive Radicals

Just as several alkyl substituents increasingly stabilize a radical center (Table 1.2), so do two phenyl substituents. The diphenylmethyl radical ("benzhydryl radical") is therefore more stable than the benzyl radical. The triphenylmethyl radical ("trityl radical") is even more stable because of the three phenyl substituents. They actually stabilize the trityl radical to such an extent that it forms by homolysis from the so-called "Gomberg hydrocarbon" even at room temperature (Figure 1.8). Although this reaction is reversible, the trityl radical is present in equilibrium quantities of about 2 mol%.

Starting from the structure of the trityl radical, radicals were designed that can be obtained as isolable, stable radicals" (Figure 1.9). There are two reasons why these radicals are so stable. For one thing, they are exceptionally well resonance-stabilized. In addition, their dimerization to valence-saturated species has a considerably reduced driving force. In the case of the trityl radical, for example, dimerization leads to the Gomberg hydrocarbon in which an aromatic sextet is lost. The trityl radical cannot dimerize giving hexaphenylethane, because severe van der Waals repulsions between the substituents would occur. There are also stable N- or O-centered radicals. The driving force for their dimerization is small because relatively weak N—N or O—O bonds would be formed.

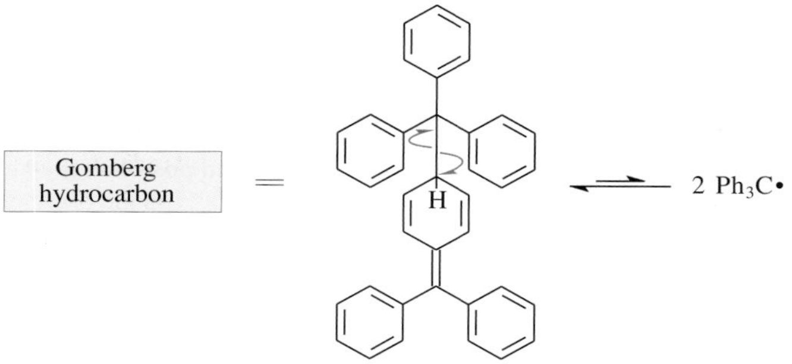

Fig. 1.8. Reversible formation reaction of the triphenylmethyl radical. The equilibrium lies on the side of the Gomberg hydrocarbon.

The destabilization of the dimerization product of a radical is often more important for the existence of stable radicals than optimum resonance stabilization. This is shown by comparison of the trityl radical derivatives **A** and **B** (Figure 1.9). In radical **A**, the inclusion of the radical center in the polycycle makes optimum resonance stabilization possible because the dihedral angle χ between the $2p_z$ AO at the central atom and the neighboring π orbitals of the three surrounding benzene rings is exactly 0°. And yet radical **A** dimerizes! In contrast, the trityl radical derivative **B** is distorted like a propeller, to minimize the interaction between the methoxy substituents on the adjacent rings. The $2p_z$ AO at the central atom of radical **B** and the π orbitals of the surrounding benzene rings therefore form a dihedral angle χ of a little more than 45°. The resonance stabilization of radical **B** is therefore only one half as great—$\cos^2 45° = 0.50$—as that of radical **A**. In spite of this, radical **B** does not dimerize at all!

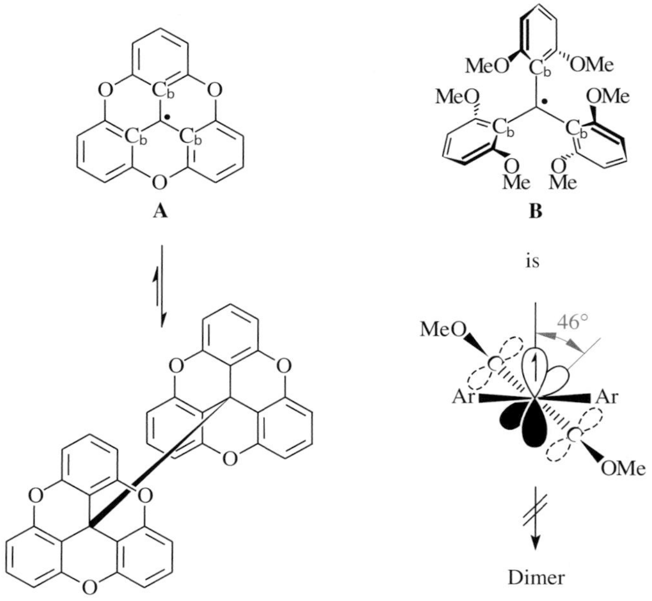

Fig. 1.9. Comparison of the trityl radical derivatives **A** and **B**; **A** dimerizes, **B** does not.

1.3 Relative Rates of Analogous Radical Reactions

In Section 1.2.1 we discussed the stabilities of reactive radicals. It is interesting that they make an evaluation of the relative rates of formation of these radicals possible. This follows from the **Bell–Evans–Polanyi principle** (Section 1.3.1) or the **Hammond postulate** (Section 1.3.2).

1.3.1 The Bell–Evans–Polanyi Principle

In thermolyses of aliphatic azo compounds, two alkyl radicals and one equivalent of N_2 are produced according to the reaction at the bottom of Figure 1.10. A whole series of such reactions was carried out, and their reaction enthalpies ΔH_r, were determined. They were all endothermic reactions (ΔH_r has a positive sign). Each substrate was thermolyzed at several different temperatures T and the associated rate constants k_r were determined. The temperature dependence of the k_r values for each individual reaction was analyzed by using the **Eyring equation** (Equation 1.1).

$$k_r = \frac{k_B \cdot T}{h} \exp\left(-\frac{\Delta G^{\ddagger}}{RT}\right) = \frac{k_B \cdot T}{h} \exp\left(-\frac{\Delta H^{\ddagger}}{RT}\right) \exp\left(+\frac{\Delta S^{\ddagger}}{R}\right) \tag{1.1}$$

k_B: Boltzmann constant (3.298×10^{-24} cal/K)
T: absolute temperature (K)
h: Planck's constant (1.583×10^{-34} cal s)
$\Delta G^{\ddagger}$: Gibbs (free) energy of activation (kcal/mol)
$\Delta H^{\ddagger}$: enthalpy of activation (kcal/mol)
$\Delta S^{\ddagger}$: entropy of activation (cal mol^{-1} K^{-1})
R: gas constant (1.986 cal mol^{-1} K^{-1})

Equation 1.1 becomes Equation 1.2 after (a) dividing by T, (b) taking the logarithm, and (c) differentiating with respect to T.

$$\Delta H^{\ddagger} = RT \frac{\partial \ln\left(\frac{k_r}{T}\right)}{\partial T} \tag{1.2}$$

With Equation 1.2 it was possible to calculate the activation enthalpy $\Delta H^{\ddagger}$ for each individual reaction.

The pairs of values $\Delta H_r/\Delta H^{\ddagger}$, which were now available for each thermolysis, were plotted on the diagram in Figure 1.10, with the enthalpy change ΔH on the vertical axis and the reaction progress on the horizontal axis. The horizontal axis is referred to as the **reaction coordinate** (RC). Among "practicing organic chemists" it is not accurately calibrated. What is implied is that on the reaction coordinate one has moved by x% toward the reaction product(s) when all the structural changes that are necessary *en route* from the starting material(s) to the product(s) have been x% completed.

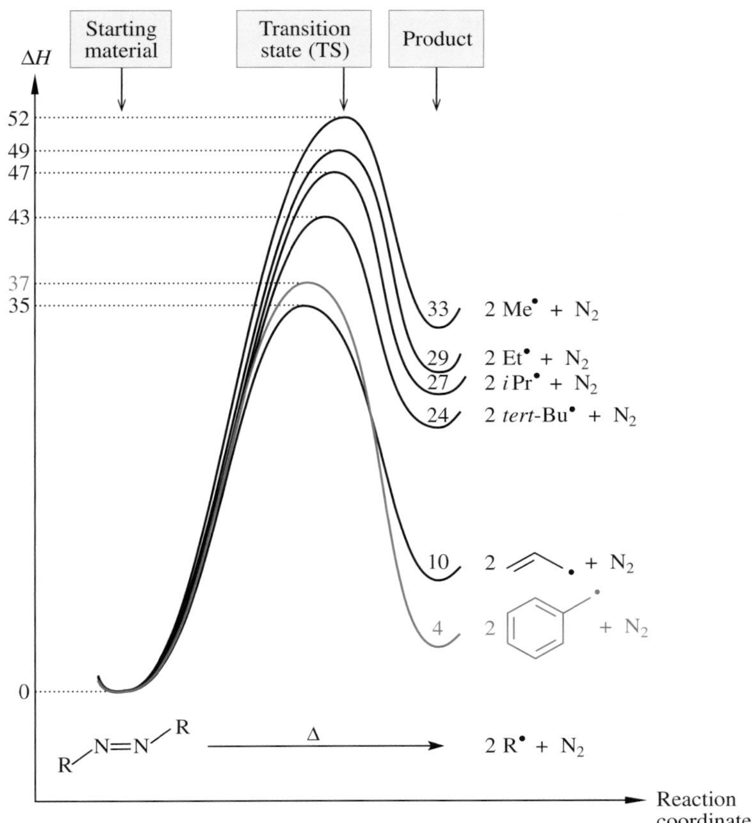

Fig. 1.10. Enthalpy change along the reaction coordinate in a series of thermolyses of aliphatic azo compounds. All thermolyses in this series except the one highlighted in color follow the Bell–Evans–Polanyi principle.

For five out of the six reactions investigated, Figure 1.10 shows a decrease in the activation enthalpy $\Delta H^\ddagger$ with increasingly negative reaction enthalpy ΔH_r. Only for the sixth reaction—drawn in red in Figure 1.10—is this not true. Accordingly, except for this one reaction $\Delta H^\ddagger$ and ΔH_r are proportional for this series of radical-producing thermolyses. This proportionality is known as the Bell–Evans–Polanyi principle and is described by Equation 1.3.

$$\Delta H^\ddagger = \text{const.} + \text{const.}'\, \Delta H_r \tag{1.3}$$

The thermolyses presented in this chapter are *one* example of a series of analogous reactions. The Bell–Evans–Polanyi relationship of Equation 1.3 also holds for many other series of analogous reactions. The general principle that can be extracted from Equation 1.3 is that, at least for a reaction series, the more exothermic the enthalpy of reaction, the faster it will be. But this doesn't mean that all reactions that are exothermic are fast, so be careful.

1.3.2 The Hammond Postulate

In many series of analogous reactions a second proportionality is found experimentally, namely, between the free energy change (ΔG_r; a thermodynamic quantity) and the free energy of activation ($\Delta G^{\ddagger}$, a kinetic quantity). In a series of analogous reactions, a third parameter besides $\Delta H^{\ddagger}$ and $\Delta G^{\ddagger}$ no doubt also depends on the ΔH_r and ΔG_r values, namely, the structure of the transition state. This relationship is generally assumed or postulated, and only in a few cases has it been supported by calculations (albeit usually only in the form of the so-called "transition structures"; they are likely to resemble the structures of the transition state, however). This relationship is therefore not stated as a law or a principle but as a postulate, the so-called Hammond postulate.

The Hammond Postulate

The Hammond postulate can be stated in several different ways. For *individual reactions* the following form of the Hammond postulate applies. In an endergonic reaction (positive ΔG_r) the transition state (TS) is similar to the *product(s)* with respect to energy and structure and is referred to as a **late transition state**. Conversely, in an exergonic reaction (negative ΔG_r), the transition state is similar to the *starting material(s)* with respect to energy and structure and is referred to as an **early transition state**.

For series of analogous reactions, this results in the following form of the Hammond postulate: in a series of *increasingly endergonic* analogous reactions the transition state is *increasingly* similar to the product(s), i.e., increasingly late. On the other hand, in a series of *increasingly exergonic* analogous reactions, the transition state is *increasingly* similar to the starting material(s), i.e., increasingly early.

What does the statement that "increasingly endergonic reactions take place via increasingly product-like transition states" mean for the special case of two irreversible endergonic analogous reactions that occur as competitive reactions? Using the Hammond postulate, the outcome of this competition can often be predicted. The energy of the competing transition states should be ordered in the same way as the energy of the potential reaction products. This means that the more stable reaction product is formed via the lower-energy transition state. It is therefore produced more rapidly than the less stable reaction product.

The form of the Hammond postulate just presented is very important in the analysis of the selectivity of many of the reactions we will discuss in this book in connection with chemoselectivity (Section 1.7.2; also see Section 3.2.2), stereoselectivity (Section 3.2.2), diastereoselectivity (Section 3.2.2), enantioselectivity (Section 3.2.2), and regioselectivity (Section 1.7.2).

Selectivity

Selectivity means that one of several reaction products is formed preferentially or exclusively. In the simplest case, for example, reaction product 1 is formed at the expense of reaction product 2. Selectivities of this type are usually the result of a *kinetically controlled reaction process*, or "**kinetic control**." This means that they are usually not the consequence of an equilibrium being established under the reaction conditions between the alternative reaction products 1 and 2. In this latter case one would have a *thermodynamically controlled reaction process*, or "**thermodynamic control**."

Hammond Postulate and Kinetically Determined Selectivities

For reactions under kinetic control, the Hammond postulate now states that:

- If the reactions leading to the alternative reaction products are one step, the most stable product is produced most rapidly, that is, more selectively. This type of selectivity is called product development control.
- If these reactions are two-step, the product that is derived from the more stable intermediate is produced more rapidly, that is, more selectively.
- If these reactions are more than two-step, one must identify the least stable intermediate in each of the alternative pathways. Of these high-energy intermediates, the least energy-rich is formed most rapidly and leads to a product that, therefore, is then formed more selectively.

The selectivity in all three cases is therefore due to "**product development control.**"

1.4 Radical Substitution Reactions: Chain Reactions

Radical substitution reactions can be written in the following form:

$$R_{sp^3}-X \xrightarrow[\text{radical initiator (cat.)}]{\text{reagent}} R_{sp^3}-Y$$

All radical substitution reactions are chain reactions. Every chain reaction starts with an initiation reaction. In one or more steps, this reaction converts a valence-saturated compound into a radical, which is sometimes called an **initiating radical** (the reaction arrow with the circle means that the reaction takes place through several intermediates, which are not shown here):

Substrate (R_{sp^3}–X) and/or reagent and/or radical initiator (Section 1.5)	$\xrightarrow[\text{(1 or more steps)}]{\circ}$	Initiating radical (from substrate) (i.e., R_{sp^2}•) or initiating radical (from reagent)

The initiating radical is the radical that initiates a sequence of two, three, or more so-called **propagation steps**:

$$\text{Initiating radical (from substrate)} + \text{reagent} \xrightarrow{k_{prop,1}} \ldots$$
$$\ldots \xrightarrow{k_{prop,n}} \ldots$$
$$\ldots \xrightarrow{k_{prop,\varpi}} \text{initiating radical (from substrate)} + \ldots$$

$$\Sigma_{\text{propagation steps}}: R_{sp^3}-X + \text{reagent} \longrightarrow R_{sp^3}-Y + \text{by-product(s)}$$

Depending on whether the initiating radical comes from the substrate or the reagent, the propagation steps must be formulated as above or as follows:

$$\text{Initiating radical (from reagent)} + R_{sp^3}-X \xrightarrow{k_{\text{prop},1}} \ldots$$
$$\ldots \xrightarrow{k_{\text{prop},n}} \ldots$$
$$\ldots \xrightarrow{k_{\text{prop},\varpi}} \text{initiating radical (from reagent)} + \ldots$$

$$\Sigma_{\text{propagation steps}}: R_{sp^3}-X + \text{reagent} \longrightarrow R_{sp^3}-Y + \text{by-product(s)}$$

As the reaction equations show, the last propagation step supplies the initiating radical consumed in the first propagation step. From this you also see that the mass conversion of a chain reaction is described by an equation that results from the propagation steps alone: they are added up, and species that occur on both sides of the reaction arrow are dropped.

If the radical intermediates of the propagation steps did nothing other than always enter into the next propagation step of the chain again, even a single initiating radical could initiate a complete starting material(s) → product(s) conversion. However, radical intermediates may also react with each other or with different radicals. This removes them from the process, and the chain reaction comes to a stop.

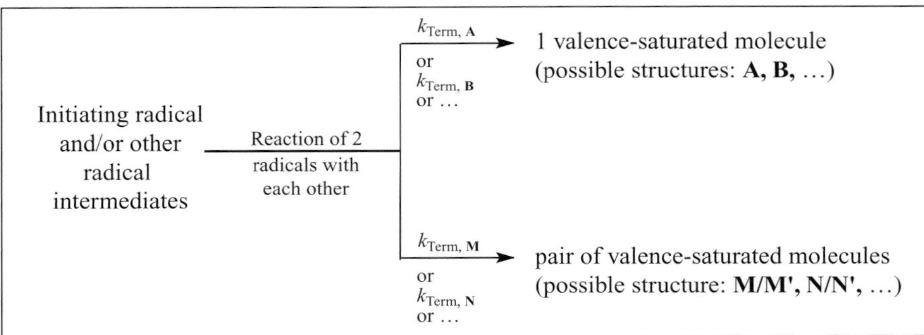

Reactions of the latter type therefore represent **termination steps** of the radical chain. A continuation of the starting material(s) → product(s) conversion becomes possible again only when a new initiating radical is made available via the starting reaction(s). Thus, for radical substitutions via chain reactions to take place *completely*, new initiating radicals must be produced continuously.

The ratio of the rates $v_{\text{prop}} = k_{\text{prop}}$ [R•][reagent] and $v_{\text{term}} = k_{\text{term}}$ [R•][R'•] of the propagation and the termination steps determines how many times the sequence of the propagation steps is run through before a termination step ends the conversion of starting material(s) to product(s). The rates of the propagation steps (v_{prop} in the second- and third-to-last boxes of the present section) are greater than those of the termination steps (v_{term} in the fourth box), frequently by several orders of magnitude. An initiating radical can therefore initiate from 1 000 to 100 000 passes through the propagation steps of the chain.

How does this order of the rates $v_{prop} \gg v_{term}$ come about? As high-energy species, radical intermediates react exergonically with most reaction partners. According to the Hammond postulate, they do this very rapidly. Radicals actually often react with the first reaction partner they encounter. Their average lifetime is therefore very short. The probability of a termination step in which *two* such short-lived radicals meet is consequently low.

There is a great diversity of initiating and propagation steps for radical substitution reactions. Bond homolyses, fragmentations, atom abstraction reactions, and addition reactions to C=C double bonds are among the possibilities. All of these reactions can be observed with substituted alkylmercury(II) hydrides as starting materials. For this reason, we will examine these reactions as the first radical reactions in Section 1.6.

1.5 Radical Initiators

Only for some of the radical reactions discussed in Sections 1.6—1.10 is the initiating radical produced immediately from the starting material or the reagent. In all other radical substitution reactions another substance, the **radical initiator**, added in a substoichiometric amount, is responsible for producing the initiating radical.

Radical initiators are thermally labile compounds, which decompose into radicals upon moderate heating or photolysis. These radicals initiate the actual radical chain through the formation of the initiating radical. The most frequently used radical initiators are azobis-isobutyronitrile (AIBN) and dibenzoyl peroxide (Figure 1.11). AIBN has has a half-life of 1 h at 80 °C, and dibenzoyl peroxide a half-life of 1 h at 95 °C.

Azobisisobutyronitrile (AIBN) as radical initiator:

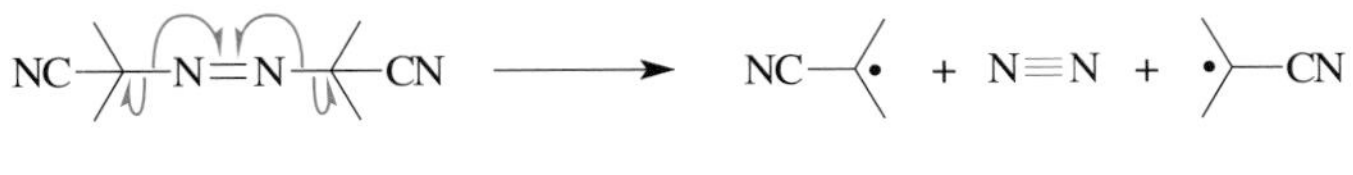

NC–C(CH₃)₂–N=N–C(CH₃)₂–CN ⟶ NC–C(CH₃)₂• + N≡N + •C(CH₃)₂–CN

Dibenzoyl peroxide as radical initiator:

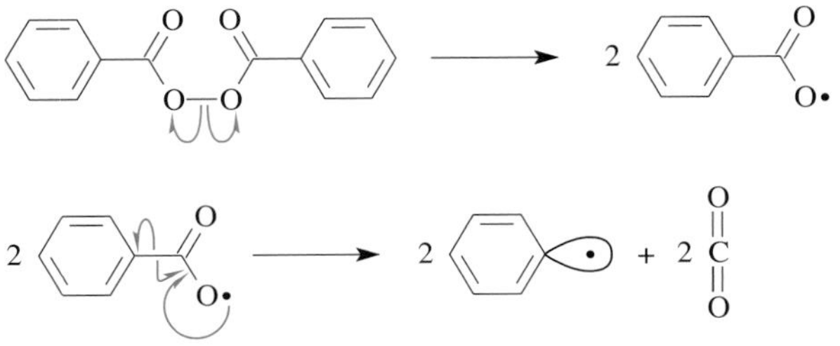

PhC(=O)O–OC(=O)Ph ⟶ 2 PhC(=O)O•

2 PhC(=O)O• ⟶ 2 Ph• + 2 O=C=O

Fig. 1.11. Radical initiators and their mode of action (in the "arrow formalism" for showing reaction mechanisms used in organic chemistry, arrows with half-heads show where *single* electrons are shifted, whereas arrows with full heads show where electron *pairs* are shifted).

Side Note 1.1. Decomposition of Ozone in the Upper Stratosphere

Certain undesired reactions can take place via radical intermediates. Examples are the autoxidation of ethers (see Figure 1.38) or the decomposition of ozone in the upper stratosphere. This decomposition is initiated by, among other things, the **chlorofluorohydrocarbons** ("HCFCs"), which form chlorine radicals under the influence of the short-wave UV light from the sun (Figure 1.12). They function as initiating radicals for the decomposition of ozone, which takes place via a radical chain. However, this does not involve a radical substitution reaction.

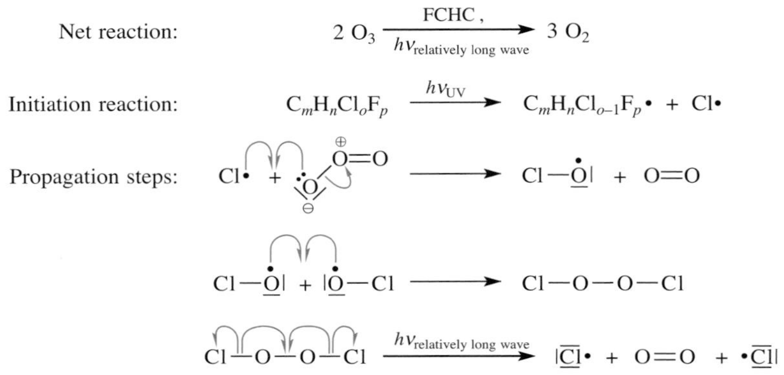

Net reaction: $2\,O_3 \xrightarrow[h\nu_{\text{relatively long wave}}]{\text{FCHC},} 3\,O_2$

Initiation reaction: $C_mH_nCl_oF_p \xrightarrow{h\nu_{UV}} C_mH_nCl_{o-1}F_p\bullet + Cl\bullet$

Propagation steps: $Cl\bullet + O_3 \longrightarrow Cl{-}O\bullet + O{=}O$

$Cl{-}O\bullet + \bullet O{-}Cl \longrightarrow Cl{-}O{-}O{-}Cl$

$Cl{-}O{-}O{-}Cl \xrightarrow{h\nu_{\text{relatively long wave}}} Cl\bullet + O{=}O + \bullet Cl$

Fig. 1.12. FCHC-initiated decomposition of stratospheric ozone.

1.6 Radical Chemistry of Alkylmercury(II) Hydrides

Alkyl derivatives of mercury in the oxidation state + 2 are produced during the solvomercuration of alkenes (the first part of the reaction sequence in Figure 1.13). Oxymercuration provides (β-hydroxyalkyl)mercury(II) carboxylates, while alkoxymercuration gives (β-alkoxyalkyl)mercury(II) carboxylates. These compounds can be reduced with $NaBH_4$ to the corresponding mercury(II) hydrides. A ligand exchange takes place at the mercury: a carboxylate substituent is replaced by hydrogen. The β-oxygenated alkylmercury(II) hydrides obtained in this way are so unstable that they immediately react further. These reactions take place via radical intermediates. The latter can be transformed into various kinds of products by adjusting the reaction conditions appropriately. The most important radical reactions of alkylmercury(II) hydrides are fragmentation to an alcohol (Figure 1.14), addition to a C=C

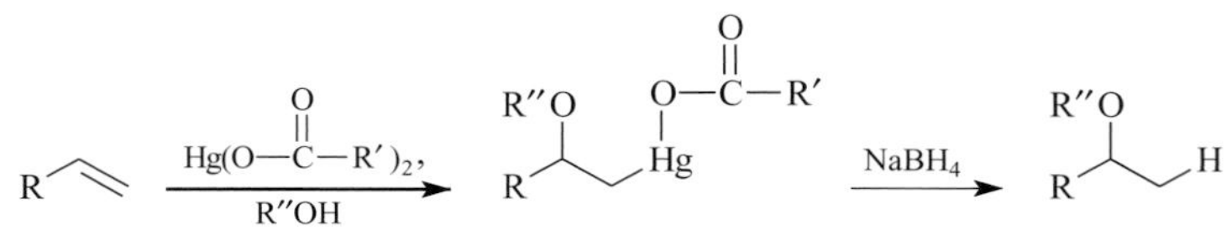

$R{-}CH{=}CH_2 \xrightarrow[R''OH]{Hg(O{-}C(=O){-}R')_2,} R{-}CH(OR''){-}CH_2{-}Hg{-}O{-}C(=O){-}R' \xrightarrow{NaBH_4} R{-}CH(OR''){-}CH_2{-}H$

Fig. 1.13. Net reaction (a) for the hydration of alkenes ($R' = CH_3$, $R'' = H$) or (b) for the addition of alcohol to alkenes ($R' = CF_3$, R'' = alkyl) via the reaction sequence (1) solvomercuration of the alkene (for mechanism, see Figure 3.48; regioselectivity: Figure 3.49); (2) reduction of the alkylmercury compound obtained (for mechanism, see Figure 1.14).

double bond (Figure 1.15), and oxidation to a glycol derivative (Figure 1.16). The mechanisms for these reactions will be discussed below.

When (β-hydroxyalkyl)mercury(II) acetates are treated with $NaBH_4$ and no additional reagent, they first form (β-hydroxyalkyl)mercury(II) hydrides. These react via the chain process shown in Figure 1.14 to give a mercury-free alcohol. Overall, a substitution reaction R—Hg(OAc) → R—H takes place. The initiation step for the chain reaction participating in this transformation is a homolysis of the C—Hg bond. This takes place rapidly at room temperature and produces the radical •Hg—H and a β-hydroxylated alkyl radical. As the initiating radical, it starts the first of the two propagation steps. This first step is an atom transfer reaction, more specifically, a hydrogen atom transfer reaction. The second propagation step involves a radical fragmentation. This is the decomposition of a radical into a radical with a lower molecular weight and at least one valence-saturated compound (in this case, elemental mercury). The net reaction equation is obtained according to Section 1.4 by adding up the two propagation steps.

Ionic prereaction:

Initiation step:

Propagation steps:

Termination steps, for example:

Net equation:
Σ Propagation steps:

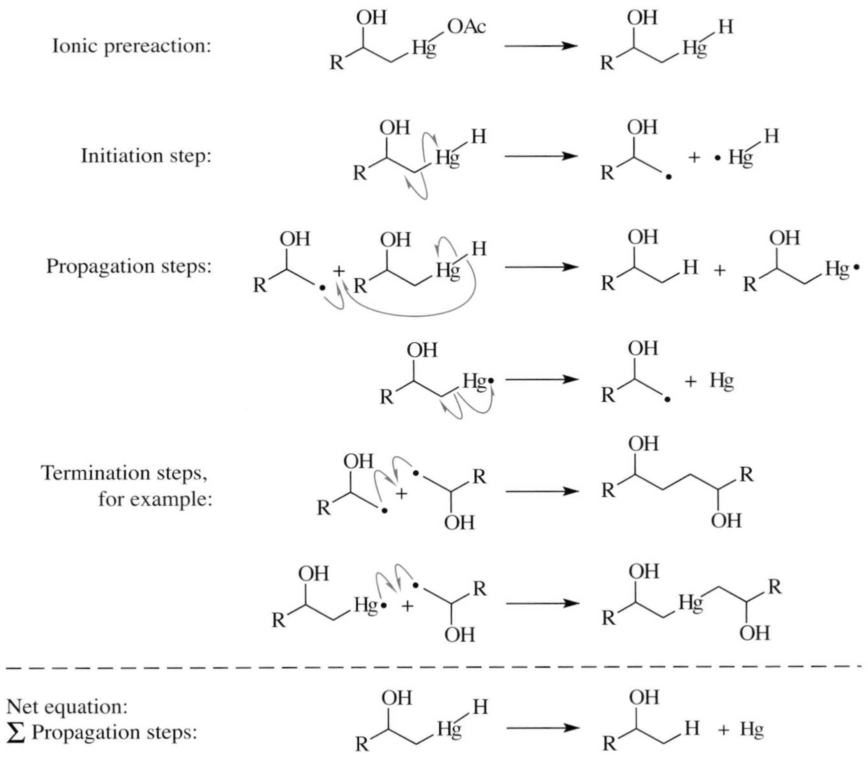

Fig. 1.14. $NaBH_4$, reduction of (β-hydroxyalkyl)mercury(II) acetates to alcohols and radical fragmentation of (β-hydroxyalkyl)mercury (II) hydrides. According to the terminology used in Figure 1.2 it is a "substitution by fragmentation." The fate of the radical H–Hg• of the initiation reaction is unknown. This, however, does not limit our understanding of the overall reaction since under no circumstances is this radical H–Hg• a chain carrier.

These propagation steps are repeated many times while the organic mercury compound is consumed and alcohol and elemental mercury are released. This process is interrupted only by termination steps (Figure 1.14). Thus, for example, two mercury-free radicals can combine to form one dimer, or a mercury-free and a mercury-containing radical can combine to form a dialkylmercury compound.

(β-Hydroxyalkyl) mercury(II) acetates and $NaBH_4$ react to form carbon-centered radicals through the reaction steps shown in Figure 1.14. When methyl acrylate is present in the reaction mixture, these radicals can add to the C=C double bond of the ester (Figure 1.15). The addition takes place via a reaction chain, which comprises three propagation steps.

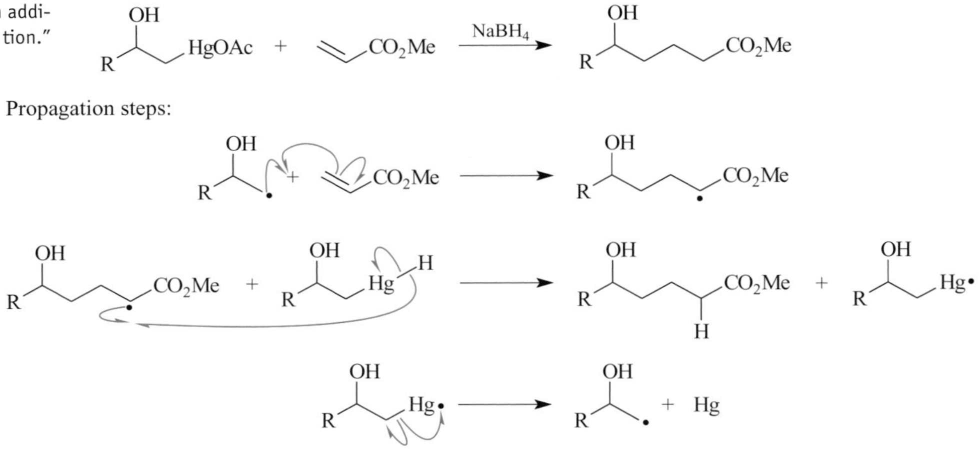

Fig. 1.15. $NaBH_4$-mediated addition of (β-hydroxyalkyl)mercury(II) acetates to an acceptor-substituted alkene. In terms of Figure 1.2 it is a "substitution reaction including a fragmentation and an addition."

Propagation steps:

The radicals produced during the decomposition of alkylmercury(II) hydrides can also be added to molecular oxygen (Figure 1.16). A hydroperoxide is first produced in a chain reaction, which again comprises three propagation steps. However, it is unstable in the presence of $NaBH_4$ and is reduced to an alcohol.

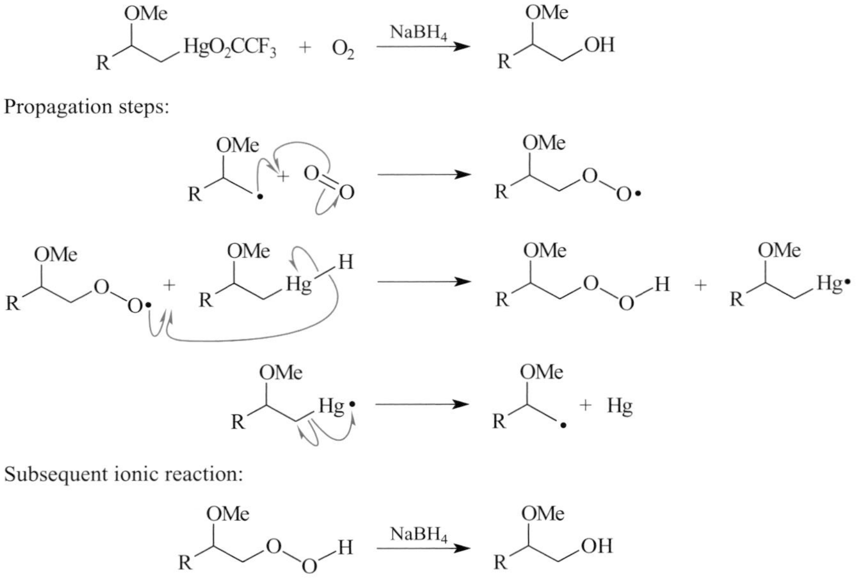

Propagation steps:

Subsequent ionic reaction:

Fig. 1.16. $NaBH_4$-induced air oxidation of a (β-alkoxyalkyl)mercury (II) trifluoroacetate (see Figure 3.39) to a glycol derivative. In terms of Figure 1.2 it is "substitution reaction including a fragmentation and an addition."

1.7 Radical Halogenation of Hydrocarbons

Many hydrocarbons can be halogenated with elemental chlorine or bromine while being heated and/or irradiated together:

$$\gt C_{sp^3}\text{—H} + Cl_2(Br_2) \xrightarrow{h\nu \text{ or } \Delta} \gt C_{sp^3}-\text{Cl(Br)} + \text{HCl (HBr)}$$

The result is simple or multiple halogenations.

1.7.1 Simple and Multiple Chlorinations

You should already be familiar with the mechanism for the thermal chlorination of methane. We will use Figure 1.17 to review briefly the net equation, the initiation step, and the propagation steps of the monochlorination of methane. Figure 1.18 shows the energy profile of the propagation steps of this reaction.

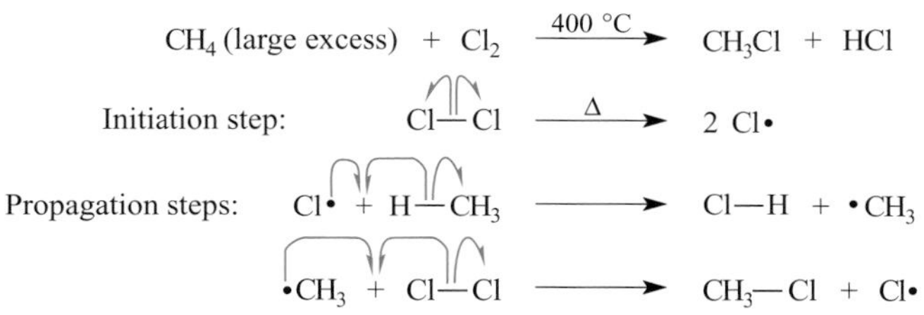

CH4 (large excess) + Cl2 400 °C → CH3Cl + HCl
Initiation step: Cl—Cl Δ → 2 Cl•
Propagation steps: Cl• + H—CH3 → Cl—H + •CH3
•CH3 + Cl—Cl → CH3—Cl + Cl•

Fig. 1.17. Mechanism for monochlorination of methane with Cl_2.

Starting materials
TS1
I1
TS2
Products
E
+1
−26
Cl2 + CH4 + Cl•
Cl2 + •CH3 + HCl
Cl• + CH3Cl + HCl
Reaction coordinate

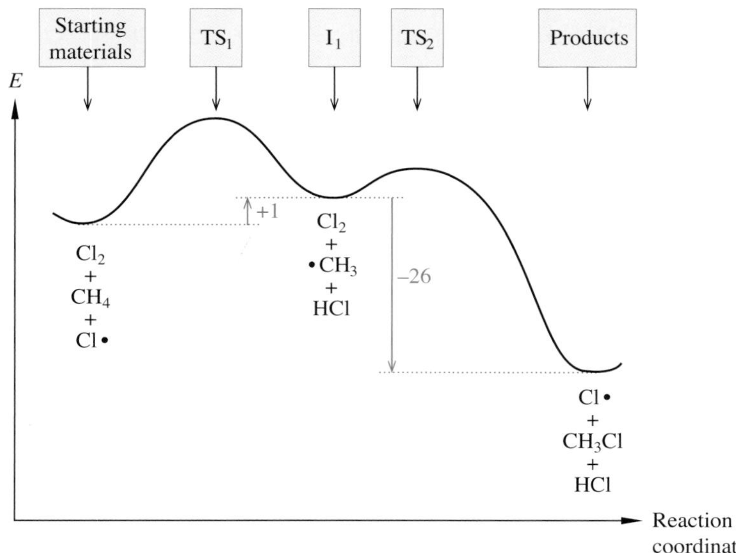

Fig. 1.18. Energy profile of the propagation steps of the monochlorination of methane with Cl_2 (enthalpies in kcal/mol).

In the energy profile, each of the two propagation steps is represented as a transition from one energy minimum over an energy maximum into a new energy minimum. Energy minima in an energy profile characterize either long-lived species [starting material(s), product(s)] or short-lived intermediates. On the other hand, energy maxima in an energy profile (transition states) are snapshots of the geometry of a molecular system, whose lifetime corresponds to the duration of a molecular vibration (approx. 10^{-13} s). These are general concepts.

A chemical transformation that takes place via exactly one transition state is called an **elementary reaction.** This holds regardless of whether it leads to a short-lived intermediate or to a product that can be isolated. According to the definition, an *n*-step reaction consists of a sequence of *n* elementary reactions. It takes place via *n* transition states and $(n - 1)$ intermediates.

In the reaction of a 1:1 mixture of methane and chlorine one does not obtain the monochlorination product selectively, but a 46:23:21:9:1 mixture of unreacted methane, mono-, di-, tri-, and tetrachloromethane. Thus, all conceivable multiple chlorination products are also produced. Multiple chlorinations, like monochlorinations, occur as radical chain substitutions. They are based on completely analogous propagation steps (Figure 1.19).

According to Figure 1.20, *analogous* propagation steps possess the same heat of reaction, independent of the degree of chlorination. With the help of Hammond's postulate, one concludes from this that the associated free activation energies should also be independent of the degree of chlorination. This means that the monochlorination of methane and each of the sub-

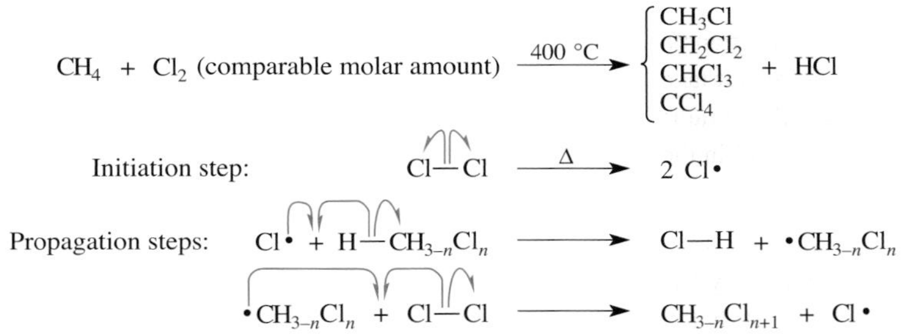

$$CH_4 + Cl_2 \text{ (comparable molar amount)} \xrightarrow{400\ °C} \left\{ \begin{array}{l} CH_3Cl \\ CH_2Cl_2 \\ CHCl_3 \\ CCl_4 \end{array} \right. + HCl$$

Initiation step: $Cl{-}Cl \xrightarrow{\Delta} 2\ Cl\bullet$

Propagation steps: $Cl\bullet + H{-}CH_{3-n}Cl_n \longrightarrow Cl{-}H + \bullet CH_{3-n}Cl_n$

$\bullet CH_{3-n}Cl_n + Cl{-}Cl \longrightarrow CH_{3-n}Cl_{n+1} + Cl\bullet$

Fig. 1.19. Mechanism for the polychlorination of methane.

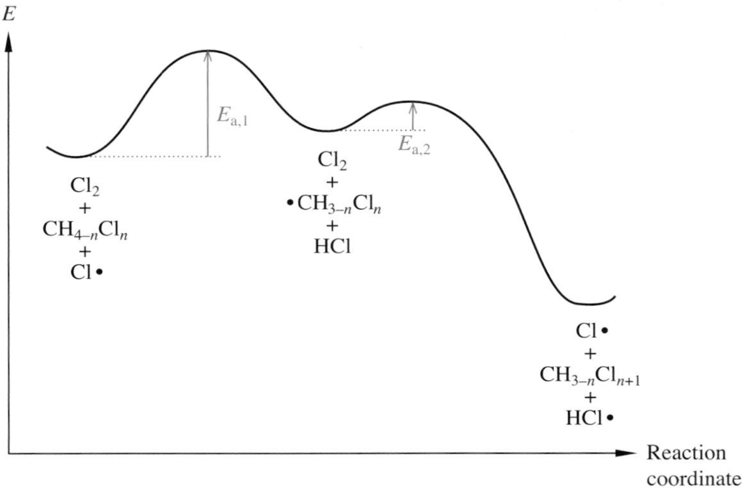

Fig. 1.20. Energy profile of the propagation steps of the polychlorinations $CH_3Cl \rightarrow CH_2Cl_2$, $CH_2Cl_2 \rightarrow CHCl_3$, and $CHCl_3 \rightarrow CCl_4$ of methane (n = 1–3 in the diagram), and of the monochlorination $CH_4 \rightarrow CH_3Cl$ (n = 0 in the diagram).

sequent multiple chlorinations should take place with one and the same rate constant. This is essentially the case. The relative chlorination rates for $CH_{4-n}Cl_n$ in the temperature range in question are 1 ($n = 0$), 3 ($n = 1$), 2 ($n = 2$), and 0.7 ($n = 3$).

If only methyl chloride is needed, it can be produced essentially free of multiple chlorination products only if a large excess of methane is reacted with chlorine. In this case, there is always more unreacted methane available for more monochlorination to occur than there is methyl chloride available for a second chlorination.

Another preparatively valuable multiple chlorination is the photochemical perchlorination of methyl chloroformate, which leads to diphosgene:

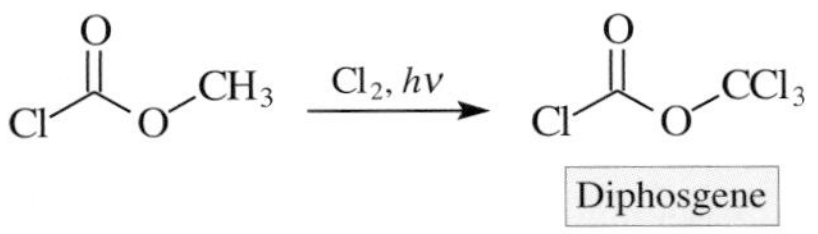

$$\mathrm{Cl{-}C({=}O){-}O{-}CH_3} \xrightarrow{Cl_2,\ h\nu} \mathrm{Cl{-}C({=}O){-}O{-}CCl_3}$$

Diphosgene

1.7.2 Regioselectivity of Radical Chlorinations

Clean, regioselective, monochlorinations can be achieved with hydrocarbons that react via resonance-stabilized radicals.

Regioselectivity

A given molecular transformation, for example, the reaction C—H → C—Cl, is called **regioselective** when it takes place preferentially or exclusively at one place on a substrate. Resonance-stabilized radicals are produced regioselectively as a consequence of product development control in the radical-forming step.

In the industrial synthesis of benzyl chloride (Figure 1.21), only the H atoms in the benzyl position are replaced by Cl because the reaction takes place via resonance-stabilized benzyl radicals (cf. Table 1.1, bottom line) as intermediates. At a reaction temperature of 100 °C, the first H atom in the benzyl position is substituted a little less than 10 times faster (⟶ benzyl chloride) than the second (⟶ benzal chloride) and this is again 10 times faster than the third (⟶ benzotrichloride).

In the reaction example of Figure 1.22, the industrial synthesis of allyl chloride, only an H atom in the allylic position is substituted. Its precursor is a resonance-stabilized (Table 1.1, center line) allylic radical.

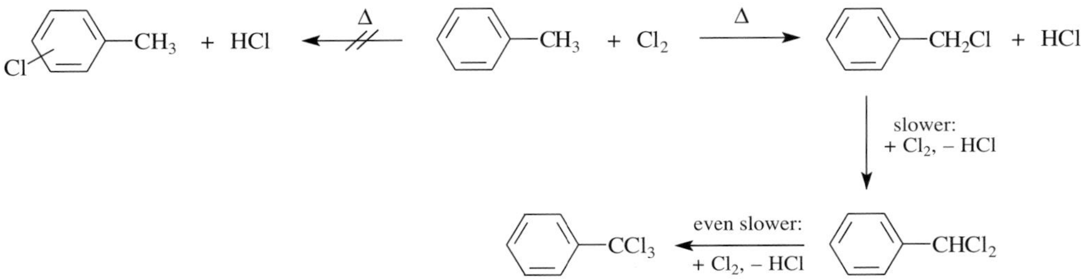

Fig. 1.21. Industrial synthesis of benzyl chloride.

$$\mathrm{Cl{-}C_6H_4{-}CH_3} + \mathrm{HCl} \;\xleftarrow[\;/\!/\;]{\Delta}\; \mathrm{C_6H_5{-}CH_3} + \mathrm{Cl_2} \xrightarrow{\Delta} \mathrm{C_6H_5{-}CH_2Cl} + \mathrm{HCl}$$

$$\mathrm{C_6H_5{-}CH_2Cl} \xrightarrow[\text{+ Cl}_2,\ -\ \text{HCl}]{\text{slower:}} \mathrm{C_6H_5{-}CHCl_2} \xrightarrow[\text{+ Cl}_2,\ -\ \text{HCl}]{\text{even slower:}} \mathrm{C_6H_5{-}CCl_3}$$

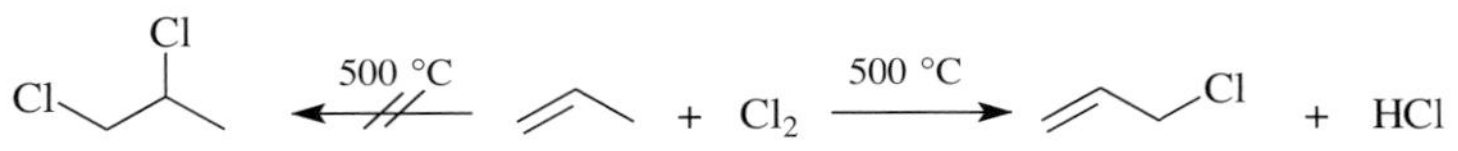

Fig. 1.22. Industrial synthesis of allyl chloride.

Cl … Cl ⟵(500 °C, ✗) propene + Cl_2 ⟶(500 °C) allyl–Cl + HCl

Incidentally, this reaction of chlorine with propene is also chemoselective.

Chemoselectivity

Reactions in which the reagent effects preferentially or exclusively one out of several types of possible transformations are **chemoselective.**

In the present case, the only transformation that takes place is C—H → C—Cl, that is, a substitution, and not a transformation C=C + Cl_2 → Cl—C—C—Cl, which would be an addition.

The monochlorination of isopentane gives all four conceivable monochlorination products (Table 1.4). These monochlorination products are obtained with relative yields of 22% (substitution of C_{tert}—H), 33% (substitution of C_{sec}—H), 30 and 15% (in each case substitution of C_{prim}—H). This is clearly not a regioselective reaction.

Two factors are responsible for this result. The first one is a statistical factor. Isopentane contains a single H atom that is part of a C_{tert}—H bond. There are two H atoms that are part of C_{sec}—H bonds, and 6 + 3 = 9 H atoms as part of C_{prim}—H bonds. If each H atom were sub-

Table 1.4. Regioselectivity of Radical Chlorination of Isopentane

Isopentane $\xrightarrow{Cl_2,\ \Delta}$	Cl-substituted (tert) +	Cl-substituted (sec) +	Cl-substituted (prim) +	Cl-substituted (prim) +	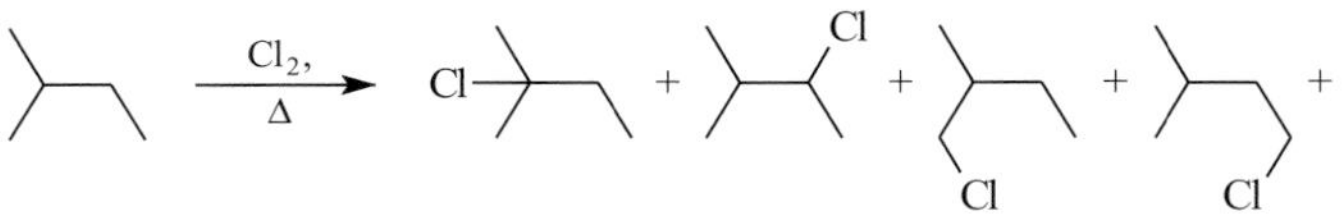 multiple chlorinated compounds
The relative yields of the above monochlorination products are...	... 22%	33%	30%	15%	
In order to produce the above compounds in the individual case...	... 1	2	6	3 ...	
...H atoms were available for the substitution. Yields on a per-H-atom basis were...	... 22%	16.5%	5%	5% ...	
... for the monochlorination product. In other words: $k_{C-H \rightarrow C-Cl,\ rel}$ in the position concerned is ...	... 4.4	3.3	≡ 1	≡ 1	
..., that is, generally for ...	... C_{tert}—H	C_{sec}—H	C_{prim}—H		

stituted at the same rate, the cited monochlorides would be produced in a ratio of 1:2:6:3. This would correspond to relative yields of 8, 17, 50, and 25%.

The discrepancy from the experimental values is due to the fact that H atoms bound to different types of C atoms are replaced by chlorine at different rates. The substitution of C_{tert}—H takes place via a tertiary radical. The substitution of C_{sec}—H takes place via the somewhat less stable secondary radical, and the substitution of C_{prim}—H takes place via even less stable primary radicals (for the stability of radicals, see Table 1.2). According to Hammond's postulate, the rate of formation of these radicals should decrease as the radical's stability decreases. Hydrogen atoms bound to C_{tert} should thus be substituted more rapidly than H atoms bound to C_{sec}, and these should in turn be substituted by Cl more rapidly than H atoms bound to C_{prim}. As the analysis of the regioselectivity of the monochlorination of isopentane carried out by means of Table 1.4 shows, the relative chlorination rates of C_{tert}—H, C_{sec}—H, and C_{prim}—H are 4.4:3.3:1, in agreement with this expectation.

1.7.3 Regioselectivity of Radical Brominations Compared to Chlorinations

In sharp contrast to chlorine, bromine and isopentane form monosubstitution products with pronounced regioselectivity (Table 1.5). The predominant monobromination product is pro-

Table 1.5. Regioselectivity of Radical Bromination of Isopentane

$\xrightarrow{Br_2,\ \Delta}$	Br	+ Br	+ Br	+ Br + multiple brominated compounds
The relative yields of the above monobromination products are...	... 92.2%	7.38%	0.28%	0.14%
In order to produce the above compounds in the individual case...	... 1	2	6	3 ...
...H atoms were available for the substitution. Yields on a per-H-atom basis were...	... 92.2%	3.69%	0.047%	0.047% ...
... for the monobromination product above. In other words: $k_{C-H \rightarrow C-Br,\ rel}$ in the position concerned is ...	... 2000	79	≡ 1	≡ 1
..., that is, generally for ...	... C_{tert}—H	C_{sec}—H	C_{prim}—H	

duced in 92.2% relative yield through the substitution of C_{tert}—H. The second most abundant monobromination product (7.4% relative yield) comes from the substitution of C_{sec}—H. The two monobromination products in which a primary H atom is replaced by Br occur only in trace quantities. The analysis of these regioselectivities illustrated in Table 1.5 gives relative rates of 2000, 79, and 1 for the bromination of C_{tert}—H, C_{sec}—H, and C_{prim}—H, respectively.

The low regioselectivity of the radical chain chlorination in Table 1.4 and the high regioselectivity of the analogous radical chain bromination in Table 1.5 are typical: bromine is generally considerably more suitable than chlorine for the regioselective halogenation of saturated hydrocarbons. In the following we will explain mechanistically why the regioselectivity for chlorination is so much lower than for bromination.

How the enthalpy ΔH of the substrate/reagent pair R—H/Cl• changes when R• and H—Cl are produced from it is plotted for four radical chlorinations in Figure 1.23 (left). These afford carbon radicals, the methyl, a primary, a secondary, and a tertiary radical. The reaction enthalpies ΔH_r for all four reactions are known and are plotted in the figure. Only the methyl radical is formed slightly endothermically ($\Delta H_r = +2.0$ kcal/mol). The primary radical, which is more stable by 4.3 ± 0.7 kcal/mol (cf. Table 1.2), is formed exothermically with $\Delta H_r = -2.3$ kcal/mol. The formation of the more stable secondary and tertiary radicals are more exothermic by –4.7 and –7.6 kcal/mol, respectively.

Hammond's postulate can be applied to this series of the selectivity-determining steps of the radical chlorination. They all take place via early transition states, i.e. via transition states that are similar to the starting materials. The more stable the resulting radical, the more similar the transition state is to the *starting materials*. The stability differences between these radicals are therefore manifested only to a very small extent as stability differences between the transition states that lead to them. All transition states are therefore very similar in energy and thus the different reaction rates are very similar. This means that the regioselectivity of the radical chlorination of saturated hydrocarbons is generally low.

In radical brominations, the energy profiles of the selectivity-determining step are completely different from what they are in analogous chlorinations. This is shown on the right side of Figure 1.23 and is rationalized as follows. The abstraction of a C-bound H atom by Cl atoms leads to the formation of an H—Cl bond with a bond energy of 103.1 kcal/mol. In contrast, the abstraction of a C-bound H atom by Br atoms leads to the formation of a H–Br bond with a bond energy of 87.4 kcal/mol, significantly weaker than the H-Cl bond. Accordingly, even the most stable radical considered in Figure 1.23, the tertiary radical, is formed endothermically in radical brominations ($\Delta H_r = +\ 8.1$ kcal/mol). From the secondary through the primary to the methyl radical increasingly less stable radicals are produced in the analogous brominations of this figure. They are therefore formed increasingly endothermically and consequently probably also increasingly endergonically. According to Hammond's postulate, the selectivity-determining step of radical brominations thus proceeds via late, that is, product-like, transition states. Consequently, the substituent effects on the free energy changes of the selectivity-determining step manifest as substituent effects on the free energies of the respective transition states. This results in different radicals being formed at very different reaction rates. The regioselectivity of radical brominations is consequently considerably higher than the regioselectivity of analogous chlorinations.

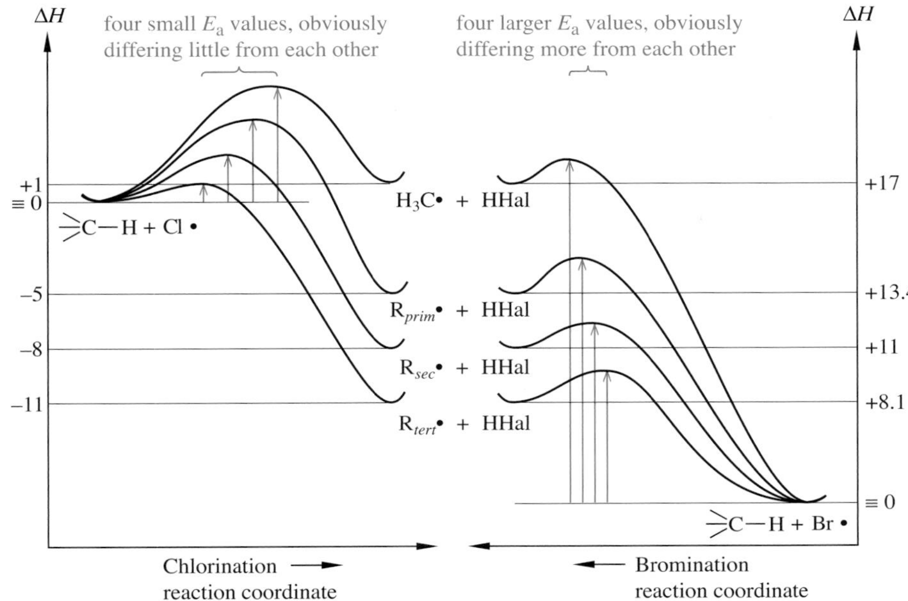

Fig. 1.23. Thermochemical analysis of that propagation step of radical chlorination (left) and bromination (right) of alkanes that determines the regioselectivity of the overall reaction. The ΔH_r values were determined experimentally; the values for the activation enthalpies ($\Delta H^{\ddagger}$) are estimates.

At the end of Section 1.7.4 we will talk about an additional aspect of Figure 1.23. To understand this aspect, however, we must first determine the rate law according to which radical halogenations take place.

1.7.4 Rate Law for Radical Halogenations; Reactivity/Selectivity Principle and the Road to Perdition

A simplified reaction scheme for the kinetic analysis of radical chain halogenations can be formulated as follows:

$$\mathrm{Hal_2} \underset{k_2}{\overset{k_1}{\rightleftharpoons}} 2\,\mathrm{Hal}\bullet \quad \text{with } K_{\mathrm{Dis}} = \frac{k_1}{k_2}$$

$$2\,\mathrm{Hal}\bullet + \mathrm{RH} \xrightarrow{k_3} \mathrm{HalH} + \mathrm{R}\bullet$$

$$\mathrm{R}\bullet + \mathrm{Hal_2} \xrightarrow{k_4} \mathrm{R\,Hal} + \mathrm{Hal}\bullet$$

Let us assume that only the reaction steps listed in this scheme participate in the radical chain halogenations of hydrocarbons and ignore termination reactions. According to this scheme, the thermolysis of halogen molecules gives halogen atoms with the rate constant k_1. These can recombine with the rate constant k_2 to form the halogen molecule again. The halogen atoms can also participate as the initiating radical in the first propagation step, which takes place with the rate constant k_3. The second and last propagation step follows with the rate constant k_4.

Explicit termination steps do not have to be considered in this approximate kinetic analysis. A termination step has already been *implicitly* considered as the reversal of the starting reaction (rate constant k_2). As soon as all halogen atoms have been converted back into halogen molecules, the chain reaction stops.

The rate law for the halogenation reaction shown above is derived step by step in Equations 1.4–1.8. We will learn to set up derivations of this type in Section 2.4.1. There we will use a much simpler example. We will not discuss Bodenstein's steady-state approximation used in Equations 1.6 and 1.7 in more detail until later (Section 2.5.1). What will be explained there and in the derivation of additional rate laws in this book is sufficient to enable you to follow the derivation of Equations 1.4—1.8 in detail in a second pass through this book (and you should make several passes through the book to ensure you understand the concepts).

$$\frac{d[\text{R Hal}]}{dt} = k_4[\text{R}\bullet][\text{Hal}_2] \quad (1.4)$$

$$\frac{d[\text{R}\bullet]}{dt} = k_3[\text{Hal}\bullet][\text{RH}] - k_4[\text{R}\bullet][\text{Hal}_2]$$
$$= 0 \text{ in the framework of Bodenstein's steady-state approximatio} \quad (1.5)$$

$$\frac{d[\text{Hal}\bullet]}{dt} = k_1[\text{Hal}_2] - k_2[\text{Hal}\bullet]^2$$
$$= 0 \text{ in the framework of Bodenstein's steady-state approximation} \quad (1.6)$$

$$\text{Eq.}(1.6) \Rightarrow [\text{Hal}\cdot] = \sqrt{\frac{k_1}{k_2}[\text{Hal}_2]}$$
$$= \sqrt{K_{\text{dis}}[\text{Hal}_2]} \quad (1.7)$$

$$\text{Eq.}(1.7) \text{ in Eq. } (1.5) \Rightarrow$$
$$k_4\ [\text{R}\cdot][\text{Hal}_2] = k_3\ [\text{RH}]\ \sqrt{K_{\text{dis}}[\text{Hal}_2]} \quad (1.8)$$

$$\text{Eq.}(1.8) \text{ in Eq. } (1.4) \Rightarrow \frac{d[\text{R Hak}]}{dt} = k_3\ [\text{RH}]\sqrt{K_{\text{dis}}[\text{Hal}_2]}$$

At this stage, it is sufficient to know the result of this derivation, which is given as Equation 1.9:

$$\text{Gross reaction rate} = k_3\ [\text{RH}]\sqrt{K_{\text{dis}}[\text{Hal}_2]} \quad (1.9)$$

It says that the substitution product R—X is produced at a rate that is determined by two constants and two concentration terms. For given initial concentrations of the substrate R—H and the halogen and for a given reaction temperature, the rate of formation of the substitution product is *directly proportional* to the rate constant k_3, k_3 being the rate constant of the propagation step in which the radical R• is produced from the hydrocarbon R—H.

Let's look at the energy profiles in Figure 1.23. They represent the step of chlorination (left side) and bromination (right side) that determines the regioselectivity and takes place with the rate constant k_3. According to Section 1.7.3, this step is faster for chlorination than for bromination.

If we look at the reaction scheme we set up at the beginning of this section, then this means that

$$k_3 \text{ (chlorination)} > k_3 \text{ (bromination)}$$

Also, according to Equation 1.9, the *overall reaction* "radical chlorination" takes place on a given substrate considerably faster than the overall reaction "radical bromination." If we consider this and the observation from Section 1.7.3, which states that radical chlorinations on a given substrate proceed with considerably lower regioselectivity than radical brominations, we have a good example of the so-called reactivity/selectivity principle. This states that more reactive reagents and reactants are less selective than less reactive ones. So selectivity becomes a measure of reactivity and vice versa. However, the selectivity-determining step of radical chlorination reactions of hydrocarbons takes place near the diffusion-controlled limit. Bromination is considerably slower. Read on.

Reactivity/Selectivity Principle: The Road to Perdition?

A highly reactive reagent generally reacts with lower selectivity than a less reactive reagent. It sounds beautiful. The problem with this principle is that it is considered by many to be completely useless and that the preceding example was actually an exception to what is the reality in many other reactions.

The reactivity-selectivity principle seems to be intuitively sound. But data indicate otherwise. The work of Mayr and others has shown that there are many reactions in which selectivity is not compromised at the expense of reactivity. Rates increase, but relative reactivities, that is, selectivities, remain constant. The example discussed above is an exception and conceptually very tempting. It teaches us the lesson that our models about chemistry must be continually questioned and refined.

The reactivity-selectivity principle holds when one of the reference reactions being considered occurs close to the diffusion-controlled limit. At or near this limit, reactions are very fast and not selective. Below this point, selectivity reveals itself. Compare two analogous reactions that differ substantially in rate, neither of which occurs near the diffusion-controlled limit and each will exhibit similar selectivities, though the rate by which these reactions occur will differ.

1.7.5 Chemoselectivity of Radical Brominations

Let us go back to radical brominations (cf. Section 1.7.3). The bromination of alkyl aromatics takes place completely regioselectively: only the benzylic position is brominated. The intermediates are the most stable radicals that are available from alkyl aromatics, namely, benzylic radicals. Refluxing *ortho*-xylene reacts with 2 equiv. of bromine to give one monosubstitution

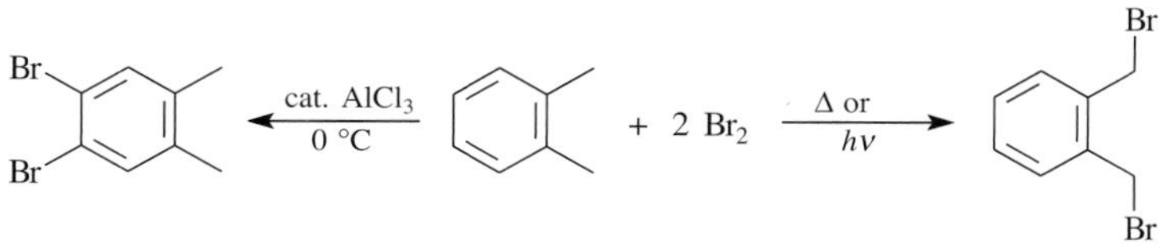

Br Br ←(cat. $AlCl_3$, 0 °C)— + 2 Br_2 —(Δ or $h\nu$)→ Br Br

Fig. 1.24. Competing chemoselectivities during the reaction of bromine with *ortho*-xylene by a polar mechanism (left) and a radical mechanism (right).

per benzylic position. The same transformation occurs when the reactants are irradiated at room temperature in a 1:2 ratio (Figure 1.24, right).

Starting from the same reagents, one can also effect a double substitution *on the aromatic ring* (Figure 1.24, left). However, the mechanism is completely different (Figure 5.13 and following figures). This substitution takes place under reaction conditions in which no radical intermediates are formed. (Further discussion of this process will be presented in Section 5.2.1.)

Hydrogen atoms in the benzylic position can be replaced by elemental bromine as shown. This is not true for hydrogen atoms in the allylic position. The alkene reacts rapidly with molecular bromine via addition and allylic bromination is not observed (Figure 1.25, left). A chemoselective allylic bromination of alkenes succeeds only according to the **Wohl-Ziegler process** (Figure 1.25, right), that is, with *N*-bromosuccinimide (NBS).

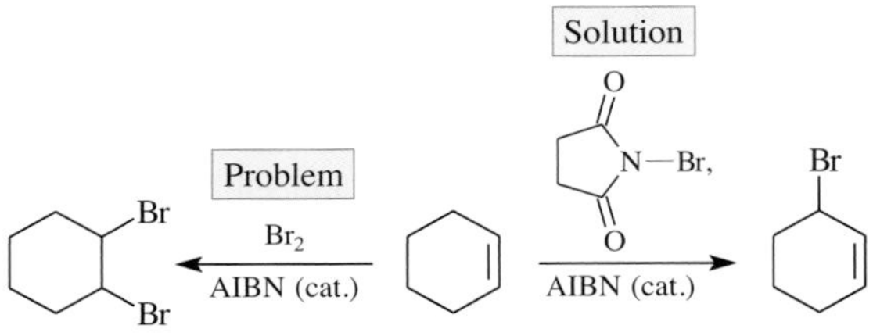

Problem

Br_2

AIBN (cat.)

Solution

N—Br,

AIBN (cat.)

Fig. 1.25. Bromine addition and bromine substitution on cyclohexene.

Figure 1.26 gives a mechanistic analysis of this reaction. NBS is used in a stoichiometric amount, and the radical initiator AIBN (cf. Figure 1.11) is used in a sub-stoichiometric amount, generally a few percent of the alkene to be brominated. Initiation involves several reactions, which lead to Br• as the initiating radical. Figure 1.26 shows one of several possible starting reaction sequences. Next follow three propagation steps. The second propagation step—something new in comparison to the reactions discussed before—is an ionic reaction between NBS and HBr. This produces succinimide along with the elemental bromine, which is required for the third propagation step. Yes, in the radical bromination using NBS, bromine, not NBS, reacts with the carbon radical to form the product!

In the first propagation step of the Wohl–Ziegler bromination, the bromine atom abstracts a hydrogen atom from the allylic position of the alkene and thereby initiates a **substitution.** This is not the only reaction mode conceivable under these conditions. As an alternative, the bromine atom could react with the C=C double bond and thereby start a radical **addition** to it (Figure 1.27). Such an addition is indeed observed when cyclohexene is reacted with a Br_2/AIBN mixture.

The difference is that in the Wohl-Ziegler process there is always a much lower Br_2 concentration than in the reaction of cyclohexene with bromine itself. Figure 1.27 shows *qualitatively* how the Br_2 concentration controls whether the combined effect of Br/Br_2 on cyclohexene is an addition or a substitution. The critical factor is that the addition takes place via a reversible step and the substitution does not. During the addition, a bromocyclohexyl radical forms from cyclohexene and Br in an equilibrium reaction. This radical is intercepted by forming dibromocyclohexane only when a *high* concentration of Br_2 is present. However, if the concentration of Br_2 is *low*, this reaction does not take place. The bromocyclohexyl radical is then produced only in an unproductive equilibrium reaction. In this case, the irreversible substitution therefore determines the course of the reaction.

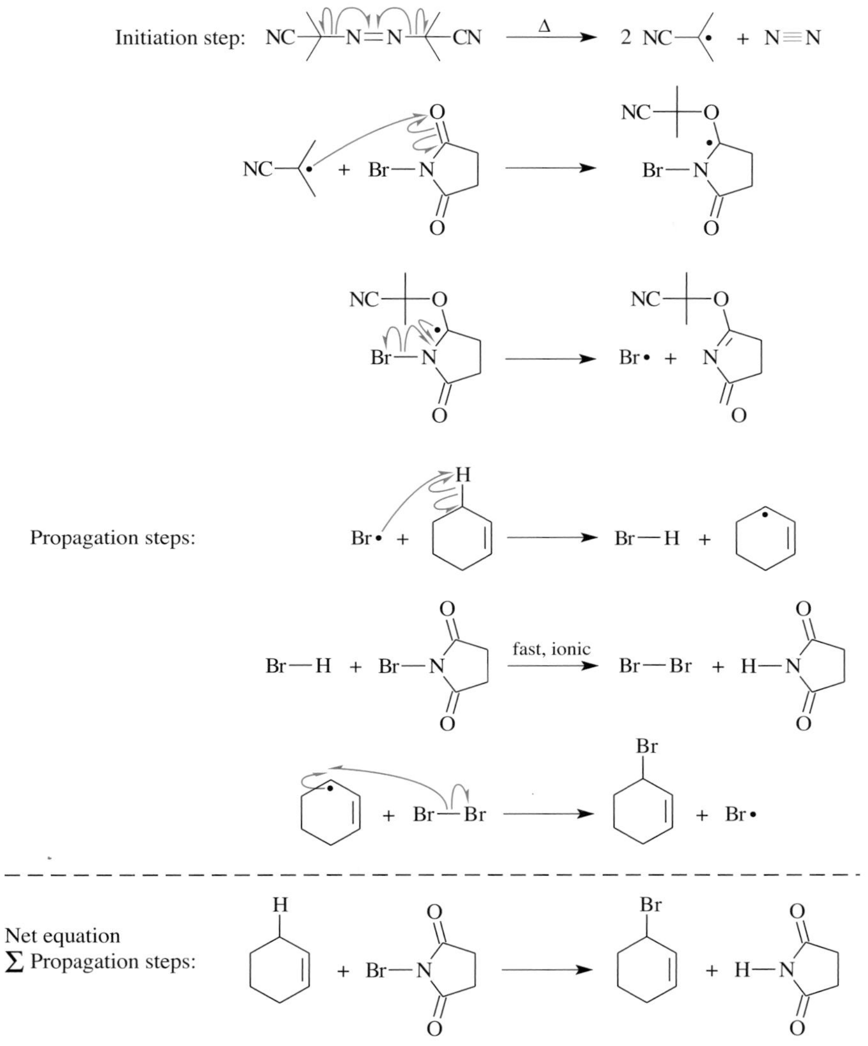

Initiation step:

Propagation steps:

fast, ionic

Net equation
Σ Propagation steps:

Fig. 1.26. Mechanism for the allylic bromination of cyclohexene according to the Wohl-Ziegler process.

Figure 1.28 gives a *quantitative* analysis of the outcome of this competition. The ratio of the rate of formation of the substitution product to the rate of formation of the addition product—which equals the ratio of the yield of the substitution product to the yield of the addition product—is inversely proportional to the concentration of Br_2.

An allyl radical can be brominated at both termini of the radical. This is why *two* allyl bromides can result from the Wohl–Ziegler bromination of an alkene if the allyl radical intermediate is unsymmetrical (examples see Figures 1.29–1.31). Even more than two allyl bromides may form. This happens if the substrate possesses constitutionally different allylic H atoms, and if, as a result thereof, several constitutionally isomeric allyl radicals form and react with bromine without selectivity.

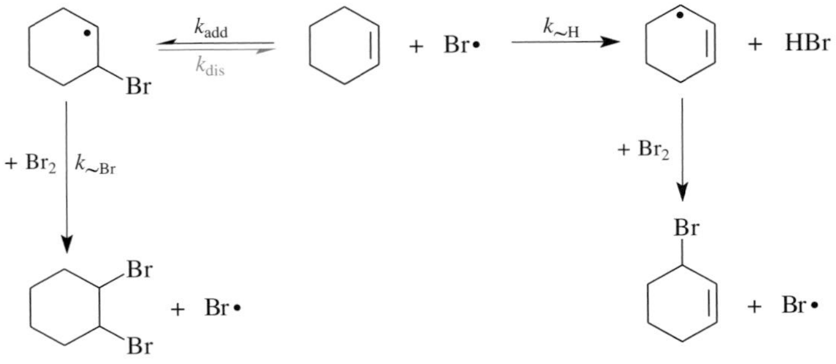

Fig. 1.27. Reaction scheme for the action of Br•/Br_2 on cyclohexene and the kinetic analysis of the resulting competition between allylic substitution (right) and addition (left) (in $k_{\sim x}$, ~X means homolytic cleavage of a bond to atom X).

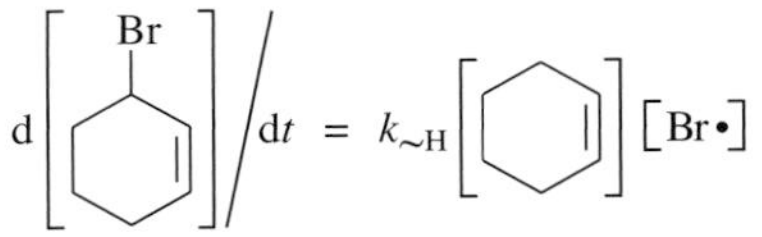

$$\mathrm{d}[\text{3-bromocyclohexene}]/\mathrm{d}t = k_{\sim H}[\text{cyclohexene}][\mathrm{Br}\bullet] \quad (1.10)$$

because $k_{\sim H}$ describes the rate-determining step of the allylic substitution

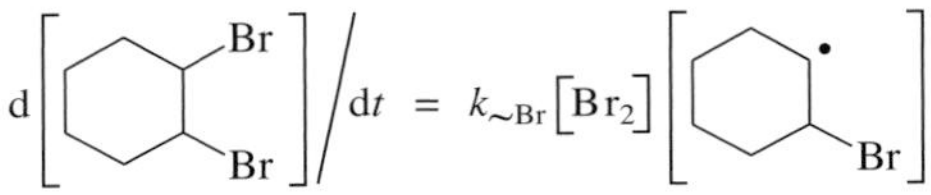

$$\mathrm{d}[\text{1,2-dibromocyclohexane}]/\mathrm{d}t = k_{\sim Br}[\mathrm{Br_2}][\text{2-bromocyclohexyl radical}] \quad (1.11)$$

because $k_{\sim Br}$ describes the rate-determining step of the addition reaction

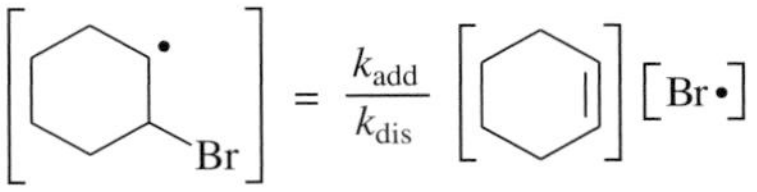

$$[\text{2-bromocyclohexyl radical}] = \frac{k_{add}}{k_{dis}}[\text{cyclohexene}][\mathrm{Br}\bullet] \quad (1.12)$$

because the equilibrium condition is met

Equation (1.12) in Equation (1.11) ⇒

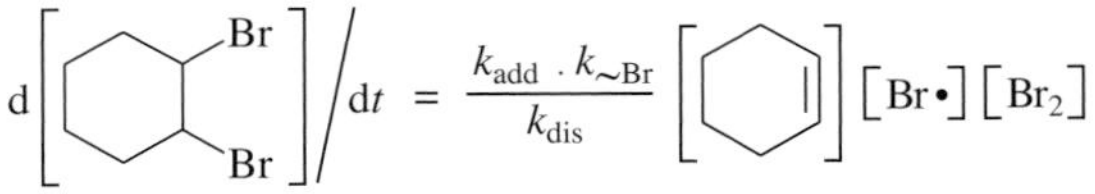

$$\mathrm{d}[\text{1,2-dibromocyclohexane}]/\mathrm{d}t = \frac{k_{add}\cdot k_{\sim Br}}{k_{dis}}[\text{cyclohexene}][\mathrm{Br}\bullet][\mathrm{Br_2}] \quad (1.13)$$

Divide Equation (1.10) by Equation (1.13) ⇒

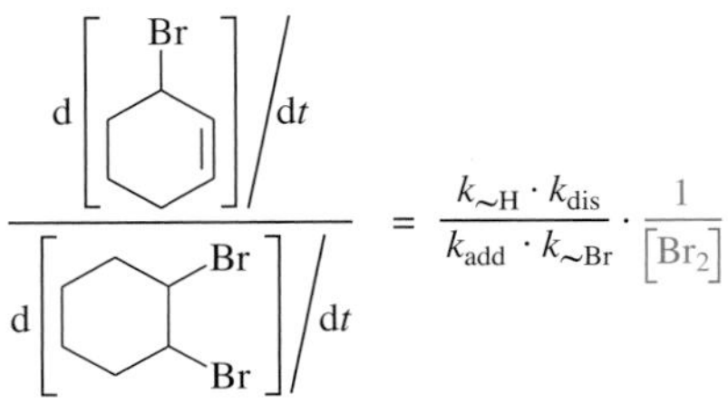

$$\frac{\mathrm{d}[\text{3-bromocyclohexene}]/\mathrm{d}t}{\mathrm{d}[\text{1,2-dibromocyclohexane}]/\mathrm{d}t} = \frac{k_{\sim H}\cdot k_{dis}}{k_{add}\cdot k_{\sim Br}}\cdot\frac{1}{[\mathrm{Br_2}]} \quad (1.14)$$

Fig. 1.28. Derivation of the kinetic expression for the chemoselectivity of allylic substitution versus bromine addition in the Br•/Br_2/cyclohexene system. The rate constants are defined as in Figure 1.27.

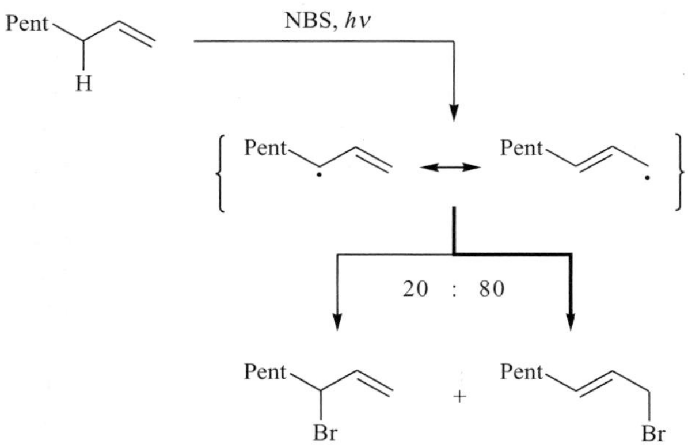

Pent H NBS, $h\nu$ Pent Pent 20 : 80 Pent Br + Pent Br

Fig 1.29. With Wohl–Ziegler brominations unsymmetric allyl radicals can basically react with the bromine atom at any of their nonequivalent ends. The product in which the bromine is localized on the less alkylated C atom and in which the higher alkylated C=C bond is present will form preferentially.

The bromination depicted in Figure 1.29 proceeds via an unsymmetrical allyl radical. This radical preferentially (80:20) reacts to yield the bromination product with the more highly substituted C=C double bond. As this reaction proceeds under kinetic control, the selectivity is based on product development control: the more stable (since higher alkylated) alkene is formed more rapidly than the isomer.

The bromination shown in Figure 1.30 also proceeds via an unsymmetrical allyl radical intermediate, and the bromination product with the aryl-substituted C=C double bond is formed exclusively. Since this process, too, is under kinetic control, the selectivity is again due to product development control: the alkene isomer, which is strongly favored because of conjugation, is formed so much faster than the non-conjugated isomer that the latter is not observed at all.

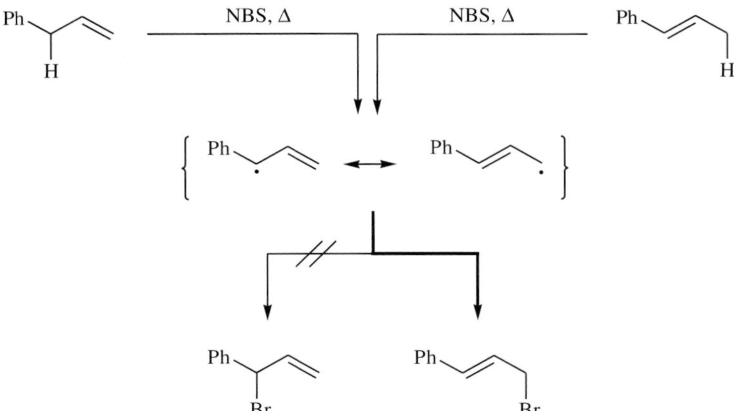

Fig. 1.30. In some cases the allyl radical intermediate of Wohl-Ziegler brominations is available from alkene double bond isomers, which can profitably be used when one of the substrates is more easily accessible or cheaper than its isomer.

Figure 1.30 illustrates that the allyl radical intermediate of several Wohl-Ziegler brominations can be accessed from isomeric alkene substrates as starting materials. This is worth considering when one substrate is more easily accessible or cheaper than its isomer. The price of allyl benzene, for example, is just a fraction of what has to be paid for 1-phenylpropene and would thus be preferred for the synthesis shown in Figure 1.30.

Side Note 1.2. The Rate of the Wohl–Ziegler Bromination Is More Dependent on a Polar Effect than on Product Development Control

Surely you would have correctly predicted the result of the Wohl–Ziegler bromination of crotonic acid methyl ester (Formula **D** in Figure 1.31): bromocrotonic ester (**G**) is formed exclusively, and no bromovinyl acetic acid ester (**F**), which is the result of product development control and the resonance stabilization (4–5 kcal/mol) observed with conjugated esters. You would certainly also expect that the allylic bromination of vinyl acetic acid ester (**C**) would lead to the same 100:0 ratio of the bromides **G** and **F** as the allyl bromination of crotonic ester (**D**). Under Wohl–Ziegler conditions, however, vinyl acetic acid ester (**C**) does not undergo an allylic substitution, but an addition according to the net equation **C** + 2 NBS → **B** + 2 succinimide, which is most unusual for NBS. The failure of the allylic substitution with the ester **C** is all the more surprising since **C** is a nonconjugated ester with a monosubstituted C=C double bond and is therefore less stable than the isomeric ester **D**, which is conjugated and pos-

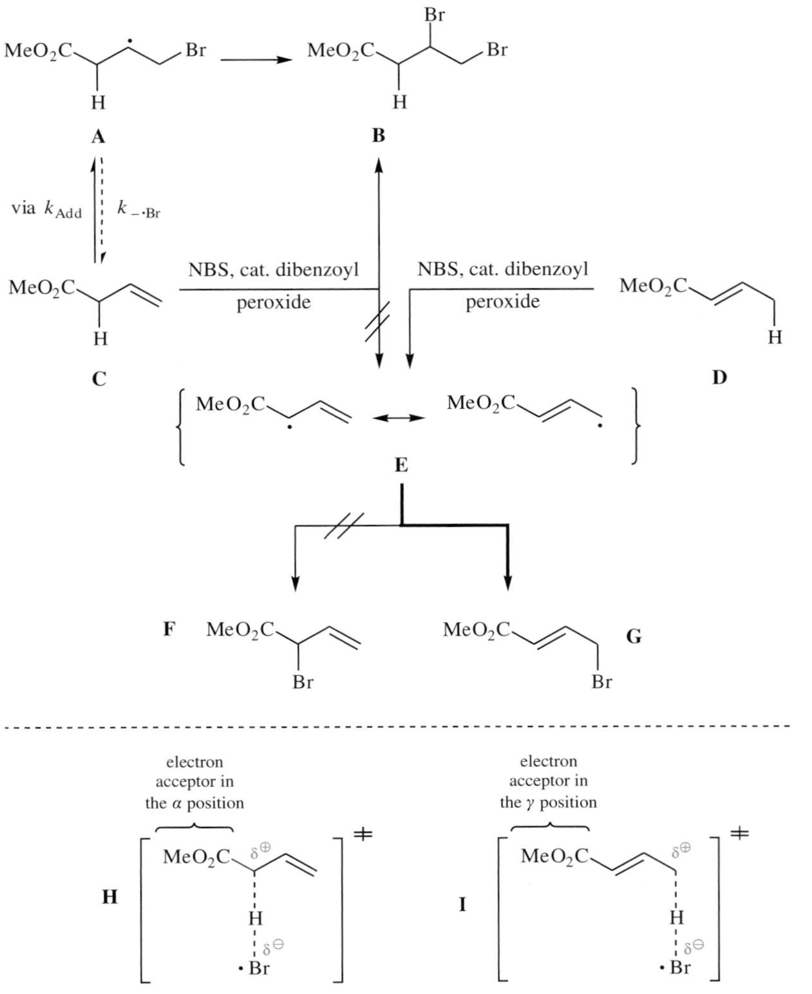

Fig. 1.31. The Wohl-Ziegler bromination of crotonic acid methyl ester **D** exclusively supplies the bromocrotonic ester **G**. However, vinyl acetic acid ester **C** and NBS exclusively yield the dibromo addition product **B** under Wohl–Ziegler conditions. Here, NBS acts as a Br_2 source for addition rather than substitution. The fact that vinyl acetic acid ester and NBS do not react likewise to yield the bromocrotonic ester **G** is due to an electronic effect discussed in the text.

sesses a disubstituted C=C double bond. The formation of the allylic radical intermediate **E**, which is actually obtained from the ester **D**, would be less endergonic if it originated from **C**. According to the Hammond postulate this reaction would have had to be even faster than the one starting from **D**. Instead, **C** turns out to be does not afford **E** at all. Why?

The rate analysis of radical reactions will only be complete if electronic effects are also considered. In the transition state of the hydrogen transfer from the substrate to a bromine atom, partial charges develop during the course of the reaction. This is reflected in the transition states **H** and **I** in terms of a polarization $^{\delta\oplus}C\cdots\cdots H\cdots\cdots Br^{\delta\ominus}$. A partial positive charge developing on the carbon atom suggests that substituents that destabilize this charge will slow the reaction. The effect of the CO_2Me group is negligible when it is removed from that reaction center as in transition state **I**. However, transition state **H** is energetically inaccessible and this allows other reactions to compete with the radical bromination.

This concept can be generalized. Radical substitution reactions that take place via *electrophilic* radicals ($Y_{electrophilic}$) will proceed through transition states in which positive charge develops at the carbon atom involved in the C-H abstraction. These are destabilized by neighboring electron acceptors, which decelerates or prevents this kind of substitution. The transition states of such substitution reactions are stabilized by substituents on the reacting carbon atom that stabilized positive charge, electron-donating substituents. In the extreme case, they can even occur without any explicit initiation. Such an effect is likely to contribute to the fact that ethers easily undergo autoxidation (Figures 1.37 and 1.38).

So what have we learned here? Both **C** and **D** lead to the same intermediate **E** upon hydrogen abstraction, but **H** is much higher in energy than **I**, allowing a new pathway to complete for **C** to the exclusion of **H**. Product development control does not work here and it should be known that, like many of our models in organic chemistry, it has its limits.

1.7.6 Radical Chain Chlorination Using Sulfuryl Chloride

In university laboratories, radical chlorinations are not usually conducted with the gaseous and highly aggressive Cl_2, but with the more easy-to-handle liquid sulfuryl chloride (SO_2Cl_2). An application example is given in Figure 1.32, which shows the chlorination of *tert*-butyl benzene. As expected, this reaction proceeds as a regioselective substitution at the C_{sp3}–H- and not at the C_{sp2}–H bonds. The mechanism of this substitution is somewhat more complex than the chlorination mechanism using Cl_2 (Figures 1.17 and 1.19): it includes three instead of two propagation steps (Figure 1.32), and the chain initiation comprises a three-step sequence instead of a single reaction.

In the first propagation step of the sulfuryl chloride reaction $\bullet SO_2Cl$ reacts as the initiating radical (Figure 1.32). It generates a C radical and reacts to furnish chlorosulfuric acid, which in the second propagation step decomposes to hydrogen chloride and sulfur dioxide. In the third propagation step, the C radical abstracts a chlorine atom from a sulfuryl chloride molecule. In this way, the substitution product and a new initiating radical $\bullet SO_2Cl$ are formed. As mentioned above, $\bullet SO_2Cl$ is the initiating radical for chlorinations with sulfuryl chloride (Figure 1.32) and Cl• for chlorinations with chlorine gas (Figures 1.17 and 1.19). As the product of a radical chlorination only depends on which C–H bond of the substrate reacts with the

proceeds via 3 propagation steps:

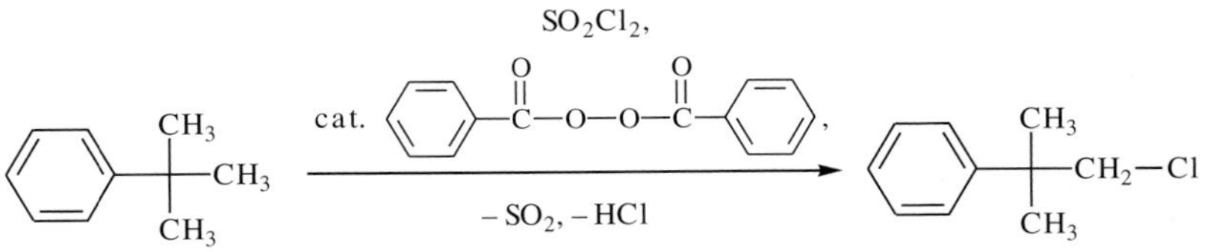

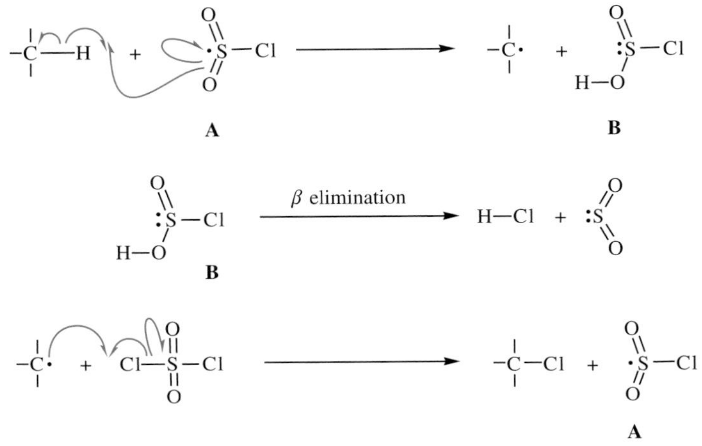

Fig. 1.32. Sulfuryl chloride (SO_2Cl_2) acts as a chlorinating reagent. Here, the chlorination of *tert*-butylbenzene is shown. To initiate the chain reaction catalytic amounts of dibenzoyl peroxide are used. Dibenzoyl peroxide undergoes multistep fragmentation to give the phenyl radical (cf. Figure 1.37), which abstracts a Cl atom from sulfuryl chloride and thus generates the initiating radical $\bullet SO_2Cl$.

initiating radical, different regio- and/or chemoselectivities can be achieved, depending on the chlorinating agent used. The potential differences will always reflect that $\bullet SO_2Cl$ is a less reactive initiating radical than Cl•. This might largely be due to the fact that it demands more space (steric hindrance) and, secondarily, that the unpaired electron is not localized on the S atom, but delocalized over the two O atoms (reactivity/selectivity principle). Different chemoselectivities can be observed, for example, with the polychlorination of toluene (compare Figure 1.33 to Figure 1.21). With an excess of sulfuryl chloride — irrespective of how large it is—toluene can (only) be chlorinated to benzal chloride ($PhCHCl_2$), whereas an excess of chlorine converts toluene into benzotrichloride.

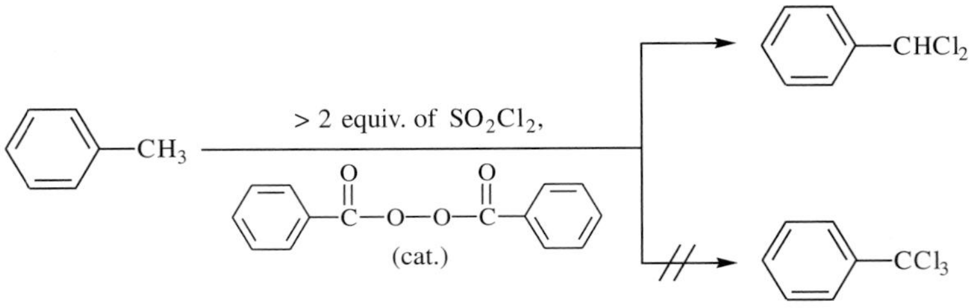

Fig. 1.33. An excess of sulfuryl chloride is used to chlorinate toluene to furnish benzal chloride ($PhCHCl_2$) (while benzotrichloride = $PhCCl_3$ is obtained with an excess of chlorine).

A dependence of the regioselectivity on the reagent used can be observed, for example, with the monochlorination of adamantane (Figure 1.34). SO_2Cl_2 and adamantane yield *one* monochlorination product, whereas Cl_2 and adamantane furnish *both monochlorination products*. However, the SO_2Cl_2 experiment brings a surprise as the monochlorination product is the *tert*-adamantyl chloride **A** in Figure 1.34. **A** must have been formed from the *tert*-adamantyl radical (**D**), which is by 2.5 kcal/mol less stable than the *sec*-adamantyl radical (**C**). Why does $\bullet SO_2Cl$ selectively form the *less* stable radical? When introducing the reactivity/selectivity principle in Section 1.7.4 reasons were given why the opposite is true, namely that less reactive and thus more selectively reacting radicals — $\bullet SO_2Cl$ *is* one — lead to the highly regioselective formation of the *most stable* C radical *possible*!

Side Note 1.3. Pyramidalized Bridgehead Radicals—and/or Less Substituted Planar Radicals as Intermediates of Radical Chlorinations

The preferential formation of **D** instead of **C** when $\bullet SO_2Cl$ is employed proves that a steric effect prevails here, upstaging the electronically based stability sequence **D** < **C**. Being a sterically demanding radical, $\bullet SO_2Cl$ can abstract only an H atom from adamantane that is acces-

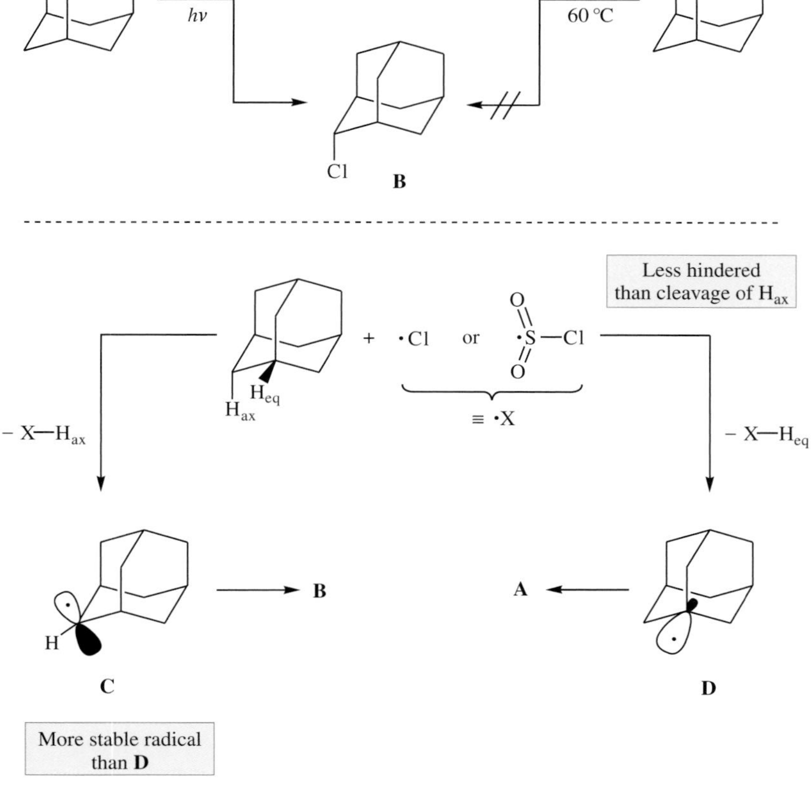

Fig. 1.34. Chlorination of adamantane to furnish *tert*-adamantyl chloride (**A**) and *sec*-adamantyl chloride (**B**), using differents reagents. In the reaction with Cl_2 the **A**:**B** selectivity amounts to 68:32 in carbon disulphide and 39:61 in carbon tetrachloride (each at 25°C). In the reaction with SO_2Cl_2 in sulfolan the **A**:**B** selectivity is > 97.5:2.5 (at 60°C). The formation of *sec*-adamantyl chloride (**B**) proceeds via the secondary radical **C**, *tert*-adamantyl chloride (**A**) is formed via the tertiary radical **D**.

sible to it, that is, is unhindered trajectory. Adamantane contains four configuratively fixed cyclohexane chairs. It is a common experience that the equatorial substituents in cyclohexane chairs are more easily accessible—they are on the convex face of the chair (for a definition of this term see Figure 10.11) — than the axial substituents, since the latter are on the concave face. As is emphasized in the lower part of Figure 1.34, the C_{tert}–H bonds are the equatorial, that is, reactive, C–H bonds of adamantane. *They* react with by $\bullet SO_2Cl$, leading to the formation of *tert*-adamantyl radicals (**D**) and, subsequently, of *tert*-adamantyl chloride (**A**). In contrast, the C_{sec}–H bonds of adamantane are axial C–H bonds and as such too hindered to react with $\bullet SO_2Cl$. This is why no *sec*-adamantyl radical (**C**) is formed and thus no *sec*-adamantyl chloride (**B**) either.

In order to fully understand Figure 1.34, we still need to account for the following order of stability: *tert*- < *sec*-adamantyl radical. It contradicts the rule that tertiary radicals are more stable than secondary ones. The latter statement holds true only if the respective radicals are able to adopt their preferential geometry, that is, if according to Section 1.1.1 they possess a planar or partially pyramidalized radical center. This is warranted with the *sec*-adamantyl radical (**C**); it is planar. The *tert*-adamantyl radical (**D**), though, is a bridgehead radical. A bridgehead radical is a tertiary radical, which — as a ring-linkage site — is clamped into a bi- or polycyclic ring system like into a corset. The rigid periphery forces such radical centers into a stereostructure that would be avoided by radicals of a mobile conformation. For this reason, the rigid *tert*-adamantyl radical (**D**) must display 80% of the pyramidalization of an ideal tetrahedral center, while the mobile *tert*-butyl radical prefers 40% pyramidalization. Despite its tertiary nature such a "forcefully pyramidalized" radical **D** becomes more energy-rich than the merely secondary but non-bridgehead radical **C**.

1.8 Autoxidations

Reactions of compounds with oxygen with the development of flames are called combustions. In addition, flameless reactions of organic compounds with oxygen are known. They are referred to as autoxidations. Of the autoxidations, only those that take place via sufficiently stable radical intermediates can afford pure compounds in good yields. Preparatively valuable autoxidations are therefore limited to substitution reactions of hydrogen atoms that are bound to tertiary, allylic, or benzylic carbon atoms. An example can be found in Figure 1.35. *Unintentional autoxidations* can unfortunately occur at the oxygenated carbon of ethers such as diethyl ether or tetrahydrofuran (THF) (Figure 1.36).

Side Note 1.4. An Industrially Important Autoxidation

The most important autoxidation used industrially is the synthesis of cumene hydroperoxide from cumene and air (i.e., "diluted" oxygen) (Figure 1.37). It is initiated by catalytic amounts of dibenzoyl peroxide as the radical initiator (cf. Figure 1.11). The cumyl radical is produced

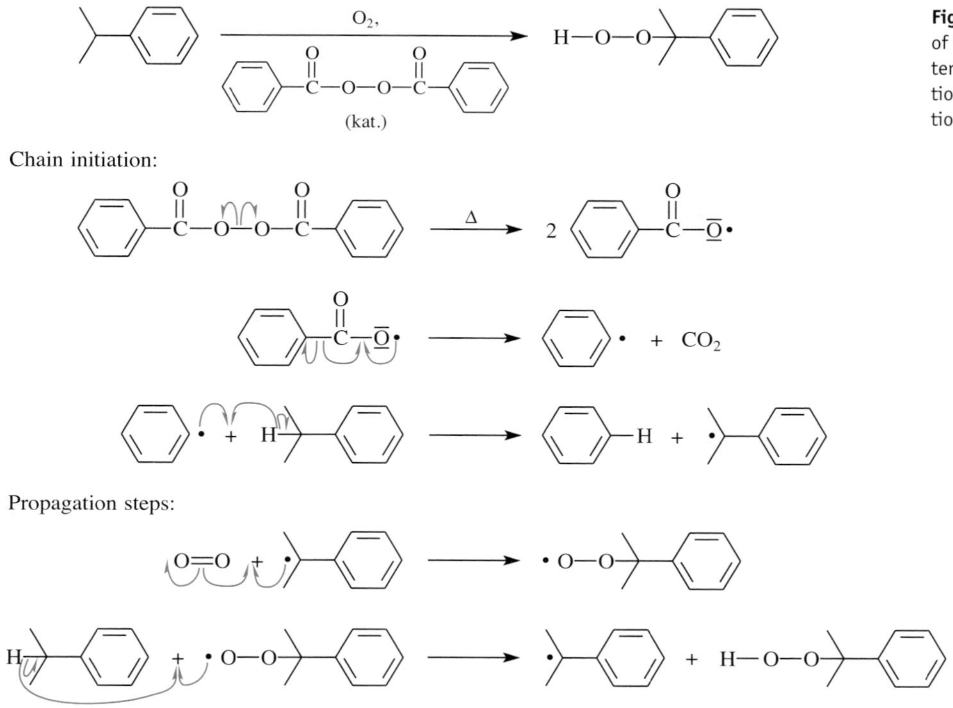

O_2,

(kat.)

Chain initiation:

Propagation steps:

Fig. 1.35. Industrial synthesis of cumene hydroperoxide. In terms of Figure 1.2, this reaction is classified as a "substitution including an addition."

as the initiating radical from a sequence of three starting reactions. It is a tertiary radical, which is additionally stabilized by resonance. The cumyl radical is consumed in the first propagation step of this autoxidation and is regenerated in the second propagation step. These steps alternate until there is a chain termination. The autoxidation product, cumene hydroperoxide, is a useful reagent, but is more important as the starting material in the cumene hydroperoxide rearrangement (see Section 14.4.1), which produces phenol and acetone. Worldwide, 90% of each of these industrially important chemicals is synthesized by this process.

Two ethers that are frequently used as solvents are relatively easy to autoxidize — unfortunately, because this reaction is not carried out intentionally. Diethyl ether and THF form hydroperoxides through a substitution reaction in the α position to the oxygen atom (Figure 1.36). These hydroperoxides are fairly stable in dilute solution. However, they are highly explosive in more concentrated solutions or in pure form. Concentration of a solution of an ether hydroperoxide on a rotatory evaporator can lead to a really big explosion…be careful to use pure ethers.

What makes the α position of these ethers autoxidizable? One reason is that the intermediates in this process are α-oxygenated radicals, which are stabilized by the free electron pairs on the heteroatom (see Table 1.3 and Figure 1.7). However, this factor is probably not *critical*

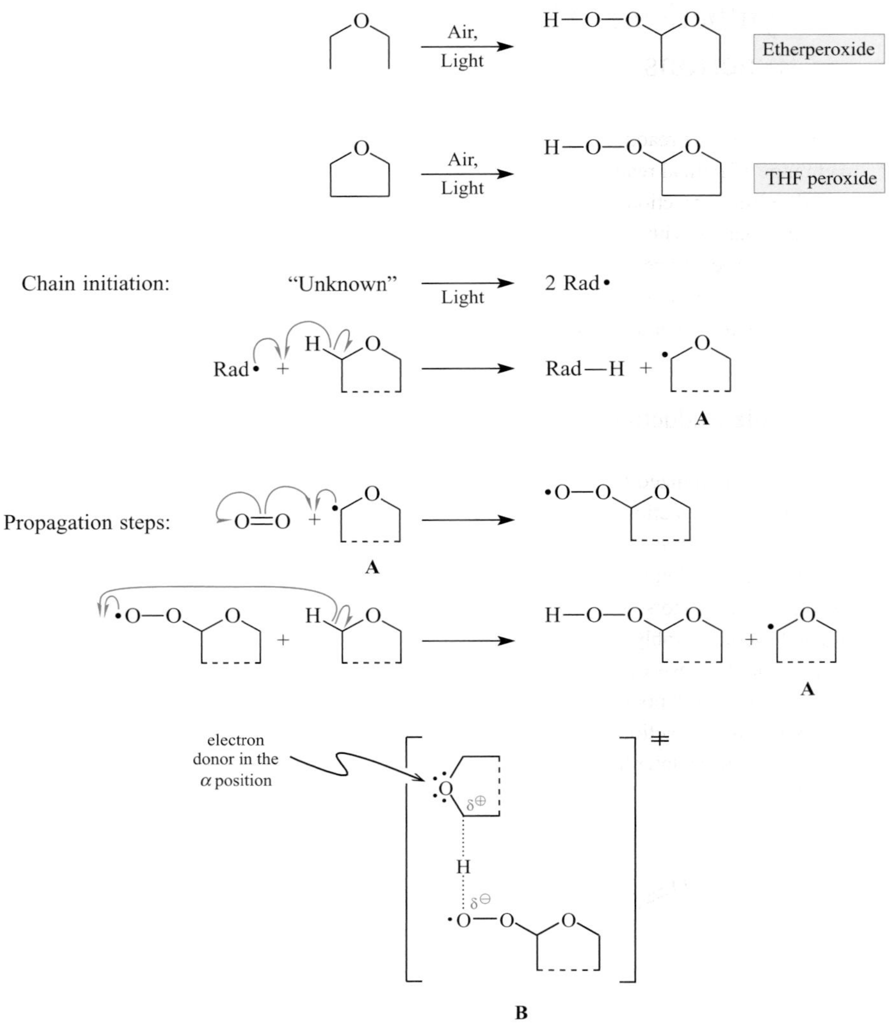

Air, Light → Etherperoxide

Air, Light → THF peroxide

Chain initiation: "Unknown" → (Light) 2 Rad•

Rad• + H-THF → Rad—H + A

Propagation steps: O=O + A → •O—O-THF

•O—O-THF + H-THF → H—O—O-THF + A

electron donor in the α position

B

Fig. 1.36. Autoxidation of diethyl ether and THF: net equations (top) and mechanism (bottom). In terms of Figure 1.2 it is a "substitution including an addition." The mechanism is given in the middle, the electronic effect in the rate- and selectivity-determining step at the bottom.

for the easy autoxidizability of the cited ethers. Rather, a related electronic effect is important. As shown in transition state **B** of the H atom transfer (Figure 1.36), positive charge character builds on the carbon undergoing hydrogen abstraction. $^{\delta\oplus}$C······±-H······$^{\delta\ominus}$OOR. The partial positive charge is stabilized through the oxygen directly bound to it. Related to this is the fact that overlap of the lone pairs of electrons of the oxygen with the unoccupied orbitals (σ*) of the adjacent C-H bond(s) weaken that bond, and make hydrogen abstraction easier.

1.9 Synthetically Useful Radical Substitution Reactions

Although some of the reactions we have examined are certainly synthetically useful, there are a broad range of radical reactions that extend well beyond things like simple benzylic bromination. All of these reactions proceed by essentially the same mechanism: a radical chain process. Though details with respect to how a chain might initiate and propagate certainly differ among different reactions, at the most basic level there is no difference. Recognizing this is a key to understanding not just radical reactions, but all organic chemistry. There are lots of reagents, but many common mechanisms.

1.9.1 Simple Reductions

A series of functionalized hydrocarbons can be defunctionalized (reduced) with the help of radical substitution reactions. The functional group is then replaced by a hydrogen atom. The groups that can be removed in this way include iodide, bromide, and some sulfur-containing alcohol derivatives. Compounds with an easily cleaved hydrogen–element (H-X) bond serve as hydrogen atom donors. The standard reagent is Bu_3Sn–H. A substitute that does not contain tin and is considerably less toxic is $(Me_3Si)_3Si$—H. Reactions of this type are usually carried out for the synthesis of a hydrocarbon or to produce a hydrocarbon-like substructure. Figure 1.37 illustrates the possibilities of this method using a deiodination and a debromination as examples. These reactions represent general synthetic methods for obtaining cyclic esters or ethers. We will see later how easy it is to prepare halides like those shown from alkene precursors (Figure 3.47).

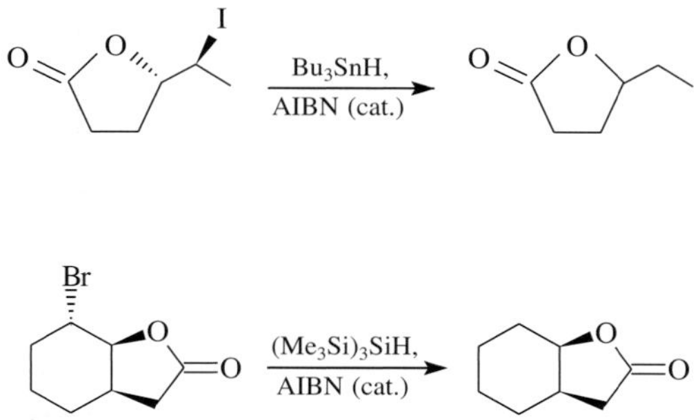

Bu_3SnH, AIBN (cat.)

$(Me_3Si)_3SiH$, AIBN (cat.)

Fig. 1.37. Dehalogenations through radical substitution reactions. (Both substrates are racemic when they are prepared according to Figure 3.47.)

Both in radical reductions effected with Bu_3SnH and in those carried out with $(Me_3Si)_3SiH$, the radical formation is initiated by the radical initiator AIBN (Figure 1.38). The initiation sequence begins with the decomposition of AIBN, which is triggered by heating or by irradiation with light, into the cyanated isopropyl radical. In the second step of the initiation sequence, the cyanated isopropyl radical produces the respective initiating radical; that is, it converts Bu_3SnH into Bu_3Sn• and $(Me_3Si)_3SiH$ into $(Me_3Si)_3Si$•. The initiating radical gets the actual reaction chain going, which in each case consists of two propagation steps.

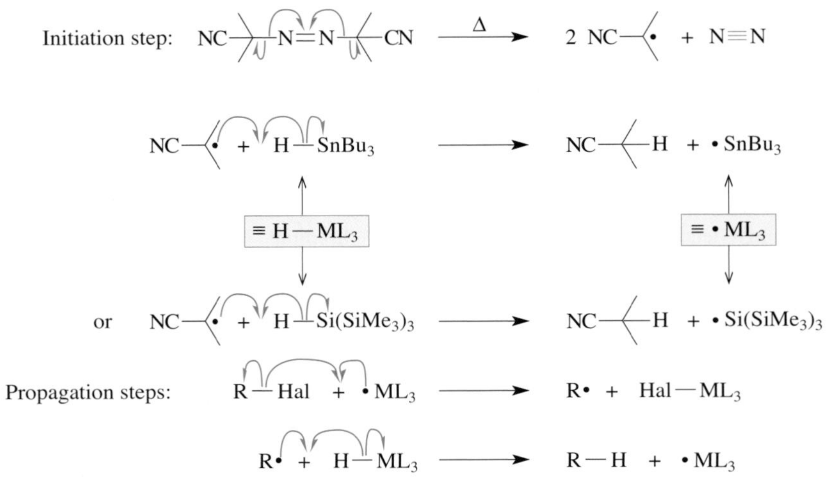

Initiation step: NC–C(Me)₂–N=N–C(Me)₂–CN →(Δ) 2 NC–C(Me)₂• + N≡N

NC–C(Me)₂• + H–SnBu₃ → NC–C(Me)₂–H + •SnBu₃

≡ H–ML₃ ≡ •ML₃

or NC–C(Me)₂• + H–Si(SiMe₃)₃ → NC–C(Me)₂–H + •Si(SiMe₃)₃

Propagation steps: R–Hal + •ML₃ → R• + Hal–ML₃

R• + H–ML₃ → R–H + •ML₃

Fig. 1.38. Mechanism of the radical dehalogenations of Figure 1.37.

Both Bu_3SnH and $(Me_3Si)_3SiH$ are able to reduce alkyl iodides or bromides but not alcohols. However, in the Barton–McCombie reaction, they reduce certain alcohol *derivatives*, namely, ones that contain a C=S double bond (e. g., thiocarboxylic esters or thiocarbonic esters). Figure 1.39 shows how the OH group of cholesterol can be removed by means of a Barton–McCombie reaction. The C=S-containing alcohol derivative used there is a xanthate.

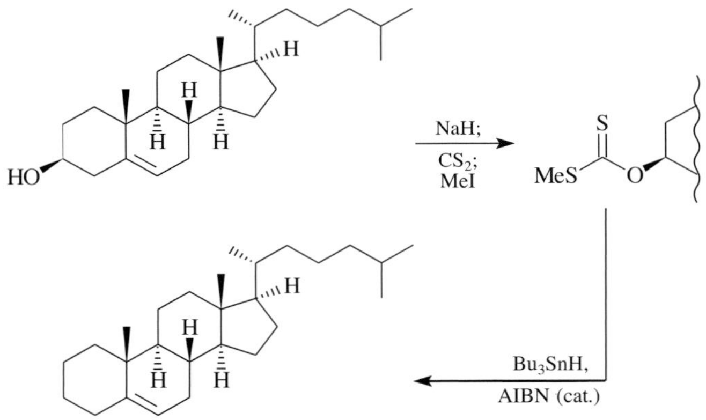

NaH; CS₂; MeI

MeS–C(=S)–O–

Bu₃SnH, AIBN (cat.)

Fig. 1.39. Reduction of an alcohol by means of the radical substitution reaction of Barton and McCombie. In terms of Figure 1.2 a "substitution including a fragmentation" occurs.

The starting sequence of this reduction is identical to the one that was used in Figure 1.39 and leads to Bu_3Sn• radicals. These radicals enter into the actual reaction chain, which consists of three propagation steps (Figure 1.40). The Bu_3Sn radical is a thiophile; that is, it likes to combine with sulfur. Thus in the first propagation step, the tin radical bonds to the sulfur atom of the xanthate. The second propagation step is a radical fragmentation.

Other C=S-containing esters derived from alcohols can also be defunctionalized in a similar fashion. Imidazolylthiocarbonic esters, which in contrast to xanthates can be synthesized under neutral conditions, are one example. This is important for the reduction of base-sensitive alcohols. Figure 1.41 shows a reaction of this type. If the corresponding xanthate was pre-

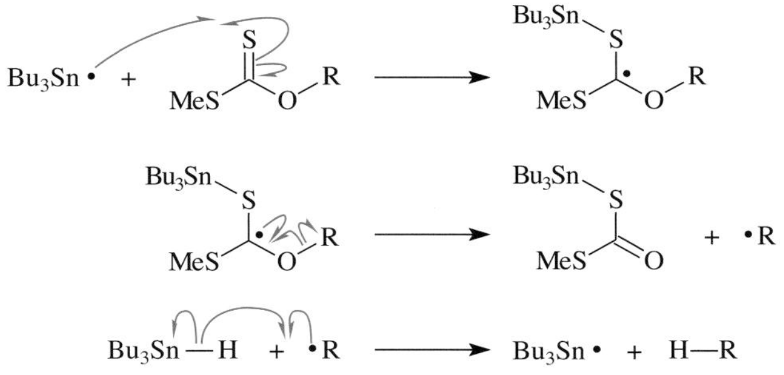

Fig. 1.40 Propagation steps of the Barton–McCombie reaction in Figure 1.39.

pared through consecutive reactions with NaH, CS_2, and MeI, this compound would lose its *cis*-configuration. This would occur because of the presence of the keto group. It would undergo a reversible deprotonation giving the enolate (Figure 1.41). The latter would be reprotonated in part—actually even preferentially — to form the *trans* isomer.

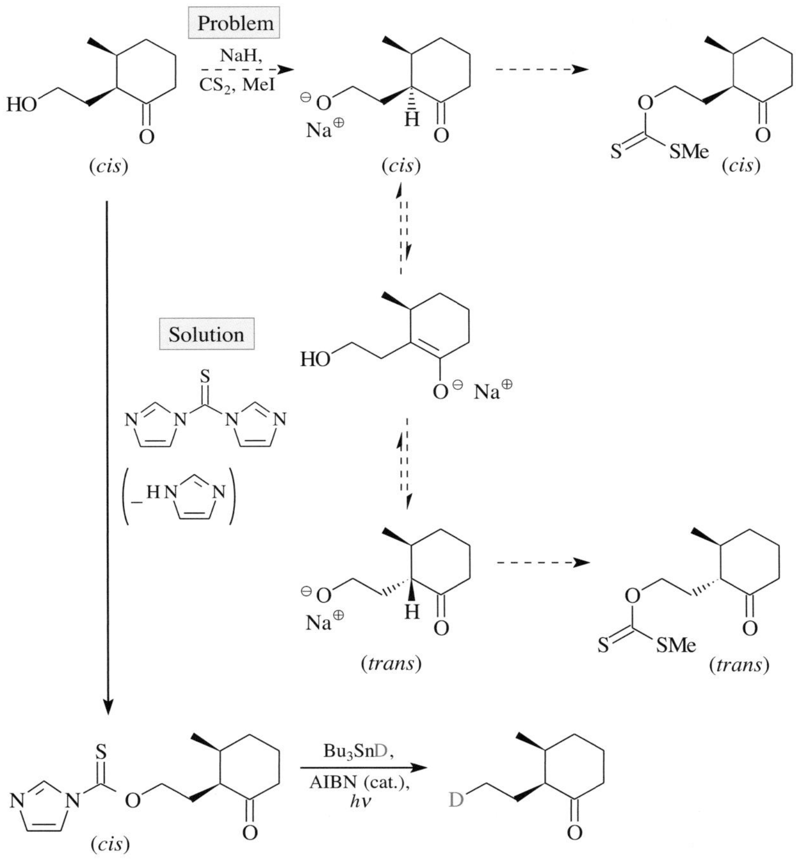

Fig. 1.41. Deoxygenation/deuteration of alcohols via a radical substitution reaction.

The starting sequence of the Barton–McCombie reaction in Figure 1.41 is again identical to the one in Figure 1.39. However, because now Bu_3SnD is the reducing agent instead of Bu_3SnH, D is incorporated into the product instead of H.

1.9.2 Formation of 5-Hexenyl Radicals: Competing Cyclopentane Formation

The reductions discussed in Section 1.9.1, considered in a different way, are also reactions for producing radicals that might engage in chemistry beyond H atom abstraction. Certain radicals ordinarily cannot be reduced to the corresponding hydrocarbon: the 5-hexenyl radical and its derivatives. These radicals often cyclize (irreversibly) before they abstract a hydrogen atom from the reducing agent. By this cyclization, the cyclopentylmethyl radical or a derivative of it is produced selectively. An isomeric cyclohexyl radical or the corresponding derivative is practically never obtained. Often only the cyclized radical abstracts a hydrogen atom from the reducing agent. Figure 1.42 gives an example of a reaction of this type in the form of a cyclopentane annulation.

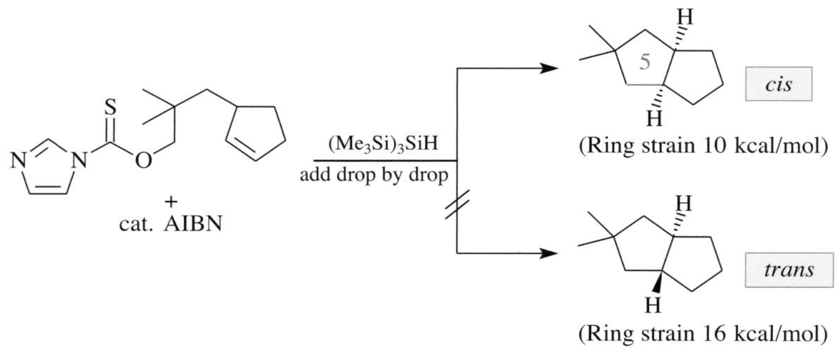

Fig. 1.42. Cyclization of a 5-hexenyl radical intermediate from a Barton–McCombie defunctionalization as a method for cyclopentane annulation (cyclizing defunctionalization, in terms of Figure 1.2 this means a "substitution including a fragmentation and an addition").

The precursor to the cyclizable radical is again (cf. Figure 1.43) a thiocarbonic acid imidazolide, as shown in Figure 1.44. The reducing agent is $(Me_3Si)_3SiH$, but one also could have used Bu_3SnH. AIBN functions as the radical initiator. After the usual initiation sequence (Figure 1.40, formula lines 1 and 3) the initiating radical $(Me_3Si)_3Si\bullet$ is available, and a sequence of four propagation steps is completed (Figure 1.43). In the second propagation step, the 5-hexenyl radical is produced, and in the third step it is cyclized. The cyclization stereoselectively leads to a *cis*- instead of a *trans*-annulated bicyclic system. The reason for this is that the preferentially formed *cis*-radical has less (by approx. 6 kcal/mol) ring strain than its *trans*-isomer. Yet, the *cis*-selectivity is due to kinetic rather than thermodynamic control. Therefore, it is a consequence of product-development control.

It is interesting that one can also use the same reagents to reduce the same thiocarbonic acid derivative *without cyclization* (Figure 1.44). To do this, one simply changes the sequence in which the reagents are added according to Figure 1.44. There is no cyclization when the substrate/AIBN mixture is added dropwise to an excess of the reducing agent. A relatively concentrated solution of the reducing agent is then always available as a reaction partner for the substrate and the radicals derived from it. According to Figure 1.44, on the other hand,

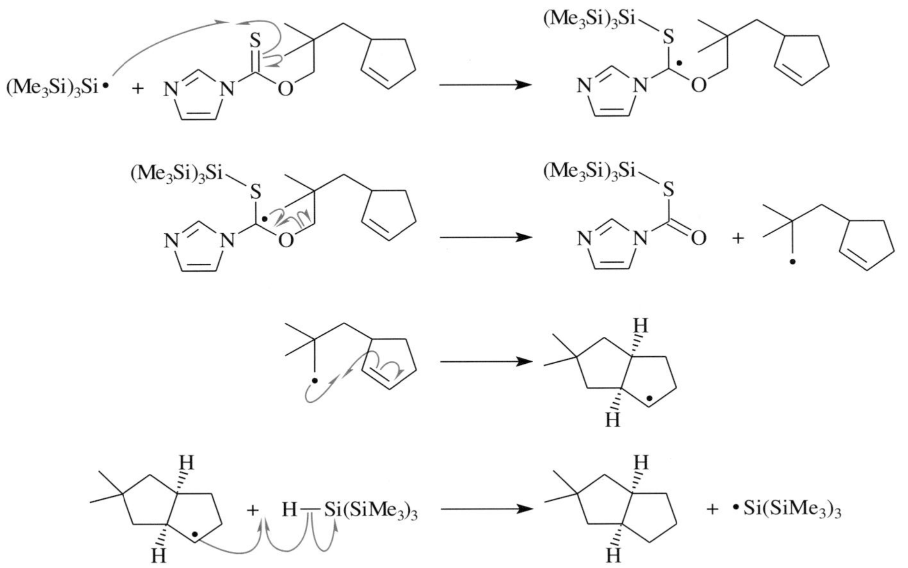

Fig. 1.43. Propagation steps in the radical substitution/cyclization of Figure 1.42.

cyclization predominates when the reducing agent/AIBN mixture is added to the substrate dropwise over several hours. In this, way the substrate and the derived radicals are exposed to only an extremely dilute solution of the reducing agent during the entire reaction.

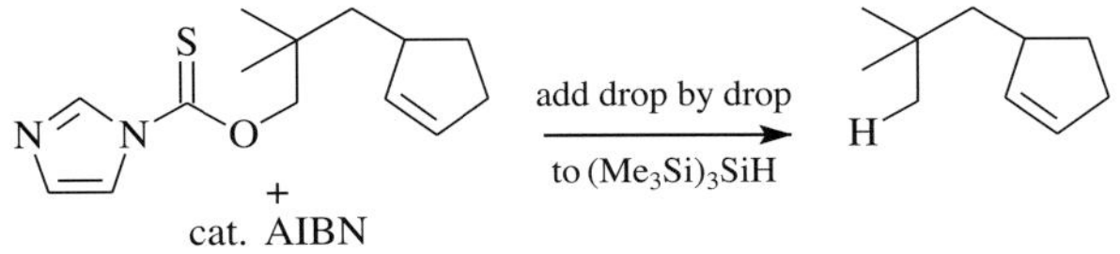

Fig. 1.44. Cyclization-free defunctionalization of the thiocarbonic acid derivative from Figure 1.42.

According to Figure 1.45, the question whether cyclization takes place or not is decided as soon as the C—O bond of the substrate is cleaved and radical **A** is present. Radical **A** either cyclizes (at a rate that is equal to the product of the rate constant k_{cycl} and the concentration of **A**) or it reacts with the silane. In the latter case, the rate is the product of the rate constant $k_{\sim\text{H}\rightarrow prim}$, the concentration of the radical **A**, and the concentration of the silane. Let us assume that the concentration of the silane does not change during the reduction. This assumption is not correct, but sufficiently accurate for the present purpose. Then the ratio of the rates of the two alternative reactions of the hexenyl radical **A** is equal to the yield ratio of the cyclized and the non-cyclized reaction products.

$$\frac{\text{\% Cyclization to the diquinane}}{\text{\% Reduction to cyclopentene}} = \frac{k_{\text{cycl}}}{k_{\sim H \rightarrow prim}} \cdot \frac{1}{[(Me_3Si)_3SiH]_{\text{quasistationary}}} \quad (1.15)$$

According to Equation 1.15, the yield ratio of the two reduction products depends on a single variable, namely, the concentration of the reducing agent. This is in the denominator of Equation 1.15, which means two things: (1) when the radical intermediate **A** encounters a small

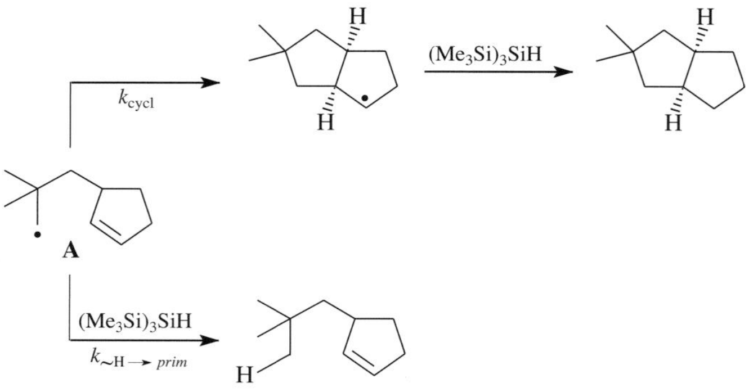

Fig. 1.45. Reaction scheme for the kinetic analysis of the complementary chemoselectivity of the cyclizing defunctionalization (top reaction) of Figure 1.42 and the noncyclizing defunctionalization (bottom reaction) of Figure 1.44.

amount of reducing agent, the cyclized product is produced preferentially (see Figure 1.42) and (2) when the same radical encounters a large amount of reducing agent, the cyclopentane is preferentially formed (see Figure 1.44). The most important lesson, one that can be applied to many other reaction processes that are under kinetic control, is that it is all about relative rates. Understanding kinetics of reactions, even just qualitatively, can help you design reactions or solve reaction problems.

1.10 Diazene Fragmentations as Novel Alkane Syntheses

One problem with the reduction of alcohols according to the procedure developed by Barton and McCombie (Figure 1.39–1.42) is the reducing agent: the standard reducing agent Bu_3SnH is toxic, and the alternative reagent, $(Me_3Si)_3SiH$, is very expensive. This sparked researchers' ambition to develop other methods for the transformation R–OH → R–H. For example, the interesting four-component reaction R_{prim}–OH + **A** + **B** + **C** → R_{prim}–H of Figure 1.46 can be used to reduce primary alcohols. This process involves many individual events, some of which shall be highlighted as follows:

1. The sulfonyl hydrazide **D** is obtained through a Mukaiyama redox condensation while consuming compounds **A**–**C** (cf. Figure 2.38).
2. β-Elimination (for the term see Section 4.1.1) of *ortho*-nitrobenzene sulfinic acid (**F**) leads from **D** to the N=N double bond of compound **E**. Because of the functional group C–N=N–H **E** is called a diazene. Diazenes assume a middle position between azo compounds, which contain a C–N=N–C group, and the NH compound diimide (sometimes called diimine; H–N=N–H).
3. Like diimide, diazenes are unstable at room temperature. They decompose while releasing elemental nitrogen. The decomposition mechanism is a radical chain reaction.

4. The starting reaction and the propagation steps of the fragmentation R_{prim}–N=N–H → R_{prim}–H + N≡N in Figure 1.46 proceed in complete analogy to the corresponding steps of the demercuration RCH(OH)–Hg–H → RCH(OH)–H + Hg in Figure 1.14.

But **diazenes** R–CH_2–N=N–H do not only occur as the intermediates **E** of the deoxygenation of primary alcohols according to the procedure outlined in Figure 1.46. They are also formed in the course of the two-step aldehyde → alkane transformation discussed in Figure 17.70. There, the diazene forming step is R–CH=N–NH–Ts + $NaBH_4$(MeOH) → R–CH_2–N=N–H. The latter species subsequently proceeds along the same route as **E** as shown in Figure 1.46.

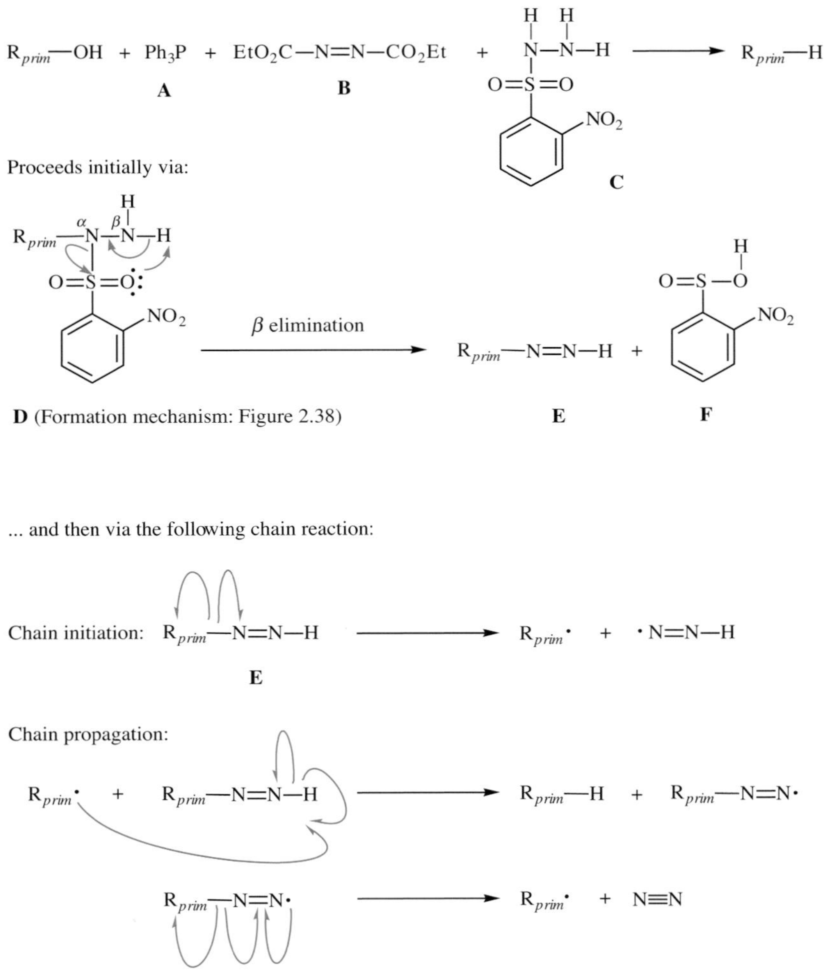

R_{prim}—OH + Ph_3P + EtO_2C—N=N—CO_2Et + C → R_{prim}—H

A **B** **C**

Proceeds initially via:

D (Formation mechanism: Figure 2.38) —β elimination→ R_{prim}—N=N—H (**E**) + **F**

... and then via the following chain reaction:

Chain initiation: R_{prim}—N=N—H (**E**) → $R_{prim}\cdot$ + $\cdot$N=N—H

Chain propagation:

$R_{prim}\cdot$ + R_{prim}—N=N—H → R_{prim}—H + R_{prim}—N=N$\cdot$

R_{prim}—N=N$\cdot$ → $R_{prim}\cdot$ + N≡N

Fig. 1.46. One-step defunctionalizations of primary alcohols. The diazenes R_{prim}–N=N–H (**E**) formed as intermediates decompose in a chain reaction, which, according to Figure 1.2, corresponds to a "substitution reaction by fragmentation." Apart from this radical decomposition (further examples see Figure 1.49 and 17.70), diazenes are also able to release nitrogen via (carb)anionic intermediates (examples see Figures 17.67–17.69) or in a single step (example see Figure 17.71).

Side Note 1.5. An (Almost) Universal C,C Bond-Forming Alkane Synthesis

"Wherever you go, there you are". Those possessing the wisdom of Yogi Berra might have recognized that the generation of a diazene from the decomposition reactions in Figure 1.46 and 17.70 as being equivalent to the generation of the corresponding alkane—and so it is! This insight is at the bottom of Myers' synthesis of alkanes from aldehydes (Figure 1.47). It involves three steps: the formation of a hydrazone (→ **B**), a hydrazone silylation (→ **C**) and an RLi-addition followed by protonation (→ sulfonyl hydrazine **H**). As soon as the latter species is formed, the same reactions that in Figure 1.46 relate to the intermediate **D** are started without any further active intervention of the chemist. These include the β-elimination **H** → **E** + **G**, the desilylation **E** → **D** → **F** and, finally, the "radical substitution by fragmentation" **F** → **I**.

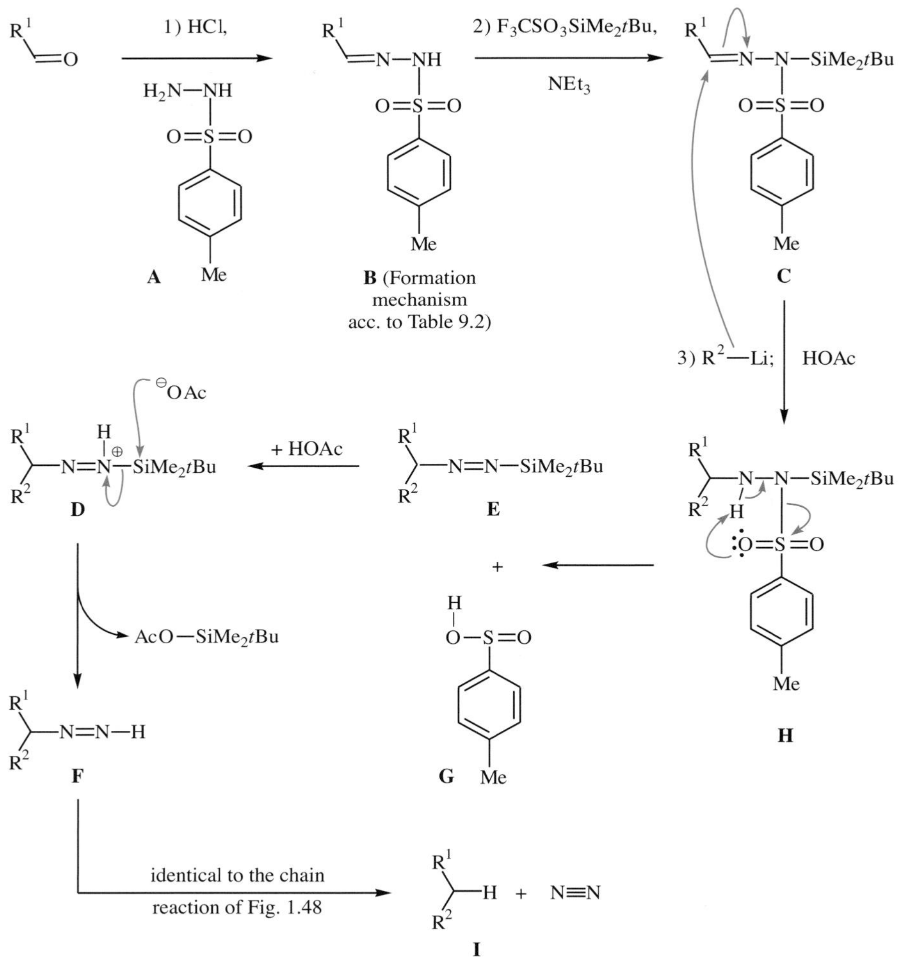

Fig. 1.47. Myers' synthesis of alkanes from aldehydes via the sulfonylated hydrazones **B** and their silyl derivatives **C**. This procedure provides a skeleton-generating three-step synthesis of alkanes from aldehydes. In terms of Figure 1.2, the chain reaction **F** → **I** involved is a "substitution reaction by fragmentation." (The sulfonylated hydrazide anion, which here results from the addition of R_3–Li to the sulfonyl hydrazone **C**, does—at this reaction's temperature of −78 °C—not release toluene sulfinate. Compare the different behavior of the sulfonylated hydrazide anion **C** in the reaction of Figure 17.69, where a mesitylene sulfinate residue is eliminated at a temperature that is higher by 160 °C).

You should be able to draw a mechanism for any radical chain reaction. You should recognize at least some radical initiators and appreciate when a radical reaction is taking place. More difficult, but no less important, is to begin to appreciate the occurrence of radicals in reactions where you might not expect them. For example, treatment of benzophenone with LDA should result in no reaction, right? There are no protons available to form an enolate. But the reaction results in the formation of the corresponding alcohol. LDA can reduce benzophenone. How is this possible? Electron transfer. An electron is transferred from the LDA to benzophenone to form a ketyl radical. Hydrogen abstraction by the ketyl from LDA gives the alkoxide and a new species also capable of electron transfer.

So radical intermediates can be accessed from full valence compounds. Keep electron transfer in mind when thinking about possible reaction mechanisms for a process.

References

B. Giese, "C-Radicals: General Introduction," in *Methoden Org. Chem.* (Houben-Weyl) 4th ed. **1952**, *C-Radicals* (M. Regitz, B. Giese, Eds.), Bd. E19a, 1, Georg Thieme Verlag, Stuttgart, **1989**.

W. B. Motherwell, D. Crich, "Free-Radical Chain Reactions in Organic Chemistry," Academic Press, San Diego, CA, **1991**.

J. E. Leffler, "An Introduction to Free Radicals," Wiley, New York, **1993**.

M. J. Perkins, "Radical Chemistry," Ellis Horwood, London, **1994**.

J. Fossey, D. Lefort, J. Sorba, "Free Radicals in Organic Chemistry," Wiley, Chichester, U.K., **1995**.

Z. B. Alfassi (Ed.), "General Aspects of the Chemistry of Radicals," Wiley, Chichester, U. K., **1999**.

Z. B. Alfassi, "The Chemistry of N-Centered Radicals," Wiley, New York, **1998**.

J. Hartung, T. Gottwald, K. Spehar, "Selectivity in the Chemistry of Oxygen-Centered Radicals—The Formation of Carbon-Oxygen Bonds," *Synthesis* **2002**, 1469–1498.

Z. B. Alfassi (Ed.), "S-Centered Radicals," Wiley, Chichester, U. K., **1999**.

P. P. Power, "Persistent and Stable Radicals of the Heavier Main Group Elements and Related Species," *Chem. Rev.* **2003**, *103,* 789–809.

A. F. Parsons, "An Introduction to Free Radical Chemistry," Blackwell Science, Oxford, **2000**.

1.1

R. S. Nyholm, R. J. Gillespie, "Inorganic Stereochemistry," *Q. Rev. Chem. Soc.* **1957**, *11*, 339.

R. J. Gillespie, "Electron-Pair Repulsions and Molecular Shape," *Angew. Chem. Int. Ed. Engl.* **1967**, *6*, 819–830.

R. J. Gillespie, "Electron-Pair Repulsion Model for Molecular Geometry," *J. Chem. Educ.* **1970**, *47*, 18.

R. J. Gillespie, "Molecular Geometry," Van Nostrand Reinhold Co., London, **1972**.

R. J. Gillespie, "A Defense of the Valence Shell Electron Pair Repulsion (VSEPR) Model," *J. Chem. Educ.* **1974**, *51*, 367.

R. J. Gillespie, "Molekülgeometrie," Verlag Chemie, Weinheim, **1975**.

R. J. Gillespie, E. A. Robinson, "Electron Domains and the VSEPR Model of Molecular Geometry," *Angew. Chem. Int. Ed. Engl.* **1996**, *35*, 495–514.

R. Ahlrichs, "Gillespie- und Pauling-Modell – ein Vergleich," *Chem. unserer Zeit*, **1980**, *14*, 18–24.

M. Kaupp, "'Non-VSEPR' Structures and Bonding in d^0 Systems," *Angew. Chem. Int. Ed. Engl.* **2001**, *40*, 3534–3565.

1.2

C. Rüchardt, H.-D. Beckhaus, "Steric and Electronic Substituent Effects on the Carbon-Carbon Bond," *Top. Curr. Chem.* **1985**, *130*, 1–22.

D. D. M. Wayner, D. Griller, "Free Radical Thermochemistry," in *Adv. Free Radical Chem.* (D. D. Tanner, Ed.), Vol. 1, Jai Press, Inc., Greenwich, CT, **1990**.

D. Gutman, "The Controversial Heat of Formation of the *tert*-C_4H_9 Radical and the Tertiary Carbon-Hydrogen Bond Energy," *Acc. Chem. Res.* **1990**, *23*, 375–380.

J. C. Walton, "Bridgehead Radicals," *Chem. Soc. Rev.* **1992**, *21*, 105–112.

W. Tsang, "Heats of Formation of Organic Free Radicals by Kinetic Methods," in: "Energetics of Organic Free Radicals," J. A. M. Simoes, A. Greenberg, J. F. Liebman (Eds.), Blackie Academic & Professional, Glasgow, **1996**, 22–58.

A. Studer, S. Amrein, "Tin Hydride Substitutes in Reductive Radical Chain Reactions," *Synthesis* **2002**, 835–849.

1.3

G. S. Hammond, "A Correlation of Reaction Rates," *J. Am. Chem. Soc.* **1955**, *77*, 334–338.

D. Farcasiu, "The Use and Misuse of the Hammond Postulate," *J. Chem. Educ.* **1975**, *52*, 76–79.

B. Giese, "Basis and Limitations of the Reactivity-Selectivity Principle," *Angew. Chem. Int. Ed. Engl.* **1977**, *16*, 125–136.

H. Mayr, A. R. Ofial, "The Reactivity-Selectivity Principle: An Imperishable Myth In Organic Chemistry," *Angew. Chem. Int. Ed. Engl.* **2006**, *45*, 1844 - 1854.

E. Grunwald, "Reaction Coordinates and Structure/Energy Relationships," *Progr. Phys. Org. Chem.* **1990**, *17*, 55–105.

A. L. J. Beckwith, "The Pursuit of Selectivity in Radical Reactions," *Chem. Soc. Rev.* **1993**, *22*, 143–161.

1.5

J. O. Metzger, "Generation of Radicals," in *Methoden Org. Chem.* (Houben-Weyl) 4th ed. **1952**, *C-Radicals* (M. Regitz, B. Giese, Eds.), Vol. E19a, 60, Georg Thieme Verlag, Stuttgart, **1989**.

H. Sidebottom, J. Franklin, "The Atmospheric Fate and Impact of Hydrochlorofluorocarbons and Chlorinated Solvents," *Pure Appl. Chem.* **1996**, *68*, 1757–1769.

M. J. Molina, "Polar Ozone Depletion (Nobel Lecture)," *Angew. Chem. Int. Ed. Engl.* **1996**, *35*, 1778–1785.

F. S. Rowland, "Stratospheric Ozone Depletion by Chlorofluorocarbons (Nobel Lecture)," *Angew. Chem. Int. Ed. Engl.* **1996**, *35*, 1786–1798.

1.6

R. C. Larock, "Organomercury Compounds in Organic Synthesis," *Angew. Chem. Int. Ed. Engl.* **1978**, *17*, 27–37.

G. A. Russell, "Free Radical Chain Reactions Involving Alkyl- and Alkenylmercurials," *Acc. Chem. Res.* **1989**, *22*, 1–8.

G. A. Russell, "Free Radical Reactions Involving Saturated and Unsaturated Alkylmercurials," in *Advances in Free Radical Chemistry* (D. D. Tanner, Ed.), **1990**, *1*, Jai Press, Greenwich, CT.

1.7

J. O. Metzger, "Reactions of Radicals with Formation of C,Halogen-Bond," in *Methoden Org. Chem.* (Houben-Weyl) 4th ed. **1952**–, *C-Radicals* (M. Regitz, B. Giese, Eds.), Vol. E19a, 268, Georg Thieme Verlag, Stuttgart, **1989**.

K. U. Ingold, J. Lusztyk, K. D. Raner, "The Unusual and the Unexpected in an Old Reaction. The Photochlorination of Alkanes with Molecular Chlorine in Solution," *Acc. Chem. Res.* **1990**, *23*, 219–225.

H. A. Michelsen, "The Reaction of Cl with CH_4: A Connection Between Kinetics and Dynamics," *Acc. Chem. Res.* **2001**, *34*, 331–337.

1.8

J. O. Metzger, "Reactions of Radicals with Formation of C,O-Bond," in *Methoden Org. Chem.* (Houben-Weyl) 4th ed. **1952**–, *C-Radicals* (M. Regitz, B. Giese, Eds.), Vol. E19a, 383, Georg Thieme Verlag, Stuttgart, **1989**.

W. W. Pritzkow, V. Y. Suprun, "Reactivity of Hydrocarbons and Their Individual C-H Bonds in Respect to Oxidation Processes Including Peroxy Radicals," *Russ. Chem. Rev.* **1996**, *65*, 497–503.

Z. Alfassi, "Peroxy Radicals," Wiley, New York, **1997**.

1.9

J. O. Metzger, "Reactions of Radicals with Formation of C,H-Bond," in *Methoden Org. Chem.* (Houben-Weyl) 4th ed. **1952**–, *C-Radicals* (M. Regitz, B. Giese, Eds.), Vol. E19a, 147, Georg Thieme Verlag, Stuttgart, **1989**.

S. W. McCombie, "Reduction of Saturated Alcohols and Amines to Alkanes," in *Comprehensive Organic Synthesis* (B. M. Trost, I. Fleming, Eds.), Vol. 8, 811, Pergamon Press, Oxford, **1991**.

A. Ghosez, B. Giese, T. Göbel, H. Zipse, "Formation of C-H Bonds via Radical Reactions," in *Stereoselective Synthesis* (Houben-Weyl) 4th ed. **1996**, (G. Helmchen, R. W. Hoffmann, J. Mulzer, E. Schaumann, Eds.), **1996**, Vol. E 21 (Workbench Edition), *7*, 3913–3944, Georg Thieme Verlag, Stuttgart.

C. Chatgilialoglu, "Organosilanes as Radical-Based Reducing Agents in Synthesis," *Acc. Chem. Res.* **1992**, *25*, 180–194.

V. Ponec, "Selective De-Oxygenation of Organic Compounds," *Rec. Trav. Chim. Pays-Bas* **1996**, *115*, 451–455.

S. Z. Zard, "On the Trail of Xanthates: Some New Chemistry from an Old Functional Group," *Angew. Chem. Int. Ed. Engl.* **1997**, *36*, 672–685.

C. Chatgilialoglu, M. Newcomb, "Hydrogen Donor Abilities of the Group 14 Hydrides," *Adv. Organomet. Chem.* **1999**, *44*, 67–112.

P. A. Baguley, J. C. Walton, "Flight from the Tyranny of Tin: The Quest for Practical Radical Sources Free from Metal Encumbrances," *Angew. Chem. Int. Ed. Engl.* **1998**, *37*, 3072–3082.

A. Studer, S. Amrein, "Tin Hydride Substitutes in Reductive Radical Chain Reactions," *Synthesis* **2002**, 835–849.

Further Reading

A. Ghosez, B. Giese, H. Zipse, W. Mehl, "Reactions of Radicals with Formation of a C,C-Bond," in *Methoden Org. Chem.* (Houben-Weyl) 4th ed. **1952**–, *C-Radicals* (M. Regitz, B. Giese, Eds.), Vol. E19a, 533, Georg Thieme Verlag, Stuttgart, **1989**.

M. Braun, "Radical Reactions for Carbon-Carbon Bond Formation," in *Organic Synthesis Highlights* (J. Mulzer, H.-J. Altenbach, M. Braun, K. Krohn, H.-U. Reißig, Eds.), VCH, Weinheim, New York, etc., **1991**, 126–130.

B. Giese, B. Kopping, T. Göbel, J. Dickhaut, G. Thoma, K. J. Kulicke, F. Trach, "Radical Cyclization Reactions," *Org. React.* **1996**, *48*, 301–856.

D. P. Curran, "The Design and Application of Free Radical Chain Reactions in Organic Synthesis," *Synthesis* **1988**, 489.

T. V. RajanBabu, "Stereochemistry of Intramolecular Free-Radical Cyclization Reactions," *Acc. Chem. Res.* **1991**, *24*, 139–145.

D. P. Curran, N. A. Porter, B. Giese (Eds.), "Stereochemistry of Radical Reactions: Concepts, Guidelines, and Synthetic Applications," VCH, Weinheim, Germany, **1995**.

C. P. Jasperse, D. P. Curran, T. L. Fevig, "Radical Reactions in Natural Product Synthesis," *Chem. Rev.* **1991**, *91*, 1237–1286.

G. Mehta, A. Srikrishna, “Synthesis of Polyquinane Natural Products: An Update,” *Chem. Rev.* **1997**, *97*, 671–720.

V. K. Singh, B. Thomas, “Recent Developments in General Methodologies for the Synthesis of Linear Triquinanes,” *Tetrahedron* **1998**, *54*, 3647–3692.

S. Handa, G. Pattenden, “Free Radical-Mediated Macrocyclizations and Transannular Cyclizations in Synthesis,” *Contemp. Org. Synth.* **1997**, *4*, 196–215.

A. J. McCarroll, J. C. Walton, “Programming Organic Molecules: Design and Management of Organic Syntheses Through Free-Radical Cascade Processes,” *Angew. Chem. Int. Ed. Engl.* **2001**, *40*, 2224–2248.

G. Descotes, “Radical Functionalization of the Anomeric Center of Carbohydrates and Synthetic Applications,” in *Carbohydrates* (H. Ogura, A. Hasegawa, T. Suami, Eds.), 89, Kodansha Ltd, Tokyo, Japan, **1992**.

C. Walling, E. S. Huyser, “Free Radical Addition to Olefins to Form Carbon-Carbon Bonds,” *Org. React.* **1963**, *13*, 91–149.

B. Giese, T. Göbel, B. Kopping, H. Zipse, “Formation of C–C Bonds by Addition of Free Radicals to Olefinic Double Bonds,” in *Stereoselective Synthesis* (Houben-Weyl) 4th ed. **1996**, (G. Helmchen, R. W. Hoffmann, J. Mulzer, E. Schaumann, Eds.), **1996**, Vol. E 21 (Workbench Edition), *4*, 2203–2287, Georg Thieme Verlag, Stuttgart.

L. Yet, “Free Radicals in the Synthesis of Medium-Sized Rings,” *Tetrahedron* **1999**, *55*, 9349–9403.

H. Ishibashi, T. Sato, M. Ikeda, “5-Endo-Trig Radical Cyclizations,” *Synthesis* **2002**, 695–713.

B. K. Banik, “Tributyltin Hydride Induced Intramolecular Aryl Radical Cyclizations: Synthesis of Biologically Interesting Organic Compounds,” *Curr. Org. Chem.* **1999**, *3*, 469–496.

H. Fischer, L. Radom, “Factors Controlling the Addition of Carbon-Centered Radicals to Alkenes—An Experimental and Theoretical Perspective,” *Angew. Chem. Int. Ed. Engl.* **2001**, *40*, 1340–1371.

B. Giese, “The Stereoselectivity of Intermolecular Free Radical Reactions,” *Angew. Chem. Int. Ed. Engl.* **1989**, *28*, 969–980.

P. Renaud, M. Gerster, “Use of Lewis Acids in Free Radical Reactions,” *Angew. Chem. Int. Ed. Engl.* **1998**, *37*, 2562–2579.

M. P. Sibi, T. R. Ternes, “Stereoselective Radical Reactions,” in *Modern Carbonyl Chemistry*, Ed. by J. Otera, Wiley-VCH, Weinheim, **2000**, 507–538.

F. W. Stacey, J. F. Harris, Jr, “Formation of Carbon-Heteroatom Bonds by Free Radical Chain Additions to Carbon-Carbon Multiple Bonds,” *Org. React.* **1963**, *13*, 150–376.

O. Touster, “The Nitrosation of Aliphatic Carbon Atoms,” *Org. React.* **1953**, *7*, 327–377.

C. V. Wilson, “The Reaction of Halogens with Silver Salts of Carboxylic Acids,” *Org. React.* **1957**, *9*, 332–387.

R. A. Sheldon, J. K. Kochi, “Oxidative Decarboxylation of Acids by Lead Tetraacetate,” *Org. React.* **1972**, *19*, 279–421.

M. Quaedvlieg, “Methoden zur Herstellung und Umwandlung von aliphatischen Sulfonsäuren und ihren Derivaten,” in *Houben-Weyl, Methoden der Org. Chem.*, 4th ed., Vol. 9, Thieme, Stuttgart, **1955**, 345–405; here: “Sulfoxidation,” 366–367 and “Sulfochlorierung,” 391–392.

H. Ramloch, G. Täuber, “Verfahren der Großchemie: Die Sulfoxidation,” *Chemie in unserer Zeit* **1979**, *13*, 157–162.

Nucleophilic Substitution Reactions at the Saturated C Atom

2

The book began with chemistry that is by no means uncommon, but the bulk of organic chemistry has and continues to be based on two electron processes. Substitution reactions comprise one class of reactions involving such chemistry. Addition reactions comprise the other. You should have been introduced to both by now.

2.1 Nucleophiles and Electrophiles; Leaving Groups

Although 4 out of 5 medical doctors probably would not agree, organic chemistry is comparatively simple to learn because most organic chemical reactions follow a single pattern. This pattern is

$$\text{nucleophile} + \text{electrophile} \xrightarrow[\text{pair shifts(s)}]{\text{valence electron}} \text{product(s)}$$

A nucleophile is a Lewis base that donates a pair of electrons to the electrophile, a Lewis acid, to create a bond.

Electrophiles and Nucleophiles

As **electron pair donors**, nucleophiles must either contain an electron pair that is easily available because it is nonbonding or a bonding electron pair that can be donated and thus be made available to the reaction partner, the electron acceptor. Thus, nucleophiles are usually anions or neutral species but not cations. In this book, nucleophile is abbreviated as "$Nu^{\ominus}$," regardless of charge.

Electrophiles are Lewis acids, **electron pair acceptors.** They therefore either have an empty orbital in the valence electron shell of one of their constituent atoms or they have their valence shell filled but contain an atom that can leave with an pair of electrons, that is, a leaving group. The atom that bears the leaving group is formally the electrophilic atom. Other electrophiles can become hypervalent. They can exceed normal bonding constraints that would be associated with bonding schemes like the octet rule. Third row (and below) elements can do this by taking advantage of empty "d" orbitals. For the systems we will discuss, electrophiles are cations or neutral compounds, but not anions, and we won't dig any deeper than that. In this book electrophile is abbreviated as "$E^{\oplus}$," regardless of charge.

Bruckner R (author), Harmata M (editor) In: *Organic Mechanisms – Reactions, Stereochemistry and Synthesis*
Chapter DOI: 10.1007/978-3-642-03651-4_2,

For most organic chemical reactions, the pattern just discussed can be written briefly as follows:

$$\text{Nu:}^{\ominus} + \text{E}^{\oplus} \longrightarrow \text{Nu—E} \quad (+\text{ by-products})$$

It looks simple. It is simple. It is also profound and a huge amount of organic chemistry is based on this process. In this chapter, we deal with nucleophilic substitution reactions at the saturated, that is, the sp^3-hybridized C atom (abbreviated "S_N reactions"). In these reactions, nucleophiles react with carbon compounds that have leaving groups, alkylating agents. They have the structure $(R_{3-n}H_n)C_{sp3}$–X. The group X is displaced by the nucleophile according to the equation

$$\text{Nu:}^{\ominus} + -\overset{|}{\underset{|}{C}}_{sp^3}-\text{X} \xrightarrow{k} \text{Nu}-\overset{|}{\underset{|}{C}}_{sp^3}- + \text{:X}^{\ominus}$$

as $X^{\ominus}$. Consequently, both the bound group X and the departing entity $X^{\ominus}$ are called **leaving groups.**

2.2 Good and Poor Nucleophiles

What is the difference between a good nucleophile and a poor one? What about a good electrophile and a poor one? These questions are kinetic in nature in that they ask how fast a particular species might react. Something with high nucleophilicity will react faster with a given electrophile than something with low nucleophilicity. The same statement can be made of electrophiles. What are the guidelines that allow us to make reasonable guesses about answers to the questions in specific cases? That's what we are after here.

Answers to these questions can experimentally obtained via pairs of S_N reactions, which are carried out as **competition experiments**. In a competition experiment, two reagents react simultaneously with one substrate (or vice versa). Two reaction products can then be produced. The main product is the compound that results from the *more reactive* (or or more nucleophilic/electrophilic, as the case may be) reaction partner.

For example, if two nucleophiles are reacted with some alkylating agent, the nucleophile that reacts to form the main product is then the "better" nucleophile. From the investigation of a large number of competition experiments of this type, gradations of the nucleophilicity exist that are essentially independent of the substrate.

Why are the differences in nucleophilicity relatively the same over many different substitution reactions? Nucleophilicity describes the ability of the nucleophile to make an electron pair available to the electrophile (i.e., the alkylating agent). With this as the basic idea, the experimentally observable nucleophilicity trends can be interpreted as follows.

Nucleophilicity Rankings

- Within a group of nucleophiles that react with electrophile with the same atom, the nucleophilicity decreases with *decreasing basicity of the nucleophile* (Figure 2.1). Decreasing basicity is equivalent to decreasing affinity of an electron pair for a *proton*, which to a certain extent, is a model electrophile for the electrophiles since both alkylating agents and protons are Lewis acids.

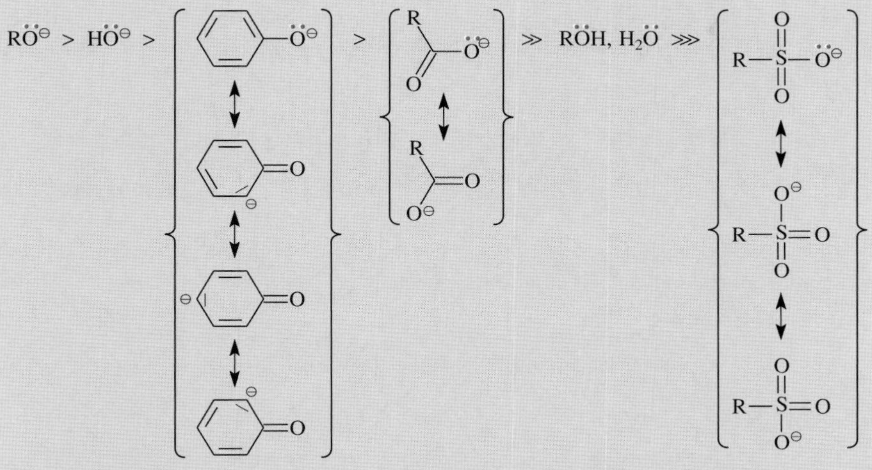

$$RO^{\ominus} > HO^{\ominus} > \{PhO^{\ominus}\} > \{RCO_2^{\ominus}\} \gg ROH, H_2O \ggg \{RSO_3^{\ominus}\}$$

Fig. 2.1. Nucleophilicity of O nucleophiles with different basicities.

- The nucleophilicity of a given nucleophilic center is increased by attached heteroatoms—i.e., heteroatoms in the α-position—with free electron pairs (the so-called α-effect):

$$HO{-}O^{\ominus} > H{-}O^{\ominus}$$

$$H_2N{-}NH_2 > H{-}NH_2$$

The reason for this is the unavoidable overlap of the orbitals that contain the free electron pairs at the nucleophilic center and its neighboring atom. This raises the energy of the HOMO of the molecule, making it a better nucleophile.

- *Nucleophilicity decreases with increasing electronegativity of the reacting atom.* This is always true in both comparisons of atoms that belong to the same row of the periodic table of the elements

$$R_2N^{\ominus} \gg RO^{\ominus} \gg F^{\ominus} \qquad RS^{\ominus} \gg Cl^{\ominus}$$

$$Et_3N \gg Et_2O$$

- and in comparisons of atoms from the same column of the periodic table:

$$RS^{\ominus} > RO^{\ominus} \qquad I^{\ominus} > Br^{\ominus} > Cl^{\ominus} \gg F^{\ominus}$$

$$RSH > ROH$$

- Solvation of anionic nucleophiles is poorer in dipolar aprotic than in dipolar protic solvents. The reason is that dipolar aprotic solvents cannot form a hydrogen bond to such nucleophiles. Consequently, anionic nucleophiles are not stabilized in dipolar aprotic solvents and therefore are better nucleophiles. The higher the charge density in the anionic nucleophile and the higher the demand for stabilization of its charge, the stronger the effect. Therefore, it is especially the small anionic nucleophiles that exhibit an increased nucleophilic reactivity when changing from a dipolar protic to a dipolar aprotic solvent:

$$\text{Me—I} + \text{Cl}^{\ominus} \xrightarrow{k_{S_N2}} \text{Me—Cl} + \text{I}^{\ominus}$$

	in MeOH	in $H-C(=O)-NH_2$	in $H-C(=O)-NMe_2$	in $Me-C(=O)-NMe_2$
$k_{S_N2,\,rel}$	1.0	12.5	120,000	7,400,000

- this causes a complete reversal of the nucleophilicities of the halide ions in a dipolar aprotic instead of a dipolar protic environment:

k_{S_N2} in *protic* polar solvent → increase

$F^{\ominus}$ $Cl^{\ominus}$ $Br^{\ominus}$ $I^{\ominus}$

k_{S_N2} in *aprotic* polar solvent → decrease

An important lesson from this is that the idea of nucleophilicity in the real world of organic reactions is not easy to pigeonhole. Polarizability is important, but basicity is also very important and can be influenced by solvation. Values of the pKa of a given compound vary as a function of solvent, and so does basicity. You can make a species, anions in particular, more reactive by putting them in solvents that don't solvate them very well. Dipolar aprotic solvents interact nicely with cations, but not so well with anions. Polar protic solvents (e. g., water, alcohols) can hydrogen bond to anions, diminishing their basicity and literally blocking them sterically.

Side Note 2.1. Ether Cleavages

The increase in reactivity of nucleophiles by using dipolar aprotic solvents that was just described can be exploited when, for example, aromatic ethers are cleaved with thiolate or chloride ions instead of with boron tribromide (BBr_3, a Lewis acid, is quite an aggressive reagent (Figure 2.2). Aryl methyl ethers are demethylated ("deprotected") to furnish the corresponding phenols using these reagents. When performed with chloride or thiolate anions, the reaction medium is essentially neutral and groups sensitive to acids like BBr_3 can survive.

The demethylation of the arylmethyl ethers **A** in Figure 2.2 is run in dimethyl formamide (DMF; structural formula see Figure 2.17); the fourfold demethylation of the aryl tetramethyl ether **B** is performed in molten salt and the monodemethylation of the aryl dimethyl ethers C in DMF. The selectivity of the latter reaction deserves comment. Compound **C** and lithium

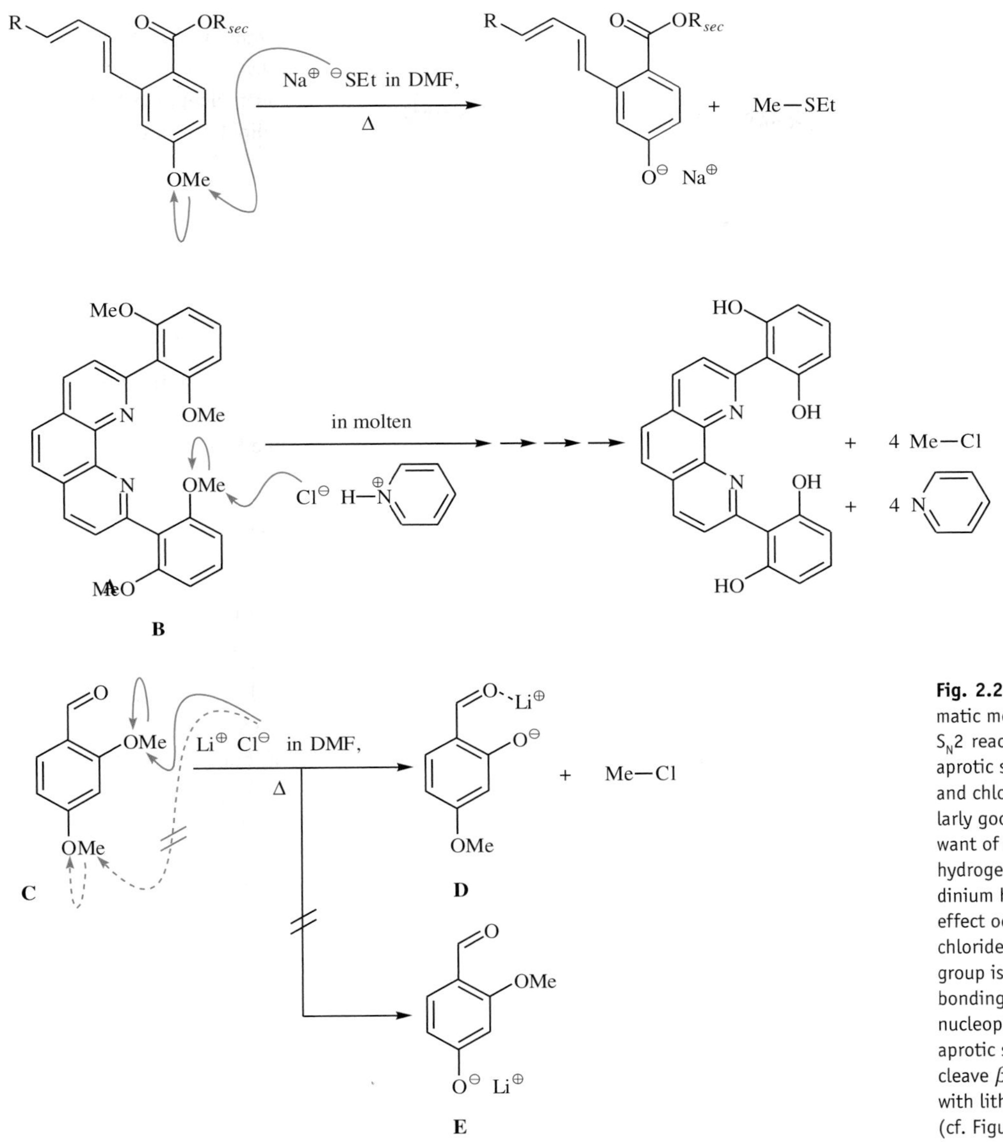

Fig. 2.2. Cleavage of aromatic methyl ether using S_N2 reactions. In the dipolar aprotic solvent DMF, thiolate and chloride ions are particularly good nucleophiles for want of solvation through hydrogen bonding. In pyridinium hydrochloride a similar effect occurs because for each chloride only one $N^{\delta\ominus}-H^{\delta\oplus}$ group is available for hydrogen bonding. The same increase in nucleophilicity in a dipolar aprotic solvent is used to cleave β-ketomethyl esters with lithium iodide in DMF (cf. Figure 13.29).

chloride exclusively form the lithium phenolate **D** and not the isomeric lithium phenolate **E**. This might be due to a chelation effect: bidentate chelation of the lithium ion will selectively activate the methoxy group that reacts.

2.3 Leaving Groups: Good, Bad and Ugly

In Figure 2.3, substructures have been listed in the order of their suitability as leaving groups in S_N reactions. Substrates with good leaving groups are listed on top and substrates with increasingly poor leaving groups follow. At the bottom of Figure 2.3 are substrates whose functional group is an extremely poor leaving group. The effect of nucleophiles on substrates of the latter type can result in a reaction, but it is not an S_N reaction. For example, a nucleophile might remove an acidic proton from the carbon that bears a certain functional group instead of replacing the functional group. Attempted S_N reactions with ammonium salts, nitro compounds, sulfoxides, sulfones, sulfonium salts, phosphonic acid esters, phosphine oxides, and phosphonium salts often fail as a result of such a deprotonation. Another reaction competing with the substitution of a functional group by a nucleophile is a reaction on the functional group by the nucleophile. For example, S_N reactions of nitriles, phosphonic acid esters, and phosphonium salts often fail because of this problem.

Alcohols, ethers, and carboxylic acid esters occupy an intermediate position. These compounds as such—except for the special cases shown in Figure 2.3—do not particpate in any S_N reactions with nucleophiles. The reason for this is that poor leaving groups would have to leave ($^{\ominus}OH$, $^{\ominus}OR$, $^{\ominus}O_2CR$; see below for details). However, these compounds can enter into S_N reactions with nucleophiles when they are activated as oxonium ions, for example, via a reversible protonation, via bonding of a Lewis acid (LA in Figure 2.4), or via a phosphorylation. Thus, upon attack by the nucleophile, better leaving groups (e. g., HOH, HOR, HO_2CR, $O{=}PPh_3$) can be released.

Only special carboxylic acid esters, aryl ethers and special ethers like epoxides enter into S_N reactions as such, that is, without derivatization in some fashion (Figure 2.4). In carboxylic acid esters of secondary and tertiary alcohols, the carboxylate group $^{\ominus}O_2CR$ can become a leaving group, namely in solvolyses. With epoxides as the substrate, an alkoxide ion is also an acceptable leaving group. Its depature, which is actually bad because of the high basicity, is coupled with the release of part of the 26 kcal/mol epoxide ring strain. Product development control therefore makes this reaction path feasible.

What makes a leaving group good or bad in substrates that react with nucleophiles as alkylating agents? The Hammond postulate implies that a good leaving group is a stabilized species, not a high-energy species. Therefore, good leaving groups are usually weak bases. Another way of thinking about it: A strong base reacts rapidly with protons (electrophiles) in an energetically favorable process, the reverse of which is necessarily energetically unfavorable. By analogy we can conclude that a mixture of a strongly basic leaving group with the product of an S_N reaction is also relatively high in energy. Very basic leaving groups are produced relatively slowly.

The suitability of halide ions as leaving groups is predicted correctly based on this reasoning alone, where $I^{\ominus} > Br^{\ominus} > Cl^{\ominus} > F^{\ominus}$. The **trifluoromethanesulfonate anion (triflate anion)** $F_3C{-}SO_3^{\ominus}$ is a far better leaving group than the ***p*-toluenesulfonate anion (tosylate anion)** $Me{-}C_6H_4{-}SO_3^{\ominus}$ or the **methanesulfonate anion (mesylate anion)** $H_3C{-}SO_3^{\ominus}$. The pKa values for the corresponding conjugate acids of these leaving groups are –13, –6 and –2.5, respectively. Leaving group ability is correlated with the pKa of the conjugate acid of the leaving group, at least to a first approximation.

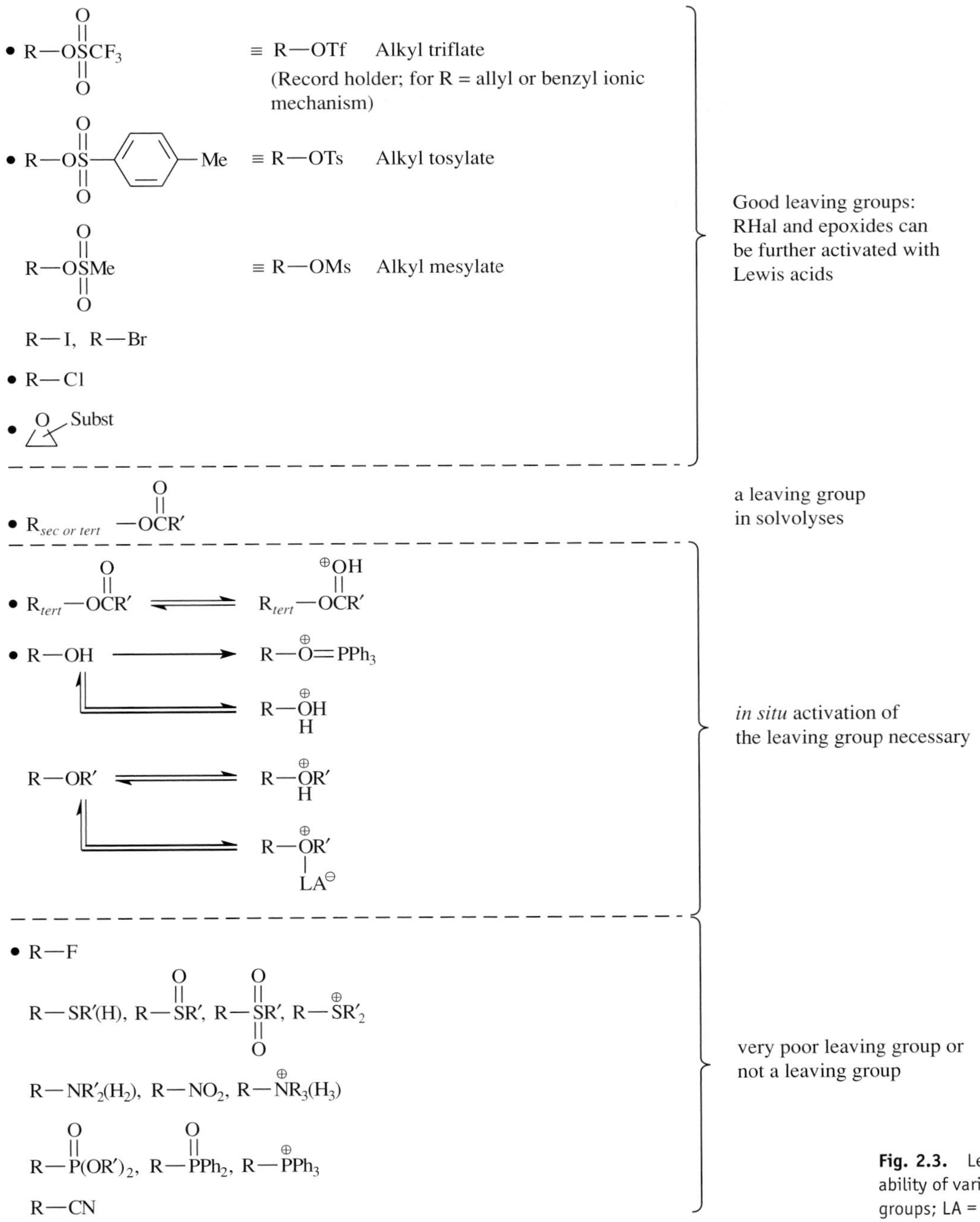

- $R{-}OSO_2CF_3$ ≡ R—OTf Alkyl triflate (Record holder; for R = allyl or benzyl ionic mechanism)
- $R{-}OSO_2C_6H_4Me$ ≡ R—OTs Alkyl tosylate
- $R{-}OSO_2Me$ ≡ R—OMs Alkyl mesylate
- R—I, R—Br
- R—Cl
- Epoxide (O, Subst)

Good leaving groups: RHal and epoxides can be further activated with Lewis acids

- $R_{sec\ or\ tert}{-}OC(=O)R'$

a leaving group in solvolyses

- $R_{tert}{-}OC(=O)R' \rightleftharpoons R_{tert}{-}OC(=\overset{\oplus}{O}H)R'$
- $R{-}OH \longrightarrow R{-}\overset{\oplus}{O}{=}PPh_3$; $R{-}OH \rightleftharpoons R{-}\overset{\oplus}{O}H_2$
- $R{-}OR' \rightleftharpoons R{-}\overset{\oplus}{O}(H)R'$; $R{-}OR' \rightleftharpoons R{-}\overset{\oplus}{O}(LA^{\ominus})R'$

in situ activation of the leaving group necessary

- R—F
- $R{-}SR'(H)$, $R{-}S(=O)R'$, $R{-}SO_2R'$, $R{-}\overset{\oplus}{S}R'_2$
- $R{-}NR'_2(H_2)$, $R{-}NO_2$, $R{-}\overset{\oplus}{N}R_3(H_3)$
- $R{-}P(=O)(OR')_2$, $R{-}P(=O)Ph_2$, $R{-}\overset{\oplus}{P}Ph_3$
- R—CN

very poor leaving group or not a leaving group

Fig. 2.3. Leaving-group ability of various functional groups; LA = Lewis acid.

Finally, HOH and ROH can leave *protonated* alcohols or ethers as leaving groups, but neither the $OH^{\ominus}$ group (from alcohols) nor the $OR^{\ominus}$ group (from ethers, except for epoxides, see above) can leave. Remember, the conjugate acid of water is the hydronium ion, a strong acid (pKa = –1.7)

Another consideration is that lower the bond energy of the bond between the C atom and the leaving group, the better the leaving group. This again follows from the Hammond postulate. For this reason as well, the suitability of the halide ions as leaving groups is predicted as $I^{\ominus} > Br^{\ominus} > Cl^{\ominus} > F^{\ominus}$.

2.4 S_N2 Reactions: Kinetic and Stereochemical Analysis—Substituent Effects on Reactivity

2.4.1 Energy Profile and Rate Law for S_N2 Reactions: Reaction Order

An S_N2 reaction refers to an S_N reaction

$$\text{Nu:}^{\ominus} + \text{R—X} \xrightarrow{k} \text{Nu—R} + \text{:X}^{\ominus}$$

in which the nucleophile and the alkylating agent are converted into the substitution product in one step, that is, via one transition state (Figure 2.4).

Do you remember the definition of an elementary reaction (Section 1.7.1)? The S_N2 reaction *is* such an elementary reaction. Recognizing this is a prerequisite for deriving the rate law for the S_N2 reaction, because the rate law for any elementary reaction can be obtained *directly from the reaction equation*.

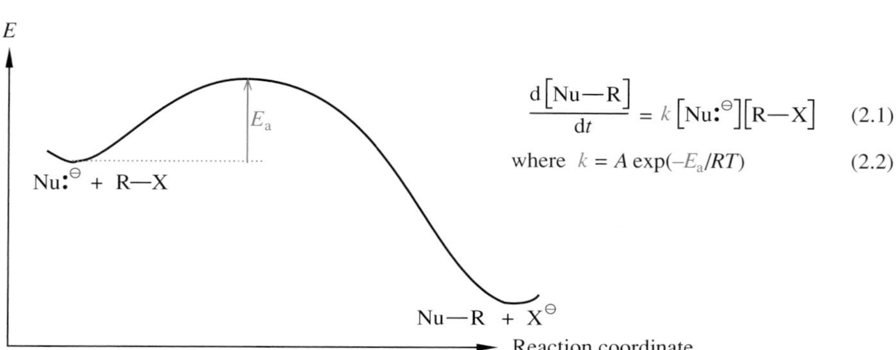

Fig. 2.4. Energy profile and rate law for S_N2 reactions.

Side Note 2.2. Rate Laws

Rate laws establish a relationship between

- the change in the concentration of a product, an intermediate, or a starting material as a function of time,
- and the concentrations of the starting material(s) and possibly the catalyst,
- and the rate constants of the elementary reactions involved in the overall reaction.

The overall reaction rate is the rate of product formation $d[\text{product}]/dt$ or a starting material consumption rate $-d[\text{starting material}]/dt$. The following applies unless the stoichiometry requires an additional multiplier:

$$\frac{d[\text{final product}]}{dt} = -\frac{-d[\text{starting material(s)}]}{dt} \qquad (2.3)$$

With the help of the rate laws that describe the elementary reactions involved, it is possible to derive Equation 2.4:

$$\frac{d[\text{final product}]}{dt} = f\{[\text{starting material(s) and (if needed) catalyst}], \text{rate constants}_{\text{ER}}\} \qquad (2.4)$$

where the subscript ER refers to the elementary reactions participating in the overall reaction. If the right-hand side of Equation 2.4 does not contain any sums or differences, the sum of the powers of the concentration(s) of the starting material(s) in this expression is called the order (m) of the reaction. It is also said that the **reaction** is **of the *m*th order**. A reaction of order $m = 1$ is a first-order reaction, or **unimolecular**. A reaction of order $m = 2$ (or 3) is a second- (or third-) order reaction or a **bimolecular** (or **trimolecular**) **reaction**. A reaction of order m, where m is not an integer, is a reaction of a mixed order. The rate laws for elementary reactions are especially easy to set up. The recipe for this is

$$\frac{d[\text{product of an elementary reaction}]}{dt} = -\frac{d[\text{starting material(s) of the elementary reaction}]}{dt}$$

The rate of product formation or starting material consumption is equal to the product of the rate constant k for this elementary reaction and the concentration of *all* starting materials involved, including any catalyst. It turns out that all elementary reactions are either first- or second-order reactions.

Because the reactions we consider in this section are single-step and therefore elementary reactions, the rate law specified in Section 2.3 as Equation 2.1 is obtained for the rate of formation of the substitution product Nu—R in Figure 2.4. It says that these reactions are bimolecular substitutions. They are consequently referred to as **S_N2 reactions.** The bimolecularity makes it possible to distinguish between this type of substitution and S_N1 reactions, which we will examine in Section 2.5: nucleophile concentration affects the rate of an S_N2 reaction, but not an S_N1 reaction.

The rate constant of each elementary reaction is related to its activation energy E_a by the Arrhenius equation. This of course also holds for the rate constant of S_N2 reactions (see Equation 2.2 in Figure 2.4).

2.4.2 Stereochemistry of S_N2 Substitutions

In sophomore organic chemistry you learned that S_N2 reactions take place stereoselectively. Let us consider Figure 2.5 as an example: the displacement by potassium acetate on the *trans*-tosylate **A** gives only the cyclohexyl acetate *cis*-**B.** No *trans*-isomer is formed. In the starting material, the leaving group is equatorial and the C—H bond at the reacting C atom is axial. In the substitution product *cis*-**B**, the acetate is axial and the adjacent H atom is equatorial. A 100% inversion of the configuration has taken place in this S_N2 reaction. This is also true for all other S_N2 reactions investigated stereochemically.

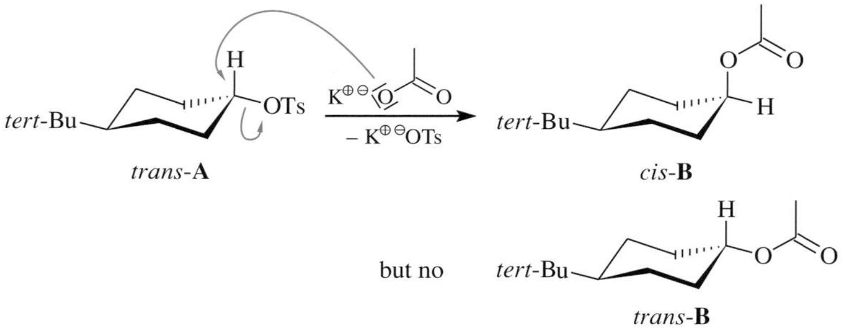

Fig. 2.5. Proof of the inversion of configuration at the reacting C atom in an S_N2 reaction.

The reason for the inversion of configuration is that S_N2 reactions take place with a backside entry by the nucleophile on the bond between the C atom and the leaving group. In the transition state of the S_N2 reaction, the reacting C atom has five bonds. The three substituents at the reacting C atom not participating in the S_N reaction and this C atom itself are for a short time located in one plane:

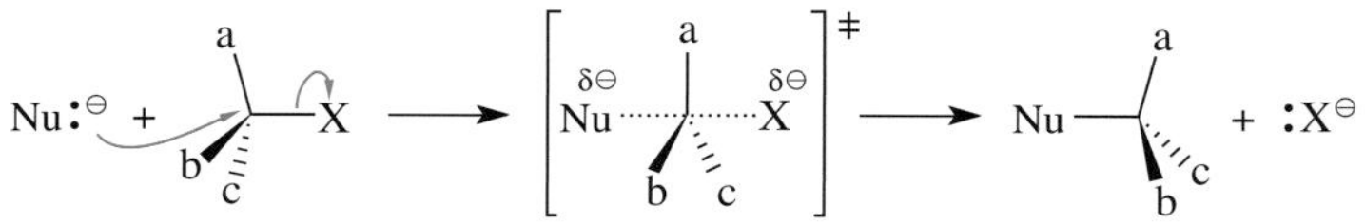

The S_N2 mechanism is sometimes referred to as an "umbrella mechanism." The nucleophile enters in the direction of the umbrella handle and displaces the leaving group, which was originally lying above the tip of the umbrella. The geometry of the transition state corresponds to the geometry of the umbrella, which is just flipping over. The geometry of the substitution product corresponds to the geometry of the flipped-over umbrella. The former nucleophile is located at the handle of the flipped-over umbrella. Can you see that in your mind's eye?

2.4.3 A Refined Transition State Model for the S_N2 Reaction; Crossover Experiment and Endocyclic Restriction Test

Figure 2.6 shows several methylations, which in each case take place as one-step S_N reactions. The nucleophile is in each case a sulfonyl anion; a methyl (arenesulfonate) reacts as electrophile. These classic experiments were carried out to clarify whether these methylations take place inter- or intramolecularly.

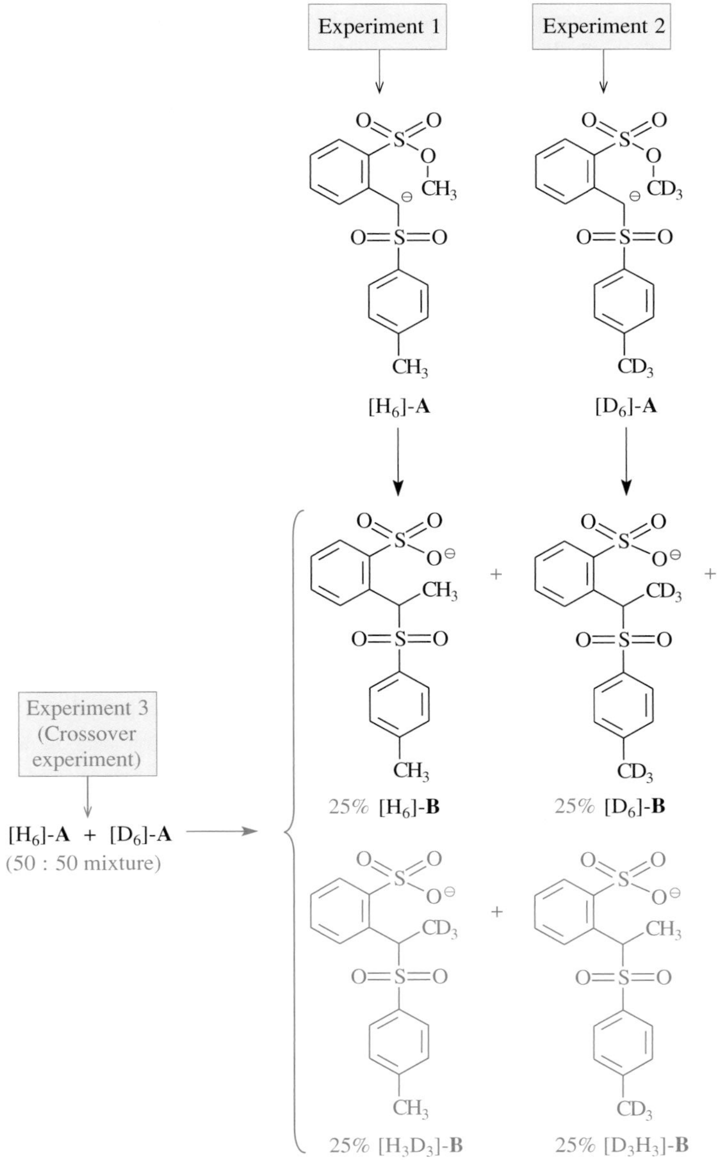

Experiment 1

Experiment 2

[H6]-A

[D6]-A

Experiment 3 (Crossover experiment)

[H6]-A + [D6]-A (50 : 50 mixture)

25% [H6]-B

25% [D6]-B

25% [H3D3]-B

25% [D3H3]-B

Fig. 2.6. Determination of the mechanism of one-step S_N reactions on methyl (arenesulfonates): intra- or intermolecularity.

In experiment 1, the perprotio-sulfonyl anion $[H_6]$-**A** reacts to form the methylated perprotio-sulfone $[H_6]$-**B.** It is not known whether this is the result of an intra- or an intermolecular S_N reaction. In experiment 2 of Figure 2.6, the sulfonyl anion $[D_6]$-**A**, which is perdeuterated in both methyl groups, reacts to form the hexadeuterated methylsulfone $[D_6]$-**B.** Even this result does not clarify whether the methylation is intra- or intermolecular. An explanation is not provided until the third experiment in Figure 2.6, a so-called **crossover experiment.**

The purpose of every crossover experiment is to determine whether reactions take place intra- or intermolecularly. In a crossover experiment two substrates differing from each other by a *double* substituent variation are reacted as a *mixture*. This substrate mixture is subjected to the same reaction conditions in the crossover experiment that the two individual substrates had been exposed to in separate experiments. This double substituent variation allows one to determine the origin of the reaction products from their structures, i.e., from which parts of which starting materials they were formed (see below for details).

The product mixture is then analyzed. There are two possible outcomes. It can contain nothing other than the two products that were already obtained in the individual experiments. In this case, each substrate would have reacted only with itself. This is possible only for an intramolecular reaction. The product mixture of a crossover experiment could alternatively consist of four compounds. Two of them would not have arisen from the individual experiments. They could have been produced only by "crossover reactions" *between* the two components of the mixture. A crossover reaction of this type can only be intermolecular.

In the third crossover experiment of Figure 2.6, a 1:1 mixture of the sulfonyl anions $[H_6]$-**A** and $[D_6]$-**A** was methylated. The result did *not* correspond to the sum of the individual reactions. Besides a 1:1 mixture of the methylsulfone $[H_6]$-**B** obtained in experiment 1 and the methylsulfone $[D_6]$-**B** obtained in experiment 2, a 1:1 mixture of the two crossover products $[H_3D_3]$-**B** and $[D_3H_3]$-**B** was isolated, in the same yield. The fact that both $[H_3D_3]$-**B** and $[D_3H_3]$-**B** occurred proves that the methylation was intermolecular. The crossover product $[H_3D_3]$-**B** can only have been produced because a CD_3 group was transferred from the sulfonyl anion $[D_6]$-**A** to the deuterium-free sulfonyl anion $[H_6]$-**A.** The crossover product $[D_3H_3]$-**B** can only have been produced because a CH_3 group was transferred from the sulfonyl anion $[H_6]$-**A** to the deuterium-containing sulfonyl anion $[D_6]$-**A.**

The *intramolecular* methylation of the substrate of Figure 2.6, which was *not observed*, would have had to take place through a six-membered cyclic transition state. In other cases, cyclic, six-membered transition states of intramolecular reactions are so favored that intermolecular reactions usually do not occur. Why then is a cyclic transition state not able to compete in the S_N reactions in Figure 2.6?

The conformational degrees of freedom of cyclic transition states are considerably limited or "restricted" relative to the conformational degrees of freedom of noncyclic transition states (cf. the fewer conformational degrees of freedom of cyclohexane vs. *n*-hexane). Mechanistic investigations of this type are therefore also referred to as **endocyclic restriction tests.** They suggest certain transition state geometries in a very simple way because of this conformational restriction. The endocyclic restriction imposes limitations of the conceivable geometries on cyclic transition states. These geometries do not comprise all the possibilities that could be realized for intermolecular reactions proceeding through acyclic transition states.

In the S_N reactions of Figure 2.6, the endocyclic restriction would therefore impose a geometry in an intramolecular substitution that is energetically disfavored relative to the geometry that can be obtained in an intermolecular substitution. As shown on the right in Figure 2.7, in an intramolecular substitution the reason for this is that the nucleophile, the reacting C atom, and the leaving group cannot lie on a common axis. However, such a geometry can be realized in an intermolecular reaction (see Figure 2.7, left). From this, one concludes that in an S_N2 reaction the approach path of the nucleophile must be *collinear* with the bond between the reacting C atom and the leaving group.

This approach path is preferred in order to achieve a transition state with optimum bonding interactions. Let us assume that in the transition state, the distance between the nucleophile and the reacting C atom and between the leaving group and this C atom are exactly the same. The geometry of the transition state would then correspond precisely to the geometry of an umbrella that is just flipping over. The reacting C atom would be—as shown in Figure 2.7 (left)—sp^2-hybridized and at the center of a trigonal bipyramid. The nucleophile and the leaving group would be bound to this C atom via (partial) σ bonds. Both would result from overlap with one lobe of the $2p_z$ AO. *For this reason*, a linear arrangement of the nucleophile, the attacked C atom, and the leaving group is preferred in the transition state of S_N reactions. Thus, S_N2 reactions proceed with a certain *stereoelectronic* requirement, a well-defined orientation of electron pairs within the transition state. Specifically, this is a 180° angle

Fig. 2.7. Illustration of the intermolecular course of the S_N reactions of Figure 2.6; energy profiles and associated transition state geometries.

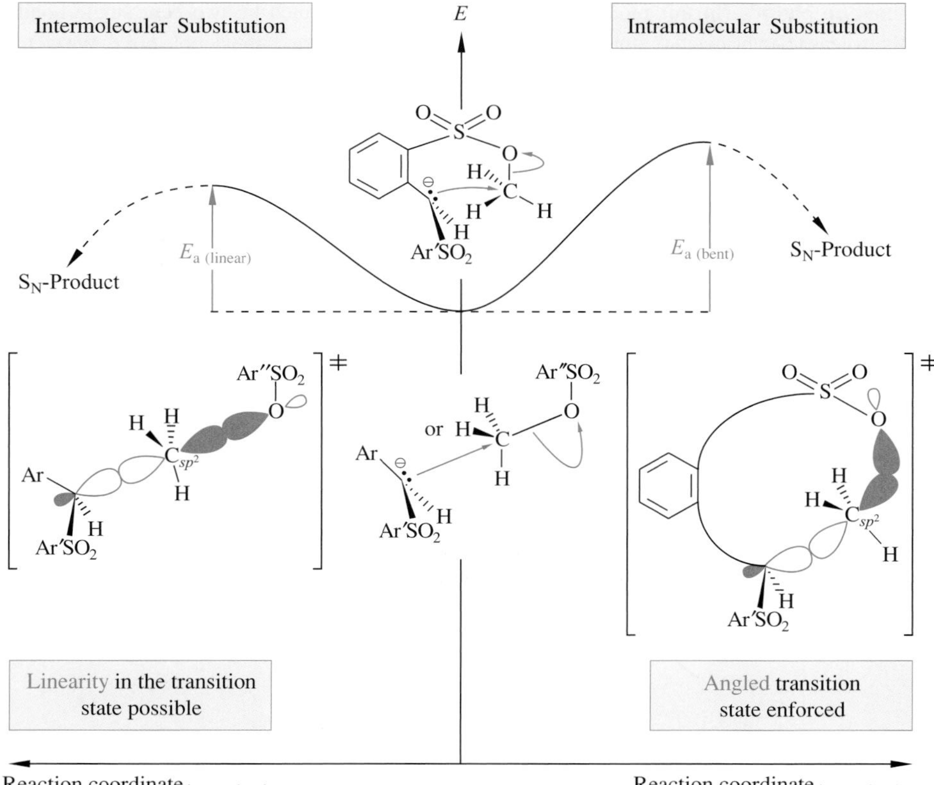

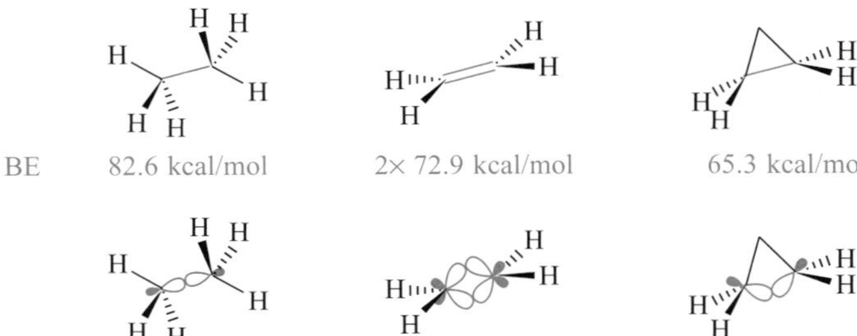

Fig. 2.8. Bond enthalpy (BE) of linear (left) and bent (middle and right; cf. explanatory text) C—C bonds.

between the pair of electrons in the bond to the leaving group and the pair of electrons coming from the nucleophile and forming the new bond.

Figure 2.7 (right) shows that in a bent transition state of the S_N reaction neither the nucleophile nor the leaving group can form similarly stable σ bonds by overlap with the $2p_z$ AO of the reacting C atom. Because the orbital lobes under consideration are not parallel, both the Nu··· C_{sp^2} and the C_{sp^2}···leaving group bonds would be bent. **Bent bonds** are weaker than linear bonds because of the smaller orbital overlap. This is known from the special case of bent C—C bonds (Figure 2.8) as encountered, for example, in the very strained C—C bonds of cyclopropane. Bent bonds are also used in the "banana bond" model to describe the C=C double bond in olefins. (In this model, the double bond is represented by two bent single bonds between sp^3-hybridized C atoms. Don't worry about this picture of the double bond. We won't monkey around with this concept too much.) Both types of bent C—C bonds are less stable than linear C—C bonds, such as in ethane.

2.4.4 Substituent Effects on S_N2 Reactivity

How substituents in the alkylating agent influence the rate constants of S_N2 reactions can be explained by means of the transition state model developed in Section 2.4.3. This model makes it possible to understand both the steric and the electronic substituent effects.

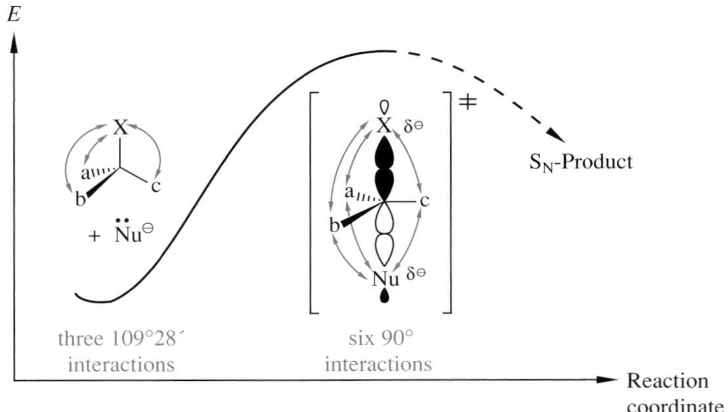

Fig. 2.9. Steric effects on S_N2 reactivity: substituent compression in the transition state.

When an alkylating agent is approached by a nucleophile, the steric interactions become larger in the vicinity of the reacting C atom (Figure 2.9). The spectator substituents come closer to the leaving group X. The bond angle between these substituents and the leaving group decreases from. the tetrahedral angle of ~109° to about 90°. However, the reacting nucleophile approaches the spectator substituents until the bond angle that separates it from them is also approximately 90°. The resulting increased steric interactions destabilize the transition state. Consequently, the activation energy E_a increases and the rate constant k_{S_N2} decreases. It should be noted that this destabilization is not compensated for by the simultaneous increase in the bond angle *between* the inert substituents from the tetrahedral angle to approximately 120°.

This has three consequences:

Tendencies and Rules

- First, the S_N2 reactivity of an alkylating agent decreases with an **increasing number** of the alkyl substituents at the electrophilic C atom. In other words, α-branching at the C atom of the alkylating agent reduces its S_N2 reactivity. This reduces the reactivity so much that tertiary C atoms don't react via the S_N2 mechanism in intermolecular processes:

	Nu⊖ + Me–X	Et–X	*i*Pr–X	*tert*-Bu–X
$k_{S_N2,\,rel}$ =	30	1	0.025	tiny

  Generally stated, for S_N2 reactivity we have $k(\text{Me—X}) > k(\text{R}_{prim}\text{—X}) >> k(\text{R}_{sec}\text{—X})$; $k(\text{R}_{tert}\text{—X}) \approx 0$.
- Second, the S_N2 reactivity of an alkylating agent decreases with **increasing size** of the alkyl substituents at the reacting C atom. In other words, β-branching in the alkylating agent reduces its S_N2 reactivity. This reduces the reactivity so much that a C atom (neopentyl) with a quaternary C atom in the β-position does not react at all via an S_N2 mechanism:

	Nu⊖ + $MeCH_2$–X	$EtCH_2$–X	*i*PrCH_2–X	*tert*-BuCH_2–X
$k_{S_N2,\,rel}$ =	1	0.4	0.03	tiny

  Generally stated, for S_N2 reactivity we have $k(\text{MeCH}_2\text{—X}) > k(\text{R}_{prim}\text{CH}_2\text{—X}) >> k(\text{R}_{sec}\text{CH}_2\text{—X})$; $k(\text{R}_{tert}\text{CH}_2\text{—X}) \approx 0$ (unit: $1\ mol^{-1}\ s^{-1}$).
- Third, there is a substituent effect on the S_N2 reactivity that results from a change in the geometry at the attacked C atom. This effect occurs with S_N reactions with cyclic alkylating agents. Here, an increase of the bond angle between the reaction center and the two adjacent ring carbon atoms to the 120° of the transition-state geometry is only strain-free in five-membered rings (cf. Figure 2.10); therefore, cyclopentylation occurs as fast as isopropylation:

	*i*Pr–Br	cyclopropyl–Br	cyclobutyl–Br	cyclopentyl–Br	cyclohexyl–Br
$k_{S_N2,\,rel}$ =	1.0	< 0.0001	0.008	1.6	0.01

- Cyclohexylations are different. In such cases, ring strain is built up when a bond angle of 120° is reached in the transition state, leading to a severely reduced S_N2 reactivity. S_N2 cyclobutylations proceed even more slowly. Intermolecular S_N2 cyclopropylations are very uncommon. The ring strain associated with achieving an sp^2-hybridized carbon atom in a three-membered ring is just too large and it slows S_N2 reactions of cyclopropyl halides and related compounds to a virtual halt.

In S_N2 reactions there is also a rate-increasing electronic substituent effect. It is due to facilitation of the rehybridization of the reacting C atom from sp^3 to sp^2 (Figure 2.10). This effect is exerted by unsaturated substituents bound to the reacting C atom. These include substituents, such as alkenyl, aryl, or the C=O double bond of ketones or esters. When it is not prevented by the occurrence of strain, the π-electron system of these substituents can line up in the transition state parallel to the $2p_z$ atomic orbital of the reacting C atom. This orbital thereby becomes part of a *delocalized* π-electron system. Consequently, there is a reduction in energy and a corresponding increase in the S_N2 reaction rate. Allyl and benzyl halides are therefore as good electrophiles as methyl iodide:

$Nu^{\ominus}$ +	$MeCH_2{-}X$	$vinylCH_2{-}X$	$PhCH_2{-}X$
$k_{S_N2,\,rel}$ =	1	40	120

Because of the substituent effect just described, allyl and benzyl halides generally react with nucleophiles according to an S_N2 mechanism. This occurs even though the S_N1 reactivity of allyl and benzyl halides is *higher* than that of nonconjugated alkylating agents (see Section 2.5.4).

α-Halogenated ketones and α-halogenated carboxylic acid esters also react with nucleophiles according to the S_N2 mechanism. However, for them the alternative of an S_N1 mechanism is not possible. This is because it would have to take place via a carbenium ion, which would be extremely destabilized by the strongly electron-withdrawing acyl or alkoxyacyl substituent.

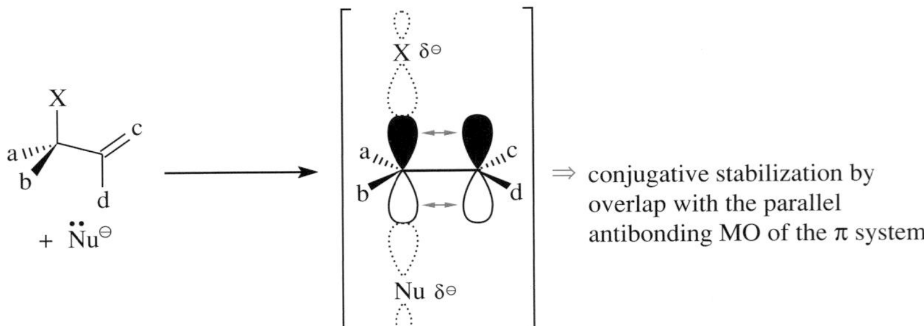

Fig. 2.10. Electronic effects on S_N2 reactivity: conjugative stabilization of the transition state by suitably aligned unsaturated substituents.

2.5 S_N1 Reactions: Kinetic and Stereochemical Analysis; Substituent Effects on Reactivity

2.5.1 Energy Profile and Rate Law of S_N1 Reactions; Steady State Approximation

Substitution reactions according to the S_N1 mechanism take place in two steps (Figure 2.11). In the first and slower step, heterolysis of the bond between the C atom and the leaving group takes place. A carbenium ion is produced as a high-energy, and consequently short-lived, intermediate. In a considerably faster second step, it reacts with the nucleophile to form the substitution product Nu—R.

In the S_N1 mechanism, the nucleophile does not actively react with the alkylating agent. The reaction mechanism consists of the alkylating agent first dissociating into a carbenium ion and the leaving group. *Only then* does the nucleophile change from a "spectator" into an active participant that intercepts the carbenium ion to form the substitution product.

What does the rate law for the substitution mechanism of Figure 2.11 look like? The rate of formation of the substitution product Nu—R in the second step can be written as Equation 2.5 because this step represents an elementary reaction. Here, as in Figure 2.11, k_{het} and k_{assoc} designate the rate constants for the heterolysis and the nucleophilic reaction, respectively.

However, Equation 2.5 cannot be correlated with experimentally determined data. The reason for this is that the concentration of the carbenium ion intermediate appears in it. This concentration is extremely small during the entire reaction and consequently cannot be measured. However, it cannot be set equal to zero, either. In that case, Equation 2.5 would mean that the rate of product formation is also equal to zero and thus that the reaction does not take place at all. Accordingly, we must have a better approximation, one that is based on the following consideration:

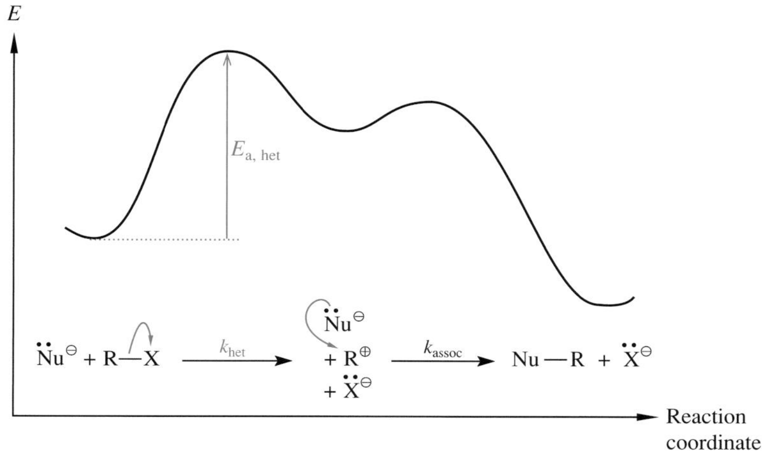

Fig. 2.11 Mechanism and energy profile of S_N1 reactions: $E_{a,\,het}$ designates the activation energy of the heterolysis, k_{het} and k_{assoc} designate the rate constants for the heterolysis and the nucleophilic reaction on the carbenium ion, respectively.

$$\frac{d[\text{Nu—R}]}{dt} = k_{\text{assoc}}[\text{R}^{\oplus}][\text{Nu}^{\ominus}] \tag{2.5}$$

$$\frac{d[\text{R}^{\oplus}]}{dt} = 0 \text{ within the limits of the Bodenstein approximation} \tag{2.6}$$

$$= k_{\text{het}}[\text{R—X}] - k_{\text{attack}}[\text{R}^{\oplus}][\text{Nu}^{\ominus}] \tag{2.7}$$

$$\Rightarrow [\text{R}^{\oplus}] = \frac{k_{\text{het}}}{k_{\text{assoc}}} \bullet \frac{[\text{R—X}]}{[\text{Nu}^{\ominus}]} \tag{2.8}$$

Equation 2.8 in Equation 2.5 ⇒

$$\frac{d[\text{Nu—R}]}{dt} = k_{\text{het}} \bullet [\text{R—X}] \tag{2.9}$$

The Steady State Approximation

The concentration of an intermediate in a multistep reaction is always very low when it reacts faster than it is produced. If this concentration is set equal to zero in the derivation of the rate law, unreasonable results may be obtained. In such a case, one resorts to a different approximation. The *change of the concentration* of this intermediate as a *function of time* is set equal to zero. This is equivalent to saying that the concentration of the intermediate during the reaction takes a value slightly different from zero. This value can be considered to be invariant with time, i.e., steady. Consequently this approximation is called the **steady state approximation**.

Equipped with this principle, let us now continue the derivation of the rate law for S_N reactions. The approximation [carbenium ion] = 0 must be replaced by Equation 2.6. Let us now set the left-hand side of Equation 2.6, the change of the carbenium ion concentration with time, equal to the difference between the rate of formation of the carbenium ion and its consumption. Because the formation and consumption of the carbenium ion are elementary reactions, Equation 2.7 can be set up straightforwardly. Now we set the right-hand sides of Equations 2.6 and 2.7 equal and solve for the concentration of the carbenium ion to get Equation 2.8. With this equation, it is possible to rewrite the previously unusable Equation 2.5 as Equation 2.9. The only concentration term that appears in Equation 2.9 is the concentration of the alkylating agent. In contrast to the carbenium ion concentration, it can be readily measured.

The rate law of Equation 2.9 identifies the S_N reactions of Figure 2.11 as unimolecular reactions. That is why they are referred to as **S_N1 reactions**. The rate of product formation thus depends only on the concentration of the alkylating agent and not on the concentration of the nucleophile. This is the key experimental criterion for distinguishing the S_N1 from the S_N2 mechanism.

From Equation 2.9 we can also derive the following: the S_N1 product is produced with the rate constant k_{het} of the first reaction step. Thus the rate of product formation does not depend on the rate constant k_{attack} of the second reaction step. In a multistep reaction, a particular step may be solely responsible for the rate of product formation. This is referred to as the **rate-determining step**. In the S_N1 reaction, this step is the heterolysis of the alkylating agent. The energy profile of Figure 2.11 shows that the rate-determining step of a multistep sequence is the step in which the highest activation barrier must be overcome. This is a general concept.

Equation 2.9 can also be interpreted as follows. Regardless of which nucleophile enters into an S_N1 reaction with a given alkylating agent, the substitution product is produced with the same rate. Figure 2.12 illustrates this using as an example S_N1 reactions of two different nucleophiles with $Ph_2CH—Cl$. Both take place via the benzhydryl cation $Ph_2CH^{\oplus}$. In the first

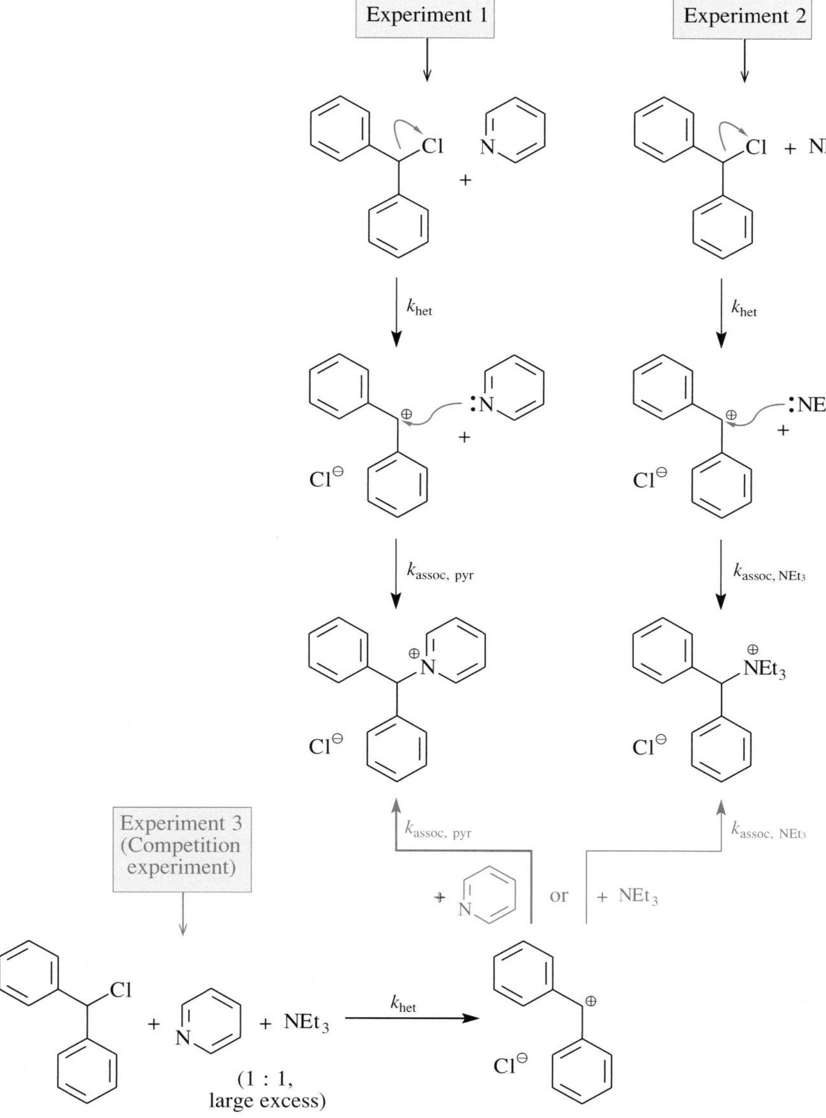

Fig. 2.12. Rate and selectivity of S_N1 reactions. Both reactions 1 and 2 take place via the benzhydryl cation. Both reactions (with pyridine and triethylamine) take place with the same rate constant. The higher nucleophilicity of pyridine does not become noticeable until experiment 3, the "competition experiment": The pyridinium salt is by far the major product.

experiment, this cation is intercepted by pyridine as the nucleophile to form a pyridinium salt. In the second experiment, it is intercepted by triethylamine to form a tetraalkylammonium salt. Both reactions take place with the same rate, represented by the heterolysis constant k_{het} of benzhydryl chloride. The fact that pyridine is a better nucleophile than triethylamine therefore is not apparent under these conditions.

The situation is different in experiment 3 of Figure 2.12. There, both S_N1 reactions are carried out as competitive reactions: benzhydryl chloride heterolyzes (just as fast as before) in the presence of equal amounts of both amines. The benzhydryl cation is now intercepted faster by the more nucleophilic pyridine than by the less nucleophilic triethylamine. The major product will be determined by the relative rates of the reaction of each nucleophile with the benzhydryl cation. The major product in this case is the pyridinium salt.

2.5.2 Stereochemistry of S_N1 Reactions; Ion Pairs

What can be said about the stereochemistry of S_N1 reactions? In the carbenium ion intermediates $R^1R^2R^3C^{\oplus}$, the positively charged C atom has a trigonal planar geometry (cf. Figure 1.3). These intermediates are therefore achiral, if the substituents R^i themselves do not contain stereogenic centers.

When in carbenium ions of this type $R^1 \neq R^2 \neq R^3$, these carbenium ions react with achiral nucleophiles to form chiral substitution products $R^1R^2R^3C—Nu$. These must be produced as a 1:1 mixture of both enantiomers (i.e., as a racemic mixture). Achiral reaction partners alone can never form an optically active product. But in apparent contradiction to what has just been explained, *not all* S_N1 reactions that start from enantiomerically pure alkylating agents and take place via achiral carbenium ions produce a racemic substitution product. Let us consider, for example, Figure 2.13.

Figure 2.13 shows an S_N1 reaction with optically pure (*R*)-2-bromooctane carried out as a solvolysis. By **solvolysis** we mean an S_N1 reaction performed in a polar solvent that also functions as the nucleophile. The solvolysis reaction in Figure 2.13 takes place in a water/ethanol mixture. In the rate-determining step, a secondary carbenium ion is produced. This ion must be planar and therefore achiral. However, it is highly reactive. Consequently, it reacts so quickly with the solvent that at this point in time it has still not completely "separated" from the bromide ion that was released when it was formed. In other words, the reacting carbenium ion is still almost in contact with this bromide ion. It exists as part of a so-called **contact ion pair** $R^1R^2HC^{\oplus}\cdots LBr^{\ominus}$.

The contact ion pair $R^1R^2HC^{\oplus}\cdots LBr^{\ominus}$, in contrast to a free carbenium ion $R^1R^2HC^{\oplus}$, is *chiral*. Starting from enantiomerically pure (*R*)-2-bromooctane, the contact ion pair first produced is also a pure enantiomer. In this ion pair, the bromide ion adjacent to the carbenium ion center partially protects one side of the carbenium ion from the reaction by the nucleophile. Consequently, the nucleophile preferentially reacts from the side that lies opposite the bromide ion. Thus, the solvolysis product in which the configuration at the reacting C atom has been inverted is the major product. To a minor extent the solvolysis product with retention of configuration at the reacting C atom is formed.

It was actually found that 83% of the solvolysis product was formed with inversion of configuration and 17% with retention. This result is equivalent to the occurrence of 66% inver-

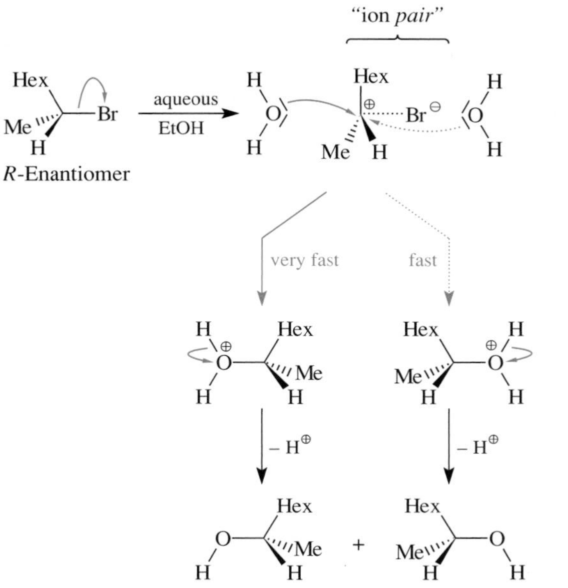

83% *S*-Enantiomer + 17% *R*-Enantiomer, or:
66% *S*-Enantiomer + 34% racemic mixture

Fig. 2.13. Stereochemistry of an S_N1 reaction that takes place via a contact ion pair. The reaction proceeds with 66% inversion of configuration and 34% racemization.

sion of configuration and 34% racemization. We can therefore generalize and state that when in an S_N1 reaction the nucleophile reacts with the contact ion pair, the reaction proceeds with *partial* inversion of configuration.

On the other hand, when the nucleophile reacts with the carbenium ion after it has separated from the leaving group, the reaction takes place with *complete* racemization. This is the case with more stable and consequently longer-lived carbenium ions. For example, the α-methyl benzyl cation, which is produced in the rate-determining step of the solvolysis of *R*-phenethyl bromide in a water/ethanol mixture, is such a cation (Figure 2.14). As in the solvolysis of Figure 2.13, the nucleophile is a water/ethanol mixture.

The following should be noted for the sake of completeness. The bromide ion in Figure 2.14 moves far enough away from the α-methylbenzyl cation intermediate that it allows the solvent to react on both sides of the carbenium ion with equal probability. However, the bromide ion does not move away from the carbenium ion to an arbitrary distance. The electrostatic attraction between oppositely charged particles holds the carbenium ion and the bromide ion together at a distance large enough for solvent molecules to fit in between. This is the so-called **solvent-separated ion pair**.

2.5.3 Solvent Effects on S_N1 Reactivity

In contrast to the S_N2 mechanism (Section 2.4.3), the structure of the transition state of the rate-determining step in the S_N1 mechanism cannot be depicted in a simple way. As an aid, we

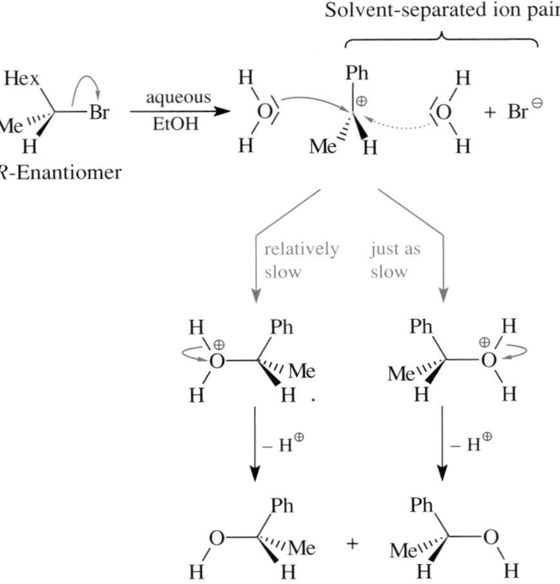

50% *S*-Enantiomer + 50% *R*-Enantiomer, thus:
0% *S*-Enantiomer + 100% racemic mixture

Fig. 2.14. Stereochemistry of an S_N1 reaction that takes place via a solvent-separated ion pair. The reaction proceeds with 0% inversion and 100% racemization.

can use a transition state model. According to the Hammond postulate for an endothermic reaction like carbenium ion formation, the transition state should be late, or product-like, so a suitable model for the transition state of the rate-determining step of an S_N1 reaction is the corresponding carbenium ion. The rate constant k_{het} of S_N1 reactions should therefore be greater the more stable the carbenium ion produced in the heterolysis:

$$R{-}X \xrightarrow{k_{Het}} R^{\oplus} + {:}X^{\ominus}$$

As a measure of the rate of formation of this carbenium ion, one can take the free energy of heterolysis of the alkylating agent from which it arises. Low free energies of heterolysis are related to the formation of a stable carbenium ion and thus high S_N1 reactivity. Because S_N1 reactions are carried out in solution, the free energies of heterolysis in solution are the suitable stability measure. These values are not available, but they can be calculated from other experimentally available data (Table 2.1) by means of the thermodynamic cycle in Figure 2.15.

Heterolyses of alkyl bromides in the gas phase always require more energy than the corresponding homolyses (cf. Table 2.1). In fact, the ΔG values of these heterolyses are considerably more positive than the ΔH values of the homolyses (even if we take into consideration that ΔG and ΔH values cannot be compared directly with each other). The reason for this is that extra energy is required for separating the charges. As Table 2.1 shows, this situation is reversed in polar solvents such as water. There the heterolyses require less energy than the homolyses.

Tab. 2.1 Free Energy Values from Gas Phase Studies (Lines 1–3). Free Energies of Heterolysis in Water (Line 4) Calculated Therefrom According to Figure 2.15*

$R{-}Br \longrightarrow R^{\oplus} + Br^{\ominus}$	Me	Et	*i*Pr	*tert*-Bu	$PhCH_2$
$\frac{\Delta G_{het,\ g}}{kcal/mol}$	+214	+179	+157	+140	+141
$\frac{\Delta G_{hyd\ R^{\oplus})}}{kcal/mol}$	–96	–78	–59	–54	–59
$\frac{\Delta G_{hyd\ Br^{\ominus})}}{kcal/mol}$	–72	–72	–72	–72	–72
$\frac{\Delta G_{het,\ H_2O}}{kcal/mol}$	+47	+30	+27	+14	+11
$\left(vgl.\ \frac{\Delta H_{hom,\ any\ medium}}{kcal/mol}\right)$	(+71)	(+68)	(+69)	(+63)	(+51)
$\tau_{1/2,\ Het,\ 298\ K}$	$\geq 10^{16}$ yr	$\geq 10^5$ yr	≥ 220 yr	≥ 0.7 s	≥ 0.007 s

*From these energies and using the Eyring equation (Equation 1.1), one can calculate minimum half-lives for the pertinent heterolyses in water (line 6). These are minimum values because the $\Delta G_{Het,\ H_2O}$ values were used for $\Delta G^{\ddagger}$ in the Eyring equation, whereas actually $\Delta G^{\ddagger} > {}^{*}\Delta G_{Het,\ H_2O}$ ΔG_{hyd} = free energy of hydration, ΔG_{het} = free energy of heterolysis, and ΔH_{hom} = enthalpy of homolysis.

You see the reason for this solvent effect when you look at lines 2 and 3 in Table 2.1. Ions are stabilized considerably by solvation. Heterolyses of alkylating agents, and consequently S_N1 reactions, therefore, succeed only in highly solvating media. These include the **dipolar protic** (i.e., hydrogen-bond donating) **solvents** like methanol, ethanol, acetic acid, and aqueous acetone as well as the **dipolar aprotic** (i.e., non-hydrogen-bond donating) **solvents** acetone, acetonitrile, DMF, NMP, DMSO, and DMPU and HMPA (Figure 2.16).

It is important to note that just because a reaction occurs with an species that could give rise to a reasonably stable cation and the solvent is polar does not mean that an S_N1 reaction is necessarily taking place.

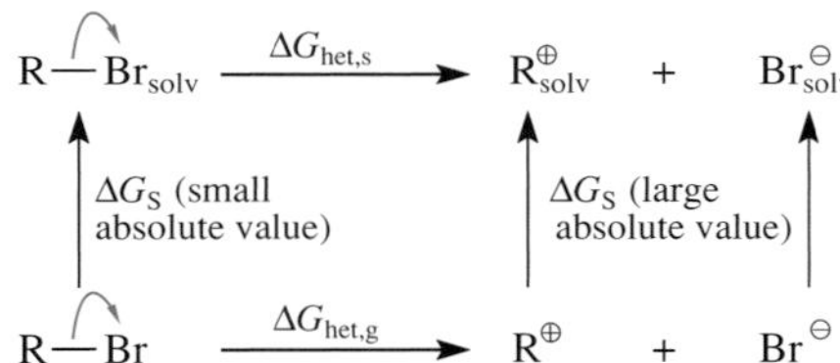

Fig. 2.15. Thermodynamic cycle for calculating the free energy of heterolysis of alkyl bromides in solution ($\Delta G_{het,\ s}$) from free energies in the gas phase: ΔG_s refers to the free energy of solvation, $\Delta G_{het,\ g}$ refers to the free energy of heterolysis in the gas phase.

DMF = **Dim**ethyl**f**ormamide: H—C(=O)—NMe_2

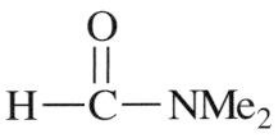

NMP = ***N*-M**ethyl**p**yrrolidone:

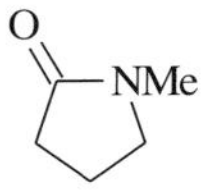

DMSO = **Dim**ethyl**s**ulf**o**xide: Me—S(=O)—Me

DMPU = *N,N*′-**Dim**ethyl-*N,N*′-**p**ropylene **u**rea:

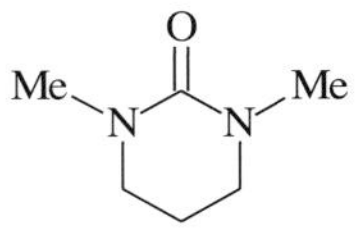

HMPA = **H**exa**m**ethyl **p**hosphoric acid tri**a**mide: $(Me_2N)_3P{=}O$

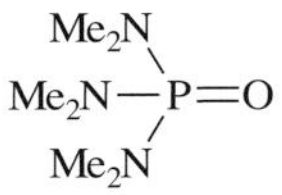

Fig. 2.16. Polar aprotic solvents and the abbreviations that can be used for them.

2.5.4 Substituent Effects on S_N1 Reactivity

The stabilities of five carbenium ions formed in the heterolyses of alkyl bromides in water (as an example of a polar solvent) can be compared by means of the associated free energy values ΔG_{het}, H_2O (Table 2.1, row 4). From these free energy values we get the following order of stability in polar solvents: $CH_3^{\oplus} < MeCH_2^{\oplus} < Me_2CH^{\oplus} < Me_3C^{\oplus} <$ Ph—$CH_2^{\oplus}$. One can thus draw two conclusions: (1) a phenyl substituent stabilizes a carbenium ion center, and (2) alkyl substituents also stabilize a carbenium ion center. The stability order just cited is similar to the stability order for the same carbenium ions in the gas phase (Table 2.1, line 1). However, in the gas phase considerably larger energy differences occur. They correspond to the inherent stability differences (i.e., without solvation effects).

We observed quite similar substituent effects in Section 1.2.1 in connection with the stability of radicals of type $R^1R^2R^3C•$. These effects were interpreted both in VB and in MO terms. For the carbenium ions, completely analogous explanations apply, which are shown in Tables 2.2 (effect of conjugating groups) and 2.3 (effect of alkyl groups). These tables use res-

Tab. 2.2 Stabilization of a Trivalent Carbenium Ion Center by Conjugating Substituents: Experimental Findings and Their Explanation by Means of Resonance Theory

	Stabilization of ⊕	VB formulation
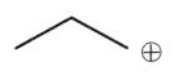 propyl cation	↓ increases ↓	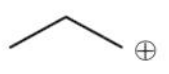 propyl cation
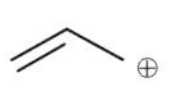 allyl cation		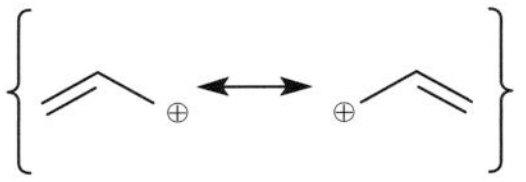 {allyl cation ⟷ allyl cation}
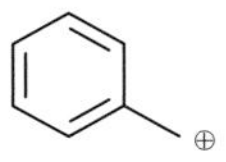 benzyl cation		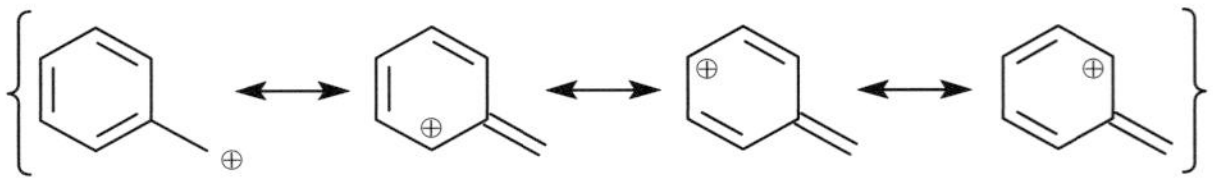 {benzyl cation ⟷ ⟷ ⟷ resonance structures with ⊕ in ring}

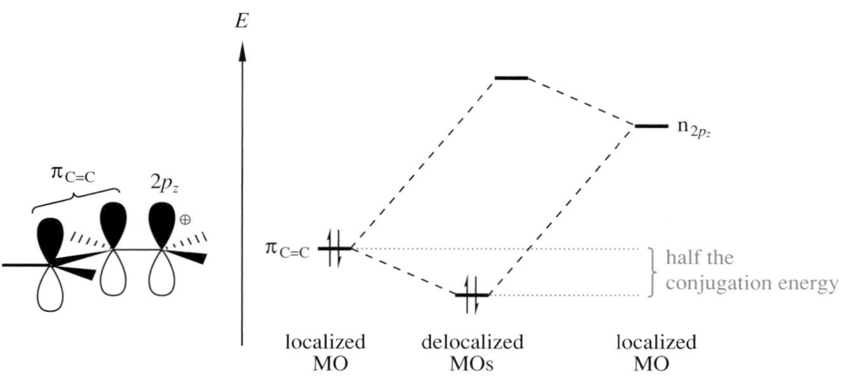

Fig. 2.17. MO interactions responsible for the stabilization of trivalent carbenium ion centers by suitably oriented unsaturated substituents ("conjugation").

onance theory for the explanation. Figures 2.17 and 2.18 explain the same phenyl- and alkyl group effects by means of MO theory. Conjugating substituents, which are electron-rich, stabilize a neighboring carbenium ion center via a resonance effect, and alkyl substituents do the same via an inductive effect. Hyperconjugation plays a role as well.

The very large inherent differences in the stability of carbenium ions are reduced in solution—because of a solvent effect—but they are not eliminated. This solvent effect arises because of the dependence of the free energy of solvation $\Delta G_{hyd(R^\oplus)}$ on the structure of the carbenium ions (Table 2.1, row 2). This energy becomes less negative going from $Me^\oplus$ to Ph—$CH_2^\oplus$, as well as in the series $Me^\oplus \rightarrow Et^\oplus \rightarrow iPr^\oplus \rightarrow$ *tert*-$Bu^\oplus$. The reason for this is hindrance of solvation. It increases with increasing size or number of the substituents at the carbenium ion center.

Allyl halides heterolyze just as easily as benzyl halides because they also produce a resonance-stabilized carbenium ion. Even faster heterolyses are possible when the charge of the resulting carbenium ion can be delocalized by more than one unsaturated substituent and can thereby be stabilized especially well. This explains the remarkably high S_N1 reactivities of the benzhydryl halides (via the benzhydryl cation) and especially of the triphenylmethyl halides (via the trityl cation):

	$PhCH_2$–X	Ph_2CH–X	Ph_3C–X
$k_{S_N1,\,rel}$	1	1000	10^8

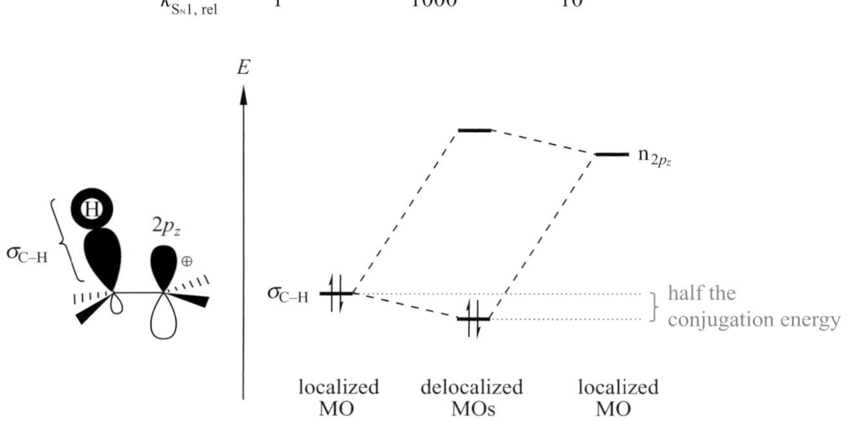

Fig. 2.18. MO interactions responsible for the stabilization of trivalent carbenium ion centers by suitably oriented C—H bonds in the β position ("hyperconjugation").

Tab. 2.3 Stabilization of Trivalent Carbenium Ion Centers by Methyl Substituents: Experimental Findings and Their Explanation by Means of Resonance Theory

	Stabilization of ⊕	VB formulation
$H_3C^{\oplus}$	increases (↓)	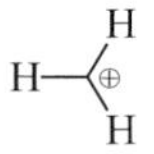 $H_2C^{\oplus}$–H
$H_3C{-}H_2C^{\oplus}$		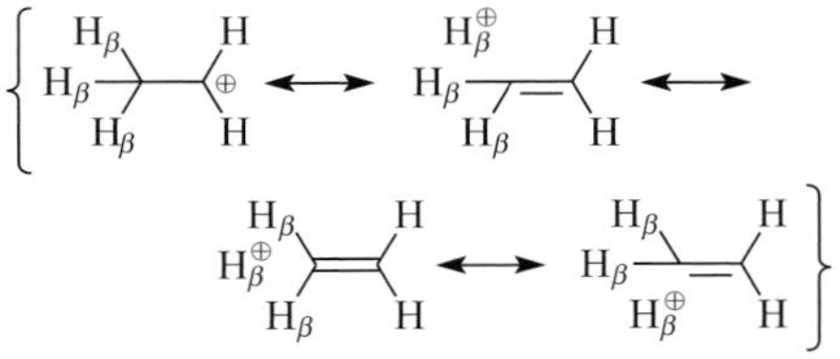 { $(H_\beta)_3C{-}CH_2^{\oplus}$ ⟷ $H_\beta^{\oplus}\ (H_\beta)_2C{=}CH_2$ ⟷ $H_\beta^{\oplus}\ (H_\beta)_2C{=}CH_2$ ⟷ $H_\beta^{\oplus}\ (H_\beta)_2C{=}CH_2$ } i.e. 1 no-bond resonance form per H_β
$(H_3C)_2HC^{\oplus}$		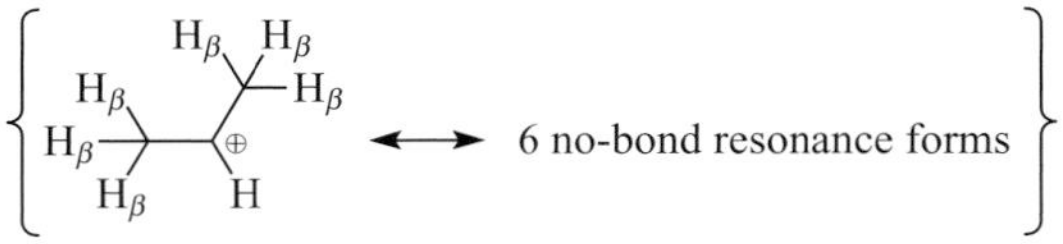 { $((H_\beta)_3C)_2CH^{\oplus}$ ⟷ 6 no-bond resonance forms }
$(H_3C)_3C^{\oplus}$		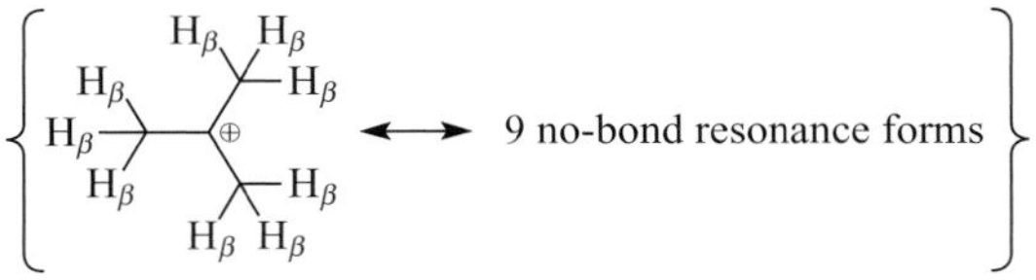 { $((H_\beta)_3C)_3C^{\oplus}$ ⟷ 9 no-bond resonance forms }

Side Note 2.3. The Structures of Alkyl- and Aryl-substituted Carbenium Ions

The fact that one or more resonance form(s) for the carbenium ions in Tables 2.2 and 2.3 can be drawn, does not only account for the stabilities already discussed, but also for structural factors. First, the C=C double bonds that in the resonance forms of a tertiary carbenium ion may be drawn from the central C atom to each of its three neighboring C atoms result in a trigonal planar coordination (the same conclusion is arrived at with VSEPR theory, see Section 1.1.1). Second, these C=C double bonds imply that the $C^{\oplus}$–C bonds of carbenium ions are shorter than the C–C bonds in neutral analogous compounds. Third, wherever a C=C double bond occurs in the resonance form of one of the carbenium ions in Table 2.3, this is accompanied by the disappearance of a C–H_β bond. Therefore, the C–H_β bonds in carbenium ions are expected to be longer than in neutral analogous compounds. The same is true of C–C bonds in the same relative postion. All this means is that hyperconjugation (double bond/no bond resonance) can occur with both C-H and C-C bonds. Actually, it can occur, in principle, with a variety of sigma bonds. The C-Si bond is one of the most important of these and such hyperconjugation contributes significantly to the stability of β-silyl carbenium ions, the so-called β-silyl effect.

Hyperconjugation using C-H and C-C bonds is illustrated by X-ray crystal structure analyses of the *tert*-butyl and *tert*-adamantyl cations (Formula **A** and **B** in Figure 2.19, respectively):

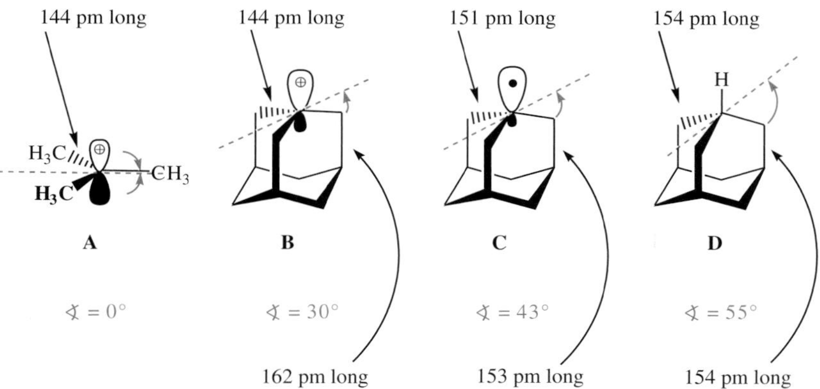

Fig. 2.19. C^{β}-$C^{\oplus}$ bond-length reduction in the stabilization of carbenium ions through hyperconjugation. The planar carbenium ion center experiences the highest hyperconjugative stabilization leading to a large C^{β}-$C^{\oplus}$ bond length reduction.

- The cationic center is planar unless it is incorporated into a (poly)cyclic ring system and extant pyramidalization becomes inevitable.
- The $C^{\oplus}$–C bonds of the adamantyl cation are by 10 pm shorter than the C–C bonds in the neutral adamantane (Formula **D** in Figure 2.19).
- The C–C bonds at the β position in the adamantyl cation are by 8 pm longer than in the adamantane.

The four resonance forms of the benzylic cations (bottom of Table 2.2) allow for the prediction of structural details. They are confirmed, for example, by crystal structure analysis of the cumyl cation (Figure 2.20):

- The three substituents of the benzylic carbon atom and the phenyl ring are coplanar.
- The $C^{\oplus}$–C_{ipso} bond is shortened.
- The C_{ipso}–C_{ortho} bond is elongated.
- The C_{ortho}–C_{meta} bond is shortened.
- The C_{meta}–C_{para} bond is elongated.

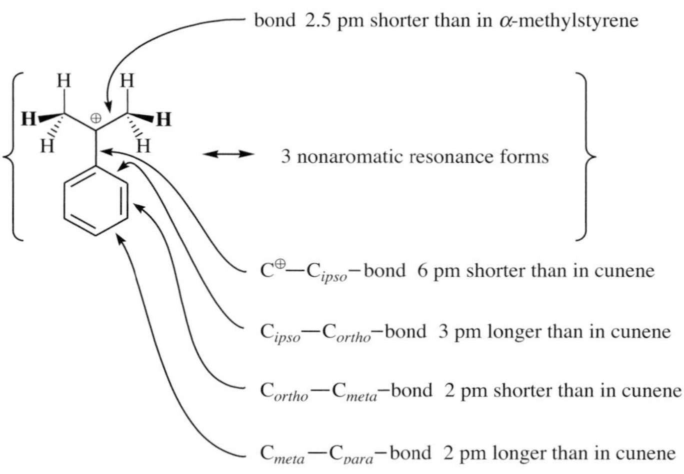

Fig. 2.20. C,C bond-length reductions and elongations in the cumyl cation (compared to cunene) confirming the stabilization of the carbenium ion center through conjugation, and H_3C^{β}-$C^{\oplus}$ bond-length reduction (compared to α-methylstyrene) due to the additional stabilization of the carbenium ion center caused by hyperconjugation (cf. Figure 2.19).

The $C^{\oplus}$–CH_3 bonds in the cumyl cation are also shortened, but only slightly, by 2.5 pm. Clearly, this falls short of the truncation observed with the $C^{\oplus}$–CH_3 bonds of the *tert*-butyl cation (Figure 2.19). The bulk of the stabilization of the cation is performed by the phenyl ring via resonance, mitigating the need for extensive stabilization via C-H hyperconjugation.

Side Note 2.4. *para*-Methoxylated Trityl Cations in Nucleotide Synthesis

One methoxy group in the *para*-position of the trityl cation contributes to further stabilization due to its pi electron-donating ability. The same is true for each additional methoxy group in the *para*-position of the remaining phenyl rings. The resulting increase in the S_N1 reactivity of multiply *para*-methoxylated trityl ethers is used in nucleotide synthesis (Figure 2.21), where these ethers serve as acid-labile protecting groups. The more stable the resulting trityl cation (i.e., the more methoxy groups it contains in *para*-positions), the faster it is formed under acidic conditions.

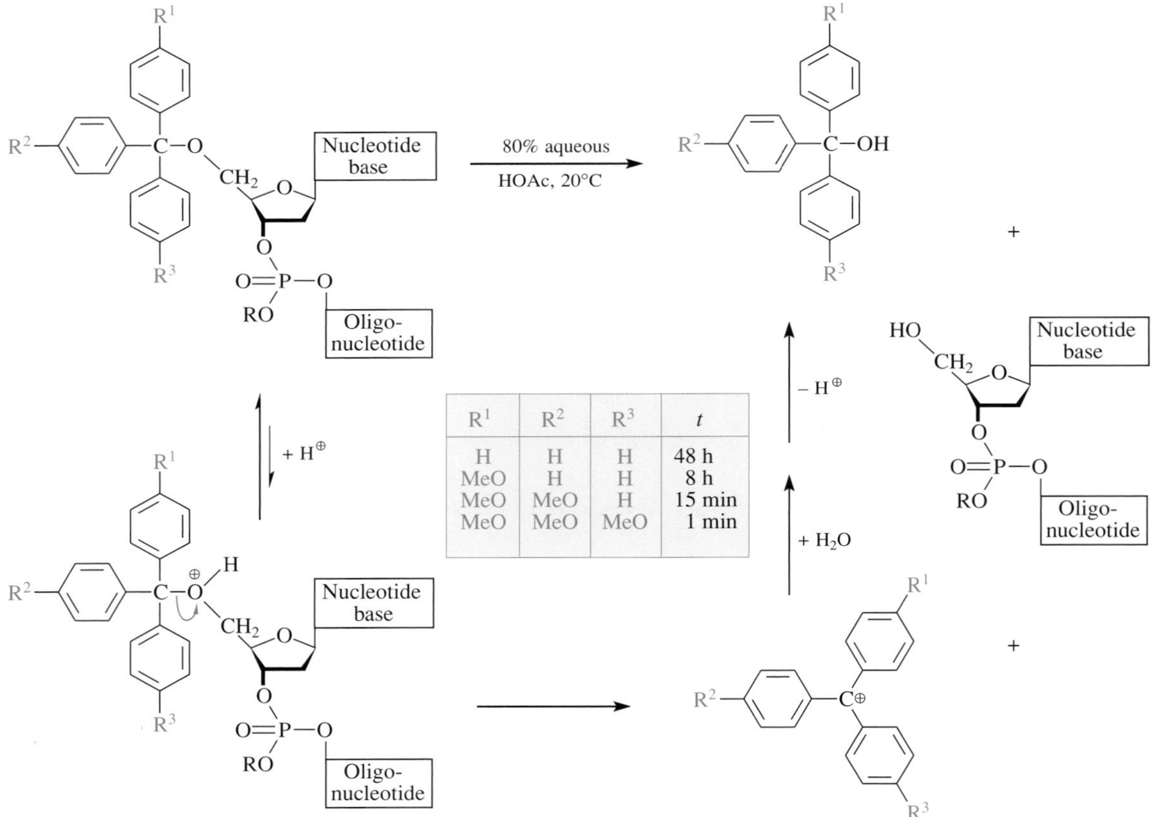

Fig. 2.21. Acid catalysed S_N1 substitutions of trityl ethers to trityl alcohols, using deprotection procedures from nucleotide synthesis as an example. The table in the center indicates the time (t) it takes to completely cleave the respective trityl groups.

Finally, two or three amino or dimethylamino groups in the *para*-position stabilize trityl cations so efficiently that the associated nonionized neutral compounds are no longer able to exist at all but heterolyze quantitatively to salts. These salts include, for example, the well-known triphenylmethane dyes malachite green and crystal violet:

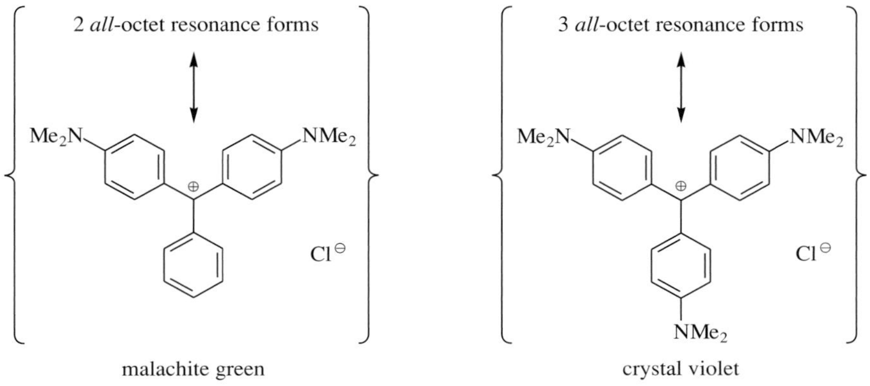

2 *all*-octet resonance forms

malachite green

3 *all*-octet resonance forms

crystal violet

Only one other substituent is capable of stabilizing trityl cations more effectively than amino or dimethylamino groups: namely, the oxyanion substituent $O^{\ominus}$. The stabilization of the trityl cation form of the dianion of the indicator dye phenolphthalein depends on the presence of two such substituents:

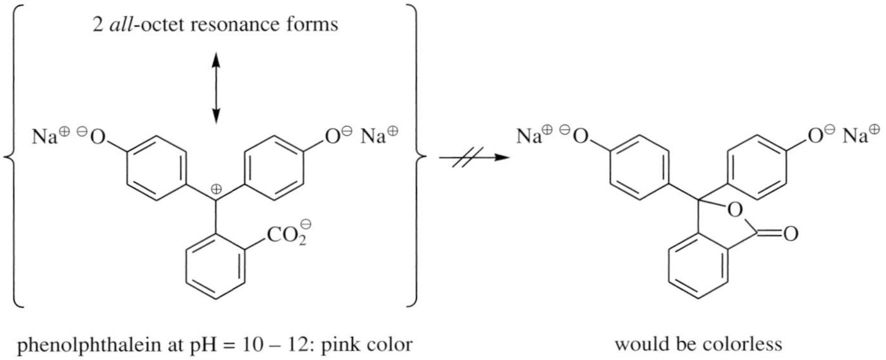

2 *all*-octet resonance forms

phenolphthalein at pH = 10 – 12: pink color

would be colorless

Crystal structure analyses of trityl cations have always shown that their benzylic cation substructures deviate considerably from the planarity that might be expected (cf. Figure 2.20). It is found that the aryl rings are distorted like a propeller (Figure 2.22). The cause of this distortion is similar to the one observed in trityl radicals (cf. Figure 1.9): Here, it is the *ortho* hydrogen atoms of adjacent aryl residues that avoid each other.

Why then are trityl cations still more stable than benzylic cations? An aryl residue that is rotated out of the nodal plane of the $2p_z$ AO of the benzylic cation center by an angle χ provides resonance stabilization that is decreased by $\cos^2\chi$-fold. Three aryl residues in a trityl cation can thus provide up to $3 \times \cos^2(30°) = 2.25$ times more resonance stabilization than *one*

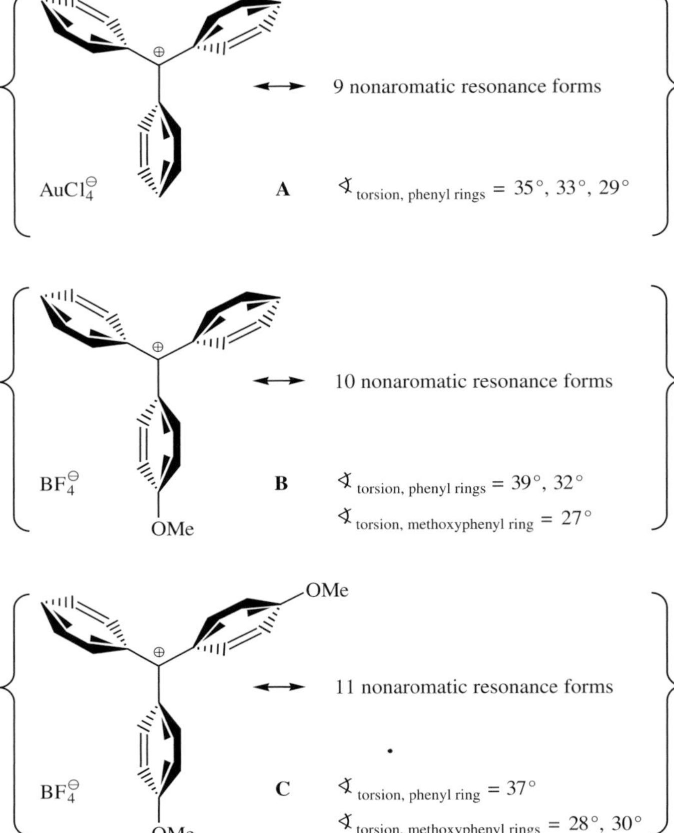

Fig. 2.22. Non-planarity of trityl cations: compromise between maximum benzylic resonance (planarizing effect) and minimum aryl/aryl repulsion (twisting effect).

aryl residue in a planar benzylic cation assuming that the distortion from planarity of the aryl residues in trityl cations amounts to an average 30°. An additive net stabilization would thus actually be predictable. Incidentally, in the trityl cations of **B** and **C** in Figure 2.22 the (*para*-methoxyphenyl) residues are associated with smaller χ values than the phenyl residues. This ensures that the cations are most effectively stabilized by the better donor—a MeO substituted phenyl residue.

2.6 When Do S_N Reactions at Saturated C Atoms Take Place According to the S_N1 Mechanism and When Do They Take Place According to the S_N2 Mechanism?

From Sections 2.4.4 and 2.5.4, the following rules of thumb can be derived:

Rules of Thumb

S_N1 reactions are always observed
- in substitutions on R_{tert}—X, Ar_2HC—X, and Ar_3C—X;
- in substitutions on substituted and unsubstituted benzyl and allyl triflates;
- in substitutions on R_{sec}—X when poor nucleophiles are used (e. g., in solvolyses);
- in substitutions on R_{sec}—X that are carried out in the presence of strong Lewis acids such as in the substitution by aromatics ("Friedel–Crafts alkylation," see Figure 5.25);
- but almost never in substitutions on R_{prim}—X (exception: R_{prim}—N$^{\oplus}$ N).

S_N2 reactions always take place
- (almost) in substitutions in sterically unhindered benzyl and allyl positions (exception: benzyl and allyl triflates react by S_N1);
- in substitutions in MeX and R_{prim}—X;
- in substitutions in R_{sec}—X, provided a reasonably good nucleophile is used.

These are guidelines. Nucleophilic substitution reactions proceed via a mechanistic continuum from S_N2 to S_N1. And that doesn't include processes that proceed by SET (single electron transfer)! Life is neither easy nor fair.

2.7 Getting by with Help from Friends, or a Least Neighbors: Neighboring Group Participation

2.7.1 Conditions for and Features of S_N Reactions with Neighboring Group Participation

A leaving group in an alkylating agent can be displaced not only by a nucleophile added to the reaction mixture but also by one in the alkylating agent itself. This holds for alkylating agents that contain a nucleophilic electron pair at a suitable distance from the leaving group. The structural element on which this electron pair is localized is called a neighboring group. It displaces the leaving group stereoselectively through a backside reaction. This process thus corresponds to that of an S_N2 reaction. But because the substitution through the neighboring group takes place intramolecularly, it represents a unimolecular process. In spite of this, organic chemists, who want to emphasize the mechanistic relationship and not the rate law, classify substitution reactions with neighboring group participation as S_N2 reactions.

Because of this neighboring group participation, a cyclic and possibly strained (depending on the ring size) intermediate **A** is formed from the alkylating agent:

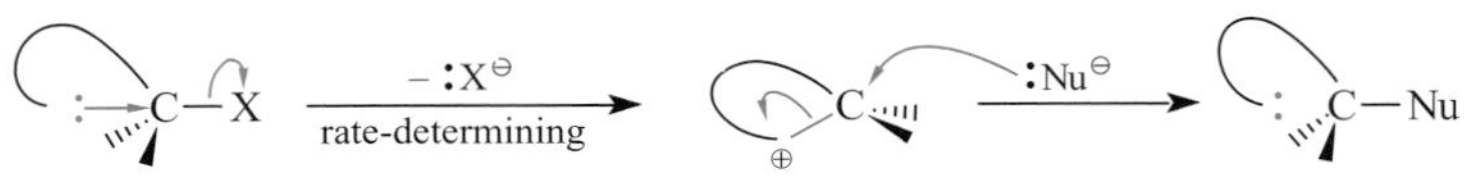

$- :X^{\ominus}$ rate-determining $:Nu^{\ominus}$ C–X C–Nu

A (three- or five-membered) **B**

This intermediate generally contains a positively charged center, which represents a new leaving group. This group is displaced in a second step by the external nucleophile through another backside reaction. This step is clearly an S_N2 reaction. In the reaction product **B**, the external nucleophile occupies the same position the leaving group X originally had. Reactions of this type thus take place with complete retention of configuration at the reacting C atom. This distinguishes them both from substitutions via the S_N2 mechanism and from substitutions according to the S_N1 mechanism. Two inversions equal retention.

For an S_N reaction to take place with neighboring group participation, a neighboring group not only must be present, but it must be sufficiently reactive. The reaction of the "neighboring group" must take place faster than the reaction with the external nucleophile. Otherwise, the latter would initiate a normal S_N2 reaction. In addition, the neighboring group must actively displace the leaving group before this group leaves the alkylating agent on its own. Otherwise, the external nucleophile would enter via an S_N1 mechanism. Therefore, reactions with neighboring group participation take place more rapidly than comparable reactions without such participation. This fact can be used to distinguish between S_N reactions with neighboring group participation and normal S_N1 and S_N2 reactions.

The nucleophilic electron pairs of the neighboring group can be nonbonding, or they can be in a π bond or, in special cases, in a σ bond (Figure 2.23). Generally they can displace the leaving group only when this produces a three- or five-membered cyclic intermediate. The formation of rings of a different size is almost always too slow for neighboring group participations to compete with reactions that proceed by simple S_N1 or S_N2 mechanisms.

The "range" of possible neighboring group participations mentioned above reflects the usual ring closure rates of five-membered ≥ three-membered > six-membered >> other ring sizes.

In general, one speaks of a neighboring group effect only when the cyclic and usually positively charged intermediate cannot be isolated but is subject to an S_N2-like ring opening by the

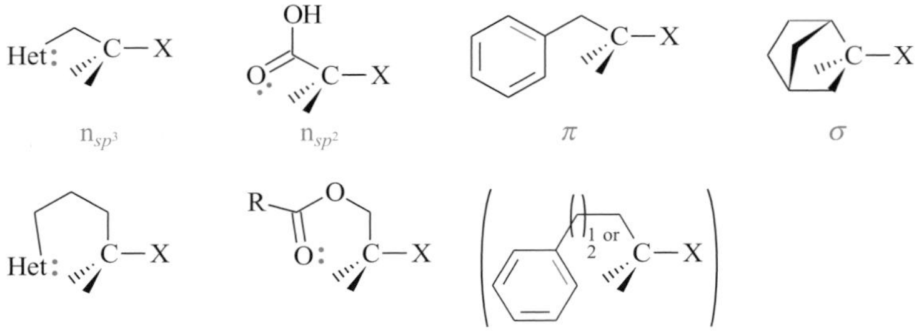

Fig. 2.23. Alkylating agents with structural elements that can act as neighboring groups in S_N reactions. (In the example listed in parentheses, the neighboring group participation initiates an alkylation of the aromatic compound, in which an external nucleophile is not participating at all.)

Het: C–X n_{sp^3} OH O: C–X n_{sp^2} C–X π C–X σ

Het: C–X R O O: C–X 1 or 2 C–X

external nucleophile. This ring opening always happens for three-membered ring intermediates. There it profits kinetically from the considerable reduction in ring strain. In the case of five-membered intermediates, the situation is different. They are not only subject to reaction with the external nucleophile, but they can also react to form *isolable* five-membered rings by elimination of a cation—usually a proton.

2.7.2 Increased Rate through Neighboring Group Participation

The sulfur-containing dichloride "mustard gas," a high-boiling liquid used in World War I as a combat gas in the form of an aerosol, hydrolyzes much more rapidly to give HCl and a diol (Figure 2.24) than its sulfur-free analog 1,5-dichloropentane. Therefore, mustard gas released HCl especially efficiently into the lungs of soldiers who had inhaled it (and thus killed them very painfully). The reason for the higher rate of hydrolysis of mustard gas is a neighboring group effect. It is due to the availability of a free electron pair in a nonbonding orbital on the sulfur atom.

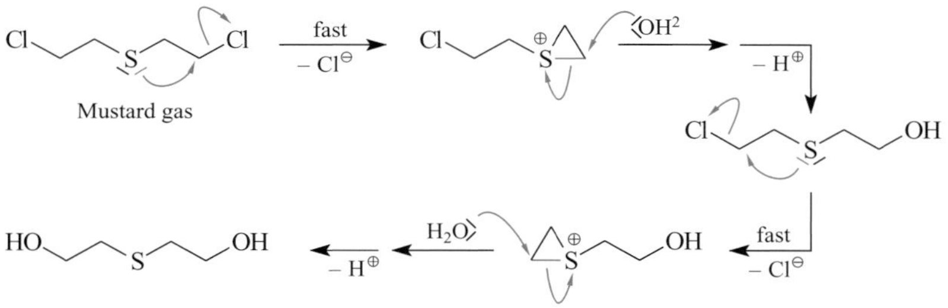

Fig. 2.24. Acceleration of the hydrolysis of mustard gas by neighboring group participation.

Phenethyl tosylate solvolyzes in CF_3CO_2H orders of magnitude faster than ethyl tosylate (Figure 2.25). Because the neighboring phenyl ring can make a π-electron pair available, a phenonium ion intermediate is formed. Phenonium ions are derivatives of the spirooctadienyl

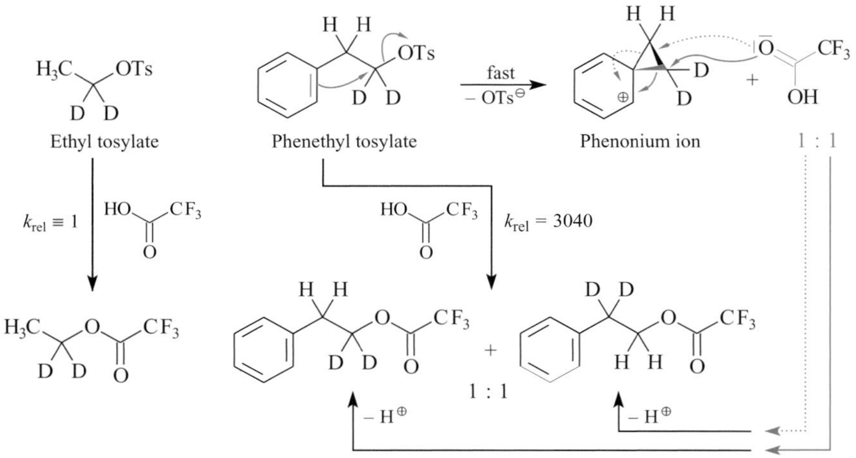

Fig. 2.25. Acceleration of the trifluoroacetolysis of phenethyl tosylate by neighboring group participation.

cation shown. They are therefore related to the sigma complexes of electrophilic aromatic substitution (see Chapter 5). The intermediacy of a phenonium ion in the solvolysis of phenethyl tosylate was proven by isotopic labeling. A deuterium label located exclusively α to the leaving group in the tosylate was scrambled in the reaction product—it emerged half in α-position and half in β-position.

2.7.3 Stereoselectivity through Neighboring Group Participation

A carboxylic ester can participate in an S_N reaction as a neighboring group. In this case, a nonbonding electron pair on the double-bonded oxygen displaces the leaving group in the first reaction step. Let us consider, for example, the glycosyl bromide **B** in Figure 2.26. There a bromide ion is displaced by an acetate group in the β-position. The departure of the bromide ion is facilitated by reaction with Ag(I) ions and converted into insoluble AgBr. The five-membered ring of the resonance-stabilized cation ion **E** is formed. At this stage, an axially oriented C—O bond has emerged from the equatorial C—Br bond of substrate **B** via an inversion of the configuration. In the second reaction step, this C—O bond is broken by the solvent methanol via another inversion of the configuration at the reacting C atom. In the resulting substitution product **C**, the methoxy group is equatorial just like the bromine in the starting material **B**. The saponification of the acetate groups of this compound leads to isomerically

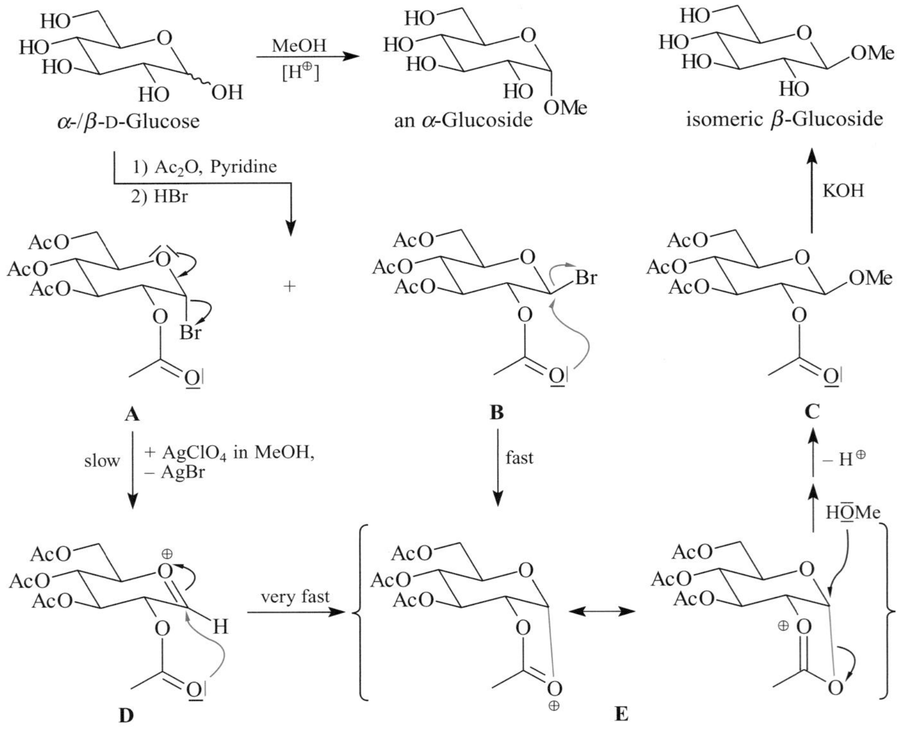

Fig. 2.26. Stereoselectivity due to neighboring group participation in glycoside syntheses.

pure β-D-methylglucopyranoside (Figure 2.26, top right). This is of preparative interest because the direct acetalization of glucose leads to the diastereomeric α-methylglucoside (Figure 2.26).

Bromide **A**, is a stereoisomer of bromide **B**. It is interesting that in methanol in the presence of silver cations, **A** stereoselectively produces the same substitution product **C** as produced by **B**. In glycosyl bromide **A**, however, the C—Br bond can*not* be broken with the help of the acetate group in the β-position. In that case, a *trans*-annulated and consequently strained cationic intermediate would be produced. Glycosyl bromide **A** therefore reacts without neighboring group participation and consequently more slowly than its diastereomer **B.** The first step is a normal S_N1 reaction with heterolysis of the C—Br bond to form the cation **D.** However, once formed, the acetate group in the β-position still exerts a neighboring group effect. With its nucleophilic electron pair, it closes the ring to the cyclic cation **E** that was produced in a *single* step from the diastereomeric glycosyl bromide **B.** The subsequent ring opening of **E** by methanol gives the methyl glucoside **C**, regardless of whether it was derived from **A** or **B.**

S_N reactions of enantiomerically pure diazotized α-amino acids are also preparatively important (Figure 2.27). There the carboxyl group acts as a neighboring group. It has two effects. First, it slows down the S_N1-like decomposition of the diazonium salt, which for normal aliphatic diazonium salts takes place extremely fast to form carbocations. But heterolysis of a diazotized α-amino acid is slowed down. This is because the carbenium ion produced would be strongly destabilized by the carboxyl group, an electron acceptor. In addition, the carboxylic acid group enters actively into the reaction as a neighboring group and displaces the leaving group (N_2) with the nucleophilic, free electron pair of its double-bonded O atom. A highly strained protonated three-membered ring lactone is produced, in which the C—O_2C bond reacts from the back side with the external nucleophile. The nucleophile therefore assumes the position of the original amino group, i.e., appears with retention of the configuration. As in Figure 2.27, H_2O is the entering nucleophile when the diazotization is carried out with aqueous H_2SO_4 and $NaNO_2$. But if the diazotization is carried out with aqueous HBr and

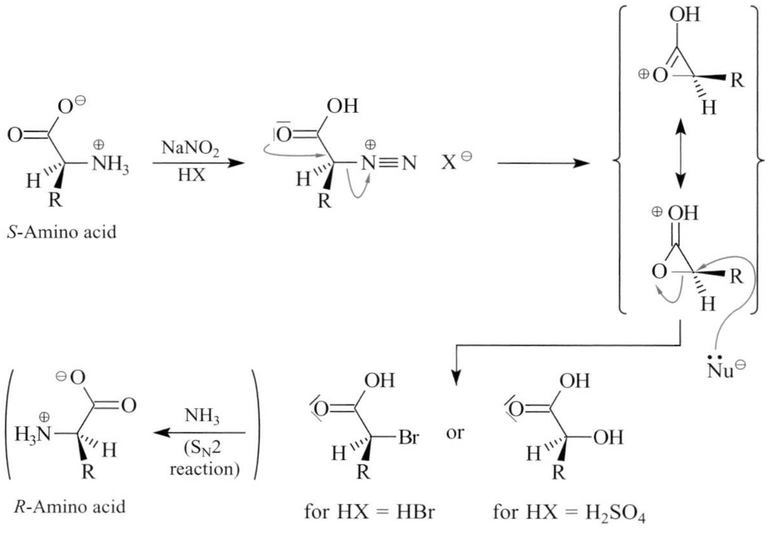

Fig. 2.27. Stereoselectivity due to neighboring group participation in the synthesis of α-functionalized carboxylic acids.

$NaNO_2$, then a bromide ion acts as the external nucleophile. These chemoselectivities are explained by the nucleophilicity order $HSO_4^{\ominus} < H_2O < Br^{\ominus}$ (cf. Section 2.2).

A bonding electron pair that is contained in a C—C single bond normally is not a neighboring group. However, one famous exception is a particular bonding electron pair in the norbornane ring system (Figure 2.28). It is fixed in precisely such a way that it can interact with a reaction center on C2.

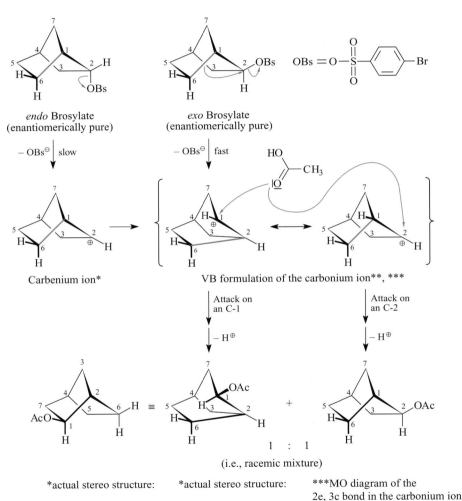

endo Brosylate (enantiomerically pure)

exo Brosylate (enantiomerically pure)

OBs = O–S($=O$)$_2$–C_6H_4–Br

– $OBs^{\ominus}$ slow

– $OBs^{\ominus}$ fast

Carbenium ion*

VB formulation of the carbonium ion**, ***

Attack on an C-1, – $H^{\oplus}$

Attack on an C-2, – $H^{\oplus}$

1 : 1

(i.e., racemic mixture)

*actual stereo structure:

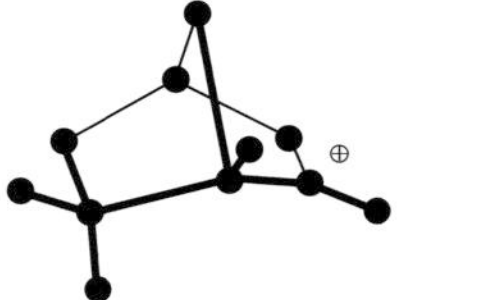

*actual stereo structure:

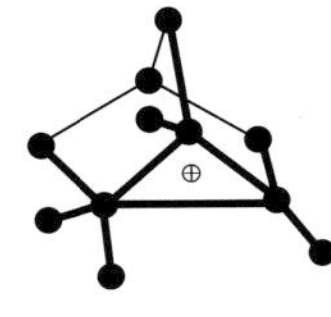

***MO diagram of the 2e, 3c bond in the carbonium ion

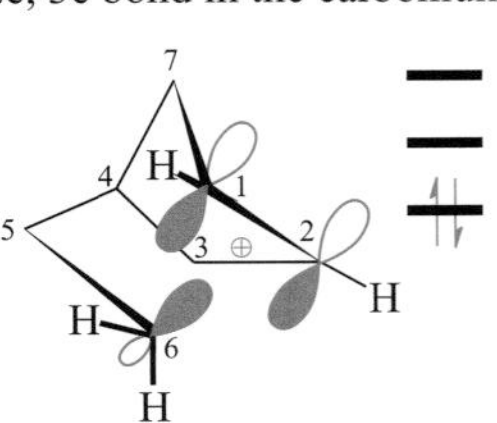

Fig. 2.28. Stereoselectivity by neighboring group participation in the acetolysis of norbornane derivatives; 2e, 3c, two-electron, three-center.

This electron pair speeds up the departure of a suitably oriented leaving group. It also determines the orientation of the nucleophile in the reaction product. The formulas in Figure 2.28 describe both of these aspects better than many words.

Nonetheless, Figure 2.28 requires two comments: (1) the six-membered ring substructure of all other molecules including the cations possesses a boat conformation and not a twist-boat conformation, as the two-dimensional structures of the figure imply. If they had been drawn with the correct boat conformation, the crucial orbital interaction could not have been drawn with so few C—C bonds crossing each other. The ball-and-stick stereoformulas at the bottom of Figure 2.28 show the actual geometry of both *cations*. (2) The carbenium ion shown in Figure 2.24 is an isomer of, and less stable than, the carbonium ion shown there. It has slightly longer C1–C6 and C1–C2 bonds, and no C2–C6 bond (the C2/C6 separation is 2.5 Å in the carbenium ion and 1.8 Å in the carbonium ion). Whether the acetolysis of the *endo*-norbornyl brosylate takes place via the fully formed carbenium ion or follows a reaction path that leads to the carbonium ion without actually reaching the carbenium ion has not been clarified experimentally.

For many decades chemists had been interested in whether the positively charged intermediate of the S_N reaction in Figure 2.28 was a carbenium or a carbonium ion. Also the existence of a rapidly equilibrating mixture of two carbenium ions was considered. It is now known with certainty that this intermediate is a **carbonium ion**; it is known as a **nonclassical carbocation**. In this carbonium ion, there is a bond between the centers C1, C2, and C6, that consists of two sp^2 AOs and one sp^3 AO (see MO diagram, lower right, Figure 2.28). It accommodates two electrons. There are many examples of nonclassical carbocations.

2.8 S_Ni Reactions

A common method for the transformation of alcohols into alkyl chlorides is the reaction with thionyl chloride. The advantages of this method include the lower price of the chlorinating agent and the fact that only gases (sulfur dioxide, hydrogen chloride) occur as by-products in stoichiometric amounts, thus facilitating workup. Moreover, a primary OH group may react selectively as shown in the example in Figure 2.29. The two secondary and therefore slightly more hindered OH groups remain untouched without requiring protection.

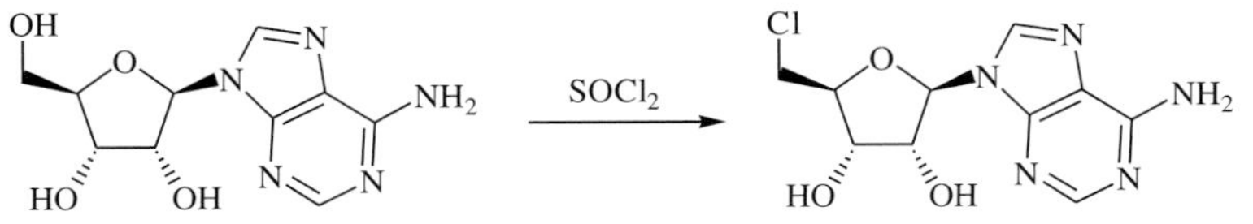

Fig. 2.29. Selective exchange of an OH group for a Cl residue by reaction with thionyl chloride.

The term "S_Ni reaction" was coined during mechanistic investigations into this transformation with the substrates shown in Figure 2.30. Using the *R* enantiomer as the substrate, it was possible to elucidate the steric course of the substitution. Upon reaction with thionyl chloride in

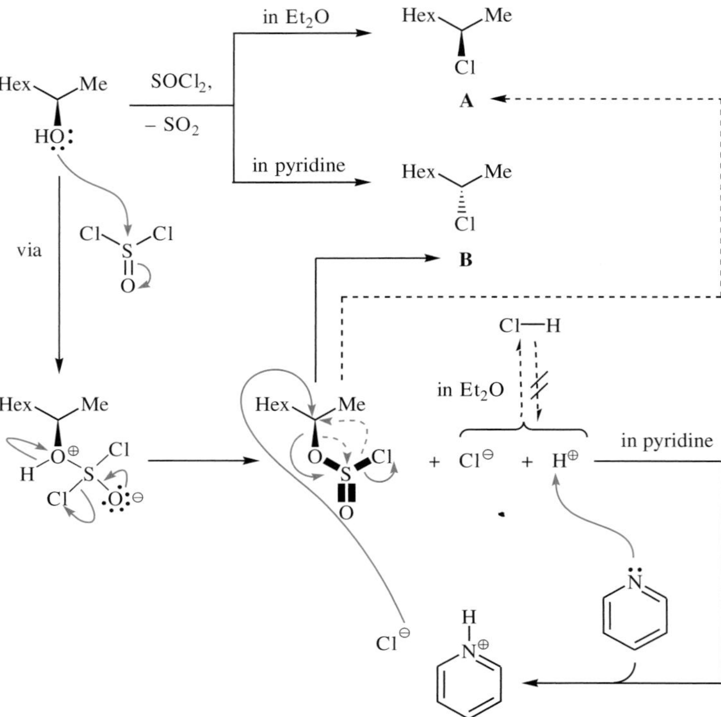

Fig. 2.30. Stereoselective exchange of an OH group for a Cl residue by reaction with thionyl chloride. Depending on the solvent, either complete (100%) retention of configuration (S_Ni reaction; → **A**) or complete (100%) inversion of configuration (S_N2 reaction; **B**) can be observed.

pyridine the chloride **B** was formed with complete *inversion* of configuration. Amazingly, the reaction of the same alcohol and the same reagent in diethyl ether also proceeded selectively, but yielded the chloride **A**, which is the mirror image of **B**. **A** is the result of a substitution proceeding with complete *retention* of configuration.

The bottom part in Figure 2.30 makes all the difference. Irrespective of whether pyridine or diethyl ether is used as the solvent, a short-lived alkyl chlorosulfite is formed initially. In addition, a proton and a chloride ion are generated. These ions react with pyridine to give pyridinium hydrochloride; in diethyl ether they form an H–Cl molecule. Pyridinium hydrochloride contains nucleophilic chloride ions, a solvated H–Cl molecule, however, does not. The chloride ions of the pyridinium hydrochloride react with the alkyl chlorosulfite with inversion of configuration. It is an S_N2 reaction; no surprises. In diethyl ether, however, the analogous attack fails from lack of free chloride ions. In this solvent the alkyl chlorosulfite undergoes a unimolecular decomposition. The C–O bond and the S–Cl bond are simultaneously cleaved, and the Cl atom, together with the previously bonding electron pair, moves from S to C before any contact ion pair can be formed from a secondary carbenium ion and the chlorosulfite anion. This means that an "**i**nternal" S_N reaction has occurred, for which the abbreviation "S_N**i** reaction" is used, as mentioned above.

2.9 Preparatively Useful S_N2 Reactions: Alkylations

The three-part Figure 2.31, with a list of synthetically important reactions, shows the various nucleophiles that can be alkylated via S_N2 reactions. This list refers to alkylations with alkyl

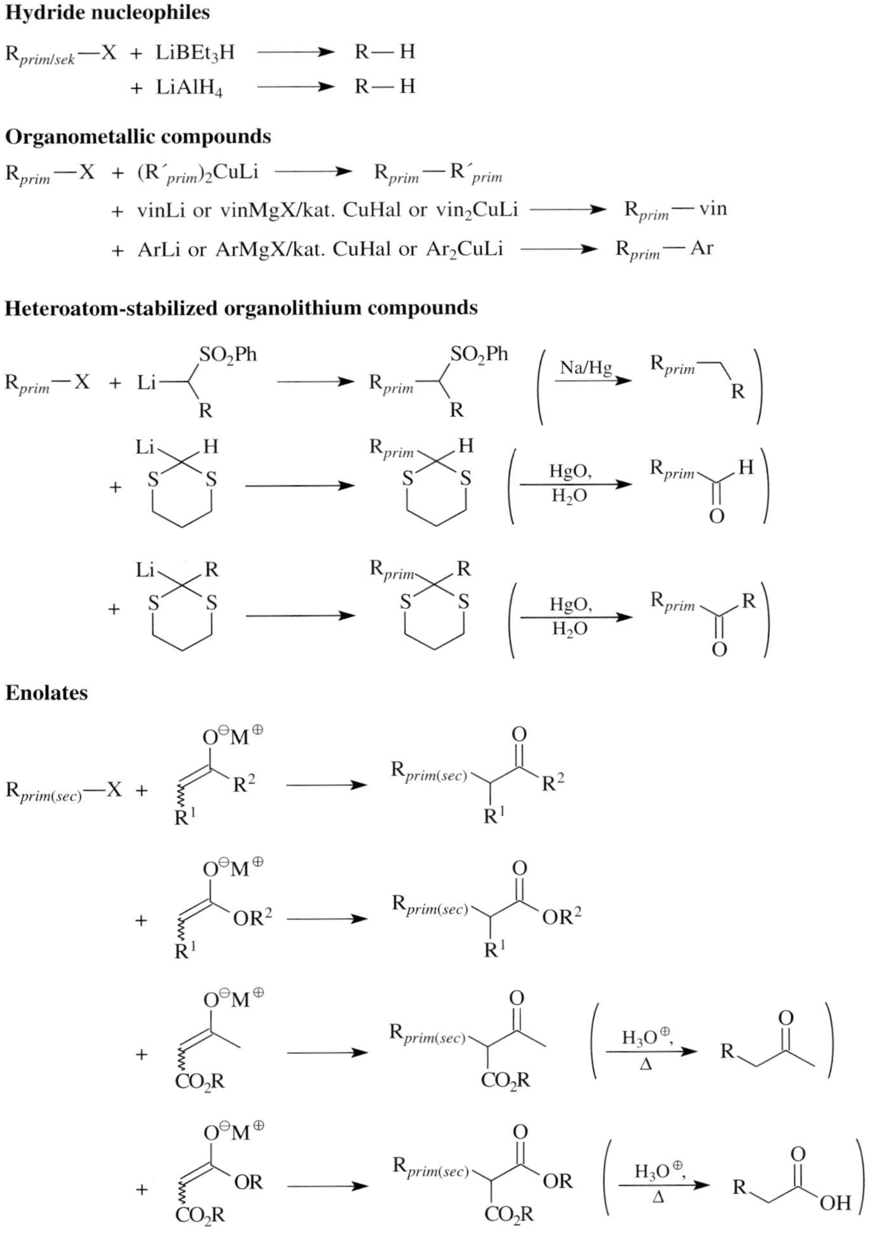

Hydride nucleophiles

$R_{prim/sek}—X$ + $LiBEt_3H$ ⟶ R—H

+ $LiAlH_4$ ⟶ R—H

Organometallic compounds

$R_{prim}—X$ + $(R'_{prim})_2CuLi$ ⟶ $R_{prim}—R'_{prim}$

+ vinLi or vinMgX/kat. CuHal or vin_2CuLi ⟶ R_{prim}—vin

+ ArLi or ArMgX/kat. CuHal or Ar_2CuLi ⟶ R_{prim}—Ar

Heteroatom-stabilized organolithium compounds

Enolates

Fig. 2.31. Preparatively important S_N2 reactions.

Further C-nucleophiles

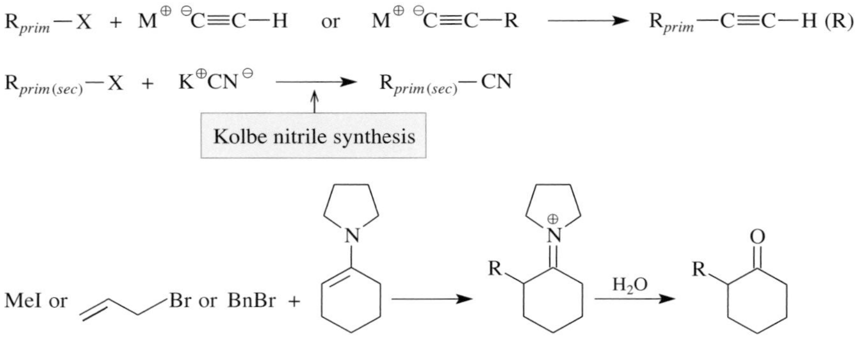

$R_{prim}-X + M^{\oplus}\ {}^{\ominus}C\equiv C-H$ or $M^{\oplus}\ {}^{\ominus}C\equiv C-R \longrightarrow R_{prim}-C\equiv C-H\ (R)$

$R_{prim(sec)}-X + K^{\oplus}CN^{\ominus} \longrightarrow R_{prim(sec)}-CN$

Kolbe nitrile synthesis

MeI or allyl-Br or BnBr + 1-(pyrrolidin-1-yl)cyclohexene ⟶ 2-R-cyclohexylidene-pyrrolidinium (N⊕) $\xrightarrow{H_2O}$ 2-R-cyclohexanone

N-nucleophiles

$R-X$ + $K^{\oplus}$ ${}^{\ominus}$N-phthalimide ⟶ R–N-phthalimide $\left(\xrightarrow{N_2H_4} R-NH_2 + \text{phthalhydrazide}\right)$

Gabriel synthesis of primary amines

$+ Na^{\oplus}N_3^{\ominus} \longrightarrow R-N_3 \left(\xrightarrow[H_2,\ Pd/C]{LiAlH_4 \text{ or } PPh_3 \text{ followed by } H_2O \text{ or}} R-NH_2\right)$

$+ K^{\oplus}NCO^{\ominus} \longrightarrow R-N=C=O$

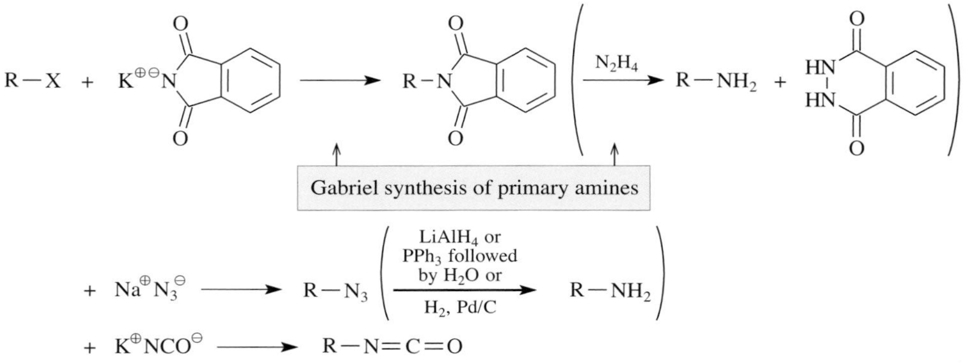

P-nucleophiles (⟶ precursors for Wittig or Horner-Wadsworth-Emmons reaction)

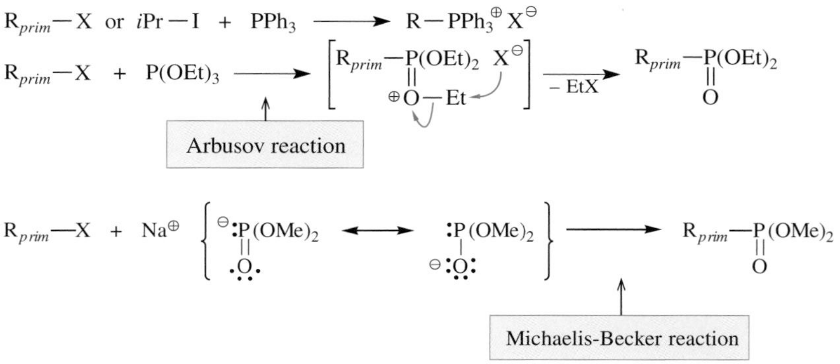

$R_{prim}-X$ or $iPr-I + PPh_3 \longrightarrow R-PPh_3^{\oplus}X^{\ominus}$

$R_{prim}-X + P(OEt)_3 \longrightarrow \left[R_{prim}-P(OEt)_2(={}^{\oplus}O-Et)\ X^{\ominus}\right] \xrightarrow{-\ EtX} R_{prim}-P(=O)(OEt)_2$

Arbusov reaction

$R_{prim}-X + Na^{\oplus} \left\{ {}^{\ominus}:P(=O)(OMe)_2 \longleftrightarrow :P(OMe)_2-O:^{\ominus} \right\} \longrightarrow R_{prim}-P(=O)(OMe)_2$

Michaelis-Becker reaction

Fig. 2.31. (continued)

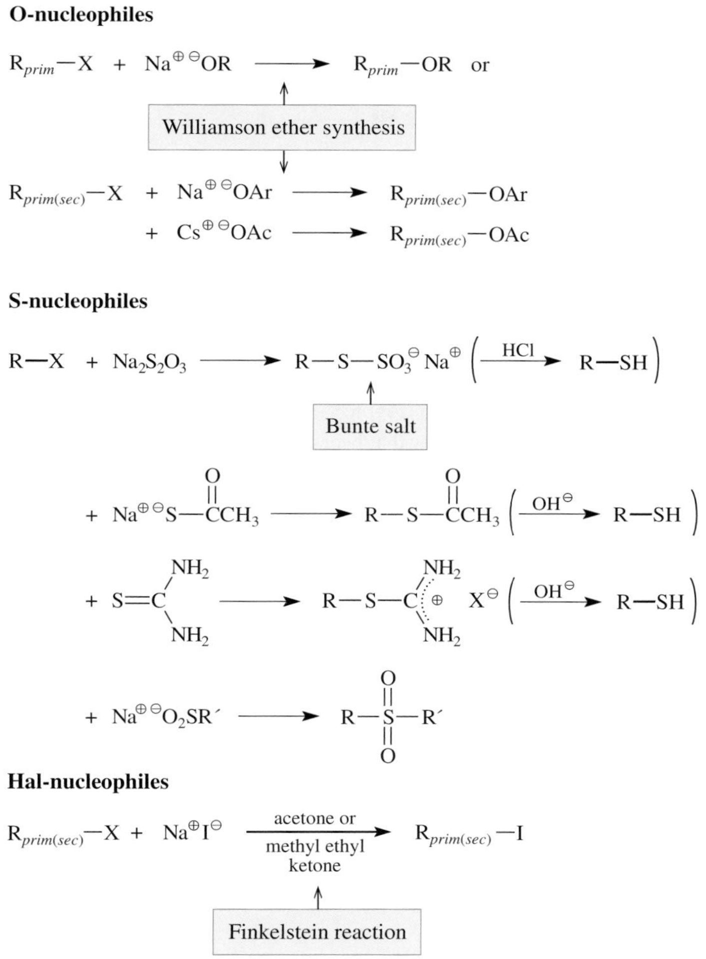

O-nucleophiles

$$R_{prim}\text{—}X + Na^{\oplus}\,{}^{\ominus}OR \longrightarrow R_{prim}\text{—}OR \quad \text{or}$$

Williamson ether synthesis

$$R_{prim(sec)}\text{—}X + Na^{\oplus}\,{}^{\ominus}OAr \longrightarrow R_{prim(sec)}\text{—}OAr$$

$$+ Cs^{\oplus}\,{}^{\ominus}OAc \longrightarrow R_{prim(sec)}\text{—}OAc$$

S-nucleophiles

$$R\text{—}X + Na_2S_2O_3 \longrightarrow R\text{—}S\text{—}SO_3^{\ominus}\,Na^{\oplus} \left(\xrightarrow{HCl} R\text{—}SH\right)$$

Bunte salt

$$+ Na^{\oplus}\,{}^{\ominus}S\text{—}C(=O)CH_3 \longrightarrow R\text{—}S\text{—}C(=O)CH_3 \left(\xrightarrow{OH^{\ominus}} R\text{—}SH\right)$$

$$+ S{=}C(NH_2)_2 \longrightarrow R\text{—}S\text{—}C^{\oplus}(NH_2)_2\; X^{\ominus} \left(\xrightarrow{OH^{\ominus}} R\text{—}SH\right)$$

$$+ Na^{\oplus}\,{}^{\ominus}O_2SR' \longrightarrow R\text{—}S(=O)_2\text{—}R'$$

Hal-nucleophiles

$$R_{prim(sec)}\text{—}X + Na^{\oplus}I^{\ominus} \xrightarrow[\text{methyl ethyl ketone}]{\text{acetone or}} R_{prim(sec)}\text{—}I$$

Finkelstein reaction

Fig. 2.31. (continued)

halides, sulfonates and sulfates. Relevant alkylations using alcohols are displayed in Figures 2.33–2.37, and important alkylations with ethylene oxide ("hydroxy alkylations") are depicted in Figure 2.38 at the end of the chapter.

Whereas all the alkylations in Figure 2.31 take place in basic or neutral solutions, carboxylic acids can be directly methylated with diazomethane (Figure 2.32). The actual nucleophile (the carboxylate ion) and the actual methylating agent (H_3C—$N^{\oplus}$ N) are produced from the reaction partners by proton transfer.

Several S_N reactions of alcohols are initiated by positively charged phosphorus(V) compounds (Figures 2.33–2.37). In these cases, alcohols are treated with phosphonium salts $Ph_3P^{\oplus}$—X that contain a leaving group X. This leaving group is displaced by the O atom of the OH group of the "alkylating agent." In this way this O atom is bound to the P atom and becomes part of a good leaving group, $Ph_3P{=}O$. If

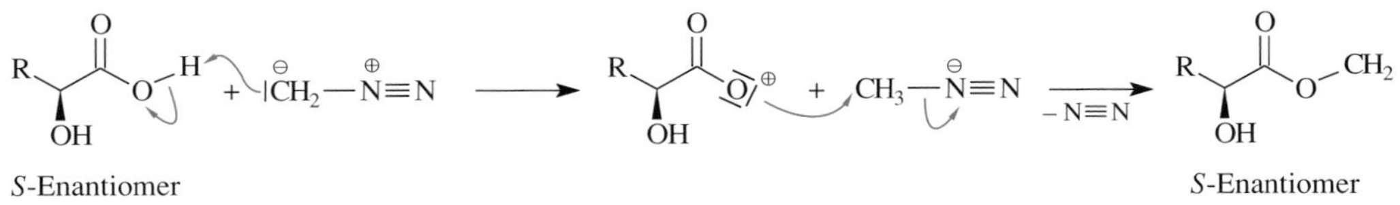

Fig. 2.32 Preparation of methyl esters from carboxylic acids and diazomethane.

- the species $Ph_3P^{\oplus}$—X, which reacts with by the OH group, is produced from Ph_3P and EtO_2C—N=N—CO_2Et, and
- the nucleophile is a carboxylic acid or the carboxylate ion formed from it *in situ*, and
- in addition, the OH group is located at a stereocenter,

then the reaction is called a **Mitsunobu inversion.** With its help it is possible to invert the configuration of a stereocenter bearing an OH group (Figure 2.33). The enantiomers of optically pure alcohols, which contain a single stereocenter that bears an OH group, are easily accessible through a Mitsunobu inversion process.

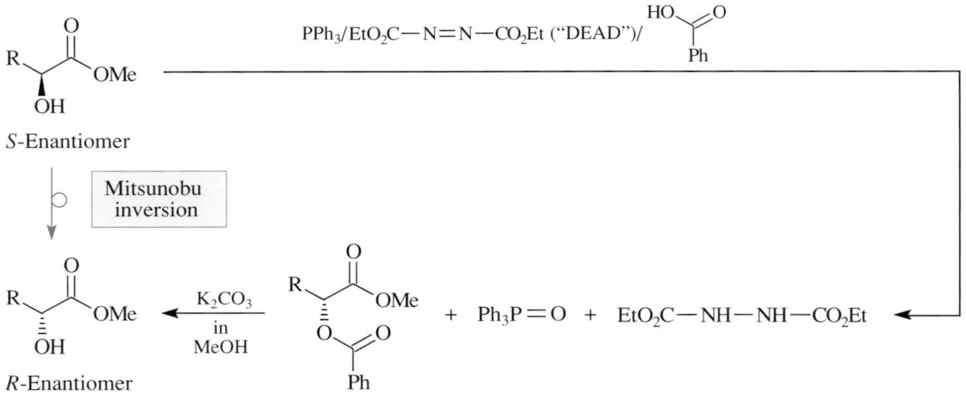

Fig. 2.33. Mitsunobu inversion: a typical substrate, the reagents, and products (possible preparation of the substrate: Figure 2.27). "DEAD" stands for diethylazodicarboxylate.

Figure 2.34 shows the mechanism of this reaction. A key intermediate is the alkylated phosphine oxide **A**, with which the carboxylate ion reacts to displace the leaving group O=PPh_3. Figure 2.34 also shows that this carboxylate ion results from the deprotonation of the carboxylic acid used by the intermediate carbamate anion **B**. Nucleophiles that can be deprotonated by **B** analogously, i.e., quantitatively, are also alkylated under Mitsunobu-like conditions (see Figure 2.36). In contrast, nucleophiles that are too weakly acidic cannot undergo Mitsunobu alkylation. Thus, for example, there are Mitsunobu etherifications of phenols, but not of alcohols.

If the Mitsunobu inversion is carried out intramolecularly (i.e., in a hydroxycarboxylic acid), a lactone is produced with inversion of the configuration at the OH-bearing stereocenter (Figure 2.35). This lactonization is stereochemically complementary to the paths via activated hydroxycarboxylic acids, which lead to lactones with retention of the configuration at the OH-bearing C atom (Section 6.4.2).

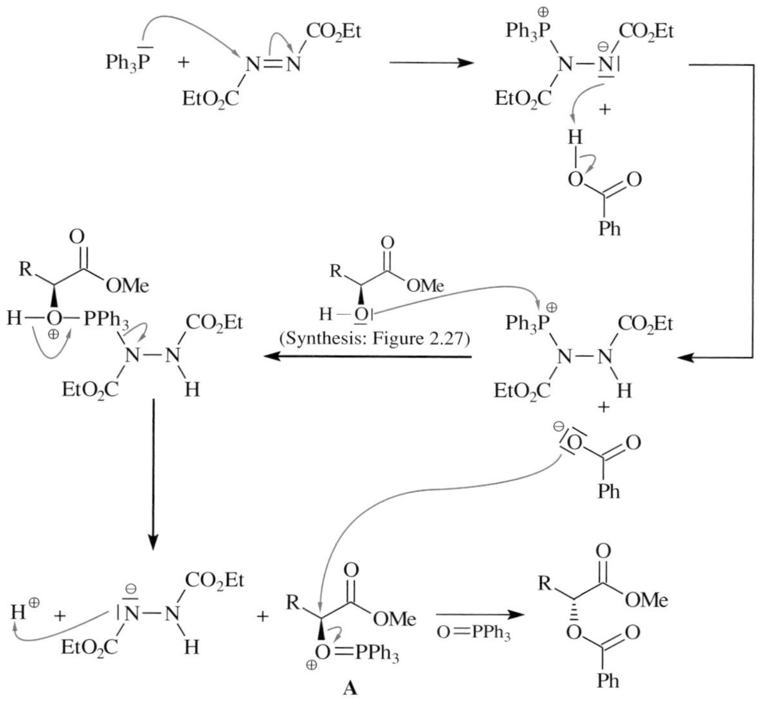

Fig. 2.34. Mechanism of the Mitsunobu inversion in Figure 2.33.

Let us finally consider the reactions shown in Figure 2.36. There the oxophilic species $Ph_3P^{\oplus}$—X are produced from Ph_3P and reagents of the form X—Y. While one half of such a reagent is transferred as $X^{\oplus}$ to Ph_3P, an anion $Y^{\ominus}$ is produced from the other half of the molecule. The phosphonium ion now converts the OH group of the alcohol—through a nucleophilic reaction of the oxygen at the phosphorus—to the same leaving group $Ph_3P{=}O$ that was present during the Mitsunobu inversion. The group X, which was bound to the phosphorus, is released in the form of a nucleophile $X^{\ominus}$, which ultimately displaces the leaving group $Ph_3P{=}O$. Consequently, from three molecules of starting materials (alcohol, Ph_3P, X—Y) one substitution product, one equivalent of $Ph_3P{=}O$, and one equivalent of H—Y are produced.

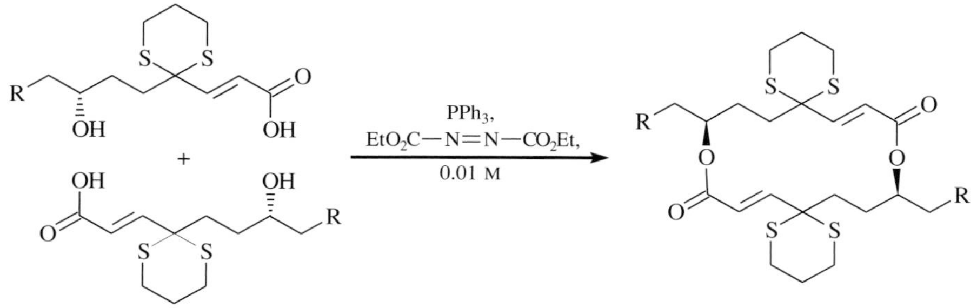

Fig. 2.35. Stereoselective lactonization by means of a Mitsunobu inversion.

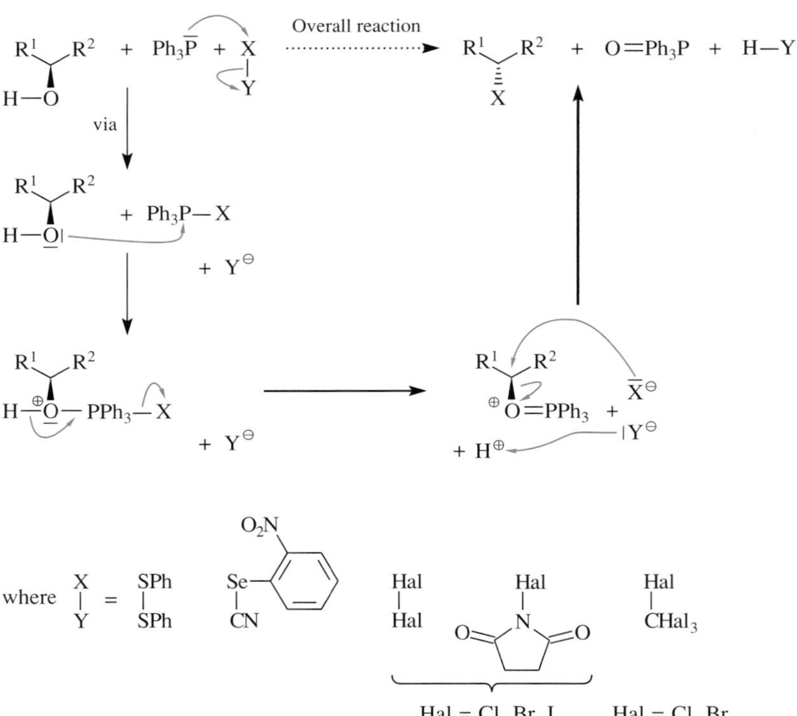

Fig. 2.36. Preparatively important redox condensations according to Mukaiyama, and related reactions.

At the same time the phosphorus, which was originally present in the oxidation state +3, is oxidized to P(V). This type of chemistry was extensively developed by Mukaiyama.

Side Note 2.5. Deoxygenation of Primary Alcohols under Tin-free Conditions

The caption of Figure 2.37 specifies a reagent mix that can be used the deoxygenation of primary alcohols. It involves a complex scheme of reactions, the last part of which has already been discussed in Figure 1.48. Figure 2.37 focuses on the initial reactions. The intermediates **B**, **C** and **E** explain how the alkylated triphenylphosphine oxide **H** is derived from the substrate triphenylphosphine and DEAD. Our first encounter with this intermediate was the analysis of the Mitsunobu inversion (Figure 2.34). Both reactions are accompanied by a carbamate anion. This carbamate anion (intermediate **I** in Figure 2.37) deprotonates the sulfonylhydrazone **A** to give the corresponding sulfonylhydrazide anion **K** in just the same way that it deprotonates benzoic acid to form the corresponding benzoate anion in Figure 2.34. Only then does the real S_N2 substitution take place: In Figure 2.37 the sulfonylhydrazine **F** is produced, whereas the Mitsunobu inversion leads to the benzoate of the inverted alcohol.

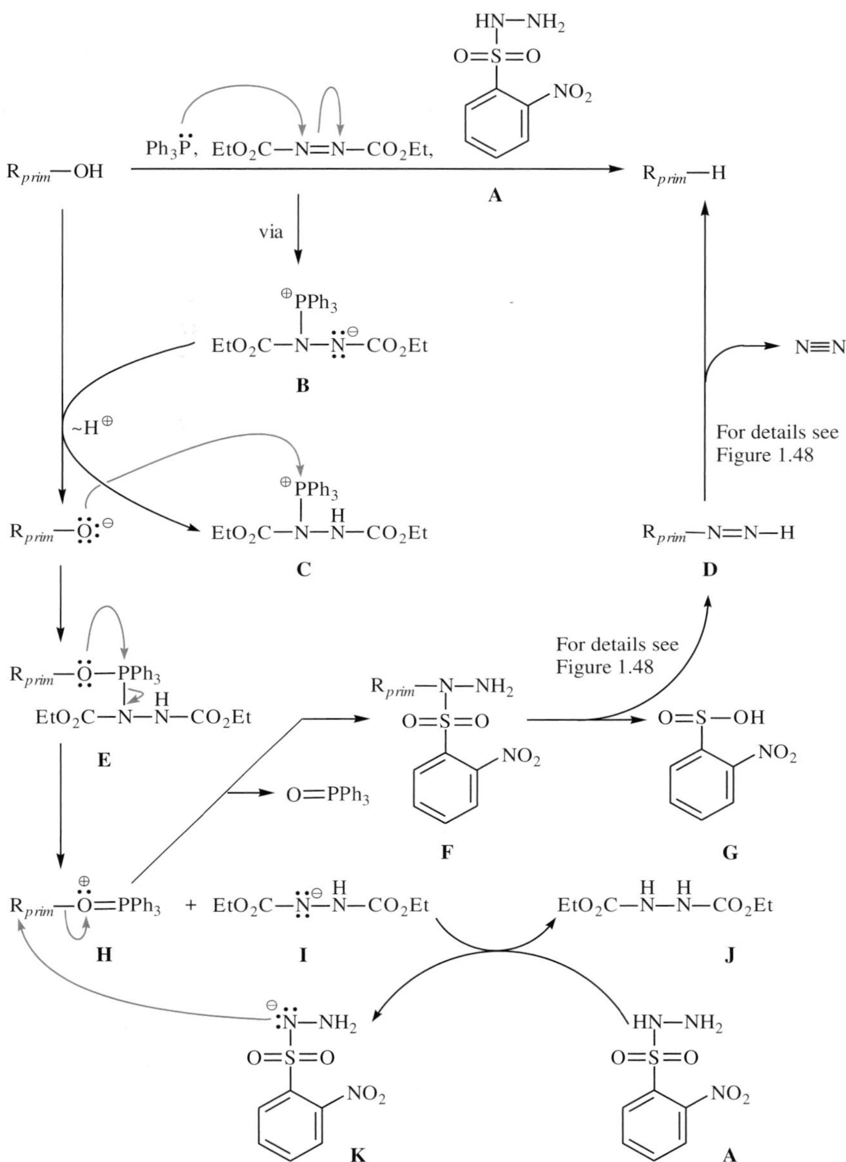

Fig. 2.37. One-step defunctionalization of primary alcohols to give alkanes: mechanistic details of the initial step, i. e. the Mitsunobu reaction.

Finally, many important S_N2 reactions with ethylene oxide as the alkylating agent are being employed in industrial-scale applications, as they deliver solvents, intermediates or polyethylene glycol (Figure 2.38). As can be seen, two examples require only catalytic amount of nucleophile: namely, the formation of 1,4-dioxane (proceeding in the presence of catalytic amounts of sulfuric acid) and the formation of polyethylene glycol (proceeding in the pres-

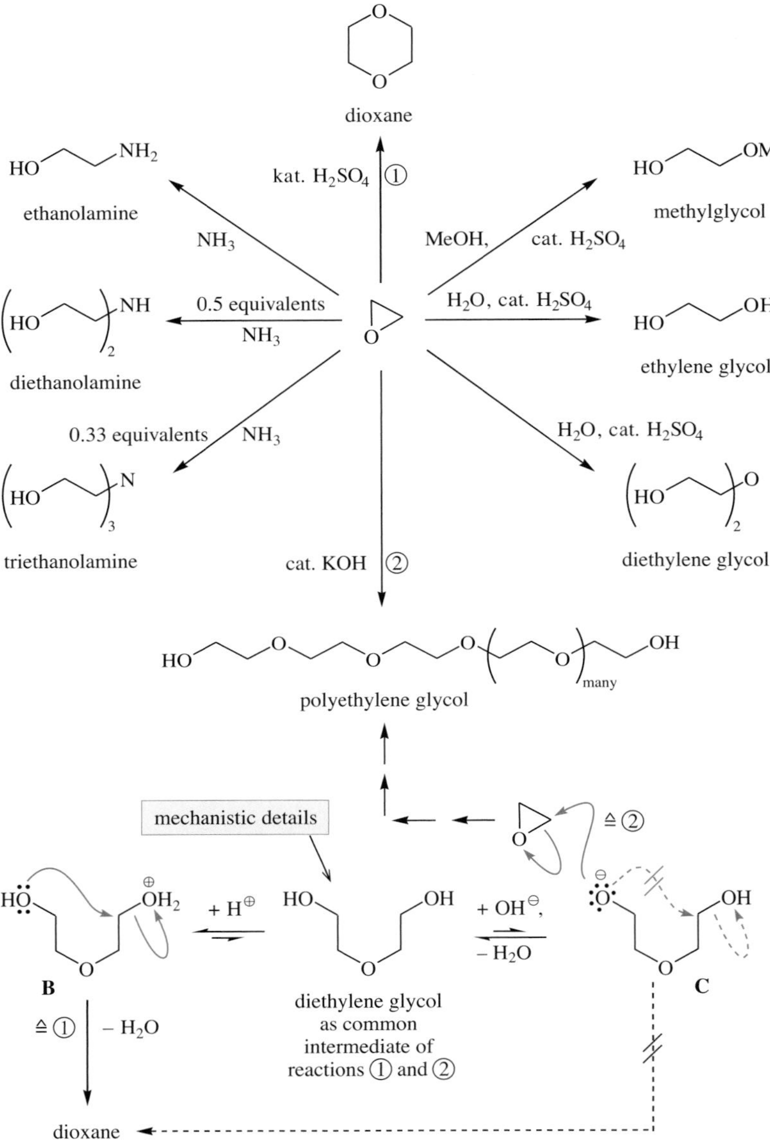

Fig. 2.38. Preparatively important SN_2 reactions with ethylene oxide.

ence of catalytic amounts of potassium hydroxide solution). Ethylene oxide and epoxides in general are good electrophiles. What about aziridines? Find out for yourself, you life-long-learner!

References

G. L. Edwards, "One or More CC Bond(s) Formed by Substitution: Substitution of Halogen," in *Comprehensive Organic Functional Group Transformations* (A. R. Katritzky, O. Meth-Cohn, C. W. Rees, Eds.), Vol. 1, 105, Elsevier Science, Oxford, U. K., **1995**.

2.1

H. Mayr, M. Patz, "Scales of Nucleophilicity and Electrophilicity: A System for Ordering Polar Organic and Organometallic Reactions," *Angew. Chem. Int. Ed. Engl.* **1994**, *33*, 938–958.

2.2

N. J. Fina, J. O. Edwards, "The Alpha Effect. A Review," *Int. J. Chem. Kinet.* **1973**, *5*, 1–26.

A. P. Grekov, V. Y. Veselov, "The β-Effect in the Chemistry of Organic Compounds," *Russ. Chem. Rev.* **1978**, *47*, 631–648

E. Buncel, S. Hoz, "The β-Effect: A Critical Evaluation of the Phenomenon and Its Origin," *Isr. J. Chem.* **1985**, *26*, 313–319.

2.3

M. Klessinger, "Polarity of Covalent Bonds," *Angew. Chem. Int. Ed. Engl.* **1970**, *9*, 500–512.

2.4

P. Beak, "Determinations of Transition-State Geometries by the Endocyclic Restriction Test: Mechanism of Substitution at Nonstereogenic Atoms," *Acc. Chem. Res.* **1992**, *25*, 215–222

P. Beak, "Mechanisms of Reactions at Nonstereogenic Heteroatoms: Evaluation of Reaction Geometries by the Endocyclic Restriction Test," *Pure Appl. Chem.* **1993**, *65*, 611–615.

2.5

T. Baer, C.-Y. Ng, I. Powis (Eds.), "The Structure, Energetics and Dynamics of Organic Ions," Wiley, Chichester, U. K., **1996**.

P. Buzek, P. v. Ragué Schleyer, St. Sieber, "Strukturen von Carbokationen," *Chem. unserer Zeit*, **1992**, *26*, 116–128.

G. A. Olah, "Carbocations and Electrophilic Reactions," *Angew. Chem. Int. Ed. Engl.* **1973**, *12*, 173–212.

H. Schwarz, "Pyramidal Carbocations," *Angew. Chem. Int. Ed. Engl.* **1981**, *20*, 991–1003.

G. A. Olah, "My Search for Carbocations and Their Role in Chemistry (Nobel Lecture)," *Angew. Chem. Int. Ed. Engl.* **1995**, *34*, 1393–1405.

P. von R. Schleyer, R. E. Leone, "Degenerate Carbonium Ions," *Angew. Chem. Int. Ed. Engl.* **1970**, *9*, 860–890.

C. A. Grob, "The Norbornyl Cation: Prototype of a 1,3-Bridged Carbocation," *Angew. Chem. Int. Ed. Engl.* **1982**, *21*, 87–96.

M. Saunders, H. A. Jiménez-Vázquez, "Recent Studies of Carbocations," *Chem. Rev.* **1991**, *91*, 375–397.

V. D. Nefedov, E. N. Sinotova, V. P. Lebedev, "Vinyl Cations," *Russ. Chem. Rev.* **1992**, *61*, 283–296.

T. T. Tidwell, "Destabilized Carbocations," *Angew. Chem. Int. Ed. Engl.* **1984**, *23*, 20.

J.-L. M. Abboud, I. Alkorta, J. Z. Davalos, P. Muller, E. Quintanilla, "Thermodynamic Stabilities of Carbocations," *Adv. Phys. Org. Chem.* **2002**, *37,* 57–135.

K. Okamoto, "Generation and Ion-Pair Structures of Unstable Carbocation Intermediates in Solvolytic Reactions", in *Advances in Carbocation Chemistry* (X. Creary, Ed.) **1989**, *1*, JAI Press, Greenwich, CT.

J. P. Richard, T. L. Amyes, M. M. Toteva, "Formation and Stability of Carbocations and Carbanions in Water and Intrinsic Barriers to Their Reactions," *Acc. Chem. Res.* **2001**, *34,* 981–988.

P. E. Dietze, "Nucleophilic Substitution and Solvolysis of Simple Secondary Carbon Substrates," in *Advances in Carbocation Chemistry* (J. M. Coxon, Ed.) **1995**, *2*, JAI, Greenwich, CT.

R. F. Langler, "Ionic Reactions of Sulfonic Acid Esters," *Sulfur Rep.* **1996**, *19*, 1–59.

H. Normant, "Hexamethylphosphoramide," *Angew. Chem. Int. Ed. Engl.* **1967**, *6*, 1046–1067.

M. S. Shchepinov, V. A. Korshun, "Recent Applications of Bifunctional Trityl Groups," *Chem. Soc. Rev.* **2003**, *32,* 170–180.

2.6

A. R. Katritzky, B. E. Brycki, "Nucleophilic Substitution at Saturated Carbon Atoms. Mechanisms and Mechanistic Borderlines: Evidence from Studies with Neutral Leaving Groups," *J. Phys. Org. Chem.* **1988**, *1*, 1–20.

J. P. Richard, "Simple Relationships Between Carbocation Lifetime and the Mechanism for Nucleophilic Substitution at Saturated Carbon," in *Advances in Carbocation Chemistry* (X. Creary, Ed.) **1989**, *1*, JAI Press, Greenwich, CT.

A. R. Katritzky, B. E. Brycki, "The Mechanisms of Nucleophilic Substitution in Aliphatic Compounds," *Chem. Soc. Rev.* **1990**, *19*, 83–105.

H. Lund, K. Daasbjerg, T. Lund, S. U. Pedersen, "On Electron Transfer in Aliphatic Nucleophilic Substitution," *Acc. Chem. Res.* **1995**, *28*, 313–319.

2.7

B. Capon, *Neighboring group participation*, Plenum Press, New York :, **1976**.

G. M. Kramer, C. G. Scouten, "The 2-Norbornyl Carbonium Ion Stabilizing Conditions: An Assessment of Structural Probes", in *Advances in Carbocation Chemistry* (X. Creary, Ed.) **1989**, *1*, JAI Press, Greenwich, CT.

2.9

J. M. Klunder, G. H. Posner, "Alkylations of Nonstabilized Carbanions," in *Comprehensive Organic Synthesis* (B. M. Trost, I. Fleming, Eds.), Vol. 3, 207, Pergamon Press, Oxford, **1991**.

H. Ahlbrecht, "Formation of C–C Bonds by Alkylation of σ-Type Organometallic Compounds," in *Stereoselective Synthesis* (Houben-Weyl) 4th ed. **1996**, (G. Helmchen, R. W. Hoffmann, J. Mulzer, E. Schaumann, Eds.), **1996**, Vol. E 21 (Workbench Edition), *2*, 645–663, Georg Thieme Verlag, Stuttgart.

D. W. Knight, "Alkylations of Vinyl Carbanions," in *Comprehensive Organic Synthesis* (B. M. Trost, I. Fleming, Eds.), Vol. 3, 241, Pergamon Press, Oxford, **1991**.

H. Ahlbrecht, "Formation of C–C Bonds by Alkylation of π-Type Organometallic Compounds," in *Stereoselective Synthesis* (Houben-Weyl) 4th ed. **1996**, (G. Helmchen, R. W. Hoffmann, J. Mulzer, E. Schaumann, Eds.), **1996**, Vol. E 21 (Workbench Edition), *2*, 664–696, Georg Thieme Verlag, Stuttgart.

P. J. Garratt, "Alkylations of Alkynyl Carbanions," in *Comprehensive Organic Synthesis* (B. M. Trost, I. Fleming, Eds.), Vol. 3, 271, Pergamon Press, Oxford, **1991**.

G. H. Posner, "Substitution Reactions Using Organocopper Reagents," *Org. React.* **1975**, *22*, 253–400.

B. H. Lipshutz, S. Sengupta, "Organocopper Reagents: Substitution, Conjugate Addition, Carbo/Metallocupration, and Other Reactions," *Org. React.* **1992**, *41*, 135–631.

M. V. Bhatt, S. U. Kulkarni, "Cleavage of Ethers," *Synthesis* **1983**, 249–282.

C. Bonini, G. Righi, "Regio- and Chemoselective Synthesis of Halohydrins by Cleavage of Oxiranes with Metal Halides," *Synthesis* **1994**, 225–238.

M. S. Gibson, R. W. Bradshaw, "The Gabriel Synthesis of Primary Amines," *Angew. Chem. Int. Ed. Engl.* **1968**, *7*, 919–930.

U. Raguarsson, L. Grehn, "Novel Gabriel Reagents," *Acc. Chem. Res.* **1991**, *24*, 285–289.

T. H. Black, „The Preparation and Reactions of Diazomethane," *Aldrichimica Acta* **1983**, *16*, 3–10.

R. Appel, "Tertiary Phosphane/Tetrachloromethane, a Versatile Reagent for Chlorination, Dehydration, and P-N Linkage," *Angew. Chem. Int. Ed. Engl.* **1975**, *14*, 801–811.

B. R. Castro, "Replacement of Alcoholic Hydroxy Groups by Halogens and Other Nucleophiles via Oxyphosphonium Intermediates," *Org. React.* **1983**, *29*, 1–162.

T. Mukaiyama, "Oxidation-Reduction Condensation," *Angew. Chem. Int. Ed. Engl.* **1976**, *15*, 94–103.

D. L. Hughes, "Progress in the Mitsunobu Reaction. A Review," *Org. Prep. Proced. Int.* **1996**, *28*, 127–164.

D. L. Hughes, "The Mitsunobu Reaction," *Org. React.* **1992**, *42*, 335–656.

C. Simon, S. Hosztafi, S. Makleit, "Application of the Mitsunobu Reaction in the Field of Alkaloids," *J. Heterocycl. Chem.* **1997**, *34*, 349–365.

A. Krief, A.-M. Laval, "*Ortho*-Nitrophenyl Selenocyanate, a Valuable Reagent in Organic Synthesis: Application to One of the Most Powerful Routes to Terminal Olefins from Primary-Alcohols (The Grieco-Sharpless Olefination Reaction) and to the Regioselective Isomerization of Allyl Alcohols)," *Bull. Soc. Chim. Fr.* **1997**, *134*, 869–874.

Further Reading

R. M. Magid, "Nucleophilic and Organometallic Displacement Reactions of Allylic Compounds: Stereo- and Regiochemistry," *Tetrahedron* **1980**, *36*, 1901–1930.

Y. Yamamoto, "Formation of C–C Bonds by Reactions Involving Olefinic Double Bonds, Vinylogous Substitution Reactions," in *Methoden Org. Chem.* (Houben-Weyl) 4th ed. **1952**–, *Stereoselective Synthesis* (G. Helmchen, R. W. Hoffmann, J. Mulzer, E. Schaumann, Eds.), Vol. E21b, 2011, Georg Thieme Verlag, Stuttgart, **1995**.

L. A. Paquette, C. J. M. Stirling, "The Intramolecular S_N' Reaction," *Tetrahedron* **1992**, *48*, 7383–7423.

R. A. Rossi, A. B. Pierini, A. B. Penenory, "Nucleophilic Substitution Reactions by Electron Transfer," *Chem. Rev.* **2003**, *103,* 71–167.

R. M. Hanson, "The Synthetic Methodology of Nonracemic Glycidol and Related 2,3-Epoxy Alcohols," *Chem. Rev.* **1991**, *91*, 437–476.

P. C. A. Pena, S. M. Roberts, "The Chemistry of Epoxy Alcohols," *Curr. Org. Chem.* **2003**, *7,* 555–571.

B. B. Lohray, "Cyclic Sulfites and Cyclic Sulfates: Epoxide Like Synthons," *Synthesis* **1992**, 1035–1052.

J. B. Sweeney, "Aziridines: Epoxides' Ugly Cousins?," *Chem. Soc. Rev.* **2002**, *31,* 247–258.

S. C. Eyley, "The Aliphatic Friedel-Crafts Reaction," in *Comprehensive Organic Synthesis* (B. M. Trost, I. Fleming, Eds.), Vol. 2, 707, Pergamon Press, Oxford, **1991**.

Y. Ono, "Dimethyl Carbonate for Environmentally Benign Reactions," *Pure Appl. Chem.* **1996**, *68*, 367–376.

C. M. Sharts, W. A. Sheppard, "Modern Methods to Prepare Monofluoroaliphatic Compounds," *Org. React.* **1974**, *21*, 125–406.

J. E. McMurry, "Ester Cleavages via S_N2-Type Dealkylation," *Org. React.* **1976**, *24*, 187–224.

S. K. Taylor, "Reactions of Epoxides with Ester, Ketone and Amide Enolates", *Tetrahedron* **2000**, *56*, 1149–1163.

M. Schelhaas, H. Waldmann, "Protecting Group Strategies in Organic Synthesis§, *Angew. Chem. Int. Ed. Engl.* **1996**, *35*, 2056–2083.

K. Jarowicki, P. Kocienski, "Protecting Groups§, *Contemp. Org. Synth.* **1996**, *3*, 397–431.

Electrophilic Additions to the C=C Double Bond

3

Alkenes contain a C=C double bond. The C=C double bond can be described with two different models. According to the most commonly used model, a C=C double bond consists of a σ- and a π-bond. The bond energy of the σ-bond is 83 kcal/mol, about 20 kcal/mol higher than the π-bond (63 kcal/mol). The higher stability of σ bonds in comparison to π bonds is due to the difference in the overlap between the atomic orbitals (AOs) that form these bonds. Sigma bonds are produced by the overlap of two sp^n atomic orbitals ($n = 1, 2, 3$), which is quite effective because it is *frontal*. Pi bonds are based on the overlap of $2p_z$ atomic orbitals, which is not as good because it is lateral or parallel.

Upon exposure to a suitable reagent, the relatively weak C=C double bond of an alkene will be broken and a relatively stable C—C single bond will remain. From the point of view of our bonding model, this means that the reactions of alkenes are addition reactions, in which the C—C π-bond is converted into σ-bonds to two new substituents, a and b.

$$R^1R^2C_{sp^3} \overset{\pi}{=} C_{sp^3}R^3R^4 + \text{Reagent(s)} \xrightarrow{\text{Addition}} a \overset{\sigma}{-} C_{sp^3}(R^1)(R^2) - C_{sp^3}(R^3)(R^4) \overset{\sigma}{-} b$$

Suitable reagents have the structure a—b in which the two groups are atoms in the most simple case. However, they may also be more complex. The fragments a and b, which are added to the olefinic C=C double bond, do not necessarily constitute the *entire* reagent nor must they originate from a single reagent. It does not matter with respect to the fundamental bonding changes that occur in the reaction, though mechanisms may differ. Read on.

In addition reactions to C=C double bonds, two new sp^3-hybridized C atoms are produced. Each of them can be a stereogenic center (stereocenter) if properly substituted. If stereocenters are produced, their configuration must be specified. The first question is which *absolute* configuration is produced at any new stereocenter. Then the question arises about the configuration of the new stereocenters *relative* (a) to each other or (b) to additional stereocenters the molecule contained before the addition reaction occurred.

Stereochemical aspects are therefore an important part of the chemistry of electrophilic addition reactions to the C=C double bond. We will therefore investigate them in detail in this chapter. In fact, the content of Chapter 3 is arranged according to the stereochemical characteristics of the (addition) reactions.

Bruckner R (author), Harmata M (editor) In: *Organic Mechanisms – Reactions, Stereochemistry and Synthesis*
Chapter DOI: 10.1007/978-3-642-03651-4_3, © Springer-Verlag Berlin Heidelberg 2010

3.1 The Concept of *cis*- and *trans*-Addition

In a number of additions of two fragments a and b to a C=C double bond, a new stereocenter is produced at *both* reacting C atoms (Figure 3.1). For describing the configurations of these new stereocenters *relative* to each other the following nomenclature is used. If the stereostructure of the addition product arises because fragments a and b have added to the C=C double bond of the substrate from the same side, a *cis*-addition has taken place. Conversely, if the stereostructure of the addition product is obtained because fragments a and b were added to the C=C double bond of the substrate from opposite sides, we have a *trans*-addition. The terms *cis*- and *trans*-addition are also used when the addition products are acyclic. We need to learn more about stereochemical concepts, so go the next section.

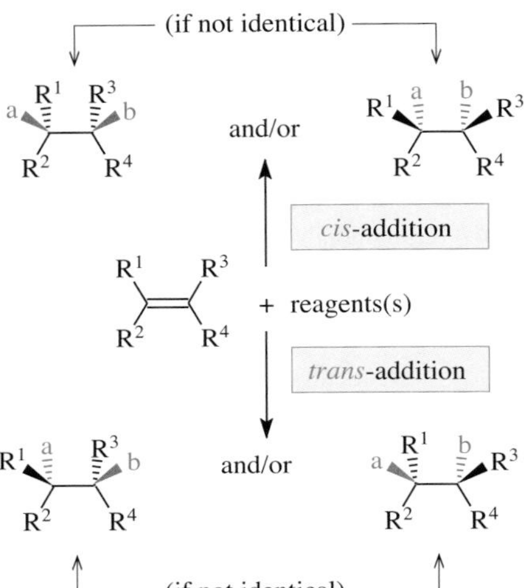

Fig. 3.1. The *cis*- and *trans*-additions of two fragments a and b to C=C double bonds.

3.2 Vocabulary of Stereochemistry and Stereoselective Synthesis I

3.2.1 Isomerism, Diastereomers/Enantiomers, Chirality

Two molecules that have the same empirical formula can exhibit four different degrees of relationship, which are shown in Figure 3.2.

Molecules of the same empirical formula are either identical or isomers. Isomers either differ in the connectivity of their constituent atoms—this then involves **constitutional isomers** (structural isomers)—or they do not differ in this way; then they are **stereoisomers.** Stereoiso-

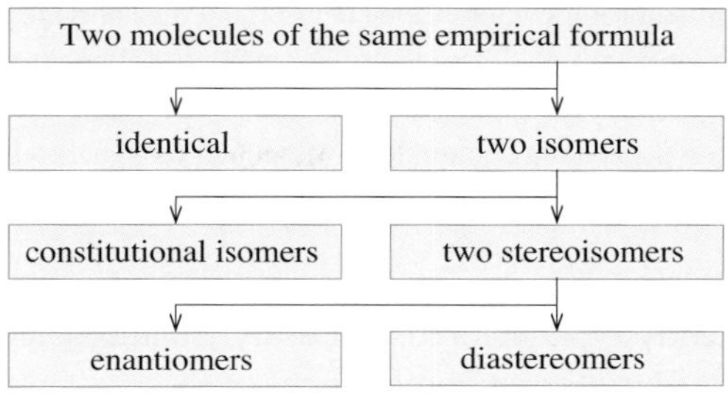

Fig. 3.2. Possible relationships between molecules of the same empirical formula.

mers that are related to each other as an image and a nonsuperimposable mirror image are **enantiomers**; otherwise they are **diastereomers.** Diastereomers that contain several stereocenters but differ in the configuration of only one stereocenter are called **epimers.**

According to the foregoing analysis, conformers such as *gauche-* and *anti-*butane or chair and twist-boat cyclohexane would be considered to be diastereomers of each other. However, under most conditions these conformers interconvert so rapidly that butane and cyclohexane are considered to be single species and not mixtures of stereoisomers. When we have to write chemistry books for people living on the outer planets of the solar system, we might have to modify these concepts.

Let us also visualize the important concept of **chirality.** Only molecules that differ from their mirror image have enantiomers. Molecules of this type are called **chiral.** For anything to be chiral, that is, non-superimposable on its mirror—there is a necessary and sufficient condition: *it is the absence of an intramolecular rotation/reflection axis*. What's that, you say? See the next paragraph.

The most frequently occurring rotation/reflection axis in organic chemistry is S_1, the intramolecular mirror plane. Its presence makes *cis*-1,2-dibromocyclohexane (structure **A** in Figure 3.3) as well as *meso*-2,3-dibromosuccinic acid (structure **B**) achiral. This is true even though there are two stereocenters in each of these compounds. But why is the dibromocyclohexane dicarboxylic acid **C** (Figure 3.3) achiral? The answer is that it contains an S_2 axis, an inversion center, which occurs rarely in organic chemistry. This compound is thus achiral although it contains four stereocenters.

According to the foregoing definition, chirality occurs only in molecules that do not have a *rotation/reflection axis*. However, if the molecule only has an *axis of rotation*, it is chiral. For

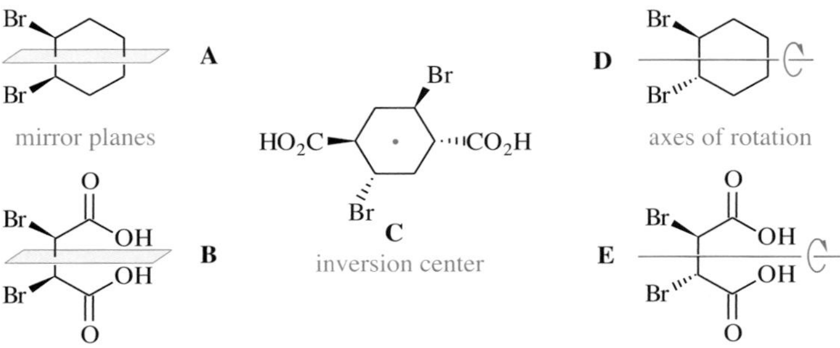

Fig. 3.3. Molecules that contain several stereocenters as well as one (**A–C**) or no (**D, E**) rotation/reflection axis and consequently are achiral (**A–C**) or chiral (**D, E**).

example, both *trans*-1,2-dibromocyclohexane (**D** in Figure 3.3) and the dibromosuccinic acid **E** have a twofold axis of rotation (C_2) as the only symmetry element. In spite of this, these compounds *are* chiral because the presence of an axis of rotation, in contrast to the presence of a rotation/reflection axis, is not a enough for a molecular to be achiral. By the way, if you're not using molecular models while going through this, you should be.

3.2.2 Chemoselectivity, Diastereoselectivity/Enantioselectivity, Stereospecificity/Stereoconvergence

Preparatively useful reactions should give a *single* product or at least a *major* product and not a mixture of many products. If they give one product exclusively or preferentially, they are called selective reactions. One usually specifies the type of selectivity involved, depending on whether the conceivable side products are absent (highly selective reaction) or are formed in small amounts (moderately selective reaction) (Figure 3.4).

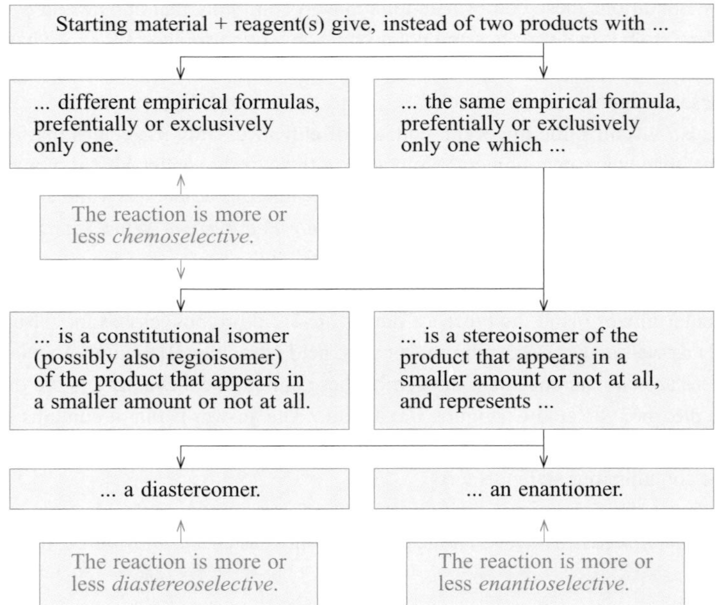

Fig. 3.4. Chemo-, diastereo-, and enantioselectivity.

A reaction that takes place preferentially or exclusively at one functional group among several takes place with **chemoselectivity.** A reaction that preferentially gives one of several conceivable stereoisomers is referred to as moderately or highly **stereoselective**, depending on the extent of stereoselectivity. When these conceivable stereoisomers are diastereomers, we have diastereoselective reaction and the occurrence of **diastereoselectivity.** When the conceivable stereoisomers are enantiomers, we have an enantioselective reaction, or **enantioselectivity.**

The objective of **stereoselective synthesis** is to produce compounds as pure diastereomers (diastereomerically pure) and/or pure enantiomers (enantiomerically pure). To be able to eval-

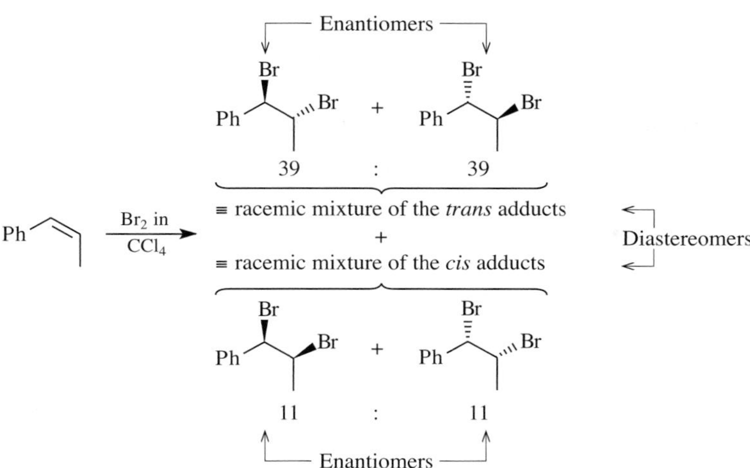

Fig. 3.5. An unusually detailed representation of a moderately diastereoselective reaction. (Figure 3.6 gives a short form of the same reaction.)

uate the success of efforts of this type, one needs quantitative measures for diastereoselectivity and enantioselectivity.

The stereochemical quality of diastereoselective reactions is expressed as a numeric ratio *ds* = *x:y (:z:...)* (from **dias**tereoselectivity). It is the ratio of the yields of the two (or more) diastereomers formed normalized to 100. "Normalized to 100" means that we must have $x + y$ $(+ z + \ldots) = 100$. Figures 3.5 and 3.6 illustrate the use of the measure *ds* as applied to a moderately diastereoselective bromination. The complete description of the stereochemical result is shown in Figure 3.5, and the common short form is shown in Figure 3.6.

Of the dibromides obtained, 78% is the *trans*-adduct, which is produced as a racemic mixture (i.e., a 1:1 mixture) of the two enantiomers, each of which is formed in 39% yield. The remaining dibromide (22%) is the *cis*-adduct, which is also produced as a racemic mixture. The diastereoselectivity with which the *trans*-(vs. the *cis*-) addition product is produced is consequently *ds* = 78:22.

In *all* reactions between achiral reaction partners, chiral products occur as racemic mixtures. In general, equations for such reactions are written so that only different diastereomers are depicted, though it is *understood* that these diastereomers occur as racemic mixtures. Figure 3.6 is the shorthand documentation of what Figure 3.5 presented in much more detail. The same shorthand notations will be used in this chapter and in the remainder of the book.

The quality of enantioselective reactions can be numerically expressed using the **enantiomer ratio** or the **enantiomeric excess** (*ee*). The former is equal to the ratio of the yield of enantiomers normalized to 100. The latter is the difference of this ratio, expressed as a per-

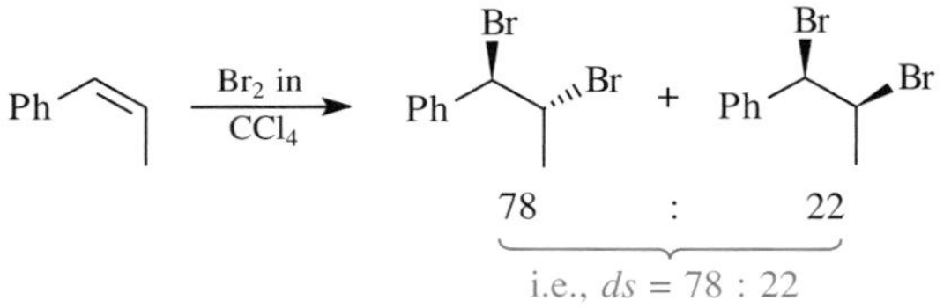

Ph — Br_2 in CCl_4 → Ph(Br)(Br) + Ph(Br)(Br)

78 : 22

i.e., *ds* = 78 : 22

Fig. 3.6. Short way of writing the stereochemical result of the bromination of Figure 3.5.

centage. For example, in the Sharpless epoxidation of allylic alcohol (see Figure 3.7) *S*- and *R*-glycidol are formed in a ratio of 19:1. For a total glycidol yield standardized to 100%, the *S*-glycidol fraction (95% yield) thus exceeds the *R*-glycidol fraction (5% yield) by 90%. Consequently, *S*-glycidol is produced with an enantiomer ratio of 95:5 and an *ee* of 90%.

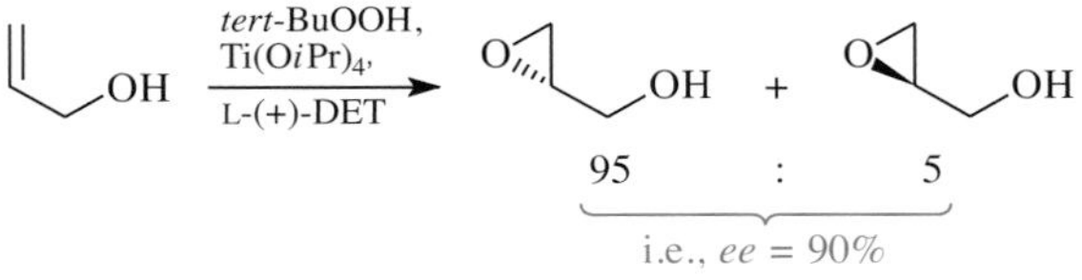

Fig. 3.7. Definition of the enantiomeric excess *ee* using the Sharpless epoxidation of allylic alcohol as an example. The chiral auxiliary is tartaric acid diethyl ester (diethyl tartrate, DET).

tert-BuOOH, $Ti(OiPr)_4$, L-(+)-DET

95 : 5

i.e., *ee* = 90%

The concept of **stereospecificity** has been introduced to characterize the stereochemical course of *pairs* of highly stereoselective reactions. Another term, **stereoconvergence**, is also very useful in this connection.

Definition: Stereospecific Reaction

If in a *pair* of analogous reactions *one* starting material is converted with complete stereoselectively into a *first* product and a *stereoisomer* of the starting material is converted with complete stereoselectively into a *stereoisomer* of the first product, these reactions are called **stereospecific.** An example of this can be found in Figure 3.8. The S_N2 reaction of 2 enantiomers is stereospecific too.

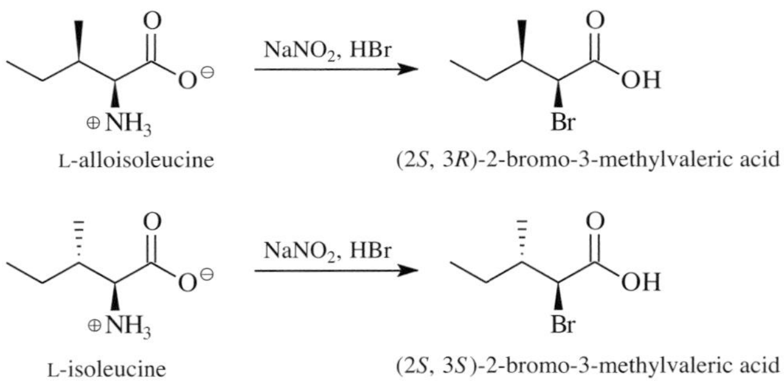

Fig. 3.8. A pair of stereospecific reactions (using the type of reaction in Figure 2.28 as an example).

$NaNO_2$, HBr

L-alloisoleucine → (2*S*, 3*R*)-2-bromo-3-methylvaleric acid

$NaNO_2$, HBr

L-isoleucine → (2*S*, 3*S*)-2-bromo-3-methylvaleric acid

Definition: Stereoconvergent Reaction

It is possible that in a pair of analogous reactions, *one* starting material is converted highly stereoselectively into one product, and a *stereoisomer* of the starting material is converted highly stereoselectively into *the same* product. Reactions of this type can be called **stereoconvergent** (for an example, see Figure 3.9).

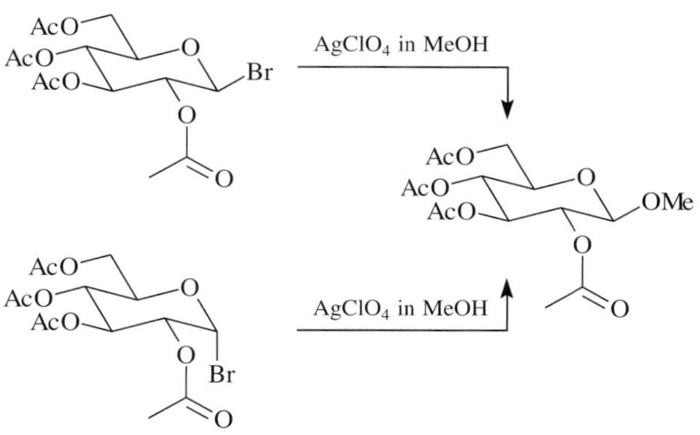

AcO AcO AcO O Br O O AgClO$_4$ in MeOH AcO AcO AcO O OMe O O AcO AcO AcO O O Br O AgClO$_4$ in MeOH

Fig. 3.9. A pair of stereo-convergent reactions (using the glucoside syntheses from Figure 2.27 as an example).

3.3 Electrophilic Additions that Take Place Diastereoselectively as *cis*-Additions

All one-step additions to C=C double bonds are mechanistically required to take place *cis*-selectively (*ds* > 99:1) (Sections 3.3.1–3.3.3). In addition, the heterogeneously catalyzed hydrogenation of alkenes also usually takes place with very high *cis*-selectivity, in spite of its being a multistep reaction (Section 3.3.4).

3.3.1 A Cycloaddition Forming Three-Membered Rings

Cycloadditions are ring-forming addition reactions in which the product, the so-called cycloadduct, possesses an empirical formula that corresponds to the sum of the empirical formulas of the starting material and the reagent. All one-step cycloadditions take place with *cis*-selectivity. Three-, four-, five-, or six-membered can be produced by cycloadditions to alkenes (Figure 3.10).

You will become familiar with selected cycloadditions that lead to four-, five-, or six-membered rings in Chapter 15. *Here* we only discuss the addition of dichlorocarbene (Cl_2C) to olefins as an example of a *cis*-addition of the cycloaddition type.

Dichlorocarbene cannot be isolated, but it can be produced in the presence of an alkene with which it rapidly reacts. The best dichlorocarbene precursor is the anion $Cl_3C^{\ominus}$, which easily eliminates a chloride ion. This anion is obtained from Cl_3CH and fairly strong bases like potassium *tert*-butoxide (KO-*tert*-Bu), KOH or NaOH. The deprotonation of Cl_3CH is run very efficiently with a solution of KO-*tert*-Bu in THF. Alternatively, when a concentrated aqueous solution of NaOH or KOH is vigorously stirred with a chloroform solution of the alkene (which is not miscible with the first) there is only moderate conversion into the corre-

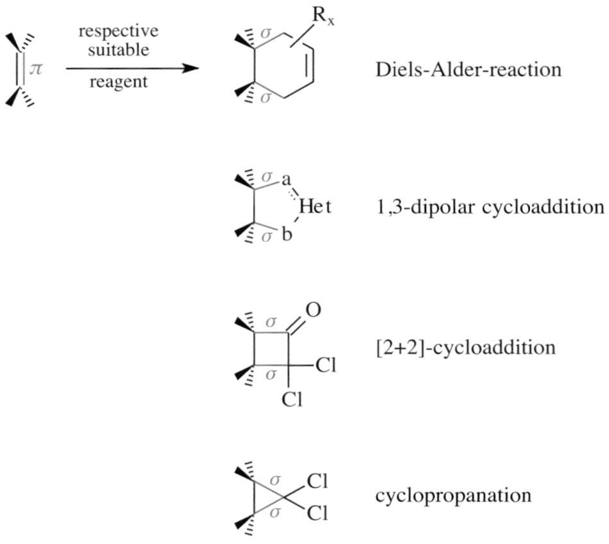

Fig. 3.10. *cis*-Selective cycloadditions.

sponding dichlorocyclopropane: the solvation of ions such as $K^{\oplus}$, $Na^{\oplus}$ or $OH^{\ominus}$ in the organic phase would be far too low for them to efficiently migrate from the aqueous phase into Cl_3CH. Therefore, the deprotonation reaction $Cl_3CH + OH^{\ominus} \rightarrow Cl_3C^{\ominus}$ is limited in such a way that no more than a monomolecular layer is produced at the interface between the chloroform and the aqueous phase. This limitation can be overcome using phase transfer catalysis.

Side Note 3.1. The Principles of Phase Transfer Catalysis

The migration of **inorganic anions** into organic solvents can often be facilitated by a **phase transfer catalyst**. Tetraalkylammonium salts can serve as phase transfer catalysts. Such ammonium salts are characterized by the following special features: on the one hand, their cationic moiety is so hydrophobic that it cannot be solvated by water. Further, it is so lipophilic that solvation can be readily achieved by organic solvents. Consequently, in two-phase systems consisting of chloroform and water, tetraalkylammonium cations tend to prefer the organic phase. Because of Coulombic attraction they "drag along" equivalent amounts of anions. The crucial factor for the catalyst function of tetraalkylammonium salts is that only a small fraction of the anions that are "dragged along" still consist of the original anions; most of them have been exchanged for the anions that are required for the reaction to be catalyzed.

Figure 3.11 illustrates a scenario where $OH^{\ominus}$ ions are transported from the aqueous into the chloroform phase by tetraalkylammonium cations. There, the tetraalkylammonium hydroxide is the base and is available for deprotonation in the entire chloroform phase—a process that was previously limited to just the interface. The $Cl_3C^{\ominus}$ so formed could undergo fragmentation to dichlorocarbene, which could then add to the alkene to be cyclopropanated. This scenario provides a plausible explanation of the reaction mechanism for dichlorocyclopropanations, which in practice are usually performed under phase-transfer catalysis (cf. Figure 3.13 for an example).

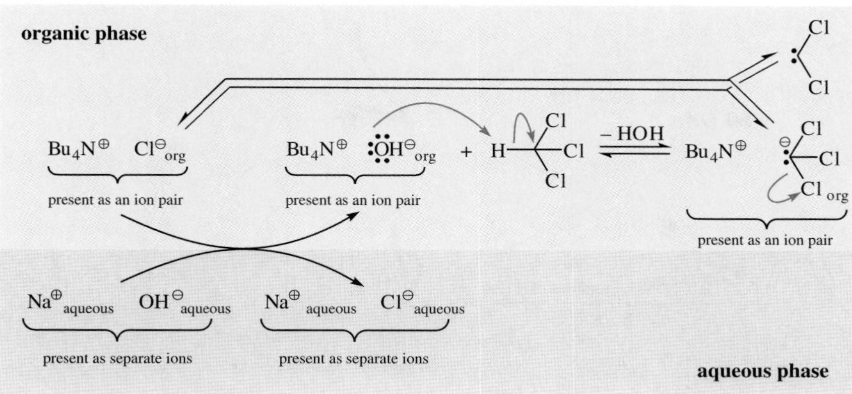

Fig. 3.11. Plausible, but incorrect mechanism of the phase-transfer catalyzed dichlorocyclopropanation of alkenes as often encountered in (close!) analogy to the $Bu_4N^{\oplus}Cl^{\ominus}$ catalyzed Kolbe nitrile synthesis in the two-phase system CH_2Cl_2/aqueous solution of NaCN.

As a reminder that just because something is reasonable does not mean it is true, detailed studies on phase-transfer catalyzed dichlorocyclopropanations suggest the mechanism in Figure 3.11 must be replaced with the mechanism given in Figure 3.12. It is presently assumed that even in the presence of the phase transfer catalyst the deprotonation of chloroform is also limited to the interface between chloroform/aqueous NaOH. This means that its efficiency is as low as described above. But as soon as $Cl_3C^{\ominus}$ ions have been produced they "pair" with the tetraalkylammonium cations. The resulting ion pairs leave the interface (where new $Cl_3C^{\ominus}$ ions are then formed) and disseminate into the entire chloroform phase where chloride abstraction forming the dichlorocarbene and the three-ring formation occur.

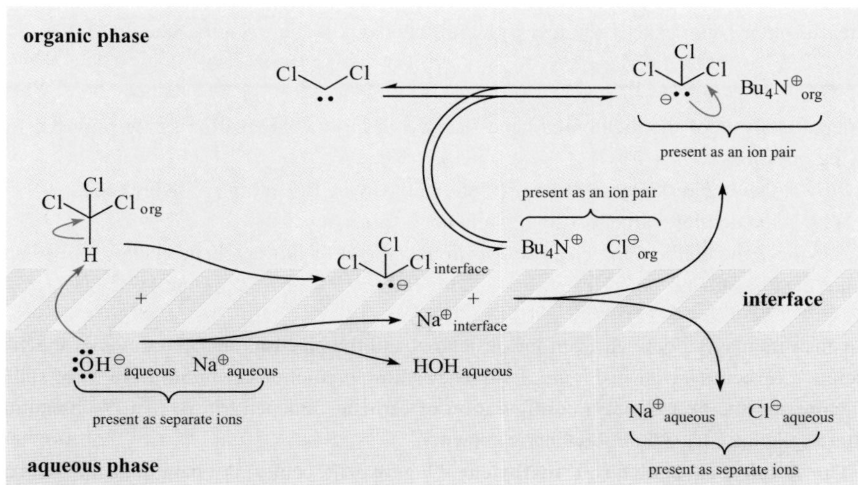

Fig. 3.12. Correct mechanism of the phase-transfer catalyzed dichlorocyclopropanation of alkenes.

The dichlorocarbene produced in this way *in* chloroform adds to alkenes not only stereoselectively but also stereospecifically (Figure 3.13). Doubly chlorinated cyclopropanes are produced. These are not only of preparative interest as such, but they can also be used as starting

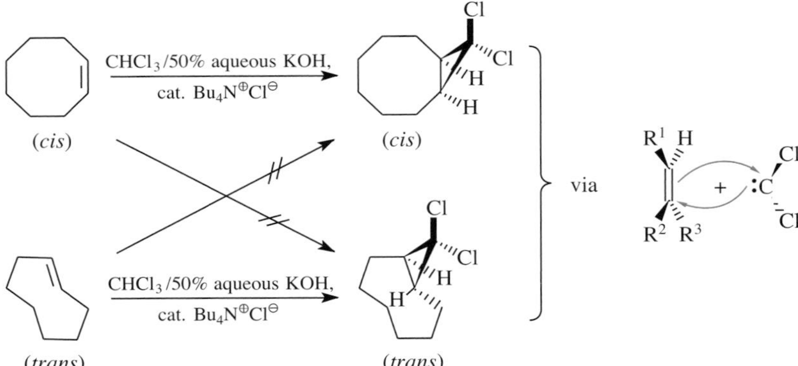

Fig. 3.13. Two reactions that demonstrate the stereospecificity of the *cis*-addition of dichlorocarbene to alkenes.

materials for the synthesis of chlorine-free cyclopropanes. Chlorine can successfully be removed, for example, with tributyltin hydride (Bu_3SnH) via the mechanism discussed in Section 1.10.1 for bromides and iodides. Being able to prepare dichlorocarbene as conveniently as described makes the two-step process attractive for producing chlorine-free cyclopropanes.

The right portion of Figure 3.13 shows the transition state geometry of dichlorocyclopropanations, which is determined in part by the electron configuration of the dichlorocarbene. The C atom of dichlorocarbene is sp^2-hybridized. The two nonbonding electrons are spin-paired—that is, dichlorocarbene is a singlet species—and occupy an sp^2 AO; the $2p_z$ AO is vacant. The C–C bonds of the three-membered ring are formed as follows: due to its occupied sp^2 AO the carbene carbon atom provides the alkene with an electron pair; *at the same time* (one-step process!) it accepts—due to its vacant $2p_z$ AO—an electron pair from the alkene. As both the sp^2 and the $2p_z$ AO interact with their reaction partner in the transition state, the latter is not exactly aligned with either the axis of the sp^2 AO or the $2p_z$ AO.

Side Note 3.2. Carbenes

In-depth analysis of the dichlorocarbene formula in Figure 3.13 will possibly raise the following questions:

- Why is the dichlorocarbene structure bent, i.e., why is its C atom sp^2-hybridized?
- Why is the dichlorocarbene a singlet and not a triplet species?
- Why does the nonbonding electron pair in the singlet dichlorocarbene occupy the sp^2 and not the $2p_z$ AO?

The answers may be derived from the structures and the electron configurations of the five reference carbenes given in Figure 3.14. They allow predictions regarding the most stable structure and the best electron configuration of carbenes independent of their "substitution pattern" (incl. CCl_2) if the following is known:

The simplest carbene CH_2 ("methylene," Figure 3.14, center) is a bent molecule with an H,C,H bond angle of 135° and has a triplet ground state. The singlet CH_2 is less stable by 8 kcal/mol. Its free electron pair occupies the sp^2 AO (because in this orbital it is nearer to the nucleus and therefore more stabilized than in the $2p_z$ AO), and the H,C,H bond angle amounts to 105° (*two* electrons in the sp^2 AO as compared to *one* electron in the sp^2 AO of triplet CH_2; cf. the discussion of VSEPR theory in Section 1.1.1).

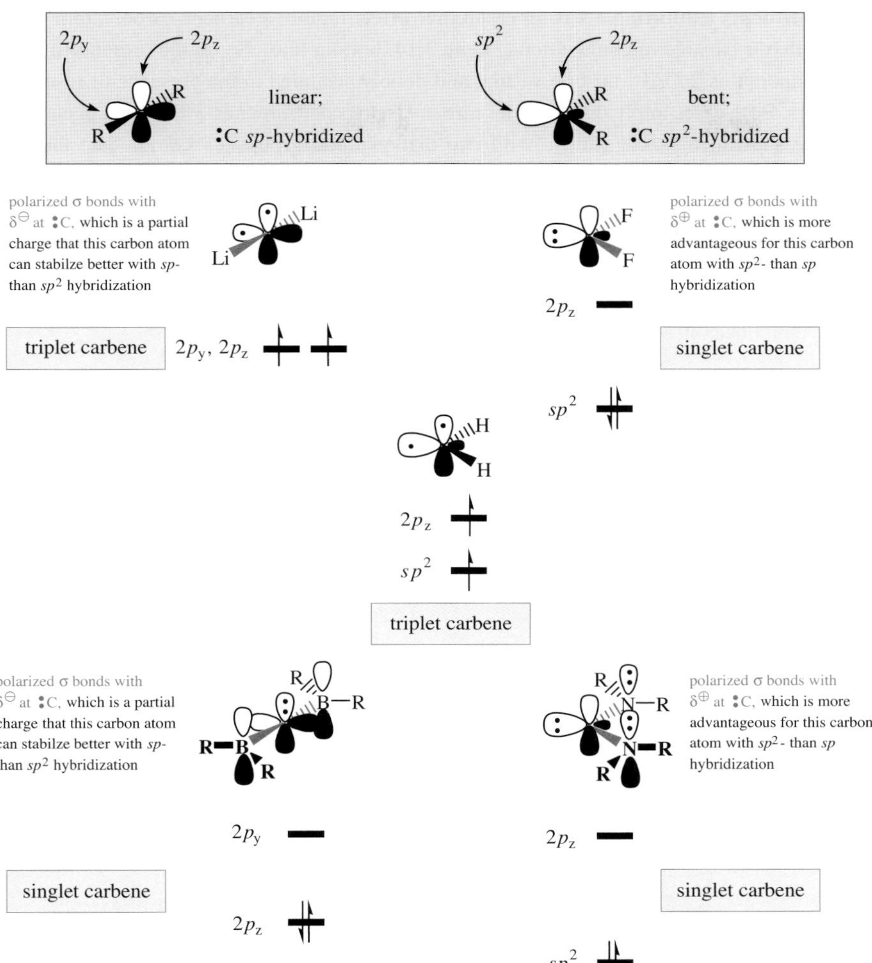

Fig. 3.14. Preferential geometries, bonding and electron configurations of carbenes. Here, the parent compound "methylene" (center) and four disubstituted carbenes are shown, with the substituents mainly exerting a single electronic effect: Li (+I effect, no resonance effect), F (−I effect, no resonance effect), R_2B (−M effect, no inductive effect) and R_2N (+M effect, no inductive effect).

Substituents exhibiting a strong inductive effect generate partial charges at the carbene carbon atom (Figure 3.14, top line). So, carbenes with strong inductively electron-withdrawing substituents—e. g., with two fluorine atoms—have a C atom with a charge deficit and thus increased electronegativity. Such a C atom will place lone electrons in an orbital that interacts with the nucleus. The two nonbonding electrons therefore prefer an sp^2 AO. This means the carbene is bent and singlet (Figure 3.14, top right). Carbenes with strong inductively electron-donating substituents—e. g., those with two lithium atoms—have a carbene carbon atom that is unusually rich in electrons. In order to minimize Coulomb repulsion forces within the electron shell the nonbonding electrons are bound as far away as possible from the carbene carbon atom and also from each other. This is achieved by accommodating the electrons within the two 2*p* AOs of a carbene with *sp*-hybridization and thus linear structure. According to Hund's Rule no spin pairing occurs under these conditions, i.e. the carbene is triplet (Figure 3.14, top left).

Substituents exhibiting a strong resonance effect require a carbene carbon atom with a $2p_z$ AO that is capable of conjugation (Figure 3.14, bottom line). So, in the presence of resonance donors this $2p_z$ AO must be vacant, and doubly occupied in the presence of resonance acceptors. Both cases imply that a singlet carbene is present. Carbenes with strong pi electron-donating substituents—e. g., those with two dialkylamino groups—accommodate the nonbonding electron pair in the most stabilizing atomic orbital possible, which is an AO with partial *s* character, namely an sp^2 AO. The latter occurs with a bent structure (Figure 3.14, bottom right). Carbenes with strong pi electron-withdrawing substituents—e. g., those with two dialkylboryl groups—accommodate the electron deficiency by placing electrons in the orbital best suited for overlap with the empty p orbitals on boron, an $2p_y$ AO, to produce a singlet carbene with a linear structure (Figure 3.14, bottom left).

3.3.2 Additions to C=C Double Bonds That Are Related to Cycloadditions and Also Form Three-Membered Rings

In contrast to the dichlorocyclopropanations from Section 3.3.1, the reactions discussed in this section are not cycloadditions in a strict sense. The reason is that the empirical formula of the addition products presented here is not equal to the sum of the empirical formulas of the reaction partners.

The addition of the **Simmons–Smith reagent** to alkenes leads to chlorine-free cyclopropanes in a single step (Figure 3.16). This addition also runs stereoselectively and stereospecifically. The Simmons–Smith reagent (Figure 3.15, Formula **A**) is generated from diiodomethane and a so-called Zn/Cu couple, which in turn is prepared from an excess of Zn dust and catalytic amounts of $CuSO_4$, CuCl or $Cu(OAc)_2$. Formally, this reaction resembles the

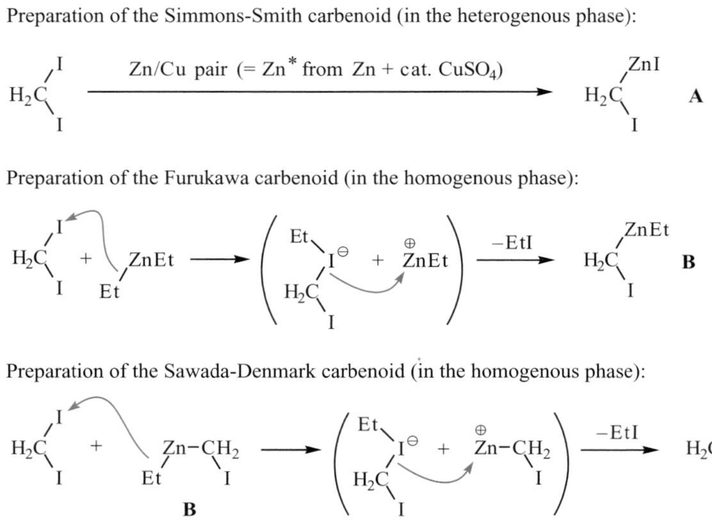

Preparation of the Simmons-Smith carbenoid (in the heterogenous phase):

H_2CI_2 —[Zn/Cu pair (= Zn* from Zn + cat. $CuSO_4$)]→ ICH_2ZnI **A**

Preparation of the Furukawa carbenoid (in the homogenous phase):

H_2CI_2 + $ZnEt_2$ → ($EtI^{\ominus}CH_2I$ + $^{\oplus}ZnEt$) —[−EtI]→ ICH_2ZnEt **B**

Preparation of the Sawada-Denmark carbenoid (in the homogenous phase):

H_2CI_2 + $EtZnCH_2I$ (**B**) → ($EtI^{\ominus}CH_2I$ + $^{\oplus}ZnCH_2I$) —[−EtI]→ ICH_2ZnCH_2I **C**

Fig. 3.15. Zinc carbenoids: methods of preparation, stoichiometries and details of their nomenclature.

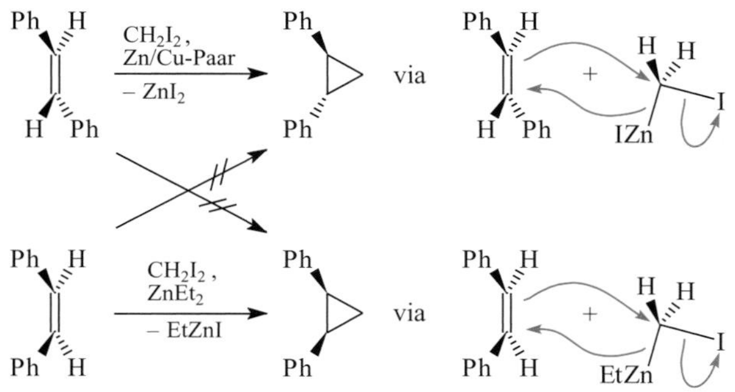

Fig. 3.16. Two reactions that demonstrate the stereospecificity of *cis*-cyclopropanations with the Simmons–Smith reagent. In the first reaction the zinc carbenoid is produced according to the original method, and in the second it is produced by the Furukawa variant.

formation of methyl magnesium iodide from methyl iodide and activated magnesium. Related "Simmons-Smith reagents" include zinc compounds **B** ("Furukawa reagent") and **C** ("Sawada reagent" or "Sawada-Denmark reagent") in Figure 3.15. The formation of the reagents **B** and **C** are based on iodine/zinc exchange reactions that you may have never come across before. In Section 5.3.2 you will be introduced to iodine/magnesium, iodine/lithium and bromine/lithium exchange reactions proceeding similarly to the iodine/zinc exchange reactions of Figure 3.15.

The three Simmons–Smith reagents **A**, **B** and **C** in Figure 3.15 would be ordinary organic zinc compounds with a rather covalent C–Zn bond if the carbon that binds the zinc were not also bound to iodine, a good leaving group. Because of this arrangement, both the C—I bond and the adjacent C—Zn bond are highly elongated. As a result of this distortion, the geometry of the Simmons–Smith reagents **A**–**C** is similar to the geometry of the carbene complexes of ZnI_2 or ethyl zinc iodide (EtZnI) or (iodomethyl)zinc iodide (**I-H_2CZnI**). The fact that these carbene complexes are not the same as a mixture of free carbene (CH_2) and ZnI_2 or EtZnI or I-H_2CZnI is expressed by designating the Simmons–Smith reagents as **carbenoids.** These carbenoids have the ability to transfer a CH_2 group to alkenes.

A $C(CO_2Me)_2$ unit can add to C=C double bonds by means of another carbenoid, namely a rhodium–carbene complex **A** (**B** is a resonance form, Figure 3.17). Again, these additions are

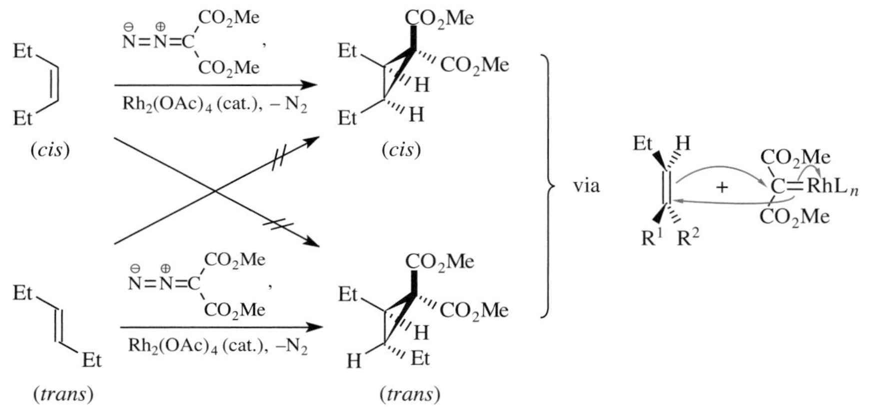

Fig. 3.17. Two reactions that demonstrate the stereospecificity of Rh-catalyzed *cis*-cyclopropanations of electron-rich alkenes. — The zwitterionic resonance form **A** turns out to be a better presentation of the electrophilic character of rhodium–carbene complexes than the (formally) charge-free resonance form **B** or the zwitter-ionic resonance form (not shown here) with the opposite charge distribution (⊖ adjacent to the CO_2Me groups, ⊕ on Rh): rhodium–carbene complexes preferentially react with electron-rich alkenes.

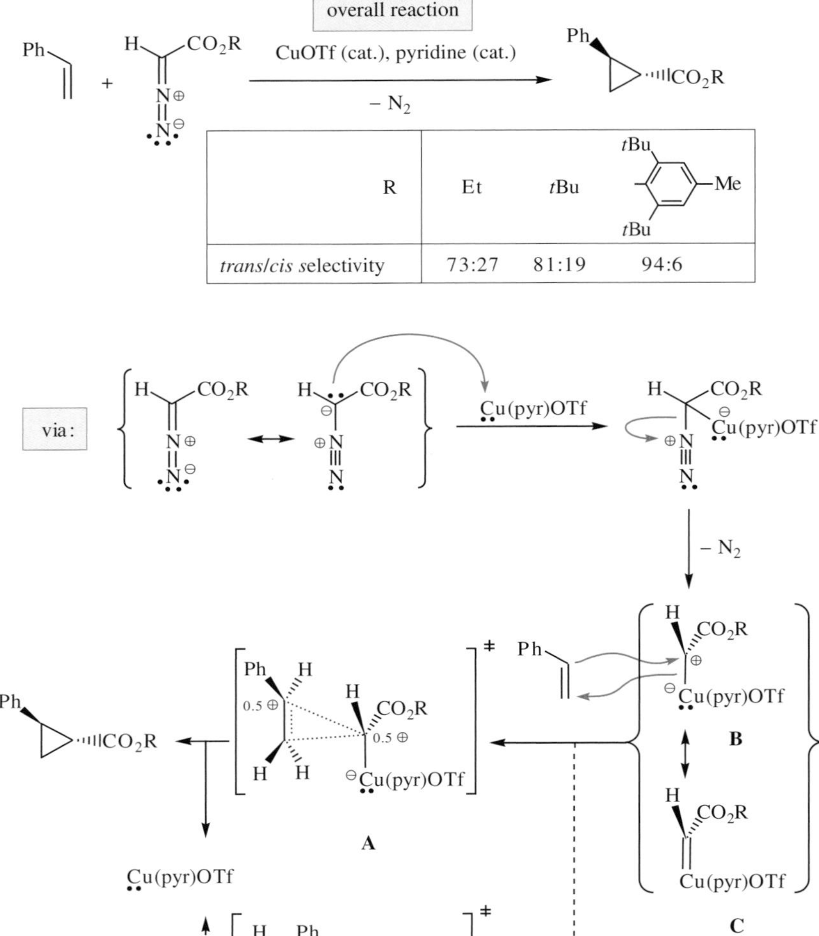

R	Et	tBu	tBu, tBu, Me-substituted aryl
trans/cis selectivity	73:27	81:19	94:6

Fig. 3.18. Mechanistic details on the transition-metal catalyzed (here: Cu-catalyzed) cyclopropanation of styrene as a prototypical electron-rich alkene. The more bulky the substituent R of the ester group CO_2R, the stronger is the preference of transition state **A** over **D** and hence the larger the portion of the *trans*-cyclopropane carboxylic acid ester in the product mixture.—The zwitterionic resonance form **B** turns out to be a better presentation of the electrophilic character of copper–carbene complexes than the (formally) charge-free resonance form **C** or the zwitterionic resonance form (not shown here) with the opposite charge distribution ($^{\ominus}\alpha$ to the CO_2R substituent, $^{\oplus}$ on Cu): copper–carbene complexes preferentially react with electron-rich alkenes.

not only stereoselective but also stereospecific. The rhodium–carbene complex mentioned forms via the reaction of diazomalonic ester (for a preparation, see Figure 15.42) and dimeric rhodium(II)acetate with elimination of molecular nitrogen. The detailed mechanism of the formation of carbenoid like **A** is shown in Figure 3.18 for the formation of the copper–carbene complex **B**, generated from diazoacetic ester and copper(I) trifluoromethane sulfonate.

Diazoacetic esters and a number of transition metal salts form similar transition metal–carbene complexes while eliminating N_2. These allow for the addition of a $CH(CO_2R)$ group to

C=C double bonds. Here, a stereochemical aspect has to be considered that does not occur with carbenoid additions originating from diazomalonic ester: usually, *one* alkene and *one* transition metal–carbene complex can generate *two* cyclopropanes. Figure 3.18 illustrates this observation with a series of cyclopropanations of styrene with copper–carbene complexes **B**, which contain a bulky ester residue R. The resulting phenylcyclopropane carboxylic acid esters are obtained as mixtures of more *trans*- than *cis*-isomers. Depending on R, the proportion of the *trans*-cyclopropane ester increases from 73% to 81% and finally to 94%, indicating that transition state **A**, leading to the *trans*-cyclopropane carboxylic acid ester, is preferred over transition state **D**, leading to the *cis*-cyclopropane carboxylic acid ester—and the bulkier the ester residue R, the stronger this preference. In the transition state **A**, the residue R avoids the phenyl ring of styrene; in transition state **D** it does not.

Another one-step addition reaction to C=C double bonds that forms three-membered rings is the epoxidation of alkenes with **percarboxylic acids** (Figure 3.19). Most often, ***meta*-*chloroperbenzoic acid*** (MCPBA) is used for epoxidations. Magnesium monoperoxyphthalate (MMPP) has become an alternative. Imidopercarboxylic acids are used to epoxidize olefins as well. Their use (for this purpose) is *mandatory* when the substrate contains a ketonic C=O double bond in addition to the C=C double bond. In compounds of this type, percarboxylic acids preferentially cause a Baeyer–Villiger oxidation of the ketone (see Section 14.4.2), whereas imidopercarboxylic acids selectively effect epoxidations (for an example see Figure 14.35).

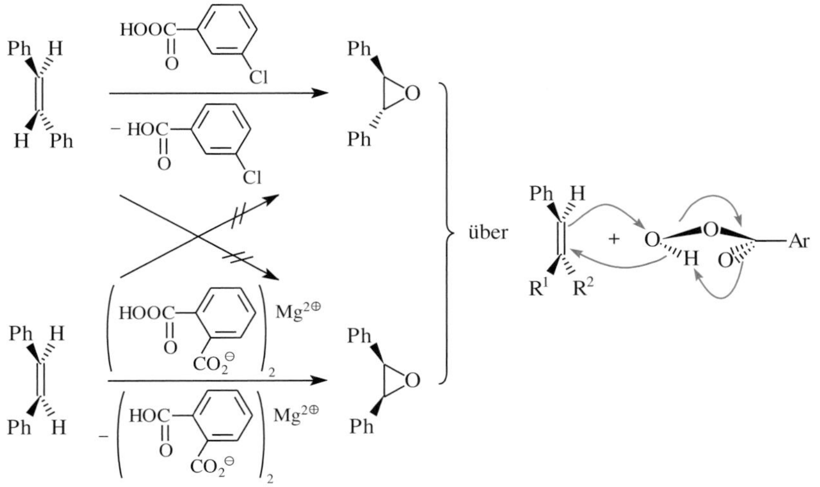

HOOC (C=O), Cl; – HOC (C=O), Cl; Ph H / H Ph; Ph, O, Ph; über; Ph H, R^1 R^2, + O, O, H, O, Ar; HOOC, $CO_2^{\ominus}$, $Mg^{2\oplus}$, 2; Ph H / Ph H; – HOC, $CO_2^{\ominus}$, $Mg^{2\oplus}$, 2

Fig. 3.19. Stereospecific *cis*-epoxidations of alkenes with percarboxylic acids.

In the transition state of the epoxidation of alkenes with a percarboxylic acid the C=C axis of the alkene is rotated out of the plane of the percarboxylic acid group by 90° ("spiro transition state"). In this process, four electron pairs are shifted simultaneously shifted. This very special transition state geometry make peracid oxidations of C=C double bonds largely insensitive to steric hindrance. The epoxidation given in Figure 3.20 provides an impressive example.

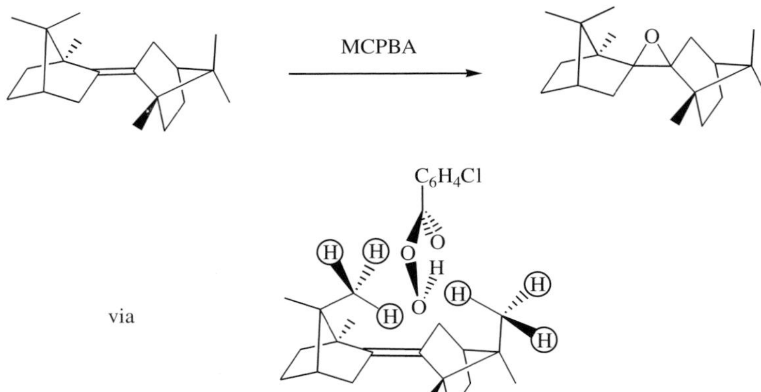

Fig. 3.20. Due to the transition state geometry shown there is hardly any steric hindrance in *cis*-epoxidations with percarboxylic acids—as is impressively demonstrated by the example given here.

3.3.3 *cis*-Hydration of Alkenes via the Hydroboration/Oxidation/Hydrolysis Reaction Sequence

Boranes

In monoborane (BH_3), monoalkylboranes RBH_2, or dialkylboranes R_2BH there is only an electron sextet at the boron atom. In comparison to the more stable electron octet, the boron atom thus lacks two valence electrons. It "obtains" them by bonding with a suitable electron pair donor. When no better donor is available, the bonding electron pair of the B—H bond of a second borane molecule acts as the donor so that a "two-electron, three-center bond" is produced. Under these conditions, boranes are consequently present as dimers: "BH_3," for example, as B_2H_6. Still, small fractions of the monomers appear as minor components in the dissociation equilibrium of the dimer: B_2H_6, for example, thus contains some BH_3.

The Lewis bases Me_2S or THF are better electron pair donors than the boranes themselves. Therefore, they convert dimeric boranes into Me_2S or THF complexes of the monomers such as Me_2S-BH_3 or THF-BH_3. This type of complex also dissociates to a small extent and thus contains a small equilibrium concentration of free monomeric boranes.

Only *monomeric* boranes can undergo a *cis*-addition to alkenes. In principle, *every* B—H bond of a monomeric borane can be involved in this kind of reaction. It effects a one-step addition of formerly joined fragments BR_2 and H to the C=C double bond. This addition is therefore called the **hydroboration** of a C=C double bond.

Since BH_3 contains three B—H bonds, it can add to as many as three C=C double bonds. Monoalkylboranes RBH_2, with their two B—H bonds, add to as many as two C=C double bonds. Dialkylboranes R_2BH naturally can add to only one. The steric hindrance of a B—H bond increases in the series $BH_3 < RBH_2 < R_2BH$. Therefore its reactivity decreases. Consequently, BH_3 does not necessarily react to produce trialkylboranes (but may do so as shown shortly: see Figure 3.22). This failure to produce trialkylboranes is especially common with sterically demanding alkenes, which may therefore only lead to dialkyl- or even to monoalkylboranes (examples: Figure 3.21). The compound 9-BBN (9-borabicyclo[3.3.1]nonane), which is produced from 1,5-cyclooctadiene and B_2H_6, illustrates the selective formation of a

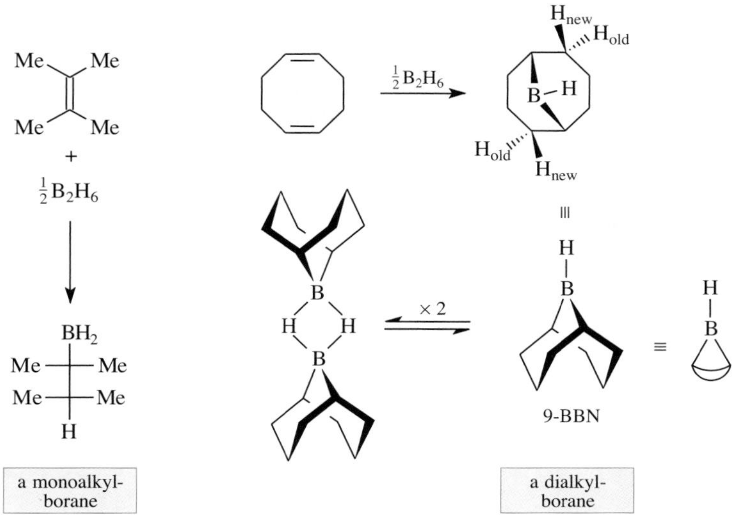

Fig. 3.21. Synthesis of monoalkyl- and dialkylboranes through incomplete hydroboration of multiply substituted C=C double bonds.

dialkylborane; 9-BBN itself is a valuable hydroborating agent (synthesis applications are given later: Figures 3.24–3.27 and Table 3.1).

Hydration of Cyclohexene

Hydroboration of alkenes is very important in organic chemistry, mainly because of a secondary reaction, which is often carried out after it. In this reaction, the initially obtained trialkylboranes are first oxidized with sodium hydroxide solution/H_2O_2 to boric acid trialkyl esters (trialkyl borates), which are then hydrolyzed to alcohol and sodium borate. Thus, through the reaction sequence of (1) hydroboration and (2) NaOH/H_2O_2 treatment, one achieves an H_2O addition to C=C double bonds. The reaction example in Figure 3.22 illustrates this starting from Me_2S -BH_3 or THF-BH_3 and cyclohexene.

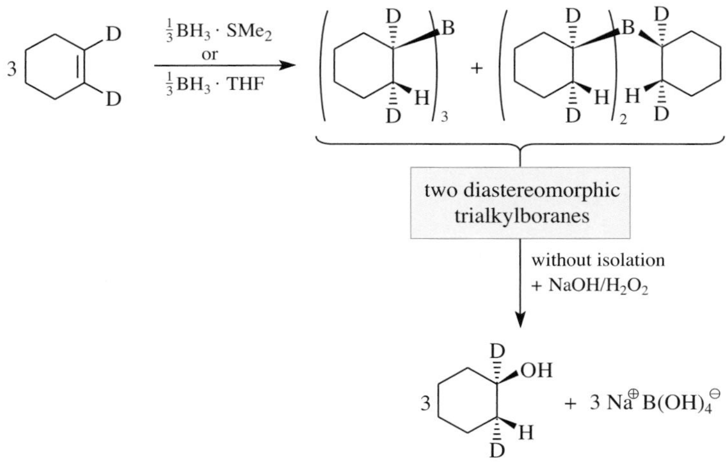

Fig. 3.22. *cis*-Hydration of an alkene via the reaction sequence hydroboration/oxidation/hydrolysis. (The chiral molecules occur as racemates.)

Because the C=C double bond of the cyclohexene used in Figure 3.22 is labeled with deuterium, it is possible to follow the stereochemistry of the whole reaction sequence. First there is a *cis*-selective hydroboration. Two diastereomeric, racemic trialkylboranes are produced. Without isolation, these are oxidized/hydrolyzed with sodium hydroxide solution/H_2O_2. The reaction product is the sterically homogeneous but, of course, racemic dideuteriocyclohexanol. The stereochemistry of the product proves the *cis*-selectivity of this hydration.

Three types of reactions participate in the reaction sequence of Figure 3.22 (cf. Figure 3.23):

1. the hydroboration of the C=C double bond to form the trialkylboranes **A** and **B**,
2. their oxidation to the trialkylborates **D** and **E**, respectively,
3. their hydrolyses to the dideuteriocyclohexanol **C.**

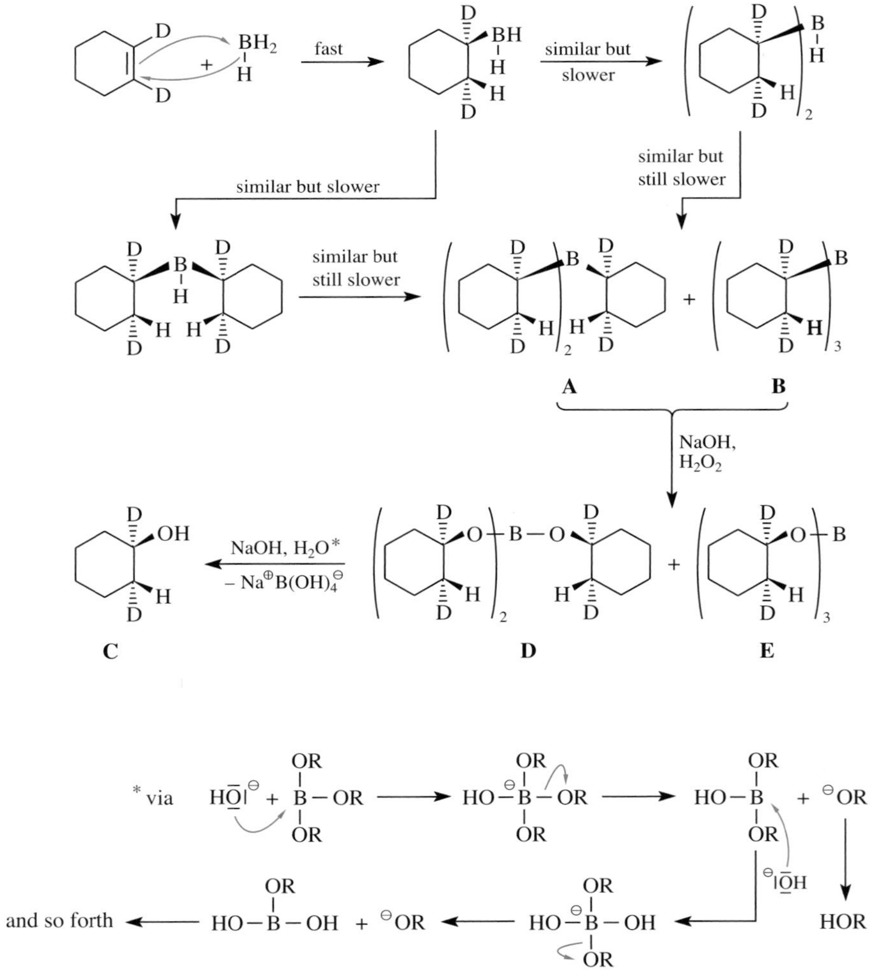

Fig. 3.23. Mechanistic details of the hydroboration/oxidation/hydrolysis sequence of Figure 3.22. (The chiral molecules occur as racemates.)

The mechanism of reaction 1 is detailed in Figure 3.23 (top), and the mechanism of reaction 3 is detailed in the same figure at the bottom. We will discuss step 2 later (Figure 14.40). The upper part of Figure 3.23 illustrates the details of the consecutive hydroboration of C=C double bonds by BH_3, mono-, and dialkylboranes, respectively. The second and the third hydroboration steps take place without any diastereoselectivity with respect to the stereocenters, which are already contained in the reacting mono- or dialkylborane. The isotopomeric trialkylboranes **A** and **B** are therefore produced in the statistical (i.e., 3:1) ratio. During the oxidation of these trialkylboranes with NaOH/H_2O_2, a 3:1 mixture of the isotopomeric trialkylborates **D** and **E** is produced. In this step, the C—B bonds are converted to C—O bonds with retention of configuration. During the subsequent alkaline hydrolysis of the borates **D** and **E** (Figure 3.23, bottom), the three O—B bonds are broken one after the other while the O—C bonds remain intact. In each case the intermediate is an anion with a tetracoordinate B atom (i.e., a borate complex). Because the sodium borate is produced in an alkaline rather than acidic solution, the hydrolysis is irreversible.

Regioselective Hydroboration of Unsymmetrical Alkenes

The hydroboration of alkenes, in which the $C_\alpha{=}C_\beta$ is not symmetrically substituted can lead to constitutionally isomeric trialkylboranes. This is because the new C—B bond can form either at the C_α or at the C_β of the $C_\alpha{=}C_\beta$ double bond. In the oxidation/hydrolysis sequence that follows, constitutionally isomeric alcohols are produced. In one of them, the OH group binds to C_α and in the other it binds to C_β. If only one constitutional isomer of the trialkylborane and consequently only one constitutional isomer of the alcohol is to be produced, the hydroboration step must take place regioselectively. Whether regioselectivity occurs is determined by steric and electronic effects.

The two reactions in Figure 3.24 prove that the regioselectivity of the hydroboration of unsymmetrical alkenes is influenced by steric effects. The substrate is an alkene whose reacting centers C_α and C_β differ only in the size of the alkyl fragment bound to them, Me or *i*Pr (while they do not differ with respect to the small positive partial charges located there; see below). The upper hydroboration of this alkene shown in Figure 3.24 takes place with BH_3 itself. It exhibits no regioselectivity at all. This highly reactive and sterically undemanding reagent forms C_α—B bonds as rapidly as C_β—B bonds. The monoalkylboranes thus formed become the reagents for the second hydroboration step. In their reaction with the next $C_\alpha{=}C_\beta$ double bond, they are subject to a very small steric hindrance. They therefore react with the less hindered $C\alpha$ center with a very slight preference. The dialkylboranes thus formed are the most hindered hydroborating agents in this mixture and react with a slight preference for C_α. Nonetheless, these steric effects are small. After oxidation/hydrolysis, one therefore finds only a 57:43 regioselectivity in favor of the product hydroxylated at C_α.

The bottom half of Figure 3.24 shows an almost perfect solution to this regioselectivity problem using 9-BBN. The B—H bond of 9-BBN is much less readily accessible than the B—H bond in BH_3 and the latter's resulting primary products. Thus, the hydroboration with 9-BBN takes place not only much more slowly than the one with BH_3, but also with much higher regioselectivity. The less hindered hydroboration product and the alcohol derived from it are produced with a regioselectivity of 99.8:0.2.

Unsymmetric alkenes, which carry more alkyl substituents at the C_β center than at the C_α center, are also hydroborated by the unhindered BH_3 with considerable regioselectivity

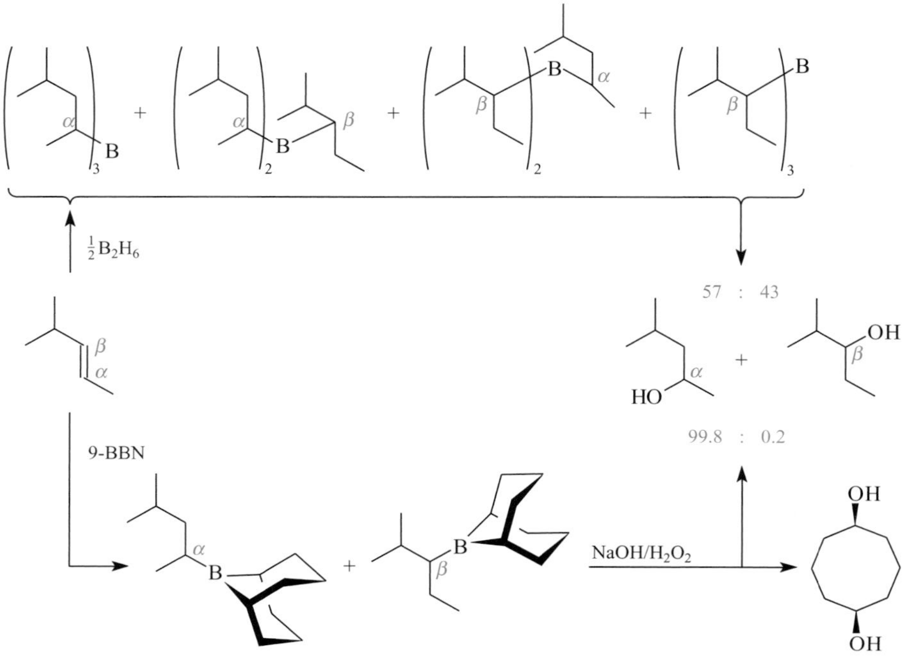

Fig. 3.24. Regioselectivity of the hydroboration of an α,β-dialkylated ethylene.

(Table 3.1). After oxidative workup, one isolates the alcohol in the α-position almost exclusively. According to what has already been stated, as the more bulky reagent, 9-BBN reacts with more sensitivity to steric effects than BH_3 and its secondary products. It therefore makes possible alkene hydrations with almost perfect regiocontrol.

The regioselectivities in Table 3.1 are probably also caused by an electronic effect, which acts in the same direction as the steric effect. The electronic effect can be explained in various ways: for example, by considering a resonance formula of the type

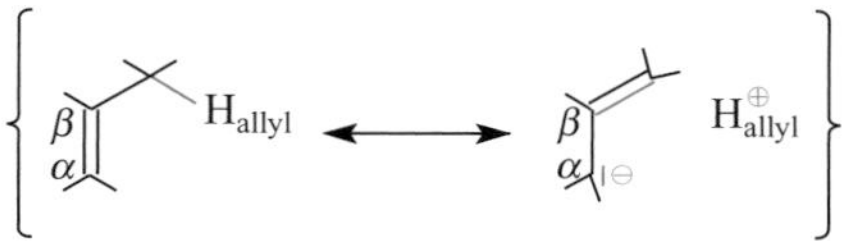

Tab. 3.1 Regioselectivity for the Hydration (via Hydroboration/Oxidation/Hydrolysis) of Ethylene Derivatives that Contain More Alkyl Substituents at C_β Than at C_α

Fraction of α-alcohol in the hydroboration/ oxidation of ...	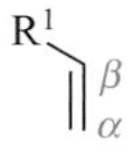 R^1, β, α	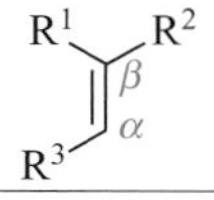 R^1, R^2, R^3, β, α	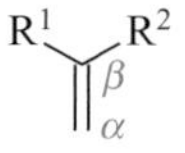 R^1, R^2, β, α
... with B_2H_6:	94%	100%	99%
... with 9-BBN:	99.9%		99.8%

for each H atom in the allylic position. When more H atoms are allylic next to C_β than to C_α—precisely what we have for the alkenes in Table 3.1—a negative partial charge appears at C_α relative to C_β. Because the B atom in all boranes is an *electrophilic* center, it preferentially reacts at the more electron-rich C_α.

Addition reactions to unsymmetrical alkenes in which a reagent with the structure H—X transfers the H atom to the less-substituted C atom and the "X group" to the more substituted C atom give, according to an old nomenclature, a so-called **Markovnikov addition product.** On the other hand, addition reactions of reagents H—X in which the H atom is transferred to the more substituted C atom of an unsymmetrically substituted C=C double bond and the "X group" is transferred to the less substituted C atom lead to a so-called **anti-Markovnikov addition product.**

The hydroboration of unsymmetrical alkenes thus gives monoalkylboranes (addition of H—BH_2), dialkylboranes (addition of H—BHR), or trialkylboranes (addition of H—BR_2), which are typical anti-Markovnikov products. Therefore, the reaction sequence hydroboration/oxidation/hydrolysis brings about the anti-Markovnikov addition of H_2O to unsymmetrically substituted alkenes.

The hydroboration/oxidation/hydrolysis of trisubstituted alkenes also takes place as a *cis*-addition. The reaction equation from Figure 3.25 shows this using 1-methylcyclo-hexene as

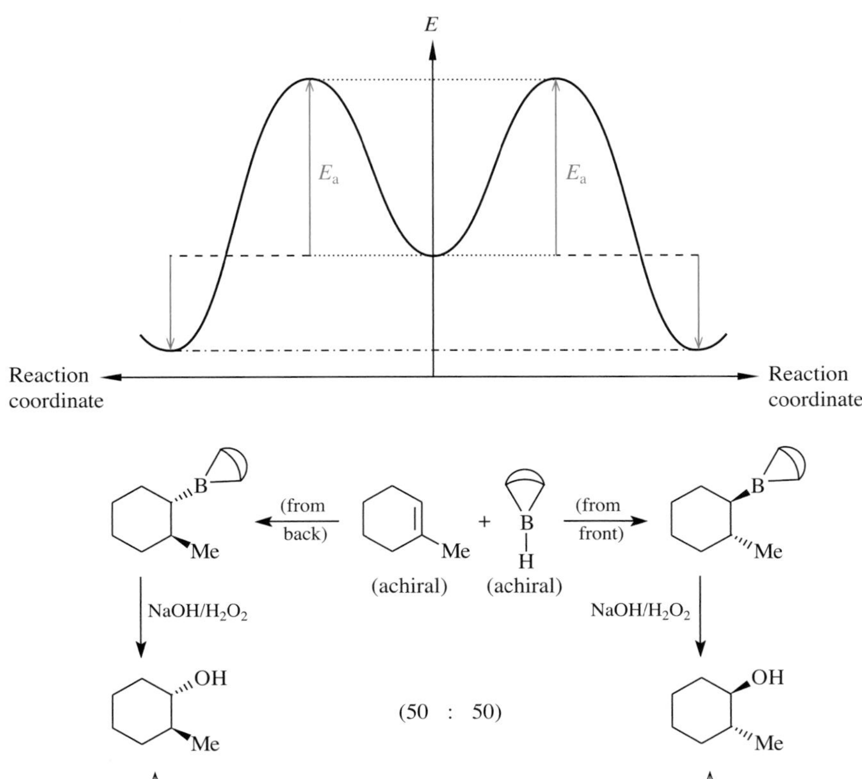

Fig. 3.25. *cis*-Selective hydration of an achiral trisubstituted alkene.

an example. 9-BBN adds to both sides of its C=C double bond at the same reaction rate. In the energy profile, this fact means that the activation barriers for both modes of addition are equally high. The reason for this equality is that the associated transition states are enantiotopic. They thus have the same energy—just like enantiomers.

Stereoselective Hydration of Unsymmetrical Alkenes and Substrate Control of the Stereoselectivity

Let us now consider the analogous hydroborations/oxidations of chiral derivatives of 1-methylcyclohexene, namely of racemic 3-ethyl-1-methylcyclohexene (Figure 3.26) and of enantiomerically pure α-pinene (Figure 3.27).

The stereochemical result is no longer characterized solely by the fact that the newly formed stereocenters have a uniform configuration *relative to each other*. This was the only type of stereocontrol possible in the reference reaction 9-BBN + 1-methylcyclohexene (Figure 3.25). In the hydroborations of the cited *chiral* alkenes with 9-BBN, an additional question arises. What is the relationship between the new stereocenters and the stereocenter(s) already present in the alkene? When a uniform relationship between the old and the new stereocenters arises, a type of diastereoselectivity not mentioned previously is present. It is called **induced or relative diastereoselectivity.** It is based on the fact that the substituents on the stereocenter(s) of the chiral alkene hinder one face of the chiral alkene more than the other. This is an example of what is called **substrate control of stereoselectivity.** Accordingly, in the hydroborations/oxidations of Figures 3.26 and 3.27, 9-BBN does *not* add to the top and the bottom sides of the alkenes with the same reaction rate. The transition states of the two modes of addition are not equivalent with respect to energy. The reason for this inequality is that the associated transition states are diastereotopic. They thus have different energies—just diastereomers.

When 9-BBN reacts with 3-ethyl-1-methylcyclohexene from the side on which the ethyl group is located (Figure 3.25), the transition state involved is higher in energy than the transition state of the 9-BBN addition to methylcyclohexene. This is due to repulsion between the reagent and the ethyl group (Figure 3.26). On the other hand, the reaction of 9-BBN from the other side of 3-ethyl-1-methylcyclohexene is not sterically hindered. Therefore, the corresponding transition state has approximately the same energy as that of the 9-BBN addition to methylcyclohexene. This circumstance makes the “left” transition state in Figure 3.26 the more energetically favorable, and it can therefore be passed through faster. As a consequence, one obtains a single, diastereomerically pure trialkylborane, or after oxidation and hydrolysis, a single alcohol. Because the alkene is used as a racemic mixture, this alcohol is also racemic. Had the starting material been a single enantiomer, so would the product.

Figure 3.27 shows reaction equations and the energy relationships of the hydroboration of enantiomerically pure α-pinene with 9-BBN. The reagent approaches only the side of the C=C double bond that lies opposite the isopropylidene bridge. The addition is thus completely diastereoselective. Moreover, the trialkylborane obtained is a pure enantiomer, since the starting material is a pure enantiomer. It is used as Alpine-Borane® for the enantioselective reduction of carbonyl compounds (Section 10.4).

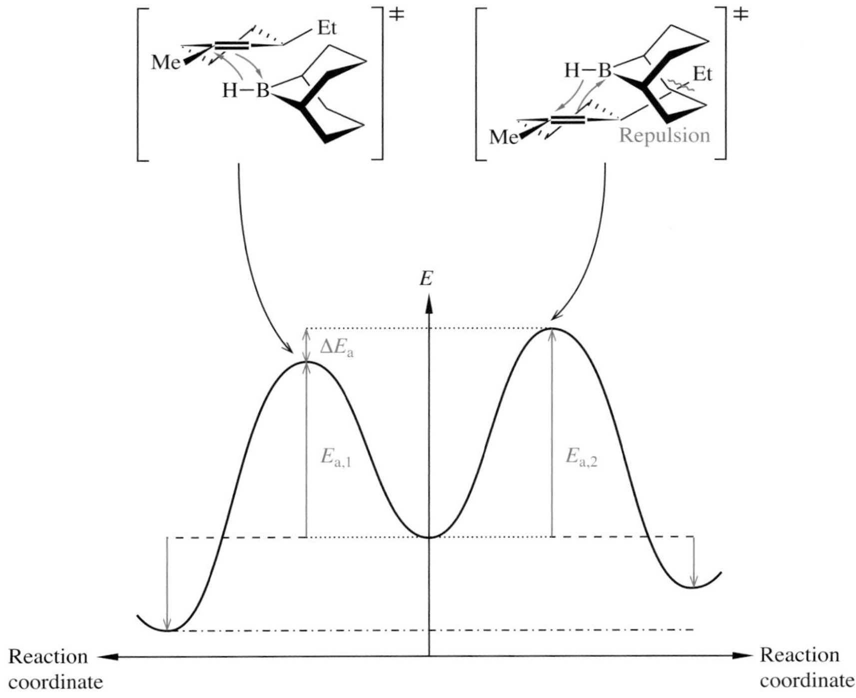

Et, B, Me
k_1 (from back)
Et, Me + B–H
k_2 (from front)
Et, B, Me

A (+ 50% *ent*-**A**)
(+ 50% of the enantiomer)
B (+ 50% *ent*-**B**)

$NaOH/H_2O_2$
$NaOH/H_2O_2$

Et, OH, Me
trans,trans (+ 50% of the enantiomer)

Et, OH, Me
cis,trans (+ 50% of the enantiomer)

Fig. 3.26. *cis*-Selective hydration of a chiral, racemic, trisubstituted alkene with induced diastereoselectivity; ΔE_a corresponds to the extent of the substrate control of diastereoselectivity.

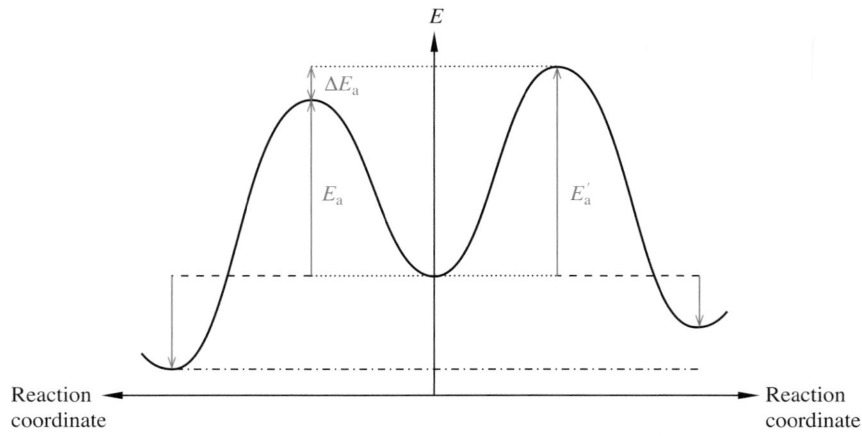

(from back) (from front)

(pure enantiomer) (pure enantiomer) (pure enantiomer)

$NaOH/H_2O_2$ $NaOH/H_2O_2$

(the only product)

cis,trans (pure enantiomer) *trans,trans* (pure enantiomer)

Fig. 3.27. *cis*-Selective hydration of a chiral, enantiomerically pure, trisubstituted alkene with induced diastereoselectivity; ΔE_a corresponds to the degree of substrate control of diastereoselectivity.

3.3.4 Heterogeneous Hydrogenation

There is no known one-step addition of molecular hydrogen to C=C double bonds (hydrogenation).

The addition of hydrogen to alkenes is made possible only with noble metal catalysts. They allow for a multistep, low-energy reaction pathway. The noble metal catalyst may be soluble in the reaction mixture; in that case we have a homogeneous hydrogenation (cf. Sec-

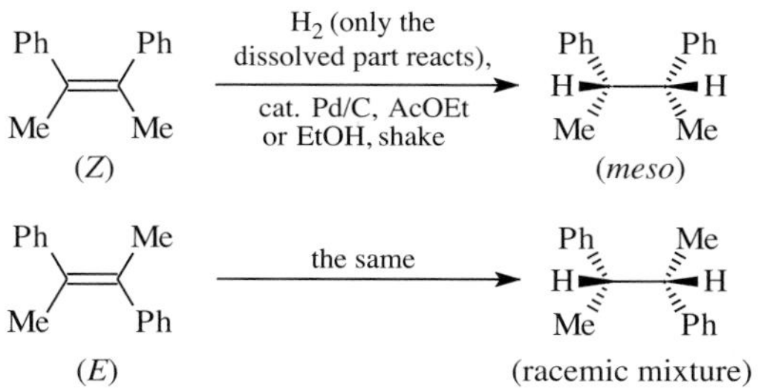

Ph Ph / Me Me (*Z*) —H_2 (only the dissolved part reacts), cat. Pd/C, AcOEt or EtOH, shake→ Ph, H, Me / Ph, H, Me (*meso*)

Ph Me / Me Ph (*E*) —the same→ Ph, H, Me / Me, H, Ph (racemic mixture)

Fig. 3.28. Stereoselectivity and stereospecificity of a pair of heterogeneously catalyzed hydrogenations.

tion 17.4.7). But the catalyst may also be insoluble; then we deal with a heterogeneous hydrogenation.

Heterogeneous hydrogenations are by far the most common. They take place as surface reactions (see Section 17.4.7 for a discussion of the mechanism). Their rate is higher, the greater the surface of the catalyst. Therefore, the catalyst is precipitated on a carrier with a high specific surface such that it is very finely distributed. Suitable carriers are activated carbon or aluminum oxide. To bring the alkene, the catalyst, and the hydrogen, which is dissolved continuously from the gas phase, into intimate contact, the reaction mixture is either shaken or stirred. This procedure makes it possible to carry out heterogeneous hydrogenations at room temperature and under normal or often only slightly increased pressure. Most heterogeneous hydrogenations take place with high *cis*-selectivity. This implies that stereoisomeric alkenes are frequently hydrogenated with stereospecificity (Figure 3.28). For steric reasons, monosubstituted alkenes react more rapidly with hydrogen than 1,1-disubstituted alkenes, which react faster than *cis*-disubstituted alkenes, which react faster than *trans*-disubstituted alkenes, which react faster than trisubstituted alkenes, which in turn react faster than tetrasubstituted alkenes:

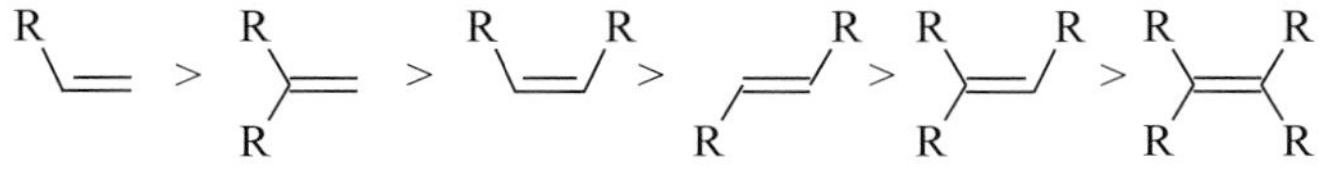

R > R, R > R R > R R > R R, R > R R, R R

Therefore, in polyenes that contain differently substituted C=C double bonds, it is often possible to chemoselectively hydrogenate the least hindered C=C double bond:

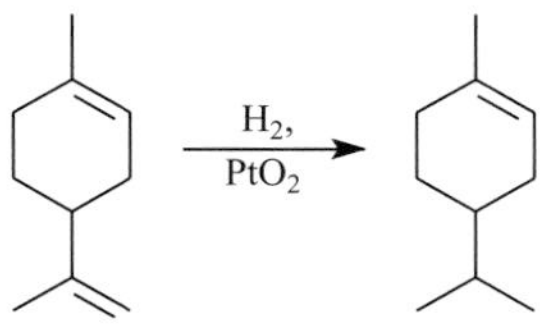

H_2, PtO_2

3.4 Enantioselective *cis*-Additions to C=C Double Bonds

In the discussion of the hydroboration in Figure 3.26, you saw that one principle of stereoselective synthesis is the use of *substrate control* of stereoselectivity. But there are quite a few problems in stereoselective synthesis that cannot be solved in this way. Let us illustrate these problems by means of two hydroborations from Section 3.3.3:

- 9-BBN effects the hydration of the C=C double bond of 1-methylcyclohexene according to Figure 3.25 in such a way that after the oxidative workup, *racemic* 2-methyl-1-cyclohexanol is obtained. This brings up the question: Is an *enantioselective* H_2O addition to the same alkene possible? The answer is yes, but only with the help of *reagent control* of stereoselectivity (cf. Section 3.4.2).
- 9-BBN reacts with the C=C double bond of 3-ethyl-1-methylcyclohexene according to Figure 3.26 exclusively from the side that lies opposite the ethyl group at the stereocenter. Consequently, after oxidation and hydrolysis, a *trans,trans*-configured alcohol is produced. The question that arises is: Can this diastereoselectivity be reversed in favor of the *cis,trans*-isomer? The answer is possibly, but, if so, only by using *reagent control* of stereoselectivity (cf. Section 3.4.4).

Before these questions are answered, we want to consider a basic concept for stereoselective synthesis—topicity—in the following subsection.

3.4.1 Vocabulary of Stereochemistry and Stereoselective Synthesis II: Topicity, Asymmetric Synthesis

The "faces" on both sides of a C=C, C=O, or C=N double bond can have different geometrical relationships to each other (Figure 3.29). If they can be converted into each other by means of a rotation of 180° about a two-fold intramolecular axis of rotation, then they are **homotopic.** If they cannot be superimposed on each other in this way, they are **stereoheterotopic.** Stereoheterotopic faces of a double bond are **enantiotopic** if they can be related to each other by a reflection; otherwise they are **diastereotopic.** Let us illustrate this with examples. The faces of the C=C double bond of cyclohexene are homotopic, those in 1-methylcyclohexene are enantiotopic, and those in 3-ethyl-1-methylcyclohexene are diastereotopic.

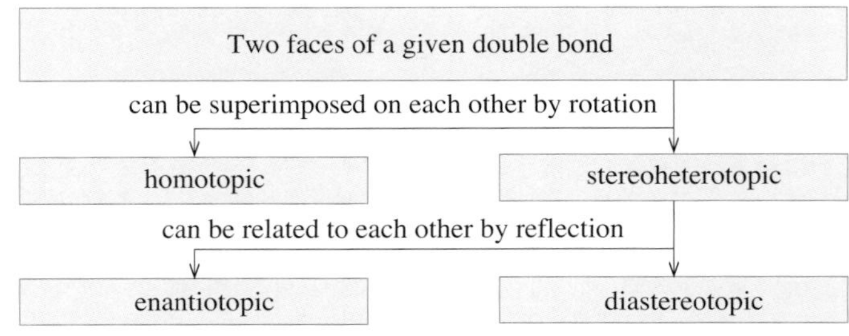

Fig. 3.29. Topicities of molecular faces (**A**) on the sides of different double bonds or (**B**) on either side of a specific double bond of an organic compound.

Let us consider an addition reaction to one of the cited C=X double bonds during which at least one stereocenter is produced. The topicity of the faces of the C=X double bond in question allows one to predict whether such an addition can, in principle, take place stereoselectively. This is because depending on the topicity of the faces of the reacting C=X double bond, the transition states that result from the reaction with reagents of one kind or another (see below) from one or the other face are enantiotopic or diastereotopic.

If the transition states are enantiotopic, they have the same energy. The addition proceeds through each of them to the same extent and we thus obtain a 50:50 ratio of enantiomers. This means that there is no enantioselectivity, i.e., *ee* = 0%.). If the transition states are diastereotopic, they may have different energies. The addition preferentially takes place via the lower energy transition state and preferentially results in more of one stereoisomer. It is thus diastereoselective (i.e., *ds* ≠ 50:50) or enantioselective (i.e., *ee* ≠ 0%).

When all these results are analyzed accurately, it is seen that addition reactions to the C=X double bond

- of achiral compounds (i.e., to C=X double bonds with homotopic or enantiotopic faces) cannot take place enantioselectively; they always give racemic mixtures (e. g., hydroboration in Figure 3.25);
- of chiral compounds (i.e., to C=X double bonds with diastereotopic faces) can take place diastereoselectively with achiral reagents. In this case, we have substrate control of stereoselectivity (e. g., hydroborations in Figures 3.26 and 3.27); or
- of achiral compounds (i.e., to C=X double bonds with homotopic or enantiotopic faces) with chiral reagents can lead to enantiomerically enriched or enantiomerically pure compounds.

An enantioselective addition of the latter type or, in general, the successful conversion of achiral starting materials into enantiomerically enriched or enantiomerically pure products is referred to as **asymmetric synthesis** (for examples, see Sections 3.4.2, 3.4.6, 10.4, 10.5.2, 17.4.7).

3.4.2 Asymmetric Hydroboration of Achiral Alkenes

The conclusion drawn from Section 3.4.1 for the hydroborations to be discussed here is this: an addition reaction of an enantiomerically pure chiral reagent to a C=X double bond with enantiotopic faces can take place via two transition states that are diastereotopic and thus generally different from one another in energy. In agreement with this statement, there *are* diastereoselective additions of enantiomerically pure mono- or dialkylboranes to C=C double bonds that possess enantiotopic faces. Consequently, when one subsequently oxidizes all C—B bonds to C—OH bonds, one has realized an enantioselective hydration of the respective alkene.

An especially efficient reagent of this type is the boron-containing five-membered ring compound shown in Figure 3.30. Since this reagent is quite difficult to synthesize, it has not been used much in asymmetric synthesis. Nonetheless, this reagent will be presented here simply because it is particularly easy to see with which face of a C=C double bond it will react.

In the structure shown in Figure 3.30, the top side reaction of this borane on the C=C double bond of 1-methylcyclohexene prevails kinetically over the bottom side reaction. This

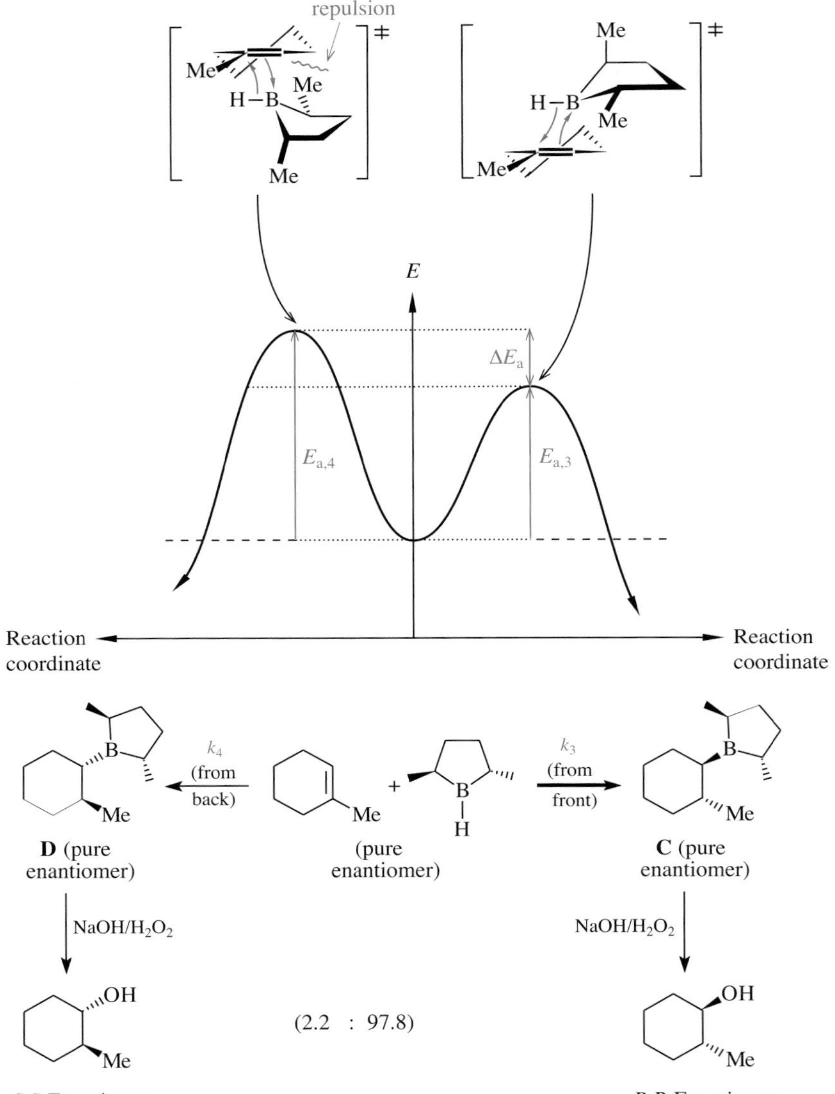

Fig. 3.30. Asymmetric hydration of an achiral alkene via hydroboration/oxidation/hydro lysis; ΔE_a corresponds to the extent of reagent control of diastereoselectivity.

is because only the top side reaction of the borane avoids steric interactions between the methyl substituents on the borane and the six-membered ring. In other words, the *reagent* determines the face to which it adds. We thus have **reagent control of stereoselectivity.** As a result, the mixture of the diastereomeric trialkylboranes **C** and **D**, both of which are pure enantiomers, is produced with a high diastereoselectivity (*ds* = 97.8:2.2). After the normal NaOH/H_2O_2 treatment, they give a 97.8:2.2 mixture of the enantiomeric *trans*-2-methylcyclohexanols. The 1*R*,2*R* alcohol will therefore have an *ee* value of 97.8% – 2.2% = 95.6%.

The *S,S* enantiomer of this alcohol is obtained with the same *ee* value of 95.6% when the enantiomer of the borane shown in Figure 3.30 is used for the hydroboration of 1-methylcyclohexene. The first problem we ran into in the introduction to Section 3.4 is solved!

3.4.3 Thought Experiment I on the Hydroboration of Chiral Alkenes with Chiral Boranes: Mutual Kinetic Resolution

During the addition of a racemic chiral dialkylborane to a racemic chiral alkene a maximum of four diastereomeric racemic trialkylboranes can be produced. Figure 3.31 illustrates this using the example of the hydroboration of 3-ethyl-1-methylcyclohexene with the cyclic borane from Figure 3.30. This hydroboration, however, has not been carried out experimentally. This should not prevent us from considering what would happen if it were performed.

Our earlier statements on substrate and reagent control of stereoselectivity during hydroborations are incorporated in Figure 3.31. Because of the obvious analogies between the old and the new reactions, the following can be predicted about the product distribution shown:

- As the main product we expect the racemic trialkylborane **E**; the substrate and the reagent controls work together to promote its formation.
- The racemic trialkylborane **H** should be produced in trace quantities only. Its formation is disfavored by both the reagent and the substrate control.
- As minor products we expect the racemic trialkylboranes **F** and/or **G**; **F** is favored by reagent and disfavored by substrate control of stereoselectivity, whereas for **G** it is exactly the opposite.

We thus summarize: the yield ratios of the conceivable hydroboration products E, F, G, and H should be much:little:little:none. One enantiomer of **E** comes from the reaction of the *S*-alkene with the *S,S*-borane; the other enantiomer of **E** comes from the reaction of the *R*-alkene with the *R,R*-borane. Thus, each enantiomer of the reagent has preferentially reacted with *one* enantiomer of the substrate. The diastereoselectivity of this reaction thus corresponds to a *mutual kinetic resolution*.

The condition for the occurrence of a mutual kinetic resolution is therefore that considerable substrate control of stereoselectivity and considerable reagent control of stereoselectivity occur simultaneously.

From the diastereoselectivities in Figure 3.31 one expects the following for the rate constants: $k_5 > k_6$, $k_7 > k_g$ (which implies $k_5 >> k_8$). We can't say whether k_6 or k_7 is greater.

For the discussion in Sections 3.4.4 and 3.4.5, we will assume (!) that $k_6 > k_7$; that is, the reagent control of stereoselectivity is more effective than the substrate control of stereoselectivity. The justification for this assumption is simply that it makes additional thought experiments possible. These are useful for explaining interesting phenomena associated with stereoselective synthesis, which are known from other reactions. Because the thought experiments are much easier to understand than many of the actual experiments, their presentation is given preference for introducing concepts.

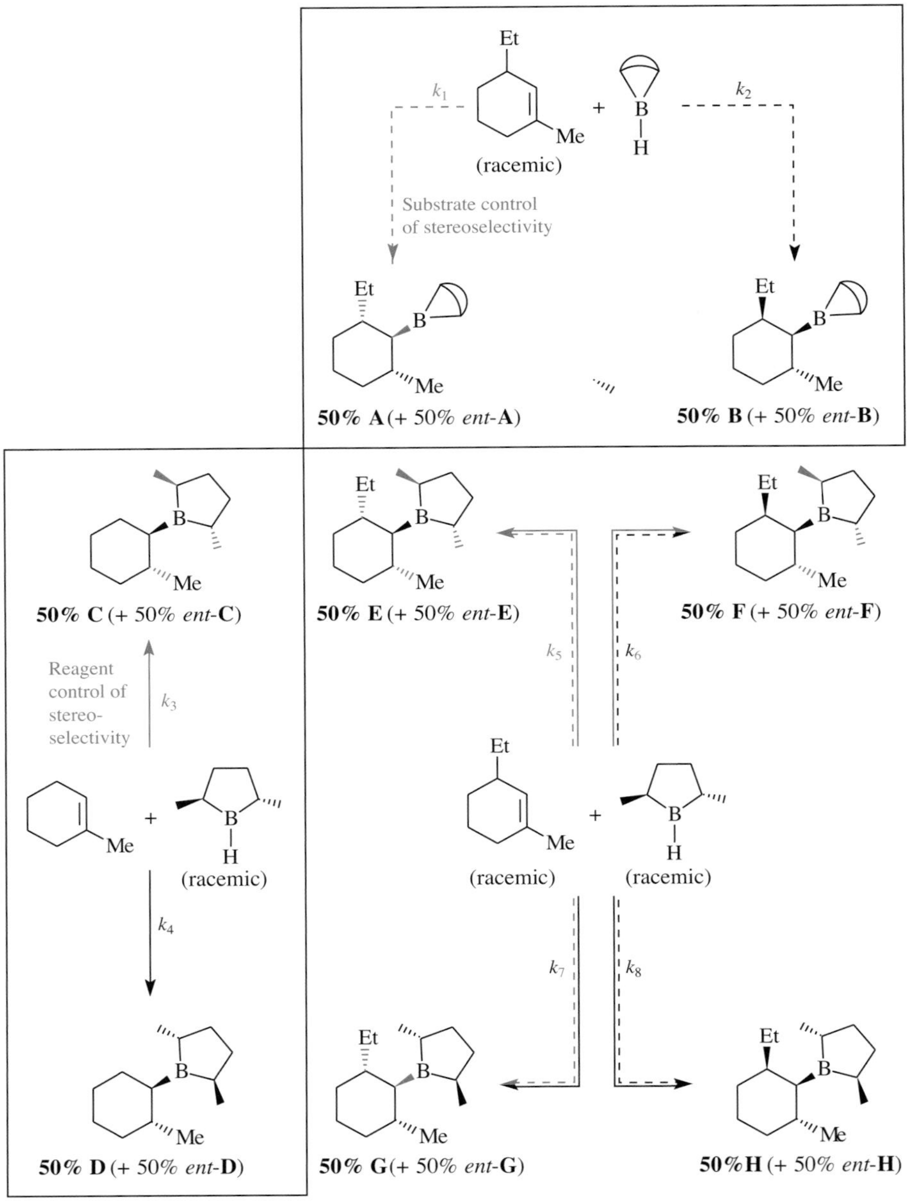

Fig. 3.31. Thought experiment I: products from the addition of a racemic chiral dialkylborane to a racemic chiral alkene. Rectangular boxes: previously discussed reference reactions for the effect of substrate control (top box: reaction from Figure 3.26) or reagent control of stereoselectivity [leftmost box: reaction from Figure 3.30 (rewritten for racemic instead of enantiomerically pure reagent)]. Solid reaction arrows, reagent control of stereoselectivity; dashed reaction arrows, substrate control of stereoselectivity; red reaction arrows (kinetically favored reactions), reactions proceeding with substrate control (solid lines) or reagent control (dashed lines) of stereoselectivity; black reaction arrows (kinetically disfavored reactions), reactions proceeding opposite to substrate control (solid lines) or reagent control (dashed lines) of stereoselectivity.

3.4.4 Thought Experiments II and III on the Hydroboration of Chiral Alkenes with Chiral Boranes: Reagent Control of Diastereoselectivity, Matched/Mismatched Pairs, Double Stereodifferentiation

At the beginning of Section 3.4, we wondered whether 3-ethyl-1-methylcyclohexene could also be hydroborated/oxidized/hydrolyzed to furnish the *cis,trans*-configured alcohol. There is a solution (Figure 3.32) if two requirements are fulfilled. First, we must rely on the assumption made in Section 3.4.3 that this alkene reacts with the cyclic borane in such a way that the reagent control of stereoselectivity exceeds the substrate control of the stereoselectivity. Second, both the alkene and the borane must be used in enantiomerically pure form.

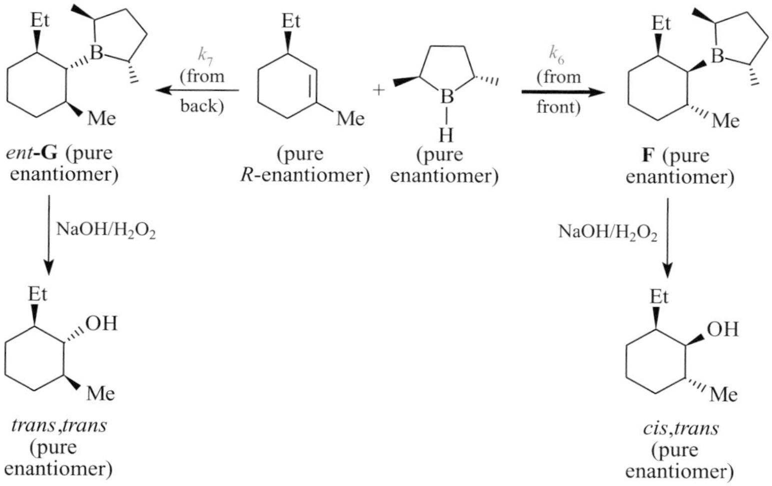

Et, B, Me, k_7 (from back), Et, Me, +, B, H, k_6 (from front), Et, B, Me

ent-**G** (pure enantiomer) (pure *R*-enantiomer) (pure enantiomer) **F** (pure enantiomer)

$NaOH/H_2O_2$ $NaOH/H_2O_2$

Et, OH, Me Et, OH, Me

trans,trans (pure enantiomer) *cis,trans* (pure enantiomer)

Fig. 3.32. Thought experiment II: reagent control of stereoselectivity as a method for imposing on the substrate a diastereoselectivity that is alien to it (mismatched pair situation).

Figure 3.31 already contains all the information that is necessary to deduce the product distribution of the reaction in Figure 3.32. The reaction partners in Figure 3.32 can provide nothing but the *cis,trans*-adduct as the pure enantiomer **F** and the *trans,trans*-adduct as the pure enantiomer *ent*-**G.** The adduct **F** is produced with the rate constant k_6 and the adduct *ent*-**G** is produced with the rate constant $k_{7.}$ After oxidative cleavage and hydrolysis with $NaOH/H_2O_2$, the adduct **F** results in the *cis,trans*-alcohol. The diastereoselectivity of this reaction is given by $ds = (k_6/k_7)$:1. When $k_6 > k_7$, the *cis,trans*-alcohol is produced in excess. If on the other hand, in contrast to our assumption, k_7 were to exceed k_6, the *cis,trans*-alcohol would still be produced although only in less than 50% yield. It would be present in the product mixture—as before—in a (k_6/k_7):1 ratio, but now only as a minor component.

If, as in the reaction example in Figure 3.32, during the addition to enantiomerically pure chiral alkenes, substrate and reagent control of diastereoselectivity act in opposite directions, we have a so-called **mismatched pair**. For obvious reasons it reacts with relatively little diastereoselectivity and also relatively slowly. Side reactions and, as a consequence, reduced yields are not unusual in this type of reaction. However, there are cases in which "mismatched" paris still give rise to highly diastereoselective reactions, just not as high as the matched pair.

Conversely, the addition of enantiomerically pure chiral dialkylboranes to enantiomerically pure chiral alkenes can also take place in such a way that substrate control and reagent control of diastereoselectivity act in the same direction. Then we have a **matched pair.** It reacts faster than the corresponding **mismatched pair** and with especially high diastereoselectivity. This approach to stereoselective synthesis is also referred to as **double stereodifferentiation.**

Thought experiment III in Figure 3.33 provides an example of how such a double stereodifferentiation can be used to increase stereoselectivity. The two competing hydroborations take place with $ds = (k_5/k_8){:}1$, as a comparison with the rate constants in Figure 3.31 reveal. After oxidation/hydrolysis, the *trans,trans*-alcohol is obtained with the *cis,trans*-alcohol in the same ratio $(k_5/k_8){:}1$. According to the results from Section 3.4.3, it was certain that $k_5 > k_8$. The diastereoselectivity of the thought experiment in Figure 3.33 should be considerably higher than the one in Figure 3.32, since the substrate and reagent are matched.

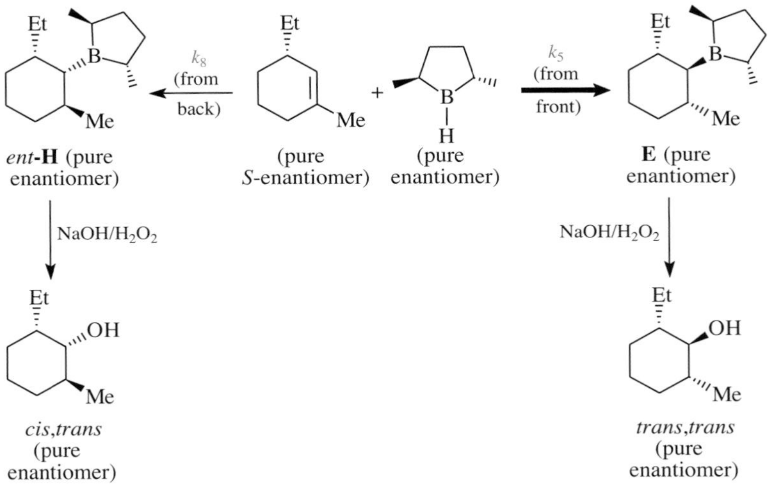

Et B Me *ent*-**H** (pure enantiomer) ← k_8 (from back) Et Me (pure *S*-enantiomer) + B H (pure enantiomer) → k_5 (from front) Et B Me **E** (pure enantiomer)

$NaOH/H_2O_2$ $NaOH/H_2O_2$

Et OH Me *cis,trans* (pure enantiomer) Et OH Me *trans,trans* (pure enantiomer)

Fig. 3.33. Thought experiment III: reagent control of stereoselectivity as a method for enhancing the substrate control of stereoselectivity (matched pair situation).

3.4.5 Thought Experiment IV on the Hydroboration of Chiral Olefins with Chiral Dialkylboranes: Kinetic Resolution

Figure 3.32 showed the reaction of our enantiomerically pure chiral cyclic dialkylborane with (*R*)-3-ethyl-1-methylcyclohexene. It took place relatively slowly with the rate constant k_6. The reaction of the same dialkylborane with the isomeric *S*-alkene was shown in Figure 3.33. It took place considerably faster with the rate constant k_5. The combination of the two reactions is shown in Figure 3.34. There the same enantiomerically pure borane is reacted simultaneously with both alkene enantiomers (i.e., the racemate). What is happening? In the *first* moment of the reaction the *R*- and the *S*-alkene react in the ratio k_6 (small)/k_5 (big). The **matched pair** thus reacts faster than the **mismatched pair**. This means that at low conversions (≤ 50%) the trialkylborane produced is essentially derived from the *S*-alkene only. It has the stereostructure **E**. Therefore, relative to the main by-product **F**, compound **E** is produced with a diastereoselectivity of almost k_5/k_6. However, this selectivity decreases as the reaction progresses because the reaction mixture is depleted of the more reactive *S*-alkene. When the reaction is almost complete, the less reactive *R*-alkene has been enriched to a ratio k_5/k_6. At this point it has an *ee* value of almost:

$$\frac{k_5 - k_6}{k_5 + k_6} \cdot 100\%$$

Of the greatest preparative interest is a point in time at which there is approximately 50% conversion in the reaction. Only then can the yield of the trialkylborane of one enantiomeric series or the yield of the alkene of the other enantiomeric series reach its theoretical maximum value of 50%. In other words, in the ideal case, at a conversion of exactly 50%, the alkene enantiomer of the **matched pair** has reacted completely (owing to the high rate constant k_5), and the alkene enantiomer from the **mismatched pair** has not yet reacted (owing to the much smaller rate constant k_6). If the hydroboration were stopped at this time by adding H_2O_2/NaOH, the following desirable result would be obtained: 50% of the alcohol derived exclusively from the *S*-alkene and 50% of exclusively *R*-configured unreacted alkene would be isolated. The reaction in Figure 3.34 would thus have been a kinetic resolution. The kinetic requirements for the success of a kinetic resolution are obviously more severe than the requirements for the success of a mutual kinetic resolution (Section 3.4.3). For the latter, both substrate and reagent control of stereoselectivity must be high. For a good kinetic resolution to occur, it is additionally required that one type of stereocontrol clearly dominate the other type (i.e., $k_5 \geq k_6$).

Note that these concepts are general ones and not limited to the specific reactions used as examples to present them.

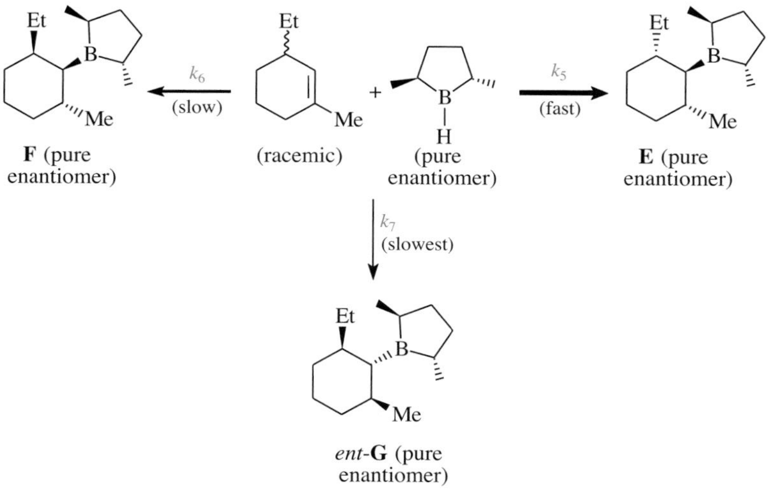

Fig. 3.34. Thought experiment IV: kinetic resolution of a racemic chiral alkene through reaction with $\leq$ 0.5 equiv. of an enantiomerically pure chiral dialkylborane.

3.4.6 Catalytic Asymmetric Synthesis: Sharpless Oxidations of Allylic alcohols

Asymmetric syntheses can be carried out even more easily and elegantly than by reacting achiral substrates with enantiomerically pure chiral reagents if one allows the substrate to react with an enantiomerically pure species formed *in situ* from an achiral reagent and an enantiomerically pure chiral additive. The exclusive reaction of this species on the substrate implies that the reagent itself reacts substantially slower with the substrate than its adduct with the chiral additive. If high stereoselectivity is observed, it is exclusively due to the presence of the additive. The chiral additive speeds up the reaction. This is an example of **ligand accelerated asymmetric catalysis**.

This type of **additive** (or ligand) **control of stereoselectivity** has three advantages. First of all, after the reaction has been completed, the chiral additive can be separated from the product with physical methods, for example, chromatographically. In the second place, the chiral additive is therefore also easier to recover than if it had to be first liberated from the product by means of a chemical reaction. The third advantage of additive control of enantioselectivity is that the enantiomerically pure chiral additive does not necessarily have to be used in stoichiometric amounts; catalytic amounts may be sufficient. This type of **catalytic asymmetric synthesis**, especially on an industrial scale, is important and will continue to be so.

The most important catalytic asymmetric syntheses include addition reactions to C=C double bonds. One of the best known is the Sharpless epoxidations. Sharpless epoxidations cannot be carried out on all alkenes but only on primary or secondary allylic alcohols. Even with this limitation, the process has seen a great deal of application.

There are two reasons for this. First, the Sharpless epoxidation can be applied to almost *all* primary and secondary allylic alcohols. Second, it makes trifunctional compounds accessible in the form of enantiomerically pure α,β-epoxy alcohols. These can react with a wide variety of nucleophiles to produce enantiomerically pure "second-generation" products. Further transformations can lead to other enantiomerically pure species that ultimately may bear little structural resemblance to the starting α,β-epoxy alcohols.

Sharpless epoxidations are discussed in the plural because primary (Figures 3.35 and 3.38) and secondary allylic alcohols (Figure 3.39) are reacted in different ways. Primary allylic alcohols are reacted to completion. Secondary allylic alcohols—if they are racemic—are usually reacted only to 50% conversion (the reason for this will become clear shortly). The oxidation agent is always a hydroperoxide, usually *tert*-BuOOH. The chiral additive used is 6–12 mol% of an enantiomerically pure dialkyl ester of tartaric acid, usually the diethyl ester (**d**iethyl**t**ar-

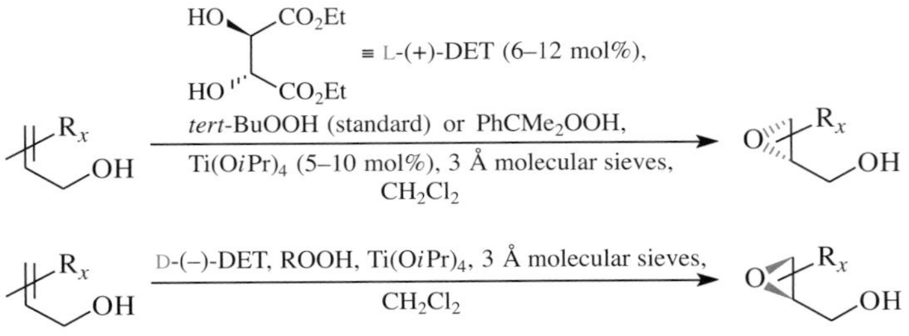

HO, CO_2Et / HO, CO_2Et ≡ L-(+)-DET (6–12 mol%), *tert*-BuOOH (standard) or $PhCMe_2OOH$, $Ti(O\textit{i}Pr)_4$ (5–10 mol%), 3 Å molecular sieves, CH_2Cl_2

D-(–)-DET, ROOH, $Ti(O\textit{i}Pr)_4$, 3 Å molecular sieves, CH_2Cl_2

R_x, OH, O

Fig. 3.35. Enantioselective Sharpless epoxidation of achiral primary allylic alcohols. If the substrates are drawn as shown, the direction of the reaction of the complexes derived for L-(+) and D-(–)-DET can be remembered with the following mnemonic: **L**, from **l**ower face; **D**, **d**oesn't react from **d**ownface.

trate, DET). Titanium tetraisopropoxide (Ti(O*i*Pr)$_4$) is added so that oxidizing agent, chiral ligand, and substrate can assemble to form an enantiomerically pure chiral complex. No epoxidation at all takes place in the absence of Ti(O*i*Pr)$_4$.

Figure 3.36 indicates the structure of the catalytically active species in Sharpless epoxidations. It is generated *in situ* from the mixture of the cited reagents and L-diethyltartrate as a chiral additive. There is a subtle effect preferring the stereostructure of the catalytically active species (Figure 3.36, top) over a very similar alternative (Figure 3.36, bottom): the *tert*-Bu-O-Ti substructure is located on that side of the 5-ring/4-ring/5-ring midplane that only contains *one* CO_2Et group in the β-position (with the numbering starting from the Ti atom of this substructure). The allyl-O-Ti substructure occupies the opposite side of the plane.

O-transferring intermediate in the presence of L-(+)-DET is ...

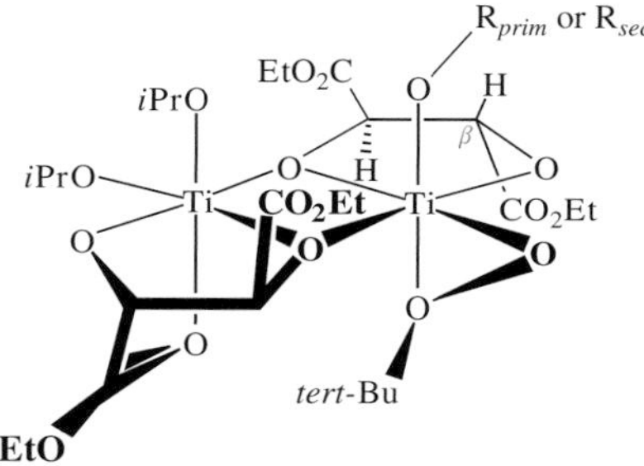

... instead of the more sterically hindered diastereomer

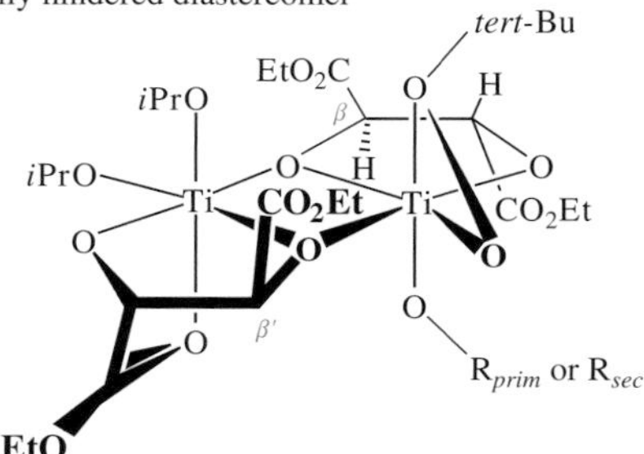

Fig 3.36. Mechanistic details of Sharpless epoxidations, part I: the actual oxidant is a stereouniform *tert*-butyl hydroperoxide complex of a titanium tartrate "dimer" with the least hindrance possible.

The upper half of Figure 3.37 shows the catalytically active species in the transition state of the Sharpless epoxidation of a primary allylic alcohol (cf. Figure 3.36) that has the same substitution pattern as the substrate of the epoxidation in Figure 3.35 (top). The curved arrows indicate the transfer of oxygen from the *tert*-butyl hydroperoxide substructure to the C=C double bond. For clarity, the resulting epoxy alcohol is shown in two different orientations: first according to its position in the transition state, and second according to the formula given in Figure 3.35. In the lower half of Figure 3.37, the transition state of Sharpless epoxidations of primary allylic alcohols is presented in mirror-image form. Here we obtain the enantiomer of the epoxy alcohol resulting from the reaction given in the lower part of Figure 3.35.

Achiral primary allylic alcohols undergo *enantioselective* epoxidation (cf. Figure 3.35), whereas—*chiral primary* allylic alcohols undergo *diastereoselective* oxidation. So the reagent

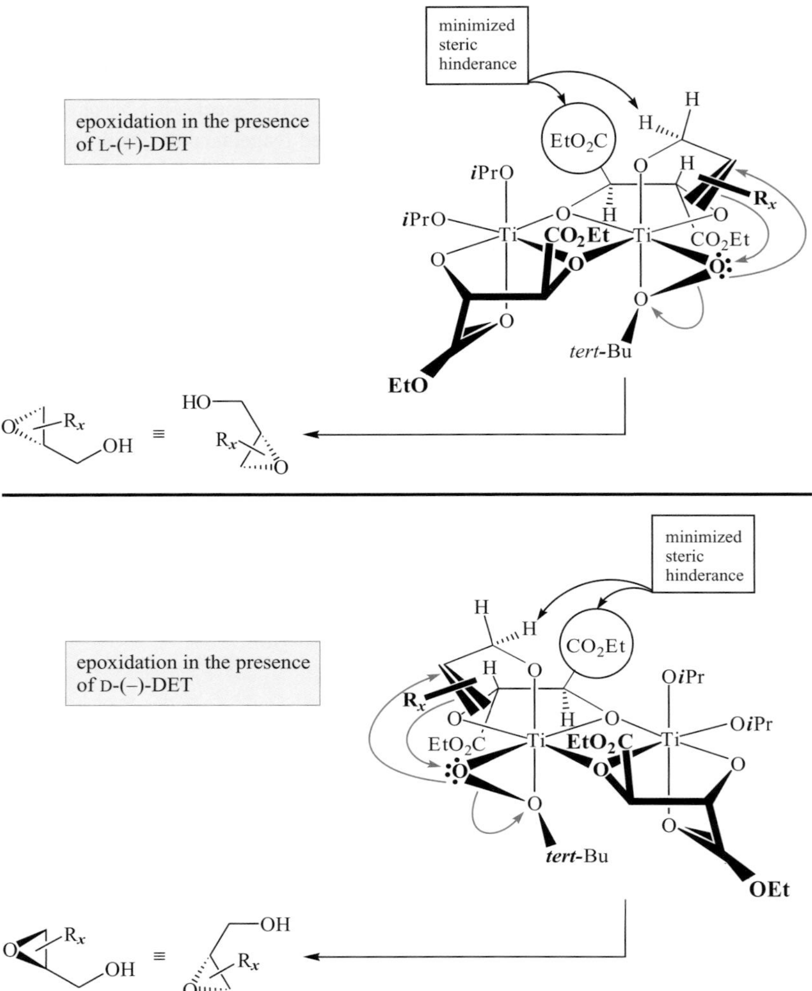

Fig. 3.37. Mechanistic details of Sharpless epoxidations, part II: preferred transition state of enantioselective epoxidations of achiral primary allylic alcohols in the presence of L-(+)-DET (*top*) or D-(−)-DET (*bottom*).

control of stereoselectivity is much higher than the substrate control (third line in Figure 3.38). Thus the diastereoselectivity in the ***mismatched* pair** (middle line in Figure 3.38) is only slightly smaller than in the ***matched* pair** (top line in Figure 3.38).

Racemic chiral secondary allylic alcohols can be subjected to a kinetic resolution by means of the Sharpless epoxidation (Figure 3.39). The reagent mixture reacts with both enantiomers of the allylic alcohol—they may be considered as α-substituted crotyl alcohols—with very different rates. The unreactive enantiomer is therefore isolated with enantiomer excesses close to *ee* = 100% in almost 50% yield at approximately 50% conversion. The other enantiomer is the reactive enantiomer. Its epoxidation proceeds much faster (i.e., almost quantitatively) at 50% conversion. The epoxide obtained can also be isolated and, due to its enantiomeric excess, used synthetically.

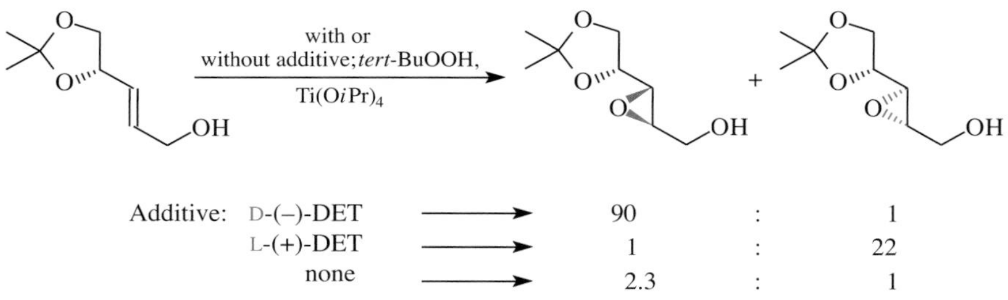

with or without additive; *tert*-BuOOH, $Ti(OiPr)_4$

Additive:				
D-(–)-DET	→	90	:	1
L-(+)-DET	→	1	:	22
none	→	2.3	:	1

Fig. 3.38. Diastereoselective Sharpless epoxidation of chiral primary allylic alcohols (for the preparation of the substrate see Figure 17.63)

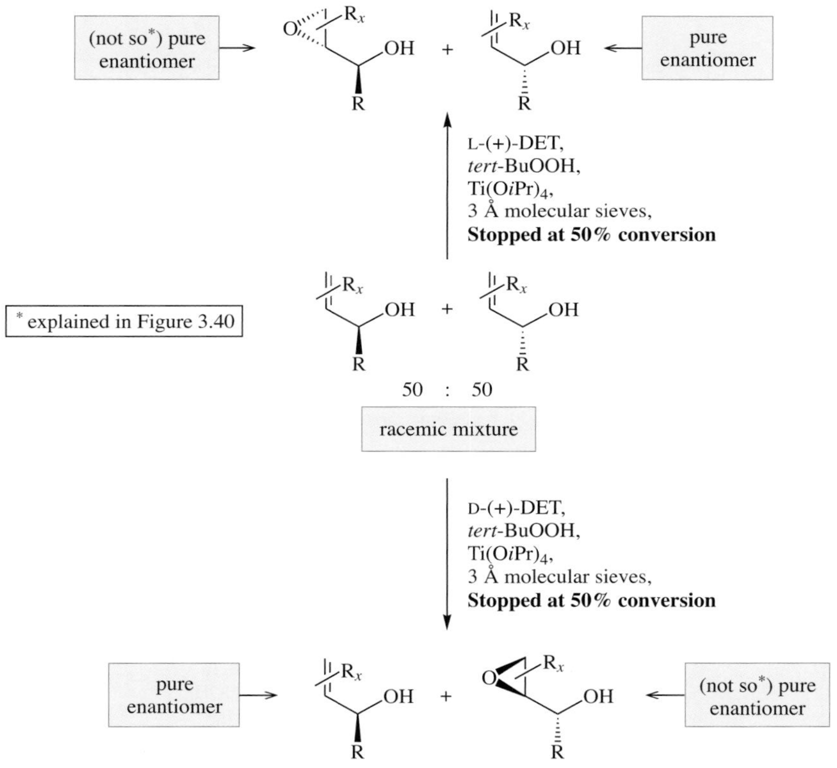

(not so*) pure enantiomer → ... + ... ← pure enantiomer

L-(+)-DET, *tert*-BuOOH, $Ti(OiPr)_4$, 3 Å molecular sieves, **Stopped at 50% conversion**

* explained in Figure 3.40

50 : 50

racemic mixture

D-(+)-DET, *tert*-BuOOH, $Ti(OiPr)_4$, 3 Å molecular sieves, **Stopped at 50% conversion**

pure enantiomer → ... + ... ← (not so*) pure enantiomer

Fig. 3.39. Sharpless epoxidations of chiral racemic secondary allylic alcohols; if they are stopped at (a good) 50% conversion they become kinetic resolutions. The unreacted allylic alcohol is obtained as enantiomerically pure(st) material.

Side Note 3.3.
***ee* Values at a Closer Glance: Kinetic Resolution with Sharpless Epoxidations**

In the kinetic resolution of secondary allylic alcohols by means of the Sharpless epoxidation (Figure 3.39), the *ee* value of the epoxy alcohol formed does not quite reach the *ee* value of the unreacted allylic alcohol. The reason might be deduced from the epoxidation exemplified in Figure 3.40. Until 50% conversion is reached it is almost only the reactive enantiomer of the α-substituted crotyl alcohol that undergoes epoxidation. This is not only a very rapid, but a highly diastereoselective reaction as well, yielding a 98:2 mixture of the *anti*- and *syn*-epoxy alcohol. Epoxidation of the unreactive enantiomer of the α-substituted crotyl alcohol is also observed until 50 % conversion is reached, but only to a very minor extent. The result is most clearly revealed at a 100% conversion, when a 38:62 mixture of the *anti*- and *syn*-epoxy alcohol is obtained. Relating to the resolution of the racemate in Figure 3.39, the latter finding implies that the *anti*-configured epoxy alcohol at 50% conversion mainly arises from the reactive enantiomer of the substrate, though a small amount arises from the unreactive enantiomer. In other words, because the *anti*-configured epoxy alcohol in Figure 3.39 originates from both enantiomers of the allylic alcohol, its enantiomeric excess is necessarily smaller than the enantiomeric excess of the re-isolated allylic alcohol.

The enantiomer of the allylic alcohol in Figures 3.39 and 3.40, which is reactive in the presence of L-(+)-DET, can occupy the same *strain-free* position in the transition state of the Sharpless epoxidation as the one that in Figure 3.37 was presented as the transition state of

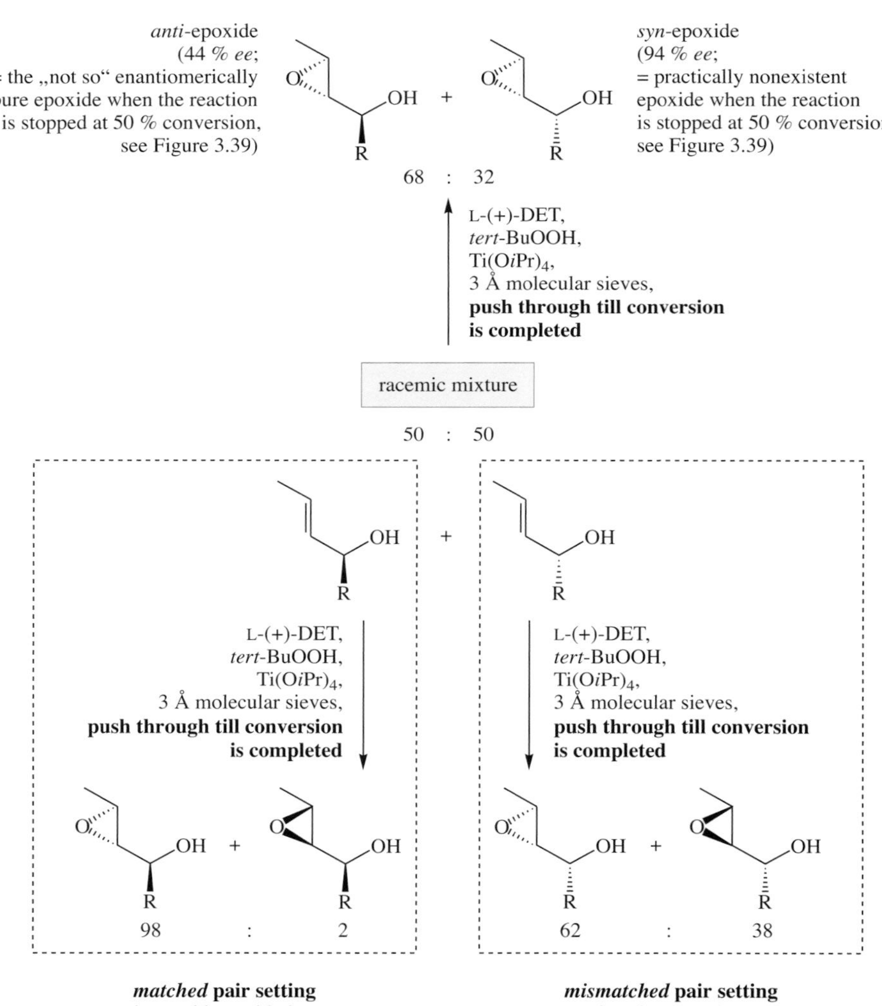

Fig. 3.40. Sharpless epoxidations of chiral racemic secondary allylic alcohols (in contrast to the upper half of Figure 3.39) driven to a 100% conversion; divided into 50% rapid and also highly diastereoselective epoxidation of the **matched pair**, and 50% slow epoxidation in the **mismatched pair** where hardly any diastereocontrol occurs.

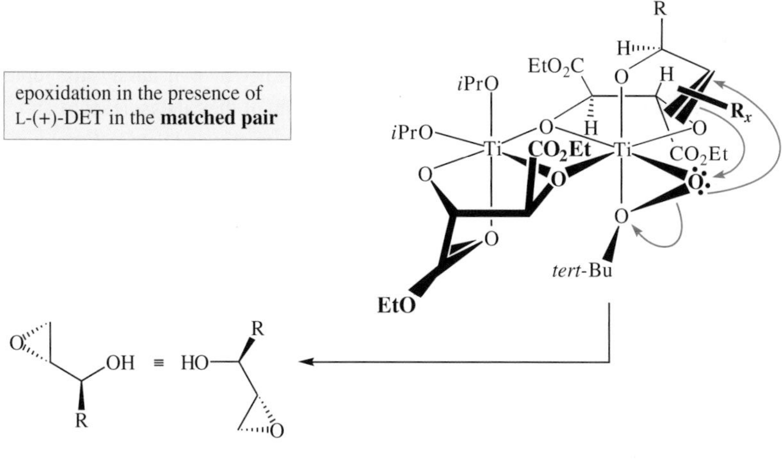

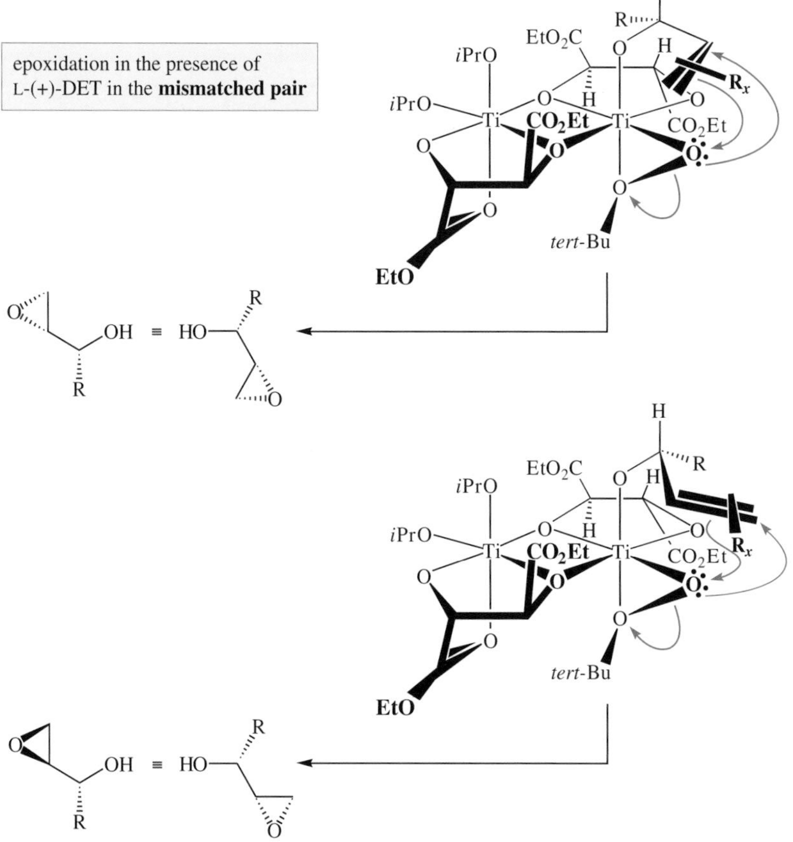

Fig. 3.41. Mechanistic details of Sharpless epoxidations, part III: epoxidations of chiral racemic secondary allylic alcohols in the presence of L-(+)-DET and their diastereoselectivities. Transition state of the **matched pair** (*top*), transition states of the **mismatched pair** (*bottom*).

the Sharpless epoxidation of primary allylic alcohols (Figure 3.41, top). The allylic alcohol enantiomer in Figures 3.39 and 3.40, which is unreactive in the presence of L-(+)-DET, destabilizes the transition state of its Sharpless epoxidation either by van der Waals' repulsion between the α substituent of the substrate and an ester group of the catalyst (Figure 3.41, center) or by an untoward steric interaction between the α substituent and the C=C double bond (Figure 3.41, bottom). These arguments account for the configuration-dependent rate of the Sharpless epoxidation of chiral secondary allylic alcohols.

At the same time the curved arrows in Figure 3.41 explain the kind of diastereoselectivity observed with the oxygen transfer from *tert*-butyl hydroperoxide to the respective allylic alcohol. The first stereoformulas of the resulting epoxy alcohol each emulate the transition state geometry. The second stereoformulas of the resulting epoxy alcohol each ensure comparability to the formulas for the products of Figures 3.39 and 3.40.

There are many other examples of highly efficient catalytic asymmetric syntheses. These include the asymmetric dihydroxylation of alkenes and certain homogeneous catalyzed hydrogenations. The latter will be discussed in the context of redox reactions in Sections 17.3.2 and 17.4.7. Further examples for catalytic asymmetric syntheses also mentioned in this book are the proline-catalyzed cyclohexenone annulations in Figure 12.19.

3.5 Additions that Take Place Diastereoselectively as *trans*-Additions (Additions via Onium Intermediates)

trans-Additions to alkenes are two-step reactions. They occur only for reagents that for certain reasons are unable to form the *final* σ-bonds in a one-step addition. Instead they form two *preliminary* σ-bonds (Figure 3.42), which are incorporated into a three-membered heterocyclic intermediate. This intermediate carries a positive formal charge on the heteroatom. Thus it contains an "onium" center there. The fact that this intermediate contains two new σ-bonds provides a better energy balance for breaking the π-bond of the alkene than if an acyclic intermediate were produced, which, of course, would contain only one new σ-bond. During the formation of the cyclic intermediate, however, it is necessary to compensate for the energy that (a) is required to separate the charges that result when the onium center is formed and (b) must be invested as ring strain. Therefore the first step of *trans*-additions to the C=C double bond, i.e., the one that leads to the onium ion is endothermic. The Hammond postulate thereby identifies it as the rate-determining step of the overall reaction (Figure 3.42).

The onium centers of the three-membered intermediates considered in this section are $Cl^{\oplus}$, $Br^{\oplus}$, $I^{\oplus}$, and $Hg^{\oplus}$ (OAc). In each of these onium ions the ring strain is lower than in an epoxide. This is because the C—$Cl^{\oplus}$, C—$Br^{\oplus}$, C—$I^{\oplus}$, and C—$Hg^{\oplus}$ (OAc) bonds are considerably longer than the C—O bonds in an epoxide. The reason for this is that the heteroatoms of these onium ions have larger atomic radii than oxygen. The C—C bond within the ring of each onium ion is approximately just as long as the C—C bond in an epoxide ring. The ratio of C-

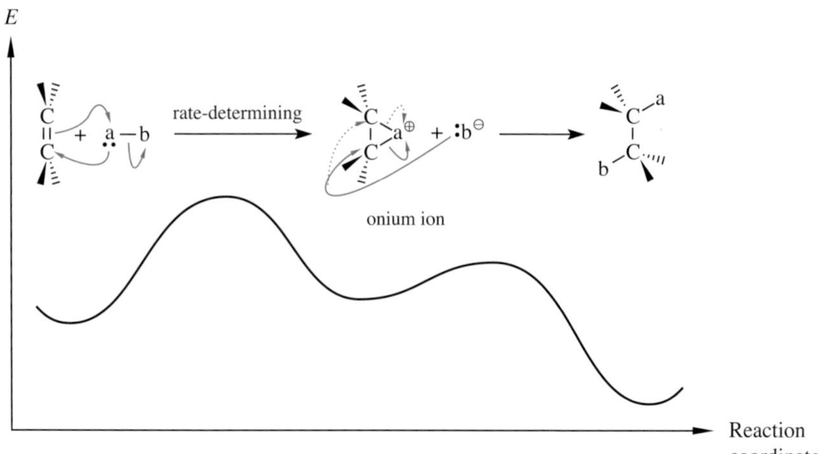

Fig. 3.42. Mechanism and energy profile of *trans*-selective additions to C=C double bonds. The first step, the formation of the onium ion, is rate-determining. The second step corresponds to the S_N2 opening of an epoxide.

heteroatom bond *length* to the endocyclic C—C bond *length* now defines the bond *angle* in these heterocycles. In an epoxide, the C—C—O and C—O—C angles measure almost exactly 60° and cause the well-known 26 kcal/mol ring strain. On the other hand, in symmetrically substituted onium ions derived from $Cl^{\oplus}$, $Br^{\oplus}$, $I^{\oplus}$, or $Hg^{\oplus}$ (OAc), the C—C—$Het^{\oplus}$ bond angles must be greater than 60° and the C—$Het^{\oplus}$—C bond angles must be smaller:

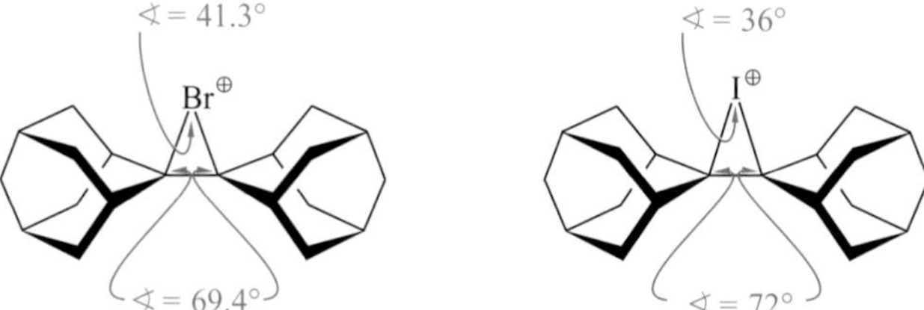

The increase in the C—C—$Het^{\oplus}$ angle reduces the angular strain and thus increases the stability. On the other hand, the simultaneous decrease in the C—$Het^{\oplus}$—C bond angle hardly costs any energy. According to the VSEPR theory, the larger an atom is, the smaller the bond angle can be between bonds emanating from that atom. Consequently, for atoms such as Cl, Br, I, or Hg, relative to C and O, a smaller bond angle does not result in a significantly increased strain.

In the second step of *trans*-additions, the onium intermediates react with a nucleophile. Such a nucleophile can be either an anion released from the reagent in the first step or a better nucleophile present in the reaction mixture. It opens the onium intermediate in an S_N2 reaction with inversion. In this way, the *trans* addition product is produced stereoselectively (Figure 3.42).

3.5.1 Addition of Halogens

The *trans*-addition of bromine to C=C double bonds follows the onium mechanism generally formulated in Figure 3.42. Cyclohexene reacts stereoselectively to give racemic *trans*-1,2-dibromocyclohexane:

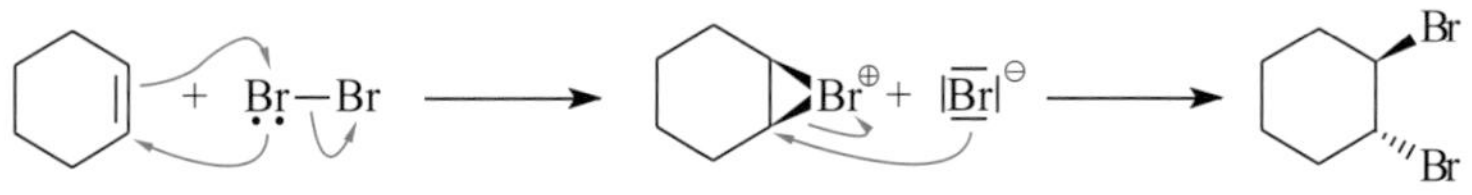

trans-Selectivity is also exhibited by the additions of bromine to fumaric or maleic acid, which follow the same mechanism:

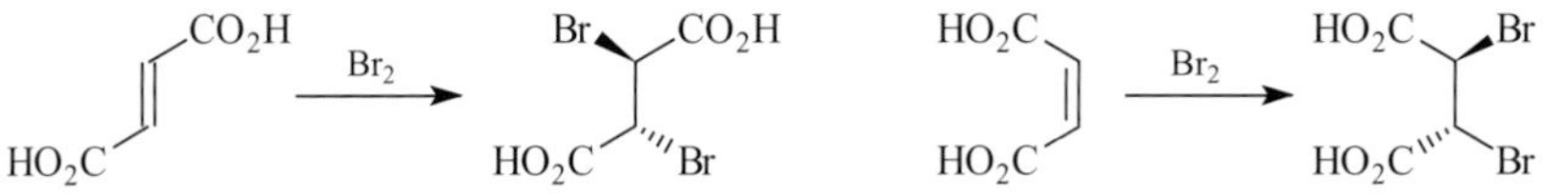

(*meso*)

(racemic mixture)

This is supported by the fact that the reaction with fumaric acid gives *meso*-dibromosuccinic acid, whereas maleic acid gives the *chiral* dibromosuccinic acid, of course as a racemic mixture. Note that these reactions are stereospecific.

Instead of a bromonium ion, in certain cases an isomeric acyclic and sufficiently stable cation can occur as an intermediate of the bromine addition to alkenes. This holds true for bromine-containing benzyl cations. Therefore, the bromine addition to β-methyl styrene shown in Figures 3.5 and 3.6 takes place *without* stereocontrol.

The addition of Cl_2 to C=C double bonds is *trans*-selective only when it takes place via three-membered chloronium ions. But it can also take place without stereocontrol, namely, when carbenium ion intermediates appear instead of chloronium ions. This is observed in Cl_2 additions that can occur via benzyl or *tert*-alkyl cations.

In general, an addition of I_2 to C=C double bonds is thermodynamically impossible, although an iodonium ion can still form.

3.5.2 The Formation of Halohydrins; Halolactonization and Haloetherification

Halohydrins are β-halogenated alcohols. They can be obtained in H_2O-containing solvents from alkenes and reagents, which transfer Hal$^\oplus$ ions. *N*-Bromosuccinimide (transfers Br$^\oplus$; Figures 3.43 and 3.44 as well as 3.47), chloramine-T (transfers Cl$^\oplus$; Figure 3.46), and elemental iodine (transfers I$^\oplus$; Figure 3.47) have this ability. Bromonium and chloronium ions react with H_2O via an S_N2 mechanism. This furnishes the protonated bromo- or chlorohydrins, which are subsequently deprotonated.

Instead of H_2O, COOH or OH groups of the substrate located at a suitable distance can also open the halonium ion intermediate through a nucleophilic backside reaction. In this way, cyclic halohydrin derivatives are produced (Figure 3.47). They are referred to as halolactones or haloethers.

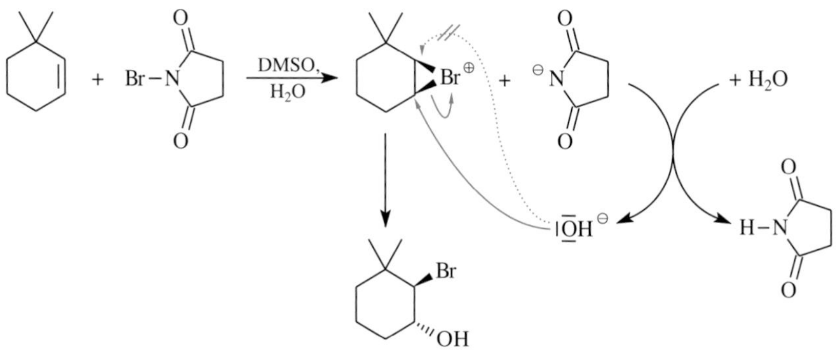

Fig. 3.43. Stereo- and regioselective formation of a bromohydrin from a 1,2-disubstituted alkene.

With NBS in aqueous DMSO, cyclohexene gives racemic *trans*-2-bromo-1-cyclohexanol. This stereochemical result means that we have a *trans*-addition. In the analogous bromohydrin formation from 3,3-dimethylcyclohexene, the analogous dimethylated 2-bromo-1-cyclohexanol is also produced *trans*-selectively as well as regioselectively (Figure 3.43). In the bromonium ion intermediate, the H_2O molecule does not react at the hindered neopentyl center. This is as expected from the rules for S_N2 reactivity (Section 2.4.4), and results in high regioselectivity.

As shown in Figure 3.44, trisubstituted alkenes are also converted to bromohydrins by NBS in an aqueous organic solvent. This reaction takes place via a *trans*-addition and therefore must take place via a bromonium ion. Once again, the reaction is also regioselective. At first glance, the regioselectivity of this reaction might seem surprising: The H_2O molecule reacts with the bromonium ion intermediate at the tertiary instead of at the secondary C atom. One might have expected the backside (S_N2) reaction of the nucleophile to be directed at the secondary C atom of the bromonium ion (cf. Section 2.4.4). However, this is not the case. The bromonium ion is very distorted: The C_{sec}—$Br^{\oplus}$ bond at 1.9 Å is considerably shorter than and consequently *considerably stronger* than the C_{tert}—$Br^{\oplus}$ bond. The latter is stretched to 2.6 Å and thereby weakened. This distortion reduces the ring strain of the bromonium ion. The distortion becomes possible because the stretching of the C_{tert}—$Br^{\oplus}$ bond produces a partial

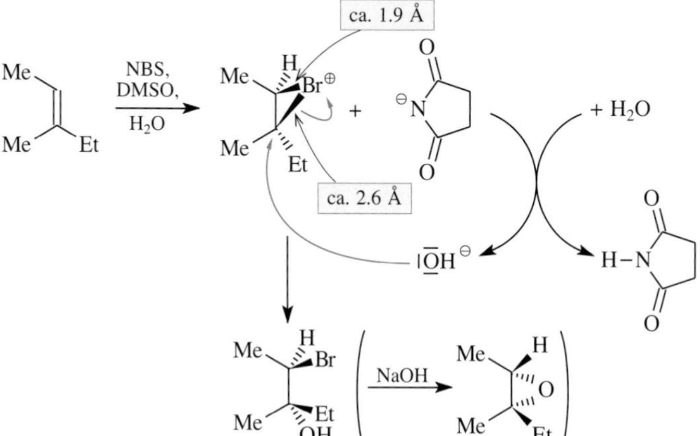

Fig. 3.44. Stereo- and regioselective formation of a bromohydrin from a trisubstituted ethylene; conversion to an epoxide.

positive charge on the tertiary C atom, which is stabilized by the alkyl substituents located there. In this bromonium ion, the bromine atom has *almost* separated from the tertiary ring C atom (but only almost, because otherwise the result would not be 100% *trans*-addition).

Side Note 3.4. The Fürst-Plattner Rule

When cyclohexene is transformed into its bromohydrin, regiocontrol of the ring-opening of the onium ion may be exerted —in addition to the factors already discussed (factor 1: see Figure 3.43; factor 2: see Figure 3.44)—by a third factor. An example is given in Figure 3.45.

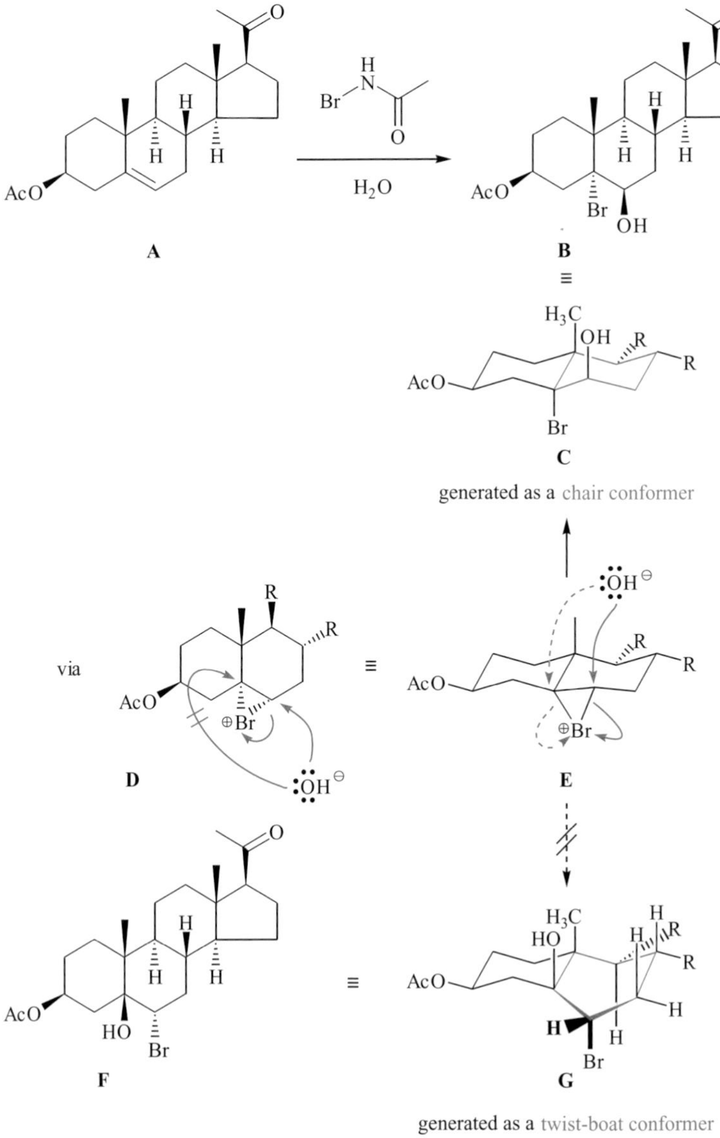

Fig. 3.45. Diastereo- and regioselective formation of a bromohydrin from an asymmetrically substituted cyclohexene.

There the bromonium ion **D** is opened with the opposite regioselectivity than the one shown in Figure 3.44. The bromonium ions that derive from a cyclohexene and, like **D,** prefer a half-chair conformation (here: **E**) are ring-opened according to the so-called Fürst-Plattner rule: the hydroxide ion reacts at the C atom where this reaction—at least initially—leads to a cyclohexane chair conformer (**C** in Figure 3.45) such that both the new C–OH bond and the remaining C–Br bond adopt an axial orientation. The opposite regioselectivity would lead to a cyclohexane twist-boat conformer as the primary ring-opening product. This is considerably more strained and its formation therefore (Hammond postulate!) kinetically disfavored (**G** in Figure 3.45).

The Fürst-Plattner rule is more comprehensive and thus of greater significance than it might appear at first. Indeed, it applies to the preferential course of all nucleophilic ring-openings of any "cyclohex-annulated" three-membered heterocycles. Under (the usual) kinetic control, the regioselectivity that is observed primarily leads to a cyclohexane chair conformer with a *trans*-diaxially oriented nucleophile and leaving group.

Alkenes and chloramine-T in aqueous acetone form chlorohydrins through another *trans*-addition (Figure 3.46).

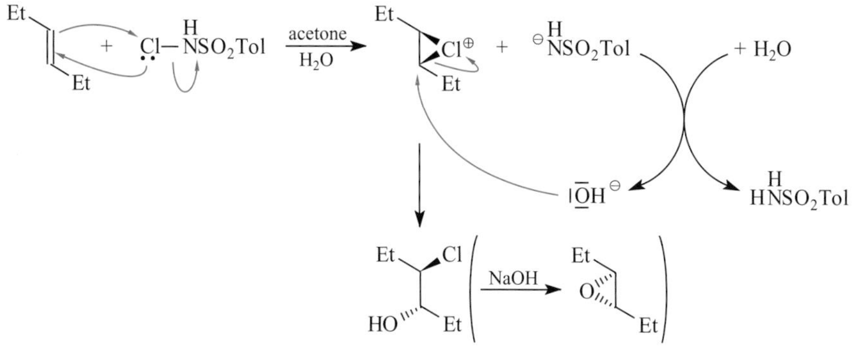

Et, Et, H, Cl–NSO$_2$Tol, acetone, H_2O, Et, Cl$^\oplus$, Et, $^\ominus$H NSO$_2$Tol, + H_2O, $|\overline{O}H^\ominus$, H HNSO$_2$Tol, Et, Cl, HO, Et, NaOH, Et, O, Et

Fig. 3.46. Stereoselective formation of a chlorohydrin; conversion to an epoxide.

Halohydrin formation is not only important in and of itself. Bromohydrins (see Figure 3.44) as well as chlorohydrins (Figure 3.46) can be cyclized stereoselectively to epoxides by deprotonation with NaH or NaOH via an intramolecular S_N2 reaction. In this way, epoxides can be obtained from alkenes in two steps. Though this route is longer than single-step epoxidation of alkenes with percarboxylic acids (Figure 3.19), it has merits: it avoids the use of reagents that in the extreme case might explode. In addition, this process is stereochemically complementary to a peracid epoxidation.

Similar addition mechanisms explain the so-called halolactonization and the related haloetherification (Figure 3.47). With the help of these reactions one can produce halogenated five- and six-membered ring lactones or ethers stereoselectively. Dehalogenation afterward is possible (Figure 1.38).

The formation of the halonium ion intermediate in halolactonizations and haloetherifications is a reversible step. Therefore, initially, comparable amounts of the diastereomeric bromonium ions **B** and *iso*-**B** are produced from the unsaturated alcohol **A** of Figure 3.47.

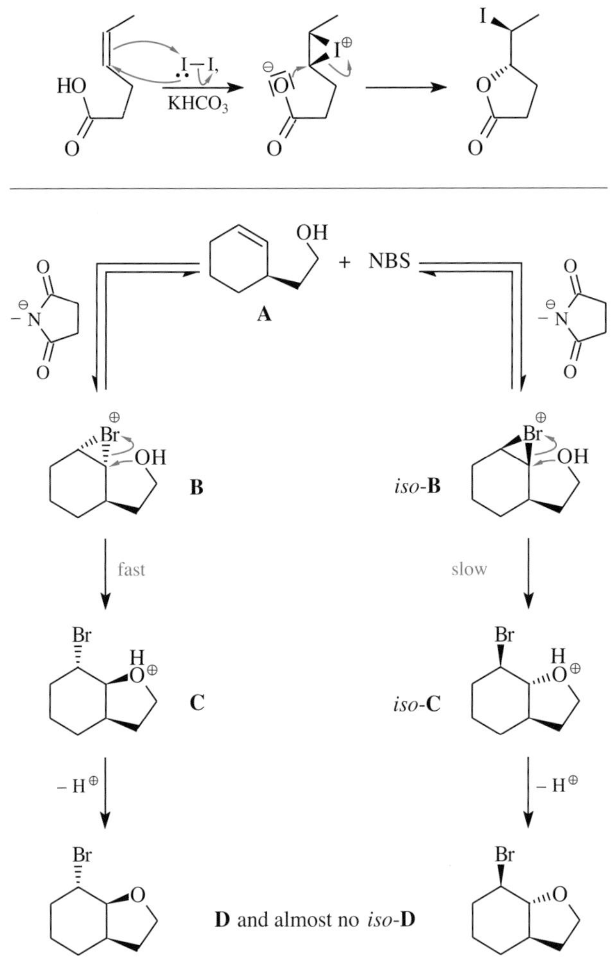

Fig. 3.47. Stereoselective iodolactonization (*top*) and stereoselective bromoetherification (*bottom*). See Figure 1.39 for the dehalogenations of the iodolactone and of the bromoether shown.

However, essentially only one of them—namely, **B**—undergoes a nucleophilic backside attack by the hydroxy group. The brominated tetrahydrofuran **D** is produced via the oxonium ion **C.** An analogous intramolecular backside reaction by the alcoholic OH group in the bromonium ion *iso*-**B** is energetically disfavored and hardly observed. The result is that the bromonium ion *iso*-**B** can revert to the starting material **A,** whereby the overall reaction takes place almost exclusively via the more reactive bromonium ion **B.**

3.5.3 Solvomercuration of Alkenes: Hydration of C=C Double Bonds through Subsequent Reduction

Mercury(II) salts add to C=C double bonds (Figure 3.48) in nucleophilic solvents via the the onium mechanism of Figure 3.42. However, the heterocyclic primary product is not called an onium, but rather a mercurinium ion. Its ring opening in an H_2O-containing solvent gives a

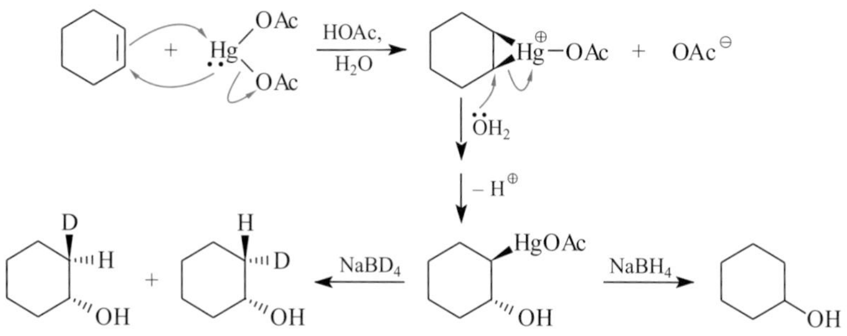

Fig. 3.48. Hydration of a symmetric alkene through solvomercuration/reduction.

trans-configured alcohol, which can then be demercurated with $NaBH_4$. The hydration product of the original olefin is obtained and is, in our case, cyclohexanol.

The mechanism of this defunctionalization was discussed in connection with Figure 1.14. It took place via approximately planar radical intermediates. This is why in the reduction of alkyl mercury(II) acetates, the C—Hg bond converts to a C—H bond without stereocontrol. The stereochemical integrity of the mercury-bearing stereocenter is thus lost. When the mercurated alcohol in Figure 3.48 is reduced with $NaBD_4$ rather than $NaBH_4$, the deuterated cyclohexanol is therefore produced as a mixture of diastereomers.

Unsymmetrically substituted C=C double bonds are hydrated according to the same mechanism (Figure 3.49). The regioselectivity is high, and the explanation for this is that the mercurinium ion intermediate is distorted in the same way as the bromonium ion in Figure 3.44. The H_2O preferentially breaks the stretched and therefore weakened C_{sec}—$Hg^{\oplus}$ bond by a backside attack and does not affect the shorter and therefore more stable C_{prim}—$Hg^{\oplus}$ bond.

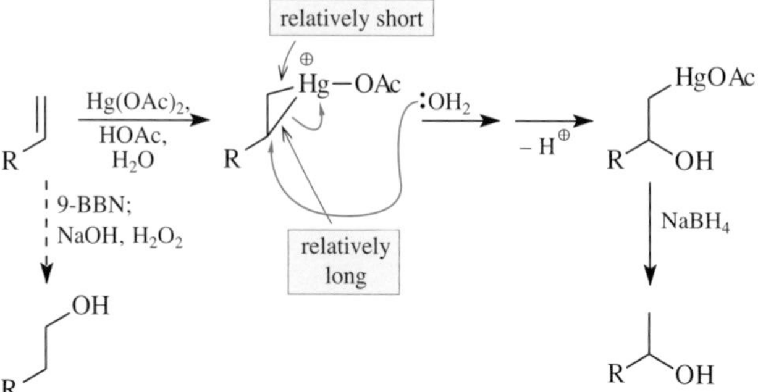

Fig. 3.49. Regioselective hydration of an unsymmetric alkene via solvomercuration/reduction. The regioselectivity of the solvomercuration/reduction sequence is complementary to that of hydroboration/oxidation/hydrolysis.

From the Hg-containing alcohol in Figure 3.49 and $NaBH_4$, one can obtain the Hg-free alcohol. The overall result is a hydration of the C=C double bond. According to the nomenclature of Section 3.3.3, its regioselectivity corresponds to a Markovnikov addition. It is complementary to the regioselectivity of the reaction sequence (1) reaction with 9-BBN, (2) reaction with H_2O_2/NaOH (Table 3.1). The latter sequence would have converted the same alkene regioselectively into the primary instead of the secondary alcohol.

Besides H_2O, simple alcohols or acetic acid can also be added to alkenes by solvomercuration/reduction. Figure 3.50 shows MeOH addition as an example. The regioselectivities of this reaction and of the H_2O addition in Figure 3.49 are identical.

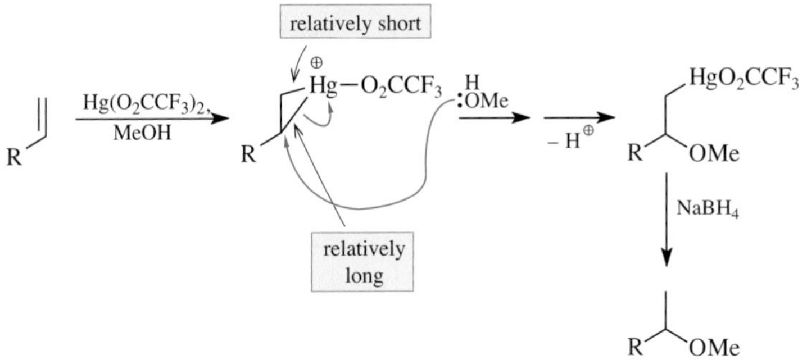

Fig. 3.50. Regioselective methanol addition to an asymmetric alkene via solvomercuration/reduction.

3.6 Additions that Take Place or Can Take Place without Stereocontrol Depending on the Mechanism

3.6.1 Additions via Carbenium Ion Intermediates

Carbenium ions become intermediates in two-step additions to alkenes when they are more stable than the corresponding cyclic onium intermediates. With many electrophiles (e. g., Br_2 and Cl_2), this occasionally is the case. With others, such as protons and carbenium ions, it is always the case (Figure 3.51).

In Section 3.5.1, it was mentioned that Br_2 and Cl_2 form resonance-stabilized benzyl cation intermediates with styrene derivatives and that *gem*-dialkylated alkenes react with Br_2 but not

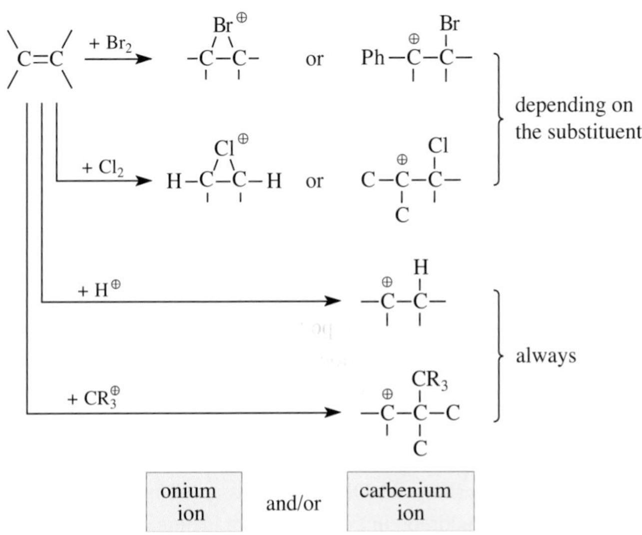

Fig. 3.51. Competition between onium and carbenium ion intermediates in electrophilic additions to C=C double bonds.

with Cl_2 via halonium ions. As C—Cl bonds are shorter than C—Br bonds, chloronium ions presumably have a higher ring strain than bromonium ions. Accordingly, a β-chlorinated tertiary carbenium ion is more stable than the isomeric chloronium ion, but a β-brominated tertiary carbenium ion is less stable than the isomeric bromonium ion.

Protons and carbenium ions *always* add to C=C double bonds via carbenium ion intermediates simply because no energetically favorable onium ions are available from a reaction with these electrophiles. An onium intermediate formed by the reaction of a proton would contain a divalent, positively charged H atom. An onium intermediate produced by the reaction of a carbenium ion would be a carbonium ion and would thus contain a pentavalent, positively charged C atom. Species of this type are generally rare , but an excellent example is the norbornyl cation (Figure 2.29).

Preparatively, it is important that mineral acids, carboxylic acids, and *tert*-carbenium ions can be added to alkenes via carbenium ion intermediates. Because of their relatively low stability, primary carbenium ions form more slowly in the course of such reactions than the more stable secondary carbenium ions, and these form more slowly than the even more stable tertiary carbenium ions (Hammond postulate!). Therefore, mineral and carboxylic acids add to unsymmetrical alkenes regioselectively to give Markovnikov products (see Section 3.3.3 for an explanation of this term). In addition, these electrophiles add most rapidly to those alkenes from which tertiary carbenium ion intermediates can be derived.

An example of an addition of a mineral acid to an alkene that takes place via a tertiary carbenium ion is the formation of a tertiary alkyl bromide from a 1,1-dialkylethylene and HBr (Figure 3.52).

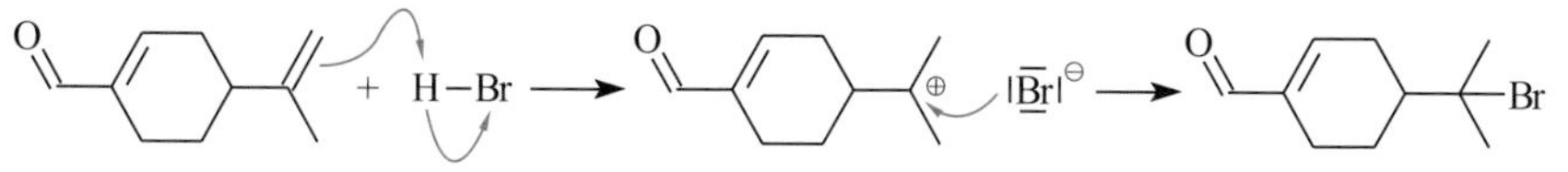

O + H−Br → O ⊕ |Br|⊖ → O −Br

Fig. 3.52. Addition of a mineral acid to an alkene.

Figure 3.53 shows an addition of a carboxylic acid to isobutene, which takes place via the *tert*-butyl cation. This reaction is a method for forming *tert*-butyl esters. Because the acid shown in Figure 3.53 is a β-hydroxycarboxylic acid whose alcohol group adds to an additional isobutene molecule, this also shows an addition of a primary alcohol to isobutene, which takes place via the *tert*-butyl cation. Because neither an ordinary carboxylic acid nor, of course, an alcohol is sufficiently acidic to protonate the alkene to give a carbenium ion, catalytic amounts of a mineral or sulfonic acid are also required here.

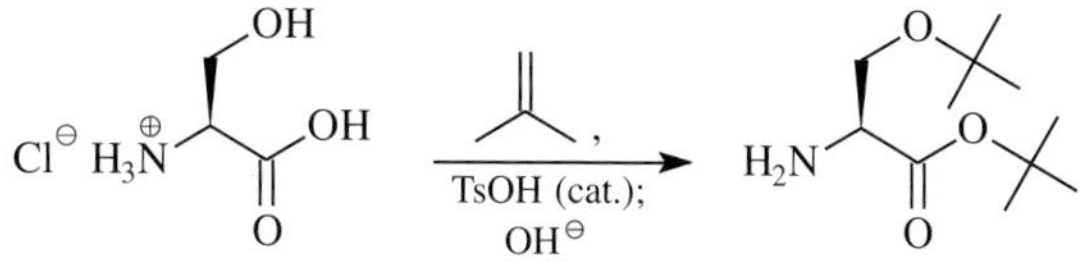

OH; $Cl^{\ominus}$ $H_3\overset{\oplus}{N}$; OH; O → (isobutene), TsOH (cat.); $OH^{\ominus}$ → O; H_2N; O; O

Fig. 3.53. Addition of a carboxylic acid and an alcohol to an alkene.

One of the rare cases of an intermolecular carbenium ion addition to an alkene without polymer formation occurs in the industrial synthesis of isooctane (Figure 3.54).

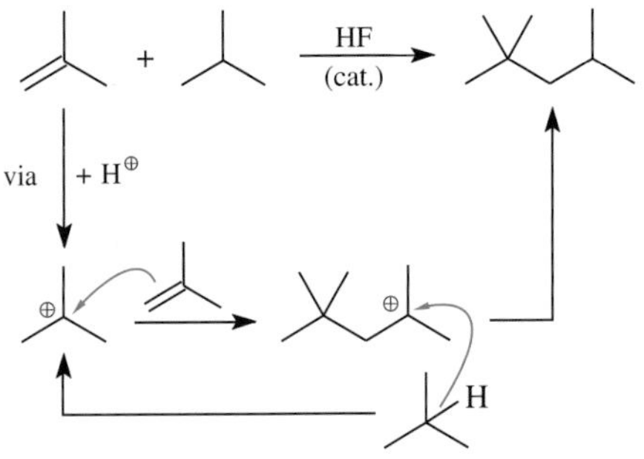

HF (cat.)

via + $H^{\oplus}$

H

Fig. 3.54. Carbenium ion additions to isobutene as key steps in the cationic polymerization of isobutene. The dashed arrow corresponds to the overall reaction.

Carbenium ion additions to C=C double bonds are of greater preparative importance when they take place intramolecularly as ring closure reactions (see Figure 3.55).

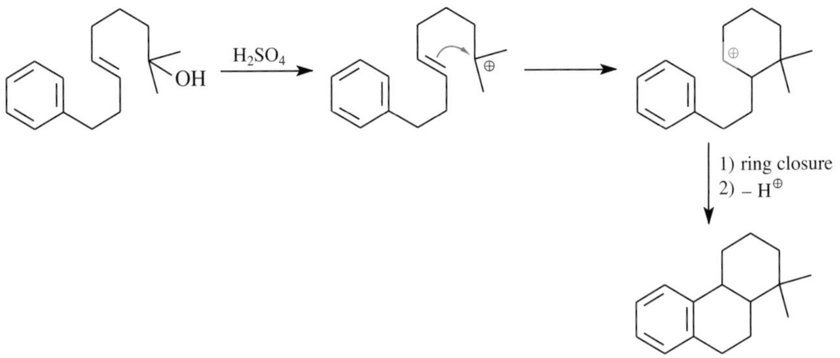

OH H_2SO_4

1) ring closure
2) – $H^{\oplus}$

Fig. 3.55. Ring closure reaction by addition of a carbenium ion to a C=C double bond (for the mechanism of the second ring closure, see Section 5.2.5).

In polyenes even tandem additions are possible. The best known and the most impressive example is the biosynthesis of steroid structures from squalene or squalene epoxide (Figures 14.12 and 14.13). The corresponding biomimetic syntheses of the steroid structure are simply beautiful.

3.6.2 Additions via "Carbanion" Intermediates

Nucleophiles can be added to acceptor-substituted alkenes. In that case, enolates and other "stabilized carbanions" occur as intermediates. Reactions of this type are discussed in this book only in connection with 1,4-additions of organometallic compounds (Section 10.6), or enolates (Section 13.6) to α,β-unsaturated carbonyl and carboxyl compounds.

References

G. Melloni, G. Modena, U. Tonellato, "Relative Reactivities of C–C Double and Triple Bonds Towards Electrophiles," *Acc. Chem. Res.* **1981**, *14*, 227.

H. Mayr, B. Kempf, A. R. Ofial, "π-Nucleophilicity in Carbon-Carbon Bond-Forming Reactions," *Acc. Chem. Rev.* **2003**, *36,* 66–77.

3.2

G. Helmchen, "Nomenclature and Vocabulary of Organic Stereochemisty," in *Stereoselective Synthesis* (Houben-Weyl) 4th ed. **1996**, (G. Helmchen, R. W. Hoffmann, J. Mulzer, E. Schaumann, Eds.), **1996**, Vol. E 21 (Workbench Edition), *1*, 1–74, Georg Thieme Verlag, Stuttgart.

K. Mislow, "Molecular Chirality," *Top. Stereochem.* **1999**, *22*, 1–82.

3.3

G. Poli, C. Scolastico, "Formation of C-O Bonds by 1,2-Dihydroxylation of Olefinic Double Bonds," in *Methoden Org. Chem.* (Houben-Weyl) 4th ed. **1952**–, *Stereoselective Synthesis* (G. Helmchen, R. W. Hoffmann, J. Mulzer, E. Schaumann, Eds.), Vol. E21e, 4547–4599, Georg Thieme Verlag, Stuttgart, **1995**.

H.-U. Reissig, "Formation of C–C Bonds by [2+1] Cycloadditions," in Stereoselective Synthesis (Houben-Weyl) 4th ed. **1996**, (G. Helmchen, R. W. Hoffmann, J. Mulzer, E. Schaumann, Eds.), **1996**, Vol. E 21 (Workbench Edition), *5*, 3179–3270, Georg Thieme Verlag, Stuttgart.

W. Sander, G. Bucher, S. Wierlacher, "Carbenes in Matrixes: Spectroscopy, Structure, and Reactivity," *Chem. Rev.* **1993**, *93*, 1583–1621.

W. E. Parham, E. E. Schweizer, "Halocyclopropanes from Halocarbenes," *Org. React.* **1963**, *13*, 55–90.

R. R. Kostikov, A. P. Molchanov, A. F. Khlebnikov, "Halogen-Containing Carbenes," *Russ. Chem. Rev.* **1989**, *58*, 654–666.

M. G. Banwell, M. E. Reum, "gem-Dihalocyclopropanes in Chemical Synthesis" in *Advances in Strain in Organic Chemistry* (B. Halton, Ed.) **1991**, *1*, Jai Press, Greenwich, CT.

A. J. Arduengo, R. Krafczyk, "Auf der Suche nach stabilen Carbenen," *Chem. unserer Zeit*, **1998**, *32*, 6–14.

D. Bourissou, O. Guerret, F. P. Gabba, G. Bertrand, "Stable Carbenes," *Chem. Rev.* **2000**, *100*, 39–91.

G. Boche, J. C. W. Lohrenz, "The Electrophilic Nature of Carbenoids, Nitrenoids, and Oxenoids," *Chem. Rev.* **2001**, *101*, 697–756.

E. V. Dehmlow, S. S. Dehmlow, "Phase Transfer Catalysis," 3rd ed., VCH, New York, **1993**.

Y. Goldberg, "Phase Transfer Catalysis. Selected Problems and Applications," Gordon and Bresch, Philadelphia, PA, **1992**.

H. E. Simmons, T. L. Cairns, S. A. Vladuchick, C. M. Hoiness, "Cyclopropanes from Unsaturated Compounds, Methylene Iodide, and Zinc-Copper Couple," *Org. React.* **1973**, *20*, 1–131.

W. B. Motherwell, C. J. Nutley, „The Role of Zinc Carbenoids in Organic Synthesis," *Contemporary Organic Synthesis* **1994**, *1*, 219.

S. D. Burke, P. A. Grieco, "Intramolecular Reactions of Diazocarbonyl Compounds," *Org. React.* **1979**, *26*, 361–475.

J. Adams, D. M. Spero, "Rhodium(II) Catalyzed Reactions of Diazo-Carbonyl Compounds," *Tetrahedron* **1991**, *47*, 1765–1808.

T. Ye, M. A. McKervey, "Organic Synthesis with α-Diazo Carbonyl Compounds," *Chem. Rev.* **1994**, *94*, 1091–1160.

A. Padwa, D. J. Austin, "Ligand Effects on the Chemoselectivity of Transition Metal Catalyzed Reactions of α-Diazo Carbonyl Compounds," *Angew. Chem. Int. Ed. Engl.* **1994**, *33*, 801–811.

M. P. Doyle, M. A. McKervey, T. Ye, "Reactions and Syntheses with α-Diazocarbonyl Compounds," Wiley, New York, **1997**.

H. M. L. Davies, S. A. Panaro, "Effect of Rhodium Carbenoid Structure on Cyclopropanation Chemoselectivity," *Tetrahedron* **2000**, *56*, 4871–4880.

M. P. Doyle, "Asymmetric Addition and Insertion Reactions of Catalytically-Generated Metal Carbenes," in *Catalytic Asymmetric Synthesis*, Ed.: I. Ojima, Wiley-VCH, New York, 2nd ed., **2000**, 191–228.

J. Aube, "Epoxidation and Related Processes," in *Comprehensive Organic Synthesis* (B. M. Trost, I. Fleming, Eds.), Vol. 1, 843, Pergamon Press, Oxford, **1991**.

A. S. Rao, "Addition Reactions with Formation of Carbon-Oxygen Bonds: (i) General Methods of Epoxidation," in *Comprehensive Organic Synthesis* (B. M. Trost, I. Fleming, Eds.), Vol. 7, 357, Pergamon Press, Oxford, **1991**.

R. Schwesinger, J. Willaredt, T. Bauer, A. C. Oehlschlager, "Formation of C–O Bonds by Epoxidation of Olefinic Double Bonds," in *Methoden Org. Chem.* (Houben-Weyl) 4th ed. **1952**–, *Stereoselective Synthesis* (G. Helmchen, R. W. Hoffmann, J. Mulzer, E. Schaumann, Eds.), Vol. E21e, 4599, Georg Thieme Verlag, Stuttgart, **1995**.

D. Swern, "Epoxidation and Hydroxylation of Ethylenic Compounds with Organic Peracids," *Org. React.* **1953**, *7*, 378–433.

H. Heaney, "Oxidation Reactions Using Magnesium Monoperphthalate and Urea Hydrogen Peroxide," *Aldrichimica Acta* **1993**, *26*, 35–45.

K. Smith, A. Pelter, "Hydroboration of C=C and Alkynes," in *Comprehensive Organic Synthesis* (B. M. Trost, I. Fleming, Eds.), Vol. 8, 703, Pergamon Press, Oxford, **1991**.

M. Zaidlewicz, "Formation of C–O Bonds by Hydroboration of Olefinic Double Bonds Followed by Oxidation," in *Stereoselective Synthesis* (Houben-Weyl) 4th ed. **1996**, (G. Helmchen, R. W. Hoffmann, J. Mulzer, E. Schaumann, Eds.), **1996**, Vol. E 21 (Workbench Edition), *8*, 4519–4530, Georg Thieme Verlag, Stuttgart.

G. Zweifel, H. C. Brown, "Hydration of Olefins, Dienes, and Acetylenes via Hydroboration," *Org. React.* **1963**, *13*, 1–54.

B. Carboni, M. Vaultier, "Useful Synthetic Transformations Via Organoboranes. 1. Amination Reactions," *Bull. Soc. Chim. Fr.* **1995**, *132*, 1003–1008.

J. V. B. Kanth, "Borane-Amine Complexes for Hydroboration," *Aldrichim. Acta* **2002**, *35,* 57–66.

J. S. Siegel, "Heterogeneous Catalytic Hydrogenation of C=C and Alkynes," in *Comprehensive Organic Synthesis* (B. M. Trost, I. Fleming, Eds.), Vol. 8, 417, Pergamon Press, Oxford, **1991**.

U. Kazmaier, J. M. Brown, A. Pfaltz, P. K. Matzinger, H. G. W. Leuenberger, "Formation of C–H Bonds by Reduction of Olefinic Double Bonds: Hydrogenation," in *Methoden Org. Chem.* (Houben-Weyl) 4th ed. **1952**–, *Stereoselective Synthesis* (G. Helmchen, R. W. Hoffmann, J. Mulzer, E. Schaumann, Eds.), Vol. E21d, 4239, Georg Thieme Verlag, Stuttgart, **1995**.

B. H. Lipshutz, "Development of Nickel-on-Charcoal as a ‚Dirt-Cheap' Heterogeneous Catalyst: A Personal Account," *Adv. Synth. Catal.* **2001**, *343,* 313–326.

3.4

A. Pelter, K. Smith, H. C. Brown, "Borane Reagents," Academic Press, London, **1988**.

H. C. Brown and B. Singaram, "Development of a Simple Procedure for Synthesis of Pure Enantiomers via Chiral Organoboranes," *Acc. Chem. Res.* **1988**, *21*, 287.

H. C. Brown, P. V. Ramachandra, "Asymmetric Syntheses via Chiral Organoboranes Based on α-Pinene," in *Advances in Asymmetric Synthesis* (A. Hassner, Ed.), **1995**, *1*, JAI, Greenwich, CT.

H. B. Kagan, J. C. Fiaud, "Kinetic Resolution," *Top. Stereochem.* **1988**, *18*, 249.

D. E. J. E. Robinson, S. D. Bull, "Kinetic Resolution Strategies using Non-Enzymatic Catalysts," *Tetrahedron: Asymmetry* **2003**, *14,* 1407–1446.

H. B. Kagan, "Various Aspects of the Reaction of a Chiral Catalyst or Reagent with a Racemic or Enantiopure Substrate," *Tetrahedron* **2001**, *57*, 2449–2468.

S. Masamune, W. Choy, J. S. Petersen, L. R. Sita, "Double Asymmetric Synthesis and a New Strategy for Stereochemical Control in Organic Synthesis", *Angew. Chem. Int. Ed. Engl.* **1985**, *24*, 1–30.

A. Pfenniger, "Asymmetric Epoxidation of Allylic Alcohols: The Sharpless Epoxidation", *Synthesis* **1986**, 89.

R. A. Johnson, K. B. Sharpless, "Addition Reactions with Formation of Carbon-Oxygen Bonds: (ii) Asymmetric Methods of Epoxidation," in *Comprehensive Organic Synthesis* (B. M. Trost, I. Fleming, Eds.), Vol. 7, 389, Pergamon Press, Oxford, **1991**.

R. A. Johnson, K. B. Sharpless, "Asymmetric Oxidation: Catalytic Asymmetric Epoxidation of Allylic Alcohols," in *Catalytic Asymmetric Synthesis* (I. Ojima, Ed.), 103, VCH, New York, **1993**.

D. J. Berrisford, C. Bolm, K. B. Sharpless, "Ligand-Accelerated Catalysis," *Angew. Chem. Int. Ed. Engl.* **1995**, *34*, 1059–1070.

T. Katsuki, V. S. Martin, "Asymmetric Epoxidation of Allylic Alcohols: The Katsuki-Sharpless Epoxidation Reaction," *Org. React.* **1996**, *48*, 1–299.

Roy A. Johnson , K. Barry Sharpless, "Catalytic Asymmetric Epoxidation of Allylic Alcohols," in *Catalytic Asymmetric Synthesis*, Ed.: I. Ojima, Wiley-VCH, New York, 2nd ed., **2000**, 231–286.

R. M. Hanson, "The Synthetic Methodology of Nonracemic Glycidol and Related 2,3-Epoxy Alcohols," *Chem. Rev.* **1991**, *91*, 437–476.

P. C. A. Pena, S. M. Roberts, "The Chemistry of Epoxy Alcohols," *Curr. Org. Chem.* **2003**, *7*, 555–571.

3.5

M. F. Ruasse, "Bromonium Ions or β-Bromocarbocations in Olefin Bromination. A Kinetic Approach to Product Selectivities," *Acc. Chem. Res.* **1990**, *23*, 87–93.

M.-F. Ruasse, "Electrophilic Bromination of Carbon-Carbon Double Bonds: Structure, Solvent and Mechanism," *Adv. Phys. Org. Chem.* **1993**, *28*, 207.

R. S. Brown, "Investigation of the Early Steps in Electrophilic Bromination Through the Study of the Reaction with Sterically Encumbered Olefins," *Acc. Chem. Res.* **1997**, *30*, 131–138.

S. Torii, T. Inokuchi, "Addition Reactions with Formation of Carbon-Halogen Bonds," in *Comprehensive Organic Synthesis* (B. M. Trost, I. Fleming, Eds.), Vol. 7, 527, Pergamon Press, Oxford, **1991**.

J. Mulzer, "Halolactonization: The Career of a Reaction," in *Organic Synthesis Highlights* (J. Mulzer, H.-J. Altenbach, M. Braun, K. Krohn, H.-U. Reißig, Eds.), VCH, Weinheim, New York, etc., **1991**, 158–164.

G. Cardillo, "Lactonization," in *Stereoselective Synthesis* (Houben-Weyl) 4th ed. **1996**, (G. Helmchen, R. W. Hoffmann, J. Mulzer, E. Schaumann, Eds.), **1996**, Vol. E 21 (Workbench Edition), *8*, 4704–4759, Georg Thieme Verlag, Stuttgart.

M. Orena, "Cycloetherification," in *Stereoselective Synthesis* (Houben-Weyl) 4th ed. **1996**, (G. Helmchen, R. W. Hoffmann, J. Mulzer, E. Schaumann, Eds.), **1996**, Vol. E 21 (Workbench Edition), *8*, 4760–4817, Georg Thieme Verlag, Stuttgart.

A. N. Mirskova, T. I. Drozdova, G. G. Levkovskaya, M. G. Voronkov, "Reactions of N-Chloramines and N-Haloamides with Unsaturated Compounds," *Russ. Chem. Rev.* **1989**, *58*, 250–271.

I. V. Koval, "N-Halo Reagents. Preparation of Chlorosulfonamide Sodium Salts and Their Application to Organic Synthesis," *Russ. J. Org. Chem.* **1999**, *35*, 475–499.

E. Block, A. L. Schwan, "Electrophilic Addition of X-Y Reagents to Alkenes and Alkynes," in *Comprehensive Organic Synthesis* (B. M. Trost, I. Fleming, Eds.), Vol. 4, 329, Pergamon Press, Oxford, **1991**.

T. Hosokawa, S. Murahashi, "New Aspects of Oxypalladation of Alkenes," *Acc. Chem. Res.* **1990**, *23*, 49–54.

J. M. Takacs, X.-t. Jiang, "The Wacker Reaction and Related Alkene Oxidation Reactions," *Curr. Org. Chem.* **2003**, *7*, 369–396.

3.6

R. C. Larock, W. W. Leong, "Addition of H–X Reagents to Alkenes and Alkynes," in *Comprehensive Organic Synthesis* (B. M. Trost, I. Fleming, Eds.), Vol. 4, 269, Pergamon Press, Oxford, **1991**.

U. Nubbemeyer, "Formation of C–C Bonds by Addition of Carbenium Ions to Olefinic Double Bonds and Allylic Systems," in *Stereoselective Synthesis* (Houben-Weyl) 4th ed. **1996**, (G. Helmchen, R. W. Hoffmann, J. Mulzer, E. Schaumann, Eds.), **1996**, Vol. E 21 (Workbench Edition), *4*, 2288–2370, Georg Thieme Verlag, Stuttgart.

H. Mayr, "Carbon-Carbon Bond Formation by Addition of Carbenium Ions to Alkenes: Kinetics and Mechanism," *Angew. Chem. Int. Ed. Engl.* **1990**, *29*, 1371–1384.

R. Bohlmann, "The Folding of Squalene: an Old Problem has New Results," *Angew. Chem. Int. Ed. Engl.* **1992**, *31*, 582–584.

I. Abe, M. Rohmer, G. D. Prestwich, "Enzymatic Cyclization of Squalene and Oxidosqualene to Sterols and Triterpenes," *Chem. Rev.* **1993**, *93*, 2189–2206.

S. R. Angle, H. L. Mattson-Arnaiz, "The Formation of Carbon-Carbon Bonds via Benzylic-Cation-Initiated Cyclization Reactions," in *Advances in Carbocation Chemistry* (J. M. Coxon, Ed.) **1995**, *2*, JAI, Greenwich, CT.

D. Schinzer, "Electrophilic Cyclizations to Heterocycles: Iminium Systems," in *Organic Synthesis Highlights II* (H. Waldmann, Ed.), VCH, Weinheim, New York, etc., **1995**, 167–172.

D. Schinzer, "Electrophilic Cyclizations to Heterocycles: Oxonium Systems," in *Organic Synthesis Highlights II* (H. Waldmann, Ed.), VCH, Weinheim, New York, etc., **1995**, 173–179.

D. Schinzer, "Electrophilic Cyclizations to Heterocycles: Sulfonium Systems," in *Organic Synthesis Highlights II* (H. Waldmann, Ed.), VCH, Weinheim, New York, etc., **1995**, 181–185.

C. F. Bernasconi, "Nucleophilic Addition to Olefins. Kinetics and Mechanism," *Tetrahedron* **1989**, *45*, 4017–4090.

Further Reading

K. B. Wilberg, "Bent Bonds in Organic Compounds," *Acc. Chem. Res.* **1996**, *29*, 229–234.

T. Arai, K. Tokumaru, "Present Status of the Photoisomerization About Ethylenic Bonds," *Adv. Photochem.* **1996**, *20,* 1–57.

B. B. Lohray, "Recent Advances in the Asymmetric Dihydroxylation of Alkenes," *Tetrahedron: Asymmetry* **1992**, *3*, 1317–1349.

V. K. Singh, A. DattaGupta, G. Sekar, "Catalytic Enantioselective Cyclopropanation of Olefins Using Carbenoid Chemistry," *Synthesis* **1997**, 137–149.

I. P. Beletskaya, A. Pelter, "Hydroborations Catalyzed by Transition Metal Complexes," *Tetrahedron* **1997**, *53*, 4957–5026.

V. K. Khristov, K. M. Angelov, A. A. Petrov, "1,3-Alkadienes and their Derivatives in Reactions with Electrophilic Reagents," *Russ. Chem. Rev.* **1991**, *60*, 39–56.

A. Hirsch, "Addition Reactions of Buckminsterfullerene," *Synthesis* **1995**, 895–913.

β-Eliminations

4

4.1 Concepts of Elimination Reactions

4.1.1 The Concepts of α,β- and 1,n-Elimination

Reactions in which two atoms or atom groups X and Y are removed from a compound are referred to as **eliminations** (Figure 4.1). In many elimination reactions, X and Y are removed

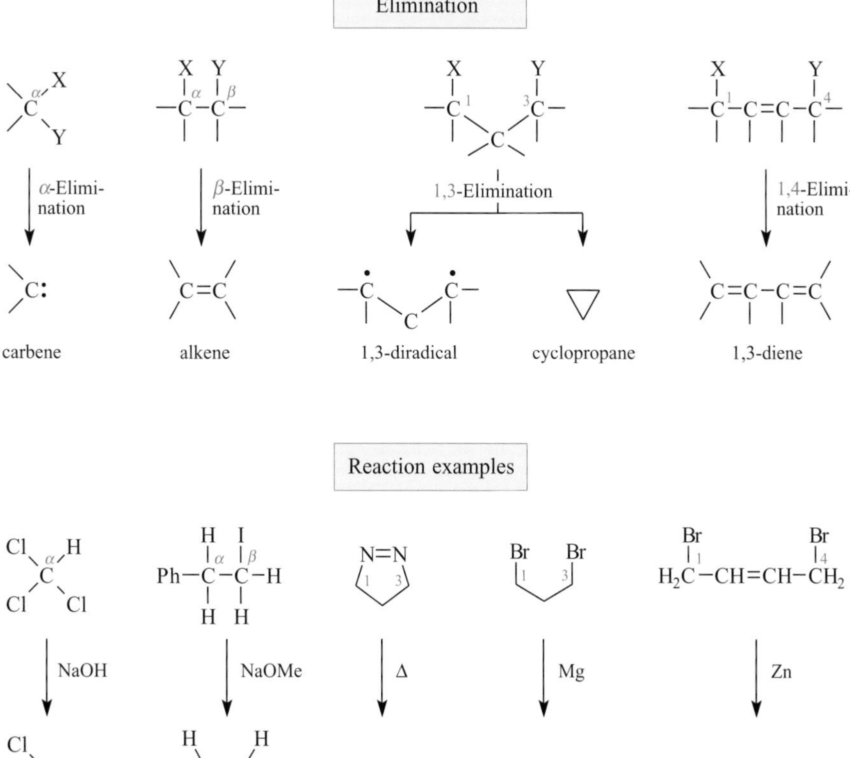

Fig. 4.1. 1,n-Eliminations (n = 1–4) of two atoms or groups X and Y, which are bound to sp^3-hybridized C atoms.

Bruckner R (author), Harmata M (editor) In: *Organic Mechanisms – Reactions, Stereochemistry and Synthesis*
Chapter DOI: 10.1007/978-3-642-03651-4_4, © Springer-Verlag Berlin Heidelberg 2010

in such a way that they do not become constituents of same molecule. In other eliminations, they become attached to one another such that they leave as a molecule of type X—Y or X=Y or as X≡Y. The atoms or groups X and Y can be bound to C atoms and/or to heteroatoms in the substrate. These atoms can be sp^3- or sp^2-hybridized.

Depending on the distance between the atoms or groups X and Y removed from the substrate, their elimination has a distinct designation. If X and Y are geminal, their removal is an ***α*-elimination**. If they are vicinal, it is a ***β*-elimination**. If X and Y are separated from each other by *n* atoms, their removal is called **1,*n*-elimination**, i.e., 1,3-, 1,4-elimination, and so on (Figure 4.1).

Some α-eliminations have already been discussed, like the formation of dichlorocarbene from chloroform and base. Others will be presented in certain contexts later. 1,3-Eliminations are mentioned in the preparation of 1,3-dipoles such as diazoalkanes or *α*-diazoketones and nitrile oxides (Chapter 15). Chapter 4 is limited to a discussion of the most important eliminations, which are the alkene-forming, *β*-eliminations. Note that *β*-Eliminations in which at least one of the leaving groups is removed from a heteroatom are considered to be oxidations. Eliminations of this type are therefore not treated here but in the redox chapter (mainly in Section 17.3.1).

4.1.2 The Terms *syn*- and *anti*-Elimination

In various eliminations the mechanism implies a well-defined stereorelationship between the eliminated atoms or groups X and Y and the plane of the resulting C=C double bond (Figure 4.2). For example, X and Y may leave into the same half-space flanking this double bond. Their removal is then called a ***syn*- or *cis-elimination.*** Please note that this is a mechanistic descriptor and doesn't say anything about the configuration of the product alkene. There are other eliminations where group X leaves the substrate in the direction of one half-space and group Y leaves in the direction of the other half-space, both flanking the C=C double bond produced. These are so-called ***anti*- or *trans*-eliminations.** The third possibility concerns eliminations in which there is no need for an unambiguous spatial relationship between the groups X and Y to be removed and the plane of the resulting double bond.

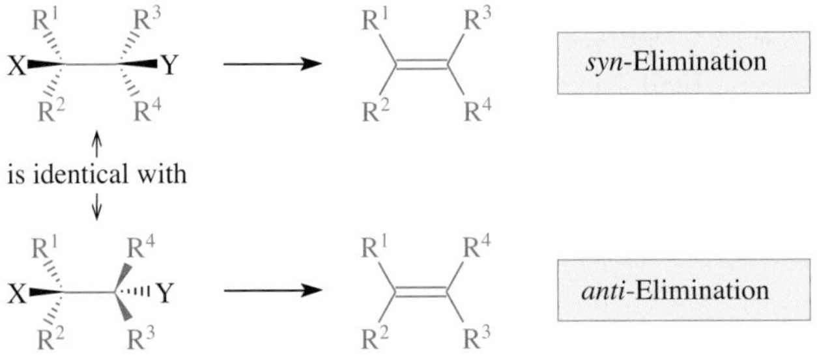

R1 R3 X Y R2 R4 → R1 R3 R2 R4 *syn*-Elimination

is identical with

R1 R4 X Y R2 R3 → R1 R4 R2 R3 *anti*-Elimination

Fig. 4.2. *Syn*- (= *cis*-) and *anti*- (= *trans*-) eliminations.

4.1.3 When Are *syn*- and *anti*-Selective Eliminations Stereoselective?

An elimination in which a *cis*, a *trans*-, an *E*-, or a *Z*-double bond is produced can be stereoselective, but they need not be. They are necessarily stereoselective when in the substrate X and Y are bound to stereocenters. Examples of stereoselectivities of this type are provided by the eliminations in Figure 4.2 for $X \neq R^1 \neq R^2$ and $Y \neq R^3 \neq R^4$. On the other hand, stereoselectivity is not guaranteed when X and Y are removed from substrates possessing the structure X—CR^{1}R^2—CY$_2$—R^3, that is, from substrates in which X is bound to a stereocenter but Y is not.

This last situation includes substrates that possess the structure Het—CR^{1}R^2—CH$_2$—R$_3$. If R^1 is different from R^2, H/Het-eliminations from such substrates *may* be stereoselective (quite independent of the elimination mechanism). As an illustration, we want to discuss three representative examples: eliminations from Het—C(Ph)H—CH$_2$—R (Figure 4.3), from Het—C(Ph)Me—CH$_2$—R (Figure 4.5), and from Het—C(Et)Me—CH$_2$—R (Figure 4.6).

β-Eliminations from the benzyl derivatives of Figure 4.3 take place preferentially to form the *trans*- and only to a lesser extent the *cis*-alkene. This is independent of whether the elimination mechanism is *syn*-selective or *anti*-selective or neither.

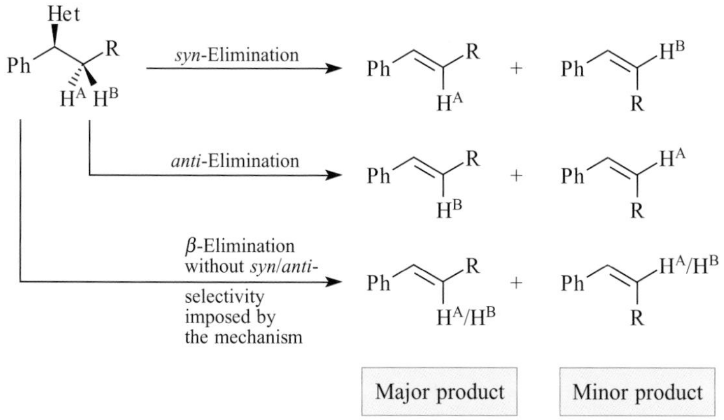

Het
Ph R
H^A H^B
syn-Elimination
Ph R H^A + Ph H^B R
anti-Elimination
Ph R H^B + Ph H^A R
β-Elimination without *syn*/*anti*-selectivity imposed by the mechanism
Ph R H^A/H^B + Ph H^A/H^B R
Major product
Minor product

Fig. 4.3. Mechanism-independent occurrence of considerable stereocontrol in *β*-eliminations of Het and H from a substrate Het—C(Ph)H—CH$_2$—R.

This *trans*-selectivity ultimately results from the fact that *trans*-alkenes are more stable than their *cis*-isomers. This energy difference is especially pronounced for the alkenes in Figure 4.3 because they are styrene derivatives. Styrenes with one alkyl group in the *trans*-position on the alkenyl C=C double bond enjoy the approximately 3 kcal/mol styrene resonance stabilization. This is lost in *cis*-styrenes because in that case the phenyl ring is rotated out of the plane of the alkenic C=C double bond to avoid the *cis*-alkyl substituent. However, the *trans*-selectivity documented in Figure 4.3 is not a consequence of thermodynamic control. This could occur only for a reversible elimination or if the alkenes could interconvert under the reaction conditions in some other way. Under the conditions of Figure 4.3, alkenes are almost always produced irreversibly and without the possibility of a subsequent *cis*/*trans*-isomerization. Therefore, the observed *trans*-selectivity is the result of kinetic control.

Figure 4.4 explains how this takes place. Whether a *cis*- or a *trans*-alkene is formed is determined in the C=C-forming step of each elimination mechanism. For a given mechanism

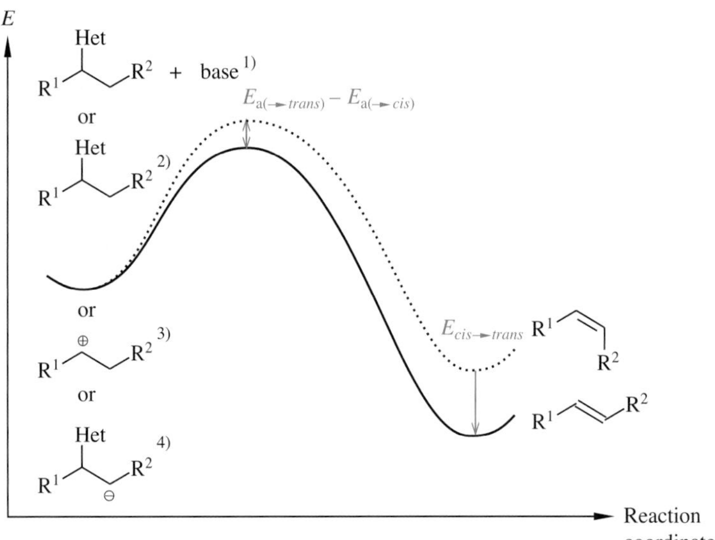

Fig. 4.4. Energy profile of the C=C-forming step of the four mechanisms according to which the β-eliminations of Figure 4.3 can take place in principle as a function of the chemical nature of the substituent Het and the reaction conditions. The conceivable starting materials for this step are, depending on the mechanism, the four species depicted on the left, where [1] is for E2 elimination, [2] is for β-elimination via a cyclic transition state, [3] is for El elimination, and [4] is for $E1_{cb}$ elimination.

the immediate precursor of each of these alkenes is the same species. This is converted either to the *cis*-alkene via one transition state or to the *trans*-alkene via another transition state. Because acyclic *cis*-alkenes have a higher energy than *trans*-alkenes, according to the Hammond postulate the *cis*-transition state should have a higher energy than the *trans*-transition state. Of course, alkene formation takes place preferentially via the lower energy pathway, and thus it leads to the *trans*-isomer. This is another example of product development control.

All β-eliminations from the benzyl derivative in Figure 4.5 exhibit *E*-stereoselectivity. This is true regardless of whether the elimination is *syn*- or *anti*-selective or neither. The reason for the preferred formation of the *E* product is product development control. This comes about because there is a significant energy difference between the isomeric elimination prod-

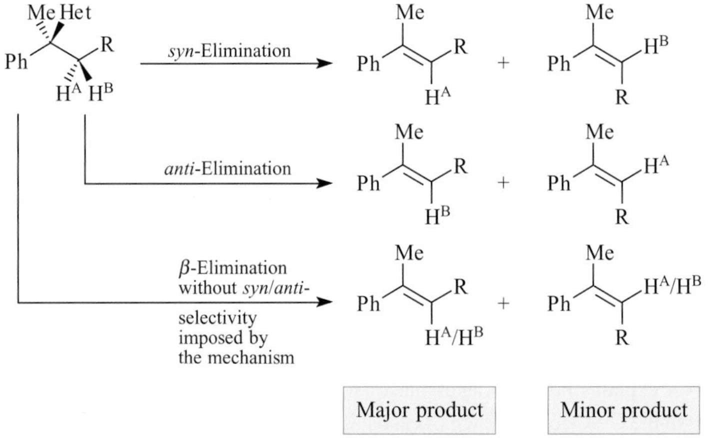

Fig. 4.5. Mechanism-independent occurrence of some stereocontrol in β-eliminations of Het and H from a substrate Het—C(Ph)Me—CH_2—R. Only eliminations in which the removed H atom does not come from the methyl group are considered.

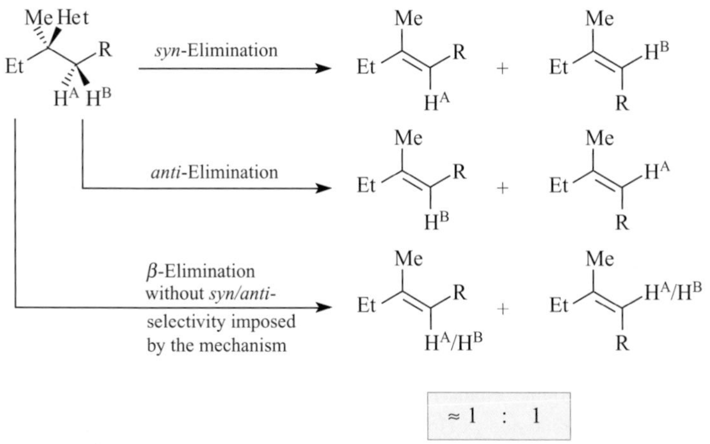

Fig. 4.6. Mechanism-independent absence of stereocontrol in β-eliminations of Het and H from a substrate Het—C(Et)Me—CH_2—R. Only eliminations in which the eliminated H atom comes neither from the ethyl group nor from the methyl group are discussed.

ucts due to the presence (in *E*-isomers) or absence (in *Z*-isomers) of styrene resonance stabilization.

The basis for the occurrence of product development control vanishes if there is only a marginal energy difference between *E,Z*-isomeric trisubstituted alkenes, as for those shown in Figure 4.6. Therefore, the corresponding β-eliminations proceed without any stereocontrol.

If we consider what has just been said from a different point of view, we arrive at the following conclusion:

Rule of Thumb for Stereocontrol in Eliminations

Eliminations of H/Het from substrates in which the leaving group (Het) is bound to a stereocenter and the H atom is not, may be suited for the stereoselective synthesis of *trans*- or of *E* alkenes. However, even this kind of stereocontrol is limited to alkenes that are *considerably* more stable than their *cis*- or *Z* isomers, respectively. On the other hand, eliminations of this type are never suitable for the synthesis of *cis*- or *Z* alkenes.

4.1.4 Formation of Regioisomeric Alkenes by β-Elimination: Saytzeff and Hofmann Product(s)

Substrates in which the leaving group "Het" is bound to a primary C atom can form just a single alkene by the β-elimination of H/Het:

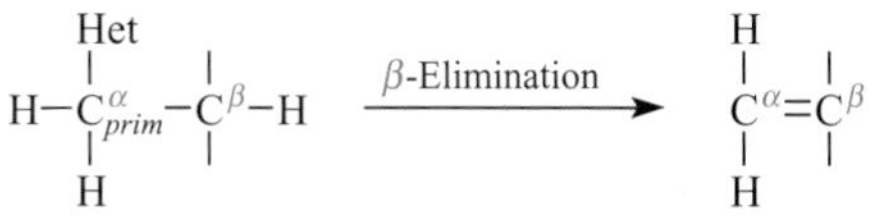

$$\text{H}-\text{C}^{\alpha}_{prim}(\text{Het})(\text{H})-\text{C}^{\beta}-\text{H} \xrightarrow{\beta\text{-Elimination}} (\text{H})(\text{H})\text{C}^{\alpha}=\text{C}^{\beta}$$

On the other hand, in β-eliminations where the leaving group Het is bound to a secondary or a tertiary C atom, a neighboring H atom can be removed from up to two (Figure 4.7) or even up to three (Figure 4.8) constitutionally different positions. Up to as many as two or three con-

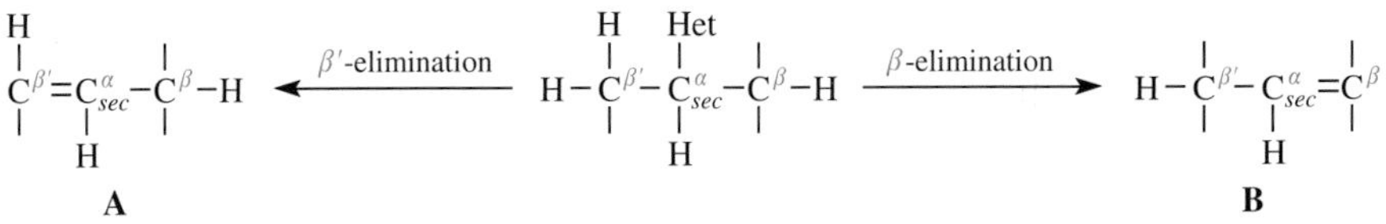

$$\overset{H}{\underset{|}{\overset{|}{C^{\beta'}}}}=\underset{\underset{H}{|}}{\overset{|}{C^{\alpha}_{sec}}}-\underset{|}{\overset{|}{C^{\beta}}}-H \xleftarrow{\beta'\text{-elimination}} H-\underset{|}{\overset{\overset{H}{|}}{C^{\beta'}}}-\underset{\underset{H}{|}}{\overset{\overset{Het}{|}}{C^{\alpha}_{sec}}}-\underset{|}{\overset{|}{C^{\beta}}}-H \xrightarrow{\beta\text{-elimination}} H-\underset{|}{\overset{|}{C^{\beta'}}}-\underset{\underset{H}{|}}{C^{\alpha}_{sec}}=\underset{|}{\overset{|}{C^{\beta}}}$$

A **B**

Fig. 4.7. Regioselectivity of the elimination of H/Het from R_{sec}—Het. When $C^{\beta'}$ has fewer alkyl substituents than C^{β} the constitutionally isomeric products are Hofmann product (**A**) and Saytzeff product (**B**).

stitutionally isomeric alkenes can be produced. In spite of this, it may be possible to form a single alkene selectively. In such a case the elimination takes place only in one particular direction. We say that it takes place **regioselectively** or that it gives only one of the **regioisomeric** alkenes.

If the β-elimination of H/Het from R_{sec}—Het can, in principle, afford regioisomeric alkenes whose C=C double bonds (Figure 4.7) contain a different number of alkyl substituents, they are differentiated as Hofmann and Saytzeff products: the Hofmann product is the alkene with the less alkylated double bond, and the Saytzeff product is the alkene with the more alkylated double bond. Because C=C double bonds are stabilized by alkyl substituents, a Hofmann product is, in general, less stable than its Saytzeff isomer. Accordingly, eliminations of H/Het from R_{sec}—Het, which exhibit product development control, furnish a Saytzeff product with some regioselectivity.

Alternatively, the (two) regioisomeric elimination products of H/Het from R_{sec}—Het could also have the same number of alkyl substituents on their C=C double bonds. In such a case, the regioselective production of one alkene at the expense of the other is generally not possible because the basis for product development control is missing, namely, a clear stability difference between the regioisomers.

Tertiary substrates R_{tert}—Het can give a maximum of three regioisomeric alkenes by β-elimination of H/Het and achieving regiocontrol can become an unsolvable problem. For example, it is impossible to eliminate H/Het regioselectively by any method from the substrate shown in Figure 4.8. Moreover, the discussion in Section 4.1.3 showed that none of the three regioisomers would be obtained stereoselectively. Consequently, none of the six alkenes can be produced selectively from the substrate specified in Figure 4.8.

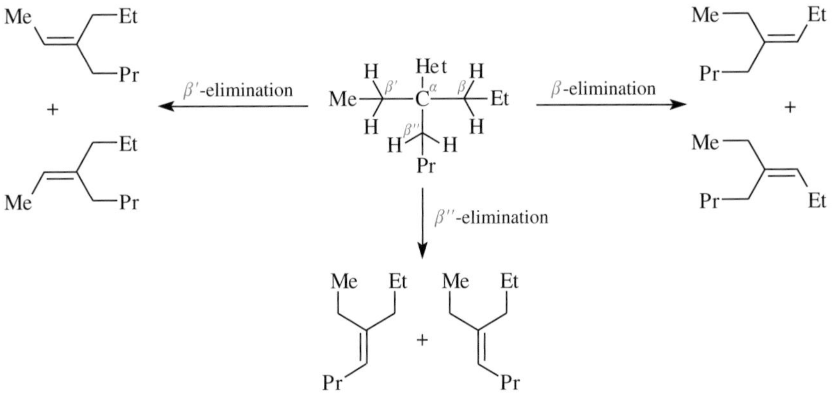

Fig. 4.8. An unsolvable regio- and stereoselectivity problem: β-elimination of H/Het from EtPrBuC—Het.

4.1.5 The Synthetic Value of Het¹/Het² in Comparison to H/Het-Eliminations

If one wants to obtain alkenes by an H/Het-elimination from R—Het, there are evidently unpleasant limitations with respect to regiocontrol (Section 4.1.4) and stereocontrol (Section 4.1.3). Most of these limitations disappear when the same alkene is synthesized by a β-elimination of Het¹/Het²:

$$-C^{\beta'}-C^{\alpha}(\text{Het}^1)(-C^{\beta''}-)-C^{\beta}(\text{Het}^2)- \xrightarrow{\beta\text{-Elimination}} -C^{\beta'}-C^{\alpha}(-C^{\beta''}-)=C^{\beta}$$

Rigorous regiocontrol in β-elimination of Het¹/Het² is available because these functional groups occur in the substrate only *once* in the β-position to each other.

In addition, many Het¹/Het²-eliminations have the advantage that their mechanism allows for *syn*-selectivity (Sections 4.7.2–4.7.3) or *anti*-selectivity (Section 4.7.2). When such a β-elimination is carried out starting from a substrate in which both Het¹ and Het² are bound to a stereocenter, 100% stereoselectivity is observed.

Figure 4.9 shows Het¹/Het²-eliminations in the same carbon framework in which the H/Het-eliminations of Figure 4.8 led to a mixture of six isomeric alkenes. In Figure 4.9, every Het¹/Het²-elimination leads to a single alkene isomer. In this way, each component of the

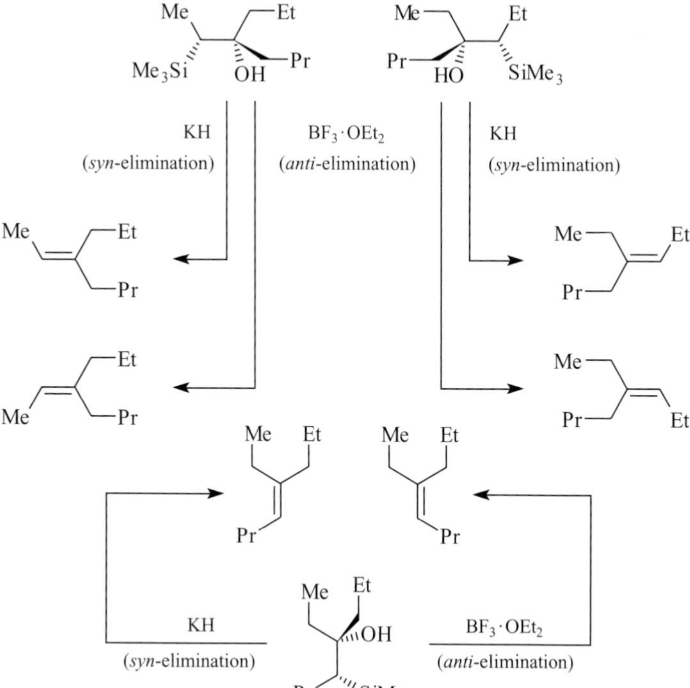

Fig. 4.9. A solution to the regio- and stereoselectivity problems depicted in Figure 4.8 with the help of the Peterson elimination (for the mechanism, see Figure 4.38) as an example of either a *syn*- or an *anti*-selective Het¹/Het²-elimination, depending on the reaction conditions.

alkene mixture in Figure 4.8 can be obtained as a pure isomer! Still, one difficulty must not be ignored: it is often more laborious to prepare the substrate for a Het^1/Het^2-elimination than for an H/Het-elimination. If one compares the alkene precursors from Figure 4.9 with those from Figure 4.8, one realizes that clearly.

4.2 β-Eliminations of H/Het via Cyclic Transition States

The thermal decomposition (pyrolysis) of alkylaryl selenoxides (**selenoxide pyrolysis**) to an alkene and an aryl selenic acid Ar—Se—OH often takes place even at room temperature (Figure 4.10). This reaction is one of the mildest methods for introducing a C=C double bond by means of a β-elimination. The mechanism is described by the simultaneous shift of three electron pairs in a five-membered cyclic transition state. One of these electron pairs becomes a nonbonding electron pair on the selenium atom in the selenic acid product. The Se atom is consequently reduced in the course of the pyrolysis.

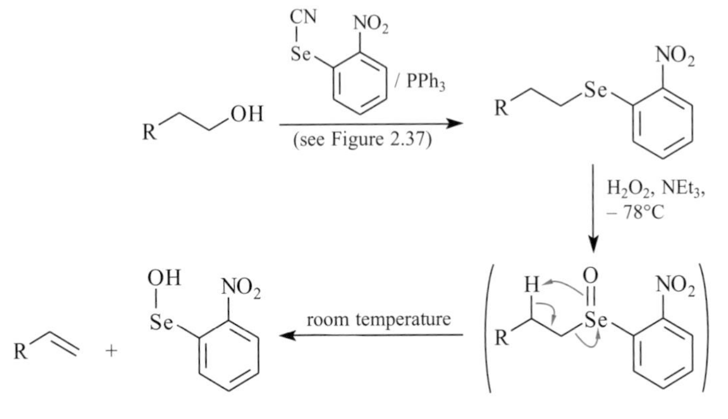

CN NO$_2$ Se / PPh_3 (see Figure 2.37) R OH → R Se NO$_2$; H_2O_2, NEt_3, – 78°C ; room temperature ; R + OH Se NO$_2$

Fig. 4.10. Selenoxide pyrolysis for the dehydration of primary alcohols.

Conveniently, the required alkylaryl selenoxide does not have to be isolated. Instead, it is produced in solution by a low-temperature oxidation of the corresponding alkylaryl selenide and eliminated in the same synthetic operation by thawing or gently heating the reaction mixture.

The reaction from Figure 4.11 proves that a selenoxide pyrolysis is a *syn*-elimination. The cyclohexylphenyl selenoxide shown reacts regioselectively to produce the less stable Hofmann product **(D).** The Saytzeff product **(E)** is not produced at all, although it is more stable than **D.** This observation shows that the transition state **C** of an *anti*-elimination (→ **E**) has a higher energy than the transition states **A** or **A'** of the conceivable *syn*-eliminations leading to **D.** With the *anti*-transition state **C** being so disfavored, the Hofmann product **D** should, of course, not form via transition state **B** either, because in that case it would also originate from an *anti*-elimination.

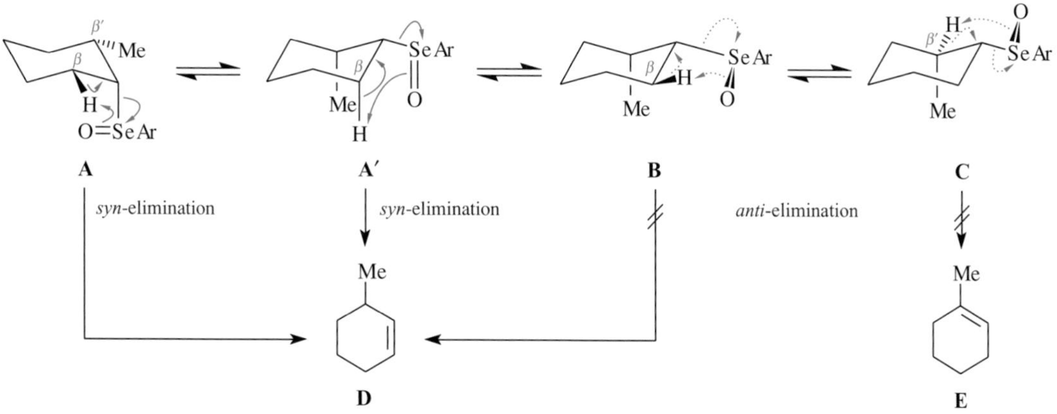

Fig. 4.11. Regiocontrol in a selenoxide pyrolysis based on its mechanism-imposed *syn*-selectivity.

Certain alkylaryl selenides can be prepared by the electrophilic selenylation of enolates (Figure 4.12; also see Table 13.5). With a subsequent H_2O_2 oxidation to produce the selenoxide followed by the elimination of Ph—Se—OH, one proceeds in a total of two synthetic steps from a carbonyl or carboxyl compound to its α,β-unsaturated analogue.

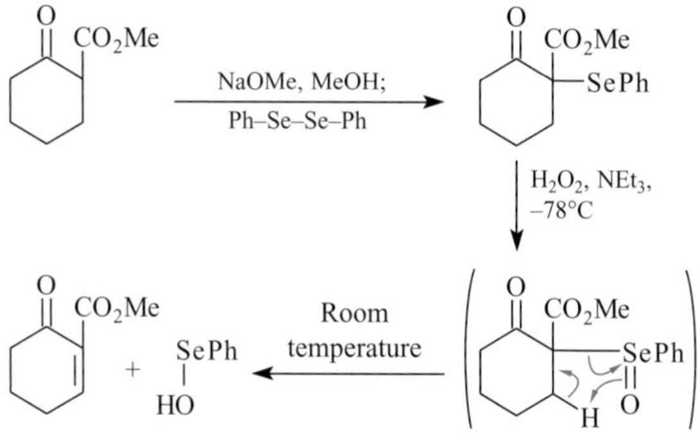

Fig. 4.12. "Dehydrogenation" of carbonyl or carboxyl compounds via their α-phenylselenyl derivatives.

The sulfur analog of the selenoxide pyrolysis is also known. In this **sulfoxide pyrolysis** the C–S bond is broken. The C-S bond is stronger than the C-Se bond and this explains why sulfoxides must typically be pyrolyzed at 200 °C to achieve elimination. Figure 4.13 shows the transformation of protected L-methionine into the corresponding sulfoxide, which then undergoes sulfoxide pyrolysis. This two-step sequence provides an elegant access to the nonnatural amino acid L-vinyl glycine.

Fig. 4.13. Sulfoxide pyrolysis for the formation of the C=C double bond of protected L-vinyl glycine, a non-proteinogenic amino acid.

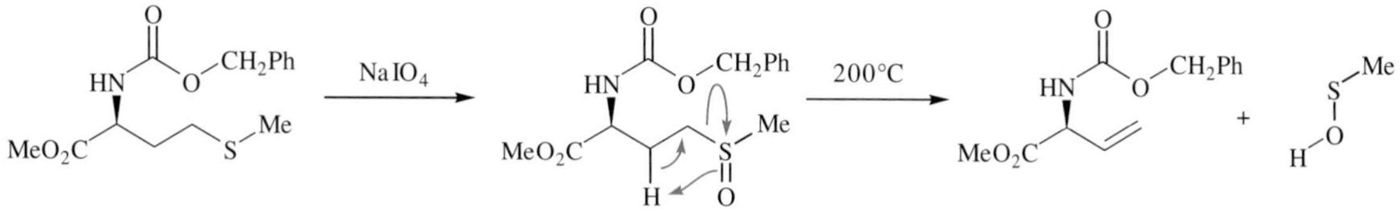

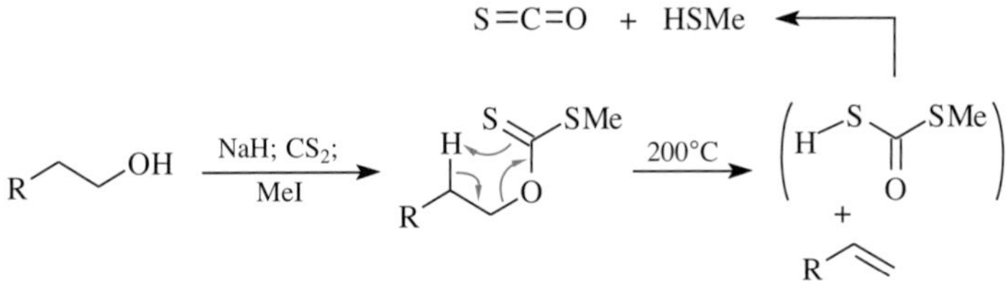

Fig. 4.14. The Chugaev reaction for the dehydration of alcohols. (The decomposition of dithiocarbonic acid methylester—here given in brackets—to form carbon oxysulfide and methane thiol is outlined in Section 8.1 near Figure 8.4.)

In the **Chugaev reaction** (Figure 4.14), *O*-alkyl-*S*-methyl xanthates are pyrolyzed to an alkene, carbon oxysulfide, and methanethiol at 200 °C.

In contrast to selenoxide pyrolysis, xanthate pyrolysis takes place via a *six-membered* transition state. As in selenoxide and sulfoxide pyrolysis, in xanthate pyrolysis three electron pairs are shifted at the same time. However, the additional atom in the ring provides a stereochemical flexibility which the selenoxide pyrolysis does not have: The Chugaev reaction is not necessarily *syn*-selective (Figure 4.15). Still, the favored transition state is the one in which the leaving group and the *syn*-H atom in the β-position leave the substrate, that is, a *syn*-transition state. However, in the transition state of the Chugaev reaction all bonds of the partial structure C^{α}···O··· C(SMe) ··· S··· H··· C^{β} are long enough that the leaving group and an *anti*-H atom in the β-position can leave the substrate, jointly as a thiocarboxylic acid half-ester.

Figure 4.15 illustrates the competition of such *syn*- vs. *anti*-pathways starting from two differently substituted cyclohexyl xanthates. The one with the substituent R = isopropyl contains a so-called **conformational lock**, at least to a certain extent. This term refers to a substituent that fixes a molecule predominantly in a single conformation while this molecule could assume several preferred conformations in the absence of this substituent. For example, an α-branched alkyl group like an isopropyl group (cf. Figure 4.15) or, even better, a *tert*-butyl group acts as a conformational lock on the cyclohexane ring. This group preferentially fixes

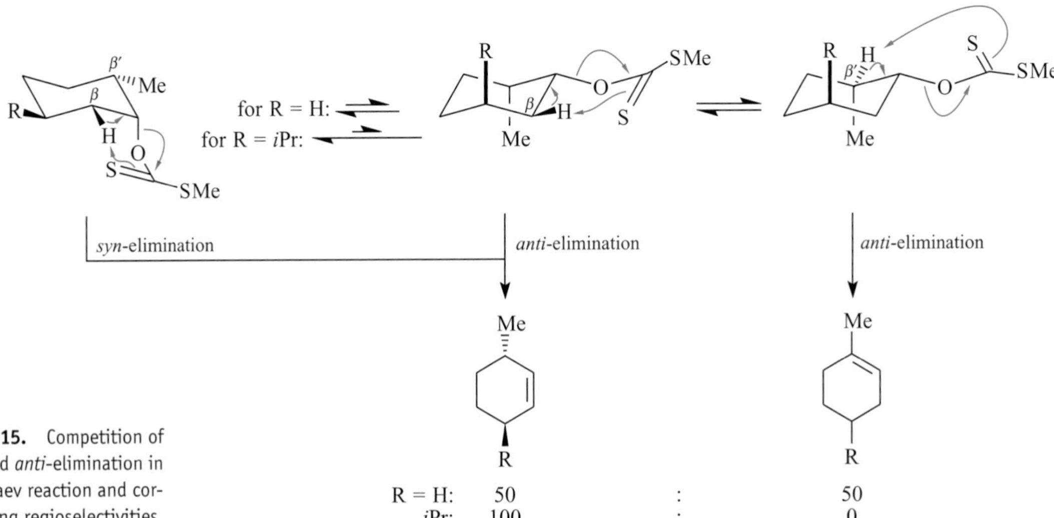

Fig. 4.15. Competition of *syn*- and *anti*-elimination in the Chugaev reaction and corresponding regioselectivities.

cyclohexane in the chair conformation in which the group is oriented equatorially. Due to the presence of this conformational lock R = isopropyl, the xanthate group of the second substrate of Figure 4.15 is oriented axially. The *syn* H atom in the B position is oriented perfectly for intramolecular abstraction. This steric course explains why the thermodynamically less stable Hofmann product is produced regioselectively in the second elimination in Figure 4.15 (R = isopropyl). When the xanthate can be equatorial (R = H), it cannot engage in a concerted, cyclic elimination—the product would be a trans-cyclohexene—and another mechanism must intervene to produce the Saytzeff product observed in the first elimination.

4.3 β-Eliminations of H/Het via Acyclic Transition States: The Mechanistic Alternatives

Let us now turn to β-eliminations that take place via acyclic transition states. There three elimination mechanisms (Figure 4.16) depending on the order in which the C—H and the C—Het bonds of the substrate are broken. If both bonds are broken at the same time, it is a one-step E2 elimination. When first one and then the other bond is broken, we have two-step eliminations. These can take place via the E1 or the $E1_{cb}$ mechanism. In the E1 mechanism, the C—Het bond is broken first and the C—H bond is broken second. Conversely, in the $E1_{cb}$ mechanism the C—H bond is broken first, by deprotonation with a base. In this way the conjugate base (cb) of the substrate is produced. Subsequently, the C—Het bond breaks.

According to the definition given above, E2 eliminations are one-step eliminations. Still, in an E2 transition state the C—H bond can be broken to a different extent than the C—Het bond. If the C—H bond is broken to a greater extent than the C—Het bond, we have an E2 elimination with an $E1_{cb}$-like distortion of the transition state geometry. Such transition states exhibit characteristic partial charges. In the $E1_{cb}$-like distorted E2 transition state, a

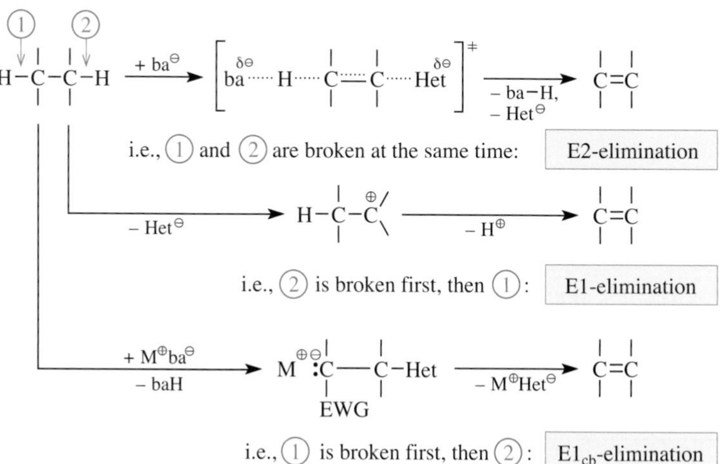

Fig. 4.16. The three mechanisms of H/Het-elimination via acyclic transition states: ba⊖: the attacking base; baH: the protonated base; EWG: electron-withdrawing group

small negative charge develops on the C atom connected to the reacting H atom. E2 eliminations of *poor* leaving groups with a *strong* base tend to proceed via an $E1_{cb}$-like transition state. In the E1-like distorted E2 transition state, the C atom linked to the leaving group carries a small positive charge. E2 eliminations of *good* leaving groups with *weak* bases are candidates for E1-like transition states.

4.4 E2 Eliminations of H/Het and the E2/S_N2 Competition

The rate of formation of the alkene in E2 eliminations is described by Equation 4.1. It shows the **bi**molecularity of this reaction, which is responsible for the short-hand notation of its mechanism. Typical substrates for E2 eliminations are alkyl halides and sulfonates:

$$\text{H–C–C–Y} + X^{\ominus} \xleftarrow[\text{(one-step)}]{k_{S_N2}} \text{H–C–C–X} + \text{Y:}^{\ominus} \xrightarrow[\text{(one-step)}]{k_{E2}} \text{C=C} + \text{H–Y} + X^{\ominus}$$

The removal of H—Hal or H—O_3S—R from these substrates is effected by bases $Y^{\ominus}$. They make their electron pair available to the H atom located in the β-position to the leaving group. This allows the H atom to be transferred to the base as a proton.

In principle, the bases $Y^{\ominus}$ are also nucleophiles, and, hence, they can react with the same alkyl halides and sulfonates via the S_N2 mechanism. The point of reaction is the C atom that bears the leaving group. In order to carry out E2 eliminations chemoselectively, competing S_N2 reactions must be excluded. To understand the outcome of the competition (E2 elimination vs. S_N2 reaction), it is analyzed kinetically with Equations 4.1–4.3.

$$\frac{d\,[\text{elimination product}]}{dt} = k_{E2}\,[RX][Y^-] \tag{4.1}$$

$$\frac{d\,[\text{substitution product}]}{dt} = k_{E_N2}\,[RX][Y^-] \tag{4.2}$$

Division of Eq. 4.1 by Eq. 4.2 yields

$$\frac{d\,[\text{elimination product}]}{d\,[\text{substitution product}]} = \frac{k_{E2}}{k_{S_N2}} = \frac{\%\ \text{elimination product}}{\%\ \text{substitution product}} \tag{4.3}$$

According to Equation 4.3, the yield ratio of E2 to S_N2 product equals the ratio of the rate constants k_{E2}:k_{SN2}. This ratio depends on the substrate structure, the nature of the added base, and stereoelectronic factors. The individual influences will be investigated in more detail in Sections 4.4.1–4.4.3.

4.4.1 Substrate Effects on the E2/S_N2 Competition

Tables 4.1 and 4.2 summarize the typical substrate effects on the chemoselectivity of E2 vs. S_N2 reactions. These substrate effects are so pronounced because NaOEt was used as base. As a reasonably strong base and a quite good nucleophile, NaOEt is able to convert a fair portion of many elimination substrates into S_N2 products (see Section 4.4.2).

Table 4.1 gives the chemoselectivity of E2 eliminations from representative bromides of the type R_{prim}—Br, R_{sec}—Br, and R_{tert}—Br. The fraction of E2 product increases in this sequence from 1 to 79 and to 100% and allows for the following generalization: E2 eliminations with sterically unhindered bases can be carried out chemoselectively (i.e., without a competing S_N2 reaction) only starting from tertiary alkyl halides and sulfonates. To obtain an E2 product from primary alkyl halides and sulfonates *at all* or to obtain an E2 product from secondary alkyl halides and sulfonates *exclusively*, one must change the base (see Section 4.4.2).

Table 4.1 also allows one to identify reasons for the chemoselectivities listed therein. According to Equation 4.3, they equal the ratio k_{E2}:k_{S_N2} of the two competing reactions. According to Table 4.1, this ratio increases rapidly in the sequence R_{prim}—Br, R_{sec}—Br, and R_{tert}—Br because the k_{S_N2} values become smaller and smaller in this order (for the reason, see Section 2.4.4) while the k_{E2} values become larger and larger.

There are two reasons for this increase in the k_{E2} values. The first reason is a statistical factor: the number of H atoms that are located in β-position(s) relative to the leaving group and can be eliminated together with it is 3, 6, and 9, respectively, in the three bromides discussed.

Tab. 4.1 Effect of Alkyl Groups in the α Position of the Substrate on the Result of the E2/S_N2 Competition

$$\text{R—Br} \xrightarrow[\text{in EtOH, 55°C}]{\geq 1\ \text{M NaOEt}} \text{R—OEt} + \text{Alkene}$$

Substrate	k_{S_N2} [10^{-5} l mol^{-1} s^{-1}]	k_{E2} [10^{-5} l mol^{-1} s^{-1}]	k_{E2} (per β-H) [10^{-5} l mol^{-1} s^{-1}]	Alkene formed	Alkene fraction
—Br	118	1.2	0.4		1%
—Br	2.1	7.6	1.3		79%
—Br	≪ 2.1	79	8.8		100%

Tab. 4.2 Effect of Alkyl Groups in the β Position in the Substrate on the Result of the E2/S_N2 Competition

$$\text{R—Br} \xrightarrow[\text{in EtOH, 55°C}]{\text{0.1 M NaOEt}} \text{R—OEt} + \text{Alkene}$$

Substrate	k_{S_N2} [10^{-5} l mol^{-1} s^{-1}]	k_{E2} [10^{-5} l mol^{-1} s^{-1}]	k_{E2} (per β-H) [10^{-5} l mol^{-1} s^{-1}]	Alkene formed	Alkene fraction
Br	172	1.6	0.53		1%
Br	54.7	5.3	2.7		9%
Br	5.8	8.5	8.5		60%

One can adjust the gross k_{E2} values of Table 4.1 for this factor by converting them to k_{E2} values per *single* H atom in a β-position. However, these numbers—"k_{E2} (per β-H atom)" in Table 4.1—still increase in the series R_{prim}—Br → R_{sec}—Br → R_{tert}—Br. This second effect is due to the fact that the E2 eliminations considered lead to C=C double bonds with an increasing number of alkyl substituents: EtBr results to an unsubstituted, *i*PrBr to a monosubstituted, and *tert*-BuBr to a disubstituted alkene. The stability of alkenes, as is well known, increases with the degree of alkylation. A certain fraction of this stability increase becomes noticeable in the transition state of E2 eliminations in the form of product development control: The rate constant k_{E2} (per β-H atom) is therefore smallest for the most unstable E2 product (ethene) and largest for the most stable E2 product (isobutene).

Table 4.2 compiles the chemoselectivities of E2 eliminations starting from Me—CH_2Br and the simplest bromides of types R_{prim}—CH_2Br and R_{sec}—CH_2Br. It shows how an increasing number of β-substituents influences the E2/S_N2 competition: the more numerous they are the greater is the fraction of E2 product. The reasons for this finding are the same as the reasons given in the discussion of Table 4.1 for the influence of the increasing number of α-substituents on the result of the E2/S_N2 competition. All of the corresponding quantities— k_{S_N2}–, gross k_{E2} and k_{E2} (per β-H atom) values—are contained in Table 4.2.

4.4.2 Base Effects on the E2/S_N2 Competition

Chemoselective E2 eliminations can be carried out with sterically hindered, sufficiently strong bases. Their bulkiness causes them to react with an H atom at the periphery of the molecule rather than at a C atom deep within the molecule. These bases are therefore called **nonnucleophilic bases.** The weaker nonnucleophilic bases include the bicyclic amidines DBN (diazabicyclononene) and DBU (diazabicycloundecene). These can be used to carry out chemoselective E2 eliminations even starting from primary and secondary alkyl halides and sulfonates (Figure 4.17).

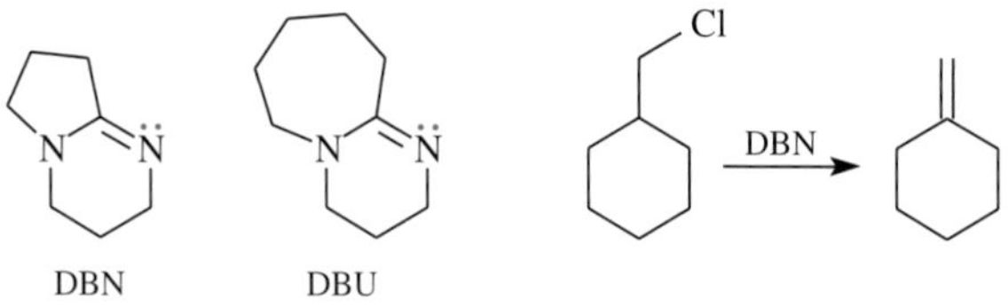

Fig. 4.17. Relatively weak nonnucleophilic bases; use in a chemoselective E2 elimination from a primary alkyl halide.

β-Eliminations of epoxides lead to allyl alcohols. For this reaction to take place, the strongly basic bulky lithium dialkylamides LDA (lithium diisopropylamide), LTMP (lithium tetramethylpiperidide) or LiHMDS (lithium hexamethyldisilazide) shown in Figure 4.18 are used. As for the amidine bases shown in Figure 4.17, the bulkiness of these amides guarantees that they are nonnucleophilic. They react, for example, with epoxides in chemoselective E2 reactions even when the epoxide contains a primary C atom that easily reacts with nucleophiles (see, e. g., Figure 4.18).

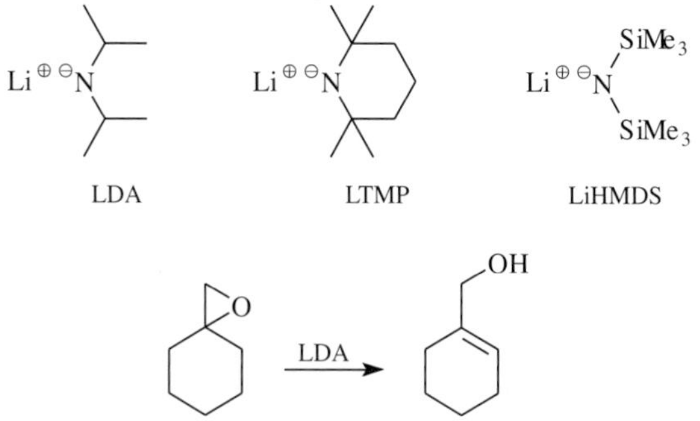

Fig. 4.18. Strong non-nucleophilic bases; use in a chemoselective E2 elimination from an epoxide.

4.4.3 A Stereoelectronic Effect on the E2/S_N2 Competition

(4-*tert*-Butylcyclohexyl)trimethylammonium iodide and potassium-*tert*-butoxide (KO*t*Bu) can undergo both an E2 and an S_N2 reaction (Figure 4.19). However, the reactivity and chemoselectivity differ drastically depending on whether the *cis*- or the *trans*-configured cyclohexane derivative is the substrate. The *cis*-isomer gives a 90:10 mixture of E2 and S_N2 products in a fast reaction. The *trans*-isomer reacts much more slowly and gives only the substitution product. According to Equation 4.3, these findings mean that $k_{E2,cis} : k_{S_N2,\,cis} =$ 90:10 and $k_{E2,trans} : k_{S_N2,\,trans} < 1{:}99$. A plausible assumption is that for the respective substitution reactions $k_{S_N2,\,cis} \approx k_{S_N2,\,trans}$. This, in turn, means that the opposite chemoselectivities of the two reactions in Figure 4.19 arise almost exclusively from the fact that $k_{E2,cis} >> k_{E2,trans}$.

How this gradation of the k_{E2} values comes about is easily understood by considering Figure 4.20 and realizing the following: in the most stable transition state of an E2 elimination the half-broken $^{\alpha}$C··· Het bond and the likewise half-broken $^{\beta}$C··· H bond are oriented *parallel* (Figure 4.20). This is because the hybridization change at $^{\alpha}$C and $^{\beta}$C from sp^3- to sp^2-hybridization has started in the transition state. As a result, $2p_z$-like AOs are being formed as $^{\alpha}$C and $^{\alpha}$C and their increasingly effective overlap establishes the π bond. In other words,

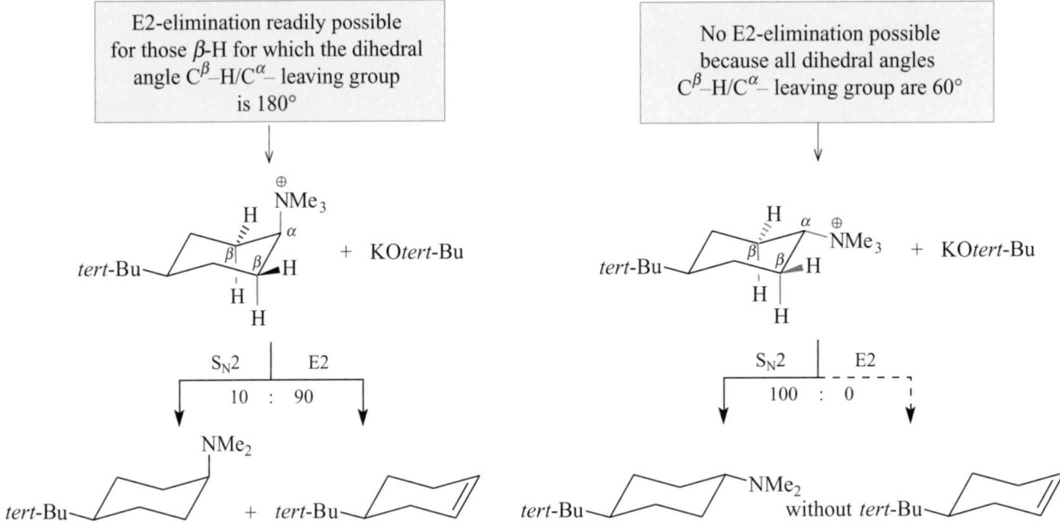

Fig. 4.19. E2/S_N2 competition in cyclohexanes with poor leaving groups locked in axial (left reaction) or equatorial (right reaction) position. Note that S_N2 reactions involve an N-bound methyl group instead of the N-bound cyclohexyl ring. The reason is the minimization of steric interactions in the S_N2 transition state (cf. "Tendencies and Rules" in Section 2.4.4)

coplanar $^{\alpha}$C··· Het and $^{\beta}$C··· H bonds stabilize the transition state of an E2 elimination through a π-like interaction. Interactions of this type are possible in the most stable transition state of *syn*-eliminations as well. When a substrate can assume both of these transition state geometries, the *anti*-elimination is always preferred (cf. the discussion in Figure 4.26). This is a consequence of the lower steric hindrance of the *anti*-transition state, in which the substituents at $^{\alpha}$C vs. $^{\beta}$C are nearly staggered. The sterically more hindered *syn*-transition state has a nearly eclipsed structure. In addition, there are reasons to believe that orbital overlap is better in the transition state of an *anti*-elimination.

Let us return to the finding that $k_{E2,\,cis} > k_{E2,\,trans}$ in the E2 eliminations from the isomeric tetraalkylammonium salts in Figure 4.19. The transition state geometries of these reactions are derived from the chair conformers shown in Figure 4.19. The stereostructure is fixed in each

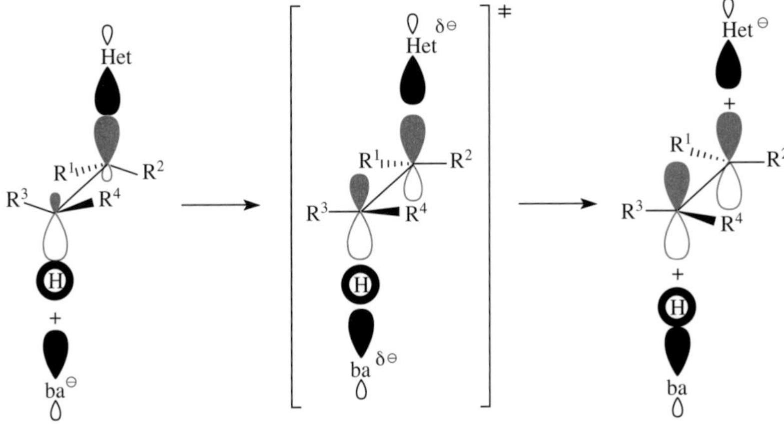

Fig. 4.20. π-Like bonding interactions in the transition state of *anti*-selective E2 eliminations; ba$^{\ominus}$ refers to the reacting base.

case by the equatorially disposed *tert*-butyl group, which acts as a conformational lock. The much higher elimination rate of the *cis* isomer is due to its ability to undergo an *anti*-elimination, because the $^{\beta}C—H$ and $C—NMe_3^+$ bonds exhibit a dihedral angle of 180°; they are antiperiplanar. In contrast, in the *trans*-configured ammonium salt, the corresponding dihedral angle is 60°, which is suitable neither for a *syn*- nor for an *anti*-elimination. The *trans*-substrate and KO*t*Bu can therefore react only via the S_N2 mode.

4.4.4 The Regioselectivity of E2 Eliminations

When E2 eliminations from secondary or tertiary substrates can give regioisomeric alkenes that differ from each other as Saytzeff and Hofmann products (i.e., in the degree of alkylation of their C=C bond), the Saytzeff product usually predominates because of product development control (see Section 4.1.4). The proportion of Hofmann product can be increased, however, by using bulky (Figure 4.21) and/or strong bases (Figure 4.22) or equipping the *starting* materials with poorer leaving groups than halides or sulfonates (Figure 4.23). On the other hand, *completely* Hofmann-selective E2 eliminations can be achieved in cyclic systems if the requirement of *anti*-selectivity gives rise to such a regiocontrol (Figures 4.24 and 4.25). In the following we want to explore this in detail.

Figure 4.21 shows how the bulkiness of the base influences the regioselectivity of the elimination of HBr from *tert*-amyl bromide. The sterically undemanding base $EtO^{\ominus}$ can react with the H atoms in all β-positions to the leaving group, irrespective of whether the H atom is bound to a primary or a secondary C atom. Thus, regioselectivity derives from product development control *alone*: the thermodynamically more stable Saytzeff product is formed preferentially, though not exclusively.

The sterically hindered base *tert*-$BuO^{\ominus}$ reacts with *tert*-amyl bromide in a different manner than $EtO^{\ominus}$ and in such a way that steric interactions are minimized (Figure 4.21). This makes the substructure C_{prim}—H a more suitable point of reaction than the substructure C_{sec}—H. If a C_{tert}—H were present, it would not react at all. As a consequence, the Hofmann product is produced predominantly, but not exclusively.

Figure 4.22 demonstrates related base effects towards an α-bromo carboxylic ester, with the regioselectivities being almost perfect as compared to the ones in Figures 4.21. Quinoline eliminates HBr from α-bromo isovaleric acid ester with Saytzeff selectivity. In contrast, KO-*tert*-Bu effects Hofmann-selective HBr elimination. The selectivity of the *tert*-$BuO^{\ominus}$ reaction

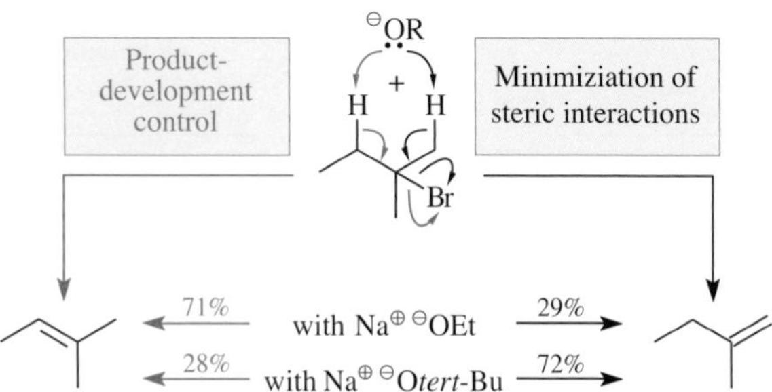

Fig. 4.21. Steric base effects on the Saytzeff/Hofmann selectivity of an E2 elimination. The small base $EtO^{\ominus}$ can attack the H atoms in both positions β to the leaving group, i.e., it does not matter whether the H atom is bound to a primary or secondary C atom. The regioselectivity therefore results *only* from product development control: the thermodynamically more stable.

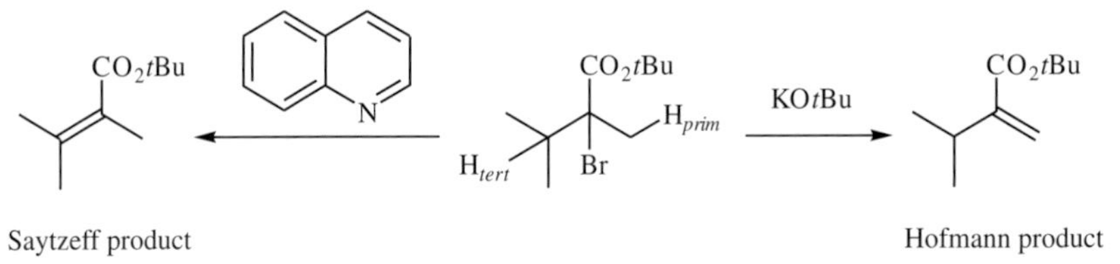

Fig. 4.22. Saytzeff product is produced preferentially although not exclusively. The reaction of the sterically hindered base tBuO⊖ is directed preferentially toward the more readily accessible H atoms at the primary C atom. Therefore, it mainly provides the Hofmann product.

depends on the steric effect shown in Figure 4.21. In Figure 4.22, this effect *increases* because the tertiary H atom, which would have to be eliminated on the way to the Saytzeff product, is virtually invulnerable towards *tert*-BuO⊖ due to the adjacent bulky $C(CH_3)Br(CO_2$*tert*-Bu). The reaction involving quinoline proceeds with much more Saytzeff selectivity than the reaction with EtO⊖ (cf. in Figure 4.21). This finding appears to be attributable to particularly strong product development control effects, which occur because quinoline, a very weak base, induces β elimination via a very late transition state.

Without changing the base, one can reverse the normal Saytzeff preference of the E2 elimination into a preference for the Hofmann elimination through a variation of the leaving group. This is especially true for eliminations from tetraalkylammonium salts R—NMe_3⊕I⊖ and trialkylsulfonium salts R—SMe_2⊕I⊖. Their positively charged leaving groups primarily acidify the H atoms in the α-position, but do the same to a lesser extent with the H atoms in the β-position. The latter effect is decisive. In addition, NMe_3 and SMe_2 are just as poor leaving groups in eliminations as in S_N reactions (cf. Section 2.3). According to Section 4.3, both propensities favor an E2 elimination via an $E1_{cb}$-like transition state. Obviously, a β-H atom participating in such an elimination must be sufficiently acidified. As is well known, C,H acidity decreases in the series C_{prim}—H > C_{sec}—H > C_{tert}—H. Therefore, in $E1_{cb}$-like E2 eliminations, the H atom is preferentially removed from the least alkylated C atom β to the leaving group. This regioselectivity corresponds to a preference for the Hofmann product, as shown in the bottom example in Figure 4.23. In addition, a steric effect probably acts in the same direction: the NMe_2^+ or SMe_2^+ groups in the substrate are relatively large because they are branched. Therefore, they allow a reacting base more ready access to the structural element C^{β}_{prim}—H relative to C^{β}_{sec}—H and to the structural element C^{β}_{sec}—H relative to C^{β}_{tert}—H.

Many cyclohexane derivatives undergo—in apparent contradiction to what has been said before—*perfectly* Hofmann-selective E2 eliminations even when they contain good leaving

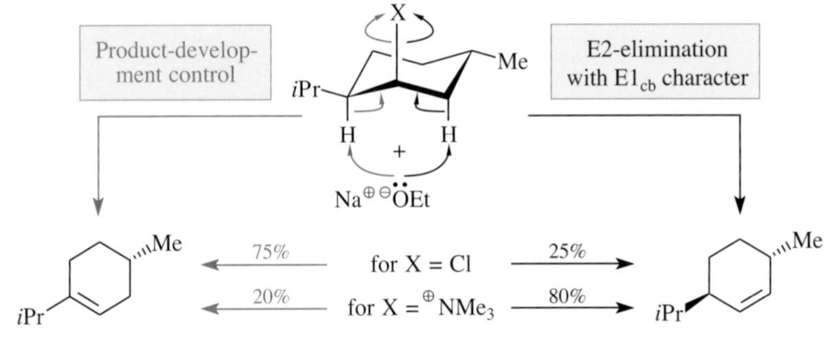

Fig. 4.23. Leaving-group effect on the Saytzeff/Hofmann selectivity of an E2 elimination. Poor, positively charged leaving groups react preferentially to give the Hofmann product via an E2 elimination with $E1_{cb}$ character.

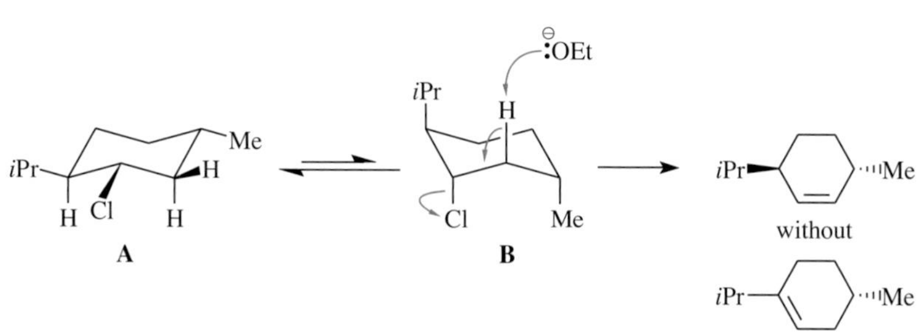

Fig. 4.24. *anti*-Selectivity of an E2 elimination as a reason for Hofmann selectivity.

groups and even when sterically modest bases are used. Figure 4.24 demonstrates this using menthyl chloride as an example. In substrates of this type, a stereoelectronic effect, which has already been discussed in another connection (Section 4.4.3), takes over regiocontrol: many E2 eliminations exhibit pronounced *anti*-selectivity. The only H/Cl elimination from menthyl chloride that is *anti*-selective leads to the Hofmann product. It takes place through the all-axial conformer **B** of substrate **A**, even though this conformer is present in only small equilibrium quantities.

Side Note 4.1: Regioselectivity of HBr Elimination from *trans*-1,2-Dibromocyclohexane

Due to the same compulsory *anti*-selectivity the potassium alkoxide-mediated HBr eliminations from *trans*-1,2-dibromocyclohexane ultimately results in the formation of 1,3-cyclohexadiene rather than 1-bromocyclohexene (Figure 4.25). The reason is that an *anti*-selective elimination occurs initially. *trans*-1,2-Dibromocyclohexane—except in very polar solvents—prefers a chair conformation with axial C–Br bonds (as it is only in this conformation that the

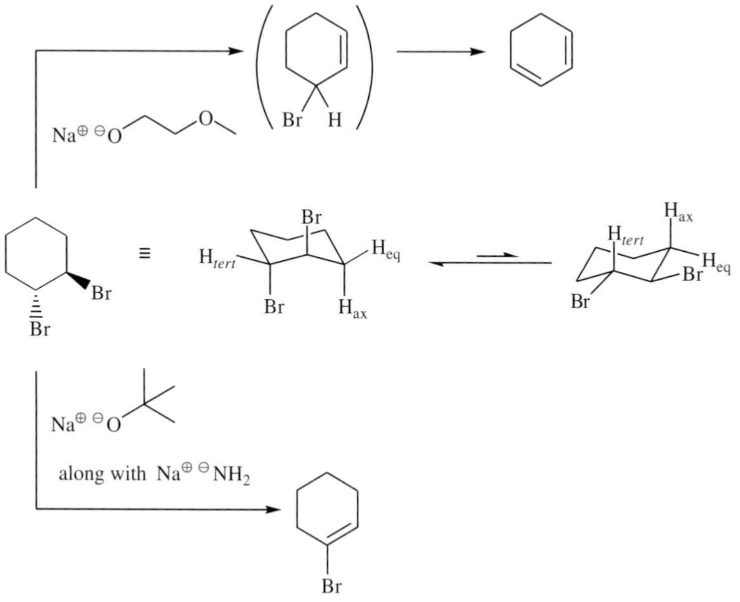

Fig. 4.25. *anti*-Selectivity of an E2 elimination as a reason for the chemoselective formation of a 1,3-diene (*top*) instead of a bromoalkene. Note: Starting from *trans*-1,2-dibromocyclohexane 1-bromocyclohexene can be obtained by HBr elimination (*bottom*). This can be achieved, e.g., by using hot pyridine in a sealed tube (not shown in the figure) or $NaNH_2$/NaO*tert*-Bu in THF (shown). The first method presumably benefits from a change in stereoelectronic control to product development control of regioselectivity (cf. the reaction with quinoline in Figure 4.22). The second method could be based on the preference of a six-membered cyclic and thus *syn*-transition state with minimal $Na^{\oplus}$/$ba^{\ominus}$ distance.

$C^{\delta\oplus}$–$Br^{\delta\ominus}$ bond dipoles are oriented antiparallel and thus able to avoid repulsion). If this chair conformer eliminates HBr, the dihedral angle between the $^{\alpha}$C–H and the $^{\beta}$C–Br that are to be broken on the way to 3-bromocyclohexene measures a favorable 180°. the dihedral angle between the H and Br atoms to be eliminated to afford 1-bromocyclohexene is an unfavorable 60°. As long as the reaction temperature is not too low and/or the potassium alkoxide not too undemanding sterically so that ether formation by the S_N2 mechanism becomes competitive, the 3-bromocyclohexene eliminates a second equivalent of HBr in another 1,2-elimination, forming 1,3-cyclohexadiene.

4.4.5 The Stereoselectivity of E2 Eliminations

Section 4.1.3 supplied essential information on the stereoselectivity of eliminations. We have learned that much depends on whether these eliminations—according to their mechanism—proceed with *anti*-selectivity, *syn*-selectivity or without *syn,anti*-selectivity. A more detailed discussion of the *syn,anti*-selectivity of E2 eliminations is worthwhile. Let us then revisit the *β*-eliminations of H and X (Het) bound to stereocenters in the substrate that we have already considered at the beginning of Section 4.1.3. With these eliminations stereoselectivity is 100% guaranteed as long as they are run as clean *anti*- or clean *syn*-eliminations. Let's take a look at Figure 4.26 and once again at Figure 4.25 in Side Note 4.1.

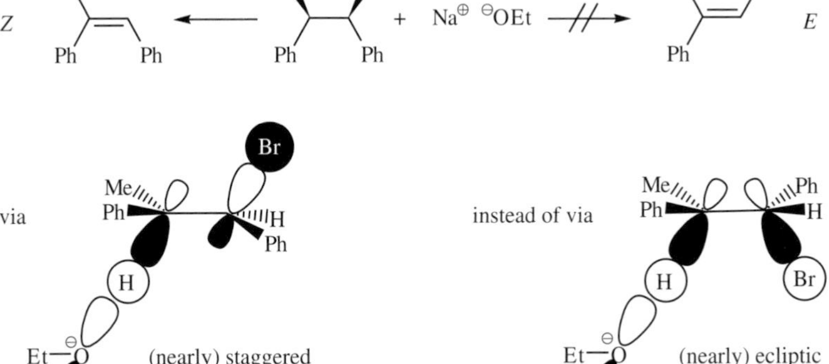

Fig. 4.26. *anti*-Selectivity of an E2 elimination as a reason for the stereoselective formation of a *Z*- instead of an *E*-configured alkene.

E2 eliminations as in Figure 4.26 usually proceed with *anti*- instead of *syn*-selectivity. This is true although both eliminations benefit from transition states with *π*-like bonding interactions between the MOs of coplanar, breaking $^{\alpha}$C-H- and $^{\beta}$C-X bonds (also cf. the discussion of Figure 4.20). The crucial factor is a steric effect that favors the *anti*- over the *syn*-transition state, since the spectator substituents at $^{\alpha}$C and $^{\beta}$C adopt a nearly staggered structure in the *anti*-transition state, while they are nearly eclipsed in the *syn*-transition state.

Side Note 4.2: Controllability of HBr Elimination from Dibromobutyric Acid Ester

The stereoselectivity of the HBr elimination from the dibromobutyric acid ethyl ester in Figure 4.27—proceeding by the E2 mechanism—is also consistent with the above observations even if, depending on the reaction conditions, different stereoselectivities are obtained: if the reaction is run with potassium carbonate in acetone, *anti*-elimination occurs. This selectivity may be deduced from the *E* configuration of the resulting bromocrotonic ester. Surprisingly, the same compound can also undergo a *syn*-selective bimolecular HBr elimination when the reaction is conducted in pentane and triethylamine is used as the base. The resulting bromocrotonic ester adopts *Z* configuration, which is only possible with *syn*-elimination.

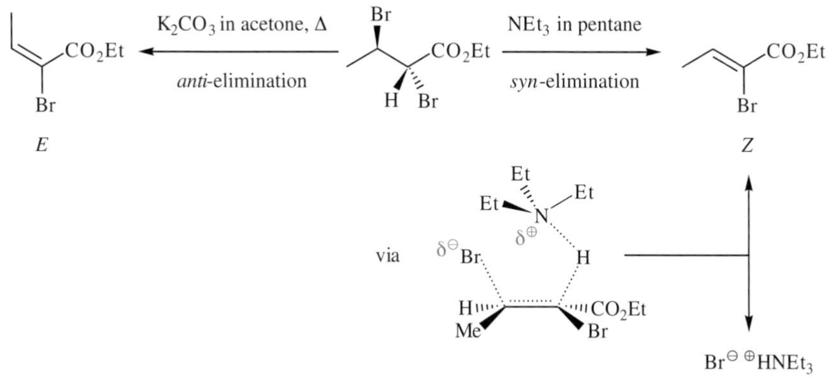

Fig. 4.27. *anti*-Selectivity of an E2 elimination of HBr as a reason for the exclusive formation of *E*-configured α-bromocrotonic acid ethyl ester, *syn*-selectivity of an E2 elimination of HBr as a reason for the preferential formation of a *Z*-configured α-bromocrotonic acid ethyl ester.

The mechanism of this preparatively valuable switch of the elimination stereochemistry is not understood completely. The most plausible explanation is to assume a quasi five-membered and thus *syn*-configured transition state as shown in Figure 4.27. Only this transition state ensures that in nonpolar pentane the elimination by-product triethylammonium hydrobromide is obtained as a contact ion pair instead of solvent-separated ions. In the latter case there would be virtually no solvation and they would be almost as unstable as "vacuum-separated" cation/anion pairs (cf. the discussion in Table 2.1.).

4.4.6 One-Pot Conversion of an Alcohol to an Alkene

The dehydration of alcohols to alkenes is possible when their OH group is converted into a good leaving group by protonation or by binding to a Lewis acid. However, eliminations in esters (e. g., sulfonates, dichlorophosphonates, amidosulfates) derived from the same alcohols usually take place under milder (i.e., nonacidic) conditions than the dehydrations of the alcohols themselves. Sometimes it can therefore be advantageous to perform such dehydrations via the corresponding esters. It is possible to prepare suitable esters from the alcohols for that purpose and subject them, without working them up, directly to the elimination. Through a mesylate produced in this way *in situ* it was possible, for example, to generate the highly sensitive dienediyne in Figure 4.28.

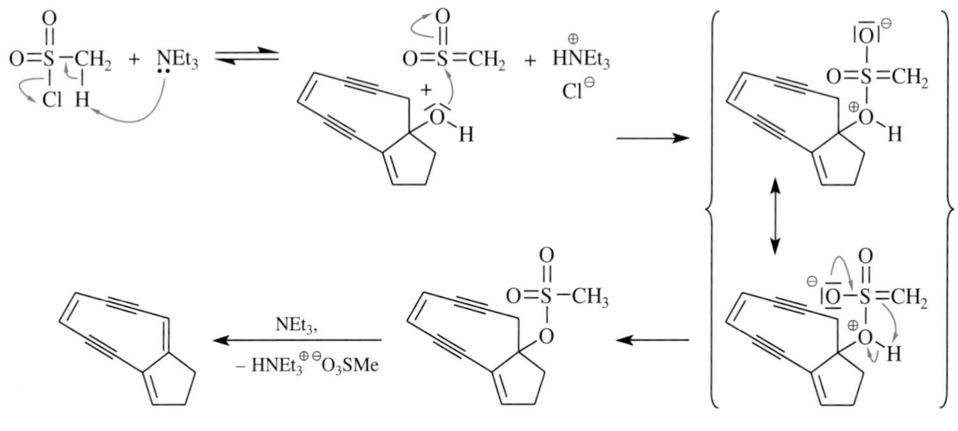

Fig. 4.28. One-pot procedure for the dehydration of alcohols: activation/elimination with MsCI/NEt$_3$.

In accord with the goal of “faster, milder, more selective” in organic synthesis, special reagents were also developed for the one-pot dehydration of particularly sensitive alcohols. One of these is the Burgess reagent (Figure 4.29). It activates the alcohol in the form of an aminosulfuric acid ester. The latter decomposes either via a cyclic transition state (i.e., by a *syn*-elimination, shown as an example in Figure 4.29), or through a NEt$_3$-induced and *anti*-selective process via an E2 mechanism, depending on the substrate.

Another special reagent for the one-pot dehydration of alcohols after an *in situ* activation is Martin’s persulfurane (Figure 4.30). It reacts with the OH group so that its O atom becomes part of a leaving group Ph$_2$S=O. For Martin’s persulfurane, the elimination mechanism depends on the exact substrate structure, too. Starting from secondary alcohols, *anti*-selective E2 eliminations take place. Tertiary alcohols can eliminate via an E2 (see example in Figure 4.30) or an E1 mechanism.

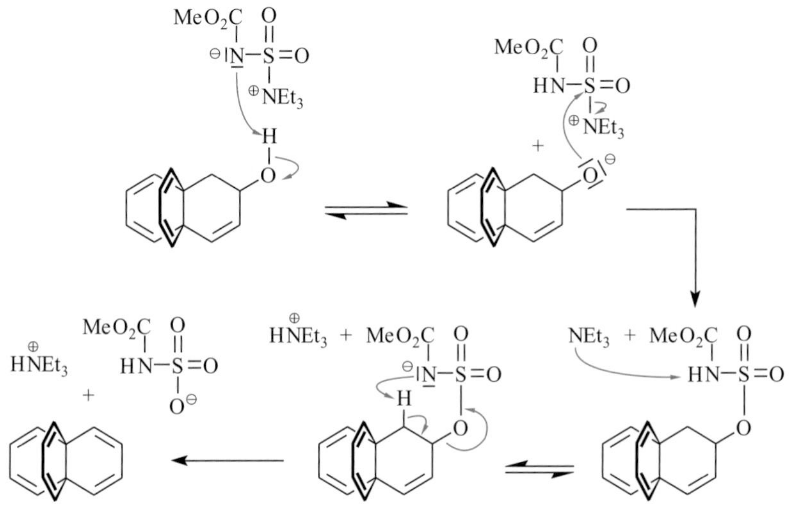

Fig. 4.29. Dehydration of an alcohol with the Burgess reagent (which in turn is prepared from O=C=N-SO$_2$-Cl by (1) addition of methanol and (2) reaction and subsequent deprotonation with triethylamine). The aminosulfuric acid ester intermediate decomposes via a cyclic transition state (*syn*-elimination).

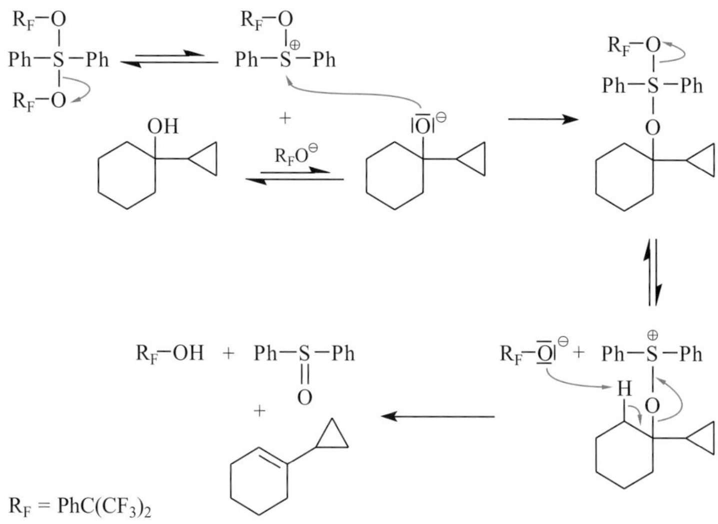

Fig. 4.30. Dehydration of an alcohol with Martin's persulfurane. The tertiary alcohol shown reacts via the E2 mechanism.

4.5 E1 Elimination of H/Het from R_{tert}—X and the E1/S_N1 Competition

E1 eliminations take place via carbenium ion intermediates. Consequently, they are observed in substrates that are prone to undergo heterolyses to a carbenium ion. E1 eliminations are also possible from substrates in which Brønsted or Lewis acids facilitate the formation of carbenium ions. In other words, E1 eliminations are possible with substrates and under conditions under which S_N1 reactions can also take place (Section 2.5).

The outcome of this competition depends only on how the carbenium ion intermediate reacts. If it encounters a poor nucleophile, the E1 product is produced. If it encounters a good nucleophile, it reacts with it to form the S_N1 product.

In general, a high reaction temperature favors the E1 at the expense of the S_N1 reaction. Only the elimination has large positive reaction entropy because a greater number of species are produced than were available at the start. Thus, the change in translational entropy ΔS increases during the reaction and the $T\Delta S$ term in $\Delta G = \Delta H - T\Delta S$ becomes more and more negative at high temperatures and thereby more and more noticeable. According to the Hammond postulate, this thermodynamic effect is felt to some extent in the transition state of the E1 elimination. Through this lowering of $\Delta G^{\ddagger}$ the elimination rate is increased.

4.5.1 Energy Profiles and Rate Laws for E1 Eliminations

tert-Butyl bromide undergoes an E1 elimination (Figure 4.31) when it is heated in a polar medium in the absence of a good nucleophile; it leads to isobutene and HBr in two steps. The

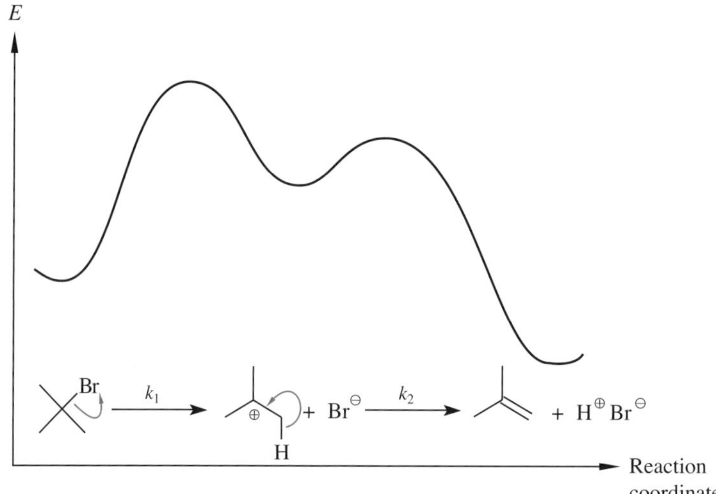

Fig. 4.31. The E1 elimination *tert*-BuBr → isobutene + HBr and its energy profile.

first step, the heterolysis to carbenium and bromide ions, is rate-determining. The second step is the deprotonation of the carbenium ion to the alkene. In Figure 4.31, and in the associated kinetic equations (Equations 4.4–4.6) it is shown as a unimolecular reaction. The possibility of a bimolecular deprotonation step, in which either the bromide ion or the solvent participates as a base, should not be ignored a priori. Taking into account Figure 4.34, a related E1 elimination in which the carbenium ion is definitely not subject to a bimolecular deprotonation by $Br^{\ominus}$, we can conclude that in the present case the deprotonation is unimolecular. By the way, from a preparative point of view, the E1 elimination in Figure 4.31 can also be used to obtain anhydrous hydrogen bromide.

Take a look at the derivation of the rate law for this E1 elimination:

$$\frac{\mathrm{d}\,[\text{alkene}]}{\mathrm{d}t} = k_2[\mathrm{R}^+] \tag{4.4}$$

$$\frac{\mathrm{d}[\mathrm{R}^+]}{\mathrm{d}t} = k_1[\mathrm{RBr}] - k_1[\mathrm{Y}^-]$$

$$= 0 \text{ because of the steady state approximation}$$

$$\Rightarrow [\mathrm{R}^+] = \frac{k_1}{k_2}[\mathrm{RBr}] \tag{4.5}$$

Inserting Eq. 4.5 into Eq. 4.4 yields

$$\frac{\mathrm{d}\,[\text{alkene}]}{\mathrm{d}t} = k_1[\mathrm{RBr}] \tag{4.6}$$

Equation 4.6 shows the **uni**molecularity of this reaction (from which the designation "E**1** elimination" is derived).

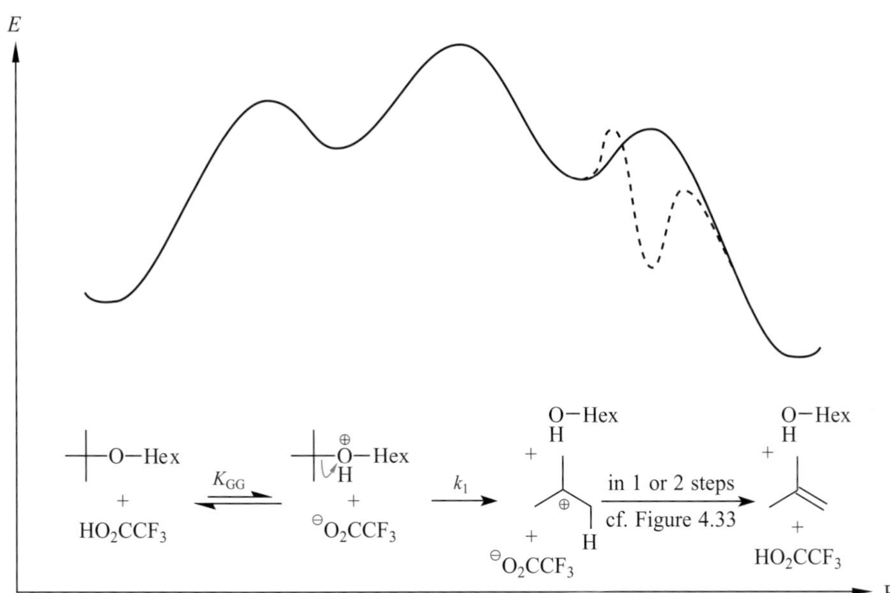

Fig. 4.32. The acid-catalyzed E1 elimination of *tert*-butyl hexyl ether to isobutene and hexanol and its energy profile. K_{eq} refers to the equilibrium constant of the acid/base reaction.

Tertiary alcohols, tertiary ethers, or carboxylic acid esters of tertiary alcohols can undergo E1 eliminations, but only in the presence of Brønsted or Lewis acids. Anyone who has prepared a tertiary alkoxide by a Grignard reaction and treated the crude reaction mixture with HCl and obtained the alkene knows that tertiary alcohols can be converted into alkenes even with dilute hydrochloric acid.

A similar E1 elimination is the cleavage of *tert*-butyl hexyl ether in trifluoroacetic acid/methylene chloride (CF_3CO_2H/CH_2Cl_2) to isobutene and hexanol (Figure 4.32). This reaction is an example of how primary or secondary alcohols can be extricated from their *tert*-butyl ether under mild conditions. The elimination in Figure 4.32 proceeds in three or four steps. In an equilibrium reaction, the ether is protonated to give the oxonium ion. The second step is rate-determining. It is a regioselective heterolysis to the *tert*-butyl cation, the more stable of the two carbenium ions that might, in principle, form. In addition, 1-hexanol is released as the leaving group. "Option **A**" in Figure 4.33 shows how this *tert*-butyl cation continues to react in a final third bimolecular step to give the alkene. "Option **B**" presents an alternative conversion of the *tert*-butyl cation to isobutene by unimolecular proton cleavage. In this case a fourth step would follow, namely the combination of the proton and the trifluoroacetate ion to yield trifluoroacetic acid. It is unclear which of these options depicts reality and it might be a case of hair-splitting.

The kinetic analysis of the E1 elimination of Figure 4.32 must be conducted for both options **A** and **B** for the further reaction of the *tert*-butyl cation:

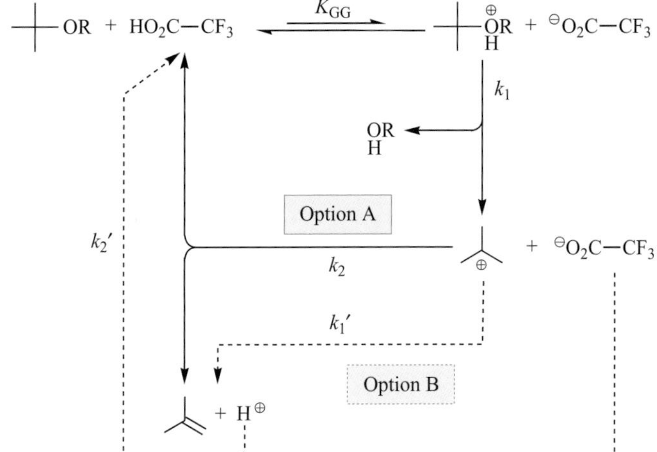

Fig. 4.33. Mechanistic details on the final phase of acid-catalyzed E1 elimination of tert-butylhexyl ether to isobutene and hexanol shown in Figure 4.32.

- Kinetic analysis of option **A**:

$$\frac{d\,[\text{Alkene}]}{dt} = k_2 \left[\text{Me}_3\text{C}^{\oplus}\right]\left[{}^{\ominus}\text{O}_2\text{C-CF}_3\right] \tag{4.7}$$

$$\frac{d\left[\text{Me}_3\text{C}^{\oplus}\right]}{dt} = k_1 \left[\text{Me}_3\text{C-}\overset{\oplus}{\text{O}}\text{R(H)}\right] - k_2 \left[\text{Me}_3\text{C}^{\oplus}\right]\left[{}^{\ominus}\text{O}_2\text{C-C}\right.$$

= 0 because of the steady state approximation

$$\Longrightarrow k_2 \left[\text{Me}_3\text{C}^{\oplus}\right]\left[{}^{\ominus}\text{O}_2\text{C-CF}_3\right] = k_1 \left[\text{Me}_3\text{C-}\overset{\oplus}{\text{O}}\text{R(H)}\right] \tag{4.8}$$

- Kinetic analysis of option **B**:

$$\frac{d\,[\text{Alkene}]}{dt} = k_1' \left[\text{Me}_3\text{C}^{\oplus}\right] \tag{4.9}$$

$$\frac{d\left[\text{Me}_3\text{C}^{\oplus}\right]}{dt} = k_1 \left[\text{Me}_3\text{C-}\overset{\oplus}{\text{O}}\text{R(H)}\right] - k_1' \left[\text{Me}_3\text{C}^{\oplus}\right]$$

= 0 because of the steady state approximation

$$\Longrightarrow k_1' \left[\text{Me}_3\text{C}^{\oplus}\right] = k_1 \left[\text{Me}_3\text{C-}\overset{\oplus}{\text{O}}\text{R(H)}\right] \tag{4.10}$$

- Inserting Eq. 4.8 into Eq. 4.7 (Option **A**) and Eq. 4.10 into Eq. 4.9 (Option **B**) leads to one and the same equation:

$$\frac{d\,[\text{Alkene}]}{dt} = k_1 \left[\mathrm{{}^{t}Bu{-}\overset{\oplus}{O}R(H)} \right] \tag{4.11}$$

- The acid/base equilibrium (protonation of the ether) is constantly maintained. This implies

$$K_{GG} = \frac{\left[\mathrm{{}^{t}Bu{-}\overset{\oplus}{O}R(H)} \right]\left[{}^{\ominus}O_2C{-}CF_3 \right]}{\left[\mathrm{{}^{t}Bu{-}OR} \right]\left[HO_2C{-}CF_3 \right]} \tag{4.12}$$

- In addition, the acid/base equilibrium of trifluoroacetic acid must be established, which is expressed in Equations 4.13 and 4.14:

$$HO_2C{-}CF_3 \overset{K_a}{\rightleftharpoons} {}^{\ominus}O_2C{-}CF_3 + H^{\oplus}$$

$$K_a = \frac{\left[{}^{\ominus}O_2C{-}CF_3 \right]\left[H^{\oplus} \right]}{\left[HO_2C{-}CF_3 \right]} \tag{4.13}$$

$$\Longrightarrow \frac{\left[{}^{\ominus}O_2C{-}CF_3 \right]}{\left[HO_2C{-}CF_3 \right]} = K_a \cdot \frac{1}{\left[H^{\oplus} \right]} \tag{4.14}$$

- Inserting Eq. 4.14 into Eq. 4.12 yields:

$$K_{GG} = \frac{\left[\mathrm{{}^{t}Bu{-}\overset{\oplus}{O}R(H)} \right]}{\left[\mathrm{{}^{t}Bu{-}OR} \right]} \cdot K_a \cdot \frac{1}{\left[H^{\oplus} \right]}$$

Solving the equation for the cationic concentration yields:

$$\left[\mathrm{{}^{t}Bu{-}\overset{\oplus}{O}R(H)} \right] = \frac{K_{GG}}{K_a} \cdot \left[H^{\oplus} \right]\left[\mathrm{{}^{t}Bu{-}OR} \right] \tag{4.15}$$

- If Eq. 4.15 is inserted into Eq. 4.11 the rate law takes the form of:

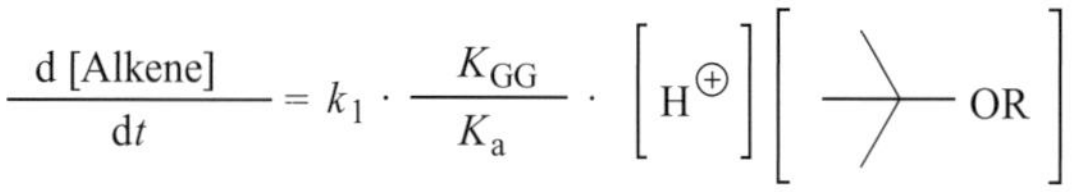

$$\frac{d\,[\text{Alkene}]}{dt} = k_1 \cdot \frac{K_{GG}}{K_a} \cdot \left[H^{\oplus}\right]\left[\text{—OR}\right] \tag{4.16}$$

- It remains to explain the meaning of the $\frac{K_{GG}}{K_a}$ ratio, which can be managed by dividing Eq. 4.12 by Eq. 4.13:

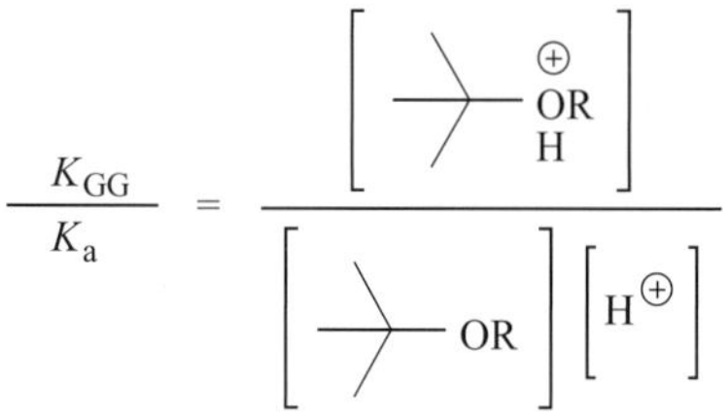

$$\frac{K_{GG}}{K_a} = \frac{\left[\text{—}\overset{\oplus}{O}RH\right]}{\left[\text{—OR}\right]\left[H^{\oplus}\right]} \tag{4.17}$$

This ratio corresponds to the equilibrium constant for the reaction

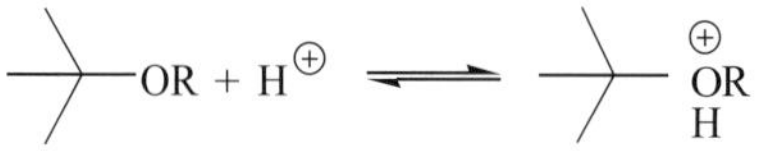

$$\text{—OR} + H^{\oplus} \rightleftharpoons \text{—}\overset{\oplus}{O}RH$$

and thus describes the basicity of the substrate. Inserting the resulting Eq. 4.17 into Eq. 4.16 the following expression is obtained for the rate law:

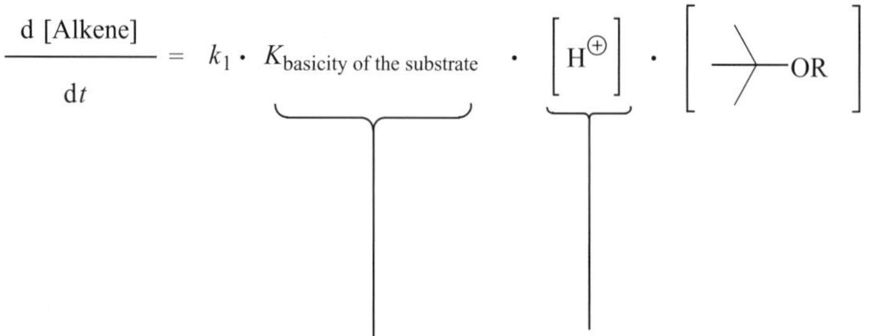

$$\frac{d\,[\text{Alkene}]}{dt} = k_1 \cdot K_{\text{basicity of the substrate}} \cdot \left[H^{\oplus}\right] \cdot \left[\text{—OR}\right] \tag{4.18}$$

i.e., *much* acid accelerates the reaction
and
strong acid accelerates the reaction

i.e. basic ether
eliminates more rapidly than
less basic ether

e.g., $\text{—}OCH_2CH_3$ more rapidly than $\text{—}OCH_2CF_3$

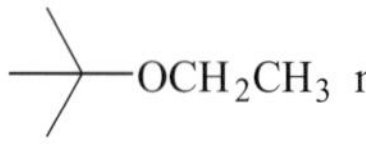

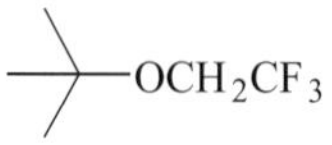

As the acid is not consumed in the course of the elimination, the concentration of protons does not vary with time. This is why the first three terms in Eq. 4.18 can be combined into a

$$\frac{d\,[\text{Alkene}]}{dt} = const. \cdot \left[\text{tert-BuOR} \right] \quad (4.19)$$

i.e. unimolecularity
(at a given concentration of the acid)

The rate law of this E1 reaction in the form of Equation 4.19 reveals unimolecularity once more. However, it hides what the more detailed form of Equation 4.18 discloses: namely, that the elimination rate increases with increasing CF_3CO_2H concentration, which means that this is a *bimolecular* E1 elimination. From Equation 4.18 we can also see that the E1 rate increases at a given concentration of the acid when a more acidic acid is used. In the end Equation 4.18 implies that the rate of ether cleavage following the E1 mechanism increases with the basicity of the substrate.

4.5.2 The Regioselectivity of E1 Eliminations

When the carbenium ion intermediate of an E1 elimination can be deprotonated to give two regioisomeric alkenes, generally both of them are produced. If these alkene isomers are Saytzeff and Hofmann products, the first one is produced preferentially because of product development control. This is illustrated in Figure 4.34 by the E1 elimination from *tert*-amyl

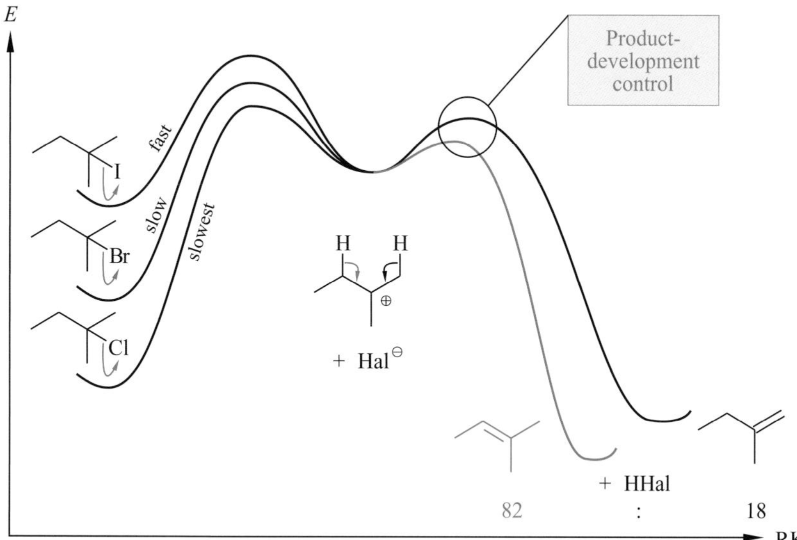

Fig. 4.34. Saytzeff preference of E1 eliminations from *tert*-amyl halides and energy profile. The Saytzeff:Hofmann preference amounts to 82:18 regardless of which halide ion accompanies the carbenium ion. This observation is explained most simply by the assumption that this halide ion is not involved in the deprotonation step forming the C=C double bond.

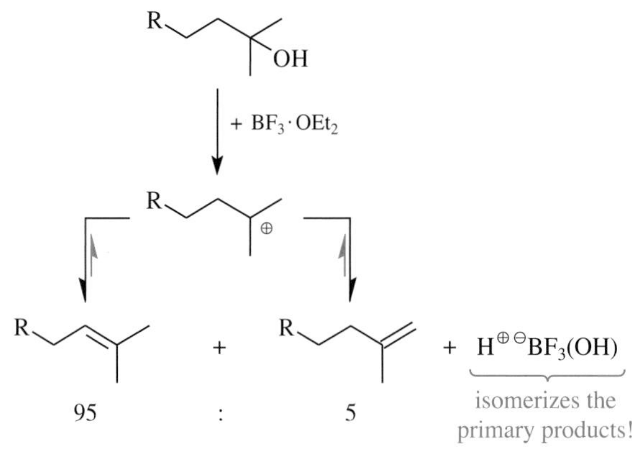

Fig. 4.35. Explanation for the particularly high Saytzeff selectivity of the E1 elimination from *tert*-amyl alcohol and its derivatives.

iodide, bromide, and chloride. The different halides form the *tert*-amyl cation and consequently the alkene, too, at *different* rates. Still, in each reaction the *same* 82:18 ratio of Saytzeff to Hofmann product is obtained. This can be explained most simply if the deprotonation of the *tert*-amyl cation proceeds unimolecularly in all cases, that is, without participation of the halide ion as a base.

It is interesting that E1 eliminations from tertiary alcohols carried out in acidic media give more Saytzeff product than E1 eliminations from the corresponding tertiary alkyl halides. As an example, compare the E1 eliminations from Figure 4.35 (amyl alcohols) with those in Figure 4.34 (amyl halides). The E1 eliminations via the—randomly—alkylated *tert*-amyl cations from Figure 4.35 give a 95:5 ratio of Saytzeff to Hofmann product. From the (unsubstituted) *tert*-amyl cations in Figure 4.34, the same compounds were produced (when we set R' = H) in the less biased ratio of 82:18.

The reason for the increased regioselectivity is that with tertiary alcohols as substrates there is no longer exclusively kinetic control. This is because the regioisomeric alkenes are no longer formed irreversibly. Instead they can be reprotonated, deprotonated again, and thereby finally equilibrated. In this way, the greatest part of the initially formed Hofmann product is converted to the more stable Saytzeff isomer. Product formation is thus the result of thermodynamic control. The fact that a greater regioselectivity results with thermodynamic control than with kinetic control and product development control is precisely what one would expect, considering the Hammond postulate.

The elimination in Figure 4.36 supports the idea that the alkenes initially formed from tertiary alcohols under E1 conditions can be reprotonated. The Saytzeff *and* the Hofmann products shown there can be protonated to provide the tertiary carbenium ion through which they were formed and also to a different tertiary carbenium ion. The consequence of this is that in the major product obtained after the final deprotonation, the C=C double bond is no longer located at the C atom that carried the OH group in the starting alcohol, but is moved one center away.

As discussed, the major product of the elimination in Figure 4.36 is produced under thermodynamic control. As such, it is surprising that it contains a trisubstituted C=C double bond, while the minor product has a tetrasubstituted one. The stability of a C=C double bond normally increases with each additional alkyl substituent. However, here an opposing effect dom-

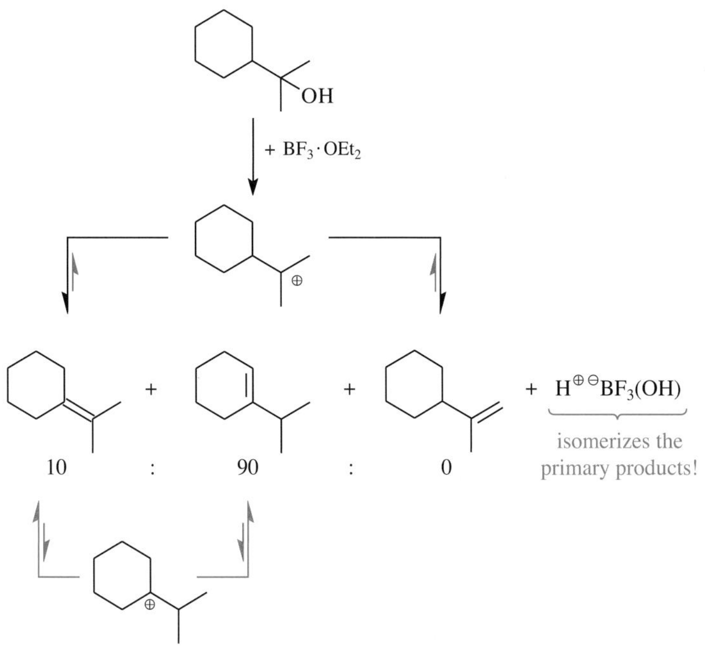

+ $BF_3 \cdot OEt_2$

10 : 90 : 0

+ $H^{\oplus}{}^{\ominus}BF_3(OH)$

isomerizes the primary products!

Fig. 4.36. E1 elimination with subsequent C=C migration.

inates. Consider the two isomers methylenecyclohexane (which contains a disubstituted C=C double bond) and 1-methylcyclohexene (which contains a trisubstituted C=C double bond) as model alkenes. Methylenecyclohexane is 2 kcal/mol higher in energy than 1-methylcyclohexene. This difference in stability is too great to be accounted for only by the difference between di-and trisubstitution of the C=C double bond. A second factor is that methylene-substituted cyclohexane has a greater ring strain than methyl-substituted cyclohexene. The same ring strain effect, possibly reinforced by greater steric hindrance, is probably responsible for the difference in stability between the alkene isomers in Figure 4.36.

4.5.3 E1 Eliminations in Protecting Group Chemistry

Acid treatment of *tert*-butyl ethers of primary and secondary alcohols produces alcohols by E1 eliminations. We saw this in the discussion of Figure 4.32. In a quite analogous way and under the same acidic conditions, *tert*-butyl esters and O-*tert*-butyl carbamates are cleaved. In accordance with that, *tert*-butyl ethers, esters, and carbamates are frequently used as protecting groups. An example of such an application in peptide synthesis is given in Figure 4.37. Of course, the elimination product (isobutene) is of no interest in this case, but the leaving group is, because it is the unprotected peptide.

Each of the E1 eliminations from the three *tert*-butyl groups shown takes place in three steps. In the first step the most basic O atom is protonated in an equilibrium reaction. Oxonium ions or carboxonium ions are produced. The second step is the heterolysis of the O-*tert*-Bu bond. Each leaving group is uncharged and therefore energetically acceptable: the *tert*-

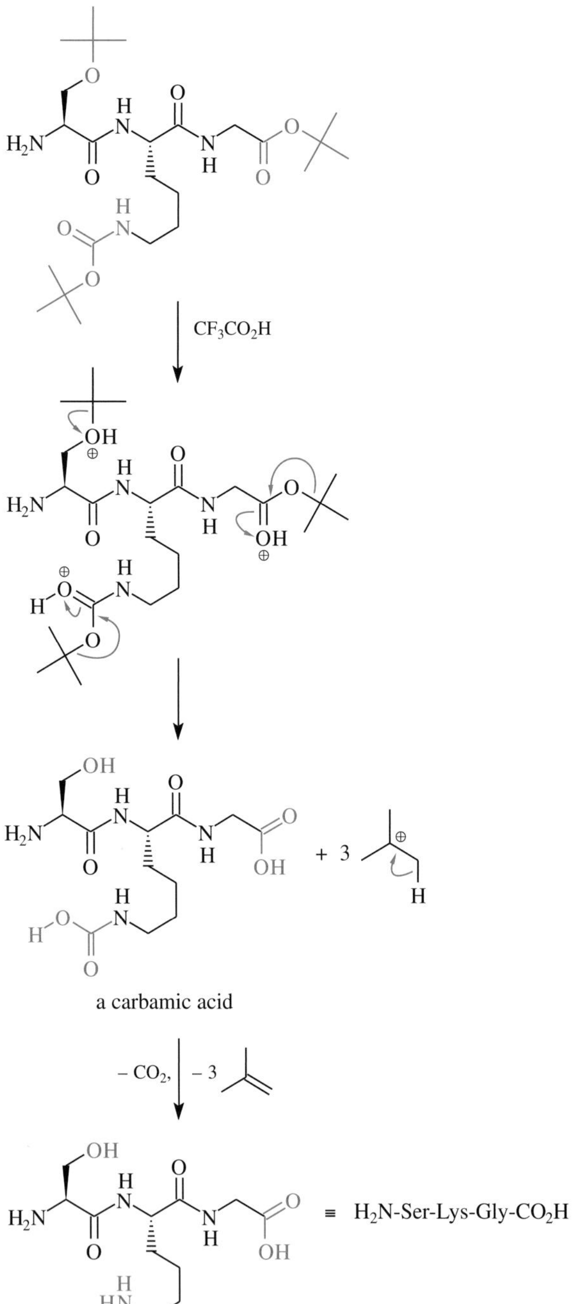

Fig. 4.37. Three E1 eliminations in the deprotection of a protected tripeptide. For the sake of brevity, a *single* formula in the second row of the scheme shows how the three *tert*-Bu—O bonds heterolyze. Of course, they are activated and broken *one after the other.* In the deprotection of the *tert*-butylated lysine side chain, the leaving group is a carbamic acid. Carbamic acids decarboxylate spontaneously (Figure 8.3, 8.5 and 8.6), which explains the final transformation. The preparation of the protected tripeptide is shown in Figure 4.41.

butyl ether of the serine side chain gives an alcohol, the *tert*-butylcarbamate of the lysine side chain gives a carbamic acid (which subsequently decarboxylates), and the *tert*-butyl ester of the glycine moiety gives a carboxylic acid. The other product of the heterolyses is the well-stabilized *tert*-butyl cation, which is deprotonated in the last step to give isobutene.

4.6 $E1_{cb}$ Eliminations

4.6.1 Unimolecular $E1_{cb}$ Eliminations: Energy Profile and Rate Law

Knoevenagel reactions (for their mechanism, see Section 13.4.2) end with H_2O being eliminated from the initially formed alcohol in the basic medium. The elimination takes place via an $E1_{cb}$ mechanism, an example of which is shown in Figure 4.38. In a fast exergonic reaction, nitroalcohol shown is deprotonated quantitatively to give a nitronate. In the following slower second reaction step, an $OH^{\ominus}$ group leaves, and a nitroalkene is produced. It is preferentially formed as the *trans*-isomer because of product development control.

The following derivation shows the rate law for this elimination as Equation 4.22. The term $[\text{nitroalcohol}]_0$ refers to the initial concentration of the nitroalcohol.

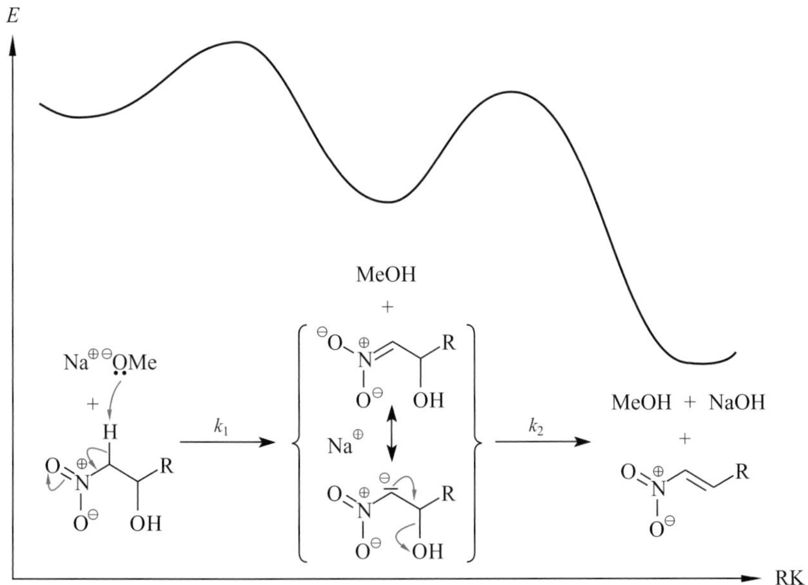

Fig. 4.38. A unimolecular $E1_{cb}$ elimination and its energy profile.

$$\frac{d\,[\text{alkene}]}{dt} = k_2[\text{nitronate}] \tag{4.20}$$

Because $k_1 \gg k_2$ and because of the stoichiometry, we have:

$$[\text{nitronate}] = [\text{nitroalcohol}]_0 - [\text{alkene}] \tag{4.21}$$

Inserting Eq. 4.21 into Eq. 4.20 yields

$$\frac{d\,[\text{alkene}]}{dt} = k_2([\text{nitroalcohol}]_0 - [\text{alkene}]) \tag{4.22}$$

This rate law contains only one concentration variable—the term $[\text{nitroalcohol}]_0$ is, of course, a constant and not a variable—and thus refers to a **uni**molecular process. This elimination is therefore designated E**1**$_{cb}$.

4.6.2 Nonunimolecular E1$_{cb}$ Eliminations: Energy Profile and Rate Law

At the end of an aldol condensation (for the mechanism, see Section 13.4.1), water is eliminated from a *β*-hydroxy carbonyl compound. This takes place according to the mechanism given in Figure 4.39. In a fast but endergonic and consequently reversible reaction, the *β*-hydroxy carbonyl compound is deprotonated to give an enolate. In the subsequent rate-determining reaction step an $OH^{\ominus}$ group is expelled, and an *α*,*β*-unsaturated carbonyl compound, the elimination product, is produced. If it were formed irreversibly, it would be produced preferentially as the *trans*-isomer because of product development control. However, in aldol con-

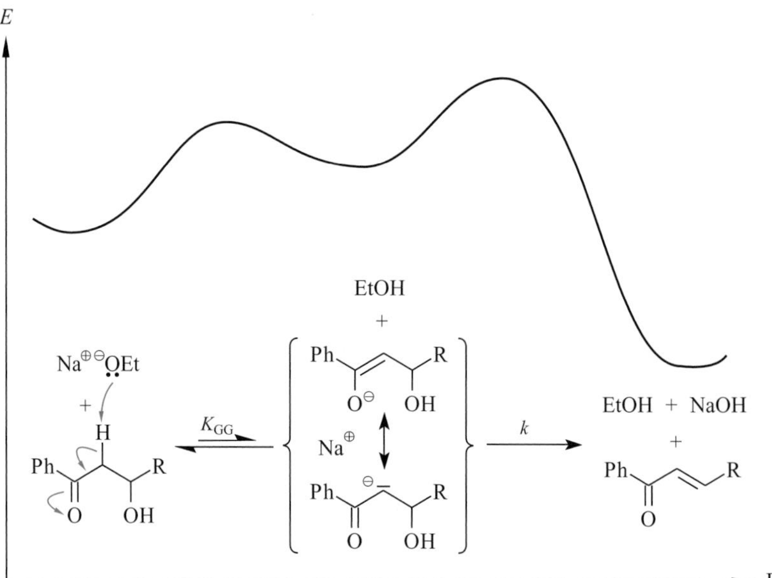

Fig. 4.39. A bimolecular E1$_{cb}$ elimination and its energy profile; K_{eq} refers to the constant of the acid/base equilibrium.

densations, this stereoselectivity is reinforced by thermodynamic control. Under the reaction conditions, any initially formed *cis*-isomer would be isomerized subsequently to the *trans*-isomer. Therefore, the *trans*-selectivity of aldol condensations is in general due to thermodynamic control.

The rate law for the $E1_{cb}$ mechanism in Figure 4.39 results as Equation 4.25 from the following derivation.

$$\frac{d\,[\text{alkene}]}{dt} = k[\text{enolate}] \tag{4.23}$$

The enolate concentration follows from the equilibrium condition

$$[\text{enolate}] = K_{eq}\,[\text{aldol}]\,[\text{NaOEt}] \tag{4.24}$$

Inserting Eq. 4.24 into Eq. 4.23 yields

$$\frac{d\,[\text{alkene}]}{dt} = k \cdot K_{eq}[\text{aldol}]\,[\text{NaOEt}] \tag{4.25}$$

Thus, this elimination takes place *bimolecularly*. Therefore, the designation $E1_{cb}$ for this mechanism is justified only if one thinks of the unimolecularity of the rate-determining step.

4.6.3 Alkene-Forming Step of the Julia-Lythgoe Olefination

α-Lithiated primary alkylphenylsulfones can be added to aldehydes (Section 11.4). A lithium alkoxide is obtained as the primary product, as with the addition of other C nucleophiles (Section 10.5). If acetic anhydride is added, a (β-acetoxyalkyl) phenylsulfone forms as a mixture of two diastereomers (*syn*-**A** and *anti*-**A**, Figure 4.40). If one treats this mixture with sodium amalgam in methanol, the result is a Het^1/Het^2 elimination. Each diastereomer of the acetoxysulfone **A** gives the same mixture of a *trans*- and a *cis*-alkene **C**. The *trans*-isomer predominates distinctly so that this so-called Julia-Lythgoe reaction has become an important synthesis for *trans*-alkene (for a detailed discussion cf. Section 11.4).

The mechanism of this elimination was elucidated in the mid-1990s. Initially, the Na/Hg_x dissolves by liberating an electron that is transferred to the solvent methanol (Figure 4.40). Ultimately, this initiates its reduction to elemental hydrogen and the simultaneous formation of sodium methanolate. While the resulting hydrogen escapes from the reaction mixture, and therefore is "harmless," the sodium methanolate remains and acts as a *base* towards the (phenylsulfonyl) acetates *syn*- and *anti*-**A** by reversibly deprotonating them to form a phenylsulfonyl-stabilized, β-acetoxylated "carbanion" **D**. This is followed by rapid elimination of an acetate ion. Whether or not the carbanion **D** is pyramidalized as shown in **E**, the intermediate adopts a conformation that minimizes steric interactions, as shown. As a result, the elimination product **B** has *E* configuration. **B** reacts *in situ* with more Na/Hg_x to give the desired alkene *trans*-**C** (for the mechanism see Figure 17.85).

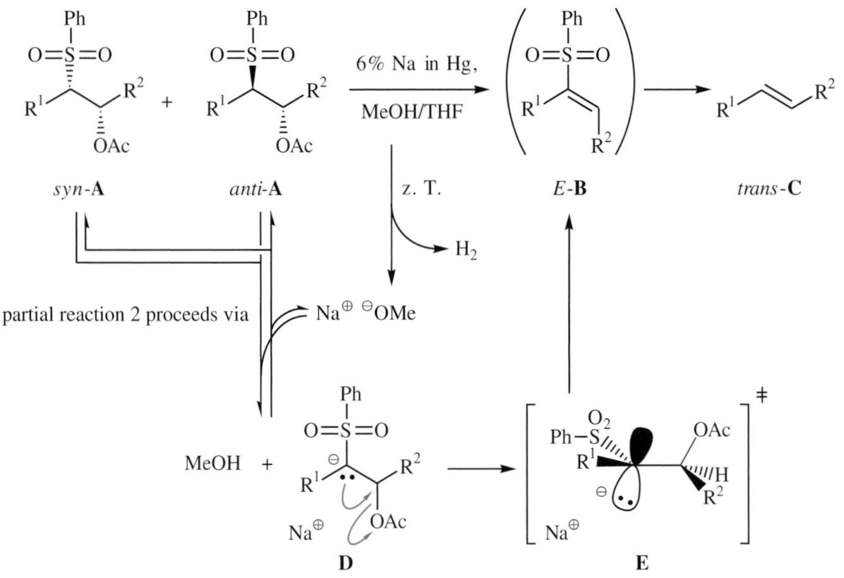

Fig. 4.40. Second step of the Julia–Lythgoe synthesis of *trans*-alkenes (first step: cf. Figure 11.22): stereoconvergent reduction of the sulfonylacetates *syn*- and *anti*-A to the uniformly configured alkene *trans*-C. The sequence starts with an E2 elimination yielding the alkenylsulfone *E*-**B** (mechanistical details as given), which is followed by the reduction to the final product (mechanistic analysis cf. Figure 17.85)

4.6.4 $E1_{cb}$ Eliminations in Protecting Group Chemistry

In peptide synthesis, functional groups in the amino acid side chains are often protected with acid-labile protecting groups (Section 4.5.3). The tripeptide in Figure 4.41 contains, for example, a serine *tert*-butyl ether and an L-lysine ε-protected as an O-*tert*-butyl carbamate. In the standard strategy of synthesizing oligopeptides from the C- to the N-terminus (cf. Section 6.4.3) the C terminus is either connected to the acid-labile Merrifield resin or to an acid-labile protecting group of lower molecular mass. For illustration purposes, Figure 4.41 shows a glycine *tert*-butyl ester as the C terminus of the mentioned tripeptide.

At the N-terminus of the growing peptide chain there is initially also a protecting group. This group must be removed as soon as the free NH_2 group is needed for condensation with the next activated amino acid. However, when the protecting group of the terminal nitrogen is removed, all side chain protecting groups as well as the C-terminal protecting group must remain intact. Because these are (only) acid-labile, the nitrogen protecting group should be base-labile. An N-bound fluorenylmethoxycarbonyl (Fmoc) group is ideal for this purpose (Figure 4.41). It can be removed from the N atom by the weakly basic reagent morpholine. This elimination takes place according to an $E1_{cb}$ mechanism. This is because the carbanion intermediate is easily produced because a fluorenyl anion (i.e., a dibenzoannulated cyclopentadienyl anion) is aromatic.

In the synthesis of oligonucleotides and oligodeoxynucleotides, the 5' ends of the nucleotide building blocks are protected with the acid-labile dimethoxytrityl group mentioned in Section 2.5.4. The phosphorous acid portion (when the synthesis takes place through phosphorous acid derivatives) or the phosphoric acid portion (which are present in all syntheses at least at the end) and the nucleotide bases are protected with base-sensitive protecting groups because these survive the acid-catalyzed removal of the dimethoxytrityl group undamaged. A

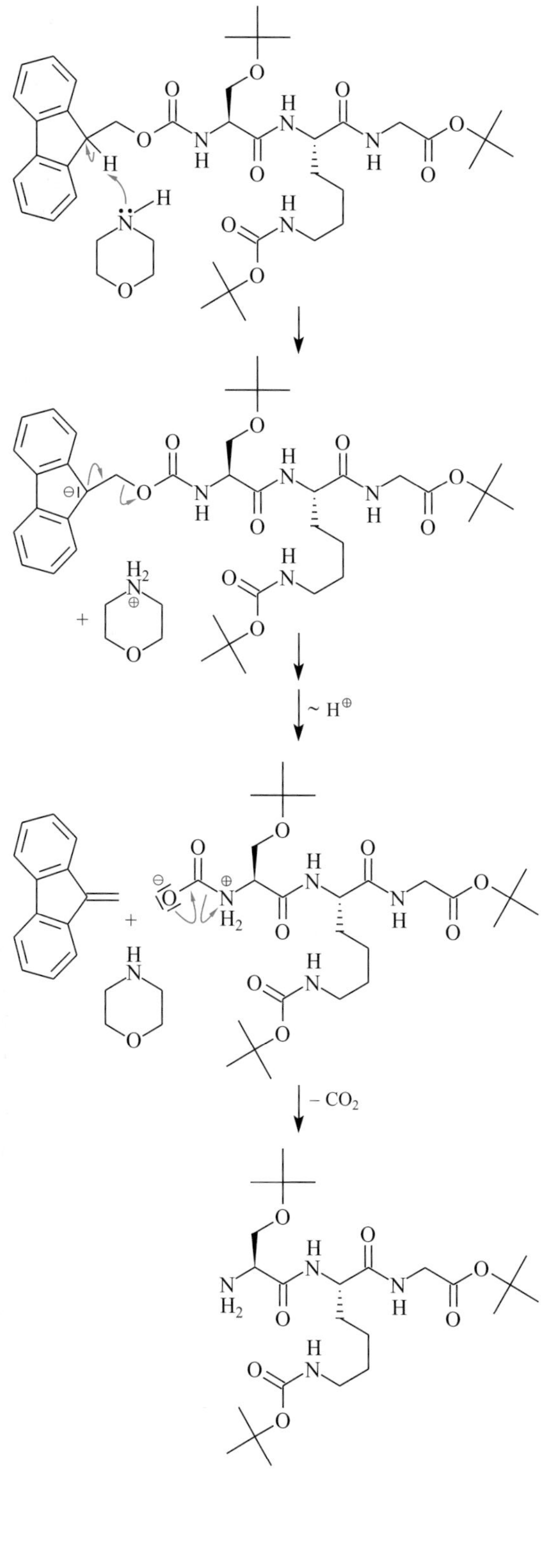

$\sim H^{\oplus}$

$- CO_2$

Fig. 4.41. Elimination of a fluorenylmethoxycarbonyl group (Fmoc group) from the terminal N atom of a protected tripeptide according to the $E1_{cb}$ mechanism.

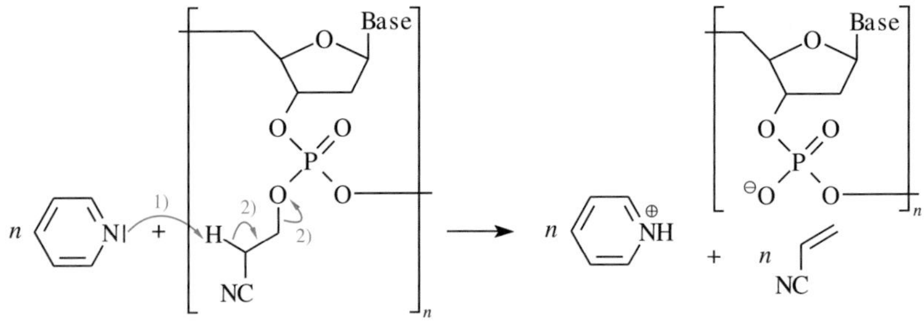

Fig. 4.42. Cleavage of a poly[(β-cyanoethyl)phosphate] according to the $E1_{cb}$ mechanism.

well-established base-labile protecting group for the phosphorus moiety (Figure 4.42) is the β-cyanoethyl ester. Its cleavage is due to an $E1_{cb}$ elimination via a cyano-stabilized "carbanion."

4.7 β-Eliminations of Het^1/Het^2

It was indicated already in Section 4.1.5 that Het^1/Het^2-eliminations offer possibilities for regio- and stereocontrol in the synthesis of alkenes that are quite different from those of H/Het-eliminations. It was also indicated that the value of these reactions for alkene synthesis depends on, among other things, how laborious it is to obtain the Het^1-and Het^2-containing substrates. Accordingly, in the next sections we will discuss such eliminations and the preparation of the elimination precursors as well.

4.7.1 Fragmentation of β-Heterosubstituted Organometallic Compounds

Haloalkanes and Li, Mg, or Zn can form organometallic compounds. How these conversions take place mechanistically will be discussed in Section 17.4.1. What is of interest here, however, is that if in organometallics of this type a leaving group is located in the β-position of the metal, the leaving group will generally be eliminated quickly together with the metal. In general, only $O^{\ominus}$- or $-N^{\ominus}C(=O)R$-groups resist this elimination; they are simply too poor as leaving groups.

A β-elimination of this type—often also referred to as fragmentation—takes place in the initiation of a Grignard reaction using ethylene bromide:

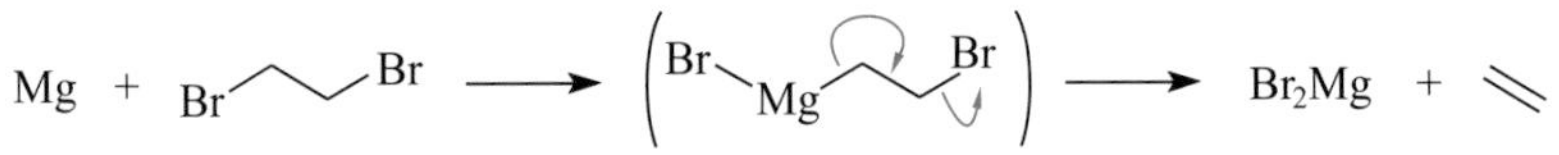

$$Mg + BrCH_2CH_2Br \longrightarrow (BrMgCH_2CH_2Br) \longrightarrow Br_2Mg + H_2C{=}CH_2$$

The purpose of this operation is not to release ethylene but to etch the Mg shavings.

A second example of this type of elimination is the generation of dichloroketene from trichloroacetyl chloride and zinc:

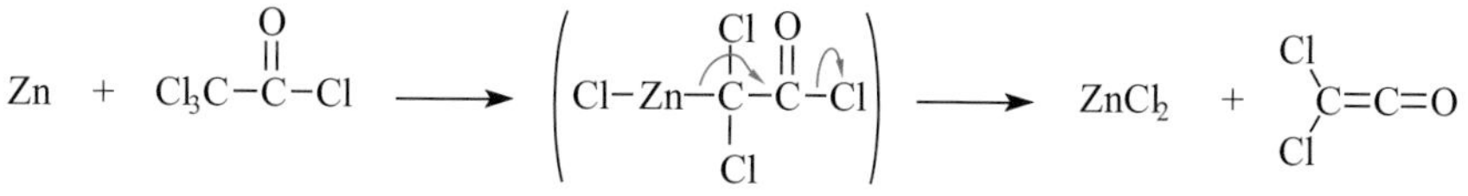

$$Zn + Cl_3C{-}C({=}O){-}Cl \longrightarrow \left(Cl{-}Zn{-}CCl_2{-}C({=}O){-}Cl \right) \longrightarrow ZnCl_2 + Cl_2C{=}C{=}O$$

You should remember that *β-heterosubstituted saturated organometallic compounds hardly ever exist.* On the contrary, they generally fragment through M⊕/Het⊖-elimination to an alkene:

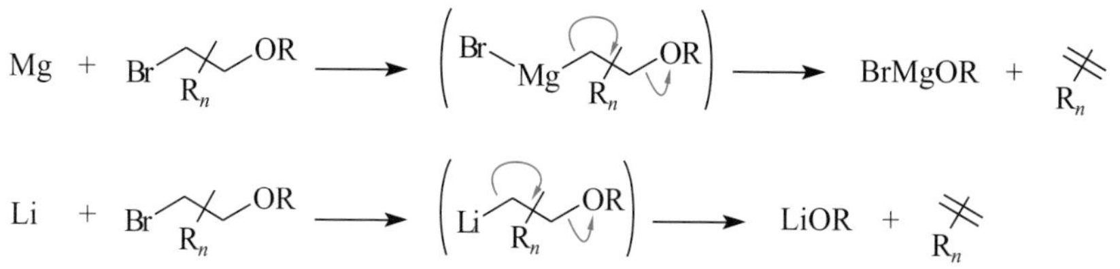

Mg + Br–CR–CR–OR ⟶ (Br–Mg–CR–CR–OR) ⟶ BrMgOR + alkene (R_n)

Li + Br–CR–CR–OR ⟶ (Li–CR–CR–OR) ⟶ LiOR + alkene (R_n)

4.7.2 Peterson Olefination

A β-hydroxysilane, like the one shown in Figure 4.43 (top, left), can be prepared stereoselectively (see Chapter 10). These compounds undergo stereoselective *anti*-elimination in the presence of acid and stereoselective *syn*-elimination in the presence of a base (Figure 4.43). Both reactions are referred to as Peterson olefinations. The stereochemical flexibility of the Peterson elimination is unmatched by any other Het¹/Het²-elimination discussed in this section.

The *acid-catalyzed* Peterson olefination is presumably an E2-elimination, that is, a one-step reaction. On the other hand, the base-induced Peterson olefination probably takes place via an intermediate. In all probability, this intermediate is a four-membered heterocycle with a pentavalent, negatively charged Si atom. This heterocycle probably decomposes by a [2+2]-cycloreversion just like the oxaphosphetane intermediate of the Wittig reaction (Section 4.7.3).

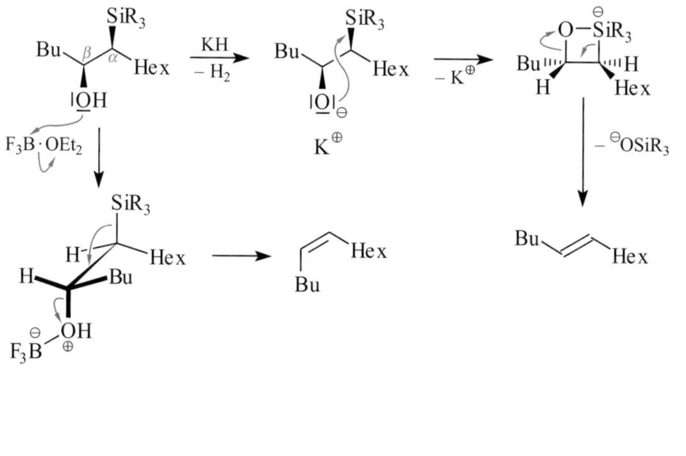

SiR₃ / Bu / β / α / Hex / OH — KH, – H₂ → Bu / SiR₃ / Hex / O⊖ K⊕ → – K⊕ → O–SiR₃⊖ / Bu / H / H / Hex → – ⊖OSiR₃ → Bu / Hex

F₃B·OEt₂ → SiR₃ / H / Hex / H / Bu / ⊖F₃B–OH⊕ → Hex / Bu

Fig. 4.43. Stereoselective Het¹/Het²-eliminations in the Peterson olefination. The base-induced Peterson olefination (top reaction) takes place as a *syn*-elimination, and the acid-catalyzed Peterson olefination takes place as an *anti*-elimination.

The same acidic or basic medium is also suitable for the stereoselective conversion of α-silylated tertiary alcohols into trisubstituted alkenes. Figure 4.9 showed an impressive series of examples of this.

4.7.3 Oxaphosphetane Fragmentation, Last Step of Wittig and Horner–Wadsworth–Emmons Reactions

According to Section 11.1.3, P-ylides and aldehydes first react in a [2+2]-cycloaddition to form a heterocycle, which is referred to as oxaphosphetane (Figure 4.44).

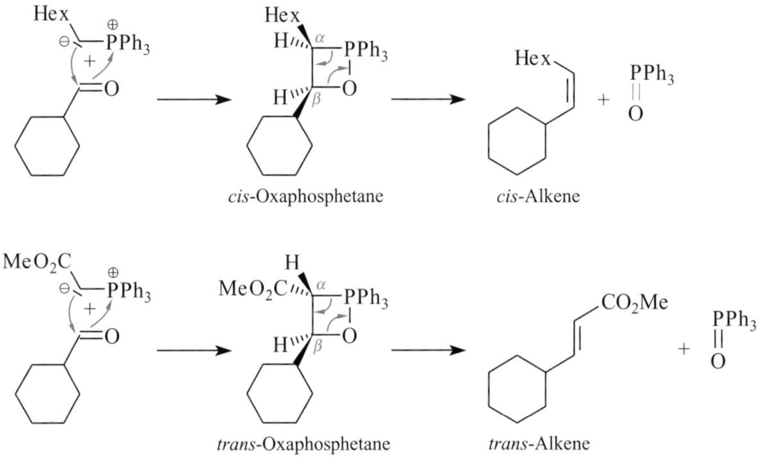

Hex ⊖ ⊕ PPh3 + O → Hex H α PPh3 H β O → Hex + PPh3 O

cis-Oxaphosphetane *cis*-Alkene

MeO2C ⊖ ⊕ PPh3 + O → H MeO2C α PPh3 H β O → CO2Me + PPh3 O

trans-Oxaphosphetane *trans*-Alkene

Fig. 4.44. *syn*-Selective eliminations from oxaphosphetanes in Wittig olefinations with unstabilized (upper row; gives *cis*-olefin) and stabilized P-ylides (bottom row; gives *trans*-olefin).

When the negative formal charge on the ylide C atom is resonance-stabilized by conjugating substituents, *trans*-configured oxaphosphetanes are produced (see Section 11.1.3). On the other hand, if the negative formal charge on the ylide C atom is not resonance-stabilized, *cis*-configured oxaphosphetanes are produced. Regardless of their stereochemistry, such oxaphosphetanes decompose rapidly and stereoselectively to give $Ph_3P{=}O$ and an alkene by a *syn*-elimination. In this way *trans*-oxaphosphetanes give conjugated *trans*-alkenes including, for example, *trans*-configured α,β-unsaturated esters. *cis*-Oxaphosphetanes lead to *cis*-alkenes.

A *syn*-elimination of $Ph_2MeP{=}O$ and simultaneous stereoselective alkene formation from an oxaphosphetane are shown in Figure 4.45 (note that this oxaphosphetane is not produced via a P-ylide). In Figure 4.45, this elimination is part of a an alkene inversion in which, via a four-step reaction sequence, an alkene such as *cis*-cyclooctene, a molecule with little strain, can be converted into its *trans*- and, in this case, more highly strained, isomer.

The first steps in this sequence are (1) *cis*-selective epoxidation, (2) ring opening of the epoxide with lithium diphenylphosphide in an S_N2 reaction, and (3) S_N2 reaction of the resulting alkyl-diarylphosphine with MeI to furnish a hydroxyphosphonium salt. In the fourth step, this salt is deprotonated to give a zwitterion or betaine with the substructure $^{\ominus}O{-}C{-}C{-}P^{\oplus}$ Ph_2Me. This zwitterion is less stable than the charge-free isomeric oxaphosphetane and consequently cyclizes to provide the latter. By way of decomposition by [2+2]-cycloreversion as known from the Wittig reaction, $Ph_2MeP{=}O$ and the stereoisomer of the starting alkene are

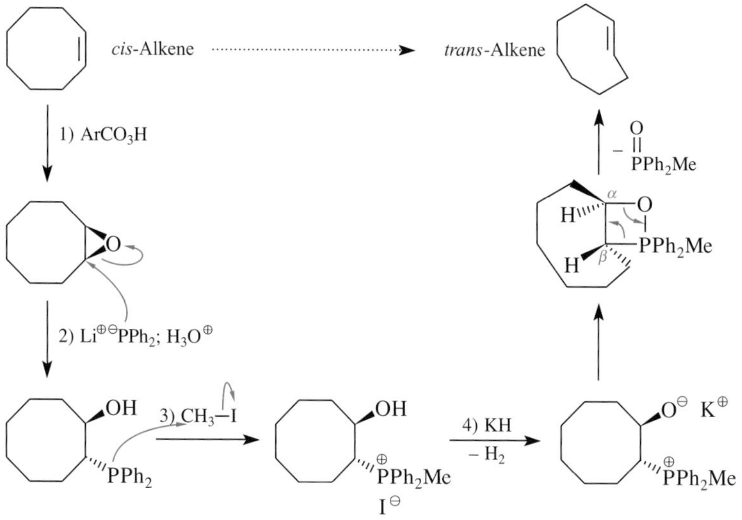

Fig. 4.45. *syn*-Selective $Ph_2MeP{=}O$ elimination as a key step in the preparation of *trans*-cyclooctene.

now produced. In the example in Figure 4.45, *trans*-cyclooctene is produced. It is even more impressive that *cis,cis*-1,5-cyclooctadiene can be converted into *trans,trans*-1,5-cyclooctadiene in a completely analogous manner.

You will learn about the reaction of α-metalated phosphonic acid esters with aldehydes in Section 11.3 in connection with the Horner–Wadsworth–Emmons reaction. This reaction also seems to give a *trans*-configured oxaphosphetane (Figure 4.46). Again, a *syn*-selective β-elimination of a compound with P=O double bond should occur. One of the elimination products is $(EtO)_2P({=}O)O^{\ominus}$. As a second product an alkene is produced that is predominantly or exclusively *trans*-configured.

In addition to the Wittig- und Horner–Wadsworth–Emmons reactions, we know a third alkene-forming reaction between carbonyl and phosphororganic compounds, i.e. the Wittig–Horner reaction. In Section 11.2, you will learn that in the course of this reaction a *syn*-elimination of $Ph_2P({=}O)O^{\ominus}$ takes places, i.e. another β-elimination of Het[1]/Het[2].

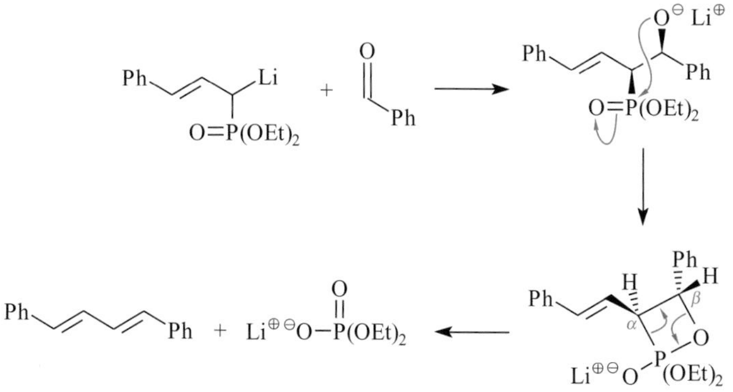

Fig. 4.46. A Het[1]/Het[2] elimination from the presumed oxaphosphetane intermediate of a Horner-Wadsworth-Emmons reaction.

References

S. E. Kelly, "Alkene Synthesis," in *Comprehensive Organic Synthesis* (B. M. Trost, I. Fleming, Eds.), Vol. 1, 729, Pergamon Press, Oxford, **1991**.

A. Krebs, J. Swienty-Busch, "Eliminations to Form Alkenes, Allenes and Alkynes and Related Reactions," in *Comprehensive Organic Synthesis* (B. M. Trost, I. Fleming, Eds.), Vol. 6, 949, Pergamon Press, Oxford, **1991**.

J. M. Percy, "One or More C=C Bond(s) by Elimination of Hydrogen, Carbon, Halogen or Oxygen Functions," in *Comprehensive Organic Functional Group Transformations* (A. R. Katritzky, O. Meth-Cohn, C. W. Rees, Eds.), Vol. 1, 553, Elsevier Science, Oxford, U. K., **1995**.

J. M. J. Williams (Ed.), "Preparation of Alkenes: A Practical Approach," Oxford University Press, Oxford, U. K., **1996**.

4.2

H. J. Reich, S. Wollowitz, "Preparation of α,β-Unsaturated Carbonyl Compounds and Nitriles by Selenoxide Elimination," *Org. React.* **1993**, *44*, 1–296.

A. Krief, A.-M. Laval, "*o*-Nitrophenyl Selenocyanate, a Valuable Reagent in Organic Synthesis: Application to One of the Most Powerful Routes to Terminal Olefins from Primary-Alcohols (The Grieco–Sharpless Olefination Reaction) and to the Regioselective Isomerization of Allyl Alcohols," *Bull. Soc. Chim. Fr.* **1997**, *134,* 869–874.

T. G. Back, "Selenoxide Eliminations," in *Organoselenium Chemistry,* (T. G. Back, Ed.), **1999**, Oxford, New York.

Y. Nishibayashi, S. Uemura, "Selenoxide Elimination and [2,3]-Sigmatropic Rearrangement," *Top. Curr. Chem.* **2000**, *208*, 201–235.

H. R. Nace, "The Preparation of Olefins by the Pyrolysis of Xanthates. The Chugaev Reaction," *Org. React.* **1962**, *12*, 57–100.

S. Z. Zard, "On the Trail of Xanthates: Some New Chemistry from an Old Functional Group," *Angew. Chem. Int. Ed. Engl.* **1997**, *36*, 672–685.

4.4

J. F. Bunnett, "The Mechanism of Bimolecular β-Elimination Reactions," *Angew. Chem. Int. Ed. Engl.* **1962**, *1*, 225–235.

J. Sicher, "The *syn* and *anti* Steric Course in Bimolecular Olefin-Forming Eliminations," *Angew. Chem. Int. Ed. Engl.* **1972**, *11*, 200–214.

S. Gronert, "Gas-Phase Studies of the Competition between Substitution and Elimination Reactions," *Acc. Chem. Res.* **2003**, *36,* 848–857.

E. Baciocchi, "Base Dependence of Transition-State Structure in Alkene-Forming E2 Reactions," *Acc. Chem. Res.* **1979**, *12*, 430.

J. K. Crandall, M. Apparu, "Base-Promoted Isomerizations of Epoxides," *Org. React.* **1983**, *29*, 345–443.

A. C. Cope, E. R. Trumbull, "Olefins from Amines: The Hofmann Elimination Reaction and Amine Oxide Pyrolysis," *Org. React.* **1960**, *11*, 317–493.

S. Khapli, S. Dey, D. Mal, "Burgess Reagent in Orgnic Synthesis," *J. Indian Inst. Sci.* **2001**, 81, 461–476.

4.5

M. F. Vinnik, P. A. Obraztsov, "The mechanism of the Dehydration of Alcohols and Hydration of Alkenes in Acid Solution," *Russ. Chem. Rev.* **1990**, *59*, 106-131.

A. Thibblin, "Mechanisms Of Solvolytic Alkene-Forming Elimination-Reactions," *Chem. Soc. Rev.* **1993**, *22*, 427.

R. F. Langler, "Ionic Reactions of Sulfonic Acid Esters," *Sulfur Rep.* **1996**, 19, 1–59.

4.6

W. C. Chan, P. D. White (Eds.), "Fmoc Solid Phase Peptide Synthesis. A Practical Approach," Oxford University Press, Oxford, **2001**.

M. Julia, "Recent Advances in Double Bond Formation," *Pure Appl. Chem.* **1985**, *57*, 763.

P. R. Blakemore, "The Modified Julia Olefination: Alkene Synthesis via the Condensation of Metallated Heteroarylalkylsulfones with Carbonyl Compounds," *J. Chem. Soc. Perkin Trans. I* **2002**, 2563–2585.

4.7

P. Kocienski, "Reductive Elimination, Vicinal Deoxygenation and Vicinal Desilylation," in *Comprehensive Organic Synthesis* (B. M. Trost, I. Fleming, Eds.), Vol. 6, 975, Pergamon Press, Oxford, **1991**.

D. J. Ager,"The Peterson Olefination Reaction," *Org. React.* **1990**, *38*, 1–223.

A. G. M. Barrett, J. M. Hill, E. M. Wallace, J. A. Flygare, "Recent Studies on the Peterson Olefination Reaction," *Synlett* **1991**, 764–770.

L. F. van Staden, D. Gravestock, D. J. Ager, "New Developments in the Peterson Olefination Reaction," *Chem. Soc. Rev.* **2002**, *31,* 195–200.

T. Kawashima, R. Okazaki, "Synthesis and Reactions of the Intermediates of the Wittig, Peterson, and their Related Reactions," *Synlett* **1996**, 600.

Further Reading

E. Block, "Olefin Synthesis via Deoxygenation of Vicinal Diols," *Org. React.* **1984**, *30*, 457–566.

H. N. C. Wong, C. C. M. Fok, T. Wong, "Stereospecific Deoxygenation of Epoxides to Olefins," *Heterocycles* **1987**, *26*, 1345.

M. Schelhaas, H. Waldmann, "Protecting Group Strategies in Organic Synthesis," *Angew. Chem. Int. Ed. Engl.* **1996**, *35*, 2056–2083.

K. Jarowicki, P. Kocienski, "Protecting Groups," *Contemp. Org. Synth.* **1996**, *3*, 397–431.

G. Köbrich, "Bredt Compounds and the Bredt Rule," *Angew. Chem. Int. Ed. Engl.* **1973**, *12*, 464–473.

Substitution Reactions on Aromatic Compounds

5

Substitution reactions on aromatic compounds are the most important methods for the preparation of aromatic compounds. Synthesizing them from nonaromatic precursors is considerably less important. Via substitution reactions, electrophiles and nucleophiles can be introduced into aromatics. A series of mechanisms is available for this. Those that are discussed in this chapter are listed in Table 5.1.

5.1 Electrophilic Aromatic Substitutions via Sigma Complexes ("Ar-SE Reactions")

The electrophilic aromatic substitution via sigma (Wheland) complexes, or the Ar-SE reaction, is the classical method for functionalizing aromatic compounds. In this section, we will focus on the mechanistic foundations as well as the preparative possibilities of this process.

5.1.1 Mechanism: Substitution of $H^{\oplus}$ vs *ipso*-Substitution

For an Ar-S_E reaction to be able to occur, first the actual electrophile must be produced from the reagent (mixture) used. Then this electrophile initiates the aromatic substitution. It takes place, independently of the chemical nature of the electrophile, according to a two-step mechanism (Figure 5.1). A third step, namely, the initial formation of a π-complex from the electrophile and the substrate, is generally of minor importance for understanding the reaction event.

In the first step of the actual Ar-S_E reaction, a substituted cyclohexadienyl cation is formed from the electrophile and the aromatic compound. This cation and its derivatives are generally referred to as a **sigma** or **Wheland complex**. Sigma complexes are described by at least three carbenium ion resonance forms (Figure 5.1). There is an additional resonance form for each substituent, which can stabilize the positive charge of the Wheland complex by a pi electron-donating (+M) effect (see Section 5.1.3). This resonance form is an *all*-octet formula.

Sigma complexes are high-energy intermediates because they do not contain the conjugated aromatic electron sextet present in the product and in the starting material. Consequently, the formation of these complexes is the rate-determining step of Ar-S_E reactions (see Figure 5.1; for the only exception see Figure 5.23). This, in turn, means that sigma complexes are excellent models for the transition state of Ar-S_E reactions.

Bruckner R (author), Harmata M (editor) In: *Organic Mechanisms – Reactions, Stereochemistry and Synthesis*
Chapter DOI: 10.1007/978-3-642-03651-4_5,

Tab. 5.1 Substitution Reactions of Aromatic Compounds: Mechanistic Alternatives*

Section	Type of substitution using a benzene derivative as an example	Substitution type also known for: naphthalene	Substitution type also known for: five-membered ring aromatic compounds	Mechanistic designation
5.1–5.2	R_x–C$_6$H$_4$–H → ($E^{\oplus}$) → R_x–C$_6$H$_4$–E	yes	yes	"Classic Ar-S_E"
5.3.2	R_x–C$_6$H$_4$–Br → (Mg or Li or BuLi; $E^{\oplus}$) → R_x–C$_6$H$_4$–E	yes	yes	Ar-S_E via organometallic compounds
5.3.2	R_x–C$_6$H$_4$–I → (Mg or Li or BuLi or *i*PrMgBr; $E^{\oplus}$) → R_x–C$_6$H$_4$–E	yes	yes	
5.3.1	R_x–C$_6$H$_3$(H)(MDG) → (*sec*-BuLi; $E^{\oplus}$) → R_x–C$_6$H$_3$(E)(DMG)	yes	yes	
5.3.3/16.4.2	R_x–C$_6$H$_4$–B(OR)$_2$ → ($E^{\oplus}$, kat. Pd(PPh$_3$)$_4$) → R_x–C$_6$H$_4$–E	yes	yes	The same; also transition metal-mediated C,C coupling
5.4	R_x–C$_6$H$_4$–$N^{\oplus}\equiv N$ → ($Nu^{\ominus}$) → R_x–C$_6$H$_4$–Nu	yes	no	Ar-S_N1
5.4	R_x–C$_6$H$_4$–$N^{\oplus}\equiv N$ → ($Nu^{\ominus}$/Cu(I)) → R_x–C$_6$H$_4$–Nu	yes	no	
5.5	EWG–C$_6$H$_4$–Hal → ($Nu^{\ominus}$) → EWG–C$_6$H$_4$–Nu	yes	no	Ar-S_N via Meisenheimer complexes
16.2	R_x–C$_6$H$_4$–Hal → ($Nu^{\ominus}$/Cu(I)) → R_x–C$_6$H$_4$–Nu	yes	no	Ar-S_N of the Ullmann type
16.3/16.4 s. a. 16.5	R_x–C$_6$H$_4$–Hal → ($Nu^{\ominus}$/Ni(0) or $Nu^{\ominus}$/Pd(0)) → R_x–C$_6$H$_4$–Nu	yes	yes	Transition metal-mediated C,C, coupling
5.6	R_x–C$_6$H$_4$–Hal → ($Nu^{\ominus}$) → R_x–C$_6$H$_4$–Nu (ortho) + R_x–C$_6$H$_4$–Nu (meta)	yes	no	Ar-S_N via arynes

*DMG, Directed Metalation Group; EWG, electron-withdrawing group.

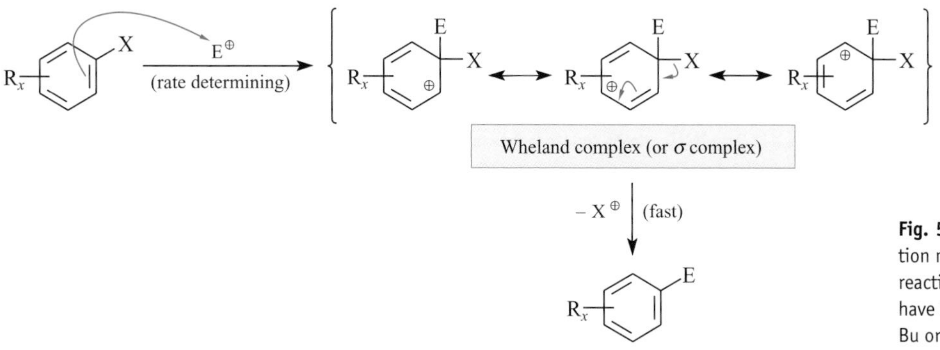

Wheland complex (or σ complex)

Fig. 5.1. The common reaction mechanism for all Ar-S_E reactions. Almost always we have X = H and rarely X = *tert*-Bu or X = SO_3H.

In the second step of the Ar-S_E reaction, an aromatic compound is regenerated by cleaving off a cation from the C atom that reacted with the electrophile. Most often the eliminated cation is a proton (Figure 5.1, X = H or $X^{\oplus} = H^{\oplus}$).

In a few cases, cations other than the proton are eliminated from the sigma complex to reconstitute the aromatic system. The *tert*-butyl cation (Figure 5.1, X = *tert*-Bu) and protonated SO_3 (Figure 5.1, X = SO_3H) are suitable for such an elimination. When the latter groups are replaced in an Ar-S_E reaction, we have the special case of an *ipso* substitution. Among other things, *ipso* substitutions play a role in the few Ar-S_E reactions that are reversible (Section 5.1.2).

Side Note 5.1. Stable Cyclohexadienyl Cations

Strictly speaking there is only indirect evidence for the occurrence of sigma complexes as short-lived intermediates of Ar-S_E reactions. Formed in the rate-determining step, sigma complexes completely explain both the reactivity and regioselectivity of most Ar-S_E reactions (cf. Section 5.1.3). In addition, the viability of short-lived sigma complexes in Ar-S_E chemistry is supported by the fact that stable sigma complexes could be isolated in certain instances.

Figure 5.2 contains X-ray crystallographic and ^{13}C-NMR spectroscopic data of the permethylated sigma complex shown. Through bond length alternation or π-charge distribution in the ring—positive partial charge at a ring C atom causes proportional deshielding—they reveal structural details exactly like the ones that would have been expected from the resonance forms in Figure 5.1.

Figure 5.3 shows the X-ray structure analysis of the protonation product of 1,3,5-tris(pyrrolidinyl)benzene. It is striking that the protonation occurs at a ring carbon atom instead of at the aniline nitrogen. The reason is that an anilinium salt would be produced upon N-protonation where the positive charge would essentially be *localized* on the ammonium nitrogen. In contrast, C-protonation as a sigma complex leads to a species where the positive charge would be *delocalized* over all three nitrogen atoms and over five of the six carbon atoms of the six-membered ring. This charge delocalization accounts for the difference in stability between the sigma complex and the unobserved N-protonated isomer. It even offsets the fact that the sigma complex—other than its N-protonated isomer—is nonaromatic.

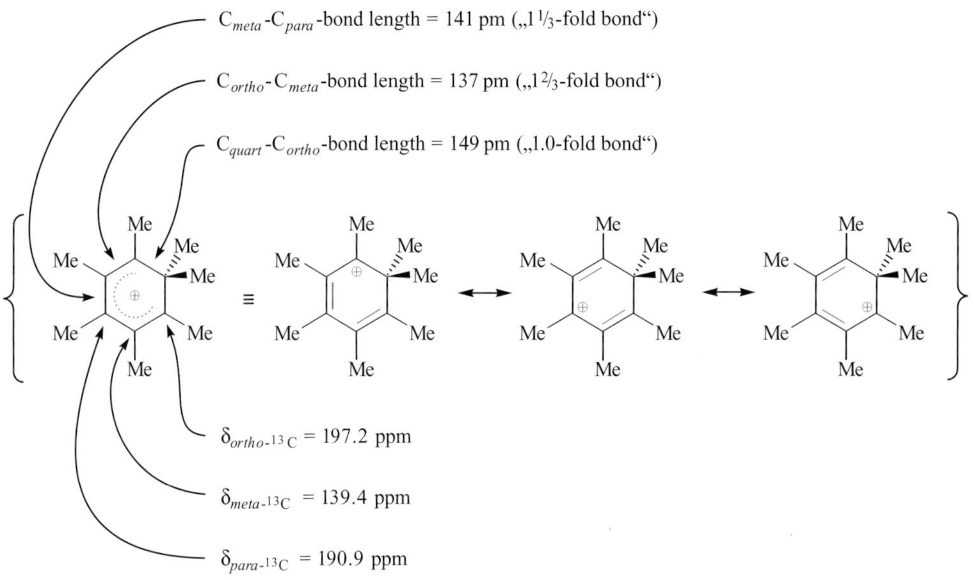

C_{meta}-C_{para}-bond length = 141 pm („1 1/3-fold bond")

C_{ortho}-C_{meta}-bond length = 137 pm („1 2/3-fold bond")

C_{quart}-C_{ortho}-bond length = 149 pm („1.0-fold bond")

$\delta_{ortho\text{-}^{13}C}$ = 197.2 ppm

$\delta_{meta\text{-}^{13}C}$ = 139.4 ppm

$\delta_{para\text{-}^{13}C}$ = 190.9 ppm

Fig. 5.2. NMR studies and geometry of a stable sigma complex. The C,C bond lengths in the six-membered ring correlate with the bond order resulting from the superimposition of the three resonance forms. The ^{13}C-NMR shifts of the sp^2-hybridized ring atoms by means of a low-field shift, i.e., an increased δ value, indicate the centers that—according to the resonance forms—bear positive partial charge.

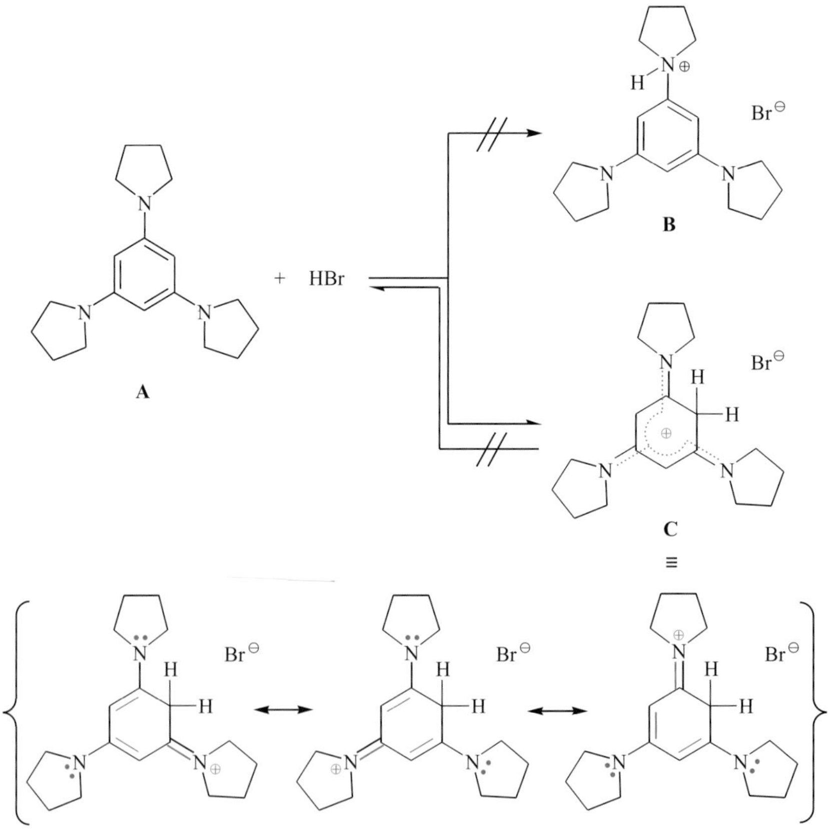

Fig. 5.3. Formation of a sigma complex (with delocalized positive charge) instead of an ammonium cation (with localized positive charge) upon protonation of 1,3,5-tris(pyrrolidinyl)benzene.

5.1.2 Thermodynamic Aspects of Ar-S_E Reactions

Substitution and Addition Compared: Heats of Reaction

As you know, Br_2 *adds* to C=C bonds (Section 3.5.1). On the other hand, Br_2 *replaces* an *sp*²-bonded H atom on the formal C=C double bond of aromatic compounds. Why is it not the other way around? That is, why do cyclohexene and Br_2 not give 1-bromocyclohexene via a substitution reaction, and why do benzene and Br_2 not give a dibromocyclohexadiene via an addition reaction?

If one compares the heats of reaction for these potentially competing reactions (Figure 5.4), one arrives at the following:

1. The substitution reaction C_{sp2}—H + Br-Br → C_{sp2}—Br + H-Br is exothermic by approximately –11 kcal/mol. It is irrelevant whether the reacting *sp*²-hybridized carbon atom is part of an alkene or an aromatic compound.
2. In an addition reaction C=C + Br—Br → Br—C—C—Br there is a drop in enthalpy of 27 kcal/mol *in the substructure shown*, that is, a reaction enthalpy of –27 kcal/mol. This enthalpy decrease equals the heat of reaction liberated when Br_2 is added to cyclohexene. However, it does not equal the heat of reaction for the addition of Br_2 to benzene.
3. When Br_2 is added to benzene, the above-mentioned –27 kcal/mol must be balanced with the simultaneous loss of the benzene conjugation, which is about +36 kcal/mol. All in all, this makes the addition of Br_2 to benzene endothermic by approximately +9 kcal/mol. Moreover, when Br_2 is added to benzene, the entropy decreases. Consequently, the addition of bromine to benzene would not only be endothermic but also endergonic. The latter means that such an addition is thermodynamically impossible.
4. The exothermic substitution reaction (see above) on benzene is also exergonic because no significant entropy change occurs. This substitution is therefore thermodynamically possible and actually takes place under suitable reaction conditions (Section 5.2.1).
5. Finally, because the **addition** of Br_2 to cyclohexene is 27 kcal/mol – 11 kcal/mol = 16 kcal/mol more exothermic than the **substitution** of Br_2 on cyclohexene, can we conclude that the first reaction also takes place more rapidly? Not necessarily! The (fictitious) substitution reaction of Br_2 on cyclohexene should be a multistep reaction and proceed via a bromonium ion formed in the first and also rate-determining reaction step. This bromonium ion has been demonstrated to be the intermediate in the known addition reaction of Br_2 to cyclohexene (Section 3.5.1). Thus, one would expect that the outcome of the competition of substitution vs. addition depends on whether the bromonium ion is converted—*in each case in an elementary reaction*—to the substitution or to the addition product. The Hammond postulate suggests that the bromonium ion undergoes the more exothermic (exergonic) reaction more rapidly. In other words, the **addition** reaction is expected to win not only thermodynamically but also kinetically.

***ipso* Substitutions and the Reversibility of Ar-S_E Reactions**

According to Section 5.1.1, electrophilic *ipso*-substitutions via sigma complexes occur, for example, when a proton reacts with the substructure C_{sp2}—*tert*-Bu or C_{sp2}—SO_3H of appropriately substituted aromatic compounds. After expulsion of a *tert*-butyl cation or an HSO_3^+ ion, an aromatic compound is obtained, which has been defunctionalized in the respective position.

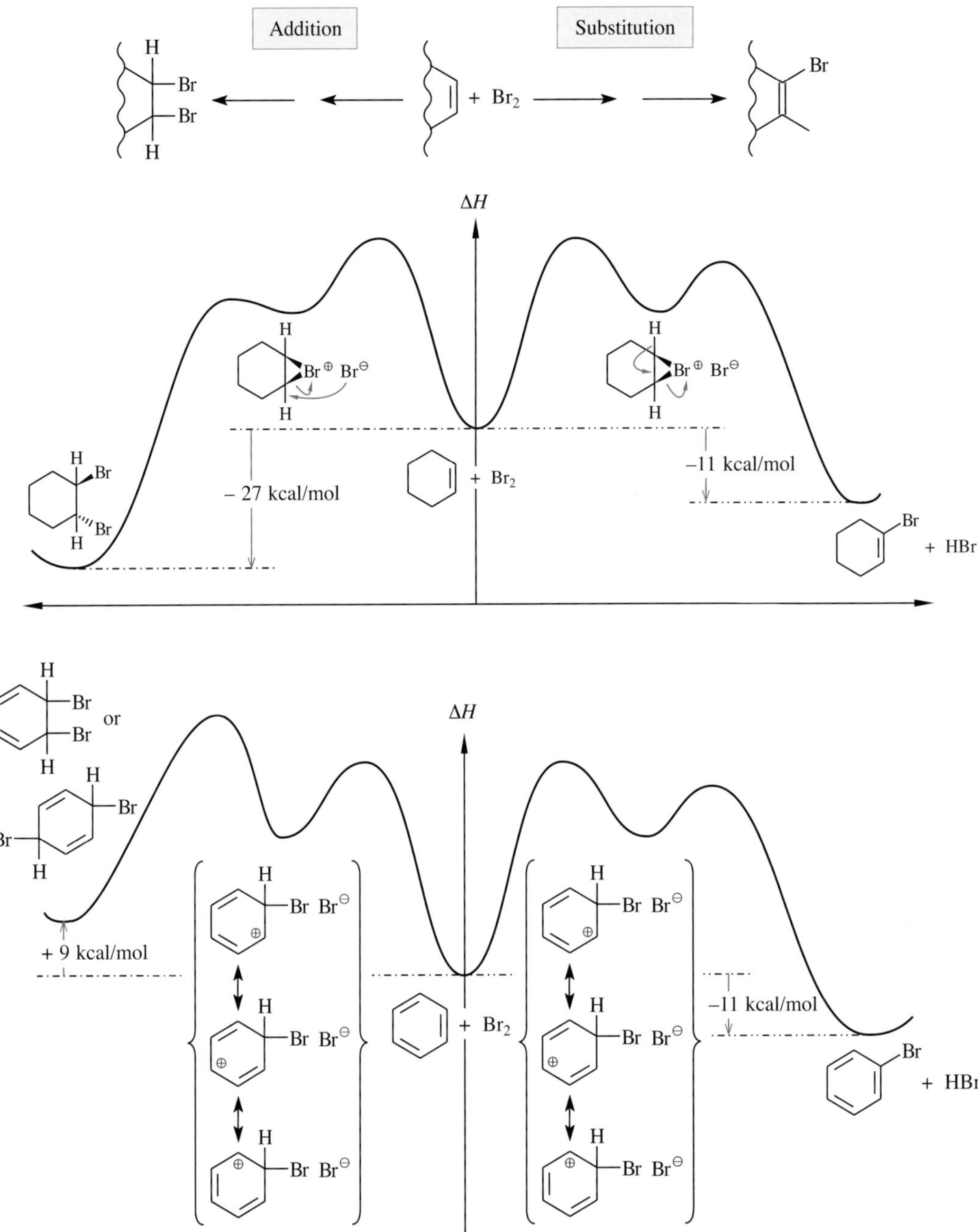

Fig. 5.4. Enthalpy profile for the electrophilic addition of Br_2 (reactions proceeding towards the left) and for the electrophilic substitution by Br_2 (reactions proceeding towards the right) of cyclohexene (top) and of benzene (bottom). *Altogether*, the facts presented here are likely to be prototypical of the chemoselectivity of all electrophilic reactions on alkenes *versus* benzenoid aromatic compounds. *In detail*, though, this need not be true: both in the alkene and the aromatic compound $\Delta H_{\text{Substitution}}$ as well as $\Delta H_{\text{addition}}$ depend on the electrophile, which is why an electrophilic dependency can in principle also be expected for $\Delta\Delta H \equiv \Delta H_{\text{substitution}} - \Delta H_{\text{addition}}$.

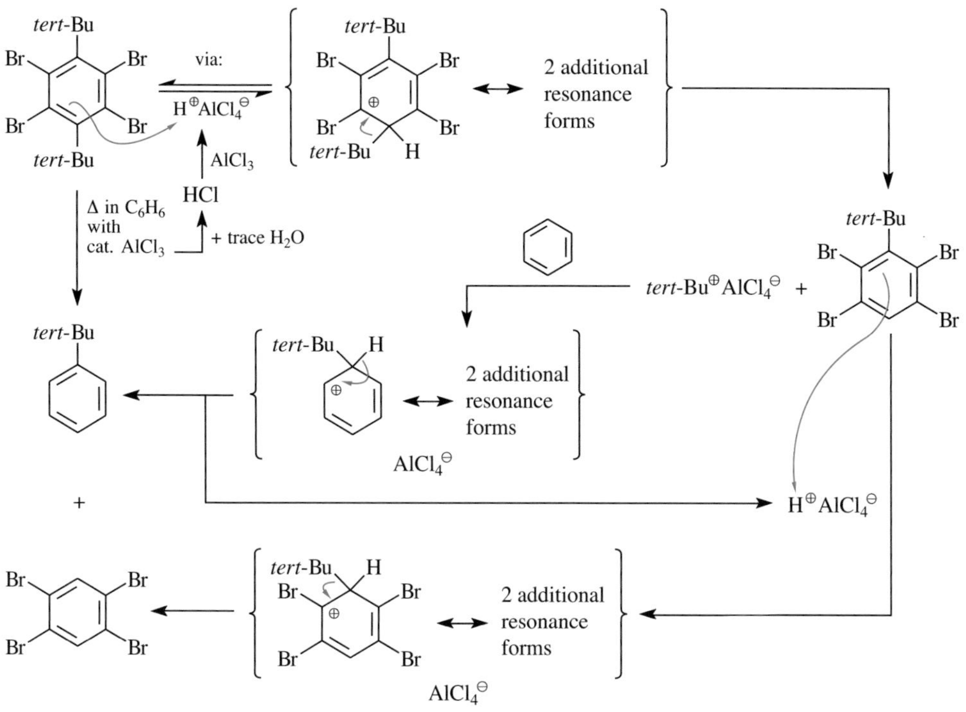

Fig. 5.5. De-*tert*-butylation via Ar—S_E reaction. The sequence consisting of *tert*-butylation and de-*tert*-butylation can in terms of a protecting group strategy be employed in the regioselective synthesis of a multiply substituted benzene derivative (cf. Figures 5.28 and 5.33).

Figure 5.5 shows an example of a de-*tert*-butylation by a reaction of this type. In fact, *both* *tert*-butyl groups of the aromatic compounds shown in the figure are removed by *ipso* substitutions. The fate of the *tert*-butyl cations released depends on whether a reactive solvent such as benzene (this case is shown in Figure 5.5) or an inert solvent (not shown) is used. The benzene is *tert*-butylated by *tert*-butyl cations in a Friedel–Crafts alkylation (cf. Section 5.2.5). From the perspective of the *tert*-butyl groups, this *ipso* substitution represents an S_N1 reaction with benzene acting as the nucleophile. The driving force for this reaction is based on thermodynamic control: the *tert*-butyl groups leave a sterically hindered substrate and enter into a sterically unhindered product. If the de-*tert*-butylation of the same substrate is carried out in an inert solvent instead of in benzene, it becomes an E1 elimination of Ar—H from the perspective of the *tert*-butyl groups: the released *tert*-butyl cations are deprotonated to give isobutene. If the latter is continuously distilled off the reaction mixture, the de-*tert*-

Fig. 5.6. De-*tert*-butylation/re-*tert*-butylation as a possibility for isomerizing *tert*-butylated aromatic compounds via Ar–S_E reactions.

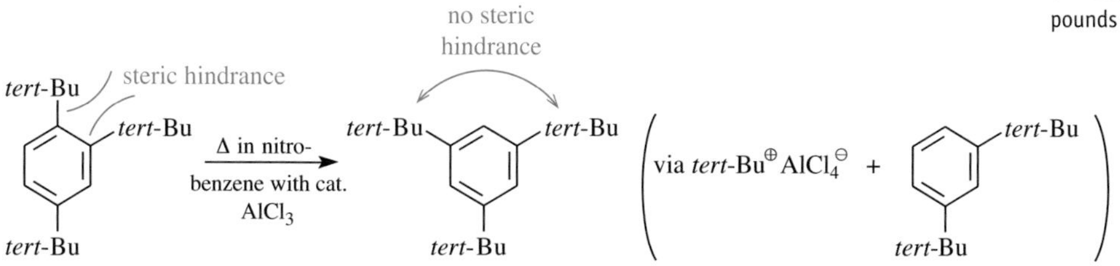

butylation equilibrium is also shifted toward the defunctionalized aromatic compounds. This represents Le Chatlier's principle in action.

The *tert*-butyl cations liberated from compounds Ar—*tert*-Bu, upon *ipso* reaction of a proton, may also react again with the aromatic compound from which they stemmed. If this course is taken, the *tert*-butyl groups are ultimately bound to the aromatic nucleus with a regioselectivity that is dictated by thermodynamic control. Figure 5.4 shows how in this way 1,2,4-tri-*tert*-butylbenzene is smoothly isomerized to give 1,3,5-tri-*tert*-butyl-benzene.

Side Note 5.2. Large-Scale Preparation of a Schäffer Acid

The other important *ipso* substitution by the Ar-S_E mechanism is the protodesulfonylation of aromatic sulfonic acids. It is used, for example, in the industrial synthesis of 2-hydroxynaphthalene-6 sulfonic acid (Schäffer acid), an important coupling component for the production of azo dyes. Schäffer acid is produced by sulfonylating 2-naphthol twice (for the mechanism, see Section 5.2.2) and then desulfonylating once. The first SO_3H group is introduced into the activated 1-position of 2-naphthol. The second SO_3H group is introduced into the 6-position of the initially formed 2-hydroxynaphthalene-1-sulfonic acid. The second step of the industrial synthesis of Schäffer acid is therefore the regioselective monodesulfonylation of the resulting 2-hydroxynaphthalene-1,6-disulfonic acid at the 1-position (Figure 5.7). Of course, the expelled electrophile HSO_3^+ must not react again with the monosulfonylated product. Therefore, it must be scavenged by another reagent. The simplest way to do this is by desulfonylation with dilute sulfuric acid. The water contained therein scavenges the HSO_3^+ cations to form HSO_4^- and $H_3O^\oplus$ or SO_4^{2-} and $H_3O^\oplus$.

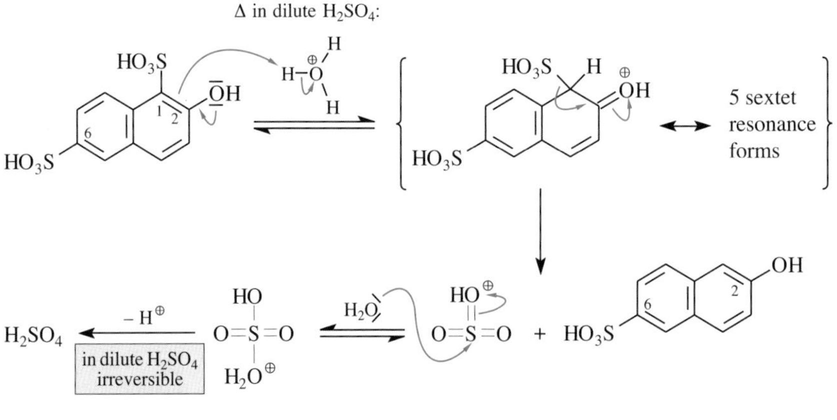

Δ in dilute H_2SO_4:

5 sextet resonance forms

in dilute H_2SO_4 irreversible

Fig. 5.7. Desulfonylation via Ar-S_E reaction (see Figure 5.10 for an explanation of this regioselectivity). The sulfonylation/desulfonylation sequence can in terms of a protecting group strategy be employed in the regioselective synthesis of a multiply substituted benzene derivative (cf. Figures 5.18 and 5.22).

If the released electrophile HSO_3^+ is *not* intercepted during the protodesulfonylation as in Figure 5.7, it reacts with the defunctionalized aromatic compound again. In this way an isomer of the original sulfonic acid may be obtained. The best-known example of such an isomerization is the conversion of naphthalene-1-sulfonic acid into naphthalene-2-sulfonic acid (Figure 5.8). Naphthalene-1-sulfonic acid is destabilized by the so-called *peri-interaction*, that is, the steric interaction between the C^8—H bond of the naphthalene and the substituent on C1. The *peri*-interaction is thus a *cis*-alkene strain. Because naphthalene-2-sulfonic acid does not

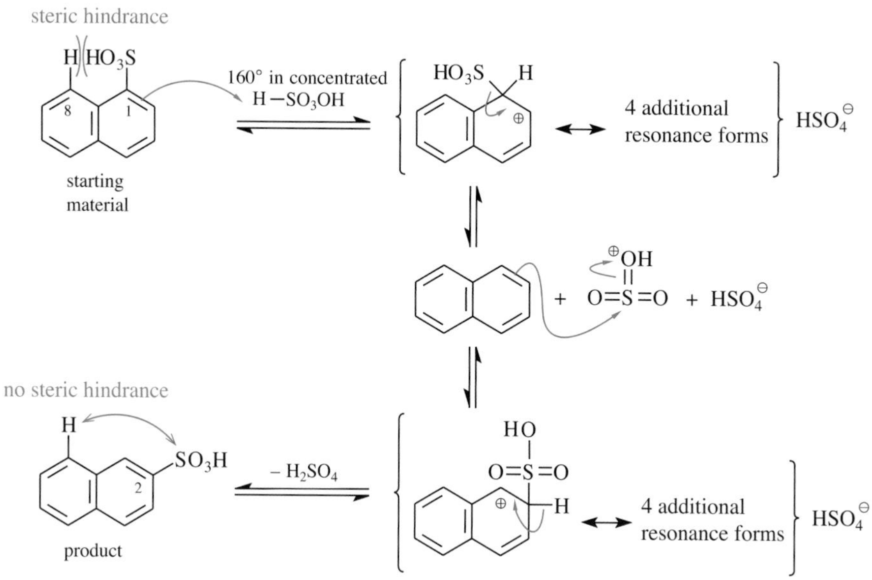

Fig. 5.8. Desulfonylation/resulfonylation as a possibility for isomerizing aromatic sulfonic acids via Ar-S_E reactions.

suffer from this interaction, it becomes the only reaction product under conditions of thermodynamic control.

Rule of Thumb

In conclusion we can make the following statement: most Ar-S_E reactions are irreversible because they have a sufficiently strong driving force and, at the same time, because they can be carried out under sufficiently mild conditions. The most important reversible Ar-S_E reactions are *tert*-alkylation and sulfonylation.

5.1.3 Kinetic Aspects of Ar-S_E Reactions: Reactivity and Regioselectivity in Reactions of Electrophiles with Substituted Benzenes

In the rate-determining step of an Ar-S_E reaction, a monosubstituted benzene and an electrophile can form three isomeric sigma complexes. By the subsequent elimination of a proton, one sigma complex gives the *ortho*-disubstituted, the second sigma complex gives the sigma *meta*-disubstituted, and the third sigma complex gives the *para*-disubstituted benzene. According to Section 5.1.1, sigma complexes are also excellent transition state models for the rate-determining step of Ar-S_E reactions. According to the Hammond postulate the more stable Wheland complexes are, the faster they form. To the extent that they form irreversibly, it is inferred that the major substitution product is the one that forms from the most stable sigma complex.

Stabilization and Destabilization of sigma Complexes through Substituent Effects

Which sigma complexes are the most stable? This is determined to a small extent by steric effects and to a considerably greater extent by electronic effects. As a carbocation, a substituted sigma complex is considerably more stable than an unsubstituted one only when it carries one or more donor substituents, and unsubstituted sigma complexes E—$C_6H_6^+$ are still considerably more stable than sigma complexes that contain one or more acceptor substituents. Therefore, donor-substituted benzenes react with electrophiles more rapidly than benzene, and acceptor-substituted benzenes react more slowly.

A more detailed analysis of the stabilizing effect of donor substituents and the destabilizing effect of acceptor substituents (both are referred to as "Subst" in the following) on sigma complexes E—$C_6H_5^+$-Subst explains, moreover, the regioselectivity of an Ar-S_E reaction on monosubstituted benzene. *Isomeric donor-containing sigma complexes and acceptor-containing sigma complexes have different stabilities.* This follows from the uneven charge distributions in the sigma complexes.

On the left side Figure 5.9 shows the "primitive model" of the charge distribution in the Wheland complex. Therein the positive charges only appear *ortho* and *para* to the reacting C atom, and in each case they equal +0.33. This charge distribution is obtained by superimposing the three resonance forms of Figure 5.1.

Fig. 5.9. Charge distribution in sigma complexes E—$C_6H_6^+$ (refined model, on the right: calculation for E = H).

The formula on the right in Figure 5.9 gives a more subtle model of the charge distribution in the sigma complex. There positive partial charges exist on all five sp^2-hybridized ring atoms. The greatest charge (+0.30) can be found in the position *para* to the reacting C atom, a somewhat smaller charge (+0.25) in the *ortho*-position, and a much smaller charge (+0.10) in the *meta*-position.

Equipped with this refined charge distribution model, it is now possible to compare (1) the stabilities of isomeric donor-substituted sigma complexes with each other, (2) the stabilities of isomeric acceptor-substituted sigma complexes with each other, and (3) the stabilities of each sigma complex already mentioned with the stability of the unsubstituted sigma complex E-C_6H_6. The results can be found in Figure 5.10 for donor-substituted and in Figure 5.11 for acceptor-substituted sigma complexes.

Every donor-substituted sigma complex E-C_6H_5-Do⊕ is more stable than the reference compound E-C_6H_6 (Figure 5.10). Regardless of its position relative to the reacting C atom, the donor turns out to be always located on a partially positively charged C atom, which it stabilizes by donating electrons. Of course, the donor provides the greatest possible stabilization when it is bound to the C atom with the greatest positive charge (+0.30), and it effects the smallest stabilization when it is bound to the C atom with the smallest positive charge (+0.10). If the donor is bound to the C atom with the +0.25 charge, the result is stabilization of intermediate magnitude. This gives the following stability order for the Wheland complexes of interest: *para*-E— E—$C_6H_5^+$—Do > *ortho*-E—C_6H_5⊕— Do > *meta*-E— $C_6H_5^+$—Do > E—$C_6H_6^+$.

Taking into account the Hammond postulate, this means two things for Ar-S_E reactions:

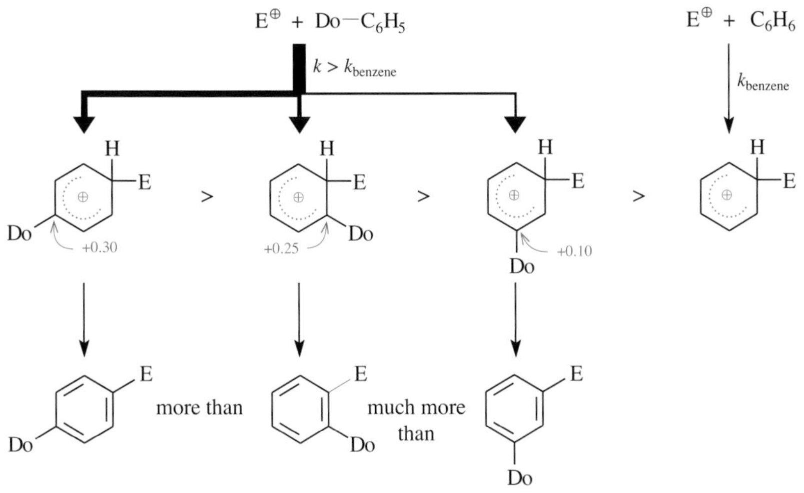

$E^{\oplus}$ + Do—C_6H_5 ($k > k_{benzene}$) — $E^{\oplus}$ + C_6H_6 ($k_{benzene}$)

Fig. 5.10. Ar-S_E reactions with donor-substituted benzenes (Do, donor substituent); comparing the regioselectivity and the reactivity with benzene. The thicknesses of the initial arrows show qualitatively to what extent the reaction takes place via the corresponding transition state.

Reactivity and Regioselectivity in Ar-S_E Reactions of Donor-Substituted Aromatic Compounds

- Each H atom of a donor-substituted aromatic compound should be substituted faster by an electrophile than an H atom in benzene.
- Donor-substituted benzenes and electrophiles should produce mixtures of *para-* and *ortho*-disubstituted aromatic compounds, in which the *para*-disubstituted product is formed in a greater amount. Only traces of the *meta*-disubstitution product are expected, even though a donor-substituted benzene is substituted faster at the *meta*-C atom than benzene itself at any of its C atoms.

Because of completely analogous considerations, every acceptor-substituted sigma complex E—C_6H_5—EWG$^{\oplus}$ is less stable than the reference compound E—$C_6H_6^+$ (Figure 5.11). From this analysis, one derives the following expectations for Ar-S_E reactions of acceptor-substituted benzenes:

Reactivity and Regioselectivity in Ar-S_E Reactions of Acceptor-Substituted Aromatic Compounds

- Each H atom of an acceptor-substituted aromatic compound should be substituted more slowly by an electrophile than an H atom in benzene.
- The substitution product should primarily be the *meta*-disubstituted aromatic compound. Among the by-products, the *ortho*-disubstituted benzene should predominate, while practically no *para*-product is expected.

Substituent Effects on Reactivity and Regioselectivity of Ar-S_E Reactions of Monosubstituted Benzenes

The vast majority of reactivities and regioselectivities observed in the reaction with electrophiles on monosubstituted benzenes (Table 5.2) are in agreement with the preceding generalizations (columns 2 and 4). The very few substituents that are not in agreement (column 3) *deactivate* the aromatic compound as do electron acceptors, but they are *para-* > *ortho-*directing as are electron donors.

Therefore, the reactivity and the regioselectivity of Ar-S_E reactions with substituted benzenes can be predicted reliably. According to what has been stated above, one only has to iden-

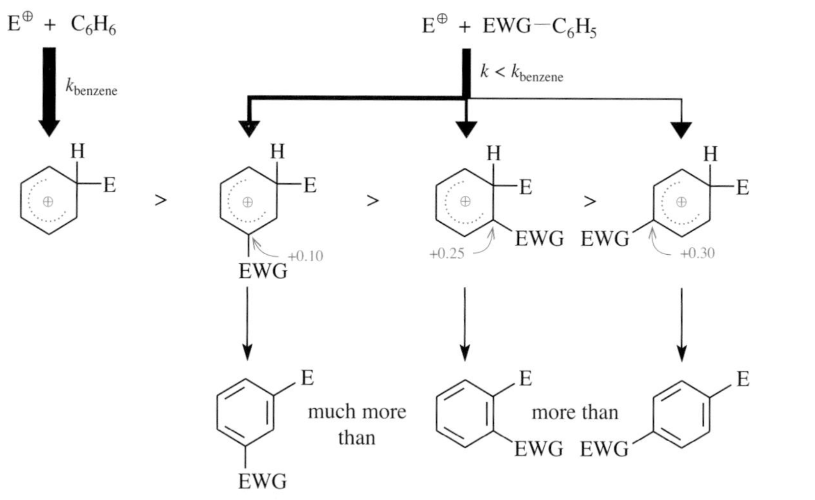

$E^{\oplus}$ + C_6H_6 ... $E^{\oplus}$ + EWG−C_6H_5; $k_{benzene}$; $k < k_{benzene}$; much more than; more than; +0.10; +0.25; +0.30

Fig. 5.11. Ar-S_E reactions with acceptor-substituted benzenes (EWG, electron-withdrawing group); comparing the regioselectivity and the reactivity with benzene. The thicknesses of the initial arrows again indicate qualitatively to what extent the reaction takes place via the respective transition states (but there is no relationship to the thicknesses of the arrows in Figure 5.10).

tify the electron donating and withdrawing substituents in the substrate. The electronic effects of the most important functional groups are listed in Table 5.3, where they are ordered semi-quantitatively: the best electron donors are on top and the best electron acceptors are at the bottom. It should be emphasized, though, that the significance of these substituent effects for the understanding of organic reactions goes far beyond the scope of this chapter.

Tab. 5.2 Experimental Results Concerning the Regioselectivity and the Reactivity of Ar-S_E Reactions of Monosubstituted Benzenes

	Do 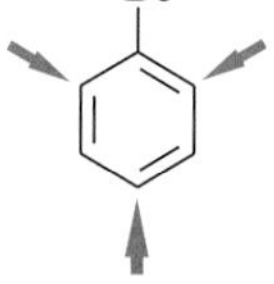	"Chameleon"-R 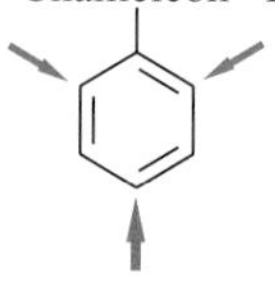	EWG 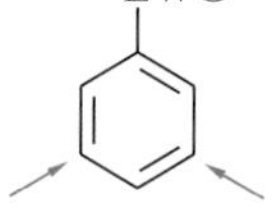
Substitution rate for each H atom of this aromatic compound compared with C_6H_5-H	greater	somewhat smaller	smaller
Regioselectivity	*para* > *ortho*	*para* > *ortho*	*meta*[1)]
The following can be introduced ...	... practically every $E^{\oplus}$	... majority of the $E^{\oplus}$	... only a good $E^{\oplus}$

[1)] For the byproducts *ortho* > *para*

Tab. 5.3 Inductive and Resonance Substituent Effects of Various Functional Groups

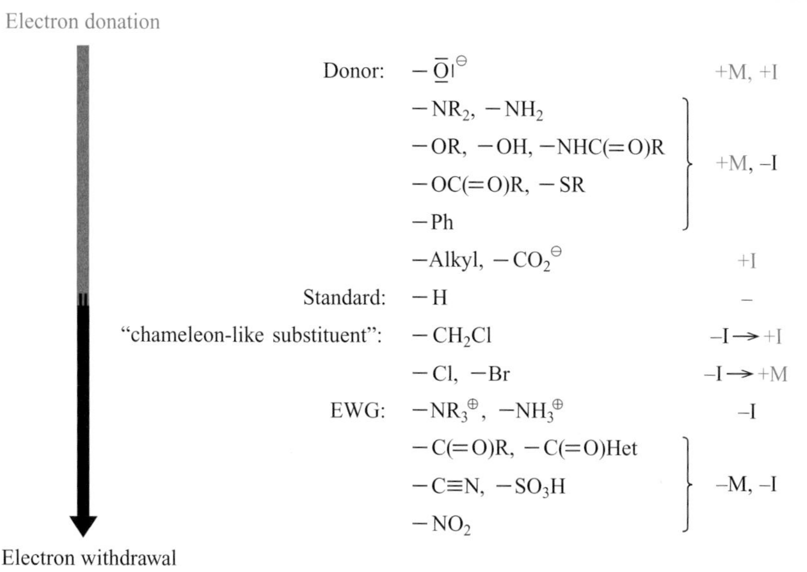

Electron donation ↓ Electron withdrawal

Donor:	$-\overline{\underline{O}}|^{\ominus}$	+M, +I
	$-NR_2$, $-NH_2$	+M, –I
	$-OR$, $-OH$, $-NHC(=O)R$	+M, –I
	$-OC(=O)R$, $-SR$	+M, –I
	$-Ph$	+M, –I
	–Alkyl, $-CO_2^{\ominus}$	+I
Standard:	$-H$	–
"chameleon-like substituent":	$-CH_2Cl$	–I → +I
	$-Cl$, $-Br$	–I → +M
EWG:	$-NR_3^{\oplus}$, $-NH_3^{\oplus}$	–I
	$-C(=O)R$, $-C(=O)Het$	–M, –I
	$-C{\equiv}N$, $-SO_3H$	–M, –I
	$-NO_2$	–M, –I

To complete this discussion, let us go back to those substituents (Table 5.2, column 3) that deactivate an aromatic compound like an electron acceptor but direct the electrophile like an electron donor in the *para*- and *ortho*-positions. The electronic effects of these substituents are obviously as variable as the color of a chameleon. These substituents destabilize each of the isomeric sigma complexes through inductive electron withdrawal. However, the extent of electron withdrawal depends on the magnitude of the positive charge in the position of this substituent. If the positive charge is small—in the *meta*-position with respect to the entering electrophile—electron withdrawal is important. Conversely, if the positive charge in the position of the substituent in question is large—*para* or *ortho* with respect to the entering electrophile—then electron withdrawal is reduced. There are two reasons for this dependence on the substituents. (1) The "chameleon-like" –I substituents Cl and—to a lesser extent—Br lend some electron density to the sigma complex through their pi electron-donating (+M) effect during a *para*- or an *ortho*-reaction of an electrophile. (2) The "chameleon-like" inductively electron withdrawing (–I) substituents CH_2Cl, Br, and I exert a diminished –I effect during a *para*- or an *ortho*-reaction of an electrophile, which results in a large positive partial charge in the α-position of the substituents, whereas they exert their full –I effect during an analogous *meta*-reaction, which gives rise to a small positive partial change in the vicinity of these substituents. It is often observed that the ability of a substituent to withdraw electron density is smaller with respect to positively charged centers than with respect to centers without a charge.

Regioselectivity for Ar-S_E Reactions of Naphthalene

Although what was stated in Section 5.1.3 is indeed completely correct, one might think that the analysis of the partial charge distribution in the aromatic substrate could allow, too, an at least qualitatively correct prediction of the reactivity and the regioselectivity in Ar-S_E reac-

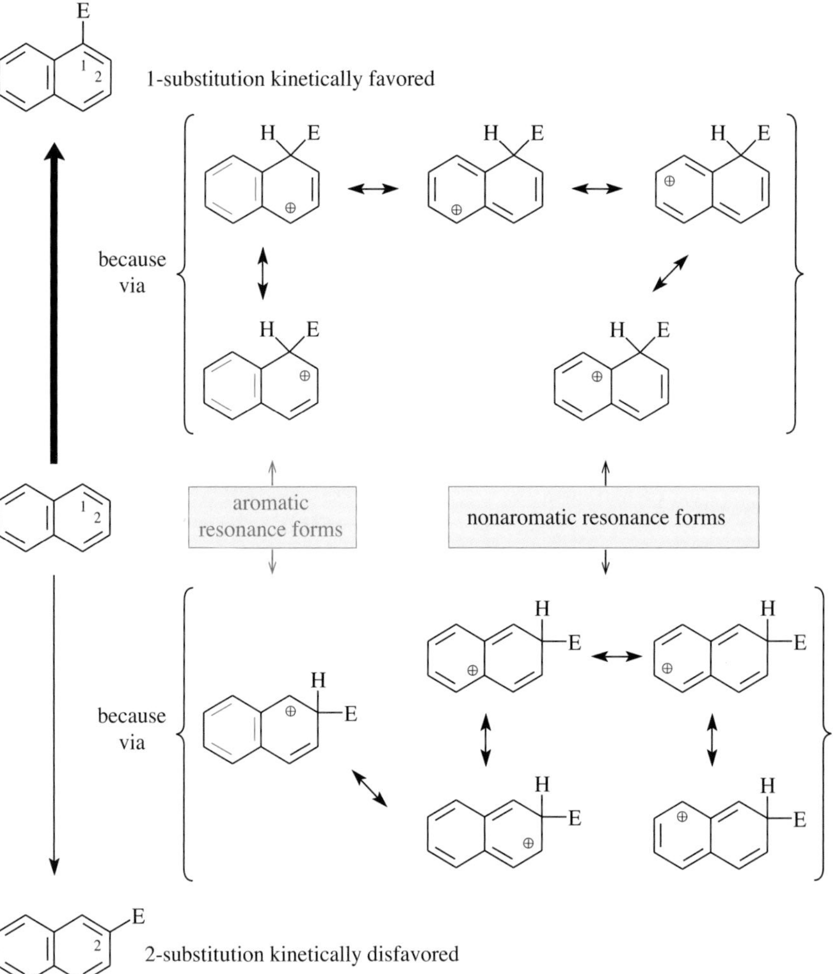

Fig. 5.12. Kinetically controlled Ar-S_E reactions of naphthalene.

tions. This opinion cannot be upheld, however, as an analysis of the rate and regioselectivity of Ar-S_E reactions with naphthalene clearly shows (Figure 5.12).

There is neither a partial positive nor a partial negative charge on the two nonequivalent positions 1 and 2 of naphthalene, which are poised for electrophilic substitution. One might consequently predict that electrophiles react with naphthalene without regiocontrol. Furthermore, this should occur with the same reaction rate with which benzene reacts. Both predictions contradict the experimental results! For example, naphthalene is brominated with a 99:1 selectivity in the 1-position in comparison to the 2-position. The bromination at C1 takes place 12,000 times faster and the bromination at C2 120 times faster than the bromination of benzene.

The regioselectivity and reactivity of Ar-S_E reactions of naphthalene are explained *correctly* by comparing the free activation enthalpies for the formation of the sigma complexes 1- E— $C_{10}H_{10}^+$ and 2-E— E— $C_{10}H_{10}^+$ from the electrophile and naphthalene and for the formation of the sigma complex E— $C_6H_6^+$ from the electrophile and benzene, respectively.

With respect to the *reactivities*, the decisive effect is the following: in the formation of either of the two isomeric sigma complexes from naphthalene, the difference between the naphthalene resonance energy (66 kcal/mol)—which is lost—and the benzene resonance energy (36 kcal/mol)—which is maintained—is about 30 kcal/mol. By contrast, the formation of a sigma complex from benzene costs the full 36 kcal/mol of the benzene resonance energy. This explains why $k_{\text{naphthalene}} > k_{\text{benzene.}}$

The *regioselectivity* of Ar-S_E reactions with naphthalene follows from the different stabilities of the sigma complex intermediate resulting from reaction at position 1 (Figure 5.12, upper half) compared with that of position 2 (Figure 5.12, lower half). For the sigma complex with the electrophile at C1, there are five resonance forms, excluding those for the intact benzene ring. In two of them, the aromaticity of one ring is retained. The latter forms are thus considerably more stable than the other three. The sigma complex with the electrophile at C2 can also be described with five resonance forms, not including one involving resonance of the intact aromatic ring. However, only one of these represents an aromatic species. The first sigma complex is thus more stable than the second. Reaction at C-1 is consequently preferred over reaction at C-2.

5.2 Ar-S_E Reactions via Sigma Complexes: Individual Reactions

5.2.1 Ar—Hal Bond Formation by Ar-S_E Reaction

Cl_2 and Br_2 react with donor-substituted aromatic compounds (aniline, acetanilide, phenol) without a catalyst. With nonactivated or deactivated aromatic compounds, these halogen molecules react only in the presence of a Lewis acid catalyst ($AlCl_3$ or $FeBr_3$). In the reaction example in Figure 5.13, as in other electrophilic substitution reactions, the following is true: if all substituents in multiply substituted aromatic compounds activate the same position (*cooperative substitution effect*), only that one position reacts. On the other hand, if these substituents activate different positions (*competitive substitution effect*), regioselectivity can be achieved only when the directing effect of one substituent predominates. Consequently, in the reaction example in Figure 5.14 ,the stronger donor methoxy, not the weaker donor methyl, determines the course of the reaction and directs the electrophile to the position *ortho to itself.*

To obtain single or multiple electrophilic brominations chemoselectively, one can simply vary the stoichiometry:

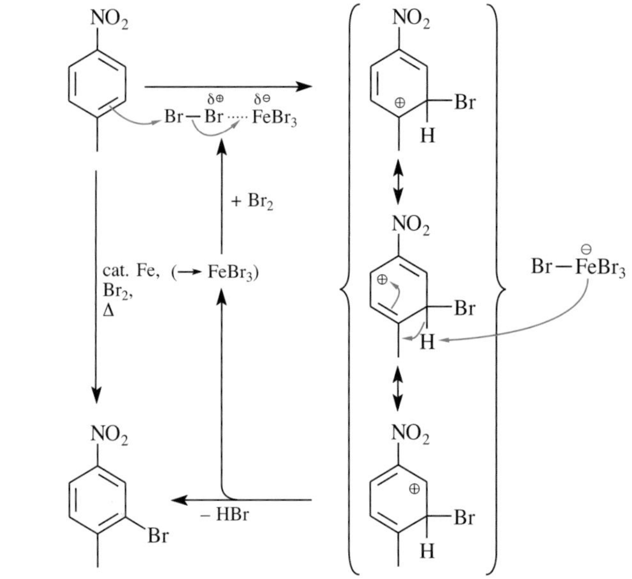

NO$_2$ δ⊕ δ⊖ Br—Br····· FeBr$_3$ + Br$_2$ cat. Fe, (→ FeBr$_3$) Br$_2$, Δ NO$_2$ Br – HBr NO$_2$ ⊕ Br H NO$_2$ ⊕ Br H NO$_2$ ⊕ Br H Br—FeBr$_3^{\ominus}$

Fig. 5.13. Regioselective bromination of a benzene derivative that contains a strong acceptor and a weak donor substituent.

Or, to the same end, different brominating reagents can be used:

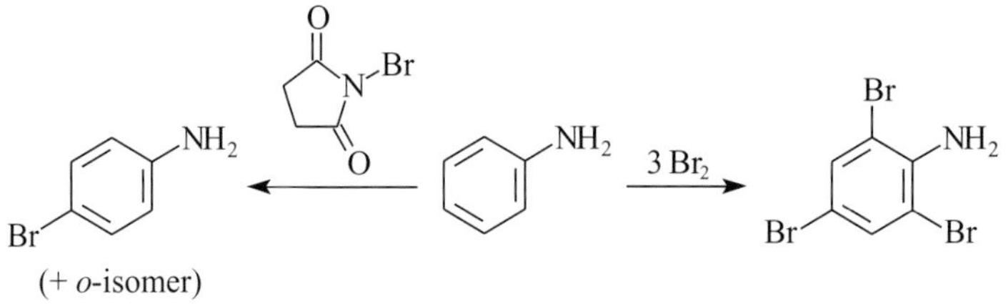

O N Br O NH$_2$ Br (+ *o*-isomer) NH$_2$ 3 Br$_2$ Br NH$_2$ Br Br

In alkylbenzenes, the *ortho-* and *para-*H atoms or the benzyl CH atoms can be replaced by Br$_2$. Chemoselective substitutions in the *ortho-* and *para-*positions succeed at low temperature in the presence of a catalyst. Figure 5.15 shows an example. Chemoselective substitutions with Br$_2$ in the benzyl position are possible using radical chemistry (see Figure 1.24), that is,

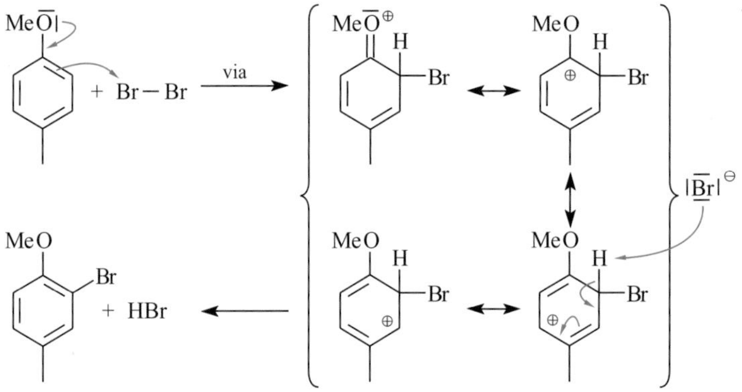

MeO + Br—Br via MeO$^{\oplus}$ H Br MeO ⊕ H Br |Br|$^{\ominus}$ MeO Br + HBr MeO H Br ⊕ MeO H ⊕ Br

Fig. 5.14. Regioselective bromination of a benzene derivative that contains two donor substituents of different strength. The stronger donor methoxy directs the electrophile to the position *ortho* to itself.

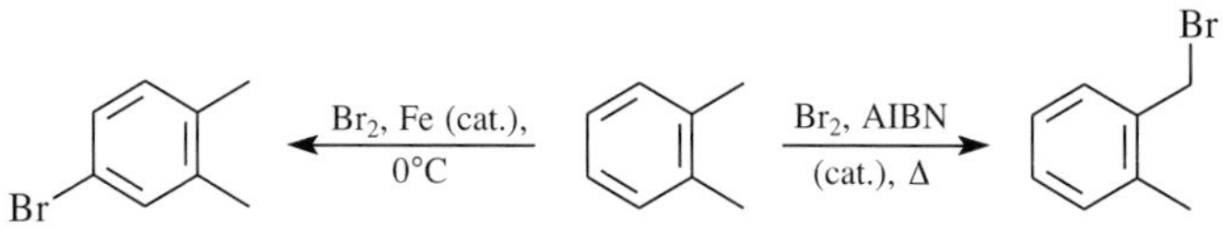

Br$_2$, Fe (cat.), 0°C; Br$_2$, AIBN (cat.), Δ

Fig. 5.15. Side chain and aromatic substitution in the reaction of *ortho*-xylene with Br_2.

by heating the reagents in the presence of a radical initiator or by irradiating them. Examples can be found in Figures 5.15 and 1.24.

When acetophenone is complexed with stoichiometric or greater amounts of $AlCl_3$, it is brominated by Br_2 in the position *meta* to the acceptor, that is, on the benzene ring.

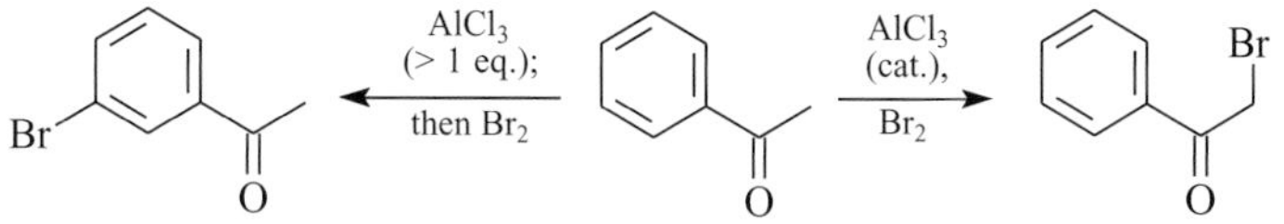

$AlCl_3$ (> 1 eq.); then Br_2; $AlCl_3$ (cat.), Br_2

On the other hand, if only catalytic amounts of $AlCl_3$ are added, the acetyl group of the acetophenone is brominated. Under these conditions the carbonyl oxygen of a fraction of acetophenone can be complexed. The bulk of the substrate still contains uncomplexed carbonyl oxygen. It allows acetophenone to equilibrate with its tautomeric enol (for details see Figure 12.5). The enol is a better nucleophile than the aromatic ring because it is brominated electrophilically without intermediate loss of aromaticity. HBr is the stoichiometric by-product of this substitution. Just like the HCl that is formed initially, it catalyzes the enolization of unreacted acetophenone and thus keeps the reaction going.

I_2 is a very weak electrophile. It is just reactive enough to react at the *para*-position of aniline (Figure 5.16, top-most example). Phenol ethers react with iodine only in the presence of

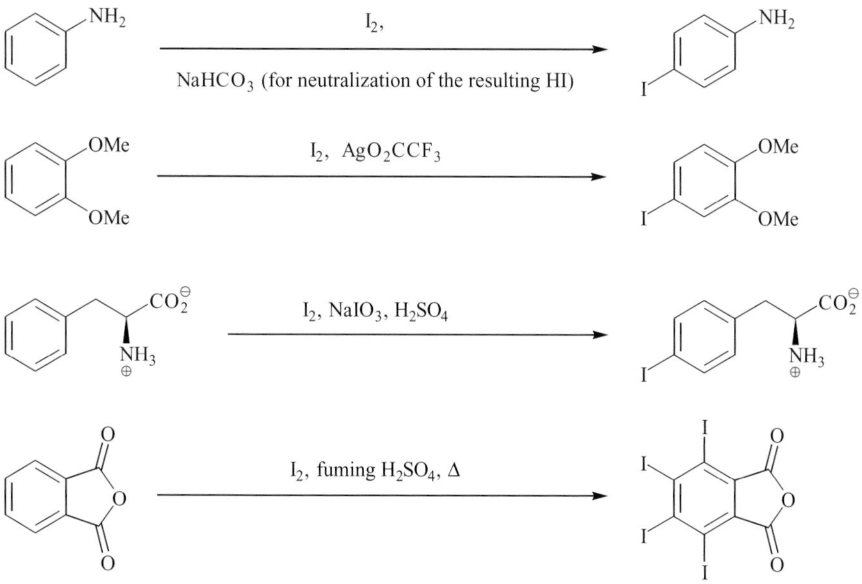

I_2, $NaHCO_3$ (for neutralization of the resulting HI)

I_2, AgO_2CCF_3

I_2, $NaIO_3$, H_2SO_4

I_2, fuming H_2SO_4, Δ

Fig. 5.16. Electrophilic iodination of electron-rich down to electron-poor—from top to bottom—benzene derivatives and typical reagents.

silver(I) salts (→ AgI + I_3^+; Figure 5.16, second example). Benzene and alkyl benzenes react with I_2 only when the iodine is activated still more strongly by oxidation with iodic or nitric acid (Figure 5.16, third example). Actually, acceptor-substituted benzenes can only undergo iodination with I_2 and fuming sulfuric acid (for an example cf. Figure 5.16, bottom). Even if the choice of the appropriate iodination agent will in the individual case also be determined by the need for providing energetic compensation for the lack of stabilization of the sigma complex intermediate, the lack of thermodynamic driving force is sufficient to explain the generally observed incapacity of elemental I_2 to enter into an aromatic ring: reactions of the type ArH + I_2 → ArI + HI exhibit $\Delta H \approx 16$ kcal/mol. This is much too endothermic to allow substitution … unless the by-product HI subsequently undergoes a highly exothermic acid-base reaction with the reaction product or an added inert amine. It is only this subsequent reaction that enables the iodination of aniline with I_2 as shown in Figure 5.16!

5.2.2 Ar—SO_3H Bond Formation by Ar-S_E Reaction

An SO_3H group can be introduced into aromatic compounds through an electrophilic substitution reaction, which is referred to as **sulfonation.** Suitable reagents are dilute, concentrated, or fuming sulfuric acid. The H_2SO_4 molecule as such probably functions as the actual electrophile only in reactions with dilute sulfuric acid. If the sulfuric acid is used concentrated or fuming, it contains much better electrophiles (Figure 5.17). One such electrophile, the protonated sulfuric acid ($H_3SO_4^{\oplus}$), is produced by autoprotolysis of sulfuric acid, and another, disulfuric acid ($H_2S_2O_7$), by condensation of two molecules of sulfuric acid. The dehydration of protonated sulfuric acid ($H_3SO_4^{\oplus}$) gives a third good electrophile ($HSO_3^{\oplus}$), which can alternatively be generated by the protonation of the SO_3 fraction of fuming sulfuric acid.

Initially, an arylamine does not react as an activated aromatic compound with sulfuric acid but as a base instead: anilinium hydrogen sulfates are produced (Figure 5.18). However, when the latter are heated, they decompose to give the starting materials reversibly. Only via the very small equilibrium amounts of free amine and free sulfuric acid does one observe the slow formation of the substitution product, an aromatic aminosulfonic acid. This aminosulfonic acid is a zwitterion, like an α-aminocarboxylic acid.

Naphthalene can be sulfonated with concentrated sulfuric acid (Figure 5.19). At 80 °C, this reaction proceeds under kinetic control and is therefore regioselective at the C1 center (cf. dis-

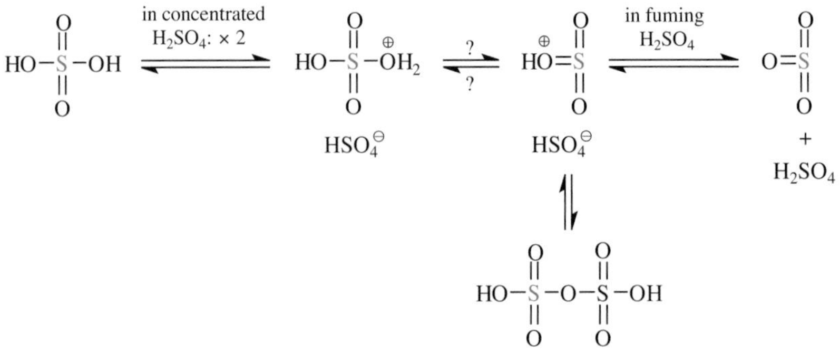

$HO-SO_2-OH$ $\xrightleftharpoons{\text{in concentrated } H_2SO_4:\ \times 2}$ $HO-SO_2-OH_2^{\oplus}$ $HSO_4^{\ominus}$ $\underset{?}{\overset{?}{\rightleftharpoons}}$ $HO=SO_2^{\oplus}$ $HSO_4^{\ominus}$ $\xrightleftharpoons{\text{in fuming } H_2SO_4}$ $O=SO_2$ + H_2SO_4

$HO=SO_2^{\oplus}$ $HSO_4^{\ominus}$ ⇅ $HO-SO_2-O-SO_2-OH$

Fig. 5.17. Effective electrophiles in sulfonations of aromatic compounds with sulfuric acids of different concentrations.

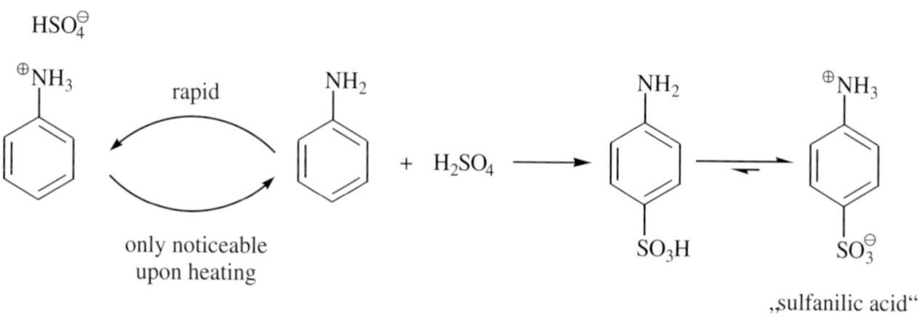

Fig. 5.18. Sulfonation of aniline. Anilinium hydrogen sulfate, which is initially formed from the substrate and concentrated sulfuric acid, reacts via the minimal equilibrium amount of free aniline; its amino groups directs the electrophile to the *para*-position.

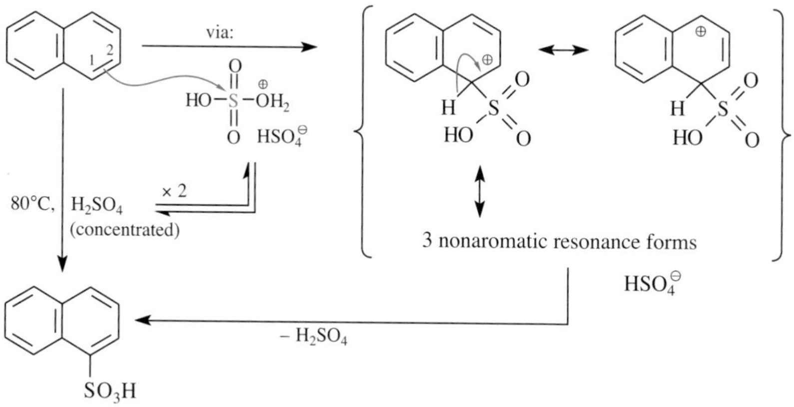

Fig. 5.19. Kinetically controlled sulfonation of naphthalene (thermodynamically controlled sulfonation of naphthalene: Figure 5.8).

cussion of Figure 5.12). At 160 °C the SO_3H group migrates into the 2-position under thermodynamic control. A mechanistic analysis of this reaction can be found in Figure 5.8.

Certain deactivated aromatic compounds can be sulfonated only with fuming rather than concentrated sulfuric acid:

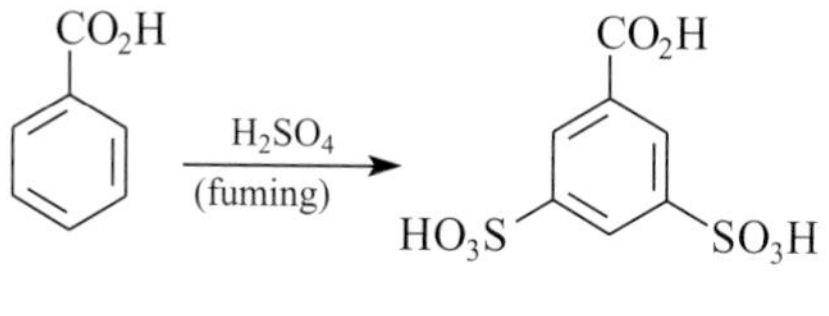

5.2.3 Ar—NO_2 Bond Formation by Ar-S_E Reaction

NO_2 groups are introduced into aromatic compounds through an Ar-S_E reaction with nitric acid **(nitration).** The HNO_3 molecule itself is such a weak electrophile that only the most electron-rich aromatic compounds will react with it (see below). Moderately activated aromatic compounds need concentrated nitric acid for nitration. Under these conditions NO_3^- and $H_2NO_3^+$ are produced in the autoprotolysis equilibrium (Figure 5.20). The NO_2^+ cation (nitronium ion), which is a considerably better electrophile than an HNO_3 molecule, is formed by eliminating a molecule of H_2O from $H_2NO_3^+$.

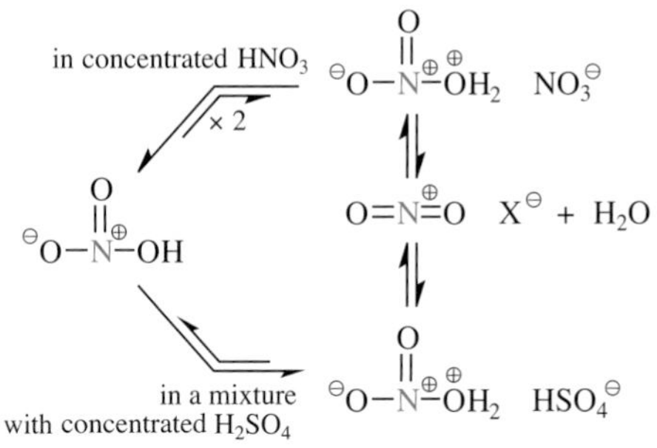

Fig. 5.20. Reactive electrophiles in the nitration of aromatic compounds with nitric acids with different concentrations and with H_2SO_4/HNO_3. (Reactive electrophiles in the nitration of (more) electron-rich aromatic compounds: cf. Figure 5.23 as well as the last equation in this section.)

in concentrated HNO_3 ×2 in a mixture with concentrated H_2SO_4

The nitronium ion initiates, for example, the following reaction:

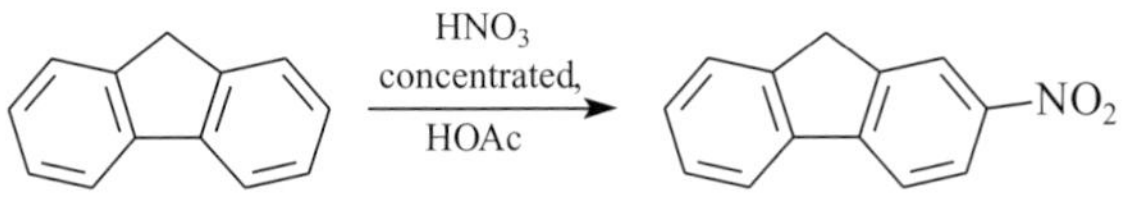

HNO_3 concentrated, HOAc

In this example, the phenyl substituent with its pi electron donating (+M) effect determines the structure of the most stable sigma complex and thus the regioselectivity. The competing inductive (+I) effect of the alkyl substituent cannot compete. This is understandable because it is not as effective at stabilizing an adjacent positive change as the phenyl group.

A higher fraction of nitronium ions is present in a mixture of concentrated sulfuric and concentrated nitric acid. In this medium, the HNO_3 molecules are protonated to a larger extent than in nitric acid. This is due to the sulfuric acid is a stronger acid than nitric acid (Figure 5.20). An HNO_3/H_2SO_4 mixture is therefore also suitable for nitrating deactivated aromatic compounds. Aromatic amines (Figure 5.21) are included in this category: In the very acidic reaction medium, they are protonated quantitatively. Thus, for example, the actual substrate of the nitration of *N, N*-dimethylaniline is **C**, in which the $HMe_2N^{\oplus}$ substituent directs the reacting nitronium ion to the *meta*-position because of its electron withdrawing inductive effect. The analogous nitration of aniline exhibits a selectivity of only 62:38 favoring the *meta*- over the *para*-reaction. This is due to (1) the lower basicity of aniline as compared to *N,N*-dimethylaniline and (2) to greater deactivating effect of an $H_3N^{\oplus}$- vis-a-vis an $HMe_2N^{\oplus}$ substituent. The implication is that the anilinium ion leads to the meta product and the free aniline leads to the para product, even though the latter will be present in extremely small amounts in the reaction mixture.

The previous argument would suggest that in even less basic anilines the reaction of the nitrating acid should be exclusively directed towards the small fraction of unprotonated amine. This would mean that anilines of this type would be nitrated in *ortho*- and/or *para*-position instead of the *meta*-position. This idea is illustrated in the nitration of sulfanilic acid (Figure 5.22), which proceeds *ortho* to the H_2N group and *meta* to the SO_3H of the free amine **B** instead of *meta* to the $H_3N^{\oplus}$ group of the anilinium cation **D**.

The actual Ar-S_E product in Figure 5.22, namely aminonitrobenzene sulfonic acid **F** (the uncharged form), which is in equilibrium with the zwitterionic form **G** (cf. Figure 5.18), will then be desulfonylated. The mechanism of this reaction resembles the one shown in Figure 5.7. So the sigma complex **E** is a key intermediate. In the end, the *ortho*-isomer **C** of

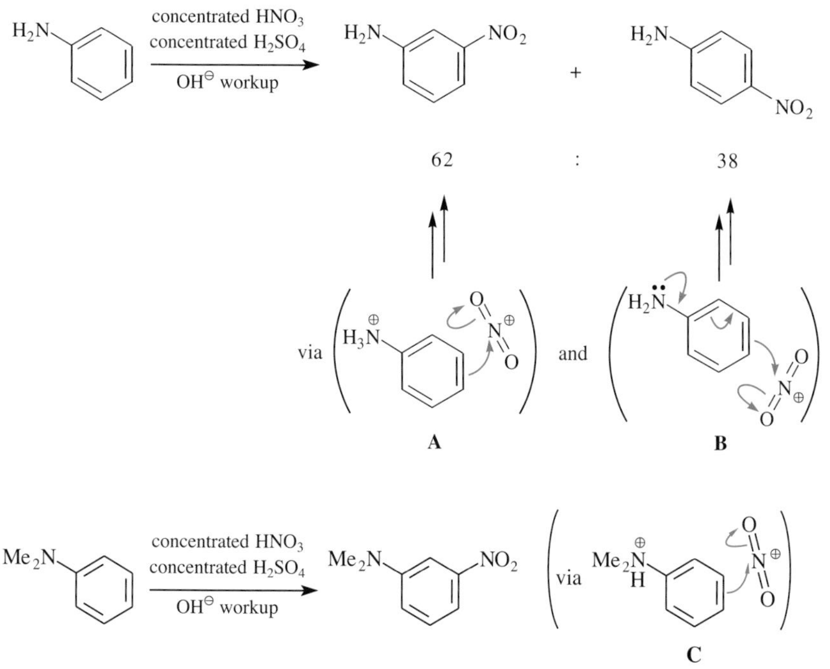

H₂N–C₆H₅ → (concentrated HNO_3, concentrated H_2SO_4, $OH^{\ominus}$ workup) → meta-nitroaniline + para-nitroaniline, 62 : 38, via A and B

Me_2N–C₆H₅ → (concentrated HNO_3, concentrated H_2SO_4, $OH^{\ominus}$ workup) → meta-nitro-N,N-dimethylaniline, via C

Fig. 5.21. Nitration of aniline (top) and of *N*,*N*-dimethylaniline (bottom). Since aniline is a somewhat weaker base than *N*,*N*-dimethyl aniline and an $H_3N^{\oplus}$ substituent is somewhat more deactivating than an $HMe_2N^{\oplus}$ substituent, both the anilinium nitrate (via **A**) and the minimal amount of free aniline (*via* **B**) react in the first reaction, whereas the second reaction only involves *N*,*N*-dimethyl anilinium nitrate (*via* **C**).

nitroaniline is formed, an isomer that is not obtained at all upon the reaction of aniline with nitric acid. Altogether, the three-step reaction sequence of aniline → sulfanilic acid → 4-amino-3-nitrobenzenesulfonic acid → *ortho*-nitroaniline in Figure 5.18/5.22 nicely demonstrates how the reversibility of an aromatic sulfonation can be exploited for regiocontrolled synthesis of multiply substituted aromatic compounds.

Still more strongly deactivated aromatic compounds are nitrated with *hot* HNO_3/H_2SO_4 or with a mixture of fuming nitric acid and concentrated sulfuric acid:

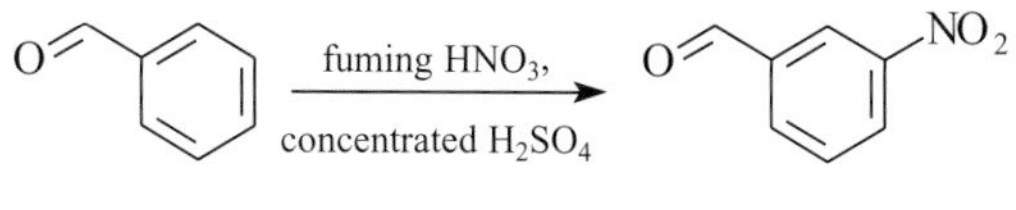

Benzaldehyde → (fuming HNO_3, concentrated H_2SO_4) → 3-nitrobenzaldehyde

The third nitro group of the explosive 2,4,6-trinitrotoluene (TNT) is introduced under similarly drastic conditions (into 2,4- or 2,6-dinitrotoluene).

A special mechanistic feature of nitrations with concentrated nitric or nitrating acid can be seen with many reasonably activated benzenes—even though they proceed via the usual sigma complexes **B** (Figure 5.23). The sigma complexes, however, no longer derive from the familiar ionic elementary reaction, but from the sequence of the three radical reactions described below, and even more unusually, these sigma complexes are not generated in the rate-determining step. Rather, the rate is determined by an electron transfer from the corresponding aromatic compounds to the nitrosyl cation ($NO^{\oplus}$). Small amounts of the latter are contained in the cited reagents that are usually employed for nitrations in organic chemistry. It is these ingredients that allow for the radical substitution mechanism of Figure 5.23! The borderline

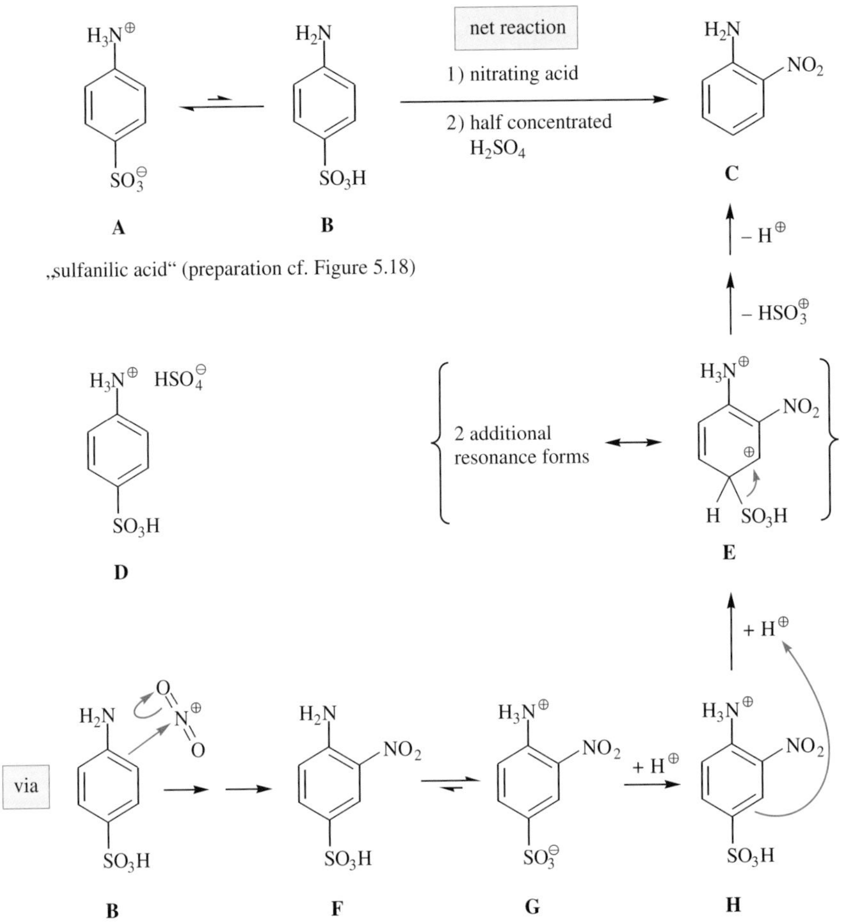

net reaction

1) nitrating acid

2) half concentrated H_2SO_4

„sulfanilic acid" (preparation cf. Figure 5.18)

2 additional resonance forms

via

Fig. 5.22. Nitration of sulfanilic acid **A** ⇌ **B** and desulfonation of the nitration product **F** ⇌ **G**. The protonation product **D** of the substrate is a benzene derivative with *two* strong electron acceptors (SO_3H and $H_3N^{\oplus}$) and therefore cannot form a sigma complex under the reacdtion conditions. The fact that an Ar-S_E reaction occurs nonetheless is due to the small equilibrium amount of the neutral form **B** of sulfanilic acid. **B** is nitrated in *ortho*-position to the electron donor (H_2N) and thus in *meta*-position to the electron acceptor (SO_3H).

between the ionic mechanism of nitration of the other examples in this section and the radical mechanism of nitration of Figure 5.23 runs between toluene on the one side and the alkoxylated or multiply alkylated benzenes on the other.

Actually, HNO_3 is the only electrophile that exists in diluted nitric acid. It is a particularly weak electrophile. This explains why HNO_3 is appropriate for the nitration of very strongly activated aromatic compounds only as shown in the following example (in diluted nitric acid an H_2N group instead of the AcNH residue would be protonated to give a deactivating substituent, whereas the AcNH group would stay unprotonated and thus remain an activating substituent):

dilute HNO_3

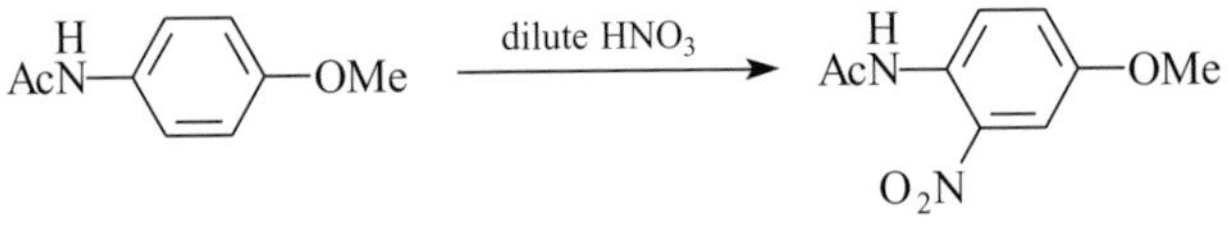

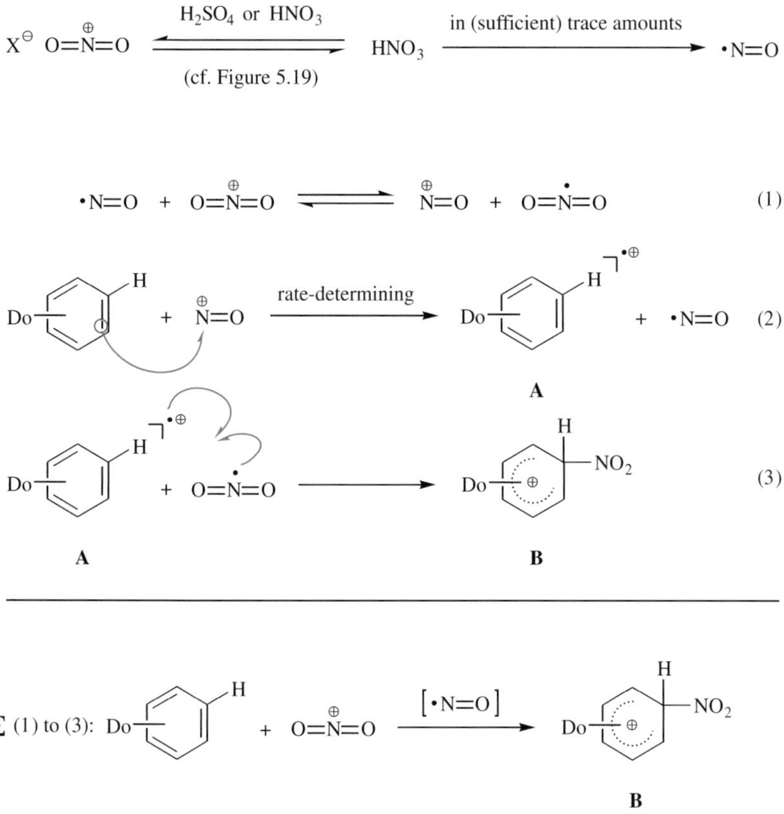

$$X^{\ominus}\ O{=}\overset{\oplus}{N}{=}O \xrightleftharpoons[\text{(cf. Figure 5.19)}]{H_2SO_4 \text{ or } HNO_3} HNO_3 \xrightarrow{\text{in (sufficient) trace amounts}} \bullet N{=}O$$

$$\bullet N{=}O + O{=}\overset{\oplus}{N}{=}O \rightleftharpoons \overset{\oplus}{N}{=}O + O{=}\overset{\bullet}{N}{=}O \quad (1)$$

$$\text{Do–C}_6\text{H}_5 + \overset{\oplus}{N}{=}O \xrightarrow{\text{rate-determining}} \mathbf{A} + \bullet N{=}O \quad (2)$$

$$\mathbf{A} + O{=}\overset{\bullet}{N}{=}O \longrightarrow \mathbf{B} \quad (3)$$

$$\Sigma \text{ (1) to (3): } \text{Do–C}_6\text{H}_5 + O{=}\overset{\oplus}{N}{=}O \xrightarrow{[\bullet N{=}O]} \mathbf{B}$$

Fig. 5.23. Mechanism of the nitration of benzene derivatives that are more electron-rich than toluene: the sigma complex **B** is not formed in a single step, as with all other Ar-S_E reactions, but in a three-step sequence.

There is also some debate over whether nitrations like the one in the last example may alternatively proceed via a nitrosyl cation ($NO^{\oplus}$) instead of a HNO_3 molecule as the electrophile. Small amounts of the nitrosyl cation occur in diluted nitric acid and would—via a Wheland complex intermediate—initially lead to a nitrosoaromatic compound (Ar–N=O) as the Ar-S_E product. (Remember: Figure 5.23 presented a different reaction mode between nitrosyl cations and—less electron-rich—aromatic compounds.) This nitrosoaromatic compound would subsequently undergo rapid oxidation by the diluted nitric acid to finally yield the nitroaromatic compound.

5.2.4 Ar—N=N Bond Formation by Ar-S_E Reaction

Aryldiazonium salts are weak electrophiles. Consequently, they undergo Ar-S_E reactions via sigma complexes (azo couplings) only with the most strongly activated aromatic compounds. Only phenolates and secondary and tertiary aromatic amines react with them. Primary aromatic amines react with diazonium salts, too, but via their N atom. Thus, triazenes, that is, compounds with the structure Ar—N=N—NH—Ar are produced. Phenol ethers or nondeprotonated phenols can react with aryldiazonium salts only when the latter are especially good electrophiles, for example, when they are activated by nitro groups in the *ortho*- or *para*-position.

On an industrial scale, azo couplings serve as a means of producing **azo dyes.** The first azo dyes included the azo compounds Orange I (Figure 5.24) and Orange II (Figure 5.25). They are obtained by azo couplings between diazotized *para*-aminobenzenesulfonic acid (sulfanilic acid) and sodium 1-naphtholate or sodium 2-naphtholate. The perfect regioselectivity of *both* couplings follows from the fact that in each case the most stable sigma complex is produced as an intermediate. The most stable sigma complex in both reactions is the only one of seven conceivable complexes for which there exists an especially stable *all-octet resonance form free of formal charges and with an aromatic ring.* For both azo couplings, this resonance form results from an Ar-S_E reaction on that ring of the naphtholate to which the $O^{\ominus}$ substituent is bound. Figure 5.24 shows as an example the most stable resonance form, which determines the regioselectivity of the formation of Orange I.

In the laboratory, azo couplings can also be carried out for another purpose. Azo compounds, such as those shown in Figures 5.24 and 5.25, can be reduced to aromatic amino com-

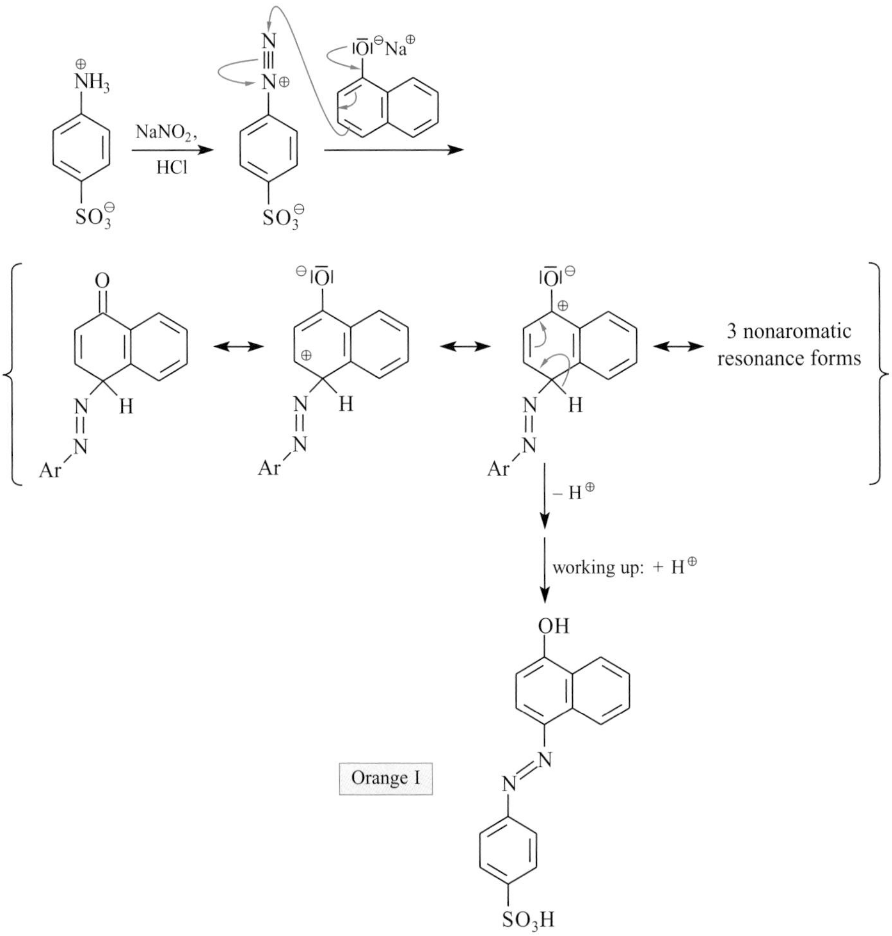

$NH_3^{\oplus}$ ... $SO_3^{\ominus}$ —NaNO₂, HCl→ $N\equiv N^{\oplus}$ ... $SO_3^{\ominus}$ + $|\bar{O}|^{\ominus}Na^{\oplus}$ →

3 nonaromatic resonance forms

$-H^{\oplus}$

working up: $+H^{\oplus}$

OH

Orange I

SO_3H

Fig. 5.24. Regioselective azo coupling between diazotized sulfanilic acid and sodium 1-naphtholate (synthesis of Orange I).

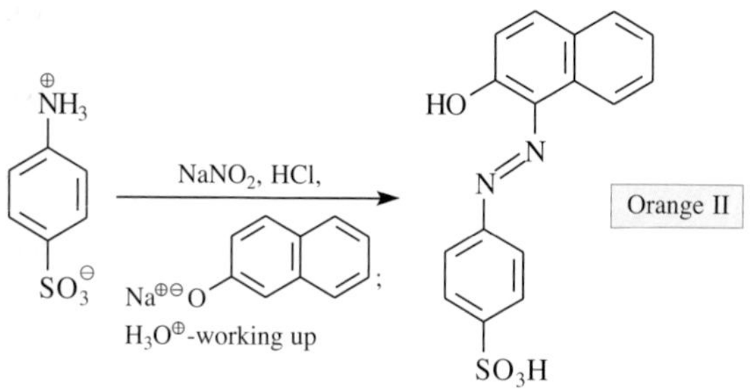

Fig. 5.25. Regioselective azo coupling between diazotized sulfanilic acid and sodium 2-naphtholate (synthesis of Orange II).

pounds. The N=N unit is then replaced by two NH_2 groups. In this case, an azo coupling would be the first step of a two-step process with which naphthols or arylamines can be functionalized with an NH_2 group (i.e., aminated). For many other aromatic compounds, the same transformation can be achieved by nitrating with HNO_3/H_2SO_4 (Section 5.2.3) and subsequently reducing the nitro group to the amino group. However, this standard process cannot be used directly with naphthols, because they are easily oxidized by the reagent mixture to give 1,4- or 1,2-naphthoquinone.

5.2.5 Ar—Alkyl Bond Formations by Ar-S_E Reaction

Suitable Electrophiles and How They React with Aromatic Compounds

Activated aromatic compounds can be alkylated via Ar-S_E reactions (Friedel–Crafts alkylation). Suitable reagents are certain

- alkyl halides or alkyl sulfonates in the presence of catalytic amounts of a Lewis acid,
- alcohols in the presence of catalytic amounts of a Brønsted acid,
- alkenes that can be converted by catalytic amounts of a Brønsted acid directly (i.e., by protonation) or indirectly (see the first reaction in Figure 3.56) into a carbenium ion.

The effective electrophiles in Friedel–Crafts alkylations are the species shown in Figure 5.26.

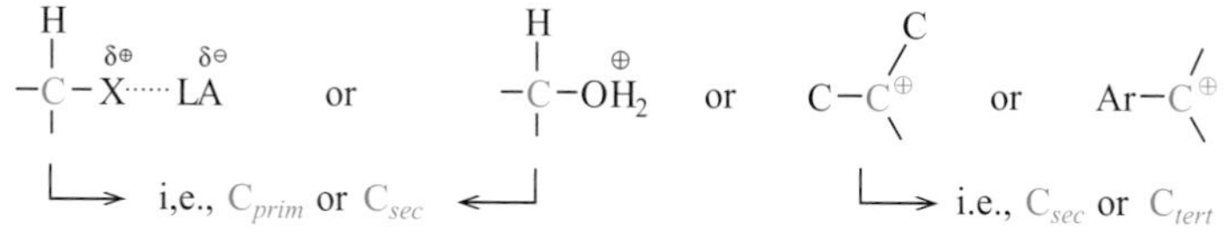

Fig. 5.26. Electrophiles that can initiate Friedel–Crafts alkylations without rearrangements; LA stands for Lewis acid.

When the reactive electrophile is a carbenium ion, a constitutionally unique alkyl group can generally be introduced through an Ar-S_E reaction only when this carbenium ion does not isomerize competitively with its reaction on the aromatic substrate. An isomerization is most reliably excluded for those Friedel–Crafts alkylations in which stable carbenium ions appear in the first step. These include *tert*-alkyl and benzyl cations (Figure 5.26, right). Suitable reagents are therefore all alkyl halides, alkyl sulfonates, alcohols, and alkenes that give either *tert*-alkyl or benzyl cations with the additives enumerated in the above list. Accordingly, the aromatic compound reacts with these species via an S_N1 reaction.

Lewis acid complexes of alkyl halides and alkyl sulfonates (Figure 5.26, left) and protonated alcohols (Figure 5.26, middle) are additional reactive electrophiles in Friedel–Crafts alkylations. The aromatic compound displaces their respective leaving group in an S_N2 process. This is in principle possible (Section 2.4.4) for primary or secondary alkylating agents and alcohols.

Alternatively, namely in the presence of a Lewis acid, secondary alkyl halides and sulfonates can react with aromatic compounds via an S_N1 mechanism. The pair of reactions in Figure 5.27 shows that the balance between S_N2 and S_N1 mechanisms can be subject to subtle substituent effects. (*S*)-2-Chlorobutane and $AlCl_3$ alkylate benzene with complete racemization. This means that the chlorobutane is substituted by the aromatic substrate through an S_N1 mechanism, which takes place via a solvent-separated ion pair (cf. Section 2.5.2). On the other hand, the mesylate of (*S*)-methyl lactate and $AlCl_3$ alkylate benzene with complete inversion of the configuration of the stereocenter. This means that the mesylate group is displaced by the aromatic compound in an S_N2 reaction. There are two reasons why these substitution mechanisms are so different and so clearly preferred in each case. On the one hand, the lactic acid derivative cannot undergo an S_N1 reaction. The carbenium ion that would be generated, $H_3C—CH^{\oplus}—CO_2Me$, would be strongly destabilized by the electron-withdrawing CO_2Me group (cf. discussion of Figure 2.29). On the other hand, S_N2 reactions in the α-position to an ester group take place especially rapidly (cf. discussion of Figure 2.11).

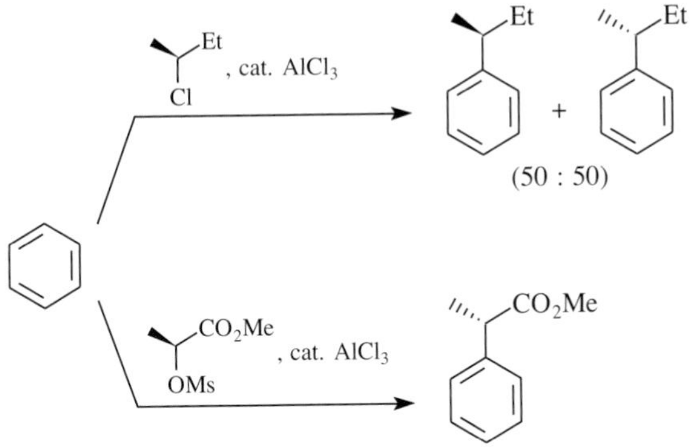

Fig. 5.27. Friedel–Crafts alkylations with secondary alkyl halides or sulfonates and $AlCl_3$: competition between S_N1 (top) and S_N2 (bottom) substitution of the electrophile.

Single or Multiple Alkylation by the Friedel–Crafts Reaction?

Friedel–Crafts alkylations differ from all other Ar-S_E reactions considered in Section 5.1 in that the reaction product is a better nucleophile than the starting material. This is because the alkyl group introduced is an *activating* substituent. Therefore, in a Friedel–Crafts alkylation we risk an overreaction of the primary product to further alkylation.

There are only three conditions under which multiple alkylations do not occur:

- First, *intermolecular* Friedel–Crafts alkylations take place as monofunctionalizations of the aromatic ring when the primary product cannot take up any additional alkyl group for steric reasons (Figure 5.28, formation of **A** and **B**).
- Second, Friedel–Crafts monoalkylations of aromatic compounds such as benzene or naphthalene, in which the introduction of a second alkyl group is not prevented by steric hindrance, can only be carried out with a trick: The reaction is performed with a large

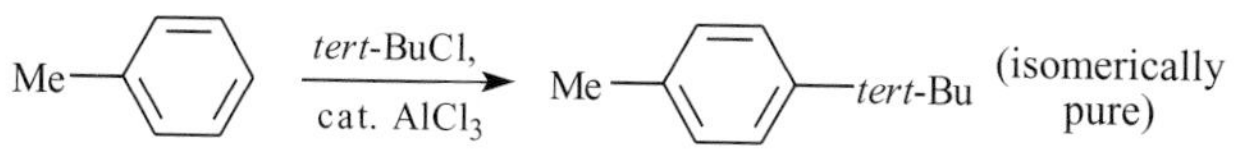

Me–C6H4 + tert-BuCl, cat. $AlCl_3$ → Me–C6H4–tert-Bu (isomerically pure)

A

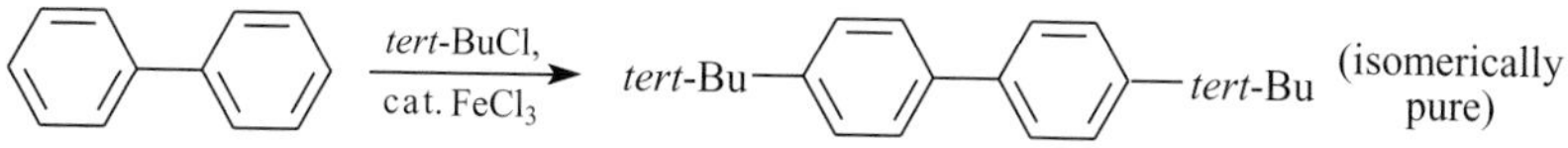

biphenyl + tert-BuCl, cat. $FeCl_3$ → tert-Bu–C6H4–C6H4–tert-Bu (isomerically pure)

B

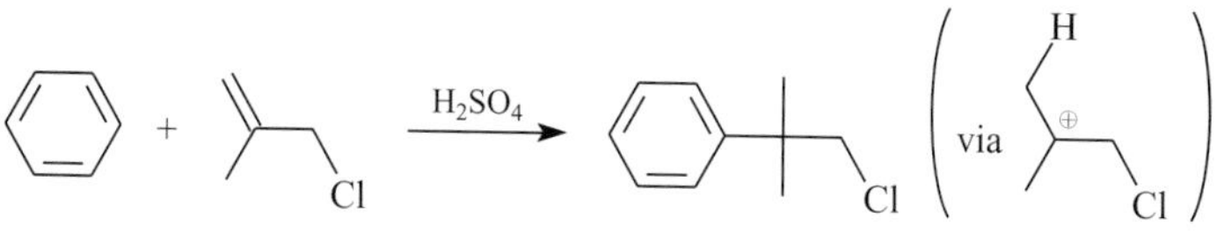

benzene + (alkene)–Cl, H_2SO_4 → **C** (via ⊕ cation)

Fig. 5.28. Friedel-Crafts alkylations with *tert*-alkyl cations as reactive electrophiles. Figure 5.33 exemplifies a way in which compound **A** can be used.

excess of the aromatic compound. There is almost no overalkylation at all. Simply for statistical reasons virtually the entire reaction is directed at the excess starting material, not the primary product. For example, the following substitution succeeds in this way (Figure 5.28, formation of **C**). This same situation occurred in the de-*tert*-butylation of Figure 5.5: mono- and not di-*tert*-butyl benzene was produced because the reaction was carried out *in* benzene.

- The *third* possibility for a selective monoalkylation is provided by *intramolecular* Friedel–Crafts alkylations. There are no multiple alkylations simply because all electrophilic centers react most rapidly intramolecularly (i.e., only once). Friedel-Crafts alkylations of this type are ring closure reactions.

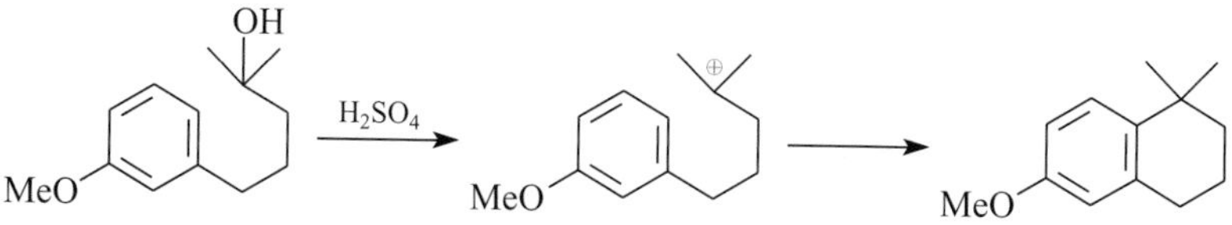

MeO–(aryl)–OH, H_2SO_4 → MeO–(aryl)–⊕ → MeO–(tetralin)

Isomerizations during Friedel–Crafts Alkylations

Intramolecularly, certain secondary carbenium ions can be introduced without isomerization in Friedel–Crafts alkylations. An example is the last step in the bicyclization of Figure 3.56.

In contrast to the intramolecular case just mentioned, in *intermolecular* Friedel–Crafts alkylations secondary carbenium ion intermediates often have sufficient time to undergo a Wagner–Meerwein rearrangement (cf. Section 14.3.1). This can lead to the formation of an unexpected alkylation product or product mixtures:

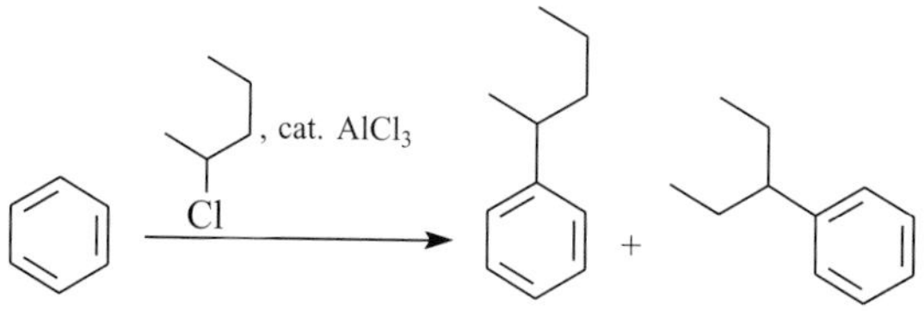

benzene + (alkyl)–Cl, cat. $AlCl_3$ → product + product

Friedel–Crafts Alkylations with Multiply Chlorinated Methanes

If $CH_2Cl_2/AlCl_3$ is used as electrophile in the Friedel–Crafts alkylation, a benzyl chloride is first produced. However, this compound is itself a Friedel–Crafts electrophile, and, in the presence of $AlCl_3$, it benzylates unconsumed starting material immediately. As a result, one has linked two aromatic rings with a CH_2 group:

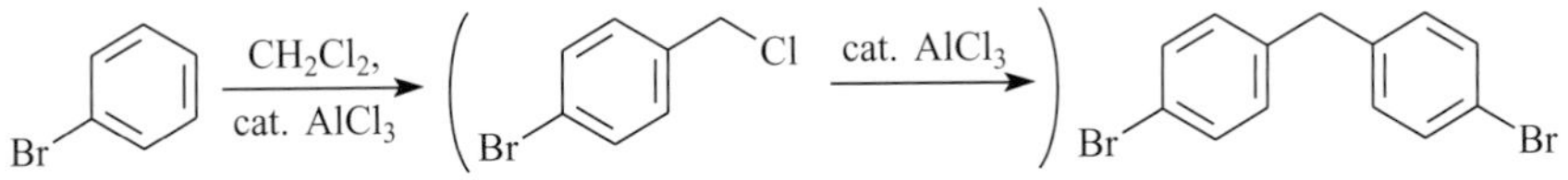

Similarly, Friedel–Crafts alkylations with $CHCl_3$ and $AlCl_3$ lead to the linking of three aromatic rings through a CH group. A related alkylation occurs with CCl_4 and $AlCl_3$. At first, three aromatic rings are linked with a C—Cl group, so that a trityl chloride is obtained. This heterolyzes in the presence of the Lewis acid and gives a trityl cation. This step may seem to set the stage for yet another Friedel–Crafts alkylation. However, the unconsumed aromatic compound is too weak a nucleophile to be able to react with the well-stabilized trityl cation.

5.2.6 Ar—C(OH) Bond Formation by Ar-S_E Reactions and Associated Secondary Reactions

Aldehydes and ketones react with aromatic compounds in the presence of Brønsted or Lewis acids. The actual electrophile is the carboxonium ion formed in an equilibrium reaction by protonation or complexation, respectively. The primary product is a substituted benzyl alcohol, which, however, is not stable and easily forms a benzyl cation. The latter continues to react further, either via an S_N1 or an E1 reaction. Thereby, the following overall functionalizations are realized: Ar—H → Ar—C-Nu or Ar—H → Ar—C=C.

If the activated aromatic compound reacts with a mixture of formalin, concentrated hydrochloric acid, and $ZnCl_2$, the result is a so-called chloromethylation (Figure 5.29). The stable reaction product is a primary benzyl chloride. This reaction is initiated by an electrophilic substitution by protonated formaldehyde; it is terminated by an S_N1 reaction in which a chloride ion acts as the nucleophile.

Phenols are such good nucleophiles that protonated carbonyl compounds functionalize two phenol molecules. The first phenol molecule reacts in an Ar-S_E reaction by the carboxonium ion formed in an equilibrium reaction. Subsequently, the second equivalent of phenol becomes the substrate of a Friedel–Crafts alkylation. The electrophile is the benzyl cation that is formed from the initially obtained benzyl alcohol and the acid. Protonated acetone is only a weak electrophile for electronic and steric reasons: it contains two electron-donating and relatively large

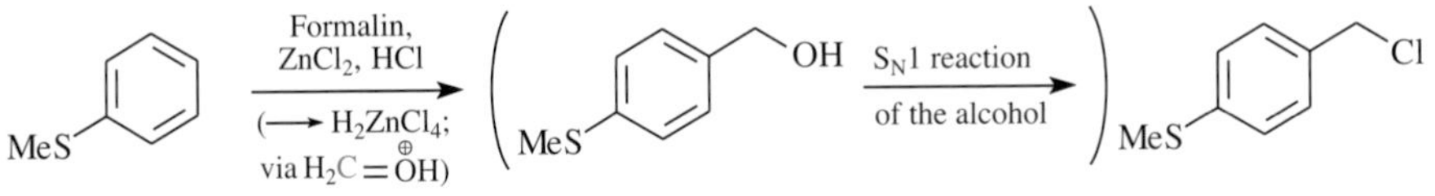

Fig. 5.29. Chloromethylation of aromatic compounds via Ar-S_E reaction.

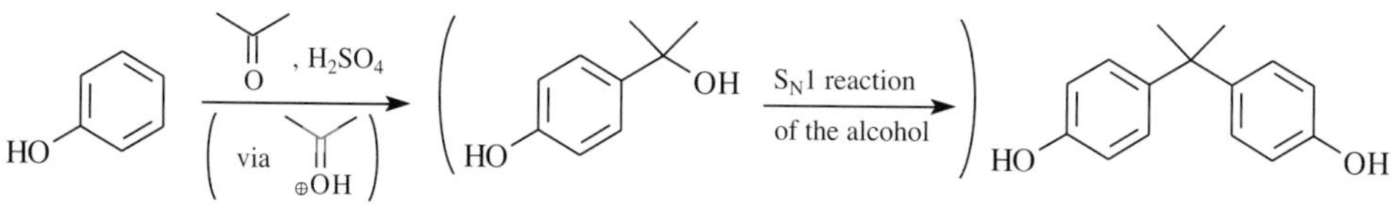

methyl groups on the electrophilic C atom. Therefore, it reacts with phenol regioselectively in the *para*- and not at all in the less favored *ortho*-position (Figure 5.30). The benzyl cation formed thereafter is a poor electrophile, too, and again for both electronic and steric reasons, it reacts with the second phenol molecule with high *para*-selectivity.

Fig. 5.30. Linking aromatic compounds with methylene units via Ar-S_E reactions—preparation of bisphenol **A**, an industrially relevant compound for the production of polycarbonates and epoxy resins.

Under comparable reaction conditions the much more reactive formaldehyde and phenol do not only give the *para*- but also the *ortho*-substituted phenol derivative. This reaction ultimately leads to the three-dimensional network of formaldehyde/phenol condensation resins such as Bakelite and to related, well-defined receptor molecules known as calixarenes.

The *intramolecular* hydroxyalkylation of aromatic compounds is a ring closure reaction. In the reaction example in Figure 5.31, it is followed by an E1 elimination, which leads to a styrene derivative.

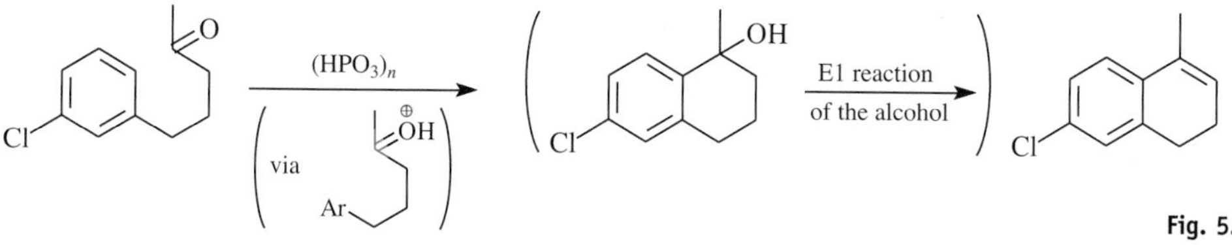

Fig. 5.31. Alkenylation of an aromatic compound by a combination of Ar-S_E and E1 reactions.

5.2.7 Ar—C(=O) Bond Formation by Ar-S_E Reaction

Bond formation between aromatic compounds and a C(=O)—$CR^1R^2R^3$ unit (R^n = H, alkyl, and/or aryl) is the domain of the Friedel–Crafts acylation. As reagents one uses:

- a carboxylic acid chloride with a stoichiometric amount of $AlCl_3$ (because the resulting ketone binds one equivalent of $AlCl_3$ in a Lewis acid complex),
- a carboxylic acid anhydride with a stoichiometric amount of $AlCl_3$,
- a carboxylic acid anhydride together with a mineral acid, or
- a carboxylic acid together with a mineral acid.

In each case the reactive electrophile is produced from these reagents in an equilibrium reaction (Figure 5.32).

The first reagent combination, carboxylic acid chloride/$AlCl_3$, reacts via the $AlCl_3$ complex **A** of the acid chloride or via the acylium tetrachloroaluminate **B** formed from it by β-elimination. A carboxylic acid anhydride and $AlCl_3$ react via analogous electrophiles, namely via the $AlCl_3$ complex **D** of the anhydride or via the acylium salt **E** formed therefore by a β-elimination. The protonated anhydride **F** and the protonated carboxylic acid **C** are the reactive electrophiles of the Friedel–Crafts acylations catalyzed by Brønsted acids.

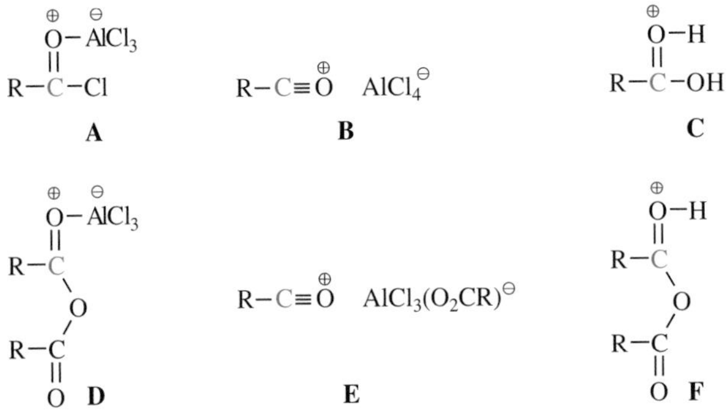

A: $R-C(=\overset{\oplus}{O}-\overset{\ominus}{AlCl_3})-Cl$

B: $R-C\equiv\overset{\oplus}{O}\quad AlCl_4^{\ominus}$

C: $R-C(=\overset{\oplus}{O}-H)-OH$

D: $R-C(=\overset{\oplus}{O}-\overset{\ominus}{AlCl_3})-O-C(=O)-R$

E: $R-C\equiv\overset{\oplus}{O}\quad AlCl_3(O_2CR)^{\ominus}$

F: $R-C(=\overset{\oplus}{O}-H)-O-C(=O)-R$

Fig. 5.32. Reactive electrophiles in the Friedel–Crafts acylation.

Importantly, a Friedel–Crafts *formylation* has *not* yet been successful. Formyl chloride and formic anhydride are not stable reagents. The mixed anhydride H—C(=O)—O—C(=O)CH_3 acts as a formylating reagent in reactions with many nucleophiles (cf. Section 6.3.3). However, in reactions with aromatic compounds under Friedel–Crafts conditions, it acts as an acetylating agent rather than as a formylating agent. Last but not least, formic acid and mineral acids proceed to react via the acylium ion $H—C\equiv O^{\oplus}$ to form carbon monoxide and water in an α-elimination. We'll solve this problem later.

Substrates of Friedel–Crafts acylations are benzene and naphthalene, as well as their halogen, alkyl, aryl, alkoxy, or acylamino derivatives. Acceptor-substituted aromatic compounds are inert. Because Friedel–Crafts acylations introduce an acceptor into the aromatic substrate, no multiple substitutions take place. This distinguishes them from Friedel–Crafts alkylations. Free OH and NH_2 groups in the aromatic compound prevent Friedel–Crafts acylations because they *themselves* are acylated. However, the *O*-acylphenols available in this way can later be rearranged with $AlCl_3$ into *ortho*-acylated isomers (**Fries rearrangement**).

Figure 5.33 presents Friedel–Crafts acylations, taking benzoylations of toluene (top line) and *para-tert*-butyl toluene (Figure 5.33, bottom) as an example. The methyl group of toluene preferentially directs the benzoyl residue into the *para*-position. The *ortho*-benzoylated toluene occurs only as a by-product. In *para-tert*-butyl toluene both the methyl- and the *tert*-butyl substituent direct the electrophile towards the *ortho*-position, since both *para*-positions are occupied and could at best react with de-*tert*-butylation, i.e., in a—sterically hindered!—*ipso*-substitution (cf. Figure 5.5). Indeed, we see reaction *ortho* to the methyl group and not *ortho* to the *tert*-butyl group. This selectivity can be ascribed to minimized steric interactions in the preferred sigma complex intermediate.

The benzoylation product of *para-tert*-butyl toluene can be de-*tert*-butylated according to the method and mechanism given in Figure 5.5, namely by treatment with ample amounts of $AlCl_3$ in benzene. The de-*tert*-butylation leads to isomerically pure *ortho*-methylbenzophenone (Figure 5.33). Upon Friedel-Crafts benzoylation of toluene this compound occurred only as a by-product only (Figure 5.33). Thus, the 3-step reaction sequence of toluene → *para-tert*-butyltoluene → 3-*tert*-butyl-6-methylbenzophenone → *ortho*-methylbenzophenone in Figure 5.28 (top) and Figure 5.33 demonstrates how the reversibility of an aromatic *tert*-butylation may be used for the regiocontrolled construction of multiply substituted aromatic compounds.

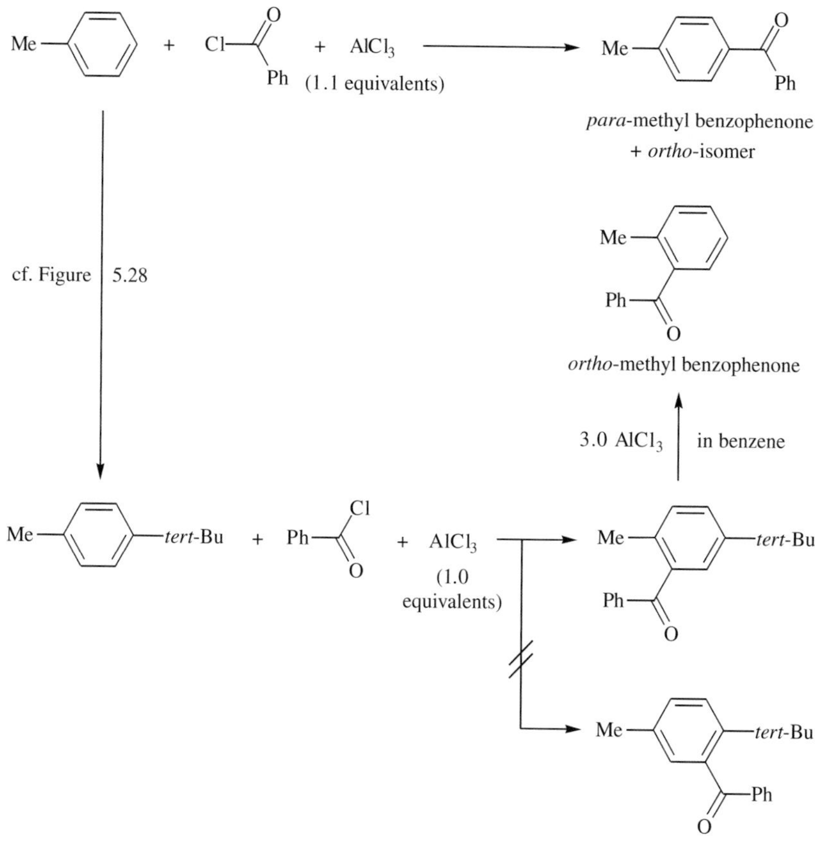

Me–C6H4 + Cl–C(=O)–Ph + $AlCl_3$ (1.1 equivalents) → para-methyl benzophenone + ortho-isomer

cf. Figure 5.28

para-methyl benzophenone
+ *ortho*-isomer

ortho-methyl benzophenone

3.0 $AlCl_3$ in benzene

Me tert-Bu + Ph–C(=O)Cl + $AlCl_3$ (1.0 equivalents)

Fig. 5.33. Friedel-Crafts alkylations of mono- or dialkyl benzenes. This figure depicts a pair of reactions, one of which mainly leads to *para*-methyl benzophenone, and the other—in combination with a subsequent de-*tert*-butylation—exclusively gives *ortho*-methyl benzophenone, respectively.

Friedel–Crafts acylations with anhydrides include the **Haworth synthesis** of naphthalenes from benzenes via the annulation of a C_4 unit (Figure 5.34). However, the two C—C(=O) bonds cannot be made in the same reaction: the acyl group that enters first deactivates the aromatic compound so that it is protected from a second reaction, even though in this case it would be intramolecular. The second step of the Haworth synthesis is therefore a Wolff-Kishner reduction of the carbonyl to a methylene group (for the mechanism, see Section 17.4.6). In the aromatic compound, which now is activated again, a second Friedel-Crafts acylation is possible as step 3. It takes place through the carboxylic acid group in the presence of polyphosphoric acid. The reacting electrophile is either the protonated carboxylic acid, a mixed carboxylic acid/phosphoric acid anhydride or an acylium ion.

Multiply hydroxylated benzene derivatives react with phthalic acid anhydride and a suitable activator, when heated, in a *double* C acylation. Evidently, under these conditions, the activating influence of the additional OH groups is important even after the first acyl group has been introduced. The resulting doubly acylated product is a 9,10-anthraquinone. Compounds of this type are important as dyes:

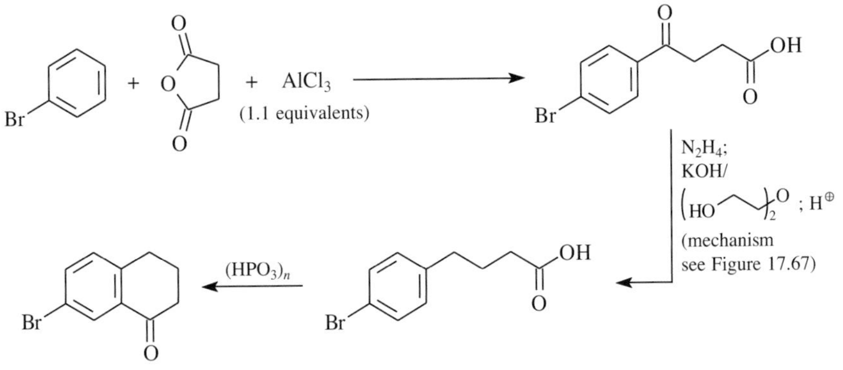

Fig. 5.34. Steps 1–3 of the five-step Haworth synthesis of substituted naphthalenes.

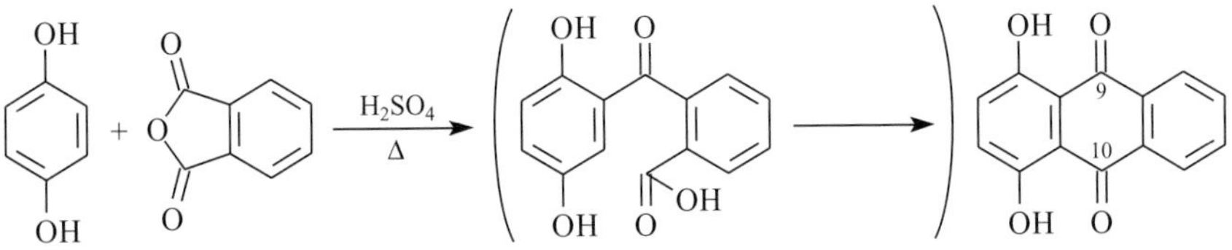

Side Note 5.3. Catalytic Friedel–Crafts Acylations

Friedel–Crafts acylations are also employed for the industrial-scale synthesis of aromatic ketones. Usually this requires Lewis acids like $AlCl_3$ and occasionally $SnCl_4$, BF_3•OEt_2 or $ZnCl_2$, but always *in (more than) stoichiometric amounts*. The demand for so much Lewis acid is due to the fact that it is bound as a 1:1-complex by the carbonyl group of the reaction product, thus precluding any further involvement of this Lewis acid in the reaction: It cannot *catalyze* the Friedel–Crafts acylation. Using *stoichiometric* instead of catalytic amounts of $AlCl_3$, $SnCl_4$, BF_3•OEt_2 or $ZnCl_2$ causes additional costs; above all, it involves much extra work for the isolation of the product and necessitates the environmentally safe disposal of considerable amounts of acidic hydrolysate.

Against this background it is important that—quite fitting in this still new millennium—the first catalytic Friedel-Crafts acylations of (still relatively electron-rich) aromatic compounds were reported (Figure 5.35). Trifluoromethane sulfonates ("triflates") of rare-earth metals, e. g., scandium(III)triflate, accomplish Friedel-Crafts acylations with amounts of as little as 1 mole percent. Something similar is true of the tris(trifluoromethanesulfonyl)-methides ("triflides") of rare-earth metals. Unlike conventional Lewis acids, the cited rare-earth metal salts can form 1:1 complexes with the ketone produced, but these are so unstable that the Lewis acid can re-enter the reaction. Whether this works analogously for the third catalytic system of Figure 5.35 is unclear.

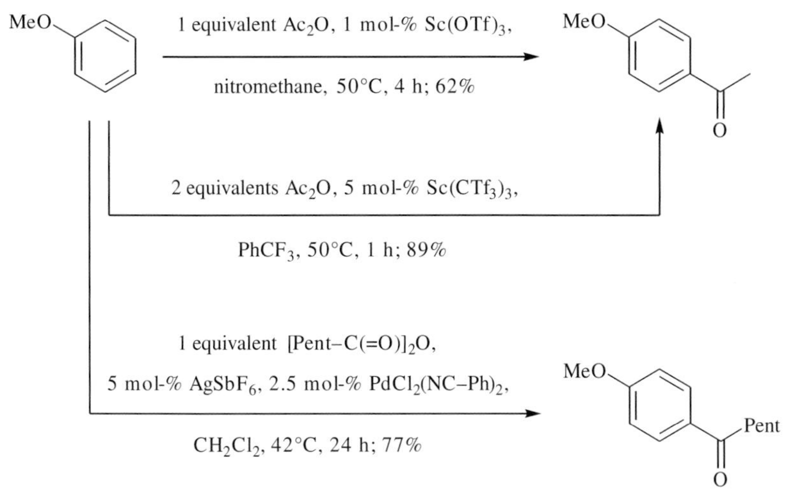

Fig. 5.35. Friedel-Crafts alkylations proceeding with only catalytic amounts of uncommon Lewis acids instead of stoichiometric amounts of aluminum trichloride of the classical procedure.

5.2.8 Ar—C(=O)H Bond Formation through Ar-S_E Reaction

As we have seen in Section 5.2.7, formylations by a Friedel-Crafts reaction are not possible. Therefore, the formation of Ar—C (=O)H bonds through Ar-S_E reactions is normally carried out using the **Vilsmeier-Haack formylation**. The reagent in the Vilsmeier-Haack formylation is a 1:1 mixture of DMF and $POCl_3$. This forms the actual electrophile, an α-chlorinated iminium ion, according to the sequence of steps shown in Figure 5.36. This species is occasionally referred to as the Vilsmeier reagent. Being an iminium ion, it is a weaker electrophile than, for example, the oxonium ion of the Friedel-Crafts acylation, because the pi electron donating effect of the NMe_2 group is greater than that of $—O—AlCl_3^-$. As one consequence, Vilsmeier-Haack formylations are possible only on especially nucleophilic aromatic com-

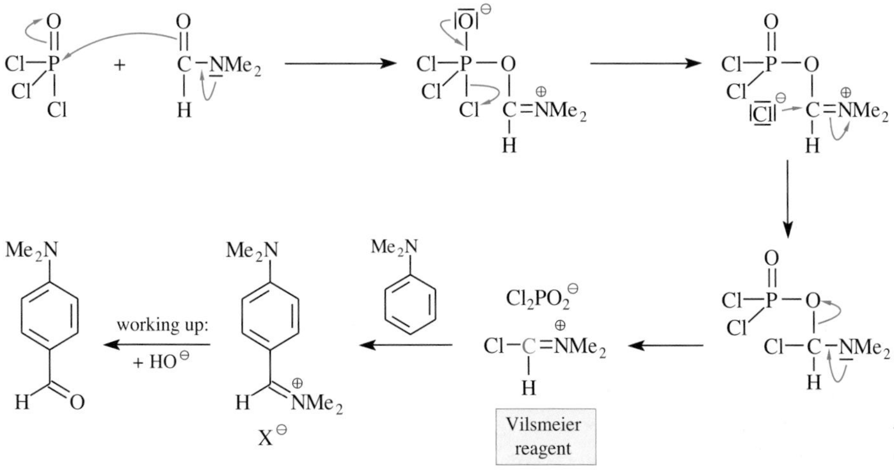

Fig. 5.36. Vilsmeier-Haack formylation ($X^- = Cl^\ominus$ or $Cl_2PO_2^-$).

pounds, i.e., on aniline, phenol, and their derivatives. After completion of the reaction, a benzylic iminium ion is present (Figure 5.36). It remains stable until it is hydrolyzed with water via an unstable *N,O*-acetal intermediate to give the desired aldehyde.

Because the iminium ion derived from *H*—C(=O)NMe_2 is a poor electrophile, it is understandable why there are in general no analogous Vilsmeier–Haack *acylations* using *R*—C(=O)NMe_2. The corresponding iminium ion *R*—C(Cl)=NMe_2^+ is a still poorer electrophile for steric and electronic reasons.

5.3 Electrophilic Substitution Reactions on Metalated Aromatic Compounds

Aryl metal compounds Ar—M (M = Li, Mg—Hal) are much better nucleophiles than the corresponding metal-free aromatic compounds Ar—H. Consequently, they react with many electrophiles that could not react with the corresponding nonmetalated aromatic compounds through an Ar-S_E reaction (Sections 5.3.1 and 5.3.2). In addition, there are methods for preparing substituted aryl metal compounds that are isomerically pure. In this way one has an especially elegant possibility for carrying out *ortho*-selective Ar-S_E reactions (Section 5.3.1), which otherwise is impossible. Finally, aryl boron compounds are the best substrates for the electrophilic hydroxylation of aromatic compounds giving phenols. Moreover, aryl boron compounds make it possible to attach unsaturated hydrocarbon substituents to aromatic rings (Section 5.3.3).

5.3.1 Electrophilic Substitution Reactions of *ortho*-Lithiated Benzene and Naphthalene Derivatives

Certain derivatives of benzene and naphthalene can be lithiated with *sec*-butyllithium (*sec*-BuLi). This reaction is regioselective. It takes place exclusively in the *ortho*-position (Directed ortho Metalation, DoM) to a so-called **D**irected-**M**etalation **G**roup (DMG), whose presence, accordingly, is a prerequisite for such a metalation. Figure 5.37 gives examples of DMGs that are bound through a C, an O, or an N atom to the aromatic compound.

All DMGs shown contain an O atom that can loosely bind the electron-deficient Li atom of *sec*-BuLi by formation of the complex **B** (Figure 5.37). It is believed that this Li atom remains complexed by the same O atom during the H/Li exchange that follows. In the corresponding transition state (**C** in Figure 5.37), the complexation of the lithium takes place in such a way that the $C_{\text{sec-butyl}}$—Li bond exhibits considerable partial charges: a positive charge on the Li atom and a negative charge in the *sec*-butyl moiety. The *sec*-butyl moiety thus becomes so basic that it can remove one of the H atoms located in the immediate vicinity of the DMG, i.e., in the *ortho*-position. Fortunately, this deprotonation does not create an energetically disfavored carbanion because, as indicated in the transition state structure **C** by the

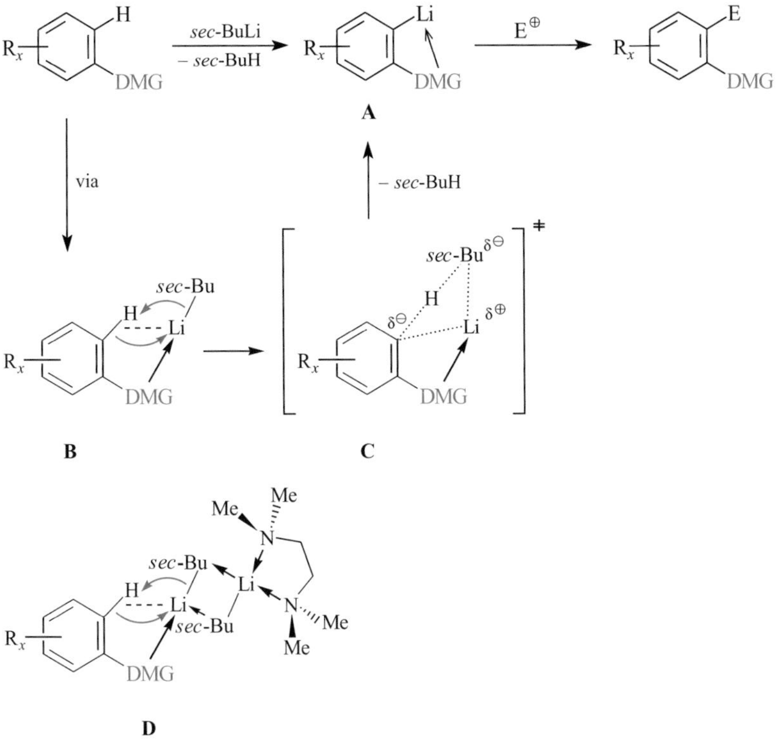

DMG is ...	... C-bound	... O-bound	... N-bound
	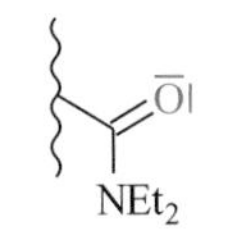 $\overline{O}$| NEt$_2$	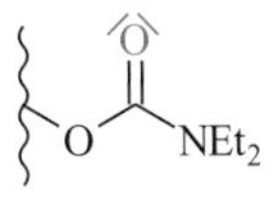 O NEt$_2$	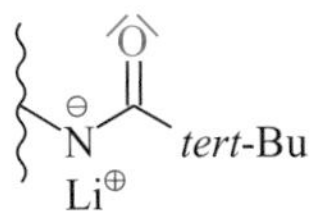 N$^{\ominus}$ Li$^{\oplus}$ *tert*-Bu
and is a ...	... amide	... carbamate	... lithioamide

Fig. 5.37. *ortho*-Selective electrophilic functionalization of aromatic compounds via a substituent-controlled lithiation. As alkyllithium compounds are oligomers (cf. Section 10.1) the presentation of the adduct **B** of the aromatic compound and *sec*-butyllithium as monomeric is likely to simplify the actual situation. Regarding the degree of oligomerization in the alkyllithium substructure, the structure of the adduct **D** of the aromatic compound and *sec*-butyllithium is the correct one. It contains one molecule $Me_2N{-}CH_2{-}CH_2{-}NMe_2$ (**t**etra**m**ethyl**e**thylen**dia**mine, TMEDA). TMEDA is frequently employed with *ortho*-lithiations. This ligand binds *sec*-butyllithium as a dimer (*sec*-butyllithium • TMEDA)$_2$ (cf. Figure 10.2, top.)

dotted bonds, concomitantly the fairly stable C—Li bond of the *ortho*-lithiated aromatic compound **A** has been established.

Strictly speaking, complex **B** and transition state **C** describe only *ortho*-lithiations proceeding under the influence of groups with a strong metalation-directing effect. Groups with a weaker metalation-directing effect do not enable *ortho*-lithiations with *sec*-BuLi alone. This is probably due to the fact that the DMG cannot break down the hexameric structure of *sec*-BuLi (cf. Figure 10.1). *ortho*-Lithiation next to weaker DMG can typically be effected if apart from *sec*-BuLi one also adds **t**etra**m**ethyl**e**thylene**dia**mine (TMEDA). This additive transforms hexameric *sec*-BuLi into a 2:2-adduct (*sec*-BuLi)$_2$TMEDA$_2$ (for the structure cf. Figure 10.2). In the latter, one molecule of TMEDA is replaced by the DMG of the aromatic compound resulting in the formation of complex **D** (Figure 5.37). Its structure corresponds exactly to the one of the previously discussed complex **B** of aromatic compounds with a stronger

DMG and thus enables *ortho*-lithiation via a transition state analogous to **C** and the formation of a lithiated aromatic structure **A**.

Any reaction Ar—H + *sec*-BuLi → Ar—Li + *sec*-BuH possesses a considerable driving force. In an DMG-controlled *ortho*-lithiation it is even greater than it would be normally. This is due to the stabilization of the *ortho*-lithiated aromatic compound **A** because of the intramolecular complexation of the Li atom by the donor oxygen of the neighboring DMG and, at least in some cases, to inductive stabilization provided by the DMG. The exclusive occurrence of *ortho*-lithiation is thus a consequence not only of the precoordination of *sec*-BuLi by the DMG but also of product development control.

Figure 5.38 shows an *ortho*-selective bromination of an aromatic compound that contains a C-bound DMG. Instead of the strong electrophile Br_2, the weak electrophile NBS is used. With respect to NBS, this reaction represents an S_N2 reaction of the organometallic on the bromine.

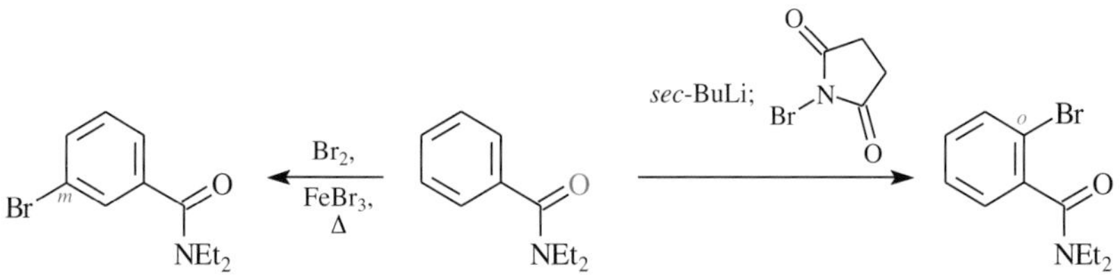

Fig. 5.38. Electrophilic functionalization *ortho* to a C-bound DMG (on the left for comparison: *meta*-selectivity of the analogous classic Ar-S_E reaction).

Figure 5.39 shows the preparation of an arylboronic acid as an example of the *ortho*-selective functionalization of an aromatic compound containing an O-bonded DMG. The $Li^+ArB(OMe)_3^-$ complex is generated and then hydrolyzed during workup.

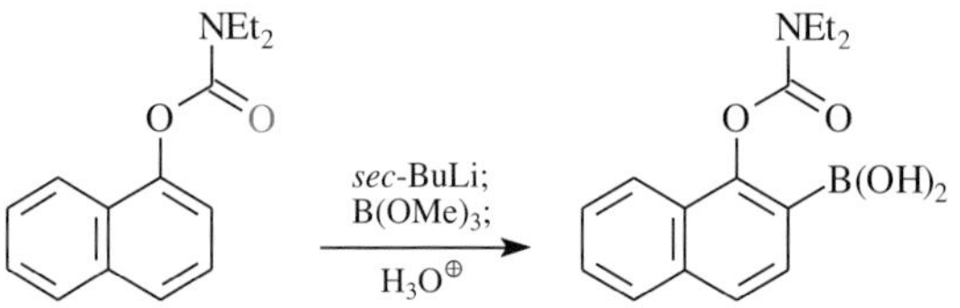

Fig. 5.39. Electrophilic functionalization *ortho* to an O-bonded DMG; preparation of an arylboronic acid.

In Figure 5.40, we see how an aromatic compound can be functionalized with complete regioselectivity exploiting an *N*-lithioamide as an N-bonded DMG. The first reaction step is an *ortho*-lithiation. It allows the introduction of an *ortho*-methyl group by quenching with MeI. The second reaction arrow illustrates that a DMG can also activate a neighboring benzylic position toward lithiation by *sec*-BuLi. The benzyllithium obtained can then be combined with the entire palette of known electrophiles (e.g., the third reaction step in Figure 5.40).

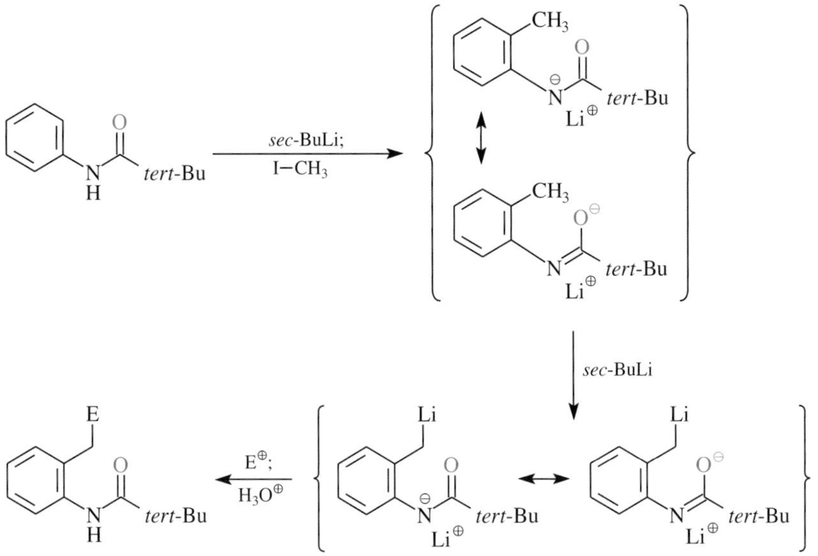

Fig. 5.40. Electrophilic functionalization of C_{sp^2}—H and CH_2—H groups in the position *ortho* to an N-bonded DMG. Here, the lithiation of the benzylic position is associated with a change from colorless to yellow, which can be used to determine the concentration of the alkyllithium solution used.

5.3.2 Electrophilic Substitution Reactions in Aryl Grignard and Aryllithium Compounds That Are Accessible from Aryl Halides

The title reactions offer a possibility for exchanging the halogen atom in aryl halides (Hal = Cl, Br, I) first with a metal (MgHal, Li) and then with an electrophile. It is generally easier to introduce bromine than chlorine or iodine into aromatic compounds. Accordingly, functionalizations of aryl bromides are the preparatively most important examples of the title reaction.

Aryl bromides can be converted to aryl Grignard or aryllithium compounds in three ways (Figure 5.41). In the first two methods, the aryl bromide is reacted with Mg shavings or with Li wire, respectively (see Section 17.4.1 for the mechanism). In the third method—which is especially convenient for small-scale preparations—aryl bromides are converted into aryllithium compounds either with 1 equivalent of butyllithium (*n*-BuLi) or with 2 equivalents of *tert*-butyllithium (*tert*-BuLi) by the so-called Br/Li exchange reaction.

The mechanism of the Br/Li exchange reaction in aryl bromides is shown in Figure 5.41. It is the same as the mechanism of the Br/Li exchange reaction in bromoalkenes (Figure 16.16). First, the aryl bromide and *n*- or *tert*-BuLi furnish the lithium salt **A**. Its anion part contains a negatively charged bromine atom with two bonds and a total of 10 valence electrons.The lithium salt **A** is called an "**-ate complex**." This –ate complex decomposes when the $Li^⊕$ ion reacts with one of its C—Br bonds as an electrophile. This reaction occurs at approximately the center of the respective C—Br bond, which is the site of the highest electron density. When the $Li^⊕$ reacts with the C—Br bond that was established during the formation of **A**, the –ate complex reverts to the starting materials. However, the preferred site of reaction by the $Li^⊕$ is the C_{aryl}-Br bond of the –ate complex **A**. In this way, **A** reacts to form the desired lithioaromatic compound and to *n*- or *tert*-butyl bromide, respectively.

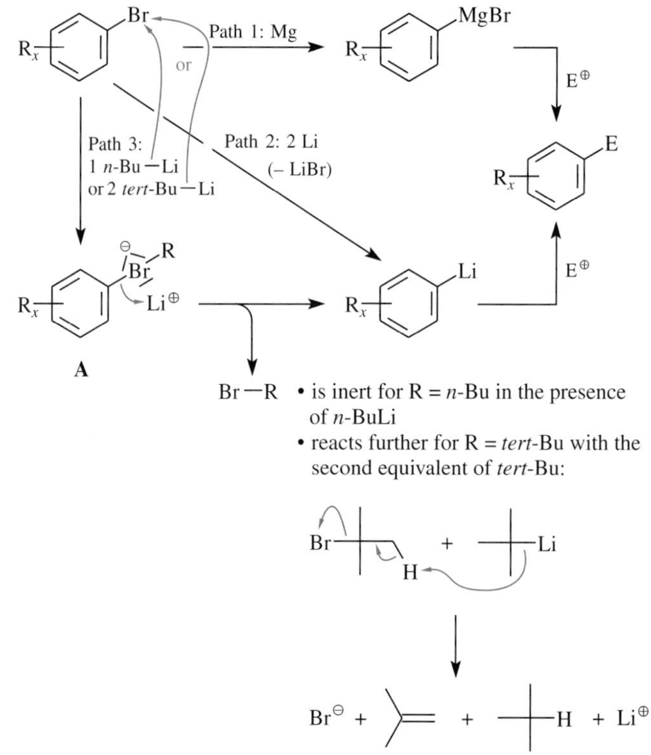

- is inert for R = *n*-Bu in the presence of *n*-BuLi
- reacts further for R = *tert*-Bu with the second equivalent of *tert*-Bu:

Fig. 5.41. Conversions of Ar—Br into Ar—MgBr or Ar—Li (for analogous conversions of Ar–I: cf. Figure 5.46.)

Overall, these reactions give rise to the following equilibrium:

It lies completely on the side of the lithio-aromatic compound and *n*- or *tert*-butyl bromide. This is because the C atom of the C—Li bond is sp^2-hybridized in the lithio-aromatic compound and therefore more electronegative than the sp^3-hybridized C atom of the C—Li bond in *n*- or *tert*-BuLi. More electronegative C atoms stabilize C—Li bonds because these are very electron-rich. Think of it as a carbanion. There is more s character in an sp^2-hybridized orbital and a pair of electrons in such an orbital should be lower in energy than in an sp^3-hybridized orbital.

If *n*-BuLi is used for the Br/Li exchange in Figure 5.41, the reaction is completed when the lithio-aromatic compound and *n*-butyl bromide have formed. It is different when this Br/Li exchange is carried out with *tert*-BuLi, where *tert*-BuLi and *tert*-butyl bromide necessarily would have to coexist during the reaction. However, *tert*-BuLi reacts very fast as a base with *tert*-butyl bromide via an E2 elimination. Therefore, Br/Li exchange reactions with *tert*-BuLi can go to completion only when 2 equivalents of the reagent are used.

The reactions of electrophiles with metalated aromatic compounds obtained in this way provide access to products that are not readily accessible through classic Ar-S_E reactions.

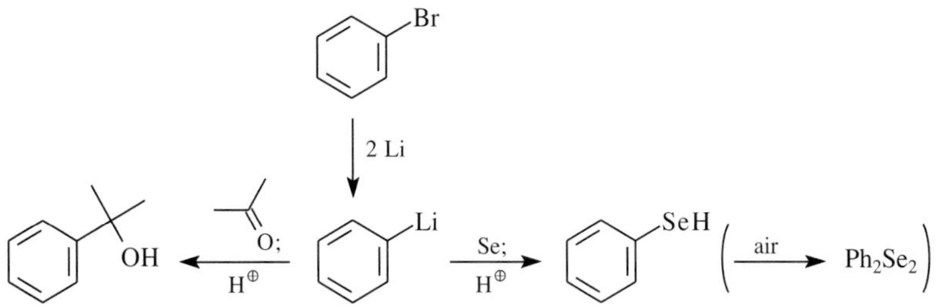

Br
2 Li
OH O; $H^{\oplus}$ Li Se; $H^{\oplus}$ SeH air Ph_2Se_2

Fig. 5.42. Electrophilic functionalization of phenyllithium. Tertiary alcohols can be synthesized from ketones. Diphenyl diselenide can be prepared from elemental selenium.

Phenyllithium, for example, unlike benzene, reacts with ketones to form tertiary alcohols or with elemental selenium to form first phenylselenol and thereafter, through oxidation, diphenyl diselenide (Figure 5.42).

There are no direct ways to introduce a CO_2H group, a single deuterium atom, or a single additional methyl group into mesitylene. However, the desired products can easily be prepared via metalated mesitylene as an intermediate. Mesitylene is first brominated on the aromatic ring, and the bromomesitylene formed is then subjected to a Li/Br exchange (Figure 5.43). The resulting mesityllithium reacts with CO_2, D_2O, and MeI to afford the target molecules.

Interestingly, it is possible to obtain *di*-Grignard compounds from many dibromoaromatic compounds. These react with 2 equivalents of an electrophile. Diphenylchlorophosphine ($ClPPh_2$ Figure 5.44), would not react with metal-free 1-1'-binaphthyl. However, with the di-Grignard compound shown it forms BINAP (2,2'-**bis**(diphenylphosphino)-1,1'-bi**nap**hthyl)—a compound that in the form of its pure enantiomer has been used with great success as a ligand in transition metal complexes for the enantioselective hydrogenation of C=C and C=O bonds (Section 17.4.7), among many other applications.

The last electrophilic substitution reaction of a Grignard compound we want to consider is a transmetalation, namely one that leads to an arylboronic ester (Figure 5.45). Arylboronic esters or their hydrolysis products, the arylboronic acids, are valuable reagents in modern aro-

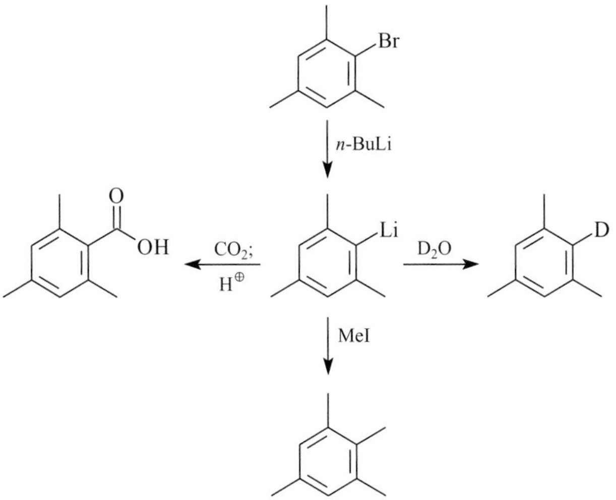

Br
n-BuLi
O OH CO_2; $H^{\oplus}$ Li D_2O D
MeI

Fig. 5.43. Electrophilic functionalization of mesityllithium. A COOH group, a deuterium atom, or a methyl group can be introduced readily.

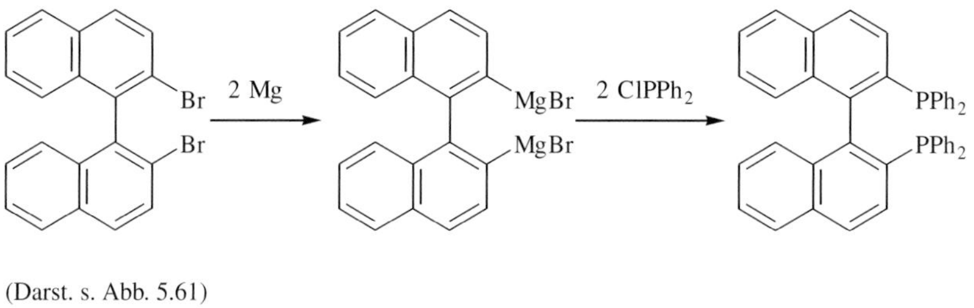

Fig. 5.44. Electrophilic functionalization of an aromatic di-Grignard compound; synthesis of *R*-BINAP. (Enantiomerically pure BINAP is one of the most important ligands in asymmetric catalysis, cf. Figures 17.76 and 17.77.)

matic chemistry. They react with a series of electrophiles that would not react with Grignard or organolithium compounds (Sections 5.3.3 and 16.4.2).

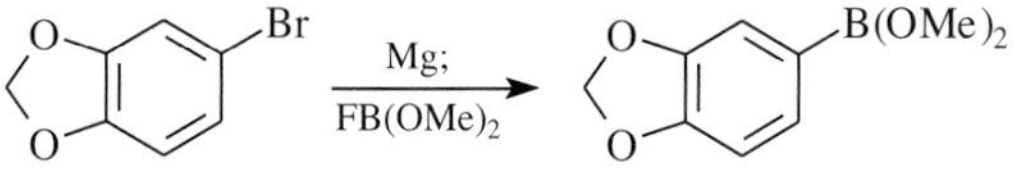

Fig. 5.45. Preparation of an arylboronic ester via a Grignard compound.

As shown in Figure 5.46, there are four methods to transform an aryl iodide into an aryllithium or aryl Grignard compound. Three of these correspond to those presented in Figure 5.41 for the transformation of aryl bromides into the respective aryl metal compounds. These are the reaction with metallic magnesium, reaction with metallic lithium and the halogen/lithium exchange upon reaction with one equivalent of *n*-butyllithium or two equivalents of *tert*-butyllithium. A special feature of aryl iodides is their transformability into aryl Grignard compounds by means of iodine/magnesium exchange (path 1 in Figure 5.46), which can be accomplished by reaction with one equivalent of isopropyl magnesium bromide. The driving force and mechanism of the I/Mg exchange are similar to the Br/Li exchange. This process actually also works for aryl bromides, but it is less efficient.

Figure 5.47 displays functionalizations of aromatic compounds that can be initiated by an I/Mg exchange, but none of the other Ar–I → Ar–M transformations in Figure 5.46:

- The temperature at which (*ortho*-bromophenyl)magnesium halide **B** would form from bromoiodobenzene (**A**) and magnesium turnings would be so high that a β-elimination to benzyne would occur (cf. Section 5.6). The lithium analog of the Grignard compound **B** would undergo the same β-elimination—irrespective of whether it was generated from bromoiodobenzene and lithium powder or from bromoiodobenzene and *n*- or *tert*-BuLi.
- The Grignard compound **D** generated from the aryl iodide **C** and magnesium turnings would suffer a polycondensation with itself because Grignard compounds are acylated by carboxylic esters. In the same way the lithium analog of **D** would lose its ester group. The ester group of **D** remains untouched only because the I/Mg exchange in the aryl iodide **C** proceeds at –30 °C, a temperature at which the ester is slow to react. At this temperature **D** is reactive enough to add to cinnamaldehyde and the resulting magnesium alkoxide undergoes and intramolecular Williamson etherification).

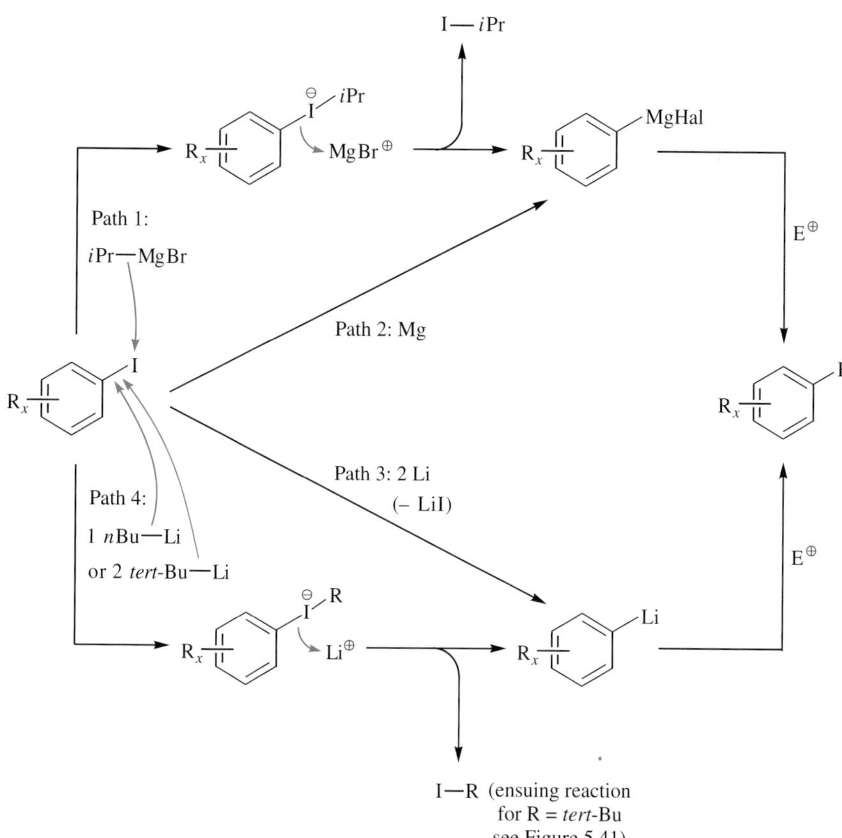

Fig. 5.46. Conversions of Ar–I in Ar–MgHal and in Ar–Li, respectively (for analogous conversions of Ar–Br: cf. Figure 5.41).

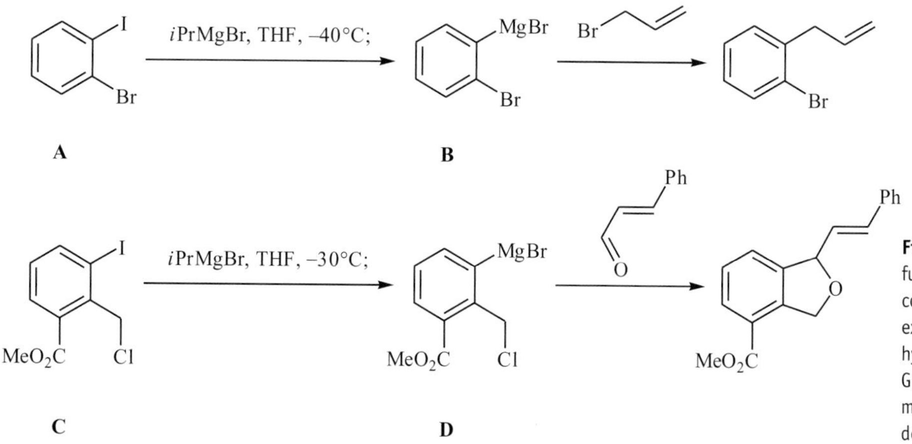

iPrMgBr, THF, –40°C;

A

B

iPrMgBr, THF, –30°C;

C

D

Fig. 5.47. Generation of functionalized aryl Grignard compounds through I/Mg exchange; alkylation and hydroxy alkylation of the Grignard reagents to yield multiply substituted benzene derivatives.

5.3.3 Electrophilic Substitutions of Arylboronic Acids and Arylboronic Esters

Arylboronic acids and their esters are nucleophiles in which the —$B(OH)_2$- or the —$B(OR)_2^{\ominus}$ group can be substituted by many electrophiles. Representative examples are shown in Figures 5.48 and 5.49.

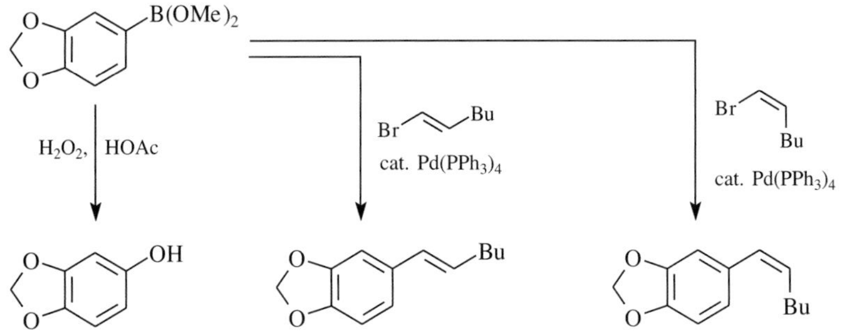

Fig. 5.48. Reactions of an arylboronic ester (for preparation, see Figure 5.45) with selected electrophiles.

Arylboronic esters (Figure 5.48) can be oxidized with H_2O_2 in acetic acid to give aryl borates, which can then be hydrolyzed to provide phenols. The mechanism for this oxidation is similar to that for the $OOH^{\ominus}$ oxidation of trialkylboranes and is discussed in Section 14.4.3. In combination with the frequently simple synthesis of arylboronic esters one can thus achieve the overall conversions Ar—H → Ar—OH or Ar—Br → Ar-OH in two or three steps.

Arylboronic esters (Figure 5.48) and arylboronic acids (Figure 5.49) can also react with unsaturated electrophiles, which cannot be introduced in one step into metal-free aromatic

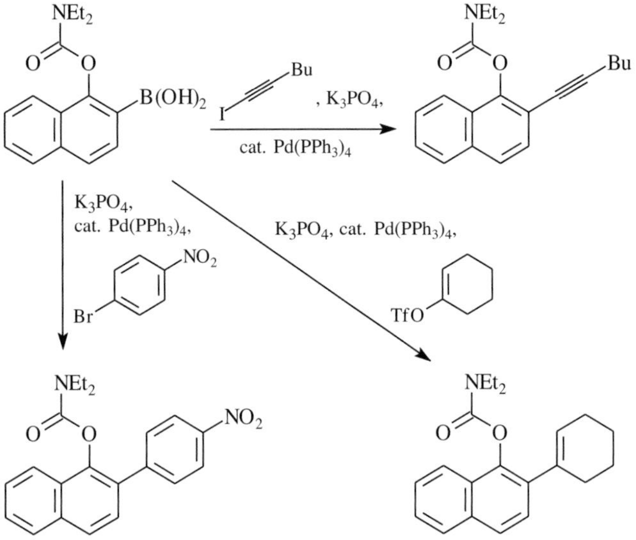

Fig. 5.49. Reactions of an arylboronic acid (for preparation, see Figure 5.39) with selected electrophiles; TfO stands for the triflate group.

compounds: these reactions are with alkenyl bromides, iodides, and triflates (with retention of the double-bond geometry!); with aryl bromides, iodides, and triflates; and with iodoalkynes. All these compounds react (only) in the presence of Pd(0) catalysts according to the mechanisms presented in Section 16.4.

5.4 Nucleophilic Substitution Reactions of Aryldiazonium Salts

Nucleophilic substitution reactions via an S_N2 reaction are not possible in aromatic compounds because the configuration at the reacting C atom cannot be inverted.

Nucleophilic substitution reactions according to the S_N1 mechanism have to take place via aryl cations. This should be exceedingly difficult because the positive charge in aryl cations is located in a low-energy sp^2 AO (Figure 5.50, middle line), and is not delocalized as one might first think. In fact, aryl cations are even less stable than alkenyl cations, which are quite unstable. However, in contrast to aryl cations, alkenyl cations can assume the favored linear geometry at the cationic center as shown in Figure 1.4. Consequently, the considerable energy expenditure required to generate an aryl cation in an S_N1 reaction can be supplied only by aryldiazonium salts. They do so successfully because the very stable N_2 molecule is produced as a second reaction product, providing a driving force. Because this nitrogen continuously

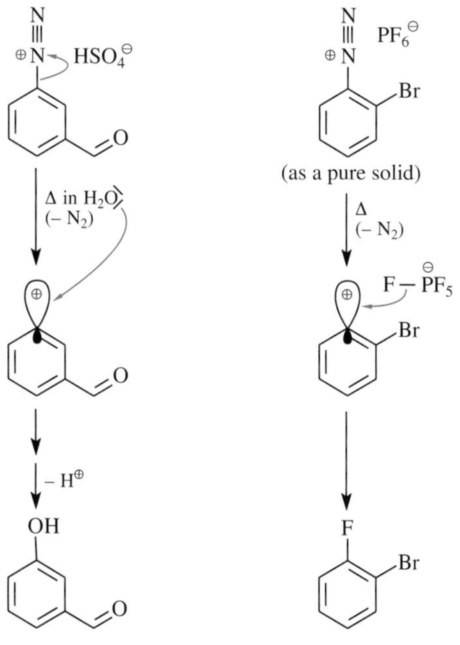

Fig. 5.50. S_N1 reactions of aryldiazonium salts I: left, hydrolysis of a diazonium salt leading to phenol; right, Schiemann reaction.

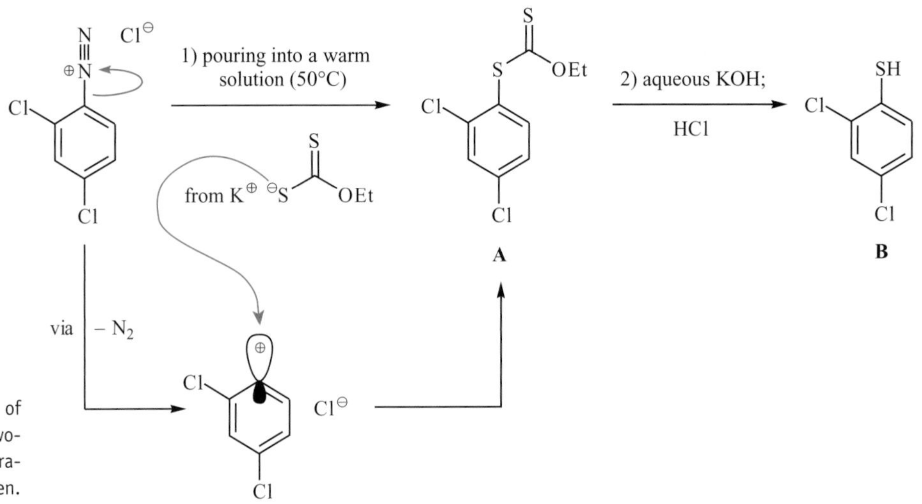

Fig. 5.51. S_N1 reaction of aryldiazonium salts II. A two-step sequence for the preparation of aromatic thiols is given.

departs from the reaction mixture, the elimination of that entity is irreversible so that the entire diazonium salt is gradually converted into the aryl cation.

Once the aryl cation is formed it reacts immediately with the best nucleophile in the reaction mixture. Being a very strong electrophile an aryl cation can even react with very weak nucleophiles. When the best nucleophile is H_2O—a weak one, indeed—a phenol is produced (**hydrolysis of diazonium salts;** Figure 5.50, left). Even when the nucleophile is a tetrafluoroborate or a hexafluorophosphate (which are *extremely* weak nucleophiles), S_N1 reactions run their course, and an aryl fluoride is obtained. This transformation is traditionally carried out by heating the dried aryldiazonium tetrafluoroborates or hexafluorophosphates ("**Schiemann reaction**," Figure 5.50, right) or—much milder—by starting with the diazotation of the respective aniline (mechanism: Figure 17.40) with sodium nitrite and the pyridine-HF complex acting in a dual capacity as supplier of both acid and fluoride.

Sulfur can be introduced into a diazonium salt by the S_N1 reaction shown in Figure 5.51. In order to prevent the reagent from effecting a double (rather than a mono-) arylation at the sulfur atom, potassium xanthogenate instead of sodium sulfide is used as the sulfur nucleophile. The resulting *S*-aryl xanthogenate **C** is hydrolyzed. In this way diarylsulfide-free aryl thiol **B** is obtained.

With other nucleophiles (overview: Figure 5.52) aryldiazonium salts react according to other mechanisms to form substitution products. These substitutions are possible because cer-

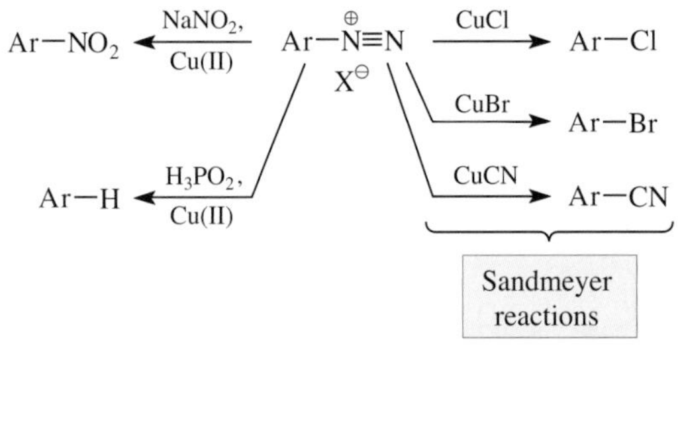

Fig. 5.52. Overview: S_N reactions of aryldiazonium salts via radicals.

tain nucleophiles reduce aryldiazonium salts Ar–N$^{\oplus}$≡N to form diazo radicals Ar—N=N•. The most common additive of this kind is a copper(I) salt or, alternatively, the addition of a copper(II) compound which is reduced to a copper(I) salt by the nucleophile *in situ*. These diazo radicals lose molecular nitrogen, leaving behind a highly reactive aryl radical that directly or indirectly reacts with the nucleophile (Nu$^{\ominus}$ = Cl$^{\ominus}$, Br$^{\ominus}$, CN$^{\ominus}$ or $NO_2^{\ominus}$, Figure 5.54; Nu$^{\ominus}$ = H_3PO_2, Figure 5.55; Nu$^{\ominus}$ = I$^{\ominus}$, Figure 5.56).

The influence of the copper salts is responsible for both the reduction of aryldiazonium salts to diazo radicals in the substitutions of Figure 5.54 and the reduction in Figure 5.55. The crucial electron originates from Cu(I), which is either added as such (reaction of aryldiazonium salts with CuCl, CuBr, CuCN or Cu_2O/H_3PO_2) or is formed *in situ* from Cu(II) by a redox reaction (reaction of aryldiazonium salts with $NaNO_2/Cu(NO_3)_2$). As soon as a diazo radical Ar–N=N• is formed, it undergoes fragmentation to Ar• + N≡N. Three of the ensuing reactions of these aryl radicals shown in Figure 5.52 are called **Sandmeyer reactions,** which give aromatic chlorides, bromides, or cyanides. Aryldiazonium salts, sodium nitrite, and Cu(II) give aromatic nitro compounds by the same mechanism (for a detailed description of the mechanism for these reactions cf. Figure 5.53).

Finally, aryl radicals Ar•, which in the presence of hypophosphoric acid (H_3PO_2) are generated from diazonium salts and Cu(I), undergo reduction to the aromatic compounds Ar–H in a radical *chain* reaction (Figure 5.54).

Thus, the reaction Ar–N$^{\oplus}$≡N + H_3PO_2 → Ar–H represents an important procedure for the defunctionalization of aromatic compounds. It is frequently employed to remove NH_2 groups from aromatic compounds if the NH_2 group was previously only introduced to accomplish a certain Ar-S_E reaction of a particular regioselectivity—*where the same reaction with the same regioselectivity would not have been possible without this NH_2 group.*

An intermediate reduction of aryldiazonium salts Ar–N$^{\oplus}$≡N to the diazo radicals Ar–N=N• also occurs when aryldiazonium salts react with KI to yield aryl iodides (Figure 5.55). Therefore, aryl radicals Ar• are obtained under these conditions, too. Their fate, however, differs from that of the aryl radicals, which are faced with nucleophiles in the presence of Cu(II) (cf. Figure 5.53) or H_3PO_2 (cf. Figure 5.54): the iodination mechanism of Figure 5.55 is a radical *chain* reaction consisting of four propagation steps.

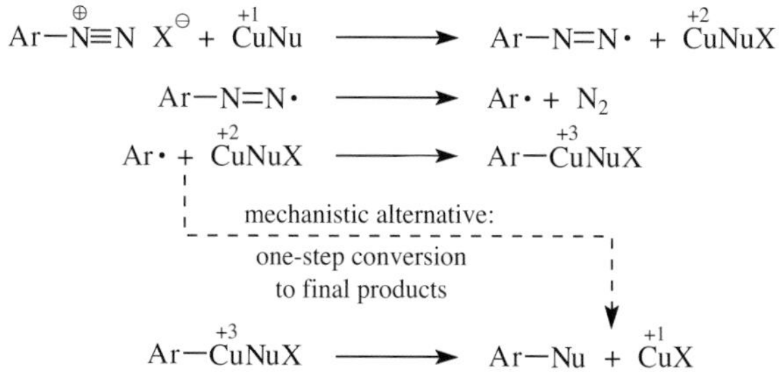

$$\text{Ar}-\overset{\oplus}{\text{N}}\equiv\text{N}\ \text{X}^{\ominus} + \overset{+1}{\text{Cu}}\text{Nu} \longrightarrow \text{Ar}-\text{N}=\text{N}\cdot + \overset{+2}{\text{Cu}}\text{NuX}$$

$$\text{Ar}-\text{N}=\text{N}\cdot \longrightarrow \text{Ar}\cdot + \text{N}_2$$

$$\text{Ar}\cdot + \overset{+2}{\text{Cu}}\text{NuX} \longrightarrow \text{Ar}-\overset{+3}{\text{Cu}}\text{NuX}$$

$$\text{Ar}-\overset{+3}{\text{Cu}}\text{NuX} \longrightarrow \text{Ar}-\text{Nu} + \overset{+1}{\text{Cu}}\text{X}$$

Fig. 5.53. Mechanistic aspects I of nucleophilic aromatic substitution reactions of aryldiazonium salts via radicals: introduction of Nu=Cl, Br, CN or NO_2 according to Figure 5.52. Following step 2 there are two alternatives: either the copper(II) salt is bound to the aryl radical (step 3) and the compound Ar–Cu(III)NuX decomposes to Cu(I)X and the substitution product Ar–Nu (step 4), or the aryl radical reacts with the copper(II) salt in a one-step radical substitution reaction yielding Cu(I)X and the substitution product Ar–Nu.

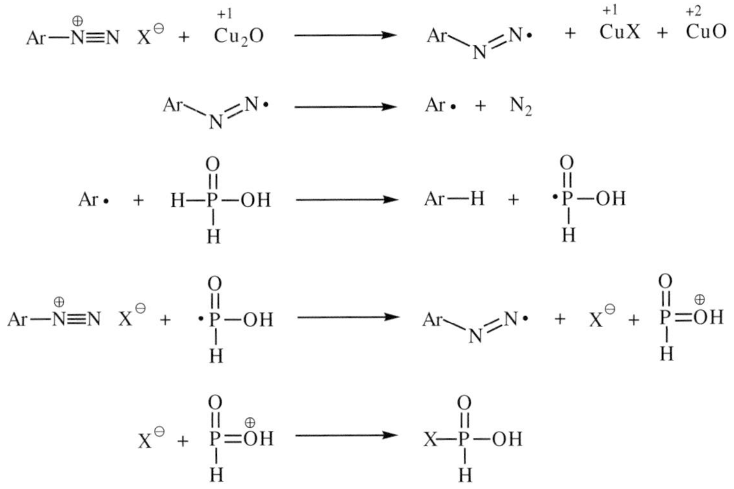

$$Ar{-}\overset{\oplus}{N}{\equiv}N\ X^{\ominus} + \overset{+1}{Cu_2O} \longrightarrow Ar{-}N{=}N\cdot + \overset{+1}{CuX} + \overset{+2}{CuO}$$

$$Ar{-}N{=}N\cdot \longrightarrow Ar\cdot + N_2$$

$$Ar\cdot + H{-}P(=O)(H){-}OH \longrightarrow Ar{-}H + \cdot P(=O)(H){-}OH$$

$$Ar{-}\overset{\oplus}{N}{\equiv}N\ X^{\ominus} + \cdot P(=O)(H){-}OH \longrightarrow Ar{-}N{=}N\cdot + X^{\ominus} + P(=O)(H){=}\overset{\oplus}{O}H$$

$$X^{\ominus} + P(=O)(H){=}\overset{\oplus}{O}H \longrightarrow X{-}P(=O)(H){-}OH$$

Fig. 5.54. Mechanistic aspects II of nucleophilic aromatic substitution reactions of aryldiazonium salts via radicals: introduction of Nu=H according to Figure 5.52.

To conclude this section, Figure 5.56 shows an elegant possibility for carrying out substitution reactions on aryldiazonium salts without using them as the substrates proper. One advantage of this method is that it allows the preparation of fluoroaromatic compounds (Figure 5.56, top) without having to isolate the potentially explosive Schiemann diazonium salts (i.e., the starting material for the reaction on the right in Figure 5.50). In addition, by the same method it is possible to prepare aryl iodides (Figure 5.56, bottom) without the risk that a nucleophile other than iodide introduced during the preparation of the diazonium salt competes with the iodide and thereby gives rise to the formation of a second substitution product.

To conduct the substitution reactions of Figure 5.56, one neutralizes an acidic solution of the aromatic diazonium salt with diethylamine. This forms the diazoamino compound **B** (called a "**triazene**"). It is isolated and subjected to the substitution reactions in an organic solvent. In each case, at first a leaving group is generated from the NEt_2 moiety of the dia-

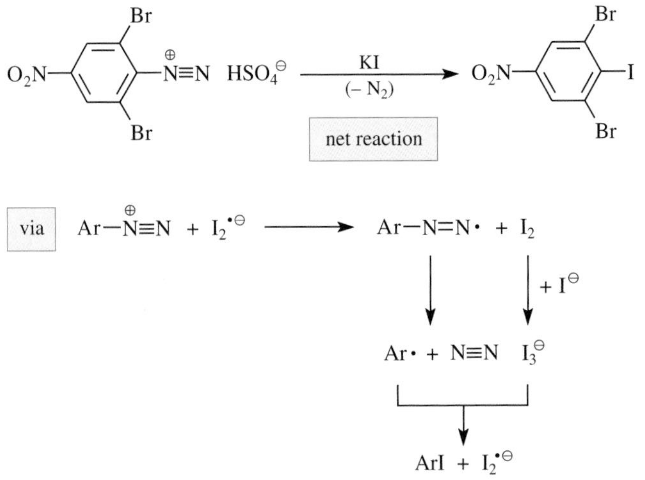

$$\text{4-}O_2N\text{-2,6-}Br_2C_6H_2{-}\overset{\oplus}{N}{\equiv}N\ HSO_4^{\ominus} \xrightarrow[(-N_2)]{KI} \text{4-}O_2N\text{-2,6-}Br_2C_6H_2{-}I$$

net reaction

via $Ar{-}\overset{\oplus}{N}{\equiv}N + I_2^{\bullet\ominus} \longrightarrow Ar{-}N{=}N\cdot + I_2$

$Ar{-}N{=}N\cdot \longrightarrow Ar\cdot + N{\equiv}N$; $I_2 + I^{\ominus} \longrightarrow I_3^{\ominus}$

$Ar\cdot + I_3^{\ominus} \longrightarrow ArI + I_2^{\bullet\ominus}$

Fig. 5.55. Mechanistic aspects III of nucleophilic aromatic substitution reactions of aryldiazonium salts via radicals: introduction of Nu = I through reaction of aryldiazonium salts with KI. In this (chain) reaction the radical $I_2\bullet^{\ominus}$—apart from its role as chain-carrying radical—plays the important role of initiating radical. The scheme shows how this radical is regenerated; the initial reaction by which it presumably forms remains to be provided, namely: (1) $Ar{-}N^{\oplus}{\equiv}N + I^{\ominus} \rightarrow Ar{-}N{=}N + I\bullet$; (2) $I\bullet + I^{\ominus} \rightarrow I_2\bullet^{\ominus}$.

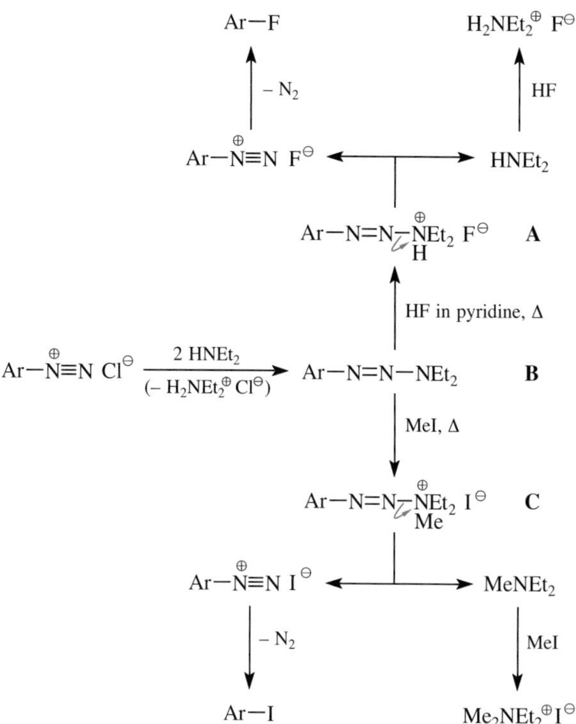

Fig. 5.56. Nucleophilic substitution reactions on a masked aryldiazonium salt. The introduction of fluorine takes place by the S_N1 mechanism in Figure 5.50, the introduction of iodine occurs by the radical mechanism of Figure 5.56.

zoamino group of compound **B**. This is achieved either by a protonation with HF (→ diazoammonium salt **A**) or by a methylation with MeI (→ diazoammonium salt **C**). The activated leaving groups are eliminated from the intermediates **A** and **C** as $HNEt_2$ or $MeNEt_2$, respectively. Aryldiazonium ions are formed, which are faced with only *one* nucleophile, namely, with $F^{\ominus}$ or $I^{\ominus}$, respectively. These reactions give Ar—F or Ar—I and they run so smoothly that the extra effort, which the isolation of intermediate product **B** requires relative to performing the corresponding direct substitutions, is often warranted.

5.5 Nucleophilic Substitution Reactions via Meisenheimer Complexes

5.5.1 Mechanism

Let us once more view the mechanism of the "**classic**" **Ar-S_E reaction** of Figure 5.1. In the rate-determining step, an *electrophile* reacts with an aromatic compound. A *carbenium ion* is produced, which is called the sigma complex. Therein a *positive* charge is delocalized over the five sp^2-hybridized centers of the former aromatic compound. The sixth center is sp^3-hybridized and linked to the electrophile (Figure 5.9). This sp^3-center also binds the substituent X, which is eliminated as a leaving group in the fast reaction step that follows. This substituent is almost always an H atom (→ elimination of $H^{\oplus}$) and only in exceptional cases a *tert*-butyl group (→ elimination of *tert*-$Bu^{\oplus}$) or a sulfonic acid group (→ elimination of $SO_3H^{\oplus}$) (Section 5.1.2).

The counterpart to this mechanism in the area of nucleophilic aromatic substitution reaction exists as the so-called **Ar-S_N reaction via a Meisenheimer complex** (Figure 5.57). In this case, a *nucleophile* reacts with the aromatic compound in the rate-determining step. A *carbanion* is produced, which is called the Meisenheimer complex (example: see Figure 5.58). A *negative* charge, which should be stabilized by electron-withdrawing substituents, is delocalized over the five sp^2-hybridized centers of the former aromatic compound. In the position *para* to the reacting C atom that has become bound to the nucleophile, a partial charge of –0.30 appears; in the *ortho*-position, a partial charge of –0.25 appears, and in the *meta*-position, a partial charge of –0.10 appears (see intermediate **A**, Figure 5.57). Corresponding to the magnitudes of these partial charges, the Meisenheimer complex is better stabilized by an electron acceptor in the *para*-position than through one in the *ortho*-position. Still, both of these provide considerably better stabilization than an electron acceptor in the *meta*-position. Actually, only when such stabilization is present can these anionic intermediates form with a preparatively useful reaction rate. The C atom that carries the former nucleophile is sp^3-hybridized in Meisenheimer intermediates and also bound to the substituent X, which is eliminated as $X^{\ominus}$in the second, fast step. X is usually Cl (→ elimination of $Cl^{\ominus}$) or, for example, in Sanger's reagent, F (→ elimination of $F^{\ominus}$; see below).

The name "Meisenheimer complex" had originally been given to carbanions of structures related to **A** (Figure 5.57) *if they could be isolated or could be maintained in solution long enough so that they could be examined spectroscopically.* The best-known Meisenheimer complex is shown in Figure 5.58. It can be prepared from trinitroanisole and NaOMe. It can be isolated because its negative charge is very well stabilized by the nitro groups located in the two *ortho*-positions and in the *para*-position of the carbanion. As another stabilizing factor, the leaving group ($MeO^{\ominus}$) is poor: as an alkoxide it is a high-energy species—so that it remains in the molecule instead of being expelled.

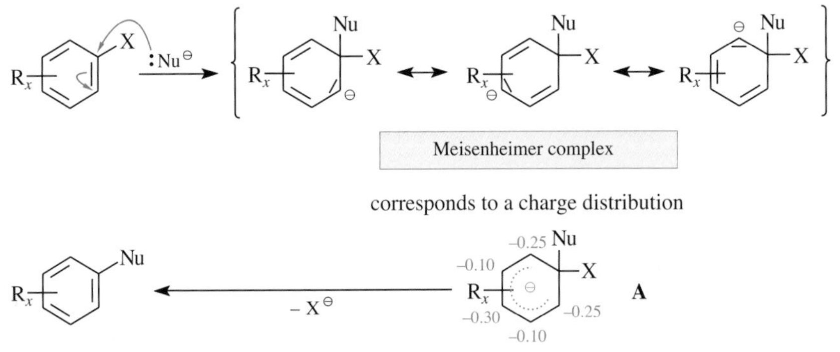

Fig. 5.57. Mechanism for the S_N reaction via Meisenheimer complexes.

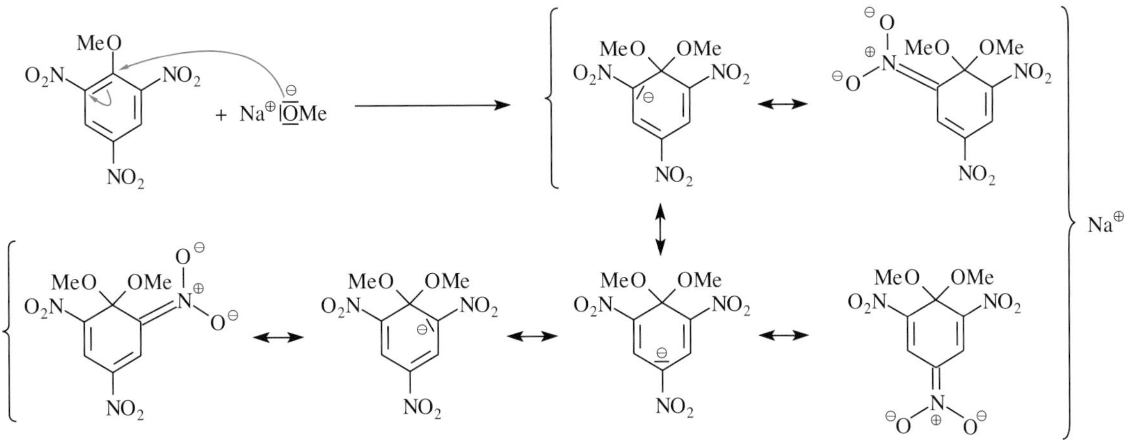

Fig. 5.58. Formation of an isolable Meisenheimer complex.

With a less comprehensive substitution by EWGs than in Figure 5.58 and/or with a better leaving group at the sp^3-C, the lifetimes of Meisenheimer complexes are considerably shorter. They then appear only as the short-lived intermediates of the Ar-S_N reactions of Figure 5.57.

5.5.2 Examples of Reactions of Preparative Interest

Two suitably positioned nitro groups make the halogen-bearing carbon atom in 2,4-dinitrohalobenzenes a favored point of reaction for nucleophilic substitution reactions. Thus, 2,4-dinitrophenyl hydrazine is produced from the reaction of 2,4-dinitrochlorobenzene with hydrazine:

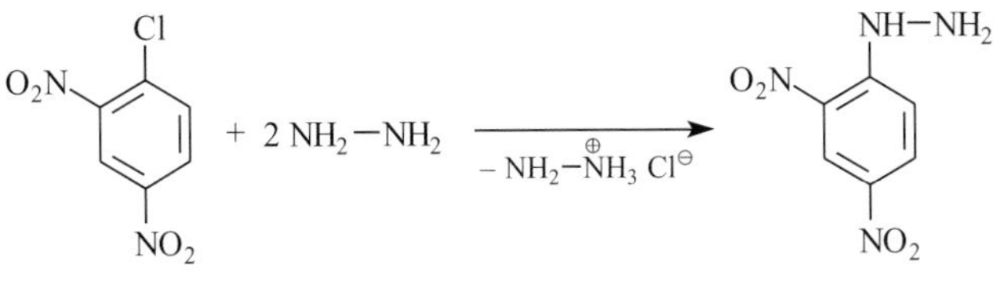

2,4-Dinitrofluorobenzene (Sanger's reagent) was used earlier in a different S_N reaction: for arylating the N atom of the N-terminal amino acid of oligopeptides. The F atom contained in this reagent strongly stabilizes the Meisenheimer complex because of its particularly great electron-withdrawing inductive (–I) effect. A chlorine atom as the leaving group would not provide as much stabilization. Consequently, as inferred from the Hammond postulate, the F atom gives Sanger's reagent a higher reactivity toward nucleophiles compared to 2,4-dinitrochlorobenzene.

The negative charge in a Meisenheimer complex, can be stabilized not only by a nitro substituent at one of the charged atoms, but also by a ring nitrogen. Therefore, pyridines, pyrimidines, and 1,3,5-triazines containing a Cl atom in the 2-, 4-, and/or 6-position, likewise enter into Ar-S_N reactions via Meisenheimer complexes very readily.

Under forcing conditions chlorine can be displaced by nucleophiles according to the mechanism described in this chapter also in compounds with fewer or weaker electron accepting substituents than hitherto mentioned. This explains the formation of tetrachlorodibenzodioxin ("dioxin") from sodium trichlorophenolate in the well-known Seveso accident:

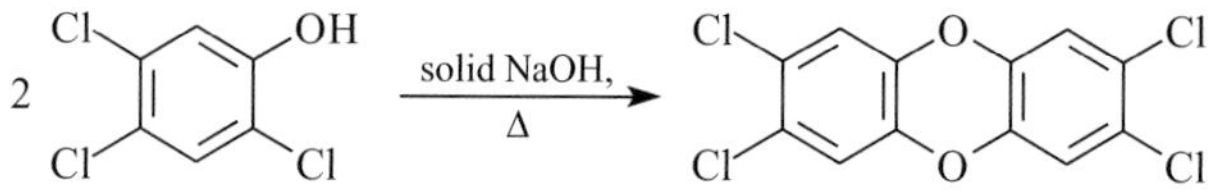

2
Cl
OH
Cl
Cl
solid NaOH,
Δ
Cl
O
Cl
Cl
O
Cl

As discussed previously, naphthalene reacts with electrophiles to form a sigma complex more rapidly than benzene does. The reason was that in this step naphthalene loses ~30 kcal/mol of aromatic stabilization, whereas benzene loses ~36 kcal/mol. For the very same reason, naphthalene derivatives react with nucleophiles faster to form Meisenheimer complexes than the analogous benzene derivatives do. In other words, naphthalene derivatives undergo an Ar-S_N reaction more readily than analogous benzene derivatives. In fact, there are naphthalenes that undergo Ar-S_N reactions even when they do not contain any electron withdrawing substituent other than the leaving group.

Let us consider as an example the synthesis of a precursor of the hydrogenation catalyst *R*-2,2'-bis(diphenylphosphino)-1,1'-binaphthyl *R*-BINAP (Figure 5.59). The substrate of this reaction is enantiomerically pure *R*-1,1'-bi-2-naphthol (*R*-BINOL). Its OH groups become

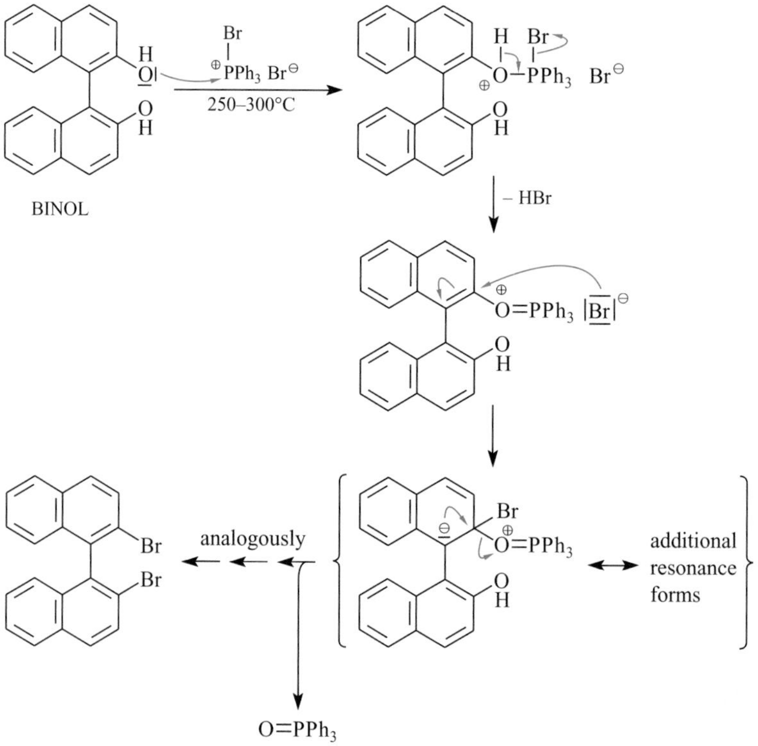

BINOL
Br
⊕PPh₃ Br⊖
250–300°C
H Br
O–PPh₃ Br⊖
– HBr
O=PPh₃ |Br|⊖
Br
O=PPh₃
additional
resonance
forms
analogously
Br
Br
O=PPh₃

Fig. 5.59. Synthesis of a precursor of the hydrogenation catalyst *R*-BINAP (for completion of its synthesis, see Figure 5.44) from *R*-BINOL. Starting from enantiomerically pure BINOL the procedure presented here leads to dibromobinaphthyl, also in racemic form—the reason is that the racemization threshold is crossed with such a high reaction temperature.

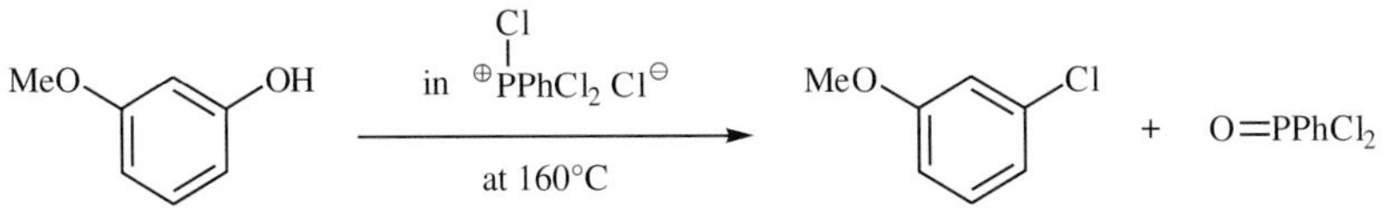

Fig. 5.60. Transformation of a phenol into a chloroaromatic compound by Ar-S_N reaction via a Meisenheimer complex.

leaving groups after activation with a phosphonium salt, which is prepared via the Mukaiyama reaction (Figure 2.36). The bromide ion contained in the reagent then acts as the nucleophile.

$PhPCl_3^{\oplus}Cl^{\ominus}$ is a stronger electrophile than $Ph_3PBr^{\oplus}Br^{\ominus}$. In $PhPCl_3^{\oplus}Cl^{\ominus}$ the phosphorus atom is linked to three chlorine atoms, each of which is more electronegative than the one bromine atom to which the P atom in $Ph_3PBr^{\oplus}Br^{\ominus}$ is bound. The superior electrophilicity of $PhPCl_3^{\oplus}Cl^{\ominus}$ explains why even the *benzene* derivative of Figure 5.60 is amenable to an OH → Cl exchange—and, what is more, at a much lower temperature than is needed for the OH → Br exchange in Figure 5.60. So mechanistically, the S_N reaction of Figure 5.60 resembles the one given in Figure 5.59 and also proceeds via a non-stabilized Meisenheimer complex.

5.6 Nucleophilic Aromatic Substitution via Arynes, *cine* Substitution

The significance of this type of reaction in preparative chemistry is limited but not negligible. We mention just the famous early phenol synthesis by Dow, which is no longer profitably carried out (Figure 5.61), but remains of mechanistic interest. The substrate of this substitution is chlorobenzene; the nucleophile is a hot aqueous solution of sodium hydroxide.

In a mechanistic investigation, a ^{14}C label was introduced at the C1 center of the substrate and in the resulting phenol the OH group was located *50% at the location and 50% next to the location* of the label. Consequently, there had been 50% *ipso*- and 50% so-called *cine*-substitution. This finding can be understood when it is assumed that in the strongly basic medium chlorobenzene first eliminates HCl to give dehydrobenzene (or benzyne). This species is an alkyne suffering from an enormous angular strain. Thus, it is so reactive that $Na^{\oplus}OH^{\ominus}$ can add to its C≡C bond. This addition, of course, takes place without regioselectivity with respect to the ^{14}C label.

The primary addition product formed is (2-hydroxyphenyl)sodium. However, this product immediately undergoes a proton transfer to give sodium phenolate, which is the conjugate base of the target molecule phenol.

The base or nucleophile in the Dow phenol process is aqueous NaOH that needs to be heated to 300° C, which is only possible at a pressure of 200 bar. This conconction is *only just* capable of deprotonating chlorobenzene to benzyne or of adding to the latter subsequently. Thus, there is no analogous synthesis of diaryl ethers from sodium phenolates and halobenzenes. And as expected, the product resulting from the Dow process did not contain more than 15% of diphenyl ether.

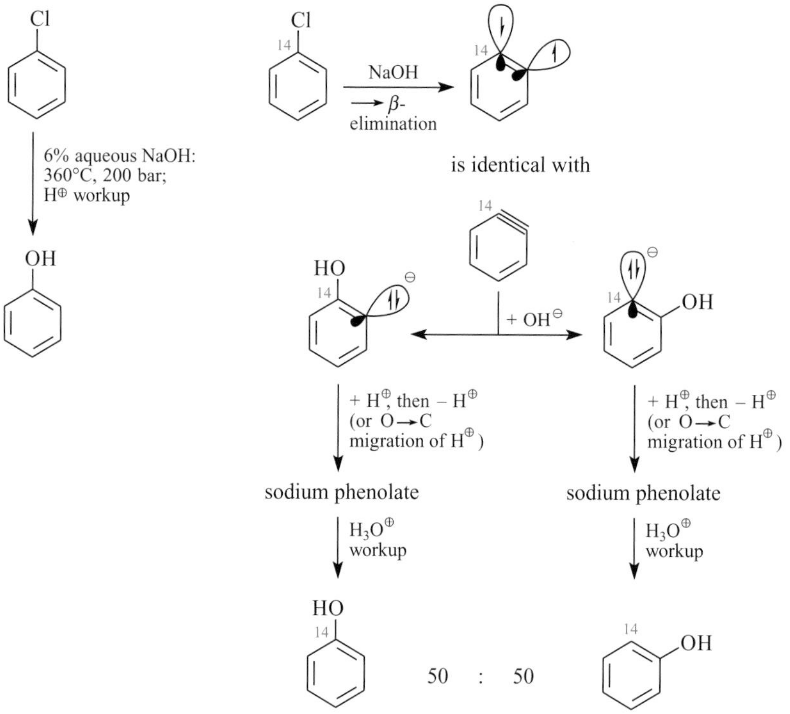

6% aqueous NaOH: 360°C, 200 bar; H⊕ workup

NaOH → β-elimination

is identical with

+ OH⊖

+ H⊕, then − H⊕ (or O→C migration of H⊕)

sodium phenolate

$H_3O^{\oplus}$ workup

50 : 50

Fig. 5.61. Phenol synthesis according to Dow: preparative (left) and mechanistic (right) aspects.

In contrast, alkylaryl ethers can be produced in analogy to the Dow process (Figure 5.62, top). Alkali metal alkoxides are even more basic/nucleophilic than alkali metal hydroxides. If, moreover, potassium alkoxide and bromo- or chlorobenzene are reacted *in DMSO*—i.e. in an aprotic dipolar solvent that does not provide the alkoxide ions with any noticeable stabilization through solvation—very good yields may already be achieved at room temperature (Figure 5.62, bottom).

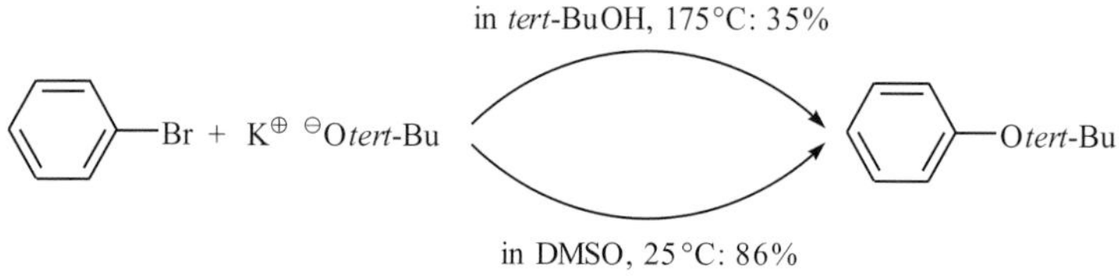

Fig. 5.62. Synthesis of an arylalkyl ether via an intermediate benzyne. Potassium-*tert*-butoxide is a much stronger base in aprotic-polar DMSO than in protic-polar *tert*-butanol, allowing for a much lower elimination temperature.

in *tert*-BuOH, 175°C: 35%

Br + K⊕ ⊖O*tert*-Bu → O*tert*-Bu

in DMSO, 25°C: 86%

Side Note 5.6. Preparation of *N,N*-Diphenylamine

Students preparing for oral examinations in organic chemistry are quite familiar with the following nerve-racking situation: questions keep coming to you that the examiner might ask and that you cannot answer spontaneously. Things get downright beastly when these questions are (seemingly) so simple that you could swear you've heard the answer before. The answers to some of these questions really require thorough and detailed knowledge. In the field of aromatic chemistry, for example, several future examinees will very likely find themselves out of

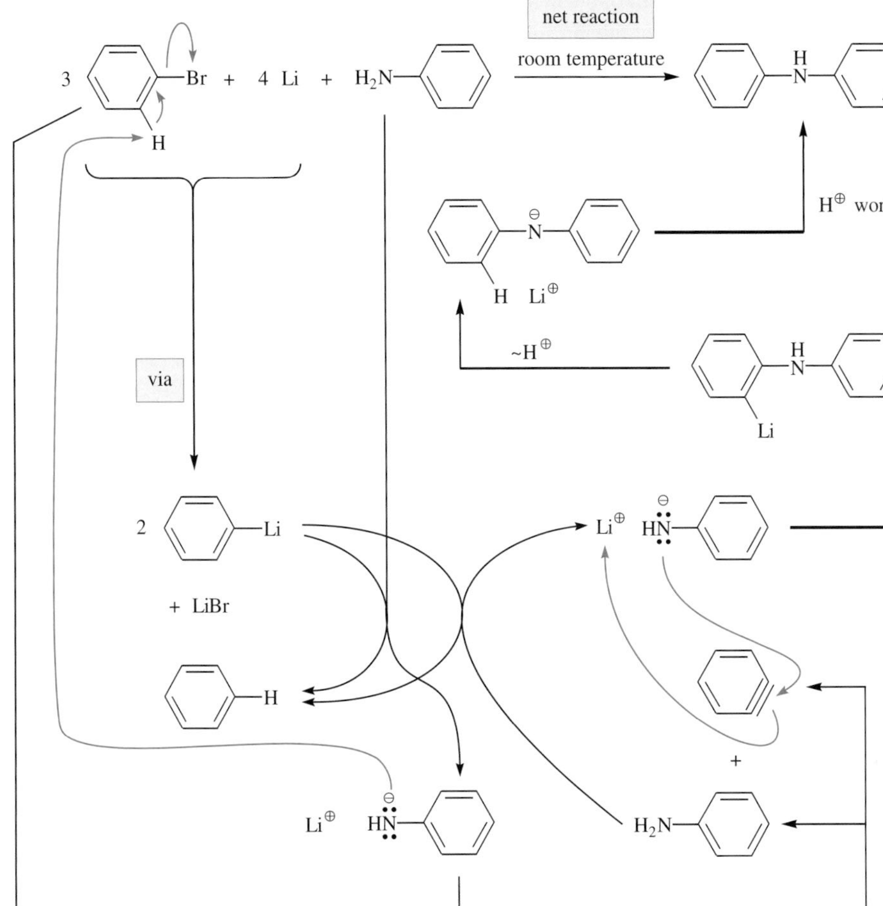

Fig. 5.63. Synthesis of diphenylamine via an intermediate benzyne. The phenyllithium that has been formed *in situ* serves to generate lithium anilide. In the synthesis presented here lithium anilide plays the same role as sodium hydroxide in the reaction given in Figure 5.61 and potassium-*tert*-butoxide in the reaction shown in Figure 5.62.

their depths when they are asked to answer the following seemingly simple questions: (1) "How can you prepare diphenyl ethers?" (2) "How can you prepare diphenyl amine?" The answer to the first question is given in Figure 16.6 and mechanistically, in fact, belongs to a totally different field of Organic Chemistry. The second question is answered in Figure 5.63, which outlines a one-pot synthesis of diphenylamine from 3 equivalents of bromobenzene, 4 equivalents of lithium and 1 equivalent of aniline ("monophenylamine").

The stoichiometry of the synthesis outlined in Figure 5.63 may well be confusing at first, but will unfold in a very simple manner when we reconstruct the details of the reaction given in this figure. The following partial reactions are involved:

- 2 equivalents of bromobenzene and 4 equivalents of lithium produce 2 equivalents of phenyllithium.
- The first equivalent of phenyllithium and aniline yield 1 equivalent of lithium anilide.
- The lithium anilide reacts with a third equivalents of bromobenzene to give 1 equivalent of benzyne. Simultaneously, 1 equivalent of aniline is regenerated.

- The second equivalent of phenyllithium and the regenerated aniline lead to the regeneration of 1 equivalent of lithium anilide.
- The latter is added to the benzyne. The initially formed "carbanion" is transformed by proton transfer to lithium-*N,N*-diphenylamide, which upon aqueous workup undergoes protonation to give diphenylamine.

References

M. Sainsbury, "Aromatic Chemistry," Oxford University Press, Oxford, U. K., **1992**.

D. T. Davies, "Aromatic Heterocyclic Chemistry," Oxford University Press, New York, **1992**.

5.1

S. W. Slayden, J. F. Liebman, "The Energetics of Aromatic Hydrocarbons: An Experimental Thermochemical Perspective," *Chem. Rev.* **2001**, *101*, 1541–1566.

R. J. K. Taylor, "Electrophilic Aromatic Substitution," Wiley, Chichester, U. K., **1990**.

H. Mayr, B. Kempf, A. R. Ofial, "π-Nucleophilicity in Carbon-Carbon Bond-Forming Reactions," *Acc. Chem. Rev.* **2003**, *36,* 66–77.

D. Lenoir, "The Electrophilic Substitution of Arenes: Is the π Complex a Key Intermediate and What is its Nature?", *Angew. Chem. Int. Ed. Engl.* **2003**, *42,* 854–857.

F. Effenberger, "1,3,5-Tris(dialkylamino)benzenes: Model Compounds for the Electrophilic Substitution and Oxidation of Aromatic Compounds," *Acc. Chem. Res.* **1989**, *22*, 27–35.

C. B. de Koning, A. L. Rousseau, W. A. L. van Otterlo, "Modern Methods for the Synthesis of Substituted Naphthalenes," *Tetrahedron* **2003**, *59*, 7–36.

K. K. Laali, "Stable Ion Studies of Protonation and Oxidation of Polycyclic Arenes," *Chem. Rev.* **1996**, *96*, 1873–1906.

A. R. Katritzky, W. Q. Fan, "Mechanisms and Rates of Electrophilic Substitution Reactions of Heterocycles," *Heterocycles* **1992**, *34*, 2179–2229.

L. I. Belen'kii, I. A. Suslov, N. D. Chuvylkin, "Substrate and Positional Selectivity in Electrophilic Substitution Reactions of Pyrrole, Furan, Thiophene, and Selenophene Derivatives," *Chem. Heterocycl. Compd.* **2003**, *39,* 36–48.

5.2

M. R. Grimmett, "Halogenation of Heterocycles: II. Six- and Seven-Membered Rings," *Adv. Heterocycl. Chem.* **1993**, *58,* 271–345.

C. M. Suter and A. W. Weston, "Direct Sulfonation of Aromatic Hydrocarbons and Their Halogen Derivatives," *Org. React.* **1946**, *3,* 141–197.

G. A. El-Hiti, "Recent Advances in the Synthesis of Sulfonic Acids," *Sulfur Rep.* **2001**, *22,* 217–250.

L. Eberson, M. P. Hartshorn, F. Radner, "Ingold's Nitration Mechanism Lives!", *Acta Chem. Scand.* **1994**, *48*, 937–950.

B. P. Cho, "Recent Progress in the Synthesis of Nitropolyarenes. A Review," *Org. Prep. Proced. Int.* **1995**, *27*, 243–272.

J. H. Ridd, "Some Unconventional Pathways in Aromatic Nitration," *Acta Chem. Scand.* **1998**, *52*, 11–22.

H. Zollinger, "Diazo Chemistry I, Aromatic and Heteroaromatic Compounds," VCH Verlagsgesellschaft, Weinheim, Germany, **1994**.

C. C. Price, "The Alkylation of Aromatic Compounds by the Friedel–Crafts Method," *Org. React.* **1946**, *3,* 1–82.

G. A. Olah, R. Krishnamurti, G. K. S. Prakash, "Friedel–Crafts Alkylations," in *Comprehensive Organic Synthesis* (B. M. Trost, I. Fleming, Eds.), Vol. 3, 293, Pergamon Press, Oxford, **1991**.

H. Heaney, "The Bimolecular Aromatic Friedel–Crafts Reaction," in *Comprehensive Organic Synthesis* (B. M. Trost, I. Fleming, Eds.), Vol. 2, 733, Pergamon Press, Oxford, **1991**.

H. Heaney, "The Intramolecular Aromatic Friedel–Crafts Reaction," in *Comprehensive Organic Synthesis* (B. M. Trost, I. Fleming, Eds.), Vol. 2, 753, Pergamon Press, Oxford, **1991**.

T. Ohwada, "Reactive Carbon Electrophiles in Friedel–Crafts Reactions," *Reviews on Heteroatom Chemistry* **1995**, *12*, 179.

R. C. Fuson and C. H. McKeever, "Chloromethylation of Aromatic Compounds," *Org. React.* **1942**, *1*, 63–90.

E. Berliner, "The Friedel and Crafts Reaction with Aliphatic Dibasic Acid Anhydrides," *Org. React.* **1949**, *5*, 229–289.

I. Hashimoto, T. Kawaji, F. D. Badea, T. Sawada, S. Mataka, M. Tashiro, G. Fukata, "Regioselectivity of Friedel-Crafts Acylation of Aromatic-Compounds with Several Cyclic Anhydrides," *Res. Chem. Intermed.* **1996**, *22*, 855–869.

A. R. Martin, "Uses of the Fries Rearrangement for the Preparation of Hydroxyaryl Ketones," *Org. Prep. Proced. Int.* **1992**, *24*, 369.

O. Meth-Cohn, S. P. Stanforth, "The Vilsmeier–Haack Reaction," in *Comprehensive Organic Synthesis* (B. M. Trost, I. Fleming, Eds.), Vol. 2, 777, Pergamon Press, Oxford, **1991**.

G. Jones, S. P. Stanforth, "The Vilsmeier Reaction of Fully Conjugated Carbocycles and Heterocycles," *Org. React.* **1997**, *49*, 1–330.

5.3

L. Brandsma, "Aryl and Hetaryl Alkali Metal Compounds," in *Methoden Org. Chem.* (Houben-Weyl) 4th ed. **1952**, *Carbanions* (M. Hanack, Ed.), Vol. E19d, 369, Georg Thieme Verlag, Stuttgart, **1993**.

H. Gilman, J. W. Morton, Jr., "The Metalation Reaction with Organolithium Compounds," *Org. React.* **1954**, *8*, 258–304.

V. Snieckus, "Regioselective Synthetic Processes Based on the Aromatic Directed Metalation Strategy," *Pure Appl. Chem.* **1990**, *62*, 671.

V. Snieckus, "The Directed Ortho Metalation Reaction. Methodology, Applications, Synthetic Links, and a Nonaromatic Ramification," *Pure Appl. Chem.* **1990**, *62*, 2047–2056.

V. Snieckus, "Directed Ortho Metalation. Tertiary Amide and O-Carbamate Directors in Synthetic Strategies for Polysubstituted Aromatics," *Chem. Rev.* **1990**, *90*, 879–933.

V. Snieckus, "Combined Directed Ortho Metalation-Cross Coupling Strategies. Design for Natural Product Synthesis," *Pure Appl. Chem.* **1994**, *66*, 2155–2158.

K. Undheim, T. Benneche, "Metalation and Metal-Assisted Bond Formation in π-Electron Deficient Heterocycles," *Acta Chem. Scand.* **1993**, *47*, 102–121.

H. W. Gschwend, H. R. Rodriguez, "Heteroatom-Facilitated Lithiations," *Org. React.* **1979**, *26*, 1–360.

R. D. Clark, A. Jahangir, "Lateral Lithiation Reactions Promoted by Heteroatomic Substituents," *Org. React.* **1995**, *47*, 1–314.

R. G. Jones, H. Gilman, "The Halogen-Metal Interconversion Reaction with Organolithium Compounds," *Org. React.* **1951**, *6*, 339–366.

W. E. Parham, C. K. Bradsher, "Aromatic Organolithium Reagents Bearing Electrophilic Groups. Preparation by Halogen-Lithium Exchange," *Acc. Chem. Res.* **1982**, *15*, 300.

N. Sotomayor, E. Lete, "Aryl and Heteroaryllithium Compounds by Metal-Halogen Exchange. Synthesis of Carbocyclic and Heterocyclic Systems," *Curr. Org. Chem.* **2003**, *7*, 275–300.

Najera, J. M. Sansano, M. Yus, "Recent Synthetic Uses of Functionalized Aromatic and Heteroaromatic Organolithium Reagents Prepared by Non-Deprotonating Methods," *Tetrahedron* **2003**, *59*, 9255–9303.

P. Knochel, W. Dohle, N. Gommermann, F. F. Kneisel, F. Kopp, T. Korn, I. Sapountzis, V. A. Vu, "Highly Functionalized Organomagnesium Reagents Prepared through Halogen-Metal Exchange," *Angew. Chem. Int. Ed. Engl.* **2003**, *42*, 4302–4320.

P. Sadimenko, "Organometallic Compounds of Furan, Thiophene, and Their Benzannulated Derivatives," *Adv. Heterocycl. Chem.* **2001**, *78,* 1–64.

A. R. Martin, Y. Yang, "Palladium Catalyzed Cross-Coupling Reactions of Organoboronic Acids with Organic Electrophiles," *Acta Chem. Scand.* **1993**, *47*, 221–230.

A. Suzuki, "New Synthetic Transformations via Organoboron Compounds," *Pure Appl. Chem.* **1994**, *66*, 213–222.

N. Miyaura, A Suzuki, "Palladium-Catalyzed Cross-Coupling Reactions of Organoboron Compounds," *Chem. Rev.* **1995**, *95*, 2457–2483.

N. Miyaura, "Synthesis of Biaryls via the Cross-Coupling Reaction of Arylboronic Acids," in *Advances in Metal-Organic Chemistry* (L. S. Liebeskind, Ed.), JAI Press, Greenwich, **1998**, *6,* 187–243.

5.4

R. K. Norris, "Nucleophilic Coupling with Aryl Radicals," in *Comprehensive Organic Synthesis* (B. M. Trost, I. Fleming, Eds.), Vol. 4, 451, Pergamon Press, Oxford, **1991**.

A. Roe, "Preparation of Aromatic Fluorine Compounds from Diazonium Fluoroborates: The Schiemann Reaction," *Org. React.* **1949**, *5*, 193–228.

N. Kornblum, "Replacement of the Aromatic Primary Amino Group by Hydrogen," *Org. React.* **1944**, *2,* 262–340.

5.5

C. Paradisi, "Arene Substitution via Nucleophilic Addition to Electron Deficient Arenes," in *Comprehensive Organic Synthesis* (B. M. Trost, I. Fleming, Eds.), Vol. 4, 423, Pergamon Press, Oxford, **1991**.

I. Gutman (Ed.), "Nucleophilic Aromatic Displacement: The Influence of the Nitro Group," VCH, New York, **1991**.

V. M. Vlasov, "Nucleophilic Substitution of the Nitro Group, Fluorine and Chlorine in Aromatic Compounds," *Russ. Chem. Rev.* **2003**, *72,* 681–764.

N. V. Alekseeva, L. N. Yakhontov, "Reactions of Pyridines, Pyrimidines, and 1,3,5-Triazines with Nucleophilic Reagents," *Russ. Chem. Rev.* **1990**, *59*, 514–530.

D. B. Kimball, M. M. Haley, "Triazenes: A Versatile Tool in Organic Synthesis," *Angew. Chem. Int. Ed. Engl.* **2002**, *41,* 3338–3351.

5.6

J. Suwinski, K. Swierczek, "*Cine-* and *Tele*-Substitution Reactions," *Tetrahedron* **2001**, *57*, 1639–1662.

Further Reading

M. Carmack, M. A. Spielman, "The Willgerodt Reaction," *Org. React.* **1946**, *3,* 83–107.

S. Sethna, R. Phadke, "The Pechmann Reaction," *Org. React.* **1953**, *7*, 1–58.

H. Wynberg, E. W. Meijer, "The Reimer-Tiemann Reaction," *Org. React.* **1982**, *28*, 1–36.

W. E. Truce, "The Gattermann Synthesis of Aldehydes," *Org. React.* **1957**, *9*, 37–72.

N. N. Crounse , "The Gattermann–Koch Reaction," *Org. React.* **1949**, *5*, 290–300.

A. H. Blatt, "The Fries Reaction", *Org. React.* **1942**, *1,* 342–369.

P. E. Spoerri, A. S. DuBois, "The Hoesch Synthesis," *Org. React.* **1949**, *5*, 387–412.

W. E. Bachmann, R. A. Hoffman, "The Preparation of Unsymmetrical Biaryls by the Diazo Reaction and the Nitrosoacetylamine Reaction," *Org. React.* **1944**, *2,* 224–261.

DeLos F. DeTar, "The Pschorr Synthesis and Related Diazonium Ring Closure Reactions," *Org. React.* **1957**, *9*, 409–462.

C. S. Rondestvedt, Jr., "Arylation of Unsaturated Compounds by Diazonium Salts," *Org. React.* **1960**, *11*, 189–260.

C. S. Rondestvedt, Jr., "Arylation of Unsaturated Compounds by Diazonium Salts (The Meerwein Arylation Reaction)," *Org. React.* **1976**, *24*, 225–259.

M. Braun, "New Aromatic Substitution Methods," in *Organic Synthesis Highlights* (J. Mulzer, H.-J. Altenbach, M. Braun, K. Krohn, H.-U. Reißig, Eds.), VCH, Weinheim, New York, **1991**, 167–173.

J. F. Bunnett, "Some Novel Concepts in Aromatic Reactivity," *Tetrahedron* **1993**, *49*, 4477.

G. A. Artamkina, S. V. Kovalenko, I. P. Beletskaya, O. A. Reutov, "Carbon-Carbon Bond Formation in Electron-Deficient Aromatic Compounds," *Russ. Chem. Rev. (Engl. Transl.)* **1990**, *59*, 750.

C. D. Hewitt, M. J. Silvester, "Fluoroaromatic Compounds: Synthesis, Reactions and Commercial Applications," *Aldrichim. Acta* **1988**, *21*, 3-10.

R. A. Abramovitch, D. H. R. Barton, J.-P. Finet, "New Methods of Arylation," *Tetrahedron* **1988**, *44*, 3039.

J. H. Clark, T. W. Bastock, D. Wails (Eds.), "Aromatic Fluorination," CRC, Boca Raton, **1996**.

L. Delaude, P. Laszlo, K. Smith, "Heightened Selectivity in Aromatic Nitrations and Chlorinations by the Use of Solid Supports and Catalysts," *Acc. Chem. Res.* **1993**, *26*, 607–613.

J. K. Kochi, "Inner-Sphere Electron Transfer in Organic Chemistry. Relevance to Electrophilic Aromatic Nitration," *Acc. Chem. Res.* **1992**, *25*, 39–47.

L. Eberson, M. P. Hartshorn, F. Radner, "Electrophilic Aromatic Nitration Via Radical Cations: Feasible or Not? ," in *Advances in Carbocation Chemistry* (J. M. Coxon, Ed.) JAI, Greenwich, CT, **1995**.

H. Ishibashi, M. Ikeda, "Recent Progress in Electrophilic Aromatic Substitution with α-Thiocarbocations," *Rev. Heteroatom Chem.* **1996**, *14*, 59–82.

H. Heaney, "The Bimolecular Aromatic Mannich Reaction," in *Comprehensive Organic Synthesis* (B. M. Trost, I. Fleming, Eds.), Vol. 2, 953, Pergamon Press, Oxford, **1991**.

P. E. Fanta, "The Ullmann Synthesis of Biaryls," *Synthesis* **1974**, *9*.

J. A. Lindley, "Copper Assisted Nucleophilic Substitution of Aryl Halogen," *Tetrahedron* **1984**, *40*, 1433.

G. P. Ellis, T. M. Romsey-Alexander, "Cyanation of Aromatic Halides," *Chem. Rev.* **1987**, *87*, 779.

I. A. Rybakova, E. N. Prilezhaeva, V. P. Litvinov, "Methods of Replacing Halogen in Aromatic Compounds by RS-Functions," *Russ. Chem. Rev. (Engl. Transl.)* **1991**, *60*, 1331.

O. N. Chupakhin, V. N. Charushin, H. C. van der Plas, "Nucleophilic Aromatic Substitution of Hydrogen," Academic Press, San Diego, CA, **1994**.

J. M. Saveant, "Mechanisms And Reactivity In Electron-Transfer-Induced Aromatic Nucleophilic Substitution—Recent Advances," *Tetrahedron* **1994**, *50*, 10117.

A. J. Belfield, G. R. Brown, A. J. Foubister, "Recent Synthetic Advances in the Nucleophilic Amination of Benzenes," *Tetrahedron* **1999**, *55*, 11399–11428.

M. Makosza, "Neue Aspekte der nucleophilen Substitution von Arenen," *Chem. unserer Zeit* **1996**, *30*, 134–140.

J. Scott Sawyer, "Recent Advances in Diaryl Ether Synthesis," *Tetrahedron* **2000**, *56*, 5045– 5065.

Nucleophilic Substitution Reactions at the Carboxyl Carbon

6

6.1 C=O-Containing Substrates and Their Reactions with Nucleophiles

C=O double bonds occur in a series of different classes of compounds:

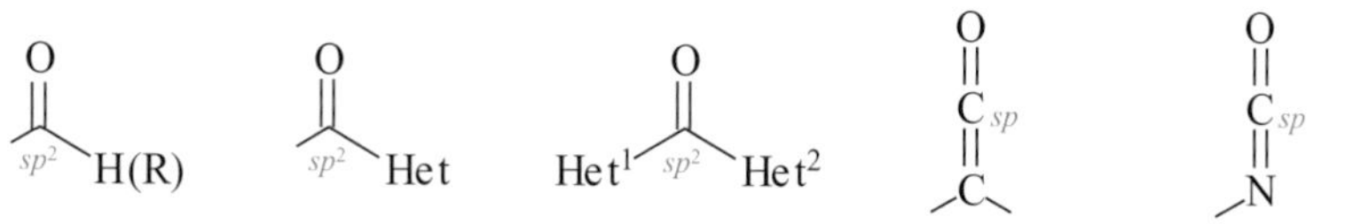

$R(H)-C_{sp^2}(=O)-H(R)$	$R-C_{sp^2}(=O)-Het$	$Het^1-C_{sp^2}(=O)-Het^2$	$O=C_{sp}=C<$	$O=C_{sp}=N-$
Aldehyde (ketone)	Carboxylic acid (derivative)	Carbonic acid derivative	Ketene	Isocyanate

In aldehydes and ketones, which together are referred to as **carbonyl compounds**, C=O double bonds are part of a carbonyl group, $C_{sp2}{=}O$. Carboxylic acids, carboxylic esters, and carboxylic amides, as well as all carboxylic acid derivatives used as acylating agents (see Section 6.3) are termed collectively as **carboxyl compounds** and are thereby distinguished from the carbonyl compounds. They contain a carboxyl group $C_{sp2}({=}O)$—Het. C=O double bonds are also part of carbonic acid derivatives Het^1—$C_{sp2}({=}O)$—Het^2. Carbonic acid derivatives contain a carboxyl carbon and a carboxyl oxygen, too. Thus, there is no difference between the nomenclatures for the C=O groups of carbonic acid derivatives and carboxylic acid derivatives. Finally, there are $C_{sp}{=}O$ double bonds; these occur in ketenes and isocyanates.

Each of the aforementioned C=O-containing compounds reacts with nucleophiles. Which kind of a reaction occurs depends almost exclusively on the nature of the substrate and hardly at all on the nucleophile:

- The typical reaction of carbonyl compounds with nucleophiles is the *addition* (Section 9.1 and Chapter 10); the C=O bond disappears:

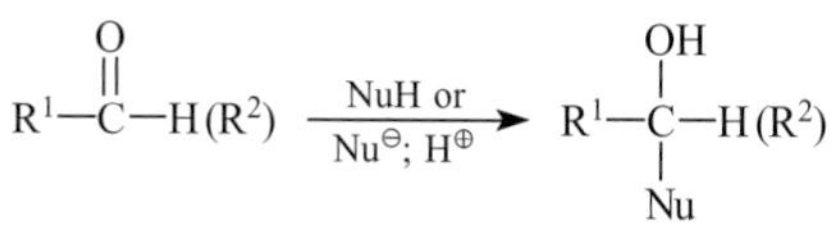

$$R^1-C(=O)-H(R^2) \xrightarrow[Nu^{\ominus};\ H^{\oplus}]{NuH\ or} R^1-C(OH)(Nu)-H(R^2)$$

- Ketenes and other C=O-containing heterocumulenes also react with nucleophiles in *addition* reactions; the C=O double bond, however, is nonetheless retained (Section 8.2):

Bruckner R (author), Harmata M (editor) In: *Organic Mechanisms – Reactions, Stereochemistry and Synthesis*
Chapter DOI: 10.1007/978-3-642-03651-4_6, © Springer-Verlag Berlin Heidelberg 2010

$$R^1R^2C{=}C{=}O \xrightarrow[\text{Nu}^{\ominus};\ \text{H}^{\oplus}]{\text{NuH or}} R^1R^2CH{-}C({=}O){-}Nu \qquad Het{=}C{=}O \xrightarrow[\text{Nu}^{\ominus};\ \text{H}^{\oplus}]{\text{NuH or}} H{-}Het{-}C({=}O){-}Nu$$

(Het = NR, O, S)

- In contrast, C=O-containing carboxylic acid and carbonic acid derivatives react with nucleophiles in *substitution* reactions. The one group or one of the two groups bound through a heteroatom to the carboxyl carbon of these substrates is substituted so that compounds **A** or **B**, respectively, are obtained.

$$R{-}C({=}O){-}Het \xrightarrow[\text{Nu}^{\ominus};\ \text{H}^{\oplus}\ (-\ \text{HetH})]{\text{NuH or}} \underset{\mathbf{A}}{R{-}C({=}O){-}Nu} \qquad Het^1{-}C({=}O){-}Het^2 \xrightarrow[\text{Nu}^{\ominus};\ \text{H}^{\oplus}\ (-\ \text{Het}^2{-}\text{H})]{\text{NuH or}} \underset{\mathbf{B}}{Het^1{-}C({=}O){-}Nu}$$

These substitution products **A** and **B** need not be the final product of the reaction of nucleophiles with carboxyl species. Sometimes they may be formed only as intermediates and continue to react with the nucleophile. Being carbonyl compounds (substitution products **A**) or carboxylic acid derivatives (substitution products **B**), they can in principle undergo, another addition or substitution reaction (see above). Thus, carboxylic acid derivatives can react with as many as two equivalents of nucleophiles, and carbonic acid derivatives can react with as many as three.

It remains to be explained how these different chemoselectivities come about. Why do nucleophiles react …

1. with aldehydes and ketones by addition and not by substitution?
2. with ketenes and other heterocumulenes by addition and not by substitution?
3. with carboxylic acids (or their derivatives) and with carbonic acid derivatives by substitution and not by addition?

Regarding question (1): in substitution reactions, aldehydes and ketones would have to eliminate a hydride ion or a carbanion, both of which are extremely poor leaving groups (cf. Section 2.3).

Regarding question (2): ketenes and other heterocumulenes contain only double-bonded substituents at the reacting C atom. However, a leaving group must be single-bonded. Consequently, the structural prerequisite for the occurrence of a substitution reaction is absent.

Regarding question (3): the addition of a nucleophile to the C=O double bond of carboxylic or carbonic acid derivatives would give products of type **C** or **D** (Figure 6.1). However, these compounds are without exception thermodynamically less stable than the corresponding substitution products **A** or **B**. The reason for this is that the three bonds in the substructure C_{sp^3}(—O—H)—Het of the addition products **C** and **D** are together less stable than the double bond in the substructure C_{sp^2}(=O) of the substitution products **A** and **B** plus the highlighted single bond in the by-product H—Het. In fact, ordinarily (Section 6.2), sub-

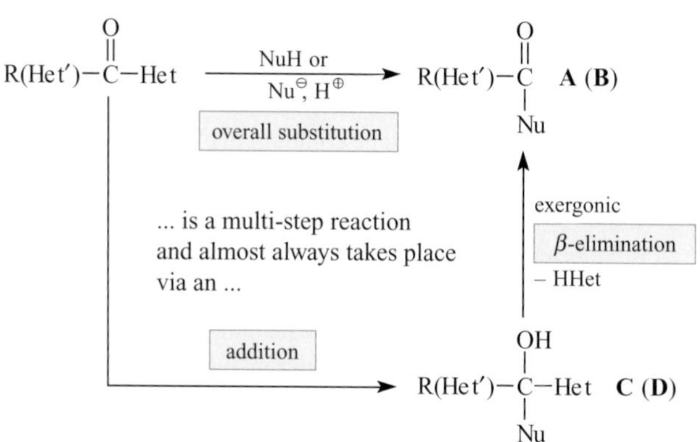

Fig. 6.1. Reactions of nucleophiles with C=O-containing carboxylic acid and carbonic acid derivatives. Substitution at the carboxyl carbon instead of addition to the acyl group.

stitution reactions on the carboxyl carbon leading ultimately to compounds of type **A** or **B** take place via neutral addition products of types **C** or **D** as *intermediates* (Figure 6.1). These addition products may be produced either continuously through reacting of the nucleophile (Figures 6.2, 6.5) or not until an aqueous workup has been carried out subsequent to the completion of the nucleophilic reaction (Figure 6.4). In both cases, once the neutral addition products **C** and **D** have formed, they decompose exergonically to furnish the substitution products **A** or **B** via rapid eliminations.

6.2 Mechanisms, Rate Laws, and Rate of Nucleophilic Substitution Reactions at the Carboxyl Carbon

When a nucleophile containing a heteroatom reacts at a carboxyl carbon S_N, reactions occur that convert carboxylic acid derivatives into other carboxylic acid derivatives, or they convert carbonic acid derivatives into other carbonic acid derivatives. When an organometallic compound is used as the nucleophile, S_N reactions at the carboxyl carbon make it possible to synthesize aldehydes (from derivatives of formic acid), ketones (from derivatives of higher carboxylic acids), or—starting from carbonic acid derivatives—carboxylic acid derivatives. Similarly, when using a hydride transfer agent as the nucleophile, S_N reactions at a carboxyl carbon allow the conversion of carboxylic acid derivatives into aldehydes.

From the perspective of the nucleophiles, these S_N reactions constitute acylations. Section 6.2 describes which acid derivatives perform such acylations rapidly as “good acylating agents” and which ones undergo them only slowly as “poor acylating agents,” and why this is so. Because of their greater synthetic importance, we will examine thoroughly only the acylations with *carboxylic acid derivatives*. Using the principles learned in this context, you can easily derive the acylating abilities of *carbonic acid derivatives*.

6.2.1 Mechanism and Rate Laws of S_N Reactions at the Carboxyl Carbon

Most S_N reactions of carboxylic acids and their derivatives follow one of three mechanisms (Figures 6.2, 6.4, and 6.5). A key intermediate common to all of them is the species in which the nucleophile is linked with the former carboxyl carbon for the first time. In this intermediate, the reacting carbon atom is tetrasubstituted and thus tetrahedrally coordinated. This species is therefore referred to as the **tetrahedral intermediate** (abbreviated as "Tet. Intermed." in the following equations). Depending on the nature of the reaction partners, the tetrahedral intermediate can be negatively charged, neutral, or even positively charged.

The tetrahedral intermediate is a *high-energy* intermediate. Therefore, independently of its charge and also independently of the detailed formation mechanism, it is formed via a *late* transition state. It also reacts further via an *early* transition state. Both properties follow from the Hammond postulate. Whether the transition state of the *formation* of the tetrahedral intermediate has a higher or a lower energy than the transition state of the subsequent *reaction* of the tetrahedral intermediate determines whether this intermediate is formed in an irreversible or in a reversible reaction, respectively. Yet, in any case, *the tetrahedral intermediate is a transition state model of the rate-determining step of the vast majority of S_N reactions at the carboxyl carbon*. In the following sections, we will support this statement by formal kinetic analyses of the most important substitution mechanisms.

S_N Reactions at the Carboxyl Carbon in Nonacidic Protic Media

Figure 6.2 shows the standard mechanism of substitution reactions carried out on carboxylic acid derivatives in neutral or basic solutions. The tetrahedral intermediate—formed in the rate-determining step—can be converted to the substitution product via two different routes. The shorter route consists of a single step: the leaving group X is eliminated with a rate constant k_{elim}. In this way the substitution product is formed in a total of two steps. The longer route to the same substitution product is realized when the tetrahedral intermediate is protonated. To what extent this occurs depends, according to Equation 6.1, on the pH value and on the equilibrium constant K_{eq} defined in the middle of Figure 6.2:

$$[\text{Protonated Tet. Intermed.}] = [\text{Tet. Intermed.}] \cdot K_{\text{eq}} \cdot 10^{-\text{pH}} \tag{6.1}$$

The protonated tetrahedral intermediate can eject the leaving group X with a rate constant k'_{elim}. Subsequently, a proton is eliminated. Thereby, the substitution product is formed in a total of four steps.

Which of the competing routes in Figure 6.2 preferentially converts the tetrahedral intermediate to the substitution product? An approximate answer to this question is possible if an equilibrium between the negatively charged and the neutral tetrahedral intermediate is established and if the rate constants of this equilibration are much greater than the rate constants k_{elim} and $k\ _{\text{elim}}$ of the reactions of the respective intermediates to give the substitution product. Under these conditions we have:

$$\frac{d[S_N\,\text{product}_{\text{two-step}}]}{dt} = k_{\text{elim}}\,[\text{Tet. Intermed.}] \tag{6.2}$$

and

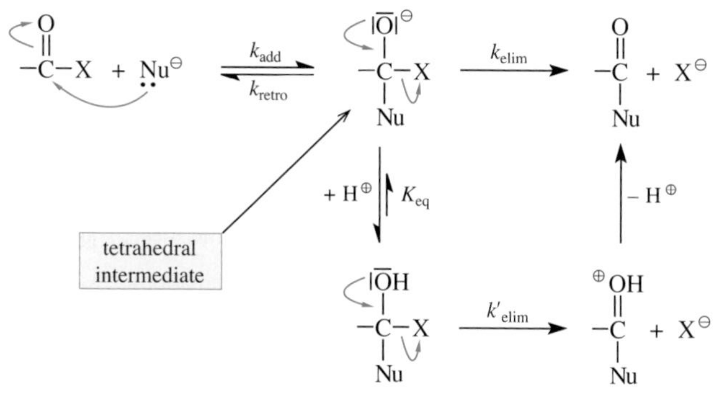

Fig. 6.2. Mechanism of S_N reactions of good nucleophiles at the carboxyl carbon: k_{add} is the rate constant of the addition of the nucleophile, k_{retro} is the rate constant of the back-reaction, and k_{elim} is the rate constant of the elimination of the leaving group; K_{eq} is the equilibrium constant for the protonation of the tetrahedral intermediate at the negatively charged oxygen atom.

$$\frac{d[S_N\text{ product}_{\text{four-step}}]}{dt} = k'_{elim}\,[\text{Protonated Tet. Intermed.}] \tag{6.3}$$

The subscripts "two-step" and "four-step" in Equations 6.2 and 6.3, respectively, refer to the rates of product formation via the two- and four-step routes of the mechanism in Figure 6.2. If one divides Equation 6.2 by Equation 6.3 and integrates subsequently, one obtains:

$$\frac{\text{Yield } S_N \text{ product}_{\text{two-step}}}{\text{Yield } S_N \text{ product}_{\text{four-step}}} = \underbrace{\frac{k_{elim}}{k'_{elim}}}_{\gg 1} \cdot \frac{10^{pH}}{K_{eq}} \tag{6.4}$$

Equation 6.4 shows the following: in strongly basic solutions, S_N products can be produced from carboxylic acid derivatives in *two* steps. An example of such a reaction is the saponification PhC(=O)OEt + KOH → PhC(=O)O$^-$K$^{\oplus}$ + EtOH. However, in approximately neutral solutions the *four-step* path to the S_N product should predominate. An example of this type of reaction is the aminolysis PhC(=O)OEt + $HNMe_2$ → PhC(=O)NMe_2 + EtOH.

Finally, we want to examine the kinetics of two-step acylations according to the mechanism in Figure 6.2. We must distinguish between two cases. Provided that the tetrahedral intermediate forms *irreversibly* and consequently in the rate-determining step, a rate law for the formation of the S_N product (Equation 6.7) is obtained from Equation 6.5.

$$\frac{d[\text{Tet. Intermed.}]}{dt} = k_{add}[-\text{C}(=\text{O})\text{X}][\text{Nu}^{\ominus}] - k_{elim}[\text{Tet. Intermed.}] \tag{6.5}$$

$$= 0 \text{ (because of the steady-state approximation)}$$

$$\Rightarrow k_{elim}[\text{Tet. Intermed.}] = k_{add}[-\text{C}(=\text{O})\text{X}][\text{Nu}^{\ominus}] \tag{6.6}$$

Inserting Equation 6.6 in Equation 6.2 ⇒

$$\frac{d[S_N\text{ product}]}{dt} = k_{add}[-\text{C}(=\text{O})\text{X}][\text{Nu}^{\ominus}] \tag{6.7}$$

In contrast, Equations 6.8–6.10 show the rate of product formation for an acylation by the two-step route in Figure 6.3, assuming that the tetrahedral intermediate is formed *reversibly*. Interestingly, it does not matter whether in the rate-determining step the tetrahedral intermediate is *formed* ($k_{retro} < k_{elim}$) or reacts further (then $k_{retro} > k_{elim}$).

$$\frac{d[\text{Tet. Intermed.}]}{dt} = k_{add}[-C(=O)X][Nu^{\ominus}] - k_{retro}[\text{Tet. Intermed.}] - k_{elim}[\text{Tet. Intermed.}]$$

$$= 0 \text{ (because of the steady-state approximation)} \tag{6.8}$$

$$\Rightarrow [\text{Tet. Intermed.}] = \frac{k_{add}}{k_{retro} + k_{elim}} [-C(=O)X][Nu^{\ominus}] \tag{6.9}$$

Inserting Eq. 6.9 in Eq. 6.2 ⇒

$$\frac{d[S_N \text{ product}]}{dt} = k_{add} \frac{1}{1 + k_{retro}/k_{elim}} [-C(=O)X][Nu^{\ominus}] \tag{6.10}$$

The *irreversible* formation of the tetrahedral intermediate allowing for the derivation of rate law 6.7 from Equation 6.5 may alternatively be regarded as a borderline case of a *reversible* formation of the tetrahedral intermediate (here rate law 6.10 applies)—namely in the case the rate constant k_{retro} approaches zero. Accordingly, Equation 6.10 for $k_{retro} = 0$ transforms into Equation 6.7.

Equations 6.7 and 6.10 are indistinguishable in the kinetic experiment because the *experimental* rate law would have the following form in both cases:

$$\frac{d[S_N \text{ product}]}{dt} = \text{const.} [-C(=O)X][Nu^{\ominus}] \tag{6.11}$$

This means that from the kinetic analysis one could conclude only that the S_N product is produced in a *bimolecular* reaction. There is no information as to whether the tetrahedral intermediate forms irreversibly or reversibly.

What does one expect regarding the irreversibility or reversibility of the formation of the tetrahedral intermediate in Figure 6.2? Answering this question is tantamount to knowing the ratio in which the tetrahedral intermediate ejects the group X (with the rate constant k_{elim}) or the nucleophile (with the rate constant k_{retro}). The outcome of this competition depends on whether group X or the nucleophile is the better leaving group. When X is a good leaving group, the tetrahedral intermediate is therefore expected to form irreversibly. This should be the case for all *good* acylating agents. When X is a poor leaving group, the nucleophile may be a better one so that it reacts with the acylating agent to give the tetrahedral intermediate in a reversible reaction. The alkaline hydrolysis of amides is an excellent example of this kind of substitution. Finally, when X and the nucleophile have similar leaving group abilities, k_{retro} and k_{elim} are of comparable magnitudes. Then, the tetrahedral intermediate decomposes partly into the starting materials and partly into the products. A good example of this kind of mechanism is the alkaline hydrolysis of carboxylic acid esters. In that case, the reversibility of the formation of the tetrahedral intermediate was proven by performing the hydrolysis with ^{18}O-labeled

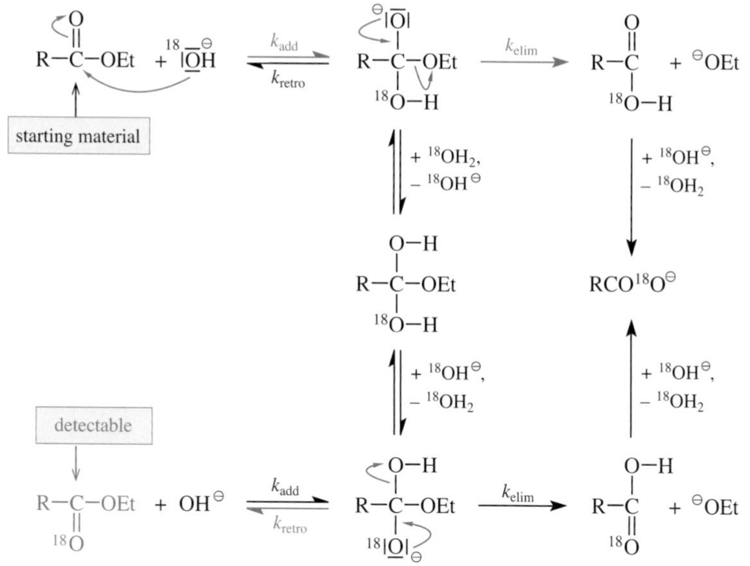

starting material

detectable

Fig. 6.3. Alkaline hydrolysis of carboxylic esters according to the mechanism of Figure 6.2: proof of the reversibility of the formation of the tetrahedral intermediate. In the alkaline hydrolysis of ethyl *para*-methylbenzoate in H_2O, for example, the ratio k_{retro}/k_{elim} is at least 0.13 (but certainly not much more).

NaOH (Figure 6.3): The unreacted ester that was recovered had incorporated ^{18}O atoms in its C=O double bond.

S_N Reactions at the Carboxyl Carbon via a Stable Tetrahedral Intermediate

A variant of the substitution mechanism of Figure 6.2 is shown in Figure 6.4. The tetrahedral intermediate is produced in an irreversible step and does not react further until it is worked up with aqueous acid. *Overall*, the substitution product is produced according to the four-step route of Figure 6.2. But in contrast to its standard course, two separate operations are required for the gross substitution to take place: first, the nucleophile must be added; second, $H_3O^{\oplus}$ must be added.

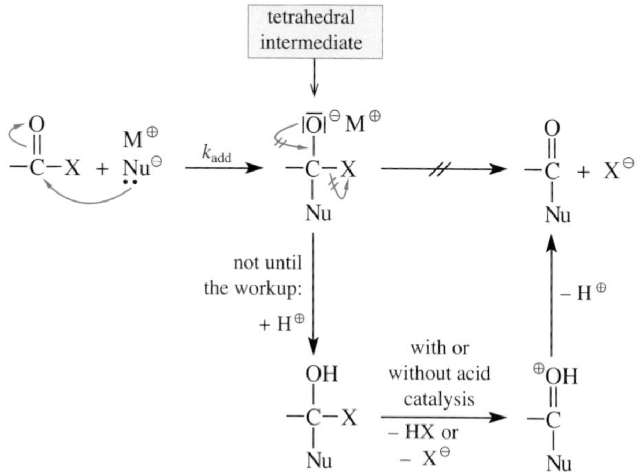

tetrahedral intermediate

not until the workup: $+ H^{\oplus}$

with or without acid catalysis $- HX$ or $- X^{\ominus}$

$- H^{\oplus}$

Fig. 6.4. Mechanism of S_N reactions at the carboxyl carbon via a stable tetrahedral intermediate.

The reactivity of carboxylic acid derivatives that react with nucleophiles according to the mechanism in Figure 6.4 cannot be measured via the rate of formation of the substitution product. Instead, the decrease in the concentration of the starting material serves as a measure of the reactivity.

$$\frac{\mathrm{d}[-\mathrm{C}(=\mathrm{O})\mathrm{X}]}{\mathrm{d}t} = -k_{\mathrm{add}}[-\mathrm{C}(=\mathrm{O})\mathrm{X}][\mathrm{Nu}^{\ominus}] \tag{6.12}$$

Many substitution reactions on the carboxyl carbon undertaken by hydride donors or organometallic reagents take place according to the mechanism of Figure 6.4 (cf. Section 6.5). For the reaction between phenyllithium and *N,N*-dimethylbenzamide, the tetrahedral intermediate could even by crystallized and characterized by X-ray analysis.

Proton-Catalyzed S_N Reactions at the Carboxyl Carbon

Figure 6.5 shows the third important mechanism of S_N reactions on the carboxyl carbon. It relates to proton-catalyzed substitution reactions of weak nucleophiles with weak acylating agents. When weak acylating agents are protonated in fast equilibrium reactions at the carboxyl oxygen, they turn into considerably more reactive acylating agents, carboxonium ions. Even catalytic amounts of acid can increase the reaction rate due to this effect because even a small equilibrium fraction of the highly reaction carboxonium ion can react easily with the nucleophile. Because protonation of the starting material continues to supply an equilibrium amount of the carboxonium ion, the entire acylating agent gradually reacts via this intermediate to give the S_N product. The proton-catalyzed esterifications of carboxylic acids or the acid-catalyzed hydrolysis of amides exemplify the substitution mechanism of Figure 6.5.

The rate law for S_N reactions at the carboxyl carbon according to the mechanism shown in Figure 6.5 can be derived as follows:

$$\frac{\mathrm{d}[\mathrm{S_N\ product}]}{\mathrm{d}t} = k_{\mathrm{elim}}[\text{Tet. Intermed.}] \tag{6.13}$$

$$\frac{\mathrm{d}[\text{Tet. Intermed.}]}{\mathrm{d}t} = k_{\mathrm{add}}[-\mathrm{C}(=\overset{\oplus}{\mathrm{O}}\mathrm{H})\mathrm{X}][\mathrm{Nu}^{\ominus}] - k_{\mathrm{retro}}[\text{Tet. Intermed.}] - k_{\mathrm{elim}}[\text{Tet. Intermed.}]$$

$$= 0 \text{ (because of steady-state approximation)}$$

$$\Rightarrow [\text{Tet. Intermed.}] = \frac{k_{\mathrm{add}}}{k_{\mathrm{retro}} + k_{\mathrm{elim}}}[-\mathrm{C}(=\overset{\oplus}{\mathrm{O}}\mathrm{H})\mathrm{X}][\mathrm{Nu}^{\ominus}] \tag{6.14}$$

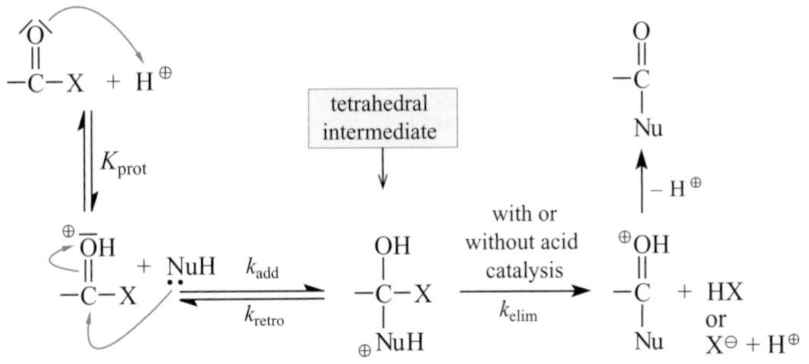

Fig. 6.5. Mechanism of proton-catalyzed S_N reactions at the carboxyl carbon; K_{prot} is the equilibrium constant of the protonation of the weak acylating agent used.

For the initiating protonation equilibrium we have:

$$[-C(=\overset{\oplus}{O}H)X] = K_{prot}\,[-C(=O)X][H^{\oplus}] \tag{6.15}$$

Eq. 6.14 and Eq. 6.15 inserted in Eq. 6.13 ⇒

$$\frac{d[S_N\,product]}{dt} = k_{add}\,\frac{1}{1 + k_{retro}/k_{elim}} \cdot K_{prot}[-C(=O)X][H^{\oplus}][Nu^{\ominus}] \tag{6.16}$$

The initial Equation 6.13 reflects that the second-to-last reaction step in Figure 6.5 is substantially slower than the last step, which is loss of a proton. Equation 6.13 is simplified using Equation 6.14. Equation 6.14 requires the knowledge of the concentration of the carboxonium ions, which are the acylating reagents in this mechanism. Equation 6.15 provides this concentration via the equilibrium constant K_{prot} of the reaction that forms the ion (Figure 6.5). Equations 6.13–6.15 allow for the derivation of the rate law of the S_N reaction according to this mechanism, which is given as Equation 6.16. This equation is the rate law of a trimolecular reaction: The rate of the reaction is proportional to the concentrations of the acylating agent, the nucleophile, and the protons.

The Rate-Determining Step of the Most Important S_N Reactions at the Carboxyl Carbon
Let us summarize: The rate laws for S_N reactions at the carboxyl carbon exhibit an important common feature regardless of whether the substitution mechanism is that of Figure 6.2 (rate laws: Equation 6.7 or 6.10), Figure 6.4 (rate law: Equation 6.12), or Figure 6.5 (rate law: Equation 6.16):

Rule of Thumb

If your head is spinning from those equations, take heart. This simple rule will get you far: The larger the rate constant k_{add} for the formation of the tetrahedral intermediate, the faster an acylating agent reacts with nucleophiles.

Therefore, *independent of the substitution mechanism*, the reactivity of a series of acylating agents with respect to a given nucleophile is characterized by the fact that the most reactive acylating agent exhibits the smallest energy difference between the acylating agent and the derived tetrahedral intermediate. This energy difference becomes small if the acylating agent R—C(=O)(—X) is high in energy (relatively destabilized) and/or the derived tetrahedral intermediate R—C(—$O^{\ominus}$)(—Nu)(—X) or R—C(—OH)(—Nu)(—X) is low in energy (relatively stabilized):

Rule of Thumb

- The acylating agent R—C(=O)—X is generally higher in energy, the lower the resonance stabilization of its C=O double bond by the substituent X. This effect is examined in detail in Section 6.2.2.
- The tetrahedral intermediate is generally lower in energy the more it is stabilized by a electron withdrawing inductive (–I) effect of the leaving group X or by an anomeric effect. This will be discussed in detail in Section 6.2.3.

6.2.2 S_N Reactions at the Carboxyl Carbon: The Influence of Resonance Stabilization of the Reacting C=O Double Bond on the Reactivity of the Acylating Agent

Table 6.1 lists acylating agents that can react with nucleophiles *without prior protonation*, that is, according to the mechanism of Figure 6.2 or according to the mechanism of Figure 6.4. They are arranged from the top to the bottom in order of decreasing resonance stabilization of the C=O double bond. It was briefly indicated in the preceding section that carboxylic acid derivatives R—C(=O)—X lose that part of their resonance stabilization that the X group had provided to the C=O double bond of the substrate when they react with the nucleophile to form the tetrahedral intermediate. This explains why the acylating agents of Table 6.1 are concomitantly arranged in the order of increasing reactivity.

The carboxylate ion (Table 6.1, entry 1) has the greatest resonance stabilization at approximately 30 kcal/mol. As expected, it is the weakest acylating agent of all and reacts only organolithium compounds as nucleophiles. Amides are also significantly resonance-stabilized (stabilization ≈ 22 kcal/mol; entry 2). Accordingly, they are rather poor acylating agents, too. Nevertheless they react not only with organolithium compounds but also with Grignard reagents and hydride donors and, under harsher conditions, with NaOH or amines. In carboxylic acids and carboxylic esters, the C=O double bond exhibits a resonance stabilization of approximately 14 kcal/mol (entry 3). Both compounds are therefore considerably more reactive than amides with respect to nucleophiles. However, it must be kept in mind that this is true for carboxylic acids only in the absence of a base. Bases, of course, deprotonate them to yield carboxylate ions. The decrease in resonance stabilization of the carboxyl group in the acylating agents carboxylate > carboxylic amide > carboxylic (ester) specified earlier is caused by the decrease in the pi electron donating (+M) effect of the substituent on the carboxyl carbon in the order $O^{\ominus}$ > NR_2, NRH, NH_2 > OAlk, OH.

All other carboxylic acid derivatives in Table 6.1, in which the leaving group is bound to the carboxyl carbon through an O atom, are increasingly better acylating agents than carboxylic acid alkyl esters (entry 3) in the order carboxylic acid phenyl ester (entry 4) < acyl isourea (entry 7) < mixed carboxylic acid/carbonic acid anhydride (entry 8) < carboxylic acid anhydride (entry 9) ≤ mixed carboxylic acid anhydride (entry 10).

The reason for this increase in reactivity is the decreasing resonance stabilization of the reacting C=O double bond by the free electron pair on the *neighboring* O atom. In the aforementioned series of compounds, this electron pair is available to an increasingly limited extent for stabilizing the reacting C=O double bond. This is because this lone pair also provides resonance stabilization to a second adjacent C=Het double bond. Note that the resonance stabilization of that second C=Het double bond is fully retained in the tetrahedral intermediate of the acylating agent. Consequently, the existence of *this* resonance does not lead to any decrease in the reactivity of the acylating agent. The demand on the lone pair of the single-bonded O atom by the second C=Het double bond of the acylating agents under consideration is naturally more pronounced, the greater the pi electron withdrawing (–M) effect of the second C=Het group. The –M effect increases in the order C=C (as part of an aromatic compound) < —C(=NAlk) – NHAlk < —C(=O)—OR < —C(=O)—R. Consequently, in the corresponding acylating agents RC(=O)—X the +M effect of the carboxyl substituent X

Tab. 6.1 Acylating Agents in the Order of Decreasing Resonance Stabilization of the Reacting C=O Double Bond*

(1) provides 30 kcal/mol stabilization

(2) provides 22 kcal/mol stabilization

R' = Alk, H

(3) provides 14 kcal/mol stabilization

R' = Alk, H

(4)

(5)

(6) additional aromatic resonance forms

(7) • X = NAlk, Y = NHAlk: DCC activation (cf. Figure 6.15)

(8) • X = O, Y = OR: $ClCO_2iBu$ activation (cf. Figure 6.14)

(9) • R' = R: inevitably 50% of the acylating agent does not go into the product but is lost as a leaving group

(10) • R' ≠ R: as much as 100% of the contained RC(=O) can be incorporated into the product (cf. Figure 6.14)

Tab. 6.1 (continued)

(11)

(12)

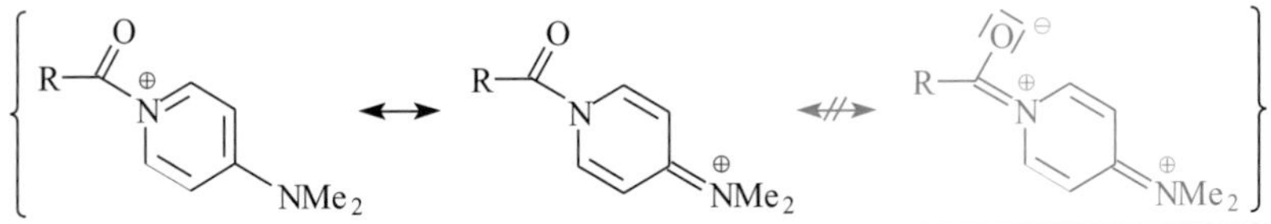

* Resonance forms drawn black contribute to the overall stabilization of the acylating agent but not to the stabilization of the C=O double bond, which reacts with the nucleophile.

decreases in the order —O—Ar > —O—C(=NAlk)—NAlk > —O—C(=O)—OR > —O—C(=O)—R. The ease of acylation increases accordingly.

Thioesters (entry 5 of Table 6.1) are quite good acylating agents. Thus, they react with nucleophiles considerably faster than their oxa analogs, the carboxylic acid alkyl esters (entry 3). This difference in reactivity is due to the fact that the pi electron donating ability of an —S—R group is smaller than that of an —O—R group. Sulfur, as an element of the second long period of the periodic table, is less capable of forming stable p_π,p_π double bonds than oxygen.

In acid chlorides, the Cl atom, which likewise is an element of the second long period of the periodic table, is not able to stabilize the neighboring C=O group by resonance at all (Table 6.1, entry 11). The main reason for this is that chlorine has only a negligible ability to form stable p_π,p_π double bonds. Because of its greater electronegativity, the electron donor capability of chlorine is even lower than that of sulfur. Acid chlorides are consequently among the strongest acylating agents.

The 6th rank in terms of acylation reactivity that is attributed to the acyl imidazolides in Table 6.1 (entry 10) is also plausible. In the acyl imidazolides, the "free" electron pair of the acylated N atom is essentially unavailable for stabilization of the C=O double bond by resonance because it is part of the π-electron sextet, which makes the imidazole ring an aromatic compound. This is why acyl imidazolides, in contrast to "normal" amides (entry 2 in Table 6.1) can act as acylating agents. Nevertheless, acyl imidazolides do not have the same acylation capacity as acylpyridinium salts because the aromatic stabilization of five-membered aromatic compounds—and thus of imidazole—is considerably smaller than that of six-membered aromatic systems (e. g., pyridine). This means that the resonance form of the acyl imidazolides printed red in Table 6.1 contributes to the stabilization of the C=O double bond. For a similar reason, there is no resonance stabilization of the C=O double bond in *N*-acylpyridinium salts: in the corresponding resonance form, the aromatic sextet of the pyridine would be destroyed in exchange for a much less stable quinoid structure.

Carboxylic amides, carboxylic esters, and carboxylic acids react with acid-stable heteroatom nucleophiles in a neutral solution much more slowly via the mechanism of Figure 6.2 than in an acidic solution via the mechanism of Figure 6.5. In an acidic solution, their carboxonium ion derivatives, which result from the reversible protonation of the carboxyl oxygen, act as precursors of the tetrahedral intermediate. According to the discussion earlier in

Tab. 6.2 Energy Gain through Resonance in Nonprotonated and Protonated Carboxylic Acid Derivatives

	Resonance forms	Stabilization	
(1)	R–C(=O)–NR'$_2$ ↔ R–C(–O$^{\ominus}$)=N$^{\oplus}$R'$_2$	provides 22 kcal/mol stabilization	R' = Alk, H
(2)	R–C(=O$^{\oplus}$H)–NR'$_2$ ↔ R–C(–OH)=N$^{\oplus}$R'$_2$	provides >22 kcal/mol stabilization	
(3)	R–C(=O)–OR' ↔ R–C(–OH)=O$^{\oplus}$R'	provides 14 kcal/mol stabilization	R' = Alk, H
(4)	R–C(=O$^{\oplus}$H)–OR' ↔ R–C(–OH)=O$^{\oplus}$R'	provides >14 kcal/mol stabilization	

this section, this might at first surprise you. The energy profiles of Figure 6.6 solve the apparent contradiction. The protonated forms of the acylating agents in question are in fact higher in energy than the corresponding nonprotonated forms, the higher resonance stabilization of the former notwithstanding. Consequently, in all the systems mentioned only a small fraction of the amide, ester, or acid present is protonated. Actually, the reason for the effectiveness of

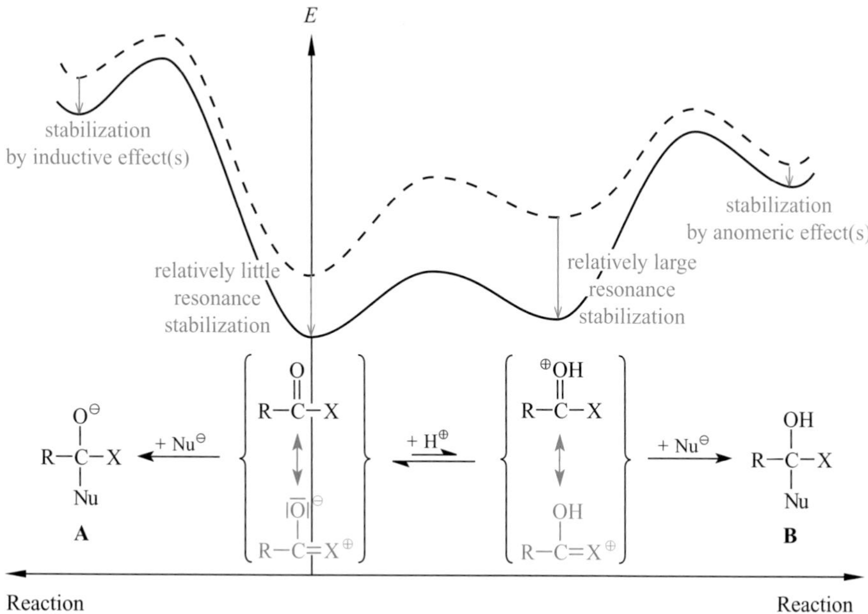

Fig. 6.6. Energy profiles for forming the tetrahedral intermediate from carboxylic amides, carboxylic esters, and carboxylic acids according to the mechanism of Figure 6.2 (the reaction coordinate goes to the left) or according to the mechanism of Figure 6.5 (the reaction coordinate goes to the right); solid curves, actual energy profiles taking stabilizing electronic effects into consideration; dashed curves, fictitious energy profiles for reactions that take place in the absence of stabilizing electronic effects.

proton catalysis of these substitutions is quite different. In the presence of excess protons, the tetrahedral intermediate is similar to an alcohol (reaction **B** in Figure 6.6), while in the absence of excess protons (i.e., in basic or neutral solutions), it is similar to an alkoxide (reaction **A** in Figure 6.6). Being strong bases, alkoxides are high-energy species in comparison to their conjugate acids, the alcohols. Accordingly, S_N reactions in acidic solutions make more stable tetrahedral intermediates **B** available than is the case in nonacidic solutions, where they would have the structure **A**.

6.2.3 S_N Reactions at the Carboxyl Carbon: The Influence of the Stabilization of the Tetrahedral Intermediate on the Reactivity

According to Section 6.2.1, the tetrahedral intermediate is the transition state model of the rate-determining step for any of the most important substitution mechanisms at the carboxyl carbon. In S_N reactions that take place according to the mechanisms of Figures 6.2 or 6.4, this tetrahedral intermediate is an alkoxide (**A** in Figure 6.7), and for those that take place according to Figure 6.5 it is an alcohol (**B** in Figure 6.7). According to the Hammond postulate, the formation of these intermediates should be favored kinetically when they experience a stabilizing substituent effect. If the reactivity of a variety of acylating agents toward a reference nucleophile were compared, any rate difference, which would be attributable to differential stabilization of the respective intermediates **A** or **B**, would be due only to a substituent effect of the leaving group X of the acylating agent. As it turns out, the nature of this substituent effect depends on whether the stabilization of alkoxide **A** or of alcohol **B** is involved.

In intermediate **A** of Figure 6.7, which is an alkoxide, the negative charge on the alkoxide oxygen must be stabilized. The leaving group X does this by its electron withdrawing inductive (–I) effect, and the greater this effect the better the leaving group's stability. The greatest –I effects are exerted by X = pyridinium and X = Cl. Therefore, *N*-acylpyridinium salts and carboxylic chlorides react with nucleophiles via especially well-stabilized tetrahedral intermediates **A**. Another effect, a stereoelectronic one, is also involved in the stabilization of tetrahedral intermediates.

The stability of intermediate **B** of Figure 6.7, which is an alcohol, is less influenced by the leaving group X and in any event not primarily through its –I effect (because it is no longer important to delocalize the excess charge of an anionic center). Nonetheless, the substituent X may even stabilize a *neutral* tetrahedral intermediate **B**. It does so through a stereoelectronic effect. This effect is referred to as the **anomeric effect**, because it is very important in sugar chemistry. Anomeric effects can occur in compounds that contain the structural element :Het^1—C_{sp^3}—Het^2. The substituents "Het" either must be halogen atoms or groups bound to the central C atom through an O- or an N atom. In addition, there is one more condition for the

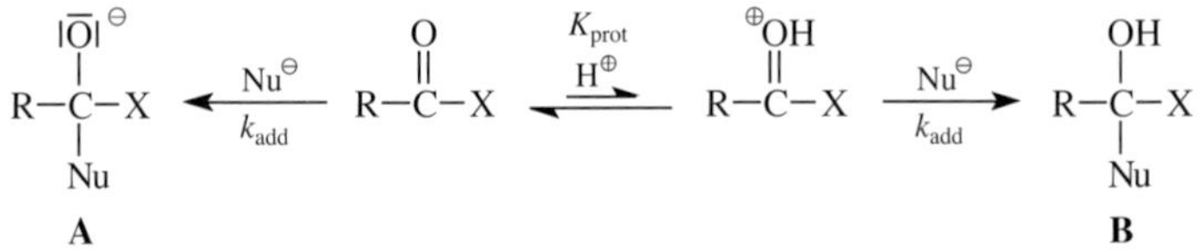

Fig. 6.7. Comparison of the substituent effects on the tetrahedral intermediates in S_N reactions at the carboxyl carbon.

$$\underset{\mathbf{A}}{R-C(O^{\ominus})(Nu)-X} \xleftarrow[k_{add}]{Nu^{\ominus}} R-C(=O)-X \underset{}{\overset{K_{prot},\ H^{\oplus}}{\rightleftharpoons}} R-C(=\overset{\oplus}{O}H)-X \xrightarrow[k_{add}]{Nu^{\ominus}} \underset{\mathbf{B}}{R-C(OH)(Nu)-X}$$

occurrence of an anomeric effect: The group :Het[1] must be oriented in such a way that the indicated free electron pair occupies an orbital that is oriented *antiperiplanar* to the C—Het[2] bond.

The stabilization of such a substructure :Het[1]—$C_{sp}{}^{3}$—Het[2] **A** can be rationalized both with the VB theory and with the MO model (Figure 6.8). On the one hand, it is possible to formulate a *no-bond* resonance form **B** for substructure **A**. In this substructure, a positive charge is localized on the substituent Het[1] and a negative charge on the substituent Het[2]. The stability of this resonance form increases with an increasing pi electron donating +M ability of the substituent Het[1] and with increasing electronegativity of the substituent Het[2]. The more stable the no-bond resonance form **B** is, the more resonance stabilization it contributes to intermediate **A**.

In MO theory, the mentioned conformer **A** of the :Het[1]—C—Het containing compound allows for overlap between the atomic orbital on the substituent Het[1], which accommodates the lone pair, and the $\sigma^*_{\text{C-Het}^2}$—$\text{Het}_2$ orbital (see formula **C** in Figure 6.8). This overlap lowers the energy of the atomic orbital (diagram in Figure 6.8, top right). As we know, this is more effective the narrower the energy-gap between the overlapping orbitals. Nonbonding electron pairs have a higher energy at nitrogen than at oxygen, and at oxygen than at fluorine. Conversely, the energy of the $\sigma^*_{\text{C-Het}^2}$—Het[2] decreases in the order Het[2] = NR_2, OR and F. Accordingly, for this group of compounds, calculations showed that the greatest anomeric effect occurs in the substructure :NR_2—C_{sp^3}—F and the smallest in the substructure :F—C_{sp^3}—NR_2. (Only theory is able to separate the last effects from each other; in fact, these anomeric effects only occur together and can therefore only be observed as a sum effect.)

The lower part of Figure 6.8 shows the application of our considerations of the general substructure :Het[1]—C_{sp3}—Het[2] to the specific substructure HÖ—C_{sp3}—X of the tetrahedral intermediate **B** of the S_N reactions of Figure 6.7. This allows us to state the following: suitable leaving groups X can stabilize intermediate **B** through an anomeric effect. This stabilization increases with increasing electronegativity of the leaving group X.

In other words: the higher the electronegativity of the leaving group X in the acylating agent R—C(=O)—X, the better stabilized is the tetrahedral intermediate of an S_N reaction at the carboxyl carbon. Whether this tetrahedral intermediate happens to be an alkoxide and is

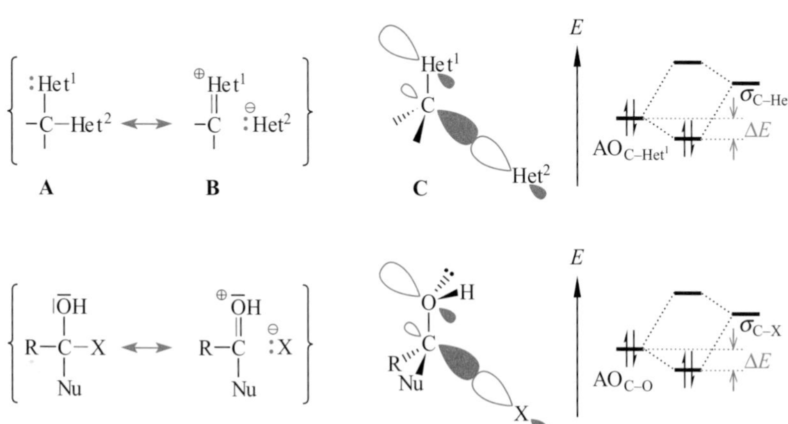

Fig. 6.8. VB explanation (left) and MO explanation (right) of the stabilization of a structure element :Het[1]—C—Het[2] (upper line) and of a suitable conformer of intermediate **B** of Figure 6.7 (bottom line) through the "anomeric effect."

stabilized inductively or whether it happens to be an alcohol and is stabilized through an anomeric effect plays a role only in the *magnitude* of the stabilization.

The observations in Sections 6.2.2 and 6.2.3 can be summarized as follows: strongly electronegative leaving groups X make the acylating agent R—C(=O)—X reactive because they provide little to no resonance stabilization to the C=O double bond of the acylating agent, and because they stabilize the tetrahedral intermediate inductively or through an anomeric effect. We thus note that:

Rule of Thumb

Carboxylic acid derivatives with a very electronegative leaving group X are good acylating agents, whereas carboxylic acid derivatives with a leaving group X of low electronegativity are poor acylating agents.

6.3 Activation of Carboxylic Acids and of Carboxylic Acid Derivatives

The conversion of a carboxylic acid into a carboxylic acid derivative, which is a more reactive acylating agent, is called "carboxylic acid activation." One can also convert an already existing carboxylic acid derivative into a more reactive one by "activating" it. Three methods are suitable for realizing such activations. One can activate carboxylic acids and some carboxylic acid derivatives through equilibrium reactions, in which, however, only part of the starting material is activated, namely, as much as is dictated by the respective equilibrium constant (Section 6.3.1). On the other hand, carboxylic acids can be converted into more reactive acylating agents *quantitatively*. One can distinguish between quantitative activations in which the acylating agent obtained must be isolated (Section 6.3.2) and quantitative activations that are effected *in situ* (Section 6.3.3).

6.3.1 Activation of Carboxylic Acids and Carboxylic Acid Derivatives in Equilibrium Reactions

Primary, secondary, and tertiary carboxylic amides, carboxylic esters, and carboxylic acids are protonated by mineral acids or sulfonic acids at the carboxyl oxygen to a small extent (Figure 6.9). This corresponds to the activation discussed in Section 6.2.3. This activation is used in acid hydrolyses of amides and esters, in esterifications of carboxylic acids and in Friedel–Crafts acylations of aromatic compounds with carboxylic acids.

In the Friedel–Crafts acylation, carboxylic acid chlorides and carboxylic acid anhydrides are activated with stoichiometric amounts of $AlCl_3$ (Section 5.2.7). However, this activation is only possible in the presence of very weak nucleophiles such as aromatic compounds. Stronger nucleophiles would react with the $AlCl_3$ instead of the carboxylic acid derivative. If one wants to acylate such stronger nucleophiles—for example, alcohols or amines—with car-

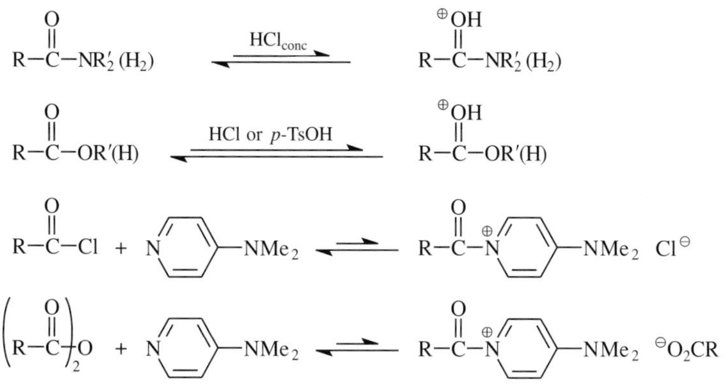

$$R{-}C({=}O){-}NR'_2\,(H_2) \xrightleftharpoons{HCl_{conc}} R{-}C({=}\overset{\oplus}{O}H){-}NR'_2\,(H_2)$$

$$R{-}C({=}O){-}OR'(H) \xrightleftharpoons{HCl\ \text{or}\ p\text{-TsOH}} R{-}C({=}\overset{\oplus}{O}H){-}OR'(H)$$

$$R{-}C({=}O){-}Cl + N(C_5H_4){-}NMe_2 \rightleftharpoons R{-}C({=}O){-}\overset{\oplus}{N}(C_5H_4){-}NMe_2\ Cl^{\ominus}$$

$$(R{-}C({=}O))_2O + N(C_5H_4){-}NMe_2 \rightleftharpoons R{-}C({=}O){-}\overset{\oplus}{N}(C_5H_4){-}NMe_2\ {}^{\ominus}O_2CR$$

Fig. 6.9. Examples of the activation of carboxylic acid derivatives in equilibrium reactions.

boxylic acid chlorides or with carboxylic acid anhydrides, and one wishes to speed up the reaction, these acylating agents can be activated by adding catalytic amounts of *para*-(dimethylamino)pyridine ("Steglich's catalyst"). Then, the acylpyridinium chlorides or carboxylates of Figure 6.9 form in equilibrium reactions via mechanism shown in Figure 6.2. They are far more reactive acylating agents (cf. discussion of Table 6.1).

6.3.2 Conversion of Carboxylic Acids into Isolable Acylating Agents

The most frequently used strong acylating agents are acid chlorides. They can be prepared from carboxylic acids especially easily with $SOCl_2$ or with oxalyl chloride (Figure 6.10). In fact, if carboxylic acids are treated with one of these reagents, only gaseous by-products are produced: SO_2 and HCl from $SOCl_2$, and CO_2, CO, and HCl from oxalyl chloride. PCl_3, $POCl_3$, or PCl_5 do not provide this advantage, although they can also be used to convert carboxylic acids to acid chlorides.

The carboxylic acid activation with $SOCl_2$ or with oxalyl chloride starts with the formation of the respective mixed anhydride (Figure 6.10). Carboxylic acids and $SOCl_2$ give a carboxylic acid/chlorosulfinic acid anhydride **A**. Carboxylic acids and oxalyl chloride furnish the carboxylic acid/chlorooxalic acid anhydride **C**, following the mechanism of Figure 6.2, that is, by a nucleophilic reaction of the carboxylic acid on the carboxyl carbon of oxalyl chloride. The anhydride **A** is probably formed in an analogous S_N reaction, i.e., one which the carboxylic acid undertakes at the S atom of $SOCl_2$.

In the formation of anhydrides **A** and **C**, one equivalent of HCl is released. It reacts with the activated carboxylic carbon atom of these anhydrides in an S_N reaction. The acid chloride is formed via the tetrahedral intermediates **B** or **D**, respectively, according to the mechanism of Figure 6.2. At the same time the leaving group Cl—S(=O)—O$^{\ominus}$ or Cl—C(=O)—C(=O)—O$^{\ominus}$, respectively, is liberated. Both leaving groups are short-lived and fragment immediately. to afford the gaseous by-products SO_2 and HCl or CO_2, CO, and HCl, respectively.

The conversion of carboxylic acids and $SOCl_2$ into acid chlorides is frequently catalyzed by DMF. The mechanism of this catalysis is shown in Figure 6.11. It is likely that $SOCl_2$ and DMF first react to give the Vilsmeier–Haack reagent **A**. It differs from the reactive interme-

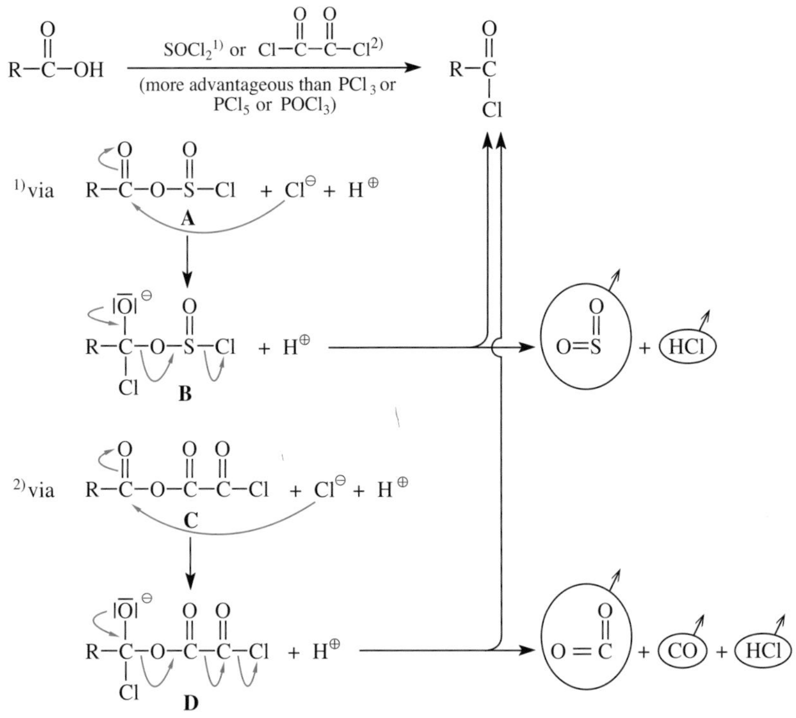

Fig. 6.10. Conversion of carboxylic acids to carboxylic acid chlorides with thionyl chloride or oxalyl chloride.

diate of the Vilsmeier–Haack formylation (Figure 5.36) only insofar as the cation is associated here with a chloride ion, rather than a dichlorophosphate ion. Now, the carboxylic acid reacts with the imminium carbon of intermediate **A** in an S_N reaction in which the Cl atom is displaced. This reaction takes place analogously to the mechanism shown in Figure 6.2. The sub-

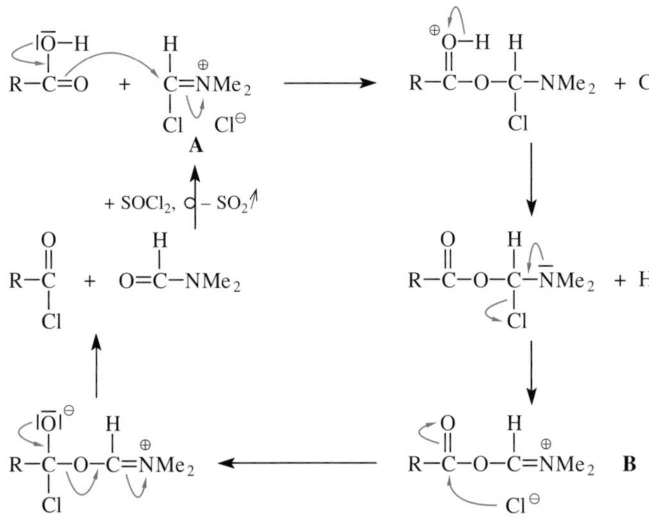

Fig. 6.11. Mechanism of the DMF-catalyzed conversion of carboxylic acids and $SOCl_2$ into acid chlorides.

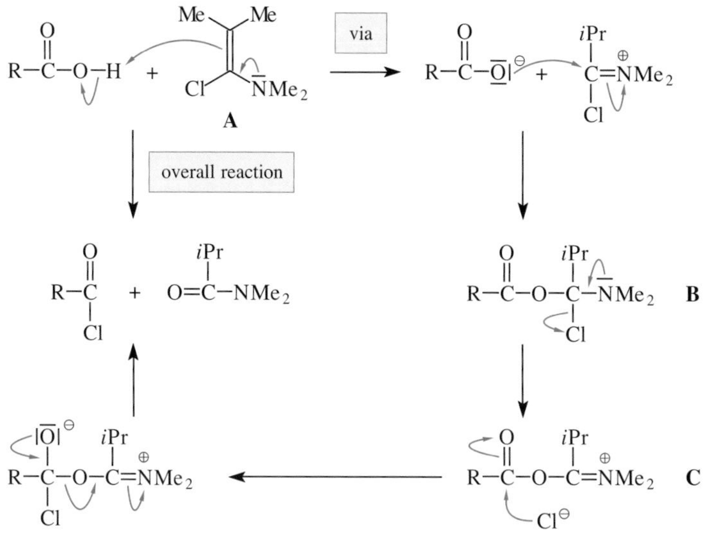

Fig. 6.12. Acid-free preparation of carboxylic chloride from carboxylic acids and a chloroenamine.

stitution product is the *N*-methylated mixed anhydride **B** of a carboxylic acid and an imidoformic acid. This mixed anhydride **B** finally acylates the released chloride ion to yield the desired acid chloride. At the same time the catalyst DMF is regenerated.

Figure 6.12 shows that carboxylic acids can also be converted into acid chlorides *without* releasing HCl. This is possible when carboxylic acids are treated with the chloroenamine **A**. First the carboxylic acid adds to the C=C double bond of this reagent electrophilically (see Figures 3.51 and 3.53). Then, the addition product **B** dissociates completely to give the ion pair **C**. This constitutes the isopropyl analog of the Vilsmeier–Haack intermediate **B** of the DMF-catalyzed carboxylic chloride synthesis of Figure 6.11. The new Vilsmeier–Haack intermediate reacts exactly like the old one (cf. previous discussion): The chloride ion undertakes an S_N reaction at the carboxyl carbon. This produces the desired acid chloride and isobutyric *N,N*-dimethylamide.

Another carboxylic acid activation in a neutral environment together with all mechanistic details is shown in Figure 6.13: carboxylic acids and carbonyldiimidazole (**A**) react to form the reactive carboxylic acid imidazolide **B**.

Carboxylic acids can also be activated by converting them to their anhydrides. For this purpose they are dehydrated with concentrated sulfuric acid, phosphorus pentoxide, or 0.5 equivalents of $SOCl_2$ (1 equivalent of $SOCl_2$ reacts with carboxylic acids to form acid chlorides rather than anhydrides). However, carboxylic anhydrides cannot transfer more than 50% of the original carboxylic acid to a nucleophile. The other 50% is released—depending on the pH value—either as the carboxylic acid or as a carboxylate ion; and is therefore lost. Consequently, in laboratory chemistry, the conversion of carboxylic acids into anhydrides is not as relevant as carboxylic acid activation. Nonetheless, acetic anhydride is an important acetylating agent because it is commercially available and inexpensive.

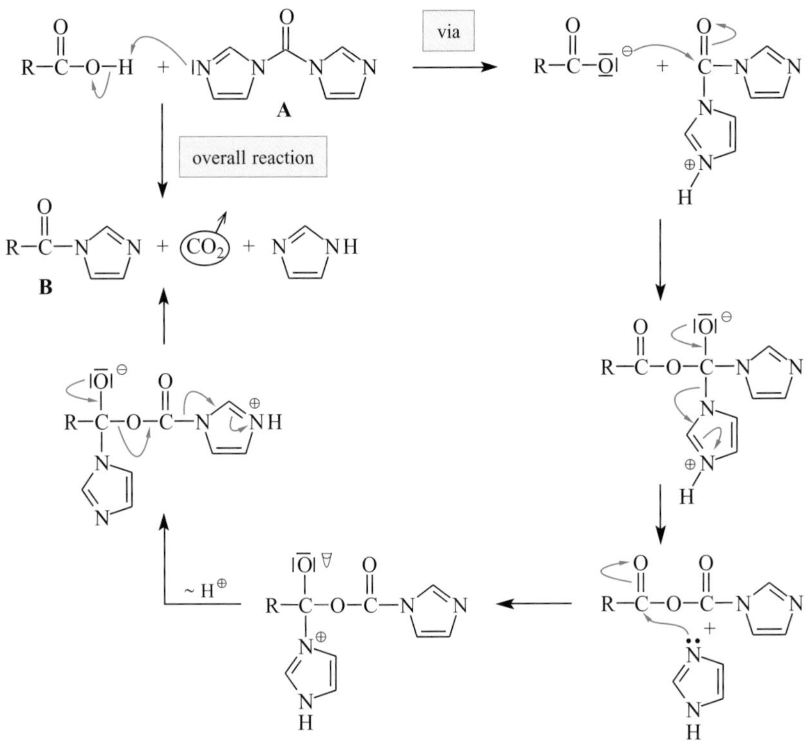

Fig. 6.13. Acid-free activation of carboxylic acids as carboxylic acid imidazolides.

6.3.3 Complete in Situ Activation of Carboxylic Acids

As can be seen from Table 6.1, a number of mixed anhydrides are good acylating agents. Some mixed anhydrides are commercially available. However, most such acylating agents are prepared *in situ* from the carboxylic acid and a suitable reagent. Four of these mixed anhydrides will be discussed in more detail in the following.

The acylation of carboxylic acids with 2,4,6-trichlorobenzoyl chloride gives mixed anhydrides **A** (Figure 6.14). Triethylamine must be present in this reaction to scavenge the released HCl. Anhydrides **A** contain two different acyl groups. In principle, both of them could react with a nucleophile. However, one observes the chemoselective reaction on the acyl group that originates from the *acid* used. This is because the carboxyl group located next to the aromatic ring is sterically hindered. In the most stable conformation the Cl atoms in the *ortho*-positions lie in the half-spaces above and below the proximal C=O double bond and therefore block the approach of the nucleophile toward that part of the molecule.

Carboxylic acids can be activated *in situ* as mixed anhydrides **B** (Figure 6.14) that are mixed anhydrides of a carboxylic acid and a carbonic acid half ester. As can be seen from Table 6.1, in anhydrides of this type the C=O double bond of the carboxylic acid moiety is stabilized less by resonance than the C=O double bond of the carbonic acid moiety. Therefore, a nucleophile chemoselectively reacts with the carboxyl carbon of the carboxylic and not the carbonic acid ester moiety.

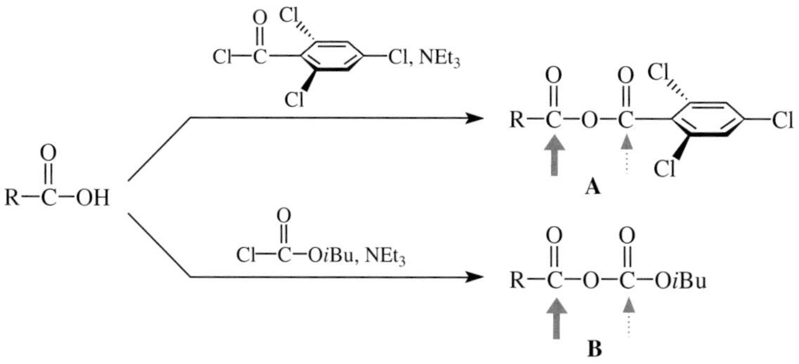

Fig. 6.14. *In situ* activation of carboxylic acids as mixed anhydrides.

Whereas the mixed anhydrides **A** of Figure 6.14 acylate amines and alcohols, the mixed anhydrides **B** are suitable for acylating amines but unsuitable for acylating alcohols. During the course of an acylation, the leaving group *i*BuO—C(=O)—O$^\ominus$ is liberated. It is unstable and fragments to give CO_2 and isobutanol. This isobutanol, can compete with the starting alcohol for the remaining anhydride **B**. In contrast, an amine as the original reaction partner of the mixed anhydride **B** remains the best nucleophile even in the presence of isobutanol. Therefore, aminolyses of **B** succeed without problems.

In peptide synthesis, the *in situ* activation of carboxylic acids with dicyclohexylcarbodiimide (DCC; compound **A** in Figure 6.15) is very important. By adding the carboxylic acid to the C=N double bond of this reagent, one obtains compounds of type **B**, so-called *O*-acyl isoureas. These constitute diaza analogs of the mixed anhydrides **B** of Figure 6.14. As can therefore be expected, *O*-acyl isoureas react with good nucleophiles with the same regioselectivity as their oxygen analogs: at the carboxyl carbon of the carboxylic acid moiety.

Poor nucleophiles react with acyl isoureas **B** so slowly that the latter start to decompose. In some sense they acylate themselves. The N atom designated with the positional number 3 intramolecularly substitutes the *O*-bound leaving group that is attached to the carboxyl carbon Cl'. A four-membered cyclic tetrahedral intermediate is formed. When the Cl'-Ol bond in this intermediate opens up, the *N*-acyl urea **E** is produced. Because compound **E** is an amide derivative it is no longer an acylating agent (cf. Section 6.2).

The "deactivation" of the *O*-acyl isoureas **B** in Figure 6.15 must be prevented when a poor nucleophile is to be acylated. In such a case, **B** is treated with a mixture of the poor nucleophile and an auxiliary nucleophile that must be a good nucleophile a good leaving group. The latter undergoes a substitution at the carboxyl carbon of the carboxylic acid moiety of **B.** However, in contrast to the acyl urea **E**, which is inert, the substitution product now obtained is still an acylating agent. Gratifyingly, it is a long-lived derivative of the originally used carboxylic acid, yet sufficiently reactive and indeed a so-called "active ester." An "active ester" is an ester that is a better acylating agent than an alkyl ester. As Table 6.1 shows, for example, phenyl esters are also "active esters." Compared with that species, Figure 6.15 shows two esters that are even more reactive, namely, the perfluorophenyl ester **C** and the hydroxybenzotriazole ester **D**. These active esters retain some of the reactivity of the acyl isourea **B**. In contrast to **B**, however, they are stable long enough even for a poor nucleophile to become acylated. The *in situ* activation of carboxylic acids to compounds of type **B, C**, or **D** is used in oligopeptide synthesis for activating *N*-protected α-amino acids (amino acid activation → **B** → **C**: Figure 6.32; amino acid activation → **B** → **D**: Figure 6.31).

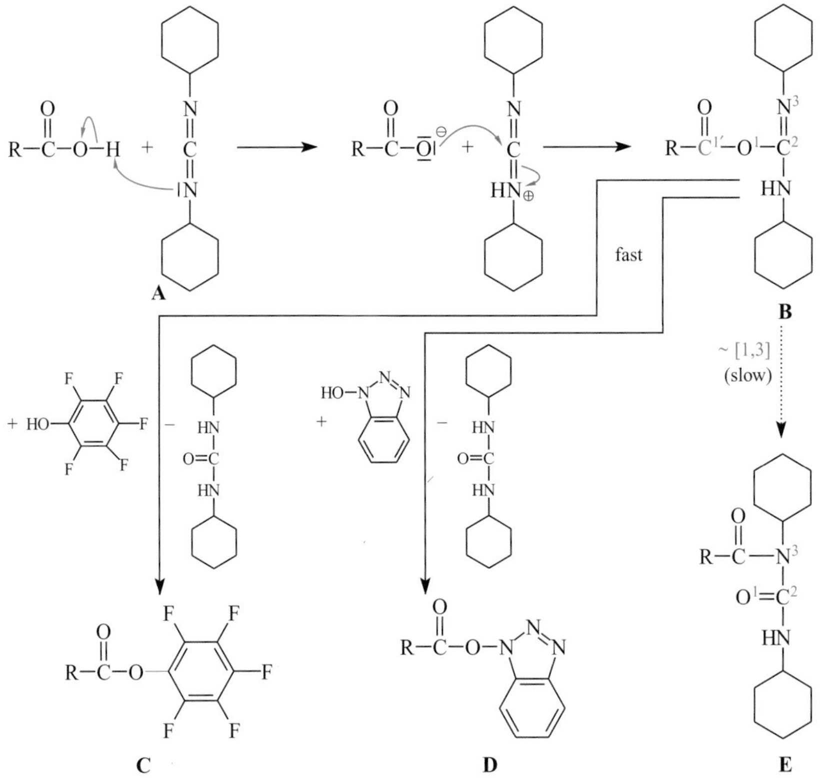

Fig. 6.15. Carboxylic acid activation with DCC. ~[1,3] means the intramolecular substitution of the oxygen atom O[1] by the N atom "3" via a cyclic four-membered tetrahedral intermediate. From the point of view of the heteroatoms, this S_N reaction corresponds to a migration of the acyl group R–C=O from the oxygen to the nitrogen. (Examples for amino acid activations in the form of the pentafluorophenyl ester **C** or the benzotriazolyl ester **D** are given in Figure 6.32 (oligopeptide synthesis) and Figure 6.31 (dipeptide synthesis), respectively.

The last but one *in situ* procedure for activating carboxylic acids is shown in Figure 6.16. There, the α-chlorinated *N*-methylpyridinium iodide **A** reacts with the carboxylic acid by an S_N reaction at a pyridine carbon. This leads to the pyridinium salt **C**, presumably via the Meisenheimer complex **B** and its deprotonation product **D** as intermediates. The activated carboxylic acid **C** is not only an aryl ester but one in which the aryl group is positively charged. This charge keeps the single-bonded O atom of this species *completely* from providing any resonance stabilization by its +M effect to the C=O double bond (cf. discussion of Table 6.1).

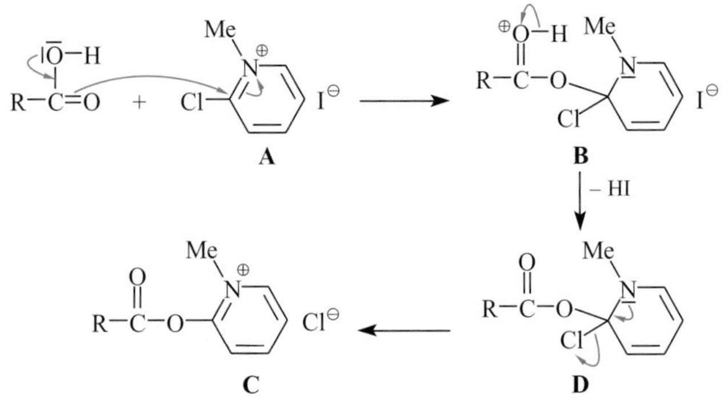

Fig. 6.16. *In situ* activation of carboxylic acids according to the procedure of Mukaiyama.

Side Note 6.1.
***In situ* Activation of a Carboxylic Acid through *O*-Silylation**

The problem with any attempt to acylate amines with carboxylic acids to furnish amides is that an acid/base reaction takes place. The resulting ammonium carboxylate is—as far as the formation of the targeted amide is concerned—dead! At best, heating up to several hundred degrees Celsius might just coerce the system into the desired direction. Although this is in line with the procedure applied in the industrial-scale synthesis of nylon-6,6 from hexamethylene diammoniumadipate, less robust amides than nylon cannot withstand these harsh reaction conditions. It can be reliably predicted that accelerated heating of, for example, the ammoniumcarboxylate mixture **B** ⇌ iso'-**B** in Figure 6.17—despite the kinetically advantageous intramolecularity—would not generate caprolactam in a reasonable yield.

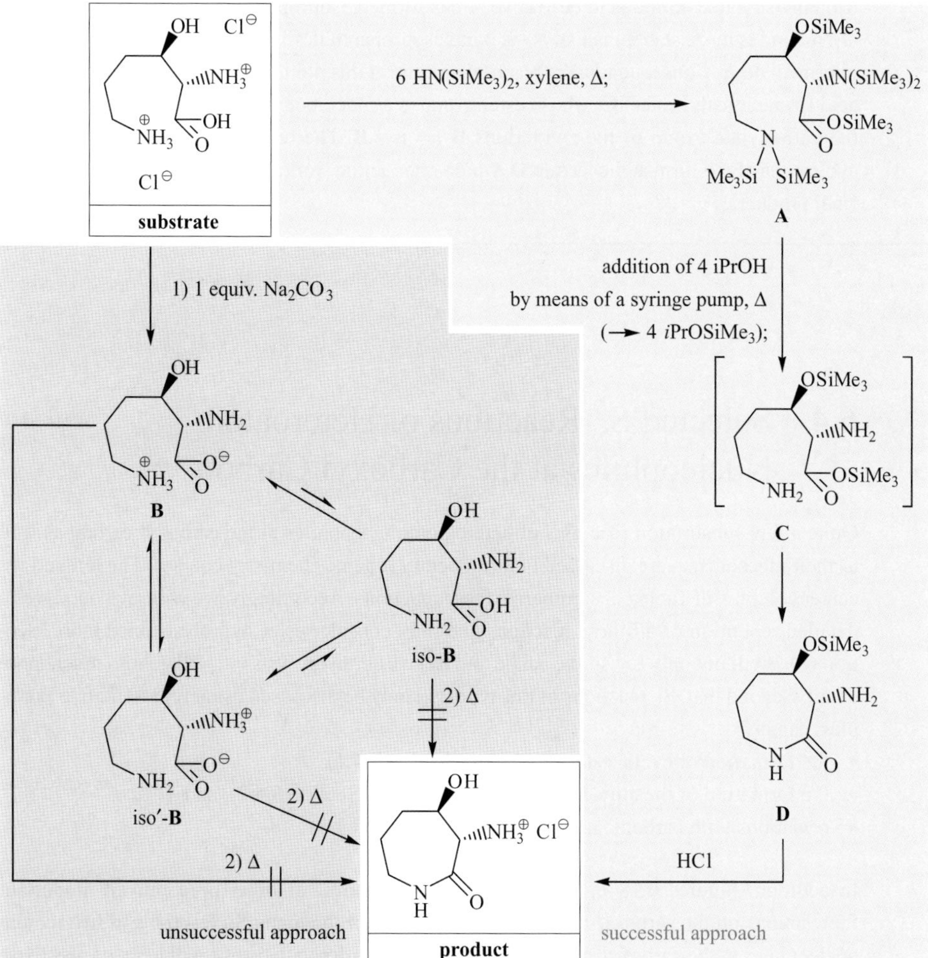

Figure 6.17. An ε-aminocaproic acid can*not* be directly transformed (area shaded grey) into the corresponding ε-aminocaprolactam; nonetheless, the transformation can be performed indirectly (area shaded red): the intermediate formation of the *O*-silylated amino acid ester **C** makes the crucial difference because it precludes the deactivation of the substrate as an ammonium carboxylate.

The formation of the desired lactam from the ammoniumcarboxylate mixture in Figure 6.17 was managed by using the following strategy. First, the dihydrochloride of the corresponding diaminohydroxy carboxylic acid was heated with $H–N(SiMe_3)_2$. This reagent (**hex**a**m**ethyl**di**silazane, HMDS) transfers trimethylsilyl groups onto *O*- and *N* nucleophiles. Such a transfer of the trimethylsilyl group onto an *O* nucleophile has a driving force because the resulting O–Si bond is more stable than the N–Si bond in HMDS. In contrast, the transfer of a trimethylsilyl group from HMDS onto an *N* nucleophile is thermodynamically neutral. Nevertheless, this transfer is fully realized in the reaction given in Figure 6.17. The reason is that $HN(SiMe_3)_2$ is transformed into NH_3 via $H_2N(SiMe_3)$ and that these two products are volatile enough to escape from the reaction mixture. Le Chatelier's principle works again! When all *O*- and *N* nucleophiles have been trimethylsilylated, i.e., when the six-fold trimethylsilylated amino acid derivative **A** has formed, isopropanol is added slowly. Since an O–Si bond is more stable than an N–Si bond, isopropanol desilylates all nitrogen atoms of the intermediate **A**. Consequently, the reaction mixture at this point contains the amino carboxylic acid (trimethylsilyl ester) **C**, whose ester group is a much better acylating agent than the initial carboxylate group of the zwitterions **B** $\rightleftharpoons$ iso'-**B**. The result is that the amino ester **C** cyclizes easily to furnish the lactam **D**, which after acidic workup gives the desired desilylated final product.

6.4 Selected S_N Reactions of Heteroatom Nucleophiles at the Carboxyl Carbon

Quite a few substitution reactions of heteroatom nucleophiles at the carboxyl carbon as well as their mechanisms are discussed in introductory organic chemistry courses. The left and the center columns of Table 6.3 summarize these reactions. Accordingly, we will save ourselves a detailed repetition of all these reactions and only consider ester hydrolysis once more. Section 6.4.1 will not only revisit the acidic and basic ester hydrolysis but will go into much more detail. Beyond that, S_N reactions of this type will only be discussed using representative examples, namely:

- the formation of cyclic esters (lactones; Section 6.4.1),
- the formation of the amide bond of oligopeptides (Section 6.4.2), and
- acylations with carbonic acid derivatives (Section 6.4.3).

In addition, Figures 6.18–6.21 briefly present a handful of other preparatively important S_N reactions on the carboxyl carbon. In other words: of the many S_N reactions of heteroatom nucleophiles at the carboxyl carbon of carboxylic acid (derivative)s we will here discuss those listed in the right column of Table 6.3.

The Figures 6.18 and 6.19 show S_N reactions with H_2O_2 at the carboxyl carbon of carboxylic acid chlorides and one carboxylic acid anhydride. They are carried out in basic solution in order to utilize the higher reactivity of the $HOO^{\ominus}$ ion. All these reactions take place

Table 6.3. Preparatively Important S_N Reactions of Heteroatom Nucleophiles at the Carboxyl Carbon of Carboxylic Acids and Their Derivatives

Standard $Nu^{\ominus}$	**Acylations of this nucleophile you should already be familiar with**	**Acylations of this nucleophile part of which are discussed *here* as prototypical examples**
H_2O or $OH^{\ominus}$	Hydrolysis of esters and amides	Ester hydrolysis (Section 6.4.1)
ROH or $RO^{\ominus}$	Esterification of carboxylic acids; transesterification giving polyethyleneterephthalate (Dacron®)	Lactonization of hydroxy acids (Section 6.4.2)
RCO_2H or $RCO_2^{\ominus}$	Formation of anhydrides; carboxylic acid activation (Section 6.3.3)	–
NH_3 or RNH_2 or R_2NH	Amide formation from carboxylic acid derivatives (mild) or from carboxylic acids (Δ; technical synthesis of nylon-6,6); transamidation [caprolactame → nylon-6 (perlon)]	Peptide synthesis (Section 6.4.3)
Special $Nu^{\ominus}$	**Acylations of this nucleophile you might have heard about in your introductory lecture to organic chemistry**	
$HOO^{\ominus}$	–	Preparation of MCPBA and MMPP (Figure 6.18)
H_2O_2	Dioxetane formation with oxalic acid diphenyl ester (→ CO_2 + chemoluminiscence)	–
$RC(=O)OO^{\ominus}$	–	Preparation of dibenzoyl peroxide (Figure 6.19)
Alkoholic OH group of α-hydroxycarboxylic acid	Glycolic acid → lactide	–
NH_2 group of α-amino acids	Glycine → diketopiperazine	–
$H_2N—NH_2$	–	2nd step of the Gabriel synthesis (Figure 6.20)
NHMe(OMe)	–	Preparation of a Weinreb amide (Figure 6.21)

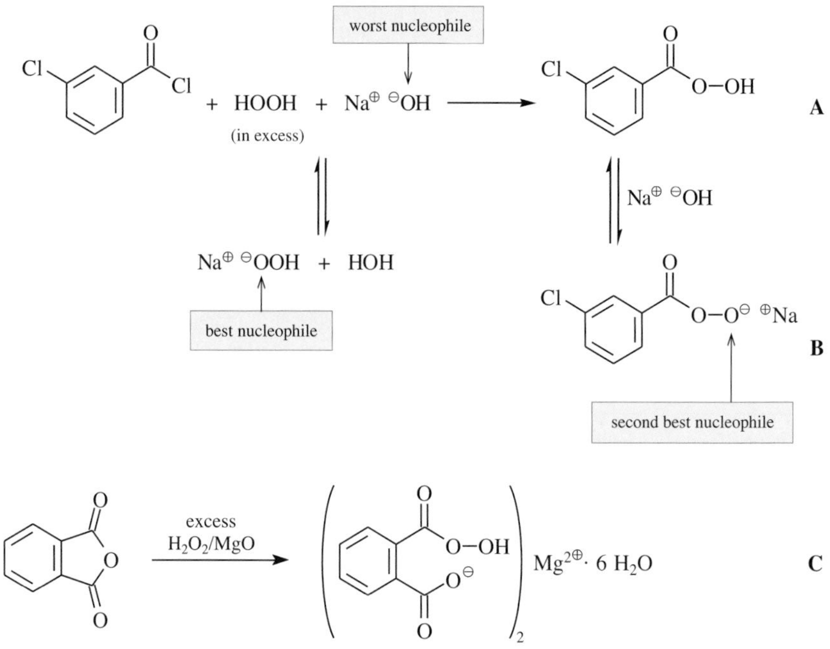

Fig. 6.18. S_N reactions with H_2O_2 at the carboxyl carbon. Syntheses of *meta*-chloroperbenzoic acid **(A)**, magnesium monoperoxophthalate hexahydrate **(B)**, and dibenzoyl peroxide **(C)** by acylation of one of the two O atoms of the nucleophile.

via the mechanism of Figure 6.2. By using these reactions it is possible—as a function of the structure of the acylating agent and the ratio of the reaction partners—to obtain the reagents *meta*-chloroperbenzoic acid (**A** in Figure 6.18;, magnesium monoperoxophthalate hexahydrate (**C** in Figure 6.18; and dibenzoyl peroxide. Dibenzoyl peroxide is produced through two such substitution reactions: The $HOO^{\ominus}$ ion is the nucleophile in the first reaction, and the Ph—C(=O)—O—$O^{\ominus}$ ion (**B**) is the nucleophile in the second reaction.

Two successive S_N reactions at two carboxyl carbons occur in the second step of the Gabriel synthesis of primary alkyl amines (Figure 6.20; first step: Figure 2.32). Hydrazine breaks both C(=O)—N bonds of the *N*-alkylphthalimide precursor **A.** The first bond cleavage is faster than the second because the first acylating agent (**A**) is an imide and the second acylating agent **(C)** is an amide, amides are comparatively inert toward nucleophiles. Still, under the conditions of Figure 6.20 even the amide **C** behaves as an acylating agent (giving the diacylhydrazide **B**). The reason for this relatively fast reaction is that it is *intramolecular*. Intramolecular reactions via a five- or a six-membered transition state are always much faster than analogous intermolecular reactions. Therefore, the *N*-alkylated phthalimide intermediates of the Gabriel synthesis are cleaved with hydrazine because the second acylation is intramolecular and it can thus take place rapidly. If one were to take NH_3 instead of hydrazine, this would have to cleave the second C(=O)—N bond in an *intermolecular* S_N reaction, which would be impossible under the same conditions.

It remains to be clarified why the 1,2-diacylated hydrazine **B** and not the 1,1-diacylated hydrazine **D** is formed in the hydrazinolysis of Figure 6.20. In the intermediate **C** the only nucleophilic electron pair resides in the NH_2 group and not in the NH group because the electron pair of the NH group is involved in the hydrazide resonance **C** → **C'** and is therefore not

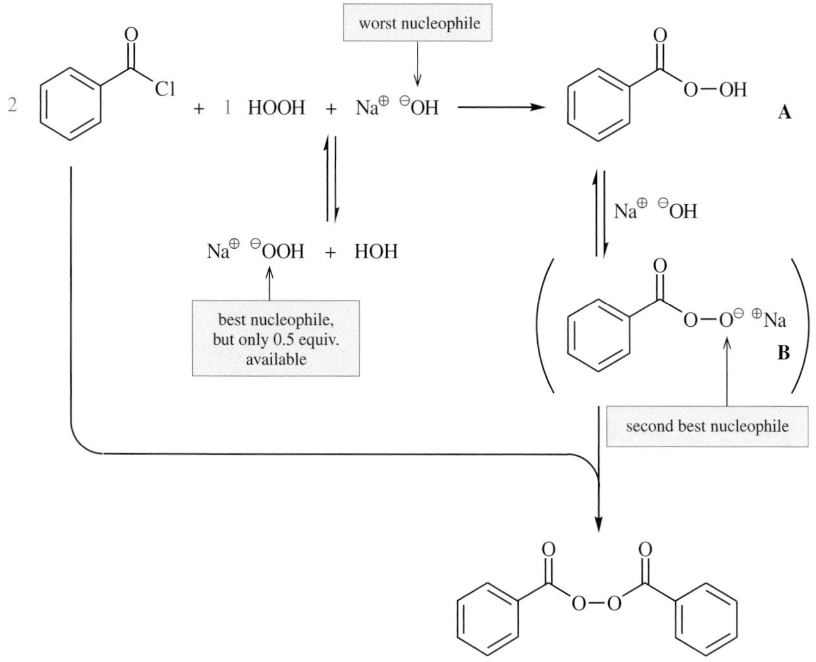

Fig. 6.19. S_N reactions with H_2O_2 at the carboxyl carbon, part II. Synthesis of dibenzoyl peroxide through acylation of both O atoms of the nucleophile.

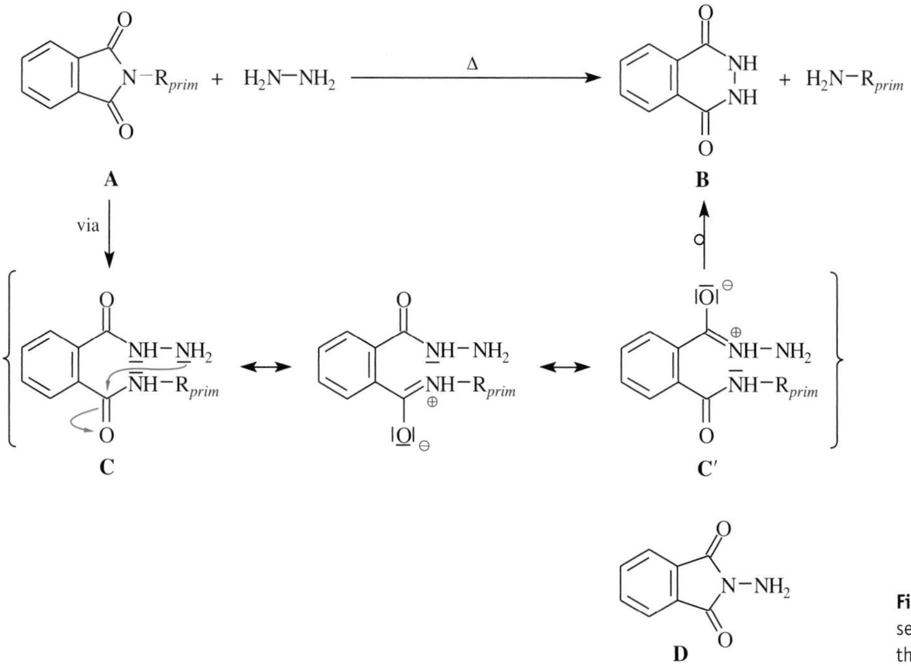

Fig. 6.20. Mechanism of the second step of the Gabriel synthesis of primary alkyl amines.

Table 6.4. Preparatively Important S_N Reactions of Heteroatom Nucleophiles at the Carboxyl Carbon of Carbonic Acid Derivatives

$Nu^{\ominus}$	**Acylations of this nucleophile you might have heard about in your introductory lecture to organic chemistry**	**Acylations of this nucleophile discussed *here* as prototypical examples**
ArOH	Polycarbonate synthesis from bisphenol **A** and phosgene	–
ROH	–	Preparation of Z-chloride (Figure 6.34) and of Boc_2O (Figure 6.35)
RCO_2H	Carboxylic acid activation with $(imidazole)_2C{=}O$ (Figure 6.13) or *i*BuO—C(=O)–Cl (Figure 6.14)	–
NH_3	Industrial-scale transformation of ammonium carbamate (generated *in situ*) into urea (also cf. Figure 8.15)	–
HO_2C—CHR—NH_2	–	Protection of amino acids as benzyl carbamates (Figure 6.36) or *tert*-butyl carbamate (Figure 6.37)
RNH_2	Conversion of aromatic diamines with phosgene into diisocyanates (for the preparation of polyurethanes)	Chlorine-free formation of diisocyanates (Figure 6.38)
RNH_2 or R_2NH	–	Formation of substituted urea from carbonic acid esters (Figure 6.39)
H_2N—C(=O)—NH_2	Synthesis of barbituric acid from urea and malonic ester; preparation of uracil (or thymine) from urea and formyl acetic acid ester (or from α-formyl propionic acid ester)	–

really available. The hydrazide resonance is about as important as the amide resonance (ca. 22 kcal/mol according to Table 6.2).

The last of these special examples of S_N reactions of heteroatom nucleophiles at the carboxyl carbon of a carboxylic acid derivative is given in Figure 6.21. There, the free carboxyl group of the aspartic acid derivative **A** is activated according to the *in situ* procedure of Figure 6.14 as a mixed carbonic acid/carboxylic acid anhydride **B** that is then treated with *N,O*-dimethylhydroxyl amine. This reagent is an N nucleophile, which is thus acylated to give the

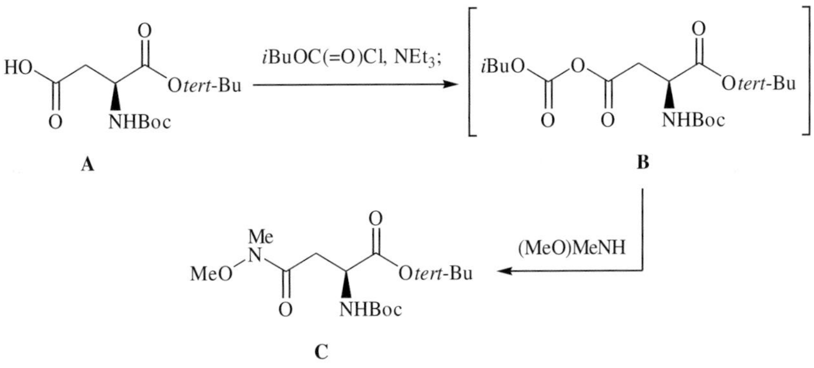

Fig. 6.21. *In situ* activation of a carboxylic acid—i.e., the side chain carboxyl group of protected L-aspartic acid—as the mixed anhydride (**B**) and its aminolysis to a Weinreb amide. How this Weinreb amide acylates an organolithium compound is shown in Figure 6.44. The acylation of an *H* nucleophile by a second Weinreb amide is presented in Figure 6.42 and the acylation of a di(ketone enolate) by a third Weinreb amide in Figure 13.64. Figure 6.50 also shows how Weinreb amides of carboxylic acids can be obtained by C,C bond formation.

amide **C**. Carboxylic acid amides derived from *N,O*-dimethylhydroxyl amine are known as **Weinreb amides**. They play an important role in synthesis as selective acylating agents for H- and C nucleophiles as can be seen from the corresponding examples (Figure 6.42 and Figure 6.44, respectively).

Many heteroatom nucleophiles also undergo S_N reactions at the carboxyl carbon of carbonic acid derivatives. In Table 6.4 these nucleophiles are listed in the left column. The middle column summarizes their S_N reactions with carbonic acid derivatives that you may have come across in your introductory organic chemistry lectures. The right column of the table refers to S_N reactions that will be taken as examples and discussed in Section 6.4.4.

6.4.1 Hydrolysis and Alcoholysis of Esters

The hydrolysis of carboxylic esters can in principle take place either as **carboxyl-O cleavage**, i.e., as an S_N reaction at the carboxyl carbon—

$$R^1{-}C({=}O){-}O{-}R^2 \xrightarrow{+\,H_2O} R^1{-}COOH + HO{-}R^2$$

or as **alkyl-O cleavage** (this variant *does not* represent an S_N reaction at the carboxyl carbon):

$$R{-}C({=}O){-}O{-}Alk \xrightarrow{+\,H_2O} R{-}COOH + HO{-}Alk$$

The hydrolysis is generally not carried out at pH 7 but is either **acid-catalyzed or base-mediated** (i.e., as a so-called saponification). Base-*catalyzed* ester hydrolyses do not exist! The carboxylic acid produced protonates a full equivalent of base and thus consumes it.

Driving Force of Base-Mediated and Acid-Catalyzed Ester Hydrolyses

Base-mediated ester hydrolyses have a high driving force. This is because of the acid/base reaction between the carboxylic acid formed in the reaction, and the base used as the reagent. The resonance stabilization of the carboxylate is approximately 30 kcal/mol, which means a gain of about 16 kcal/mol compared to the starting material, the carboxylic ester (resonance stabilization 14 kcal/mol according to Table 6.1). Accordingly, the hydrolysis "equilibrium" lies completely on the side of the carboxylate.

Acid-catalyzed ester hydrolyses lack a comparable contribution to the driving force: The starting material and the product, the ester and the carboxylic acid, possess resonance stabilizations of the same magnitude of 14 kcal/mol each (Table 6.1). For this reason, acid-catalyzed ester hydrolyses can go to completion only when one starting material (H_2O) is used in great excess and the hydrolysis equilibrium is thereby shifted unilaterally to the product side. Le Chatelier's principle again. Five- and six-membered *cyclic* esters can be saponified only in basic media. In acidic solutions they are often spontaneously formed again (see Figure 6.27 and the explanation given).

The mechanisms of ester hydrolysis are distinguished with abbreviations of the type "medium$_{\text{designation as carboxyl-O or as alkyl-O cleavage}}$ reaction order." The medium of **a**cid-catalyzed hydrolyses is labeled **"A,"** and the medium of **b**ase-mediated hydrolyses is labeled **"B."** A carboxyl-O cleavage is labeled with **"AC,"** (for **ac**yl-O cleavage), and an **al**kyl-O cleavage is labeled with **"AL."** The possible reaction orders of ester hydrolyses are 1 and 2. If all permutations of the cited characteristics were to occur, there would be eight hydrolysis mechanisms: The $A_{AC}1$, $A_{AC}2$, $B_{AL}1$, and $A_{AL}2$ mechanisms in acidic solutions and the $B_{AC}1$, $B_{AC}2$, $B_{AL}1$, and $B_{AL}2$ mechanisms in basic solutions. However, only three of these mechanisms are of importance: the $A_{AC}2$ mechanism (Figure 6.22), the $A_{AL}1$ mechanism (Figure 6.23), and the $B_{AC}2$ mechanism (Figure 6.24).

The $A_{AC}2$ mechanism (Figure 6.22) of ester hydrolysis represents an S_N reaction at the carboxyl carbon, which follows the general mechanism of Figure 6.5. Acid-catalyzed hydrolyses of carboxylic esters that are derived from primary or from secondary alcohols take place according to the $A_{AC}2$ mechanism. The reverse reactions of these hydrolyses follow the same mechanism, namely, the acid-catalyzed esterifications of carboxylic acids with alcohols. In the esterifications, the same intermediates are formed as during hydrolysis, but in the opposite order.

As you have already seen, in the system carboxylic ester + $H_2O \rightleftharpoons$ carboxylic acid + alcohol, the equilibrium constant is in general only slightly different from 1. *Complete* reactions

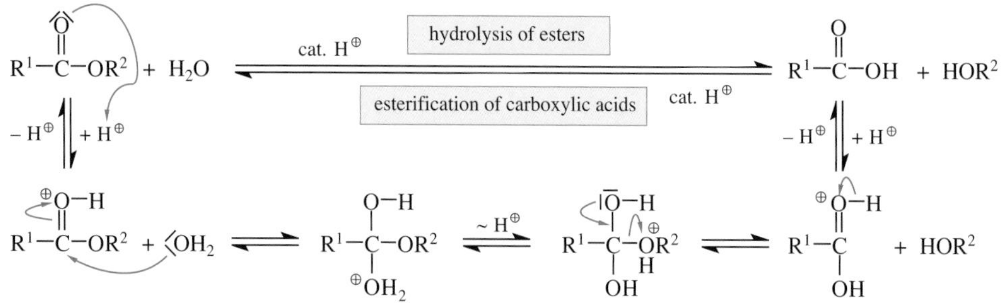

Fig. 6.22. $A_{AC}2$ mechanism of the acid-catalyzed hydrolysis of carboxylic esters (read from left to right); $A_{AC}2$ mechanism of the Fischer esterification of carboxylic acids (read from right to left). ~H⊕ means migration of a proton.

hydrolysis of esters

cat. H⊕

esterification of carboxylic acids

cat. H⊕

− H⊕ + H⊕

~ H⊕

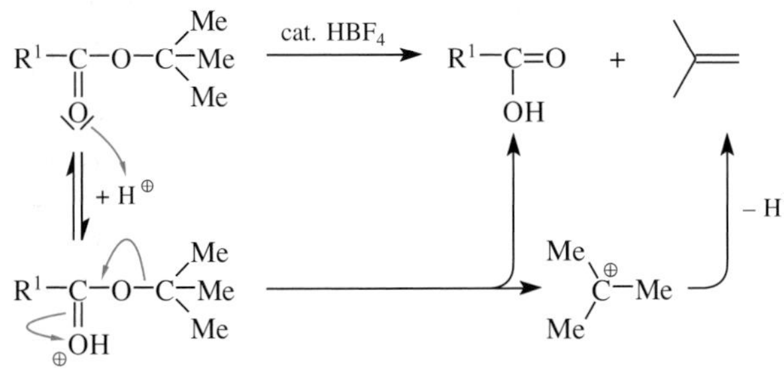

Fig. 6.23. $A_{AL}1$ mechanism of the acidic cleavage of *tert*-alkyl esters.

in both directions are therefore only possible under suitably adjusted reaction conditions that are applications of Le Chatelier's principle. Complete $A_{AC}2$ *hydrolyses* of carboxylic esters can be carried out with a large excess of water. Complete $A_{AC}2$ *esterifications* succeed when a large excess of the alcohol is used. For this purpose it is best to use the alcohol as the solvent. However, when the alcohol involved is difficult to obtain or expensive, this procedure cannot be used because the alcohol is affordable only in a stoichiometric amount. Its complete esterification by a carboxylic acid is then still possible, provided that the released water is removed. That can be done by continuously distilling it away azeotropically with a solvent such as cyclohexane. By removing one of the reaction products, the equilibrium is shifted toward this side, which is also the side of the desired ester.

In acidic media, carboxylic esters of tertiary alcohols are not cleaved via the $A_{AC}2$ mechanism (Figure 6.22) but via the $A_{AL}1$ mechanism (Figure 6.23). However, this cleavage would probably not be a "hydrolysis" even if the reaction mixture contained water. *This* mechanism for ester cleavage does not belong in Chapter 6 at all! It was already discussed in Section 4.5.3 (Figure 4.37) as the E1 elimination of carboxylic acids from *tert*-alkyl carboxylates.

However, you should know that esters of certain other alcohols can undergo acidic hydrolysis according to the $A_{AL}1$ mechanism. It only requires that a sufficiently stabilized *carbenium ion* is left after carboxylic acid removal. For example, the trifluoroacetolysis **E** → **F** in the last step of the solid phase-peptide synthesis in Figure 6.32 follows an $A_{AL}1$ mechanism. The resulting *carbenium ion*, a *para*-alkoxybenzylic cation, benefits from a considerable resonance stabilization. *This* reaction also does not belong in Chapter 6. Rather, it is an S_N1 reaction that is similar to the detritylations in Figure 2.22.

Carboxylic esters of any alcohol are saponified quantitatively (see above) in basic solution according to the $B_{AC}2$ mechanism (Figure 6.24). The $BA_{AC}2$ mechanism is an S_N reaction at

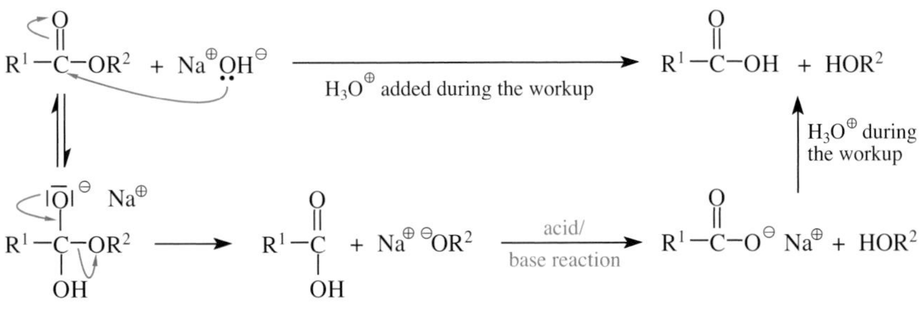

Fig. 6.24. $B_{AC}2$ mechanism of the basic hydrolysis of carboxylic esters.

the carboxyl carbon that also proceeds according to the general mechanism of Figure 6.2. The reversibility of the formation of the tetrahedral intermediate in such hydrolyses was proven with the isotope labeling experiment of Figure 6.3.

In a $B_{AC}2$ saponification, the C—O bond of the released alcohol is not formed, but is already contained in the ester. Therefore, if the C atom of this C—O bond is a stereocenter, its configuration is completely retained. This is used in the $B_{AC}2$ hydrolysis of esters with the substructure —C(=O)—O—$CR^1R^2R^3$ to stereoselectively obtain the corresponding alcohols. An application thereof is the hydrolysis of the following lactone, whose preparation as a pure enantiomer can be found in Figure 14.34:

NaOH, H_2O → $O^{\ominus}$ $Na^{\oplus}$ OH → acidify → OH OH

Transesterifications in basic solutions can also follow the $B_{AC}2$ mechanism. The reactions also can release the corresponding alcohols with retention of configuration from sterically uniform esters with the substructure —C(=O)—O—$CR^1R^2R^3$. This kind of reaction is used, for example, in the second step of a Mitsunobu inversion, such as the following, which you have already seen in Figure 2.34:

OMe, OH → $EtO_2C-N=N-CO_2Et$, PPh_3, $Ph-CO_2H$ → OMe, O, O, Ph → K_2CO_3 in MeOH (→ $K^{\oplus}$ $^{\ominus}OMe$) → OMe, O, H + MeO_2CPh

Side Note 6.2. Formation of β-Ketoesters: Transesterifications (or not)

"Diketenes" (Formula **A** in Figure 6.25) are esters in which the strain in the four-membered ring prevents the ring oxygen from stabilizing the C=O double bond through a +M effect that is as large as in open-chain esters. This effect alone would suffice to turn diketenes into reactive acylating agents. But there is another supporting strain effect: when the nucleophilic reaction (here) of an alcohol leads to the formation of the tetrahedral intermediate **B**, the carbon atom undergoing reaction is rehybridized from sp^2 to sp^3. This reduces the valence angle it desires from 120° to 109°28', relieving the Baeyer strain in the four-membered ring. This, in turn, leads to an increase in the tetrahedron formation rate according to the Hammond postulate. The bottom line is that the diketene can acylate alcohols even without the *commonly required bases or acids*. When the tetrahedral intermediate **B** decomposes, the leaving group is an enolate. Irrespective of whether its structure is **D** or **E,** the β-ketoester **F** is formed upon a proton shift to the enolate carbon. If the diketene is regarded as an enolic ester, i.e., as a compound with the substructure C=C–C(=O)–O–C, its reaction to the ketoester **F** may be conceived as a transesterification which is why it is discussed *here*.

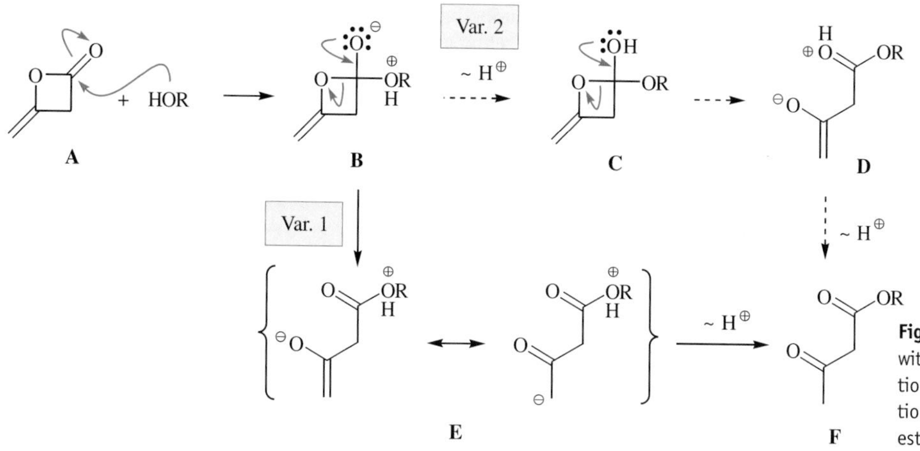

Fig. 6.25. Acylation of alcohols with a diketene (**A**; preparation: Section 15.4). The reaction product is the acetoacetic ester **F**.

Meldrum's acid is destabilized by dipole-dipole repulsions as explained in the discussion of Figure 13.7. The same applies to the (hydroxyalkylidene) Meldrum's acid **A** (Figure 6.26). When its carboxyl carbon reacts with an alcohol to form tetrahedral intermediate **B**, the dipole-dipole repulsion mentioned above is offset. In addition, the oxyanion entity of this tetrahedral intermediate could be stabilized by the neighboring enol substructure via hydrogen bond formation. Regardless of how these factors are weighted, it is clear that (hydroxyalkylidene) Meldrum's acids **A** react with alcohols in the absence of *acids and bases* and via the tetrahedral intermediate **B**. A proton shift then occurs to furnish the neutral tetrahedral intermediate **C**. Upon transformation of this tetrahedral intermediate into the ester group, the leaving group decomposes into acetone and the carboxylate **F** (mechanism: analogous to Figure 9.2). The latter equilibrates with the β-keto carboxylic acid **E**. Under the reaction condi-

Fig. 6.26. Acylation of alcohols with an acyl Meldrum's acid (**A**). The reaction product obtained after *in situ* decarboxylation is a β-keto carboxylic acid ester **D**.

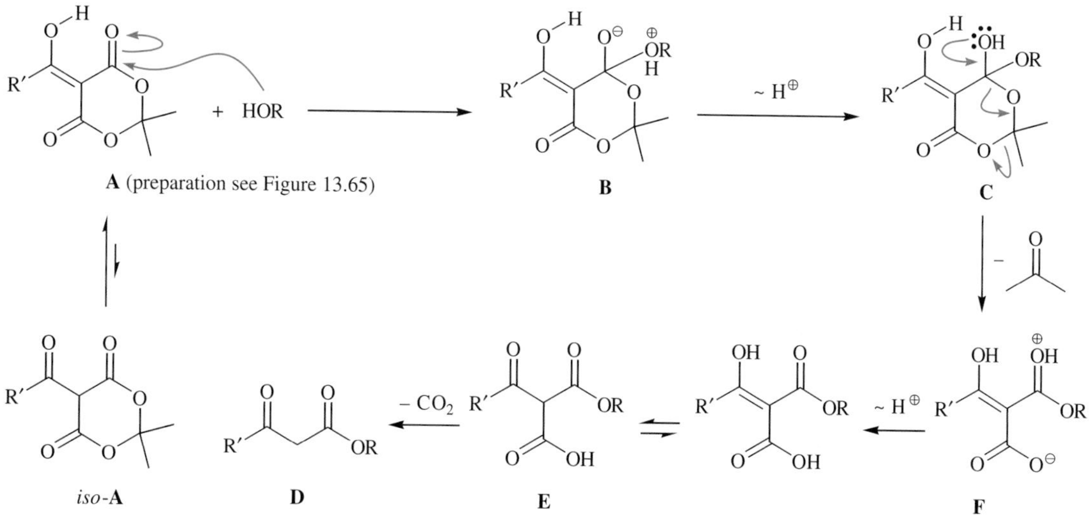

tions, the functional groups of this compound participate in a decarboxylation (mechanism: analogous to Figure 13.27 or 13.37). This decarboxylation leads to the formation of the β-ketoester **D** of the alcohol employed. The procedure described provides a very efficient method for the preparation of β-ketoesters. The initial reaction **A** + ROH $\rightarrow$ **E** + acetone identifies this procedure as a transesterification.

The following third method for the preparation of β-ketoesters also pretends to be a transesterification:

Δ in the presence of HOR

(Prep. see Fig. 9.21)

However, this is not true. Actually, it is a two-step process starting with a reversal of a Diels–Alder reaction during which a ketene is formed. In a second step the added alcohol reacts with the ketene (mechanism: Section 8.2).

According to what was generally discussed at the beginning of Section 6.2.1, the tetrahedral intermediate is also the best transition state model of the rate-determining step of the saponification of esters according to the $B_{AC}2$ mechanism. Knowing that, the substrate dependence of the saponification rate of esters is easily understood. As can be seen from Table 6.5, the saponification rate decreases sharply with increasing size of the acyl substituent because a bulky acyl substituent experiences more steric hindrance in the tetrahedral intermediate than in the starting material: in the tetrahedral intermediate, it has three vicinal O atoms compared with two in the starting material, and the three O atoms are no longer so far removed because the C—C—O bond angle has decreased from ~120° to ~109°.

Rate effects of the type listed in Table 6.5 make it possible to carry out chemoselective monohydrolyses of sterically differentiated diesters, for example:

$k_{rel} = 0.011$ *tert*-Bu Me $k_{rel} = 1$ $OH^{\ominus}$ *tert*-Bu HO

Table 6.6 shows that saponifications according to the $B_{AC}2$ mechanism are also slowed down when the esters are derived from sterically demanding alcohols. However, this structural variation takes place at a greater distance from the reaction center than the structural variation of Table 6.5. The substituent effects in Table 6.6 are therefore smaller.

Tab. 6.5 Substituent Effects on the Rate of the $B_{AC}2$ Saponification of Different Ethyl Esters

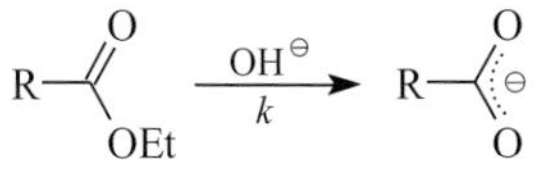

R	k_{rel}
Me	≡ 1.0
Et	0.47
*i*Pr	0.10
tert-Bu	0.011

Increase in the steric interactions in the rate-determining step:

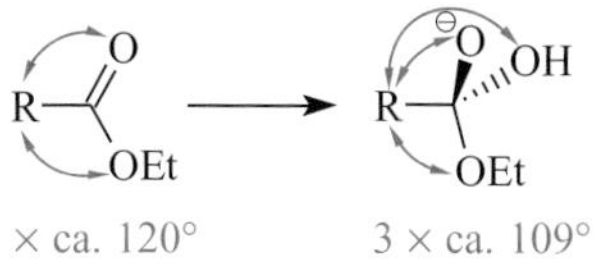

Tab. 6.6 Substituent Effects on the Rate of the $B_{AC}2$ Saponification of Different Acetic Esters

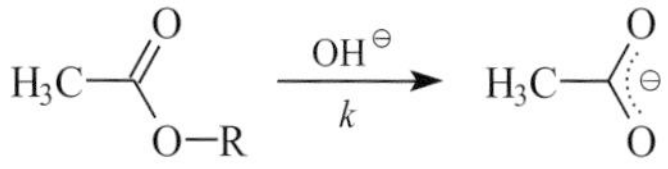

R	k_{rel}
Et	≡ 1.0
*i*Pr	0.70
tert-Bu	0.18
CEt_3	0.031

Occurrence of *syn*-pentane interactions in the rate-determining step:

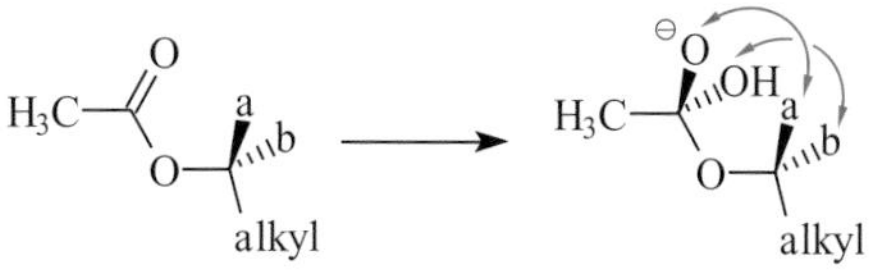

6.4.2 Lactone Formation from Hydroxycarboxylic Acids

γ- and δ-Hydroxycarboxylic acids esterify very easily intramolecularly in the presence of catalytic amounts of acid. They are thereby converted to five-membered γ-lactones or six-membered δ-lactones (Figure 6.27). These lactonizations often take place so easily that they can't be avoided. In these cases it seems that the carboxylic acid moiety of the substrate effects acid catalysis by its own acidity.

The high *rate* of the lactonizations in Figure 6.27 is a consequence of the less negative than usual activation entropies, from which intramolecular reactions that proceed via three, five-,

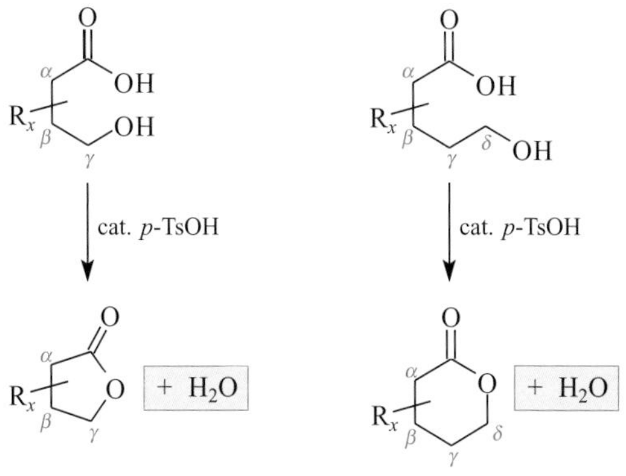

Fig. 6.27. Spontaneous lactonizations according to the $A_{AC}2$ mechanism of Figure 6.22.

or six-membered transition states always profit. The high lactonization *tendency* stems from an increase in entropy, from which intermolecular esterifications do not profit: only during lactonization does the number of molecules double (two molecules are produced from one). This entropy contribution to the driving force requires that the equilibrium constant of the reaction γ- or δ-hydroxycarboxylic acid $\rightleftharpoons$ γ- or δ-lactone + H_2O be *greater than* 1 rather than *approximately* 1, as for intermolecular esterifications (Section 6.4.1).

Lactonizations of hydroxycarboxylic acids in which the OH and the CO_2H groups are separated from each other by six to ten C atoms have a reduced driving force or are even endergonic. These lactonizations lead to **"medium-sized ring lactones."** They are—similar to medium-sized ring hydrocarbons—destabilized by eclipsing and transannular interactions. The formation of **"large-ring lactones"** with 14 or more ring members is free of such disadvantages. Still, these lactones are not obtained from the corresponding hydroxycarboxylic acids by simple acidification for kinetic reasons: the two parts of the molecule that must react in these lactonizations are so far away from each other that they encounter each other less frequently than the OH and the CO_2H group of γ- or δ-hydroxycarboxylic acids do (Figure 6.27). Stated more accurately, the activation entropy of the formation of large-ring lactones from hydroxycarboxylic acids is quite negative. This is because only a few of the many conformers of the starting material are capable of forming the tetrahedral intermediate with nothing more than a reaction of the OH group on the carboxyl group. All other conformers must first be converted into a cyclizable conformer through numerous rotations about the various C—C single bonds. Because of this difficulty, a reaction of the OH group on the carboxyl group of a nearby but different molecule is more probable: almost every conformer of the hydroxycarboxylic acid is able to do that. Consequently, when long-chain ω-hydroxycarboxylic acids are heated up in the presence of an acid intermolecular esterifications occur instead of a lactonization. The former continue in an uncontrolled fashion beyond the stage of a monoesterification and give an ester/oligoester/polyester mixture.

Large-ring lactones are available in good yields from ω-hydroxycarboxylic acids only through a combination of two measures. First of all, the carboxylic acid moiety must be activated. This ensures that the highest possible percentage of the (still improbable) encounters between the alcoholic OH group and the carboxyl carbon of the same molecule lead to a successful reaction. In addition, one must make sure that the OH group is not acylated intermolecularly, that is, by an activated neighbor molecule. To this end, the hydroxycarboxylic acid is activated in a very dilute solution. This is based on the following consideration: the rate of formation of the tetrahedral intermediate that will deliver the lactone is as follows:

$$\frac{d\,[\text{tetrahedral precursor of lactone}]}{dt} = k_{\text{lactonization}}[\text{activated } \omega\text{-hydroxycarboxylic acid}] \tag{6.17}$$

On the other hand, the rate of formation of the undesired acyclic (mono/oligo/poly)esters is:

$$\frac{d\,[\text{tetrahedral precursor of acyclic ester}]}{dt} = k_{\text{acyclic ester}}\,[\text{activated } \omega\text{-hydroxycarboxylic acid}]^2 \tag{6.18}$$

If one divides Equation 6.17 by Equation 6.18, the expression on the left side of the equals sign corresponds approximately to the yield ratio of the lactone and the mixture of the acyclic esters, oligo- and polyesters. Taking this into account, one obtains as a new equation the following:

$$\frac{\text{yield of lactone}}{\text{yield of acyclic ester}} = \frac{k_{\text{lactonization}}}{k_{\text{acyclic ester}}} \cdot \frac{1}{[\text{activated } \omega\text{-hydroxycarboxylic acid}]} \tag{6.19}$$

From Equation 6.19 it follows that the lower the concentration of the activated acid, the higher the selectivity with which the lactone is produced versus the acyclic ester, oligo- and polyesters. Macrolactones are therefore produced in very dilute (< 1 μmol/L) solutions.

For work on a 1-mole scale one would thus have to use a 1.000-liter flask to activate and then lactonize the entire ω-hydroxycarboxylic acid. Of course, it is much more practical to work in a smaller reaction vessel. However, one must *also* not exceed the mentioned concentration limit of < 1 μmol/L. Therefore, one can introduce only as much of the carboxylic acid in this smaller reaction vessel at a time so that its concentration does not exceed 1 μmol/L. Subsequently, one would have to activate this amount of acid and would then have to wait until it is lactonized. After that additional acid would have to be added and then activated, and so on. A more practical alternative is shown in Side Note 6.3.

Macrolactonizations are generally realized through an acyl activation (Figure 6.28; the alternative of an OH activation during macrolactonizations was shown in Figure 2.36). In the *in situ* activation process with trichlorobenzoyl chloride, the mixed anhydride **A** (Figure 6.28, X stands for O—C(=O)—$C_6H_2Cl_3$) is formed as shown in a more generalized manner in Figure 6.16. Alternatively, the *in situ* activation process from Figure 6.16 with the *N*-methylpyridinium halide gives the imminium analogue **A** (Figure 6.28, X stands for 1,2-dihydro-*N*-methylpyrid-2-yl) of a mixed anhydride.

Side Note 6.3. Continuous Process for Preparing Macrolactones

The process for preparing macrolactones described in the text is impractical. Instead of this process one uses a continuous method: with a syringe pump a solution of the hydroxycarboxylic acid is added very slowly—that is, in the course of hours or days—into a small flask, which contains ≥ 1 equivalent of the activator and, if necessary, just enough triethylamine to neutralize any released HCl. The rate at which the acid is added is regulated such that it is equal to or smaller than the lactonization rate. This is called pseudo high dilution. At the end of the reaction the lactone solution can be relatively concentrated, e. g., 10 mmol/L, at least 10,000 times more concentrated than without the use of this trick.

Another level of refinement regarding *in situ* acyl group activations is reached when the activated hydroxycarboxylic acid **A** is converted into the corresponding *N*-acylpyridinium salt **B** with *para*-(dimethylamino)pyridine, (Figure 6.9). Under these conditions macrocyclizations routinely succeed in yields well above 50%.

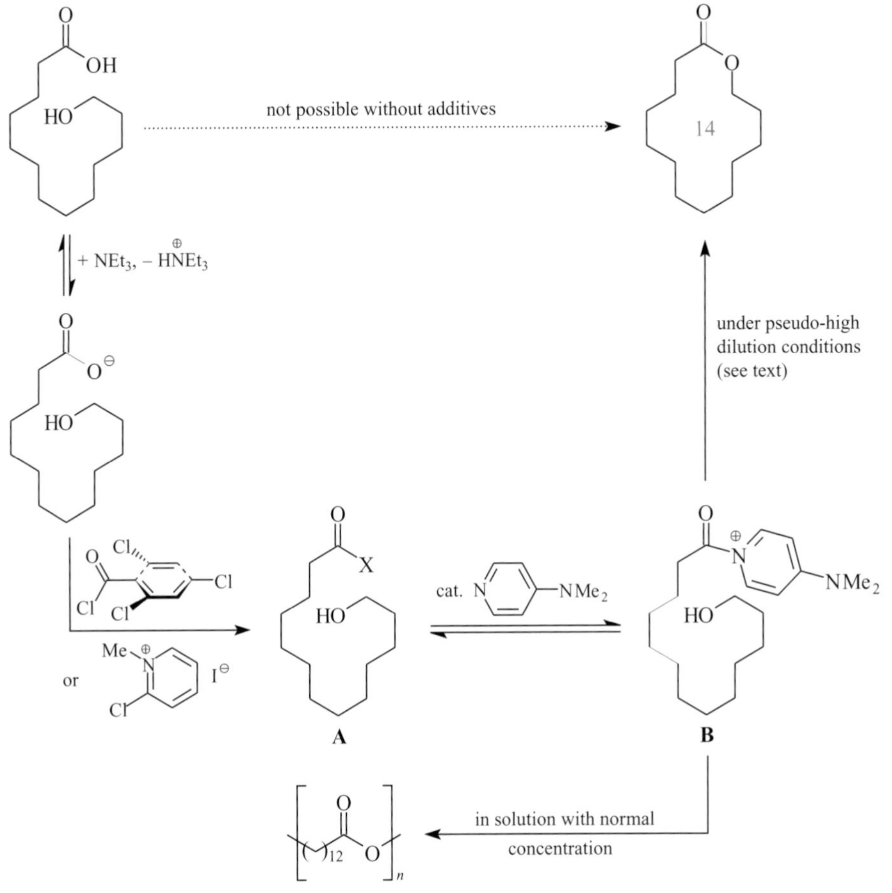

not possible without additives

$+ NEt_3, - \overset{\oplus}{H}NEt_3$

under pseudo-high dilution conditions (see text)

cat.

or

A

B

in solution with normal concentration

Fig. 6.28. Possibilities for macrolactonization.

6.4.3 Forming Peptide Bonds

An amino acid protected only at the N atom and a different amino acid in which only the CO_2H group is protected do not react with each other to form a peptide bond. On the contrary, they form an ammonium carboxylate in a fast acid/base reaction (Figure 6.29). Ammonium carboxylates can in principle be converted into amides by strong heating. Thus, for example, in the industrial synthesis of nylon-6,6, the diammonium dicarboxylate obtained from glutamic acid and hexamethylenediamine is converted to the polyamide at 300 °C. However, this method is not suitable for peptide synthesis because there would be too many undesired side reactions.

To combine the amino acid from Figure 6.29, which is only protected at the N atom, and the other amino acid from Figure 6.29, in which only the CO_2H group is protected, into a dipeptide, the CO_2H group of the N-protected amino acid must be activated.This is not easy via an acid chloride, the standard acylating agent in many organic reactions. This is because if an amino acid chloride is treated with an amino acid ester with a free NH_2 group, the desired peptide formation is only part of what happens. There is often a disastrous side reaction (Figure 6.30). The amino acid ester with the free NH_2 group can also react as a base. Hence, the

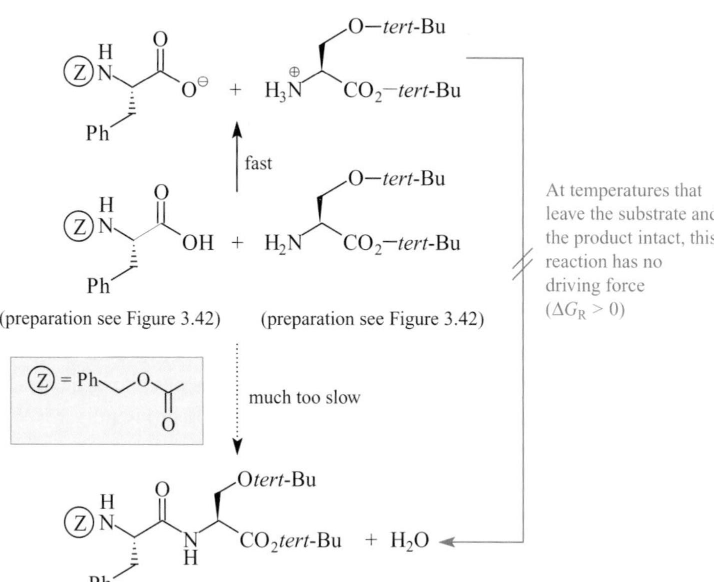

Fig. 6.29. Impossibility of preparing a dipeptide from an N-protected amino acid and an amino acid ester with a free NH_2 group.

amino acid chloride is deprotonated—reversibly—to the enolate. This is possible because this enolate is stabilized by the combined electron withdrawal by the Cl atom and by the protected amino group. The reprotonation of this enolate takes place without stereocontrol. It is not clear whether this protonation takes place at the C atom or at the O atom (whereupon the resulting enol would have to be protonated once more, now at the C atom). Be this as it may, what is important is that in this way the configurational integrity at the methine carbon of the activated amino acid is lost. However, we should note that this limitation can be overcome by judicious choice of reaction conditions.

Let us analyze once more with a different emphasis what has just been said: An N-protected amino acid chloride can be deprotonated by an amino acid ester with a free NH_2 group because the enolate produced is stabilized, among other things, by the electron-withdrawing inductive effect of the Cl atom. This immediately suggests a solution to circumvent the described dilemma: Activating the N-protected amino acid requires a derivative in which the

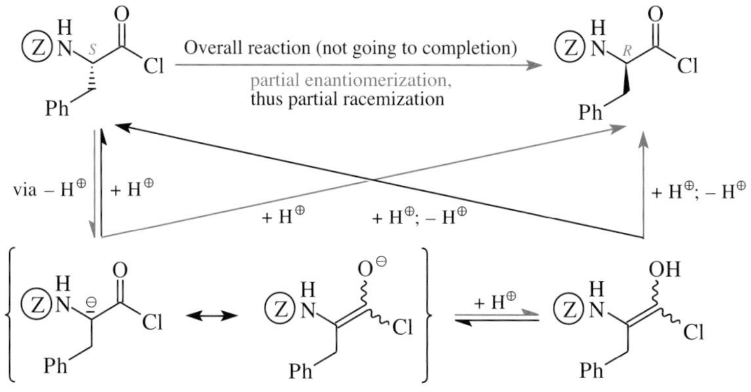

Fig. 6.30. Mechanism of the stereoisomerization of an N-protected amino acid chloride under the conditions of a peptide synthesis. The red reaction arrows stand for the steps of the racemization.

substituent at the carboxyl carbon is less electron-withdrawing than a Cl atom. In contrast to the Cl atom, a substituent of this type would not help enolate formation. Hence, one needs carboxyl substituents that activate the amino acid sufficiently but do not acidify it too much.

Acyl isoureas are extremely well suited for this purpose. Figure 6.31 shows a corresponding L-phenylalanine derivative as compound **A.** Substances such as these are prepared via the mechanism of Figure 6.15 *in situ*, by first adding dicyclohexylcarbodiimide to the N-protected L-phenylalanine. Subsequently, an amino acid ester with a free NH_2 group is added. This group is acylated without destroying the stereochemical integrity of the phenylalanine component **A** (upper part of Figure 6.31). The only drawback that remains is that the yield in this example leaves a great deal to be desired—unless the acyl isourea is used in large excess.

In its original form the DCC procedure for peptide synthesis has one basic disadvantage, which you have already learned about as side reaction **B** → **E** in Figure 6.15: When acyl isoureas are exposed to a poor nucleophile, they rearrange to form unreactive *N*-acyl ureas. These mismatched reactivities characterize the situation encountered in the peptide syntheses in Figures 6.31 and 6.32. The example given in Figure 6.31 uses *O,O*-di-*tert*-butylserine as a

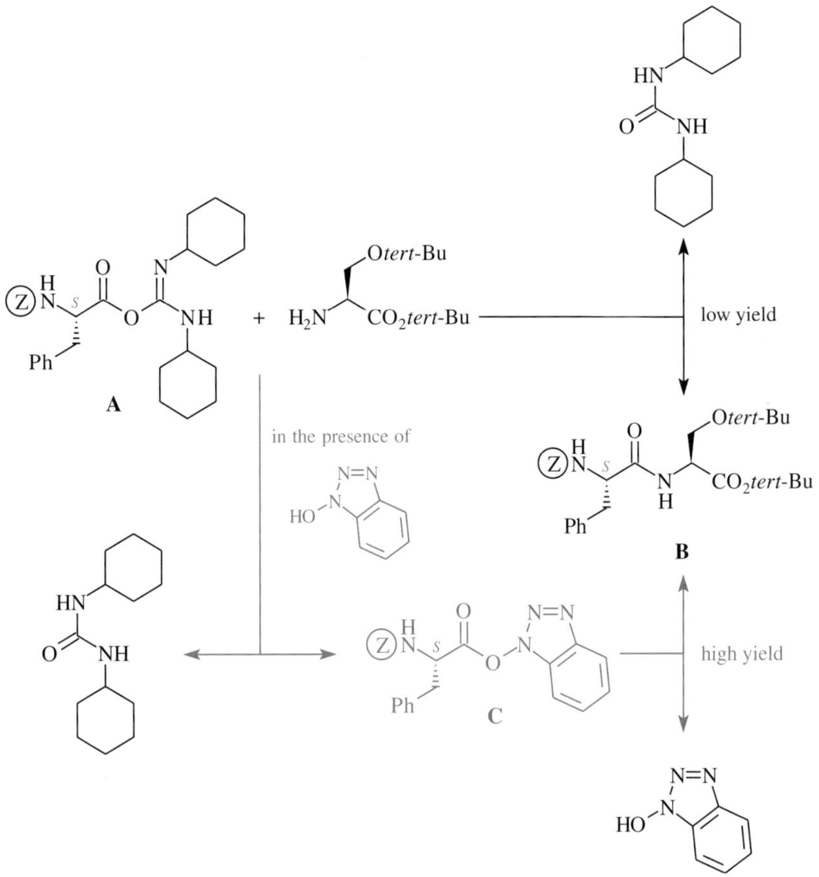

Otert-Bu
H2N CO2tert-Bu
low yield
A
in the presence of
B
C
high yield
Ph

Fig. 6.31. Stereoisomerization-free synthesis of a dipeptide according to the original DCC procedure (above) or a modified DCC procedure (below) (Z = benzyloxycarbonyl; preparation of the protected carboxylic acid underlying compound **A:** Figure 6.36).

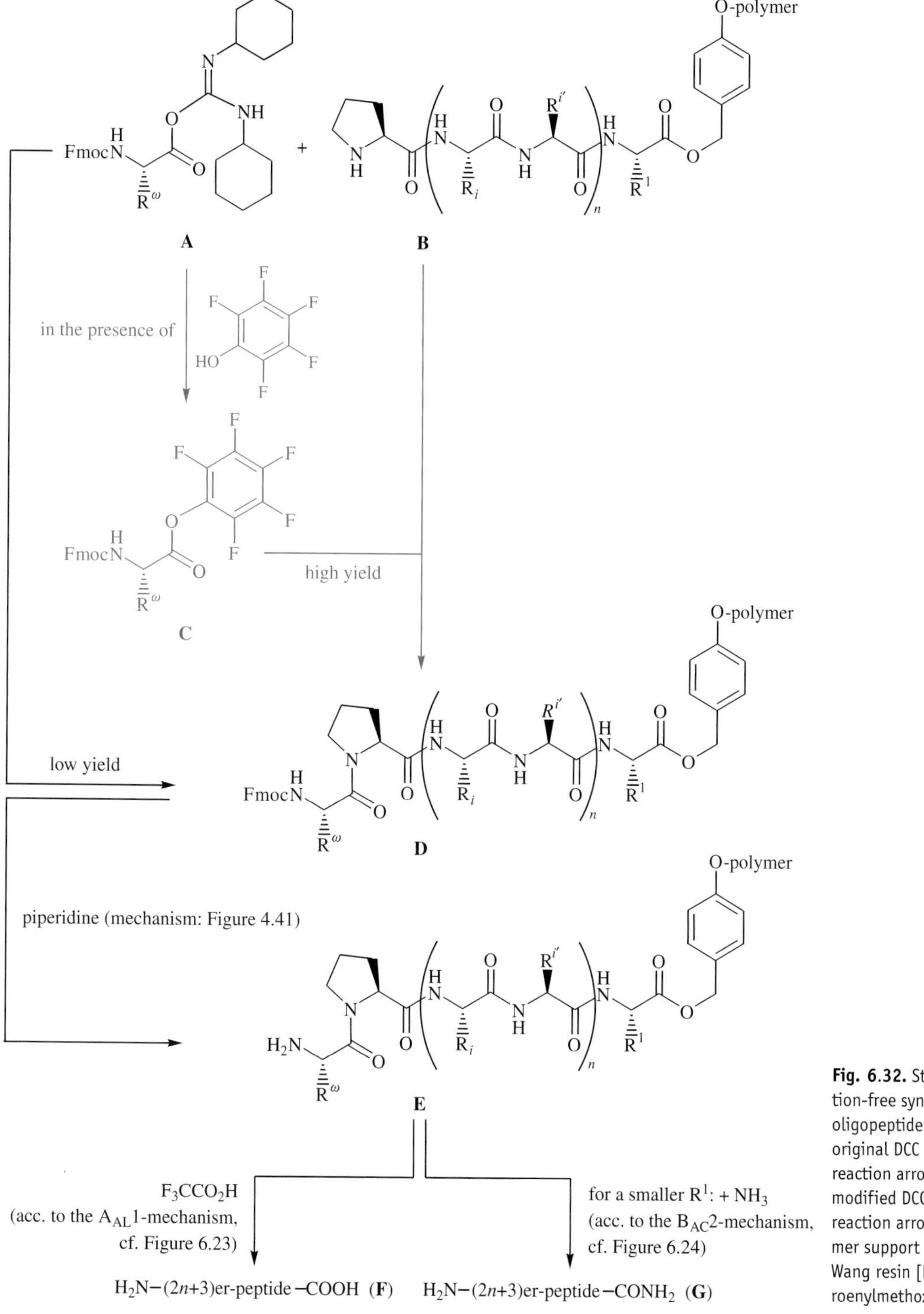

Fig. 6.32. Stereoisomerization-free synthesis of an oligopeptide following the original DCC procedure (black reaction arrow top left) and a modified DCC procedure (red reaction arrows) using a polymer support known as the Wang resin [Fmoc = (fluoroenylmethoxy) carbonyl].

nucleophile. The electron-withdrawing inductive effect of its ethereal *tert*-butoxy group makes the NH_2 group noticeably less nucleophilic than the NH_2 group in the esters of all other amino acids. In the example given in Figure 6.32, the *N*-terminal proline of an oligopeptide acts as nucleophile. For steric reasons its NH group is markedly less nucleophilic than the NH_2 group of most of the other proteinogenic amino acids. In connection with Figure 6.15, a general method was already presented allowing for the incorporation of the serine derivative in Figure 6.31 and the polypeptidyl proline in Figure 6.32 into the dipeptide or oligopeptide without forming an inert urea derivative and thus suffering yield losses. A specific solution can be found in the lower half of Figure 6.26. The unstable DCC adduct **A** is transacylated with *N*-hydroxybenzotriazole to give the stable amino acid derivative **C**. This is converted by the serine derivative into dipeptide **B** via an almost quantitative S_N reaction. A conceptually analogous, but in this particular case different, solution to the second synthetic problem is presented in Figure 6.32. The unstable DCC adduct **A** is transacylated with pentafluorophenol to give the stable amino acid derivative **C**, which is then converted into the elongated oligopeptide **D** via an S_N reaction.

Side Note 6.4. Solid-Phase Synthesis of Polypeptides

It appears worthwhile to appreciate the formation of the peptide bond **A** + **B** → **D** in the overall context of Figure 6.32 where it represents the last step of an oligopeptide synthesis in which—in principle—every one of the amino acids of **B** has been incorporated in the same way. Immediately thereafter the terminal N atom was relieved of the Fmoc group and then the next peptide bond was formed. Cleavage of the Fmoc group was conducted with piperidine according to the mechanism analyzed in Figure 4.41. The oligopeptide synthesis in Figure 6.32 ends with the Fmoc cleavage **D** → **E**, which is followed by either an aminolysis of the ester bond of the oligopeptide **E** (following the $B_{AC}2$ mechanism in Figure 6.24 and leading to the oligopeptide **G** as a *C*-terminal amide) or, alternatively, by cleavage of the ester bond of the oligopeptide **E** with trifluoroacetic acid. This latter reaction follows the $A_{AL}1$ mechanism outlined in Figure 6.23 and affords the oligopeptide **F** with a *C*-terminal carboxylic acid group.

6.4.4 S_N Reactions of Heteroatom Nucleophiles with Carbonic Acid Derivatives

Almost all carbonic acid derivatives of the type $Het^1–C(=O)–Het^2$ or, even more generally, $Het^1–C(=Het^2)–Het^3$ can undergo S_N reactions with heteroatom nucleophiles at the carboxyl carbon to produce other carbonic acid derivatives. A short survey of the preparative possibilities for these substitution reactions is given in Figure 6.33. Each of these reactions can be understood mechanistically in analogy to what you have learned in Sections 6.2–6.4.3 about S_N reactions at the carboxyl carbon of carboxylic acid derivatives. Moreover, the cross references placed in the figure mostly refer to explicitly presented mechanisms. With this information you yourself should think up suitable reagents and appropriate mechanisms for the reaction arrows that have not been labeled.

Figure 6.34 shows the SN reaction of two equivalents of benzyl alcohol (formula **B**) with $Cl_3C—O—C(=O)—Cl$ (diphosgene, formula **C**; preparation: Section 1.7.1). It is preferable to work with diphosgene, which is a liquid, rather than with phosgene Cl—C(=O)—Cl, which is a gas, because both reagents are very poisonous. As the substitution product one obtains chloroformic acid benzyl ester **A**, which is also called "Z-chloride" and used to protect the amino group of α-amino acids (cf. Figure 6.36). The reaction of fluorenyl methanol and

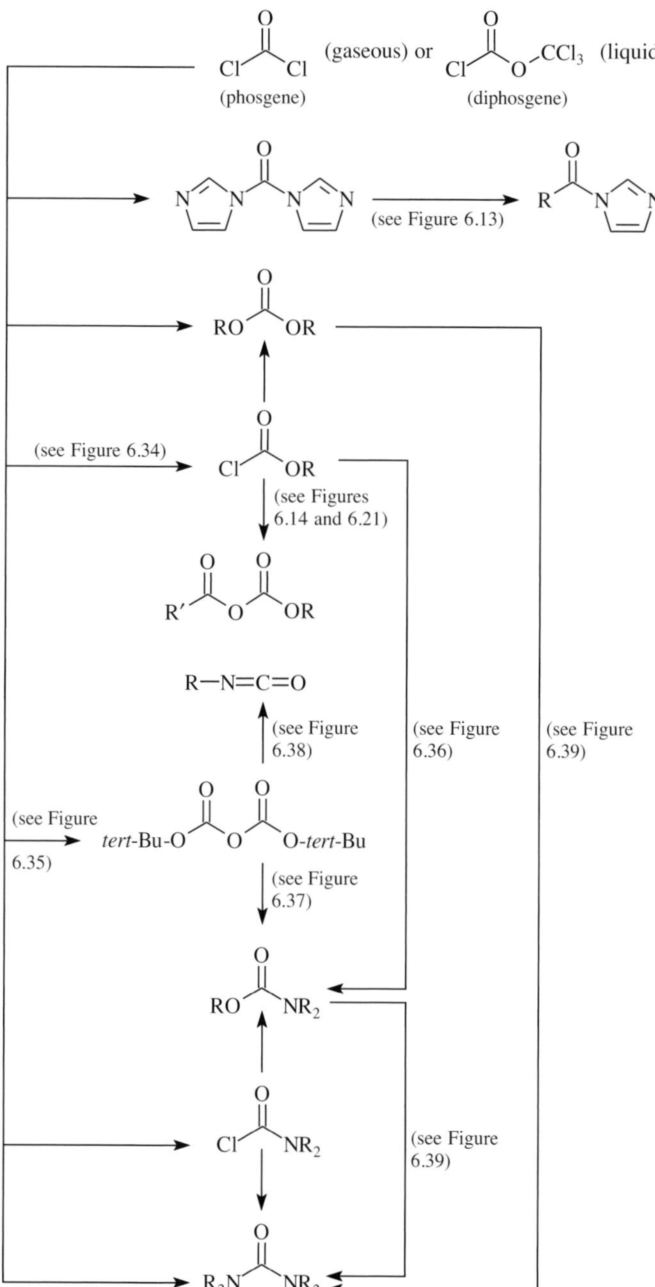

Fig. 6.33. Carbonic acid derivatives and their interconversions (plus cross-reference to a chlorine-free option for the synthesis of isocyanates).

diphosgene proceeds in perfect analogy to afford (fluorenylmethoxy)carbonyl chloride, providing the amino group of α-amino acids with the base-sensitive Fmoc protecting group (structural formula: Figure 4.41; synthesis application: Figure 6.32). Also in analogy to Figure 6.34, isobutanol and diphosgene react to give isobutyl chloroformate. The latter is used for the *in situ* activation of carboxylic acids (Figure 6.14). Finally, methyl chloroformate can be obtained from methanol and diphosgene—also in analogy to Figure 6.34. In the last reaction scheme of the present chapter it is involved in a C_1-lengthening carboxylic acid ester synthesis. In contrast to the other aforementioned chloroformates, *t*-butyl chloroformate is unknown. It is unstable and decomposes via the mechanism shown in Figure 8.7).

As you have already learned, the *tert*-butoxycarbonyl group (Boc group) is an important N-protecting group used in the synthesis of peptides. In Figure 4.37, we have seen how it can be cleaved off by the E1 mechanism. The Boc group can be attached to α-amino acids by reaction with di-*tert*-butyldicarbonate (Figure 6.37). The structure of this reagent, also known as "Boc anhydride" or "Boc$_2$O," is shown in Figure 6.35 (structural formula **H**). It can be obtained in a two-step process from potassium *tert*-butanolate, carbon dioxide and phosgene. In the first step, the potassium salt **A** of the carbonic acid *tert*-butylester is produced according to the formation mechanism of Stiles' reagent (Section 8.2). Then phosgene is introduced

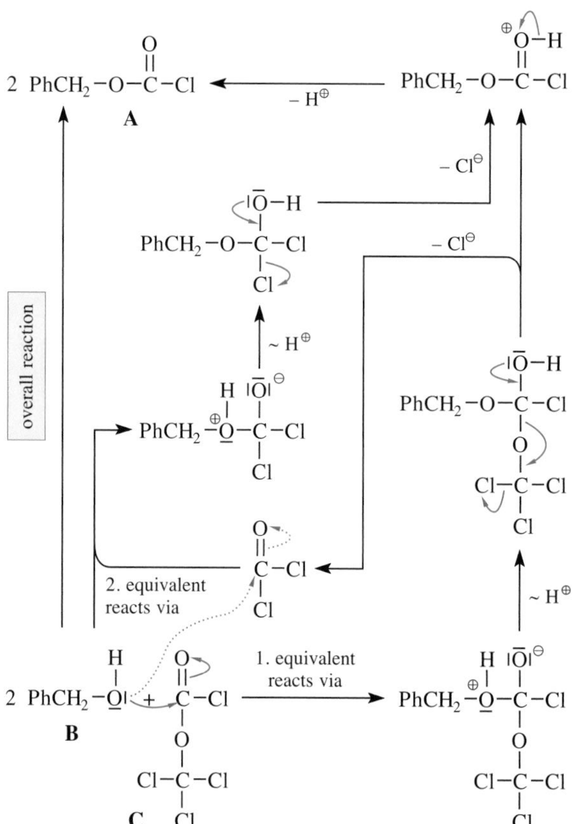

Fig. 6.34. Mechanism for obtaining a chlorocarbonate through an S_N reaction of an alcohol with diphosgene. The first equivalent of the alcohol reacts with diphosgene itself; the second equivalent reacts with the phosgene formed in the third step.

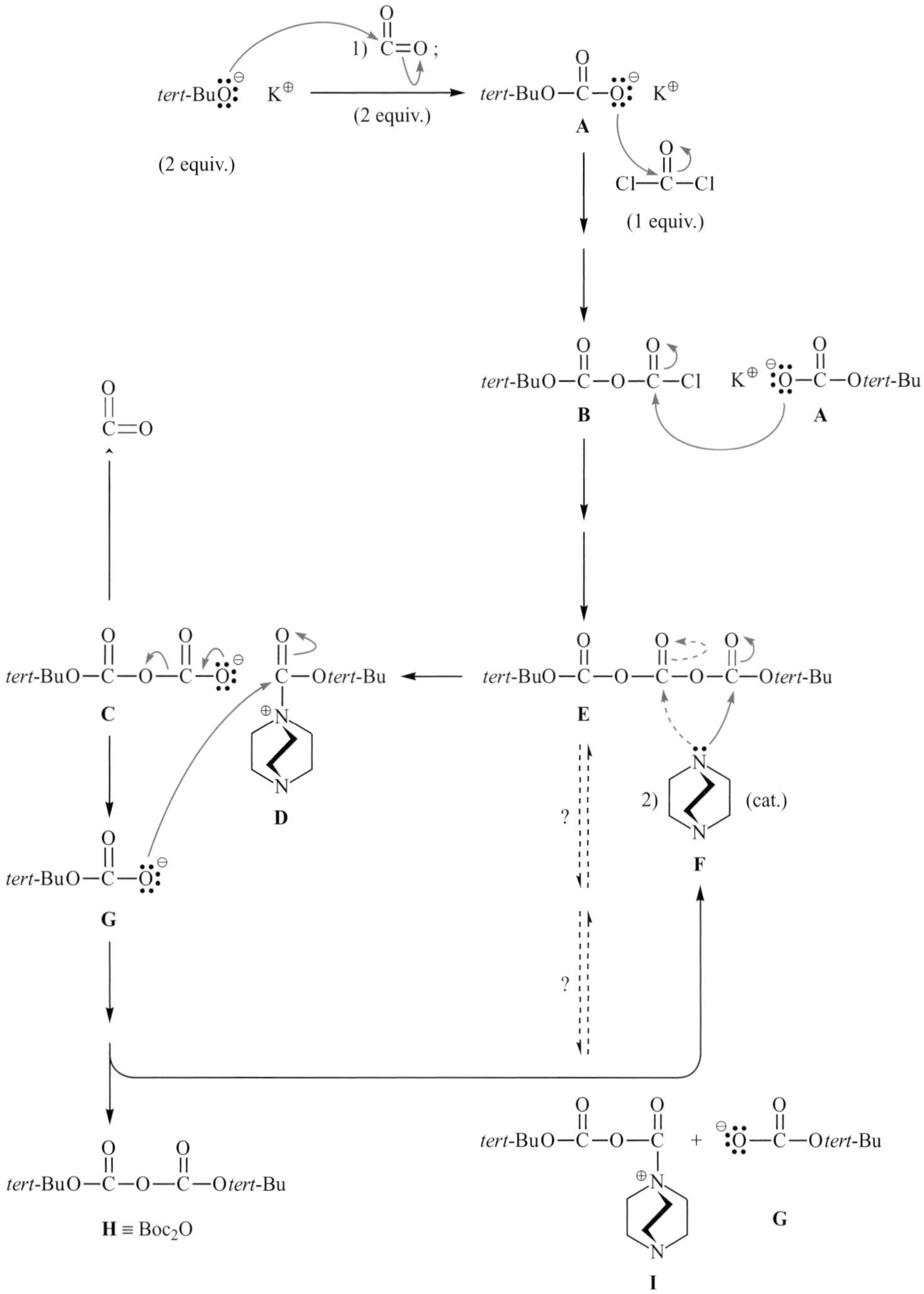

tert-BuO⁻ K⁺ (2 equiv.)
1) CO_2 (2 equiv.)
tert-BuO–C(=O)–O⁻ K⁺ **A**
Cl–C(=O)–Cl (1 equiv.)
tert-BuO–C(=O)–O–C(=O)–Cl **B**
K⁺ ⁻O–C(=O)–Otert-Bu **A**
tert-BuO–C(=O)–O–C(=O)–O–C(=O)–Otert-Bu **E**
2) **F** (cat.)
?
?
tert-BuO–C(=O)–O–C(=O)–N⁺ **I** + ⁻O–C(=O)–Otert-Bu **G**
tert-BuO–C(=O)–O–C(=O)–O⁻ **C**
C(=O)–Otert-Bu **D**
O=C=O
tert-BuO–C(=O)–O⁻ **G**
tert-BuO–C(=O)–O–C(=O)–Otert-Bu
H ≡ Boc_2O

Fig. 6.35. Preparation of the dicarbonate ("pyrocarbonate") **H** ("Boc anhydride," "Boc_2O") from potassium *tert*-butoxide, carbon dioxide and phosgene. The tricarbonate **E** is isolated as an intermediate. Only upon addition of a catalytic amount of the nucleophilic amine **F** does **E** react to afford Boc_2O, while the analogous ethyl derivative of the tricarbonate does so spontaneously. Boc_2O applications are shown in Figures 6.37 and 6.38, an application of the analogous diethyldicarbonate is given in Figure 13.62.

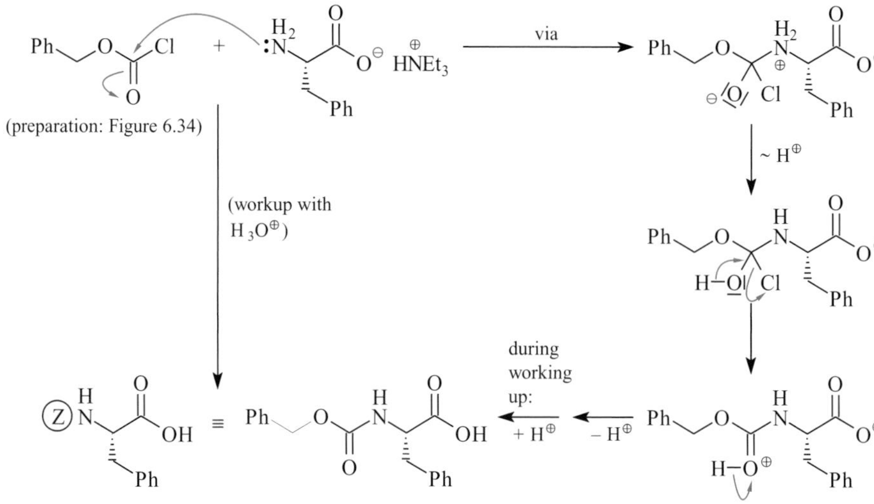

(preparation: Figure 6.34)

(workup with $H_3O^{\oplus}$)

during working up:

Fig. 6.36. Mechanism of the preparation of a Z-protected amino acid by acylation of an amino acid with Z-chloride.

and reacts twice with **A** according to the $B_{AC}2$ mechanism. The result is the formation of di-*tert*-butyltricarbonate (**E**). In a second step, this compound is treated with catalytic amounts of a tertiary amine that is acylated in an equilibrium reaction. However, the resulting acyliminium ions **D** and **I** are extremely potent acylating agents (cf. Table 6.1, entry 12), which is why they release their counter ions **C** and **G**, respectively, for a short time only and then recapture them to re-form **E**. During this process, the anion **C** decomposes into the anion **G** and CO_2. The carbon dioxide escapes from the reaction mixture thus rendering the formation of the anion **G** and the acyliminium ion **D** from the tricarbonate irreversible. Finally, **G** and **D** yield Boc_2O (**H**) and at the same time release the catalyst (the amine).

The nucleophilic nitrogen of α-amino acids undergoes an S_N reaction at the carboxyl carbon of Z-chloride according to the $B_{AC}2$ mechanism. A benzyloxycarbonyl-protected amino acid is produced. Figure 6.36 illustrates this method using the preparation of Z-protected phenylalanine as an example. Z-protected amino acids are standard building blocks for peptide synthesis (cf. Figure 6.31).

Figure 6.37 illustrates how the nitrogen of α-amino acid (derivative)s can be protected as a *tert*-butyl carbamate. *O,O*-di-*tert*-butyl serine serves as an example. Boc anhydride is the derivatization reagent of choice. It is used to protect the amino acid via the $B_{AC}2$ mechanism that you have already encountered on many other occasions. Figure 6.37 requires just one explanation: the leaving group is the same *tert*-butyl carbonate anion that you saw in Figure 6.35 as its stable potassium salt. In Figure 6.37, however, it is forms a carbonic acid half ester, which decomposes to furnish *tert*-butanol and carbon dioxide.

The protection of primary amines using Boc anhydride to give *tert*-butyl carbamates is so fast that nobody saw a reason for using Steglich's catalyst to accelerate this reaction (cf. Figure 6.9, bottom). When Knoelker in 1995 investigated the combined influence of Boc_2O and DMAP on amines the result was not only a surprise, but at the same time worth patenting: *instead of a tert-butyl carbamate an isocyanate was formed* (Figure 6.38). Amazingly, the *tert*-butyl carbamate is not a precursor of the isocyanate obtained. The transformation in Fig-

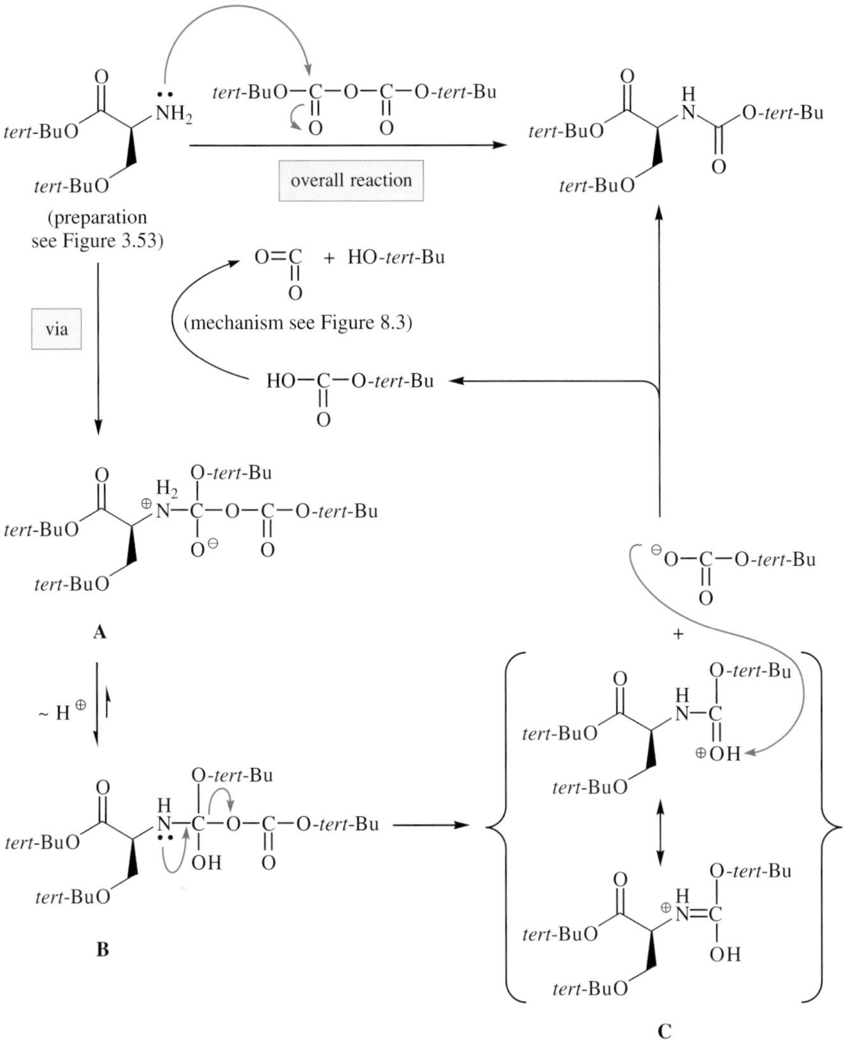

Fig. 6.37. Mechanism of the preparation of a Boc-protected amino acid by acylation of an amino acid with *tert*-butyl dicarbonate. The chemoselectivity of the reaction shown here differs from the example given in Figure 6.38 where the primary amine is treated with both *tert*-butyl dicarbonate and 4-(dimethylamino)pyridine.

ure 6.38 allows for the mild laboratory-scale isocyanation of primary amines. It further affords the technical-scale preparation of absolutely chlorine-free diisocyanates. For reasons of toxicity, this finding is interesting with respect to the subsequent reaction of diisocyanates with diols or triols that the corresponding polyurethanes (see Side Note 8.1).

As the last example of an S_N reaction at the carboxyl carbon of a carbonic acid derivative, consider the synthesis of dicyclohexylurea in Figure 6.39. In this synthesis, two equivalents of cyclohexylamine replace the two methoxy groups of dimethyl carbonate. Dicyclohexylurea can be converted into the carbodiimide dicyclohexylcarbodiimide (DCC) by treatment with tosyl chloride and triethylamine. The urea is dehydrated. The mechanism of this reaction is identical to the mechanism that is presented in Figure 8.9 for the similar preparation of a different carbodiimide.

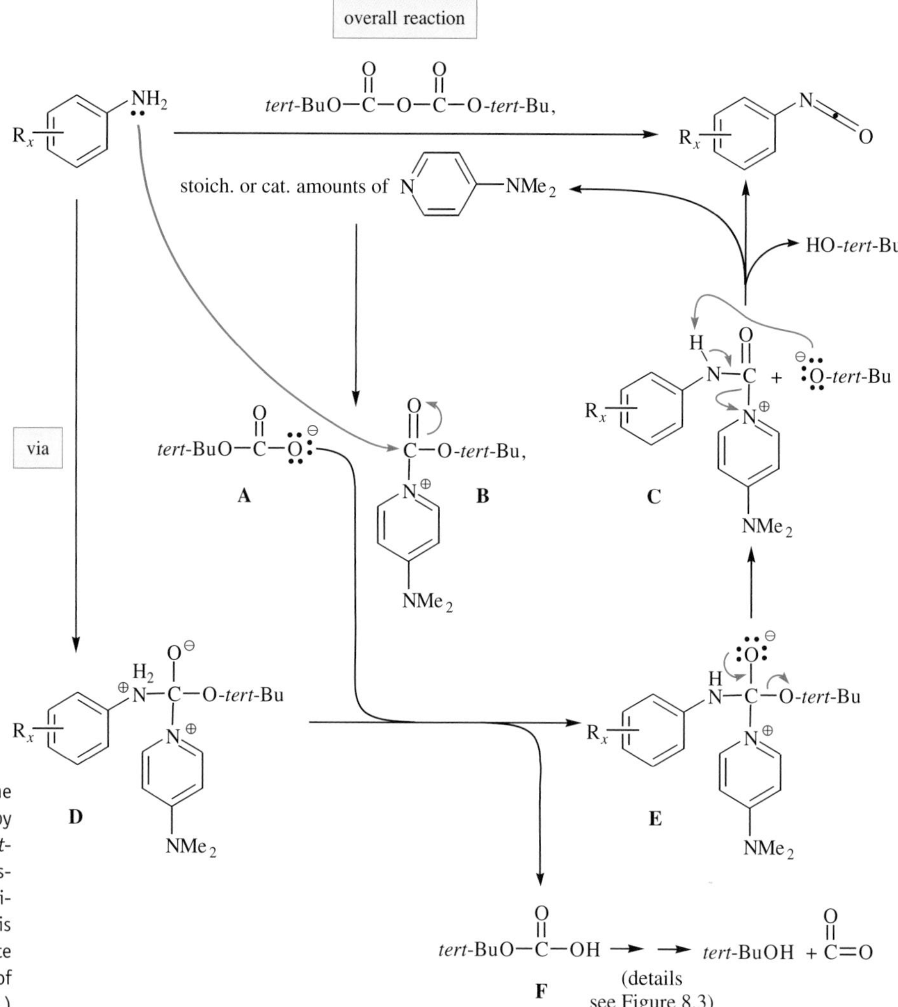

Fig. 6.38. Mechanism of the preparation of an isocyanate by acylation of anilines with *tert*-butyl dicarbonate in the presence of 4-(dimethylamino)pyridine. In the absence of this additive a *tert*-butyl carbamate would be produced instead of an isocyanate (cf. Figure 6.37.)

6.5 S_N Reactions of Hydride Donors, Organometallics, and Heteroatom-Stabilized "Carbanions" on the Carboxyl Carbon

6.5.1 When Do Pure Acylations Succeed with Carboxylic Acid (Derivative)s, and When Are Alcohols Produced?

Many hydride donors, organometallic compounds, and heteroatom-stabilized "carbanions" react with carboxylic acids and their derivatives. However, the corresponding substitution

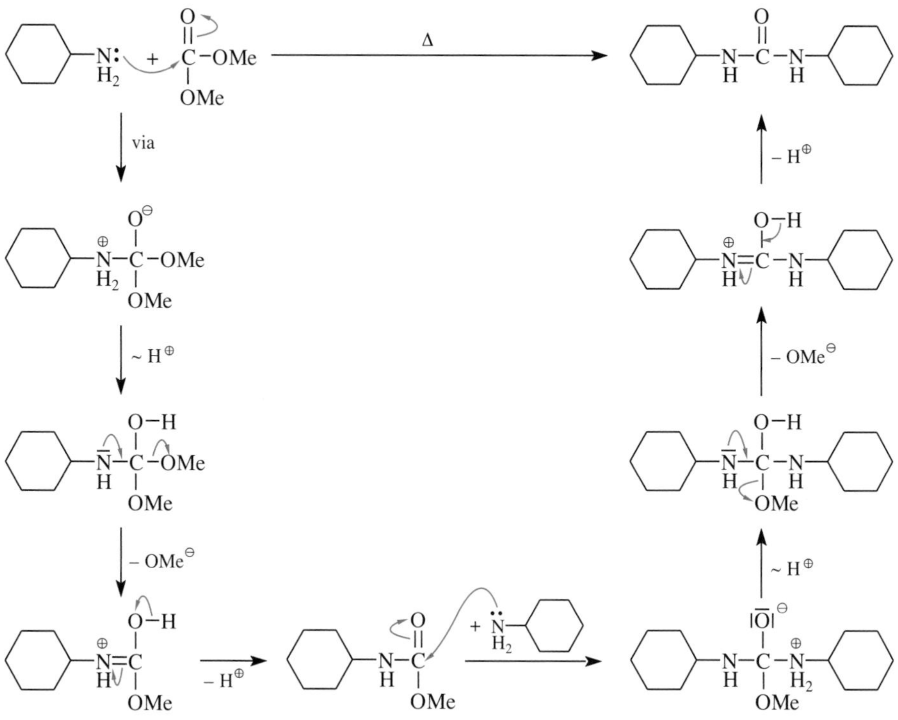

Fig. 6.39. Acylation of cyclohexylamine with dimethyl carbonate—a possibility for synthesizing the DCC precursor dicyclohexylurea.

product, i.e., the acylation product (**C** in Figure 6.40), can only be *isolated* using very specific reagent/substrate combinations. In the other cases the result is an "overreaction" with the nucleophile. It occurs as soon as the acylation product **C** forms via the mechanism of Figure 6.2 at a time when the nucleophile is still present. Because this acylation product **C** is an aldehyde or a ketone, it is able to add any remaining nucleophile (cf. Chapter 10) furnishing alkoxides **D**. Upon aqueous workup, these provide secondary (R = H) or tertiary (R ≠ H) alcohols **F**. These compounds contain two identical substituents at the carbinol carbon, both of which originate from the nucleophile.

Most S_N reactions of hydride donors, organometallic compounds, and heteroatom-stabilized "carbanions" at the carboxyl carbon follow the mechanism shown in Figure 6.2. Thus, the substitution products, i.e., the aldehydes and ketones **C**, form *in the presence of the nucleophiles.* Thus, when the nucleophile and the acylating agent are used in a 2:1 ratio, alcohols **F** are *always* produced.

However, if one of the aforementioned nucleophiles and the acylating agent are reacted in a 1:1 ratio, it is under certain circumstances possible to stop the reaction chemoselectively at the stage of the carbonyl compound **C**.

From Figure 6.40 you can see that these particular conditions are fulfilled if the rate-determining step of the acylation—i.e., the formation of the tetrahedral intermediate **B**—is considerably faster than the further reaction of the carbonyl compound **C** giving the alkoxide **D.** In more quantitative terms, it would hence be required that the rate of formation $d[\mathbf{B}]/dt$ must

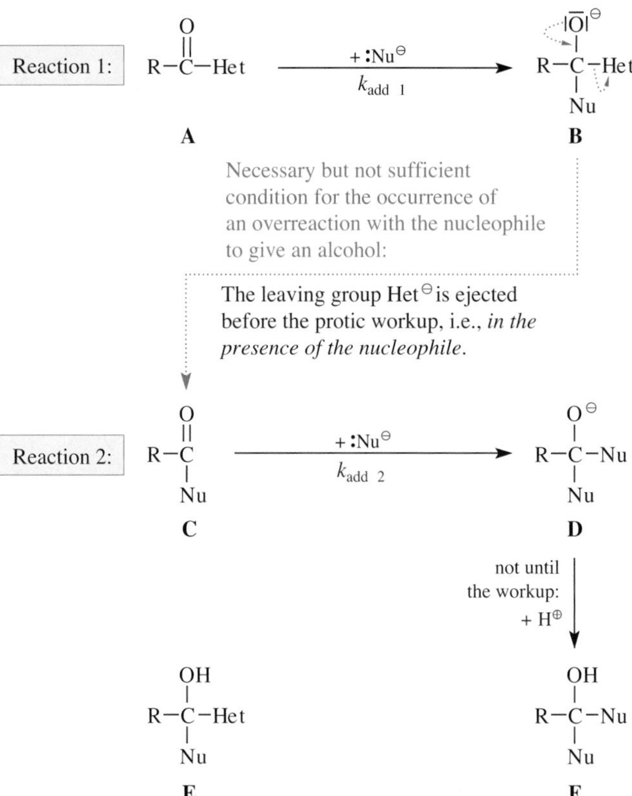

Fig. 6.40. On the chemoselectivity of the reactions of hydride donors, organometallic compounds, and heteroatom-stabilized "carbanions" with acylating agents ($k_{add\ 1}$ refers to the rate constant of the addition of the nucleophile to the carboxyl carbon, and $k_{add\ 2}$ refers to the rate constant of the addition of the nucleophile to the carbonyl carbon).

be greater than the rate of formation $d[\mathbf{D}]/dt$. Thus, in order for acylations to occur chemoselectively, the following must hold:

$$k_{add\ 1}[\mathbf{A}][\mathrm{Nu}^-] \gg k_{add\ 2}[\mathbf{C}][\mathrm{Nu}^-] \tag{6.20}$$

This is ensured during the entire course of the reaction only if

$$k_{add\ 1} \gg k_{add\ 2} \tag{6.21}$$

For the reaction of hydride donors, organometallic compounds and heteroatom-stabilized "carbanions" with acylating agents or carbonyl compounds one encounters a universal reactivity order RC(=O)Cl > RC(=O)H > R_2C=O > RC(=O)OR > RC(=O)NR_2. It applies to both good and poor nucleophiles, but—in agreement with the reactivity/selectivity principle (Section 1.7.4)—the reactivity differences are far larger for poor nucleophiles.

Accordingly, if one wants to react a nucleophile and carboxylic acid derivative to produce a carbonyl compound in a chemoselective fashion via the mechanism of Figure 6.2, then one should use acid chlorides or comparably strongly activated carboxylic acid derivatives. In addition, the reaction must be carried out with the weakest possible nucleophile because only such a nucleophile reacts *considerably* faster with the activated carboxylic acid derivative than with the product carbonyl compound. The nucleophile must react "considerably" faster with the carboxylic acid derivative because at 95% conversion there is almost twenty times more carbonyl compound present than carboxylic acid derivative, but even at this stage the carboxylic acid derivative must be the preferred reaction partner of the nucleophile.

To make as *much* carboxylic acid derivative as possible available to the nucleophile at all stages of the reaction, the nucleophile is added dropwise to the carboxylic acid derivative and not the other way around. In Figure 6.41, the approach to chemoselective acylations of hydride donors and organometallic compounds, which we have just described, is labeled as "strategy 2" and compared to two other strategies, which we will discuss in a moment.

Chemoselective S_N reactions of nucleophiles with carboxylic acid derivatives are guaranteed to take place without the risk of an overreaction when the substitution mechanism of Figure 6.4 applies. This is because as long as the nucleophile is present, only one reaction step is possible: the formation of the negatively charged tetrahedral intermediate. Figure 6.40 summarizes this addition in the top line as "Reaction 1" (→ **B**).

There may be two reasons why such a tetrahedral intermediate **B** does not fragment into the carbonyl compound **C** before the aqueous workup. For one thing, the "Het" group in it may be too poor a leaving group. For another thing, the "Het" group may be bound through a suitable metal ion to the alkoxide oxygen of the tetrahedral intermediate and stabilizing the intermediate.

The S_N reaction under consideration is not terminated until water, a dilute acid, or a dilute base is added to the crude reaction mixture. The tetrahedral intermediate **B** is then protonated to give the compound **E.** Through an E1 elimination, it liberates the carbonyl compound **C** (cf. discussion of Figure 6.4). Fortunately, at this point in time no overreaction of this aldehyde with the nucleophile can take place because the nucleophile has been destroyed during the aqueous workup by protonation or hydrolysis. In Figure 6.41, this process for chemoselective acylation of hydride donors, organometallic compounds, and heteroatom-stabilized "carbanions" has been included as "strategy 1."

As acylation "strategy 3," Figure 6.41 outlines a circumstance with which certain heteroatom-stabilized "carbanions" can be acylated chemoselectively. It is interesting to note that such a chemoselectivity exists although in this mechanism the tetrahedral intermediate does proceed to give the carbonyl compound while the nucleophile is still present. However, the heteroatom substituent, whose presence is required to make this reaction possible, now exerts a significant electron withdrawal. It makes the carbonyl compound obtained much more acidic than, for example, a simple ketone (see Section 13.1.2). Thus, this carbonyl compound is rapidly deprotonated by a second equivalent of the reagent (which is required for the success of this reaction) whereupon the enolate of the carbonyl compound is produced. This enolate can obviously not react further with any nucleophile still present. During the aqueous workup the desired carbonyl compound is formed again by protonation of the enolate. At the same time any remaining nucleophile is destroyed.

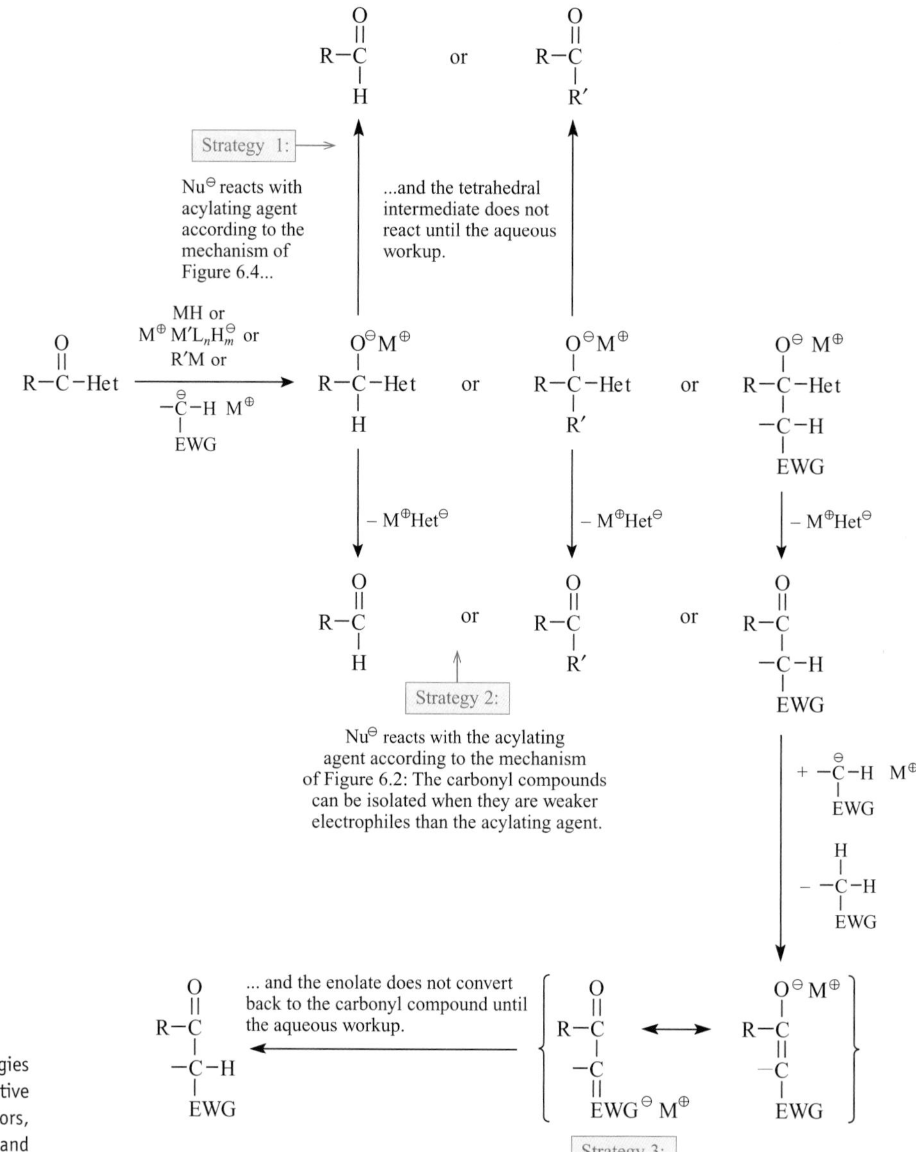

Fig. 6.41. Three strategies for the chemoselective acylation of hydride donors, organometallics and heteroatom-stabilized "carbanions" with carboxylic acid derivatives.

6.5.2 Acylation of Hydride Donors: Reduction of Carboxylic Acid Derivatives to Aldehydes

Chemoselective S_N reactions at the carboxyl carbon in which hydride donors function as nucleophiles can be carried out using "strategy 1" or "strategy 2" from Figure 6.41.

The substrates *par excellence* for "strategy 1" are tertiary amides and especially the so-called **Weinreb amides**, i.e. amides derived from *N,O*-dimethylhydroxylamine. Compound **A** in Figure 6.42 is an example. Tertiary amides or Weinreb amides and diisobutylaluminum hydride (DIBAL) furnish stable tetrahedral intermediates. In the tetrahedral intermediate, obtained from "normal" tertiary amides the $R^1R^2N^{\ominus}$ group is a particularly poor leaving group. The same qualification applies to the $(MeO)MeN^{\ominus}$ group in the tetrahedral intermediates (**B** in Figure 6.42) derived from Weinreb amides. In addition, the Weinreb intermediates **B** are stabilized by the chelation indicated in Figure 6.42 (this is why Weinreb amides contain the methoxy substituent!). During the aqueous workup these tetrahedral intermediates are hydrolyzed to give an aldehyde. Weinreb amides can be reduced to aldehydes not only with DIBAL but also with $LiAlH_4$. In general, this is not true for other amides; with $LiAlH_4$ they are reduced to form the corresponding amines (see Figures 17.65, 17.66).

Interestingly, substrate **C** in Figure 6.42 selectively reacts with by DIBAL at the carboxyl carbon of the Weinreb amide. Obviously, its reactivity is much higher than that of the carboxyl carbon of the ester group. According to Table 6.1, "normal" carboxylic acid amides are weaker acylating agents than carboxylic acid ester. The fact that Weinreb amides acylate faster than carboxylic acid esters is due, among other things, to the smaller pi electron-donating (+M) effect of their NMe(OMe) group vis-a-vis an NR^1R^2 group, resulting in a decreased resonance stabilization of the adjacent C=O double bond. In addition, a particularly large stabilization of the corresponding tetrahedral intermediates through chelation can be observed in the presence of suitable cations (see above).

Esters contain a better leaving group than "normal" amides or Weinreb amides. Esters also typically lack a heteroatom that could be incorporated in a five-membered chelate. Therefore, tetrahedral intermediates formed from esters and hydride-donating reagents decompose con-

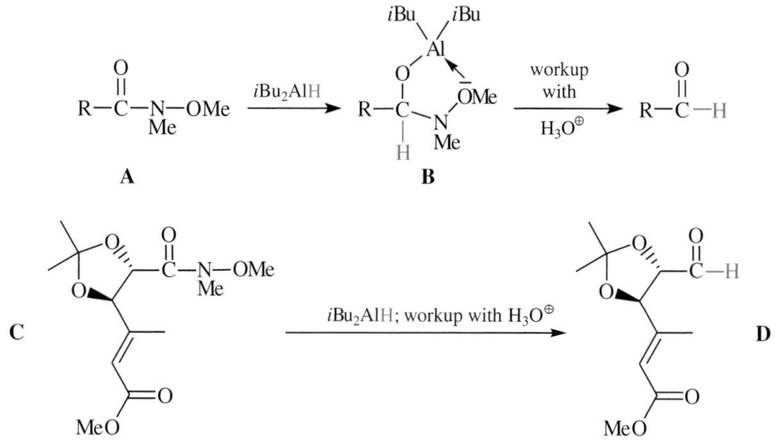

A: R–C(=O)–N(Me)–OMe —iBu_2AlH→ B: R–C(H)(O–Al(iBu)₂)–N(Me)–OMe (chelate) —workup with $H_3O^{\oplus}$→ R–C(=O)–H

C —iBu_2AlH; workup with $H_3O^{\oplus}$→ D

Fig. 6.42. Preparation of Weinreb amides through S_N reactions at the carboxyl carbon. Chemoselective reduction of Weinreb amides to aldehydes.

siderably more readily than those resulting from amides or Weinreb amides. Only from DIBAL and esters—and even then only in noncoordinating solvents—is it possible to obtain tetrahedral intermediates that do not react until the aqueous workup and can then be hydrolyzed to an aldehyde (see Figure 17.61). In coordinating solvents, DIBAL and esters give alcohols through an overreaction with the aldehydes that are now rapidly formed *in situ.* The reaction of $LiAlH_4$ and esters (Figure 17.60) always proceeds to alcohols through such an overreaction (see Figure 17.61).

Following "strategy 2" from Figure 6.41, chemoselective S_N reactions of hydride donors with carboxylic acid derivatives also succeed starting from acid chlorides (Figure 6.43). For the reasons mentioned further above, *weakly* nucleophilic hydride donors are used for this purpose. The hydride donor is added dropwise *to* the acid chloride in order to achieve success—and *not* the other way round. These reaction conditions preclude that the hydride donor is ever faced with the resulting aldehyde alone without any unreacted carboxylic acid chloride being present—a situation in which the hydride donor could undergo further reaction with the aldehyde.

$Na^{\oplus}\ BH_4^{\ominus}$, low temperature
or
$Cu(PPh_3)_2^{\oplus}\ BH_4^{\ominus}$
or
$Li^{\oplus}\ AlH(O\text{-}tert\text{-}Bu)_3^{\ominus}$, low temperature

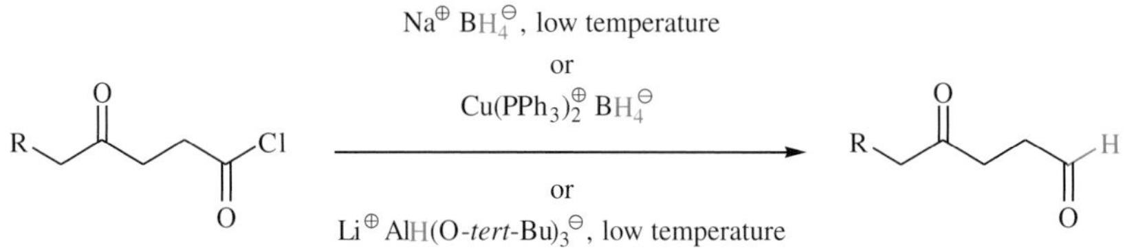

Fig. 6.43. Chemoselective reduction of carboxylic acid chloride to furnish an aldehyde; the keto group of the substrate is compatible with these reaction conditions, too.

6.5.3 Acylation of Organometallic Compounds and Heteroatom-Stabilized "Carbanions" With Carboxylic Acid (Derivative)s: Synthesis of Ketones

Tertiary amides in general and Weinreb amides in particular react according to "strategy 1" of Figure 6.41 not only with hydride donors, but also with organolithium and Grignard compounds (reactions leading to **A** or to **B**, Figure 6.44). The aqueous workup of these intermediates leads to pure acylation products. In this way, DMF or the Weinreb amide of formic acid and organometallic compounds give aldehydes (not shown). In the same way, tertiary amides or Weinreb amides of all higher monocarboxylic acids and organometallic compounds form ketones. In the lower half of Figure 6.44 this is illustrated with a somewhat more complex example, namely the reaction of the Weinreb amide **D** with the *ortho*-lithio derivative of anisole (**C**). This is another example of the superior electrophilicity of Weinreb amides as compared to carboxylic acid esters, since only the C(=O)NMe(OMe) group reacts and not the C(=O)O-*tert*-Bu group. The same grading of reactivity has already been encountered and justified in Figure 6.42. The carbamoyl residue of the substrate **D** in Figure 6.44 is deprotonated by lithioanisole and thus protected *in situ* from a competing nucleophilic reaction. This is why *two* equivalents of lithioanisole are required in order to proceed to completion.

The same reaction mechanism—again corresponding to "strategy 1" of Figure 6.41—explains why carboxylic acids and two equivalents of an organolithium compound react selectively to form ketones (Figure 6.45, substrate on the left). The first equivalent of the reagent

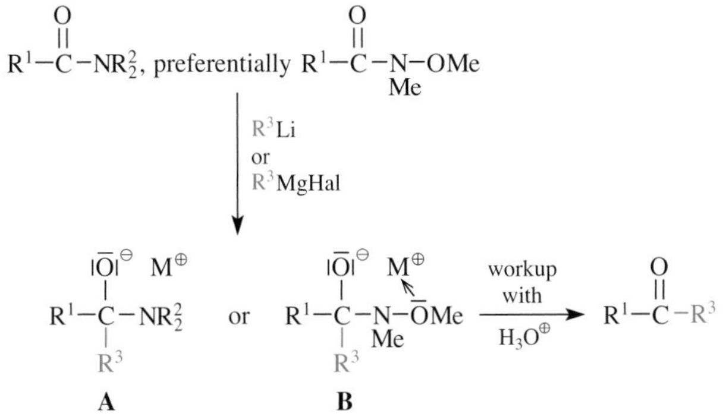

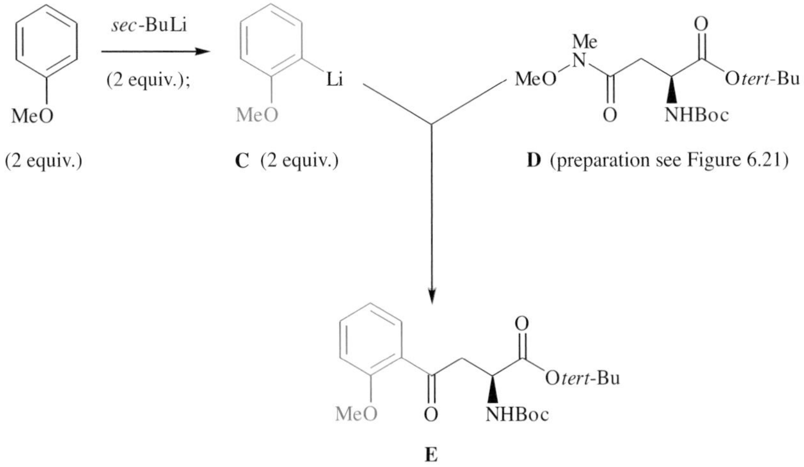

Fig. 6.44. Chemoselective acylation of an organometallic compound **C** with the Weinreb amide **D** of Figure 6.21: a general approach to ketones.

deprotonates the substrate to give a lithium carboxylate **A**. According to Table 6.1, carboxylates are the weakest of all acylating agents. Actually, lithium carboxylates acylate only a single type of nucleophile, namely organolithium compounds. Therefore, the second equivalent of the organolithium compound reacts with the carboxylate forming the tetrahedral intermediate **B**. Because this species cannot possibly eliminate Li_2O, it does not react until the aqueous workup. It is then protonated to give the ketone hydrate **C** from which the finally isolated ketone is produced by rapid loss of water (cf. Section 9.1.1).

Side Note 6.5. Reaction of Grignard Compounds with Carboxylic Acids

In times (long gone) of wet chemical instead of NMR- or MS analysis the so-called Zerevitinoff reaction was used to determine, e. g., the molecular mass of carboxylic acids (Figure 6.46, left column). This reaction exploited the fact that the carboxylic acids can be deprotonated by methylmagnesium chloride and was based on a volumetric determination of the liberated methane. According to this method, the excess of methylmagnesium chloride was not able to add to the magnesium carboxylate **A** formed. Methyllithium would have reacted, as you have just seen in the context of the reactions given in Figure 6.45. This difference in reactivity is

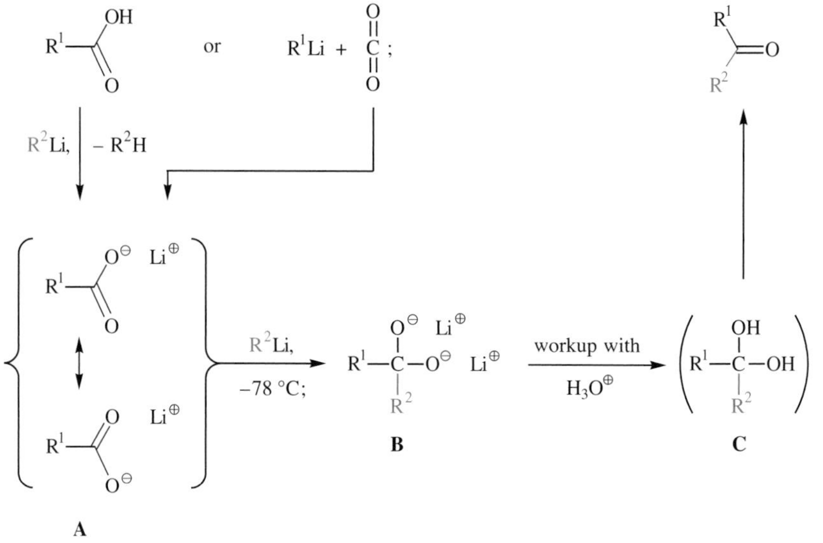

Fig. 6.45. Chemoselective acylation of organolithium compounds with lithium-carboxylates (**A**). In order to generate the substrates the choice is between the deprotonation of the corresponding carboxylic acid and the addition of an organolithium compound to carbon dioxide, i.e. via C,C bond formation.

consistent with the experience (cf. the beginning of Section 10.5.1) that Grignard compounds are poorer nucleophiles than organolithium compounds.

The middle and the right column of Figure 6.46, however, demonstrate that Grignard compounds can nevertheless add to especially reactive magnesium carboxylates in exactly the same way as is known for organolithium compounds according to Figure 6.45. So, Grignard compounds form tetrahedral intermediates of type **D** upon reaction with the sterically most undemanding magnesium carboxylate, i.e., magnesium formate **B**. These intermediates cannot release BrMgOMgBr and therefore do not react until aqueous workup. This affords an aldehyde. On the other hand, the magnesium salt **C** of the trifluoroacetic acid can react with Grignard compounds to give the tetrahedral intermediates **E**, whose two negative formal charges are significantly stabilized by the strong electron-withdrawing inductive –I effect of the CF_3 group. Aqueous workup then produces a trifluoromethyl ketone via a ketone hydrate intermediate. The reaction shown on the right side of Figure 6.46 thus offers an interesting option for introducing CF_3 groups into organic compounds. These groups can, for example, markedly modify the properties of pharmacologically active agents.

You are already familiar with “strategy 2” of Figure 6.41. This strategy also allows for the chemoselective acylation of weakly nucleophilic organometallic compounds, such as the ones shown in Figure 6.47, with acid chlorides to afford various ketones. Suitable organometallic compounds are Gilman cuprates (R_2CuLi; preparation: Figure 10.43, left), Knochel cuprates [$R_{funct}Cu(CN)ZnX$; preparation: Figure 10.43, right; R_{funct} means a functionalized group], or lithium acetylides.

In Figure 6.41, the reaction of certain heteroatom-stabilized “carbanions” with carboxylic acid derivatives is presented as “strategy 3” of Figure 6.32 for achieving chemoselective acylations. This strategy can be used to convert esters into β-ketophosphonic acid esters with

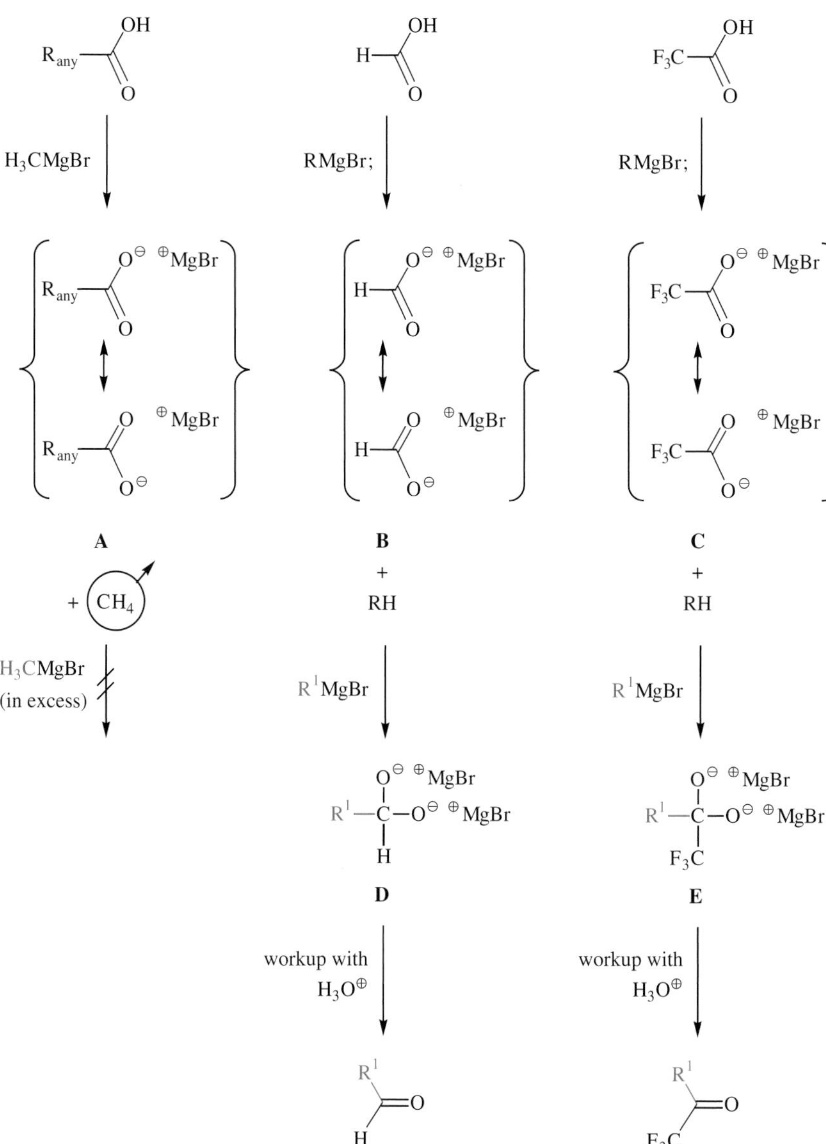

Fig. 6.46. Acylation of Grignard compounds with magnesium carboxylates: usually impossible (with magnesium carboxylate **A**), but feasible with sterically favored magnesium formate (**B**) or electronically favored magnesium(trifluoracetate) (**C**). These reactions allow the construction of aldehydes and trifluoromethylketones, respectively.

sodium phosphonates (Figure 6.48) or to acylate the sodium salt of DMSO (dimethyl sulfoxide, Figure 6.49). The last reaction can, for example, be part of a two-step synthesis of methyl ketones from carboxylic esters, which does not require the use of organometallic compounds.

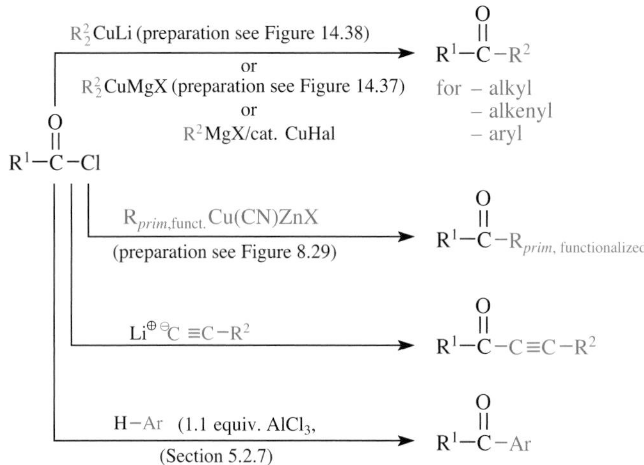

Fig. 6.47. Top three reactions: chemoselective acylations of weakly nucleophilic organometallic compounds with carboxylic acid chlorides. Bottom reaction (as a reminder): chemoselective acylation of an aromatic compound with an activated carboxylic chloride (Friedel–Crafts acylation, cf. Section 5.2.7).

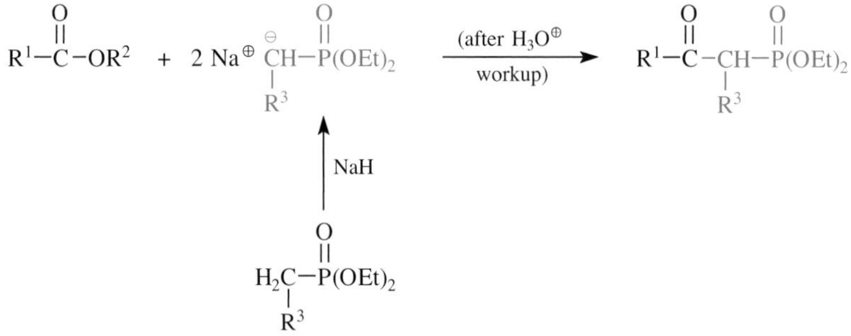

$$R^1{-}C({=}O){-}OR^2 + 2\,Na^{\oplus}\ {}^{\ominus}CH(R^3){-}P({=}O)(OEt)_2 \xrightarrow[\text{workup)}]{\text{(after } H_3O^{\oplus}} R^1{-}C({=}O){-}CH(R^3){-}P({=}O)(OEt)_2$$

$$H_2C(R^3){-}P({=}O)(OEt)_2 \xrightarrow{NaH} Na^{\oplus}\ {}^{\ominus}CH(R^3){-}P({=}O)(OEt)_2$$

Fig. 6.48. Preparation of Horner–Wadsworth–Emmons reagents (synthetic applications: Section 11.3) by chemoselective acylation of a phosphonatestabilized "carbanion" with an ester.

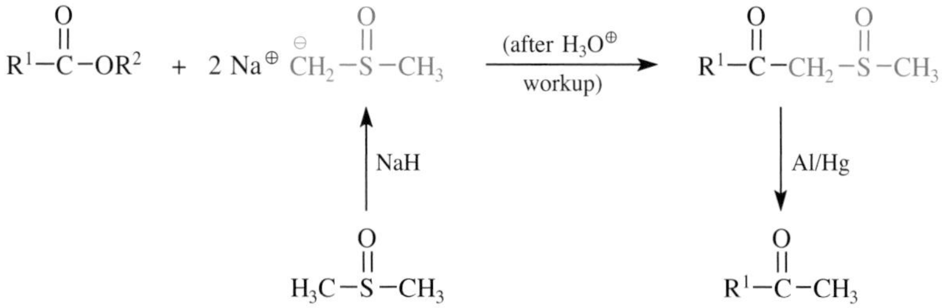

$$R^1{-}C({=}O){-}OR^2 + 2\,Na^{\oplus}\ {}^{\ominus}CH_2{-}S({=}O){-}CH_3 \xrightarrow[\text{workup)}]{\text{(after } H_3O^{\oplus}} R^1{-}C({=}O){-}CH_2{-}S({=}O){-}CH_3$$

$$H_3C{-}S({=}O){-}CH_3 \xrightarrow{NaH} Na^{\oplus}\ {}^{\ominus}CH_2{-}S({=}O){-}CH_3$$

$$R^1{-}C({=}O){-}CH_2{-}S({=}O){-}CH_3 \xrightarrow{Al/Hg} R^1{-}C({=}O){-}CH_3$$

Fig. 6.49. Preparation of methyl ketones by (1) chemoselective acylation of a sulfinylstabilized "carbanion," (2) reduction (mechanism: analogous to Figure 17.48).

6.5.4 Acylation of Organometallic Compounds and Heteroatom-Stabilized "Carbanions" with Carbonic Acid Derivatives: Synthesis of Carboxylic Acid Derivatives

Figure 6.50 displays acylations of C nucleophiles with NMe(OMe) derivatives of carbonic instead of carboxylic acids. The discussion of acylation reactions with NMe(OMe) derivatives of carboxylic acids ("Weinreb amides") in Figures 6.42 and 6.44 revealed that the NMe(OMe) group has two effects: first, it increases the reactivity and second, it is responsible for the occurrence of clean acylations. Against this background we will leave you to your own studies of a "picture without words," namely Figure 6.50. Convince yourself that the approaches **A** → **C**, **D** → **C** and **E** → **C** to Weinreb amides outlined in this figure work and find out why alternative ketone syntheses **D** → **H** and **E** → **H** are also possible!

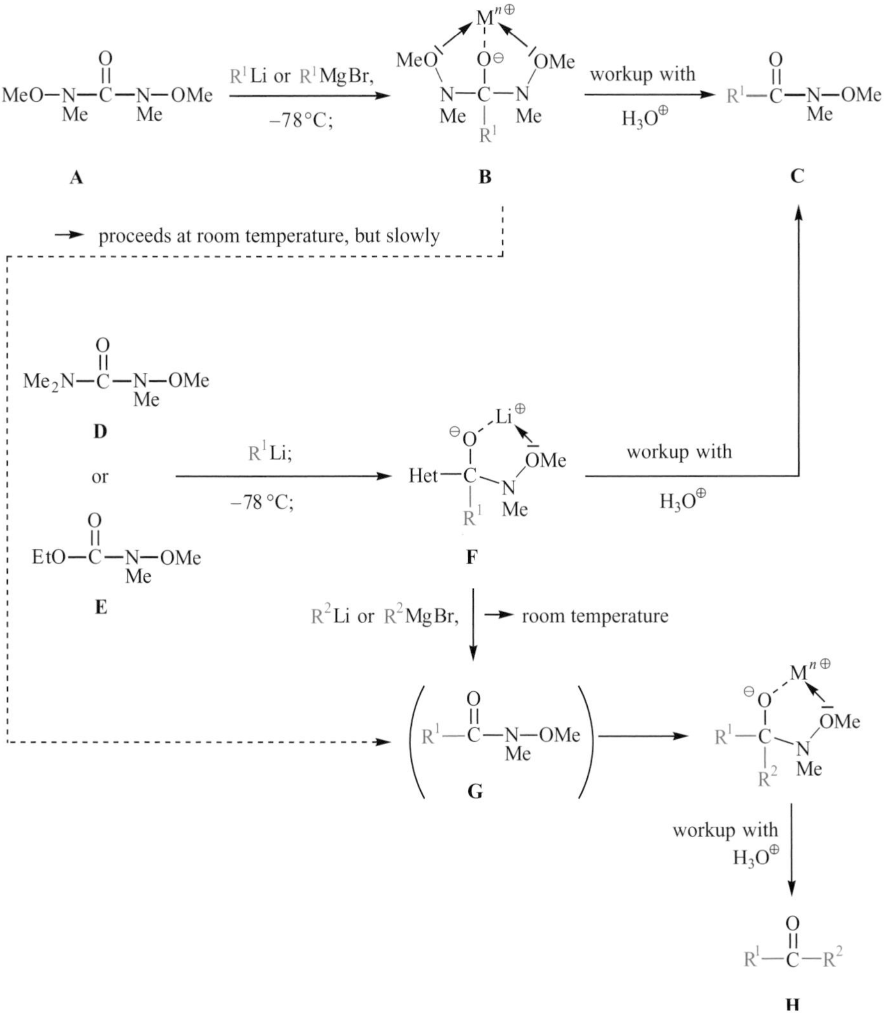

MeO–N(Me)–C(=O)–N(Me)–OMe **A** —[R^1Li or R^1MgBr, –78°C;]→ **B** (tetrahedral intermediate chelated by $M^{n\oplus}$) —[workup with $H_3O^{\oplus}$]→ R^1–C(=O)–N(Me)–OMe **C**

→ proceeds at room temperature, but slowly

Me_2N–C(=O)–N(Me)–OMe **D** or EtO–C(=O)–N(Me)–OMe **E** —[R^1Li; –78 °C;]→ Het–C(R^1)(O$^{\ominus}$···$Li^{\oplus}$)–N(Me)OMe **F** —[workup with $H_3O^{\oplus}$]→ **C**

F —[R^2Li or R^2MgBr, → room temperature]→ (R^1–C(=O)–N(Me)–OMe **G**) → R^1–C(R^2)(O$^{\ominus}$···$M^{n\oplus}$)–N(Me)OMe —[workup with $H_3O^{\oplus}$]→ R^1–C(=O)–R^2 **H**

Fig. 6.50. Acylation of organometallic compounds with various Weinreb amides (**A**, **D** and **E**) of carbonic acid. At low temperatures the Weinreb amides **C** of carboxylic acids are formed, higher temperatures lead to the formation of the ketones **H** with two different substituents. (The formation of these ketones is beyond the scope of Section 6.5.4 since intermediates **B** and **F** undergo further reactions before they are protonated by acidic workup.)

The Weinreb amide syntheses in Figure 6.50 proceeding via the stable tetrahedral intermediates **B** and **F** are chemoselective S_N reactions at the carboxyl carbon atom of carbon acid derivatives that are based on strategy 1 of the chemistry of carboxylic acid derivatives outlined in Figure 6.41. Strategy 2 of the chemistry of carboxylic acid derivatives in Figure 6.41 also has a counterpart in carbon acid derivatives, as is demonstrated by a chemoselective acylation of an organolithium compound with chloroformic acid methyl ester in this chapter's final example:

Zutropfen in Überschuss von Cl–C(=O)OMe

Oct–≡–Li → Oct–≡–C(=O)OMe

References

6.1

R. Sustmann, H.-G. Korth, "Carboxylic Acids," in *Methoden Org. Chem.* (Houben-Weyl) 4th ed. **1952**–, *Carboxylic Acids and Carboxylic Acid Derivatives* (J. Falbe, Ed.), Vol. E5, 193, Georg Thieme Verlag, Stuttgart, **1985**.

R. Sustmann, H.-G. Korth, "Carboxylic Acid Salts," in *Methoden Org. Chem.* (Houben-Weyl) 4th ed. **1952**–, *Carboxylic Acids and Carboxylic Acid Derivatives* (J. Falbe, Ed.), Vol. E5, 470, Georg Thieme Verlag, Stuttgart, **1985**.

R. Sustmann, H. G. Korth, "Carboxylic Acid Chlorides," in *Methoden Org. Chem.* (Houben-Weyl) 4th ed. **1952**–, *Carboxylic Acids and Carboxylic Acid Derivatives* (J. Falbe, Ed.), Vol. E5, 587, Georg Thieme Verlag, Stuttgart, **1985**.

M. A. Ogliaruso, J. F. Wolfe, "Carbocylic Acids," in *Comprehensive Organic Functional Group Transformations* (A. R. Katritzky, O. Meth-Cohn, C. W. Rees, Eds.), Vol. 5, 23, Elsevier Science, Oxford, U. K., **1995**.

6.2

R. S. Brown, A. J. Bennet, H. Slebocka-Tilk, "Recent Perspectives Concerning the Mechanism of $H_3O^{\oplus}$- and $OH^{\ominus}$-Promoted Amide Hydrolysis," *Acc. Chem. Res.* **1992**, *25*, 481–488.

E. Juaristi, G. Cuevas, "Recent Studies on the Anomeric Effect," *Tetrahedron* **1992**, *48*, 5019–5087.

A. J. Kirby (Ed.), "Stereoelectronic Effects," Oxford University Press, Oxford, U. K., **1996**.

C. L. Perrin, "Is There Stereoelectronic Control in Formation and Cleavage of Tetrahedral Intermediates?," *Acc. Chem. Res.* **2002**, *35*, 28–34.

6.3

K. B. Wiberg, "The Interaction of Carbonyl Groups with Substituents," *Acc. Chem. Res.* **1999**, *32*, 922–929.

R. Sustmann, "Synthesis of Acid Halides, Anhydrides and Related Compounds," in *Comprehensive Organic Synthesis* (B. M. Trost, I. Fleming, Eds.), Vol. 6, 301, Pergamon Press, Oxford, **1991**.

P. Strazzolini, A. G. Giumanini, S. Cauci, "Acetic Formic Anhydride," *Tetrahedron* **1990**, *46*, 1081–1118.

A. A. Bakibayev, V. V. Shtrykova, "Isoureas: Synthesis, Properties, and Applications," *Russ. Chem. Rev.* **1995**, *64*, 929–938.

A. R. Katritzky, X. Lan, J. Z. Yang, O. V. Denisko, "Properties and Synthetic Utility of N-Substituted Benzotriazoles," *Chem. Rev.* **1998**, *98*, 409–548.

G. Höfle, W. Steglich, H. Vorbrüggen, "4-Dialkylaminopyridines as Highly Active Acylation Catalysts," *Angew. Chem. Int. Ed. EnGl.* **1978**, *17*, 569–583.

U. Ragnarsson, L. Grehn, "Novel Amine Chemistry Based on DMAP-Catalyzed Acylation," *Acc. Chem. Res.* **1998**, *31*, 494–501.

R. Murugan, E. F. V. Scriven, "Applications of Dialkylaminopyridine (DMAP) Catalysts in Organic Synthesis," *Aldrichim. Acta* **2003**, *36,* 21–27.

A. R. Katritzky, S. A. Belyakov, "Benzotriazole-Based Intermediates: Reagents for Efficient Organic Synthesis," *Aldrichim. Acta*, **1998**, *31*, 35–45.

V. F. Pozdnev, "Activation of Carboxylic Acids by Pyrocarbonates. Scope and Limitations," *Org. Prep. Proced. Int.* **1998**, *30*, 631–655.

H. A. Staab, "Syntheses Using Heterocyclic Amides (Azolides)," *Angew. Chem. Int. Ed. EnGl.* **1962**, *1*, 351–367.

H. A. Staab, H. Bauer, K. M. Schneider, "Azolides in Organic Synthesis and Biochemistry," Wiley, New York, **1998**.

F. Albericio, R. Chinchilla, D. J. Dodsworth, C. Najera, "New Trends in Peptide Coupling Reagents," *Org. Prep. Proced. Int.* **2001**, *33*, 203–303.

6.4

U. Ragnarsson, L. Grehn, "Novel Gabriel Reagents," *Acc. Chem. Res.* **1991**, *24*, 285.

C. Salomon, E. G. Mata, "Recent Developments in Chemical Deprotection of Ester Functional Groups," *Tetrahedron* **1993**, *49*, 3691.

J. Mulzer, "Synthesis of Esters, Activated Esters and Lactones," in *Comprehensive Organic Synthesis* (B. M. Trost, I. Fleming, Eds.), Vol. 6, 323, Pergamon Press, Oxford, **1991**.

R. Sustmann, H. G. Korth, "Protecting Groups for Carboxylic Acids," in *Methoden Org. Chem.* (Houben-Weyl) 4th ed. **1952**–, *Carboxylic Acids and Carboxylic Acid Derivatives* (J. Falbe, Ed.), Vol. E5, 496, Georg Thieme Verlag, Stuttgart, **1985**.

E. Haslam, "Recent Developments in Methods for the Esterification and Protection of the Carboxyl Group," *Tetrahedron* **1980**, *36*, 2409.

J. Otera, "Transesterification," *Chem. Rev.* **1993**, *93*, 1449–1470.

N. F. Albertson, "Synthesis of Peptides with Mixed Anhydrides," *Org. React.* **1962**, *12*, 157–355.

R. C. Sheppard, "Peptide Synthesis," in *Comprehensive Organic Chemistry* (E. Haslam, Ed.), **1979**, *5* (Biological Compounds), 321–366, Pergamon, Oxford, U.K.

J. Jones, "The Chemical Synthesis of Peptides," Clarendon Press, Oxford, U. K., **1991**.

J. Jones, "Amino Acid and Peptide Synthesis (Oxford Chemistry Primers. 7)," Oxford University Press, Oxford, U. K., **1992**.

G. A. Grant, "Synthetic Peptides: A User's Guide," Freeman, New York, **1992**.

K.-H. Altmann, M. Mutter, "Die chemische Synthese von Peptiden und Proteinen," *Chem. unserer Zeit*, **1993**, *27*, 274–286.

M. Bodanszky, "Peptide Chemistry. A Practical Textbook," 2nd ed., Springer Verlag, Berlin, **1993**.

M. Bodanszky, "Principles of Peptide Synthesis,"2nd ed., Springer Verlag, Berlin, **1993**.

C. Basava, G. M. Anantharamaiah (Eds.), "Peptides: Design, Synthesis and Biological Activity," Birkhaeuser, Boston, **1994**.

M. Bodanszky, A. Bodanszky, "The Practice of Peptide Synthesis,"2nd ed., Springer Verlag, Heidelberg, **1994**.

L. A. Carpino, M. Beyermann, H. Wenschuh, M. Bienert, "Peptide Synthesis via Amino Acid Halides," *Acc. Chem. Res.* **1996**, *29*, 268-274.

G. Jung, A. G. Beck-Sickinger, "Multiple Peptide Synthesis Methods and their Applications," *Angew. Chem., Int. Ed. Engl.* **1992**, *31*, 367.

P. Lloyd-Williams, F. Albericio, E. Giralt, "Convergent Solid-Phase Peptide Synthesis," *Tetrahedron* **1993**, *49*, 11065–11133.

Y. Okada, "Synthesis of Peptides by Solution Methods," *Curr. Org. Chem.* **2001**, *5*, 1–43.

S. Aimoto, "Contemporary Methods for Peptide and Protein Synthesis," *Curr. Org. Chem.* **2001**, *5*, 45–87.

C. Najera, "From α-Amino Acids to Peptides: All You Need for the Journey," *Synlett* **2002**, 1388–1403.

W. C. Chan, P. D. White (Eds.), "FMOC Solid-Phase Peptide Synthesis: A Practical Approach," Oxford University Press, Oxford, U. K., **2000**.

T. A. Ryan, "Phosgene and Related Compounds," Elsevier Science, New York, **1996**.

L. Cotarca, P. Delogu, A. Nardelli, V. Sunjic, "Bis(trichloromethyl) Carbonate in Organic Synthesis," *Synthesis* **1996**, 553–576.

C. Agami, F. Couty, "The Reactivity of the N-Boc Protecting Group: An Underrated Feature," *Tetrahedron* **2002**, *58,* 2701–2724.

6.5

M. P. Sibi, "Chemistry of N-Methoxy-N-Methylamides. Applications in Synthesis," *Org. Prep. Proced. Int.* **1993**, *25*, 15–40.

J. L. Romine, "Bis-Protected Hydroxylamines as Reagents in Organic Synthesis. A Review," *Org. Prep. Proced. Int.* **1996**, *28*, 249–288.

G. Benz, K.-D. Gundermann, A. Ingendoh, L. Schwandt, "Preparation of Aldehydes by Reduction," in *Methoden Org. Chem.* (Houben-Weyl) 4th ed., **1952**–, *Aldehydes* (J. Falbe, Ed.), Vol. E3, 418, Georg Thieme Verlag, Stuttgart, **1983**.

E. Mosettig, "The Synthesis of Aldehydes from Carboxylic Acids," *Org. React.* **1954**, *8*, 218–257.

J. S. Cha, "Recent Developments in the Synthesis of Aldehydes by Reduction of Carboxylic Acids and their Derivatives with Metal Hydrides," *Org. Prep. Proced. Int.* **1989**, *21*, 451– 477.

R. A. W. Johnstone, "Reduction of Carboxylic Acids to Aldehydes by Metal Hydrides," in *Comprehensive Organic Synthesis* (B. M. Trost, I. Fleming, Eds.), Vol. 8, 259, Pergamon Press, Oxford, **1991**.

A. P. Davis, "Reduction of Carboxylic Acids to Aldehydes by Other Methods," in *Comprehensive Organic Synthesis* (B. M. Trost, I. Fleming, Eds.), Vol. 8, 283, Pergamon Press, Oxford, **1991**.

B. T. O'Neill, "Nucleophilic Addition to Carboxylic Acid Derivatives," in *Comprehensive Organic Synthesis* (B. M. Trost, I. Fleming, Eds.), Vol. 1, 397, Pergamon Press, Oxford, **1991**.

D. A. Shirley, "The Synthesis of Ketones from Acid Halides and Organometallic Compounds of Magnesium, Zinc, and Cadmium," *Org. React.* **1954**, *8*, 28–58.

M. J. Jorgenson, "Preparation of Ketones from the Reaction of Organolithium Reagents with Carboxylic Acids," *Org. React.* **1970**, *18*, 1–98.

R. K. Dieter, "Reaction of Acyl Chlorides with Organometallic Reagents: A Banquet Table of Metals for Ketone Synthesis," *Tetrahedron* **1999**, *55*, 4177–4236.

W. E. Bachmann, W. S. Struve, "The Arndt-Eistert Reaction," *Org. React.* **1942**, *1*, 38–62.

Further Reading

M. Al-Talib, H. Tashtoush, "Recent Advances in the Use of Acylium Salts in Organic Synthesis," *Org. Prep. Proced. Int.* **1990**, *22,* 1–36.

S. Patai, (Ed.), "The Chemistry of Ketenes, Allenes, and Related Compounds," Wiley, New York, **1980**.

H. R. Seikaly, T. T. Tidwell, "Addition Reactions of Ketenes," *Tetrahedron* **1986**, *42*, 2587.

P. W. Raynolds, "Ketene," in *Acetic Acid and Its Derivatives* (V. H. Agreda, J. R. Zoeller, Eds.), 161, Marcel Dekker, New York, **1993**.

A.-A. G. Shaikh, S. Sivaram, "Organic Carbonates," *Chem. Rev.* **1996**, *96*, 951–976.

Y. Ono, "Dimethyl Carbonate for Environmentally Benign Reactions," *Pure Appl. Chem.* **1996**, *68*, 367–376.

P. Tundo, M. Selva, "The Chemistry of Dimethyl Carbonate," *Acc. Chem. Res.* **2002**, *35,* 706–716.

V. F. Pozdnev, "Activation of Carboxylic Acids by Pyrocarbonates. Scope and Limitations. A Review," *Org. Prep. Proced. Int.* **1998**, *30*, 631–655.

7 Carboxylic Compounds, Nitriles, and Their Interconversion

The nitriles R–C≡N are anhydrides of the primary amides of carboxylic acids R–C(=O)–NH_2. As the bond enthalpy values show the dehydration of R–C(=O)–NH_2 → R–C≡N + H_2O is associated with a enthalpy change of +19 kcal/mol. The dehydration of carboxylic acids (R–C(=O)–OH → R–C(=O)–O–C(=O)–R + H_2O), leads to the formation of carboxylic acid anhydrides. This process is only half as endothermic (ΔH = +9 kcal/mol), due to the fact that the dehydration of an amide is accompanied by a loss of resonance energy of as much as 22 kcal/mol (Table 6.1, entry 2), whereas the loss from carboxylic acid is about 14 kcal/mol (Table 6.1, entries 3 and 9). The thermal formation of nitriles R–C(=O)–NH_2 → R–C≡N thus requires even higher temperatures than the thermal formation of anhydrides 2 R–C(=O)–OH → R–C(=O)–O–C(=O)–R. Actually, the dehydration of carboxylic acid amides producing the corresponding nitriles is not usually performed by heating (example: Figure 7.6), but under the influence of dehydrating reagents (examples: Figures 7.2–7.4; also cf. Figure 7.7).

The reversal of the endothermic dehydration R–C(=O)–NH_2 → R–C≡N + H_2O is the exothermic hydration R–C≡N + H_2O → R–C(=O)–NH_2 (see Figure 7.1 above). This correlation between nitriles and carboxylic acid derivatives highlighted by this pair of reactions continues, as illustrated in Figure 7.1, starting with the vertical reaction arrow. Both HO–H and heteroatom nucleophiles Het^1–H may be added to the C≡N bond of nitriles. The resulting adduct is a carboxylic acid derivative **A** and/or a tautomer **B**. **A** is the derivative of an imidic acid R–C(NH)–OH, whereas **B** is either a carboxylic acid amide (Het^1 = O), or a thiocarboxylic acid amide (Het^1 = S), or an amidine (Het^1 = NR). The carboxylic acid derivatives **A** and/or **B** in Figure 7.1 may act as an acylating agent towards the nucleophile from which

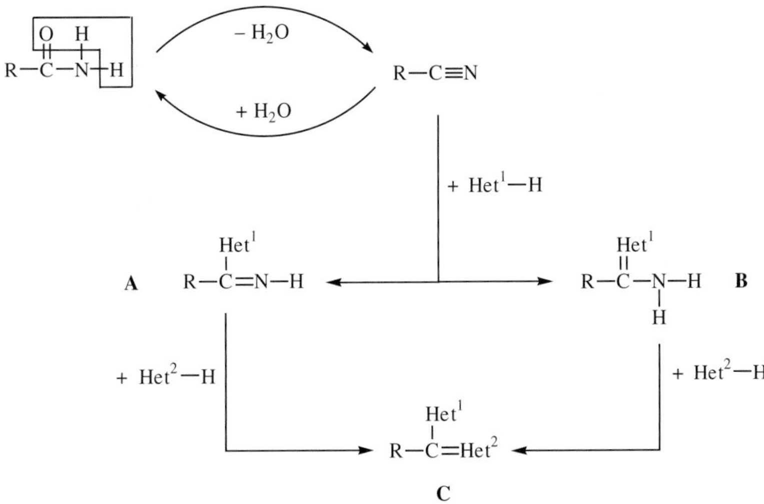

Fig. 7.1 Survey of the mutual interconversion of carboxylic acid derivatives, especially primary amides of carboxylic acids, and nitriles.

Bruckner R (author), Harmata M (editor) In: *Organic Mechanisms – Reactions, Stereochemistry and Synthesis*
Chapter DOI: 10.1007/978-3-642-03651-4_7, © Springer-Verlag Berlin Heidelberg 2010

A and/or **B** emerged by addition to the nitrile. In such cases, a new carboxylic acid derivative **C** is produced. On the other hand, **A** and/or **B** may also be used to specifically induce acylation of another heteroatom nucleophile (Het^2–H), which also gives a new carboxylic acid derivative **C**. The mechanisms underlying these "acylations" producing **C** are identical or similar to the ones introduced in Chapter 6. Note that we refer here to both R–C(=O) and the analogous R–C(=S) and R–C(=NR) transformations. Though the latter two are not discussed at length, their structures are shown and it should be realized that their chemistry is often mechanistically analogous to systems we have discussed (Figure 7.1).

As you may have suspected, the correlation between nitriles and carboxylic acid derivatives suggests that the chemistry of the former should be taught from this perspective. We will pursue this approach in this chapter.

7.1 Preparation of Nitriles from Carboxylic Acid(Derivative)s

This section presents five variations of the dehydration of primary amides to nitriles outlined in Figure 7.1. We start with three "chemical" dehydrations using phosphorus pentoxide (Figure 7.2), trifluoroacetic acid anhydride (Figure 7.3) and trifluoromethanesulfonic acid anhydride (Figure 7.4), followed by two "thermal" dehydrations. In the latter two processes, the primary amide (Figure 7.6) and acylurea (Figure 7.7) that are dehydrated are generated *in situ* from the corresponding carboxylic acid and ammonia or urea, respectively.

Figure 7.2 illustrates the phosphorus pentoxide-mediated dehydration of a primary amide to a nitrile, using the transformation of nicotine amide (**A**) into nicotine nitrile (**B**) as an example. The reaction of phosphorus pentoxide at the carboxyl oxygen furnishes the partially ring-opened iminium ion **E** (simplified as: **F**) via the polycyclic iminium ion **C**. **E** is deprotonated to give the mixed anhydride **G** from imidic acid and phosphoric acid. **Imidic acids** are characterized by the functional group R–C(=NH)–OH. This anhydride is transformed into the nitrile **B** by an E1 elimination via the intermediate nitrilium salt **D**. **Nitrilium salts** are *N*-protonated or *N*-alkylated nitriles.

The reaction of carboxylic acid derivatives of the type R–C(=O)–Het with electrophiles (abbreviated as $E^{\oplus}$) such as phosphorus pentoxide generally occurs at their doubly bound oxygen atom and almost never at their singly bound heteroatom. A resonance-stabilized cation R–C(=$O^{\oplus}$–E)–Het R–C(–O–E)=$Het^{\oplus}$ is formed only in the first case. You have already come across this *O*-selectivity of electrophilic reactions on carboxylic acid derivatives in earlier chapters of this book, though without this special emphasis. These included protonations (Figures 6.5, 6.7, 6.9, 6.22, 6.23; also see Figure 5.32), complexations with Lewis acids (Figure 5.32) as well as *O*-functionalizations by $POCl_3$ (Figure 5.36), the Vilsmeier reagent (Figure 6.11) or 2-chloro-*N*-methylpyridinium iodide (Figure 6.16). The carboxyl oxygen of carboxylic acids is also targeted by the carboxonium ion **D** in Figure 9.15.

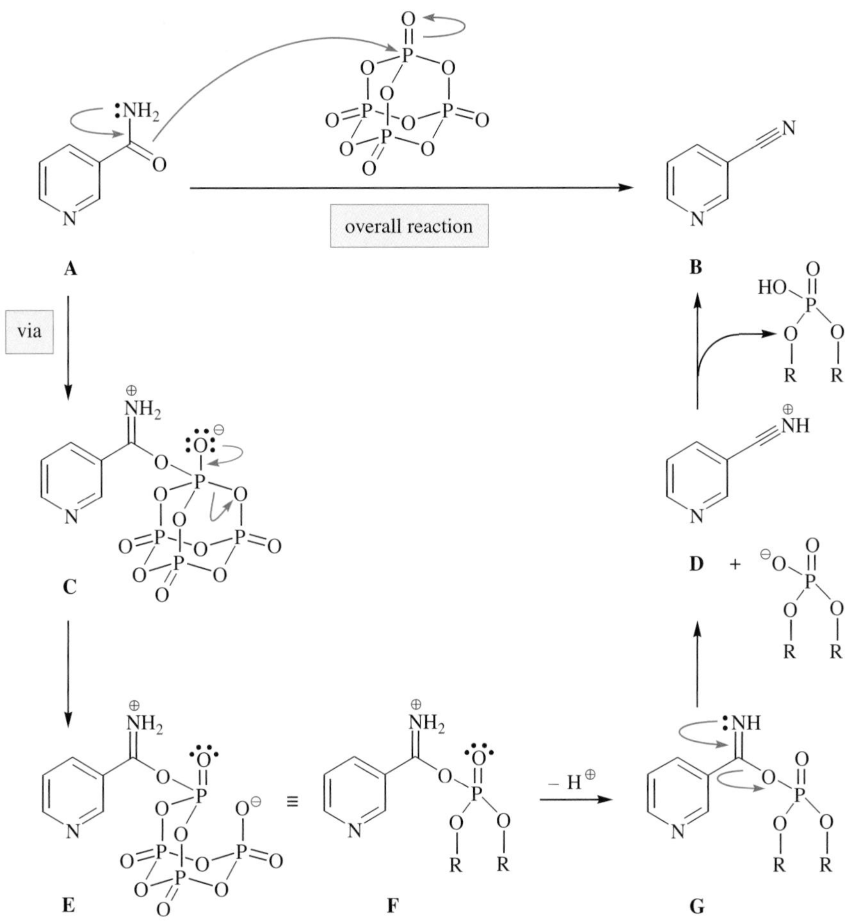

Fig. 7.2. Phosphorus pentoxide-mediated dehydration of nicotinic acid amide (**A**) to nicotinic nitrile (**B**) with the reagent forming polyphosphoric acid $(HPO_3)_x$.

Consequently, all "chemical" dehydrations of a primary amide are initiated by an *O*-selective functionalization, including dehydrations with phosphorus pentoxide (Figure 7.2), trifluoroacetic acid anhydride (Figure 7.3) or trifluoromethanesulfonic acid anhydride (Figure 7.4). One other fact makes learning easier: *for the key intermediate* **G** *of the amide dehydration in Figure 7.2 there is an exact structural analog in each of the alternative methods discussed in this section.* So, the imidic acid/phosphoric acid anhydride **G** in Figure 7.2 corresponds to the imidic acid/trifluoroacetic acid anhydride **F** in Figure 7.3, the imidic acid/trifluoromethanesulfonic acid anhydride **F** in Figure 7.4 and the amide-substituted imidic acid **E** in Figure 7.6. In addition, there is another parallel feature: *all these intermediates are transformed into the nitrile by an E1 elimination via a nitrilium cation*. These recurring basic principles should help you understand and reproduce the details of the reactions given in Figures 7.2–7.4, 7.5 and 7.7. Also, these principles should enable you to write down the mechanisms of the "chemical" dehydrations of primary amides to nitriles using $POCl_3$, PCl_5 or $SOCl_2$, and indeed of any other reagent used to convert primary amides to nitriles. Let's give a cheer for inductive reasoning.

When amides are dehydrated using phosphorus pentoxide as a reagent (Figure 7.2) the solidification of the reaction mixture caused by the resulting polyphosphoric acid is often

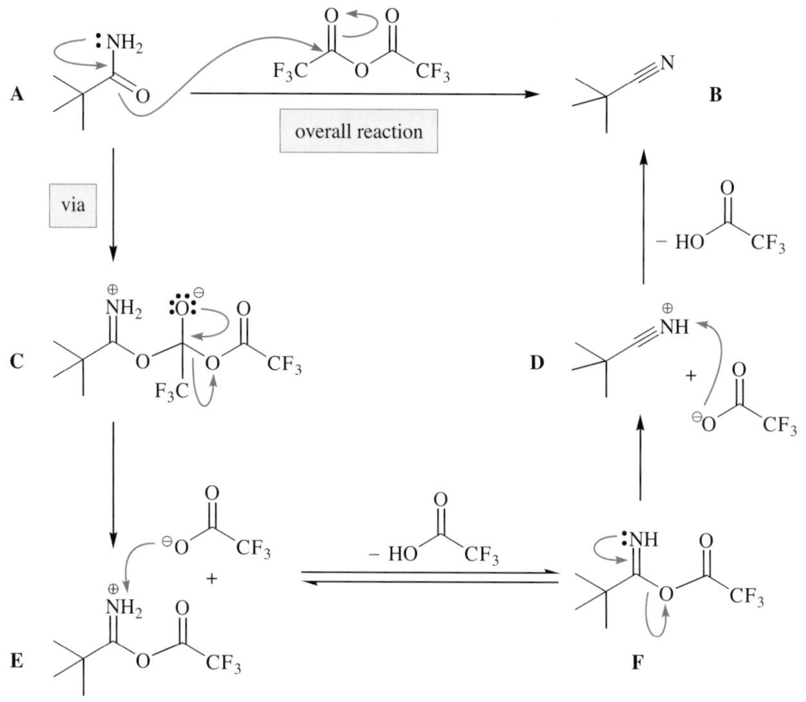

Fig. 7.3. Trifluoroacetic acid anhydride-mediated dehydration of pivalic acid amide (**A**) to pivalic acid nitrile (**B**) with the reagent forming trifluoroacetic acid $F_3C–CO_2H$.

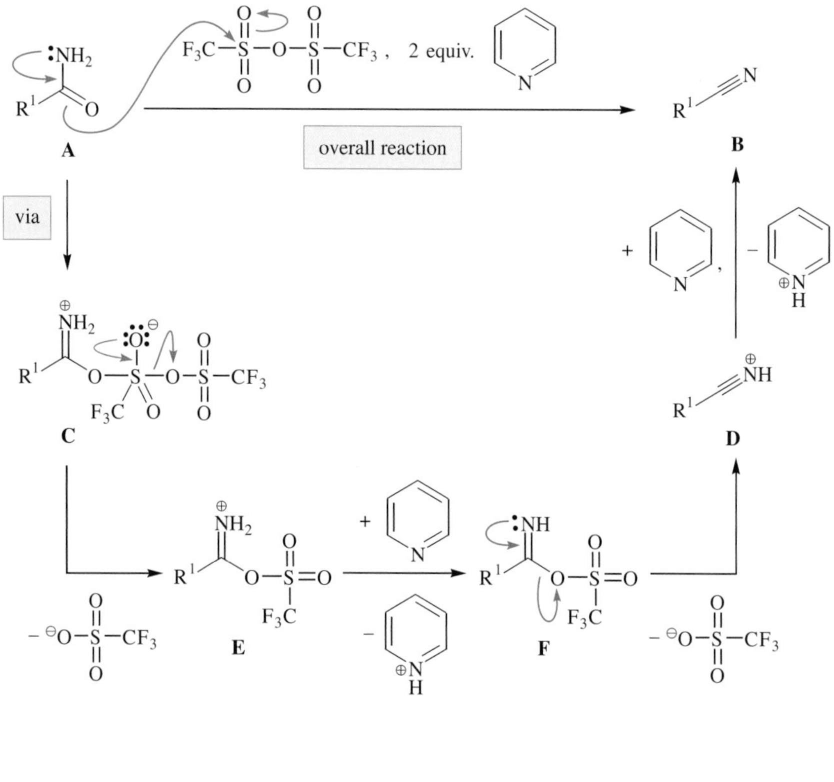

Fig. 7.4. Trifluoromethanesulfonic acid anhydride/pyridine-mediated dehydration of primary carboxylic acid amides (**A**) to give the corresponding nitriles (**B**) with the reagents converting into pyridinium trifluormethanesulfonate pyrH⊕ $F_3C–SO_3^{\ominus}$.

quite annoying. The dehydration is more comfortable and much easier to work up when trifluoroacetic acid anhydride is used, as is illustrated by the example of pivalic acid amide shown in Figure 7.3. In addition to the desired nitrile **B** trifluoroacetic acid is formed. The latter is volatile and can therefore be removed by distillation at the end of the reaction. Until the mixed anhydride **F** is formed, all steps of this reaction mechanistically resemble the formation of the analog anhydride **G** in Figure 7.2. The remaining steps also proceed quite similarly, i.e., as an E1 elimination: first, a trifluoroacetate ion and then a proton is released.

Figure 7.4 illustrates the dehydration of primary amides **A** to nitriles **B** using trifluoromethanesulfonic acid anhydride. All intermediates correspond to the ones discussed above. When the mixed anhydrides **F** finally release trifluoromethanesulfonic acid in two steps by the already familiar E1 mechanism, the formation of the nitrile **B** is completed.

Side Note 7.1. Activation of Tertiary Amides to Make Acylating Agents

Trifluoromethanesulfonic acid anhydride binds to the carboxyl oxygen of both tertiary amides (**A** in Figure 7.5) and of the primary amides just discussed (**A** in Figure 7.4). However, the

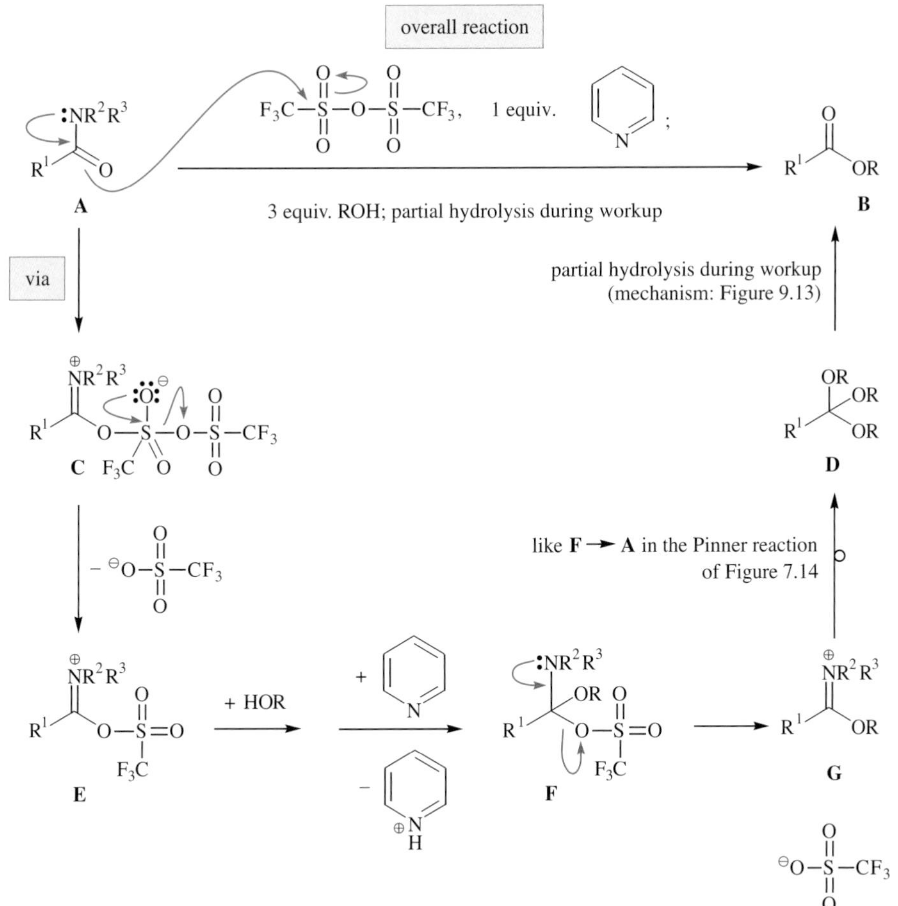

Fig. 7.5. Application of the reaction principle underlying Figure 7.4 for the conversion of tertiary carboxylic acid amides into acylating agents for alcohols. Very mild workup conditions lead to the orthoesters **D**, while the normal carboxylic acid esters **B** are obtained with aqueous standard workup.

iminium salts **E** resulting from the tertiary amides in Figure 7.5 cannot be deprotonated—rendering them stable until they react with an alcohol. The first equivalent of the alcohol is imidoylated to the iminium ion **G** via tetrahedral intermediate (**F**). This intermediate is converted into compound **D**. This reaction proceeds in several steps just as **F** → **A** in the Pinner reaction (see Figure 7.14). **D**, a so-called orthoester, can either be isolated as such or hydrolyzed to give a normal carboxylic acid ester **B** under mild conditions (mechanism: Figure 9.13). In the latter case, the overall process is an esterification of an alcohol by a tertiary amide.

The "chemical" amide dehydrations of Figures 7.2–7.4 are unsuitable for industrial-scale applications. The occurrence of stoichiometric amounts of acidic side products are a big problem when it comes down to cheap and ecologically sound production processes. For industrial applications, it is therefore important that the dehydration of primary amides to nitriles can also be performed under purely thermal conditions. Since this transformation is strongly endothermic (see above) high yields can only be achieved at really high temperatures, in accord with LeChatelier's principle. Figure 7.6 illustrates how such a thermolysis can be employed to transform adipic acid diamide (**D**) into adipic acid dinitrile (**C**). Success is guaranteed only by a tautomerization of the amide **D** to the amide-substituted imidic acid **E** that proceeds to just a minor extent. Due to the very high temperature the imidic acid **E** is subject to the same E1 elimination that we came across in Figures 7.2–7.4, the only difference being that here the hydroxide ion acts as a leaving group. This is a relatively poor leaving group that

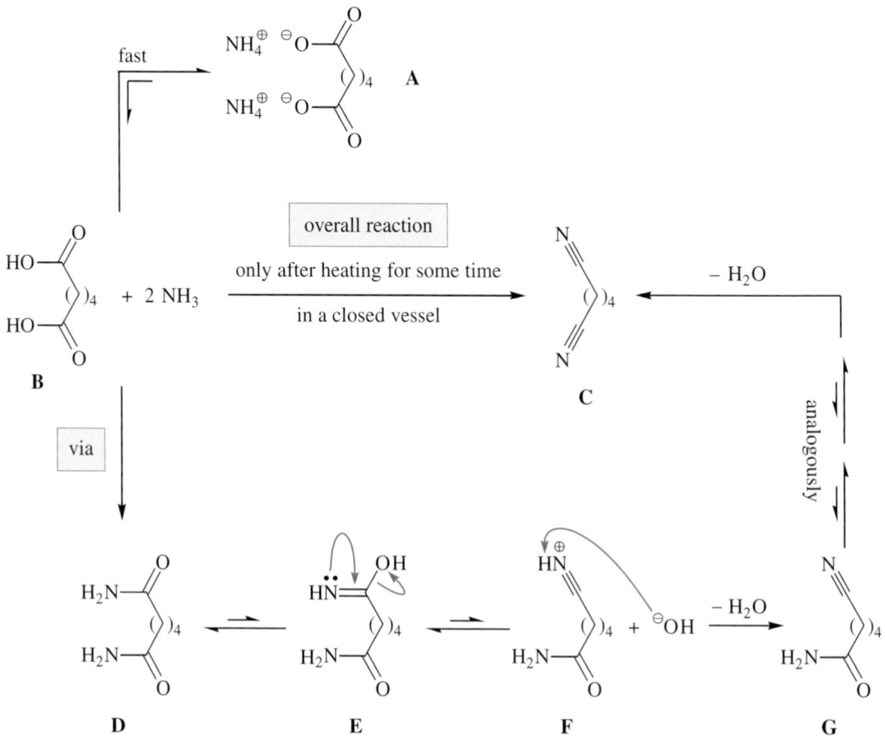

fast; A; overall reaction; only after heating for some time; in a closed vessel; $+ 2\ NH_3$; $- H_2O$; B; C; via; analogously; D; E; F; G

Fig. 7.6. Industrial-scale processing: condensation of adipic acid (**B**) and ammonia to adipic acid diamide (**D**) and its thermal dehydration to adipic acid dinitrile (**C**). On a very large scale, adipic acid dinitrile is hydrated to give 1,6-diaminohexane ("hexamethylenediamine"), which is further processed to furnish nylon-6,6 ("nylon").

at low temperatures cannot be eliminated nearly as well as the polyphosphate, the trifluoroacetate or the trifluoromethanesulfonate in the nitrile formations of Figure 7.2–7.4. At the much higher reaction temperature in Figure 7.6, the elimination works. In this way, the amide-substituted nitrile **G** is formed via the amide-substituted nitrilium cation **F**. The dehydration of the remaining amide group proceeds as previously described for the first amide group.

Actually, the technical synthesis of adipic acid dinitrile does not *begin with*, but *proceeds via* adipic acid diamide (**D**). In fact, it starts with the bis(ammoniumcarboxylate) **A** obtained from adipic acid (**B**) and two equivalents of ammonia. Upon heating in a closed reactor small amounts of the carboxylic acid and ammonia are regenerated from the salt **A** and react according to the $B_{AC}2$ mechanism (Figure 6.24) to give diamide **D**.

The synthesis of adipic acid dinitrile outlined in Figure 7.7 proceeds under milder conditions than the one presented in Figure 7.6. Here, the starting material includes adipic acid (**B**) and urea (**C**). Since urea can only be protonated by mineral acids, the formation of the ammonium salt **A** is impossible. This means that the entire reactor space is available for the initial

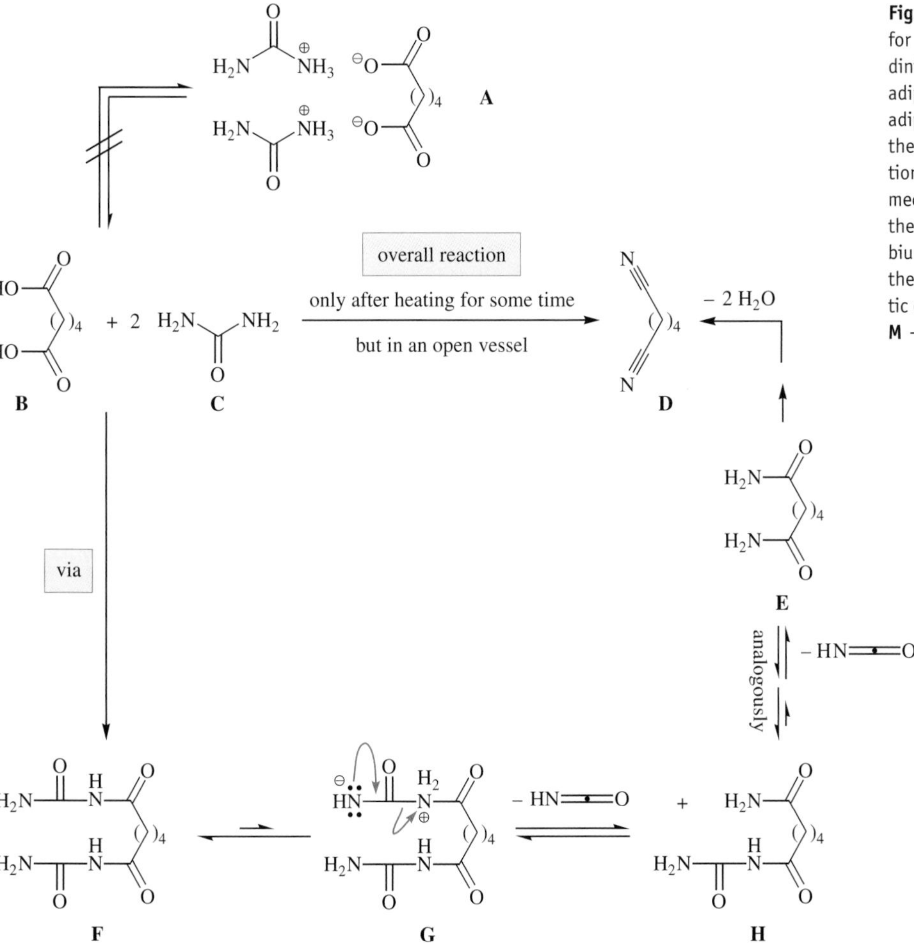

Fig. 7.7. Laboratory method for the synthesis of adipic acid dinitrile (**D**). Condensation of adipic acid (**B**) and urea to give adipic acid diamide (**E**); and its thermal dehydration. The reaction **F** → **H**—as well as its mechanism—corresponds to the reversal of the formation of biuret that takes place upon the heating of urea (mechanistic details: conversion of **M** → **O** in Figure 8.3).

acylation **B** + **C** → **F**, and not just a small equilibrium fraction as described in the procedure of Figure 7.6. The acylation product, bisacylurea **F**, is in equilibrium with the zwitterion **G**. Such zwitterions are familiar from many carbonic acid derivatives, and we know that they initiate their decomposition into a heterocumulene and a heteroatom nucleophile (see Figure 8.3). In accordance with these experiences the zwitterion **G** in Figure 7.7 decomposes into isocyanic acid (a heterocumulene) and the amide **H** (a heteroatom nucleophile). When the acylurea unit still present in **H** similarly expels isocyanic acid for the second time, adipic acid diamide (**E**) is formed. At this point, we have arrived at the reaction channel of Figure 7.6 from which the remaining path to adipic acid dinitrile can be followed.

7.2 Transformation of Nitriles and Heteroatom Nucleophiles to Carboxylic Acid (Derivative)s

The survey in Figure 7.1 has already shown that nucleophiles can add to nitriles and thus produce carboxylic acid derivatives. Water as a nucleophile either undergoes clean addition to nitriles, and thus produces primary amides (corresponding to a "partial hydrolysis of the nitrile"), or it continues with hydrolyzing the amides *in situ* to give carboxylic acid or carboxylate, which would amount to a "total hydrolysis of the nitrile":

$$R{-}C{\equiv}N \xrightarrow[\text{relatively fast}]{+\,H_2O,\ \text{cat. amounts of } H^{\oplus} \text{ or cat. amounts of } OH^{\ominus}} R{-}\overset{O}{\overset{\|}{C}}{-}NH_2 \xrightarrow[\text{relatively slowly}]{+\,H_2O,\ \text{cat. amounts of } H^{\oplus} \text{ or cat. amounts of } OH^{\ominus}} R{-}\overset{O}{\overset{\|}{C}}{-}OH$$

$$R{-}C{\equiv}N \xrightarrow[\text{relatively slowly}]{+\,2\,H_2O,\ \text{cat. amounts of } H^{\oplus} \text{ or cat. amounts of } OH^{\ominus}} R{-}\overset{O}{\overset{\|}{C}}{-}OH$$

The partial hydrolysis of a nitrile is impossible under neutral conditions, but can be managed under both acidic and basic conditions. This pH dependency is even more notable with total nitrile hydrolysis, which proceeds much more slowly than the partial hydrolysis. The reason is that the hydrolysis amide → carboxylic acid or amide → carboxylate can only be performed under *strongly* acidic or *strongly* basic conditions. Thus the partial hydrolyses nitrile → amide can be conducted chemoselectively without leading to an over-reaction. If the total nitrile hydrolysis is targeted, the choice generally is between a stepwise approach or a one-pot reaction. The latter is common (Figure 7.9) but not always possible (Figure 7.8). The reaction **B** → **C** in Figure 7.8 gives an example of a partial nitrile hydrolysis under acidic conditions.

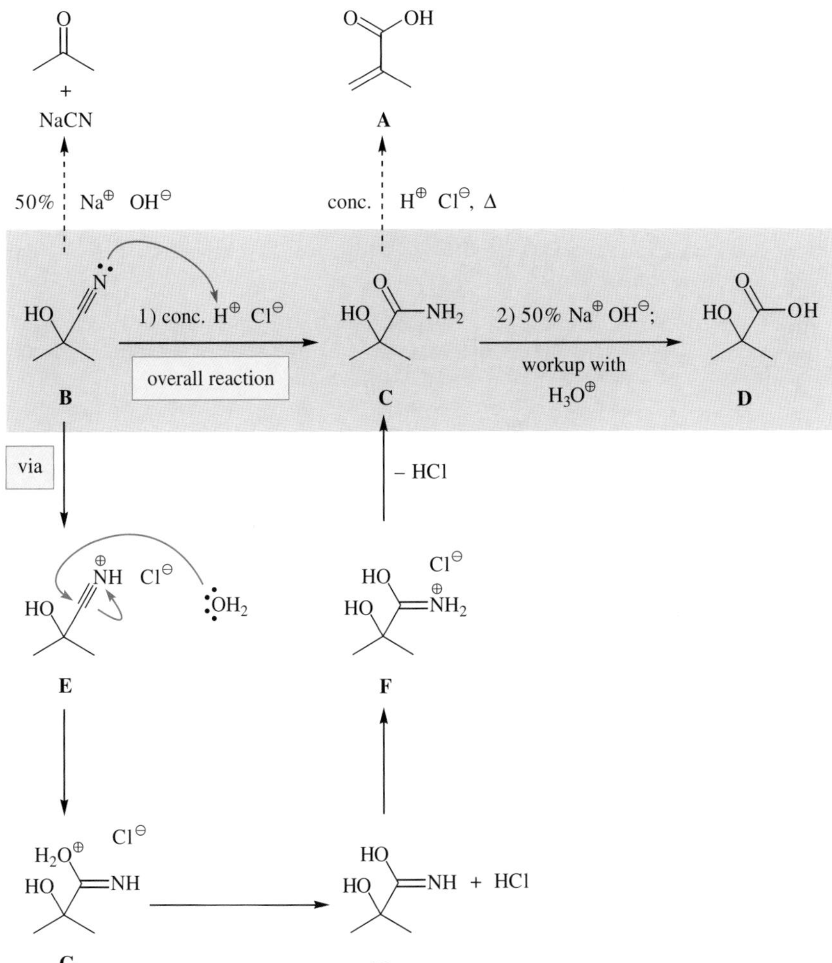

Fig. 7.8. Partial hydrolysis of a nitrile to a primary carboxylic acid amide under acidic conditions initiating the two-step total hydrolysis of acetone cyanohydrin (**B**). Grey background: overall reaction; above the grey area: competing reactions with inappropriate choice of pH; below the grey area: mechanistic details.

To get the complete picture, Figure 7.9 shows both a partial nitrile hydrolysis under basic conditions (transformation **A** → **B)** and a total nitrile hydrolysis under basic conditions (transformation **A** → **(B)** → **C)**.

In the partial nitrile hydrolysis of Figure 7.8, acetone cyanohydrin (**B**; preparation: Section 9.1.3) is reacted with hydrochloric acid to form the corresponding amide **C**. Initially, the nitrile nitrogen is protonated to the nitrilium cation **E** where the electrophilicity of the nitrile carbon atom is high enough to allow the addition of water. By deprotonation the primary addition product **G** is transformed into imidic acid **H,** which itself is protonated (→ **F**) and deprotonated to afford the isolable amide **C**. The partial hydrolysis of acetone cyanohydrin (**B**) is impossible under basic conditions: At high pH values acetone cyanohydrin (**B**)—in a reversal of its formation reaction—decomposes into acetone and sodium cyanide (see Figure 9.9). If the hydrolysis of the nitrile in Figure 7.8 is to be completed, the amide **C** must be worked up and hydrolyzed to the carboxylate with concentrated sodium hydroxide via the $B_{AC}2$ mecha-

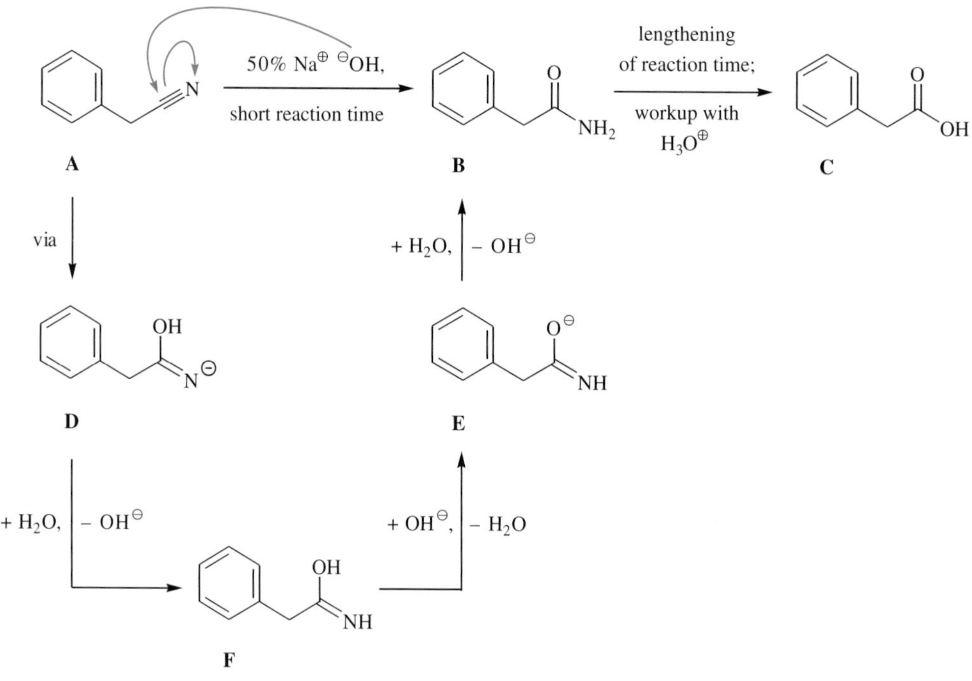

Fig. 7.9. Partial hydrolysis of phenylacetonitrile (**A**) under basic conditions to give the primary carboxylic acid amide **B**, and the possibility of hydrolyzing the latter to yield phenylacetic acid (**C**).

nism (Figure 6.24). Protonation gives the desired carboxylic acid **D**. Hydrolysis of the amide **C** under acidic conditions fails because of the E2 elimination of water so that methacrylic acid (**A**) would be produced instead of the acid-sensitive hydroxybutyric acid (**D**).

The total hydrolysis of phenylacetonitrile (Formula **A** in Figure 7.9) to phenylacetic acid (**C**) can be performed in one or two steps and both under acidic (following the mechanism in Figure 7.8) and basic conditions. In the latter case, the nucleophilic addition of a hydroxide ion to the nitrile carbon atom is the rate-determining step. This is how the imidic acid anion **D** is formed. Protonation (→ **F**), deprotonation (→ **E**) and reprotonation yield the amide **B**, which one can isolate or further hydrolyse under harsher conditions via the usual $B_{AC}2$ mechanism (cf. Figure 6.24).

The base-mediated partial hydrolysis of a nitrile → a primary amide is frequently accelerated by adding hydrogen peroxide. The underlying mechanism is shown in Figure 7.10, using the hydrolysis of benzonitrile (**A**) to benzamide (**B**) as an example. Hydrogen peroxide and the hydroxide ions are in equilibrium with water and with hydroperoxide ions. Due to the α-effect (Section 2.2), a hydroperoxide ion is a better nucleophile than a hydroxide ion. This is why the addition of the hydroperoxide ion to the C≡N triple bond (→ **D**) proceeds at a lower temperature than that of the hydroxide ion. Protonation of **D** yields the perimidic acid **E**. The name of this compound could lead us to ignore the fact that it would be deprotonated to an anion with localized charge, which is not particularly attractive from the energetic point of view. For this reason, the concentration of the nonionized perimidic acid **E** in the deprotonation equilibrium stays large enough to enable the nucleophilic replacement of the OOH- by the OH group. This substitution proceeds according to the $B_{AC}2$ mechanism via the tetrahedral intermediates **F** and **C** and yields the amide (**B**).

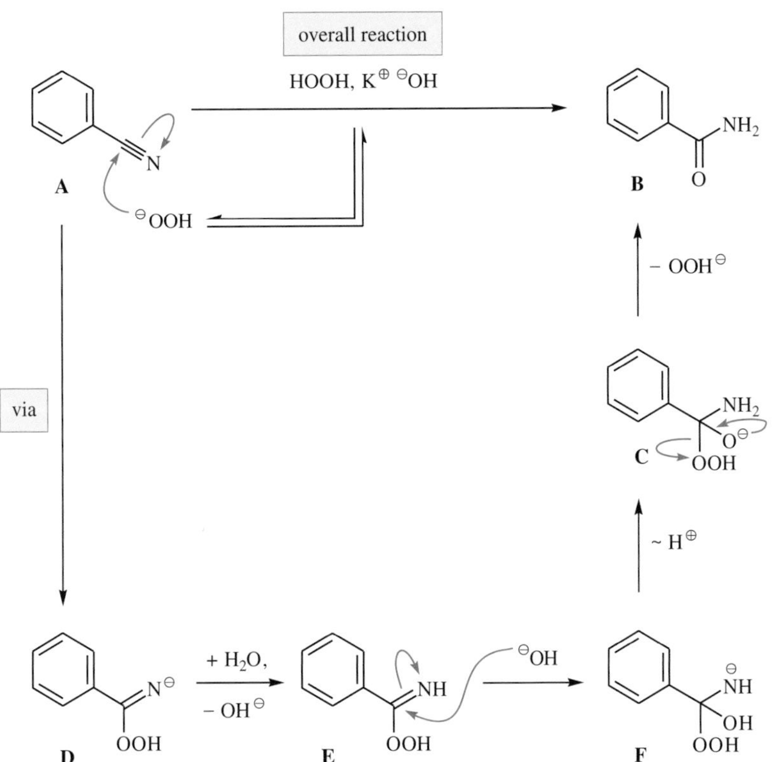

Fig. 7.10. Partial benzonitrile hydrolysis of (**A**) accelerated by hydrogen peroxide under basic conditions. The intermediate **E**, a perimidic acid, is as suitable for the epoxidation of standard alkenes as MCPBA or MMPP (cf. Section 3.19), but unlike the latter two is also capable of epoxidizing keto alkenes (which would be oxidized by MCPBA or MMPP in the ketonic substructure in a Baeyer–Villiger reaction; cf. Section 14.35).

Under especially mild conditions the total hydrolysis R–C≡N → R–CO_2H can be managed within the scope of the so-called **Bucherer modification of the Strecker synthesis of *α*-amino acids**. In Figure 7.11, the synthesis of racemic methionine (**D**) is given as an example. The substrates include *β*-(methylmercapto)propionaldehyde (**A**), ammonium hydrogen carbonate and sodium cyanide. In the first step of this reaction, they produce the heterocyclic amino acid derivative **B**, a so-called **hydantoin,** which in the second step undergoes hydrolysis with sodium hydroxide or barium hydroxide. A relatively fast hydrolysis occurs at the amide instead of the urea substructure of the heterocycle. This is a consequence of the weaker resonance stabilization of a C=O double bond that bears only one instead of two nitrogen substituents. The product of this hydrolysis is the anion **C** of a so-called **hydantoin acid**. The subsequent hydrolysis of its urea group proceeds more slowly, leaving *d,l*-methionine as a carboxylate (**E**) or—upon neutralization—as a betaine (**D**).

Figure 7.12 outlines how hydantoin is produced during the synthesis of methionine depicted in Figure 7.11. Initially, the aminonitrile **B** is formed. With its amino group it adds to carbon dioxide, which—under the weakly basic reaction conditions—is followed by the formation of the carbamate ion **D**. In a fully analogous fashion, ammonia reacts with carbon dioxide to give ammonium carbamate (cf. Section 8.2). The intermediate **D** in Figure 7.12 permits the intramolecular and thus kinetically favored—i.e., fast—addition of the oxyanion to the C≡N triple bond. The addition product is the negatively charged heterocycle **F**, which in

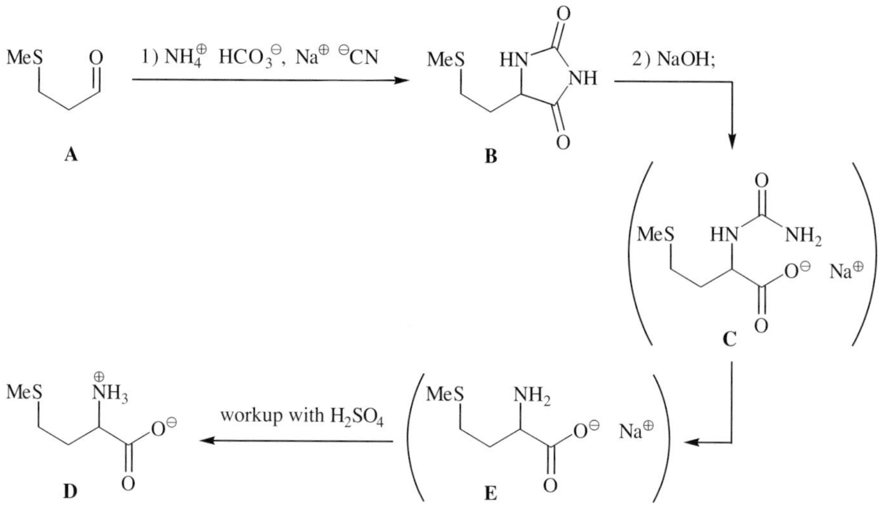

MeS O 1) $NH_4^{\oplus}$ $HCO_3^{\ominus}$, $Na^{\oplus}$ $^{\ominus}CN$ MeS HN O NH O 2) NaOH;

A B

MeS HN O NH_2 $O^{\ominus}$ $Na^{\oplus}$ O

C

MeS $\overset{\oplus}{N}H_3$ $O^{\ominus}$ O workup with H_2SO_4 MeS NH_2 $O^{\ominus}$ $Na^{\oplus}$ O

D E

Fig. 7.11. Synthesis of racemic methionine by means of the Bucherer modification of the Strecker synthesis. The first step of the reaction does not stop at the stage the step of the α-aminonitrile but yields a hydantoin (**B**; mechanistic details: Figure 7.12). The second step—via the anion **C** of a hydantoin acid—leads to the formation of the anion **E** of methionine, which can be protonated during workup to yield the uncharged methionine (**D**).

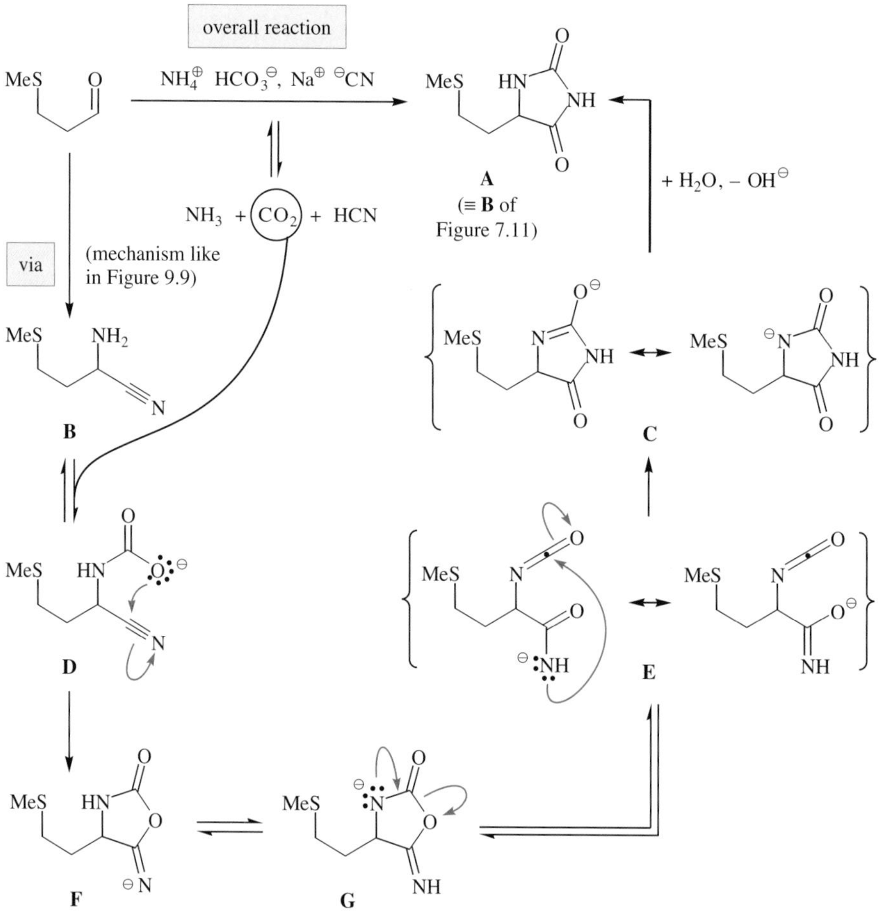

overall reaction

MeS O $NH_4^{\oplus}$ $HCO_3^{\ominus}$, $Na^{\oplus}$ $^{\ominus}CN$ MeS HN O NH O

A (≡ **B** of Figure 7.11)

NH_3 + CO_2 + HCN

+ H_2O, – $OH^{\ominus}$

via (mechanism like in Figure 9.9)

MeS NH_2 N

B

C

MeS HN O $O^{\ominus}$ N

D

E

MeS HN O O $^{\ominus}N$

F

MeS N O O NH

G

Fig. 7.12. Mechanism for the formation of hydantoin from 3-(methylthio)propionaldehyde, ammonium hydrogen carbonate and sodium cyanide, the initial reaction of the Strecker/Bucherer synthesis of methionine according to Figure 7.11.

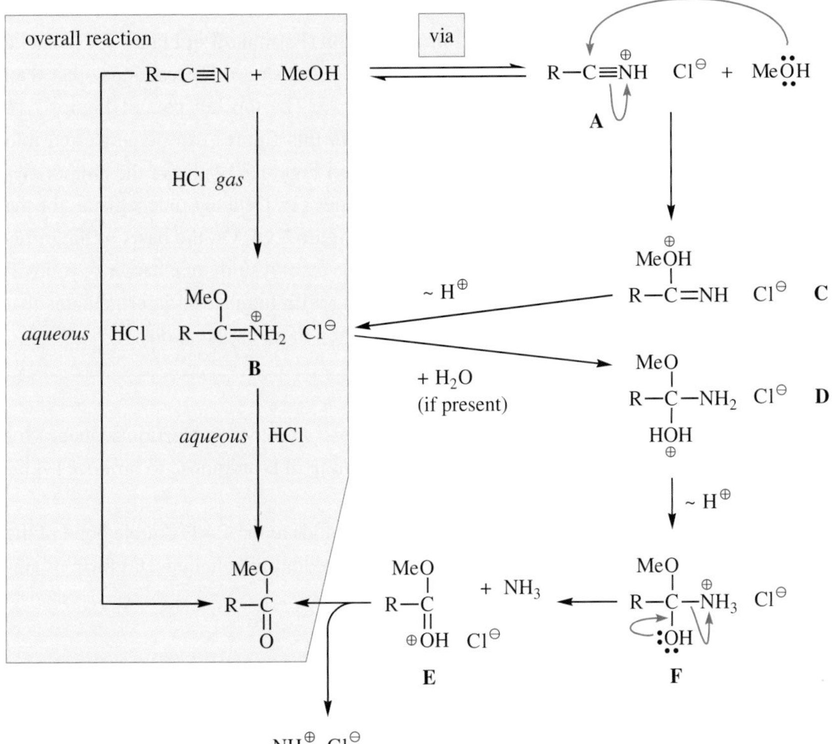

Fig. 7.13. Mechanism of the Pinner alcoholysis of nitriles—via the (possibly) isolable imidic acid ester hydrochloride (**B**)—to carboxylic acid esters.

an equilibrium reaction is converted to the negatively charged heterocycle **G**. *The latter is a carbonic acid derivative that decomposes to the ring-opened product* ***E***. *This* compound contains a *heterocumulene (isocyanate) and a heteroatom nucleophile*. The isocyanate group in the intermediate **E** of Figure 7.12 reacts very rapidly with an amide anion, which adds to it as an *N*-nucleophile. In this way, the hydantoin anion **C** is formed and finally undergoes protonation to give the neutral hydantoin **A**.

If the acidic or basic hydrolysis of a nitrile requires harsher reaction conditions than those compatible with a sensitive substrate and if there are reasons—such as the explosiveness of hydrogen peroxide—to refrain from accelerating the reaction by adding hydrogen peroxide as in Figure 7.10, the remaining alternative is the **Pinner reaction** (Figure 7.13, left side). It usually starts with the introduction of HCl gas into an alcoholic solution of the nitrile, followed by the precipitation of an imidic acid ester hydrochloride (**B**) whose C=N double bond is hydrolyzed upon subsequent treatment with aqueous HCl, yielding the corresponding carboxylic acid ester. The one-step Pinner reaction does not proceed as mildly. In that case, an alcoholic solution of the nitrile is treated with aqueous HCl and yields the same carboxylic acid ester as the two-step reaction.

The right side of Figure 7.13 shows the details of the different variants of the Pinner reaction described above. The addition of methanol to the C≡N triple bond follows the mechanism of the nitrile hydrolysis under acidic conditions (cf. Figure 7.8).

Side Note 7.2. Chemistry of Imidic Acid Ester Hydrochlorides: Preparation of Orthoesters and Amidines

The imidic acid ester hydrochlorides of the Pinner reaction (Formula **B** in Figure 7.13) can be isolated as crystalline solids and subsequently employed as substrates for reactions other than the synthesis of the carboxylic acid esters in Figure 7.13. The top left part of Figure 7.14 demonstrates that imidic acid ester hydrochlorides (**F** in this figure) can be converted into orthoesters **A** using dry methanol. The left bottom part in Figure 7.14 shows the ammonolysis of imidic acid ester hydrochlorides leading to amidines (**I**). Detailed mechanisms for the two transformations are presented on the right side of Figure 7.14. On the basis of the information so far presented in Chapter 7 and, in particular, the careful study of Chapter 6, you will by now probably have gained enough mastery in dealing with tetrahedral intermediates that you are able to comprehend these mechanisms without any more explanations.

Figure 7.15 displays the classic **Kiliani–Fischer synthesis**, a three-step reaction sequence for the homologation of aldoses. You can see the C_1-lengthening of D-arabinose to produce D-glucose and D-mannose.

In the first step of the Kiliani–Fischer synthesis, HCN adds to the C=O double bond of the tetrahydroxy aldehyde, which is in equilibrium with the predominant hemiacetal form of ara-

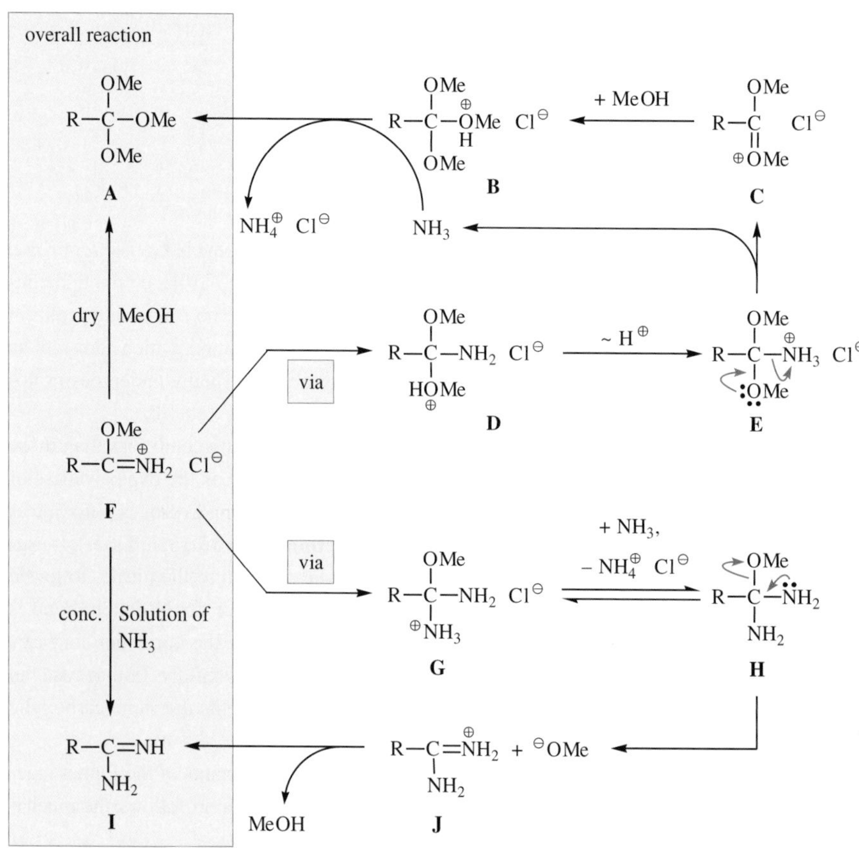

Abb. 7.14. Imidic acid ester hydrochlorides (**F**)—to be prepared from nitriles, hydrogen chloride (gas) and methanol acc. to Fig. 7.13—, their transformability into orthoester (**A**) or amidine (**I**) and the corresponding reaction mechanisms.

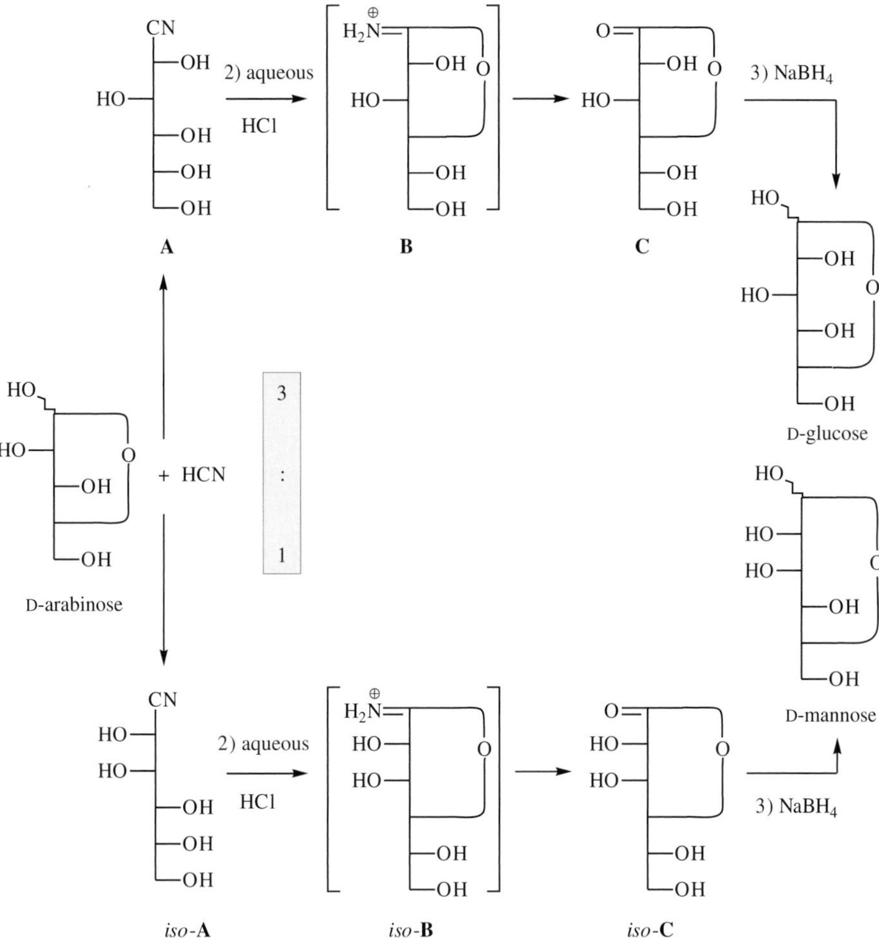

binose (cf. Section 9.1.2). This leads to the formation of a cyanohydrin according to the mechanism in Section 9.1.3. Two isomeric cyanohydrins are formed. Specifically, a 3:1 mixture of the diastereoismers cyanohydrin **A** and *iso*-**A** is obtained.

Abb. 7.15. Kiliani-Fischer method for the homologation of aldoses to give the diastereomeric mixture of the next higher aldose. Three-step syntheses of this kind (here: D-arabinose → D-glucose + D-mannose) played a central role when about a hundred years ago Emil Fischer managed to determine the configuration of aldoses by chemical (!) methods.

The second step of the Kiliani–Fischer synthesis (Figure 7.15) corresponds to the Pinner reaction of a nitrile with *aqueous* hydrochloric acid in terms of both reaction conditions and course of reaction (cf. Figure 7.13). In both cases the substrate is protonated at the C≡N group to give a nitrilium ion, and in both reactions this is followed by the nucleophilic addition of an alcoholic OH group, which is intermolecular in the Pinner reaction and intramolecular in the Kiliani–Fischer synthesis. The latter leads to the exclusive formation of a five-membered imidic acid ester hydrochloride, demonstrating the general kinetic advantage of the formation of five-membered rings over the analogous formation of six-membered rings. Of course, an ~3:1 mixture of the imidic acid ester hydrochlorides **B** and *iso*-**B** is also produced when a 3:1 mixture of the diastereomorphic pentahydroxynitriles **A** and *iso*-**A** is employed in the Kiliani–Fischer synthesis. Under Pinner *and* Kiliani–Fischer conditions, the hydrolysis of the

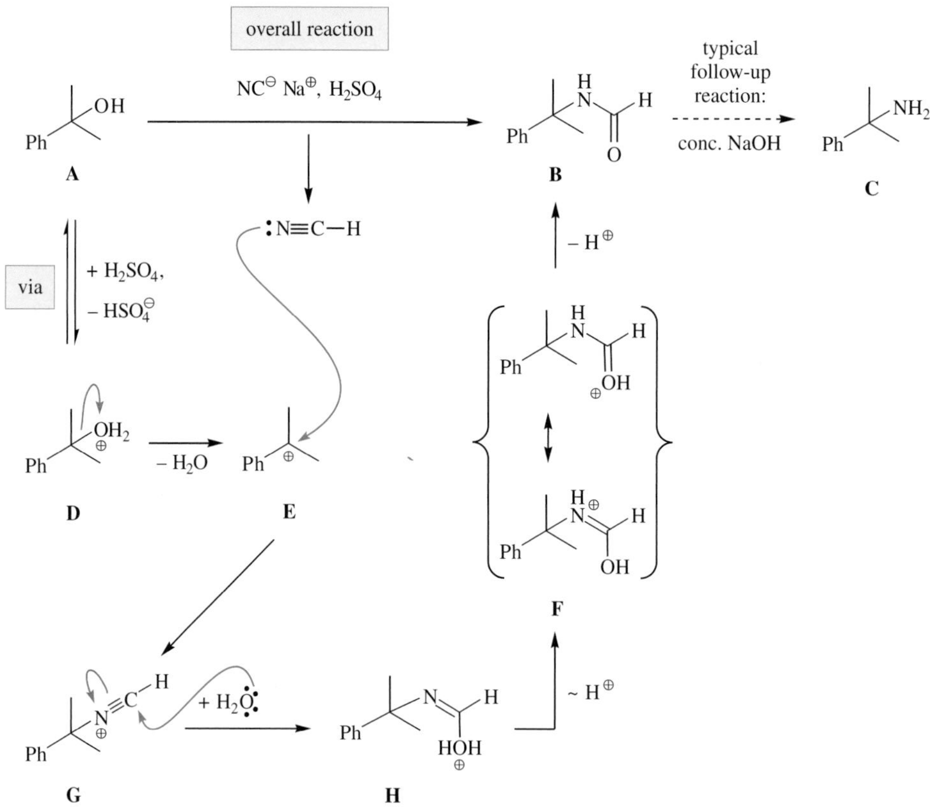

Abb. 7.16. Ritter reaction part I (cf. Fig. 7.17): S_N1 reaction between a tertiary alcohol and a nucleophile containing C≡N, with hydrogen cyanide acting as the nucleophile. Under reaction conditions the C≡N group is *tert*-alkylated at the N atom and hydrated at the nitrile nitrogen leading to the formation of an *N*-*tert*-alkylformamide (**B**).

$C{=}N^{\oplus}H_2$ double bond to the C=O double bond of the product carboxylic acid ester (Figure 7.13) or lactone (Figure 7.15, Formula **C** and *iso*-**C**) follows.

The third step of the Kiliani–Fischer synthesis in Figure 7.13 is a reduction of the lactones **C** and *iso*-**C** using sodium borohydride. Normally, sodium borohydride does not reduce γ-lactones. But since the lactones in Figure 7.15 are substituted by an electronegative group, they exhibit above-average electrophilicity. The resulting tetrahedral intermediates do not decompose until all carboxyl groups have reacted. This stability is, among other things, due to the fact that only the breakdown of the tetrahedral intermediate leads to a single species and does not possess the entropic driving force of the breakdown of tetrahedral intermediates derived from esters. Finally, the protonation of the tetrahedral intermediate furnishes the new aldoses: D-glucose as the main product and D-mannose as a by-product—both in their hemiacetal form.

Nitriles as organic species and hydrocyanic acid as their inorganic parent compound can be alkylated at the nitrogen atom by alcohols in the presence of strong acids. This can be accomplished with alcohols that readily form carbenium ions (Figure 7.16). Alcohols reacting with the nitrile nitrogen atom in an S_N2 reaction (Figure 7.17) are also suitable. Both reactions afford an *N*-alkyl nitrilium ion (Formula **G** in Figure 7.16, **B** in Figure 7.17). In an addition reaction, the latter accepts the equivalent of water that upon alkylation had been expelled by the corresponding alcohol. Via the resulting cation intermediates **H** (Figure 7.16) or **C** (Fig-

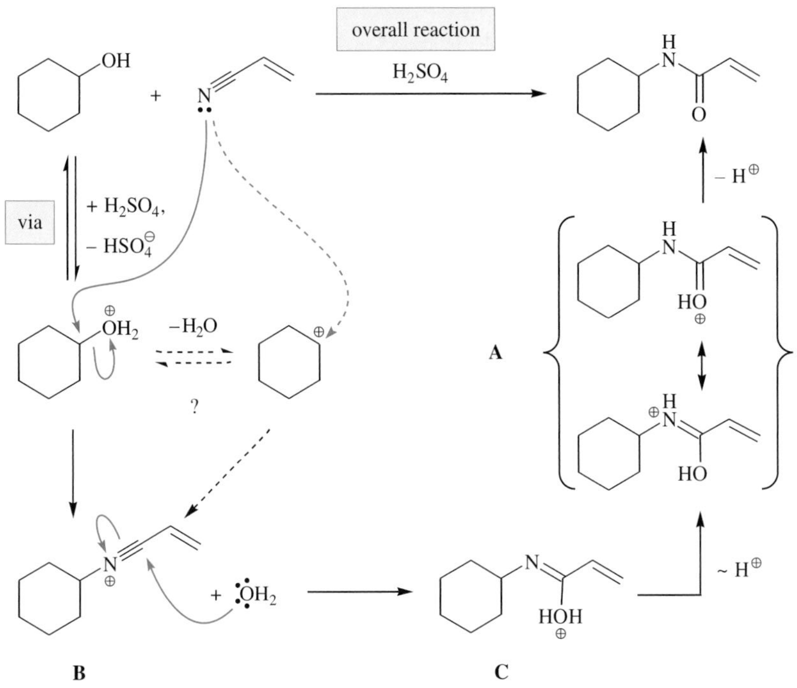

Abb. 7.17. Ritter reaction part II (cf. Fig. 7.16): S_N2 reaction between a secondary alcohol and a nucleophile containing C≡N; with acrylic nitrile acting as the nucleophile. Under reaction conditions the C≡N group is *sec*-alkylated at the N atom and hydrated at the nitrile nitrogen leading to the formation of an *N-sec*-alkylacrylic amide.

ure 7.17), a proton shift affords the protonated imidic acids **F** (Figure 7.16) and **A** (Figure 7.17), respectively. Their deprotonation to secondary carboxylic acid amides give the final products of these so-called **Ritter reactions**. Starting from HCN, the result is *N*-alkylformamides (formula **B** in Figure 7.16). Starting from organic nitriles *N*-alkylamides of higher carboxylic acids are generated (Figure 7.17).

Ritter reactions may take different courses. Figure 7.16 indicates an important one because the cumylation/hydration of HCN producing *N*-cumylformamide (**B**) is followed by the hydrolysis to cumylamine (**C**). This ensuing reaction completes the **Ritter synthesis of a primary *tert*-alkylamine**. Figure 7.17 shows the cyclohexylation/hydration of acrylonitrile. The resulting acrylamide is a synthetic target of technical relevance—not so much because its being an amide, but rather because it contains a C=C double bond that allows the acrylamide to be polymerized to a polyacrylic acid amide.

References

7.1

G. Tennant, "Imines, Nitrones, Nitriles and Isocyanides," in *Comprehensive Organic Chemistry* (I. O. Sutherland, Ed.), Vol. 2 (Nitrogen Compounds, Carboxylic Acids, Phosphorus Compounds), Pergamon Press, Oxford, U.K., **1979**, 385-590.

7.2

D. G. Neilson, "Imidates Including Cyclic Imidates," in: *The Chemistry of Amidines and Imidates* (S. Patai, Ed.), John Wiley & Sons, London, **1975**, 385–489.

I. D. Gridnev, N. A Gridneva, "Interaction of Nitriles with Electrophile Reagents," *Russ. Chem. Rev.* **1995**, *64*, 1021-1034.

L. I. Krimen, D. J. Cota, "The Ritter Reaction," *Org. React.* **1969**, *17*, 213–325.

Further Reading

S. M. Sherif, A. W. Erian, "The Chemistry of Trichloroacetonitrile," *Heterocycles* **1996**, *43*, 1083–1118.

Carbonic Acid Derivatives and Heterocumulenes and Their Interconversion

8

Heterocumulenes (**B**) are compounds with a substructure $Het^1{=}C{=}Het^2$, in which at least one of the groups "Het" is an O atom, an S atom, or an NR group (Figure 8.1). They can be regarded as anhydrides of the carbonic acid derivatives **A**. In fact, heterocumulenes are typically generated by eliminating water from such carbonic acid derivatives. Sometimes this occurs spontaneously, especially at elevated temperatures, but generally only upon treatment with a dehydrating reagent.

The reverse reaction, i.e., the addition of water to the heterocumulene **B** with formation of carbonic acid derivatives **A**, is also known. With very few exceptions this reaction occurs spontaneously with catalysis by an acid or base. The hydration of a heterocumulene is a special example of an entire class of heterocumulene reactions, namely the addition of heteroatom nucleophiles $H{-}Het^3$ (you have already come across such a reaction briefly in Section 6.1.1 in the context of the DCC activation of carboxylic acids). Occasionally, this addition

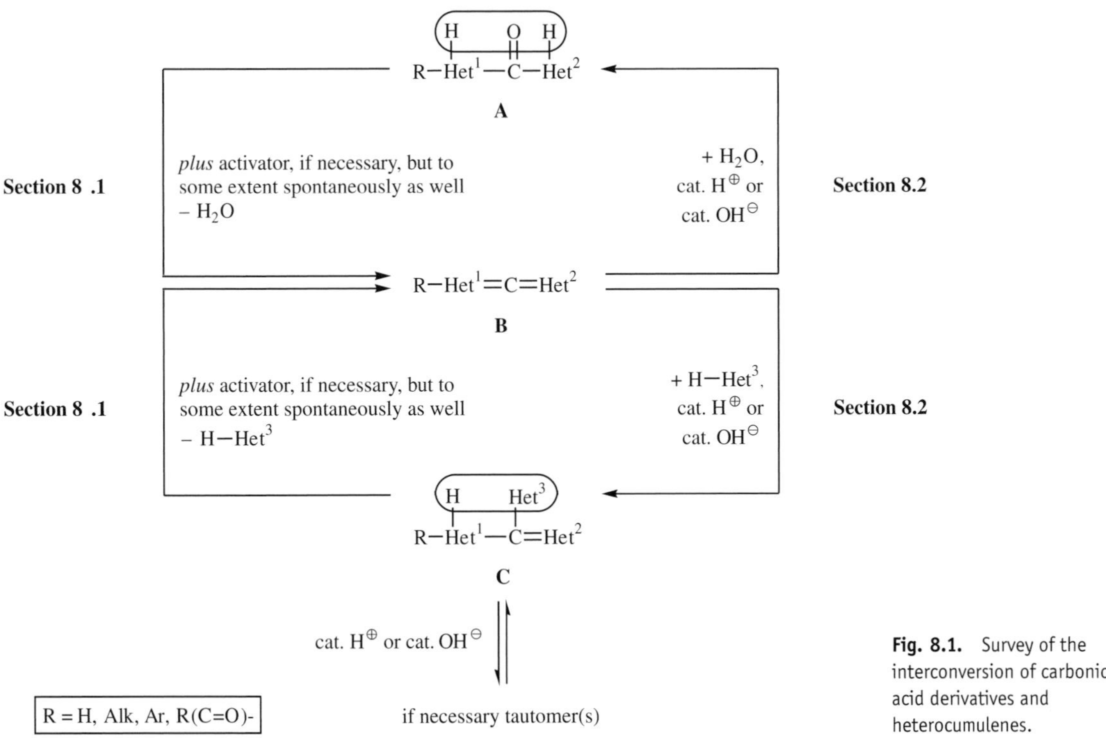

Fig. 8.1. Survey of the interconversion of carbonic acid derivatives and heterocumulenes.

Bruckner R (author), Harmata M (editor) In: *Organic Mechanisms – Reactions, Stereochemistry and Synthesis*
Chapter DOI: 10.1007/978-3-642-03651-4_8,

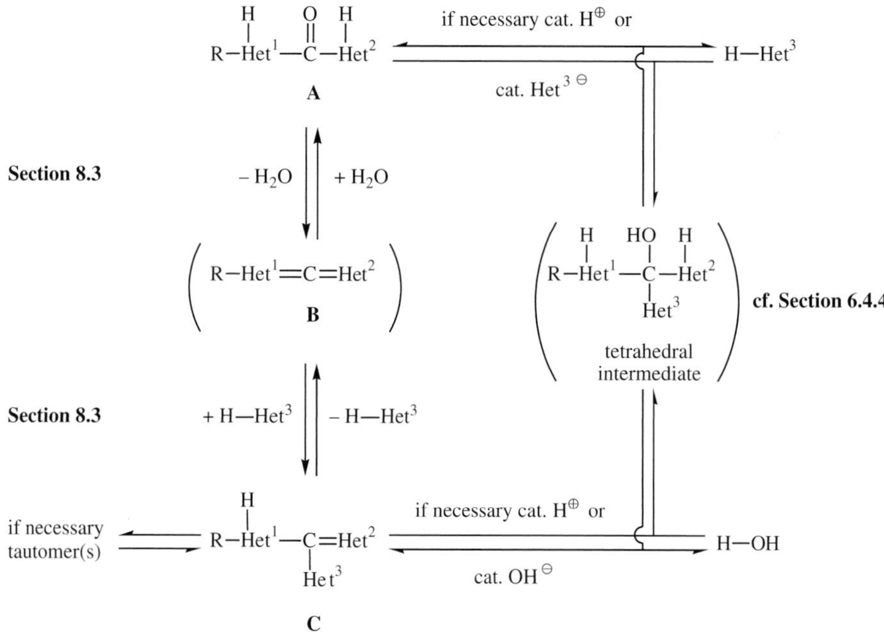

Fig. 8.2. Interconversion of carbonic acid derivatives. Elimination/addition mechanism (via a heterocumulene) and addition/elimination mechanism (via a tetrahedral intermediate).

to heterocumulenes **B** takes place spontaneously, but usually is fast enough only under acidic or basic conditions. If only the addition of a heteroatom nucleophile H–Het^3 to a heterocumulene **B** occurs, a carbonic acid derivative with the structure **C** or a tautomer of it is formed (Figure 8.1). An overreaction may occur, though, if the heteroatom nucleophile is acylated by the initially formed carbonic acid derivative. The acylation of heteroatom nucleophiles by carbonic acid derivatives has been discussed in Section 6.4.4. The reaction at the bottom left-hand side of Figure 8.1 demonstrates that not only the carbonic acid derivatives **A** act as precursors of the heterocumulenes **B** (their role as such having already been mentioned), but also the carbonic acid derivatives **C**, whose conversion into **B** proceeds via β-elimination of H and Het^3.

Because of the interconversion of heterocumulenes and carbonic acid derivatives, it is in this context useful to discuss their addition and elimination reactions—as suggested by this chapter's title.

Section 8.3 addresses the aspect of "Carbonic Acid Derivatives and Their Interconversion." Figure 8.2 shows how this subject is related to the chemistry of heterocumulenes depicted in Figure 8.1. Carbonic acid derivatives **A** and heteroatom nucleophiles H–Het^3 can react to furnish the carbonic acid derivatives **C** (see Section 6.4.4). On the other hand, they can react, via the elimination of water, to give the heterocumulene **B** and subsequently add the heteroatom nucleophile H–Het^3. This path also leads to the formation of carbonic acid derivatives **C** and is described in Section 8.3.

8.1 Preparation of Heterocumulenes from Carbonic Acid (Derivatives)

You may have heard already about carbonic acid molecules being unstable—except in very apolar solvents or in the gas phase. Why is this so? You will find the explanation in the topmost line of Figure 8.3. Through proton migration carbonic acid (**B**) is in equilibrium with the isomeric zwitterion **C**. This is true, provided that **C** is sufficiently solvated—which is not the case in very apolar media in which, *therefore*, the carbonic acid is stable. The zwitterion **C** eliminates water, much like the second step of an $E1_{cb}$ elimination, and affords CO_2. It is the

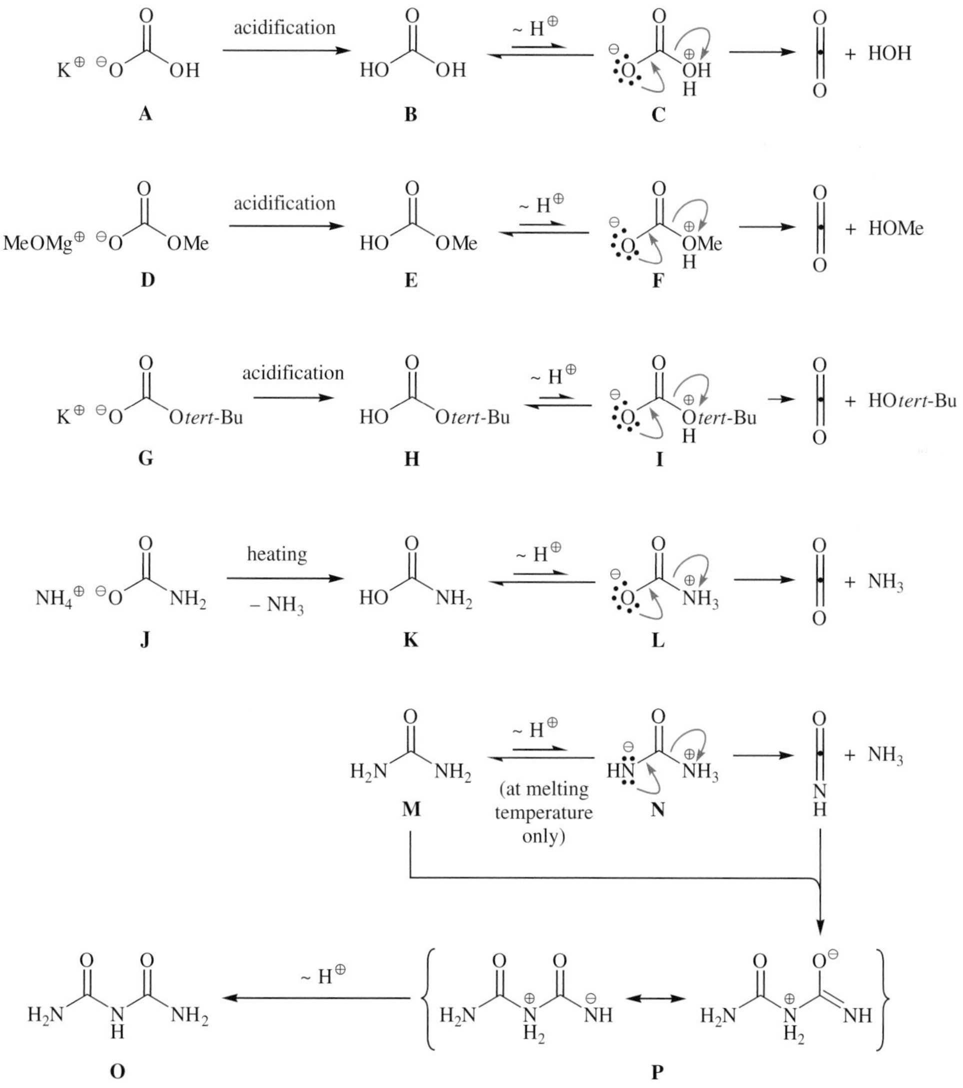

Fig. 8.3. Zwitterion mechanism of the decomposition of neutral carbonic acid (derivatives) in a heterocumulene and a hetereonucleophile—examples: carbonic acid (**B**), carbonic acid monomethylester (**E**), carbonic acid mono-*tert*-butylester (**H**), carbamic acid (**K**), and urea (**M**).

amphoteric character of carbonic acid, i.e., the simultaneous presence of an acidic OH group and a protonizable O atom, which is responsible for the instability of carbonic acid.

Experience suggests the crucial zwitterion formation **B** → **C** might not be easy since the charge separation that needs to occur might require too much energy. The following factors will convince you otherwise. (1) The C=O double bond of the zwitterion **C** benefits from a kind of carboxylate resonance, whereas in the neutral carbonic acid (**B**) only the smaller carbonic acid resonance occurs. The relevant numerical values are not exactly known, but the difference in resonance between the carboxylate (30 kcal/mol; Table 6.1, line 1) and the carbonic acid (14 kcal/mol; Table 6.1, line 3) supports the above statement. (2) Reading the top-most line in Figure 8.3 from right to left shows how carbon dioxide "chemically" dissolves in water (at the same time—and to a much larger extent—carbon dioxide also dissolves physically, of course): First, the zwitterion **C** is exactly what is formed (and subsequently transprotonated to give carbonic acid, which in turn largely dissociates into protons and hydrogen carbonate ions). So **C** is an energetically easily accessible species.

Figure 8.3 relates the decomposition of free carbonic acid (**B**) via an isomeric zwitterion (**C**) into the heterocumulene carbon dioxide and the heteroatom nucleophile water (top-most line) with four entirely analogous decomposition reactions. The scheme shows (from top to bottom):

- the decomposition of carbonic acid methyl ester (**E**) observed with the acidification of Stiles' reagent [(methoxymagnesium)monomethylcarbonate, **D**;
- the decomposition of carbonic acid *tert*-butyl ester (**H**) occurring with the acidification of potassium *tert*-butylcarbonate **G**;
- the decomposition of carbamic acid (**K**) as, e. g., results from the acid/base reaction of ammonium carbamate (**J**). Ammonium carbamate is the key intermediate in the technical urea synthesis (Figure 8.15). On the one hand, carbamic acid may decompose, as is shown here, via the zwitterion **L** into carbon dioxide and ammonia, or via the isomeric zwitterion **E** of Figure 8.15 into isocyanic acid and water.
- At the bottom, the decomposition of urea (**M**) upon heating is shown ("urea melt"). Here, the urea reacts via the zwitterion **N** to form isocyanic acid and ammonia, with the latter of the two escaping from the reaction mixture. While up to this point there is a strict analogy with the other reactions in Figure 8.3, a follow-up reaction occurs in this case leading to biuret (**O**) via the zwitterion **P**. Being a nucleophilic addition of the unreacted urea (as a heteroatom nucleophile) to the resulting isocyanic acid (a heterocumulene) it topically belongs to Section 8.2.

That urea (**M**) decomposes at much higher temperatures than carbonic acid (**B**), the carbonic acid halfesters **E** and **H** or carbamic acid (**K**) is based on the fact that only urea reacts via a zwitterion **N** in which the negative charge, ignoring resonance, is located on an N atom, whereas in all other zwitterions (**C**, **F**, **I** and **L**) the negative charge is stabilized by the more electronegative oxygen atom.

In the presence of potassium hydroxide, cellulose adds to carbon disulfide (Figure 8.4). In this way potassium xanthate **A** is produced. It is soluble in water, but restores the water-insoluble cellulose upon addition of acid. The primary protonation product is the dithiocarbonic acid *O*-cellulose ester **B**. **B** reacts just like the unstable carbonic acid derivatives in Figure 8.3, namely via a zwitterion (**C**) and its decomposition into cellulose (a heteroatom nucleophile)

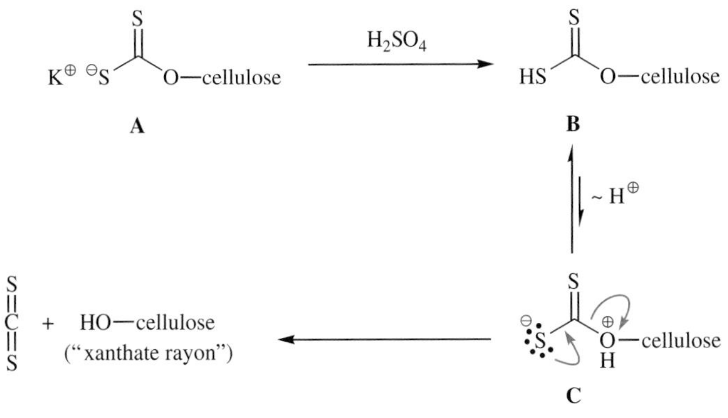

Fig. 8.4. The dithiocarbonic acid ester **B** of cellulose decomposes into carbon disulfide and cellulose according to a zwitterion mechanism. The cellulose thus obtained is called xanthate rayon and can be spun.

and carbon disulfide (a heterocumulene). This reaction provides the basis of a method for the production of artificial silk ("xanthate rayon"): first impure cellulose is made soluble through addition to CS_2, and then pure precipitates of cellulose are obtained.

In Figure 4.14, we learned about the Chugaev elimination in connection with the synthesis of alkenes. The second (primary) product of this reaction is the dithiocarbonic acid *S*-methyl ester (**A**). It equilibrates with the zwitterion **B**, which decomposes into carbon oxysulfide (a heterocumulene) and methanethiol (a heteroatom nucleophile).

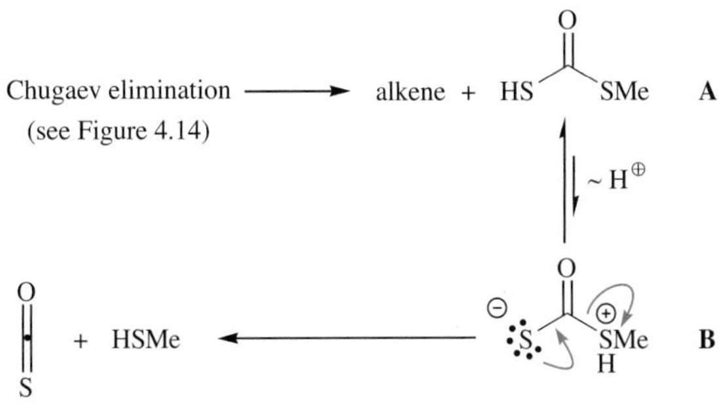

In order to better understand the various reaction conditions of all the transformations of carbonic acid derivatives into heterocumulenes that are still to be presented in this section, it is useful to take a closer look at the previously described unstable zwitterions (**C**, **F**, **I**, **L** and **N** in Figure 8.3, **C** in Figure 8.4, **B** in the reaction aequation above the previous paragraph): They are all carbonic acid derivatives in which a first heteroatom substituent has been turned into a good leaving group by protonation and a second heteroatom substituent has been made a good donor by deprotonation. This is why the leaving group can be replaced. This allows for the following general conclusion: carbonic acid derivatives react to afford heterocumulenes and heteroatom nucleophiles if they dispose of a good leaving group—right from the start or after protonation or another activation—and/or if they—innately or after anion formation contain a good donor substituent. There are six permutations of these properties, five of which are among the transformations of carbonic acid derivatives into heterocumulenes presented in Section 8.1:

1. right from the start a good leaving group is available and, at the same time, a good donor is innately present: Figure 8.8, decomposition of **A**, **C** and **E;**
2. right from the start a good leaving group is available and, at the same time, a good donor is present after anion formation: Figure 8.7, decomposition of **B**; Figure 8.10, decomposition of **D;**
3. a good leaving group is available after protonation or a good donor is innately present: Figure 8.5, decomposition of **D**; Figure 8.5, decomposition of **B**; Figure 8.10, decomposition of **A;**
4. a good leaving group is available after protonation and, at the same time, a good donor is present after anion formation: decomposition of the zwitterions mentioned above; Figure 8.11, decomposition of the zwitterion **H;**
5. a good leaving group is available after a different kind of activation and, at the same time, a good donor is innately present: Figure 8.9, decomposition of **C**; Figure 8.11, decomposition of **D;**
6. a good leaving group is available after a different kind of activation and, at the same time, a good donor is present after anion formation: unknown.

N-aryl or *N*-alkyl carbamic acids (**C** in Figure 8.5, **B** in Figure 8.6) are formed in the course of the hydrochloric hydrolysis of aryl or alkyl isocyanates (Figure 8.5) or upon cleavage of *N*-Boc groups with trifluoroacetic acid (Figure 8.6; cf. Figure 4.37). *N*-substituted carbamic acids are as unstable as the unsubstituted carbamic acid **K** of Figure 8.3. When a carbamic acid, as mentioned, is generated under acidic instead of neutral conditions, it does not decompose via a zwitterion in the form of **K** → **L** in Figure 8.3, but via a cation as the intermediate. The latter is formed through protonation at the nitrogen atom (→ **D** in Figure 8.5). The primary decomposition products are protonated carbon dioxide and an aniline or amine, respectively. An acid/base reaction between these components affords the final decomposition products, i.e., carbon dioxide and an anilinium or ammonium salt. If the neutral anilines or amines are to be released, the workup has to be conducted with sodium hydroxide.

By the way, aromatic and aliphatic isocyanates do not only undergo acidic but—even if much more slowly—neutral hydrolysis. Again, the first hydrolysis product is an unstable *N*-aryl or *N*-alkyl carbamic acid. It decomposes under these conditions, though not according

Fig. 8.5. Proton-induced decomposition of *N*-substituted carbamic acid **C** into carbon dioxide and a primary ammonium hydrochloride—a reaction occuring in the course of acidic hydrolysis of organic isocyanates.

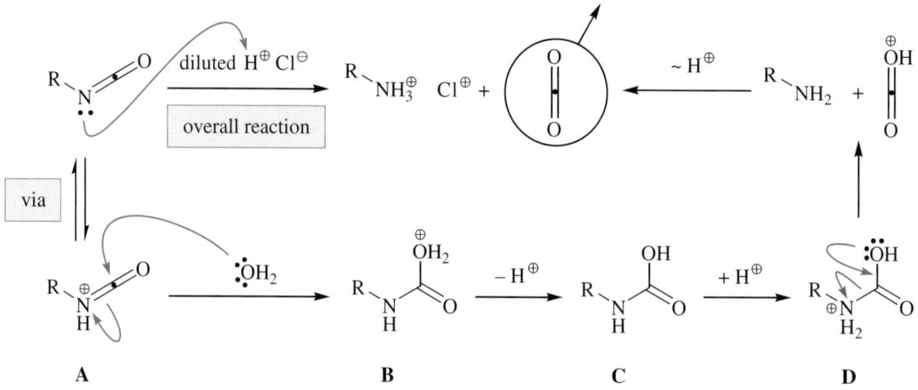

diluted $H^{\oplus} Cl^{\ominus}$

overall reaction

via

$-H^{\oplus}$

$+H^{\oplus}$

$\sim H^{\oplus}$

A **B** **C** **D**

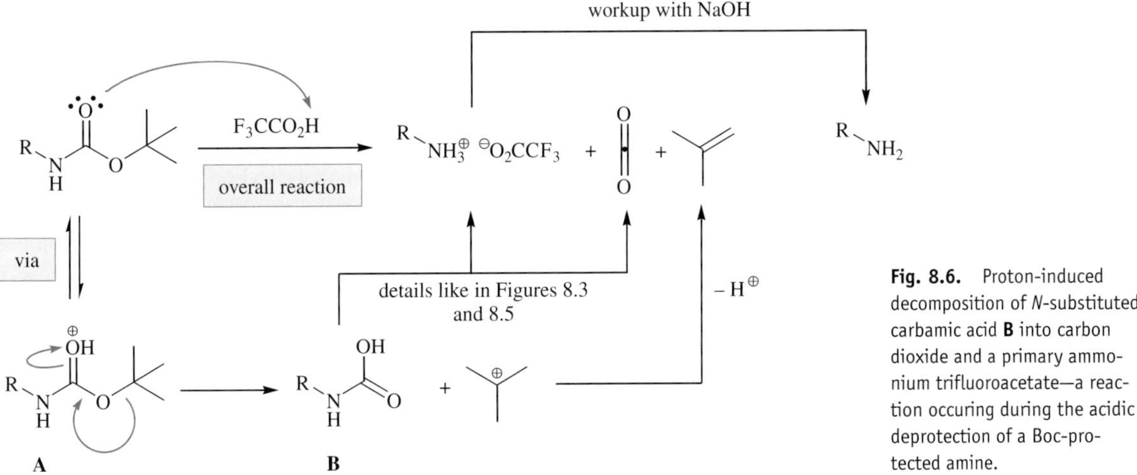

Fig. 8.6. Proton-induced decomposition of *N*-substituted carbamic acid **B** into carbon dioxide and a primary ammonium trifluoroacetate—a reaction occuring during the acidic deprotection of a Boc-protected amine.

to the cation mechanism of Figures 8.5 and 8.6, but the zwitterion mechanism in Figure 8.3. The reaction products are both the underlying aniline or amine and carbon dioxide. The formation of this gas is indispensable for the hydrolytically initiated generation of poylurethane foam formulations (Side Note 8.1), and this is what caused the tank containing methylisocyanate to burst open during the Bhopal calamity (page 354). The latter aspects will be discussed in more detail in Section 8.2 in the context of addition reactions to heterocumulenes.

Formula **B** in Figure 8.7 presents an anionic carbonic acid derivative decomposing into a heterocumulene (carbon dioxide) and a heteroatom nucleophile (the chloride ion). The driving force of this decomposition has an interesting impact. First, chloroformic acid (**A**)—though not being able to decompose according to the zwitterion mechanism of Figure 8.3—is not sta-

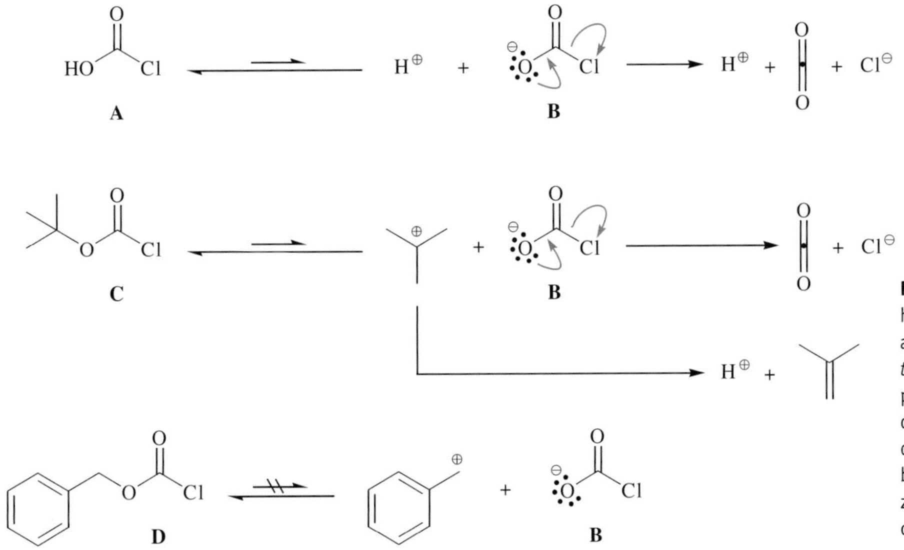

Fig. 8.7. Spontaneous heterolyses of chloroformic acid (**A**) and chloroformic acid *tert*-butylester (**C**) to give ion pairs where the anion subsequently decomposes to carbon dioxide and a chloride ion. Viability of chloroformic acid benzylester (**D**) due to the absence of an analogous heterolysis.

ble, but is in an acid/base equilibrium with a proton and the anion **B** which—according to what was outlined at the beginning of this paragraph—decomposes easily. Second, *tert*-butyl chloroformate (**C**) is not a stable carbonic acid derivative either. The reason is that this compound undergoes heterolysis as easily as a *tert*-butyl halide. In this way it equilibrates with the *tert*-butyl cation and the chlorocarbonate anion **B**, which subsequently decomposes. Since a benzyl cation is somewhat less stable than a *tert*-butyl cation, benzyl chloroformate (**D**), unlike (**C**), is not prone to undergo heterolysis. In fact, it is so stable that it is commercially available.

With the halide ion, halocarbonic acid derivatives possess such a good leaving group that they decompose upon mild heating into a protonated heterocumulene and a halide ion even without any further activation. Figure 8.8 illustrates this kind of reaction using as an example chloroformic acid amides **A** (*in situ* generation through acylation of primary amines with phosgene), chlorothiocarbonic acid amides **C** (*in situ* generation through acylation of primary amines by thiophosgene) as well as bromoamidines **E** (*in situ* generation through the addition of primary amines to the C≡N triple bond of bromocyane). Their instability renders compounds such as **A**, **C** and **E** short-lived intermediates in high-yield, one-step syntheses of isocyanates, isothiocyanates or cyanoamides from primary amines and phosgene or thiophosgene or cyanic bromide, respectively.

Dehydration of urea to carbodiimide (HN=C=NH) or its more stable tautomer cyanoamide (H_2N–C≡N) cannot be achieved via the zwitterion mechanism of Figure 8.3. It would have to proceed via the zwitterion HN$^{\ominus}$–C (=NH)–$OH_2^{\oplus}$, in which—for electronegativity reasons—both formal charges would occupy worse positions than in the zwitterion O$^{\ominus}$–C (=NH)–$NH_3^{\oplus}$, which is an intermediate in the transformation of urea into biuret upon heat-

Fig. 8.8. Spontaneous heterolyses of *N*-substituted carbamoyl chlorides (**A**), *N*-substituted thiocarbamoyl chlorides (**B**) and guanyl bromides (**E**) to give an ion pair whose cationic substructure is subsequently deprotonated to a heterocumulene.

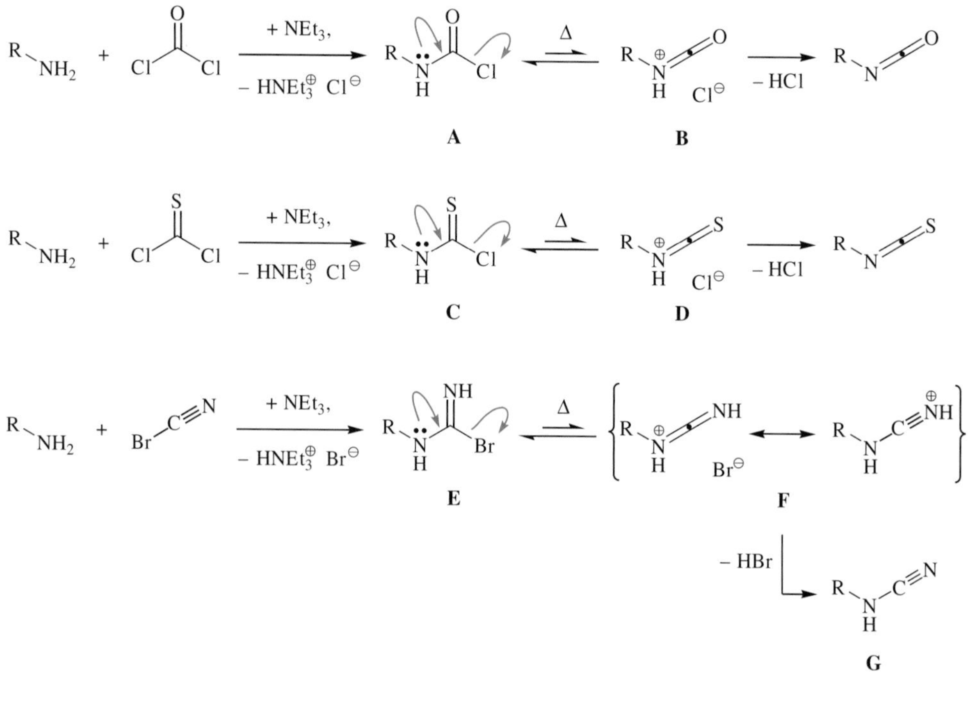

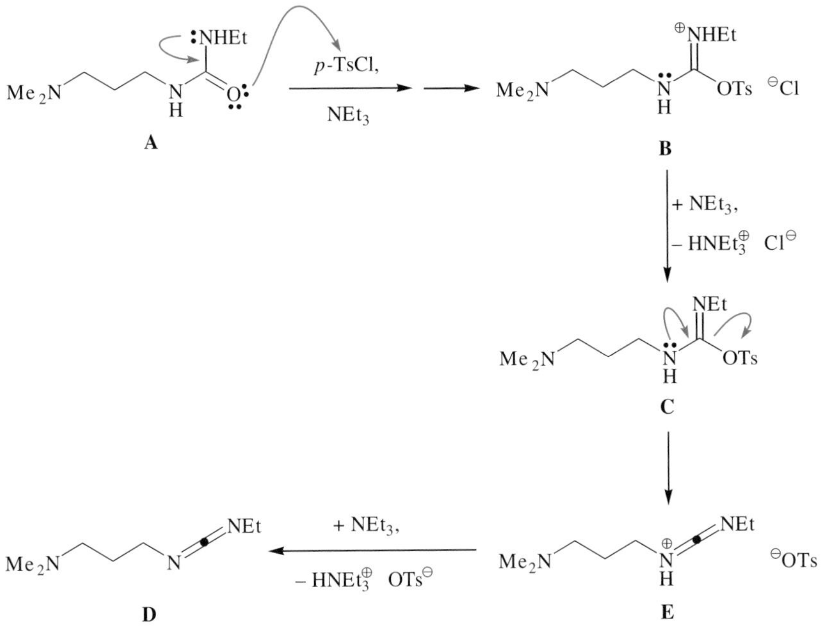

Fig. 8.9. *O*-Sulfonylation of the asymmetric *N,N'*-dialkyl urea **A** with tosyl chloride/triethylamine and subsequent. deprotonation to the unsymmetrical *N,N'*-dialkylcarbodiimide **D**.

ing (Figure 8.3, bottom). If, however, urea could be converted, e. g., into the carbonic acid derivative H_2N–C (=NH)–OTs, HOTs would certainly be easily eliminated. The result would be an indirect dehydration of urea. This is of no relevance as far as the parent compound is concerned; but it is very important with the *N,N'*-dialkyl urea **A** (Figure 8.9). Treatment of the substrate with tosyl chloride and triethylamine sequentially leads to an *O*-tosylation (→ **B**), a deprotonation (→ **C**), a *β*-elimination to give the cationic heterocumulene **E** and another deprotonation (→ **D**). Just like dicyclohexylcarbodiimide (DCC), this carbodiimide is used for the *in situ* activation of carboxylic acids (cf. Figure 6.15, 6.31 and 6.32).

To complete this series of heterocumulene syntheses from carbonic acid derivatives, we here present the acid- and base-catalyzed alcohol elimination from carbamic acid esters (**C**; Figure 8.10) and the condensation of aniline and carbon disulfide to phenylisothiocyanate (**F**;

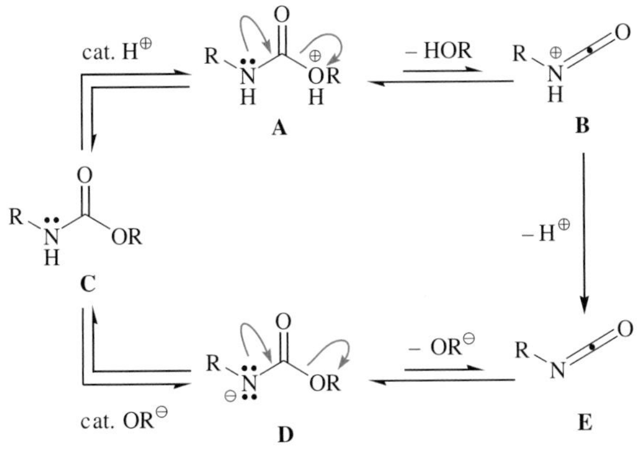

Fig. 8.10. Mechanism of the acid- and base-catalyzed *β*-elimination of an alcohol from *N,O*-disubstituted carbamate **C**—a method for the preparation of organic isocyanates.

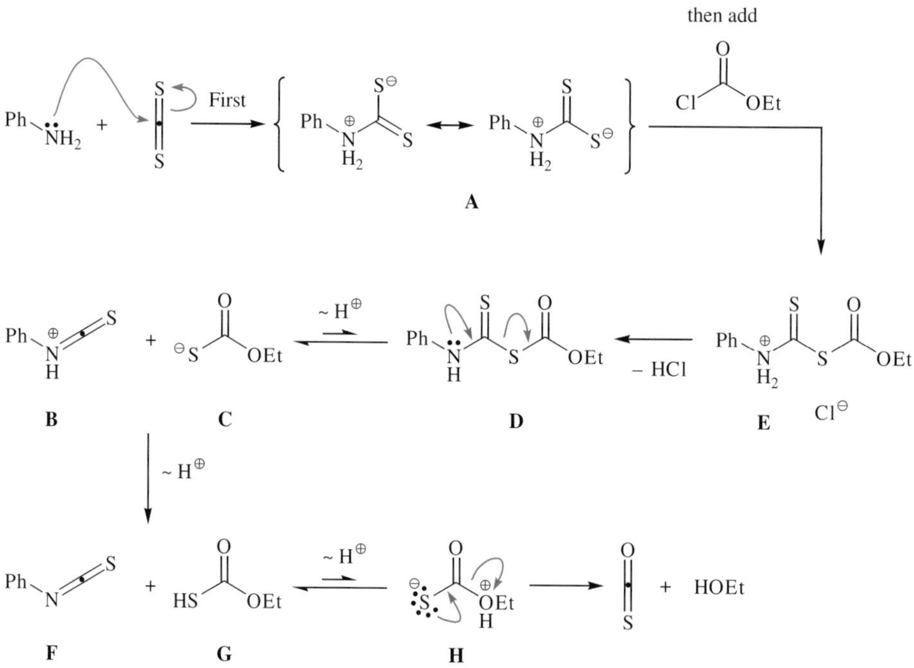

Fig. 8.11. Tandem reaction consisting of three single reactions mutually transforming heterocumulenes and heteroatom nucleophiles in a one-pot synthesis of an isothiocyanate: (1) uncatalyzed addition reaction of heteroatom nucleophile (aniline) + heterocumulene (carbon disulfide) → carbonic acid derivative (**A**); (2) heterolysis-initiated β-elimination of the carbonic acid derivative (**D**) → heterocumulene (**F**; = phenylisothiocyanate) + heteroatom nucleophile (thiocarbonic acid *O*-ethylester); (3) decomposition of a carbonic acid derivative (**D**) to a heterocumulene (carbon oxysulfide) and a heteroatom nucleophile (ethanol) via the zwitterion **H**.

Figure 8.11) via the intermediacy of an activated dithiocarbamate **D**. Phenylisothiocyanate enables the Edman degradation of peptides (Figure 8.14).

8.2 Transformation of Heterocumulenes and Heteroatom Nucleophiles into Carbonic Acid Derivatives

The reaction by the heteroatom nucleophile occurs at the *sp*-hybridized C atom in the middle of the functional group of the heterocumulene. Both the *neutral* heterocumulene or its protonated form **B** (Figure 8.12), react in this way. In each case, the rate-determining step of the addition reaction is the formation of an intermediate, in which the reacted C atom is sp^2-hybridized and thus trigonal planar. When the nucleophile adds to the neutral heterocumulene, this intermediate is represented by the resonance form **A.** If the addition involves the protonated heterocumulene **B**, the trigonal planar intermediate has the formula **C.** While intermediate **A** is then converted to a charge-free addition product through a proton *shift*, intermediate **C** provides the same product through a proton *loss*. The charge-free addition product may have an opportunity to stabilize further through tautomerism. Whether such tautomerism takes place and which isomer it favors depends on the heteroatoms in the primary addition product.

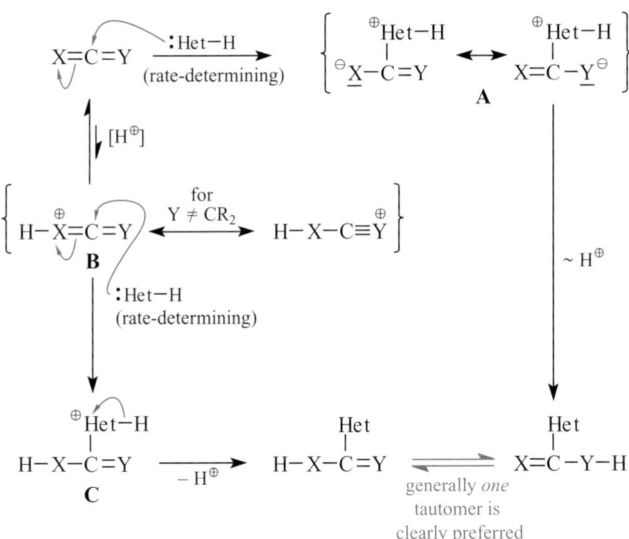

Fig. 8.12. Mechanism of the uncatalyzed addition (starting top left and proceeding clockwise) and the acid-catalyzed addition (starting top left and then proceeding counterclockwise) of heteroatom nucleophiles to heterocumulenes.

Additions to Ketenes

Ketenes are extremely powerful acylating agents for heteroatom nucleophiles. They react in each case according to the uncatalyzed mechanism of Figure 8.12. In this manner, ketenes can add

- H_2O to give carboxylic acids,
- ROH to give carboxylic esters,
- NH_3 to give primary carboxylic amides,
- RNH_2 to give secondary carboxylic amides, and
- R^1R^2NH to give tertiary carboxylic amides.

With very few exceptions ketenes cannot be isolated pure at room temperature (cf. Section 15.4). Consequently, they are prepared *in situ* and in the presence of the heteroatom nucleophile. The Wolff rearrangement of α-diazoketones is often used for this purpose (Section 14.3.2). α-Diazoketones can be obtained, for example, by the reaction between a carboxylic acid chloride and diazomethane (Figure 8.13; see also Figure 14.27) or by treating the sodium enolate of α-formylketones with tosyl azide (Figure 14.29).

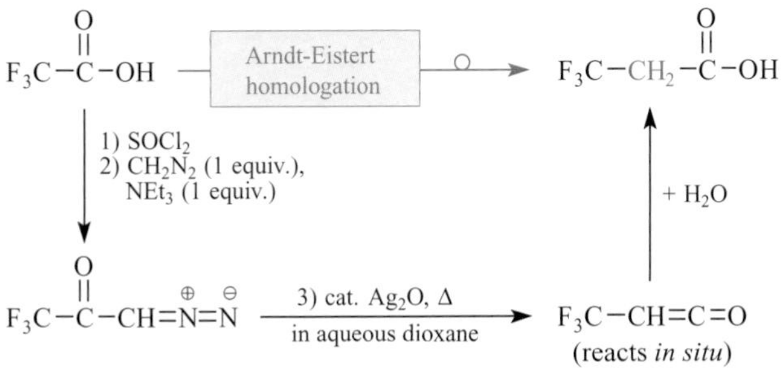

$F_3C{-}C(=O){-}OH$ — Arndt-Eistert homologation → $F_3C{-}CH_2{-}C(=O){-}OH$

1) $SOCl_2$
2) CH_2N_2 (1 equiv.), NEt_3 (1 equiv.)

$F_3C{-}C(=O){-}CH{=}\overset{\oplus}{N}{=}\overset{\ominus}{N}$ — 3) cat. Ag_2O, Δ in aqueous dioxane → $F_3C{-}CH{=}C{=}O$ (reacts *in situ*)

$+ H_2O$

Fig. 8.13. Arndt–Eistert homologation of carboxylic acids—addition of H_2O to a ketene in part 2 of the third reaction of this three-step reaction sequence.

Figure 8.13 shows a reaction sequence in which a carboxylic acid ultimately serves as the ketene precursor. For this purpose, it is converted with $SOCl_2$ into the corresponding acid chloride. This acid chloride is allowed to react with diazomethane via the mechanism with which we are familiar from Figure 6.2. The reagent is acylated at the C atom whereby an α-diazoketone is produced. The desired ketene is finally obtained from this diazoketone through a Wolff rearrangement (mechanism: Figure 14.26) the first part of the third step of this sequence. In part 2 of the third step, the heteroatom nucleophile H_2O, which was present before the rearrangement occurred, adds to this ketene. In this way, we again obtain a carboxylic acid. The newly obtained carboxylic acid is one CH_2 group longer than the carboxylic acid one started with. Therefore, the three-step reaction sequence of Figure 8.13 is a method for preparing the next homolog of a carboxylic acid—the so-called **Arndt–Eistert reaction**.

Of course, one can also carry out Wolff rearrangements in the presence of nucleophiles other than H_2O. Ketenes are then produced in *their* presence, and the addition products of *these* nucleophiles are therefore isolated. Figure 14.28 shows how one can add an alcohol to a ketene in this way. An alcohol addition to a ketene is also involved in the β-ketoester synthesis discussed in Side Note 6.2.

Additions to Carbon Dioxide

As you know water adds to CO_2—but only to the extent in which the addition product, i.e., the unstable carbonic acid (**B** in Figure 8.3) is converted into the stable hydrogen carbonate anion (like **A** in Figure 8.3). $CaCO_3$-saturated water adds to CO_2 to a larger extent than pure water because the resulting carbonic acid reacts with the basic carbonate ions in an acid/base reaction to produce the hydrogen carbonate anion. If such an initially $CaCO_3$-containing solution is provided with more CO_2 and is furthermore in contact with crystalline $CaCO_3$, which dissolves under formation of $CaCO_3$-saturated water, the addition reaction $H_2O + CO_2 + CaCO_3 \rightarrow Ca(HCO_3)_2$ runs to completion. When you admire stalactites in caves, think of this reaction and its reverse and how beautiful deposits of calcium carbonate are created by simple chemistry applied over long periods of time.

The addition of alcohols to CO_2 can also be achieved if the respective addition products—unstable carbonic acid half-esters (e. g., **E** or **H** in Figure 8.3)—are converted into the corresponding stable anions (like **D** or **G** in Figure 8.3). This is why in such addition reactions it is the alkoxides instead of the alcohols that are reacted with CO_2. In this way the addition of magnesium methoxide to CO_2 leads to the formation of (methoxymagnesium)monomethylcarbonate (the Stiles reagent, see Figure 8.3; synthesis application: Figure 13.63),

Me–O⊖ ⊕MgOMe + O=C=O ⟶ {Me–O–C(=O⊖...)–O ⟷ Me–O–C(=O)–O⊖} ⊕MgOMe

and the addition of potassium *tert*-butoxide to CO_2 yields potassium mono-*tert*-butyl carbonate:

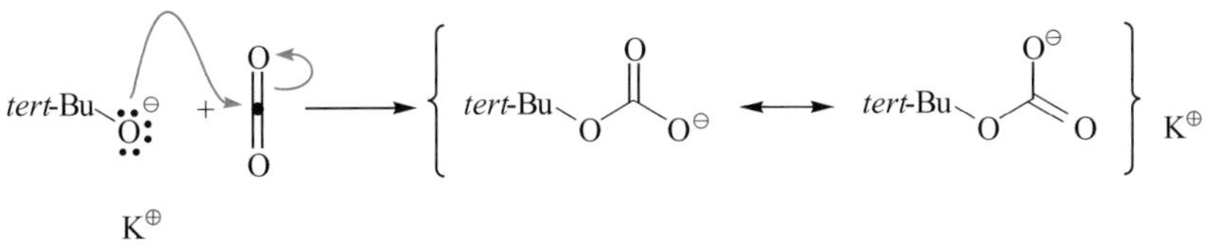

Potassium mono-*tert*-butyl carbonate is employed in the preparation of Boc_2O according to the procedure outlined in Figure 6.35.

NH_3 can also be added to CO_2 if the addition product, i.e., the unstable carbamic acid (top right of the figure), is transformed into the stable carbamic acid anion ("carbamate ion" at the bottom of the figure):

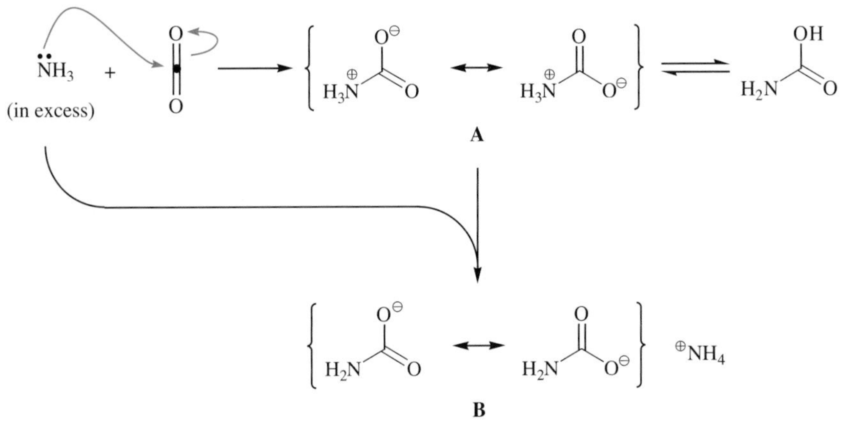

When two equivalents of ammonia are reacted with carbon dioxide the actual addition step is immediately followed by the transfer of protons onto ammonia. Incidentally, this is how the resulting ammonium carbamate (Formula **B** in the figure) is produced on a very large scale, for the synthesis of urea (Figure 8.15).

The second step of the urea synthesis follows after the reaction conditions have been changed dramatically: a temperature of 135 °C and a positive pressure of 40 bar cause the conversion of ammonium carbamate (**B**), a carbonic acid derivative, into urea, another carbonic acid derivative. The mechanism of this second part of the urea synthesis will be discussed in Section 8.3.

Additions to Other Symmetric Heterocumulenes

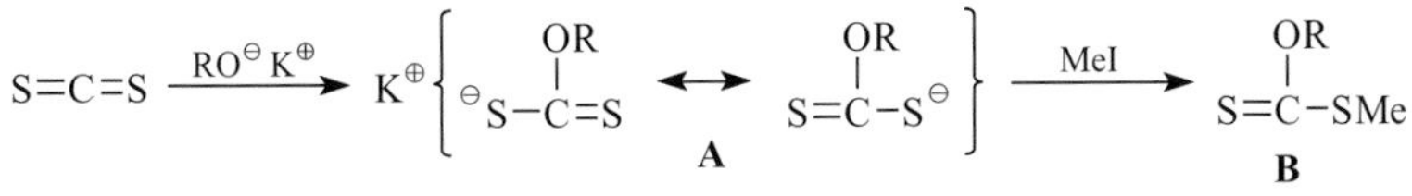

Carbon disulfide is the dithio derivative of CO_2. It is only a weak electrophile. Actually, it is so unreactive that in many reactions it can be used as a solvent. Consequently, only good nucleophiles can add to the C=S double bond of carbon disulfide. For example, alkali metal alkoxides add to carbon disulfide forming alkali metal xanthates **A** (Figure 7.4). If acidified they would produce an *O*-alkyl ester of dithiocarbonic acid which in the condensed phase is unstable in pure form—just like its cellulose analog, i.e., compound **B** in Figure 8.4. Hence, the protonation product of **A** would spontaneously decompose into the corresponding alcohol and CS_2. Their esters **B,** however, are *stable* derivatives of alkali metal xanthates **A.** They are referred to as xanthates. They are obtained by an alkylation (almost always by a methylation) of the alkali metal xanthates **A.** You have already learned about synthesis applications of xanthates in Figures 1.41, 4.14, and 4.15.

Carbodiimides are diaza derivatives of CO_2. It is also possible to add heteroatom nucleophiles to them. This is especially relevant when the heteroatom nucleophile is a carboxylic acid and when the carbodiimide is dicyclohexylcarbodiimide:

$$\text{Cy}-N{=}C{=}N-\text{Cy} \xrightarrow{HO-C(=O)-R} \text{Cy}-NH-C(-O-C(=O)-R){=}N-\text{Cy}$$

The addition of carboxylic acids to dicyclohexylcarbodiimide was mentioned in the context of Figures 6.15 and 6.26, but there we looked at it only from the point of view of *activating* a carboxylic acid. This addition follows the proton-catalyzed mechanism of Figure 8.12.

Although the carbodiimide **D** of Figure 8.9 is much more expensive than DCC, it is used for the activation of carboxylic acids—for a practical reason. Upon acylation of a heteroatom nucleophile with the DCC adduct of a carboxylic acid dicyclohexylurea (**A**) is formed along with the desired carboxylic acid derivative:

$$\underset{\text{DCC}}{\text{Cy}-N{=}C{=}N-\text{Cy}} \longrightarrow \underset{\mathbf{A}}{\text{Cy}-HN-C(=O)-NH-\text{Cy}} \qquad \underset{\mathbf{B}}{Me_2N-CH_2CH_2CH_2-N{=}C{=}NEt} \longrightarrow \underset{\mathbf{C}}{Et_2N-CH_2CH_2CH_2-HN-C(=O)-HNEt}$$

The separation of this dicyclohexylurea urea is relatively laborious, that is, it has to be conducted by chromatography or crystallization. After the activation of carboxylic acid with carbodiimide **B** and subsequent acylation of a heteroatom nucleophile, one certainly obtains a urea as a side product, too. It has the structure **C** and thus is an amine, the separation of which can, however, be accomplished easily by extraction with aqueous hydrochloric acid.

Additions to Isocyanic Acid and to Isocyanates

Wöhler's urea synthesis from the year 1828

$$O{=}C{=}N^{\ominus}\,NH_4^{\oplus} \xrightarrow{\Delta} \left(O{=}C{=}N-H + NH_3 \longrightarrow\right) O{=}C(-NH_2)-NH_2$$

and an analogous preparation of semicarbazide

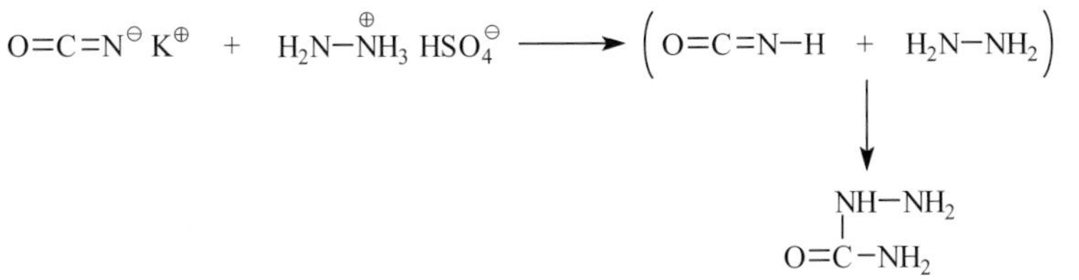

$$O{=}C{=}N^{\ominus}K^{\oplus} + H_2N{-}\overset{\oplus}{N}H_3\ HSO_4^{\ominus} \longrightarrow (O{=}C{=}N{-}H + H_2N{-}NH_2) \longrightarrow O{=}C(NH_2){-}NH{-}NH_2$$

are addition reactions to the C=N double bond of isocyanic acid (H—N=C=O). These reactions belong to the oldest organic chemical syntheses and take place via the uncatalyzed addition mechanism of Figure 8.12.

At a slightly elevated temperature, alcohols add to the C=N double bond of isocyanates according to the same mechanism. The addition products are called carbamic acid esters or **urethanes**:

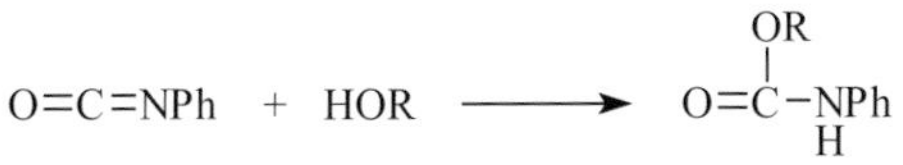

$$O{=}C{=}NPh + HOR \longrightarrow O{=}C(OR){-}NHPh$$

In classical organic analysis liquid alcohols were converted into the often easily crystallizable ***N*-phenyl urethanes**. These were characterized by their melting points and thus distinguished from each other. By comparing these melting points with the tabulated melting point of *N*-phenyl urethanes previously characterized, it was also possible to identify the alcohols used.

Side Note 8.1. Synthesis of Polyurethanes

Just as alcohols add to isocyanates, diols add to diisocyanates. In a polyaddition reaction **polyurethanes** are then produced, like the one shown here:

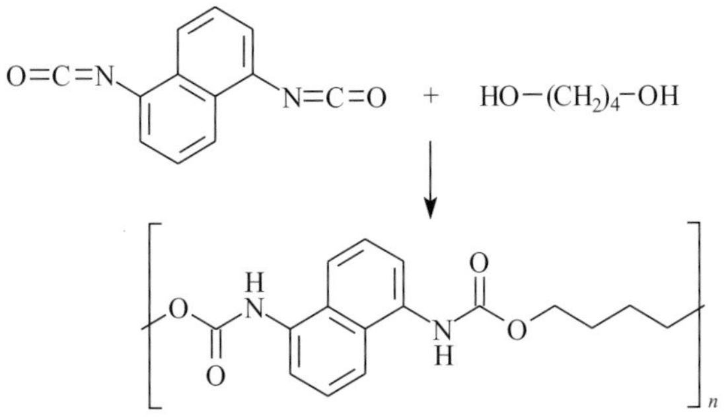

$$O{=}C{=}N{-}C_{10}H_6{-}N{=}C{=}O + HO{-}(CH_2)_4{-}OH \longrightarrow [{-}O{-}C(=O){-}NH{-}C_{10}H_6{-}NH{-}C(=O){-}O{-}(CH_2)_4{-}]_n$$

Polyurethanes are important synthetic macromolecules. They are manufactured, for example, in the form of foams. Such a foam is obtained during the formation and solidification of the polyurethane when a gas escapes from the reaction mixture, expanding the material. An elegant possibility for generating such a gas uniformly distributed everywhere in the reaction medium is as follows: besides the diol, one adds a small amount of H_2O to the diisocyanate. H_2O also adds to the C=N double bond of the diisocyanate. According to the uncatalyzed addition mechanism of Figure 8.12, this produces an *N*-arylated free carbamic acid Ar—NH—C(=O)—OH. However, such a compound decomposes easily in perfect analogy to the decomposition reaction **K** → **L** → $CO_2 + NH_3$ in Figure 8.3. The products are a primary aryl amine and *gaseous carbon dioxide* with the latter expanding the polyurethane to give a foam.

The sensitivity of isocyanates toward hydrolysis to give primary amines and CO_2 achieved sad fame with the largest chemical accident in history. In 1984 in Bhopal, India, water penetrated into or was conducted into a large tank full of methyl isocyanate. The result was a massive hydrolysis of the following type:

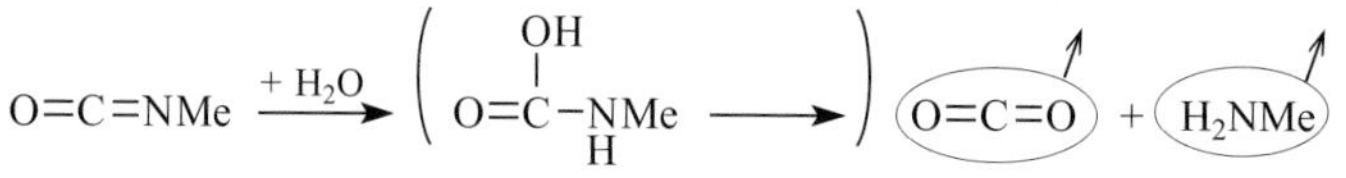

$$O{=}C{=}NMe \xrightarrow{+\,H_2O} \left(O{=}C(OH){-}NHMe \longrightarrow \right) O{=}C{=}O\uparrow + H_2NMe\uparrow$$

The released gases burst the tank, and about 40 tons of methyl isocyanate was released. It killed thousands of people and left tens of thousands with chronic ailments.

The toxicity of methyl isocyanate comes as a surprise if its high sensitivity to hydrolysis is considered. But actually, it may reach its site of action in the human body largely undecomposed. This is due to the reversible addition of another nucleophile to this heterocumulene. The tripeptide gluthathione, which is supposed to protect the body against oxidizing agents, adds to the C=N double bond of the isocyanate by means of its thiol group whereby the thiocarbamate is formed. When the latter decomposes in a reversal to its formation reaction, it releases the intact toxic methyl isocyanate.

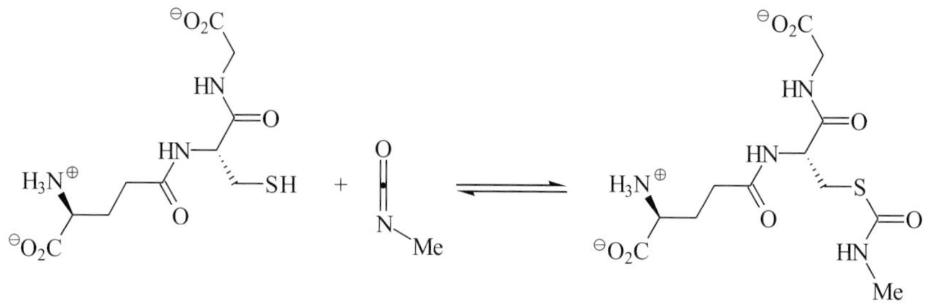

Primary and secondary amines also add to the C=N double bond of isocyanates via the mechanism of Figure 8.12. These reactions produce ureas:

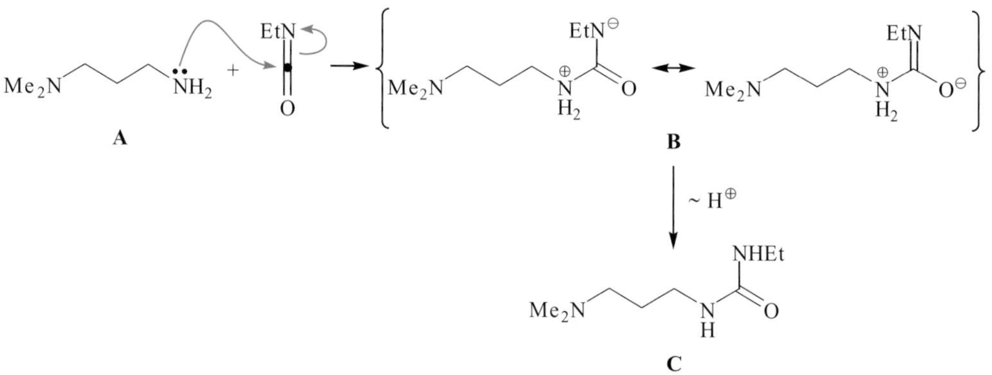

Ureas obtained in this way can contain different substituents on the two N atoms, depending on whether the amine and isocyanate had the same substituents or not.

How an *unsymmetrical* urea can be prepared from a primary amine with a (diethylamino)propyl substituent (**A**) and ethyl isocyanate is illustrated using the example of compound **C**. This urea is the starting material for preparing a carbodiimide (see Figure 8.9), which activates carboxylic acids towards heteroatom nucleophiles.

Additions to Isothiocyanates

Just as amines add to the C=N double bond of isocyanates giving ureas (Figure 8.9), they add to the C=N double bond of isothiocyanates, producing thioureas. An important example in this regard is the reaction **A** + Ph–N=C=S → **B** of the three-step Edman degradation of oligopeptides (Figure 8.14).

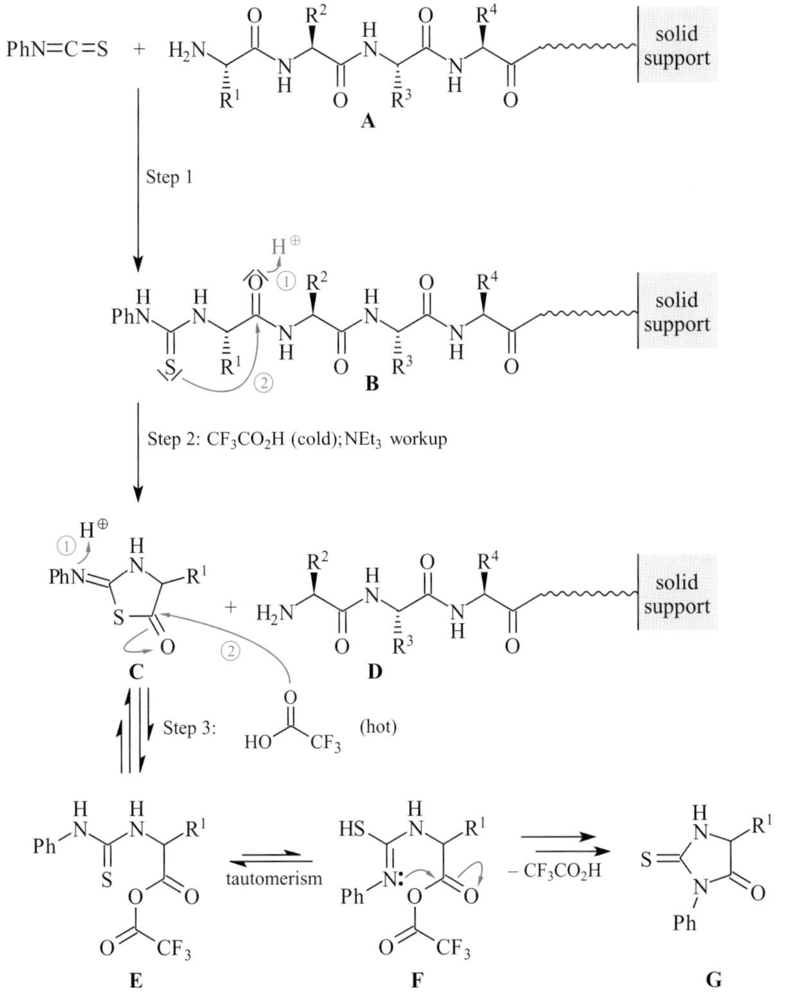

Step 1

Step 2: CF_3CO_2H (cold); NEt_3 workup

Step 3: CF_3CO_2H (hot)

tautomerism

$-\ CF_3CO_2H$

Fig. 8.14. Edman degradation of polymer-bound oligopeptides. Here, the three-step reaction sequence is shown.

Side Note 8.2. Sequence Determination of Oligopeptides with the Edman Degradation

The Edman degradation is used to determine the amino acid sequence of oligopeptides starting from the N terminus. The oligopeptide to be sequenced is linked to a solid support in the form of a derivative **A** (formula in Figure 8.14). The advantage of working on such a solid support is the following. After the first cycle of the Edman degradation, one will have obtained a mixture of oligopeptide **D,** which has lost the originally present N-terminal amino acid, and of the heterocycle **G,** into which the amino acid removed has been incorporated. Because oligopeptide **D** is polymer-bound it is separable from heterocycle **G** by filtration. This makes the isolation of **D** particularly easy and thus allows it to be *quickly* subjected to the next cycle of the Edman degradation.

Step 1 of the Edman degradation is the addition of the NH_2 group of the N-terminal amino acid to the C=N double bond of phenyl isothiocyanate. Step 2 (**B** → **C**) is an intramolecular S_N reaction of an S nucleophile on the carboxyl carbon of a protonated amide. It follows the substitution mechanism shown in Figure 6.5. The substitution product **C** is a heterocyclic derivative of the N-terminal amino acid. The other reaction product, the oligopeptide **D,** which has been shortened by one amino acid, is ejected as the leaving group. Next, the new oligopeptide **D** is degraded via another cycled. It then loses *its* N-terminal amino acid—that is, the second amino acid of the original oligopeptide **A** counting from its N-terminus—in the form of an analogous heterocycle, and so on.

The analytical objective is to determine amino acids used to form heterocycles of type **C,** one of them being released per pass through the Edman degradation. However, this determination is not easy, as it turns out. Therefore, the Edman degradation contains a third step, in which the heterocycles **C** are isomerized. This takes place through the sequence acylation (formation of **E**), tautomerism (formation of **F**), and acylation (formation of **G**), shown in Figure 8.14 at the bottom. The reaction **C** → **E** is an S_N reaction on the carboxyl carbon, which follows the mechanism of Figure 6.5. The reaction **F** → **G** is also an S_N reaction on the carboxyl carbon, but it follows the mechanism of Figure 6.2. With the heterocycle **G**—a so-called thiohydantoin—one has obtained a compound in which the heterocyle, and its constituent amino acid, can be identified very easily by a chromatographic comparison with authentic reference compounds. Thiohydantoins are the sulfur analogs of the hydantoins you have already encountered with the Bucherer modification of Strecker's amino acid synthesis (Formula **B** in Figure 7.11).

8.3 Interconversions of Carbonic Acid Derivatives via Heterocumulenes as Intermediates

The most important example of the reactions in this section's title is the second reaction of the urea synthesis (**B**) (Figure 8.15). It starts from ammonium carbamate (**A**) generated *in situ*. In a reversible acid/base reaction, **A**—to a minor extent—is transformed into NH_3 and the unsubstituted carbamic acid **C**. In a reversal of its formation reaction the latter decomposes—

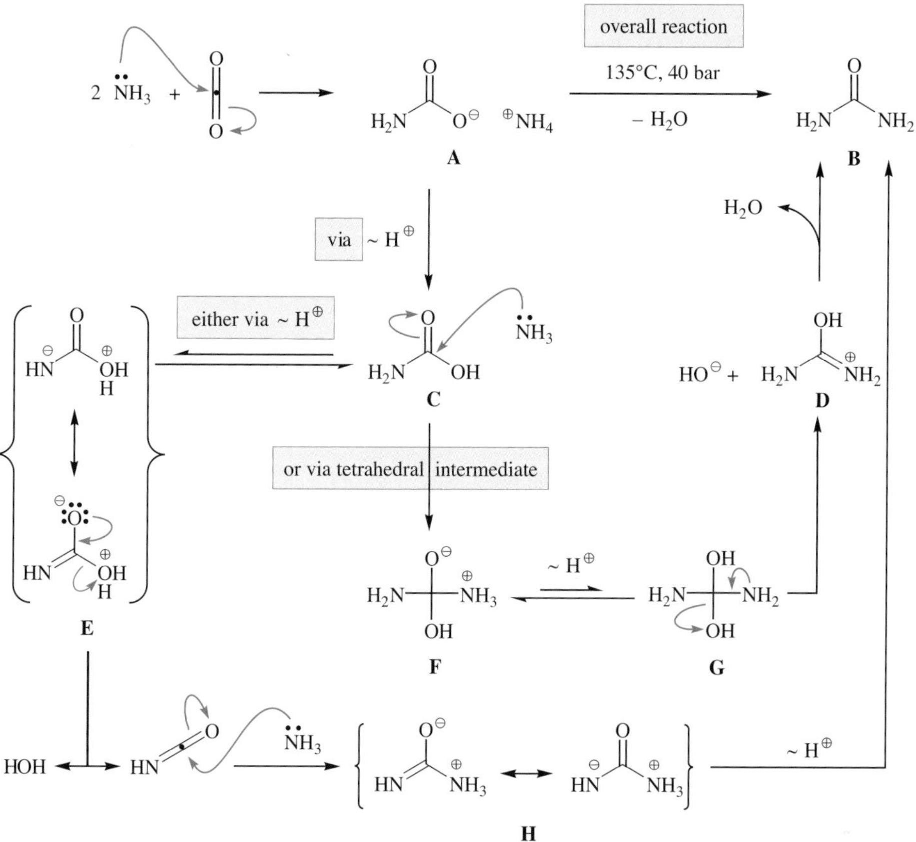

according to the third line in Figure 8.3—back into CO_2 and NH_3, but according to the same figure and Section 8.2 also continuously regenerates from these constituents.

In forward direction, the carbamic acid **C** enters into the reaction in Figure 8.15 through the acylation of NH_3. It is not known whether this acylation whether it starts with the formation of the tetrahedral intermediate **F**, followed by formation of **G** and **D**) to ultimately give urea (**B**). or if the carbamic acid is transformed into the zwitterion (**E**). Admittedly, because of its being both an amide anion and an oxonium ion, it would make a less stable zwitterion than the zwitterion **L** in Figure 8.3. The latter is the intermediate in the decomposition of carbamic acid into carbon dioxide and ammonia and—advantageously—is both a carboxylate anion and an ammonium ion. Nevertheless, *if* the zwitterion **E** of Figure 8.15 is formed, it could decompose into a heteroatom nucleophile (in Figure 8.15: water) and a heterocumulene (in Figure 8.15: isocyanic acid). The second equivalent of ammonia in the reaction mixture would undergo a nucleophilic addition to this isocyanic acid and first lead to the zwitterion **H** and then, after a proton shift, to urea (**B**).

While the mechanism of the transformation of carbonic acid derivative (**C**) → carbonic acid derivative (**B**) in Figure 8.15 is uncertain, it is well-known that two transformations of a carbonic acid derivative into another carbonic acid derivative, which have already been discussed from different perspectives, proceed via a heterocumulene intermediate. You have

Fig. 8.15. Synthesis of urea. The overall reaction is shaded red. The mechanism of the initial formation of ammonium carbamate was discussed using the (unnumbered) formula on p. 351. Here we present the mechanism of the subsequent conversion of ammonium carbamate into the final product.

already come across the first example of this heterocumulene mechanism in Figure 8.3 in the formation of biuret (**O**). The second example of such a carbonic acid derivative → carbonic acid derivative transformation via a heterocumulene—was part of the Bucherer modification of the Strecker synthesis of methionine (Figure 7.12) in the form of an isomerization of the heterocyclic carbonic acid derivatives **G** to the heterocyclic carbonic acid derivative **C** via the isocyanate **E**.

References

8.1

W. E. Hanford, J. C Sauer, "Preparation of Ketenes and Ketene Dimers," *Org. React.* **1946**, *3,* 108–140.

H. R. Seikaly, T. T. Tidwell, "Addition Reactions of Ketenes," *Tetrahedron* **1986**, *42*, 2587.

H. Ulrich, "Chemistry and Technology of Isocyanates," Wiley, Chichester, U. K., **1996**.

M. V. Vovk, L. I. Samarsi, "N-Functionalized carbodiimides," *Russ. Chem. Rev* **1992**, *61*, 297–305.

C. Agami, F. Couty, "The Reactivity of the N-Boc Protecting Group: An Underrated Feature," *Tetrahedron* **2002**, *58,* 2701–2724.

8.2

D. Belli Dell'Amico, F. Calderazzo, L. Labella, F. Marchetti, G. Pampaloni, "Converting Carbon Dioxide into Carbamato Derivatives," *Chem. Rev.* **2003**, *103,* 3857–3897.

S. Z. Zard, "On the Trail of Xanthates: Some New Chemistry from an Old Functional Group," *Angew. Chem. Int. Ed. Engl.* **1997**, *36*, 672–685.

A.-A. G. Shaikh, S. Sivaram, "Organic Carbonates," *Chem. Rev.* **1996**, *96*, 951–976.

J. P. Parrish, R. N. Salvatore, K. W. Jung, "Perspectives on Alkyl Carbonates in Organic Synthesis," *Tetrahedron* **2000**, *56*, 8207–8237.

E. Däbritz, "Syntheses and Reactions of *O,N,N'*-Trisubstituted Isoureas," *Angew. Chem. Int. Ed. Engl.* **1966**, *5*, 470–477.

V. F. Pozdnev, "Activation of Carboxylic Acids by Pyrocarbonates. Scope and Limitations," *Org. Prep. Proced. Int.* **1998**, *30*, 631–655.

Further Reading

Y. Ono, "Dimethyl Carbonate for Environmentally Benign Reactions," *Pure Appl. Chem.* **1996**, *68*, 367–376.

P. Tundo, M. Selva, "The Chemistry of Dimethyl Carbonate," *Acc. Chem. Res.* **2002**, *35,* 706–716.

L. Cotarca, P. Delogu, A. Nardelli, V. Sunjic, "Bis(trichloromethyl) Carbonate in Organic Synthesis," *Synthesis* **1996**, 553–576.

Y. I. Matveev, V. I. Gorbatenko, L. I. Samarai, "1,1-Dihaloalkyl Heterocumulenes: Synthesis and Reactions," *Tetrahedron* **1991**, *47*, 1563–1601.

N. A. Nedolya, B. A. Trofimov, A. Senning, "α,β-Unsaturated Isothiocyanates," *Sulfur Rep.* **1996**, *17*, 183–395.

J. H. Rigby, "Vinyl Isocyanates as Useful Building Blocks for Alkaloid Synthesis," *Synlett* **2000**, 1–12.

W. D. Rudorf, "Reactions of Carbon Disulfide with C-Nucleophiles," *Sulfur Rep.* **1991**, *11*, 51–141.

G. Maier, H. P. Reisenauer, R. Ruppel, "Matrix Isolation of Chalcogeno Heterocumulenes," *Sulfur Rep.* **1999**, *21*, 335–355.

A. A. Bakibayev, V. V. Shtrykova, "Isoureas: Synthesis, Properties, and Applications," *Russ. Chem. Rev.* **1995**, *64*, 929–938.

Additions of Heteroatom Nucleophiles to Carbonyl Compounds and Subsequent Reactions—Condensations of Heteroatom Nucleophiles with Carbonyl Compounds

9

In general, carbonyl compounds can react with the following heteroatom nucleophiles:

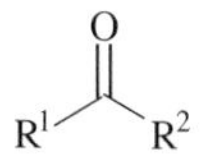

$R^1C(=O)R^2$ + H—OH, H—OR,

H—SR, H—$SO_3^{\ominus}$ $Na^{\oplus}$

H—NH_2, H—NHR including H—NH—Het and H—NH—C(=O)–, H—NR_2

R^1, R^2 = H, Alk, Ar

H—CN (a *C* nucleophile!)

These are *O*-, *S*- and *N* nucleophiles. Halide ions are not able to react as nucleophiles with carbonyl compounds, but a "pseudohalide," that is, the cyanide ion, is. The addition of the cyanide ion to aldehydes and ketones displays considerable analogies with the addition reactions of *O*-, *S*- and *N* nucleophiles and this is why Section 9.1 addresses these cyanide additions.

9.1. Additions of Heteroatom Nucleophiles or Hydrocyanic Acid to Carbonyl Compounds

Only three heteroatom nucleophiles add to a significant extent to carbonyl compounds without being followed by secondary reactions such as S_N1 reactions (Section 9.2) or E1 reactions (Section 9.3): H_2O, alcohols, and, should the substitution pattern be suitable, the carbonyl compound itself.

H_2O or alcohols as nucleophiles give low molecular weight compounds when they add to the C=O double bond of carbonyl compounds. These addition products are called aldehyde or ketone hydrates (Section 9.1.1) and hemiacetals or hemiketals (Section 9.1.2), respectively, depending on whether they result from the addition to an aldehyde or a ketone. Today, one no longer distinguishes systematically between hemiacetals and hemiketals, but the expression "hemiacetal" is frequently used to cover both.

Occasionally, aldehydes can add to themselves (Section 9.1.4). If so, cyclic oligomers or acyclic polymers are formed.

Bruckner R (author), Harmata M (editor) In: *Organic Mechanisms – Reactions, Stereochemistry and Synthesis*
Chapter DOI: 10.1007/978-3-642-03651-4_9,

9.1.1 On the Equilibrium Position of Addition Reactions of Heteroatom Nucleophiles to Carbonyl Compounds

The additions of H_2O or alcohols to the C=O double bond of carbonyl compounds as well as the oligomerizations or polymerizations of aldehydes are *reversible* reactions. Therefore, the extent of product formation is subject to thermodynamic control. The equilibrium constant of the formation of the respective addition product is influenced by steric and electronic effects.

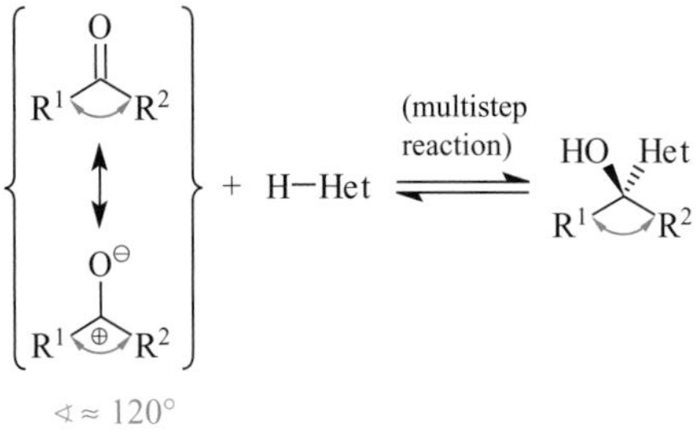

$$\left\{ R^1C(=O)R^2 \leftrightarrow R^1C^{\oplus}(O^{\ominus})R^2 \right\} + \text{H—Het} \underset{}{\overset{\text{(multistep reaction)}}{\rightleftharpoons}} R^1R^2C(OH)(\text{Het})$$

$\sphericalangle \approx 120°$

R^1	R^2	Substituent effect in starting material	Steric hindrance of the product	Consequence for equilibrium position
Alkyl	Alkyl	Stabilizes more	Significant	On starting material side
Alkyl	H	Stabilizes somewhat	Present but small	
H	H	= Reference	= Reference product	↓
EWG	H	Destabilizes clearly	Present but small	
EWG	EWG	Destabilizes much	Present but small	On product side

Fig. 9.1. Substituent effects on the equilibrium position of the addition reactions of H_2O (Het = OH), alcohols (Het = OAlkyl) and the respective carbonyl compounds [Het = $O(—CR^1R^2—O)_n—H$] to aldehydes and ketones (EWG, electron-withdrawing group).

For a given nucleophile the equilibrium lies farther on the product side the smaller the substituents R^1 and R^2 of the carbonyl compound are (Figure 9.1). Large substituents R^1 and R^2 inhibit the formation of addition products. This is because they come closer to each other in the addition product, where the bonds to R^1 and R^2 are closer than in the carbonyl compound, where these substituents are separated by a bond angle of about 120°. Formaldehyde is the sterically least hindered carbonyl compound. In H_2O this aldehyde is present completely as dihydroxymethane, and anhydrous formaldehyde is exists completely as polymer. In contrast, acetone is so sterically hindered that it does not hydrate, oligomerize, or polymerize at all.

The equilibrium position of addition reactions to carbonyl compounds is also influenced by electronic substituent effects, as was also shown in Figure 9.1:

Influence of Substituents on the Equilibrium Position of Addition Reactions to the Carbonyl Group

- Substituents with a electron-donating inductive (+I) effect (i.e., alkyl groups) stabilize the C=O double bond of aldehydes and ketones. They increase the importance of the zwitterionic resonance form by which carbonyl compounds are partly described. The driving force for the formation of addition products from carbonyl compounds therefore decreases in the order H—CH(=O) > R—CH(=O) > $R^1R^2{}_C$(=O).

- Alkenyl and aryl substituents stabilize the C=O double bond of carbonyl compounds even more than alkyl substituents. This is due to their pi electron-donating (+M) effect, which allows one to formulate additional zwitterionic resonance forms for carbonyl compounds of this type. Thus, no hydrates, hemiacetals, oligomers, or polymers can be derived from unsaturated or aromatic aldehydes.
- Electron-withdrawing substituents at the carbonyl carbon destabilize the zwitterionic resonance form of aldehydes and ketones. Thus, they deprive these compounds of the resonance stabilization, which the alkyl substituents usually present would give them. Therefore, addition reactions to acceptor-substituted C=O double bonds have an increased driving force.

The effects illustrated by Figure 9.1 are corroborated by the data in Table 9.1 for the hydration equilibria of differently substituted carbonyl compounds.

Tab. 9.1 Position of the Hydration Equilibrium in the System Carbonyl Compound/H_2O

carbonyl compound	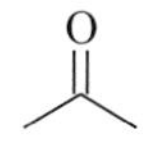 O ‖ (CH₃)C(CH₃)	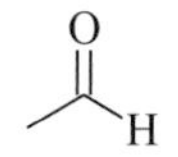 O ‖ CH₃C–H	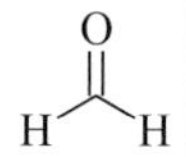 O ‖ H–C–H	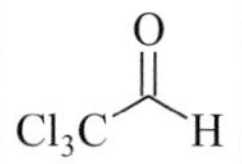 O ‖ Cl_3C–C–H	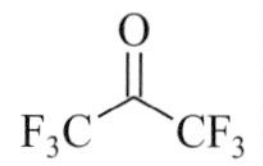 O ‖ F_3C–C–CF_3
% hydrate at equilibrium	≪ 0.1	58	100[1)]	100[2)]	100[2)]

[1)] Not isolable. [2)] Isolable.

9.1.2 Hemiacetal Formation

Structural Dependence of the Reaction

In the presence of a base or an acid as a catalyst, alcohols add to aldehydes or ketones with the formation of hemiacetals. You should already be familiar with the corresponding mechanism and its intermediates from your introductory class. Therefore it is sufficient that we briefly review it graphically by means of Figure 9.2.

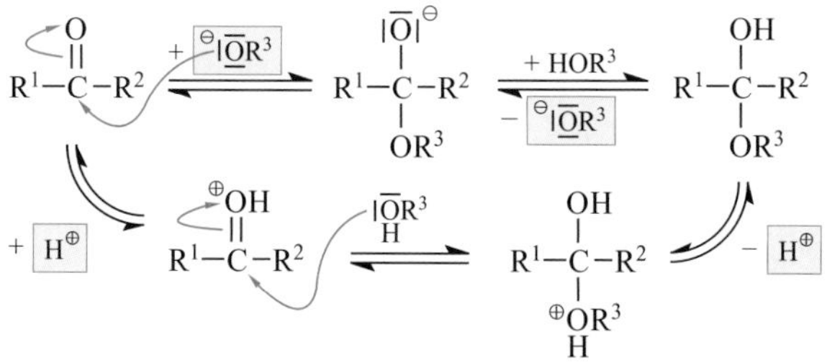

Fig. 9.2. Base-catalyzed (top) and acid-catalyzed (bottom) hemiacetal formation from carbonyl compounds and alcohols.

What you have learned in Section 9.1.1 about electronic substituent effects explains why intermolecular hemiacetal formation from the electron-deficient carbonyl compounds methyl glyoxalate **(A)** and ninhydrin **(B)** take place quantitatively:

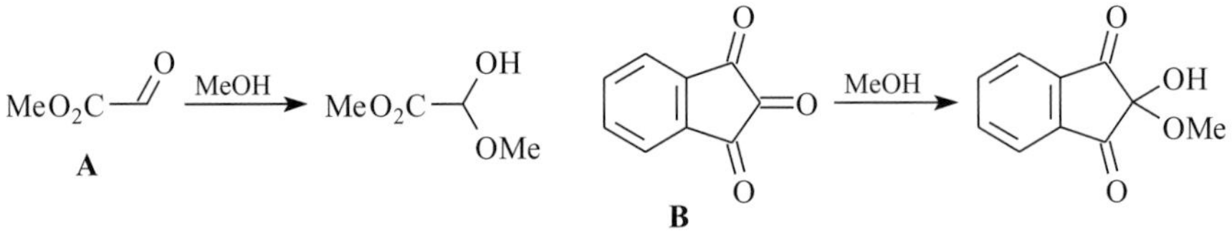

MeO_2C O MeOH MeO_2C OH OMe
A

O O O MeOH O OH OMe O
B

α-Hydroxylated aldehydes, too, are α-acceptor-substituted carbonyl compounds and as such are capable of forming relatively stable hemiacetals. However, since these aldehydes contain a hydroxy group, no additional alcohol is required in order for hemiacetals to be produced from *them*. Rather, these aldehydes form a hemiacetal through the addition of the OH group of one of their own (i.e., the OH group *of a second molecule*). Figure 9.3 shows a reaction of this type using the example of the conversion of the α-hydroxyaldehyde **C** into the hemiacetal **D.**

Compound **D** immediately reacts further, as is typical for this type of compound (Figure 9.4). The result is an intramolecular addition of remaining OH group to the remaining C=O double bond. Thus, the *bis*(hemiacetal) **B** is produced. This second hemiacetal formation is more favorable entropically than the first one: The number of molecules that can move about independently remains constant in the second hemiacetal formation (**D** → **B**), whereas this number is reduced by a factor of one-half in the first hemiacetal formation (**C** → **D**). The second reaction of Figure 9.3 therefore drives the entire reaction to the product side.

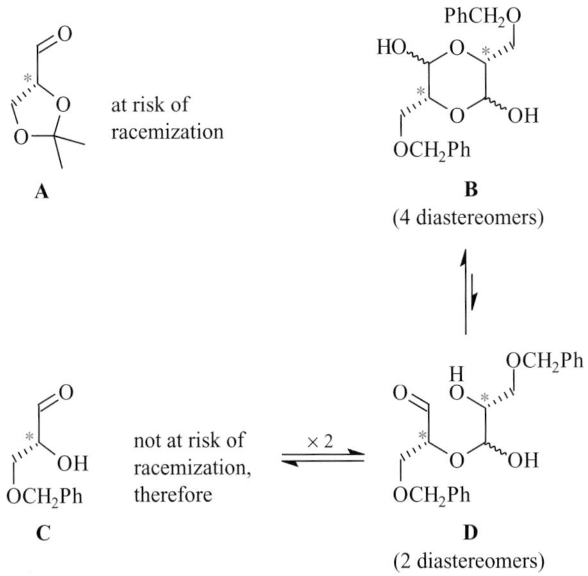

O O O
A
at risk of racemization

$PhCH_2O$ HO O O OH OCH_2Ph
B
(4 diastereomers)

O OH OCH_2Ph
C
not at risk of racemization, therefore
× 2

O H O OCH_2Ph O OH OCH_2Ph
D
(2 diastereomers)

Fig. 9.3. Formation of a bis(hemiacetal) from an α-hydroxyaldehyde by dimerization.

The quantitative formation of the bis(hemiacetal) **B** from the glyceraldehyde derivative **C,** as just discussed, led to the recent proposal to use the glyceraldehyde derivative **C** as a storage form of the glyceraldehyde derivative **A** (Figure 9.3). The latter compound cannot be stored because it is prone to racemization as its C=O group allows for ready enolate and enol formation (both of which destroy the stereochemical integrity of the stereocenter at C-α). Because compound **C** dimerizes quantitatively to give **B,** in contrast to **A, C** is not at all present as a carbonyl compound and is therefore not threatened by racemization through enolate or enol formation.

As can be seen from Figure 9.1, carbonyl compounds without electron-withdrawing α-substituents do not react *intermolecularly* with alcohols to form hemiacetals to any significant extent. However, while for such carbonyl compounds there is too little driving force for hemiacetalization to occur, the reaction is not drastically disfavored. This explains why this type of compound undergoes almost complete hemiacetal formation provided it takes place intramolecularly and leads to a nearly strain-free five- or six-membered cyclic hemiacetal—a so-called **lactol** (Figure 9.4). What makes the difference is that only in the intramolecular hemiacetal formation is no translational entropy lost (because the number of molecules moving about independently of each other does not decrease).

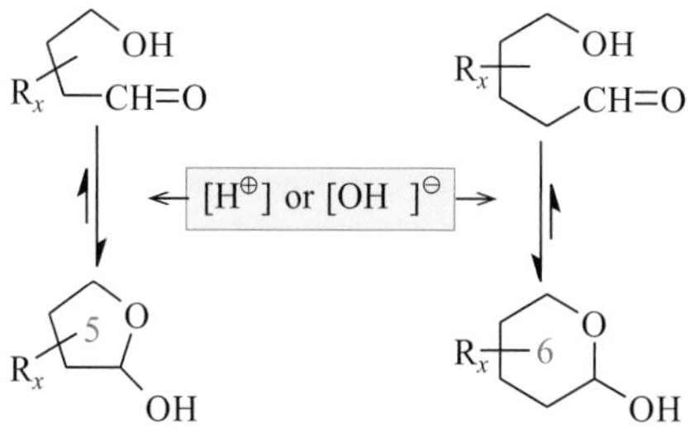

Fig. 9.4. Hemiacetal formation from γ- or δ-hydroxyaldehydes.

Five- or six-membered cyclic hemi*ketals*, which are *also* referred to as lactols, can form from γ- or δ-hydroxy*ketones*. However, hemiketals of this type are not necessarily more stable than the acyclic hydroxyketone isomers because ketones are less thermodynamically suitable to add nucleophiles than are aldehydes (Section 9.1.1).

In order to understand certain reactions of lactols with nucleophiles, you must now familiarize yourself with the **principle of microscopic reversibility.**

Principle of Microscopic Reversibility

When a molecule **A** is converted to a molecule **B** through a certain mechanism—and it does not matter whether this mechanism comprises one or more elementary reactions—a conversion of product **B** back into the starting material **A** takes place through the very same mechanism.

Lactols form, just like acyclic hemiacetals or hemiketals, according to one of the two mechanisms of Figure 9.2. From the principle of microscopic reversibility, the following is therefore concluded: the ring opening of a lactol to give a hydroxycarbonyl compound proceeds exactly via the pathways shown in Figure 9.2. Hence, both in acidic or basic solutions, a lactol gives rise to an equilibrium amount of the hydroxycarbonyl compound. Through the latter, a lactol can often react with nucleophiles.

Stereochemistry

δ-Hydroxyvaleraldehyde is achiral. When it cyclizes to give the six-membered ring lactol, which, of course, takes place under thermodynamic control, the lactol is formed as a 50:50 mixture of the two enantiomers (Figure 9.5). Each of these enantiomers in turn is present as a 78:22 mixture of two chair conformers. The conformer with the axially oriented OH group is in each case more stable than the conformer with the equatorially oriented OH group. In the carba analog (i.e., in cyclohexanol) one finds almost exactly the opposite conformational preference: at 25 °C the axial and the equatorial alcohols are present in a ratio of 27:73. This corresponds to a $\Delta G°$ value of –0.60 kcal/mol for the conversion cyclohexanol$_{\text{OH,axial}}$ → cyclohexanol$_{\text{OH,equatorial}}$. It is well-known that the equatorial conformer of a monosubstituted cyclohexane is more stable than the axial conformer due to a steric effect. The steric effect that drives the OH group in the lactols of the δ-hydroxyvaleraldehyde—i.e., in 2-hydroxytetrahydropyrans—into equatorial position should be somewhat greater than in the case of cyclohexane. This may be inferred from the fact that the $\Delta G°$ value for the conversion tetrahydropyran$_{\text{2-CH}_3\text{, axial}}$ → tetrahydropyran$_{\text{2-CH}_3\text{, equatorial}}$ is –2.9 kcal/mol. If the $\Delta G°$ value for the conversion tetrahydropyran$_{\text{2-OH, axial}}$ → tetrahydropyran$_{\text{2-OH, equatorial}}$ relates to the –2.9 kcal/mol as the $\Delta G°$ value for the conversion cyclohexane$_{\text{OH, axial}}$ → cyclohexane$_{\text{OH, equatorial}}$ (–0.60 kcal/mol) relates to the $\Delta G°$ value for the conversion cyclohexane$_{\text{CH}_3\text{, axial}}$ → cyclohexane$_{\text{CH}_3\text{, equatorial}}$ (–1.7 kcal/mol), this would mean that the $\Delta G°$ value for the tetrahydropyran$_{\text{OH, axial}}$ → tetrahydropyran$_{\text{OH, equatorial}}$ conversion is approximately (0.6/1.73) ↔ (–2.9) kcal/mol or –1.0 kcal/mol. Since the more stable lactol is *nonetheless* the 2-hydroxytetrahydropyran with the axial OH group, an opposing effect must determine the position of the conformational equilibrium. This is the anomeric effect, which is a stereoelectronic effect (Figure 6.8). In a substructure O—C$_{sp^3}$—O it contributes stabilization energy of 1–2.5 kcal/mol. The higher stability of the axially vs. equatorially hydroxylated δ-valerolactol results, in the MO picture, from an overlap between the axially oriented free electron pair on the ring O atom and the $\sigma^*_{\text{c—o}}$ orbital of the exocyclic C—OH bond.

The installment of additional hydroxy groups in the δ-hydroxyvaleraldehyde from Figure 9.5 generates the acyclic form of D-glucose (Fig. 9.6). Acyclic D-glucose is a *chiral* δ-hydroxyaldehyde. When it cyclizes to give the six-membered ring hemiacetal, a new stereocenter is formed so that two different hemiacetals can be produced. They are diastereomers and therefore not equally stable. Being formed under thermodynamic control, these hemiacetals therefore occur in a ratio different from 50:50, namely as a 63:37 mixture.

For each of these diastereotopic hemiacetals there are in principle two different chair conformers, but in each case only one chair conformer is found. The other one would contain too

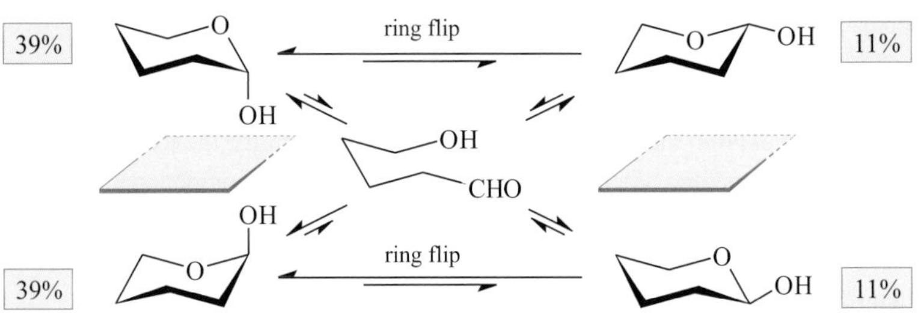

Fig. 9.5. Stereochemistry of the formation of δ-valerolactol (neat, at 38°C; ≤ 5% of the free hydroxyaldehyde is present). The top formulas represent the 78:22 mixture of the two chair conformers of one lactol enantiomer, whereas the bottom formulas represent the 78:22 mixture of the two chair conformers of the other lactol enantiomer.

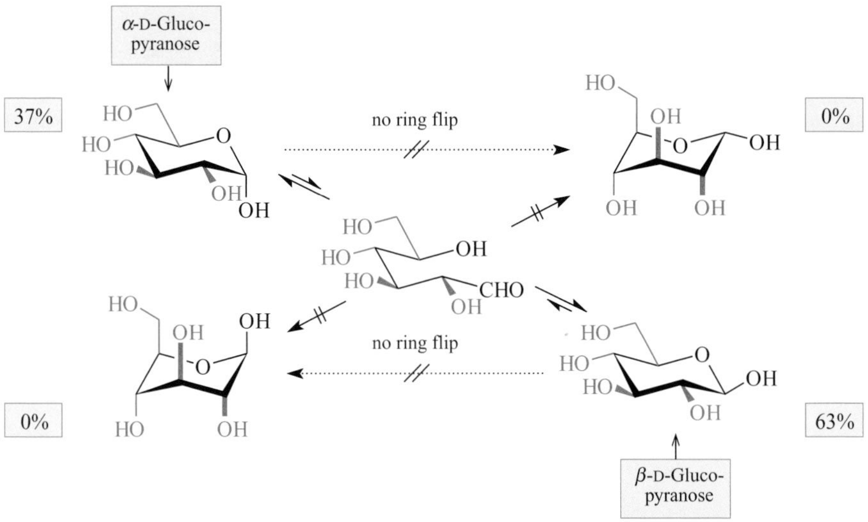

Fig. 9.6. Stereochemistry of the hemiacetal formation from D-glucose in an aqueous solution at 20 °C (in the formulas, the carbon skeleton of δ-valerolactol is black—which emphasizes the relationship to the reaction from Figure 9.5—whereas the extra substituents as well as the bonds leading to them are red).

many energetically unfavorable axial substituents. In an aqueous solution, the hemiacetal with an equatorial OH group at the newly produced stereocenter—the so-called **anomeric stereocenter**—predominates to the extent of 63%. It is the so-called β-D-glucopyranose. An axial OH group at the anomeric carbon is contained in the 37% fraction of the other hemiacetal, the α-D-glucopyranose.

The equatorial anomeric hydroxy group in β-D-glucopyranose is favored sterically but not from the anomeric effect previously described. In contrast, a stabilizing anomeric effect and a destabilizing steric effect occur in the α-D-pyranose—as was the case in the axially hydroxylated tetrahydropyran of Figure 9.5. This means that although the stability order of the underlying core structure of Figure 9.5 suggests the opposite, the α-D-glucopyranose is nevertheless the minor and the β-D-glucopyranose the major hemiacetal diastereomer of Figure 9.6.

What is the reason for this apparent discrepancy? It is a solvent effect. In *aqueous* solution (Figure 9.6), the OH group at the anomeric C atom of the glucose becomes so voluminous due to hydration that it strives for the position in which the steric interactions are as weak as possible. Thus, it moves into the equatorial position—with a $\Delta G°$ value of approximately –1.6 kcal/mol—to avoid a *gauche* interaction with the six-membered ring skeleton. (Remember that axially oriented substituents on the chair conformer of cyclohexane are subject to two *gauche* interactions with the two next-to-nearest C_{ring}—C_{ring} bonds. They therefore have a higher energy than equatorial substituents, which are not exposed to any *gauche* interactions.) This steric benefit more than compensates the absence of an anomeric stabilization, which only an axially oriented anomeric OH group would experience.

As already mentioned, δ-hydroxy*ketones* may be present in the open-chain form or as an isomeric cyclic hemiketal, depending on their substitution pattern. For example, a polyhydroxylated δ-hydroxyketone, D-fructose, is present *exclusively* in the form of hemiketals (Figure 9.7). Responsible for this equilibrium position is the fact that the carbonyl group involved contains an electron-withdrawing group in both α-positions, which favors the addition of nucleophiles (see Section 9.1.1).

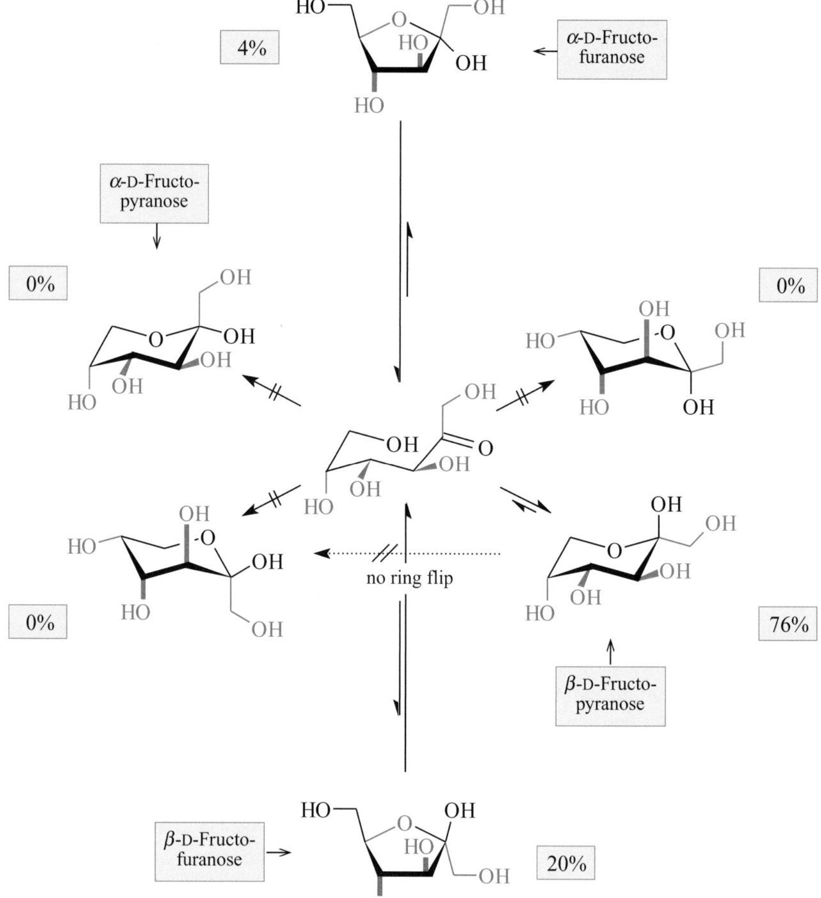

Fig. 9.7. Stereochemistry of the hemiketal formation from D-fructose in aqueous solution at 25 °C (in the formulas for the six-membered ring hemiketal the carbon skeleton of δ-valerolactol is black, which emphasizes the relationship to the reaction from Figure 9.5, whereas the extra substituents as well as the bonds leading to them are red).

In contrast to the case of D-glucose (Figure 9.6), in D-fructose there are two five-membered hemiketals in addition to one six-membered hemiketal (Figure 9.7). The six-membered ring hemiketal of D-fructose occurs as a pure β-diastereomer. The substituents on the anomeric carbon of this hemiketal are arranged in such a way that the OH group is oriented axially and the CH_2OH group equatorially. The OH group thus profits from the anomeric effect and the larger CH_2OH group from the minimization of steric hindrance.

9.1.3 Formation of Cyanohydrins and α-Aminonitriles

In the introduction to Chapter 9, it was mentioned that cyanide ions behave towards carbonyl compounds like the majority of *O*-, *S*- and *N* nucleophiles. They add to the C=O double bond. By taking up a proton the carbonyl oxygen is transformed into the OH group of a so-called **cyanohydrin**, i.e., a hydroxynitrile with geminal OH- and C≡N groups. The classical transformation of this type is is effected by treatment of a carbonyl compound with both sodium cyanide and sulfuric acid, a method that is used for the industrial-scale preparation of acetone cyanohydrin:

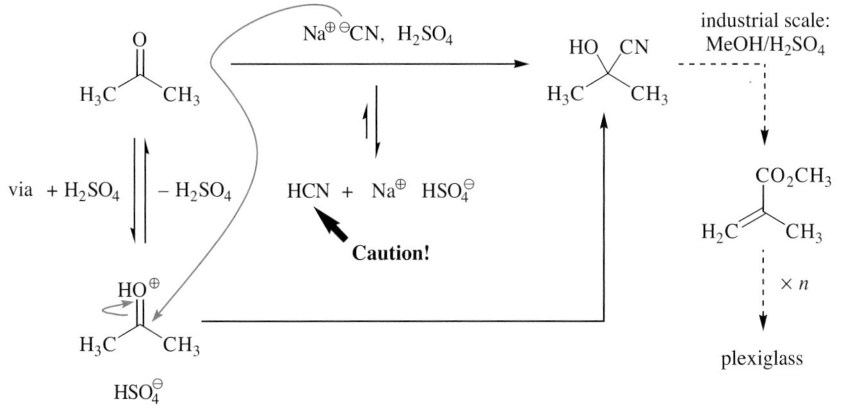

The nucleophilic reaction of the cyanide ion on the carbonyl group is facilitated by protonating the latter to a carboxonium ion. The addition of acid promotes the formation of cyanohydrins, but mainly for a thermodynamic reason. Under acidic conditions cyanohydrins equilibrate with the carbonyl compound and HCN. Under basic conditions they are in equilibrium with the same carbonyl compound and NaCN or KCN. The first reaction has a smaller equilibrium constant than the second, that is, the cyanohydrin is favored. So when cyanohydrins are formed under acidic or neutral (see Figure 9.8) instead of basic conditions, the reversal of the reaction is suppressed.

Incidentally, acetone cyanohydrin is also a reagent that may be used to transform other carbonyl compounds into cyanohydrins. An example is given in Figure 9.8. In contrast to both the above-mentioned $NaCN/H_2SO_4$ method and the $NaCN/NH_4Cl$ method (Figure 9.9), the

Fig. 9.8. Cyanide transfer from acetone cyanohydrin to 4-*tert*-butylcyclohexanone. The acetone cyanohydrin is employed as a solvent; this means it is present in great excess. Under thermodynamic control this reaction preferentially leads to the formation of a cyanohydrin with an axial nitrile group and an equatorial OH group, since a nitrile group is smaller than an OH group.

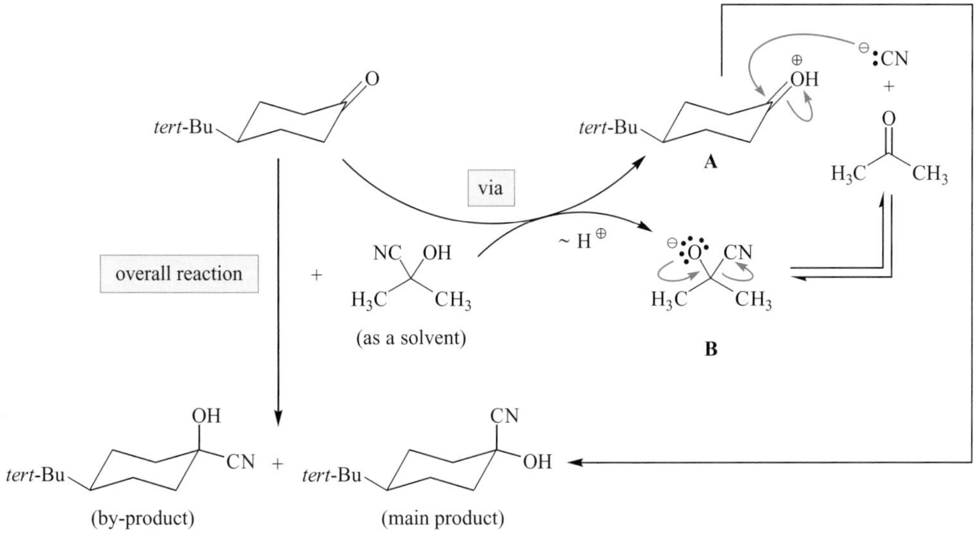

occurrence of hydrocyanic acid in the reaction mixture is negligible, which is a big advantage of this procedure. The protonation of the reaction partner to give the carboxonium ion proceeds more slowly in the procedure shown in Figure 9.8 than in the $NaCN/H_2SO_4$ approach since the proton donor is the acetone cyanohydrin itself. The latter—due to the electron-withdrawing effect of the C≡N group—is much more acidic than a normal alcohol, but less acidic than hydrocyanic acid. By the way, the success of the (trans)formation of cyanohydrin in Figure 9.8 in part relies on the fact that it is conducted in acetone cyanohydrin as the solvent, which maximizes the extent of the kinetically important protonation of the substrate to the carboxonium ion **A** and shifts the equilibrium acetone cyanohydrin + substrate ⇌ acetone + new cyanohydrin to the right.

The upper half of Figure 9.9 demonstrates that acetone can also be transformed into acetone cyanohydrin (**A**) by the combined treatment with sodium cyanide and ammonium chloride. Ammonium chloride is a weak acid. Consequently, the protonation of the (nonvolatile) sodium cyanide to the (volatile) hydrocyanic acid occurs to a lesser extent than with the $NaCN/H_2SO_4$ method (see above). However, ammonium chloride is not acidic enough to activate acetone as the carboxonium ion to the same extent as sulfuric acid. This *changes* the addition mechanism. As shown in Figure 9.9, it is the cyanide anions that react with the unactivated acetone. Only when the cyanohydrin anions have *thus* been formed does the ammonium chloride protonate them to yield the neutral cyanohydrin.

When acetone, sodium cyanide and ammonium chloride are reacted with each other for a longer time, but reversibly, an alternative reaction occurs between the continuously reformed reactants (Figure 9.9, bottom): ammonia undergoes condensation with acetone to give an iminium ion (mechanism: Figure 9.23) that irreversibly reacts with cyanide to furnish the α-aminonitrile as shown.

Fig. 9.9. Formation of cyanohydrin or α-aminonitrile from a carbonyl compound, ammonium chloride and sodium cyanide: a matter of reaction time. The formation of the cyanohydrin proceeds fast and reversibly, that of α-aminonitrile slowly and irreversibly.

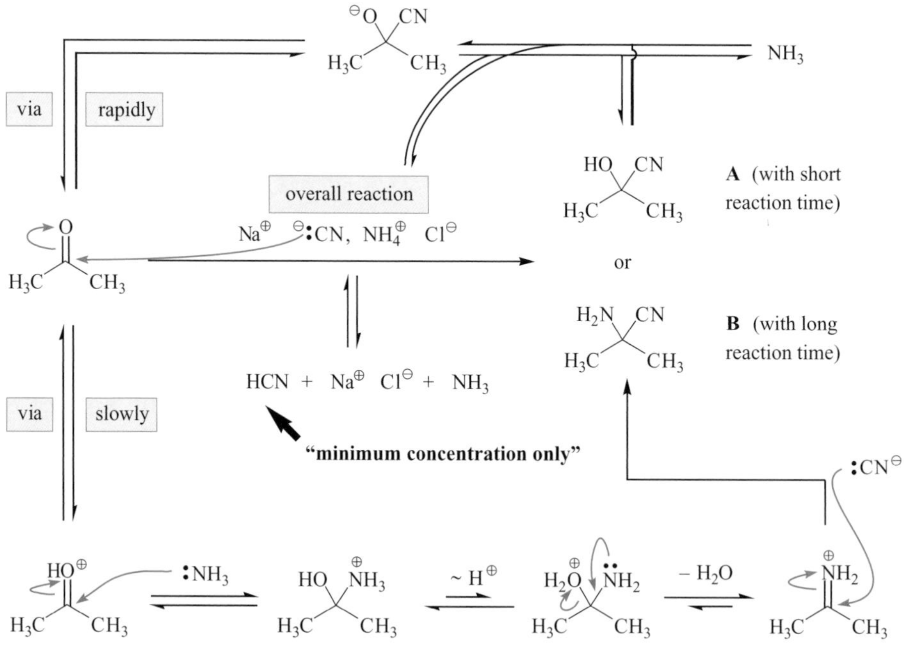

via | rapidly | overall reaction | $Na^{\oplus}$ $^{\ominus}$:CN, $NH_4^{\oplus}$ $Cl^{\ominus}$ | NH_3 | **A** (with short reaction time) | or | **B** (with long reaction time) | HCN + $Na^{\oplus}$ $Cl^{\ominus}$ + NH_3 | **"minimum concentration only"** | via | slowly | :NH_3 | ~ $H^{\oplus}$ | – H_2O | :$CN^{\ominus}$

The duration of the reaction time alone determines whether carbonyl compounds, sodium cyanide and ammonium chloride will generate a cyanohydrin (Figure 9.9, top) or an α-aminonitrile (Figure 9.9, bottom). We are already familiar with the first reaction pattern from the initial reaction of the three-step Kiliani–Fischer synthesis of aldoses (Figure 7.15). The second reaction pattern initiates the Strecker synthesis of α-amino acids, which is completed by a total hydrolysis of the C≡N group, as in the Bucherer modification discussed elsewhere (Figure 7.11).

9.1.4 Oligomerization of Aldehydes—Polymerization of Formaldehyde

Some aldehydes oligomerize in the presence of acids. The polymerization of formaldehyde (formation of **H**; see Figure 9.10) as well as that of aldehydes with strong electronegative sub-

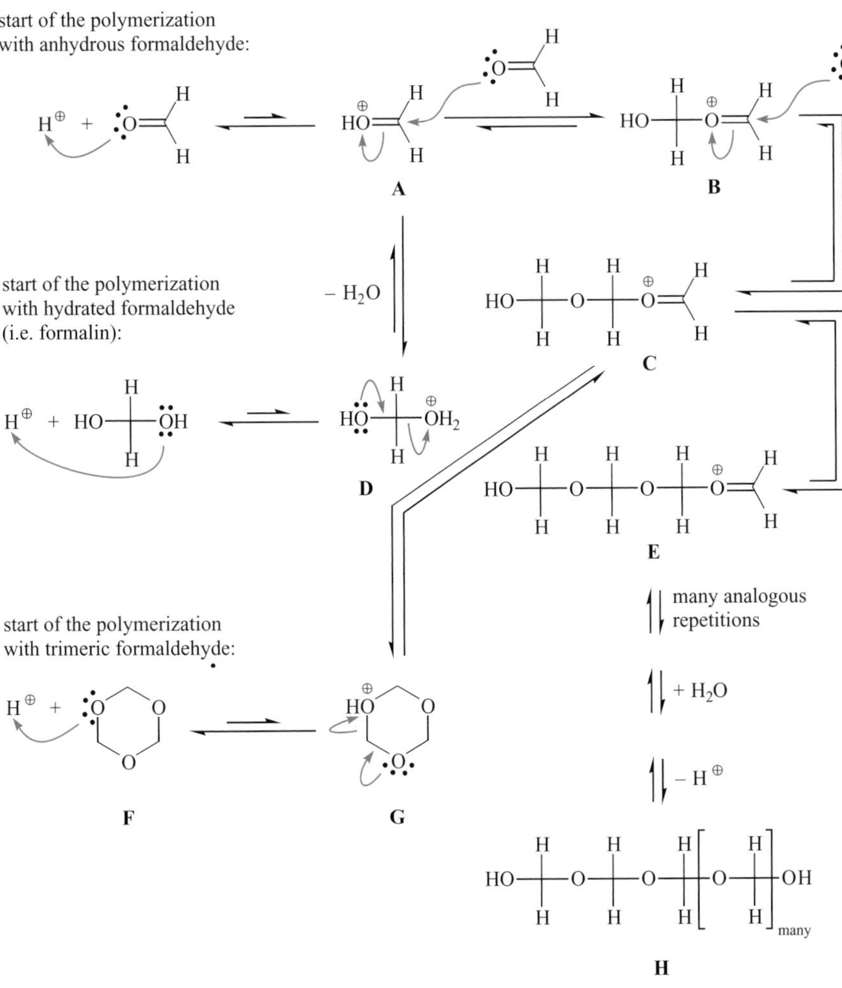

Fig. 9.10. Mechanism of the proton-catalyzed polymerization of monomeric formaldehyde (top), formalin (middle) and 1,3,5-trioxane (**F**; bottom) to give polyformaldehyde ("paraformaldehyde"; **H**), respectively. The carboxonium ions **A**, **B**, **C**, **E** etc. play a central role. They are transformed into each other through the nucleophilic reaction of a formaldehyde monomer on the respective carboxonium carbon atom. Analogously, **E** and formaldehyde continue to react to yield a high-molecular carboxonium ion that is intercepted by (traces of) water to furnish the final neutral product **H**.

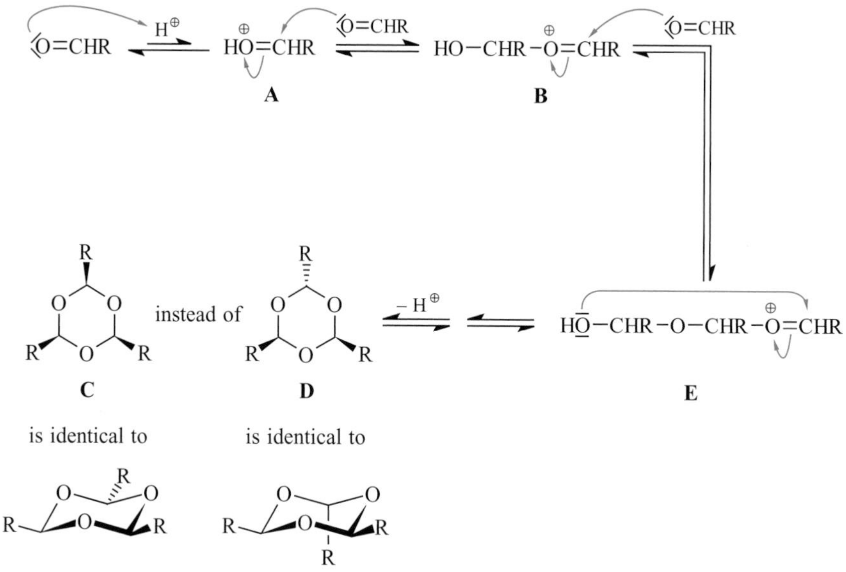

Fig. 9.11. Mechanism of the proton-catalyzed cyclotrimerization of aliphatic aldehydes.

stituents in the α-position are unavoidable reactions, even in the absence of an acid or base (example: Figure 9.10). The oligomerization of other unhindered aliphatic aldehydes is also possible, but it takes place much more slowly and usually not at neutral pH values. For example, in the presence of protons acetaldehyde and isobutyraldehyde trimerize to give the substituted trioxanes **C** (R = Me or *i*Pr, respectively) of Figure 9.11.

Figure 9.10 presents the mechanism of the polymerization of formaldehyde starting from anhydrous formaldehyde and formaldehyde hydrate. In addition, a reaction path is shown that also connects trimeric formaldehyde ("trioxane," **F**) with paraformaldehyde (**H**). In practice, though, this reaction path is only taken in the reverse direction. upon heating (entropy gain!) of paraformaldehyde in aqueous acid as a depolymerization of **H** → **F**.

The carboxonium ions of Figure 9.10 act as electrophiles in the polymerization of formaldehyde and formaldehyde hydrate. The most simple of them has the structural formula **A**, i.e., it is protonated formaldehyde from which the carboxonium ions **B**, **C**, **E** and so on are formed successively. The nucleophile causing these conversions is formaldehyde, which reacts with the cited electrophiles via its carbonyl oxygen and thus acts as a heteroatom nucleophile.

The mechanistic analysis of Figure 9.11 shows that the oligomerization of aliphatic aldehydes, in a totally analogous manner, can be considered as a reaction sequence that starts with the nucleophilic addition of *one* molecule of aldehyde to the C(=O$^{\oplus}$H) group of *another*, protonated molecule of aldehyde, namely with the reaction **A** → **B**. Again, an aldehyde acts as a nucleophile by reacting with its reaction partner using its carbonyl oxygen.

As you can see from Figures 9.10 and 9.11, all steps of the oligomerization or polymerization of aldehydes in the presence of protons are reversible. The trimerizations of the mentioned aldehydes acetaldehyde and isobutyraldehyde are therefore thermodynamically controlled. This is the reason why they take place stereoselectively and the trimers **C** (R = Me or *i*Pr) are produced diastereoselectively with a *cis*-configuration. In contrast to their *cis-, trans-, trans*-isomers **D,** the trimers **C** can occur as all equatorially substituted chair conformers.

9.2 Addition of Heteroatom Nucleophiles to Carbonyl Compounds in Combination with Subsequent S_N1 Reactions of the Primary Product: Acetalizations

9.2.1 Mechanism

The addition of one equivalent of a heteroatom nucleophile to carbonyl compounds was described in Section 9.1. However, if two or more equivalents of the heteroatom nucleophile are present, in many cases the initial *addition* products (formula **A** in Figure 9.12) react

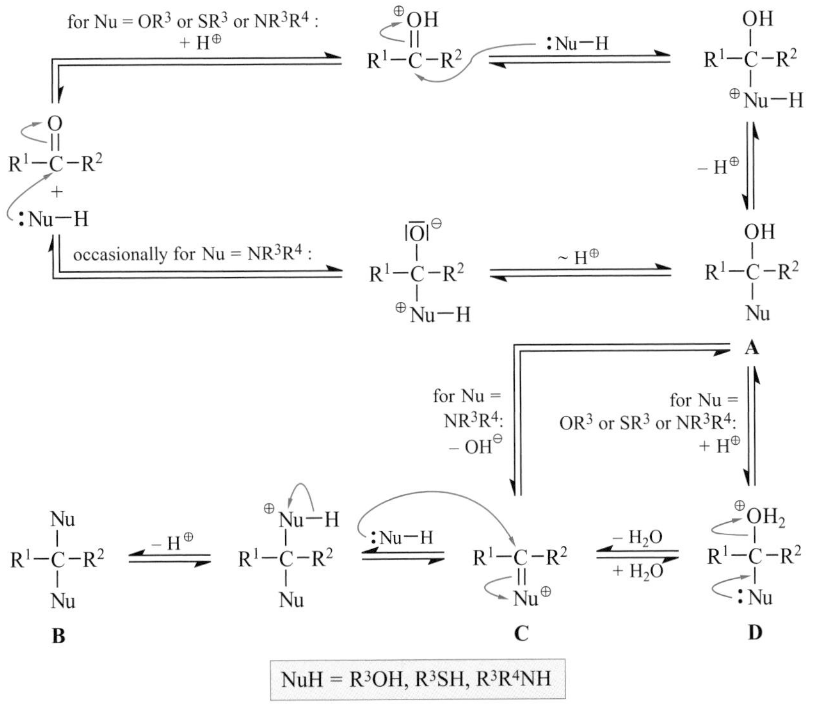

Fig. 9.12. *O,O*-, *S,S*- and *N,N*-acetalizations of carbonyl compounds.

further with the second equivalent of the nucleophile. This occurs in such a way that the OH group of **A** is replaced by that nucleophile. Thus, conversions of the type $R^1R^2C{=}O + 2HNu \rightarrow R^1R^2C(Nu)_2 + H_2O$ take place.

The product obtained (formula **B** in Figure 9.12) is, in the broadest sense of the word, an acetal. This is a collective term for compounds of the structure $R^1R^2C(Nu)_2$, where Nu can be OR^3 (so-called *O,O*-acetal), SR^3 (so-called *S,S*-acetal), or NR^3R^4 (so-called *N,N*-acetal). *O,O*- and *S,S*-acetals are isolable compounds. *N,N*-acetals can in general only be isolated when they are cyclic or substituted on both N atoms by an electron acceptor. There are also mixed acetals, compounds $R^1R^2C(Nu)_2$ with two different substituents Nu, such as $R^1R^2(OR^3)(OR^4)$ (so-called mixed *O,O*-acetal) or $R^1R^2(OR)(SR)$ (so-called *O,S*-acetal).

The mechanism for the acetalization is detailed in Figure 9.12. *O,O*-, *S,S*-, and *N,N*-acetals form via the adducts **A** that have been generated through addition of the respective nucleophiles to the carbonyl compounds in an *equilibrium(reversible) reaction*. For Nu = R^3S they can, in principle, proceed according to the same mechanism which Figure 9.12 showed for Nu = R^3O. In order for these addition products **A** to be able to react further through an S_N1 reaction, they should be able to eject H_2O as a good leaving group rather than be forced to eject OH-, a much poor leaving group. This requires that **A** be protonated in the next step, which means that one should work in the presence of acid, which in turn implies that **A** should also be generated in the presence of an acid. Interestingly, even *N* nucleophiles and carbonyl compounds react to give such addition products **A** in acidic solutions, including very acidic solutions. This is remarkable because under these conditions almost all N atoms are protonated. As a consequence, most of the *N* "nucleophile" added to the reaction mixture is no longer a nucleophile. The fact that nonetheless the overall reactivity is sufficiently high is due to two effects. First, under (strongly) acidic conditions there is a higher concentration of the carboxonium ions, which are particularly good electrophiles. Second, under (strongly) acidic conditions the addition products **A** (Nu = NR^3R^4) are protonated very efficiently to provide their conjugated acids **D** so that the iminium ions **C** ($Nu^{\oplus} = NR^3R^{4\oplus}$) are formed very rapidly *en route* to the final products.

It should be added that, in contrast to alcohols and thiols, ammonia and amines can also react with carbonyl compounds *also* under neutral or weakly basic conditions; this is possible because of their superior nucleophilicity.

The primary addition products **A** now undergo S_N1 reactions (Figure 9.12). A reversible protonation of the OH group of **A** leads to oxonium ions **D.** Water is ejected from **D** without the nucleophile getting involved. Thereby the intermediate oxocarbenium ions **C** (Nu = OR^3), thiocarbenium ions **C** (Nu = SR^3), or iminium ions **C** (Nu = NR^3R^4) are formed. These combine with the second equivalent of the nucleophile to form an *O,O*-acetal **B** (Nu = OR^3), an *S,S*-acetal **B** (Nu = SR^3), or an *N,N*-acetal (Nu = NR^3R^4).

When the hemiaminals **A** (Figure 9.12, Nu = NR^3R^4) are formed in a neutral or weakly basic solution, they also have the possibility to react further by an S_N1 reaction albeit in a different manner than just described for the corresponding hemiacetals (Nu = OR^3) and hemithioacetals (Nu = SR^3). The OH group of hemiaminals **A** is then ejected without prior protonation (i.e., simply as an $OH^{\ominus}$ ion). This is possible because an especially well-stabilized carbocation is produced at the same time, that is, the iminium ion **C** (Nu = NR^3R^4). It reacts with the second equivalent of the *N* nucleophile. Proton loss affords the *N,N*-acetal **B** (Nu = NR^3R^4).

When one follows the reaction arrows in Figure 9.12 from the bottom upward, the following important information can be noted: In an acidic water-containing solution *O,O*-acetals are hydrolyzed to give carbonyl compounds and alcohols. Such a hydrolysis consists of seven elementary reactions. First, the hemiacetal **A** (Nu = OR^3) and one equivalent of alcohol are produced from the *O,O*-acetal and water in an exact reversal of the latter's formation reaction, i.e., through a proton-catalyzed S_N1 substitution (in four steps). What follows is a three-step decomposition of this hemiacetal to the carbonyl compound and a second equivalent of the alcohol.

Acetals are thus not only produced from carbonyl compounds in acidic solutions but also can be hydrolyzed in this medium.

9.2.2 Formation of *O,O*-Acetals

The formation of *O,O*-acetals from carbonyl compounds and alcohols takes place according to the mechanism of Figure 9.12. As acidic catalysts one uses a mineral acid (HCl or H_2SO_4) or a sulfonic acid (*p*-toluenesulfonic acid, camphorsulfonic acid, or a strongly acidic cation exchange resin):

$$\text{cyclohexanone} + 2\,\text{MeOH} \underset{}{\overset{[H^{\oplus}]}{\rightleftharpoons}} \text{1,1-dimethoxycyclohexane (MeO, OMe)} + H_2O$$

The formation of the dimethylketal of cyclohexanone is an equilibrium reaction according to Figure 9.12. Complete conversion to the ketal is most simply achieved by working in methanol as a solvent and thus shifting the equilibrium to the ketal side. Whose principle is in action in this case?

One can also acetalize carbonyl compounds completely without using the alcohol in excess. This is the case when one prepares dimethyl or diethyl acetals from carbonyl compounds with the help of the orthoformic acid esters trimethyl orthoformate $HC(OCH_3)_3$ or triethyl orthoformate $HC(OC_2H_5)_3$, respectively. In order to understand these reactions, one must first clearly understand the mechanism for the hydrolysis of an orthoester to a normal ester (Figure 9.13). It corresponds nearly step by step to the mechanism of hydrolysis of *O,O*-acetals, which was detailed in Figure 9.12. The fact that the individual steps are analogous becomes very clear (see Figure 9.13) when one takes successive looks at

- the indicated acetal substructure in the substrate,
- the hemiacetal substructure in the key intermediate, and
- the C=O double bond in the hydrolysis product (i.e., in the normal ester).

The difference between the hydrolysis of an orthoester and the hydrolysis of an acetal is that protonation and C-O bond cleavage occur together in the former. In the hydrolysis of an acetal, the protonated form of the acetal is produced as an intermediate in the reaction. While we will not go into details here, the hydrolysis of the orthoester constitutes a case of what is known as general acid catalysis, while that of an acetal is specific acid catalysis.

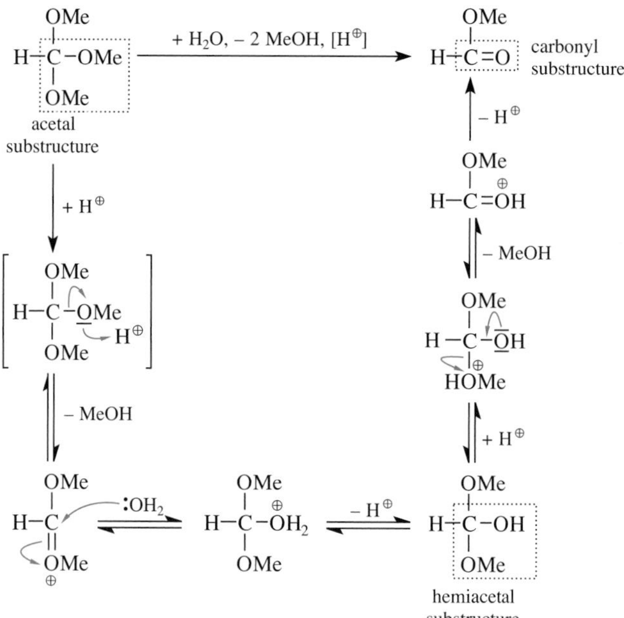

Fig. 9.13. Mechanism of the acid-catalyzed hydrolysis of orthoformic acid trimethyl ester.

The driving force for a hydrolysis orthoester → normal ester + alcohol is greater than the driving force for a hydrolysis *O,O*-acetal → carbonyl compound + alcohol. This is because the newly formed C=O double bond receives approximately 14 kcal/mol resonance stabilization in the ester (Table 6.1)—that is, starting from an orthoester—but not in the carbonyl compound.

This difference in driving force is used in the dimethyl-or diethylacetalizations of carbonyl compounds with orthoformic acid esters since these acetalizations are simply *linked* with the hydrolysis of an orthoester into a normal ester.

According to variant 1 of this process, a 1:1 mixture of the carbonyl compound and the orthoester reacts in the presence of one of the previously mentioned acid catalysts *in* the alcohol from which the orthoester is derived. Under these conditions the acetalizations take place mechanistically as shown in Figure 9.12. The H_2O released then hydrolyzes the orthoester according to the mechanism shown in Figure 9.13.

Variant 2 of the coupled acetalizations/orthoester hydrolyses succeeds completely without alcohol. Here the carbonyl compound is acetalized only with the orthoester. This method is not simply an option, but sheer necessity if the carbonyl compound that is to be acetalized is an α,β-unsaturated carbonyl compound as in the example of Figure 9.14. Otherwise, such substrates bear the risk of methanol being added to the $C\alpha{=}C\beta$ double bond before the acetalization of the C=O group takes place.

It is possible that a trace of alcohol contaminating the orthoester would make the acetalization of α,β-unsaturated carbonyl compounds via the mechanism of the preceding paragraph conceivable. But most probably the alternative acetalization mechanism of Figure 9.14 comes into play. The orthoester **A** would then first be cleaved so that MeOH would be eliminated in an equilibrium reaction This would provide the carboxonium ion **E,** which would

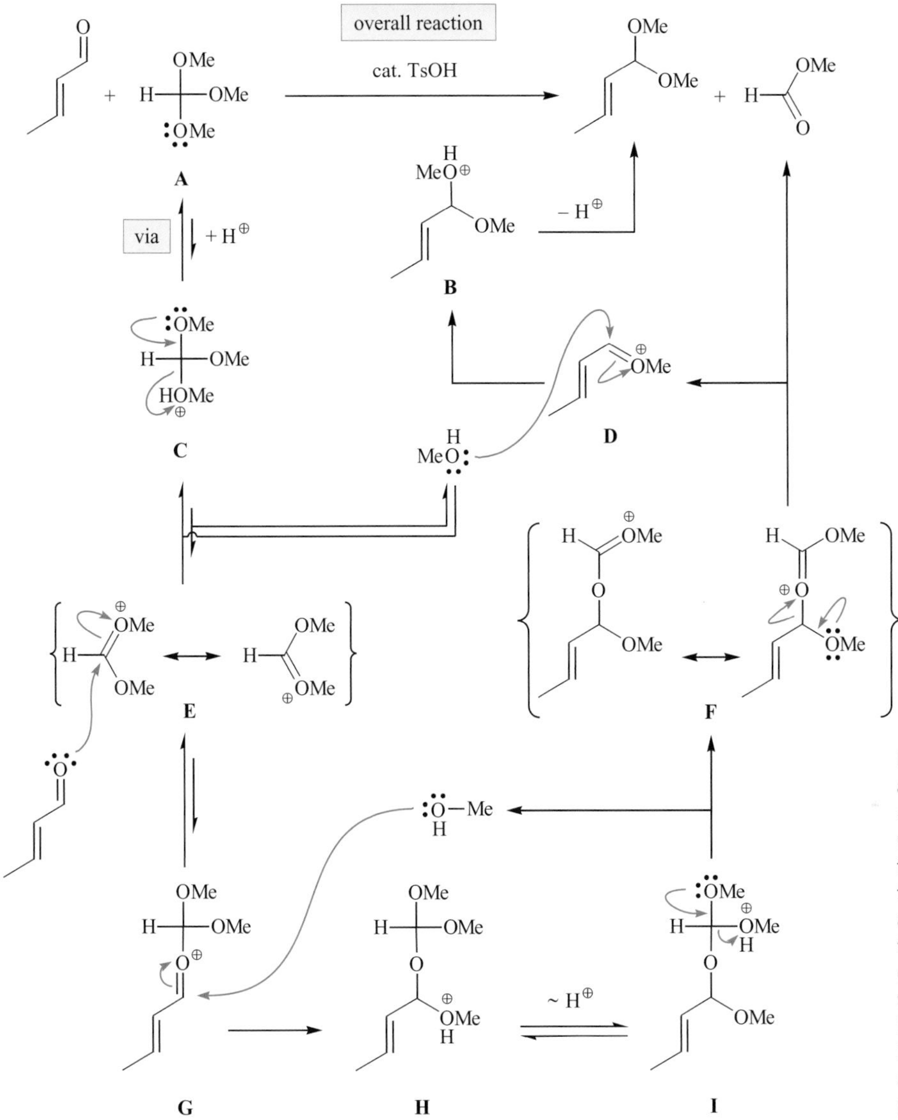

Fig. 9.14. Mechanism of the alcohol-free acetalization of an α,β-unsaturated carbonyl compound with orthoformic acid trimethyl ester. If, apart from this reagent, methanol were also present, this alcohol would first be added to the C=C double bond, then the acetal would be formed (mechanism: Figure 9.12); finally acid hydrolysis to give orthoformic acid methyl ester would occur (mechanism: Figure 9.13).

react with the carbonyl group of the substrate to give carboxonium ion **G**. When the addition of methanol to its C(=O$^{\oplus}$R) group takes place and is followed by a proton shift, the intermediate **I** is formed. The orthoester substructure of this species would be subjected to an E1 elimination leading to the elimination product HCO_2Me, and the leaving group **D** remains. **D** is a oxocarbenium ion that was already encountered as the key intermediate of the standard acetalization mechanism (Figure 9.12, Formula **C** with Nu = OR^3). This is why the two concluding steps **D** → **B** → crotonaldehyde dimethyl acetal of the acetalization in Figure 9.14 are the same as in Figure 9.12.

Side Note 9.1. Esterifications of Carboxylic Acids Following the Acetalization Mechanism of Figure 9.14

Another interpretation of the initial reactions of Figure 9.14 is that formic acid orthoester under acidic conditions produces a carboxonium ion *that can add an O nucleophile*. It thus becomes immediately clear how an orthoformate reacts with carboxylic acids under acidic conditions. This is documented in the left row of Figure 9.15. When the carboxonium ion **F** is thus obtained, two more steps lead to the formation of the carboxonium ion **H**, which is an

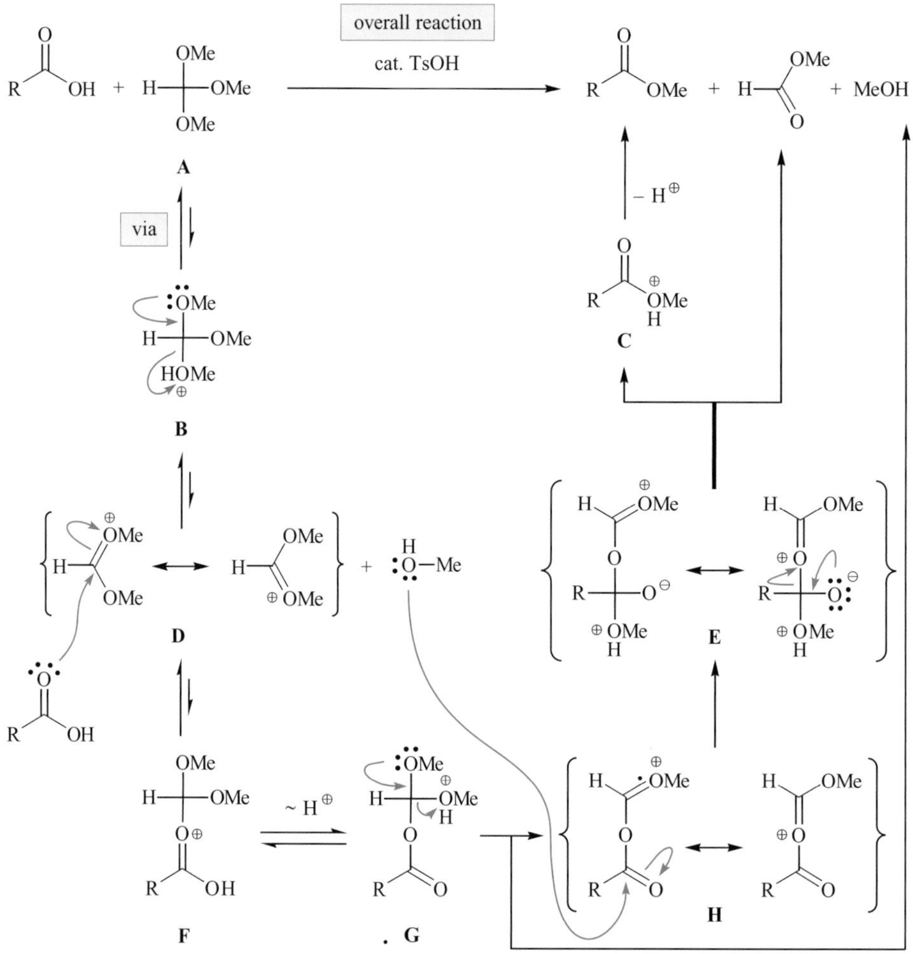

overall reaction

cat. TsOH

A B C D E F G H

via

$- H^{\oplus}$

$\sim H^{\oplus}$

Example: I  J

Fig. 9.15. Gentle esterification of a carboxylic acid I: with trimethyl orthoformate (upper part). The lower part gives an example of a chemoselective esterification according to the procedure discussed.

activated carboxylic acid. This can most easily be spotted by comparing it to the analogous structure of the iminium ion that acts as the activated carboxylic acid in the DMF-catalyzed transformation of carboxylic acids into acid chlorides (Formula **C** in Figure 6.11). The bottom line is that the carboxonium ion **H** of Figure 9.15 is also an acylating agent. As such, it acy-

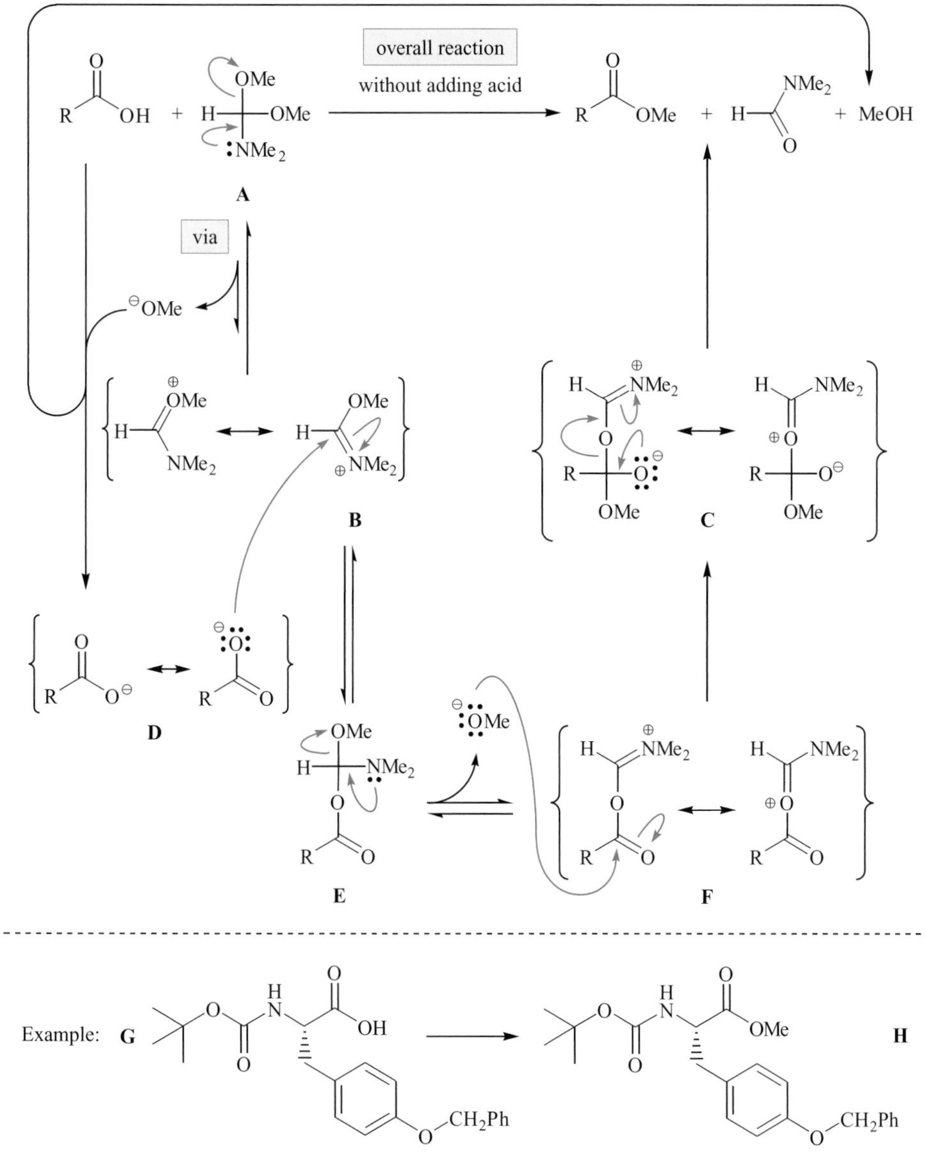

Fig. 9.16. Gentle esterification of a carboxylic acid II: with DMF acetal (**A**; upper part). The lower part gives an example of a chemoselective—since neither a cleavage of the substrate's Boc group nor a reaction on its stereocenter takes place—esterification according to the procedure discussed.

lates methanol—a stoichiometric side product of the intermediate **D**—to give the corresponding methyl ester.

In summary, Figure 9.15 presents a method for the mild esterification of carboxylic acids. For example, it allows for the chemoselective esterification **I** → **J**. Only under these reaction conditions does this sensitive substrate (**I**) not experience an addition of methanol to the C=C double bond, a C=C double bond migration or an acetalization of the aldehyde function.

The esterification of carboxylic acids with DMF acetal (Formula **A** in Figure 9.16) proceeds in line with the esterification of carboxylic acids with orthoformates (Formula **A** in Figure 9.15). The reaction conditions are even milder since no acid needs to be added. This is due to the fact that all cationic intermediates of an esterification according to Figure 9.16 are iminium ions in which the positive charge is better stabilized than by the cationic intermediates of esterifications shown in Figure 9.15, because they are carboxonium ions.

A key intermediate of the esterification in Figure 9.16 is the iminium ion **F**. It is identical to the iminium ion **B** in Figure 6.11, which represents the activated carboxylic acid in the DMF-catalyzed conversion of carboxylic acids into acid chlorides. Thus, the iminium ion **F** in Figure 9.16 is a potent acylating agent. As such, it reacts with the methoxide ion, a stoichiometric by-product of its formation reaction, via the tetrahedral intermediate **C** *to furnish the corresponding carboxylic acid methyl ester* and DMF.

Altogether, Figure 9.16 presents a method for the esterification of carboxylic acids under very mild conditions. These have been essential, for example, for the esterification **G** → **H** where an acid would have jeopardized the Boc protecting group (cf. Figure 4.37, Figure 9.16). Strictly speaking, this kind of esterification—as compared to the esterification of sensitive carboxylic acids with the explosive diazomethane—is applied much too rarely (Figure 2.33).

Acetalization equilibria of carbonyl compounds are shifted toward the side of the acetal when the reaction is carried out with one equivalent of a diol rather than with two equivalents of an alcohol (Figure 9.17). The reason for this is that the reaction entropy is no longer negative now: using the diol, two product molecules are produced from two molecules of starting material, whereas during an acetalization with mono-functional alcohols, two product molecules are produced from three molecules of starting material. Ethylene glycol and 2,2-dimethyl-1,3-propanediol are frequently used acetalization reagents because of this enhanced driving force.

When an acetalization takes place intramolecularly, the reaction entropy can even be positive because the number of particles doubles. Therefore, in the presence of protons γ, γ'-,

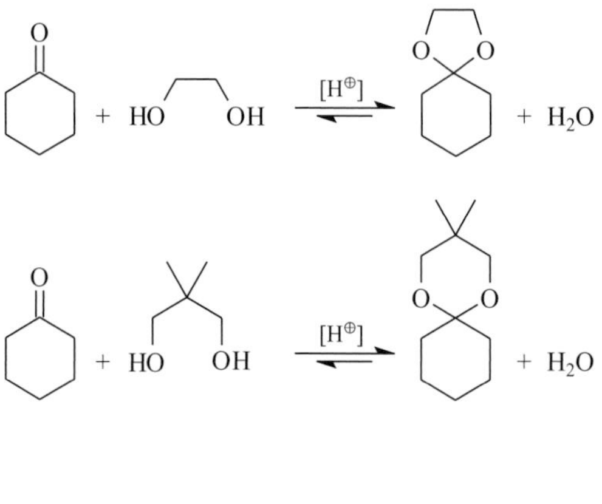

Fig. 9.17. Acetalization of a carbonyl compound with diols.

γ,δ'-, or δ, δ'-dihydroxyketones such as the following compound **A** react to give acetals of type **C** rapidly and *quantitatively*:

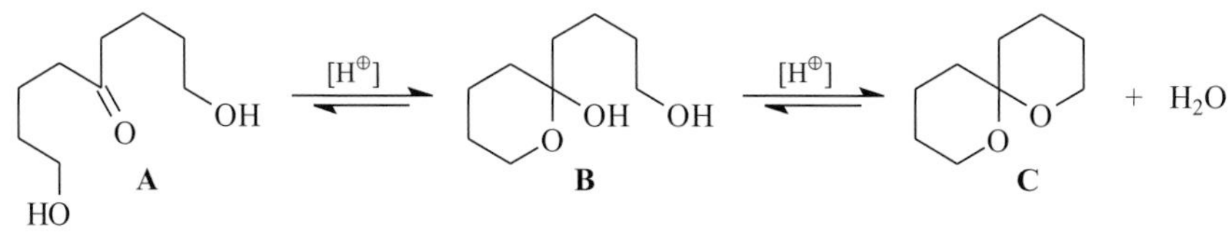

Acetals such as **C** are referred to as **spiroketals** because their acetal carbon is a spiro atom. (In a "spiro compound" two rings are connected by a single common atom that is called a "spiro atom"). The intermediates in this acetalization are lactols **B.** They resemble those lactols whose rapid formation from γ- or δ-hydroxyketones was shown in Figure 9.4. Note that because of the unfavorable reaction entropy, there is often no path back from spiroketals to the open-chain form: Usually, spiroketals cannot be hydrolyzed completely.

All acid-catalyzed acetalizations of carbonyl compounds with alcohols take place according to the mechanism of Figure 9.12—that is, through reversible steps. Therefore, they take place with thermodynamic control. Changing perspective, it also follows that acid-catalyzed acetalizations of triols (Figure 9.18) or of polyols (Figure 9.19) occur with thermodynamic control. This means that when several different acetals can in principle be produced from these reactants, the most stable acetal is produced preferentially. This circumstance can be exploited for achieving regioselective acetalizations of specific OH groups of polyalcohols.

An important polyalcohol for regioselective acetalizations is the triol of Figure 9.18 because it can be easily obtained from *S*-malic acid. This substrate contains both a 1,2-diol and a 1,3-diol. Each of these subunits can be incorporated into an acetal selectively—depending on the carbonyl compound with which the acetalization is carried out:

- With benzaldehyde the 1,3-diol moiety is acetalized exclusively. The resulting acetal **A** contains a strain-free, six-membered ring. It is present in the chair conformation and accommodates both ring substituents in the energetically favorable equatorial orientation.

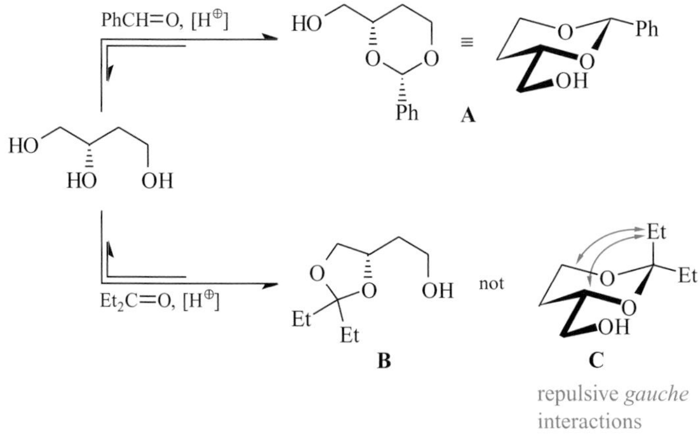

Fig. 9.18. Regioselective acetalizations of a 1,2,4-triol ("malic acid triol").

- With diethyl ketone it is possible to acetalize the 1,2-diol moiety of the malic acid triol selectively so that one obtains the five-membered acetal **B.** While, indeed, it exhibits more ring strain than the six-membered ring acetal **C,** this is compensated for because **B** suffers from considerably less Pitzer strain. The acetal isomer **C** in the most stable chair conformation would have to accommodate an axial ethyl group. However, because of their quite strong *gauche* interactions with the next-to-nearest O—C bonds of the ring skeleton, axial ethyl groups on C2 of six-membered ring acetals have almost as high a conformational energy and are therefore as unfavorable as an axial *tert*-butyl group on a cyclohexane.

Side Note 9.2. Regioselective Bisacetalization of a Pentaol

Figure 9.19 shows a biacetalization of a pentaol. There, the existence of thermodynamic control leads to the preferential production of one, namely **A,** of three possible bis(six-membered ring acetals)—**A, B,** and **C.** In the presence of catalytic amounts of *p*-toluenesulfonic acid, the pentaol from Figure 9.19 is converted into the bisacetal *in* acetone as the solvent but by a *reaction* with the dimethylacetal of acetone. From the point of view of the dimethylacetal, this reaction is therefore a transacetalization. Each of the two transacetalizations involved—remember that a bisacetal is produced—takes place as a succession of two S_N1 reactions at the acetal carbon of the dimethylacetal. Each time the nucleophile is an OH group of the pentaol.

The equilibrium of the double transacetalization of Figure 9.19 lies completely on the side of the bisacetal. There are two reasons for this. First, the dimethylacetal is used in a large excess, which shifts the equilibrium to the product side. In addition, the transacetalization is favored by an increase in the translational entropy. One molecule of pentaol and two molecules of the reagent give five product molecules.

Among the three conceivable bis(six-membered ring acetals) **A, B,** and **C** of the transacetalization of Figure 9.19, bisacetal **A** is the only one in which both six-membered rings can exist as chair conformers with three equatorial and only one axial substituent. In contrast, if

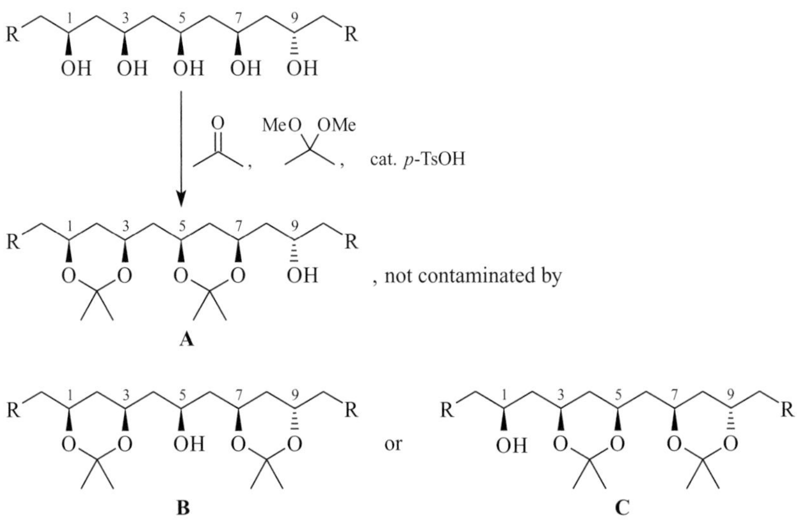

R 1 3 5 7 9 R
OH OH OH OH OH

O MeO OMe , , cat. *p*-TsOH

R 1 3 5 7 9 R
O O O O OH , not contaminated by
A

R 1 3 5 7 9 R
O O OH O O or R 1 3 5 7 9 R OH O O O O
B **C**

Fig. 9.19. Regioselective bisacetalization of a 1,3,5,7,9-pentaol.

the isomeric bis(six-membered ring acetals) **B** and **C** are also bis(chair conformers), they would contain two equatorial and two axial substituents on each six-membered ring. This is energetically worse than usual because one axial substituent would be located at the acetal carbon atom. This is especially unfavorable, as we have a ready seen with respect to the instability of the acetal **C** from Figure 9.18. Also, there is a 1,3-diaxial interaction between the two axial substituents of each six-membered ring. Thus, if the two bisacetals **B** and **C** were in fact chair conformers they would not be able to cope with so much destabilization. Consequently, they prefer to be twist-boat conformers. However, the inherently higher energy of twist-boat vs. chair conformers makes the acetals **B** and **C** still considerably less stable than acetal **A.** This is why acetal **A** is formed exclusively. Use models to prove this to yourself.

There is *one* type of carbonyl compound that cannot be converted into *O,O*-acetals of the type so far presented by treatment with alcohols and acid: γ- or δ-hydroxycarbonyl compounds. As you know from Figure 9.4, these compounds usually exist as lactols. In an acidic solution their OH group is exchanged very rapidly by an OR group via the S_N1 mechanism of Figure 9.12. This produces an *O,O*-acetal, which is derived from two different alcohols, a so-called **mixed acetal.** However, there is no further reaction delivering an open-chain nonmixed *O,O*-acetal of the γ- or δ-hydroxycarbonyl compound. This is because such a reaction would be accompanied by an unfavourable entropy loss: A single molecule would have been produced from two.

Examples of the previously mentioned formation of mixed *O,O*-acetals from lactols are given by acid-catalyzed reactions of D-glucose with MeOH (Figure 9.20). They lead to mixed *O,O*-acetals, which in this case are named methyl glycosides. The six-membered ring mixed acetal **A** is formed using HCl as a catalyst (and under thermodynamic control). The less stable five-membered ring mixed acetal **C** is produced under $FeCl_3$ catalysis. The last reaction is

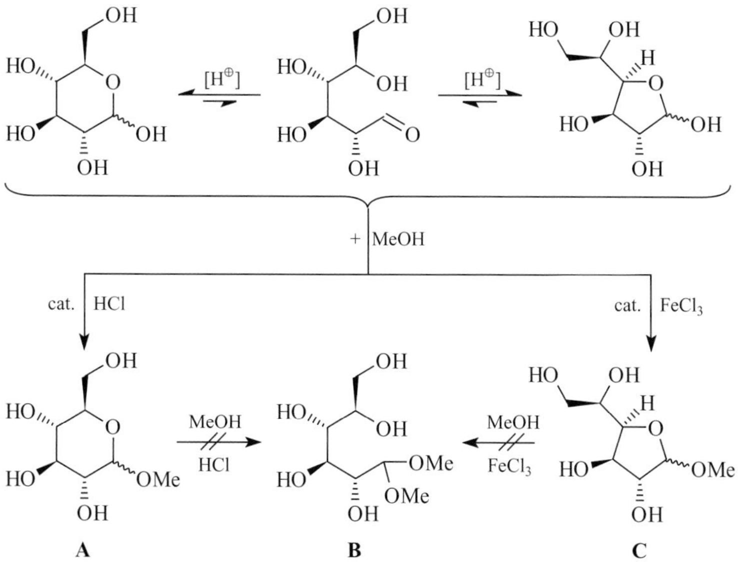

Fig. 9.20. Chemoselective formation of mixed *O,O*-acetals from D-glucose with MeOH.

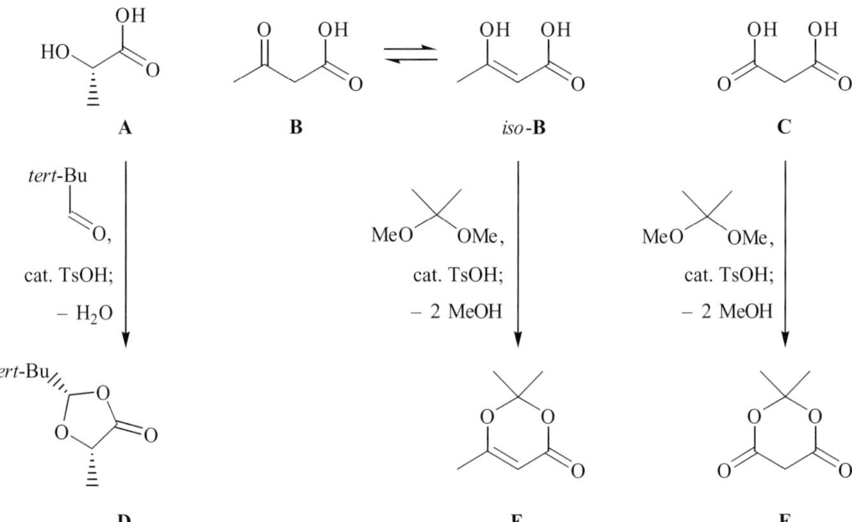

Fig. 9.21. Preparation of some analogs of *O,O*-acetals from the bisfunctional oxygen nucleophiles **A**, iso-**B** and **C**, respectively. The reactions proceed according to the mechanism in Figure 9.12.

remarkable in that this acetalization is achieved under kinetic control. The open-chain dimethyl acetal **B** never forms.

Apart from the diols examined so far there are other bis-*O* nucleophiles that also react with carbonyl compounds following the mechanisms discussed, as illustrated by Figure 9.21. The result is acetal analogs such as the compounds **D** (from hydroxy carboxylic acid **A**), **E** (from enol carboxylic acid *iso*-**B**) or **F** ("Meldrum's acid" from malonic acid). Each of these acetal analogs is used as a reagent in organic synthesis.

9.2.3 Formation of *S,S*-Acetals

In the presence of catalytic amounts of a sufficiently strong acid, thiols and carbonyl compounds form *S,S*-acetals according to the mechanism of Figure 9.12. The thermodynamic driving force for this type of *S,S*-acetal formation is greater than that for the analogous *O,O*-acetal formation (there is no clear reason for this difference). For example, although D-glucose, as shown in Figure 9.20, cannot be converted into an open-chain *O,O*-acetal, it can react to give an open-chain *S,S*-acetal:

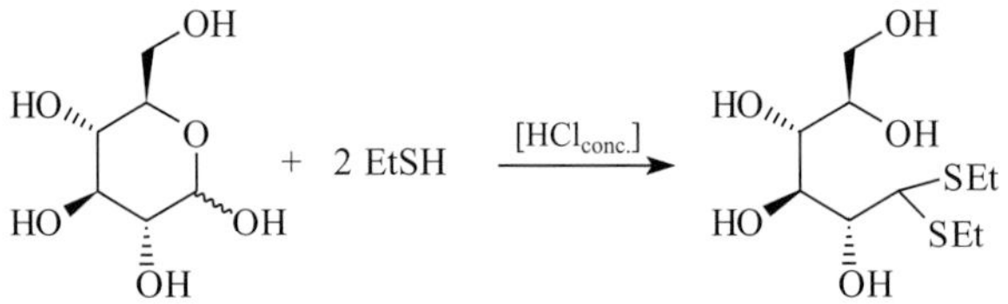

The most important *S,S*-acetals in organic chemistry are the six-membered ring *S,S*-acetals, the so-called **dithianes** (formulas **A, C,** and **F** in Figure 9.22). Dithianes are produced from

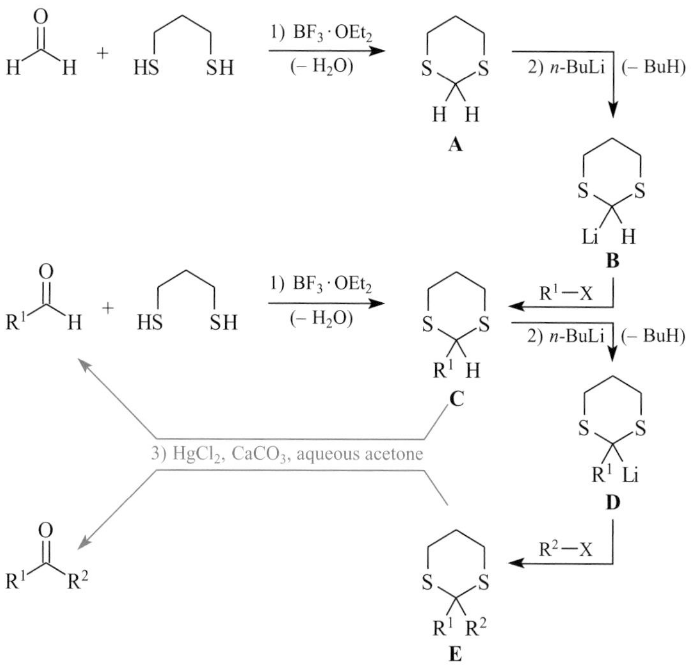

Fig. 9.22. Preparation, alkylation, and hydrolysis of dithianes.

carbonyl compounds and 1,3-propanedithiol usually in the presence of Lewis instead of Brønsted acids.

In the most simple dithiane **A** or in dithianes of type **C,** an H atom is bound to the C atom between the two S atoms. This H atom can be removed with LDA or n-BuLi and thus be replaced by a Li atom (Figure 9.22). In this way one obtains the lithiodithiane **B** or its substituted analog **D.** These compounds are good nucleophiles and can be alkylated, for example, with alkylating agents in S_N2 reactions (Figure 9.22; cf. Figure 2.32). The alkylated dithianes **C** and **E** can subsequently be hydrolyzed to carbonyl compounds. This hydrolysis is best done not simply in acids but in the presence of Hg(II) salts. Monoalkyldithianes **C** then give aldehydes and dialkyldithianes **E** give ketones. The alkyl group resulting as the carbonyl substituent of these products has been incorporated as an electrophile in the syntheses of Figure 9.22. This distinguishes this so-called **Corey–Seebach synthesis** from most aldehyde and ketone syntheses that you know.

9.2.4 Formation of *N,N*-Acetals

Ammonia, primary amines and many other compounds that contain an NH_2 group as well as secondary amines add to many carbonyl compounds only to a certain extent; that is, in equilibrium reactions (formation of hemiaminals; formula **B** in Figure 9.23). This addition is almost always followed by the elimination of an OH^- ion. As the result, one obtains an iminium ion (formula **C** in Figure 9.23).

This iminium ion **C** can be stabilized by combining with a nucleophile. As Figure 9.12 shows, when Nu = R^1R^2N, this step completes an S_N1 reaction in the initially formed addition

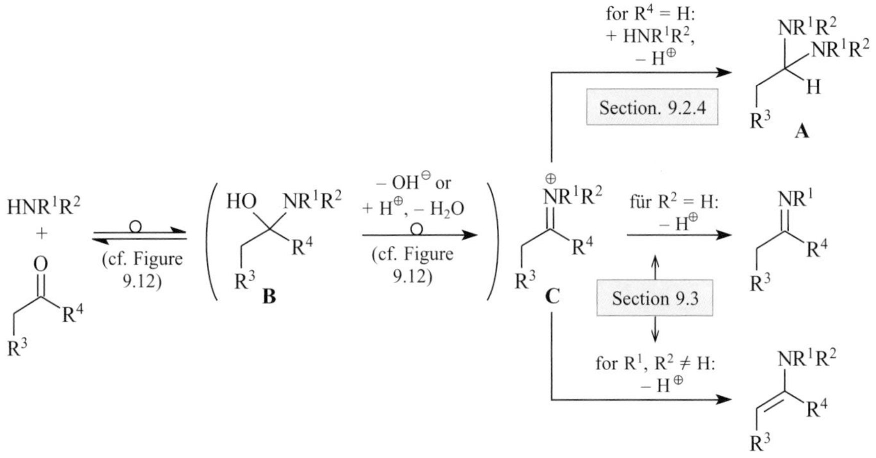

Fig. 9.23. Survey of the chemistry of iminium ions produced *in situ.*

product **B**. Thereby, an *N,N*-acetal or a derivative thereof is produced (formula **A** in Figure 9.23). We will treat this kind of reaction in more detail in Section 9.2.4. The other important reaction mode of the iminium ions **C** in Figure 9.23 is the elimination of a proton. This step would complete an E1 elimination of H_2O from the primary adduct **B**. This kind of reaction also occurs frequently and will be discussed in Section 9.3, as well.

The most important reaction examples for the formation of *N,N*-acetals involve formaldehyde because it tends more than most other carbonyl compounds to undergo additions (Section 9.1.1). With ammonia formaldehyde gives hexamethylenetetramine (Figure 9.24). This compound contains six *N,N*-acetal subunits.

With urea, formaldehyde forms two stable *N,O*-hemiacetals (Figure 9.25): a 1:1 adduct ("methylol urea") and a 1:2 adduct ("dimethylol urea"). When they are heated, both compounds are converted to *macromolecular N,N*-acetals (Figure 9.26). A three-dimensionally cross linked **urea/formaldehyde resin** is produced; it is an important plastic.

A structurally related macromolecular *N,N*-acetal is obtained from melamine (2,4,6-triamino-1,3,5-triazine) and formaldehyde. It is called **melamine/formaldehyde resin** and is likewise used as a plastic.

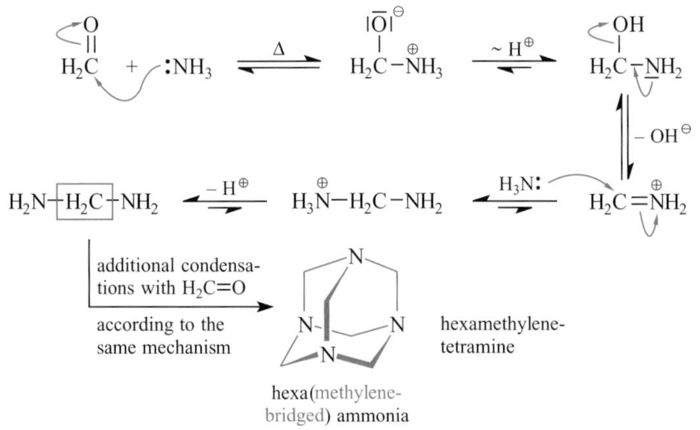

Fig. 9.24. Mechanism for the formation of hexamethylenetetramine from ammonia and formalin.

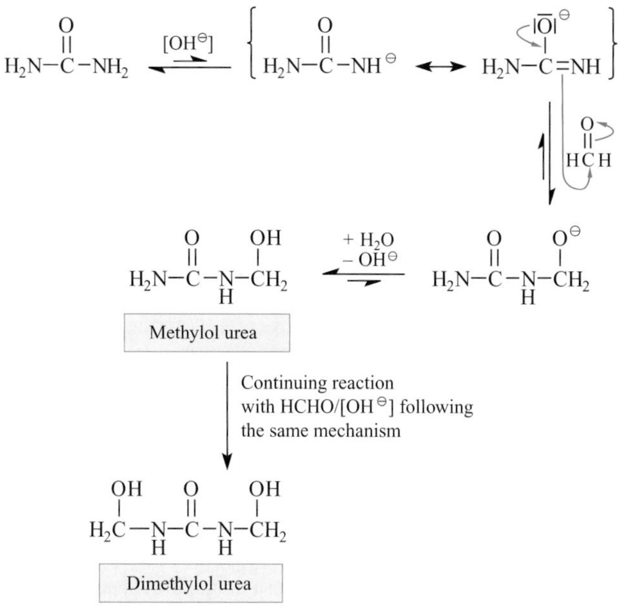

Methylol urea

Continuing reaction with HCHO/[OH⊖] following the same mechanism

Dimethylol urea

Fig. 9.25. Reaction of urea with formaldehyde at room temperature.

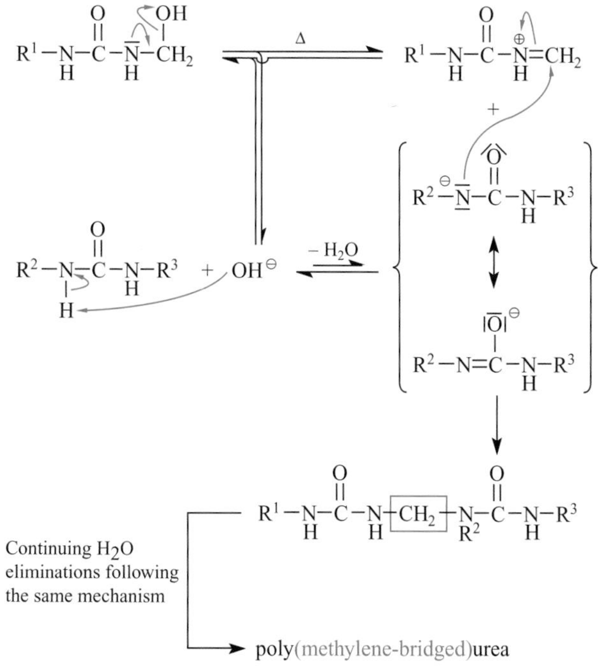

Continuing H_2O eliminations following the same mechanism

poly(methylene-bridged)urea

Fig. 9.26. Mechanism of the formation of a urea/formaldehyde resin from methylol urea (R^1 = H in formula **A**; possible preparation: Figure 9.25) or dimethylol urea (R^1 = HO—CH_2 in formula **A**; possible preparation: Figure 9.25). The substituents R^1, R^2, and R^3 represent the growing —CH_2—NH—C(=O)—NH—CH_2— chains as well as the derivatives thereof that are twice methylenated on N atoms.

9.3 Addition of Nitrogen Nucleophiles to Carbonyl Compounds in Combination with Subsequent E1 Eliminations of the Primary Product: Condensation Reactions

Iminium ions that arise from nitrogen nucleophiles and formaldehyde are more electrophilic and sterically less hindered than iminium ions obtained from higher aldehydes or from ketones. This explains why the first type of iminium ion is likely to combine with an extra molecule of the nitrogen nucleophile and in this way reacts further to give *N,N*-acetals (Section 9.2.4). The more highly substituted iminium ions are less electrophilic because of their electron-donating substituent(s) and because they are also more sterically hindered. Therefore any excess *N* nucleophile would add only slowly to iminium ions of this type. Instead such an *N* nucleophile uses a different reaction mode: It reacts with these iminium ions as a base, i.e., it deprotonates them. This takes place regioselectively: Secondary iminium ions are deprotonated at the nitrogen, whereas tertiary iminium ions are deprotonated at the β-carbon. In this way, the initially obtained addition products of *N* nucleophiles to the C=O double bond of higher aldehydes or ketones are converted to unsaturated products through an E1 elimination of H_2O. Generally, these products contain a C=N double bond. Only if this is impossible is a C=C double bond formed.

Reactions such as those just described, i.e., ones in which two molecules of starting material react to form one product molecule plus a by-product of considerably lower molecular weight are referred to as **condensation reactions.** *Here*, as in many other condensations, this by-product is H_2O. Table 9.2 summarizes the most important condensation reactions of nitrogen nucleophiles with carbonyl compounds, in which C=N double bonds are produced.

You should already be familiar with approximately half of the reactions listed in Table 9.2 from your introductory class. Moreover, you have probably tried to prepare an oxime, a phenylhydrazone, a 2,4-dinitrophenylhydrazone, or a semicarbazone. These compounds serve as crystalline derivatives with sharp and characteristic melting points for identifying aldehydes and ketones and for distinguishing them. When spectroscopic methods for structure elucidation were not available, such a means of identification was very important.

Most of the other product types in Table 9.2 are used in preparative chemistry. Aldimines, especially those that are derived from cyclohexylamine, serve as precursors to azaenolates (Figure 13.33). The same holds for the SAMP hydrazones, except that enantiomerically pure azaenolates are accessible through them (Figures 13.34 and 17.37; also see Figure 13.35). Starting from phenylhydrazones it is possible to synthesize indoles according to the Fischer procedure. Tosylhydrazones are used in two-step reductions of carbonyl compounds to hydrocarbons (Figures 17.70 and 17.71). Occasionally, semicarbazones are used similarly (Figure 17.68).

Tosylhydrazones can be silylated with (*tert*-butyldimethylsilyl)trifluoromethanesulfonate (Me_2*tert*$BuSiO_3SCF_3$) on the sulfonamide nitrogen. This is how the starting material for a broadly applicable alkane synthesis is produced (Figure 1.49). Due to their ability to undergo reductive cyanization (Figure 17.69) mesitylene sulfonyl hydrazones allow for a two-step synthesis of nitriles from ketones.

Tab. 9.2 Condensation Reactions of Nitrogen Nucleophiles with Carbonyl Compounds Through Which C=N Double Bonds are Established-Mechanism and Scope

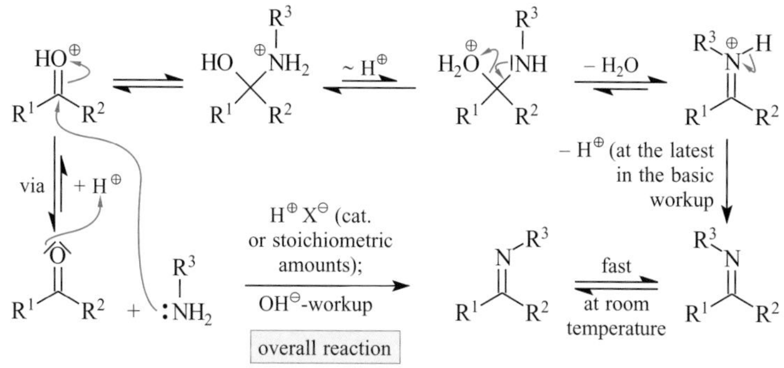

$H^{\oplus}X^{\ominus}$ (cat. or stoichiometric amounts); $OH^{\ominus}$-workup

overall reaction

$-H_2O$; $\sim H^{\oplus}$; via $+H^{\oplus}$; $-H^{\oplus}$ (at the latest in the basic workup); fast, at room temperature

R^3	Name of the N-nucleophile	Name of the condensation product
H	Ammonia	Product unstable
Alk	Primary alkylamine	For $R^1 = R^2 = H$; product unstable For $R^1 \neq H$, $R^2 = H$: aldimine For $R^1, R^2 \neq H$: ketimine
Ph	Aniline	Anil
OH	Hydroxylamine	Oxime
(pyrrolidin-1-yl with MeO-CH2 substituent)	(***S***)-*N*-**Amino**pro**l**inol methylether (SAMP)	SAMP hydrazone
NHPh	Phenylhydrazine	Phenylhydrazone
NH-2,4-(NO_2)$_2$$C_6H_3$	2,4-Dinitrophenylhydrazine	2,4-Dinitrophenylhydrazone
NHTs	Tosylhydrazide	Tosylhydrazone
NH–SO_2–mesityl	Mesithylenesulfonylhydrazide	Mesithylenesulfonylhydrazone
NH–C(=O)–NH_2	Semicarbazide	Semicarbazone

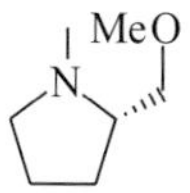

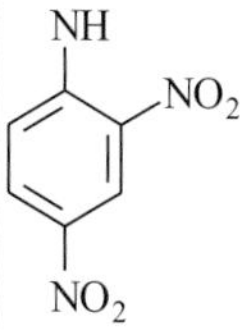

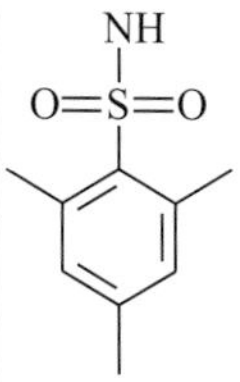

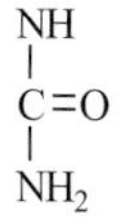

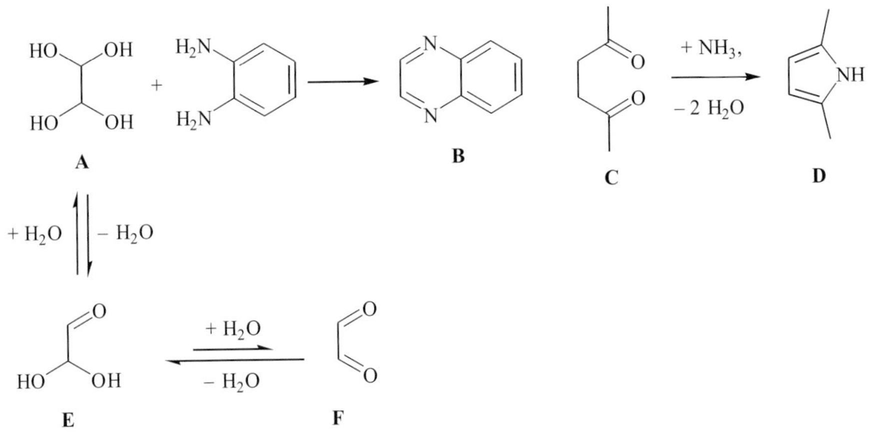

Fig. 9.27. Condensation of diamines with dicarbonyl compounds (and/or their hydrates) to give *N* heterocycles. Double imine formation yields quinoxalines (**B**), double enamine formation leads to dimethylpyrrole (**D**).

Side Note 9.3. The Ninhydrin Detection of α-Amino Acids

You have probably been introduced to a process known as the ninhydrin reaction for the detection of α-amino acids. This test reacts to any primary amines. Due to their biological relevance the most important among these amines are the proteinogenic α-amino acids—except L-proline that is proteinogenic as well, but a secondary α-amino acid. The coloring that occurs with the ninhydrin reaction is due to the formation of the anion **B** (Figure 9.28). **B** can be described by four enolate and two carbanionic resonance forms. Because of its strongly delocalized π-electron systems and the occurrence of cross-conjugation **B** is a chromophore whose high extinction coefficient at λ_{max} (the wavelength with the absorption maximum; here 570 nm) is the reason for the high sensitivity of the ninhydrin reaction.

Figure 9.28 gives the details of how the chromophore **B** is formed from ninhydrin (**A**) and a primary α-amino acid. The process starts with a condensation between the amino acid as the *N* nucleophile and the "central" carbonyl group of ninhydrin because this is the most electrophilic, the latter fact being demonstrated by the identical regioselectivity of the *hydrate formation* with ninhydrin (Section 9.1.2). The condensation product is the iminium ion **H**, which is also a carboxylic acid. A hydroxide ion is formed along with this iminium ion. This deprotonates **H** on the carboxyl group and leads to the zwitterion **I**. The latter decomposes—reminding us of the second step of an $E1_{cb}$ elimination—and yields carbon dioxide as the elimination product and **G** as the leaving group. The intermediate **G** can be described by the two resonance forms shown and two enolate resonance forms (not shown). The substructure $R_2C^{\ominus}$–$N^{\oplus}H{=}CR_2$ of the explicitly shown resonance forms is typical of a so-called azomethine ylide. Azomethine ylides are 1,3-dipoles and are fairly stable compounds (for more about other 1,3-dipoles see Section 15.5.1). Through protonation of the azomethine ylide **G** the iminium ion **E** is formed. The hydrolysis of its $C{=}N^{\oplus}HR$ double bond corresponds to the reversal of the iminium ion formation reaction outlined in Figure 9.23 and leads to an aldehyde and the primary amine **C**. Its NH_2 group again undergoes condensation with the central C=O double bond of a ninhydrin molecule to give an imine, whose deprotonation completes the formation of the chromophore **B**.

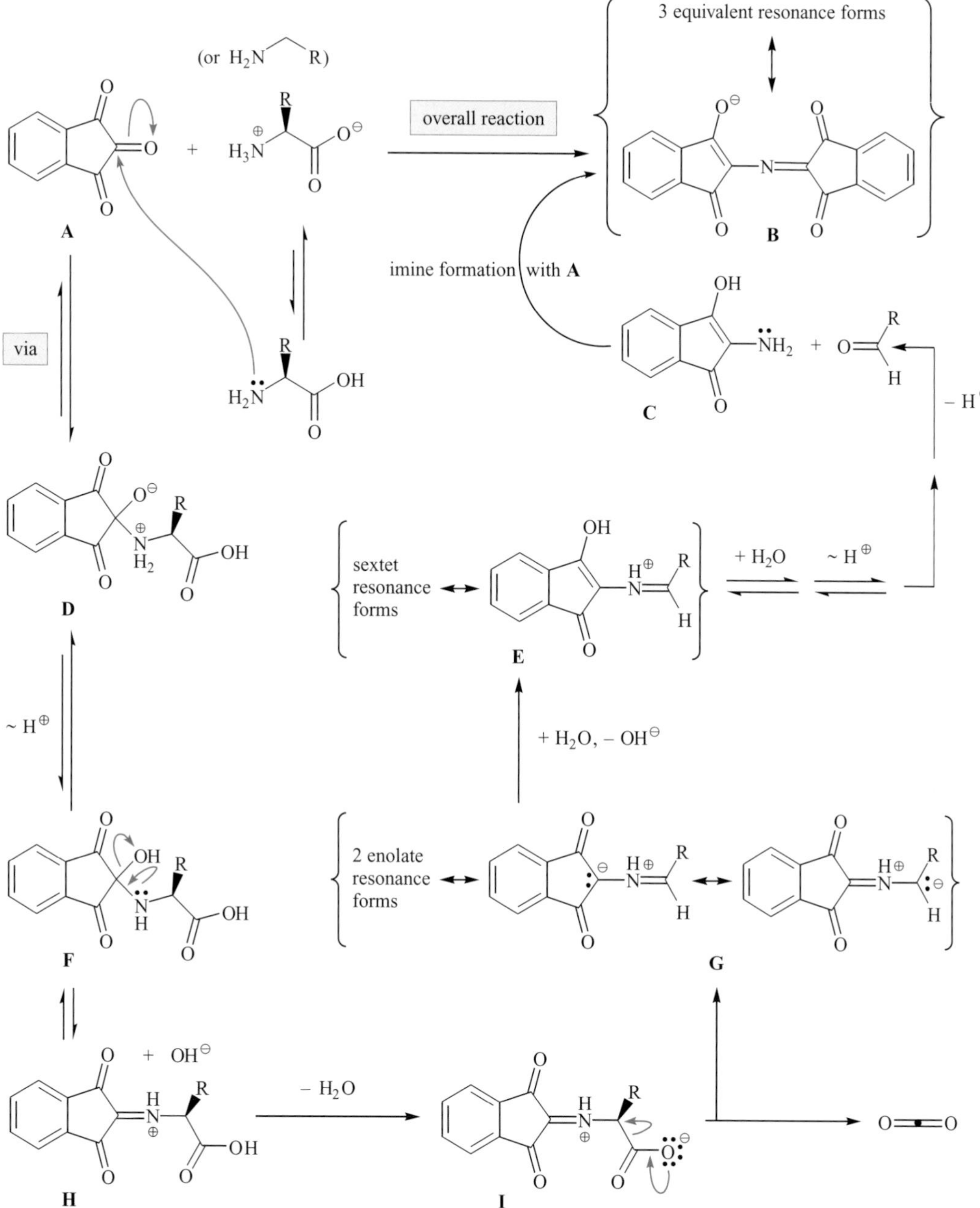

Fig. 9.28. Reactions occurring in the "Ninhydrin Test" for primary amines in general and α-amino acids in particular. The characteristic blue coloring is due to the formation of the anionic dye **B**.

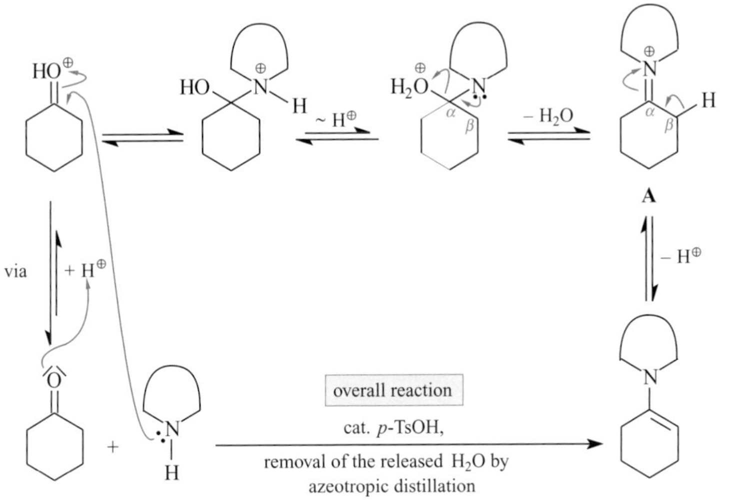

Fig. 9.29. Mechanism of the formation of enamines from secondary amines and cyclohexanone. Overall reaction in the bottom line.

Reactions proceeding *more than once in the transformation of a substrate* can be more effective than a single process If the reaction in question is a condensation of an *N* nucleophile with a carbonyl compound, the combination of the of two such reactions—allows for the synthesis of heterocycles like, the two depicted in Figure 9.27. The synthesis of quinoxaline (**B**) from glyoxal dihydrate (**A**) and *ortho*-phenylene diamine consists of two imine formations. Somewhat more complicated is the synthesis of dimethylpyrrole **D** from acetonyl acetone (**C**) and ammonia. After the formation of the first imine, an imine $\rightleftharpoons$ enamine isomerization occurs. A condensation followed by another imine → enamine isomerization leads to the product.

Secondary amines react with ketones that contain an H atom in the α-position through an addition and subsequent E1 elimination to form **enamines** (Figure 9.29). In order for enamines to be formed *at all* in the way indicated, one must add an acid catalyst. In order for them to be formed *completely*, the released water must be removed (e. g. azeotropically). The method of choice for preparing enamines is therefore to heat a solution of the carbonyl compound, the amine, and a catalytic amount of toluenesulfonic acid in cyclohexane to reflux in an apparatus connected to a Dean–Stark trap. Did someone say Le Chatelier?

The deprotonation of an iminium ion intermediate (formula **A** in Figure 9.29) to give an enamine is reversible under the usual reaction conditions. Therefore, the most stable enamine possible is produced preferentially. Figure 9.30 emphasizes this using the example of an enamine formation from α-methylcyclohexanone. The enamine with the trisubstituted double bond is produced regioselectively and not the enamine with the tetrasubstituted double bond. Since the stability of alkenes usually increases with an increasing degree of alkylation, this result is at first surprising. However, the apparent contradiction disappears when one recalls the following: The C=C double bond of enamines is stabilized considerably by the enamine resonance $C{=}C{-}NR_2 \rightarrow {}^{\ominus}C{-}C{=}NR_2^{\oplus}$. The feasibility of this resonance requires sp^2-hybridization at the nitrogen atom. Furthermore, the $2p_z$ AO at the N atom and the π-orbital of the C=C double bond must be parallel. In other words, for optimum enamine resonance to be possible the C=C double bond, the attached nitrogen and the five other atoms bound to these C and N atoms all must lie in a single plane.

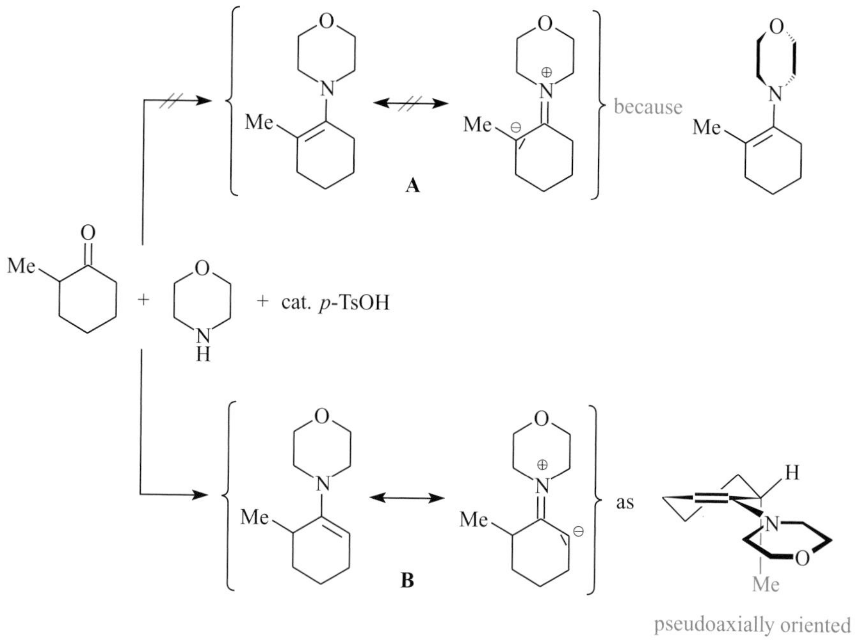

Fig. 9.30. Regioselective formation of an enamine from an asymmetrical ketone.

This is precisely what is impossible in the more highly alkylated enamine **A** of Figure 9.30. A considerable repulsion (1,3-allylic strain) between the methyl group and the heterocycle would be introduced; it would be so great that the nitrogen-containing ring would be rotated out of the plane of the C=C double bond. As a consequence, there would be little to no enamine resonance. On the other hand, the observed enamine **B** would benefit from enamine resonance. However, even this molecule must avoid the occurrence of *syn*-pentane strain between the methyl group and the heterocycle. This strain is avoided when the methyl group adopts an unusual pseudoaxial orientation.

Side Note 9.4. Preparation of Enol Ethers

The survey in Figure 9.23 shows that *N* nucleophiles can react with carbonyl compounds in the following ways: (1) An addition to the C=O double bond followed by an S_N1 reaction leads to the formation of *N,N*-acetals (details: Section 9.2.4). (2) An addition to the C=O double bond is followed by an E1 reaction by which, amongst others, enamines are formed (details: Section 9.3). (3) Imines are produced. We still need to discuss whether the reaction of *O* nucleophiles with carbonyl compounds also gives us two options—parallel to the two possibilities (1) and (2) mentioned above. According to Figure 9.12 alcohols and carbonyl compounds *always* afford *O,O*-acetals—through an addition and an S_N1 reaction (details: Section 9.2.2).

The alternative to this *O,O*-acetal formation is the sequence of addition and E1 reaction. As a matter of fact, this is familiar from the transformation of alcohols with carbonyl compounds, but only occurs in some (very rare) cases. This is illustrated by Figure 9.31 using acid-catalyzed transformations of ethanol with two *β*-diketones as an example. Here, **enol ethers**, namely 3-ethoxy-2-cyclopentene-1-one and 3-ethoxy-2-cyclohexene-1-one, respectively, are

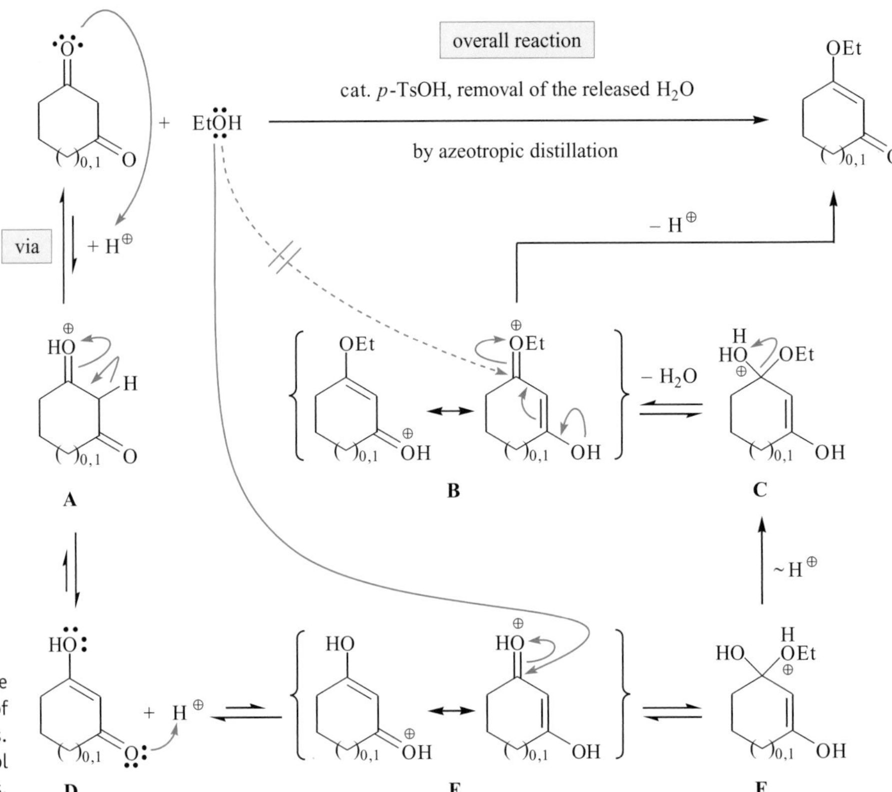

Fig. 9.31. Mechanism of the acid-catalyzed condensation of ethanol with 1,3-diketones. Synthesis of conjugated enol ethers.

produced by an addition and an E1 reaction. These enol ethers benefit from the stabilizing conjugation between their C=C- and the adjacent C=O double bond. It is this energy gain that drives the reaction towards the formation of the enol ether instead of the *O,O*-acetal. These enol ether syntheses proceed via the most stable, i.e., the conjugated, carboxonium ion (**E**) as the intermediate. The further course of this reaction follows the mechanism of the enamine

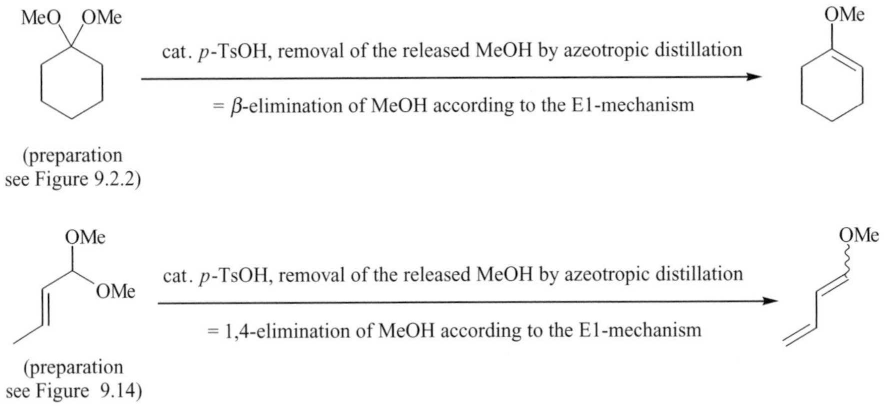

Fig. 9.32. Mechanism of the acid-catalyzed E1 elimination of methanol from *O,O*-acetals. Synthesis of an enol ether and a dienol ether, respectively.

MeO OMe

cat. p-TsOH, removal of the released MeOH by azeotropic distillation

= β-elimination of MeOH according to the E1-mechanism

OMe

(preparation see Figure 9.2.2)

OMe OMe

cat. p-TsOH, removal of the released MeOH by azeotropic distillation

= 1,4-elimination of MeOH according to the E1-mechanism

OMe

(preparation see Figure 9.14)

formation in Figure 9.23. C=O-conjugated enol ethers are employed, e.g., to produce cyclopentenones and cyclohexenones (see Figures 10.37 and 10.47).

Figure 9.32 adds the information of how enol ethers are normally produced, i.e., enol ethers with no conjugation between the C=C- and the neighboring C=O double bond: *O,O*-Acetals are subjected to an acid-catalyzed elimination of one equivalent of alcohol, via an E1 mechanism, that is, via an oxocarbenium ion intermediate that is deprotonated to give the respective enol ether (i.e., the product presented in the first line of Figure 9.32) or dienol ether (the product shown in the second line of Figure 9.32). Among other things, enol ethers are required for the Mukaiyama aldol addition (example: Figure 12.23).

References

J. K. Whitesell, "Carbonyl Group Derivatization," in *Comprehensive Organic Synthesis* (B. M. Trost, I. Fleming, Eds.), Vol. 6, 703, Pergamon Press, Oxford, U.K., **1991**.

W. E. Hanford, J. C. Sauer, "Preparation of Ketenes and Ketene Dimers," *Org. React.* **1946**, *3,* 108–140.

G. M. Coppola, H. F. Scuster, "α-Hydroxy Acids in Enantioselective Synthesis," Wiley-VCH, Weinheim, New York, **1997**.

9.1

A. Klausener, "O/O-Acetals: Special (OH/OH-Acetals, Lactols, 1,3,5-Trioxanes, OR/OEne-Acetals, OEne-Acetals etc.)," in *Methoden Org. Chem.* (Houben-Weyl) 4th ed., **1952**–, *O/O- and O/S-Acetals* (H. Hagemann, D. Klamann, Eds.), Vol. E14a/1, 591, Georg Thieme Verlag, Stuttgart, **1991**.

S. J. Angyal, "The Composition and Conformation of Sugars in Solution," *Angew. Chem. Int. Ed.* **1969**, *8*, 157–166.

K. Weissermel, E. Fischer, K. Gutweiler, H. D. Hermann, H. Cherdron, "Polymerization of Trioxane," *Angew. Chem. Int. Ed. Engl.* **1967**, *6*, 526–533.

M. M. Joullie, T. R. Thompson, "Ninhydrin and Ninhydrin Analogs. Syntheses and Applications," *Tetrahedron* **1991**, *47*, 8791–8830.

M. B. Rubin, R. Gleiter, "The Chemistry of Vicinal Polycarbonyl Compounds," *Chem. Rev.* **2000**, *100*, 1121–1164.

F. Effenberger, "Cyanohydrin Formation," in *Stereoselective Synthesis* (Houben-Weyl) 4th ed., **1996**, (G. Helmchen, R. W. Hoffmann, J. Mulzer, E. Schaumann, Hrsg.), **1996**, Bd. E21 (Workbench Edition), *3*, 1817–1821, Georg Thieme Verlag, Stuttgart.

R. J. H. Gregory, "Cyanohydrins in Nature and the Laboratory: Biology, Preparations, and Synthetic Applications," *Chem. Rev.* **1999**, *99*, 3649–3682.

Y. M. Shafran, V. A. Bakulev, V. S. Mokrushin, "Synthesis and Properties of α-Amino Nitriles," *Russ. Chem. Rev.* **1989**, *58*, 148–162.

D. Enders, J. P. Shilvock, "Some Recent Applications of α-Amino Nitrile Chemistry," *Chem. Soc. Rev.* **2000**, *29*, 359–373.

9.2

A. Klausener, "O/O-Acetals: Introduction," in *Methoden Org. Chem.* (Houben-Weyl) 4th ed., **1952**–, *O/O- and O/S-Acetals* (H. Hagemann, D. Klamann, Eds.), Vol. E14a/1, XIX, Georg Thieme Verlag, Stuttgart, **1991**.

O. Lockhoff, "Hal/O-, O/O-Acetals as Anomeric Centers of Carbohydrates," in *Methoden Org. Chem.* (Houben-Weyl) 4th ed., **1952**–, *Hal/O(S,N)-, S/S(N)-, N,N-Acetals and Hal/O-, O/O-Acetals as*

Anomeric Centers of Carbohydrates (H. Hagemann, D. Klamann, Eds.), Vol. E14a/3, 621, Georg Thieme Verlag, Stuttgart, **1992**.

F. A. J. Meskens, "Methods for the Preparation of Acetals from Alcohols or Oxiranes and Carbonyl Compounds," *Synthesis* **1981**, 501.

L. Hough, A. C. Richardson, "Monosaccharide Chemistry," in *Comprehensive Organic Chemistry* (E. Haslam, Ed.), **1979**, *5* (Biological Compounds), 687–748, Pergamon Press, Oxford, U. K.

F. Perron, K. F. Albizati, "Chemistry of Spiroketals," *Chem. Rev.* **1989**, *89*, 1617–1661.

M. A. Brimble, D. P. Furkert, "Chemistry of Bis-Spiroacetal Systems: Natural Products, Synthesis and Stereochemistry," *Curr. Org. Chem.* **2003**, *7,* 1461–1484.

S. Pawlenko, S. Lang-Fugmann, "S/S-Acetals as Anomeric Centers of Carbohydrates," in *Methoden Org. Chem.* (Houben-Weyl) 4th ed., **1952**–, *Hal/O(S,N)-, S/S(N)-, N,N-Acetals and Hal/O-, O/O-Acetals as Anomeric Centers of Carbohydrates* (H. Hagemann, D. Klamann, Eds.), Vol. E14a/3, 403, Georg Thieme Verlag, Stuttgart, **1992**.

P. C. B. Page, M. B. Van Niel, J. C. Prodger, "Synthetic Uses of the 1,3-Dithiane Grouping from 1977 to 1988," *Tetrahedron* **1989**, *45*, 7643–7677.

D. P. N. Satchell, R. S. Satchell, "Mechanisms of Hydrolysis of Thioacetals," *Chem. Soc. Rev.* **1990**, *19*, 55–81.

M. Yus, C. Najera, F. Foubelo, "The Role of 1,3-Dithianes in Natural Product Synthesis," *Tetrahedron* **2003**, *59,* 6147–6212.

9.3

H. R. Vollmer, "(4-Tolylsulfonyl)hydrazones," *Synlett* **1999**, 1844.

A. G. Cook (Ed.), "Enamines: Synthesis, Structure, and Reactions," Marcel Dekker, New York, **1988**.

Z. Rappoport (Ed.), "The Chemistry of Enamines," Wiley, Chichester, U.K., **1994**.

Further Reading

M. North, "Catalytic Asymmetric Cyanohydrin Synthesis," *Synlett* **1993**, 807–820.

F. Effenberger, "Synthesis and Reactions of Optically Active Cyanohydrins," *Angew. Chem. Int. Ed. Engl.* **1994**, *33*, 1555–1564.

L. Yet, "Recent Developments in Catalytic Asymmetric Strecker-Type Reactions," *Angew. Chem. Int. Ed.* **2001**, *40*, 875–877.

H. Groeger, "Catalytic Enantioselective Strecker Reactions and Analogous Syntheses," *Chem. Rev.* **2003**, *103*, 2795–2827.

J. K. Rasmussen, S. Heilmann, L. R. Krepski, "The Chemistry of Cyanotrimethylsilane," in *Compounds: Application to Enantioselective Synthesis of Heterocyclic Natural Products. Advances in Silicon Chemistry* (G. L. Larson, Ed.), **1991**, *1*, Jai Press, Greenwich, CT.

W. Nagata, M. Yoshioka, "Hydrocyanation of Conjugated Carbonyl Compounds," *Org. React.* **1977**, *25*, 255–476.

A. J. Kirby, "Stereoelectronic Effects on Acetal Hydrolysis," *Acc. Chem. Res.* **1984**, *17*, 305.

E. Juaristi, G. Cuevas, "Recent Studies on the Anomeric Effect," *Tetrahedron* **1992**, *48*, 5019–5087.

A. J. Kirby (Ed.), "Stereoelectronic Effects," Oxford University Press, Oxford, U. K., **1996**.

T. Y. Luh, "Regioselective C–O Bond Cleavage Reactions of Acetals," *Pure Appl. Chem.* **1996**, *68*, 635.

P. Wimmer, "O/S-Acetals," in *Methoden Org. Chem.* (Houben-Weyl) 4th ed., **1952**–, *O/O- and O/S-Acetals* (H. Hagemann, D. Klamann, Eds.), Vol. E14a/1, 785, Georg Thieme Verlag, Stuttgart, **1991**.

S. Pawlenko, S. Lang-Fugmann, "N/N-Acetals as Anomeric Centers of Carbohydrates," in *Methoden Org. Chem.* (Houben-Weyl) 4th ed., **1952**–, *Hal/O(S,N)-, S/S(N)-, N,N-Acetals and Hal/O-, O/O-Acetals as Anomeric Centers of Carbohydrates* (H. Hagemann, D. Klamann, Eds.), Vol. E14a/3, 545, Georg Thieme Verlag, Stuttgart, **1992**.

S. Pawlenko, S. Lang-Fugmann, "S/N-Acetals as Anomeric Centers of Carbohydrates," in *Methoden Org. Chem.* (Houben-Weyl) 4th ed., **1952**–, *Hal/O(S,N)-, S/S(N)-, N,N-Acetals and Hal/O-, O/O-Acetals as Anomeric Centers of Carbohydrates* (H. Hagemann, D. Klamann, Eds.), Vol. E14a/3, 483, Georg Thieme Verlag, Stuttgart, **1992**.

Corsaro, U. Chiacchio, V. Pistara, "Regeneration of Carbonyl Compounds from the Corresponding Oximes," *Synthesis* **2001**, 1903–1931.

Addition of Hydride Donors and of Organometallic Compounds to Carbonyl Compounds

10

In Section 6.5 you learned that the acylations of hydride donors or of organometallic compounds, which give aldehydes or ketones, often are followed by an unavoidable second reaction: the addition of the hydride or organometallic compound to the aldehyde or the ketone. In this chapter, we will study the intentional execution of such addition reactions.

10.1 Suitable Hydride Donors and Organometallic Compounds; the Structure of Organolithium Compounds and Grignard Reagents

The addition of a hydride donor to an aldehyde or to a ketone gives an alcohol. This addition is therefore also a redox reaction, namely, the reduction of a carbonyl compound to an alcohol. Nevertheless, this type of reaction is discussed here and not in the redox chapter (Chapter 17).

In this chapter **hydride donors** (**H nucleophiles**) are defined as reagents that ultimately transfer hydride ions onto the C atom of C=O double bonds. Such hydride donors can be subdivided into three classes. They are

- ionic, soluble hydrido complexes of B or Al ("complex hydrides"),
- covalent compounds with at least one B—H or one Al—H bond, or
- organometallic compounds which contain no M—H bond but do contain a transferable H atom at the C atom in the position β to the metal.

The first group of *H* nucleophiles includes $NaBH_4$ (which can be used in MeOH, EtOH, or HOAc), the considerably more reactive $LiAlH_4$ (with which one works in THF or ether), and alcoholysis products of these reagents such as Red-Al [$NaAlH_2(O—CH_2–CH_2—OMe)_2$] or the highly sterically hindered LiAlH(O-*tert*-Bu)$_3$. Another important hydride donor in this group is LiBH(*sec*-Bu)$_3$ (L-Selectride®), a highly sterically hindered derivative of the rarely used $LiBH_4$.

The hydride donor with a covalent M—H bond that is very frequently used for reducing carbonyl groups is *i*Bu$_2$AlH (DIBAL stands for **di**iso**b**utyl**al**uminum hydride). It can be used in ether, THF, toluene, saturated hydrocarbons, or CH_2Cl_2.

The most important organometallic compound that transfers an H atom along with the electron pair from the β-position to the carbonyl carbon is Alpine-Borane® (see Figure 10.24).

Bruckner R (author), Harmata M (editor) In: *Organic Mechanisms – Reactions, Stereochemistry and Synthesis*
Chapter DOI: 10.1007/978-3-642-03651-4_10,

It should be mentioned that certain Grignard reagents with C—H bonds in the β-position (e. g., *i*BuMgBr) can act as H nucleophiles rather than C nucleophiles (cf. Figure 10.29) with respect to the carbonyl carbon of sterically hindered ketones.

In this chapter, reagents that transfer a carbanion (in contrast to an enolate ion) to the C atom of a C=O double bond are referred to as ***C* nucleophiles.** The most important nucleophiles of this kind are organolithium compounds and Grignard reagents. Organocopper compounds transfer their carbanion moieties to the carbonyl carbon far less easily and usually not at all.

In the majority of cases, organolithium compounds and Grignard reagents can be viewed as having polarized but covalent carbon—metal bonds. Lithioalkanes, -alkenes, and -aromatics, on the one hand, and alkyl, alkenyl, and aryl magnesium halides, on the other hand, are therefore formulated with a hyphen between the metal and the neighboring C atom. Lithiated alkynes and alkynyl Grignard reagents are considered to be ionic—that is, species with carbon metal bonds similar to those in LiCN or $Mg(CN)_2$.

In covalent organolithium compounds and covalent Grignard reagents neither the lithium nor the magnesium possesses a valence electron octet. This is energetically disadvantageous. In principle, the same mechanism can be used to stabilize these metals that monomeric boranes $BH_{3-n}R_n$ use to attain a valence electron octet at the boron atom (Section 3.3.3): the formation either of oligomers or, with suitable electron pair donors, of Lewis acid/Lewis base complexes.

In the absence of Lewis bases, i.e., crystallized in pure form or dissolved in hydrocarbons, alkyllithium compounds occur as hexamers or tetramers, depending on the alkyl substituent (Figure 10.1).

For a given substituent, the degree of association drops when added Lewis bases can occupy vacant coordination sites around the lithium. An interesting effect can be observed here: the lower the temperature *T*, the more the degree of association decreases. This is due to the fact that the decrease in the degree of association of organolithium compounds paradoxically leads to a decrease in the entropy *S*. The reaction equation at the bottom of Figure 10.2 demonstrates the reason for this. The full stoichiometric equation is given in order to show that in hexane two molecules of [*tert*-butyllithium(diethyl ether)$_2$]-dimer **D** are obtained from one molecule of *tert*-butyllithium tetramer (**C**) and eight molecules of diethyl ether. The decrease

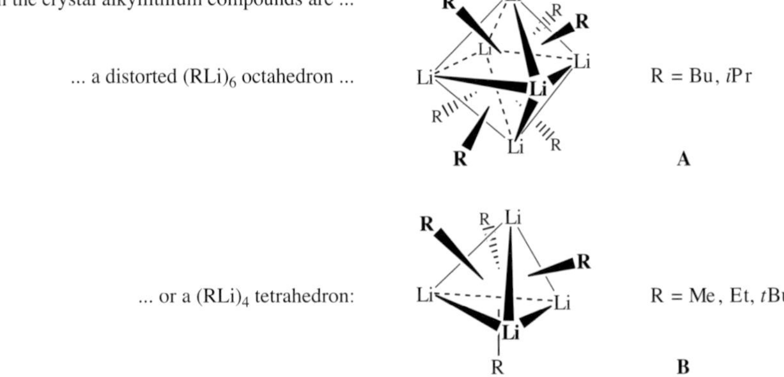
in the crystal alkyllithium compounds are ...
... a distorted $(RLi)_6$ octahedron ...
R = Bu, *i*Pr
A
... or a $(RLi)_4$ tetrahedron:
R = Me, Et, *t*Bu
B

Fig. 10.1. The structures of alkyllithium compounds in solid state.

in the number of moieties from 9 to 2 in form of the term $T\Delta S$ reduces the driving force ΔG of the deaggregation **C** → **D**. Due to the multiplication with T this effect is most strongly noticed with large T values. Because of these and similar conditions, deaggregations of oligomeric organolithium compounds only occur if their reaction enthalpy ΔH—due to the bond energy released from the Li–Lewis base—is so negative that the unfavorable entropy term is outweighed.

The preferential deaggregation of organolithium compounds at lower temperatures accounts for the fact that the latter can typically be reacted at –78°C ("dry ice temperature"): in many cases the monomeric organolithium compound is reactive, whereas the corresponding dimers, tetramers and/or hexamers are unreactive. Under these conditions a low reaction temperature not only maximizes the concentration of the monomeric organolithium compound, but the reaction rate as well (!).

Efficient deaggregations of oligomeric organolithium compounds are also made possible by adding TMEDA (for **tetra**methyl**e**thylen**dia**mine; $Me_2N–CH_2–CH_2–NMe_2$) (example: Figure 10.2, top), with the N atoms of this additive occupying two of the four lithium coordination sites. Figure 10.2 (top) illustrates this by using the conversion of *n*-butyllithium hexamer (**A**) into the (*n*-butyllithium/TMEDA) dimer **B** as an example.

Hexameric or tetrameric organolithium compounds can be further deaggregated by *dissolving* them in diethyl ether (example: Figure 10.2, bottom) or THF solution instead of just *adding* these solvents. The large excess of Lewis bases leads to a shift in the deaggregation equilibrium well beyond the dimeric step: at –100 °C phenyllithium, *sec*-butyllithium and *tert*-butyllithium are present in THF solution *mostly or partly as monomers* (cf. Formula **A**, Figure 10.3).

Pure monomeric forms can be produced if one adds 1–3 equivalents of HMPA (Figure 2.17) to the organolithium compound in diethyl ether or better still in THF solution, since the basic oxygen of this additive is an excellent electron pair donor, as illustrated in Figure 10.3 using the formation of [phenyllithium(HMPA)(THF)$_2$] monomer **B** as an example.

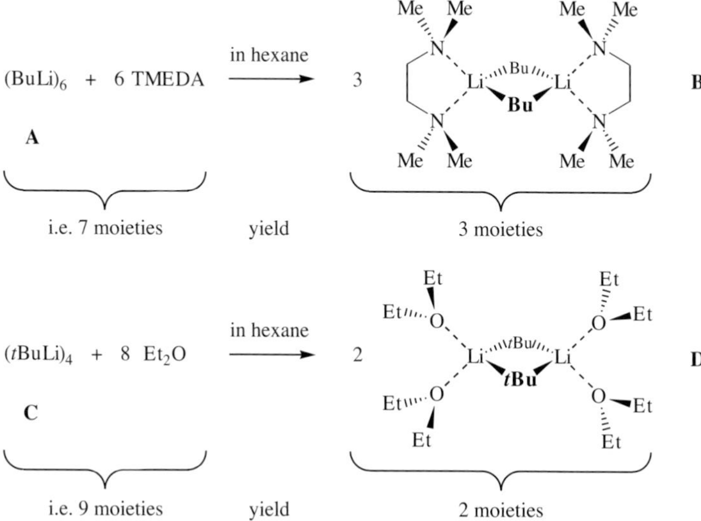

Fig. 10.2. The structures of alkyllithium compounds in solution I: Deaggregation of the alkyllithium structures, as preferred in the solid state, usually occurs when Lewis bases are added or alkyllithium compounds are dissolved in Lewis-basic solvents. By the way, this process is accompanied by a decrease of entropy, i.e., occuring more at low than at higher temperatures.

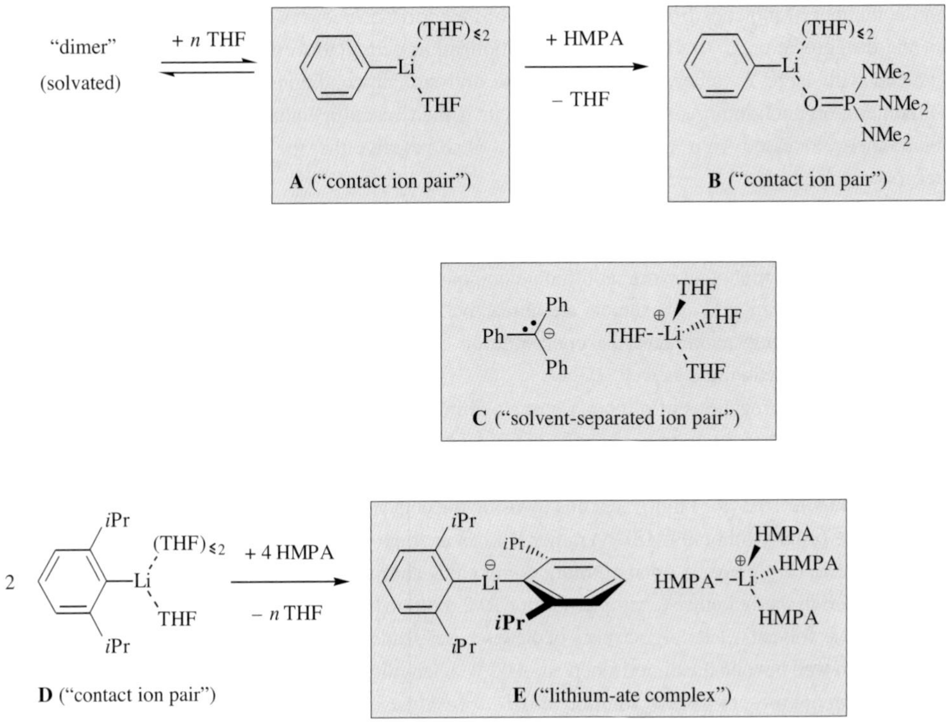

Fig. 10.3. The structures of alkyllithium compounds in solution II: contact ion pairs, solvent-separated ion pairs or lithium at-complexes. In THF solution *sec*- and *tert*-BuLi are present as contact ion pairs in the same way as phenyllithium (**A**) or (2,6-diisopropylphenyl) lithium (**D**) are.

A special nomenclature has been developed for the structures of organometallic compounds in solution. A distinction is drawn between "contact ion pairs," "solvent-separated ion pairs," and several less frequent structural types. Unfortunately, the terms "contact ion pair" and "solvent-separated ion pair" have already been used elsewhere, namely in connection with the dissociation of an alkylating agent to a carbenium ion and the leaving group in an S_N1 reaction. Figure 2.14 presented such a "contact ion pair," and Figure 2.15 a "solvent-separated ion pair. But there is a fundamental difference between the "ion pairs" in this chapter and in Chapter 2. In Chapter 2 *both* "ion pairs" (contact and solvent-separated) consist of a carbenium ion and an anion; in *both* "ion pairs" these constituents are linked via a more or less stable ionic bond. In contrast, the nomenclature used for the "ion pairs" of organometallic compounds is defined as follows: a "contact ion pair" is an organometallic compound that does not contain any ion whatsoever (!), instead they possess a *covalent*—albeit very polarized—carbon-metal bond. You have already come across these organolithium compounds that constitute such a "contact ion pair," namely in the form of the moieties **A** and **B** in Figure 10.3. The term "solvent-separated ion pair" refers to an organometallic compound that is a real pair of ions, namely a carbanion/metal cation pair whose constituents are linked to each other via a more or less stable ionic bond, but are sufficiently far apart that a solvent molecule can fit in between. For example, trityllithium in THF solution occurs as a "solvent-separated ion pair" (Formula **C** in Figure 10.3). This is made plausible by the delocalization of the negative charge due to the resonance effect of three phenyl rings.

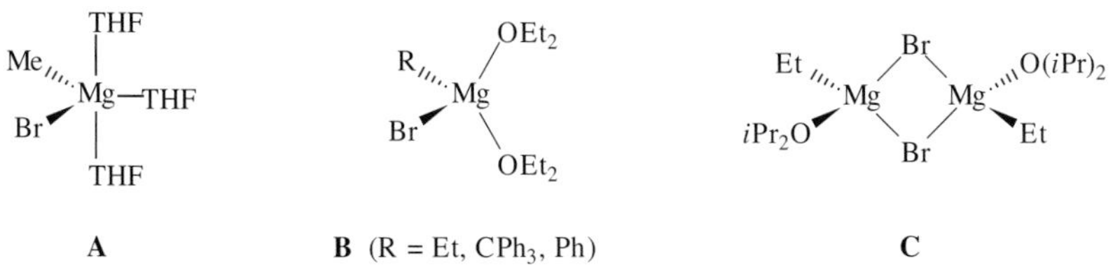

A B (R = Et, CPh_3, Ph) C

Fig. 10.4. The structures of Grignard compounds in solid state. The magnesium is trigonally bipyramidally (**A**) or tetrahedrally (**B**, **C**) surrounded by the organic residue, the halide ion and ether molecules.

Lest we think that that there is no other possibility, *one* of the complications that still makes the structural chemistry of organolithium compounds in solution a challenging research area is given in the third line of Figure 10.3. Four equivalents of the excellent donor HMPA (see above) can remove a lithium cation in the form of $Li^{\oplus}$ $(HMPA)_4$ from bis(2,6-diisopropylphenyl)lithium, which in THF is present as the contact ion pair **D**. This lithium cation is paired with a negatively charged (!) organolithium compound, forming the so-called lithium-ate complex **E**. The term -ate-complex has already been introduced in connection with the halide/metal exchange in aryl bromides and aryl iodides (Figure 5.41 and 5.46, respectively).

The structures of crystalline Grignard compounds—Figure 10.4 gives typical examples—reveal the tendency of divalent magnesium to fill its valence electron shell by binding electron pair donors, optimally ethers, to give an octet (**B**, **C**) or, in one case (**A**), even more than that. The structures presented indicate that Grignard compounds are less prone to form oligomers than organolithium compounds (see above): **A** and **B** contain a single RMgBr unit, and only **C** contains two of them.

In total, Grignard compounds in solution also contain one or two RMgBr units. The structures that are assigned to them are shaded red in Figure 10.5. For the sake of clarity they are depicted without the ether residues that are also present and bound to magnesium via their basic oxygen atom. The structures are in dynamic equilibrium. The best known of these is the so-called **Schlenk equilibrium**. It is depicted in the center of Figure 10.5 and describes the equilibration of RMgHal with R_2Mg and $MgHal_2$. It is possible that RMgHal and $MgHal_2$ do not occur separately, but are associated as shown in Figure 10.5 (top right). The same figure also shows the interconversion of various "Grignard dimers" and the intermediates that can be assumed. For some of them calculations of their stabilities have been conducted. Their stability sequence is represented by the numbers ① – ⑥. The disadvantage of these computations is that they do not take additional ether molecules into account, although it is known that they are important for the viability of Grignard compounds. Grignard reagents can be obtained from halides and magnesium but in general, this Grignard formation can be achieved only in an ethereal solvent such as diethyl ether or THF (mechanism: Figure 17.44).

Neither the mechanism for all addition reactions of hydride donors to the carbonyl carbon nor the mechanism for all addition reactions of organometallic compounds to the carbonyl carbon is known in detail. It is even doubtful whether only ionic intermediates occur. For instance, for some $LiAlH_4$ additions an electron transfer mechanism might apply. The same might be true for the addition of a number of Grignard reagents to aromatic aldehydes or ketones (see Figure 10.27). In any case, we need to assume that reactions like the $LiAlH_4$ reduction of a carbonyl compound or the addition of a Grignard compound RMgX to a carbonyl compound proceed according to one mechanism *or the other*, depending (1) on the precise structure of the reactants, including the routinely ignored counterions $Li^{\oplus}$ (with $LiAlH_4$

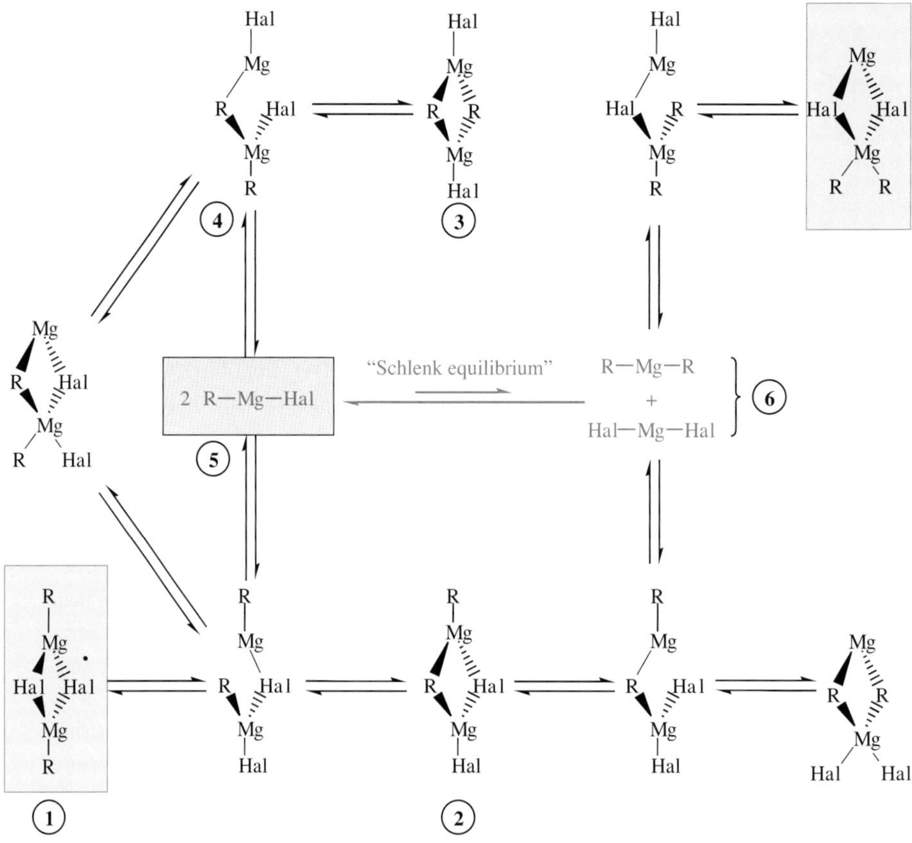

Fig. 10.5. Calculated structures of Grignard compounds in the gas phase (numbers ① to ⑥; stability given in decreasing order) or experimentally proven structures of Grignard compounds in solution (shaded pink). The Schlenk equilibrium ensures that not only RMgHal (shaded grey) occurs as a monomeric organometallic compound, but some R_2Mg as well.

reductions) or $X^{\ominus}$ (with Grignard additions) (cf. Table 10.1 and Table 10.3), (2) on the solvent and/or (3) on the stoichiometry.

Therefore, we will proceed pragmatically in this chapter: for the additions discussed, the substrates, the reagents formulated as monomers, and the tetrahedral intermediates considered established will be shown by structural formulas. How exactly the reagents are converted into these intermediates remains unknown. However, this conversion should occur in the rate-determining step.

In addition, Figures 10.27, 10.32 and 10.46 present the current view on (1) the mechanisms of the addition of Grignard reagents to carbonyl compounds, (2) the mechanism of the additions of organolithium compounds to α,β-unsaturated ketones and (3) the mechanism of the 1,4-addition of Gilman cuprates to α,β-unsaturated ketones, respectively.

10.2 Chemoselectivity of the Addition of Hydride Donors to Carbonyl Compounds

In the addition of hydride donors to aldehydes (other than formaldehyde) the tetrahedral intermediate is a primary alkoxide. In the addition to ketones it is a secondary alkoxide. When a primary alkoxide is formed, the steric hindrance is smaller. Also, when the C=O double bond of an aldehyde is broken due to the formation of the $CH(O^{\ominus}M^{\oplus})$ group of an alkoxide, less stabilization of the C=O double bond by the flanking alkyl group is lost than when the analogous transformation occurs in a ketone (cf. Table 9.1). For these two reasons aldehydes react faster with hydride donors than ketones. With a moderately reactive hydride donor such as $NaBH_4$ at low temperature one can even chemoselectively reduce an aldehyde in the presence of a ketone (Figure 10.6, left).

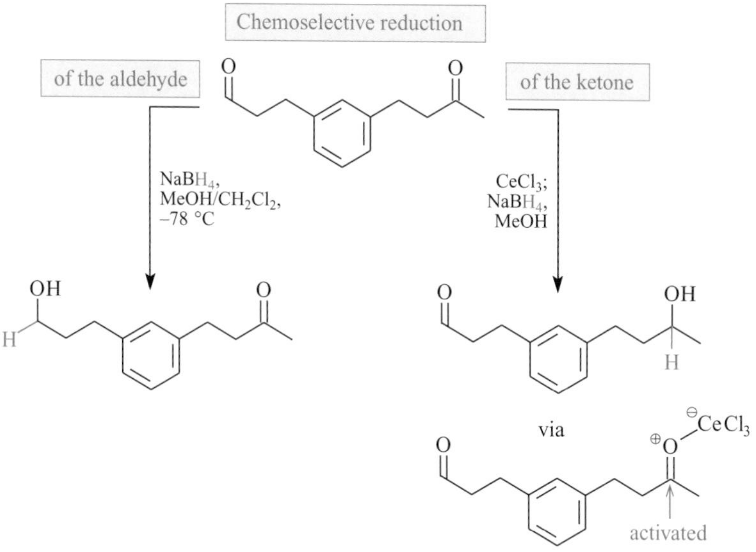

Chemoselective reduction

of the aldehyde

of the ketone

$NaBH_4$, $MeOH/CH_2Cl_2$, –78 °C

$CeCl_3$; $NaBH_4$, MeOH

via

activated

Fig. 10.6. Chemoselective carbonyl group reductions, I. On the left side a chemoselective reduction of the aldehyde takes place, whereas on the right side a chemoselective reduction of the ketone is shown.

Because of the electron-donating inductive effect of two flanking alkyl groups, the carbonyl oxygen of ketones is more basic than that of aldehydes. Therefore, ketones form more stable Lewis acid/Lewis base complexes with electrophilic metal salts than aldehydes. This is exploited in the **Luche reduction** for the chemoselective reduction of a ketone in the presence of an aldehyde (Figure 10.6, right). As the initiating step, the keto group is complexed selectively with one equivalent of $CeCl_3$. This increases its electrophilicity so that it exceeds that of the aldehyde. Therefore, the $NaBH_4$ that is now added reduces the ketonic C=O double bond preferentially.

Of two ketonic C=O double bonds, the sterically less hindered one reacts preferentially with a hydride donor: the bulkier the hydride donor, the higher the selectivity. This makes **L-Selectride®** the reagent of choice for reactions of this type (Figure 10.7, left).

Conversely, the more easily accessible C=O group of a diketone can be selectively complexed with a Lewis acid, which responds sensitively to the differential steric hindrance due

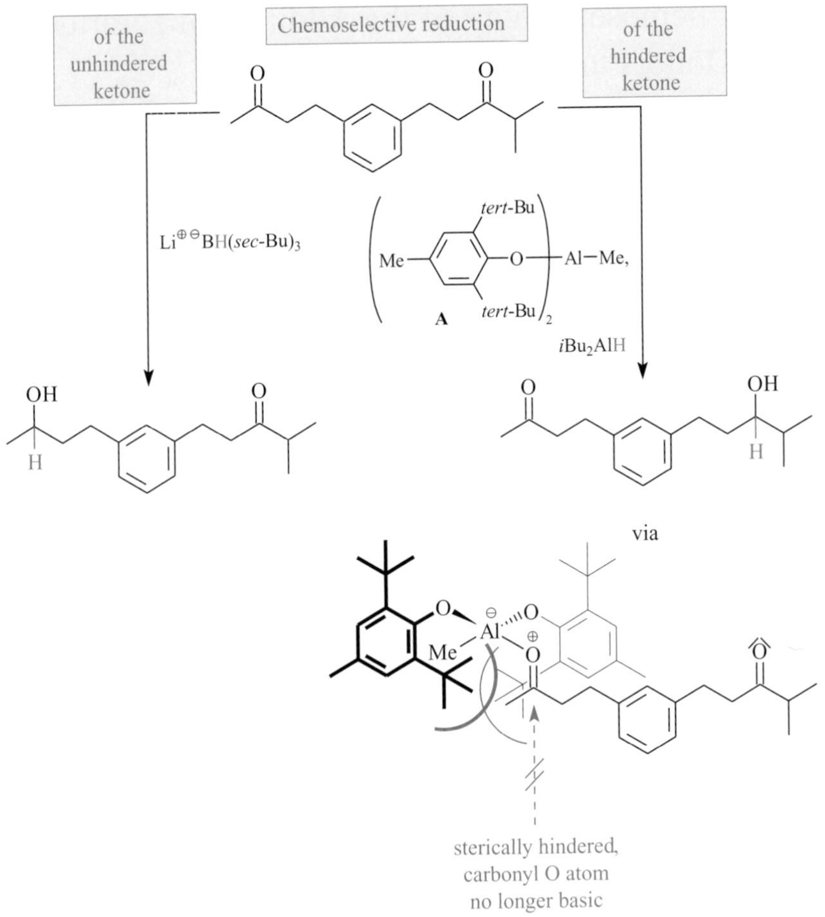

Fig. 10.7. Chemoselective carbonyl group reductions, II. A chemoselective reduction of the less hindered ketone takes place on the left side, and a chemoselective reduction of the more strongly hindered ketone takes place on the right side.

to the substituents at the two C=O groups. A suitable reagent for such a selective complexation is the aluminoxane **A** (Figure 10.7, right). After the complexation has taken place, DIBAL is added as a reducing agent. In contrast to $NaBH_4$, this reagent exhibits a certain electrophilicity because of the electron-deficient aluminum atom. Therefore, it preferentially reacts with the C=O double bond that is not complexed by the aluminoxane **A**. This is because that is the only C=O double bond that offers the DIBAL a free electron pair with which to coordinate. The result is that DIBAL reacts with the more strongly hindered ketonic C=O double bond.

An α,β-unsaturated ketone and a saturated ketone have different reactivities with respect to hydride donors (Figure 10.8). $NaBH_4$ reacts to form the tetrahedral intermediate preferentially with the nonconjugated C=O double bond (Figure 10.8, left). Product development control is responsible for this: in this addition no conjugation between the C=O and the C=C double bond is lost. Such would be the case if the addition of $NaBH_4$ occurred at the C=O double bond of the unsaturated ketone. Conversely, the unsaturated substituent, because of its resonance effect, increases the basicity of the conjugated ketonic carbonyl group. Therefore,

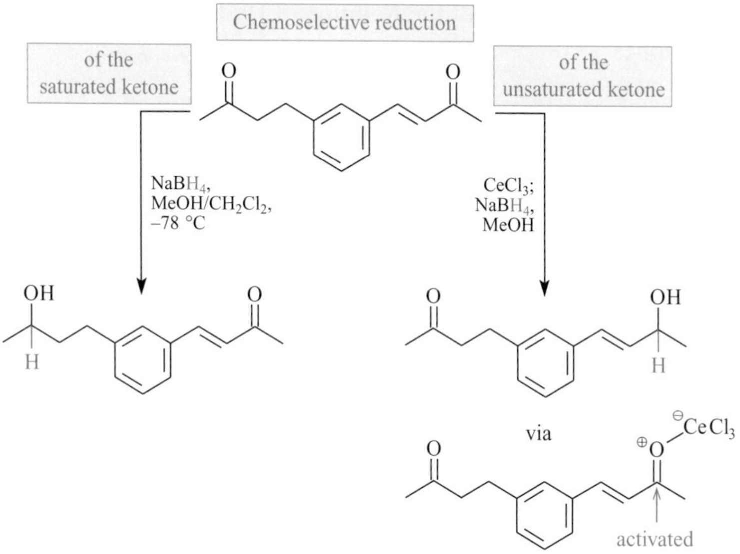

Chemoselective reduction
of the saturated ketone
of the unsaturated ketone
$NaBH_4$, $MeOH/CH_2Cl_2$, −78 °C
$CeCl_3$; $NaBH_4$, MeOH
via
activated

Fig. 10.8. Chemoselective carbonyl group reductions, III. Reduction of a saturated ketone in the presence of an unsaturated ketone (left) and reduction of an unsaturated ketone in the presence of a saturated ketone (right).

this group can be complexed preferentially with one equivalent of $CeCl_3$. Thereby, however, it becomes more electrophilic and can be reduced chemoselectively with subsequently added $NaBH_4$ in a Luche reduction (Figure 10.8, right).

The so-called 1,4-addition (for the term see Figure 10.31) of hydride donors to α,β-unsaturated ketones yielding an enolate as the primary product and, after protic workup, a saturated ketone, is also known. A hydride donor that reduces unsaturated ketones in this manner is L-Selectride®, which has been mentioned several times. An example of this type of reaction is given in the lower half of Figure 13.20.

10.3 Diastereoselectivity of the Addition of Hydride Donors to Carbonyl Compounds

When the plane of the double bond of a carbonyl compound is flanked by diastereotopic half-spaces, a stereogenic addition of a hydride can take place diastereoselectively (cf. Section 3.4.1). In Section 10.3.1, we will investigate which diastereomer is preferentially produced in such additions to the C=O double bond of cyclic ketones. In Sections 10.3.2 and 10.3.3, we will discuss which diastereomer is preferentially formed in stereogenic additions of hydride donors and acyclic chiral ketones or acyclic chiral aldehydes.

10.3.1 Diastereoselectivity of the Addition of Hydride Donors to Cyclic Ketones

In many cyclic or bicyclic molecules a stereostructure is present in which one can identify a convex and a concave side. Because reactions usually take place in such a way that the reacting reagent is exposed to the least possible steric hindrance, convex/concave substrates are generally react on their convex side.

Figure 10.9 shows an application of this principle in the diastereoselective addition of a hydride donor to a bicyclic ketone: With **L-Selectride**® [= $Li^{\oplus}\,{}^{\ominus}BH(sec\text{-}Bu)_3$] the *endo*-alcohol is produced exclusively.

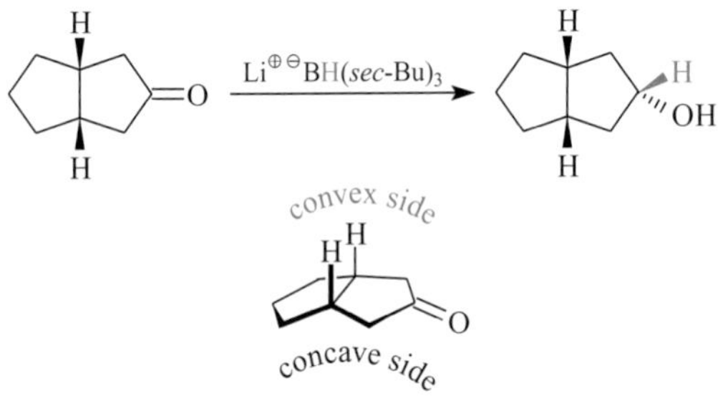

H
=O
H
$Li^{\oplus}\,{}^{\ominus}BH(sec\text{-}Bu)_3$
H
H
OH
H
convex side
H H
O
concave side

Fig. 10.9. Diastereoselective addition of a bulky hydride donor (L-Selectride) to a bicyclic ketone. The *endo*-alcohol is formed exclusively.

endo* and *exo

The stereodescriptors "*endo*" and "*exo*" are used to distinguish between the positions of a substituent beneath or outside the concave area of a bent molecule. For example, in the major product of the reduction of norbornanone with $Li^{\oplus}\,{}^{\ominus}BH\,(sec\text{-}Bu)_3$ (Formula **A** in Figure 10.10) the added H atom is oriented *exo* and the OH group obtained *endo*.

Other cyclic or bicyclic ketones do not have a convex side but only a less concave and a more concave side. Thus, a hydride donor can add to such a carbonyl group only from a concave side. Because of the steric hindrance, this normally results in a decrease in the reactivity. However, the addition of this hydride donor is still less disfavored when it takes place from the less concave (i.e., the less hindered) side. As shown in Figure 10.10 (top) by means of the comparison of two reductions of norbornanone, this effect is more noticeable for a bulky hydride donor such as L-Selectride® than for a small hydride donor such as $NaBH_4$. As can be seen from Figure 10.10 (bottom), the additions of all hydride donors to the norbornanone derivative **B** (camphor) take place with the opposite diastereoselectivity. As indicated for each substrate, the *common* selectivity-determining factor remains the principle that the reaction with hydride takes place preferentially from the less hindered side of the molecule.

In the hydride additions of Figures 10.9 and 10.10, the C=O groups are part of a bent moiety of the molecule because of the *configuration* of the substrate. *Conformational* preferences can also bring C=O groups into positions in which they have a convex side that is easier to approach and a concave side that is more difficult to approach. In 4-*tert*-butylcyclohexanone, for example, the equatorially fixed *tert*-butyl group enforces such geometry. Again, addition of the hydride donor from the convex side of this molecule is sterically favored; it corresponds to an equatorial approach (Figure 10.11). Therefore, the sterically demanding hydride donor

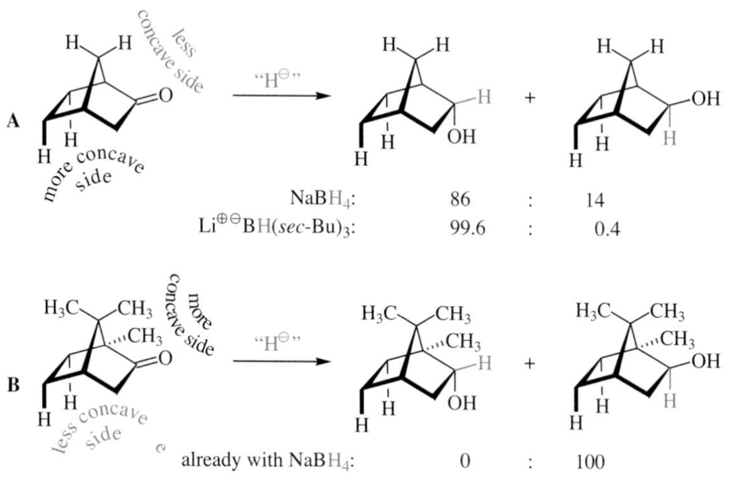

$NaBH_4$:	86	:	14
$Li^{\oplus\ominus}BH(sec\text{-}Bu)_3$:	99.6	:	0.4

already with $NaBH_4$:	0	:	100

Fig. 10.10. Addition of hydride donors to the less concave (hindered) side of the C=O double bond of norbornanone (**A**) and camphor (**B**). Since the steric differences in the diasterotopic faces of norbornanone are less pronounced than in camphor, the addition to norbornanone proceeds more rapidly and with high diastereoselectivity only if the bulky L-Selectride® instead of $NaBH_4$ serves as the hydride donor.

L-Selectride converts 4-*tert*-butylcyclohexanone into the cyclohexanol with the axial OH group by means of a preferentially equatorial approach to the extent of 93 % (Figure 10.11, top reaction).

In contrast, sterically undemanding hydride donors such as $NaBH_4$, $LiAlH_4$ or—totally unexpectedly—$LiBH_3[N(n\text{-}Pr)_2]$ reduce 4-*tert*-butylcyclohexanone preferentially/exclusively through an axial approach. This mainly produces the cyclohexanol with the equatorial OH group (Figure 10.11, second to fourth equations; for a perfectly diastereoselective reduction of 4-*tert*-butylcyclohexanone to the equatorial alcohol by a different method (see Figure 17.53). This difference results from the fact that there is also a stereoelectronic effect that influences the diastereoselectivity of the reduction of cyclohexanones. In the reaction with sterically undemanding reducing agents, this stereoelectronic effect is *fully* effective (80–99 %). However, in the reaction of such a bulky hydride donor as L-Selectride® the stereoelectronic effect is *outweighed* by the opposing steric effect discussed above.

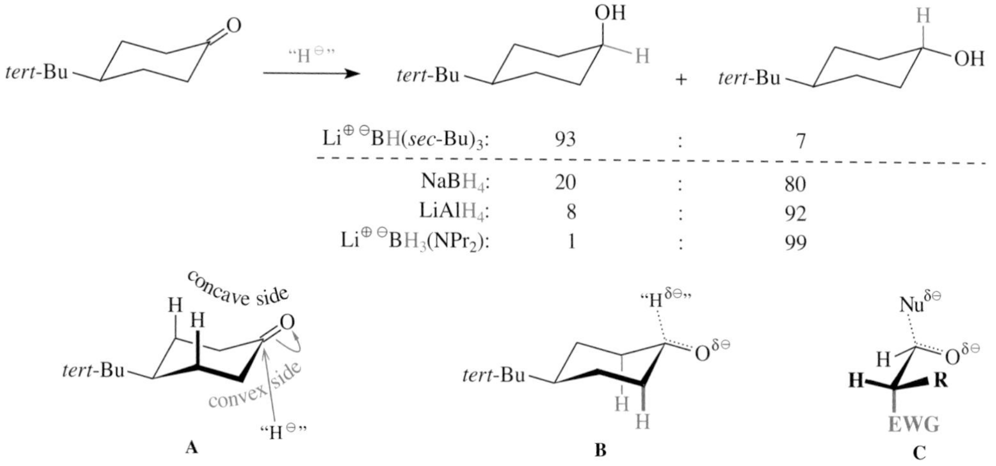

$Li^{\oplus\ominus}BH(sec\text{-}Bu)_3$:	93	:	7
$NaBH_4$:	20	:	80
$LiAlH_4$:	8	:	92
$Li^{\oplus\ominus}BH_3(NPr_2)$:	1	:	99

Fig. 10.11. Addition of various hydride donors to 4-*tert*-butylcyclohexanone. With L-Selectride® the equatorial approach (Formula **A**) is preferred, with sterically (less) demanding hydride donors the reaction proceeds axially via transition state **B** (cf. text and, particularly, Side Note 10.1). For comparison see the Felkin–Anh transition state **C** (in Figure 10.16; EWG = electron-withdrawing group).

In the explanation favored today, the reason for this stereoelectronic effect is as follows: the electronically preferred direction of approach of a hydride donor on the C=O double bond of cyclohexanone is the direction in which two of the C—H bonds at the neighboring α-positions are exactly opposite the trajectory of the approaching nucleophile. Only the axial C—H bonds in the α-positions can be in such an antiperiplanar position while the equatorial C—H bonds cannot. Moreover, these axial C—H bonds are antiperiplanar with regard to the trajectory of the hydride donor only if the nucleophile reacts via a transition state **B**, that is, axially. The "antiperiplanarity" of the two *axial* C—H bonds in the α-positions is reminiscent of the "antiperiplanarity" of the electron-withdrawing group in the α-position relative to the nucleophile in the Felkin–Anh transition state (formula **C** in Figure 10.11; cf. Figure 10.16, middle row). How the opposing C–H_{axial} bonds stabilize transition state **B** will be explained by means of Figure 10.13.

Side Note 10.1. Electronic Effects in the Reduction of Conformationally Fixed Cyclohexanones

The assumption that due to a "stereoelectronic effect" the reduction of 4-*tert*-butylcyclohexanone with sterically undemanding hydride donors proceeds diastereoselectively incited the investigation of the reduction of cyclohexanones whose stereochemical course, unlike that of 4-*tert*-butylcyclohexanone, is not subject to this competing convex/concave effect. This type of cyclohexanone includes the adamantanones in Figure 10.12. Regardless of the direction from which the hydride donor reacts with their C=O double bond, this reaction occurs axially (i.e., concave) with regard to the one cyclohexane chair substructure to which this carbonyl group belongs and equatorially (i.e., convex) with respect to the other:

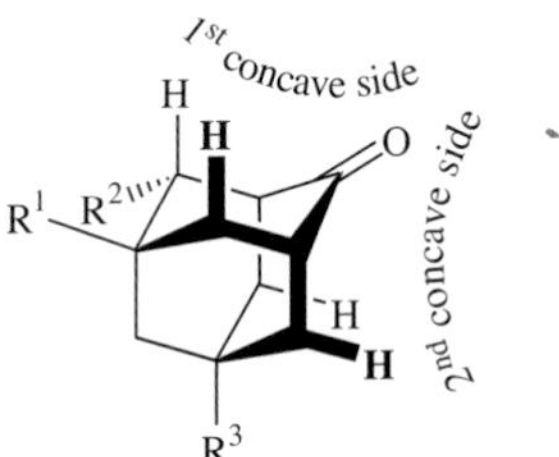

Therefore, the reduction reactions given in Figure 10.12 exhibit no asymmetric induction resulting from the skeleton. The reason for the stereochemical differentiation is substitution in position 3 with a bromine atom, in position 4 with an iodine atom, and in position 8 a nitro group and in position 8' an amino group, respectively. The numbering places the carbonyl group of the adamantanone at postion 1. With these great distances no steric effect can be operative. Electronic effects, though, have a large range in charged particles, like in the transition state of the hydride transfer where an alkoxide ion is produced. Thus, the $NaBH_4$ reductions of the adamantones mentioned proceed with diastereoselectivities of 76:24, 64:36 and 62:38, respectively. The hydride ion transfer occurs primarily on that side of the carbonyl group that is located on the side of the (more strongly) electron-withdrawing substituent. Stated differently, the hydride reacts antiperiplanar to the most electron rich bonds.

The fact that some *electronic* effect is operative in adamantanone reductions and the general preference for an axial approach in cyclohexanones initiated detailed investigations (by

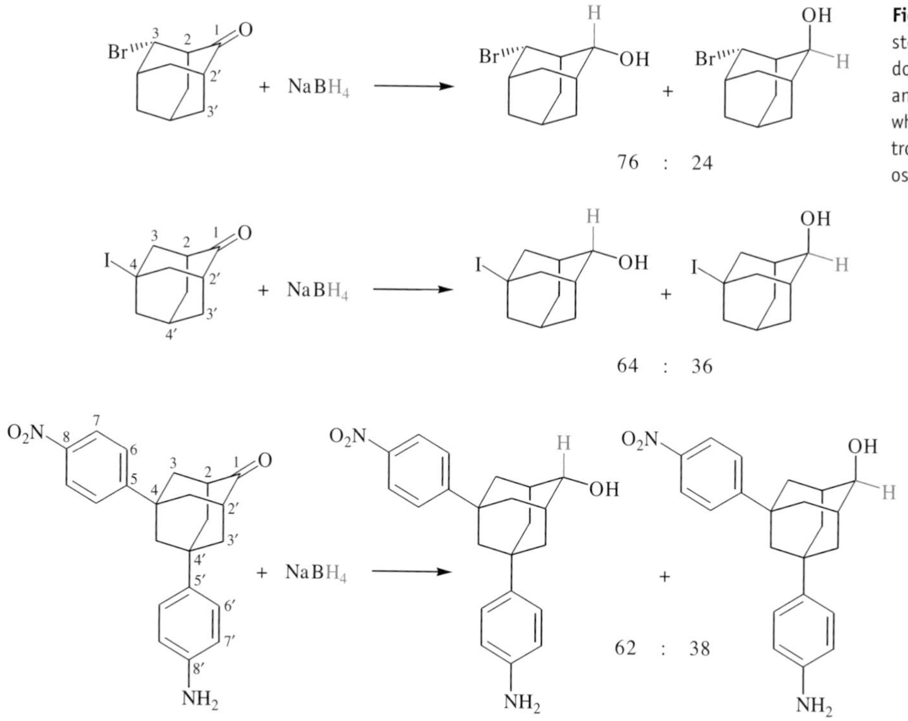

Fig. 10.12. Addition of a sterically undemanding hydride donor to tricyclic cyclohexanones ("adamantanones") where no steric but only electronic effects on the diastereoselectivity occur.

means of extensive calculations) on the origin of the effect. According to calculations, **A** in Figure 10.13 is the lowest-energy transition state of an axial hydride ion transfer to cyclohexanone. The transition state is early and the pi facial selectivity in mediated by an interaction between the axial C–H bonds next to the ketone and the π* orbital of the carbonyl group. This result in what is called orbital extension, meaning that the LUMO of the carbonyl group has a larger coefficient on the axial face of the carbonyl relative to the equatorial face. A larger coefficient leads to better bonding, stabilizing the transition state. Another possible interaction is a hyperconjugation between the axial CH bonds that the σ* orbital of the incipient CH bond. This is generally not considered important for early transition states, but may play a role in the stereochemical outcome of nucleophilic additions to carbonyl groups proceeding through late transition states. The MO model **C** of this transition state demonstrates a bonding interaction between the doubly occupied σ-MO of the resulting C–H bond and the unoccupied, parallel oriented σ*-MOs of the adjacent C–H$_{\text{axial}}$ bond. **B** is the lowest-energy transition state of an equatorial hydride ion transfer to cyclohexanone that has been calculated (Figure 10.13). It is higher in energy by 3.2 kcal/mol than transition state **A**. The MO model **D** of this transition state reveals that a bonding interaction between the doubly occupied σ-MO of the resulting C–H bond and the unoccupied, parallel oriented σ*-MOs, i.e., the σ*-MOs of the neighboring C–C bonds, may also occur. But: This interaction in transition state structure **D** results in less stabilization than the almost equal interaction in transition state **C**. From the point of view of theoretical chemistry this definitely explains the stereoelectronic preference

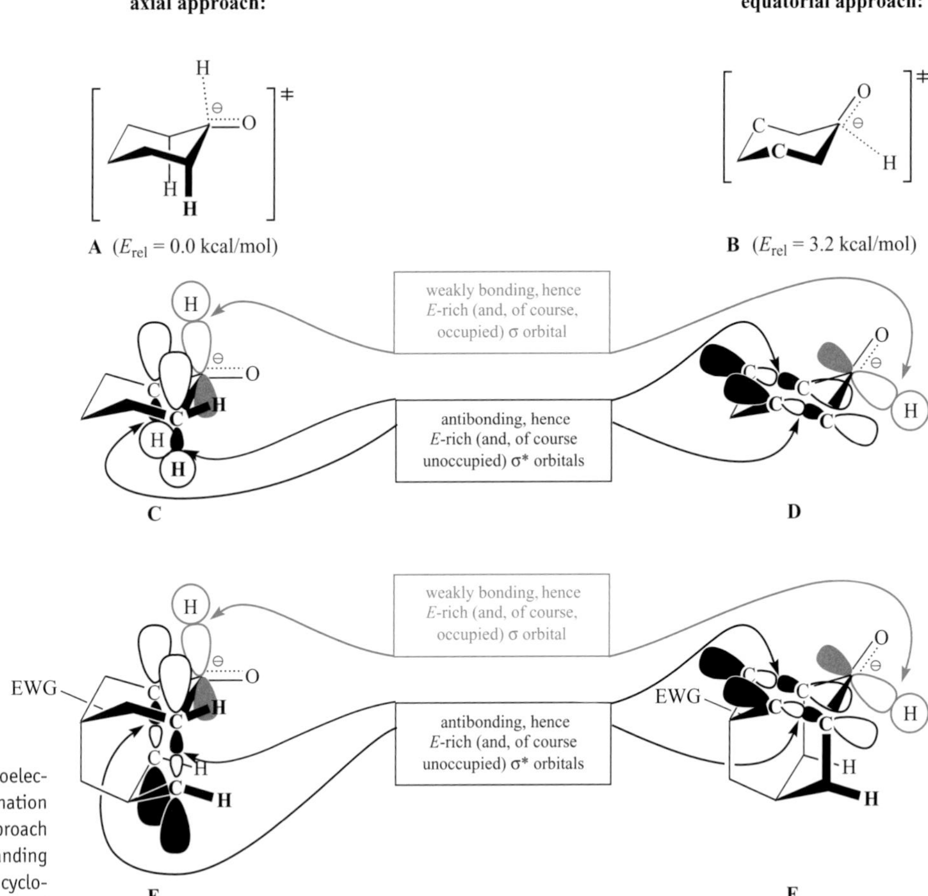

Fig. 10.13. A stereoelectronic effect as an explanation for the mostly axial approach of sterically undemanding nucleophiles with cyclohexanone (lines 1 and 2) and for the preferred *syn* approach of H-nucleophiles to acceptor-substituted adamantanones.

of the axial hydride ion approach towards *tert*-butylcyclohexanone. The only downside is that obviously there is no simple, let alone intuitive explanation for the non-equivalence of the orbital overlaps in transition states **C** and **D**.

Regarding the explanation of the "electronic effect" controlling the diastereoselectivities of the adamantone reductions in Figure 10.12 it has to be assumed on the basis of the comments on Figure 10.13 that it also represents a stereoelectronic effect. In analogy to the calculated energy difference **C** < **D**, it may be assumed that the experimental energy difference **E** < **F** is due to a non-equivalence of the corresponding orbital interactions.

10.3.2 Diastereoselectivity of the Addition of Hydride Donors to α-Chiral Acyclic Carbonyl Compounds

In this section as well as in Section 10.5.3, carbonyl compounds that contain a stereocenter in the position α to the C=O group are referred to succinctly with the term "α-chiral carbonyl compound." Because of the presence of the stereocenter, the half-spaces on both sides of the plane of the C=O double bond of these compounds are diastereotopic. In this section, we will study in detail stereogenic addition reactions of hydride donors to the C=O double bond of α-chiral carbonyl compounds. Additions of this type can take place faster from one half-space than from the other—that is, they can be diastereoselective.

Introduction: Representative Experimental Findings

If one wants to study a stereogenic addition reaction of a hydride donor to an α-chiral *aldehyde*, one must use, for instance, $LiAlD_4$ as the reducing agent (thus, a deuteride donor). In contrast, stereogenic addition reactions of hydride donors to α-chiral *ketones* require no deuterium labeling: They can be observed in "ordinary" reductions, for instance, with $LiAlH_4$. Which alcohol is preferentially produced in each case depends on, among other things, which atoms are connected to the stereocenter at C-α.

Additions of hydride donors to α-chiral carbonyl compounds that bear only hydrocarbon groups or hydrogen at C-α typically take place with the diastereoselectivities of Figure 10.14. One of the resulting diastereomers and the relative configuration of its stereocenters are referred to as the **Cram product**. The other diastereomer that results and its stereochemistry are referred to with the term **anti-Cram product**.

In contrast, the diastereoselectivities of Figure 10.15 can be observed for many additions of hydride donors to carbonyl compounds that contain a stereocenter in the α-position with an O or N atom bound to it. One of the product diastereomers and the relative configuration of its stereocenters is called the **Felkin–Anh product**. The other diastereomer and its stereo-

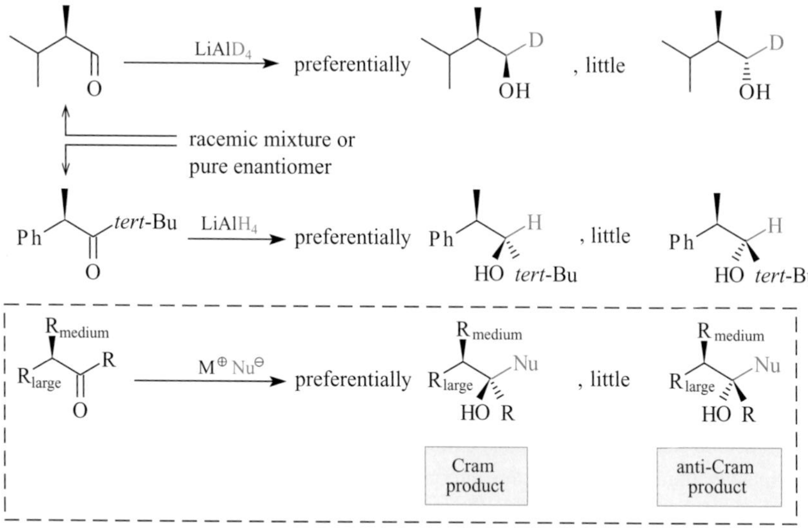

Fig. 10.14. Examples and structural requirements for the occurrence of Cram-selective additions of hydride donors to α-chiral carbonyl compounds. In the three compounds at the bottom R_{large} refers to the large and R_{medium} refers to the medium-sized substituent.

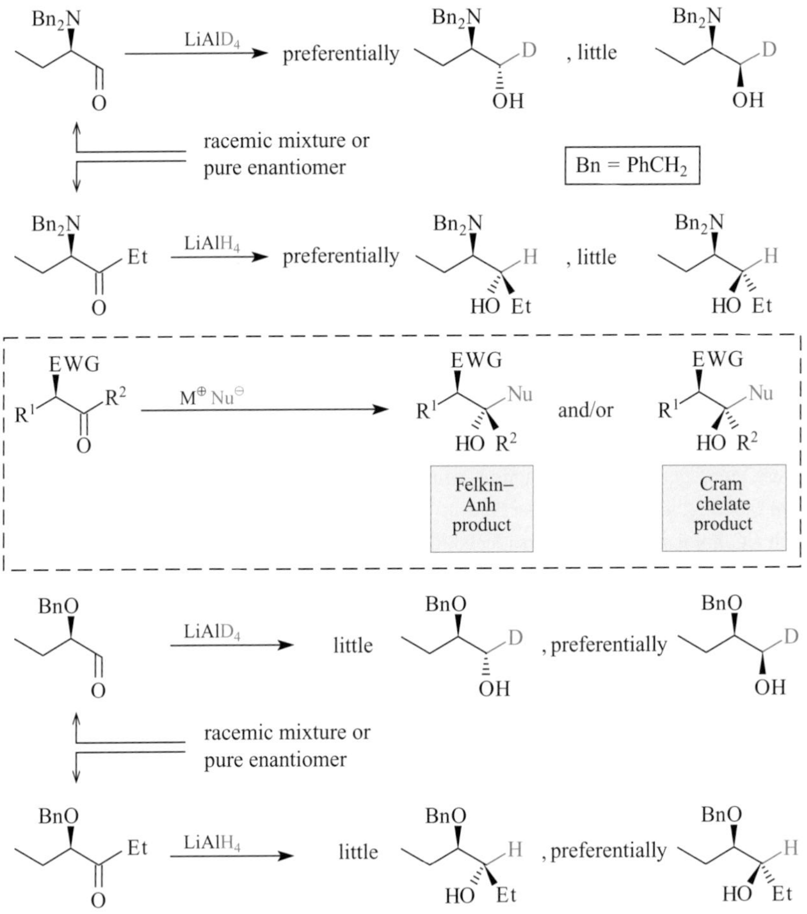

Fig. 10.15. Examples and structural requirements for the occurrence of Felkin–Anh selective (top) or chelation-controlled (bottom) additions of hydride donors to α-chiral carbonyl compounds. EWG, electron-withdrawing group.

chemistry are referred to as the so-called **Cram chelate product**. If the latter is produced preferentially, we say there is "chelation control" or that the major product is the "chelation-controlled product."

The reason for the diastereoselectivities presented in Figures 10.14 and 10.15 will be explained in the following section.

The Reason for Cram and Anti-Cram Selectivity and for Felkin–Anh and Cram Chelate Selectivity; Transition State Models

In additions of hydride donors to α-chiral carbonyl compounds, whether Cram or anti-Cram selectivity, or Felkin-Anh or Cram chelate selectivity occurs is the result of kinetic control. The rate-determining step in either of these additions is the formation of a tetrahedral intermediate. It takes place irreversibly. The tetrahedral intermediate that is accessible via the most stable transition state is produced most rapidly. However, in contrast to what is found in many other considerations in this book, this intermediate does not represent a good transition state model for its formation reaction. The reason for this "deviation" is that it is produced in an exothermic and exergonic step. Thus, according to the Hammond postulate, the tetrahedral intermediate here is formed in an early transition state. Consequently, the latter does not resemble the tetrahedral intermediate but rather the carbonyl compound and the reducing agent. Calculations confirmed this for the Cram transition state (Figure 10.16, left) and for the Felkin–Anh transition state (Figure 10.16, middle). Therefore, the existence of an early transition state is plausible also in the case of chelation-controlled addition reactions (Figure 10.16, right).

Both in the Cram (Figure 10.16, left) and Felkin–Anh transition states (Figure 10.16, middle) a stabilizing orbital overlap occurs that was already encountered in connection with the discussion of Figure 10.13: there is a—in each case bonding—overlap between the σ-MO assigned to the resulting C-Nu bond and

- the parallel oriented σ*-MO of the adjacent C-R_{large} bond in the Cram transition state and
- the parallel oriented σ*-MO of the adjacent C-EWG bond in the Felkin–Anh transition state, respectively.

The transition state geometries can be explained if we also consider the following two factors:
(1) In the Felkin–Anh transition state we find a parallel orientation of the σ_{C-EWG}*-MOs because this is lower in energy than the other neighboring σ*-MOs, i.e., the σ^*_{C-R1}-MO or the σ^*_{C-H}-MO.
(2) In the Cram transition state, the parallel orientation of the C–R_{large}*-MOs is *not* due to the fact that the resulting orbital overlap is more stable than an overlap with the parallel oriented $\sigma^*_{C-Rmedium}$-MO or σ^*_{C-H}-MO. Rather, each of these overlaps leads to a similar stabilization. But only with the first option does the reacting nucleophile face the least possible steric hindrance.

From the transition state structures in Figure 10.16 the following may be seen:

1. The additions of hydride donors to the C=O double bonds of α-chiral carbonyl compounds take place via transition states whose stereostructures do not reflect the preferred conformations of the substrates.
2. The addition of a hydride donor to an α-chiral carbonyl compound, *without* an O or N atom in the α-position, regardless of whether it is racemic or enantiomerically pure, takes place through the so-called **Cram transition state**. It is shown in Figure 10.16 (left; $Nu^{\ominus} = H^{\ominus}$) both as a Newman projection and in the sawhorse representation. In the Cram transition state, the H atom (or the smallest hydrocarbon substituent) at C-α is aligned approximately antiperiplanar to the reacting C=O double bond. The hydride donor reacts with the C=O double bond from the half-space that does *not* contain the largest hydrocarbon substituent at C-α. The nucleophile reacts with the C=O double bond at an angle of ca. 103° (i.e., slightly from the rear). In other words, in the Cram transition state the hydride donor reacts with the carbonyl carbon in a trajectory that is almost *anti* to the bond that connects C-α to the largest α-substituent.
3. The addition of a hydride donor to an α-chiral aldehyde *with* an O or an N atom in the α position or to an analogous ketone takes place through the so-called **Felkin–Anh transition state** provided that the heteroatom at C-α is *not* incorporated in a five-membered chelate ring together with the O atom of the carbonyl group. This transition state is also shown in Figure 10.16 (center; $Nu^{\ominus} = H^{\ominus}$), both as a Newman projection and in the sawhorse

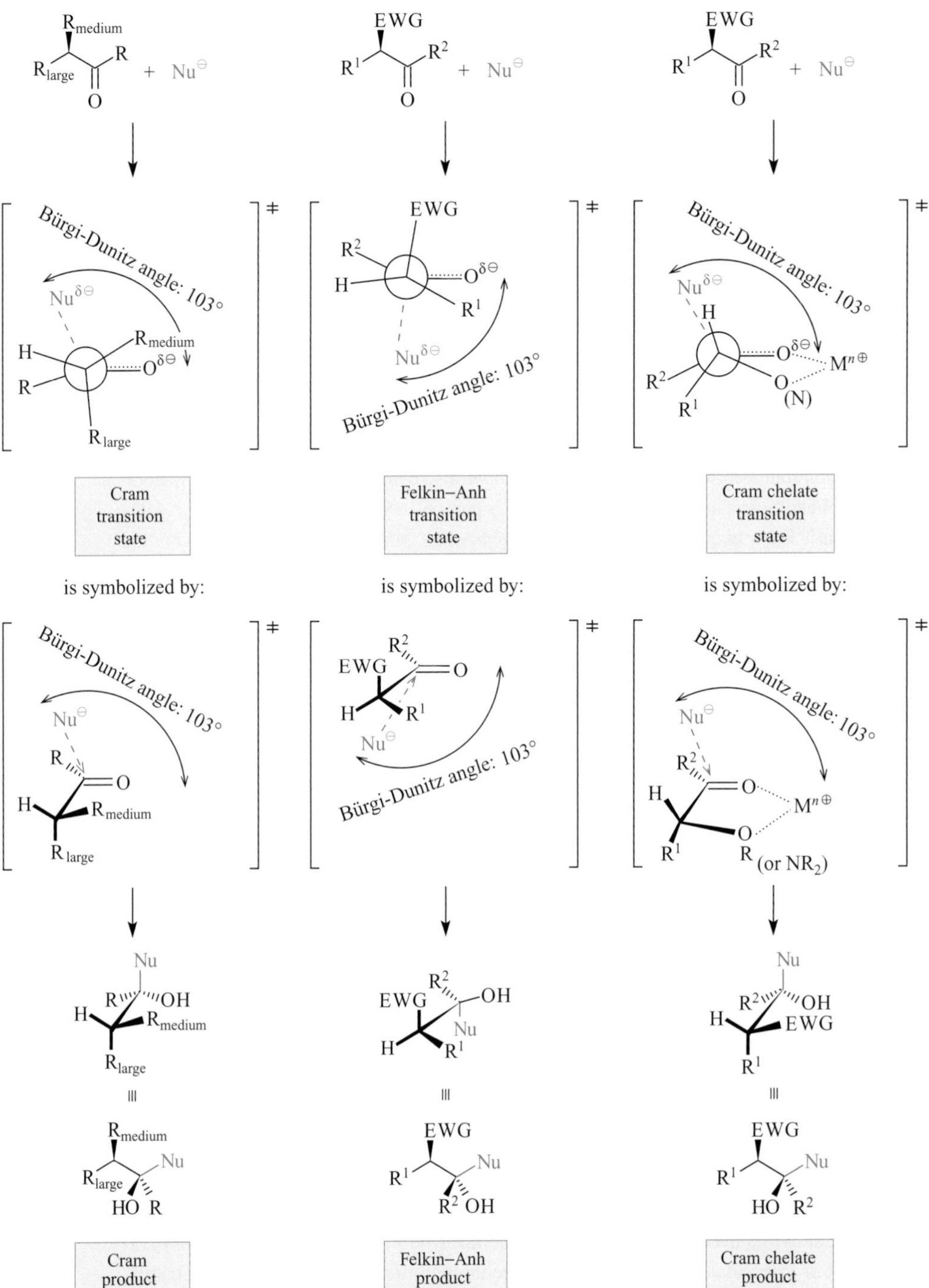

Fig. 10.16. The three transition state models for the occurrence of diastereoselectivity in the addition of hydride donors ($Nu^{\ominus} = H^{\ominus}$) or organometallic compounds ($Nu^{\ominus} = R^{\ominus}$) to α-chiral carbonyl compounds (R_{large} refers to the large substituent, R_{medium} refers to the medium-sized substituent, and EWG refers to an electron-withdrawing group).

representation. The H atom (or the second largest hydrocarbon substituent) at C-α is aligned approximately antiperiplanar to the reacting C=O double bond. The hydride donor approaches the substrate from the half-space that does *not* contain the heteroatom in the α-position. Again, the nucleophile reacts with the C=O double bond at an angle of approx. 103° (i.e., slightly from the rear). In other words, in the Felkin–Anh transition state the C—H bond being formed and the Cα—heteroatom bond are arranged almost *anti*.

4. The addition of a hydride donor to an α-chiral aldehyde with an OR or NR_2 substituent at C-α or to an analogous ketone takes place via the so-called **Cram chelate transition state** provided that the heteroatom at C-α and the O atom of the carbonyl group are incorporated in a five-membered chelate ring. This transition state is also shown in Figure 10.16 (right; $Nu^{\ominus} = H^{\ominus}$) in two different perspectives. The hydride donor approaches the carbonyl carbon from the less hindered half-space. This is the one that contains the H atom (or the second largest hydrocarbon group) in the α-position.

Side Note 10.2. Historical Explanation for the Selectivities Named after Cram

After Cram had discovered the selectivities now named after him, he proposed the transition state model for the formation of *Cram chelate* products that is still valid today. However, his explanation for the preferred formation of *Cram* products was different from current views. Cram assumed that the transition state for the addition of nucleophiles to α-alkylated carbonyl compounds was so early that he could model it with the carbonyl compound alone. His reasoning was that the preferred conformation of the free α-chiral carbonyl compound defines two sterically differently encumbered half-spaces on both sides of the plane of the C=O double bond. The nucleophile was believed to approach from the less hindered half-space.

Today, it is known that Cram was wrong about this preferred conformation (he assumed that the bulkiest Cα carbon bond was oriented *anti* to the C=O double bond; actually this Cα carbon bond prefers a *syn*-orientation). If Cram had known this conformation at that time, he would certainly not have based his explanation of Cram selectivity upon a reaction with the most stable conformer of the aldehyde. This is because, in order to establish the experimentally found product geometry, the nucleophile would have been required to approach the C=O double bond of the truly favored conformer from the sterically more hindered side. However, even at that time this error could have been avoided (see below).

Curtin-Hammett Principle

Many organic chemists have overlooked the fundamental error in Cram's old argument and are continuously inclined to commit all sorts of analogous errors in proposing "explanations" for other stereoselectivities. This type of error in reasoning is unfortunately very tempting. However, it is as easy as this: in more than 99.9% of all stereogenic reactions converting a given substrate into two diastereomers the activation barriers of the selectivity-determining steps are distinctly higher than the rotational barriers that separate the different conformers of the substrate from each other. Figure 10.17 illustrates these relationships using stereogenic addition reactions to α-methylisovaleraldehyde as an example.

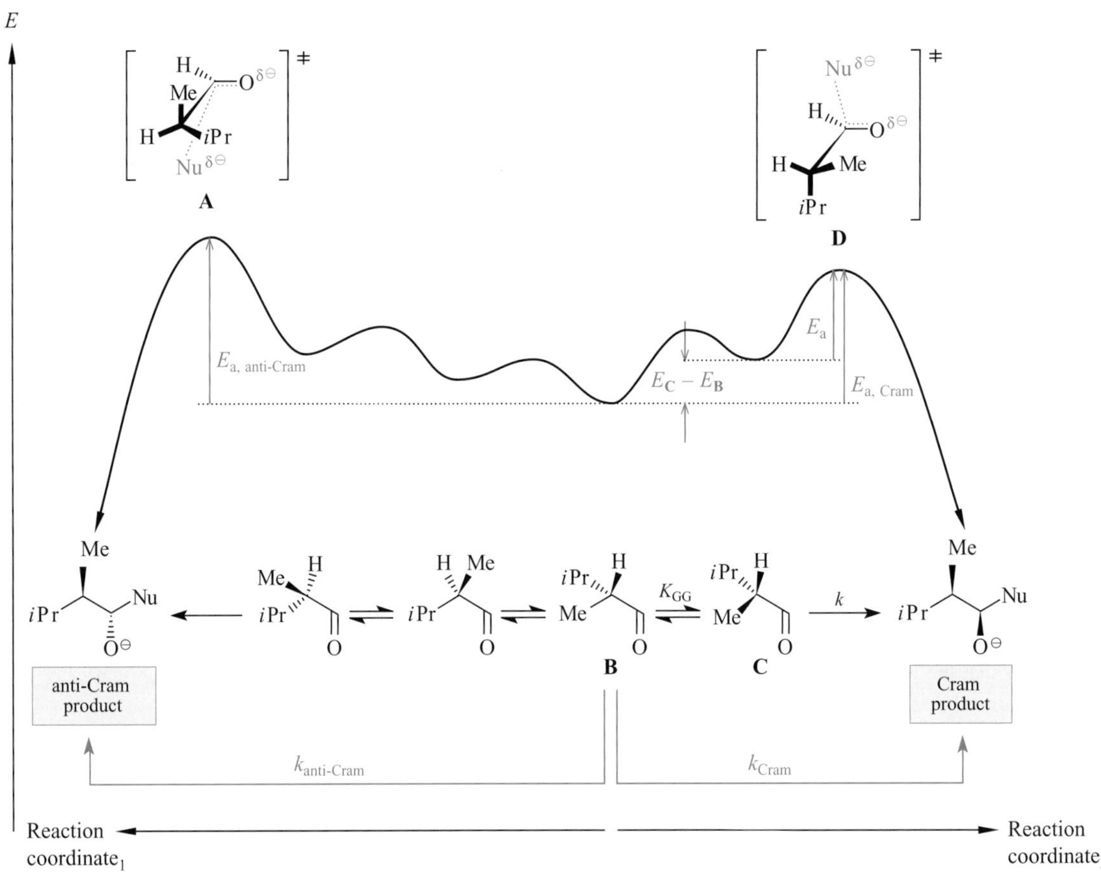

Fig. 10.17. Transition states of the selectivity-determining step of a stereogenic addition of a hydride donor to an α-chiral carbonyl compound (the energy profile would be allowed to contain additional local energy maxima provided that they do not have a higher energy than the two highest maxima shown in the figure).

Curtin-Hammett Principle

In stereogenic reactions leading from a substrate to two diastereomers, it is irrelevant which conformer of the substrate is preferred if conformational isomerization is much faster than the reaction. The favored reaction is the one that takes place via the transition state with the lowest energy and may or may not derive from the preferred conformation of the substrate. The extent of diastereoselectivity encountered in the stereogenic reaction under scrutiny results from the competition between the favored and the disfavored reaction paths; it depends exclusively on the free enthalpy difference between the competing, diastereomeric transition states.

Here, the Curtin–Hammett principle will be proven using the example of the reaction pair from Figure 10.17. The Cram product forms from a conformer **C** of the α-chiral aldehyde via the Cram transition state **D** with the rate constant k. The corresponding rate law is is shown in

Equation 10.1. However, the conformer **C** of the aldehyde also equilibrates rapidly with the more stable conformer **B.** Therefore, only a small fraction of the total aldehyde is present as **C.** The concentration of conformer **C** is to a very good approximation described by Equation 10.2 as a function of the concentration of conformer **B.** Equation 10.3 follows from the fact that virtually all of the aldehyde present adopts the conformation **B.** (If substantial amounts of other conformers were also present, one would have to introduce a proportionality factor < 1 on the right side of Equation 10.3. However, this proportionality factor would have no influence on the following derivation.)

$$\frac{\mathrm{d}[\text{Cram product}]}{\mathrm{d}t} = k[\mathbf{C}][\mathrm{Nu}^{\ominus}] \tag{10.1}$$

$$[\mathbf{C}] \approx K_{eq}\,[\mathbf{B}] \tag{10.2}$$

$$[\mathbf{B}] \approx [\text{aldehyde}] \tag{10.3}$$

Equation 10.4 is obtained by successive insertions of Equations 10.2 and 10.3 into Equation 10.1. The rate constant k of Equation 10.4 can also be written in the form of Equation 10.5 which is the Arrhenius law. In addition, one can rewrite the equilibrium constant K_{eq} of the conformer equilibrium $\mathbf{B} \rightleftharpoons \mathbf{C}$ as Equation 10.6, which relates K_{eq} to the energy difference between **C** and **B.**

$$\frac{\mathrm{d}[\text{Cram product}]}{\mathrm{d}t} = k \cdot K_{eq} \cdot [\text{aldehyde}][\mathrm{Nu}^{\ominus}] \tag{10.4}$$

$$k = A \cdot \exp\left(-\frac{E_a}{RT}\right) \tag{10.5}$$

$$K_{eq} = \exp\left(-\frac{E_c - E_B}{RT}\right) \tag{10.6}$$

We obtain Equation 10.7 by multiplying Equation 10.5 by Equation 10.6. When we now take Figure 10.17 into consideration, we can summarize the energy terms in the numerator of Equation 10.7, as shown in Equation 10.8. By using the Arrhenius relationship, we can abbreviate the exponent on the right side of Equation 10.8 as rate constant k_{Cram}. The result is Equation 10.9.

$$k \cdot K_{eq} = A \cdot \exp\left[-\frac{E_a + (E_C - E_B)}{RT}\right] \tag{10.7}$$

$$= A \cdot \exp\left(\frac{-E_{a\text{'Cram}}}{RT}\right) \tag{10.8}$$

$$= k_{\text{Cram}} \tag{10.9}$$

By inserting Equation 10.9 into Equation 10.4, we finally obtain

$$\frac{d[\text{Cram product}]}{dt} = k_{\text{Cram}} \cdot [\text{aldehyde}][\text{Nu}^{\ominus}] \tag{10.10}$$

Equation 10.11 for the rate of formation of the anti-Cram product is derived in a similar way:

$$\frac{d[\text{anti}-\text{Cram product}]}{dt} = k_{\text{anti-Cram}} \cdot [\text{aldehyde}][\text{Nu}^{\ominus}] \tag{10.11}$$

One now divides Equation 10.10 by Equation 10.11, reduces and integrates from the beginning of the reaction to the time of the work up.

In this way, and by using the Arrhenius relationship one more time, one obtains for the ratio of the yields:

$$\frac{\text{Yield of Cram product}}{\text{Yield of anti-Cram product}} = \frac{k_{\text{Cram}}}{k_{\text{anti-Cram}}} = \exp\left\{-\frac{E_{\text{a, Cram}} - E_{\text{a, anti-Cram}}}{RT}\right\} \tag{10.12}$$

Equation 10.12 states that for the diastereoselectivity of stereogenic additions of nucleophiles to α-chiral aldehydes it is neither important which conformation the substrate prefers nor is it important of how many conformers of the substrate exist.

Felkin–Anh or Cram Chelate Selectivity in the Addition of Hydride Donors to Carbonyl Compounds with an O or N Atom in the α-Position?

In order for the Cram chelate product to predominate after the addition of a hydride donor to a chiral carbonyl compound, which contains a heteroatom in the α-position, this heteroatom and part of the reagent must be able to form a five-membered ring chelate. If this is not possible, one observes Felkin–Anh selectivity (provided one observes selectivity at all). This has the following interesting consequences for synthesis.

The α-chiral ketone from Figure 10.18—the α-substituent is a benzyloxy group—is reduced to the Cram chelate product by $Zn(BH_4)_2$, a Lewis acidic reducing agent. The $Zn^{2\oplus}$ ion first bonds the benzyl and the carbonyl oxygen to a chelate. Only this species is subsequently reduced by the $BH^{\ominus}_4$ ion because a $Zn^{2\oplus}$-complexed C=O group is a better elec-

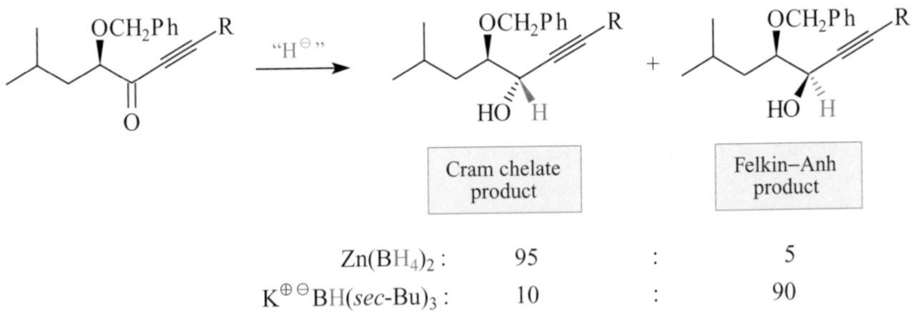

OCH₂Ph R O "H⊖" OCH₂Ph R HO H + OCH₂Ph R HO H

	Cram chelate product		Felkin–Anh product
$Zn(BH_4)_2$:	95	:	5
$K^{\oplus}\,{}^{\ominus}BH(sec\text{-}Bu)_3$:	10	:	90

Fig. 10.18. Ensuring Felkin–Anh versus Cram chelate selectivity by varying the Lewis acidity of the hydride donor.

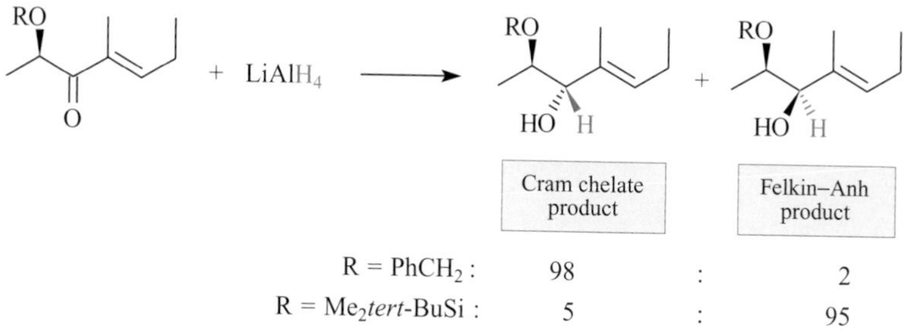

	Cram chelate product		Felkin–Anh product
R = $PhCH_2$:	98	:	2
R = Me_2*tert*-BuSi :	5	:	95

Fig. 10.19. Ensuring Felkin–Anh or Cram chelate selectivity in the addition of hydride donors to α-chiral carbonyl compounds by varying the protecting group on the stereo-directing heteroatom.

trophile than a $Zn^{2\oplus}$-free C=O group. On the other hand, the Felkin–Anh product (Figure 10.18) is produced from the same ketone and KBH(*sec*-Bu)$_3$ (**K-Selectride**®, i.e., the potassium analog of L-Selectride®). Neither the $K^{\oplus}$ ion nor the tetracoordinated boron atom of this reagent acts as a Lewis acid. Consequently, no chelate intermediate can be formed.

If, for example, the mild Lewis acid $LiAlH_4$—it is mildly acidic because its acidic moiety is $Li^{\oplus}$—is used as a hydride donor, the diastereoselectivity of the reduction of α-chiral carbonyl compounds with a heteroatom in the α-position can be controlled by a judicious choice of the protecting group on the heteroatom. Figure 10.19 shows this using the example of two differently protected α-oxygenated ketones. In the first ketone, a benzyl group is attached to the O atom at C-α. This protecting group is not bulky because it is not branched. Accordingly, it allows the incorporation of the protected α-oxygen in a five-membered ring chelate. Therefore, the reduction of this substrate with $LiAlH_4$ leads preferentially to the Cram chelate product. On the other hand, if one employs a trialkylsilyl group with branched alkyl substituents as the protecting group at the α-oxygen, it is bulky enough to prevent this formation of a chelate ring. Therefore, the hydride addition to the silyl ether from Figure 10.19 takes place via a Felkin–Anh transition state and delivers the Felkin–Anh product.

Of course, the highest diastereoselectivities in the reduction of α-chiral, α-oxygenated carbonyl compounds are expected when the reducing agent and the protecting group direct in one and the same direction; that is, when both make possible or both prevent chelation control. In the reaction examples depicted in Figure 10.20, it was possible in this way to achieve a diastereoselectivity of $> 99:1$ in the sense of Cram chelation control or optionally of $> 99:1$ in the opposite Felkin–Anh direction. Figure 10.20 puts these reductions into the context of how any glycol diastereomer can be obtained in a highly diastereoselective fashion.

10.3.3 Diastereoselectivity of the Addition of Hydride Donors to β-Chiral Acyclic Carbonyl Compounds

In this section and in Section 10.5.3, the term "β-chiral carbonyl compound" will be used as an abbreviation for carbonyl compounds that contain a stereogenic center in the position β to the C=O group. Stereogenic addition reactions of hydride donors to β-chiral carbonyl compounds are common, especially for substrates in which the stereocenter in the β-position is connected to an O atom.

β-Hydroxyketones can be reduced diastereoselectively with a hydride donor in an analogous way to that shown in Figure 10.20 for α-hydroxyketones by using chelation control. This

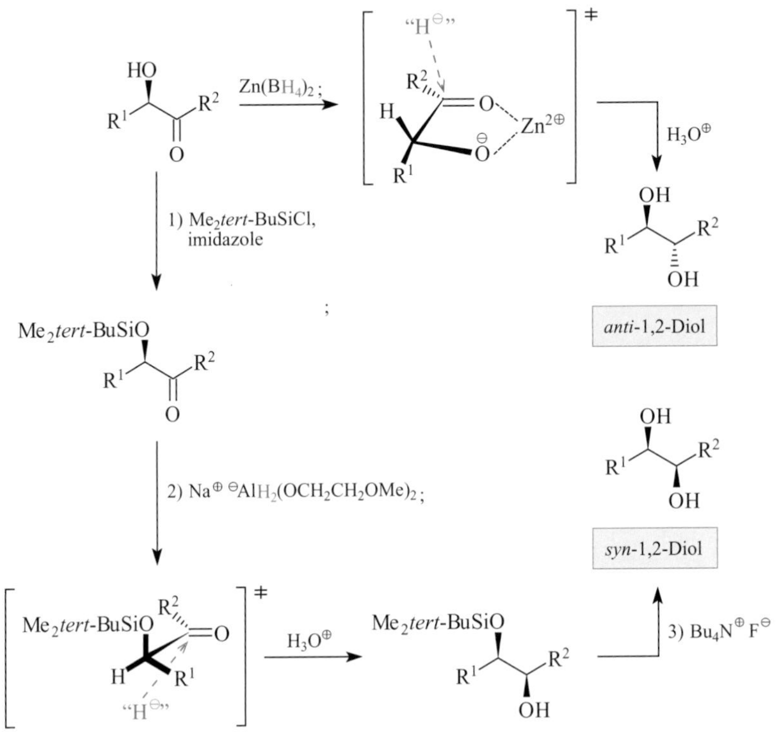

Fig. 10.20. Ensuring Felkin–Anh or Cram chelate selectivity in the addition of hydride donors to α-chiral carbonyl compounds by a combined variation of the hydride donor and the protecting group.

succeeds well when the β-hydroxyketone is fixed in a chelate complex of structure **A** through a reaction with diethylborinic acid methyl ester (Figure 10.21). In chelate **A,** the diastereotopic faces of the ketonic C=O group are not equally well accessible. Thus, $NaBH_4$ added after formation of this chelate leads to the transfer of a hydride ion from that side of the molecule that lies opposite to the pseudoequatorially oriented substituent R' in formula **B** of chelate **A.** This selectivity is not so much caused through avoiding the still quite distant R' substituent by $NaBH_4$; rather, a transition state is preferred in which the hydride ion to be transferred adopts an *axial* orientation in the resulting saturated six-membered ring while the latter assumes a *chair conformation.* Under these conditions the large borane-containing part of the reagent resides on the convex, that is, less hindered side of the least-strained conformer of the transition state.

The addition of a hydride donor to a β-hydroxyketone can also be conducted in such a way that the opposite diastereoselectivity is observed. However, the possibility previously discussed for additions to α-chiral carbonyl compounds is not applicable here. One must therefore use a different strategy as is shown in Figure 10.22, in which the OH group at the stereocenter C-β of the substrate is used *to bind the hydride donor before it reacts with the C=O double bond.* Thus, the hydridoborate **A** reacts intramolecularly. This species transfers a hydride ion to the carbonyl carbon after the latter has been protonated and thereby made more electrophilic. The hydride transfer takes place via a six-membered chair-like transition state,

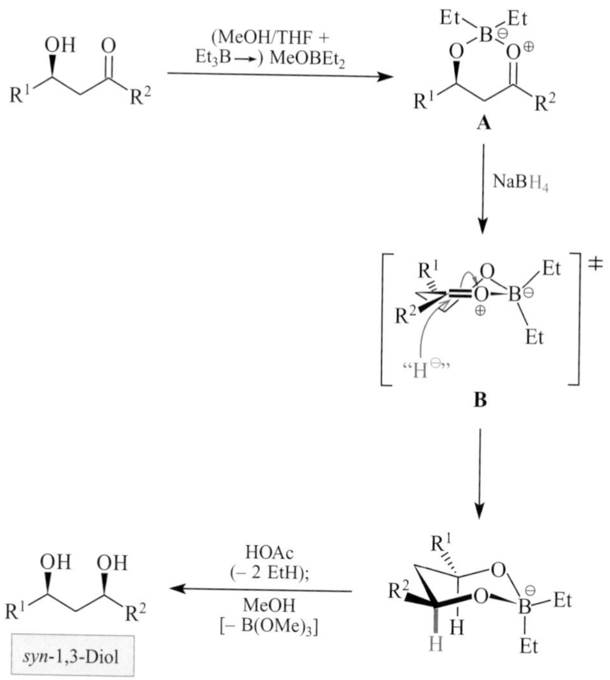

Fig. 10.21. Diastereoselective reduction of *β*-hydroxyketones to *syn*-configured 1,3-diols (**Narasaka–Prasad reduction**).

which preferentially assumes the stereostructure **B.** It is characterized by the energetically favorable pseudoequatorial arrangement of the ring substituents R^1 and R^2. This lets the double bonded oxygen atom, as the least voluminous substituent, adopt the pseudoaxial orientation shown in the figure.

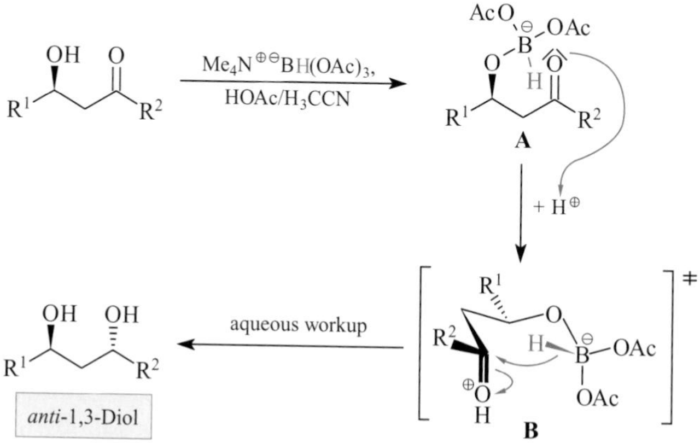

Fig. 10.22. Diastereoselective reduction of *β*-hydroxyketones to *anti*-configured 1,3-diols.

10.4 Enantioselective Addition of Hydride Donors to Carbonyl Compounds

Stereogenic additions of hydride donors to achiral deuterated aldehydes R—C(=O)D or to achiral ketones $R^1R^2C(=O)$ take place without stereocontrol using the reagents that you learned about in Section 10.3. Thus, *racemic* deuterated alcohols R—C(OH) D or *racemic* secondary alcohols $R^1R^2C(OH)H$ are produced. The reason for this is that the C=O double bond of the mentioned substrates is flanked by enantiotopic half-spaces. From these spaces achiral hydride donors *must* react with the same reaction rates (cf. Section 3.4.1). On the other hand, as you can extrapolate from Section 3.4.1, chiral and enantiomerically pure reducing agents might approach each of the half-spaces with different rates. Therefore, reducing agents of this type can in principle effect enantioselective reductions of achiral deuterated aldehydes or achiral ketones. Among the most important among these reagents are the Noyori reagent (Figure 10.23), Alpine-Borane® (Figure 10.24), Brown's chloroborane (Figure 10.25), and the oxazaborolidines by Corey and Itsuno (Figure 10.26). The first three of these reagents allow enantioselective carbonyl group reductions if they are used in stoichiometric amounts, while the fourth class of reagents make *catalytic* enantioselective carbonyl group reductions possible.

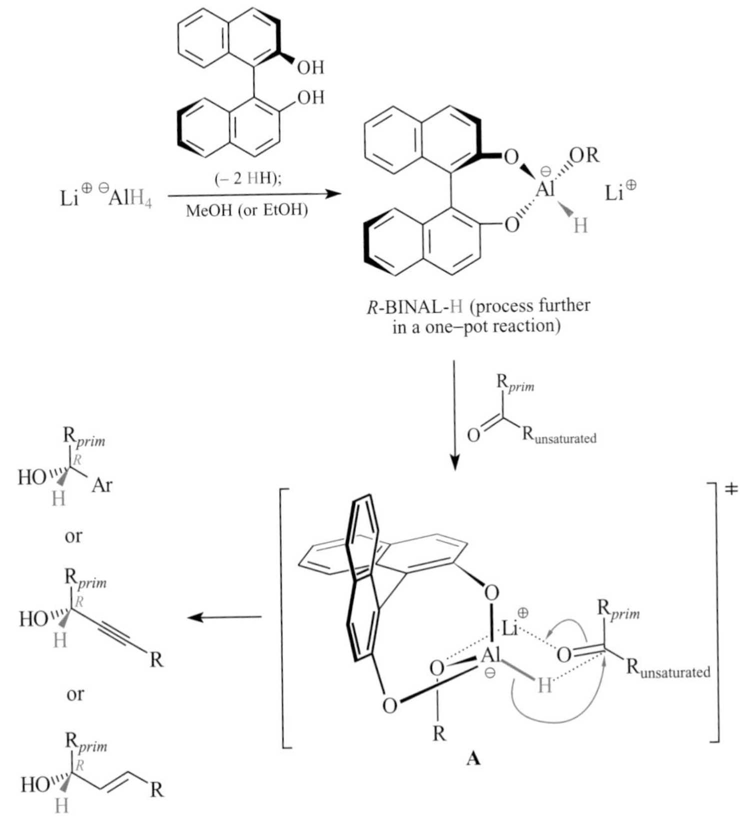

Li⊕ ⊖AlH4 —(−2 HH); MeOH (or EtOH)→ *R*-BINAL-H (process further in a one–pot reaction)

Fig. 10.23. Asymmetric carbonyl group reduction with the Noyori reagent. Note that the chirality of the reducing agent resides in the ligand but that the aluminum atom is not a stereocenter.

The hydride donor of the **Noyori reduction** of ketones is the hydrido aluminate *R*-BINAL-H shown in Figure 10.23 or its enantiomer S-BINAL-H. The new C—H bond is presumably formed via a cyclic six-membered transition state of stereostructure **A.** Unfortunately, there is no easy way to rationalize why enantioselectivity in this kind of addition is limited to substrates in which the carbonyl group is flanked by one conjugated substituent (C≡C, aryl, C═C). The suggestion that has been made is that a lone pair on the axial oxygen of the BINOL unit in the transition state undergoes a repulsive interaction with pi electrons in the unsaturated ketone if the latter is also axial.

Another set of carbonyl compounds can be reduced enantioselectively with **Alpine-Borane®** (Figure 10.24). Alpine-Borane®, just like **diisopinocampheylchloroborane (Brown's chloroborane)** (Figure 10.25), is a reducing reagent of the type H—*β*C—*α*C—ML_n. Each of these reagents transfers the H atom that is bound to the C atom in the position *β* to the boron atom to the carbonyl carbon of the substrate. The metal and its substituents L_n are transferred concomitantly to the carbonyl oxygen. At the same time, the atoms C-*β* and C-*α* become part of an alkene, which is formed as a stoichiometric by-product. Thus, both Alpine-Borane® and diisopinocampheylchloroborane release *α*-pinene in these reductions. The carbonyl compounds whose C═O double bonds are reduced by these boranes in highly enantioselective hydride transfer reactions are the same for both reagents: aryl conjugated alkynones, phenacyl halides, deuterated aromatic aldehydes, and alkyl(perfluoroalkyl)ketones.

The stereoselectivities of the carbonyl group reductions with **Alpine-Borane®** (Figure 10.24) or with **Brown's chloroborane** (Figure 10.25) are explained as shown in the for-

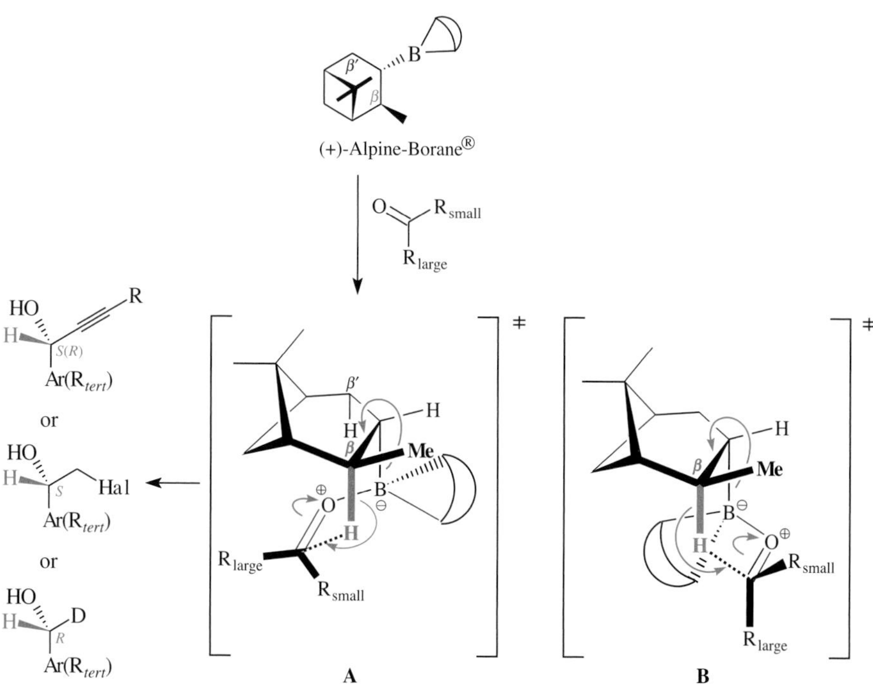

Fig. 10.24. Asymmetric carbonyl group reduction with Alpine-Borane® (preparation: Figure 3.27; for the "parachute-like" notation of the 9-BBN part of this reagent see Figure 3.21). The hydrogen atom that is in the *cis*-position to the boron atom (which applies to both *β*- and *β*′-H) and that after removal of the reducing agent leaves behind a tri- instead of a disubstituted C═C double bond (which applies to *β*-, but not *β*′-H) is transferred as a hydride equivalent. In regard to the reduction product depicted in the top row, the designation S of the configuration relates to the aryl-substituted and *R* to the R_{tert}-substituted propargylic alcohol.

mula **A**. Being Lewis acids, these reagents first bind to the carbonyl oxygen of the substrate. Then the H atom located in the position β to the boron atom is transferred to the carbonyl carbon. That the hydride ion is cleaved from the center C-β and not C-β' is due to product development control since this is the only way that a C=C double bond stabilized by three alkyl substituents can be formed in the terpene moiety. The β-hydride transfer proceeds via six-membered transition states, that are necessarily in the boat form. According to calculations, the preferred boat-like transition state **A** follows from the requirement to place both the H-transferring terpene residue and the larger substituent on the carbonyl group, i.e., R_{large}, as far away as possible from the inert ligands at the boron atom. Formerly, it was **B** that was regarded as the best boat-like transition state.

Reductions with diisopinocampheylchloroborane [$(IPC)_2BCl$] often afford higher *ee* values than reductions with Alpine-Borane® because $(IPC)_2BCl$ is the stronger Lewis acid. Therefore, in the transition state **A** (Figure 10.25), for which calculations favor the boat-like six-membered ring structure **A**, the O—B bond is stronger and, accordingly, shorter than in the Alpine-Borane® analog. This enhances the requirement that the carbonyl group substituent R_{large} adopts the orientation shown. When **B** was considered to be the best boat-like transition state of this reduction, an analogous argument was used for the orientational requirement. There is a second effect. In the transition state in which the stronger Lewis acid diisopinocam-

Fig. 10.25. Asymmetric carbonyl group reduction with diisopinocampheylchloroborane [Brown's chloroborane, $(IPC)_2BCl$]. Concerning the reduction product depicted in the top row, the designation *S* of the configuration relates to the aryl-substituted and *R* to the R_{tert}-substituted propargylic alcohol.

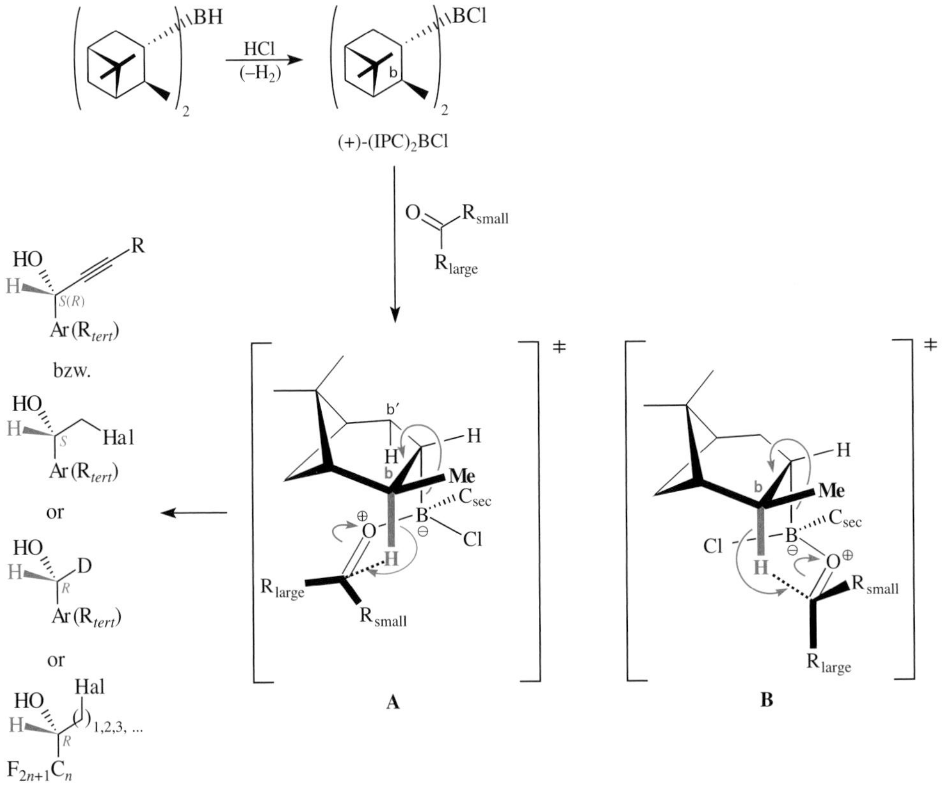

pheylchloroborane [$(IPC)_2BCl$] complexes the carbonyl oxygen, the carbonyl group is a better electrophile. Therefore, it becomes a better hydride acceptor for Brown's chloroborane than in the hydride transfer from Alpine-Borane®. Reductions with Alpine-Borane® can actually be so slow that decomposition of this reagent into α-pinene and 9-BBN takes place as a competing side reaction. The presence of this 9-BBN is problematic because it reduces the carbonyl compound competitively and of course without enantiocontrol.

The enantioselective carbonyl group reduction by the **Corey–Itsuno process** is particularly elegant (Figure 10.26). There, in contrast to the processes of Figures 10.23–10.25, one requires a stoichiometric amount only of an achiral reducing agent (BH_3), which is much cheaper than the stoichiometrically required reducing reagents of Figures 8.18–8.20. Enantiocontrol in the Corey–Itsuno reduction stems from the enantiomerically pure ligand **A.** While this *is* comparatively expensive, it can be used catalytically. That enantioselective reductions are achieved with this reagent is due to the fact that, in the absence of the chiral reagent **A**, BH_3 reacts only quite slowly with ketones of the type R_{large}—C(=O)—$R_{small.}$ However, the chiral reagent **A** forms a Lewis acid/Lewis base complex **B** with BH_3. This increases the electrophilicity of the heterocyclic boron atom such that it can now compex with a ketone. Consequently, we encounter here a new example of a *ligand-accelerated reaction*—that is, a class of reactions that also includes the Sharpless epoxidations discussed in Section 3.4.6. Additional reactions of this type are the organozinc reactions of Figures 10.39 and 10.40.

Complex B can be isolated, characterized and stored. The B atom of the complex **B** of the Corey–Itsuno reduction (Figure 10.26) is a Lewis acid. Unlike heterocycle **A**, its boron atom does not possess a neighboring lone pair of electrons and is adjacent to an atom with a formal positive charge. Therefore, the empty orbital is available for binding a carbonyl group to form the ternary complex.

Looking at all the reactions that took place for the formation of the complex **C**, the heterocycle **A** plays the role of a "molecular glue," which is possible because it represents both a Lewis acid and a Lewis base. As a result of this dual role, **A** places the electrophile (the ketone) and the hydride donor (the BH_3) in close proximity. In this way, the complex **C** makes possible a quasi-intramolecular reduction of the ketone. It takes place stereoselectively in such a way as the arrangement of the reaction partners in **C** suggests (Figure 10.26). As a bicyclic

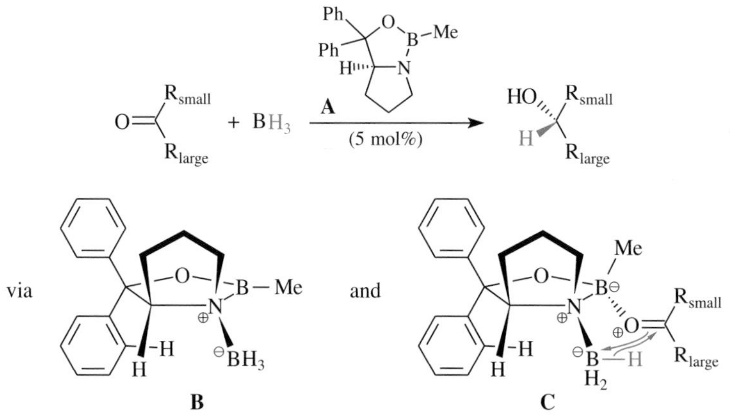

Fig. 10.26. Catalytic asymmetric carbonyl group reduction according to Corey and Itsuno.

compound with a convex and a concave side, the heterocycle binds both the ketone and the BH_3 on its less hindered convex side. The presence of the phenyl group on the convex side of the molecule ensures that the axis of the O=C bond of the coordinated ketone points away from it. Given this orientation of the ketone axis, the methyl group on the heterocyclic B atom requires the large ketone substituent R_{large} to be as far away as possible. This suffices to *unambiguously* define the face selectivity of the BH_3 addition to the C=O double bond. Consequently, Corey–Itsuno reductions exhibit high *ee* values.

10.5 Addition of Organometallic Compounds to Carbonyl Compounds

Alkyllithium and alkyl Grignard reagents, alkenyllithium and alkenyl Grignard reagents, aryllithium and aryl Grignard reagents, as well as alkynyllithium and alkynyl Grignard reagents, can be added to the carbonyl group of aldehydes and ketones. The addition of alkynyllithium compounds to sterically more demanding ketones, however, is sometimes only possible after adding one equivalent of $CeCl_3$. The addition then takes place via an organocerium compound of the composition R—C≡C—$CeCl_2$, which is formed *in situ*. Diorganozinc compounds do not add to aldehydes as such. Yet, they do so after reacting with CuCN and in the presence of a Lewis acid. Alkyl zinc iodides behave similarly under the same conditions, but the Lewis acid can be replaced by Me_3SiCl. On the other hand, organocopper compounds that do not contain an additional metal besides copper and lithium are, as a rule, not able to add to the carbonyl carbon.

10.5.1 Simple Addition Reactions of Organometallic Compounds

Similarities and Differences in the Reactions of Organolithium vs. Grignard Reagents with Carbonyl Compounds

Overall reaction equations suggest that monomeric RLi and monomeric RMgHal, respectively, act as nucleophiles in additions of organolithium compounds or Grignard reagents to carbonyl compounds. This is true of organolithium compounds (mechanism: Figure 10.32), but does not apply to Grignard compounds. Similarly, the 1,4-addition of Gilman cuprates to α,β-unsaturated ketones proceeds via dimeric instead of monomeric organometallic compounds (mechanism: Figure 10.46), but these reactions will be dealt with in more detail in Section 10.6. To begin with, we will focus on the addition mechanisms of Grignard compounds to carbonyl compounds.

According to Figure 10.5 the solution of a Grignard reagent contains—apart from the monomer—two dimers in significant concentrations and probably several others in lower concentrations. One of these "Grignard dimers" nucleophilically reacts with the carbonyl compound. The top row of Figure 10.27 presents a mechanism: in an equilibrium reaction, a "Grig-

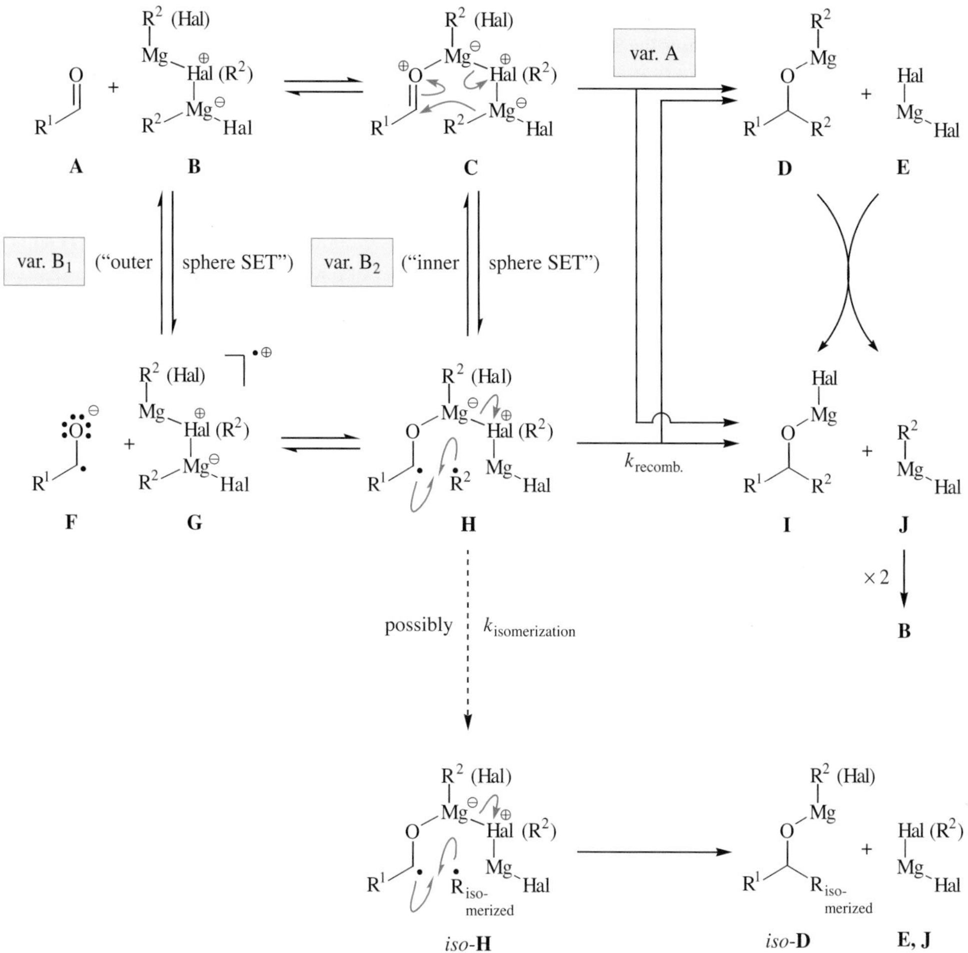

Fig. 10.27. Ionic mechanism ("variant A") and radical mechanisms ("variants B_1 or B_2") for the addition of a Grignard reagent to an aldehyde. SET = single electron transfer.

nard dimer" of the—alleged—structural type **B** and the carbonyl compound form the Lewis acid/Lewis base adduct **C**, in which three electron pairs get shifted. Directly or indirectly, this leads to the formation of magnesium alkoxide **I** and the monomeric Grignard compound **J**. The alkoxide **I** is protonated during workup and leads to the formation of the desired alcohol. The "Grignard monomer" **J**, though, dimerizes to **B** and continues to react with the carbonyl compound.

Side Note 10.3. Mechanistic Details of the Addition of Grignard Reagents to Carbonyl Compounds

"Grignard dimers" **B** can react to form the same magnesium alkoxide **I** and the same "Grignard monomer" **J** via radicals as intermediates. The center and bottom of Figure 10.27 show how this reaction might proceed. The radical-forming step is a single electron transfer (SET). Either it occurs between the separated reactants, i.e., across a larger distance, and is then termed "outer sphere SET." The latter leads to a pair of solvent-separated radical ions consisting of a ketyl radical (**F**) and a radical cation (**G**). These radical ions then form an O–Mg bond thus generating the radical pair **H**. Alternatively, the radical-providing single electron transfer takes places in the Lewis acid/Lewis base complex **C**, where a much shorter distance needs to be bridged than with the "outer sphere SET." For this process, wee use the term "inner sphere SET." This SET leads to immediate formation of the aforementioned radical pair **H**.

Basically, this radical pair **H** has two different options. First, the radical centers may form a C–C bond with each other, resulting in a magnesium alkoxide that during workup will be protonated to give the expected alcohol. Second, the component $R^2\bullet$ of the radical pair **H** could isomerize to a radical $R_{isomerized}$, which would be part of a radical pair *iso*-**H**. If the formation of the C–C bond through linkage between two radical centers does not take place prior to this step the magnesium alkoxides *iso*-**D** are generated, which are protonated during workup to give an *unexpected* alcohol.

If the addition of a Grignard reagent to a carbonyl compound leads to the formation of such an unexpected alcohol, it is clear that this reaction cannot have proceeded according to the polar mechanism outlined in the upper row of Figure 10.27. If the respective addition does not produce anything but the expected alcohol, the polar mechanism provides the easiest explanation. Nevertheless, the radical mechanism could also have applied—provided, though, that the isomerization **H** → *iso*-**H** would have been much slower than the radical annihilation reaction **H** → **I**.

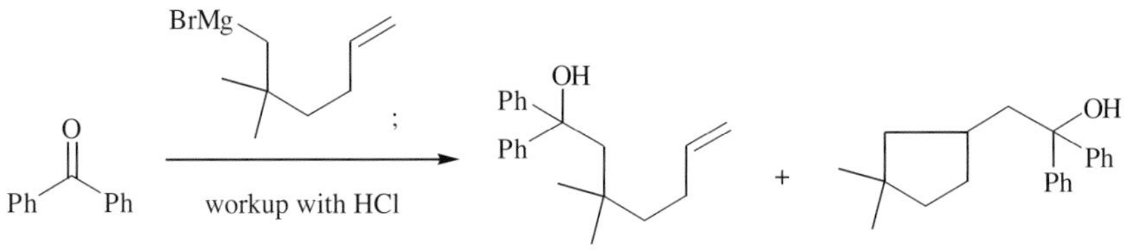

BrMg ; workup with HCl

88 : 12

A particularly stable ketyl radical is derived from benzophenone (cf. Figure 17.52). This is why additions of the Grignard reagents $R_2Mg_2Hal_2$ to *this* substrate proceed more frequently via radicals as intermediates than others. An example in which the occurrence of such a radical intermediate is documented by the typical radical cyclization 5-hexenyl → cyclopentylcarbinyl (cf. Section 1.10.2) is the following:

R. W. Hoffmann managed to provide particularly elegant evidence for the radical addition mechanism of a Grignard reagent to carbonyl compounds by means of the reaction depicted in Figure 10.28. Due to the presence of at least one conjugated aryl residue, the aromatic aldehydes and ketones given there are particularly suitable for the formation of ketyl radical intermediates. If the ketyl contained two conjugated phenyl residues instead of just one, it is more stable and thus provides the radical arising from the Grignard compounds with more time to isomerize. If the ketyl is stabilized by a perfluorophenyl- instead of a simple phenyl residue, i.e., by a stronger electron acceptor, the derived ketyl is also more stable and thus also gives the radical deriving from the Grignard compound a relatively long time to isomerize. It is exactly these nuances that the isomerizations in Figure 10.28, which are racemizations, demonstrate. These racemizations are due to the fact that secondary alkyl radicals are weakly

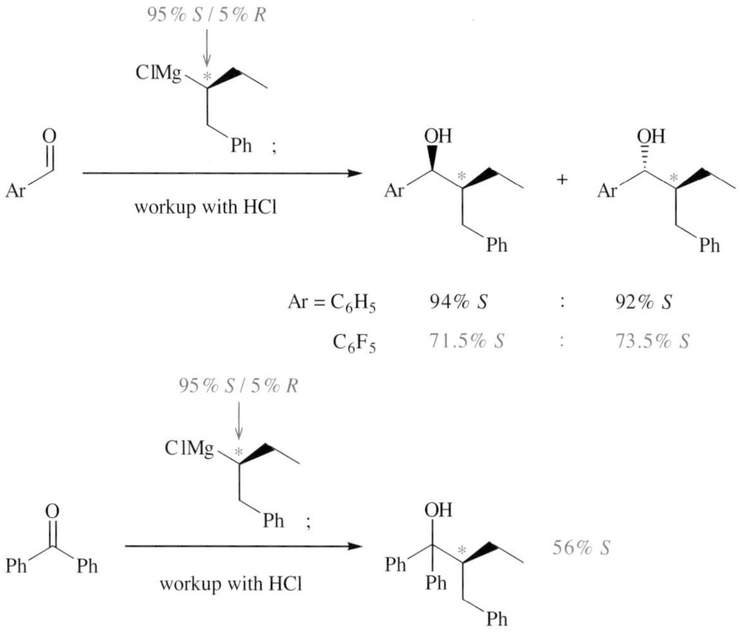

95% S / 5% R

ClMg ; workup with HCl

Ar = C_6H_5 94% S : 92% S

C_6F_5 71.5% S : 73.5% S

95% S / 5% R

ClMg ; workup with HCl

56% S

Fig. 10.28. The addition of a Grignard reagent to aromatic carbonyl compounds proceeds via SET in part, which is demonstrated by the partial racemizations observed here: the *sec*-alkyl radical that occurs as an intermediate undergoes an extremely rapid racemization. The more stable the simultaneously formed magnesium-bound ketyl radical is, the larger is the share of the racemization.

pyramidalized, but not configurationally stable (cf. Chapter 1, Section 1.1.1) and can therefore undergo isomerization through inversion of their septet center to furnish the enantiomeric pyramidalized *sec*-alkyl radical.

Often, the addition of organolithium compounds to carbonyl compounds gives the same alcohols that one would also get from the analogous Grignard reagents and the same carbonyl compounds. For example, one could prepare all the products in Figures 10.33–10.37, which result from the addition of Grignard reagents, from the addition of the analogous organolithium compounds as well.

In certain additions of aryl nucleophiles to C=O double bonds, however, it is advantageous to use organolithium instead of organomagnesium compounds:

- When one is working on a very small scale, one can prepare aryllithium compounds through a Br/Li exchange reaction (path 3 in Figure 5.41) or through an I/Li exchange (path 4 in Figure 5.41) in a much simpler way than aryl Grignard reagents.
- By means of an *ortho*-lithiation of suitably functionalized aromatics it is possible to obtain aryllithium compounds from halogen-free aromatics in a way and with a regioselectivity (Section 5.3.1) for which there is no analogy in the preparation of aryl Grignard reagents.

Some addition reactions to C=O double bonds can be carried out at all only with organolithium and not with Grignard reagents. For instance, bulky Grignard reagents do not react as C nucleophiles with sterically hindered ketones (Figure 10.29), but the analogous lithium organometallics do since they are stronger nucleophiles (Figure 10.30).

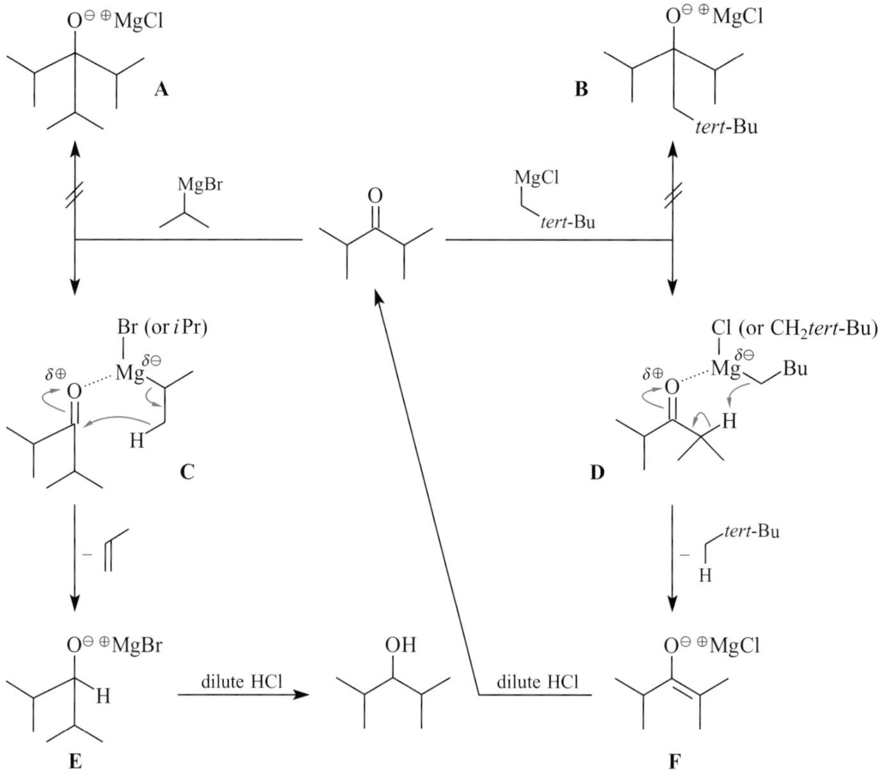

Fig. 10.29. Reduction (left) and enolate formation (right) in reactions of sterically demanding Grignard reagents with bulky ketones.

When bulky Grignard reagents contain an H atom in the position β to the magnesium, they can transfer this hydrogen to the C=O double bond in a way that is similar to that of the reduction of carbonyl compounds with Alpine-Borane® (Figure 10.24) or with diisopinocampheylchloroborane (Figure 10.25)—via a six-membered transition state. Here, it is derived from a Lewis acid/Lewis base complex of structure **C** (Figure 10.29). A magnesium alkoxide is produced, which has the structure **E.** In the aqueous workup compound **E** gives the reduction product of the ketone used and not the addition product of the Grignard reagent. This unexpected reaction is referred to as a **Grignard reduction.**

Being a *better* nucleophile, the lithium analog of the Grignard reagent in Figure 10.29 adds to the same ketone without problems (Figure 10.30, left). At first this furnishes the lithium analog of the inaccessible magnesium alkoxide **A** of Figure 10.29. Its protonation under the weakly acidic workup conditions furnishes the highly sterically hindered triisopropyl carbinol.

Neopentyl magnesium chloride also does not react with a hindered ketone such as diisopropyl ketone to form a magnesium alkoxide (formula **B** in Figure 10.29). Instead, the reactants first form a Lewis acid/Lewis base complex. This time it has the structure **D.** The neopentyl group then deprotonates the complexed ketone to give the magnesium enolate **F** via a six-membered cyclic transition state. In the aqueous workup one thus obtains only the unchanged starting ketone. Whereas diisopropylneopentyl carbinol is not accessible in *this* way, the use of the analogous lithium reagent as a C nucleophile provides help: it adds to diisopropyl ketone as desired (Figure 10.30, right).

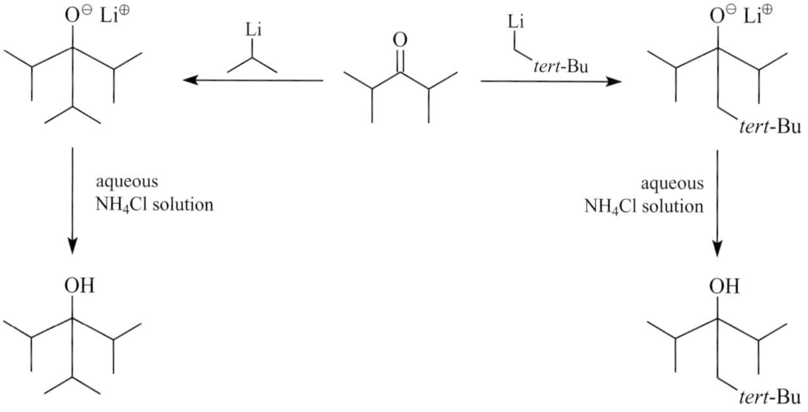

Fig. 10.30. Synthesis of sterically hindered alcohols by the reaction of sterically demanding organolithium compounds with bulky ketones. Preparation of triisopropyl carbinol (left) and diisopropylneopentyl carbinol (right).

There is one more type of addition reaction in which the use of organolithium compounds instead of Grignard reagents is a must: when the goal is to prepare an allyl alcohol through the addition of an organometallic compound to the C=O double bond of an α,β-unsaturated ketone (Figure 10.31). Grignard reagents often react with α,β-unsaturated ketones both at the carbonyl carbon and at the center Cβ. However, only in the first case is the desired addition initiated. With respect to the distance between the added organic group and the metal ion in the initially obtained alkoxide, this addition mode is called a **1,2-addition.** On the other hand, the attack of a Grignard reagent at the center C-β of an α,β-unsaturated ketone leads to an enolate. Therein, the newly added organic group and the $MgX^{\oplus}$ ion are located in positions 1 and 4. Consequently, this addition mode is called a **1,4-addition.**

A smooth 1,2-addition is almost always observed in the reaction of α,β-unsaturated ketones with organolithium compounds (Figure 10.31, bottom). Only in extreme cases such

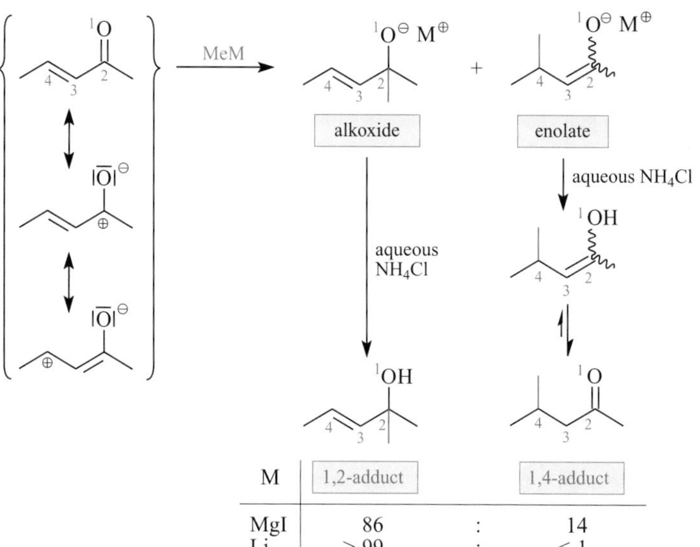

M	1,2-adduct		1,4-adduct
MgI	86	:	14
Li	> 99	:	< 1

Fig. 10.31. Competition of 1,2- and 1,4-addition in the reaction of organolithium compounds and Grignard reagents with α,β-unsaturated ketones: a phenomenological view.

as additions to α,β-unsaturated trityl ketones do *simple* alkyllithium compounds also undertake 1,4-additions. (Stabilized organolithium compounds, however, like the ones mentioned in the last paragraph of Side Note 10.4 tend to add in a 1,4-addition to normal α,β-unsaturated ketones as well.)

The intermediate formed in a 1,2-addition to an α,β-unsaturated ketone is an allyl alkoxide, whereas the intermediate of the analogous 1,4-addition is an enolate (Figure 10.31). Alkoxides are stronger bases than enolates because enols are stronger acids than alcohols. This, in turn, is due to the localization of the negative charge in the alkoxide, whereas the negative charge in the enolate is delocalized. Hence, the alkoxide intermediate of a 1,2 addition is higher in energy than the enolate intermediate of a 1,4-addition. In the reaction of Grignard reagents, which are weaker C nucleophiles than organolithium compounds, some product development control could occurs in the transition state of the addition to α,β-unsaturated ketones. This would explain why the more stable intermediate (i.e., the enolate) is obtained starting from a Grignard reagent rather than starting from an organolithium compound. If the addition of a Grignard reagent to an α,β-unsaturated ketone proceeds according to the radical mechanism of Figure 10.27, the argument that product development control occurs in the product-determining step, that is, in the radical annihilation step, holds for the same reason: namely that the enolate occurring with a 1,4-addition is more stable than the allyl alkoxide resulting from a 1,2-addition.

Selective 1,4-additions of organometallic compounds to α,β-unsaturated ketones can also be achieved starting from organocopper and organozinc reagents. We will treat this in Section 10.6.

Side Note 10.4. 1,2- as Opposed to 1,4-Selectivity in RLi Additions to α,β-Unsaturated Carbonyl Compounds

The previous paragraph provided a reasonable argument for why Grignard reagents tend to be more able to undergo 1,4-additions to α,β-unsaturated ketones than organolithium compounds. But why in general doesn't any 1,4-addition *at all* occur between organolithium compounds and α,β-unsaturated ketones? This question is addressed in Figure 10.32. Whether 1,2- and/or 1,4-selectivity occurs is attributed to a single reason, namely the structure of the reacting organolithium compound. According to the introductory passages of Chapter 10 it is a monomeric organolithium compound. Figure 10.3 presented "contact ion pairs" and "solvent-separated ion pairs" as typical structures of monomeric organolithium compounds. The upper part of Figure 10.32 provides an even finer differentiation of the latter structural type, namely into "solvent-separated, non-Lewis-acidic ion pairs" (**B**) and "solvent-separated, Lewis-acidic ion pairs" (**C**), which differ in the extent of coordinative saturation at the $Li^{\oplus}$ cation: In "non-Lewis-acidic ion pairs" the $Li^{\oplus}$ cation is connected with four donor atoms, and in "Lewis-acidic ion pairs" with three.

- The bottom part of Figure 10.32 shows how monomeric organolithium compounds as a contact ion pair (**A**), a solvent-separated, non-Lewis-acidic ion pair (**B**) or as a solvent-separated, Lewis-acidic ion pair (**C**) react with cyclohexenone as a prototypical α,β-unsaturated ketone.
- The reaction of the contact ion pair **A** with cyclohexenone proceeds in a single step. The transition state is **D**, which is structurally distinguished by the fact that the formation of a solvent-separated ion pair is avoided. The lithium separates from the organic residue $R^1R^2R^3C$ only where the latter can form a C–C bond *and* where the $Li^{\oplus}$ cation can *directly*

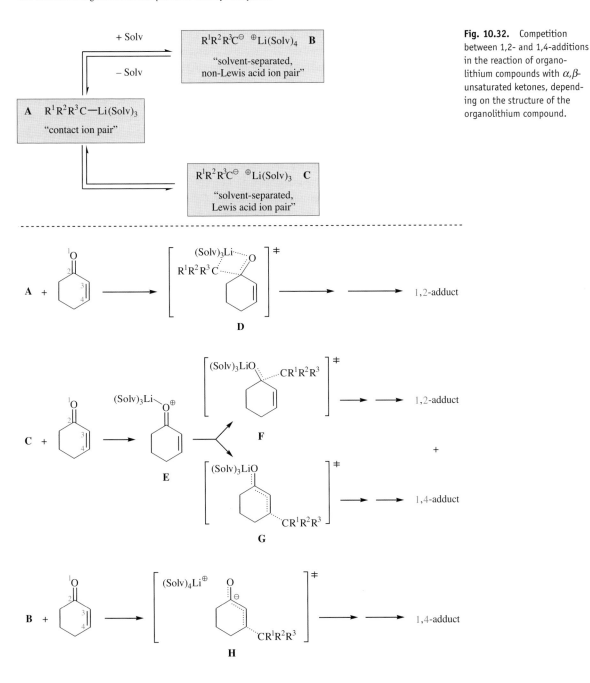

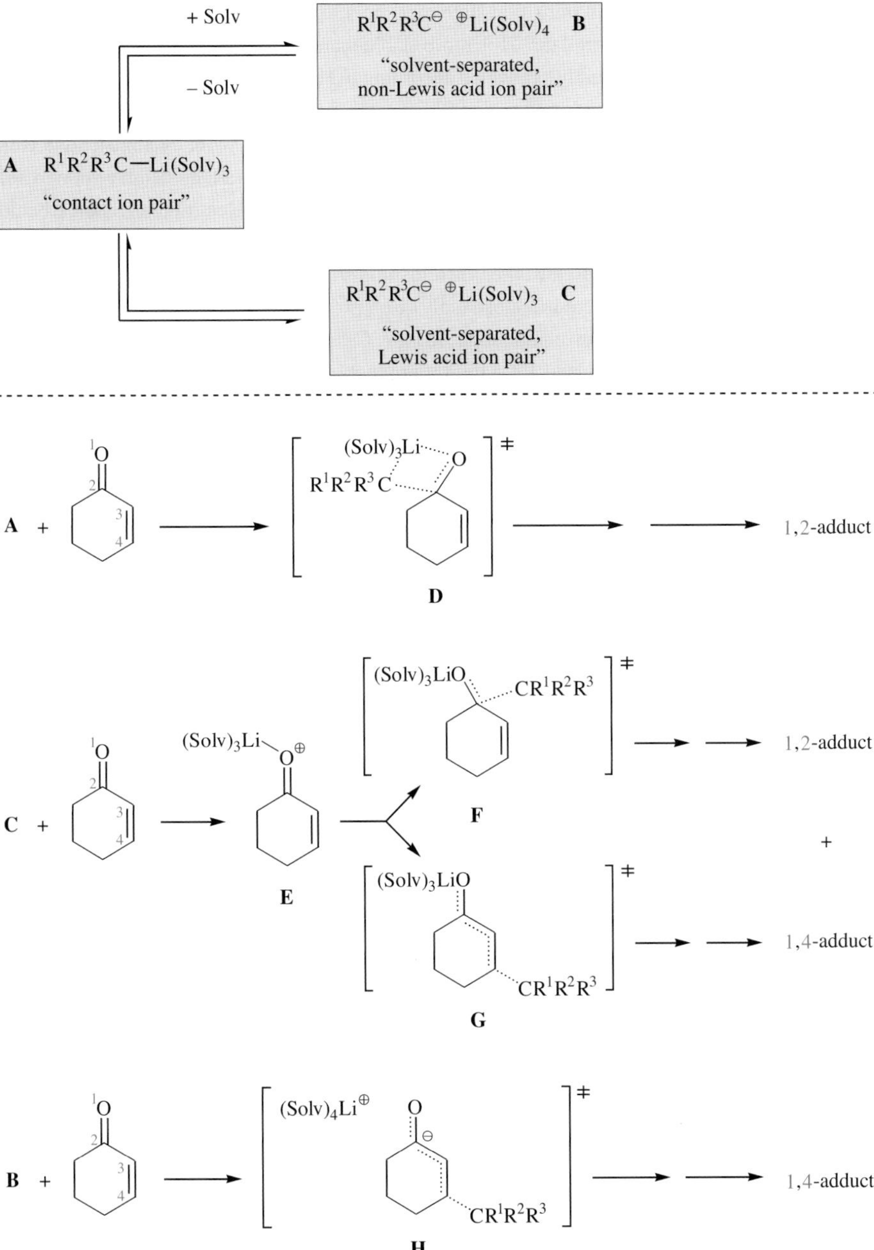

Fig. 10.32. Competition between 1,2- and 1,4-additions in the reaction of organolithium compounds with α,β-unsaturated ketones, depending on the structure of the organolithium compound.

encounter the negatively charged oxygen atom of the resulting oxyanion, which is possible only if a 1,2-addition takes place.

- The addition of the solvent-separated, non-Lewis-acidic ion pair **B** to cyclohexenone also proceeds in a single step. The coordinatively saturated $Li^{\oplus}(solv)_4$ cation is just a casual bystander. Therefore, the transition state must be a lithium-free species with the structure **H**, since being a "near-enolate" **H** is more stable than the isomeric "near-alkoxide" of the transition state of a 1,2-addition, which according to Hammond's postulate cannot compete as far as the reaction rate is concerned. This is why a direct 1,4-addition takes place.

- The reaction of the solvent-separated, Lewis-acidic ion pair **C** with cyclohexenone is the only multiple-step addition in Figure 10.32. It does not involve the cyclohexenone itself, but its $(solv)_3Li^{\oplus}$complex **E**, which, of course, is the better electrophile. The carbanion $R^1R^2R^3C^{\ominus}$ may now choose between both the 1,2- and the 1,4-addition. The former proceeds via transition state **F**, the latter via transition state **G**. These transition states are about equal in energy, since there is a close $(solv)_3Li^{\oplus}/O^{\ominus}$contact in both cases. This is why the reaction of organolithium compounds of type **C** and α,β-unsaturated ketones typically leads to a mixture of 1,2- and 1,4-addition products.

From Figure 10.32 we can thus predict the preferential mode of addition of an organolithium compound to α,β-unsaturated ketones as long or as soon as the structure of its reactive monomer is known. Alkyl-, alkenyl- and aryllithium compounds undergo addition via monomers of the structural type **A**, (triphenylmethyl)lithium via a monomer of type **B** (cf. Figure 10.3), and lithiated 1,3-dithiane (cf. Figure 9.22) belongs to the organolithium compounds that may react with α,β-unsaturated ketones via monomers of the structural type **C** (i.e., in the presence of HMPA).

Addition of Grignard Reagents to Carbonyl Compounds: The Range of Products

Grignard reagents add to the C=O double bond of formaldehyde, higher aldehydes, and sterically unhindered or moderately hindered ketones (remember, though, that combined with highly sterically hindered ketones, Grignard reagents undergo side reactions, as shown in Figure 10.29). In each case, the primary product is a magnesium alkoxide. Its protonation with dilute HCl or dilute H_2SO_4 then gives an alcohol as the final product. In the preparation of acid-sensitive alcohols—that is, of tertiary alcohols or allyl alcohols—an aqueous NH_4Cl solution or a weakly acidic phosphate buffer is used to effect the protonation (Figures 10.34 and 10.35). If in these cases one were to also use the aforementioned dilute mineral acids, one would induce acid-catalyzed secondary reactions of these acid-sensitive alcohols, for example, E1 reactions (example: Figure 10.36) or S_N1 reactions (Figure 10.37). This is because such mineral acids would protonate not only the alkoxide but also the resulting alcohol, and the latter would lead to an irreversible elimination.

The addition of Grignard reagents to formaldehyde gives primary alcohols (Figure 10.33), and the addition to higher aldehydes gives secondary alcohols (Figure 10.34).

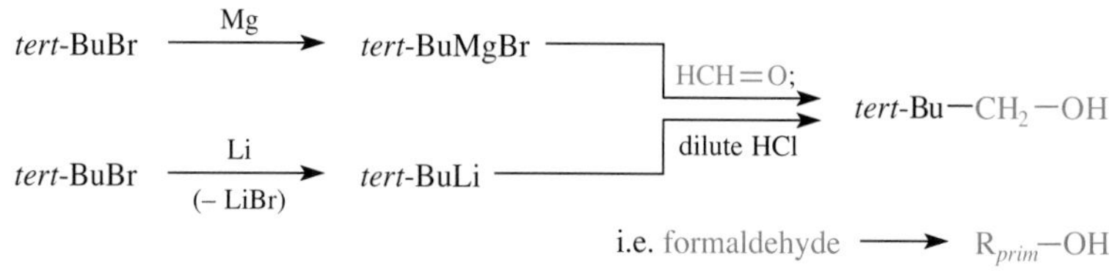

tert-BuBr $\xrightarrow{\text{Mg}}$ *tert*-BuMgBr

tert-BuBr $\xrightarrow[\text{(– LiBr)}]{\text{Li}}$ *tert*-BuLi

$\xrightarrow[\text{dilute HCl}]{\text{HCH=O;}}$ *tert*-Bu—CH_2—OH

i.e. formaldehyde ⟶ R_{prim}—OH

Fig. 10.33. Preparation of a primary alcohol from formaldehyde and a Grignard reagent or an organolithium compound.

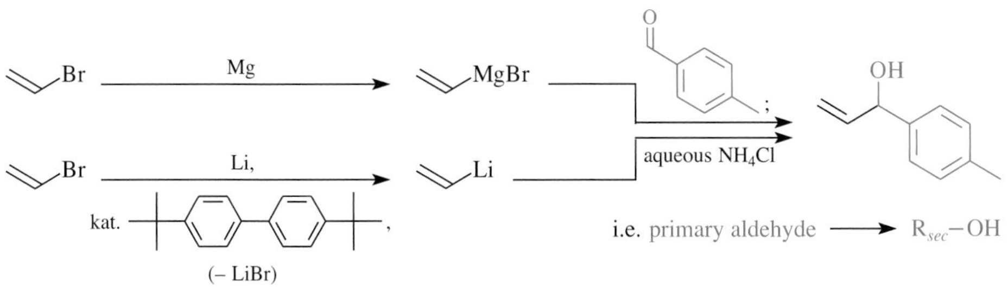

Mg
MgBr
;
OH
Li,
Li
aqueous NH_4Cl
kat.
(– LiBr)
i.e. primary aldehyde ⟶ R_{sec}–OH

Fig. 10.34. Preparation of a secondary alcohol from a higher aldehyde and a Grignard reagent or an organolithium compound.

Finally, the addition of Grignard reagents or organolithium compounds to ketones gives tertiary alcohols when one works up with a weak acid (example: Figure 10.35).

Adding a strong acid to the magnesium alkoxide or lithium alkoxide intermediate leads to further reactions of the corresponding tertiary alcohols giving, for example, E1 products instead (example: Figure 10.36; also cf. Figure 10.37).

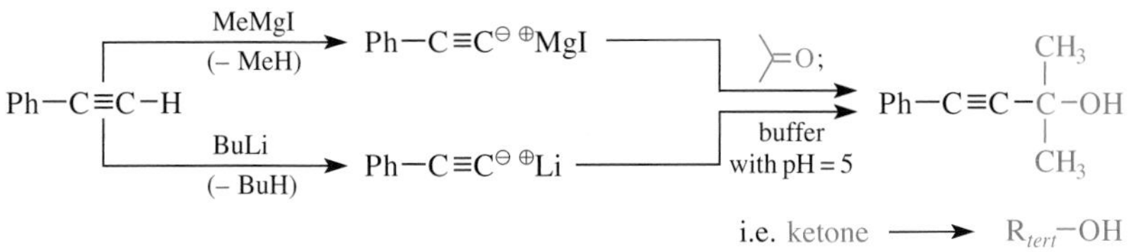

MeMgI (– MeH)
Ph–C≡C–H
Ph–C≡C$^{\ominus}$ $^{\oplus}$MgI
BuLi (– BuH)
Ph–C≡C$^{\ominus}$ $^{\oplus}$Li
=O;
buffer with pH = 5
Ph–C≡C–C(CH_3)(CH_3)–OH
i.e. ketone ⟶ R_{tert}–OH

Fig. 10.35. Preparation of a tertiary alcohol by the addition of a Grignard reagent or an organolithium compound to a ketone and subsequent workup of the reaction mixture with no more than a weak acid (not shown) or with a corresponding buffer solution (as given here).

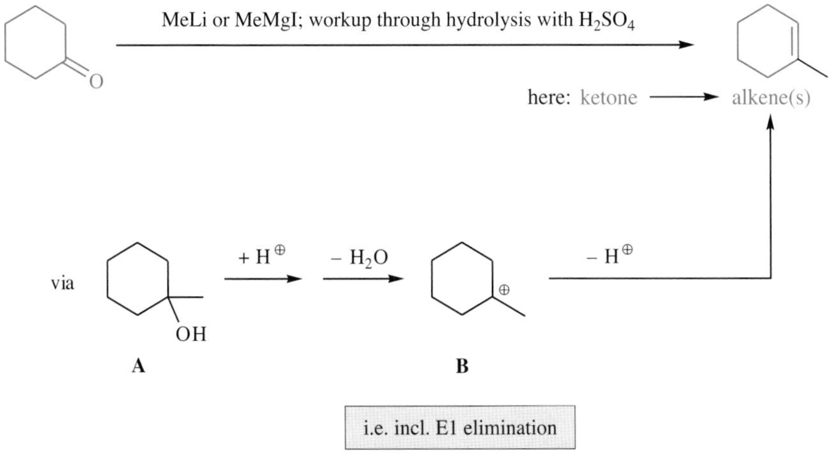

MeLi or MeMgI; workup through hydrolysis with H_2SO_4
O
here: ketone ⟶ alkene(s)
via
OH
A
+ H$^{\oplus}$
– H_2O
B
– H$^{\oplus}$
i.e. incl. E1 elimination

Fig. 10.36. Preparation of an alkene by adding a Grignard reagent or an organolithium compound to a ketone and subsequent workup with an acid that is strong enough to induce an E1 elimination.

Side Note 10.5. Preparation of 3-Substituted Cyclohexenones

The addition of an organolithium or analogous Grignard compound to 3-ethoxy-2-cyclohexene-1-one by way of 1,2-addition furnishes a tertiary alkoxide (Figure 10.37). If this is protonated during workup, the corresponding tertiary alcohol is formed (Formula **B**), which is *even more* acid-labile than the tertiary alcohol **A** in Figure 10.36, since the carbenium ion **C**, which can be generated from the initially mentioned alcohol **B** by proton-catalyzed elimination of water, is stabilized as a tertiary carbenium ion, an allylic cation and as an oxocarbenium ion, as shown in the resonance form in Figure 10.37. This cation **C** adds the heteroatom nucleophile water while forming the protonated hemiacetal **E,** which in agreement with the comments in Section 9.1.1 releases one equivalent of alcohol. Finally, the 3-substituted cyclohexenone **A** is formed. It is worthwhile to memorize this reaction as an efficient six-memberd ring synthesis.

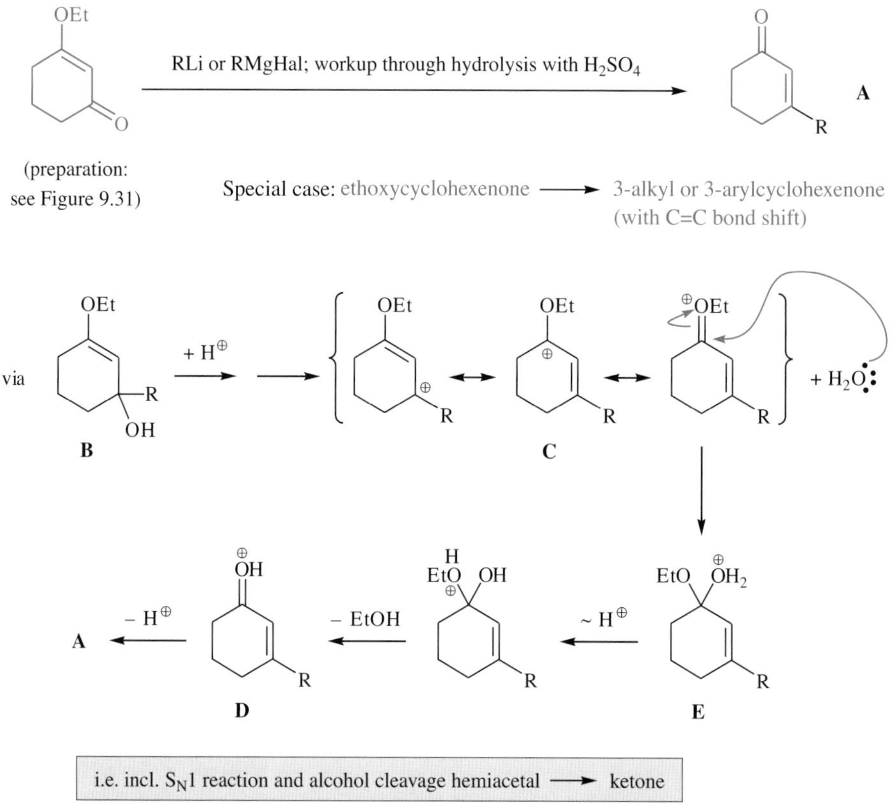

OEt
O
RLi or RMgHal; workup through hydrolysis with H_2SO_4
O
R
A
(preparation: see Figure 9.31)
Special case: ethoxycyclohexenone ⟶ 3-alkyl or 3-arylcyclohexenone (with C=C bond shift)

via
OEt R OH
B
$+ H^{\oplus}$
OEt ⊕ R ↔ OEt ⊕ R ↔ ⊕OEt R
C
$+ H_2O$
EtO ⊕OH₂ R
E
$\sim H^{\oplus}$
H EtO⊕ OH R
$- EtOH$
⊕OH R
D
$- H^{\oplus}$
A

i.e. incl. S_N1 reaction and alcohol cleavage hemiacetal ⟶ ketone

Fig. 10.37. Preparation of cyclohexenones through addition of a Grignard- or organolithium compound to 3-ethoxy-2-cyclohexenone, which is inevitably followed by an E1 elimination—even if workup of the reaction mixture is performed with a weak acid.

1,2-Addition of Knochel Cuprates to α,β-Unsaturated Aldehydes

The preparation of organolithium compounds and of Grignard reagents from halides is successful only when these reagents contain virtually no functional groups. Furthermore, these C nucleophiles can be added to the C=O double bond of a carbonyl compound only when this carbonyl compound also contains practically no additional electrophilic group(s).

Organozinc compounds exhibit a much greater compatibility with functional groups both in the reagent and in the substrate of an addition reaction. The standard preparation of organozinc compounds starts with primary alkyl iodides (see also Figure 17.45). These iodides are reduced with metallic zinc to give zinc analogs R_{FG}—Zn—I (FG, contains a functional group) of the Grignard reagents R—Mg—I. Unlike Grignard reagents, however, these alkyl zinc iodides R_{FG}—Zn—I can contain the following functional groups: CO_2R, $R^1R^2C{=}O$, C≡N, Cl, RNH, $NH_2RC({=}O)NH$, sulfoxide, sulfone, and an internal or a terminal C≡C triple bond.

Alkyl zinc iodides R_{FG}—Zn—I are poor nucleophiles. However, they are turned into good nucleophiles when they are converted into the so-called **Knochel cuprates** R_{FG}—Cu(CN)ZnHal with solubilized CuCN—that is, CuCN containing LiHal. In the presence of a Lewis acid Knochel cuprates add to aldehydes, provided these are α,β-unsaturated. With *substituted α,β*-unsaturated aldehydes a 1,2-addition can be observed, as shown in Figure 10.38. With acrolein (an unsubstituted α,β-unsaturated aldehyde) or α,β-unsaturated ketones (Fig. 10.43), however, Knochel cuprates undergo 1,4-additions.

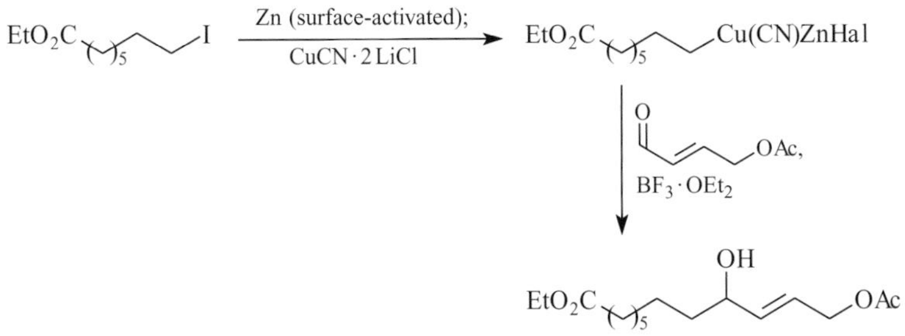

Fig. 10.38. Addition of a Knochel cuprate to an α,β-unsaturated aldehyde.

Another type of organozinc compound also adds to aldehydes in the presence of Lewis acids. The significance of this reaction is that it can also be achieved enantioselectively (Section 10.5.2).

10.5.2 Enantioselective Addition of Organozinc Compounds to Carbonyl Compounds: Chiral Amplification

Zinc-containing C nucleophiles, which tolerate the presence of diverse functional groups, are not limited to the cited Knochel cuprates obtained from alkyl iodides. The other zinc-based C nucleophiles, which can contain a comparably broad spectrum of functional groups, are dialkylzinc compounds. They are best prepared from terminal alkenes. Figure 10.39 shows in the first two reactions how this is done.

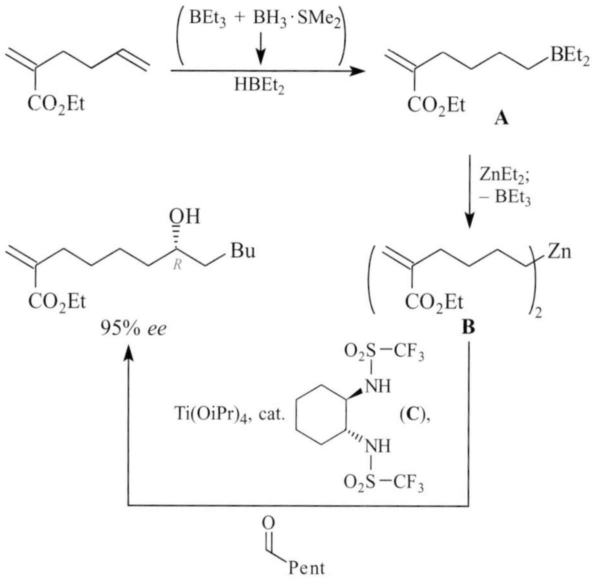

Fig. 10.39. Preparation of a dialkylzinc compound **B** and its subsequent enantioselective addition to an aldehyde.

One starts with a hydroboration with $HBEt_2$. The substrate of Figure 10.39 being a diene, this borane adds exclusively to the electron-rich C=C double bond. The C=C double bond adds the reagent regioselectively in accord with that expected for hydroboration. The trialkylborane **A** formed is then subjected to a B/Zn exchange reaction with $ZnEt_2$, an equilibrium reaction that leads to the dialkylzinc compound **B** and BEt_3. The latter is removed continuously from the equilibrium by distillation. Thereby, the equilibrium is shifted completely toward the side of the desired dialkylzinc compound **B**. Good old Le Chatelier!

Reagents of this type are suitable for performing catalytic asymmetric additions to aldehydes. For example, an enantiomerically pure Lewis acid is generated *in situ* from $Ti(O\textit{i}Pr)_4$ and the enantiomerically pure bis(sulfonamide) **C.** It catalyzes the enantioselective addition of functionalized (or unfunctionalized) dialkylzinc compounds to widely variable aldehydes. There is no detailed, substantiated rationalization of the underlying addition mechanism in this case.

A different method for the catalytic asymmetric addition of a dialkylzinc compound—Et_2Zn and aromatic aldehydes have almost always been used—is shown in Figure 10.40. With regard to stereoselective synthesis, this method has an importance that goes beyond the reaction shown there. This is because of the surprising finding that the addition product may have an *ee* of 95% even when the chiral ligand (—)-**A** is used with a relatively low enantiomeric purity of 15% *ee.* The *ee* value of the product would thus exceed the *ee* value of the chiral ligand quite considerably. This phenomenon is referred to as **chiral amplification**.

The occurrence of chiral amplification in this case is explained by the mechanism of Figure 10.40. When chiral amplification is observed one always finds—as is the case here—that two molecules of the chiral ligand become linked to each other, albeit indirectly (i.e., via another component of the reaction mixture). In other words, a "derivative of the dimer" of the chiral ligand is formed. When chiral amplification occurs, this derivative of the dimer exists

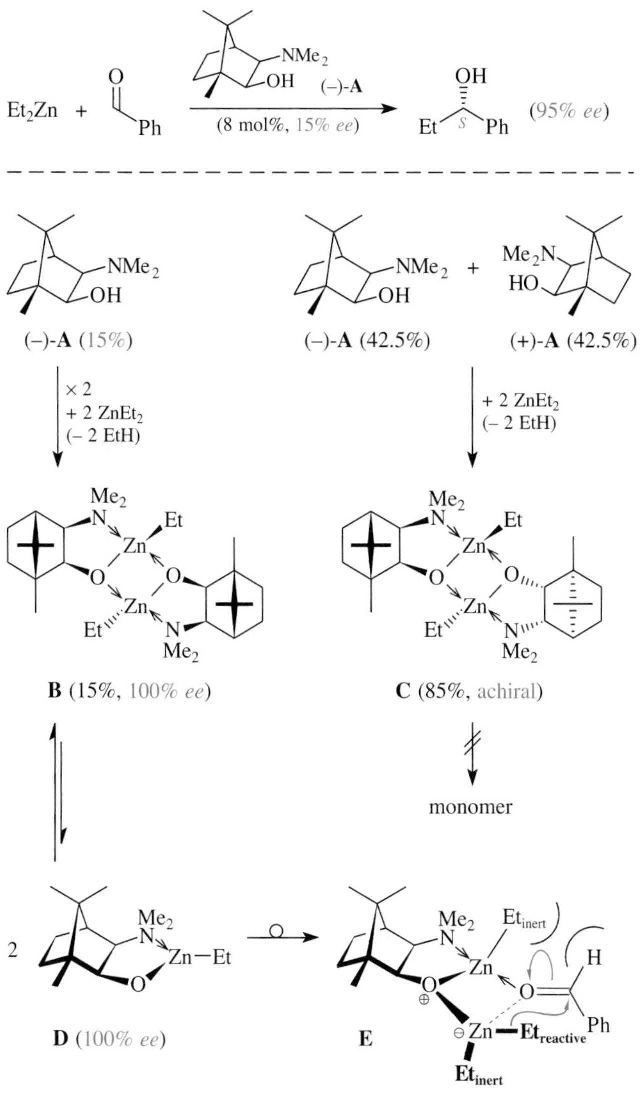

Fig. 10.40. Catalytic asymmetric addition of Et_2Zn to Ph—C(=O)H. Chiral amplification through a mutual kinetic resolution of the (auxiliary/ZnEt)$_2$ complex which is produced from two molecules of chiral aminoalcohol and diethylzinc each.

in the form of only two of its three conceivable stereoisomers. The reason for this is a mutual kinetic resolution (cf. Section 3.4.3).

In the reaction of Figure 10.40, the chiral ligand and Et_2Zn at first form as much "dimer" **C** as possible by a combination of the two enantiomers with each other. Therefore, the entire fraction of *racemic* chiral ligand **A** is used up. The remaining chiral ligand (—)-**A** is *enantiomerically 100% pure.* It reacts with additional Et_2Zn to form the "dimer" **B**. This species is less stable than "dimer" **C. B**, therefore dissociates—in contrast to **C**—reversibly to a small equilibrium fraction of the monomer **D. D** is *enantiomerically 100% pure*, just like **B**, *and represents the effective catalyst of the addition reaction that is then initiated.*

The trivalent Zn atom of the complex **D** coordinates as a Lewis acid to the carbonyl oxygen of the reaction partner benzaldehyde. Since the phenyl- and Zn-containing residue avoid each other, the phenyl group is forced into *trans*-orientation relative to the latter. At this stage a second molecule of $ZnEt_2$ gets involved. It binds to the O atom of the substructure C_{sp^3}—O—Zn—on the least hindered, i.e., the convex face of the heterocycle—and thereby creates an oxonium ion next to both the first and the second Zn atom. In the resulting complex **E**, the electrophilicity of the first Zn atom is therefore greatly increased. As a consequence, the electrophilicity of the benzaldehyde bound to this Zn atom is also greatly increased. Thus, it can then react with an ethyl group. This ethyl group is transferred from the second, negatively charged Zn atom with very high enantioselectivity. It is provided by the diethylzincate entity of the complex **E**. In the chosen projection, this happens selectively from the front side, making the reaction highly enantioselective.

10.5.3 Diastereoselective Addition of Organometallic Compounds to Carbonyl Compounds

As observed in Section 10.3.1, stereogenic addition reactions to cyclic ketones, in which one side of the carbonyl group lies in a concave region of the molecule and the other group lies in a convex region, take place preferentially from the convex side. With respect to the addition of sterically demanding hydride donors to the conformationally fixed ketone 4-*tert*-butyl cyclohexanone, this means that according to Figure 10.11 an equatorial approach is favored. However, as can also be seen from Figure 10.11, this kind of selectivity does not apply to the addition of sterically undemanding hydride donors. Because of a stereoelectronic factor, *they* preferentially approach axially.

Table 10.1 shows analogous additions of methylmagnesium bromide to the same *tert*-butylcyclohexanone. As can be seen from the complete absence of diastereoselectivity in this addition, the mentioned steric and stereoelectronic effects on diastereocontrol cancel each other out.

Tab. 10.1 Equatorial and Axial Addition of Classic Grignard Reagents and Reetz-Grignard Reagents to a Conformationally Fixed Cyclohexanone

MeMgX + *tert*-Bu–(cyclohexanone) ⟶ *tert*-Bu–(cyclohexane, axial OH, equatorial Me) + *tert*-Bu–(cyclohexane, axial Me, equatorial OH)

MeLi → MeMgX: for X ≠ Br: + MgX_2 (– LiX)

X =	axial OH : equatorial OH
Br:	49 : 51
OTf:	73 : 27
OTs:	85 : 15
O_3S–(2,4,6-trimethylphenyl):	90 : 10

Grignard prepared the alkylmagnesium halides named after him in the 19th century. The analogous alkylmagnesium *sulfonates* were not described until 90 years later by Reetz. They were prepared by transmetalation of alkyllithium compounds with magnesium sulfonates (Table 10.1, left). Reetz–Grignard reagents add with higher diastereoselectivities to 4-*tert*-butylcyclohexanone than the corresponding classic Grignard reagent (Table 10.1). The larger the sulfonate group, the more it favors an equatorial approach.

Cram selectivity occurs (Table 10.2) in the addition of methylmagnesium bromide to α-chiral aldehydes and ketones that do not contain a heteroatom at the alpha stereocenter. In agreement with the Cram transition state model of Figure 10.16 ($Nu^{\ominus} = Me^{\ominus}$), the Cram product is produced via the transition state **A.** Whether the corresponding anti-Cram product is formed via the transition state **B** or the transition state **C** has not yet been determined (by calculations). As one can see from Table 10.2, for a given substitution pattern at the stereocenter the fraction of anti-Cram product decreases with increasing size of the other, achiral substituent of the carbonyl group. In the transition state **B**, this effect would be understandable since a destabilizing interaction (1,2-allylic strain) would occur between a large substituent R at the carbonyl group and the neighboring methyl group, which is oriented almost *syn*-periplanar.

Table 10.3 shows how the Cram selectivity of the least Cram-selective addition reaction in Table 10.2 can be increased considerably by using Reetz–Grignard compounds.

Grignard reagents also add diastereoselectively to α-chiral, α-oxygenated aldehydes and ketones. The chelation-controlled product constitutes the major product (Figure 10.41). It is the diastereomer that is produced via the Cram chelate transition state of Figure 10.16

Tab. 10.2 Diastereoselectivity of the Addition of MeMgBr to α-Chiral Carbonyl Compounds

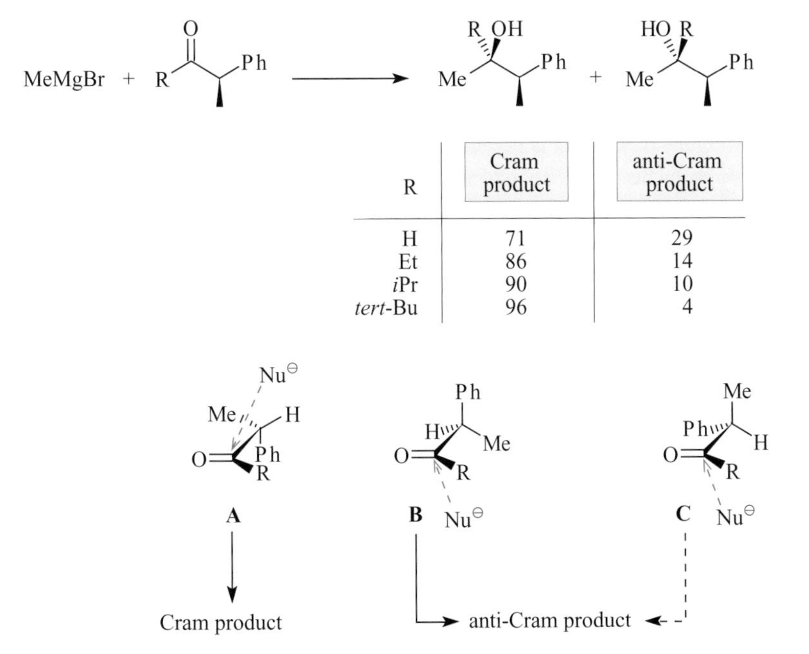

R	Cram product	anti-Cram product
H	71	29
Et	86	14
*i*Pr	90	10
tert-Bu	96	4

Tab. 10.3 Diastereoselectivity of the Addition of Classic Grignard and Reetz-Grignard Reagents to an α-Chiral Aldehyde

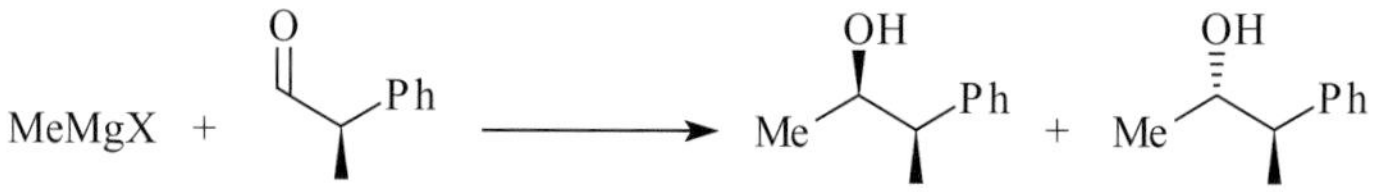

MeMgX + [aldehyde] → Cram product + anti-Cram product

X	Cram product	anti-Cram product
Cl, Br, I	ca. 70	30
OAc	91	9
OTs	92	8
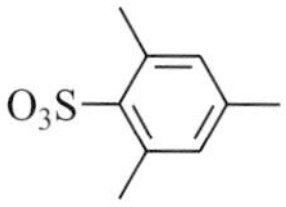 O_3S–(2,4,6-trimethylphenyl)	94	6
O_2C–*tert*-Bu	94	6

($Nu^\ominus = R^\ominus$). In fact, the Lewis acidity of the magnesium is so high that even a bulky protecting group at the O atom in the α-position of the carbonyl group does not completely prevent chelate formation. Therefore, the addition of Grignard reagents to α-oxygenated aldehydes via a Felkin–Anh transition state is unknown.

Grignard reagents and α-chiral aldehydes that contain a dialkylamino group at their stereocenter, as does compound **A** of Figure 10.42, react selectively via the Felkin-Anh transition state of Figure 10.16 ($Nu^\ominus = Ph^\ominus$). The N atom in substrate **A** is tertiary and sufficiently hindered by the two benzyl substituents that it cannot be incorporated into a chelate.

A sterically less hindered N atom is found, for example, in the α-aminated α-chiral aldehyde **B** in Figure 10.42. This is because this amine is "secondary" and bears only two nonhydrogen substituents. The N atom of aldehyde **B** is so unhindered that it can be bound firmly in a chelate by the magnesium. Consequently, the addition of a Grignard reagent to aldehyde **B** exclusively affords the chelation-controlled product.

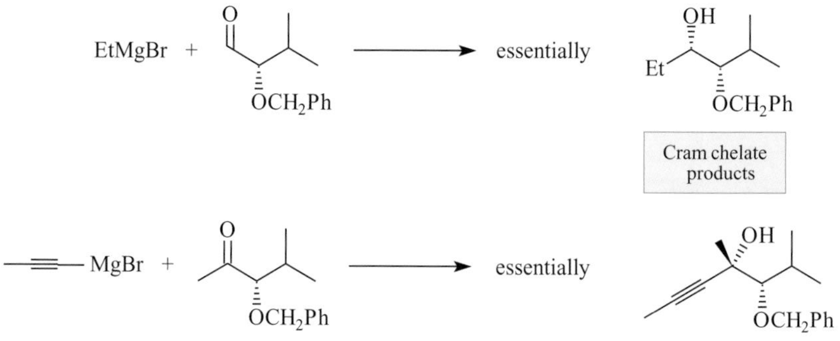

EtMgBr + [aldehyde with OCH_2Ph] → essentially [Et, OH, OCH_2Ph product]

Cram chelate products

[propynyl]–MgBr + [ketone with OCH_2Ph] → essentially [OH, OCH_2Ph product]

Fig. 10.41. Diastereoselective addition of Grignard reagents to α-chiral, α-oxygenated carbonyl compounds.

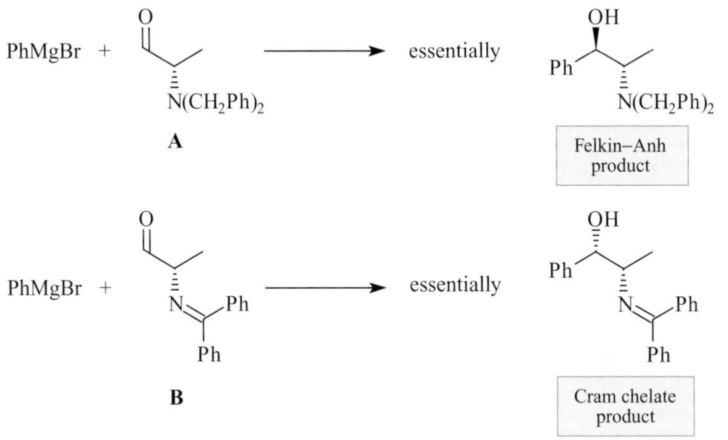

PhMgBr + A ⟶ essentially Felkin–Anh product

PhMgBr + B ⟶ essentially Cram chelate product

Fig. 10.42. Diastereoselective addition of Grignard reagents to α-chiral, α-aminated aldehydes.

Grignard reagents add to the C=O double bond of β-chiral, β-oxygenated carbonyl compounds with only little diastereoselectivity. Highly diastereoselective additions to such substrates would be expected to take place under chelation control (cf. Section 10.3.3). However, in order to obtain significant chelate formation with these substrates, one must use organometallic compounds that are stronger Lewis acids than Grignard reagents.

10.6 1,4-Additions of Organometallic Compounds to α,β-Unsaturated Ketones; Structure of Copper-Containing Organometallic Compounds

Many copper-containing organometallic compounds add to α,β-unsaturated ketones in smooth 1,4-additions (for the term "1,4-addition" see Figure 10.31). The most important ones are **Gilman cuprates** (Figure 10.43, left). For sterically hindered substrates, their rate of addition can be increased by adding Me_3SiCl to the reactants.

Grignard reagents are converted into C nucleophiles capable of undergoing 1,4-additions selectively when they are transmetalated with Cu(I) compounds to give so-called **Normant cuprates** (Figure 10.43, middle). Usually, this transmetalation is carried out with only catalytic amounts of CuI or CuBr-SMe_2. This means that Normant cuprates undergo the 1,4-addition to α,β-unsaturated ketones considerably faster than the Cu-free Grignard reagents undergo the nonselective 1,2-/1,4-addition (cf. Figure 10.31). This is another example of a ligand-accelerated reaction (see also the comments to Figure 10.26).

Knochel cuprates likewise add to α,β-unsaturated ketones with 1,4 selectivity, but only in the presence of Me_3SiCl or BF_3-OEt_2 (Figure 10.43, right). Under the same conditions, how-

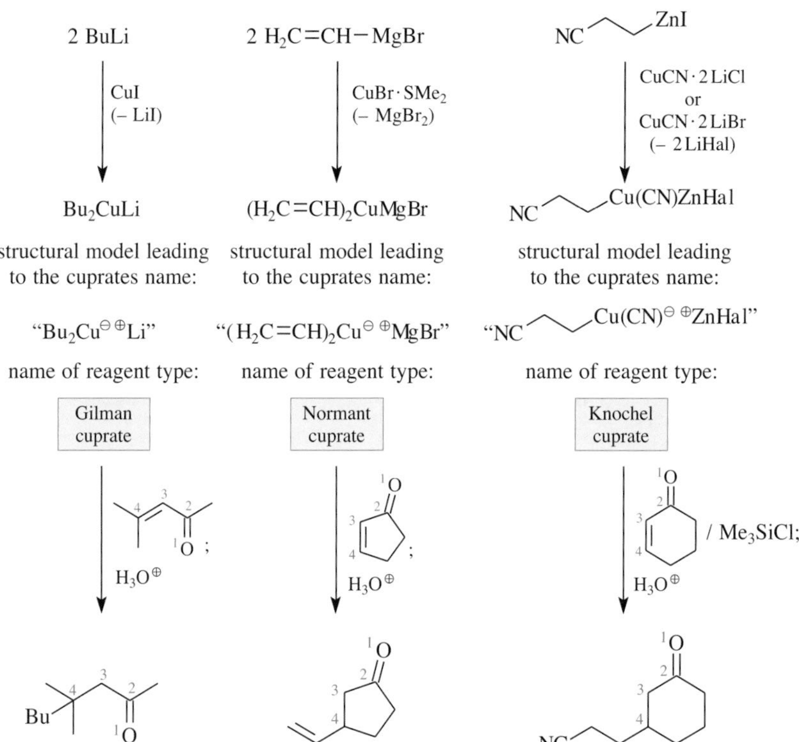

Fig. 10.43. 1,4-Addition of various organometallic compounds to α,β-unsaturated ketones.

ever, they add to α,β-unsaturated aldehydes (except acrolein) with 1,2-selectivity (Figure 10.38).

So far only little is known about the mechanisms of such 1,4-additions. To start with, it is uncertain whether they only depend on the metal used or on both metal and substrate. At the beginning of the new millennium, however, the prototype of this reaction, i.e., the 1,4-addition of Gilman cuprates to α,β-unsaturated ketones, could finally be assigned a mechanism after many years of studies and with a proper finish of crystallographic, NMR-spectroscopic, kinetic and quantum-chemical studies (Figure 10.46, see farther below).

Gilman cuprates in solution occur as an equilibrating mixture of a "contact ion pair" and a "solvent-separated ion pair" (Figure 10.44). In diethyl ether, whose oxygen atom is not very easily accessible and hence does not provide a good donor towards $Li^{\oplus}(solv)_2$ ions, Gilman cuprates are usually present as contact ion pairs. THF, whose O atom is particularly easy to access, or 1,2-dimethoxyethane (DME), with its two O atoms that are capable of chelation, are solvents that usually convert these contact ion pairs into their separated ion pair form. Figure 10.44 shows that two reactions are involved here: first, the complexation of the $Li^{\oplus}(solv)_2$ cations by the solvents mentioned to give $Li^{\oplus}(THF)_4$- and $Li^{\oplus}(DME)_2$ cations, respectively, and second, the transformation of the "Gilman dimer" into the "Gilman monomer."

Figure 10.45 also shows that lithium dialkyl cuprates in solutions containing lithium iodide or lithium cyanide—in addition to the dimeric or monomeric Gilman cuprates presented in Figure 10.44—may contain the following species: the contact ion pairs **A** and **C** in diethyl

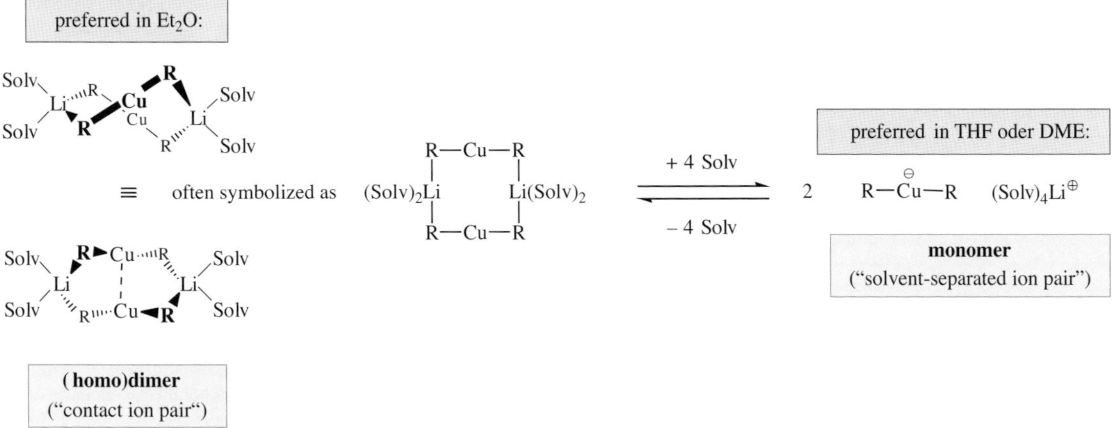

Fig. 10.44. The structures of Gilman cuprates in halide- and cyanide-free solution: dimer-to-monomer conversion. The dimer is a contact ion pair and occurs in diethyl ether. The monomer is a solvent-separated ion pair and occurs in the more solvating ethers THF (with the stoichiometry mentioned) and DME, respectively. (DME stands for 1,2-**dim**ethoxy**e**thane; since DME is a bidentate ligand, the stoichiometry given in this figure does not apply: only two DME molecules are bound per lithium).

ether solution and/or the solvent-separated ion pairs **B** and **D** in THF or DME solution. It seems as if neither **A** nor **B** nor **C** nor **D** are capable of undergoing a 1,4-addition in such LiI or LiCN-containing cuprate solutions, but only the "Gilman dimers" that are present at equilibrium (Figure 10.44). This is why iodide and cyanide ions should not occur in 1,4-additions of cuprates to α,β-unsaturated ketones.

Figure 10.46 shows the current views on the mechanism of the 1,4-addition of Gilman cuprates to cyclohexenone. The numerous details, which have been elucidated or made plausible over the past years, force us to break down the general view of this mechanism into two halves.

The first part of Figure 10.46 depicts the actual 1,4-addition up to the formation of the copper-containing enolate **G**. Initially, the ketone reacts with either the "Gilman dimer" **A** (→ **D**) or with its ring-opening product **C** (→ **F**), which in both cases happens *via a coordinatively unsaturated lithium atom*. This is why 1,4-additions of cuprates, from which lithium was removed by complexation as the $Li^{\oplus}(solv)_4$ cation like in the "Gilman monomer" of Figure 10.44, or as the $Li^{\oplus}$(crown ether) cation, usually fail. In conjunction with the statements about solvent effects in Figure 10.44, this essential role of coordinatively unsaturated lithium

A: Li–I–Li / R–Cu–R — **heterodimer** ("contact ion pair")

B: $R{-}\overset{\ominus}{Cu}{-}R \quad (Solv)_nLi{-}I{-}Li(Solv)_n^{\oplus}$ — **monomer with complex cation** ("solvent-separated ion pair")

C: Li–C≡N–Li / R–Cu–R — **heterodimer** ("contact ion pair")

D: $R{-}\overset{\ominus}{Cu}{-}R \quad (Solv)_nLi{-}C{\equiv}N{-}Li(Solv)_n^{\oplus}$ — **monomer with complex cation** ("solvent-separated ion pair")

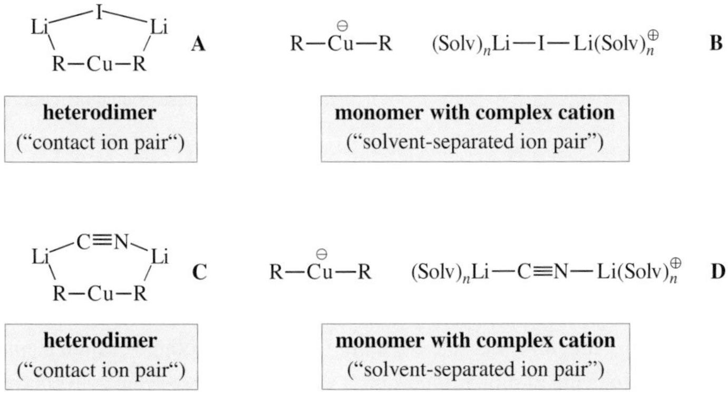

Fig. 10.45. The structures of Gilman cuprates in halide- or cyanide-containing solution: apart from the (homo-)dimer and/or the monomer in Figure 10.44, the (hetero-)dimers **A** or **C** and/or the corresponding "monomers with complex cation" **B** or **D** presented here can occur (but no "cyanocuprates" in which the bonding partner of the copper is a CN group).

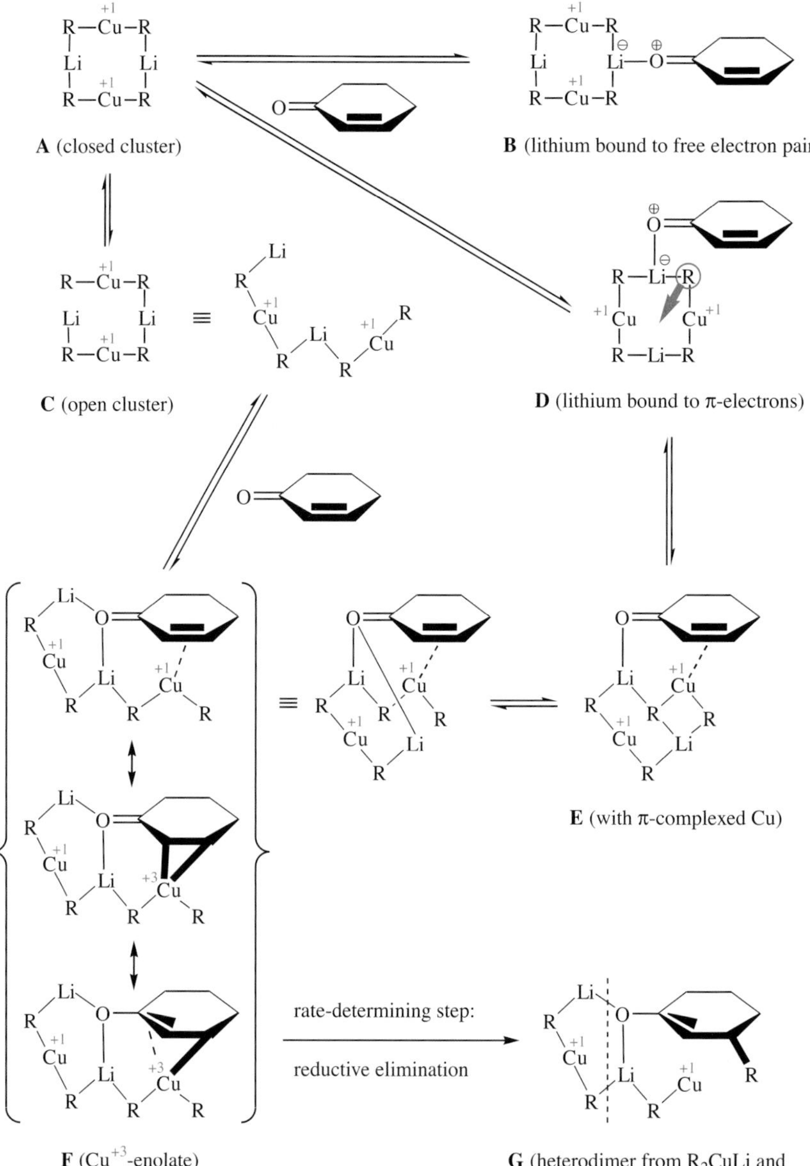

Fig. 10.46. Mechanistic possibilities for the 1,4-addition of a Gilman cuprate to an α,β-unsaturated ketone. Part 1 shows the reaction up to the rate-determining step (**F** $\rightarrow$ **G**). For the sake of greater clarity the solvation of the lithium atoms is left unconsidered here.

also explains why 1,4-additions of Gilman cuprates to enones are preferentially conducted in diethyl ether, and not in THF or DME. This also helps to understand that 1,4-additions to highly oxygenated α,β-unsaturated ketones can often only be performed if a considerable excess of cuprate is used: a great deal of reagent is consumed simply by binding to the extra oxygens atoms.

It is essential to recognize this role of lithium, with all the other reaction steps in the first part of Figure 10.46 being of minor importance. Only the last step—the conversion of the eno-

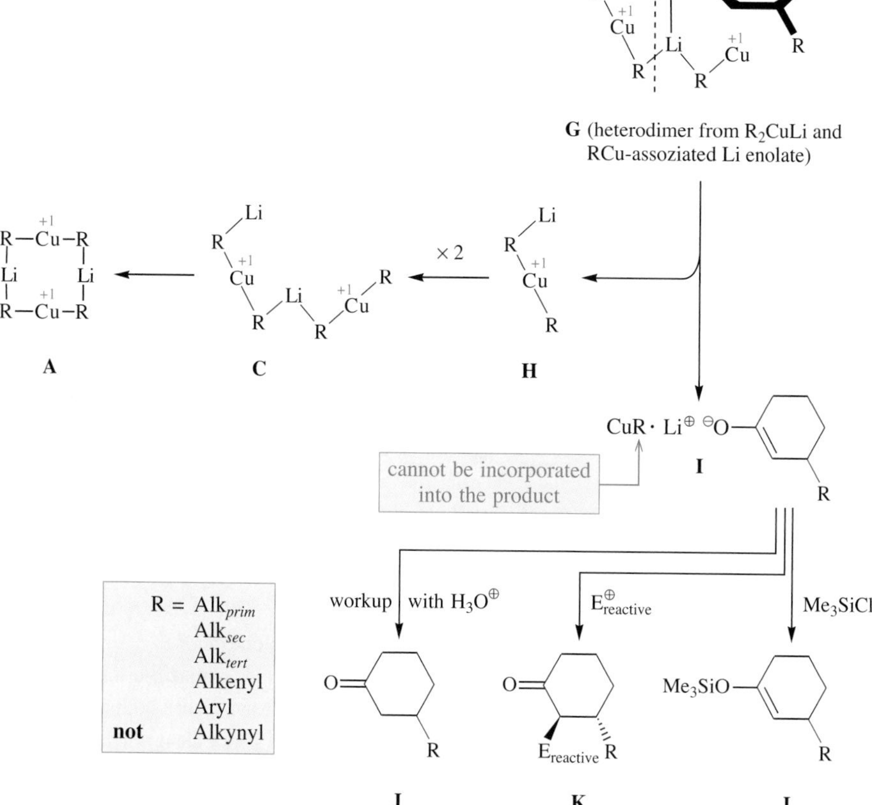

Fig. 10.46 (cont.) Mechanistic possibilities for the 1,4-addition of a Gilman cuprate to an α,β-unsaturated ketone. Part 2 shows the formation of the CuR-containing enolate (**I**) and its further reaction with typical electrophiles. The fact that all intermediates of this 1,4-addition are chiral, but generated in racemic form should not be overlooked; consequently, the same is true of the final products **J**, **K** and **L**.

late **F** with its C–Cu bond in the β-position into the enolate **G** with its C–R bond in the β-position—is really significant since the new C–C bond is formed here. The label "reductive elimination" at the corresponding arrow in Figure 10.46 will not help you unless you have mastered Chapter 16 and are able to recognize this step as a typical reaction of organotransition metal chemistry.

In the second part of Figure 10.46, the 1,4-addition of the Gilman cuprates to cyclohexenone is completed, picking up with the same copper-containing enolate **G** with which the first part of this figure ended. This enolate **G** decomposes into the RCu-associated lithium enolate **I** and the "Gilman monomer" (**H**). The latter reacts to yield the "ring-opened Gilman dimer" (**C**) and further to give the eight-membered stable "Gilman dimer" **A**. So the organocopper compounds **C** and **A**, according to the first part of Figure 10.46, are again ready to undergo the actual 1,4-addition. The lithium enolate **I** is not an ordinary lithium enolate, since it is associated with one equivalent of CuR. As you can see in Figure 10.46 (part 2), 50% of the organic residue R contained in Gilman cuprates is lost in 1,4-additions because this by-product CuR is formed and it is not reactive. This disadvantage is not seen with 1,4-additions of Normant and Knochel cuprates.

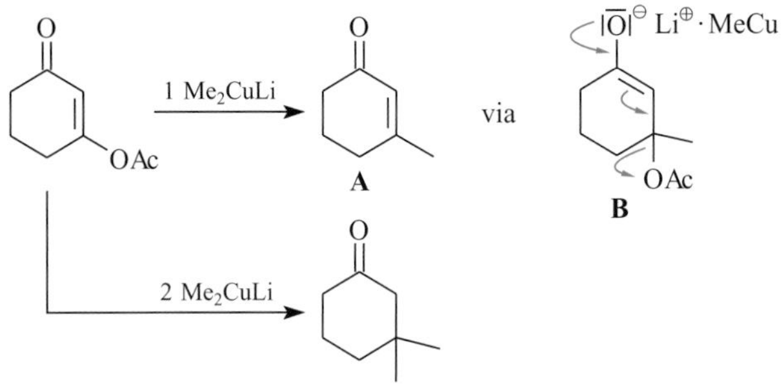

Fig. 10.47. Nucleophilic substitution I on an acceptor-substituted alkene with a leaving group in the β-position.

The enolate **I** (Figure 10.46, part 2) behaves inertly until aqueous workup is performed and it is protonated to furnish the saturated ketone **J**. However, **I** can also be further functionalized with other electrophiles if these are highly reactive. Thus, with methyl iodide or allyl bromide, CuR-containing enolates **I** often form the alkylation products **K** (example: Figure 13.30). With Me_3SiCl they reliably give the silyl enol ethers **L** (example: Figure 13.19).

The enol ester from Figure 10.47 and one equivalent of Me_2CuLi react to give the methylcyclohexenone **A.** Presumably, the enolate **B** is formed first in this reaction—according to one of the two mechanisms shown in Figure 10.46. An acetate ion is eliminated from this enolate as in the second step of an $E1_{cb}$ elimination. As an alternative, the enol ester of Figure 10.47 can also react with two equivalents of Me_2CuLi. The second equivalent adds to the enone **A**, which is formed as before, and converts it into 3,3-dimethylcyclohexanone.

Other reactions of copper-containing organometallic compounds with α,β-unsaturated carbonyl compounds or with α,β-unsaturated esters (they do not always react), both containing a leaving group at C-β, are shown in Figures 10.48 and 10.49. It is not clear whether in these reactions, a 1,4-addition takes place first and is then followed by the second part of an $E1_{cb}$ elimination or whether the organometallics substitute C_{sp2}-bound leaving groups in the respective β-positions via a Cu-catalyzed mechanism, which is similar to the mechanisms in Chapter 16.

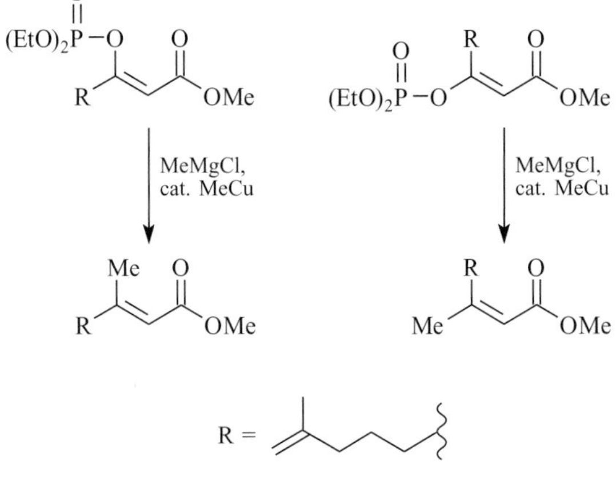

Fig. 10.48. Nucleophilic substitution II on an acceptor-substituted alkene with a leaving group in the β-position (possibility for preparing the starting materials; Figure 13.25). Stereospecifity occurs in the reaction pair shown.

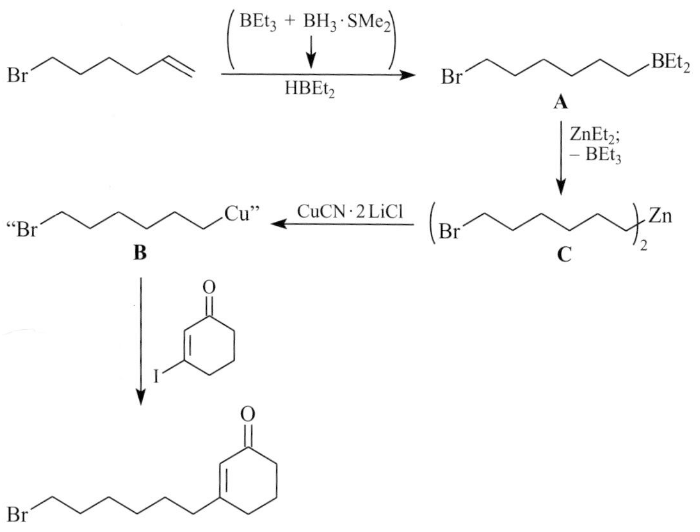

Fig. 10.49. Nucleophilic substitution III on acceptor-substituted alkenes with a leaving group in the β-position.

There is a serious limitation with respect to the structure of ketones that are accessible by 1,4-addition of Gilman or Normant cuprates to α,β-unsaturated ketones. As a result of preparing these reagents from organolithium or from Grignard reagents, only functional groups can be incorporated into these cuprates that are compatible with forming the Li- or Mg-containing precursors. This essentially limits the substrates to contain only ethers, acetals, and internal C≡C triple bonds as functional groups.

As mentioned in the discussion of Figure 10.38, in Knochel cuprates a large variety of functional groups can be present: CO_2R, RC=O, C≡N, Cl, RNH, NH_2, RC(=O)NH, sulfoxide, sulfone, and an internal or a terminal C≡C triple bond. This compatibility with functional groups makes Knochel cuprates the reagents of choice for 1,4-additions of organometallic compounds that contain *functional groups.* An example of such a reaction is shown in Figure 10.43 (right).

A second type of organometallic compound, dialkylzinc compounds, is compatible with the presence of the most varied functional groups, too. However, they only react with those α,β-unsaturated ketones that contain a leaving group in the β-position. The preparation of the required dialkylzinc compounds from alkenes was described in Figure 10.39. Figure 10.49 shows how such a species, through the intermediacy of a Cu-containing derivative of unknown composition, can be reacted with a suitable substrate.

References

10.1

H. Günther, D. Moskau, P. Bast, D. Schmalz, "Modern NMR Spectroscopy of Organolithium Compounds," *Angew. Chem. Int. Ed. Engl.* **1987**, *26*, 1212–1220.

C. Lambert, P. von R. Schleyer, "Are Polar Organometallic Compounds 'Carbanions'? The Gegenion Effect on Structure and Energies of Alkali-Metal Compounds," *Angew. Chem. Int. Ed. Engl.* **1994**, *33*, 1129–1140.

A. Basu, S. Thayumanavan, "Configurational Stability and Transfer of Stereochemical Information in the Reactions of Enantioenriched Organolithium Reagents," *Angew. Chem. Int. Ed. Engl.* **2002**, *41,* 716–738.

G. S. Silverman, "Common Methods of Grignard Reagent Preparation," in *Handbook of Grignard Reagents* (G. S. Silverman, P. E. Rakita, Eds.), Marcel Dekker Inc., New York, **1996**, 9–22.

J.-A. K. Bonesteel, "Nuclear Magnetic Resonance Analyses of Grignard Reagents," in *Handbook of Grignard Reagents* (G. S. Silverman, P. E. Rakita, Eds.), Marcel Dekker Inc., New York, **1996**, 103–116.

H. L. Uhm, "Crystal Structures of Grignard Reagents," in *Handbook of Grignard Reagents* (G. S. Silverman, P. E. Rakita, Eds.), Marcel Dekker Inc., New York, **1996**, 117–144.

K. C. Cannon, G. R. Krow, "The Composition of Grignard Reagents in Solution: The Schlenk Equilibrium and Its Effect on Reactivity," in *Handbook of Grignard Reagents* (G. S. Silverman, P. E. Rakita, Eds.), Marcel Dekker Inc., New York, **1996**, 271–290.

F. Bickelhaupt, "Structures of Organomagnesium Compounds as Revealed by X-Ray Diffraction Studies," in *Grignard Reagents—New Developments* (H. G. Richey, Jr., Ed.), John Wiley & Sons, Chichester, **2000**, 299–328.

T. S. Ertel, H. Bertagnolli, "X-ray Absorption Spectroscopy and Large Angle X-Ray Scattering of Grignard Compounds," in *Grignard Reagents—New Developments* (H. G. Richey, Jr., Ed.), John Wiley & Sons, Chichester, **2000**, 329–366.

F. Bickelhaupt, "Di- and Polyfunctional Organomagnesium Compounds," in *Grignard Reagents—New Developments* (H. G. Richey, Jr., Ed.), John Wiley & Sons, Chichester, **2000**, 367–393.

M. Westerhausen, "100 Years after Grignard: Where does the Organometallic Chemistry of the Heavy Alkaline Earth Metals Stand Today?," *Angew. Chem. Int. Ed. Engl.* **2001**, *40,* 2975–2977.

P. Knochel, W. Dohle, N. Gommermann, F. F. Kneisel, F. Kopp, T. Korn, I. Sapountzis, V. A. Vu, "Highly Functionalized Organomagnesium Reagents Prepared through Halogen-Metal Exchange," *Angew. Chem. Int. Ed. Engl.* **2003**, *42,* 4302–4320.

R. W. Hoffmann, "The Quest for Chiral Grignard Reagents," *Chem. Soc. Rev.* **2003**, *32,* 225–230.

10.2

N. Greeves, "Reduction of C=O to CHOH by Metal Hydrides," in *Comprehensive Organic Synthesis* (B. M. Trost, I. Fleming, Eds.), Vol. 8, 1, Pergamon Press, Oxford, **1991**.

A. P. Davis, M. M. Midland, L. A. Morell, "Formation of C–H Bonds by Reduction of Carbonyl Groups (C=H), Reduction of Carbonyl Groups with Metal Hydrides," in *Methoden Org. Chem.* (Houben-Weyl) 4th ed., **1952**–, *Stereoselective Synthesis* (G. Helmchen, R. W. Hoffmann, J. Mulzer, E. Schaumann, Eds.), Vol. E21d, 3988, Georg Thieme Verlag, Stuttgart, **1995**.

M. M. Midland, L. A. Morell, K. Krohn, "Formation of C–H Bonds by Reduction of Carbonyl Groups (C=O) – Reduction with Hydride Donors," in *Methoden Org. Chem.* (Houben-Weyl) 4th ed., **1952**–, *Stereoselective Synthesis* (G. Helmchen, R. W. Hoffmann, J. Mulzer, E. Schaumann, Eds.), Vol. E21d, 4082, Georg Thieme Verlag, Stuttgart, **1995**.

T. Ooi, K. Maruoka, "Exceptionally Bulky Lewis Acidic Reagent, MAD," *Rev. Heteroatom Chem.*, **1998**, *18*, 61–85.

10.3

A. S. Cieplak, "Inductive and Resonance Effects of Substituents on π-Face Selection," *Chem. Rev.* **1999**, *99*, 1265–1336.

T. Ohwada, "Orbital-Controlled Stereoselection in Sterically Unbiased Cyclic Systems," *Chem. Rev.* **1999**, *99*, 1337–1376.

W. Adcock, N. A. Trout, "Nature of the Electronic Factor Governing Diastereofacial Selectivity in Some Reactions of Rigid Saturated Model Substrates," *Chem. Rev.* **1999**, *99*, 1415–1435.

G. Mehta, J. Chandrasekhar, "Electronic Control of Facial Selection in Additions to Sterically Unbiased Ketones," *Chem. Rev.* **1999**, *99*, 1437–1468.

J. M. Coxon, R. T. Luibrand, "π-Facial Selectivity in Reactions of Carbonyls: A Computational Approach," in *Modern Carbonyl Chemistry* (J. Otera, Ed.), Wiley-VCH, Weinheim, **2000**, 155–184.

Y. Senda, "Role of the Heteroatom on Stereoselectivity in the Complex Metal Hydride Reduction of Six-membered Cyclic Ketones," *Chirality* **2002**, *14,* 110–120.

J. Mulzer, "Cram's Rule: Theme and Variations," in *Organic Synthesis Highlights* (J. Mulzer, H.-J. Altenbach, M. Braun, K. Krohn, H.-U. Reißig, Eds.), VCH, Weinheim, New York, etc., **1991**, 3–8.

E. L. Eliel, "Application of Cram's Rule: Addition of Achiral Nucleophiles to Chiral Substrates," in *Asymmetric Synthesis* (J. D. Morrison, Ed.), Vol. 2, 125, AP, New York, **1983**.

F. Vögtle, J. Franke, A. Aigner, D. Worsch, „Stereochemie in Stereobildern: Die Cramsche Regel – plastisch", *Chem. unserer Zeit,* **1984**, *18,* 203–210.

E. L. Eliel, S. V. Frye, E. R. Hortelano, X. B. Chen, "Asymmetric Synthesis and Cram's (chelate) Rule," *Pure Appl. Chem.* **1991**, *63,* 1591–1598.

A. Mengel, O. Reiser, "Around and Beyond Cram's Rule," *Chem. Rev.* **1999**, *99,* 1191–1223.

A. P. Davis, "Diastereoselective Reductions," in *Stereoselective Synthesis* (Houben-Weyl) 4th ed., **1996**, (G. Helmchen, R. W. Hoffmann, J. Mulzer, E. Schaumann, Eds.), **1996**, Vol. E21 (Workbench Edition), *7,* 3988–4048, Georg Thieme Verlag, Stuttgart.

M. T. Reetz, "New Approaches to the Use of Amino Acids as Chiral Building Blocks in Organic Synthesis," *Angew. Chem. Int. Ed. Engl.* **1991**, *30,* 1531–1546.

M. T. Reetz, "Structural, Mechanistic, and Theoretical Aspects of Chelation-Controlled Carbonyl Addition Reactions," *Acc. Chem. Res.* **1993**, *26,* 462–468.

D. Gryko, J. Chalko, J. Jurczak, "Synthesis and Reactivity of N-Protected-α-Amino Aldehydes," *Chirality* **2003**, 15, 514–541.

X. Liang, J. Andersch, M. Bols, "Garner's Aldehyde," *J. Chem. Soc. Perkin Trans. 1* **2001**, *18,* 2136–2157.

10.4

M. M. Midland, L. A. Morell, "Enantioselective Reductions," in *Stereoselective Synthesis* (Houben-Weyl) 4th ed. **1996**, (G. Helmchen, R. W. Hoffmann, J. Mulzer, E. Schaumann, Eds.), **1996**, Vol. E21 (Workbench Edition), *7,* 4049–4066, Georg Thieme Verlag, Stuttgart.

M. Nishizawa, R. Noyori, "Reduction of C=X to CHXH by Chirally Modified Hydride Reagents," in *Comprehensive Organic Synthesis* (B. M. Trost, I. Fleming, Eds.), Vol. 8, 159, Pergamon Press, Oxford, **1991**.

E. R. Grandbois, S. I. Howard, J. D. Morrison, "Reductions with Chiral Modifications of Lithium Aluminium Hydride," in *Asymmetric Synthesis* (J. D. Morrison, Ed.), Vol. 2, 71, AP, New York, **1983**.

H. Haubenstock, "Asymmetric Reductions with Chiral Complex Aluminium Hydrides and Tricoordinate Aluminium Reagents," *Top. Stereochem.* **1983**, *14,* 213.

P. Daverio, M. Zanda, "Enantioselective Reductions by Chirally Modified Alumino- and Borohydrides," *Tetrahedron: Asymmetry* **2001**, 12, 2225–2259.

M. M. Midland, "Asymmetric Reductions with Organoborane Reagents," *Chem. Rev.* **1989**, *89,* 1553–1561.

H. C. Brown, P. V. Ramachandran, "Asymmetric Reduction with Chiral Organoboranes Based on α-Pinene," *Acc. Chem. Res.* **1992**, *25,* 16–24.

R. K. Dhar, "Diisopinocamphenylchloroborane, (DIP-Chloride), an Excellent Chiral Reducing Reagent for the Synthesis of Secondary Alcohols of High Enantiomeric Purity," *Aldrichimica Acta* **1994**, *27,* 43–51.

B. B. Lohray, V. Bhushan, "Oxazaborolidines and Dioxaborolidines in Enantioselective Catalysis," *Angew. Chem. Int. Ed. Engl.* **1992**, *31,* 729–730.

S. Wallbaum, J. Martens, "Asymmetric Syntheses with Chiral Oxazaborolidines," *Tetrahedron: Asymmetry* **1992**, *3,* 1475–1504.

S. Itsuno, "Enantioselective Reduction of Ketones," *Org. React.* **1998,** 52, 395–576.

E. J. Corey, C. J. Helal, "Reduction of Carbonyl Compounds with Chiral Oxazaborolidine Catalysts: A New Paradigm for Enantioselective Catalysis and a Powerful New Synthetic Method," *Angew. Chem. Int. Ed. Engl.* **1998**, *37,* 1987–2012.

B. T. Cho, "Boron-Based Reducing Agents for the Asymmetrical Reduction of Functionalized Ketones and Ketimines," *Aldrichim. Acta* **2002**, *35*, 3–16.

V. K. Singh, "Practical and Useful Methods for the Enantioselective Reduction of Unsymmetrical Ketones," *Synthesis* **1992**, 607–617.

M. Wills, J. R. Studley, "The Asymmetric Reduction of Ketones," *Chem. Ind.* **1994**, 552–555.

10.5

G. Salem, C. L. Raston, "Preparation and Use of Grignard Reagents and Group II Organometallics in Organic Synthesis," in *The Use of Organometallic Compounds in Organic Synthesis* (F. R. Hartley, Ed.), Vol. 4, 159, Wiley, Chichester, **1987**.

R. M. Kellogg, "Reduction of C=X to CHXH by Hydride Delivery from Carbon," in *Comprehensive Organic Synthesis* (B. M. Trost, I. Fleming, Eds.), Vol. 8, 79, Pergamon Press, Oxford, **1991**.

C. Blomberg, "Mechanisms of Reactions of Grignard Reagents," in *Handbook of Grignard Reagents* (G. S. Silverman, P. E. Rakita, Eds.), Marcel Dekker Inc., New York, **1996**, 219–248.

T. Holm, I. Crossland, "Mechanistic Features of the Reactions of Organomagnesium Compounds," in *Grignard Reagents—New Developments* (H. G. Richey, Jr., Ed.), John Wiley & Sons, Chichester, **2000**, 1–26.

L. Miginiac, "Nucleophilic Addition to Carbon-Heteroatom Multiple Bonds: O, S, N, P," in *Handbook of Grignard Reagents* (G. S. Silverman, P. E. Rakita, Eds.), Marcel Dekker Inc., New York, **1996**, 361–372.

L. Miginiac, "Nucleophilic Additions to Conjugated Carbon-Heteroatom Multiple Bonds: O, S, N," in *Handbook of Grignard Reagents* (G. S. Silverman, P. E. Rakita, Eds.), Marcel Dekker Inc., New York, **1996**, 391–396

P. E. Rakita, "Safe Handling Practices of Industrial Scale Grignard Reagents," in *Handbook of Grignard Reagents* (G. S. Silverman, P. E. Rakita, Eds.), Marcel Dekker Inc., New York, **1996**, 79–88.

F. R. Busch, D. M. De Antonis, "Grignard Reagents—Industrial Applications and Strategy," in *Grignard Reagents—New Developments* (H. G. Richey, Jr., Ed.), John Wiley & Sons, Chichester, **2000**, 165–183

M. Umeno, A. Suzuki, "Alkynyl Grignard Reagents and Their Uses," in *Handbook of Grignard Reagents* (G. S. Silverman, P. E. Rakita, Eds.), Marcel Dekker Inc., New York, **1996**, 645–666.

W. Kosar, "Grignard Reagents as Bases," in *Handbook of Grignard Reagents* (G. S. Silverman, P. E. Rakita, Eds.), Marcel Dekker Inc., New York, **1996**, 441–454.

P. Knochel, M. J. Rozema, C. E. Tucker, C. Retherford, M. Furlong, S. A. Rao, "The Chemistry of Polyfunctional Organozinc and Copper Reagents," *Pure Appl. Chem.* **1992**, *64*, 361.

P. Knochel, R. D. Singer, "Preparation and Reactions of Polyfunctional Organozinc Reagents in Organic Synthesis," *Chem. Rev.* **1993**, *93*, 2117–2188.

P. Knochel, "New Preparations of Polyfunctional Dialkylzincs and Their Application in Asymmetric Synthesis," *Chemtracts: Organic Chemistry* **1995**, *8*, 205.

A. Boudier, L. O. Bromm, M. Lotz, P. Knochel, "New Applications of Polyfunctional Organometallic Compounds in Organic Synthesis," *Angew. Chem. Int. Ed. Engl.* **2000**, *39*, 4414–4435.

L. Pu, H.-B. Yu, "Catalytic Asymmetric Organozinc Additions to Carbonyl Compounds," *Chem. Rev.* **2001**, *101*, 757–824.

R. M. Devant, H.-E. Radunz, "Addition of σ-Type Organometallic Compounds," in *Stereoselective Synthesis* (Houben-Weyl) 4th ed**., 1996**, (G. Helmchen, R. W. Hoffmann, J. Mulzer, E. Schaumann, Ed.), **1996**, Vol. E21 (Workbench Edition), *2*, 1151–1334, Georg Thieme Verlag, Stuttgart.

D. Hoppe, "Formation of C–C Bonds by Addition of Benzyl-Type Organometallic Compounds to Carbonyl Groups," in *Stereoselective Synthesis* (Houben-Weyl) 4th ed**., 1996**, (G. Helmchen, R. W. Hoffmann, J. Mulzer, E. Schaumann, Eds.), **1996**, Vol. E21 (Workbench Edition), *3*, 1335–1356, Georg Thieme Verlag, Stuttgart.

D. Hoppe, W. R. Roush, E. J. Thomas, "Formation of C–C Bonds by Addition to Carbonyl Groups (C=O) – Allyl-Type Organometallic Compounds," in *Methoden Org. Chem.* (Houben-Weyl) 4th ed**., 1952–**,

Stereoselective Synthesis (G. Helmchen, R. W. Hoffmann, J. Mulzer, E. Schaumann, Eds.), Vol. E21b, 1357, Georg Thieme Verlag, Stuttgart, **1995**.

G. Solladié, "Formation of C–C Bonds by Addition to Carbonyl Groups (C=O) – Metalated Sulfoxides or Sulfoximides," in *Methoden Org. Chem.* (Houben-Weyl) 4th ed., **1952**–, *Stereoselective Synthesis* (G. Helmchen, R. W. Hoffmann, J. Mulzer, E. Schaumann, Eds.), Vol. E21b, 1793, Georg Thieme Verlag, Stuttgart, **1995**.

A. B. Sannigrahi, T. Kar, B. G. Niyogi, P. Hobza, P. v. R. Schleyer, "The Lithium Bond Reexamined," *Chem. Rev.* **1990**, *90*, 1061–1076.

W. Bauer, P. von Rague Schleyer, "Recent Results in NMR Spectroscopy of Organolithium Compounds," in *Advances in Carbanion Chemistry* (V. Snieckus, Ed.), **1992**, *1*, JAI Press, Greenwich, CT.

K. Maruyama, T. Katagiri, "Mechanism of the Grignard Reaction," *J. Phys. Org. Chem.* **1989**, *2*, 205–213.

D. M. Huryn, "Carbanions of Alkali and Alkaline Earth Cations: (ii) Selectivity of Carbonyl Addition Reactions," in *Comprehensive Organic Synthesis* (B. M. Trost, I. Fleming, Eds.), Vol. 1, 49, Pergamon Press, Oxford, **1991**.

G. S. Silverman, P. E. Rakita (Eds.), "Handbook of Grignard Reagents," [In: *Chem. Ind.* **1996**; 64] Dekker, New York, **1996**.

R. Noyori, M. Kitamura, "Enantioselective Addition of Organometallic Reagents to Ccarbonyl Compounds: Chirality Transfer, Multiplication and Amplification," *Angew. Chem. Int. Ed. Engl.* **1991**, *30*, 49.

K. Soai, S. Niwa, "Enantioselective Addition of Organozinc Reagents to Aldehydes," *Chem. Rev.* **1992**, *92*, 833–856.

P. Knochel, "Stereoselective Reactions Mediated by Functionalized Diorganozincs," *Synlett* **1995**, 393–403.

Y. L. Bennani, S. Hanessian, "*Trans*-1,2-Diaminocyclohexane Derivatives as Chiral Reagents, Scaffolds, and Ligands for Catalysis: Applications in Asymmetric Synthesis and Molecular Recognition," *Chem. Rev.* **1997**, *97*, 3161–3195.

C. Girard, H. B. Kagan, "Nonlinear Effects in Asymmetric Synthesis and Stereoselective Reactions: Ten Years of Investigation," *Angew. Chem. Int. Ed. Engl.* **1998**, *37*, 2923–2959.

K. Soai, "Rational Design of Chiral Catalysis for the Enantioselective Addition Reaction of Dialkylzincs," *Enantiomer* **1999**, *4*, 591–598.

D. R. Fenwick, H. B. Kagan, "Asymmetric Amplification," *Top. Stereochem.* **1999**, *22*, 257– 296.

D. G. Blackmond, "Kinetic Aspects of Nonlinear Effects in Asymmetric Catalysis," *Acc. Chem. Res.* **2000**, *33*, 402–411.

K. Soai, T. Shibata, "Asymmetric Amplification and Autocatalysis," in *Catalytic Asymmetric Synthesis* (I. Ojima, Ed.), Wiley-VCH, New York, 2nd ed., **2000**, 699–726.

H. B. Kagan, "Practical Consequences of Non-Linear Effects in Asymmetric Synthesis," *Adv. Synth. Catal.* **2001**, *343*, 227–233.

H. B. Kagan, "Nonlinear Effects in Asymmetric Catalysis: A Personal Account," *Synlett* **2001**, 888–899.

M. T. Reetz, "Chelation or Non-Chelation Control in Addition Reactions of Chiral α- or β-Alkoxy Carbonyl Compounds," *Angew. Chem. Int. Ed. Engl.* **1984**, *23*, 556.

M. T. Reetz, "Structural, Mechanistic, and Theoretical Aspects of Chelation-Controlled Carbonyl Addition Reactions," *Acc. Chem. Res.* **1993**, *26*, 462–468.

M. T. Reetz, "New Approaches to the Use of Amino Acids as Chiral Building Blocks in Organic Synthesis," *Angew. Chem. Int. Ed. Engl.* **1991**, *30*, 1531–1546.

M. T. Reetz, "Synthesis and Diastereoselective Reactions of N,N-Dibenzylamino Aldehydes and Related Compounds," *Chem. Rev.* **1999**, *99*, 1121–1162.

J. Jurczak, A. Golebiowski, "Optically Active N-Protected α-Amino Aldehydes in Organic Synthesis," *Chem. Rev.* **1989**, *89*, 149–164.

E. L. Eliel, "Application of Cram's Rule: Addition of Achiral Nucleophiles to Chiral Substrates," in *Asymmetric Synthesis* (J. D. Morrison, Ed.), Vol. 2, 125, AP, New York, **1983**.

F. Vögtle, J. Franke, A. Aigner, D. Worsch, „Stereochemie in Stereobildern: Die Cramsche Regel – plastisch", *Chem. unserer Zeit*, **1984**, *18*, 203–210.

R. M. Pollack, "Stereoelectronic Control in the Reactions of Ketones and their Enolates," *Tetrahedron* **1989**, *45*, 4913.

E. L. Eliel, S. V. Frye, E. R. Hortelano, X. B. Chen, "Asymmetric Synthesis and Cram's (chelate) Rule," *Pure Appl. Chem.* **1991**, *63*, 1591–1598.

A. Mengel, O. Reiser, "Around and Beyond Cram's Rule," *Chem. Rev.* **1999**, *99*, 1191–1223.

10.6

Y. Yamamoto, "Addition of Organometallic Compounds to α,β-Unsaturated Carbonyl Compounds," in *Stereoselective Synthesis* (Houben-Weyl) 4th ed., **1996**, (G. Helmchen, R. W. Hoffmann, J. Mulzer, E. Schaumann, Eds.), **1996**, Vol. E21 (Workbench Edition), *4*, 2041–2067, Georg Thieme Verlag, Stuttgart.

G. H. Posner, "Conjugate Addition Reactions of Organocopper Reagents," *Org. React.* **1972**, *19*, 1–114.

E. Erdik, "Copper(I)-Catalyzed Reactions of Organolithiums and Grignard Reagents," *Tetrahedron* **1984**, *40*, 641.

R. J. K. Taylor, "Organocopper Conjugate Addition-Enolate Trapping Reactions," *Synthesis* **1985**, 364.

M. J. Chapdelaine, M. Hulce, "Tandem Vicinal Difunctionalization: β-Addition to α,β-Unsaturated Carbonyl Substrates Followed by β-Functionalization," *Org. React.* **1990**, *38*, 225–653.

J. A. Kozlowski, "Organocuprates in the Conjugate Addition Reaction," in *Comprehensive Organic Synthesis* (B. M. Trost, I. Fleming, Eds.), Vol. 4, 169, Pergamon Press, Oxford, **1991**.

B. H. Lipshutz, S. Sengupta, "Organocopper Reagents: Substitution, Conjugate Addition, Carbo/Metallocupration, and Other Reactions," *Org. React.* **1992**, *41*, 135–631.

R. A. J. Smith, A. S. Vellekoop, "1,4-Addition Reactions of Organocuprates with α,β -Unsaturated Ketones," in *Advances in Detailed Reaction Mechanisms* (J. M. Coxon, Ed.), Vol. 3, Jai Press, Greenwich, CT, **1994**.

N. Krause, A. Gerold, "Regio- and Stereoselective Syntheses with Organocopper Reagents," *Angew. Chem. Int. Ed. Engl.* **1997**, *36*, 186–204.

M. P. Sibi, S. Manyem, "Enantioselective Conjugate Additions," *Tetrahedron* **2000**, *56*, 8033– 8061.

S. Woodward, "Decoding the 'Black Box' Reactivity that is Organocuprate Conjugate Addition Chemistry," *Chem. Soc. Rev.* **2000**, *29*, 393–401.

E.-i. Nakamura, S. Mori, "Wherefore Art Thou Copper? Structures and Reaction Mechanisms of Organocuprate Clusters in Organic Chemistry," *Angew. Chem. Int. Ed. Engl.* **2000**, *39*, 3750–3771.

F. F. Fleming, Q. Wang, "Unsaturated Nitriles: Conjugate Additions of Carbon Nucleophiles to a Recalcitrant Class of Acceptors," *Chem. Rev.* **2003**, *103*, 2035–2077.

Further Reading

Kevin C. Cannon, Grant R. Krow, "Dihalide-Derived Di-Grignard Reagents: Preparation and Reactions," in *Handbook of Grignard Reagents* (G. S. Silverman, P. E. Rakita, Eds.), Marcel Dekker Inc., New York, **1996**, 497–526.

D. A. Hunt, "Michael Addition of Organolithium Compounds. A Review," *Org. Prep. Proced. Int.* **1989**, *21*, 705-749.

G. Boche, "The Structure of Lithium Compounds of Sulfones, Sulfoximides, Sulfoxides, Thioethers and 1,3-Dithianes, Nitriles, Nitro Compounds and Hydrazones," *Angew. Chem. Int. Ed. Engl.* **1989**, *28*, 277–297.

Y. Yamamoto, "Selective Synthesis by Use of Lewis Acids in the Presence of Organocopper and Related Reagents," *Angew. Chem. Int. Ed. Engl.* **1986**, *25*, 947.

E. Nakamura, "New Tools in Synthetic Organocopper Chemistry," *Synlett* **1991**, 539.

B. H. Lipshutz, "The Evolution of Higher Order Cyanocuprates," *Synlett* **1990**, *3*, 119.

B. H. Lipshutz, "Organocopper Reagents," in *Comprehensive Organic Synthesis* (B. M. Trost, I. Fleming, Eds.), Vol. 1, 107, Pergamon Press, Oxford, **1991**.

B. H. Lipshutz, "Synthetic Procedures Involving Organocopper Reagents," in *Organometallics. A Manual* (M. Schlosser, Ed.), 283, Wiley, Chichester, **1994**.

R. J. K. Taylor, "Organocopper Chemistry: An Overview," in *Organocopper Reagents: A Practical Approach* (R. J. K. Taylor, Ed.), 1, Oxford University Press, Oxford, U. K., **1994**.

R. J. K. Taylor, J. M. Herbert, "Compilation of Organocopper Preparations," in *Organocopper Reagents: A Practical Approach* (R. J. K. Taylor, Ed.), 307, Oxford University Press, Oxford, U. K., **1994**.

R. J. K. Taylor, "Organocopper Reagents: A Practical Approach," Oxford University Press, Oxford, U. K., **1995**.

N. Krause, A. Gerold, "Regioselective and Stereoselective Syntheses with Organocopper Reagents," *Angew. Chem. Int. Ed. Engl.* **1997**, *36*, 187–204.

E. Erdik, "Use of Activation Methods for Organozinc Reagents," *Tetrahedron* **1987**, *43*, 2203.

E. Erdik, "Transition Metal Catalyzed Reactions of Organozinc Reagents," *Tetrahedron* **1992**, *48*, 9577.

E. Erdik (Ed.), "Organozinc Reagents in Organic Synthesis," CRC, Boca Raton, FL, **1996**.

Y. Tamaru, "Unique Reactivity of Functionalized Organozincs," in *Advances in Detailed Reaction Mechanism. Synthetically Useful Reactions* (J. M. Coxon, Ed.), **1995**, *4*, JAI Press, Greenwich, CT.

P. Knochel, "Organozinc, Organocadmium and Organomercury Reagents," in *Comprehensive Organic Synthesis* (B. M. Trost, I. Fleming, Eds.), Vol. 1, 211, Pergamon Press, Oxford, **1991**.

P. Knochel, M. J. Rozema, C. E. Tucker, C. Retherford, M. Furlong, S. A. Rao, "The Chemistry of Polyfunctional Organozinc and Copper Reagents," *Pure Appl. Chem.* **1992**, *64*, 361–369.

P. Knochel, "Zinc and Cadmium: A Review of the Literature 1982–1994," in *Comprehensive Organometallic Chemistry II* (E. W. Abel, F. G. A. Stone, G. Wilkinson, Eds.), Vol. 11, 159, Pergamon, Oxford, UK, **1995**.

P. Knochel, "Preparation and Application of Functionalized Organozinc Reagents," in *Active Metals* (A. Fürstner, Ed.), 191, VCH, Weinheim, Germany, **1996**.

P. Knochel, J. J. A. Perea, P. Jones, "Organozinc Mediated Reactions," *Tetrahedron*, **1998**, *54*, 8275–8319.

P. Knochel, "Carbon-Carbon Bond Formation Reactions Mediated by Organozinc Reagents," in *Metal-catalyzed Cross-coupling Reactions* (F. Diederich, P. J. Stang, Eds.), Wiley-VCH, Weinheim, **1998**, 387–416.

P. Knochel, P. Joned (Eds.), "Organozinc Reagents: A Practical Approach," Oxford University Press, Oxford, U. K., **1999**.

M. Melnik, J. Skorsepa, K. Gyoryova, C. E. Holloway, "Structural Analyses of Organozinc Compounds," *J. Organomet. Chem.* **1995**, *503*, 1.

R. Bloch, "Additions of Organometallic Reagents to C:N Bonds: Reactivity and Selectivity," *Chem. Rev.* **1998**, *98*, 1407–1438.

A. Alexakis, "Asymmetric Conjugate Addition," in *Organocopper Reagents: A Practical Approach* (R. J. K. Taylor, Ed.), 159, Oxford University Press, Oxford, U. K., **1994**.

T. Ooi, K. Maruoka, "Carbonyl-Lewis Acid Complexes," in *Modern Carbonyl Chemistry* (J. Otera, Ed.), Wiley–VCH, Weinheim, **2000**, 1–32.

S. Saito, H. Yamamoto, "Carbonyl Recognition," in *Modern Carbonyl Chemistry* (J. Otera, Ed.), Wiley-VCH, Weinheim, **2000**, 33–67.

J. M. Coxon, R. T. Luibrand, "π-Facial Selectivity in Reaction of Carbonyls: A Computational Approach," in *Modern Carbonyl Chemistry*, (J. Otera, Ed.), Wiley-VCH, Weinheim, **2000**, 155–184.

F. Effenberger, "Cyanohydrin Formation," in *Stereoselective Synthesis* (Houben–Weyl) 4th ed., **1996**, (G. Helmchen, R. W. Hoffmann, J. Mulzer, E. Schaumann, Eds.), **1996**, Vol. E21 (Workbench Edition), *3*, 1817–1821, Georg Thieme Verlag, Stuttgart.

W. Nagata, M. Yoshioka, "Hydrocyanation of Conjugated Carbonyl Compounds," *Org. React.* **1977**, *25*, 255–476.

Conversion of Phosphorus- or Sulfur-Stabilized *C* Nucleophiles with Carbonyl Compounds: Addition-induced Condensations

11

11.1 Condensation of Phosphonium Ylides with Carbonyl Compounds: Wittig Reaction

11.1.1 Bonding in Phosphonium Ylides

It is possible to remove a proton from the methyl group of a tetramethylphosphonium halide with strong bases. Thereby, a betaine (see Section 4.7.3) is produced with the structure $Me_3P^{\oplus}–CH_2^{\ominus}$. A betaine in which the positive and the negative charges are located on adjacent atoms as in $Me_3P^{\oplus}–CH_2^{\ominus}$ is called an **ylide.** The “yl” part of the name ylide refers to the *covalent bond* in the substructure $P^{\oplus}–CH_2^{\ominus}$. The “ide” part indicates that it also contains an *ionic bond.* The ylide $Me_3P^{\oplus}–CH_2^{\ominus}$ is the parent compound of the **phosphonium ylides** or ***P* ylides** $R^1R^2R^3P^{\oplus}–CR^4R^{5\ominus}$. This terminology is used to distinguish them from other ylides like **ammonium ylides** or ***N* ylides** $R^1R^2R^3N^{\oplus}–CR^4R^{5\ominus}$ and **sulfur ylides** or ***S* ylides** $R^1R^2S^{\oplus}–CR^3R^{4\ominus}$. With strong bases it is also possible to remove a proton from the methyl group of triarylmethylphosphonium halides to form the ylides $Ar_3P^{\oplus}–CR^4R^{5\ominus}$.

Figure 11.1 shows the parent compound $Me_3P^{\oplus}–CH_2^{\ominus}$ of the phosphonium ylides as **A** and the most simple triarylphosphonium ylide, $Ph_3P^{\oplus}–CH_2^{\ominus}$, as **B**. The ionic representation of the ylides in Figure 11.1 represents only one of two conceivable resonance forms of such

$$\left\{ Me_3P^{\oplus}–\bar{C}H_2^{\ominus} \longleftrightarrow Me_3P{=}CH_2 \right\} \qquad \left\{ Ph_3P^{\oplus}–\bar{C}H_2^{\ominus} \longleftrightarrow Ph_3P{=}CH_2 \right\}$$

A (parent compound of the phosphonium ylides)

B (an especially important compo phosphonium ylide)

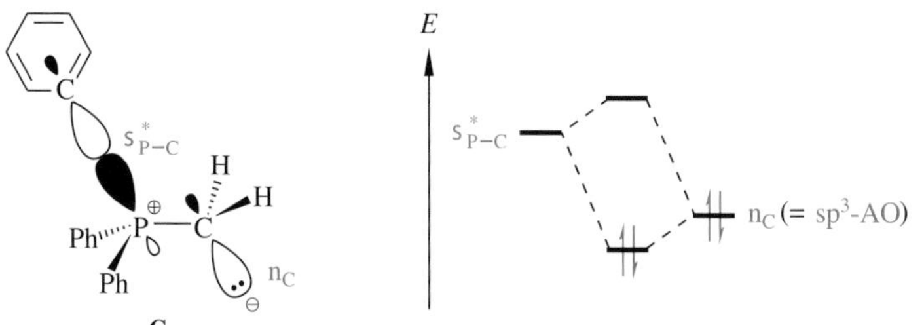

Fig. 11.1. Parent compound of phosphonium ylides (**A**) and a typical phosphonium ylide (**B**). Below: the $Ph_3P^{\oplus}–CH_2^{\ominus}$conformer, which is capable of maximum anomeric stabilization, and the corresponding MO diagram.

Bruckner R (author), Harmata M (editor) In: *Organic Mechanisms – Reactions, Stereochemistry and Synthesis*
Chapter DOI: 10.1007/978-3-642-03651-4_11, © Springer-Verlag Berlin Heidelberg 2010

species. The P atom in the center of *P* ylides may exceed its valence electron octet and share a fifth electron pair. For each *P* ylide one can therefore also write a resonance form that is free of formal charges (Figure 11.1) and is called an **ylene resonance form**. The "ene" part of the designation "ylene" refers to the double bond between the heteroatom and the deprotonated methyl group.

It should be remembered that the heteroatom under scrutiny belongs to the second full period of the periodic table of the elements. Thus, the double bond of an ylene form is not a p_π, p_π, but a d_π, p_π double bond. Though a d_π, p_π double bond is not very stable, the ylene resonance should make some contribution to the electron distribution in the ylides of Figure 11.1. One reason is that this resonance form must does not suffer from charge separation, as does the ylidic resonance form.

However, it seems more likely that the formal charge at the carbanionic center of these ylides is rather stabilized to a considerable extent by two other effects:

1. electron withdrawal due to the inductive (–I) effect of the heteroatom, and
2. stabilization of the electron pair on the carbanionic carbon atom by an anomeric effect.

The occurrence of an anomeric effect was discussed with the help of Figure 6.8 using the example of a structural element :Het^1—C—Het^2. In the ylides of Figure 11.1, the anomeric effect arises in an analogous substructure, namely, in :$CH_2^{\ominus}$–$PR_2^{\oplus}$–R. In the MO picture, this is due to an interaction between the electron pair in the non-bonding carbanionic sp^3 AO and an appropriately oriented vacant $\sigma^*_{\text{P–R}}$ orbital. Such an anomeric stabilization of an ylide is explained in Figure 11.1 using the example of the *P* ylide $CH_2^{\ominus}$–$PPh_2^{\oplus}$–Ph (**B**) (see Formula **C**). In order for this stabilization to be optimal, the crucial $P^{\oplus}$–R bond must be oriented *anti* with respect to the nonbonding sp^3 AO on the (pyramidalized) carbanionic center. This arrangement is in agreement with the findings of crystal structure analyses.

Phosphonium ylides are the crucial reagents in the **Wittig reaction,** which is a C,C-forming alkene synthesis starting from a phosphonium ylide $R^1R^2R^3P^{\oplus}$–$CR^4R^{5\ominus}$ and a carbonyl compound $R^6R^7C{=}O$. A short-lived four-membered ring intermediate is formed that decomposes to give an alkene $R^6R^7C{=}CR^4R^5$ and a phosphine oxide $R^1R^2R^3P{=}O$ (see the brief discussion in Section 4.7.3). In more than 99% of all Wittig reactions, ylides of the structure $Ph_3P^{\oplus}$–$CH^{\ominus}$–X are used (i.e., triphenylphosphonium ylides). Therein, X usually stands for H, alkyl, aryl, or CO_2-alkyl, though other substituents are possible.

11.1.2 Nomenclature and Preparation of Phosphonium Ylides

Most *P* ylides for Wittig reactions are prepared *in situ* and not isolated. Actually, they are always prepared *in situ* when the ylide $Ph_3P^{\oplus}$–$CH^{\ominus}$–X contains a substituent X which is unable to stabilize the negative formal charge of the carbanionic center. This type of *P* ylide is called a **nonstabilized ylide.** The so-called **semistabilized ylides** $Ph_3P^{\oplus}$–$CH^{\ominus}$–X, on the other hand, contain a substituent X which slightly stabilizes the carbanionic center. This type of ylide is also prepared *in situ.* The **stabilized ylides** are the third and last *P* ylide type. They carry a *strongly* electron-withdrawing substituent on the carbanionic carbon atom and are the only triphenylphosphonium ylides that are shelf-stable and can be stored. Therefore, they can be (and generally are) added as neat compounds to Wittig reactions.

Tab. 11.1 Triphenylphosphonium Ylides: Nomenclature, Preparation, and Stereoselectivity of Their Wittig Reactions

P-Ylide	$Ph_3\overset{\oplus}{P}-\overset{\ominus}{C}HAlkyl$	$Ph_3\overset{\oplus}{P}-\overset{\ominus}{C}HAryl$	$Ph_3\overset{\oplus}{P}-\overset{\ominus}{C}H-CO_2R$
Ylide type	non-stabilized ylide	semi-stabilized ylide	stabilized ylide
Ylide is prepared...	... *in situ* ...	... *in situ* ...	... in prior reactions ...
... prepared from $Ph_3\overset{\oplus}{P}-CHR\ Hal^{\ominus}$ and ...	*n*-BuLi or $Na^{\oplus\ominus}CH_2S(=O)CH_3$[1)] or $Na^{\oplus}\ NH_2^{\ominus}$[1)] or $K^{\oplus\ominus}O$*tert*-Bu[2)]	NaOEt or aqueous NaOH	aqueous NaOH
1,2 disubstituted alkenes typically result[3)]...	...with ≥ 90% *cis*-selectivity	... as *cis*-, *trans*-mixture	... with > 90% *trans*-selectivity

1) So-called salt-free Wittig reaction.
2) So-called high-temperature Wittig reaction, which takes place via the equilibrium fraction of the ylide.
3) For $Ph_3\overset{\oplus}{P}-\overset{\ominus}{C}-HAlkyl$ under conditions 1 or 2.

All *P* ylides for Wittig reactions are obtained by deprotonation of phosphonium salts. Depending on whether one wants to prepare a nonstabilized, a semistabilized, or a stabilized ylide, certain bases are especially suitable (see Table 11.1; an unusual, i.e., base-free, generation of ylides is described in Side Note 11.1). In stereogenic Wittig reactions with aldehydes, *P* ylides exhibit characteristic stereoselectivities. These depend mainly on whether the ylide involved is nonstabilized, semistabilized, or stabilized. This can also be seen in Table 11.1.

Side Note 11.1. Base-free Deprotonation of a Phosphonium Salt

Interestingly, no base is necessary for the phosphonium salt **A** and the ester-substituted C_5 aldehyde in Figure 11.2 to form vitamin A acetate (**B**) in a Wittig synthesis. Instead, the ylide formation is initiated through addition of 1,2-epoxybutane. This epoxide is ring-opened in a nucleophilic fashion by the bromide counterion of the phosphonium salt, thus forming the metal-free alkoxide **C**, which undertakes the deprotonation. In the case of the synthesis shown in Figure 11.2, working without bases in this manner is interesting if one needs to work under essentially neutral conditions. In the case at hand, both strong and weak bases might be problematic for the acetoxy group in the aldehyde. The secondary alkoxide **C**, which is generated from 1,2-epoxybutane, is more basic than NaOH or NaOEt, but more sterically hindered and thus too slow for the competing alcoholysis.

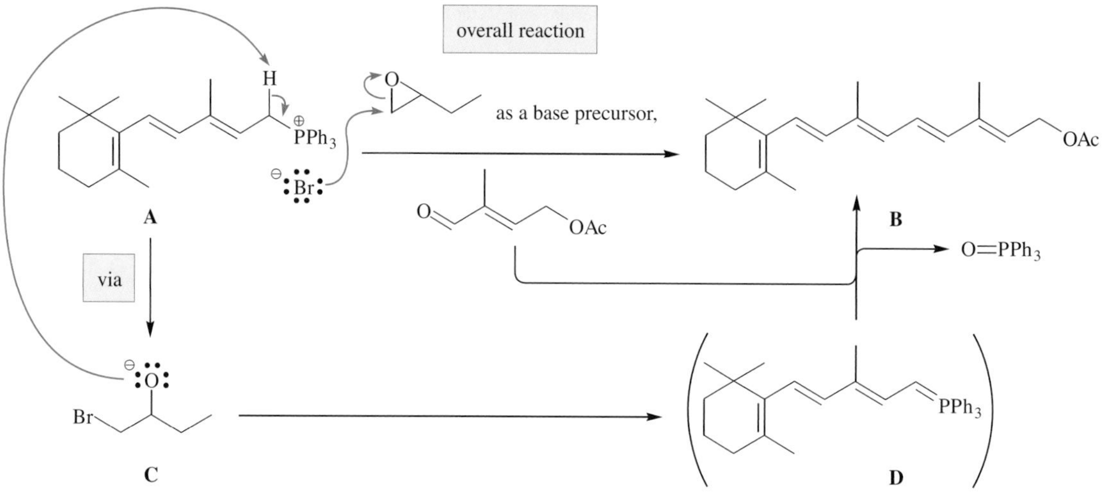

overall reaction

as a base precursor,

Fig. 11.2. Working without bases in the Wittig reaction with the (semi)stabilized phosphonium ylide **D**: *in-situ* formation of this reagent from the alkoxide **C** resulting from the S_N2 ring opening of butylenoxide through the bromide ion of the phosphonium salt **A**.

11.1.3 Mechanism of the Wittig Reaction

cis-Selective Wittig Reactions

We have a fairly detailed knowledge of the mechanism of the Wittig reaction (Figure 11.3). It starts with a one-step [2+2]-cycloaddition of the ylide to the aldehyde. This leads to a heterocycle called an oxaphosphetane. The oxaphosphetane decomposes in the second step—which is a one-step [2+2]-cycloreversion—to give triphenylphosphine oxide and an alkene. This decomposition takes place stereoselectively (cf. Figure 4.44): a *cis*-disubstituted oxaphosphetane reacts exclusively to give a *cis*-alkene, whereas a *trans*-disubstituted oxaphosphetane gives only a *trans*-alkene. The reaction is stereospecific.

The [2+2]-cycloaddition between *P* ylides and carbonyl compounds to give oxaphosphetanes can be stereogenic. It *is* stereogenic when the carbanionic C atom of the ylide bears—besides the P atom—two different substituents and when this holds true for the carbonyl group, too. The most important stereogenic oxaphosphetane syntheses of this type start from monosubstituted ylides $Ph_3P^{\oplus}$—$CH^{\ominus}$—X and from substituted aldehydes R—CH=O. We will therefore study this case in Figure 11.3.

The [2+2]-cycloaddition between the mentioned reaction partners to form an oxaphosphetane is not only stereogenic but frequently also exhibits a considerable degree of stereoselectivity. The latter is more precisely called **simple diastereoselectivity**.

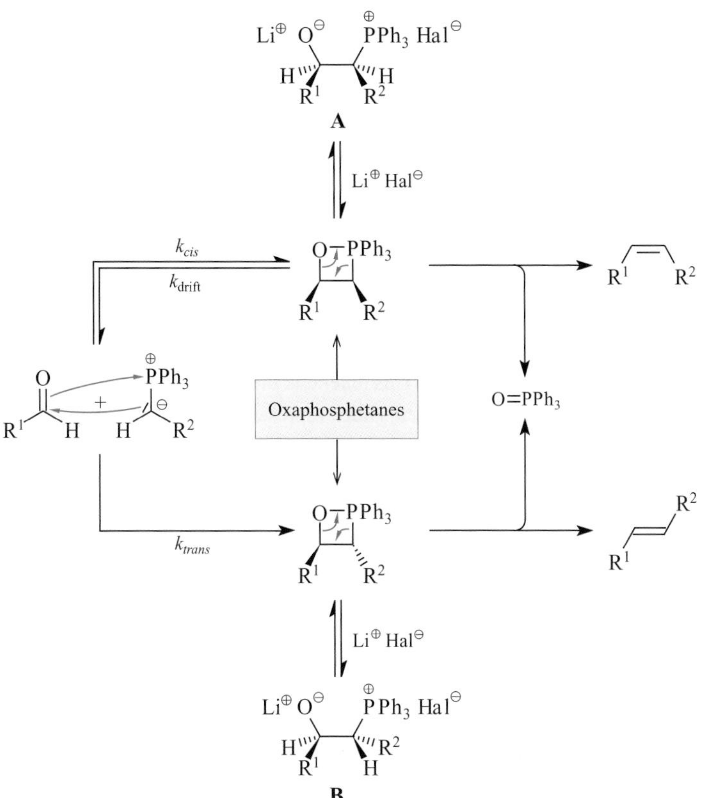

Fig. 11.3. Mechanism of the Wittig reaction. k_{cis} is the rate constant for the formation of the *cis*-oxaphosphetane, k_{trans} is the rate constant for the formation of the *trans*-oxaphosphetane, and k_{drift} is the rate constant for the isomerization of *cis*- to *trans*-configured oxaphosphetane, which is called "stereochemical drift."

Simple Diastereoselectivity

By simple diastereoselectivity we mean the occurrence of diastereoselectivity when bond formation between the C atom of a C=X double bond and the C atom of a C=Y double bond establishes a stereocenter at each of these C atoms.

The simple diastereoselectivity of the formation of oxaphosphetanes depends mainly on which ylide type from Table 11.1 is used.

Stereogenic Wittig reactions of nonstabilized ylides of the structure $Ph_3P^{\oplus}$–$CH^{\ominus}$–R^2 have been studied in-depth in many instances. They give the *cis*-configured oxaphosphetane rapidly with the rate constant k_{cis} (Figure 11.3). On the other hand, the same nonstabilized ylide produces the *trans*-oxaphosphetane slowly, with the rate constant k_{trans}. The primary product of the [2+2]-cycloaddition of a nonstabilized *P* ylide to a substituted aldehyde therefore is a *cis*-oxaphosphetane. Why this is so has not been ascertained despite the numerous suggestions about details of the mechanism that have been made. However, one rationalization suggests that with unstabilized ylides, oxaphosphetane formation proceeds through an early transition state. Minimization of steric interactions between the aldehyde alkyl substituent and a phosphorus substituent gives rise to a *cis*-oxaphosphetane.

The initially obtained *cis*-oxaphosphetane can subsequently isomerize irreversibly to a *trans*-oxaphosphetane with the rate constant $k_{drift} \times [k_{trans}/(k_{trans} + k_{cis})]$ (Figure 11.3). This isomerization is referred to as **stereochemical drift**.

The stereochemical drift in Wittig reactions of nonstabilized ylides can be suppressed if these reactions are carried out in the absence of Li salts ("salt-free"). However, Li salts are necessarily present in the reaction mixture when the respective nonstabilized ylide is prepared from a phosphonium halide and a lithium-containing base such as *n*-BuLi or LDA. In order to avoid a stereochemical drift in this case, the ylide must be prepared from a phosphonium halide and Na- or K-containing bases, for example, by using $NaNH_2$, KO-*tert*-Bu, KHMDS (K-hexamethyldisilazide, the potassium analog of LiHMDS, whose structural formula is shown in Figure 4.18). The latter deprotonation conditions initiate Wittig reactions, which in the literature are referred to as **"salt-free" Wittig reactions**.

Under salt-free conditions, the *cis*-oxaphosphetanes formed from nonstabilized ylides can be kept from participating in the stereochemical drift and left intact until they decompose to give the alkene in the terminating step. This alkene is then a pure *cis*-isomer. In other words, salt-free Wittig reactions of nonstabilized ylides represent a stereoselective synthesis of *cis*-alkenes.

However, according to the data in Figure 11.4 (first row of figures), the previous remark applies without limitation only to Wittig olefinations of *saturated* aldehydes. Indeed, α,β-unsaturated, α,β—C≡C-containing, or aromatic aldehydes have a tendency to react similarly, but the achievable *cis*-selectivities are lower. However, almost optimum *cis*-selectivities are also obtained with these unsaturated aldehydes when they are treated not with the common triphenyl-substituted *P* ylides but instead with its unusual tris(*ortho*-tolyl) analogs (Figure 11.4; second row of figures).

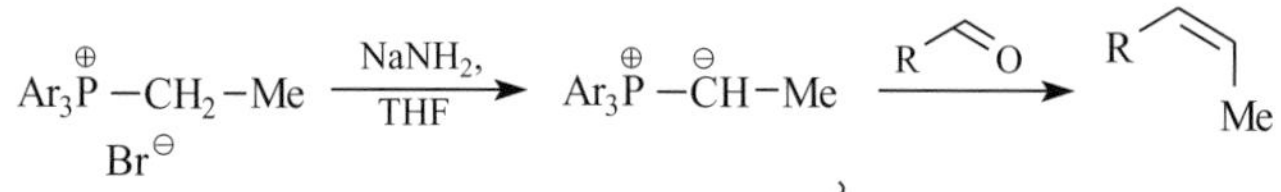

$$Ar_3\overset{\oplus}{P}-CH_2-Me\ Br^{\ominus} \xrightarrow[THF]{NaNH_2,} Ar_3\overset{\oplus}{P}-\overset{\ominus}{C}H-Me \xrightarrow{R-CHO} R-CH=CH-Me\ (cis)$$

% *cis*-alkene for ...	... R = Pent	Pr–CH=CH–	Ph	Pr–C≡C–
Ar = Ph	96	90	92	84
Ar = *o*-Tolyl	98	97	96	95

Fig. 11.4. Optimum *cis*-selectivities of Wittig olefinations of different aldehydes with nonstabilized ylides under "salt-free" conditions.

The *P* ylides react with C=O double bonds faster the more electrophilic these C=O double bonds are. One can therefore occasionally olefinate aldehydes even in the presence of ketones. If one works salt-free, this is also accomplished with *cis*-selectivity (Figure 11.5).

The C=O double bond of esters is usually not electrophilic enough to be olefinated by *P* ylides. Only formic acid esters can undergo condensation with $Ph_3P^{\oplus}–CH_2^{\ominus}$ and then they give enol ethers of the structure H—C(=CH_2)—OR. α,β-Unsaturated esters can sometimes react with *P* ylides, but this then results in a cyclopropanation:

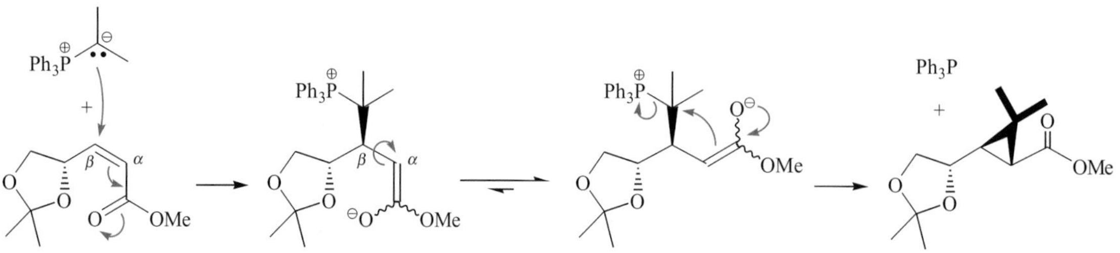

Ph₃P⊕ (ylide) + α,β-unsaturated ester (OMe) → A (Ph₃P⊕, β, α, ⊖O, OMe) ⇌ B (Ph₃P⊕, O⊖, OMe) → Ph₃P + cyclopropane ester (OMe)

A **B**

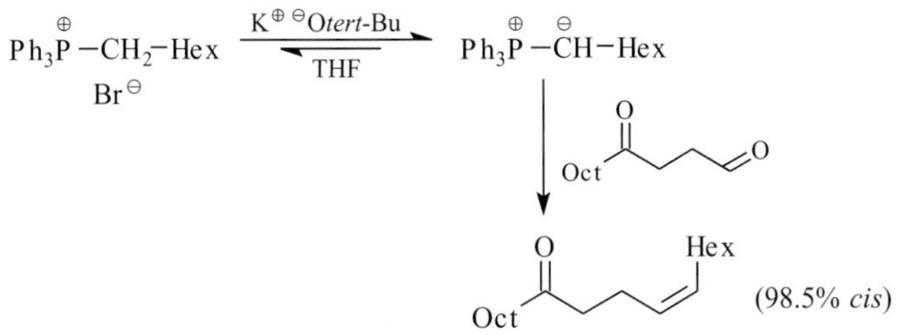

Fig. 11.5. Chemoselective and stereoselective Wittig olefination with a nonstabilized ylide.

As show above, $Ph_3P^{\oplus}$–$CMe_2^{\ominus}$ undergoes an initial 1,4-addition forming an enolate. The fact that the addition here—due to induced diastereoselectivity (for the term cf. Section 3.3.3)—generates a *uniformly* configured stereocenter is of preparative interest but irrelevant for the mechanistic understanding of cyclopropane formation. The initial Michael addition affords the enolate as conformer **A,** which undergoes rapid 180° rotation around the C^{β}–C^{α}H=C(–$O^{\ominus}$)(OMe) bond, minimizing van der Waals repulsion. The resulting enolate conformer **B** undergoes an intramolecular S_N2 reaction. The enolate carbon as a nucleophile expels the leaving group triphenylphosphane. The preferential conformation **B** of the enolate translates into a *trans*-configuration of the resulting cyclopropane.

Wittig Reactions without Stereoselectivity

If nonstabilized *P* ylides react with carbonyl compounds in the presence of Li salts, i.e., not under salt-free conditions, several changes take place in the mechanism of Figure 11.3. First, the *cis*-oxaphosphetanes are produced with a lower selectivity; that is, in contrast to the salt-free case, here the rate constant k_{cis} is not much greater than k_{trans} Secondly, some of the initially formed *cis*-oxaphosphetane has an opportunity to isomerize to give the *trans*-oxaphosphetane—that is, to undergo the stereochemical drift—before an alkene is produced. However, *cis* → *trans*-isomerization is not complete. For these reasons, the alkene ultimately obtained is a *cis/trans* mixture. Because such a mixture is normally useless, aldehydes are usually not treated with nonstabilized *P* ylides under non-salt-free conditions. On the other hand, non-salt-free Wittig reactions of nonstabilized ylides can be used without disadvantage when no stereogenic double bond is produced:

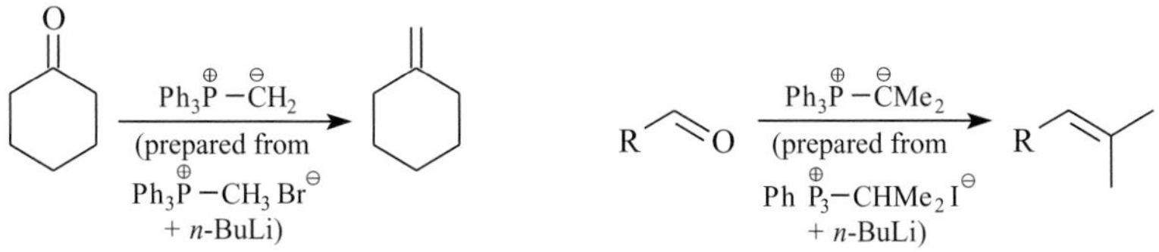

There is a third Li effect under non-salt-free conditions. At first sight, it is only relevant mechanistically. Li salts are able to induce a heterolysis of the O–P bond of oxaphosphetanes. They thereby convert oxaphosphetanes into the so-called lithiobetaines (formulas **A** and **B** in Figure 11.3). The disappearance of the ring strain and the gain in $Li^{\oplus}O^{\ominus}$ bond energy more than compensate for the energy required to break the P–O bond and to generate a cationic and an anionic charge.

Beware of thinking that the occurrence of the lithiobetaines **A** and **B** must have stereochemical implications. Until fairly recently, lithium-*free* betaines were incorrectly considered intermediates in the Wittig reaction. Today, it is known that lithium-*containing* betaines are formed in a dead-end side reaction. They must revert back to an oxaphosphetane—which occurs with retention of the configuration—before the actual Wittig reaction can continue.

trans-Selective Wittig Reactions

A different reaction mode of lithiobetaines is used in the **Schlosser variant** of the Wittig reaction. Here, too, one starts from a nonstabilized ylide and works under non-salt-free conditions. However, the Schlosser variant is an olefination which gives a pure *trans*-alkene rather than a *trans,cis* mixture. The experimental procedure looks like magic at first:

$$Ph_3\overset{\oplus}{P}-CH_2-Me\ Br^{\ominus} \xrightarrow[\text{PhLi/LiBr; HCl; KO}tert\text{-Bu}]{\text{PhLi/LiBr; RCH=O;}} R\text{-CH=CH-Me}$$

	% *trans*-alkene
for R = Pent	99
R = Ph	97

What is going on? First, the nonstabilized ylide is prepared by deprotonating the phosphonium halide with PhLi. For this purpose the deprotonating agent is best prepared according to the equation PhBr + 2 Li → PhLi + LiBr as a 1:1 mixture with LiBr. When the ylide, which then contains *two* equivalents of LiHal, is added to an aldehyde, a *cis,trans* mixture of the oxaphosphetanes is produced. However, because of the considerable excess of lithium ions, these oxaphosphetanes are immediately and completely converted to the corresponding lithiobetaine, diastereomers **A** and **C** (Figure 11.6). For kinetic and thermodynamic reasons, this ring opening of the oxaphosphetanes takes place faster and more completely, respectively, when two equivalents of LiHal are present than if only one equivalent were present. In fact, the first objective of the Schlosser olefination is to convert the starting materials into lithiobetaines as fast and completely as possible in order to minimize fragmentation of the initially formed oxaphosphetane mixture, which would result in the formation of a mixture of *cis* and *trans* alkenes.

The lithiobetaines **A** and **C** in Figure 11.6 are phosphonium salts. As such, they contain an acidic H atom in the position α to the P atom. By adding a second equivalent of the previously mentioned PhLi/LiBr reagent to the reaction mixture, deprotonation occurs. A single **oxido ylide** (**B**) is produced from each of the diastereomeric lithiobetaines **A** and **C**. At this point, exactly 1.0 equivalent of HC1 is added to protonate the oxido ylide. This produces a lithiobetaine again, but this time it is obtained diastereomerically pure as compound **A**. This species reacts very cleanly to give the alkene when it is treated with KO*tert*-Bu. This treatment establishes an equilibrium reaction in which the Li^+ ion migrates from the lithiobetaine **A** to the *tert*-BuO$^{\ominus}$ ion. This creates a betaine **D**, which lacks the stabilizing Li$^{\oplus}$ ion. However, as you have learned, lithium-free betaines are relatively unstable. Consequently, **D** collapses exergonically to the oxaphosphetane **E.** Of course, the stereocenters in **E** have the same configuration as the stereocenters in its precursor molecule **D** and the same configuration as in lithio-

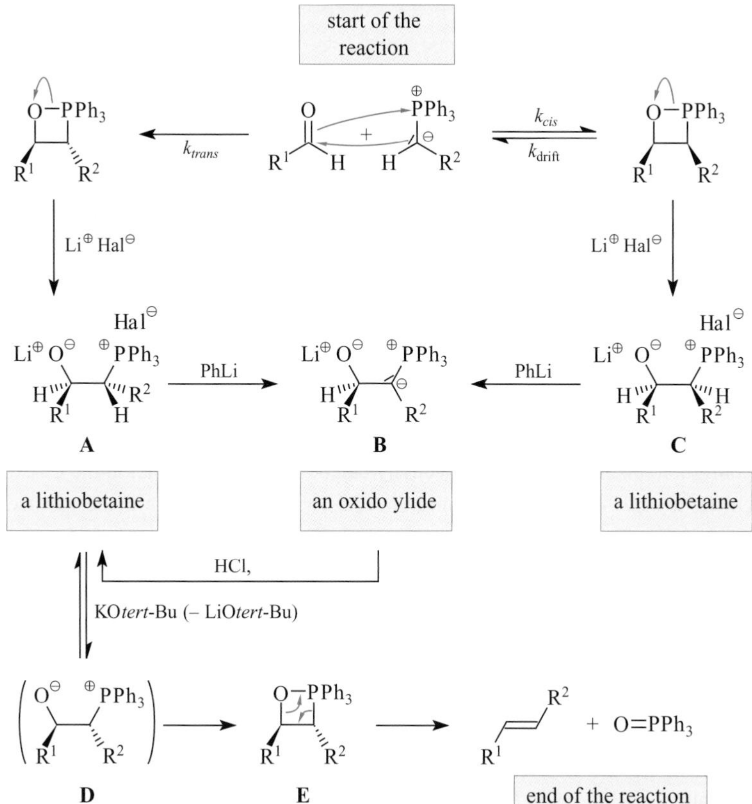

Fig. 11.6. Mechanism of the Schlosser variant of the Wittig reaction of nonstabilized ylides.

betaine **A.** The oxaphosphetane **E** is therefore uniformly *trans*. Accordingly, the decomposition of **E** gives the pure *trans*-alkene in the last step of the Schlosser variant of the Wittig olefination. Why is **B** protonated so stereoselectively?

Semistabilized ylides generally react with aldehydes to form mixtures of *cis*- and *trans*-oxaphosphetanes before the decomposition to the alkene starts. Therefore, stereogenic reactions of ylides of this type usually give alkene mixtures regardless of whether the work is carried out salt-free or not.

On the other hand, stabilized ylides react with aldehydes almost exclusively via *trans*-oxaphosphetanes. Initially, a small portion of the *cis*-isomer may still be produced. However, all the heterocyclic material isomerizes very rapidly to the *trans*-configured, four-membered ring through an especially pronounced stereochemical drift. Only after this point does the

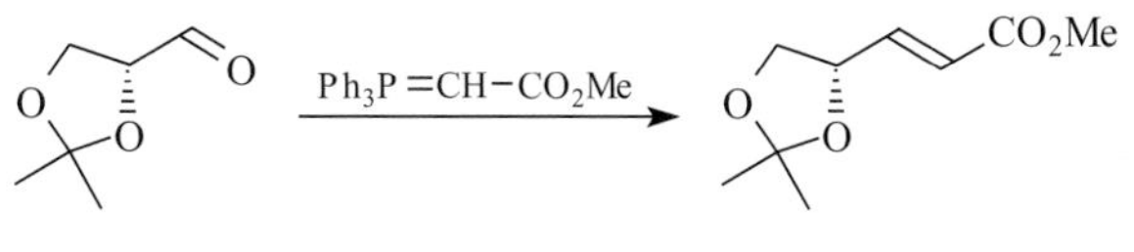

$Ph_3P{=}CH{-}CO_2Me$

Fig. 11.7. *trans*-Selective Wittig olefination of aldehydes I—Preparation of a *trans*-configured α,β-unsaturated ester (preparation of the starting material: Figure 17.24).

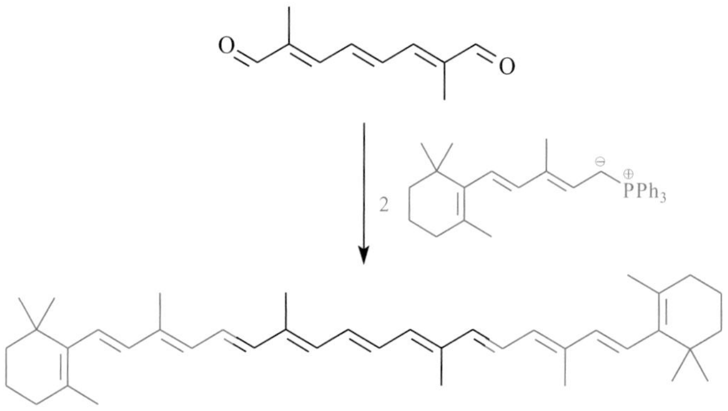

Fig. 11.8. *trans*-Selective Wittig olefination of aldehydes II—Synthesis of β-carotene from a dialdehyde. The ylide used here is already known from Figure 11.2. In a way, it is "(semi)stabilized" since it is prepared *in situ* like a semistabilized phosphonium ylide, but reacts as *trans*-selectively as a stabilized ylide.

[2+2]-cycloreversion start. It leads to triphenylphosphine oxide and an acceptor-substituted *trans*-configured alkene. This *trans*-selectivity can be used, for example, in the C_2 extension of aldehydes to *trans*-configured α,β-unsaturated esters (Figure 11.7) or in the *trans*-selective synthesis of polyenes such as β-carotene (Figure 11.8).

The same *trans*-selectivity can be observed with the condensation of aldehydes with phosphonium ylides of the type $Ph_3P^{\oplus}–CH^{\ominus}–C(=O)$–alkyl or $Ph_3P^{\oplus}–CH^{\ominus}–C(=O)–H$, which is illustrated in Figure 11.9, using as an example the preparation of the α,β-unsaturated ketone **C** and the preparation of the α,β-unsaturated aldehyde **D**, although the latter does not work that smoothly (see below).

As far as their applicability is concerned, Wittig reactions producing α,β-unsaturated carbonyl compounds (cf. Figure 11.9) are fundamentally different from Wittig reactions leading to α,β-unsaturated carboxylic acid esters (cf. Figure 11.7) or fully conjugated polyenes (cf. Figure 11.8). The products of the Wittig reaction mentioned first, i.e., carbonyl compounds, are potential *new starting materials* for a Wittig reaction. Accordingly, a Wittig reaction producing an aldehyde or a ketone cannot easily be carried out as a selective monocondensation. The question is whether the product will compete with the starting material for the reagent. Remember the Wittig reaction in Figure 11.5? There an aldehydic carbonyl group undergoes olefination by a *nonstabilized* phosphonium ylide much more rapidly than a ketonic carbonyl group. Ketones are even less reactive with stabilized ylides than they are with nonstabilized ylides. This leads to the following:

(1) Ketones are never suited for use in Wittig reactions with $Ph_3P^{\oplus}–CH^{\ominus}–C(=O)H$ since the condensation product would be an aldehyde that would more rapidly undergo a second Wittig olefination than the residual unreacted ketone for the first time.

(2) Ketones are sometimes suitable as substrates for Wittig reactions with $Ph_3P^{\oplus}–CH^{\ominus}–C(=O)Me$. This ylide would yield an α,β-unsaturated ketone as the condensation product, which should be less reactive than the starting material due to conjugation, assuming the starting ketone is not more sterically hindered than the product. Frankly, ketones are hardly ever used with stabilized ylides so this is more a thought experiment than anything else.

(3) Aldehydes selectively react with $Ph_3P^{\oplus}–CH^{\ominus}–C(=O)Me$ once because a ketone is produced and it is less reactive than the starting aldehyde.

(4) In general, aldehydes are suited for use in Wittig reactions with $Ph_3P^{\oplus}–CH^{\ominus}–C(=O)H$ only under carefully controlled reaction conditions. When the starting material is a saturated

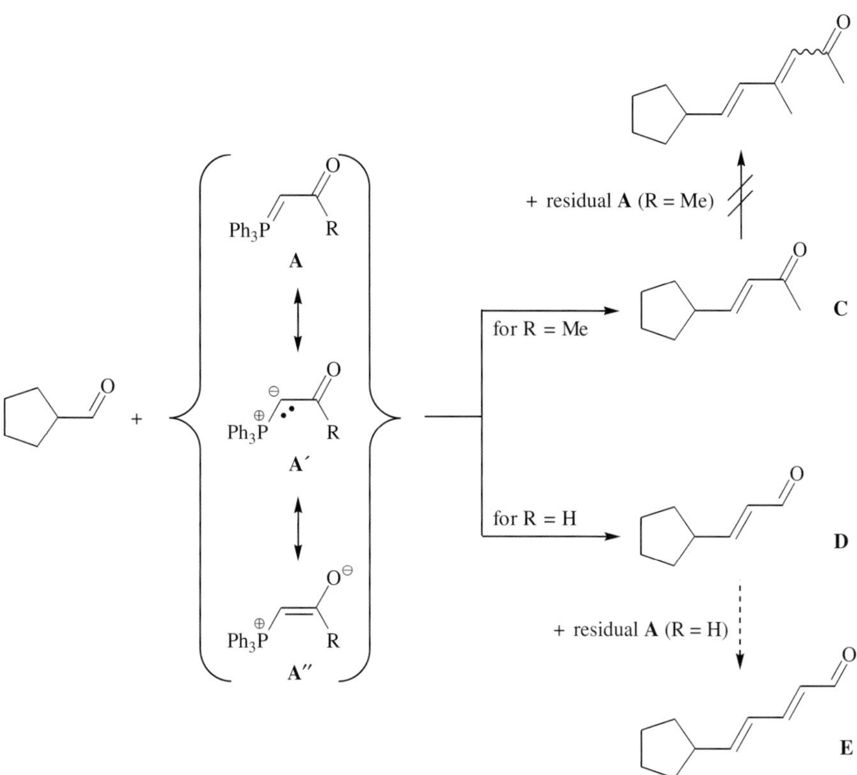

Fig. 11.9. *trans*-Selective Wittig olefination of aldehydes III—Preparation of a *trans*-configured α,β–unsaturated ketone. The analogous preparation of *trans*-configured α,β–unsaturated aldehydes can also be managed, but only when the reaction's progress is carefully monitored.

aldehyde, the electrophilicity in the resulting aldehyde usually is sufficiently reduced through the conjugation of the C=O- and the C=C double bond so that more mono- than repeated condensation takes place. Accordingly, the main product in the reaction at the bottom of Figure 11.9 is the monocondensation product **D**, with the biscondensation product **E** occurring as a minor by-product. If the starting material had been an α-oxygenated aldehyde, its electrophilicity would have exceeded the electrophilicity of the product aldehyde to such an extent that a monocondensation would have taken place *exclusively*.

11.2 Wittig–Horner Reaction

A Wittig-Horner reaction or Wittig–Horner olefination is a C,C-bond-forming alkene synthesis starting from an alkyldiphenylphosphine oxide R^1–CH_2–P(=O)Ph_2 and an aldehyde R^2–CHO. The phosphine oxide is deprotonated with *n*-BuLi in a position α to the phosphorus, and the resulting phosphorus-stabilized organolithium compound is added to the carbonyl group of the aldehyde. After oxaphosphetane formation and fragmentation, an alkene $R^1HC{=}CHR^2$ and

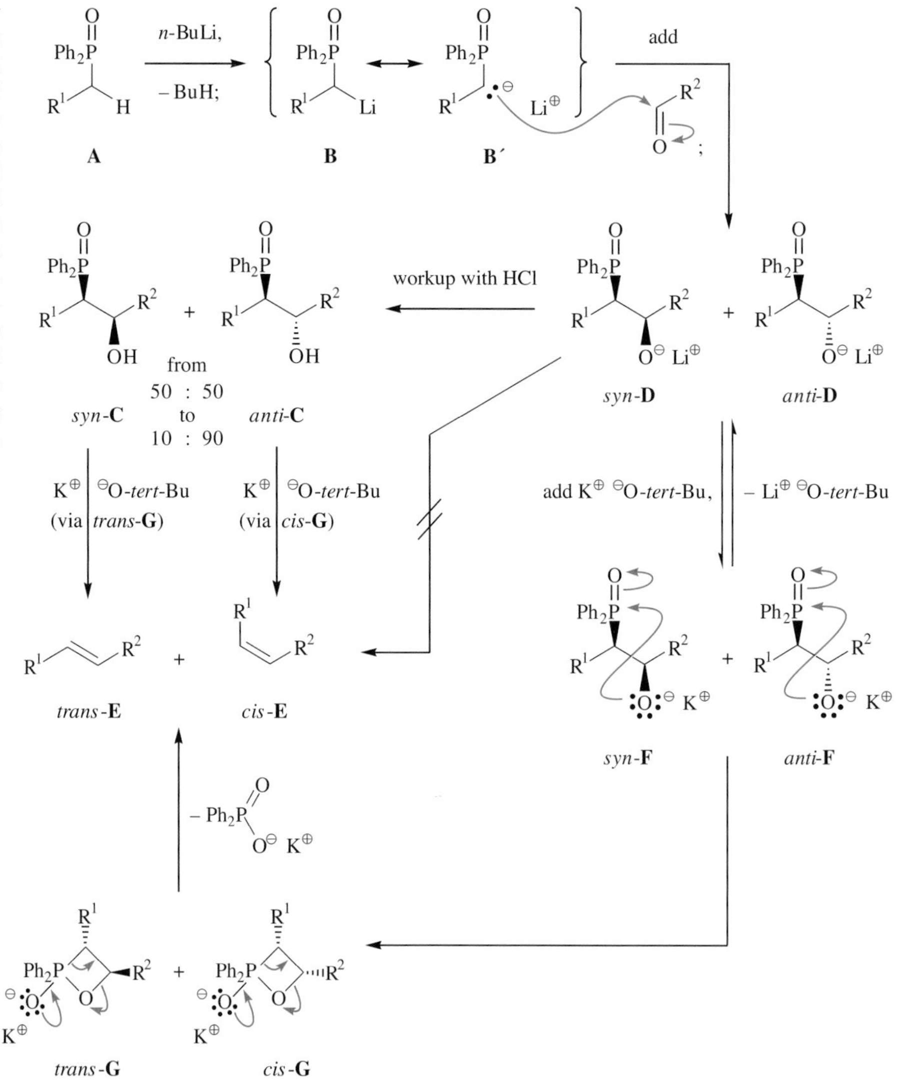

Fig. 11.10. Wittig–Horner synthesis of alkenes with stereogenic C=C double bond via condensation of lithiated phosphine oxides **B** ↔ **B′** with aldehydes. In the two-step reaction the initially formed addition products *syn*- and *anti*-**D** are protonated and the resulting alcohols *syn*- and *anti*-**C** are isolated and separated; the suitable diastereomer is then stereoselectively converted into the desired alkene isomer. The one-step reaction leads to the same alkene in a *cis,trans* mixture.

an alkali metal salt of diphenylphosphinic acid are formed. The mechanistic details are presented in Figure 11.10. As you can see, the last step of the Wittig–Horner reaction is the same as the last step of the Wittig reaction, stereospecific decomposition of an oxaphosphetane (**G**). The most noticeable mechanistic difference between the Wittig–Horner and the Wittig reaction is that the oxaphosphetane is formed in several steps instead of just a single step.

The nucleophilic addition of the α-lithiated alkyldiphenylphosphine oxide **B** to the carbonyl group of an aldehyde at the beginning of a Wittig–Horner reaction results in the phosphorylated lithium alkoxide **D**. If the alkene synthesis is carried out in a single step, the $Li^{\oplus}$ of the intermediate **D** is, without workup, reversibly replaced by $K^{\oplus}$ by adding potassium-*tert*-butoxide. In this way, the phosphorylated potassium alkoxide **F** is made available. Only in **F** is the alkoxide oxygen nucleophilic enough to add to the adjacent P=O double bond, thereby forming the oxaphosphetane **G**. This intermediate decomposes to give the alkene **E** as well as its water-soluble and thus conveniently separable accompanying product, i.e., potassium diphenylphosphinate.

The Wittig-Horner reaction performed in a *single* operation, as described, is flawed by the lack of stereocontrol. That's because the phosphorylated lithium alkoxide **D** is formed without diastereoselectivity. So the alkene **E** is also formed without diastereoselectivity—via a *syn*:*anti*-mixture of the phosphorylated potassium alkoxides **F** and a *trans*:*cis*-mixture of the oxaphosphetane **G**, all of which are essentially 1:1 mixtures of diastereomers. It is generally not easy to separate a cis alkene from its trans isomer.

Single alkene diastereomers are accessible through a Wittig–Horner reaction only if it is performed in two steps (Figure 11.10). A 1:1 mixture of the phosphorylated lithium alkoxides *syn*- and *anti*-**D** is still formed but if the mixture is protonated at this point, the resulting phosphorylated alcohol diastereomers **C** can usually be separated without difficulty. The suitable diastereomer will be deprotonated with potassium-*tert*-butoxide in the second step and then be converted into the stereouniform *trans*- or *cis*-alkene **E** via stereospecific oxaphosphetane formation and fragmentation.

Side Note 11.2. An Oblique Approach to the Wittig–Horner Olefination

Two-step Wittig–Horner reactions may lead to pure *trans*- and pure *cis*-alkenes, as shown. But this procedure comes at the expense of a lower than 50% yield since nearly half of the starting material is lost in form of the "incorrectly" configured phosphorylated alcohol **C** (see Figure 11.10). After a mixture of *syn*-C and *anti*-**C** has been obtained, as shown in Figure 11.10, we can turn to the stereocontrolled alkene synthesis in Figure 11.11. For this purpose, the *syn*-**C**/*anti*-**C** mixture needs to be oxidized to form the α-phosphorylated ketone **B**. Alternatively, the same ketone can be obtained by acylating the α-lithium derivative of the Wittig–Horner alkyldiphenylphosphine oxide **A** with an ester R^2–CO_2Et (Figure 11.11, top row). The stoichiometry of the acylation differs from the one of the formally analogous synthesis of ketophosphonate in Figure 6.48. In Figure 11.11, complete turnover is achieved with only one equivalent of the metalated phosphorus reagent, while in Figure 6.48 two equivalents are required. Why? Addition and decomposition of a tetrahedral intermediate will give a relatively acidic β-keto phosphine oxide. That should react with the organolithium via deprotonation. Well, the tetrahedral intermediate in the acylation **A** → **B** in Figure 11.11 is so stable that its break down is not very fast. This is due to the stability of the $Ph_2P{=}O/Li^{\oplus}/C{-}C{-}O^{\ominus}$ bridge in the tetrahedral intermediate of Figure 11.11, which is higher than that of a potential $(EtO)_2P{=}O/Na^{\oplus}/C{=}C{-}O^{\ominus}$ bridge in the tetrahedral intermediate of the acylation in Figure 6.48 (not shown). (Could you have predicted this? It is not likely that anyone could have.) Being α-chiral carbonyl compounds, the phosphorylated ketones **B** can be diastereoselectively reduced according to the principles outlined in Section 10.3.2: to form the *syn*-configured phosphorylated alcohols **D** under Felkin-Anh control, and to give its *anti*-diastereomers **E** under chelation control. Subsequent treatment with potassium-*tert*-butoxide leads us back to the Wittig–Horner reaction path (Figure 11.10). Starting from **D** this reaction provides the pure *trans*-alkene; starting from **E** the pure *cis*-alkene is obtained. In each case the yield is often much higher than the $<$ 50% mentioned above and it really depends on the stereoselectivity of the reduction step.

Fig. 11.11. Wittig–Horner synthesis of stereouniform alkenes via ketophosphine oxide **B**. The reaction proceeds via its Felkin-Anh-selective or chelate-controlled reduction to form the *syn*-configured hydroxyphosphine oxides **D** and the *anti*-configured hydroxyphosphine oxides **E**. **D** and **E** continue to react—after deprotonation with KO-*tert*-Bu—via a *syn*-elimination to give the *trans*- and *cis*-alkene, respectively. R^1 in the formula **A**–**C** corresponds to a primary (*prim*-alkyl) or a secondary alkyl residue (*sec*-alkyl).

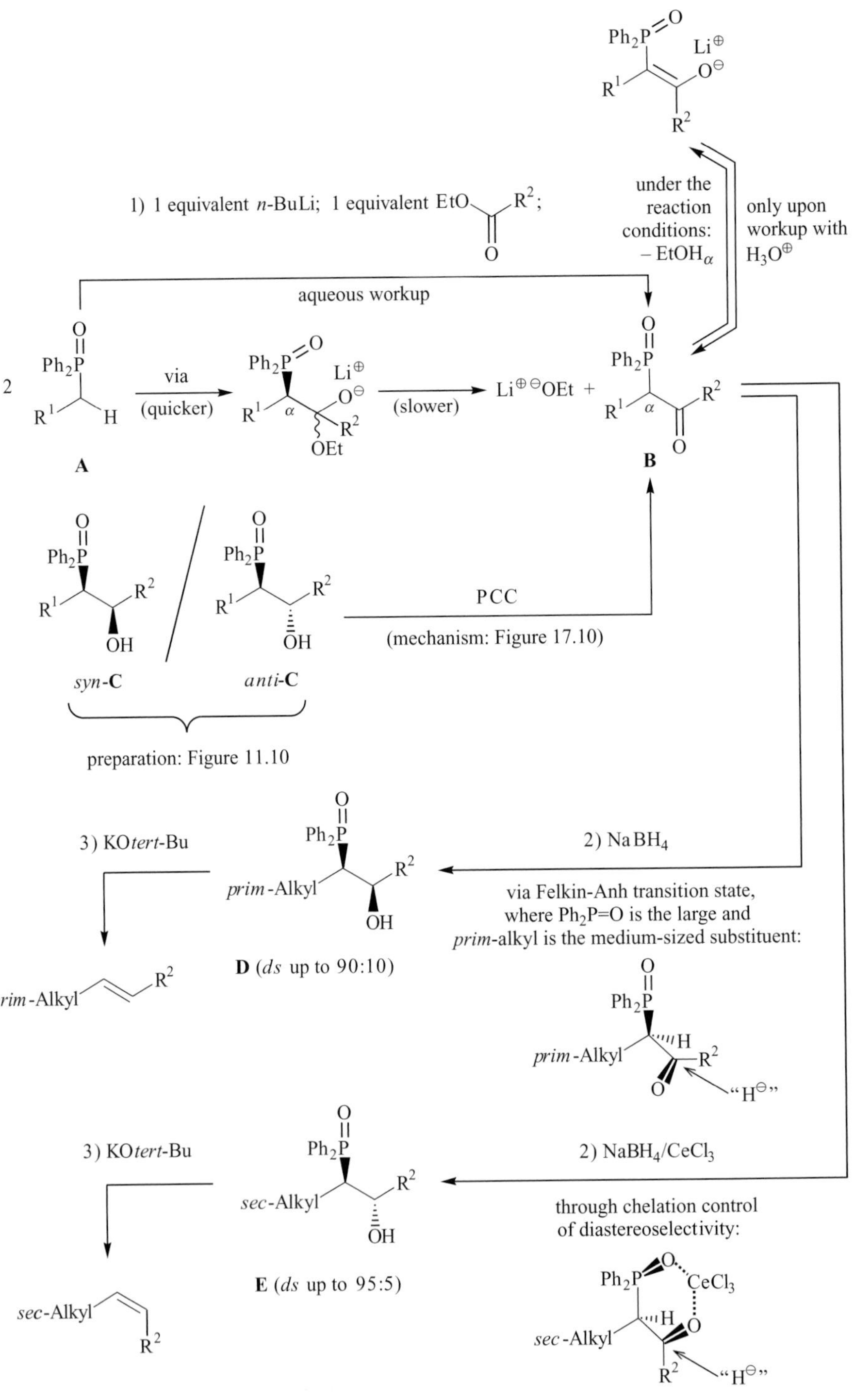

11.3 Horner–Wadsworth–Emmons Reaction

The Horner–Wadsworth–Emmons reaction represents a methodologically more important and more commonly used supplement to the Wittig reaction (cf. Section 11.1.3) than the Wittig–Horner reaction (Section 11.2).

11.3.1 Horner–Wadsworth–Emmons Reactions Between Achiral Substrates

Horner–Wadsworth–Emmons reactions are C=C-forming condensation reactions between the Li, Na, or K salt of a *β*-keto- or an *α*-(alkoxycarbonyl)phosphonic acid dialkyl ester and a carbonyl compound (see Figure 4.46). These reactions furnish *α*,*β*-unsaturated ketones or *α*,*β*-unsaturated esters, respectively, as the desired products and a phosphoric acid diester anion as a water-soluble by-product. In general, starting from aldehydes, the desired compounds are produced *trans*-selectively or, in the case of alkenes with trisubstituted C=C double bonds, *E*-selectively.

The precursors for these Horner–Wadsworth–Emmons reagents are *β*-ketophosphonic acid dialkyl esters or *α*-(alkoxycarbonyl)phosphonic acid dialkyl esters. The first type of compound, i.e., a *β*-ketophosphonic acid dialkyl ester is available, for example, by acylation of a metalated phosphonic acid ester (Figure 6.48). The second type of compound, i.e., an *α*-(alkoxycarbonyl)phosphonic acid dialkyl ester, can be conveniently obtained via the Arbuzov reaction (Figure 11.12).

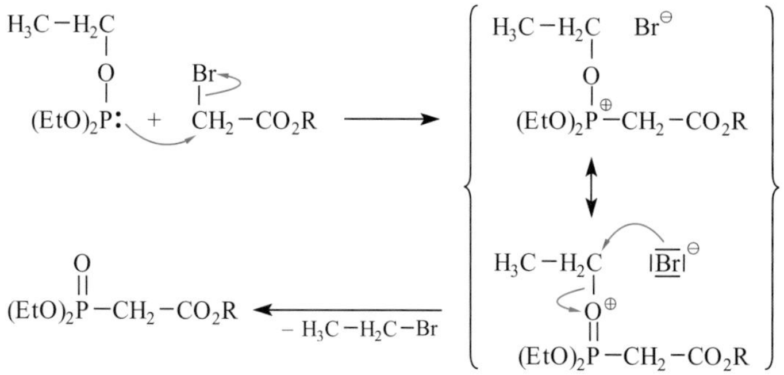

$$(EtO)_2P{:}{-}O{-}CH_2{-}CH_3 + Br{-}CH_2{-}CO_2R \longrightarrow \left\{ (EtO)_2\overset{\oplus}{P}(O{-}CH_2{-}CH_3){-}CH_2{-}CO_2R \;\; Br^{\ominus} \longleftrightarrow (EtO)_2P({=}\overset{\oplus}{O}{-}CH_2{-}CH_3){-}CH_2{-}CO_2R \;\; |\overline{Br}|^{\ominus} \right\} \xrightarrow{-\,H_3C-H_2C-Br} (EtO)_2P({=}O){-}CH_2{-}CO_2R$$

Fig. 11.12. The Arbuzov reaction, which provides the most important access to Horner–Wadsworth–Emmons reagents: a sequence of two S_N2 reactions (cf. Figure 2.32).

Condensations between aldehydes and metalated phosphonic acid dialkyl esters other than those mentioned previously are also referred to as Horner–Wadsworth–Emmons reactions. Nevertheless, in these esters, too, the carbanionic center carries a substituent with a pi electron withdrawing group, for example, an alkenyl group, a polyene or a C≡N group. The Horner–Wadsworth–Emmons reactions of these reagents are also stereoselective and form the new C=C double bond *trans*-selectively.

The mechanism of Horner–Wadsworth–Emmons reactions has not been definitively established. A contemporary rationalization is shown in Figure 11.13 for the reaction between an

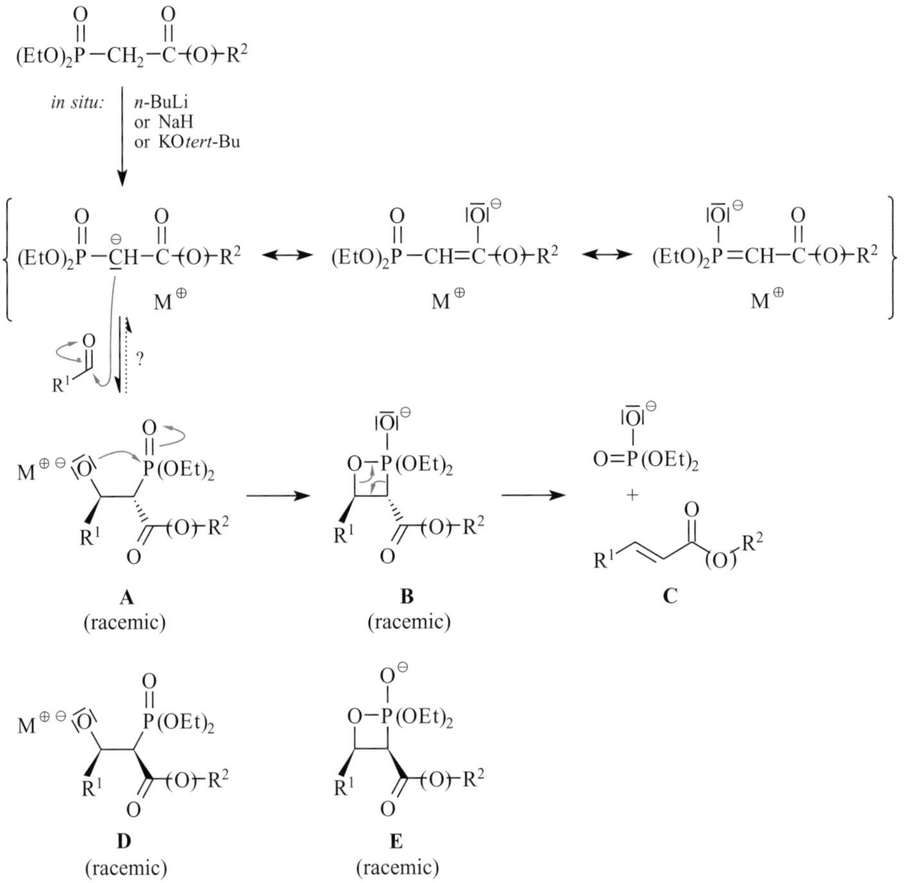

Fig. 11.13. The currently assumed mechanism for the Horner–Wadsworth–Emmons reaction.

aldehyde and a metalated phosphonic acid diester, whose carbanionic center is conjugated with a keto group or an ester function. The phosphonate ions from Figure 11.13 have a delocalized negative charge. It is not known whether they react as conformers in which the O=P—C$^{\ominus}$—C=O substructure is U-shaped and bridged through the metal ion in a six-membered chelate. In principle, this substructure could also have the shape of a sickle or a W. It is also not known for certain whether this reaction leads to the oxaphosphetane **B** in two steps through the alkoxide **A** or directly in a one-step cycloaddition. It is not even known whether the oxaphosphetane is formed in a reversible or an irreversible reaction.

The formal analogy to the Wittig reaction (Figure 11.3) would perhaps suggest a one-step formation of the oxaphosphetane **B** in Figure 11.13. On the other hand, the presumably closer analogy to the Wittig–Horner reaction (Section 11.2) would argue for a two-step formation of this oxaphosphetane. If the Horner–Wadsworth–Emmons reaction takes place analogously, the phosphonate ion from Figure 11.13 would first react with the aldehyde to form the alkoxide **A**, which would then cyclize to give the oxaphosphetane **B**. The decomposition of this heterocycle by a one-step [2 + 2]-cycloreversion (cf. Figure 4.46) would finally lead to the *trans*-alkene **C**.

Why *trans*-selectivity occurs is not known because of the lack of detailed knowledge about the mechanism. Perhaps the reason is that only the alkoxide **A** is cyclized to the more stable *trans*-oxaphosphetane shown. This is conceivable because the diastereomeric alkoxide (**D** in Figure 11.13) should cyclize comparatively slowly to the less stable *cis*-oxaphosphetane **E** if product development control were to occur in this step. It would thus be possible that both the alkoxide **A** and its diastereomer **D** form unselectively, but reversibly, from the phosphonate ion and the aldehyde. Then an irreversible cyclization of the alkoxide **A** would give the *trans*-oxaphosphetane **B**. The alkoxide **D** would also gradually be converted into the *trans*-oxaphosphetane **B** through the equilibrium **D** $\rightleftharpoons$ starting materials $\rightleftharpoons$ **A.**

According to Figure 11.13, Horner–Wadsworth–Emmons reactions always produce α,β-unsaturated *ketones* that are *trans*- or *E*-configured as the major product. In general, this also applies for similarly prepared α,β-unsaturated *esters* (Figure 11.14, left). However, an apparently innocent structural variation in the phosphonic ester moiety of metalated α-(alkoxycarbonyl)phosphonates completely reverses the stereochemistry of the formation of α,β-unsaturated esters. For example, the replacement of the $H_3C—CH_2—O$ groups by $F_3C—CH_2—O$ groups is the main feature of the **Still-Gennari variant** of the Horner–Wadsworth–Emmons reaction. It makes *cis*-substituted acrylic esters and *Z*-substituted methacrylic esters accessible (Figure 11.14, right).

The stereostructure of the alkoxide intermediate of a Horner–Wadsworth–Emmons reaction that affords the *trans*-alkene was shown in Figure 11.13 (as formula **A**). The Still–Gennari variant of this reaction (Figure 11.14) must proceed via an alkoxide with the inverse stereostructure because an alkene with the opposite configuration is produced. According to Figure 11.15, this alkoxide is a 50:50 mixture of the enantiomers **C** and *ent*-**C.** Each of these enantiomers contributes equally to the formation of the finally obtained *cis*-configured acrylic ester **D**.

The **Ando variant** is an alternative to the Still–Gennari variant of the Horner–Wadsworth–Emmons reaction. Here, phosphonates are employed that contain two aryloxy residues, for example, the *ortho*-tolyloxy residues of the phosphonate **A**. The Ar–O groups in this reaction

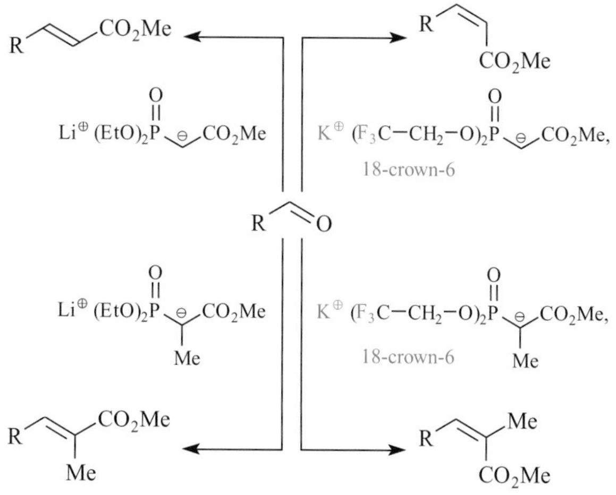

Fig. 11.14. Preparation of *trans*- or *E*-configured α,β-unsaturated esters by the Horner–Wadsworth–Emmons reaction (left) or preparation of their *cis*- or *Z*-isomers by the Still–Gennari variant of it (right). 18-Crown-6 is a so-called crown ether containing a saturated 18-membered ring that is made up from six successive $—CH_2—O—CH_2$-units. 18-Crown-6 dissociates the $K^{\oplus}$ ions of the Horner–Wadsworth–Emmons reagent by way of complexation.

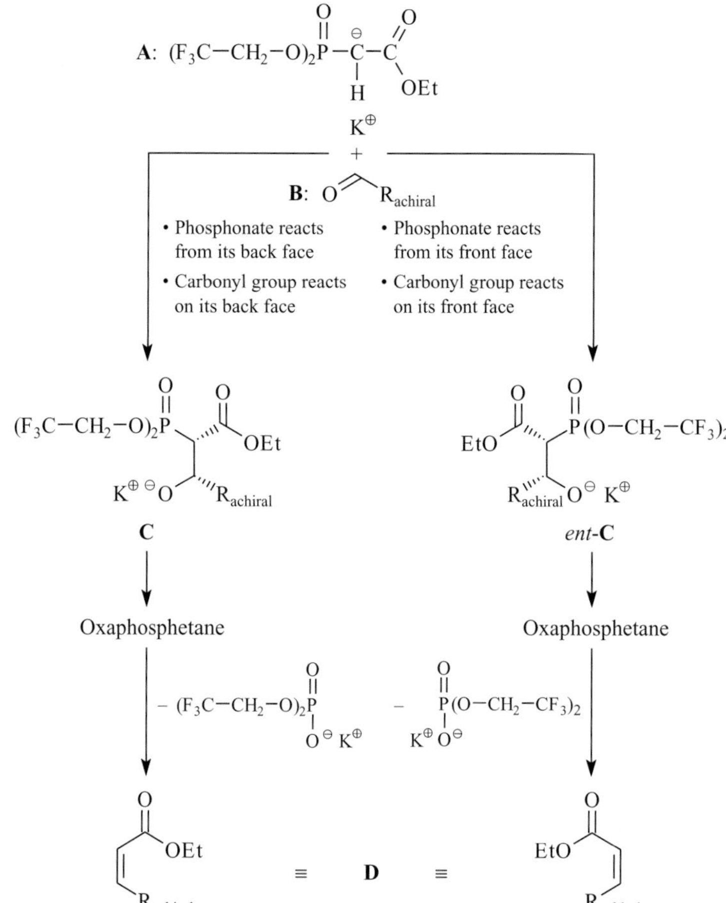

Fig. 11.15. Analysis of the overall stereoselectivity of a Still–Gennari olefination such as the one in Figure 11.13; simple diastereoselectivity of the formation of the alkoxide intermediate from the achiral phosphonate **A** and the achiral aldehyde **B.** For both reagents the terms "back face" and "front face" refer to the selected projection.

exert a similar electron-withdrawing inductive effect as the $F_3C{-}CH_2{-}O$ groups in the variant of Still and Gennari *and* influence the stereochemistry in the same way:

NaI, DBU; Hept–CHO

A → (*Z*)-Hept–CH=CH–C(=O)OEt *cis* : *trans* = 93:7

B

In the example shown the deprotonation of the phosphonate can be achieved under much milder conditions than with the usually employed sodium hydride, which by the way also applies to the standard Horner–Wadsworth–Emmons reagents. Incorporation of the two doubly bound oxygen atoms of the reagent **A** into sodium chelate **B** does the trick. Being a cation, **B** is much more acidic than **A** and can thus already be deprotonated by an amidine.

11.3.2 Horner–Wadsworth–Emmons Reactions between Chiral Substrates: A Potpourri of Stereochemical Specialties

Let us now consider the stereostructures **C**/*ent*-**C** of the two enantiomeric Still–Gennari intermediates of Figure 11.15 from another point of view. The simple diastereoselectivity (see Section 11.1.3) with which the phosphonate **A** and the aldehyde **B** must combine in order for the alkoxides **C** and *ent*-**C** to be produced is easy to figure out. If we use the formulas as written in the figure, this simple diastereoselectivity can be described as follows: the phosphonate ion **A** and the aldehyde **B** react with each other in such a way that a back $\text{face}_{\text{phosphonate}}$/back $\text{face}_{\text{aldehyde}}$ linkage (formation of alkoxide **C**) and a front $\text{face}_{\text{phosphonate}}$/front $\text{face}_{\text{aldehyde}}$ linkage (formation of alkoxide *ent*-C) take place *concurrently*.

The Still–Gennari reactions of Figures 11.16–11.21 start from similar substrates as those shown in Figure 11.15. However, at least one of them is chiral. Since each of these Still–Gennari olefinations is *cis*-selective, it is clear that, when their reactants combine to form an alkoxide intermediate, they must do so with the same type of simple diastereoselectivity that we have just unraveled for the Still–Gennari reaction of Figure 11.15: either the back faces of both reactants are linked or the front faces are linked. However, with regard to how important the back face/back face linkage is in comparison to the front face/front face linkage, these various Still–Gennari reactions will differ from one another. We will study this in each individual reaction of Figures 11.16–11.21 to see how each of these two possible linkages is energetically influenced by the steric or electronic characteristics of the substrates involved. This influence stems from the stereocenters in the *chiral* substrates of Figures 11.16–11.21. They, of course, may exert reagent control or substrate control of the diastereoselectivity.

Figure 11.16 shows what happens when the achiral aldehyde **B** from Figure 11.15 is olefinated with an enantiomerically pure chiral phosphonate ion **A**. A *cis*-alkene (formula **D**) is again formed preferentially, and in principle it can again be produced in two ways: via the alkoxide **C** or via its diastereomer *iso*-**C**. If only a simple diastereoselectivity as in Figure 11.15 were involved, this would mean that these alkoxides would be produced in a 50:50 ratio. However, there is, a reagent control of the diastereoselectivity since the phosphonate ion **A** of Figure 11.16 is chiral. The reagent control stems from the shielding of the front face of this phosphonate in its presumed reactive conformation **A** by the protruding phenyl ring. This shielding makes a front $\text{face}_{\text{phosphonate}}$ reaction virtually impossible. Therefore, this phosphonate ion can combine with aldehyde **B** only at the back face of the phosphonate. For this to occur, the simple diastereoselectivity of the Horner–Wadsworth–Emmons reaction requires a back $\text{face}_{\text{phosphonate}}$/back $\text{face}_{\text{aldehyde}}$ linkage. The alkoxide intermediate of the reaction of Figure 11.16 is thus mainly compound **C.**

We now analyze the Still–Gennari reaction of Figure 11.17. The reagents there are the enantiomerically pure chiral phosphonate **A**, with which you are familiar from Figure 11.16,

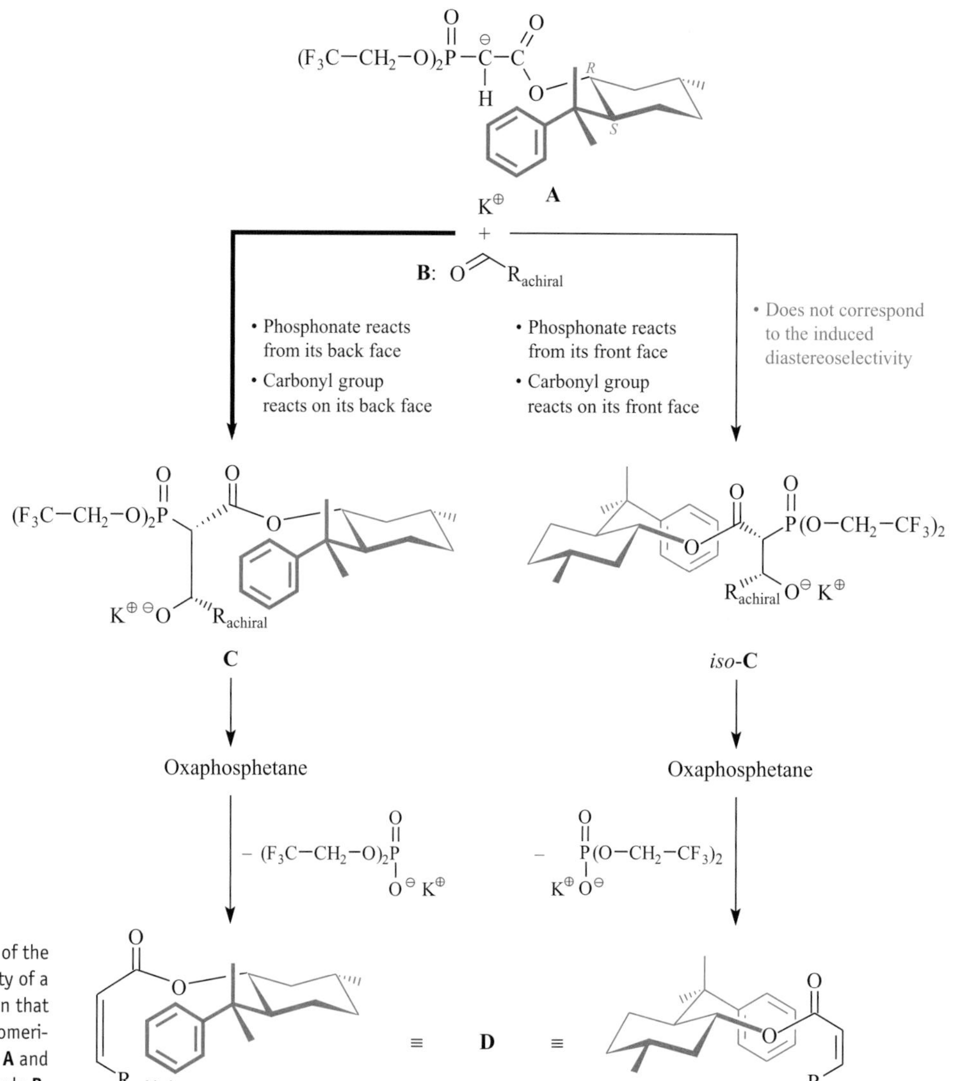

Fig. 11.16. Analysis of the simple diastereoselectivity of a Still–Gennari olefination that starts from the enantiomerically pure phosphonate **A** and the achiral aldehyde **B.**

and an enantiomerically pure α-chiral aldehyde **B**. The diastereoselectivity of the formation of the crucial alkoxide intermediate(s) is in this case determined by the interplay of three factors:

- the simple diastereoselectivity (as it already occurred in the olefinations of Figures 11.15 and 11.16);
- the reagent control of the diastereoselectivity originating from phosphonate **A** (which you know from the olefination of Figure 11.16), and
- the substrate control of the diastereoselectivity, which originates from aldehyde **B** and makes a Felkin–Anh transition state preferred in the reaction of the phosphonate ion on the C=O double bond (compare Figure 10.16, middle).

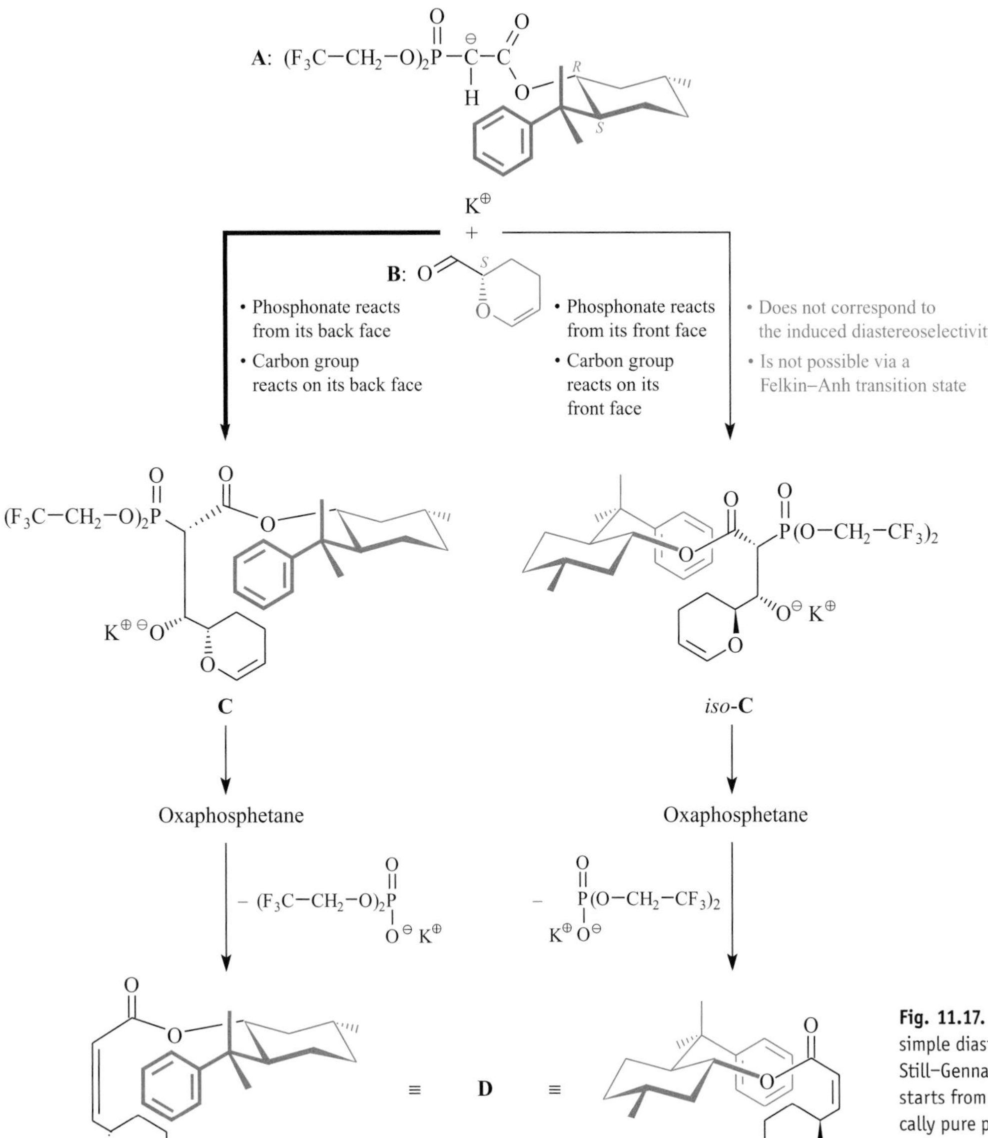

Fig. 11.17. Analysis of the simple diastereoselectivity of a Still–Gennari olefination that starts from the enantiomerically pure phosphonate **A** and the enantiomerically pure α-chiral aldehyde **B.**

When reagent control and substrate control occur in one reaction *at the same time*, this is a case of double stereodifferentiation (see section 3.4.4). In a matched substrate/reagent pair, both effects act in the same direction, whereas in a mismatched substrate/reagent pair they act against each other. In the Still–Gennari olefination of Figure 11.17, we have a case of double stereodifferentiation, and a matched substrate/reagent pair is realized. Reagent *and* substrate control cooperate and, in combination with the simple diastereoselectivity, cause the reaction to proceed practically exclusively via the alkoxide **C.**

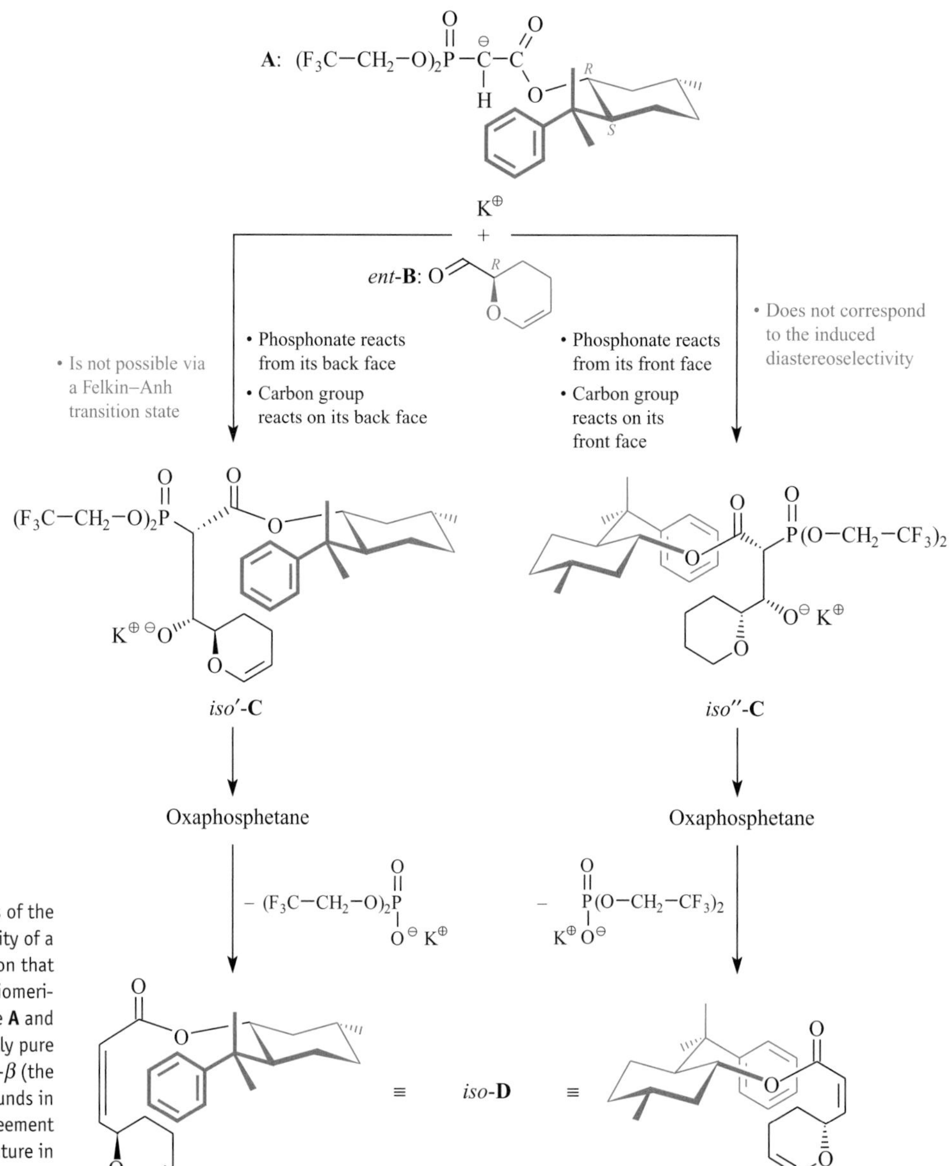

Fig. 11.18. Analysis of the simple diastereoselectivity of a Still–Gennari olefination that starts from the enantiomerically pure phosphonate **A** and the enantiomerically pure α-chiral aldehyde *ent-β* (the naming of the compounds in this figure is in agreement with the nomenclature in Figure 11.17).

The mismatched case of this Still–Gennari olefination, can be found in Figure 11.18. There, reagent control and substrate control act against each other. The substrate in Figure 11.18 is the enantiomer *ent*-**B** of the aldehyde used in Figure 11.17. Although the olefination product of Figure 11.18 must be *iso*-**D**, it is impossible to predict by which of the two conceivable routes it is formed: alkene *iso*-**D** can only be produced if either the substrate control (through formation of the alkoxide intermediate *iso*'-**C**) or the reagent control of the diastereoselectivity (through formation of the alkoxide intermediate *iso*″-**C**) is violated. It is

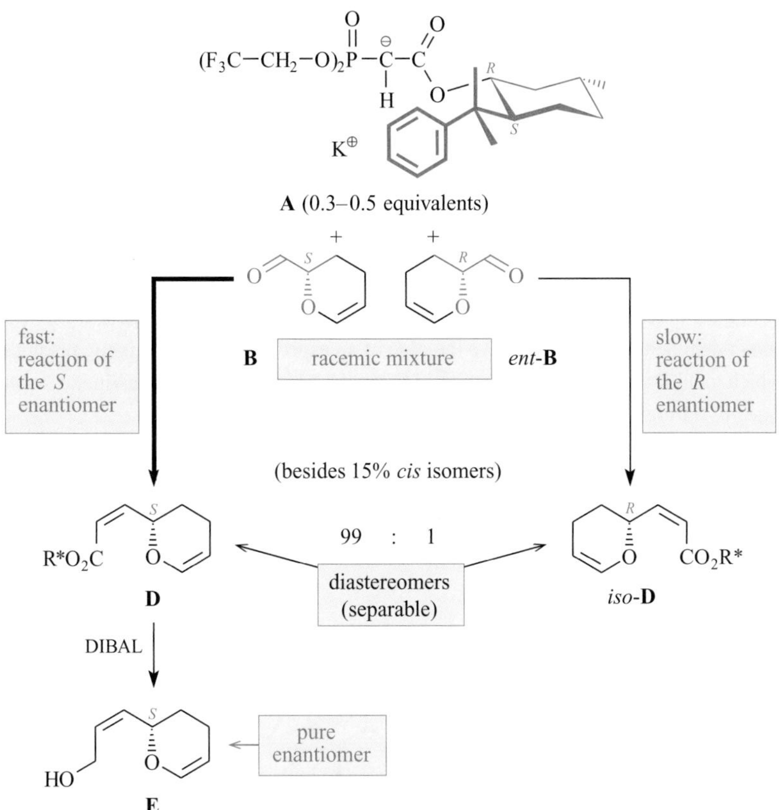

Fig. 11.19. Still-Gennari olefination of a racemic α-chiral aldehyde with an enantiomerically pure phosphonate as kinetic resolution I—Loss of the unreactive enantiomer *ent*-**B** of the aldehyde (R^* stands for the phenylmenthyl group in the Horner-Wadsworth-Emmons products; the naming of the products in this figure is in agreement with the nomenclature of Figures 11.17 and 11.18).

not known which of these alkoxide intermediates is less disfavored and is therefore formed less slowly. One thing is certain: The alkene *iso*-**D** is produced in a considerably slower reaction than the alkene **D** of Figure 11.17.

Let us express the outcomes of Figures 11.17 and 11.18 once more in a different way. The *S*-configured chiral aldehyde **B** is olefinated by the enantiomerically pure phosphonate **A** considerably faster than its *R*-enantiomer *ent*-**B.** This can be exploited when 0.5 equivalents of the enantiomerically pure phosphonate **A** are treated with a 1:1 mixture of the aldehydes **B** and *ent*-**B**—that is, treated with the racemic aldehyde (Figure 11.19). The *S*-enantiomer **B** reacts with the phosphonate **A** in a matched pair (as per Figure 11.17), hence fast and consequently virtually completely. On the other hand, the *R*-enantiomer *ent*-**B** of the aldehyde is doesn't react (almost) because it forms a mismatched pair with the phosphonate (according to Figure 11.18). Thus, the result is a kinetic resolution, and the *cis*-esters formed in Figure 11.19 are formed in a 99:1 ratio from the *S*- or *R*-aldehyde, respectively. Since these esters represent a 99:1 mixture of diastereomers—**D** and *iso*-**D**—they can be purified chromatographically to give the pure *S*-ester **D.** The chiral auxiliary can be removed by a reduction with DIBAL (method: Figure 17.63). In this way one obtains the allyl alcohol **E** *in 100% enantiomeric purity.*

A completely analogous kinetic resolution succeeds with the Still-Gennari olefination of Figure 11.20. Here the racemic substrate is a different α-chiral aldehyde. It carries a sulfonamido group at the stereocenter in the α-position. Nucleophiles also add to the carbonyl group of such an aldehyde preferentially (cf. Figure 10.42) via a Felkin–Anh transition state (Figure 10.16, middle). Thus, in the Horner–Wadsworth–Emmons reaction of this aldehyde, exactly as in the analogous one from Figure 11.19, the *S*-enantiomer is olefinated faster than the *R*-enantiomer. However, the α-chiral aldehyde in Figure 11.20 differs from the one in Figure 11.19 in that it racemizes under the reaction conditions. It is therefore not only the initially available 50% of the reactive *S*-enantiomer that can be olefinated. A major fraction of the *R*-aldehyde, which is not olefinated to a significant extent (only 13% of the *R*-alkene's *cis-iso-* and *trans-iso-***C** are produced), isomerizes to the *S*-aldehyde and is subsequently *also* converted to the *S*-alkene *cis*-**C.** The latter's yield is 73% based on the racemate but is 146% based on the *S*-aldehyde used. Here we have the case of an especially efficient kinetic resolution—one which includes the enantiomerization of the undesired antipode. Such a kinetic resolution is termed **dynamic kinetic resolution.**

The Still-Gennari olefination in Figure 11.21 is recommended to anyone who wants to enjoy a third stereochemical delicacy. The substrate is a dialdehyde that contains oxygenated stereocenters in both α-positions. Nevertheless, this aldehyde is achiral because it has a mirror plane and thus represents a *meso*-compound. *Meso*-compounds can sometimes be con-

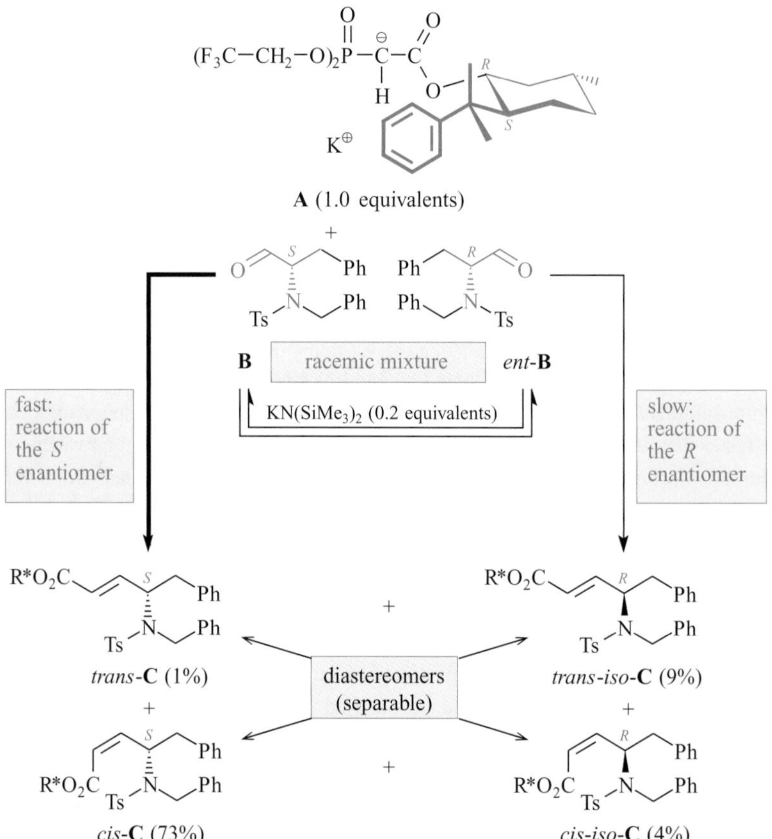

Fig. 11.20. Still-Gennari olefination of a racemic α-chiral aldehyde with an enantiomerically pure phosphonate as kinetic resolution II—Use of the unreactive enantiomer *ent*-**B** of the aldehyde (R* stands for the phenylmenthyl group in the Horner-Wadsworth-Emmons products).

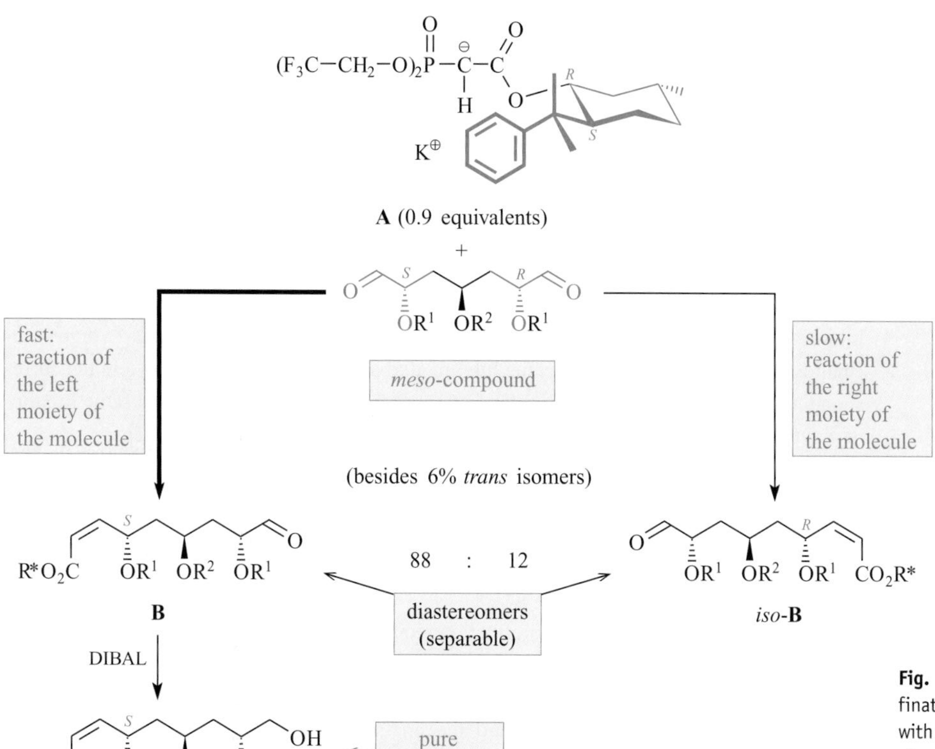

Fig. 11.21. Still-Gennari olefination of a *meso*-aldehyde with an enantionmerically pure phosphonate—Conversion of a *meso*- into an enantiomerically pure compound.

verted into enantiomerically pure chiral compounds by reacting them with enantiomerically pure reagents. Figure 11.21 shows how this can be done using a Still–Gennari olefination with the phosphonate ion **A.** The outcome of this olefination is readily understood when it is compared with the olefination in Figure 11.19. The left portion of the *meso*-dialdehyde in Figure 11.21 corresponds structurally to the *reactive* enantiomer **B** of the aldehyde in Figure 11.19. Similarly, the right portion of the *meso*-dialdehyde in Figure 11.21 corresponds structurally to the *unreactive* enantiomer *ent*-**B** of the aldehyde in Figure 11.19. We already discussed in detail why in Figure 11.19 the aldehyde enantiomer **B** is olefinated and the aldehyde enantiomer *iso*-**B** is not. These analogies perfectly rationalize why in the *meso*-dialdehyde of Figure 11.21 the carbonyl group on the left is olefinated faster than the one on the right.

The *cis*-alkene **B** is consequently the major product of the olefination of Figure 11.21. It can be separated chromatographically from the minor *cis*-alkene *iso*-**B** and also from the *trans*-alkenes formed in small amounts because these species are diastereomers. Finally, the chiral auxiliary can be removed reductively from the obtained *pure* alkene **B** (cf. the analogous reaction at the bottom of Figure 11.19). The resulting allyl alcohol **C** is 100% enantiomerically pure. Figure 11.21 thereby gives an example of the conversion of a *meso*-compound into an *enantiomerically pure* substance; a 100% yield is possible in principle in such reactions.

11.4 (Marc) Julia–Lythgoe– and (Sylvestre) Julia–Kocienski Olefination

Since the 1980s the **Julia-Lythgoe olefination** has blossomed into one of the most important C=C double-bond forming synthesis of acceptor-free *trans*-alkenes **E** (Figure 11.22). It usually consists of two or three steps. In the first step, a primary alkylphenyl sulfone **A** is lithiated with *n*-BuLi in the position α to the sulfur atom. The resulting sulfonyl-stabilized organolithium compound **B** → **B'** is then added to the carbonyl group of an aldehyde R^2–CHO. The sulfonyl-substituted lithium alkoxides **D** are formed as a mixture of diasteromers. These lithium alkoxides can be acetylated without prior workup, shortening the Julia-Lythgoe olefination to two steps. Alternatively, the lithium alkoxides **D** are protonated to yield the corresponding alcohols, the latter being esterified in a separate reaction to give the same mixture of *syn*-**C** and *anti*-**C**. In this case, the Julia-Lythgoe sequence requires three steps.

The last step of the Julia–Lythgoe olefination is an elimination, which is typically performed with sodium amalgam and starts with an $E1_{cb}$ elimination to give an alkenyl sulfone (mechanistic analysis: Figure 4.40) with its reduction to the alkene following *in situ*. Both the related mechanism and an explanation of the resulting *trans*-selectivity will be outlined later in Figure 17.85.

The fact that the Julia–Lythgoe olefination requires more than one step to prepare alkenes has generally been accepted as an inconvenient and inevitable part of the procedure developed by Marc Julia and Basil Lythgoe. This flaw kept nagging at Marc Julia's brother Sylvestre, who would not rest until he had found the one-step **(Sylvestre) Julia olefination**. The **(Sylvestre) Julia–Kocienski olefination** has become the *state-of-the-art*-variant of this olefination (Figure 11.23). It may be applied to any kind of aldehyde.

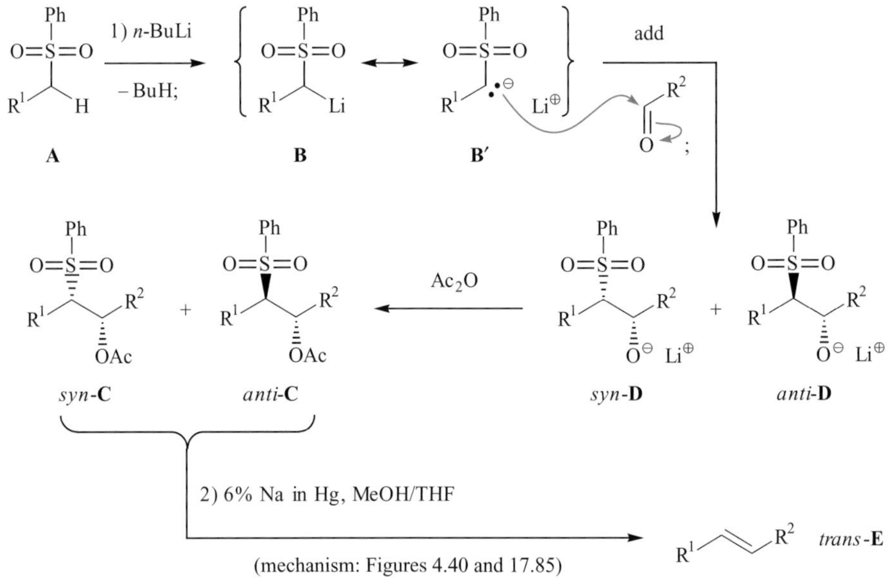

1) *n*-BuLi
– BuH;
add
A **B** **B′**
Ac_2O
syn-**C** *anti*-**C** *syn*-**D** *anti*-**D**
2) 6% Na in Hg, MeOH/THF
(mechanism: Figures 4.40 and 17.85)
trans-**E**

Fig. 11.22. Julia-Lythgoe olefination of aldehydes to form *trans*-alkenes in two steps: (1) addition of a lithium sulfone **B** ↔ **B'** to an aldehyde; *in-situ* acetylation; (2) reduction of the *syn:anti*-diastereomeric mixture of the resulting sulfonylacetates **C** with sodium amalgam.

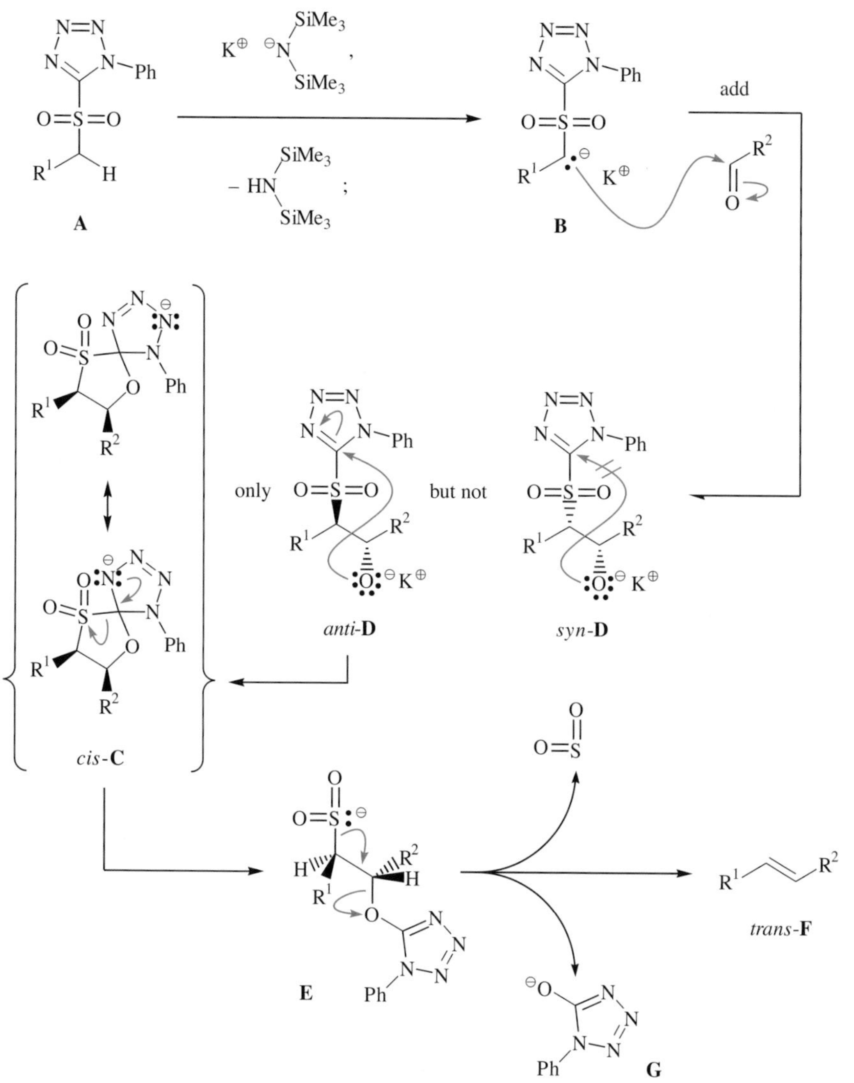

Fig. 11.23. Julia–Kocienski olefination to obtain *trans*-alkenes from aldehydes in a single step: their condensation reaction with the sulfonyl anion **B**, followed by an Ar-S_N reaction (→ **E**) and a fragmentation.

Initially the alkylaryl sulfone **A** is deprotonated with KHMDS to form the sulfonyl-substituted carbanion **B**, which adds to the carbonyl group of the aldehyde R^2–CHO. However, this addition does not furnish a diastereomeric mixture of the sulfonyl-substituted potassium alkoxides **D**, as might have been expected in analogy to the primary step of the Julia–Lythgoe olefination (Figure 11.22), Rather, the *anti*-configured alkoxide **D** is obtained exclusively. This circumstance accounts for the stereoselectivity of the net formation of the alkene since next the potassium alkoxide *anti*-**D** undergoes an intramolecular nucleophilic aromatic substitution reaction (see Section 5.5.1). It proceeds via a Meisenheimer intermediate **C** where the vicinal substituents R^1 and R^2 of the five-membered ring are oriented *cis* to each other, which is favorable since in this way they can *both* avoid the phenyl substituents of the neighboring ring. Or maybe not. It is not known whether the diastereomeric Meisenheimer intermediate occurs in which the substituents R^1 and R^2 lie on the same side of the five-membered ring as the phenyl group. The same final product would be obtained regardless, however. Subsequently, the intermediate **C** is re-aromatized regardless of its configuration, with a sulfinate ion acting as the leaving group. The resulting aromatic compound **E** undergoes fragmentation and—because this is an *anti*-selective process—the alkene **F** is formed with a *trans*-configured C=C double bond.

References

J. M. J. Williams (Ed.), "Preparation of Alkenes: A Practical Approach," Oxford University Press, Oxford, U.K., **1996.**

11.1

H. J. Bestmann, R. Zimmerman, "Synthesis of Phosphonium Ylides," in *Comprehensive Organic Synthesis* (B. M. Trost, I. Fleming, Eds.), Vol. 6, 171, Pergamon Press, Oxford, **1991**.

A. W. Johnson, "Ylides and Imines of Phosphorus," Wiley, New York, **1993**.

O. I. Kolodiazhnyi, "C-Element-Substituted Phosphorus Ylides," *Tetrahedron* **1996**, *52*, 1855–1929.

G. Wittig, „Von Diylen über Ylide zu meinem Idyll (Nobel Lecture)," *Angew. Chem.* **1980**, *92*, 671–675; *Angew. Chem. Int. Ed. Engl.* **1980**, *19* [This paper has not been translated into the English language].

A. Maercker, "The Wittig Reaction," *Org. React.* **1965**, *14*, 270–490.

H. Freyschlag, H. Grassner, A. Nürrenbach, H. Pommer, W. Reif, W. Sarnecki, "Formation and Reactivity of Phosphonium Salts in the Vitamin A Series," *Angew. Chem. Int. Ed. Engl.* **1965**, *4*, 287–291.

H. J. Bestmann, O. Vostrowski, "Selected Topics of the Wittig Reaction in the Synthesis of Natural Products," *Top. Curr. Chem.* **1983**, *109*, 85.

H. Pommer, P. C. Thieme, "Industrial Applications of the Wittig Reaction," *Top. Curr. Chem.* **1983**, *109*, 165.

B. E. Maryanoff, A. B. Reitz, "The Wittig Olefination Reaction and Modifications Involving Phosphorylstabilized Carbanions. Stereochemistry, Mechanism, and Selected Synthetic Aspects," *Chem. Rev.* **1989**, *89*, 863–927.

E. Vedejs, M. J. Peterson, "Stereochemistry and Mechanism in the Wittig Reaction," *Top. Stereochem.* **1994**, *21*, 1–167.

K. Becker, "Cycloalkenes by Intramolecular Wittig Reaction," *Tetrahedron* **1980**, *36*, 1717.

B. M. Heron, "Heterocycles from Intramolecular Wittig, Horner and Wadsworth-Emmons Reactions," *Heterocycles* **1995**, *41*, 2357.

11.2

F. R. Hartley (Ed.), "The Chemistry of Organophosphorus Compounds: Phosphine Oxides, Sulfides, Selenides, and Tellurides. The Chemistry of Functional Groups," Wiley, Chichester, U. K., **1992**.

W. S. Wadsworth, Jr., "Synthetic Applications of Phosphoryl-Stabilized Anions," *Org. React.* **1977**, *25*, 73–253.

J. Clayden, S. Warren, "Stereocontrol in Organic Synthesis Using the Diphenylphosphoryl Groupv *Angew. Chem. Int. Ed. Engl.* **1996**, *35*, 241–270.

11.3

W. S. Wadsworth, Jr., "Synthetic Applications of Phosphoryl-Stabilized Anions," *Org. React.* **1977**, *25*, 73–253.

D. F. Wiemer, "Synthesis of Nonracemic Phosphonates," *Tetrahedron* **1997**, *53*, 16609–16644.

J. Seyden-Penne, "Lithium Ccoordination by Wittig-Horner Reagents Formed by α-Carbonyl Substituted Phosphonates and Phosphine Oxide," *Bull. Soc. Chim. Fr.* **1988**, 238.

T. Rein, T. M. Pedersen, "Asymmetric Wittig Type Reactions," Synthesis **2002**, 579–594.

H. B. Kagan, J. C. Fiaud, "Kinetic Resolution," *Top. Stereochem.* **1988**, *18*, 249–330.

H. Pellissier, "Dynamic Kinetic Resolution," *Tetrahedron* **2003**, *59,* 8291–8327.

11.4

P. R. Blakemore, "The Modified Julia Olefination: Alkene Synthesis via the Condensation of Metallated Heteroarylalkylsulfones with Carbonyl Compounds," *J. Chem. Soc. Perkin Trans. I* **2002**, 2563–2585.

Further Reading

P. J. Murphy, J. Brennan, "The Wittig Olefination Reaction with Carbonyl Compounds Other than Aldehydes and Ketones," *Chem. Soc. Rev.* **1988**, *17*, 1–30.

P. J. Murphy, S. E. Lee, "Recent Synthetic Applications of the Non-Classical Wittig Reaction," *J. Chem. Soc.-Perkin Trans 1* **1999**, 3049–3066.

H. J. Cristau, "Synthetic Applications of Metalated Phosphonium Ylides," *Chem. Rev.* **1994**, *94*, 1299.

C. Yuan, S. Li, C. Li, S. Chen, W. Huang, G. Wang, C. Pan, Y. Zhang, "New Strategy for the Synthesis of Functionalized Phosphonic Acids," *Heteroatom Chem.* **1997**, *8*, 103–122.

C. Najera, M. Yus, "Desulfonylation Reactions: Recent Developments," *Tetrahedron* **1999**, *55*, 10547–10658.

The Chemistry of Enols and Enamines

12

In six chapters we have dealt with the chemistry of molecules containing C=O double bonds (Chapter 6–11). In this context we encountered

- nucleophilic substitutions (Chapter 6),
- H-Het eliminations leading to nitriles (Section 7.1) or heterocumulenes (Section 8.1, 8.3),
- nucleophilic additions (Section 7.2, 8.2, 8.3, 9.1, Chapter 10, Section 11.2–11.4),
- nucleophilic additions in combination with an S_N1 reaction (Section 9.2), an E1 elimination (Section 9.3) or an Ar-S_N reaction (Section 11.4), and
- the one-step formation of oxaphosphetanes through [2+2]-cycloadditions (Section 11.1).

Despite their diversity, the cited reactions share a common feature: they take place *on the* carbonyl or carboxyl group.

In the present—and next—chapter, the focus will once more be on the chemistry of molecules containing C=O double bonds. However, Chapters 12 and 13 will for the first time deal with reactions taking place in the position *next to* the carbonyl or carboxyl group. They are known for both neutral derivatives of carbonyl or carboxyl compounds (Chapter 12) and their conjugate bases, i.e., the enolates (Chapter 13). The neutral derivatives of carbonyl or carboxyl compounds, that allow for reactions *next to* the C=O double bond, are presented in the following scheme:

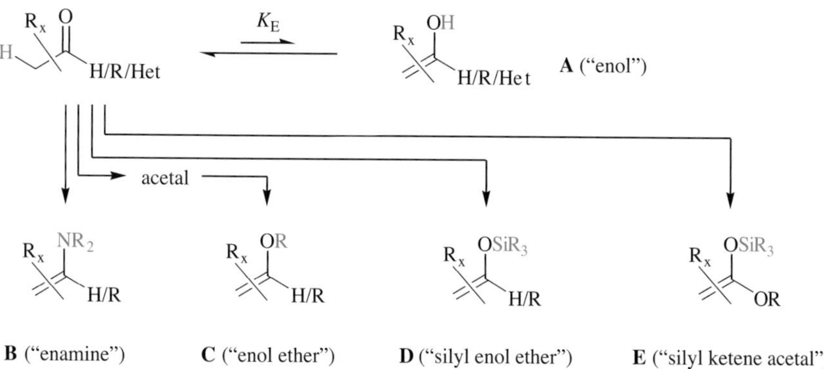

These substrates include:

- enols (**A**) that are in a dynamic equilibrium with the respective carbonyl or carboxyl compounds, that is, that occur along with them and therefore need not be prepared separately in the presence of a suitable catalyst,
- enamines (**B**) that must be prepared beforehand (or in situ) according to Section 9.3 via condensation of a secondary amine with the respective ketone or aldehyde
- enol ethers (**C**) that can be obtained from carbonyl compounds in two steps, i.e., by formation of an *O,O*-acetal (cf. Section 9.2.2) and subsequent E1 elimination (cf. Figure 9.32);

Bruckner R (author), Harmata M (editor) In: *Organic Mechanisms – Reactions, Stereochemistry and Synthesis*
Chapter DOI: 10.1007/978-3-642-03651-4_12, © Springer-Verlag Berlin Heidelberg 2010

- silyl enol ethers (**D**) prepared from carbonyl compounds by deprotonation and silylation (example: Figure 13.21), or
- silylketene acetals (**E**) that are prepared from carboxylic acid derivatives by deprotonation and silylation (example: Figure 13.22).

All the substrates **A**–**E** introduced here are alkenes that are electron-rich. This is why they react with electrophiles. Enols (**A**), enol ethers (**C**), silyl enol ethers (**D**) and silylketene acetals (**E**) react electrophiles to form oxocarbenium or carboxonium ions, whereas the reaction of enamines (**B**) with electrophiles gives iminium ions:

R_x OH, H/R/Het **A** | R_x NR_2, H/R **B** | R_x OR, H/R **C** | R_x $OSiR_3$, H/R(OR) **D (E)**

$E^{\oplus}$ | $E^{\oplus}$ | $E^{\oplus}$ | $E^{\oplus}$ $X^{\ominus}$

R_x $^{\oplus}$OH, H/R/Het, E | R_x $^{\oplus}NR_2$, H/R, E ⇄ (+ $ba^{\ominus}$, – baH; + HCl) R_x NR_2, H/R, E | R_x $^{\oplus}$OR, H/R, E | R_x $^{\oplus}OSiR_3$, H/R(OR), E

– $H^{\oplus}$ | + $H_3O^{\oplus}$ | z.B. + ROH, – $H^{\oplus}$ | – $XSiMe_3$

R_x O, H/R/Het, E | R_x O, H/R, E | R_x OR, H/R, OR, E | R_x O, H/R(OR), E

The fate of these oxocarbenium, carboxonium or iminium ions and thus the structure of the reaction product depend on the type of substrate used and the reaction partners:

- Carboxonium ions resulting from the electrophilic reaction on an enol are deprotonated at the O atom in the second reaction step. In this way an α-functionalized carbonyl or carboxyl compound is formed. This amounts to a two-step mechanism, but as far as experimental work is concerned it involves only a single-step electrophilic substitution next to the C=O double bond of a carbonyl or carboxyl compound. In Section 12.2, this type of reaction is illustrated with numerous examples.
- Iminium ions formed from an electrophile and an enamine are stable in the absence of a base. Upon aqueous workup they react via the unstable intermediate of an "imine hydrate" to furnish the corresponding α-functionalized aldehyde or ketone. Iminium ions formed in the presence of a base are deprotonated by this base to give an enamine. The latter is hardly ever isolated, but hydrolyzed upon a somewhat more aggressive workup with diluted

hydrochloric acid to form the underlying α-functionalized carbonyl compound. Regardless of whether a base is present or absent, this functionalized carbonyl compound is prepared by electrophilic substitution in two steps from the carbonyl compound corresponding to the enamine employed. The process forms the basis of a great deal of asymmetric organocatalysis initiated by chiral secondary amines like proline. We will only briefly touch upon this important topic here. Reactions based on enamines are outlined in Section 12.3.

- Carboxonium/oxocarbenium ions emerging from the reaction of an electrophile towards an enol ether are encountered less frequently. Figure 12.23 shows an important, since generalizable example (a Mukaiyama aldol condensation); Figure 12.25 contains a special example which, however, leads to an important reagent.
- Carboxonium/oxocarbenium ions that are produced from a silyl enol ether or a silylketene acetal and an electrophile are desilylated in the second step of the reaction. This also produces an α-functionalized carbonyl or carboxyl compound.

12.1 Keto-Enol Tautomerism; Enol Content of Carbonyl and Carboxyl Compounds

Enols are isomers of molecules containing C=O double bonds. Both isomers equilibrate, and even small amounts of acid or base serve as a catalyst. The mechanism of acid-catalyzed isomerizations will be presented Figure 12.4. Isomers existing in a dynamic equilibrium of this kind are referred to as **tautomers**; the respective isomerism is termed **tautomerism,** and the isomerization of tautomers is called **tautomerization**.

The electrophilic functionalization of carbonyl and carboxyl compounds *next to* the C=O double bond generally proceeds in a single step via an S_E reaction of the enol tautomer. The identification of suitable substrates for this type of reaction requires rough knowledge of the extent to which the respective enolization occurs. The enolization equilibrium constants K_E or their —more easy-to-handle—negative logarithm values, the so-called pK_E values, provide a quantitative measure thereof. These measures are defined as follows:

R$_x$ O H H/R/Het $\xrightarrow{K_E}$ R$_x$ OH H/R/Het

$$pK_E = -\lg K_E$$

pK_E Values relate to the equilibrium constants K_E as the familiar pH values relate to the acidity constants K_a.

The top row of Figure 12.1 displays the pK_E values of selected aldehydes. According to the effect of the substituents at the α-carbon atom they can be categorized into the following groups. Acetaldehyde (**A**) contains up to a hundred times less enol than higher-order aliphatic

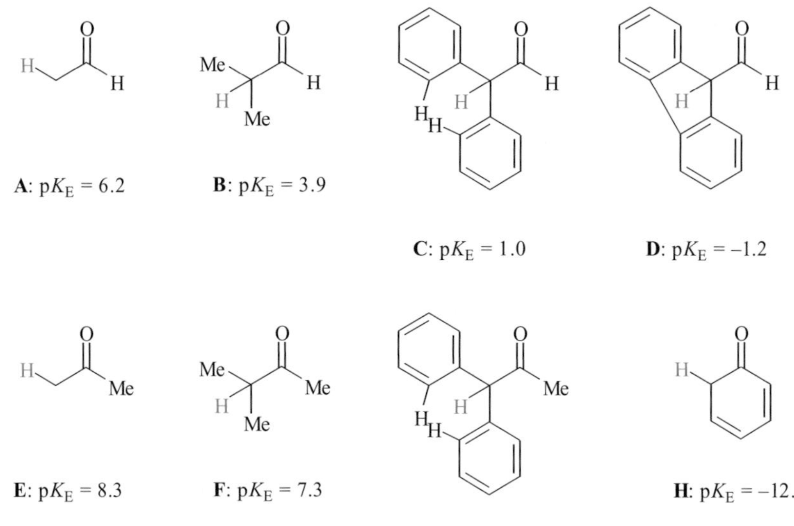

Fig. 12.1. Substituent dependency of the enol content of aldehydes and ketones. Here, the pK_E values, i.e., the negative of the base-10 logarithm of the equilibrium constants K_E of the respective tautomerization carbonyl compound $\rightleftharpoons$ enol, are used as a measure.

aldehydes and, in particular, an α,α-dialkyl-substituted aldehyde like the isobutyraldehyde (**B**), since the $C^{\alpha}{=}C$ double bond of enols is stabilized by an α-alkyl- instead of an α-hydrogen substituent in the same way as the C=C double bond of normal alkenes. *α-Alkylated* aldehydes (like, e. g., **B** in Figure 12.1), however, contain by several orders of magnitude less enol than *α-arylated* aldehydes (such as **C** in Figure 12.1). The reason is that C=C double bonds are better stabilized by conjugated substituents, for example, an aryl group, than by an alkyl group with just hyperconjugative effects. In this context an interesting detail may be observed: the enol content of the fluorenecarbaldehyde (**D**) is one hundred times larger than that of diphenylacetaldehyde (**C**). This is due to the fact that the benzene rings in the fluorene-containing enol lie exactly in the plane of the enolic C=C double bond, which allows for optimal conjugation. In the enol of the diphenylacetaldehyde, however, they avoid each other by twisting, which reduces their contribution to conjugation.

The bottom line complements Figure 12.1 by adding the pK_E values of representative ketones. The comparison of **E**–**G** reveals the same substituent effects that are familiar from the analogous aldehydes **A**–**C**: the enol content is increased by alkyl substituents in the α-position, and even more so by aryl substituents in the α-position. The ketone **H** in Figure 12.1, the nonexistent "isophenol," has by far the highest propensity to enolization of all the carbonyl compounds shown. The reason, of course, is that the tautomeric enol, phenol, is favored because of its aromaticity and thus *particularly* efficient C=C double bond stabilization.

In Figure 12.1, it is interesting to compare the pK_E values that are placed one below the other. This comparison reveals that ketones contain by 10–1.000 times less enol than the corresponding aldehydes. This is due to the fact that in comparison to an aldehyde C=O double bond, a ketonic C=O double bond is stabilized by 3 kcal/mol—a fact that is familiar from the discussion of Figure 9.1. Essentially, the price that has to be paid for the loss of the C=O

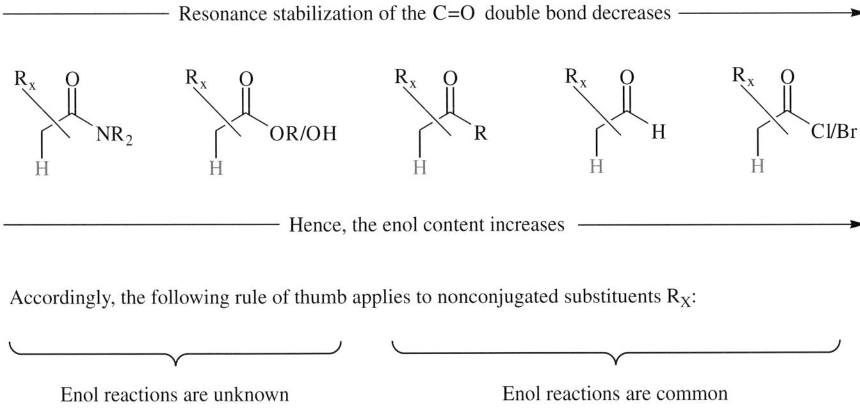

Fig. 12.2. (Mono-)carboxyl- and (mono-)carbonyl compounds, in the order of increasing enol content (for the enol contents of active-methylene compounds see Table 12.1.)

double bond during the conversion into an enol is higher for ketones by these 3 kcal/mol than for aldehydes, decreasing the equilibrium enol content for ketones relative to aldehydes.

This latter thought has an important consequence: if compounds with C=O double bonds are sorted in decreasing order of resonance stabilization of their C=O group they are at the same time sorted according to their increasing propensity to enolization. So as the resonance stabilization of the C=O double bond decreases from 22 kcal/mol to somewhere near zero in the order carboxylic acid amide > carboxylic acid ester/carboxylic acid > ketone > aldehyde > carboxylic acid chloride/-bromide, the enol content increases in this same order (Figure 12.2). These circumstances immediately explain why no enol reactions whatsoever are known of carboxylic acid amides, virtually none of normal carboxylic acid esters/carboxylic acids, but are commonly encountered with ketones, aldehydes and carboxylic acid halides.

If enolization affords a C=C double bond that is conjugated with an acceptor substituent, the propensity to undergo enolization is notably increased. Figure 12.3 illustrates this finding

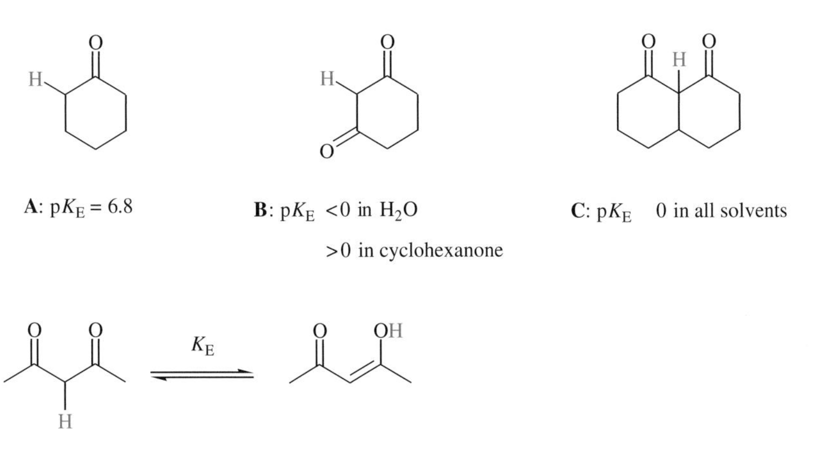

Fig. 12.3. Enol content of three representative β-diketones (**B–D**): ten-million-fold increase as compared to cyclohexanone (**A**).

with the pK_E values of the β-diketones **B**–**D**, which contain 10 million times more enol than cyclohexanone. This may advance to the point where the enol always becomes the dominant tautomer (**C**) or dominant only in certain solvents (**B**, **D**).

In some enols of β-diketones an intramolecular hydrogen bridge occurs, for example, in crystalline acetylacetone (where it is even symmetrical). But there are also β-diketones with a similarly high degree of enolization where such a hydrogen bridge cannot exist for geometrical reasons (for example, in the enol of compound **B** in Figure 12.3). Hence, a different reason than the existence of a hydrogen bridge is largely held responsible for the stabilization of a ketone enol by a keto substituent: for such an enol a zwitterionic resonance form can be drawn in which *due to the electron withdrawing resonance (–M) effect of the acceptor substituent* the anionic part is an enolate and therefore quite stable, unlike the situation when no such acceptor substituent is present:

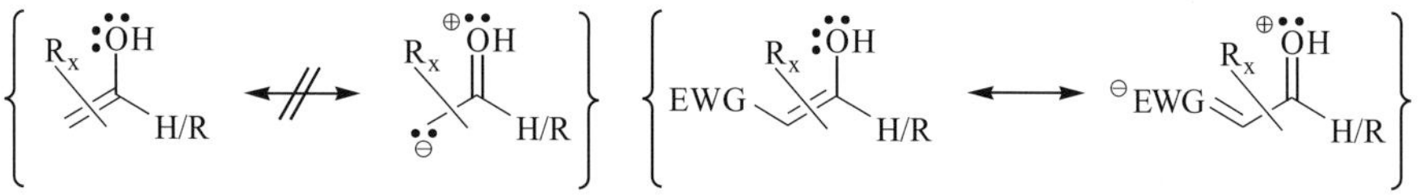

The stabilization of ketone-substituted ketone enols that has just been presented suggests that all ketone enols that are acceptor-substituted should have an enhanced enol content. All car-

Tab. 12.1 Substituent Effects on the Enol Content of Active-Methylene Compounds (for the enol content of monocarboxyl and monocarbonyl compounds cf. Figure 12.2)

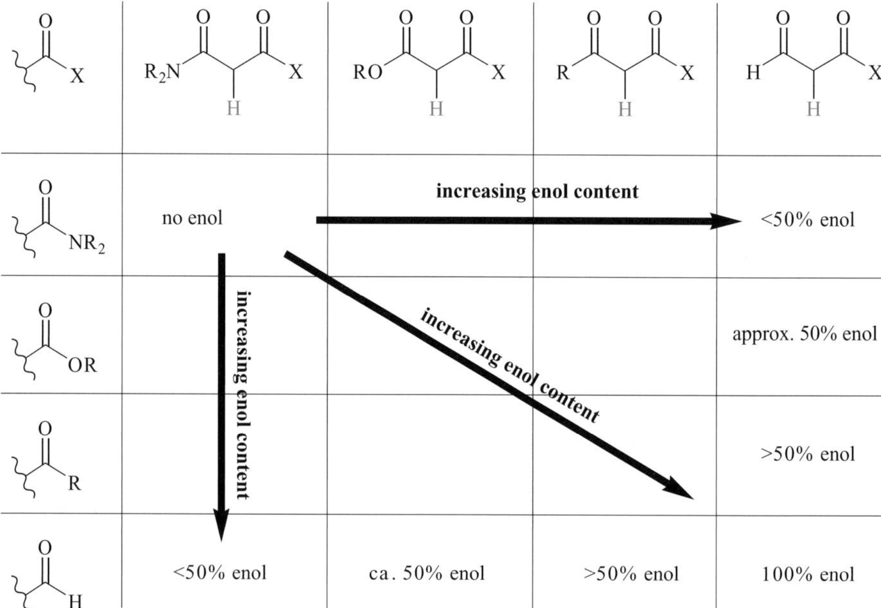

bonyl and carboxyl compounds that are generally capable of enolization should be able to benefit from such extra stabilization by a conjugated acceptor substituent. This idea, in combination with the intrinsically dissimilar enolization tendencies of classes of compounds with a C=O double bond in Figure 12.2 suggest a degree of enolization of β-acceptor-substituted carbonyl and carboxyl compounds as summarized in Table 12.1: all the substrates outside the upper left quadrant contain enough enol to react with an electrophile.

12.2 α-Functionalization of Carbonyl and Carboxyl Compounds via Tautomeric Enols

If a carbonyl or carboxyl compound that is capable of enolization, is to undergo complete reaction with an electrophile via its enol tautomer, it must be ensured that the latter is constantly resupplied from the carbonyl or carboxyl tautomer under the reaction conditions. If the tautomerization, which this process requires, occurs faster than the further reaction of the enol, the enol will be continuously available for the electrophile to react to completion—and always at the respective equilibrium concentration. But if the tautomerization is the slower reaction of the two, the enol will be depleted to less than its equilibrium concentration; in the extreme case, it may even be totally consumed, and it will take some time before new enol will have formed.

In the presence of an electrophile, tautomerization of a substrate with a C=O double bond to its enol only takes place when catalyzed by either a Brønsted- or a Lewis acid. The proton-catalyzed mechanism is shown for the ketone → enol conversion **B** → *iso*-**B** (Figure 12.4), the carboxylic acid → enol conversion **A** → **E** (Figure 12.6), the carboxylic acid bromide → enol conversion **E** → **G** (Figure 12.7) and the carboxylic acid ester → enol conversion diethylmalonate → **E** (Figure 12.9). Each of these enol formations is a two-step process consisting of the protonation to a carboxonium ion and the latter's deprotonation. The mechanism of a Lewis acid-catalyzed enolization is illustrated in Figure 12.5, exemplified by the ketone → enol conversion **A** → *iso*-**A**. Again, a protonation to a carboxonium ion and the latter's deprotonation are involved; the Lewis acid-complexed ketone acts as a proton source (see below).

Side Note 12.1. Enol Titration of Acetoacetic Ester

If the reaction between the enol and the electrophile proceeds extremely fast, the enol tautomer of a carbonyl or carboxyl compound might be consumed completely. The generation of enol becomes the rate-determining step. This situation occurs with the "enol titration of acetoacetic ester," (Figure 12.4). In this process, bromine is added to an equilibrium mixture of the ketone form (**B**) and the enol form (*iso*-**B**) of an acetoacetic ester. Bromine functionalizes the enol form via the intermediacy of the carboxonium ion **E** to form the bromoacetic ester **D**. The trick of conducting the enol *titration* is to capture the enol portion of a known amount of acetoacetic ester by adding exactly the equivalent amount of bromine. From the values for

Fig. 12.4. Enol titration of acetoacetic acid ester. The mechanism of the acid-catalyzed **B** ⇌ *iso*-**B** enolization and the mechanism of the electrophilic *iso*-**B** → **D** bromination is shown.

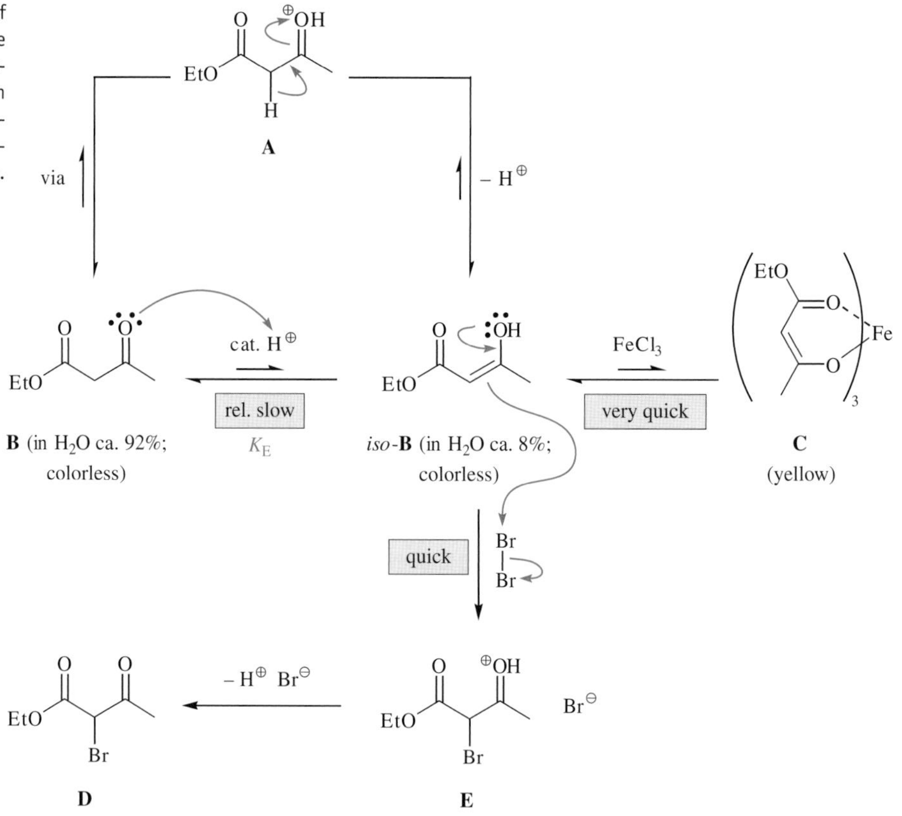

$m_{\mathbf{B}\,+\,iso\text{-}\mathbf{B}}$ (from the initial weight) and for $m_{iso\text{-}\mathbf{B}}$ (from the amount of bromine consumed) we may calculate the enolization constant K_E.

The proper amount of bromine can be determined by a series of identical parallel experiments. Iron(III) chloride is added as an indicator to the equilibrium mixtures consisting of the ketone and the enol form of the acetoacetic ester and various amounts of bromine are added. If too much bromine has been added, the reaction mixture turns brown. If too little bromine has been added, the reaction mixture remians yellow, as the added iron(III) chloride reacts with the enol *iso*-**B** extremely fast, but reversibly to form the yellow octahedral complex **C.** This way the enol tautomer is made visible. Only if the proper amount of bromine has been identified does the reaction mixture turn colorless—only for a few seconds, of course, until by HBr catalysis enough of the enol form (*iso*-**B**) has been regenerated from the ketone form (**B**) of the acetoacetic ester and complexed to give the yellow **C**.

If the enolization has enough time to proceed, the transformation in Figure 12.4 completely leads to a brominated β-ketoester. Basically, the same method can be employed to also α-brominate ketones (Figure 12.5), alkylated malonic acids (Figure 12.6), acid bromides (Figure 12.7, 12.8) and acid chlorides (not shown). The mechanistic details are detailed in the cited figures. Look at how similar they are.

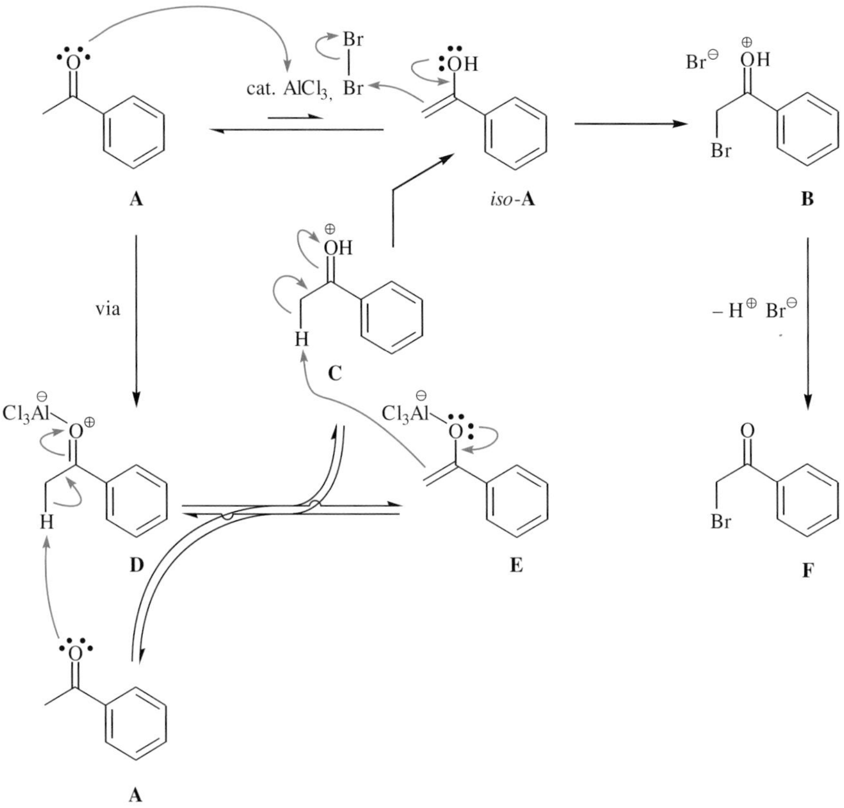

Fig. 12.5. Electrophilic side chain bromination of acetophenone (**A**). With catalytic amounts of $AlCl_3$ the acetophenone is transferred into the enol *iso*-**A**, while stoichiometric amounts of $AlCl_3$ lead to the formation of acetophenone in the form of the Lewis acid/Lewis base complex **D**. So in the presence of catalytic amounts of $AlCl_3$ the enol *iso*-**A** is brominated (→ **F**, "phenacyl bromide"); in the presence of stoichiometric amounts of $AlCl_3$, however, bromination of the aromatic moiety of **D** takes place (→ *meta*-bromoacetophenone, cf. Section 5.2.1).

In the presence of a catalytic amount of aluminum chloride, acetophenone (Formula **A** in Figure 12.5) and bromine react to give phenacyl bromide (**F**). In contrast, the same reactants and a stoichiometric amount of aluminum chloride yield *meta*-bromoacetophenone (Section 5.2.1). This difference is due to the different ratios of substrate quantities in the respective product-determining step. The acetophenone enol (*iso*-**A**) is the substrate for the formation of phenacyl bromide, and the aluminum chloride/acetophenone complex **D** is the substrate in the reaction leading to *meta*-bromoacetophenone. The acetophenone enol derives from a bimolecular reaction between the complexed (**D**) and the free acetophenone (**A**). This access is blocked, though, as soon as the *all of the* acetophenone has been complexed to form **D**—which occurs if aluminum chloride is added in stoichiometric amounts.

According to Figure 12.6, the bromination of alkylated malonic acids **A** initially furnishes α-brominated alkyl malonic acids **B**. Upon heating they decarboxylate to form the α-bromocarboxylic acids **C**. This two-step approach to **C** can be managed without using the expensive phosphorus tribromide, which would be required in the alternative single-step Hell–Volhard–Zelinsky synthesis of this compound (Figures 12.7, 12.8).

Two variants of the Hell–Volhard–Zelinsky brominations are commonly conducted. According to the first procedure, carboxylic acids **A** are reacted with one equivalent of bromine and a *catalytic* amount of phosphorus tribromide to form α-bromocarboxylic acids (Formula **B**

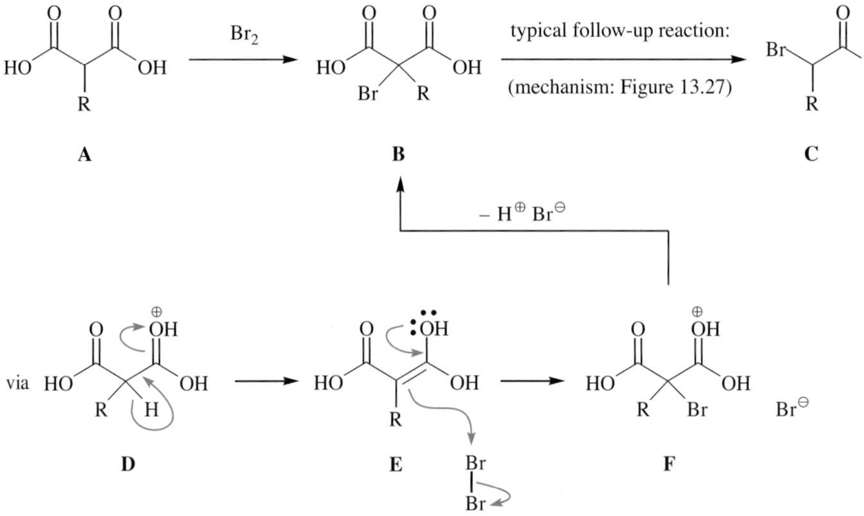

Fig. 12.6. Bromination of malonic acids or alkylated malonic acids. The figure shows the mechanisms of the acid-catalyzed enolization (alkyl) malonic acid ⇌ enol of (alkyl) malonic acid (**E**) and the actual bromination (**E** → **B**).

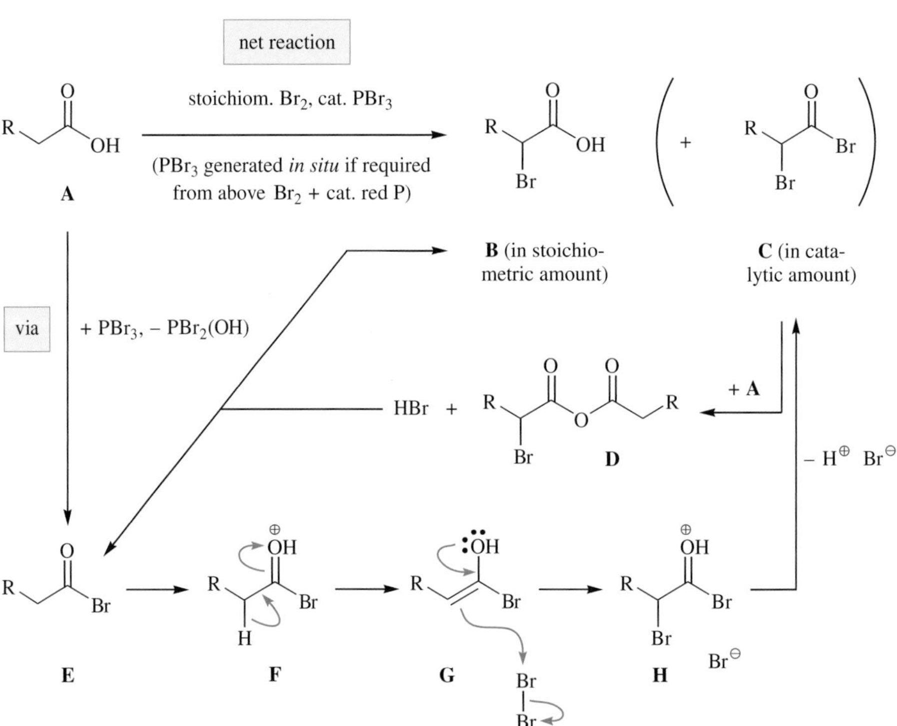

Fig. 12.7. Bromination of carboxylic acids in the presence of catalytic amounts of phosphorus tribromide ("Hell–Volhard–Zelinsky reaction I"): the reaction product is an α-bromocarboxylic acid.

in Figure 12.7). The second, more frequently employed method uses the same substrates, but stoichiometric amounts of phosphorus tribromide. In this way, an α-bromocarboxylic acid bromide is generated (Formula **B** in Figure 12.8). The course of the reaction with phosphorus tribromide in stoichiometric amounts will become immediately clear (Figure 12.8) if the effect of catalytic amounts of phosphorus tribromide is understood (Figure 12.7).

Carboxylic acid **A** in Figure 12.7 cannot form an enol since this would entail a loss of resonance stabilization of 14 kcal/mol (cf. Figure 12.3). In contrast, the carboxylic acid bromide **E** that forms from phosphorus tribromide does not have any resonance stabilization and can hence be converted into the enol **G**. The latter further reacts with bromine to give the α-bromocarboxylic acid bromide **C**. Since there is no phosphorus tribromide (which was employed in catalytic amounts only) left to react with the still unconsumed carboxylic acid **A**, the latter reacts with the small amount of α-bromocarboxylic acid bromide **C** that has formed to yield the mixed anhydride **D**. If one or the other carboxyl oxygen in this mixture is protonated by HBr in an equilibrium reaction, the resulting carboxonium ions act as acylating agents towards the concomitantly formed bromide ion—a heteroatom nucleophile! The "unproductive" acylation of this bromide ion leads back to the formation of carboxylic acid **A** and the α-bromocarboxylic acid bromide **C**. The other acylation is "productive," which means it promotes the overall reaction. This acylation yields the α-bromocarboxylic acid **B** and the new carboxylic acid bromide **E**. Via the reaction path **A** + **C** → **D** → **B**+ **E**, and the cited reaction **E** + Br_2 → **C**, the entire carboxylic acid **A** of Figure 12.7 is ultimately converted into the α-bromocarboxylic acid **B**. Some additional α-bromocarboxylic acid bromide **C** that is present at the end of the reaction gets consumed during workup.

The mechanism in Figure 12.7 implies that the *stoichiometric* use of phosphorus tribromide in the second variant of the Hell–Volhard–Zelinsky bromination leads to the *selective* formation of the α-bromocarboxylic acid bromide (Formula **B** in Figure 12.8). Often these compounds are hydrolyzed to give the bromocarboxylic acids **C**. More importantly, α-bromocarboxylic acid bromides **B** can also undergo alcoholysis and in this way provide access to α-brominated esters **D**.

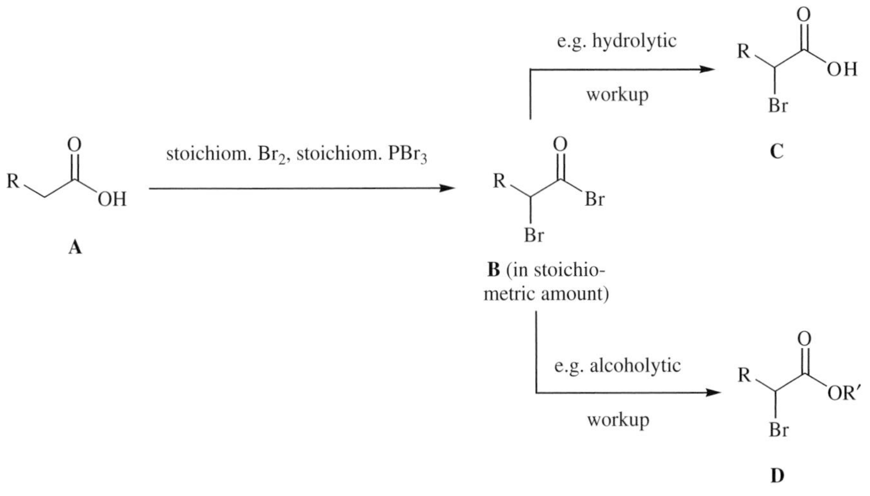

stoichiom. Br_2, stoichiom. PBr_3

A

B (in stoichiometric amount)

e.g. hydrolytic workup

C

e.g. alcoholytic workup

D

Fig. 12.8. Bromination of carboxylic acids in the presence of stoichiometric amounts of phosphorus tribromide ("Hell–Volhard–Zelinsky reaction II"). The actual reaction product is an α-bromocarboxylic acid bromide (**B**), which often undergoes further *in situ* reaction to afford an α-bromocarboxylic acid (**C**) or an α-bromocarboxylic acid ester (**D**).

Like the acid bromide generated *in situ*, an acid chloride may be α-brominated with bromine as well since an acid chloride does not lose any resonance stabilization if it tautomerizes to give the corresponding enol. In contrast to the conventional Hell–Volhard–Zelinsky procedure (Figure 12.7, 12.8), this synthesis—of α-bromocarboxylic acid chlorides—can be performed without any phosphorus tribromide whatsoever. The enolizability of *N*-protected α-amino*carboxylic acid chlorides* was already mentioned in connection with their propensity to undergo racemization, which is based on this very circumstance (cf. Figure 6.30).

If we turn our attention in enol chemistry from bromine as an electrophilic reaction partner to an *N* (Figure 12.9) or *O* electrophile (Figure 12.10), you will find that the following familiar steps are involved in the course of this reaction: (1) tautomerism to the enol; (2) reaction with the electrophile and formation of a carboxonium ion; (3) deprotonation of the latter to form a compound that contains a C=O double bond and differs from the substrate by a heteroatom substituent in the α-position. Try to locate these steps that tend to recur, like a chorus, in the two figures! You will be gratified to hear that the same steps will turn up again in connection with α-functionalizations of enols with *C* electrophiles—organic chemistry is full of patterns. Learn to recognize them.

An aqueous solution of sodium nitrite that is treated with HCl contains nitrosyl cations O=N$^{\oplus}$. These can react with the enol **E** of the malonic acid diethyl ester (cf. Figure 12.9, bottom). First, a nitroso compound (**F**) is formed, which then undergoes acid-catalyzed isomerization to give the oxime **A**. Usually, the oxime is reduced by zinc, which is dissolved in acetic acid, to yield an amine that normally undergoes *in situ* acetylation in acetic acid. In this way the (acetamido)malonic acid diethyl ester **B** is obtained as the reduction/acetylation product, which can be employed, for example, in the synthesis of amino acids (Figure 13.39).

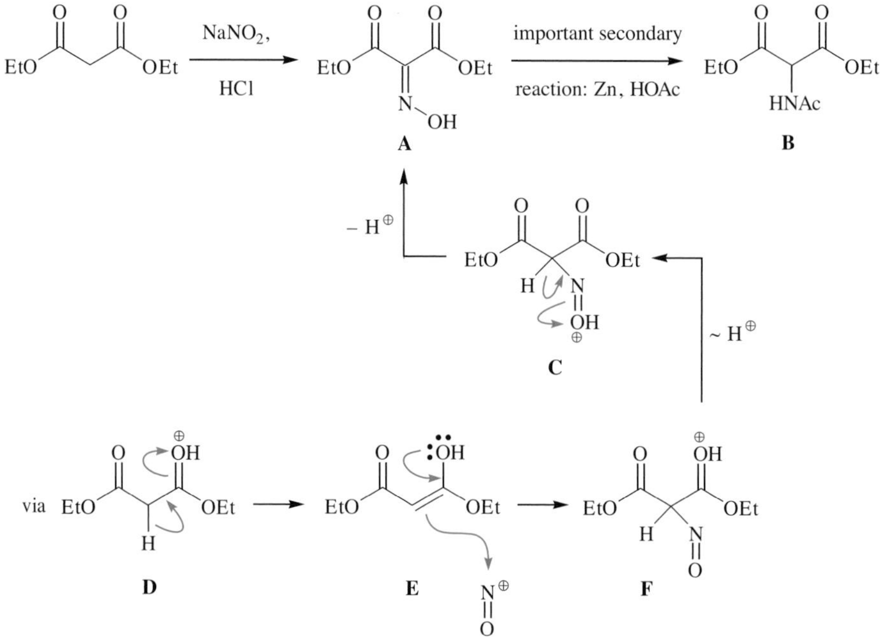

Fig. 12.9. Nitrosation of malonic acid diethyl ester. The mechanism of the acid-catalyzed enolization malonic ester ⇌ malonic ester enol (**E**) and the mechanism of the actual nitrozation (**E** → **A**) is shown here.

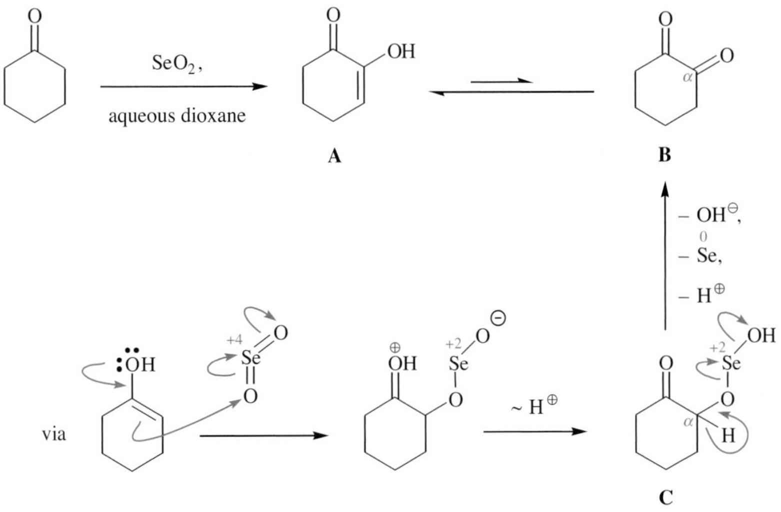

Fig. 12.10. Mechanism of the α-oxygenation of ketones in reactions with selenium dioxide: an electrophilic substitution reaction ($\rightarrow\rightarrow$ **C**) is followed by a β-elimination at the C–O single bond.

Selenium dioxide is able to α-oxygenate ketones via their enol tautomers. As is demonstrated in Figure 12.10 by the reaction of selenium dioxide with cyclohexanone, the *actual* electrophilic substitution product **C** is unstable. The latter contains selenium in the oxidation state +2 that takes the opportunity to transform into selenium in the oxidation state 0, i.e., elemental selenium, by way of the fragmentation reaction indicated. Thereby, the α-C–O single bond of the primary product **C** is transformed into the α-C=O double bond of the final product **B** (which, however, is largely present as the tautomeric enol **A**).

Section 12.2 closes with the α-functionalization of enols with *C* electrophiles (Figure 12.11–12.15). Figure 12.11 presents a *tert*-butylation of malonic ester with *tert*-butyl chlo-

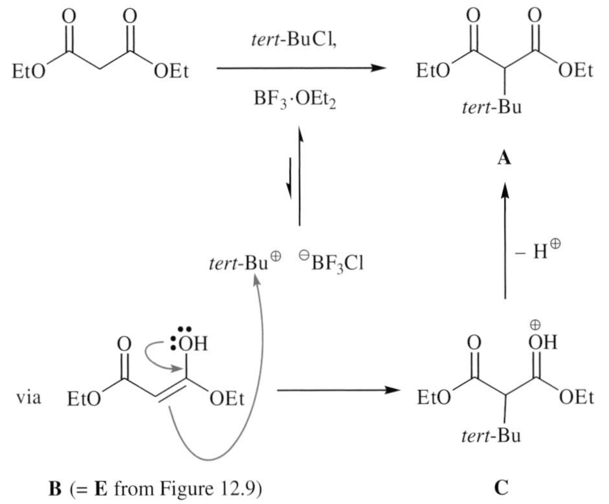

Fig. 12.11. Mechanism of the *tert*-butylation of malonic acid diethyl ester. From the point of view of *tert*-butyl chloride it is an S_N1 reaction with the malonic ester enol (**B**) acting as the nucleophile.

ride. The malonic ester enol **B** undergoes an S_N1 reaction with this alkylating agent, since the latter forms *tert*-butyl cations under the influence of the boron trifluoride etherate that is also present. In Section 13.2.2 a complementary procedure to obtain alkylated malonic esters will be introduced as the "malonic ester synthesis." There the malonic ester enolate acts as the nucleophile and the alkylation follows an S_N2 mechanism. Thus, *tert*-alkylations of malonic esters are performed via the enol and proceed via the S_N1 mechanism; *sec*- and *prim*-alkylations proceed via the enolate and follow an S_N2 mechanism.

Side Note 12.2. Cyclodehydration of Ketones to Furnish Six-membered Aromatic Rings

If you work through this book from beginning to end your impression may be that aromatic compounds can be synthesized in many ways, but that one always starts from aromatic compounds. This is the case. Nonetheless, it may occasionally be advantageous to obtain an aromatic from a non-aromatic compound. This is demonstrated in Figure 12.12 using a synthesis of mesitylene (**B**) from acetone as an example. The mechanism comprises many individual

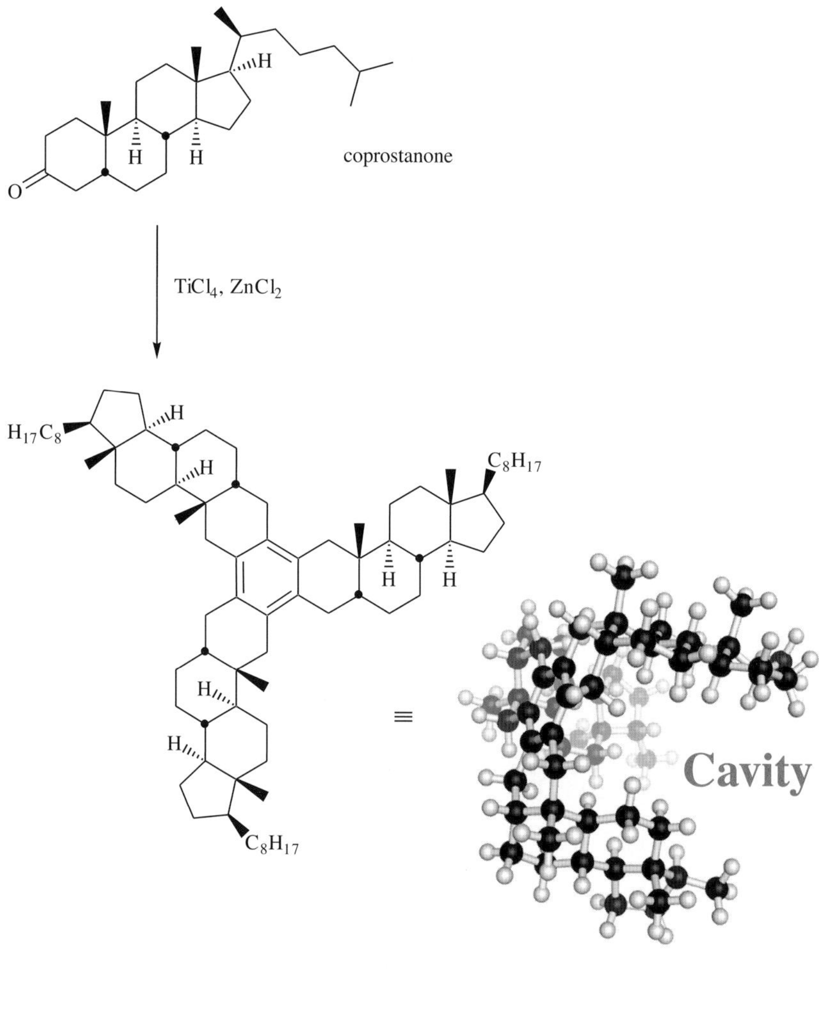

coprostanone

$TiCl_4$, $ZnCl_2$

$H_{17}C_8$ H H H H H C_8H_{17} H H C_8H_{17} ≡ Cavity

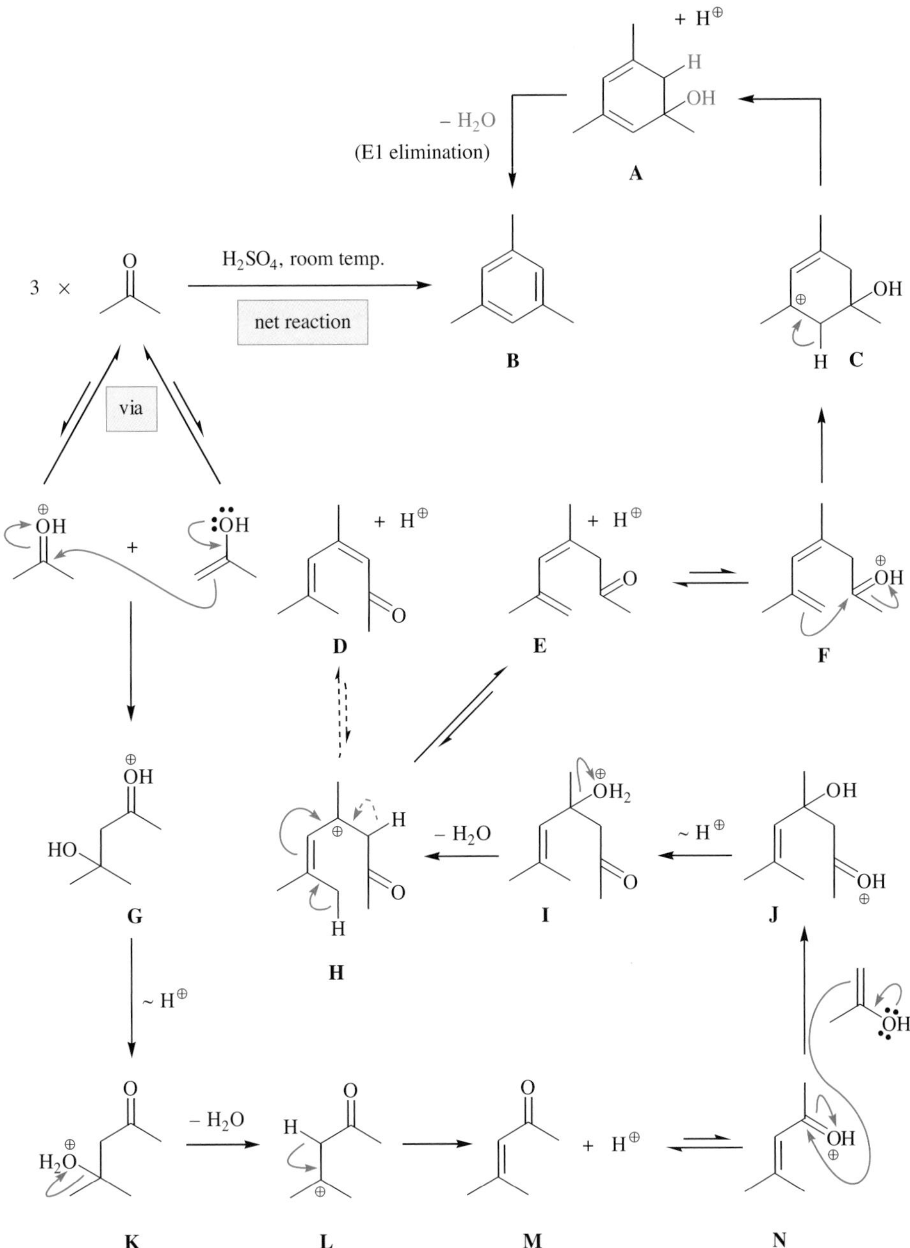

Fig. 12.12. Cyclocondensation 3 × acetone → 1,3,5-trimethylbenzene (**B**, "mesitylene") in the presence of H_2SO_4 as a Brønsted acid. If conducted intentionally, this reaction offers a large-scale access to mesitylene. It occurs as a side reaction when acetone is dried with phosphorus pentoxide—since phosphorus acid is formed and the latter also catalyzes the condensation reaction shown. (This figure conceals the fact that under acidic conditions acetone generally reacts via **M** to furnish phorone, i.e., $Me_2C{=}CH{-}C({=}O){-}CH{=}CMe_2$. That process, however, is *reversible*.)

reactions, and it should not be too difficult to track them down. In the first part the reaction starts with acetone enol acting as the nucleophile and protonated acetone acting as the electrophile, and it ends with mesityl oxide (**M**). The second part begins with the reaction of the acetone enol as the nucleophile again and protonated mesityl oxide (**N**) as the electrophile, and for the time being, ends with the dienone **D**. The latter forms a small equilibrium amount of the double bond isomer **E** under acidic conditions. Its diene moiety undergoes intramolecular hydroxyalkylation (→ six-membered ring **A**), followed by an E1 elimination that yields the aromatic compound (**B**).

Under the conditions of Figure 12.12 acetophenone yields 1,3,5-triphenylbenzene. Lewis rather than Brønsted acids initiate the cyclodehydration of ketones to give aromatic compounds in an analogous fashion. Under influence of titanium tetrachloride and zinc chloride the steroid coprostanone, for example, leads to the formation of hexasubstituted benzene: the 3×3 cyclohexane chairs that constitute the substituents of the resulting benzene ring are rigid

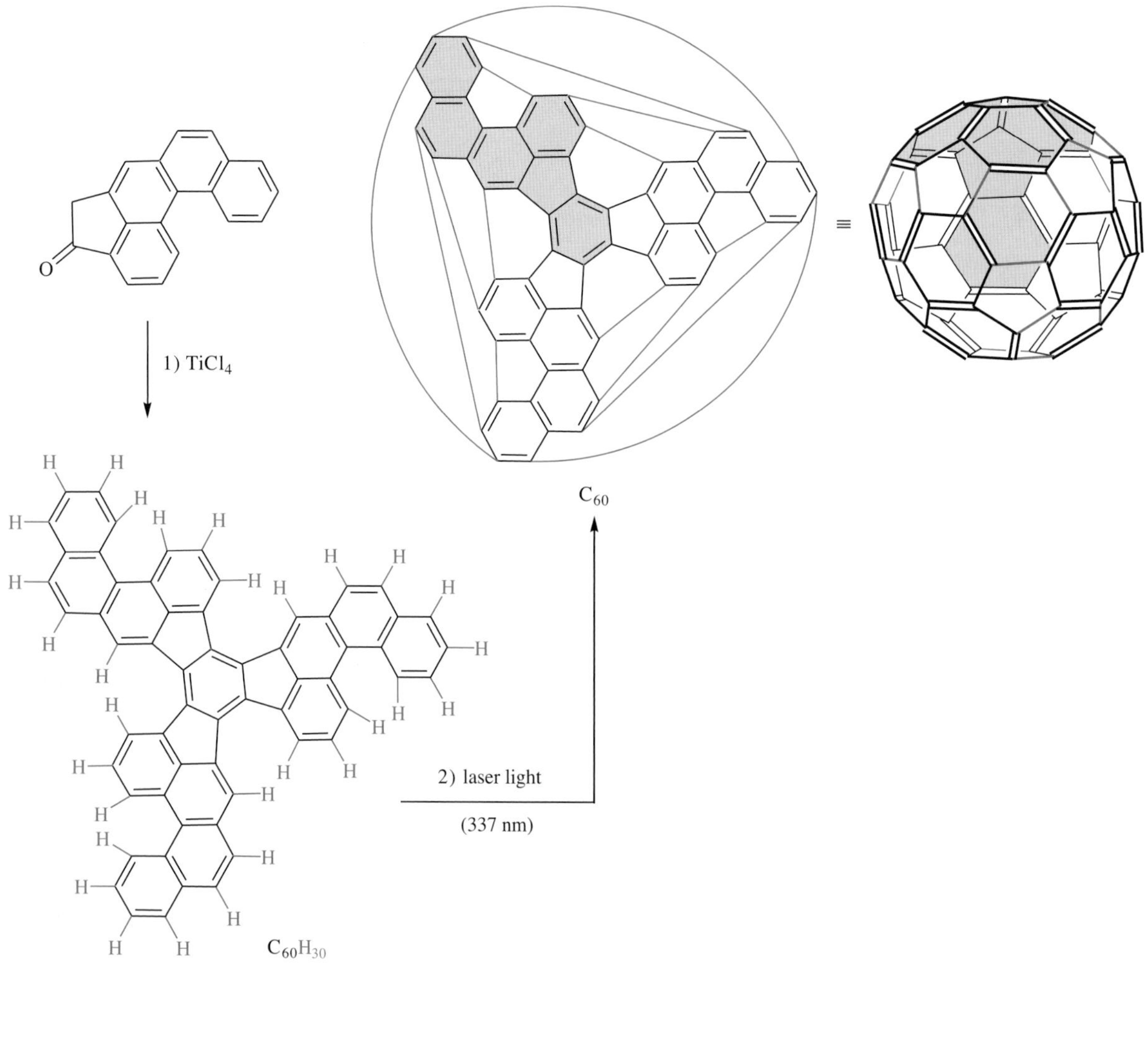

1) $TiCl_4$

$C_{60}H_{30}$

2) laser light

(337 nm)

C_{60}

≡

structural elements rendering the benzene derivative shown in the figure a "host molecule." A host molecule is a molecule with an internal cavity where "guest molecules" can be non-covalently bound—in the present case by van der Waals interactions.

The titanium tetrachloride-mediated cyclodehydration of the above aromatic ketone also forms a benzene ring carrying substituents on all six positions. These were designed to react with each other when exposed to the energetically rich photons of an ultraviolet-emitting laser. In combination with the subsequent cleavage of hydrogen, which benefits from the fact that more and more fully conjugated compounds are formed successively, Scott managed to develop the first *directed* synthesis of the fullerene C_{60} in this way.

Figure 12.13 shows that the *iso*-**A** enols of the β-diketones **A** react with an α,β-unsaturated carboxonium ion **C** that acts as a *C* electrophile. This oxocarbenium ion is formed by reversible protonation of the α,β-unsaturated methyl vinyl ketone in acetic acid. However, the oxocarbenium ion **C** in this figure does not react with the *iso*-**A** enols at its carbonyl carbon atom—as the protonated acetone in Figure 12.12 does with the enol of acetone—but at the center C-β of the conjugated C=C double bond. Accordingly, an addition reaction takes place whose regioselectivity resembles that of a 1,4-addition of an organometallic compound to an α,β-unsaturated carbonyl compound (see Section 10.6). 1,4-additions of enols (like in this case) or enolates (as in Section 13.6) to α,β-unsaturated carbonyl and carboxyl compounds are referred to as **Michael additions**.

An iminium ion acting as the *C* electrophile and a carbonyl compound react—after passing the usual intermediates from Section 12.2 shown in Figure 12.14—to form a β-aminocarbonyl compound (Figure 12.14 and 12.15). First, this compound is obtained as the corresponding hydrochloride (Formula **B** in both figures). But it can also be deprotonated to form

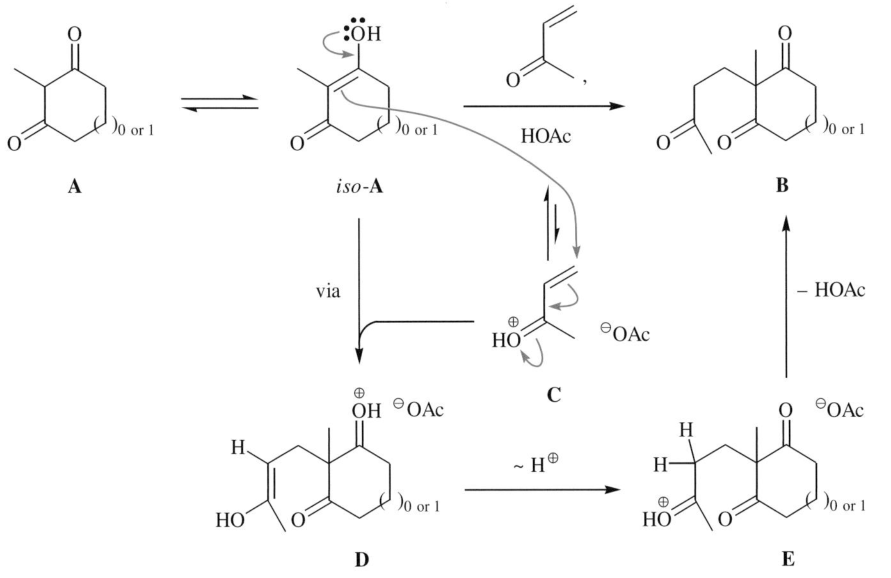

Fig. 12.13. An *acid-catalyzed* Michael addition. Unlike *base-catalyzed* Michael additions, which are also known (Section 13.6.1), a Michael addition occurring under acidic conditions cannot be followed by the formation of a six-membered ring through aldol addition, aldol condensation or—starting from other reactants than the ones shown here—acylation. Under basic conditions, however, the corresponding tandem reactions are quite common (see Section 13.6.2).

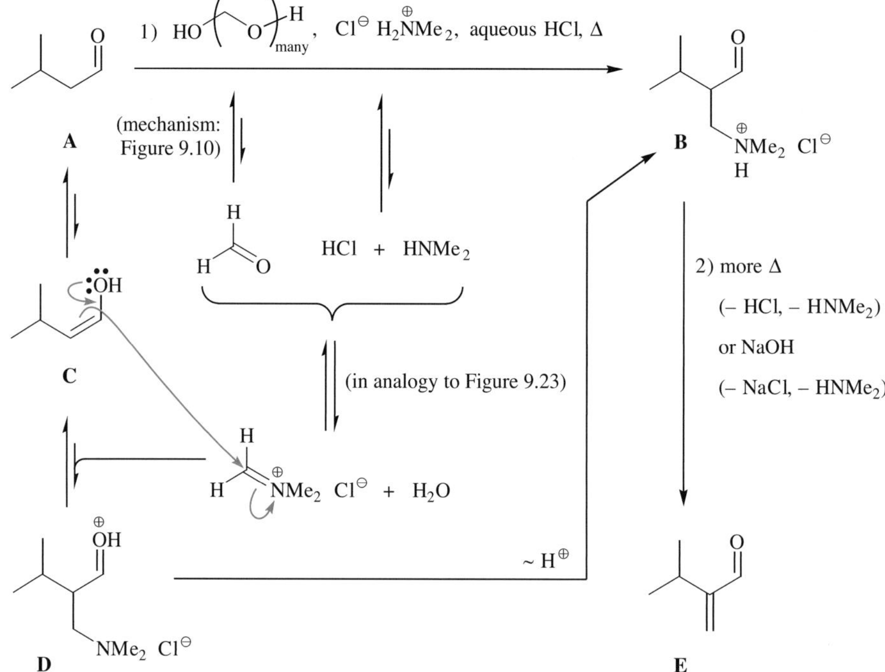

Fig. 12.14. Aminomethylation of an aldehyde via its enol (**C**) ("Mannich reaction"). This transformation can then be employed (as shown) to obtain an α-methylenated aldehyde (**E**).

the underlying amine, called the **Mannich base**. Reactions of this type are called **Mannich reactions** or **aminoalkylations**. Usually, the iminium ion involved is generated *in situ*, i.e., through condensation of a secondary amine with an aldehyde in hydrochloric acid solution (mechanism: Section 9.3).

The Mannich reaction of an aldehyde enol (example: Formula **C** in Figure 12.14) or a ketonic enol (example: Formula **C** in Figure 12.15) often proceeds beyond the hydrochloride of a β-aminocarbonyl compound or the Mannich base. The reason is that the secondary amine or its hydrochloride, which has previously been incorporated as part of the iminium ion, is relatively easy to eliminate from these two types of product. The elimination product is an α,β-unsaturated aldehyde (example: Formula **E** in Figure 12.14) or an α,β-unsaturated ketone (example: Formula **D** in Figure 12.15)—that is, an α,β-unsaturated carbo*n*yl compound. Figure 13.51 will show how the Mannich reaction of a carboxylated lactonic enol provides access to an α-methylene lactone, that is, an α,β-unsaturated carbo*x*yl compound.

By way of Mannich reaction (step 1) and β-elimination (step 2), the transformations shown in Figures 12.14 and 12.15 demonstrate how an aldol condensation (for the term see Section 13.4.1) can be conducted under acidic conditions as well. Both the enamine reaction in Figure 12.18 and the enol ether reaction in Figure 12.23 illustrate the same thing differently. Many aldol condensations, however, start from carbonyl compounds only and proceed under basic conditions. They follow a totally different mechanism (Section 13.4.1).

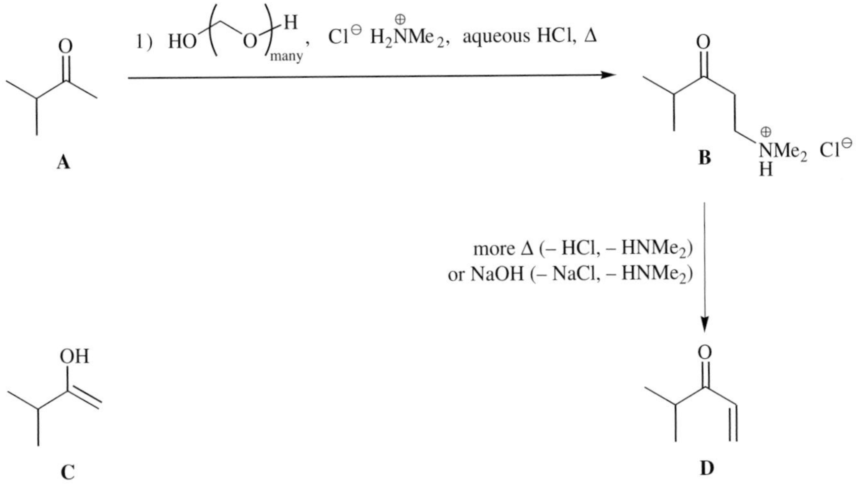

1) $HO\!-\!(\!\sim\!O)_{many}\!-\!H$, $Cl^{\ominus}$ $H_2\overset{\oplus}{N}Me_2$, aqueous HCl, Δ

A

B $\overset{\oplus}{N}HMe_2$ $Cl^{\ominus}$

more Δ (– HCl, – $HNMe_2$)
or NaOH (– NaCl, – $HNMe_2$)

C

D

Fig. 12.15. Aminomethylation of a ketone via its enol (**C**) ("Mannich reaction"). This transformation can then be employed (as shown) to obtain an α-methylenated ketone (**D**).

12.3 α-Functionalization of Ketones via Their Enamines

The enol content of simple ketones is much lower than that of β-ketoesters or β-diketones. For a number of electrophiles it is often too low. Hence, functionalizations with the respective electrophile via the enol form do not succeed in these cases. This problem can be managed, though, by converting the ketone (Formula **A** in Figure 12.16) into an enamine **D** with the aid of a condensation with a secondary amine that is in line with Figure 9.29 and the mechanism given there. Enamines are common synthetic equivalents for ketonic and aldehyde enols.

Synthetic Equivalents

A **synthetic equivalent** is an compound exhibiting the following properties: (1) a synthetic equivalent reacts in a way that the actual substrate or the actual reagent cannot. (2) After having reacted a synthetic equivalent can be modified in such a way that the same product is obtained indirectly whose direct generation—i.e., its generation from the actual substrate or reagent—was not possible.

Enamines react faster with electrophiles than the corresponding enols. The reason is that the electrophilic reaction with enamines yields an iminium ion (Formula **E** in Figure 12.16) where the positive charge is better stabilized than in an oxocarbenium ion derived from the electrophile and an enol. The fate of the iminium ion **E** thus formed relies on the reaction conditions as already outlined in the opening passages of this chapter. It can be deprotonated to give the enamine **G** (if a base is present or if it is strongly acidified by the substituent E = EWG), or it survives until hydrolysis takes place upon aqueous workup. Irrespective of whether an enamine **G** or an iminium ion **E** is present at this stage, this leads to the desired α-functionalized ketone **B** and the secondary amine or an ammonium salt thereof.

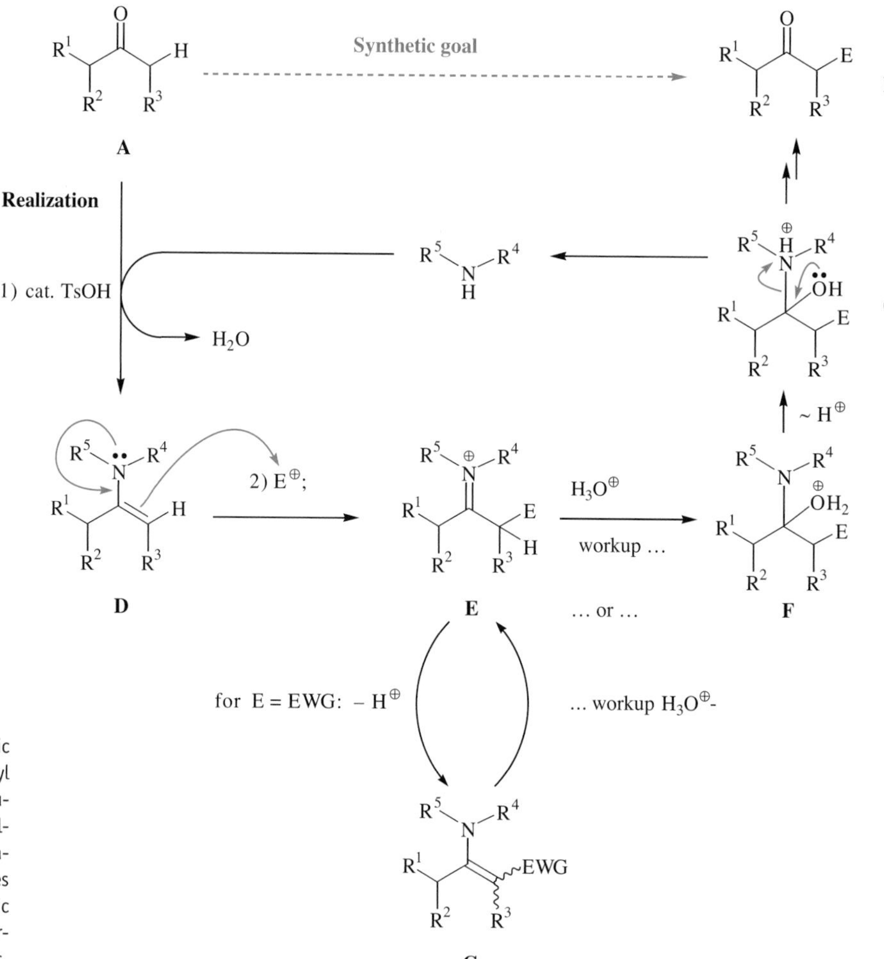

Fig. 12.16. Electrophilic functionalization of carbonyl compounds **A**, whose enol tautomer is unsuitable or unavailable at sufficient concentration, via the related enamines **D**: a survey of mechanistic details and important intermediate products.

Following the mechanism of Figure 12.16, enamines can react with good alkylating agents (Figure 12.17). First, the iminium ions **B** are formed and subsequent hydrolysis leads to the formation of the α-alkylated ketones **C**. The attempt to produce the alkylation products in Figure 12.17 from the same alkylating agents and cyclohexanone or its equilibrium fraction of enol would be bound to fail.

Under acidic conditions enamines such as compound **A** in Figure 12.18 and aldehydes undergo condensation to form the conjugated iminium ions **D**. These will be deprotonated by the concomitantly formed hydroxide ions, In this way dienamines of type **F** are formed, which will then be hydrolyzed upon acidic workup to give a carbonyl group. The generation of the α,β-unsaturated ketones **E** is thus completed. You will learn about type **E** compounds in Section 13.4.1 in connection with the so-called crossed aldol condensation products. It should be noted that it is not possible to form the same unsaturated ketone by reacting cyclopentanone or its equilibrium fraction of enol with an aliphatic aldehyde. Instead, a cyclodehydration of

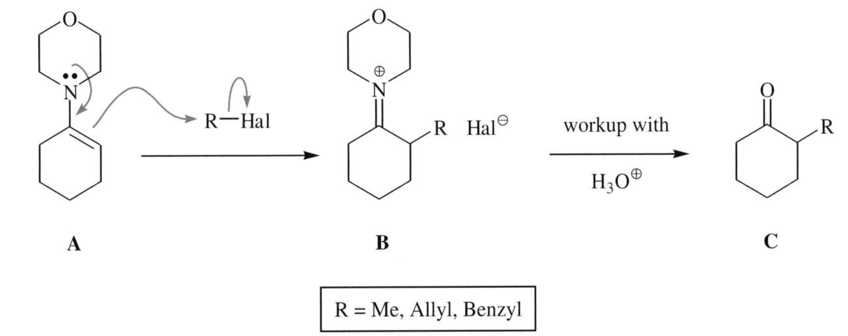
R—Hal
R Hal⊖
workup with
H3O⊕
R
A
B
C
R = Me, Allyl, Benzyl

Fig. 12.17. Alkylation of enamines. Only highly reactive alkylating agents can be employed. First, an iminium ion **B** is formed. Hydrolysis of **B** then yields an α-alkylated ketone.

the cyclopentanone in the fashion of Figure 12.12 or perhaps a trimerization of the aldehyde (cf. Figure 9.11) would occur.

Side Note 12.3. Catalytic Enantioselective Aldol Addition Using Enamine Chemistry

Do you remember Figure 12.13? 2-Methyl-1,3-cyclopentanedione and 2-methyl-1,3-cyclohexanedione in acetic acid undergo Michael additions to methyl vinyl ketone to form 1,5-diketones. Their structure reappears in Figure 12.19 as the substrate **A**. In the discussion of the formation mechanism of these 1,5-diketones you were probably not at all surprised to

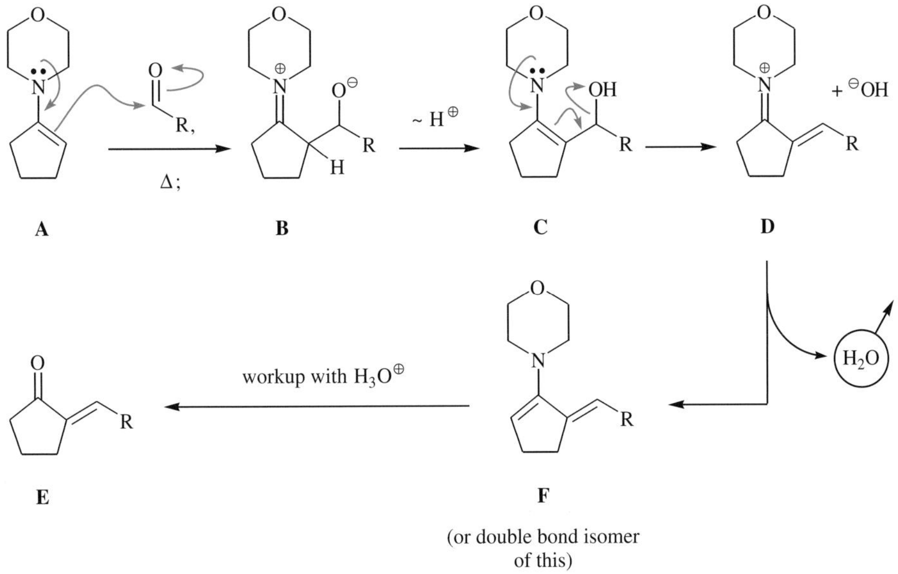
R,
Δ;
~ H⊕
+ ⊖OH
A
B
C
D
workup with H3O⊕
H2O
E
F
(or double bond isomer of this)

Fig. 12.18. Hydroxyalkylation of an enamine ($\rightarrow$ hydroxy-enamine **C**), followed by *in-situ*-dehydration ($\rightarrow$ dienamine **F**) and acidic workup ($\rightarrow$ α,β-unsaturated ketone **E**). Since the enamine **A** is produced from cyclopentanone, the figure shows the second part of a two-step reaction, which is an alternative to the base-mediated crossed aldol condensation (see Section 13.4.1).

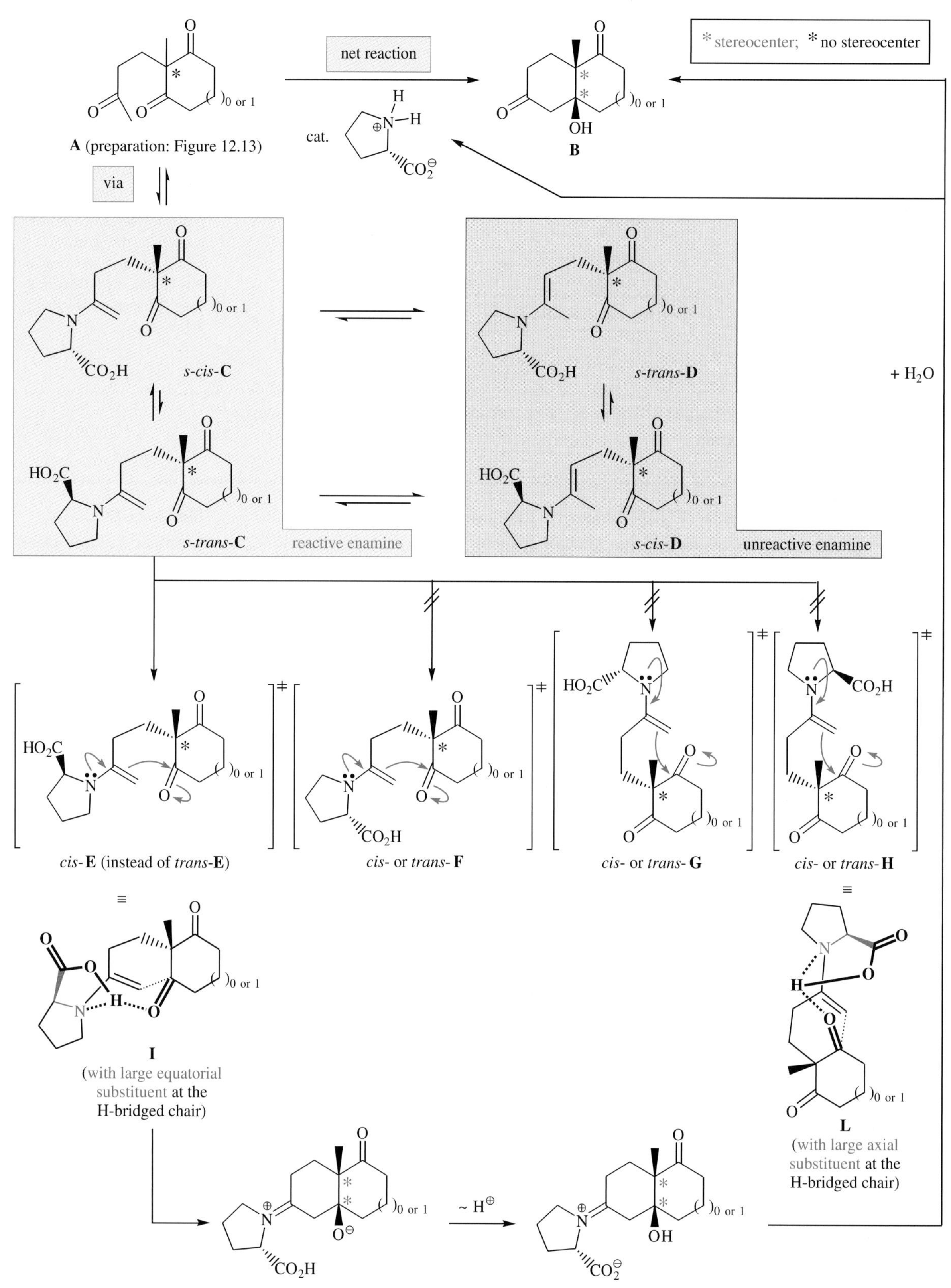

* stereocenter; * no stereocenter
net reaction
A (preparation: Figure 12.13)
B
cat.
via
s-cis-C
s-trans-C
reactive enamine
s-trans-D
s-cis-D
unreactive enamine
+ H2O
cis-E (instead of trans-E)
cis- or trans-F
cis- or trans-G
cis- or trans-H
I
(with large equatorial substituent at the H-bridged chair)
L
(with large axial substituent at the H-bridged chair)
~ H⊕
cis-J
K

Fig. 12.19. Catalytic asymmetric aldol additions in steroid synthesis. The key intermediates are the enamines **C**, and the key step is their intramolecular hydroxyalkylation (→ *cis*-**J**). However, the hydroxyiminium ions **K** are formed instead of hydroxyenamines or dienamines (as is the case with the hydroxyalkylation shown in Figure 12.18). The hydrolysis of the hydroxyiminium ions **K** furnishes the β-hydroxyketones **B** as the finally isolated products. In the stereogenic step of the aldol additions **C** → *cis*-**J** three factors contribute to stereoselectivity: (1) the newly formed ring is *cis*- instead of *trans*-annulated on the cycloalkane dione, which eliminates the transition states *trans*-**E**, *trans*-**F**, *trans*-**G** and *trans*-**H**. (2) The cycloalkane dione moiety preferentially approaches the C=C double bond of the enamine moiety from the half-space that is not occupied by the carboxyl group (the reason is that only under these conditions can the carboxyl proton electrophilically activate the carbonyl group by forming a hydrogen bond); this criterion eliminates the transition states *cis*-**F** and *cis*-**G**. (3) In order to establish this activating H bond, the proline substituent must be able to offer its CO_2H group to the reacting C=O double bond without steric hindrance. Of the two remaining transition states *cis*-**E** (cf. Stereo Formula **I**) is more stable than *cis*-**H** (cf. Stereo Formula **L**) as a result of this criterion. In conclusion, the cyclization exclusively proceeds via transition state *cis*-**E**.

learn that the reaction stopped at this stage. Basically, it would have been also conceivable that an aldol addition **A** → **B** or an aldol addition **A** → stereoisomer of **B** would have followed immediately. The absence of such aldol additions indicates that the amount of **A**-enol is not sufficient for a reaction of the respective protonated cyclohexanedione moiety acting as the *C* electrophile to proceed fast enough.

The enamine chemistry in Figure 12.19 can help to overcome this reactivity problem. The crucial enamine **C** is so electron-rich that it reacts electrophilically with a C=O group of the adjacent cycloalkane dione moiety without the need for prior protonation to an oxocarbenium ion. This is just like the electrophilic reaction of the aldehyde on the enamine **A** in Figure 12.18.

In the reaction of Figure 12.19, the alkoxide formed in this step deprotonates a carboxylic acid (*cis*-**I** → **K**), whereas in Figure 12.18 an iminium ion is deprotonated (**B** → **C**). Accordingly, different chemoselectivities are observed: Figure 12.19 shows an enamine-mediated aldol *addition*, and Figure 12.18 presents an enamine-mediated aldol *condensation*. Hydrolysis of the iminium ion **K** in Figure 12.19 leads to the formation of the aldol addition products **B** and the amine which, together with the still unconsumed substrate **A**, forms the new enamine **C**, to start the catalytic cycle anew.

The enamine-mediated aldol addition of Figure 12.19 has several special features:

- The enamine **C** as the intermediate of the primary reaction is generated *in situ* from the small amount of L-proline that is added as a catalyst. The more stable but unreactive enamine **D** is formed too, but it equilibrates with **C**.
- The iminium ion *cis*-**J** resulting from the primary reaction is hydrolyzed via **K** *in situ*, that is, no additional step is required.
- Both of these factors makie it possible for the aldol addition in Figure 12.19 to proceed via a catalytic amount of the enamine intermediate and hence a catalytic amount of proline.
- The aldol additions in Figure 12.19 are stereogenic and proceed with high (induced) diastereoselectivities: with $n = 0$ (formation of a five-membered ring) and $n = 1$ (formation of a six-membered ring) the iminium ion *cis*-**J** is obtained with ds = 96.5:3.5 and ds = 97.5:2.5, respectively. The origin of this stereocontrol is the stereocenter of L-proline. Since it is added in catalytic amounts, the examples in Figure 12.19 are catalytic asymmetric syntheses (for the term see Section 3.4.6).

- The hydrolysis of iminium ions **K** proceeds with complete retention of configuration (Figure 12.19). Hence, the resulting aldol adducts **B** for $n = 0$ (formation of a five-membered ring) and for $n = 1$ (formation of a six-membered ring) are obtained with 93% *ee* and 95% *ee*, respectively.

The unsaturated diketones that are obtained from the β-hydroxy ketones in Figure 12.19 by proton-catalyzed dehydration according to the E2 mechanism are of considerable importance:

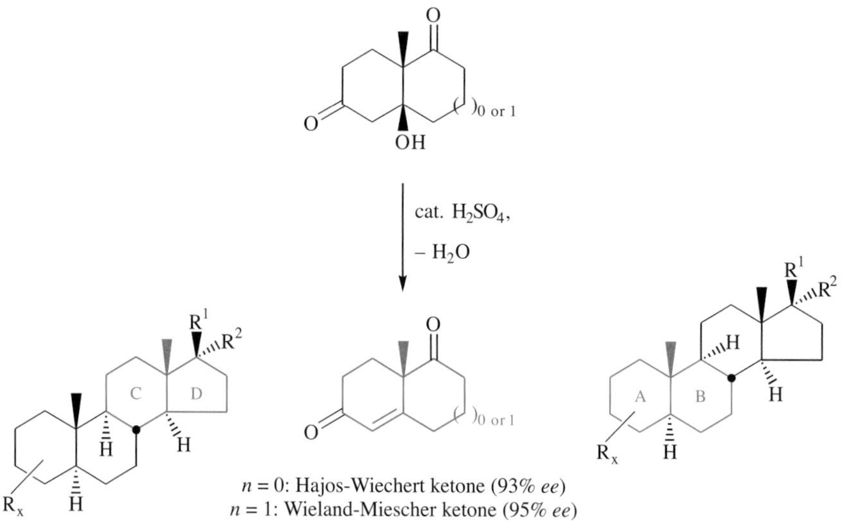

cat. H_2SO_4,
$- H_2O$

$n = 0$: Hajos-Wiechert ketone (93% *ee*)
$n = 1$: Wieland-Miescher ketone (95% *ee*)

In the six-/five-membered ring series the dehydration product is called the Hajos–Wiechert ketone, whereas in the six-/six-membered ring series it is referred to as the Wieland–Miescher ketone. Regarding its constitution and configuration the Hajos–Wiechert ketone corresponds to the C-ring/D-ring moiety of the steroid skeleton and the Wieland–Miescher ketone to the A-ring/B-ring moiety.

Figure 12.20 demonstrates that Michael acceptors can also act as electrophiles towards enamines. Aqueous workup leads to the regeneration of the carbonyl group, and the Michael addition product **C** is obtained. An acrylic acid ester would react with the enamine **A** in complete analogy to the reaction of the acrylonitrile shown here; the same holds true for the methyl vinyl ketone.

And finally, acylating agents can also be used as electrophiles to react with enamines. Following the hydrolysis of the enaminoketones (i.e., compounds with the substructure R_2N–C=C–C(=O)–R') or enaminoesters (i.e., compounds with the substructure R_2N–C=C–C(=O)–OR') the acylation products of the corresponding ketones are obtained. Figure 12.21 gives the mechanistic details of the acylation with an acid chloride, and Figure 12.22 shows the acylation with ethyl chloroformate. The first acylation yields β-diketones, the second furnishes a β-ketoester.

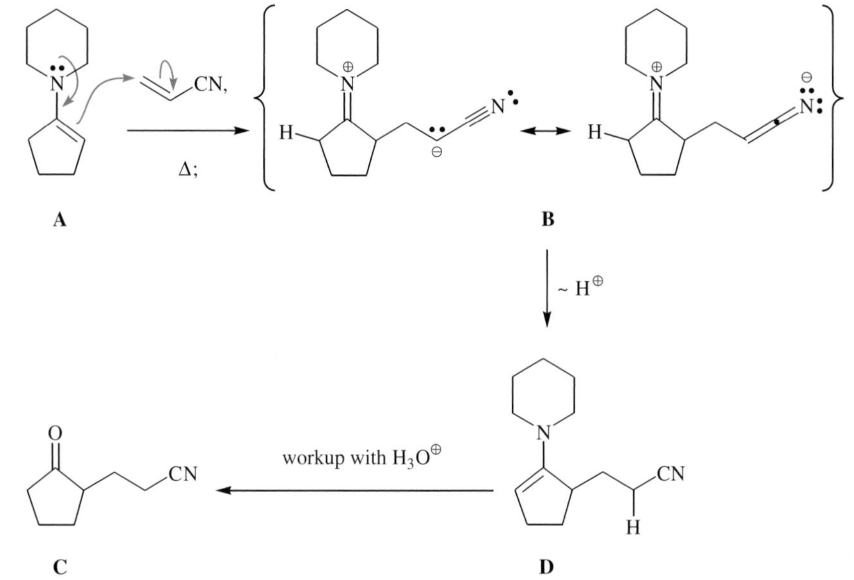

CN, Δ;

A B

~ H$^{\oplus}$

workup with $H_3O^{\oplus}$

C D

Fig. 12.20. Michael addition of an enamine (→ enamine **D**), followed by acidic workup → δ-ketonitrile **C**). Since the enamine **A** is produced from cyclopentanone, the figure shows the second part of a two-step reaction that is an alternative to the base-mediated one-step Michael addition (see Section 13.6.1).

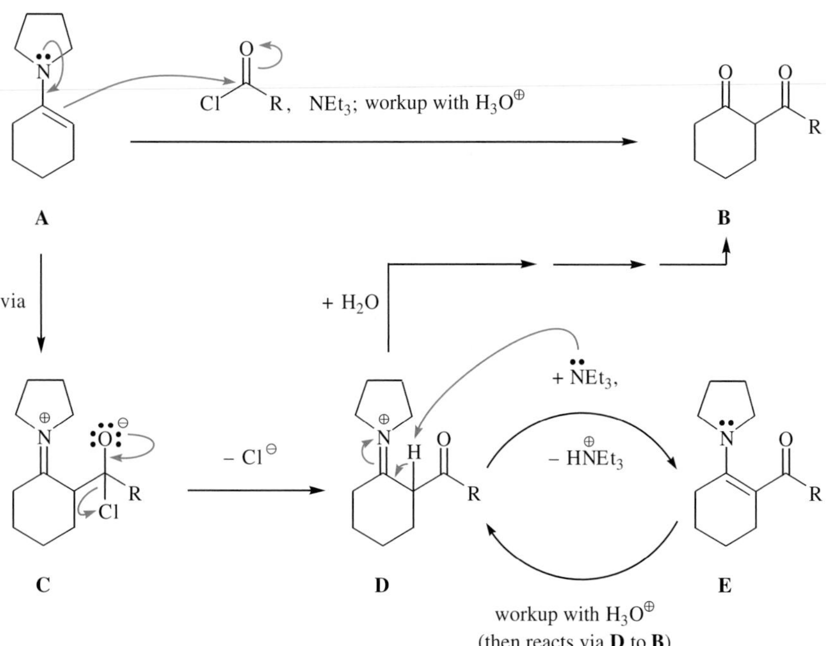

Fig. 12.21. Acylation of an enamine with an acid chloride (→ enaminoketone **E**), followed by acidic workup (→ β-diketone **B**). Since the enamine **A** is produced from cyclohexanone, the figure shows the second part of a two-step reaction which is an alternative to the one-step acylation of a ketone enolate (cf. Section 13.5.2).

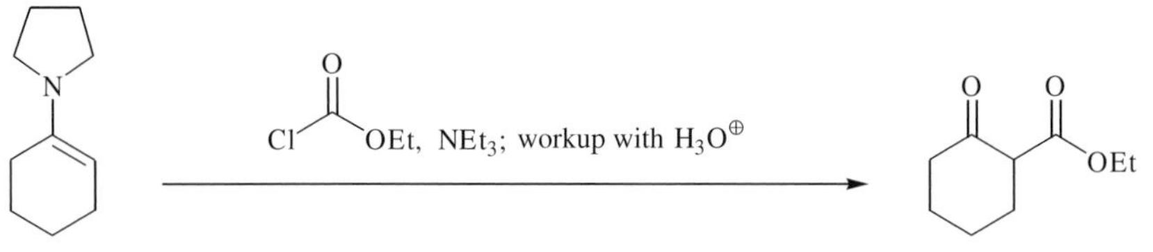

Fig. 12.22. Acylation of an enamine with ethyl chloroformate. The mechanism of this reaction corresponds to the mechanism outlined in Figure 12.21. As in Figure 12.21, the second part of a two-step reaction representing an alternative to the one-step acylation of a ketone enolate is given here.

12.4 α-Functionalization of Enol Ethers and Silyl Enol Ethers

According to Section 12.3 enamines are just *one* synthetic equivalent for enols that are not sufficiently represented in equilibrium with a carbonyl compound to allow for α-functionalizations. Enol ethers and silyl enol ethers, which are addressed in this section, are *other* synthetic equivalents for such enols. An enol ether, for example, is used as an enol equivalent for aldehyde enols, since several aldehydes do not form stable enamines. In addition, enol ethers or silyl enol ethers are usually employed as synthetic equivalents for the enols of α,β-unsaturated carbonyl compounds. The attempt to react α,β-unsaturated carbonyl compounds with secondary amines to give a dienamine is often frustrated by a competing 1,4-addition of the amine. The combination of these factors turns the dienol ether **B** of Figure 12.23 into a species for which there is no analog in enamine chemistry.

Under the influence of a Brønsted- or Lewis acid the dienol ether **B** in the example of Figure 12.23 reacts with an acetal. As far as the dienol ether is concerned the reaction that occurs corresponds to an electrophilic substitution reaction, and it corresponds to an S_N1 reaction with respect to the acetal, which is illustrated at the very bottom of Figure 12.23. Initially, cleavage of a methoxy residue from the acetal **A** in an equilibrium reaction produces the oxocarbenium ion **F**, which is the active *C* electrophile. The latter reacts with the dienol ether **B**, forming a new oxocarbenium ion **G**. **G** picks up the equivalent of methanol originating from the acetal **A** and thus delivers the methoxy-substituted acetal **C**. Why the methanol and not **B**?

The methoxy-substituted acetal **C** is isolated and is subjected to another two reactions: (1) the acetal function is hydrolyzed under acidic conditions to provide the methoxylated aldehyde **E**. (2) With a catalytic amount of sodium methoxide, the methoxylated aldehyde **E** reacts via an $E1_{cb}$ elimination, which leads to the formation of the poly-unsaturated aldehyde **D**. The substrates **A** *and* **B** of the three-step reaction sequence in Figure 12.23 derive from crotonic aldehyde ($H_3C{-}CH{=}CH{-}CH{=}O$); the respective methods of preparation are outlined in the figure. The final product, which is the highly unsaturated aldehyde **D**, displays the structure of an aldol condensation product, namely of the crotonic aldehyde mentioned above.

Following the mechanism given in Figure 12.23, the addition of an acetal to a "simple" enol ether (in contrast to the dienol ether **B** shown above) leads to a β-alkoxy acetal. This reaction is known as the **Mukaiyama aldol addition**. If this is followed by a hydrolysis (of the

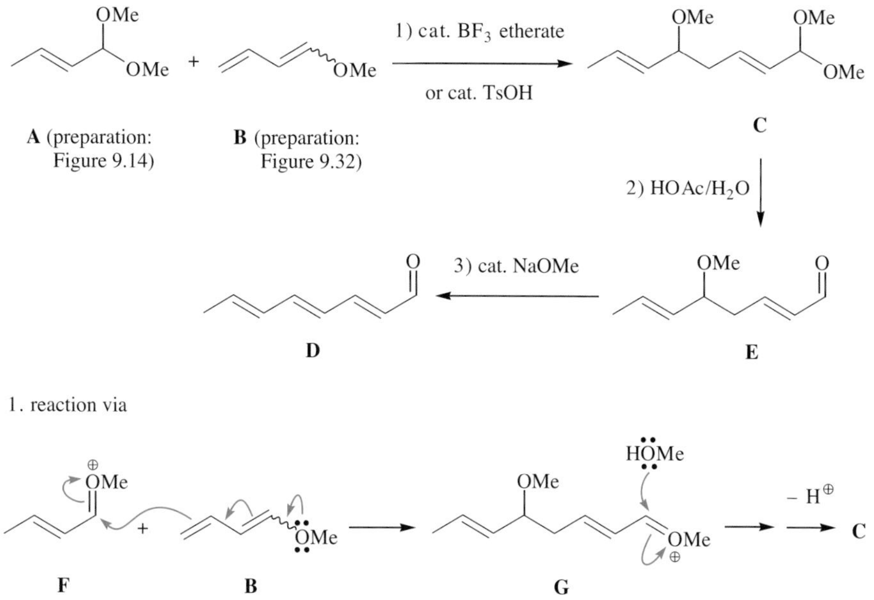

1) cat. BF_3 etherate or cat. TsOH

A (preparation: Figure 9.14) + **B** (preparation: Figure 9.32) → **C**

2) $HOAc/H_2O$ → **E**

3) cat. NaOMe → **D**

1. reaction via

F + **B** → **G** → $-H^{\oplus}$ → **C**

Fig. 12.23. A Mukaiyama aldol addition (→ **C**) and its reaction mechanism (bottom row). As shown here, this method can be exploited to obtain the poly-unsaturated aldehyde **D**. Under the conditions of the first reaction step the primary product **C**—which, like the substrate **A**, is an acetal—does not compete with **A** for still unconsumed enol ether **B**. This is due to the fact that the methoxy substituent in the oxocarbenium ion **G**, which would have to be regenerated from **C** in order to undergo further reaction with **B**, destabilizes **G** because of its electron-withdrawing inductive (–I) effect.

acetal) and a β-elimination (of one equivalent of alcohol)—in analogy to the reaction sequence of Figure 12.23—a **Mukaiyama aldol condensation** has been performed.

Modern variants of the Mukaiyama aldol addition start from silyl enol ethers, not from enol ethers, and use an aldehyde instead of the acetal as the electrophile. Mukaiyama aldol additions of this kind have been included in the C,C coupling reactions that build the basic repertoire of modern synthetic chemistry and can even be performed in a catalytic enantioselective fashion.

Side Note 12.4. Dehydrogenation of Ketones via Silyl Enol Ethers

Figure 12.24 depicts the oxidation of a silyl enol ether **A** to give an α,β-unsaturated ketone **B**. Mechanistically, three reactions must be distinguished. The first justifies why this reaction is introduced *here*. The silyl enol ether **A** is electrophilically substituted by palladium(II) chloride. The α-palladated cyclohexanone **E** is formed via the intermediary *O*-silylated oxocarbenium ion **C** and its parent compound **D**. The enol content of cyclohexanone, which is the origin of the silyl enol ether **A**, would have been too low to allow for a reaction with palladium(II) chloride. Once more, the synthetic equivalence of a silyl enol ether and a ketonic enol is the basis for success (Figure 12.24).

In the second part of the reaction of Figure 12.24, the α-palladated cyclohexanone **E** decomposes (since it is an alkylpalladium(II) compound with a *syn*-H atom in the position β to the metal) to an alkene, namely the previously mentioned *unsaturated* ketone **B**, and to H–Pd–Cl. In Chapter 16 this type of reaction will be repeatedly encountered in connection with the keyword "β-hydride elimination": as **A** → **B** in the Figures 16.13 and 16.14 and as step 7 of Figure 16.35 (part II). H–Pd–Cl decomposes according to the equation H–Pd–Cl → Pd(0) + HCl, which describes a reductive elimination. Reductive eliminations will be

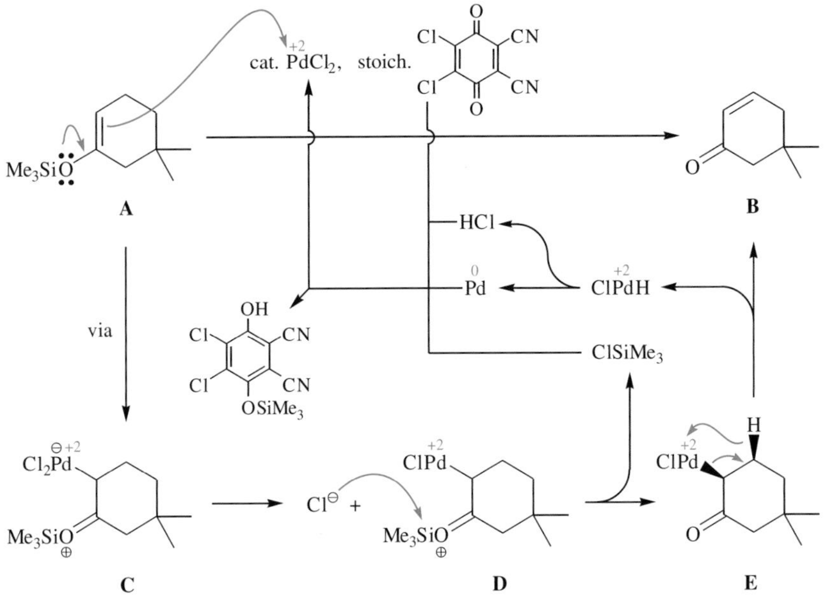

Fig. 12.24. Oxidation of a silyl enol ether (**A**) to yield an α,β-unsaturated ketone **B**. Since the substrate used here is the silylation product of the enolate **B** of Figure 13.13, this figure shows the second part of a two-step sequence that is used to convert ketones into α,β-unsaturated ketones.

investigated in more detail in Chapter 16. The third part involved in the dehydrogenation of Figure 12.24, namely the reoxidation of Pd(0) to $PdCl_2$, ensures the affordability of the overall reaction in the first place. Thus, catalytic amounts of the expensive palladium(II) chloride are sufficient to achieve full conversion of the silyl enol ether **A**. The reoxidation Pd(0) $\rightarrow$ $PdCl_2$ is performed with a stoichiometric amount of 2,3-dichloro-4,5-dicyanobenzoquinone (**DDQ**).

Figure 12.25 shows how acetals can be brominated electrophilically because of the (weakly) acidic reaction conditions. Proper acidity and electrophilicity is ensured by the use of pyridinium tribromide (**B**). This reagent is produced from pyridinium hydrobromide and one equivalent of bromine. Pyridinium tribromide is acidic enough to cleave the acetal **A** into the enol ether **G**. This cleavage succeeds by way of an E1 elimination like the one encountered in Figure 9.32 as an enol ether synthesis. The enol ether **G** reacts with the tribromide ion via the bromine-containing oxocarbenium ion **H** and the protonated acetal **D** to form the finally isolated neutral bromoacetal **C**. (The reaction can be conducted despite the unfavorable equilibrium between the acetal **A** and the enol ether **G**, since **G** continuously reacts and is thus eliminated from the equilibrium.)

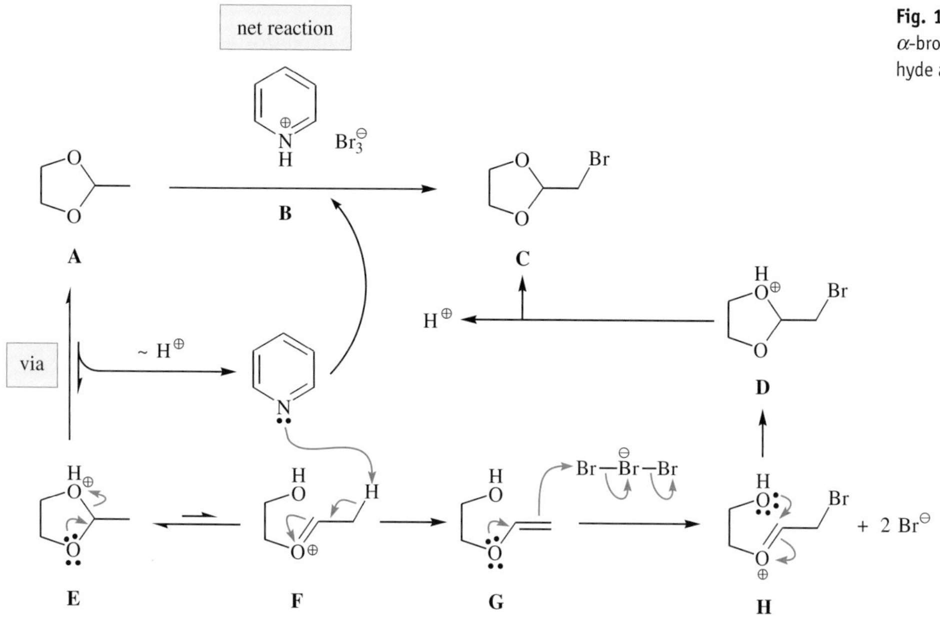

Fig. 12.25. Mechanism of the α-bromination of the acetaldehyde acetal **A**.

References

12.1

R. Brettle, "Aldehydes," in *Comprehensive Organic Chemistry* (J. F. Stoddart, Ed.), Vol. 1 (Stereochemistry, Hydrocarbons, Halo Compounds, Oxygen Compounds), Pergamon Press, U. K., **1979**, 943–1016.

T. Laird, "Aromatic Aldehydes," in *Comprehensive Organic Chemistry* (J. F. Stoddart, Ed.), Vol. 1 (Stereochemistry, Hydrocarbons, Halo Compounds, Oxygen Compounds), Pergamon Press, U. K., **1979**, 1105–1160.

A. J. Waring, "Ketones," in *Comprehensive Organic Chemistry* (J. F. Stoddart, Ed.), Vol. 1 (Stereochemistry, Hydrocarbons, Halo Compounds, Oxygen Compounds), Pergamon Press, U. K., **1979**, 1017–1104.

T. Laird, "Aromatic Ketones," in *Comprehensive Organic Chemistry* (J. F. Stoddart, Ed.), Vol. 1 (Stereochemistry, Hydrocarbons, Halo Compounds, Oxygen Compounds), Pergamon Press, U. K., **1979**, 1161–1212.

Z. Rappoport (Ed.), "The Chemistry of Enols (The Chemistry of Functional Groups)," Wiley, Chichester, U. K., **1990**.

A. J. Kresge, "Flash Photolytic Generation and Study of Reactive Species: From Enols to Ynols," *Acc. Chem. Res.* **1990**, *23*, 43–48.

Y. Chiang, A. J. Kresge, "Enols and Other Reactive Species," *Science* **1991**, *253*, 395–400.

A. J. Kresge, "Ingold Lecture: Reactive Intermediates: Carboxylic Acid Enols and Other Unstable Species," *Chem. Soc. Rev.* **1996**, *25*, 275–280.

A. J. Kresge, "Keto-Enol Tautomerism of Phenols in Aqueous Solution," *Chemtracts Org. Chem.* **2002**, *15*, 212–215.

12.2

A. Cox, "Halo Carboxylic Acids," in *Comprehensive Organic Chemistry* (I. O. Sutherland, Ed.), Vol. 2 (Nitrogen Compounds, Carboxylic Acids, Phosphorus Compounds), Pergamon Press, U. K., **1979**, 719–738.

O. Touster, "The Nitrosation of Aliphatic Carbon Atoms," *Org. React.* **1953**, *7*, 327–377.

F. F. Blicke, "The Mannich Reaction," *Org. React.* **1942**, *1,* 303–341.

H. E. Zaugg, W. B. Martin, "α-Amidoalkylations at Carbon," *Org. React.* **1965**, *14*, 52–269.

M. Tramontini, L. Angiolini, "Further Advances in the Chemistry of Mannich Bases," *Tetrahedron* **1990**, *46*, 1791–1837.

M. Tramontini, L. Angiolini, "Mannich Bases," CRC Press, Boca Raton, Florida, **1994**.

A. V. Bordunov, J. S. Bradshaw, V. N. Pastushok, R. M. Izatt, "Application of the Mannich Reaction for the Synthesis of Azamacroheterocycles," *Synlett* **1996**, 933–948.

M. Arend, B. Westermann, N. Risch, "Modern Variants of the Mannich Reaction," *Angew. Chem. Int. Ed. Engl.* **1998**, *37*, 1044–1070.

N. Rabjohn, "Selenium Dioxide Oxidation," *Org. React.* **1976**, *24*, 261–415.

12.3

Z. Rappoport (Ed.), "The Chemistry of Enamines," Wiley, Chichester, U. K., **1994**.

J. P. Adams, G. Robertson, "Imines, Enamines and Related Functional Groups," *Contemp. Org. Synth.* **1997**, *4*, 183–260.

J. P. Adams, "Imines, Enamines and Oximes," *Contemp. Org. Synth.* **1997**, *4*, 517–543.

J. P. Adams, "Imines, Enamines and Oximes," *J. Chem. Soc. Perkin Trans. 1* **2000**, 125–139.

D. C. Oare, C. H. Heathcock, "Acyclic Stereocontrol in Michael Addition Reactions of Enamines and Enol Ethers," *Top. Stereochem.* **1991**, *20,* 87–170.

C. Agami, "Mechanism of the Proline-Catalyzed Enantioselective Aldol Reaction. Recent Advances," *Bull. Soc. Chim. Fr.* **1988**, *3*, 499–507.

B. List, "Asymmetric Aminocatalysis," *Synlett* **2001**, 1675–1686.

B. List, "Proline-Catalyzed Asymmetric Reactions," *Tetrahedron* **2002**, *58,* 5573–5590.

12.4

T. Mukaiyama, "The Directed Aldol Reaction," *Org. React.* **1982**, *28*, 203–331.

A. Deagostino, C. Prandi, P. Venturello, "α,β-Unsaturated Acetals in Synthesis," *Curr. Org. Chem.* **2003**, *7,* 821–839.

Further Reading

S. K. Bur, S. F. Martin, "Vinylogous Mannich Reactions: Selectivity and Synthetic Utility," *Tetrahedron* **2001**, *57*, 3221–3242.

S. F. Martin, "Evolution of the Vinylogous Mannich Reaction as a Key Construction for Alkaloid Synthesis," *Acc. Chem. Res.* **2002**, *35,* 895–904.

C. Reichardt, "Vilsmeier-Haack-Arnold Formylations of Aliphatic Substrates with N-Chloromethylene-N,N-dimethylammonium Salts," *J. Prakt. Chem.* **1999**, *341,* 609–615.

P. Lue, J. V. Greenhill, "Enamines in Heterocyclic Synthesis," *Adv. Heterocycl. Chem.* **1997**, *67*, 207–343.

V. G. Granik, V. A. Makarov, C. Parkanyi, "Enamines as Synthons in the Synthesis of Heterocycles," *Adv. Heterocycl. Chem.* **1998**, *72,* 283–360.

A. W. Erian, "The Chemistry of β-Enaminonitriles as Versatile Reagents in Heterocyclic Synthesis," *Chem. Rev.* **1993**, *93*, 1991–2005.

S. Rajappa, "Nitroenamines: An Update," *Tetrahedron* **1999**, *55*, 7065–7114.

A.-Z. A. Elassar, A. A. El-Khair, "Recent Developments in the Chemistry of Enaminones," *Tetrahedron* **2003**, *59,* 8463–8480.

G. Casiraghi, F. Zanardi, G. Appendino, G. Rassu, "The Vinylogous Aldol Reaction: A Valuable, Yet Understated Carbon-Carbon Bond-Forming Maneuver," *Chem. Rev.* **2000**, *100*, 1929–1972.

P. I. Dalko, L. Moisan, "Enantioselective Organocatalysis," *Angew. Chem. Int. Ed. Engl.* **2001**, *40,* 3726–3748.

Chemistry of the Alkaline Earth Metal Enolates

13

Aldehydes, ketones, carboxylic esters, carboxylic amides, imines and *N,N*-disubstituted hydrazones react as electrophiles at their sp^2-hybridized carbon atoms. These compounds also become nucleophiles, if they contain an H atom in the α-position relative to their C=O or C=N bonds. This is because they can undergo tautomerization to the corresponding enol as seen in Chapter 12. They are also C,H-acidic at this position, i.e., the H atom in the α-position can be removed with a base (Figure 13.1). The deprotonation forms the conjugate bases of these substrates, which are called **enolates.** The conjugate bases of imines and hydrazones are called **aza enolates.** The reactions discussed in this chapter all proceed via enolates.

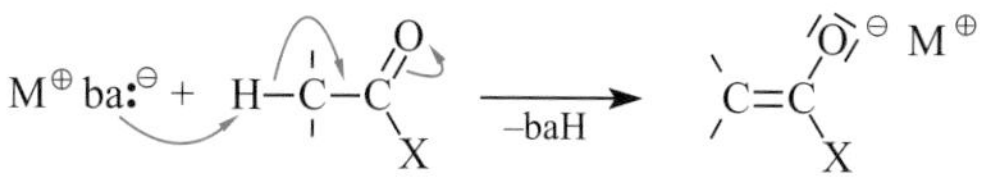

$M^{\oplus}$ ba:$^{\ominus}$ + H–C–C(=O)–X —(–baH)→ C=C(–O$^{\ominus}$ $M^{\oplus}$)–X

($M^{\oplus}$ = $Li^{\oplus}$, $Na^{\oplus}$, $K^{\oplus}$)

X	
H	Aldehyde enolate[1]
alkyl, aryl	Ketone enolate[1]
Oalkyl, Oaryl	Ester enolate
NR^1R^2	Amide enolate

[1]Also called simply enolate.

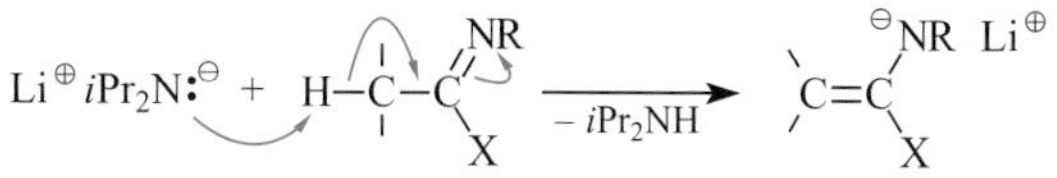

$Li^{\oplus}$ $i Pr_2N$:$^{\ominus}$ + H–C–C(=NR)–X —(– iPr_2NH)→ C=C(–$^{\ominus}$NR $Li^{\oplus}$)–X

X = H, alkyl
R = alkyl, $N(alkyl)_2$ } aza-enolate

Fig. 13.1. Formation of enolates from different C,H acids.

13.1 Basic Considerations

13.1.1 Notation and Structure of Enolates

In valence bond theory, every enolate can be described by two resonance forms. The negative formal charge is located at a C atom in one of these resonance forms and at an O or an N atom in the other resonance form. In the following, we refer to these resonance forms as the *carbanion* and the *enolate* resonance forms, respectively. Only the enolate resonance form is shown in Figure 13.1, because this resonance form has the higher weight according to reso-

Bruckner R (author), Harmata M (editor) In: *Organic Mechanisms – Reactions, Stereochemistry and Synthesis*
Chapter DOI: 10.1007/978-3-642-03651-4_13, © Springer-Verlag Berlin Heidelberg 2010

nance theory. The enolate resonance form places negative charge on the more electronegative heteroatom (O or N). These heteroatoms stabilize the negative charge better than the less electronegative C atom in the carbanion resonance form.

In Figure 13.1, the enolate structures are shown with the charge on the heteroatom and with the heteroatom in association with a metal ion. The metal ion stems from the reagent used in the enolate formation. In the majority of the reactions in Chapter 13, the enolate is generated by deprotonation of C,H acids. The commonly employed bases contain the metal ions $Li^{\oplus}$, $Na^{\oplus}$, or $K^{\oplus}$. Therefore, in Chapter 13, we will consider the chemistry of lithium, sodium, and potassium enolates.

It is known that the chemistry of enolates depends on the nature of the metal. Moreover, the metals are an integral part of the structures of enolates. Lithium enolates are most frequently employed, and in the solid state the lithium cations definitely are associated with the heteroatoms rather than with the carbanionic C atoms. Presumably the same is true in solution. The bonding between the heteroatom and the lithium may be regarded as ionic or polar covalent. However, the heteroatom is not the only bonding partner of the lithium cation irrespective of the nature of the bond between lithium and the heteroatom:

- Assuming ionic $Li^{\oplus}O^{\oplus}$ or $Li^{\oplus}NR^{\ominus}$ interactions, it may be appropriate to draw a parallel between the structures of enolates and ionic crystals of the $Li^{\oplus}O^{\oplus}$ or $Li^{\oplus}NR^{\ominus}$ types. In the latter structures, every lithium atom is coordinated by six neighboring anions.
- From the viewpoint of polar, yet covalent Li—O and Li—N bonds, lithium would be unable to reach a valence electron octet in the absence of bonding partners besides the heteroatom. The lithium thus has to surround itself by other donors in much the same way as has been seen in the case of the organolithium compounds (cf. Section 10.1).

Be this as it may, lithium attempts to bind to several bonding partners; the structural consequences for the enolates of a ketone, an ester, and an amide are shown in Figure 13.2: In contrast to the usual notation, these enolates are not monomers at all! The heteroatom that carries the negative charge in the enolate resonance form is an excellent bonding partner such that several of these heteroatoms are connected to every lithium atom. Lithium enolates often result in "tetramers" if they are crystallized in the absence of other lithium salts and in the absence of other suitable neutral donors. The lithium enolate of *tert*-butyl methyl ketone, for example, crystallizes from THF in the form shown in Figure 13.3.

"Tetramers" like the one in Figure 13.3 contain cube skeleta in which the corners are occupied in an alternating fashion by lithium and enolate oxygen atoms. Every lithium atom is surrounded by three enolate oxygen atoms, and vice versa. Every lithium atom binds a molecule of THF as its fourth ligand. It is for this reason that the term "tetramer" was used in quotation marks; the overall structure is a THF complex of the tetramer.

Figure 13.2 shows structures that contain two lithium enolates each. But again, these structures are not *pure* dimers. Both lithium atoms employ two of their coordination sites to bind to an N atom of the bidentate ligand TMEDA (see Figure 13.2 for name and structure).

"Oligomeric" enolates along with the associated neutral ligands also are called **aggregates.** Lithium enolates are likely to exist as aggregates not only in the solid state but also in solution. The neutral ligands in these aggregates can be TMEDA, DMPU (structure in Figure 2.17), HMPA (structure in Figure 2.17), THF and/or $HN(i\text{Pr})_2$. Lithium enolates may occur in such **homoaggregates**, but they also may be part of so-called **mixed aggregates.**

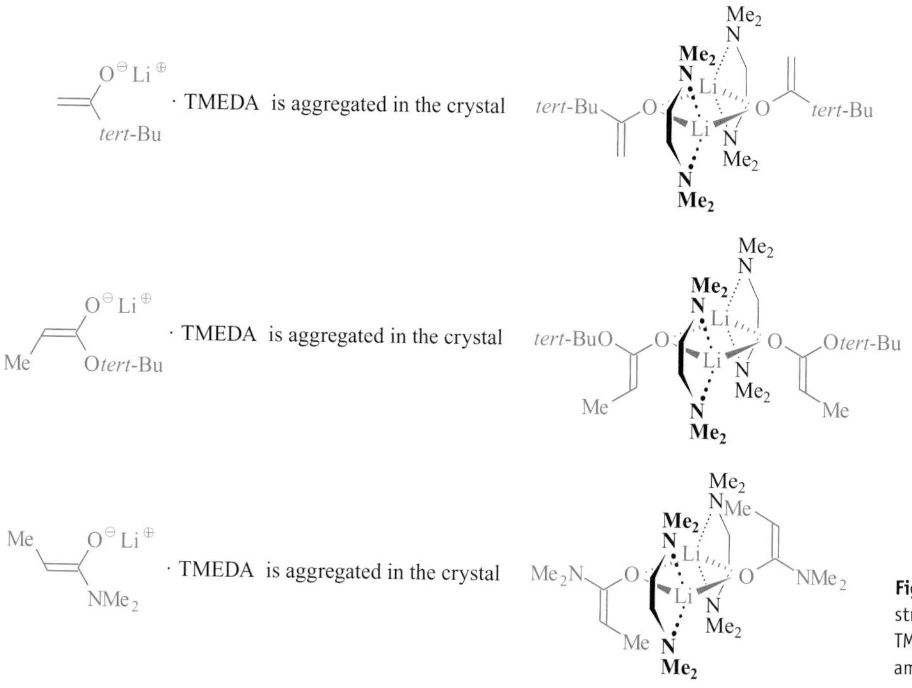

· TMEDA is aggregated in the crystal

· TMEDA is aggregated in the crystal

· TMEDA is aggregated in the crystal

Fig. 13.2. X-ray single crystal structures of lithium enolates. TMEDA, tetramethylethylenediamine.

The latter are aggregates that also contain other lithium compounds (e. g., LiHal, LiOR or LDA).

It is not known whether lithium enolates exist in solution as homoaggregates or as mixed aggregates, nor is it known whether lithium enolates react as aggregates or via other species that might be present in low concentration. But it is certain that the reactivity of lithium enolates is affected by the presence or absence of molecules that are capable of forming aggregates. However, all these insights about aggregation do not preclude focusing on the enolate monomers in discussions of the elemental aspects of enolate reactivity. Hence, in Chapter 13, all reactions are formulated in a simplified and unified format considering monomeric enolates. There also are enolates of metals other than Li, Na, or K, but these are not considered in this book. In addition, there are some metal-free enolates, the ammonium enolates, which can be generated in equilibrium reactions between amines and so-called active-methylene

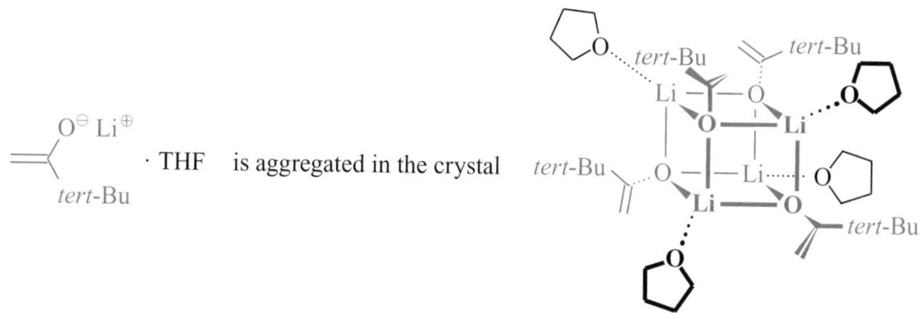

· THF is aggregated in the crystal

Fig. 13.3. X-ray crystal structure of $H_2C{=}C(O^{-}Li^{\oplus})(\textit{tert}\text{-Bu})$ THF.

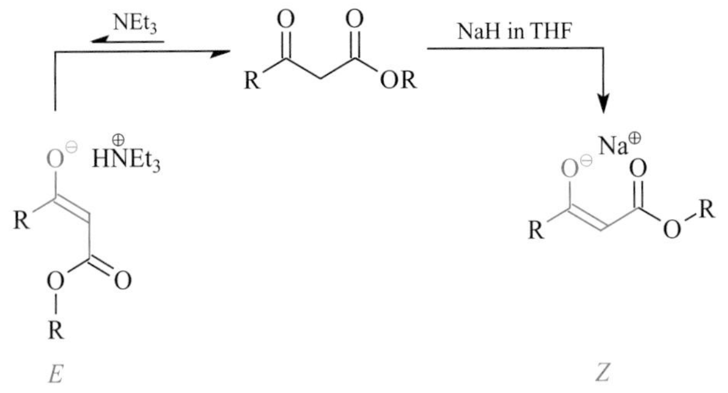

Fig. 13.4. Stereoselective deprotonation of a β-ketoester to trialkylammonium or sodium enolates. The *E*- and *Z*-enolates are formed when NEt_3 and NaH, respectively, are employed.

compounds. These are compounds that contain two geminal acceptor substituents with strong –M effects.

Starting from β-ketoesters (Figure 13.4, see Figure 13.25 for a synthetic application), or β-ketoaldehydes (Figure 13.5), ammonium enolates are formed with a stereostructure (*E*-enolates) that differs from that of the corresponding alkaline earth metal enolates (*Z*-enolates). In the latter, the lithium atom forms a bridge between the two negatively charged O atoms such that a six-membered chelate results. In contrast, the ammonium ion cannot play such a role in the ammonium enolate, and the more stable *E*-enolate is formed as the result of product development or thermodynamic control. The negatively charged O atoms are at a greater distance from each other in the *E*-enolate than in the *Z*-enolate. The greater distance between the O atoms in the *E*-enolate reduces charge-charge and dipole-dipole repulsions.

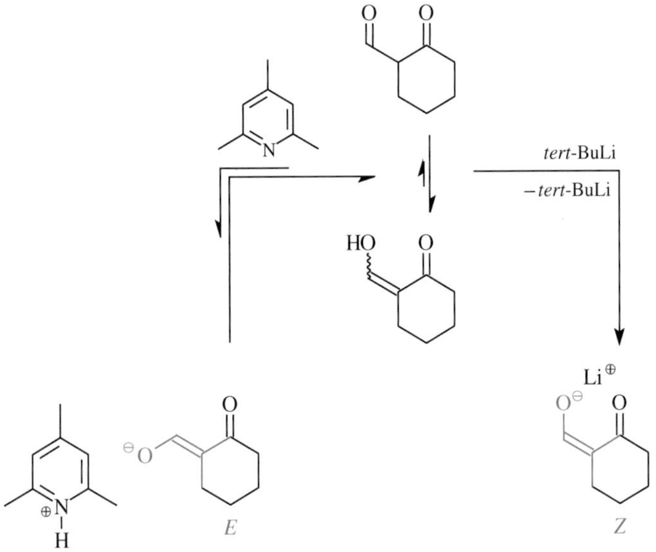

Fig. 13.5. Stereoselective deprotonation of a β-ketoaldehyde and its enol tautomer to substituted pyridinium or lithium enolates, respectively. Similar to the deprotonation of Figure 13.4, the *E*-enolate is formed when the amine is used and the *Z*-enolate is formed when the metal-containing base is employed.

13.1.2 Preparation of Enolates by Deprotonation

Suitable Bases

According to Figure 13.1, carbon-bound H atoms are acidic if they are bound to carbon atoms that are in the α-position with respect to an electron acceptor that can stabilize a negative charge via resonance (–M effect). Carbon-bound H atoms are even more acidic if they are located in the α-position of *two* such electron acceptors, which is the case in the so-called active-methylene compounds. Enolates derived from active-methylene compounds require three resonance forms for their description, and resonance forms **A** and **B** (Figure 13.6) are the more important ones. Compounds that contain an H atom in the α-position with respect to three electron acceptors are even more acidic than active-methylene compounds. However, such compounds do not play a significant role in organic chemistry.

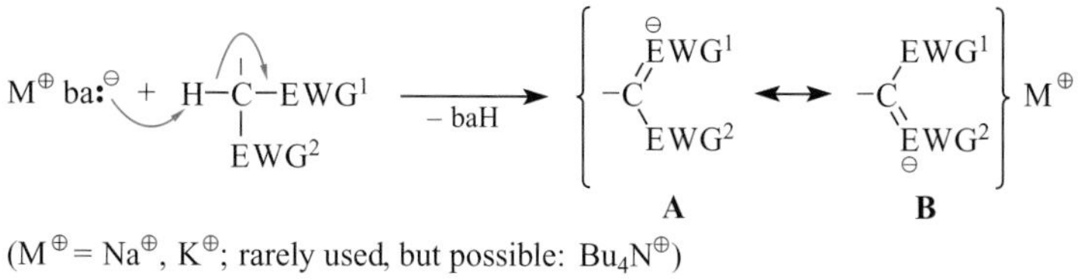

$M^{\oplus}$ ba:$^{\ominus}$ + H–C(EWG^2)–EWG^1 —(– baH)→ {A ↔ B} $M^{\oplus}$

($M^{\oplus}$ = $Na^{\oplus}$, $K^{\oplus}$; rarely used, but possible: $Bu_4N^{\oplus}$)

Fig. 13.6. Enolate formation of active-methylene compounds.

Table 13.1 lists the pK_a values of C,H-acidic compounds with a variety of electron acceptors. It shows that multiple substitution by a given acceptor enhances the acidity of the α-H atom more than monosubstitution. Table 13.1 also shows that the nitro group is the most activating substituent. One nitro group causes the same acidity of an α-H atom as do two carbonyl or two ester groups.

The acidifying effect of the remaining acceptor substituents of Table 13.1 decreases in the order —C(=O)—H > —C(=O)—alkyl > –C(=O)—O-alkyl, and the amide group —C(=O)–NR_2 is even less effective. This ordering essentially reflects substituent effects on the

Tab. 13.1 Effects of Substituents on C,H Acidity*

pK_a of ... for EWG =	H–C–EWG	H–C(EWG)–EWG
$-NO_2$	10.2	3.6
–C(=O)–H	16	5
–C(=O)–Me	19.2	9.0
–C(=O)–OMe	24.5	13.3

stability of the C=O double bond in the respective C,H-acidic compound. The resonance stabilization of these C=O double bonds drastically increases in the order R—C(=O)—H < R—C(=O)—alkyl < R—C(=O)—O—alkyl < R—C(=O)—NR_2 (cf. Table 6.1; see Section 9.1.1 for a comparison between the C=O double bonds in aldehydes and ketones).This resonance stabilization is lost completely once the α-H atom has been removed by way of deprotonation and the respective enolate has formed.

The equilibrium constant K_{eq} of the respective deprotonation equilibrium shows whether a base can deprotonate a C,H-acidic compound quantitatively, in part, or not at all:

$$(\mathrm{EWG})_{1\text{ or }2}\overset{|}{\mathrm{C}}-\mathrm{H} + \mathrm{ba}^{\ominus} \underset{}{\overset{K_{GG}}{\rightleftharpoons}} (\mathrm{EWG})_{1\text{ or }2}\overset{|}{\mathrm{C}}^{\ominus} + \mathrm{H{-}ba} \tag{13.1}$$

$$K_{GG} = \frac{K_{a,\text{CH-acid}}}{K_{a,\text{H-ba}}}$$

$$= 10^{pK_{a,\text{H-ba}} - pK_{a,\text{CH-acid}}} \tag{13.2}$$

Equation 13.1 shows that these equilibrium constants in turn depend on the acidity constants of the two weak acids involved, that is, the acidity constant $K_{a,\text{C,H acid}}$ of the C,H acid and the acidity constant $K_{a,\text{H-base}}$ of the conjugate acid (H-base) of the base (base$^{\ominus}$) employed. Equation 13.2 makes the same statement in terms of the corresponding pK_a values. From this equation it follows:

Rules of Thumb Regarding the Position of the Equilibria of C,H-Acidic Compounds

1. A C,H acid is deprotonated quantitatively (or nearly so) by an equimolar amount of base if the pK_a value of the conjugate acid of the base employed is higher than the pK_a value of the C,H acid.
2. A C,H acid is deprotonated by an equimolar amount of base only to the extent of 10%, 1%, or 0.1%, and so on, if the pK_a value of the conjugate acid of the base employed is 1, 2, 3, ... units lower than the pK_a value of the C,H acid.
3. In cases of incomplete C,H acid deprotonation, an excess of base can be employed to increase the enolate fraction. According to the principle of Le Chatelier, the base excess increases the enolate fraction by a factor that equals the square root of the number of mole equivalents of the base employed.

Side Note 13.1. Conformational Dependence of the C,H Acidity of Carboxylic Acid Esters

"For every rule, there is an exception!" or so they say. Sometimes this is also true in chemistry. It appears, for example, that the pK_a values of carboxylic acid esters vary more widely than is allowed or acknowledged by Table 13.1. This does not justify, though, the abandonment of the above mentioned "rules of thumb" for estimating the position of the deprotonation equilibrium of C,H-acidic compounds. As it were, *one* structural effect on the C,H acidity of carboxylic acid esters has so far been totally ignored, namely the effect of the conformation that the substructure C–O–C=O adopts in relation to the highlighted bond (printed in boldface), i.e., which dihedral angle occurs between the C–O and the C=O bond.

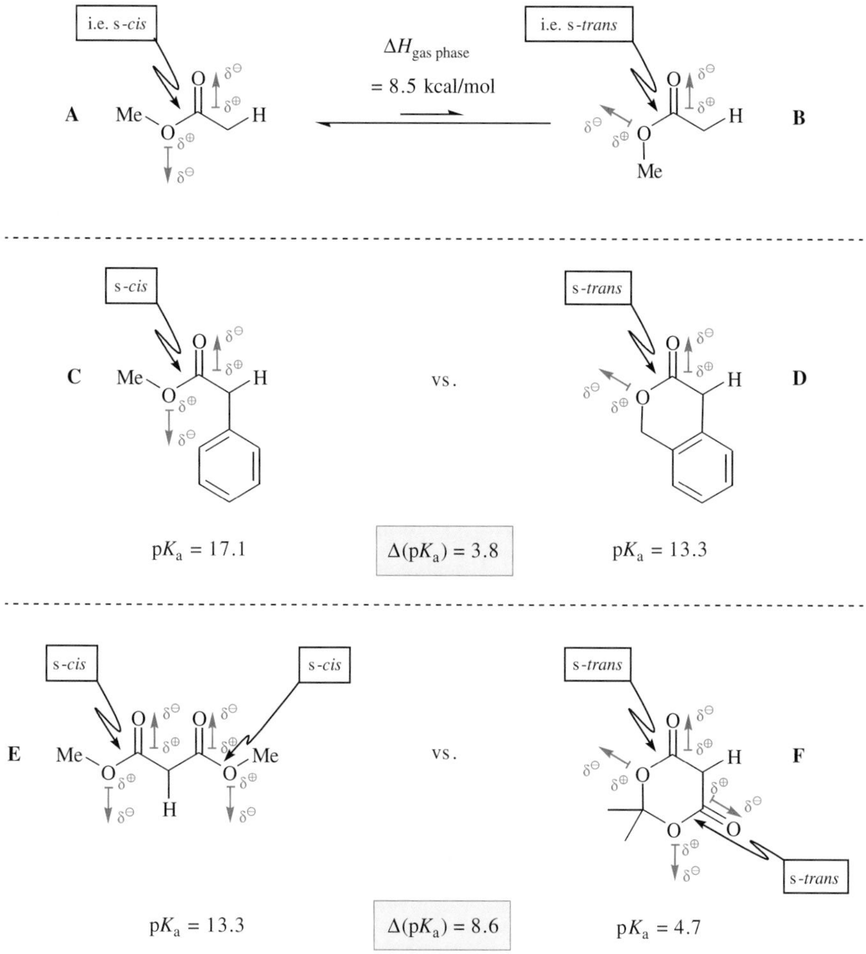

A (i.e. s-*cis*) $\Delta H_{\text{gas phase}}$ = 8.5 kcal/mol B (i.e. s-*trans*)

C (s-*cis*) vs. D (s-*trans*)

pK_a = 17.1 $\Delta(pK_a)$ = 3.8 pK_a = 13.3

E (s-*cis*, s-*cis*) vs. F (s-*trans*, s-*trans*)

pK_a = 13.3 $\Delta(pK_a)$ = 8.6 pK_a = 4.7

Fig. 13.7. Conformational dependence of the enthalpy of carboxylic acid esters and the conformational dependence of the C,H acidity of carboxylic acid esters based on it.

Using the pK_a values of the ester pairs in Figure 13.7, this conformational effect on the C,H acidity may be conceived as follows: if the dihedral angle between the C–O and the C=O bond in the C–O–C=O substructure of carboxylic acid esters is 0°—that is, if this substructure displays the so-called ***s-cis*-conformation**—the pK_a value of this ester is predicted by Table 13.1 in the correct range. If in the same substructure C–O–C=O of a carboxylic acid ester the dihedral angle between the C–O and the C=O bond measures 180°—i.e., if this ester displays the so-called ***s-trans*-conformation**—its pK_a value is underestimated in Table 13.1 by several orders of magnitude. This is why phenylacetic ester (Formula **C** in Figure 13.7; dihedral angle$_{\text{C–O/C=O}}$ = 0°) is by 4 pK units less acidic than its lactone analog **D** (dihedral angle$_{\text{C–O/C=O}}$ = 180°). The doubling of this conformational effect in comparison to malonic acid dimethyl ester (Formula **E** in Figure 13.7; both dihedral angles$_{\text{C–O/C=O}}$ = 0°) and Meldrum's acid (**F**; both dihedral angles$_{\text{C–O/C=O}}$ = 180°) explains why the former is less acidic than the latter by almost 9 pK units.

The reason for this conformational effect has to do with the pronounced endothermic character of the transformation of *s-cis-* into *s-trans*-methyl acetate ($\Delta H_R = +\,8.5$ kcal/mol; Figure 13.7). This is due to the partial dipole moments that occur in each of the conformers involved because of the polarized C=O double bond or the asymmetric distribution of the nonbonding electrons on the other oxygen atom. In the *s-cis*-conformer of methyl acetate *and all other s-cis-carboxylic acid esters* these partial moments are antiparallel, i.e., they nearly eliminate each other. However, in the *s-trans*-conformer of methyl acetate *and all other s-trans-carboxylic acid esters* these partial dipole moments are aligned. Hence, repulsion occurs between them, destabilizing the *s-trans*-ester conformer by 8–9 kcal/mol. In the enolate resulting from the deprotonation of a carboxylic acid ester, there are no similar large polar effects—irrespective of the configuration of the enolate double bond. This means that only an *s-trans-*, but not an *s-cis*-configured carboxylic acid ester will lose dipole/dipole repulsion when it reacts as a C,H acid and is converted into the enolate. *This* is what renders *s-trans*-configured esters more acidic than *s-cis*-configured esters.

With the pK_a values of the conjugate acids of the most commonly used organic bases (Table 13.3) and the pK_a values of the C,H acids compiled in Table 13.1, the foregoing statements lead to the following deductions (see also Table 13.2):

Tab. 13.2 Survey of the Deprotonation Ability of C,H-Acidic Compounds. The ease of deprotonation of C,H-acidic compounds depends on (a) the type and number of the electron-withdrawing groups in the substrate and (b) on the base employed

	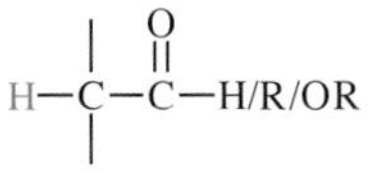 H–C–C(=O)–H/R/OR	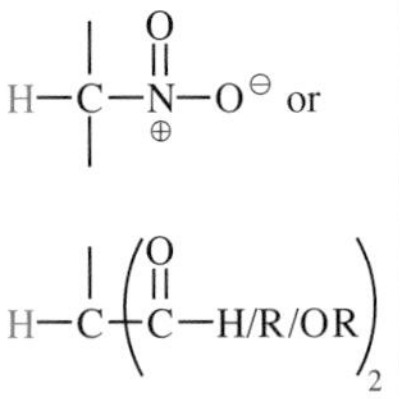 H–C–N⁺(=O)–O⁻ or H–C–(C(=O)–H/R/OR)$_2$
pK_a	18–25	4–13
+ NEt_3 (→ $H\overset{\oplus}{N}Et_3$, with a $pK_a = 10.7$)	little deprotonation	partial deprotonation
+ $OH^{\ominus}$ (→ H_2O, with a $pK_a = 15.5$) or + $OEt^{\ominus}$ (→ EtOH, with a $pK_a = 15.7$) or + O*tert*-$Bu^{\ominus}$ (→ *tert*-BuOH, with a s $pK_a = 19$)	partial deprotonation	quantitative deprotonation
+ LDA (→ $HNiPr_2$, with a $pK_a \approx 36$)	quantitative deprotonation	hardly ever employed as this would be overkill

Tab. 13.3 Basicity of Typical Reagents Employed in the Generation of Enolates via Deprotonation

Reagent	pK_a-value of the conjugate acid
$Li^{\oplus}NR_2^{\ominus}$	35 to 40
$K^{\oplus}$O*tert*-$Bu^{\ominus}$	19
$Na^{\oplus}OEt^{\ominus}$	15.7
$Na^{\oplus}OH^{\ominus}$	15.5
NEt_3	10.7

- All aldehydes, ketones, and carboxylic esters can be deprotonated quantitatively to enolates with lithium amides. The same substrates and alkoxides give only small amounts of enolate in equilibrium reactions, but even these small amounts of enolate may be large enough to allow for enolate reactions.
- For the quantitative deprotonation of nitroalkanes and active-methylene compounds, there is no need to employ the "heavy artillery" of lithium amides. Rather, it suffices to employ alkaline earth metal alkoxides or alkaline earth metal hydroxides. In addition, equilibrium reactions between these C,H acids and amines form enough enolate to initiate enolate reactions.

The foregoing classification is of fundamental significance for the understanding of enolate chemistry. For every pair of C,H acid and base, one needs to know whether the combination effects quantitative or partial enolate formation. If deprotonation is only partial, then the unreacted substrate may represent an electrophile that can react with the enolate nucleophile. In such a case, it depends on the specific circumstances whether an enolate reacts with any remaining substrate or whether it reacts only with an added *different* electrophile. The occurrence of a reaction between enolate and unreacted substrate is avoided if the C,H acid is deprotonated completely with a stoichiometric amount of a sufficiently strong base.

There is only one exception to the last statement in that aldehydes cannot be converted quantitatively into aldehyde enolates. Any attempt to achieve a quantitative deprotonation of an aldehyde—with a lithium amide, for example—necessarily leads to a situation in which some aldehyde enolate is formed while some aldehyde substrate is still present, and these species cannot coexist even at temperatures as low as that of dry ice. The aldehyde is such an excellent electrophile that it reacts much faster with the enolate than it is deprotonated by the base.

Table 13.4 allows for a comparison of the basicities of the strongest lithium-containing bases. The basicities are measured by the heats of deprotonation liberated upon mixing the reference acid isopropanol with these bases. These heats of deprotonation reveal that organolithium compounds are even stronger bases than lithium amides. Their basicities decrease from *tert*-BuLi via *sec*-BuLi and *n*-BuLi to PhLi.

Considering these heats of deprotonation, one wonders whether organolithium compounds should not be at least as suitable as lithium amides for effecting the deprotonation of carbonyl and carboxyl compounds. However, this is usually not the case, since organolithium com-

Tab. 13.4 Thermochemistry of Selected Acid/Base Reactions: Deprotonation Enthalpies (kcal/mol) for Deprotonations of *i*PrOH with Various Organolithium Compounds and Lithium Amides

tert-BuLi	–56.2	LTMP	–30.4
sek-BuLi	–52.8	LDA	–28.6
n-BuLi	–50.0	LiHMDS	–12.1
PhLi	–42.3		

pounds react almost always as nucleophiles rather than as bases. Organolithium compounds thus would add to the carbonyl carbon (Section 10.5) or engage in a substitution reaction at the carboxyl carbon (Section 6.5).

Obviously, only **nonucleophilic bases** can be employed for the formation of enolates from carbonyl and carboxyl compounds. A base is nonnucleophilic if it is very bulky. The only non-nucleophilic organolithium compounds that *deprotonate* carbonyl and carboxyl compounds are mesityllithium (2,4,6-trimethylphenyllithium) and trityllithium (triphenylmethyllithium). However, these bases do not have any significance for the generation of enolates because of the difficulties associated with their preparation and with the separation of their conjugate acid hydrocarbons.

Alkaline earth metal amides have a unique place in enolate chemistry in light of the preceding discussion. Yet, amides without steric demand—from $NaNH_2$ to $LiNEt_2$—also are usually not suitable for the formation of enolates, since their nucleophilicities exceed their basicities. On the other hand, the amides LTMP, LDA, and LiHMDS (structures in Figure 4.18) are so bulky that they can never act as nucleophiles and always deprotonate C,H acids to the respective enolates.

Table 13.4 also shows that the deprotonation of isopropanol with LiHMDS is less than half as exothermic as the deprotonations with LDA or LTMP. Hence, LiHMDS is a much weaker base than the other two amides. This is due to the ability of the $SiMe_3$ groups of LiHMDS to stabilize the negative charge in the α-position at the N atom. The mechanism of this stabilization might be the same as in the case of the isoelectronic triphenylphosphonium center in *P* ylides (Figure 11.1), that is, a combination of an inductive effect and anomeric effect. Because of its relatively low basicity, LiHMDS is employed for the preparation of enolates primarily when it is important to achieve high chemoselectivity.

Side Note 13.2. Neutral, but Superbasic: Schwesinger Bases!

Schwesinger's "P_5 base" is as basic as the strongest amide bases and also non-nucleophilic. Figure 13.8 suggests that the index "5" represents the five phosphorus atoms of this base and that there might be analogous "P_2", "P_3" and "P_4 bases." And indeed, there are. They are all collected under the generic term "Schwesinger bases." But regarding basicity, the "P_5 base" is entitled to leadership: at 34–35, the pK_a value of its conjugate acid is quite similar to the pK_a value of diisopropylamine, the conjugate acid of LDA! This is due to the extraordinarily good stabilization of the positive charge in the conjugate acid of the "P_5 base": there are 12 *all*-octet resonance forms that contribute to this stabilization (Figure 13.8) . The "P_5 base" provides access to metal-free enolates and thus to a chemistry which, in particular cases, differs markedly from the chemistry of metal enolates.

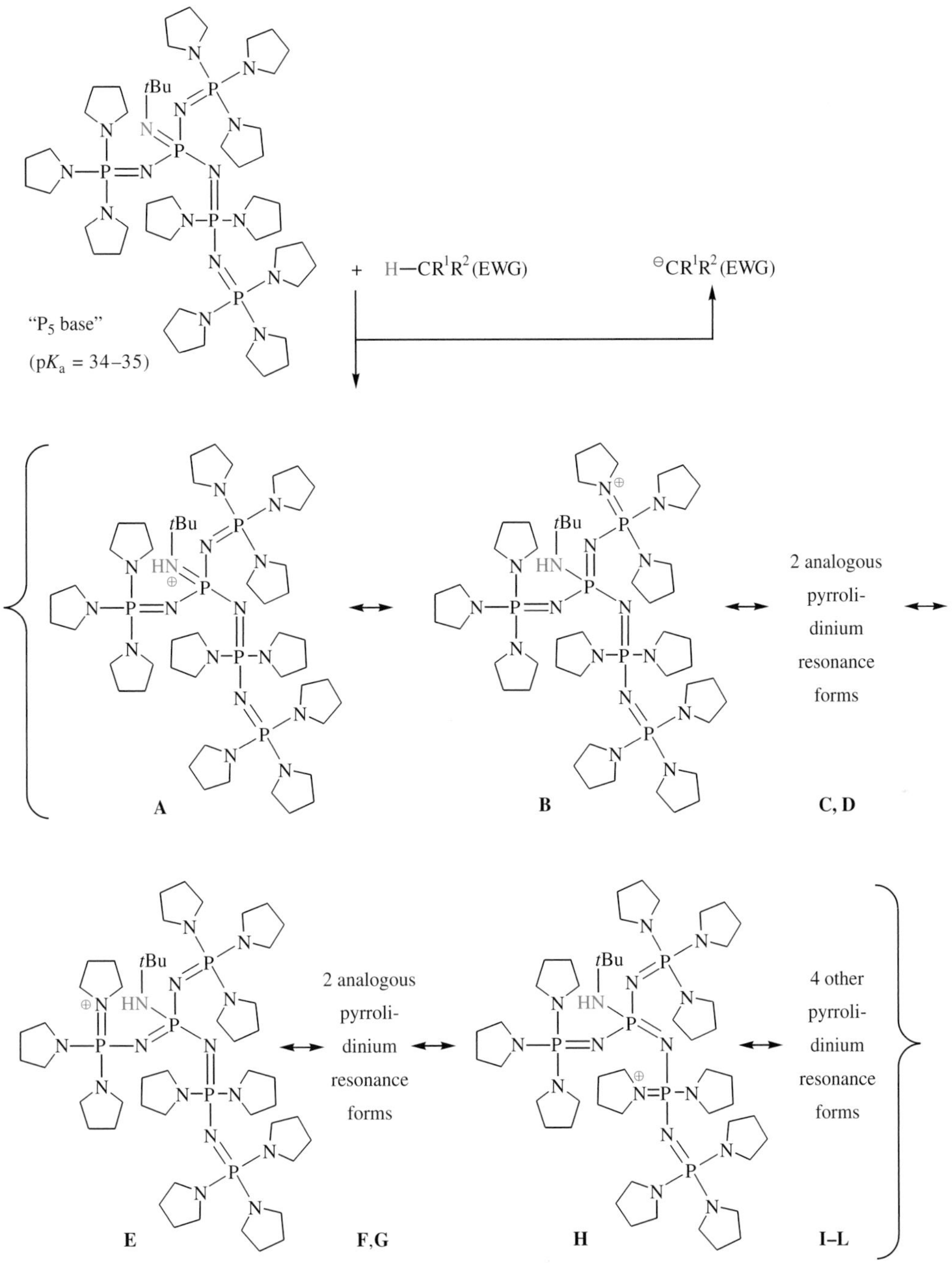

"P_5 base"

(pK_a = 34–35)

+ H—CR^1R^2(EWG)

$^{\ominus}CR^1R^2$(EWG)

tBu

A

B

2 analogous pyrroli-dinium resonance forms

C, D

E

2 analogous pyrroli-dinium resonance forms

F,G

H

4 other pyrroli-dinium resonance forms

I–L

Fig. 13.8. The strongest Schwesinger base: metal-free, but as strong as alkali metal amides!

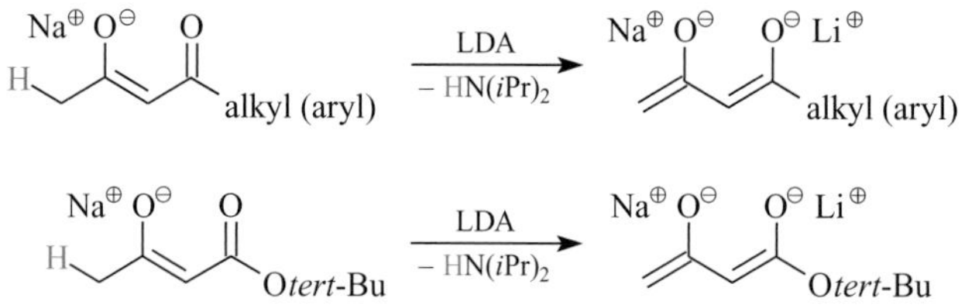

Fig. 13.9. Formation of bisenolates.

The basicity of LDA is so high that it is even possible to generate bisenolates from β-diketones and β-ketoesters (Figure 13.9). Even carboxylates can be deprotonated at the α-carbon if the strongest organic bases are employed (Figure 13.10). However, the twofold deprotonation of phenylacetic acid by ethylmagnesium bromide is not commonly used. This reaction is mentioned in Figure 13.10 only because the resulting enolate **A** acts as the nucleophile in the Ivanov reaction in Figure 13.45. The double deprotonation of malonic acid monoethyl ester actually requires only magnesium ethoxide, which may conveniently be generated *in situ*. Of course, this is due to the fact that the resulting **B** is not a true carboxylate enolate, but a carboxylate-substituted ester enolate. An example of the synthetic use of this enolate is given in Figure 13.65 (middle).

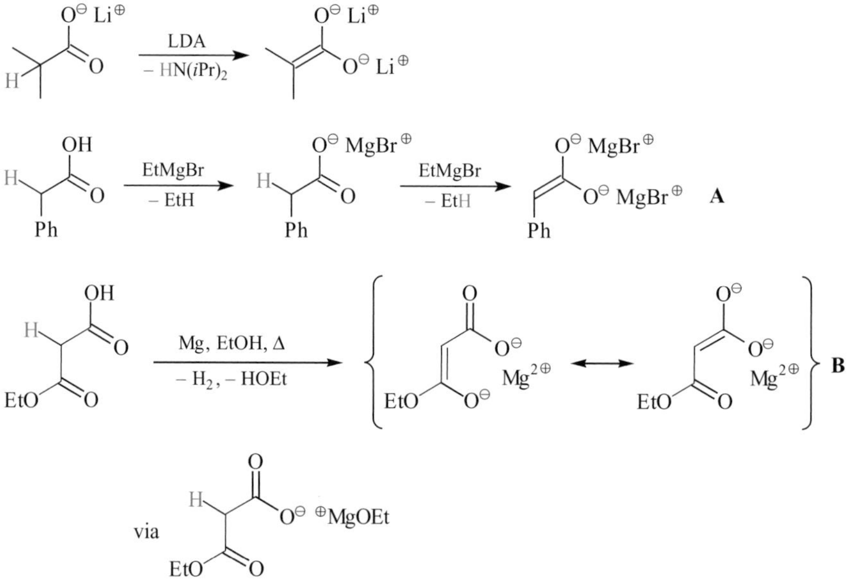

Fig. 13.10. Formation of carboxylate enolates ("carboxylic acid dianions") by reaction of carboxylic acid salts with strong bases.

Regiocontrol in the Formation of Lithium Enolates

Only one enolate can be generated from aldehydes or their aza analogs, from symmetric ketones or their aza analogs, or from carboxylic esters or carboxylic amides. For the moment we are ignoring the possibility that two stereoisomers, *E*- and *Z*-enolates, may occur for each of these enolates. On the other hand, constitutionally isomeric (regioisomeric) enolates may be derived from unsymmetrical ketones and from their aza analogs if they contain acidic H atoms at the C_α and $C_{\alpha'}$ centers. From certain unsymmetrical ketones or their aza analogs one or sometimes even both of these enolates can be generated regioselectively (see also Figure 13.35).

For example, 2-phenylcyclohexanone can be deprotonated regioselectively with LDA (Figure 13.11). This reaction is most successful at –78 °C in THF because the reaction is irreversible under these conditions as long as a small excess of LDA is employed. Hence, the reaction is kinetically controlled and proceeds via the most stable transition state. The standard transition state of *all* enolate formations from C,H acids with LDA is thought to be cyclic, six-membered, and preferentially in the chair conformation (**A** and **B** in Figure 13.11). To be as

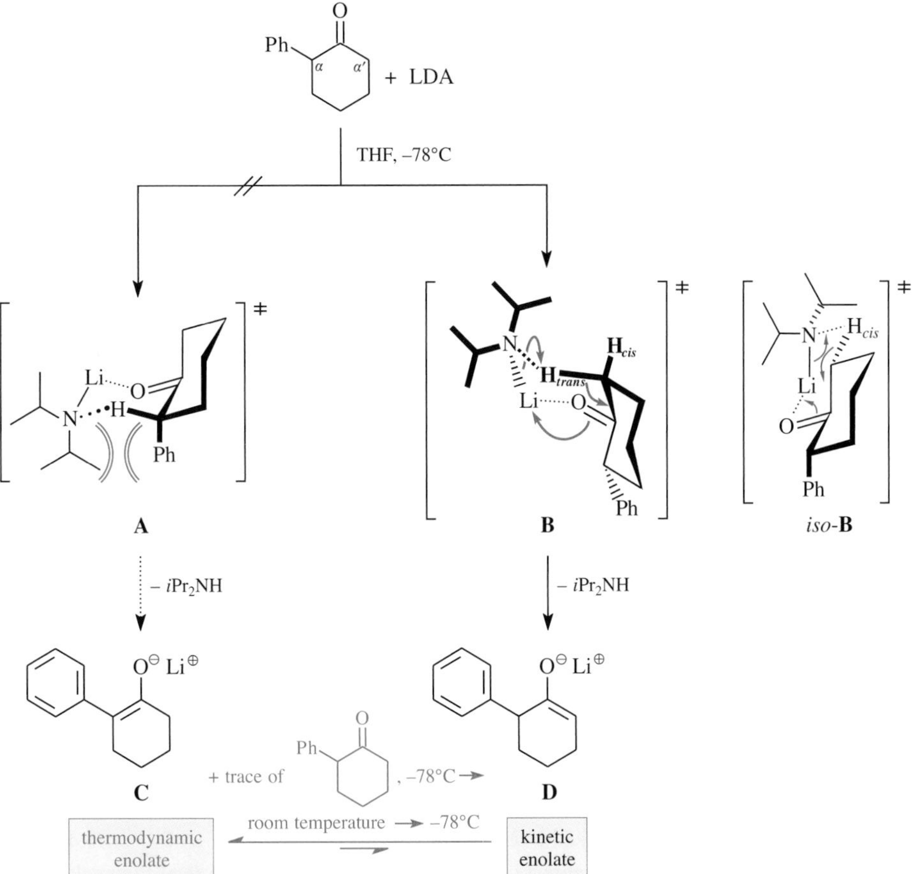

Fig. 13.11. Regioselective generation of ketone enolates, I: the effects of different substituents in the α- and α'-positions. Enolate **D** is formed in THF at –78 °C with LDA irrespective of whether a substoichiometric amount or an excess of LDA is used. However, if one employs slightly less than the stoichiometric amount of LDA (so that a trace of the neutral ketone is present), then, upon warming, the initially formed enolate **D** isomerizes quantitatively to enolate **C** with its more highly substituted C=C double bond. It should be noted that LDA removes an axially oriented α-H from the cyclohexanone; this is because only then does the resulting lone pair of electrons receive optimum stabilization by the adjacent C=O bond. With the kinetically preferred deprotonation leading to the enolate **D** the axial α'-H is transferred to the base (via transition state **B**), but not the equatorial α'-H (via transition state *iso*-**B**.)

stable as possible, this transition state should not feature any steric hindrance that can be avoided. In particular, the transition state should not contain any substituent in the six-membered ring that is parallel with the pseudo-axially oriented amide *N*-isopropyl group. Such a substituent would suffer from 1,3-diaxial repulsion because of its interaction with the isopropyl group. It follows that transition state **A** of Figure 13.11 is less stable than transition state **B.** The enolate formation thus proceeds selectively via transition state **B** and results in the **kinetic enolate D.**

The C=C double bond of **D** is not conjugated to the phenyl ring. It is therefore less stable than the regioisomeric enolate **C**, which benefits from such a conjugation. In this context, **C** is called the **thermodynamic enolate.** Because **C** is more stable than **D, C** can be generated from the kinetic enolate **D** as soon as the opportunity for isomerization is provided. The opportunity for isomerization arises if a weak acid is present that allows for the protonation of the enolate to the ketone. This can be accomplished by allowing the presence of a trace of unreacted substrate ketone, which happens if one treats the substrate ketone with a slightly less than stoichiometric amount of LDA. Under these conditions, at temperatures above –78 °C, the remaining substrate ketone reacts with the kinetic enolate **D** to yield the thermodynamic enolate **C** and newly formed substrate ketone. This occurs fast enough to effect a quantitative isomerization of **D** into **C.**

We thus reach the following interesting result. Depending on the reaction conditions, both the kinetic and the thermodynamic enolates of 2-phenylcyclohexanone can be generated with perfect regiocontrol. The same is true for many ketones that carry a different number of *alkyl* groups at the C_α and $C_{\alpha'}$ centers, but not always to the same extent. 2-Methylcyclohexanone, for example, reacts with LDA at –78 °C to yield a 99:1 mixture of kinetic and thermodynamic enolates. Under equilibrium conditions, however, a ratio of only 80:20 of the thermodynamic and the kinetic enolate, respectively, is obtained. A much more noticeable stabilization is provided by the same methyl group in the corresponding magnesium enolate. Thus, the thermodynamic enolate is produced "exclusively" if under equilibrium conditions 2-methylcyclohexanone is deprotonated with (iPr$_2$N)MgBr (Figure 13.23).

Side Note 13.3. Cyclohexanone Conformations and C,H Acidity

One structural aspect of the preferred transition state **B** (Figure 13.11) of the deprotonation of 2-phenylcyclohexanone has so far been ignored because it did not have any relation with the topic "Regiocontrol in the Formation of Lithium Enolates" discussed in that section. Yet, it shall be explicitly addressed because familiarity with this aspect will in the following subsection "Stereocontrol in the Formation of Lithium Enolates" allow us to understand why the deprotonation of carboxylic acid esters with LDA in THF containing DMPU or HMPT will lead to a *"Z"*-enolate (see Figure 13.17). Also, in the transition state **B** of Figure 13.11 the cyclohexanone depicted is deprotonated *diastereoselectively*: The LDA selectively abstracts the H atom that is in *trans*-position to the phenyl substituent and thus pseudo-axially oriented. This is due to the bonding situation in the resulting enolate **D** and, in particular, to the fact that in this transition state rehybridization from sp^3 to sp^2 begins with the carbon atom that will become the enolate carbon.

At the beginning of the reaction, an sp^3 AO originating from this carbon atom is bound to the hydrogen atom H_{trans}. At the end of the reaction, the $2p_z$ AO, which originates with the C atom under scrutiny via deprotonation, must overlap with the π^* MO of the adjacent C=O double bond, *since it is only because of this "C-H/π* $_{C=O}$ overlap" that the enolate **D** can be formed.* In the transition state **B** of the formation of **D**, the latter overlap is needed and requires the $sp^3 \rightarrow 2p_z$ AO to be as perpendicular as possible to the double bond plane of the adjacent carbonyl group. The dihedral angle between the $C_{enolate}$–H_{trans} bond and the C=O double bond in the phenylcyclohexanone of Figure 13.11 is about 117°, since the corresponding dihedral angle between the $C_{enolate}$–$H_{pseudo\text{-}axial}$ bond and the C=O double bond in the cyclohexanone itself exactly measures these 117° (Figure 13.12). In contrast, the dihedral angle between the $C_{enolate}$–H_{cis} and the C=O double bond in the phenylcyclohexanone of Figure 13.11 must be around 8°, since the corresponding dihedral angle between the $C_{enolate}$–$H_{pseudo\text{-}equatorial}$ bond and

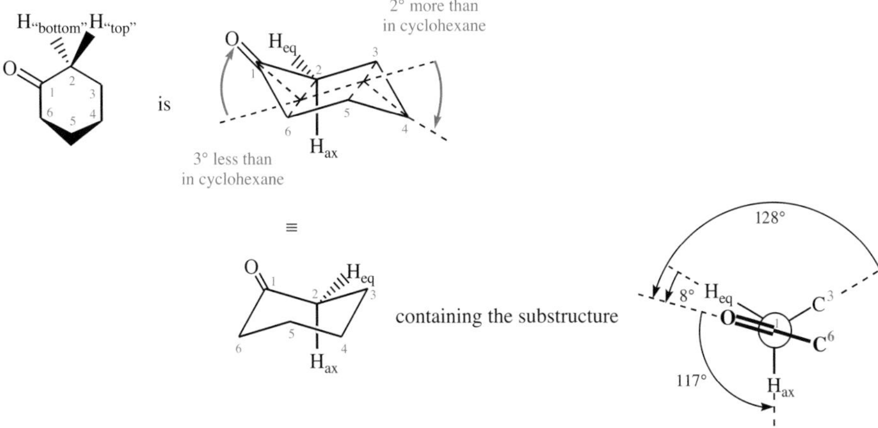

It can be deprotonated during the gas phase of OH⊖, either ...

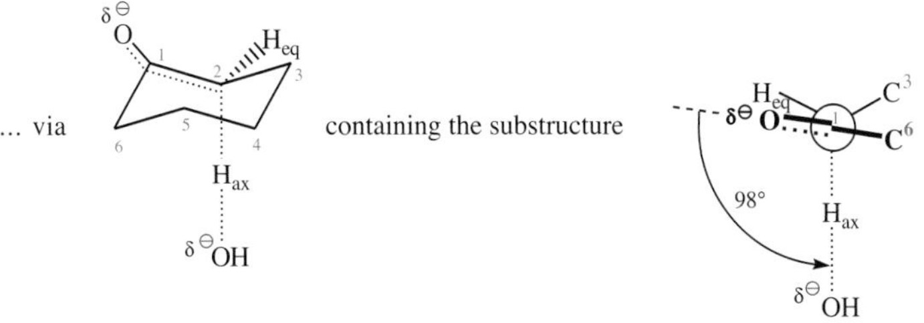

... or – with an extra activation energy of 2.8 kcal/mol through – ...

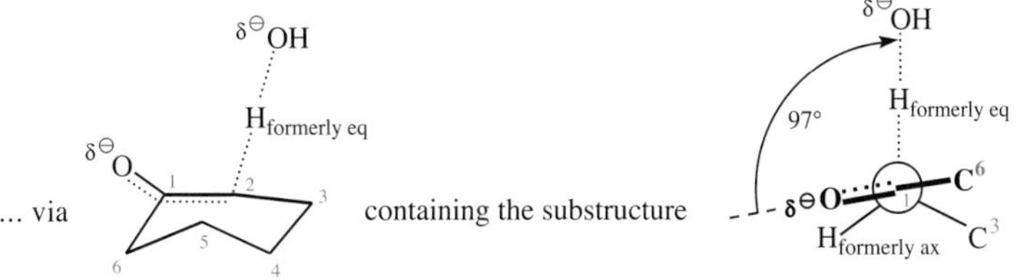

Fig. 13.12. Which H atom in the α-position to the carbonyl group of cyclohexanone(s) will be more rapidly removed by a base: the equatorial or the axial one? Answer: the H atom which is bound to the C–H bond whose σ_{C-H} orbital in the transition state of the deprotonation is *parallel* to the $2p_z$ orbitals of the $\pi^*_{C=O}$ orbital of the adjacent carbonyl group, and hence overlaps with it most effectively (explanation: see text).

the C=O double bond of cyclohexanone again measures precisely 8° (Figure 13.12). The dihedral angle of 117° sufficiently approaches the optimal dihedral angle of 90°, and 8° is sufficiently remote so that the LDA reacts with the $C_{enolate}$–H_{trans} bond of 2-phenylcyclohexanone while leaving its $C_{enolate}$–H_{cis} bond intact.

From the point of view of conformation cyclohexanones are more flexible than cyclohexanes. Therefore it can*not* be generally assumed that hydrogen atoms in the α- position of cyclohexanones can be abstracted as protons only if they are pseudo-axially oriented. Accordingly, it is incorrect to say that removing them is never possible if their orientation is pseudo-equatorial. This situation is illustrated in Figure 13.12. The most favorable transition state (shown in the second line) corresponds (see above) to the pseudo-axial deprotonation, just as expected. The most stable transition state of the pseudo-equatorial deprotonation (shown in line 3), though, is higher in energy by only 2.8 kcal/mol and may—in the presence of suitable substituents, including bridging substituents—definitely become competitive.

An unsymmetrical ketone may even lead to the generation of one enolate in a selective fashion if the lack of symmetry is caused by the substitution patterns in the β- rather than α-positions. The difference in the β-positions may be due to the number or the kind of substituents there. This point is emphasized in Figure 13.13 with cyclohexanones that contain one (**D**) or two (**A**) β-substituents. In this case, deprotonation occurs preferentially on the side opposite to the location of the extra substituent; that is, the sterically less hindered acidic H atom reacts.

In the discussion of Figure 13.11, we mentioned that LDA abstracts acidic H atoms from the α-position of C,H-acidic compounds via a cyclic and six-membered transition state and that *the carbonyl group is an integral part of this six-membered ring.* This emphasis explains why the deprotonation of conjugated ketones with LDA yields the kinetic enolate (**A** in Figure 13.14) in a regioselective fashion instead of the thermodynamic enolate (**B** in Figure 13.14).

Stereocontrol in the Formation of Lithium Enolates

We have seen that LDA forms enolates of carbonyl compounds, carboxylic esters, and carboxylic amides via cyclic and six-membered transition states with the chair conformation. This geometry of the transition state for enolate formation has consequences if a stereogenic C=C double bond is generated.

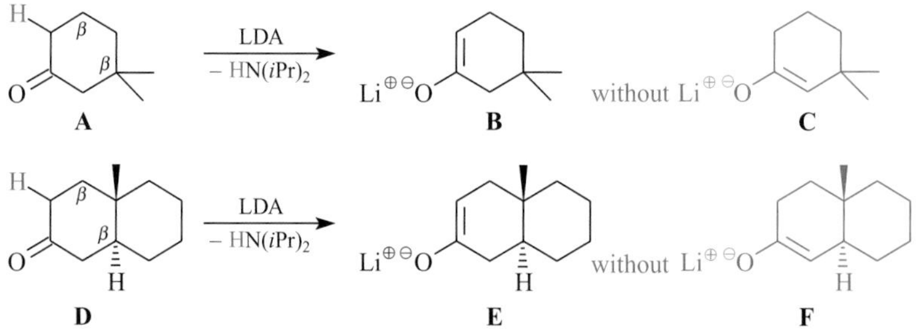

H, β, β, O, A; LDA, – HN(iPr)$_2$; $Li^{\oplus}$ $^{\ominus}$O, B; without $Li^{\oplus}$ $^{\ominus}$O, C
H, β, β, O, H, D; LDA, – HN(iPr)$_2$; $Li^{\oplus}$ $^{\ominus}$O, H, E; without $Li^{\oplus}$ $^{\ominus}$O, H, F

Fig. 13.13. Regioselective generation of ketone enolates II: the effects of different substituents in the β- and β'-positions (for the regioselective preparation of enolates **C** and **F**, see Figures 13.19 and 13.20, respectively).

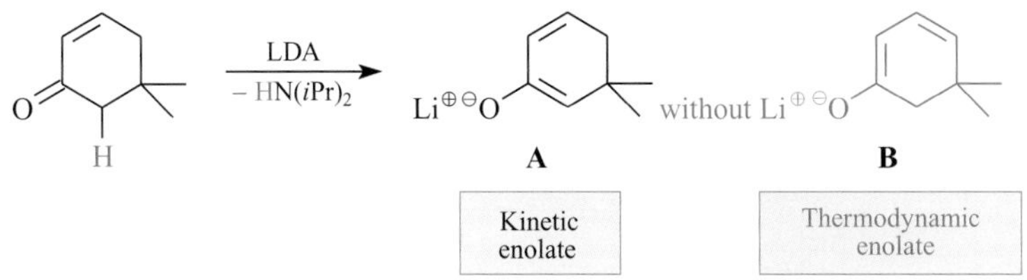

Fig. 13.14. Regioselective generation of ketone enolates III: the effects of conjugated versus nonconjugated substituents (for the regioselective preparation of enolate **B**, see Figure 13.21).

The reaction of unhindered aliphatic ketones with LDA yields more *E*- than *Z*-enolates. On the other hand, the *Z*-enolate is formed selectively if either a bulky aliphatic or a conjugating aromatic group is the inert group attached to the carbonyl carbon. Figure 13.15 shows how this *Z*-selectivity results from the differing steric demands in the potentially available transition states for deprotonation. The point is made using the example of an ethyl ketone with a sterically demanding substituent. Transition state **A** is so strongly destabilized by the 1,2-interaction indicated that deprotonation occurs exclusively via transition state **B** in spite of the repulsive 1,3-interaction in the latter.

Unhindered aliphatic ketones selectively yield *E*-enolates if they are deprotonated by a lithium amide via a transition state structure of type **A** of Figure 13.15. This occurs, for example, when the **B**-type transition state is destabilized because of the use of a base that is even more sterically demanding than LDA such as, for example, LTMP (for structure, see Figure 4.18). For example, diethyl ketone and LTMP form the *E*-enolate with *ds* = 87:13.

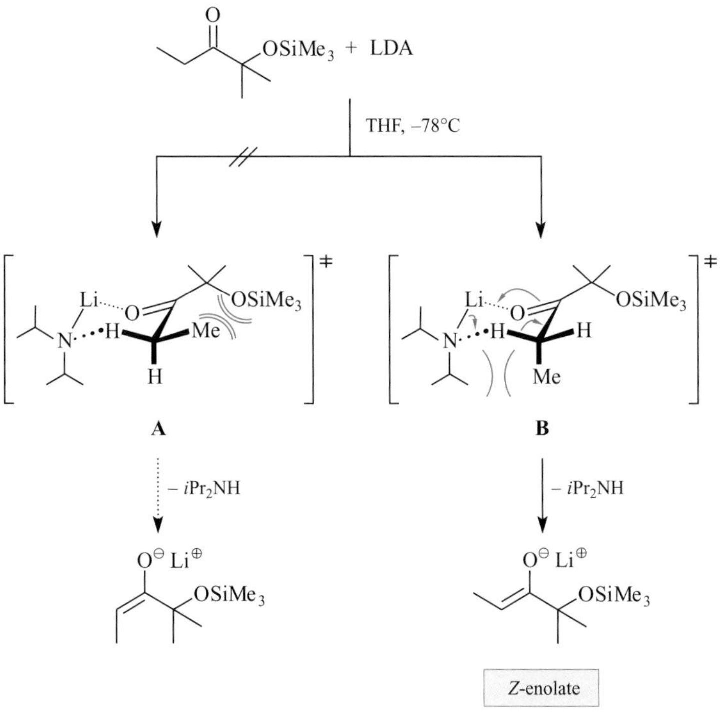

Fig. 13.15. Highly *Z*-selective generation of a ketone enolate. The transition state **A** is destabilized so strongly by 1,2-interactions that deprotonation occurs exclusively via transition state **B**. – Note: Don't let yourself be misled if it seems as though this is a deprotonation of H atoms lying almost in the plane of the C=O double bond. According to the discussion of Figure 13.12, this would be quite unfavorable due to the lack of $\sigma_{C-H}/\pi^*_{C=O}$ interaction. This is not true, though. For, if the transition states **A** and **B** displayed the same dihedral angle as a cyclohexane chair, the dihedral angle between the crucial C–H bond and the C=O bond in the transition states **A** and **B**, respectively, would amount to 60° each.

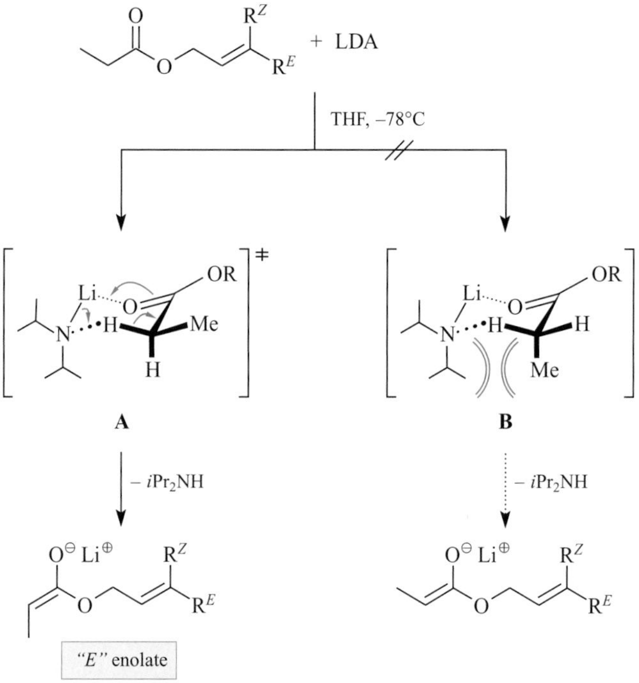

Fig. 13.16. Highly *"E"*-selective generation of ester enolates. The deprotonation of the ester occurs preferentially via the strain-free transition state **A.** – The caption of Figure 13.15 analogously applies to the dihedral angle between the C=O and the C–H-bond that reacts (**A**) or does not react (**B**) with LDA.

Deprotonation of α-alkylated acetic acid esters (e.g., the propionic acid ester of Figure 13.16) with LDA at –78 °C selectively yields the *"E"*-enolates. The quotation marks indicate that this application of the term is based on an extension of the *E/Z*-nomenclature: here, the Cahn–Ingold–Prelog priority of the $O^{\ominus}Li^{\oplus}$ substituent is considered to be higher than the priority of the OR group. The deprotonation of the ester shown in Figure 13.16 occurs via the strain-free transition state **A.** The alternative transition state **B** is destabilized by a 1,3-diaxial interaction.

"Z"-Enolates also are accessible from the same propionic acid esters with the same high stereoselectivity (Figure 13.17). This is a complete reversal of stereoselectivity in comparison to the deprotonation shown in Figure 13.16. This reversal is solely due to the change in solvent from THF to a mixture of THF and DMPU. The stereocontrol is again the result of kinetic control. In this solvent, the lithium of LDA coordinates with the oxygen atom of DMPU. Given the large excess of DMPU each lithium atom binds several equivalents of this ligand. Almost surely, in this way the lithium atom is fully coordinatively saturated by DMPU. Hence, in a mixture of THF and DMPU, LDA occurs as a "solvent-separated ion pair". This is why the deprotonation in Figure 13.17 proceeds via an acyclic transition state. We may assume that the active base is the metal-free diisopropylamide anion of the above-mentioned solvent-separated ion pair. From these general conditions, we can derive structure **A** for the transition state of the *"Z"*-selective ester deprotonation. Its conformation corresponds to the most favorable transition state **B** of the gas-phase deprotonation of propionaldehyde, which also leads to the formation of a *"Z"*-enolate. The stereostructure of both transition states **A** and **B** benefits from the C-H/$\pi^*_{C=O}$ overlap. The importance of the C-H/$\pi^*_{C=O}$ overlap for the rapid depro-

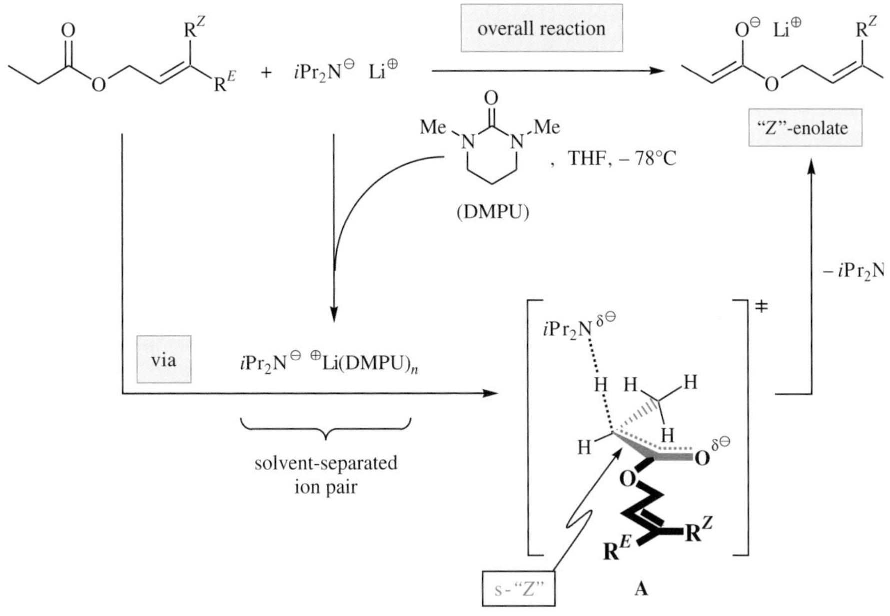

Fig. 13.17. Highly "Z"-selective generation of ester enolates in a THF/DMPU solvent mixture (DMPU, *N,N'*-dimethylpropyleneurea). The transition state **A** of *this* deprotonation with a metal-free diisopropylamide anion (in solution) corresponds to the calculated transition state **B** of the deprotonation of propionic aldehyde with a hydroxide anion (in the gas phase).

tonation of cyclohexanones was emphasized in connection with the discussion of Figure 13.12.

The generation of amide enolates by the reaction of carboxylic amides and LDA at –78 °C occurs with complete stereoselectivity and yields the *"Z"*-enolate (Figure 13.18). This is just the opposite selectivity than in the case of the generation of ester enolates under the same conditions (Figure 13.16). The NR_2 group of an amide is branched and therefore sterically more demanding than the OR group of an ester. Hence, in the transition state for deprotonation, the NR_2 group of an amide requires more space than the OR group of an ester. Consequently, the propionic acid amide of Figure 13.18 as well as all other carboxylic amides cannot react via transition states **A**. Unlike transition state **A** of the analogous ester deprotonation (Figure 13.16), the 1,2-interaction would be *prohibitively* high. Therefore, carboxylic amides are deprotonated via **B**-type transition state structures in spite of the repulsive 1,3-interaction. Remember, it is due to this repulsive 1,3-interaction that these kinds of transition state structures (**B** in Figure 13.16) are *not involved* in the deprotonation of esters.

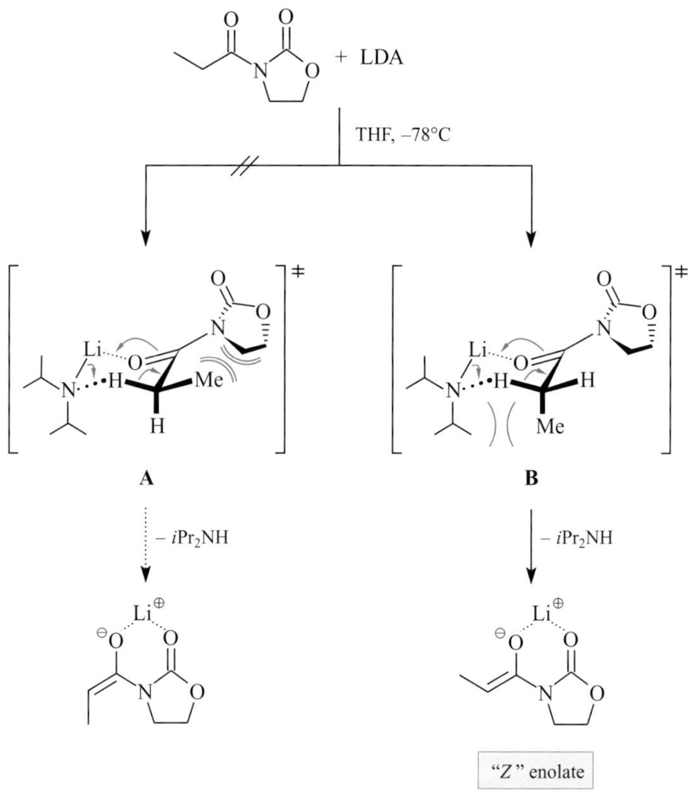

Fig. 13.18. Highly "Z"-selective generation of amide enolate.

13.1.3 Other Methods for the Generation of Enolates

Fig. 13.19. Generation of Cu-containing enolates from Gilman cuprates and their use as substrates for the preparation of Cu-free enolates via silyl enol ethers.

Figures 13.13 and 13.14 demonstrate that deprotonation might afford certain enolates with only *one* regioselectivity. However, there might be other reaction paths that lead to the other regioisomer (Figures 13.19–13.21).

One of these alternative synthetic paths consists of the addition of Gilman cuprates (for preparation, see Figure 10.43) to α,β-unsaturated ketones. In the discussion of this addition

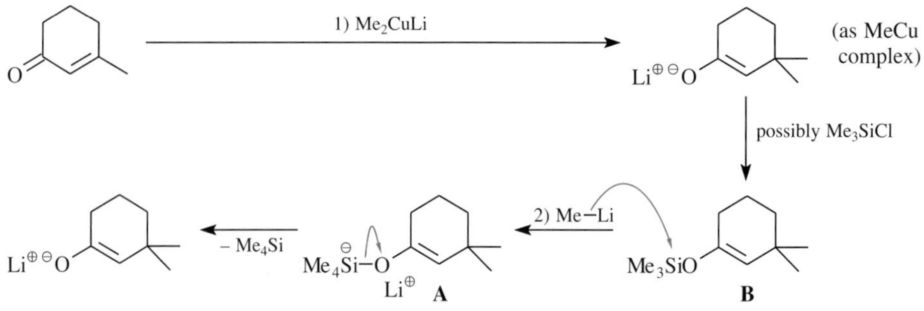

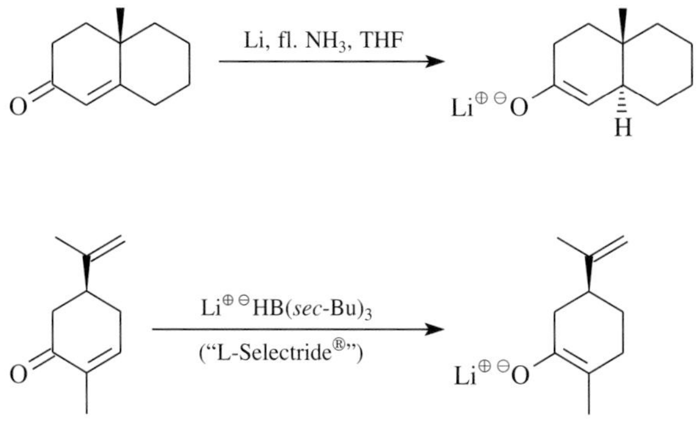

Fig. 13.20. Generation of enolates from α,β-unsaturated ketones via Birch reduction (top line) or by reduction with L-Selectride® (bottom line).

mechanism (Figure 10.46), it was pointed out that the enolates formed are associated with CuR. It is assumed that mixed aggregates are formed. As it turns out, the enolate fragments contained in such aggregates are significantly less reactive than CuR-free enolates. It is possible, however, to convert these CuR-containing enolates into the CuR-free lithium enolates. This conversion typically starts with the reaction between the CuR-containing enolate and Me_3SiCl to form a silyl enol ether such as **B** (Figure 13.19). The silyl enol ether reacts with MeLi via the silicate complex **A** (pentacoordinated Si center) to provide the Cu-free enolate.

Another way of obtaining enolates is the Birch reduction of α,β-unsaturated ketones (Figure 13.20, top; mechanism: in analogy to Figure 17.53). The transformation of α,β-unsaturated ketones into enolates using L-Selectride® proceeds via a different mechanism, namely the 1,4-addition of a hydride ion, with the result being indistinguishable, though, from the above (Figure 13.20, bottom).

Accordingly, trimethylsilyl enol ethers are enolate precursors (Figure 13.19). Fortunately, they can be prepared in many ways. For instance, silyl enol ethers are produced in the silylation of ammonium enolates. Such ammonium enolates can be generated at higher temperature by partial deprotonation of ketones with triethylamine (Figure 13.21). The incompleteness of this reaction makes this deprotonation reversible. Therefore, the regioselectivity of such deprotonations is subject to thermodynamic control and assures the preferential formation of the more stable enolate. Consequently, upon heating with Me_3SiCl and NEt_3, α,β-unsaturated ketones are deprotonated to give 1,3-dienolates **A** with the $O^{\ominus}$ substituent in the 1-position.

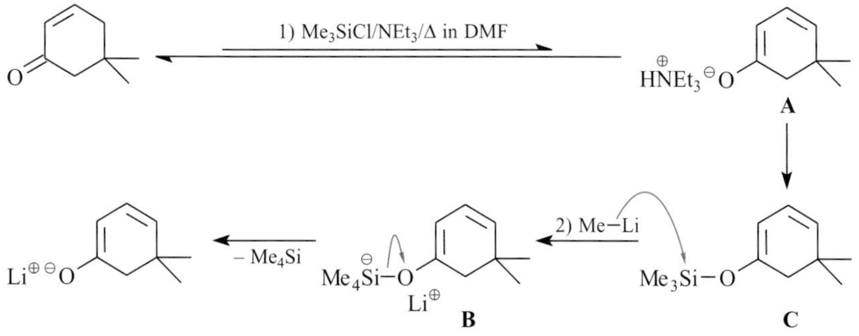

Fig. 13.21. Generation of enolates from silyl enol ethers.

This enolate is more stable than the isomeric 1,3-enolate in which the $O^{\ominus}$ substituent is in the 2-position; three resonance forms can be written for the former enolate while only two can be written for the latter. Me_3SiCl then reacts with the dienolate **A** at the oxygen atom. The dienol silyl ether **C** is obtained in this way. Silyl ether **C** reacts with MeLi—in analogy to the reaction **B** → **A** shown in Figure 13.19—via the silicate complex **B** to give the desired enolate. Note that the product enolate is not accessible by treatment of cyclohexanone with LDA (Figure 13.14).

13.1.4 Survey of Reactions between Electrophiles and Enolates and the Issue of Ambidoselectivity

Enolates and aza enolates are so-called **ambident nucleophiles.** This term describes nucleophiles with two nucleophilic centers that are in conjugation with each other. In principle, enolates and aza enolates can react with electrophiles either at the heteroatom or at the carbanionic C atom. Ambidoselectivity occurs if one of these alternative modes of reaction dominates.

Most enolates exhibit an ambidoselectivity toward electrophiles that depends on the electrophile and not on the substrate. The extent of the ambidoselectivity almost always is complete:

1) Only very few electrophiles react at the enolate *oxygen* of the enolates of aldehydes, ketones, esters, and amides, and these few electrophiles are
   - silyl chlorides (examples in Figures 13.19, 13.21, and 13.22)

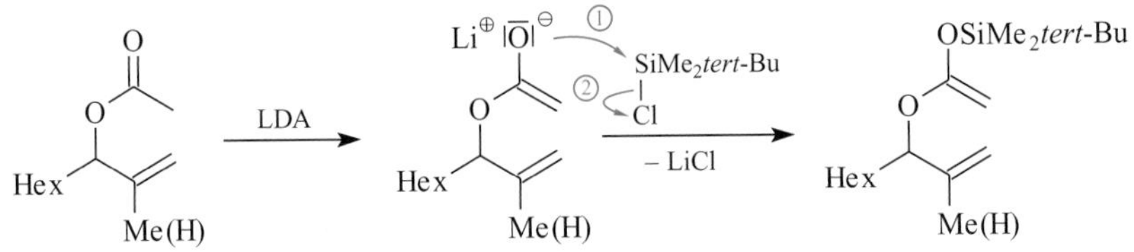

Fig. 13.22. *O*-Silylation of an ester enolate to give a silyl ketene acetal. (The formation of the silicate complex in step 1 of the reaction is plausible but has not yet been proven.)

   - Derivatives of sulfonic acids such as the *N*-phenylbisimide of trifluoromethane-sulfonic acid (examples: Figure 13.23, Side Note 13.3 and Figure 16.2). Note that alkenyl triflates are obtained in this way and these are the substrates of a variety of Pd-mediated C,C-coupling reactions (Chapter 16).
   - Derivatives of chlorophosphonic acid such as chlorophosphonic acid diamide (for example, see Figure 13.24) or chlorophosphonic acid esters (Figure 13.25).

2) Essentially all other electrophiles react with the enolate *carbon* of the enolates of aldehydes, ketones, esters, and amides (important examples are listed in Table 13.5).

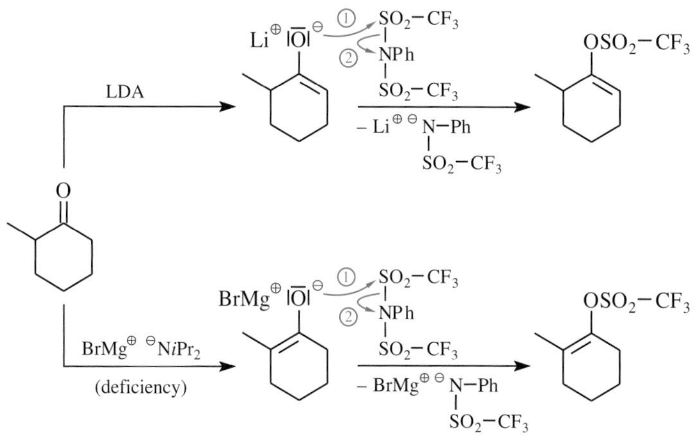

LDA
Li⊕ ⊖O
SO2−CF3
NPh
SO2−CF3
− Li⊕ ⊖N−Ph
SO2−CF3
OSO2−CF3
O
BrMg⊕ ⊖NiPr2
(deficiency)
BrMg⊕ ⊖O
− BrMg⊕ ⊖N−Ph

Fig. 13.23. *O*-Sulfonylations of regioisomeric ketone enolates to give an enol triflate (regarding the regiocontrol of the enolate formations cf. the discussion of Figure 13.11):

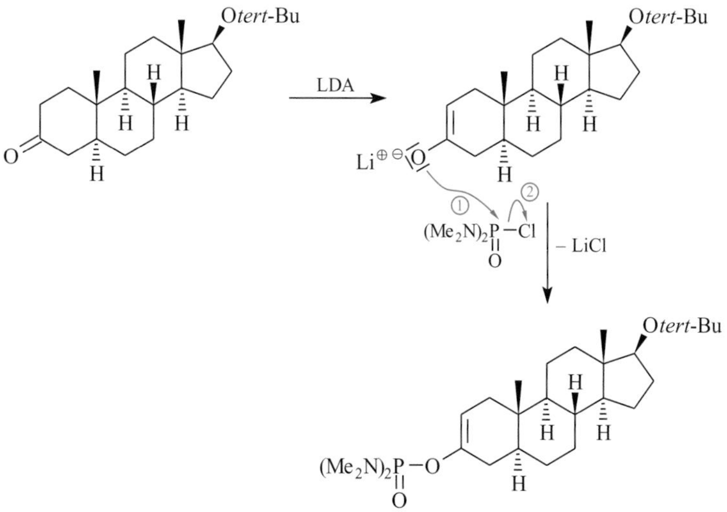

Otert-Bu
LDA
Li⊕ ⊖O
(Me2N)2P−Cl
O
− LiCl
(Me2N)2P−O

Fig. 13.24. *O*-Phosphorylation of a ketone enolate to afford an enol phosphonamide (see Figure 13.13, bottom row, regarding the regioselectivity of the enolate formation):

NaH in THF
Na⊕
Rprim
OMe
(EtO)2P−Cl
− NaCl
(EtO)2P
NEt3
O⊖ HNEt3⊕
− HNEt3⊕ Cl⊖

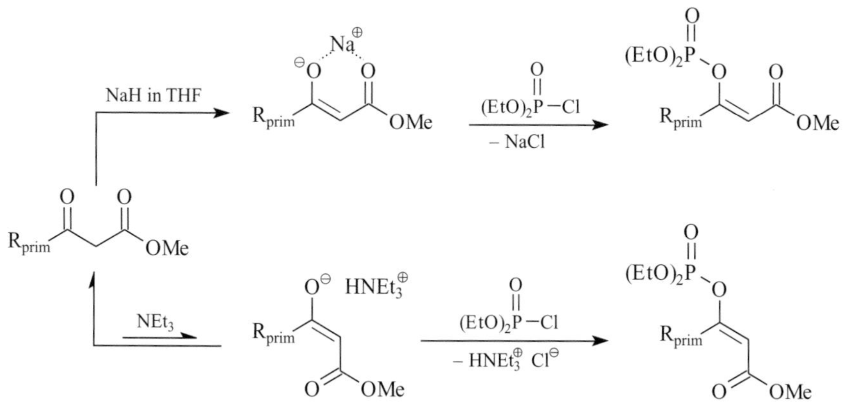

Fig. 13.25. *O*-Phosphorylation of a ketone enolate to afford an enol phosphate (see Figure 13.4 regarding the stereochemistry of the enolate formation).

Tab. 13.5 Electrophiles That React Ambidoselectively at the C Atom of Enolates Derived from Aldehydes, Ketones, Esters, and Amides

Electrophile	$M^{\oplus}\,{}^{\ominus}$O enolate (X, Subst) + $E^{\oplus}$ ↓	$M^{\oplus}\,{}^{\ominus}$O enolate (X, Subst, EWG) + $E^{\oplus}$ ↓	For details or applications in synthesis, sec …
PhSe–SePh	X–C(=O)–CH(Subst)–SePh	X–C(=O)–C(Subst)(EWG)–SePh	Fig. 4.12
Ph-oxaziridine ($PhSO_2$–N)	X–C(=O)–CH(Subst)–OH	X–C(=O)–C(Subst)(EWG)–OH	–
$N^{\ominus}=N^{\oplus}=N-SO_2$–(2,4,6-triisopropylphenyl)	X–C(=O)–CH(Subst)–N_3		–
$N^{\ominus}=N^{\oplus}=N-SO_2$–Ph	X–C(=O)–C(Subst)=$N^{\oplus}=N^{\ominus}$	X–C(=O)–C(EWG)=$N^{\oplus}=N^{\ominus}$	Fig. 4.29, Fig. 15.42
RX	X–C(=O)–CH(Subst)–R	X–C(=O)–C(Subst)(EWG)–R	Section 13.2
O=C(R¹)(R²)	X–C(=O)–CH(Subst)–C(OH)R¹R²	(X–C(=O)–CH(EWG)–C(OH)R¹R² ↓ – H_2O)	Section 13.3
	(für X = H, Alkyl oder Aryl auch ⟶ X–C(=O)–CH=CR¹R²)	X–C(=O)–C(EWG)=CR¹R²	Section 13.4
Y–C(=O)–(O)R	X–C(=O)–CH(Subst)–C(=O)–(O)R	X–C(=O)–C(Subst)(EWG)–C(=O)–(O)R	Section 13.5
CH_2=C(Subst')–EWG′	X–C(=O)–CH(Subst)–CH(Subst')–CH₂–EWG′	X–C(=O)–C(Subst)(EWG)–CH₂–CH(Subst')–EWG'	Section 13.6

13.2 Alkylation of Quantitatively Prepared Enolates and Aza-enolates; Chain-Elongating Syntheses of Carbonyl Compounds and Carboxylic Acid Derivatives

All the reactions discussed in this section are S_N2 reactions with respect to the alkylating reagent. The most suitable alkylating reagents for enolates and aza enolates are therefore the most reactive alkylating reagents (Section 2.4.4), that is, MeI, R_{prim}—X, and especially H_2C=CH—CH_2—X and Ar—CH_2—X (X = Hal, OTs, OMs). Isopropyl bromide and iodide also can alkylate enolates in some instances. Analogous compounds R_{sec}—X and R_{tert}—X either do not react with enolates at all or react via E2 eliminations to afford alkenes.

13.2.1 Chain-Elongating Syntheses of Carbonyl Compounds

Acetoacetic Ester Synthesis of Methyl Ketones

An acetoacetic ester is an active-methylene compound and it can be deprotonated (Table 13.1) with one equivalent of NaOEt in EtOH to the sodium enolate **A** (Figure 13.26). As is depicted in Figure 13.26, **A** is *monoalkylated* by butyl bromide. This is possible even though the butylated sodium enolate **B** is already present while the reaction is still under way. The sodium enolate **B** is formed in an equilibrium reaction between not-yet-butylated enolate **A** and the butylation product **C**. **B** represents a nucleophilic alternative to unreacted enolate **A.** However, the butylated enolate **B** is sterically more demanding than the nonbutylated enolate **A.** The first butylation of **A** is thus faster than the second butylation reaction, that is, the butylation of **B.** This reactivity difference is not large enough to cause 100% monobutylation and 0% dibutylation. Still, the main product is the product of monobutylation **C.** Distillation is required to separate the monobutylation product from the dibutylation product and from unreacted substrate.

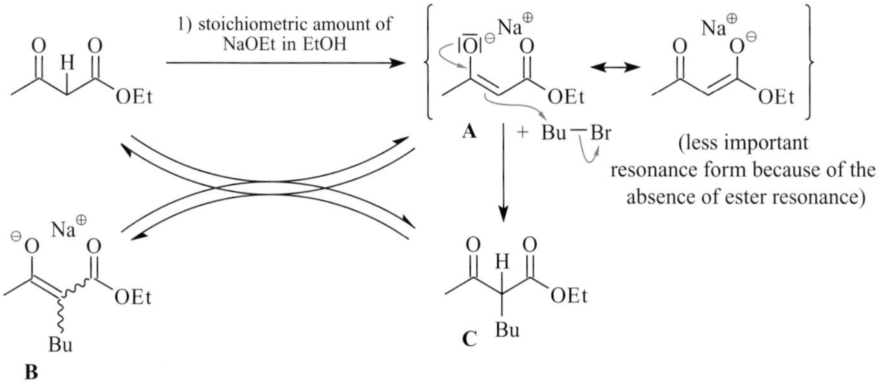

1) stoichiometric amount of NaOEt in EtOH

+ Bu—Br

(less important resonance form because of the absence of ester resonance)

Fig. 13.26. Acetoacetic ester synthesis of methyl ketones I: preparation of an alkylated acetoacetic ester.

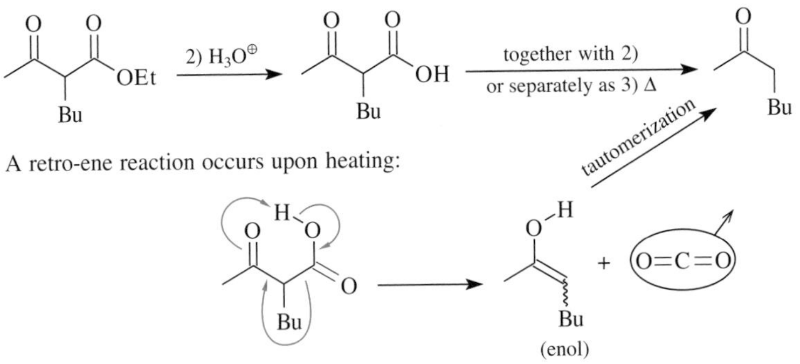

Fig. 13.27. Acetoacetic ester synthesis of methyl ketones II: hydrolysis of the alkylated acetoacetic ester/decarboxylation of the alkylated acetoacetic acid.

The butylated β-ketoester **C** of Figure 13.26 is not the final synthetic target of the *acetoacetic ester synthesis of methyl ketones.* In that context, the β-ketoester **C** is converted into the corresponding β-ketocarboxylic acid via acid-catalyzed hydrolysis (Figure 13.27; for the mechanism, see Figure 6.22). This β-ketocarboxylic acid is then heated either in the same pot or after isolation to effect decarboxylation. The β-ketocarboxylic acid decarboxylates via a cyclic six-membered transition state in which three valence electron pairs are shifted at the same time. The reaction product is an enol, which isomerizes immediately to a ketone (to phenyl methyl ketone in the specific example shown).

Other β-ketoesters can be converted into other ketones under the reaction conditions of the acetoacetic ester synthesis and with the same kinds of reactions as are shown in Figures 13.26 and 13.27. Two examples are provided in Figures 13.28 and 13.29. These β-ketoesters also are first converted into sodium enolates by reaction with NaOEt in EtOH, and the enolates are then reacted with alkylating reagents. The reaction shown in Figure 13.28 employs a β-ketoester that can be synthesized by a Claisen condensation (see Figure 13.57). The alkylating reagent employed in Figure 13.28 is bifunctional and reacts at both its reactive centers, thereby cross-linking two originally separate β-ketoester molecules. The new bis(β-ketoester) is hydrolyzed to afford a bis(β-ketocarboxylic acid), a twofold decarboxylation of which occurs subsequently according to the mechanism depicted in Figure 13.27. The reaction sequence of Figure 13.28 represents the synthesis of a diketone and illustrates the value of the acetoacetic ester synthesis to access a variety of alkyl ketones.

β-Ketoesters derived from cyclic five- or six-membered ketones are conveniently accessible via the Dieckmann condensation (for example, see Figure 13.58). Such β-ketoesters can be converted into cyclic ketones under the reaction conditions of the acetoacetic ester synthesis. Step 1 in Figure 13.29 shows how such a β-ketoester is allylated at its activated position. The allylation product **A** *could* be converted into the alkylated ketone **B** via a sequence comprising hydrolysis and decarboxylation. There exists, however, an alternative to achieve the special transformation of β-keto *methyl* esters ketones. It is shown in Figure 13.29. This alternative is based on the knowledge that good nucleophiles like the iodide ion or the phenylthiolate ion undergo an S_N2 reaction at the methyl group of the β-ketoester. The reaction is carried out at temperatures above 100 °C. The β-ketocarboxylate leaving group decarboxylates immediately under these conditions, producing the enolate of the desired ketone. This enolate is protonated to the ketone either by acidic contaminants of the solvent or later, during aqueous workup.

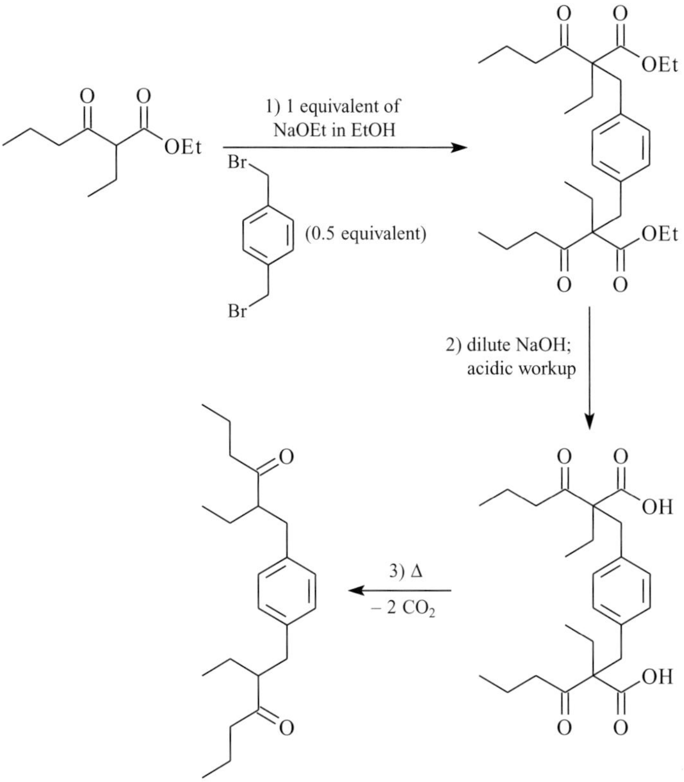

Fig. 13.28. Synthesis of complicated ketones in analogy to the acetoacetic ester synthesis I: generation of a diketone.

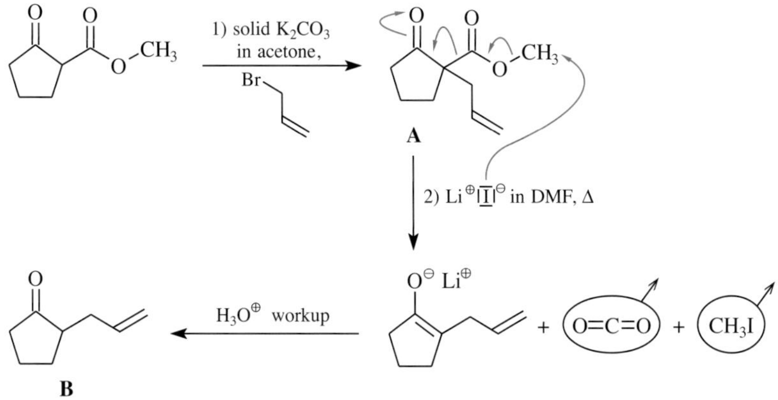

Fig. 13.29. Synthesis of complicated ketones in analogy to the acetoacetic ester synthesis II: generation of a cyclic ketone. In the first step, the β-ketoester is alkylated at its activated position. In the second step, the β-ketoester is treated with $Li^{\oplus}I^{\ominus}$. S_N2 reaction of the iodide at the methyl group generates the β-ketocarboxylate ion as the leaving group. The β-ketocarboxylate decarboxylates immediately under the reaction conditions (temperature above 100 °C) and yields the enolate of a ketone.

Alkylation of Ketone Enolates

Ester-substituted ketone enolates are stabilized, and these enolates can be alkylated (acetoacetic ester synthesis). Alkylation is, however, also possible for enolates that are not stabilized. In the case of the stabilized enolates, the alkylated ketones are formed in two or three steps, while the nonstabilized enolates afford the alkylated ketones in one step. However, the preparation of nonstabilized ketone enolates requires more aggressive reagents than the ones employed in the acetoacetic ester synthesis.

Figure 13.30 shows that even sterically hindered ketone enolates can be alkylated. The carbon atom in the β-position relative to the carbonyl carbon of an α,β-dialkylated α,β-unsaturated ketone can be converted into a quaternary C atom via 1,4-addition of an Gilman cuprate (for conceivable mechanisms, see Figure 10.46). As can be seen, a subsequent alkylation allows for the construction of another quaternary C atom in the α-position even though it is immediately adjacent to the quaternary center generated initially in the β-position.

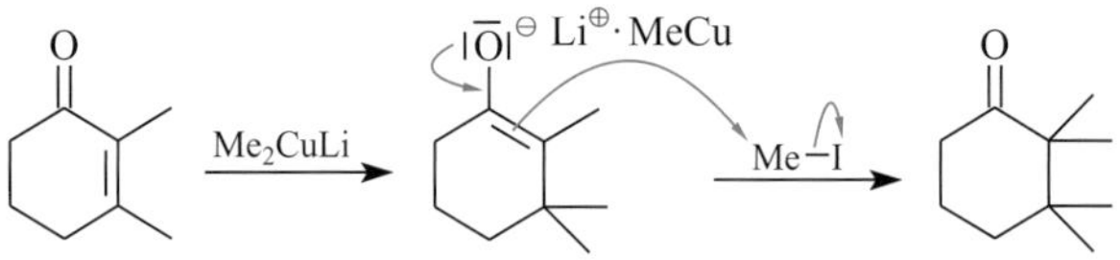

Fig. 13.30. 1,4-Addition plus enolate alkylation—a one-pot process for the α- and β-functionalization of α,β-unsaturated ketones. Two quaternary C atoms can be constructed via addition of a Gilman reagent and subsequent alkylation with MeI.

Diastereoselective alkylations of enolates may occur if the enolate is chiral, i.e., surrounded by diastereotopic half-spaces. This was discussed in Section 3.4.1. In general, it is difficult to predict the preferred side of reaction of the alkylating reagent on such enolates. For cyclic enolates the situation is relatively simple, because these enolates always react from the less-hindered side. Hence, for the methylation of the enolate in Figure 13.31, the reaction with methyl iodide occurs equatorially, that is, from the side that is opposite to the axially oriented methyl group at the bridgehead.

Bisenolates such as compound **A** derived from the acetoacetic ester in Figure 13.32 react with one equivalent of alkylating reagent in a regioselective fashion to give the enolate **C**. This could be the result of product development control, since the isomeric alkylation product

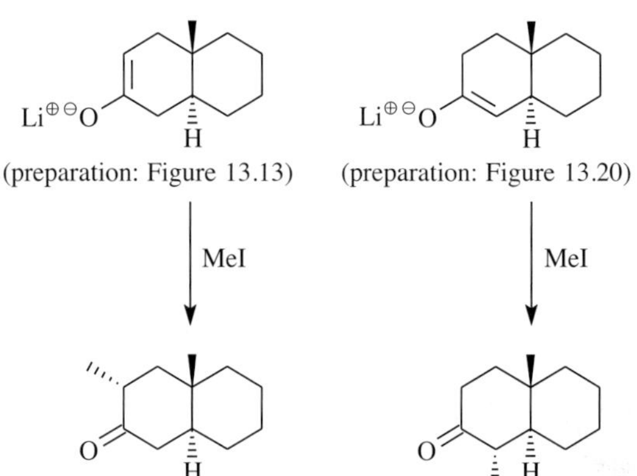

Fig. 13.31. Alkylation of a chiral ketone enolate. The reaction with methyl iodide occurs preferentially from the side opposite to the side of the axially oriented methyl group at the bridgehead carbon.

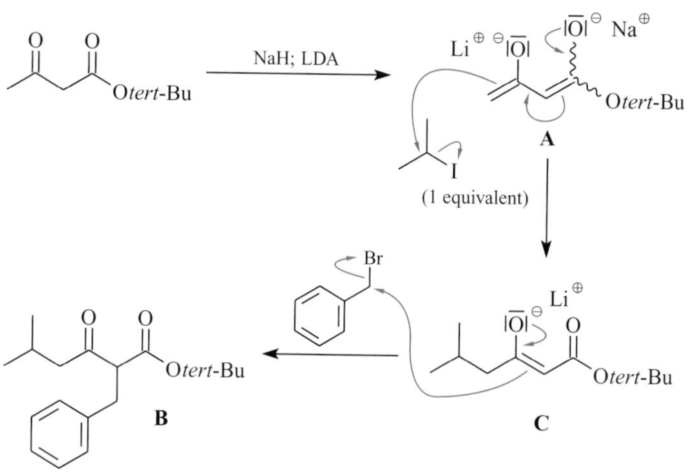

Fig. 13.32. Regiocontrolled bisalkylation of a bisenolate. The enolate **C** is formed with one equivalent of the alkylating reagent. **C** can be alkylated again, and **B** is formed in that way.

would be less stable. The enolate **C** resembles the nucleophile of the acetoacetic ester synthesis (Figure 13.26) and can be alkylated likewise. One can employ different alkylating reagents in the first and second alkylations to obtain a β-ketoester **B** with two new substituents. This product may feature a substitution pattern that could not be constructed via a Claisen condensation (Figures 13.57 and 13.59), as is true for the example presented in Figure 13.32.

Alkylation of Lithiated Aldimines and Lithiated Hydrazones

The quantitative conversion of aldehydes into enolates with lithium amides hardly ever succeeds because an aldol reaction (cf. Section 13.1.2) occurs while the deprotonation with LDA is in progress. Aldol additions also occur upon conversion of a small fraction of the aldehyde into the enolate with a weak base (Section 13.3.1). Hence, it is generally impossible to alkylate an aldehyde without the simultaneous occurrence of an aldol addition. There is only one exception: certain α-branched aldehydes can be deprotonated to their enolates in equilibrium reactions, and these enolates can be reacted with alkylating reagents to obtain tertiary aldehydes.

Since very few aldehydes can be converted into an α-alkylated aldehyde directly, there are some "detours" available. Figure 13.33 shows such a detour for the conversion of an aldehyde (without α-branching) into an α-alkylated aldehyde. First, the aldehyde is reacted with a primary amine—cyclohexyl amine is frequently used—to form the corresponding aldimine. Aldimines can be deprotonated with LDA or *sec*-BuLi to give azaenolates. The success of the deprotonation with *sec*-BuLi demonstrates that aldimines are much weaker electrophiles than aldehydes: *sec*-BuLi would immediately add to an aldehyde.

The obviously low electrophilicity of the C=N double bonds of aldimines precludes the addition of the azaenolate to remaining aldimine in the course of aldimine deprotonation. The aldimine enolate is obtained quantitatively and then reacted with the alkylating reagent. This step results cleanly in the desired product, again because of the low electrophilicity of imines: as the alkylation progresses, azaenolate and the alkylation product coexist without reacting with each other, no aldol-type reaction, no proton transfer. *All* the azaenolate is thus converted

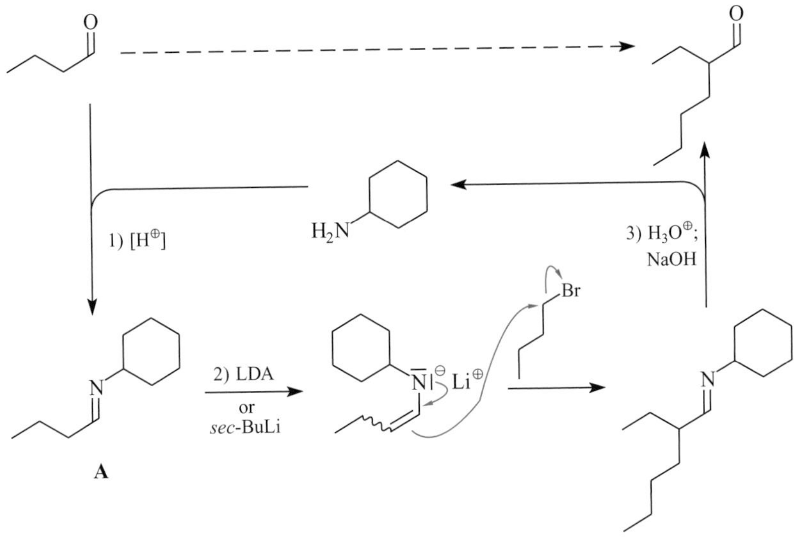

1) $[H^{\oplus}]$
H_2N
2) LDA or *sec*-BuLi
Br
Li$^{\oplus}$
3) $H_3O^{\oplus}$; NaOH
A

Fig. 13.33. α-Alkylation of an aldehyde via an imine derivative.

into the alkylated aldimine. In step 3 of the sequence of Figure 13.33, the imine is subjected to an acid-catalyzed hydrolysis, and the alkylated aldehyde results.

The formation of the alkylated aldimine in Figure 13.33 involves the generation of a stereocenter, yet without stereocontrol. The aldehyde derived from this aldimine consequently is obtained as a racemate. Figure 13.34 shows how a variation of this procedure allows for the enantioselective generation of the same aldehyde.

The "aldimine" of Figure 13.34 is a chiral and enantiomerically pure aldehydrazone **C.** This hydrazone is obtained by condensation of the aldehyde to be alkylated, and an enantiomerically pure hydrazine **A**, the *S*-proline derivative ***S*-am**ino**p**rolinol methyl ether (SAMP). The hydrazone **C** derived from aldehyde **A** is called the SAMP hydrazone, and the entire reaction sequence of Figure 13.34 is the **Enders SAMP alkylation.** The reaction of the aldehydrazone **C** with LDA results in the chemoselective formation of an azaenolate **D**, as in the case of the analogous aldimine **A** of Figure 13.33. The C=C double bond of the azaenolate **D** is *trans*-configured. This selectivity is reminiscent of the *E*-preference in the deprotonation of sterically unhindered aliphatic ketones to ketone enolates and, in fact, the origin is the same: both deprotonations occur via six-membered ring transition states with chair conformations. The transition state structure with the least steric interactions is preferred in both cases. It is the one that features the C atom in the β-position of the C,H acid in the pseudo-equatorial orientation.

The N—Li bond of azaenolate **D** lies outside the plane of the enolate. The structure created via chelation is a rigid polycyclic species. In this structure, the 4 and 5 carbons of the pyrollidine ring block one side of the azaenolate, resulting in facial selectivity during alkylation. The alkylation product **E** is formed preferentially with the *S*-configuration shown. Only traces of the *R*-configured product are formed.

The main and trace products are diastereoisomers, which can be completely separated by using chromatography. The separation affords a diastereomerically and enantiomerically pure SAMP hydrazone **E.**

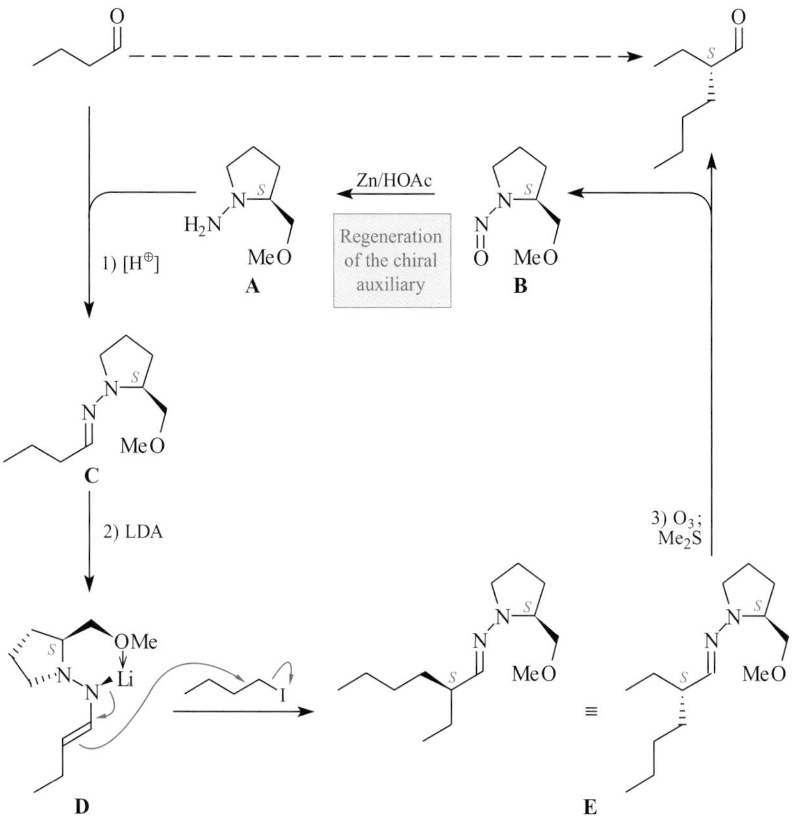

Fig. 13.34. Enders' SAMP method for the generation of enantiomerically pure α-alkylated carbonyl compounds; SAMP, (*S*)-aminoprolinol methyl ether = (*S*)-2-methoxymethyl-1-pyrrolidinamine, or **S**-1-**a**mino-2-(**m**ethoxymethyl)**p**yrrolidine (which is the name according to IUPAC rules).

In the third step of the reaction sequence depicted in Figure 13.34, the hydrazone **E** is converted into the desired sterically homogeneous aldehyde. This transformation can be achieved, for example, by ozonolysis of the C=N double bond. One of the products of ozonolysis is the desired enantiomerically pure α-butylated butanal. The other product of the ozonolysis also is valuable, since it is the nitroso derivative **B** of reagent **A.** The N=O group of **B** can be reduced to give an amino group to regenerate **A** from **B.** The possibility of recycling valuable chiral auxiliaries greatly enhances the attractiveness of any method for asymmetric synthesis.

The strategy of Figure 13.34 also is suitable for the synthesis of enantiomerically pure α-alkylated *ketones.* Figure 13.35 shows a procedure for the synthesis of the *S*-configured 6-methyl-2-cyclohexenone. The desired *S*-configuration is achieved with the help of a so-called RAMP hydrazone **C**, which is a derivative of the ***R*-amino*pro*linol** methyl ether **A.** In step 2 of the RAMP procedure, hydrazone **C** is deprotonated with LDA to give the azaenolate **D.** This deprotonation occurs with the same regioselectivity as the formation of the kinetic enolate **A** in the reaction of cyclohexenone with LDA. The common regioselectivities have the same origin. Deprotonations with LDA prefer cyclic transition state structures that are six-membered rings and include the heteroatom of the acidifying C=X double bond (X = O, N).

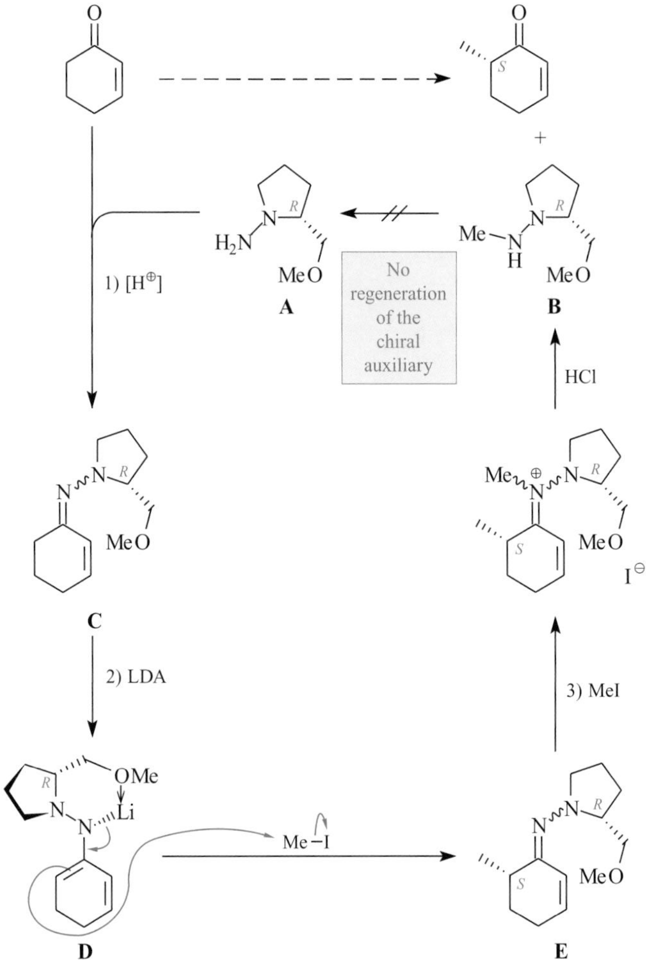

Fig. 13.35. Enders' RAMP method for the generation of enantiomerically pure α-alkylated carbonyl compounds; RAMP, (*R*)-aminoprolinol methyl ether or **R**-1-**a**mino-2-(**m**ethoxymethyl)**p**yrrolidine. While hydrazones derived from ketones, such as compound **E**, are methylated at the doubly bound nitrogen, the methylation of the corresponding aldehyde-derived hydrazones preferentially occurs at the nitrogen atom of the heterocycle.

As with the azaenolate of Figure 13.34, the azaenolate **D** in Figure 13.35 contains a chelate. As before a preferred conformation of the azaenolate and a rigid structure results in high diastereoselectivity during alkylation. The kethydrazone **E** is formed with high diastereoselectivity and, after chromatographic separation, it is obtained in 100% stereochemically pure form.

To complete the reaction sequence of Figure 13.35, the desired alkylated ketone needs to be released from the kethydrazone. Ozonolysis cannot be used in the present case. Ozonolysis would cleave not only the C=N double bond but also the C=C double bond. Another method must therefore be chosen. The kethydrazone is alkylated to give an iminium ion. The iminium ion is much more easily hydrolyzed than the hydrazone itself, and mild hydrolysis yields the desired *S*-enantiomer of 6-methyl-2-cyclohexenone. The other product of hydrolysis is a RAMP derivative. This RAMP derivative carries a methyl group at the N atom and cannot be recycled to the enantiomerically pure chiral auxiliary **A** that was employed initially.

13.2.2 Chain-Elongating Syntheses of Carboxylic Acid Derivatives

Malonic Ester Synthesis of Substituted Acetic Acids

Malonic ester syntheses are the classical analog of acetoacetic ester syntheses of methyl ketones. Neither case requires the use of an amide base for the enolate formation, and in both cases alkoxides suffice to deprotonate the substrate completely. Malonic esters are active-methylene compounds just like acetoacetic ester and its derivatives.

One equivalent of NaOEt in EtOH deprotonates diethyl malonate completely to give the sodium enolate **A** (Figure 13.36). This enolate is monoalkylated upon addition of an alkylating reagent such as BuBr, and a substituted malonic ester **C** is formed. *During* the alkylation reaction, the substituted malonic ester **C** reacts to a certain extent with some of the enolate **A**, resulting in the butylated enolate **B** and unsubstituted neutral malonic ester. It is for this reason that the reaction mixture contains *two* nucleophiles—the original enolate **A** and the butylated enolate **B.** The alkylation of **A** with butyl bromide is much faster than that of **B**, since **A** is less sterically hindered than **B.** The main product is therefore the product of monoalkylation. Distillation can be used to separate the main product from small amounts of the product of dialkylation.

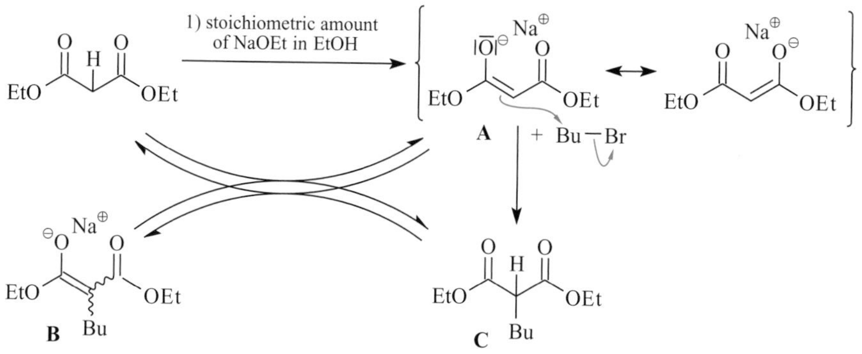

Fig. 13.36. Malonic ester synthesis of alkylated acetic acids I: preparation of alkylated malonic esters.

The butylated malonic ester **C** of Figure 13.36 is not the actual synthetic target of the *malonic ester synthesis of substituted acetic acids.* Instead, **C** is subjected to further transformations as shown in Figure 13.37. Ester **C** first is hydrolyzed with acid catalysis to afford the corresponding alkylated malonic acid (for the mechanism, see Figure 6.22). The alkylated malonic acid then is heated either directly in the hydrolysis mixture or after it has been isolated. This heating leads to decarboxylation. The mechanism of this decarboxylation resembles the mechanism of the decarboxylation of β-ketocarboxylic acids (see Figure 13.27), and it involves a cyclic, six-membered ring transition state in which three valence electron pairs are shifted at the same time. The primary products of this decomposition are carbon dioxide and the enol of the carboxylic acid. The enol immediately tautomerizes to give the carboxylic acid. This carboxylic acid—an alkylated acetic acid—represents the typical final product of a malonic ester synthesis.

Both acidic H atoms of a malonic ester can be replaced by alkyl groups. These dialkylated malonic esters are formed by successively removing the acidic protons with sodium alkoxide

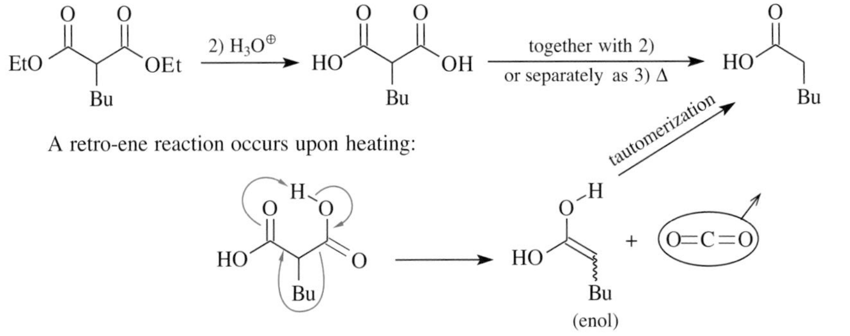

Fig. 13.37. Malonic ester synthesis of alkylated acetic acids II: hydrolysis and decarboxylation of the alkylated malonic ester.

and treatment of the enolates with an alkylating reagent. The subsequent hydrolysis and decarboxylation of these dialkylated malonic esters affords α,α-dialkylated acetic acids as another class of products accessible via the malonic ester synthesis.

If one employs *monofunctional* alkylating reagents in the alkylation of malonic esters, one obtains dialkylated acetic acids in which the two α-alkyl groups are not connected with each other. On the other hand, if one employs a *difunctional* alkylating reagent, the dialkylated acetic acid synthesized is a cycloalkane carboxylic acid. This is the case when the second alkylation occurs in an intramolecular instead of an intermolecular fashion. Over 100 years ago, Perkin employed this principle and succeeded at the synthesis of cyclopropane carboxylic acid (Figure 13.38), the first cyclopropane derivative ever made. Until that time, the synthesis of a cyclopropane was thought to be impossible because of its high Baeyer strain ("angular strain").

Figure 13.39 shows that malonic ester syntheses can also lead to acetic acid derivatives with a heteroatom in the α-position. The benzylation of (acetamido)malonic acid diethyl ester

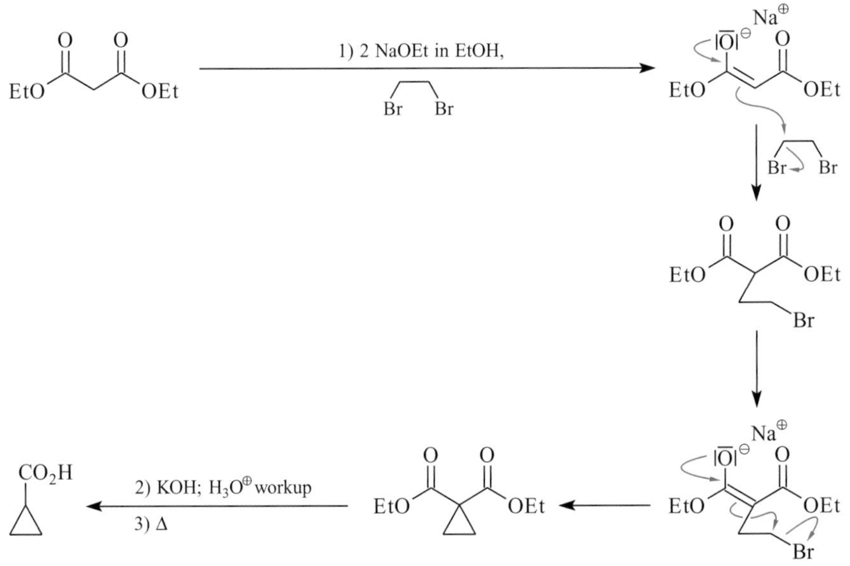

Fig. 13.38. Perkin's first cyclopropane synthesis via a malonic ester synthesis.

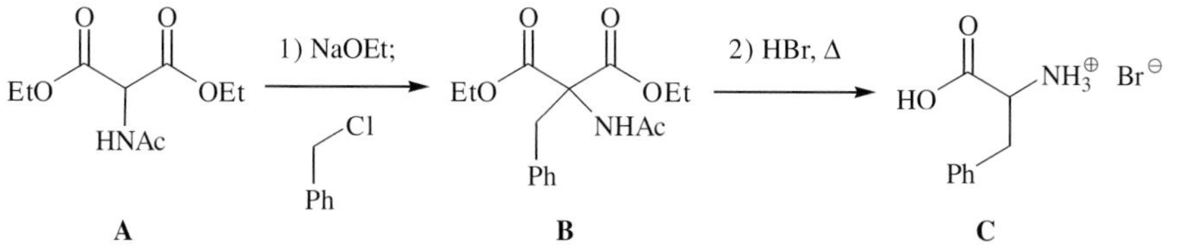

EtO / OEt / HNAc / A / (preparation: Figure 12.9) — 1) NaOEt; Cl / Ph → EtO / OEt / NHAc / Ph / B — 2) HBr, Δ → HO / $NH_3^{\oplus}$ $Br^{\ominus}$ / Ph / C

Fig. 13.39. Synthesis of an α-amino acid through malonic ester synthesis, here using the example of phenylalanine (in the form of its hydrobromide).

(**A**) produces the disubstituted malonic ester **B**. Hydrobromic acid hydrolyzes the ester functions of **B**—which, upon heating, is followed by the usual decarboxylation—as well as the acetamido group. The reaction product is the hydrobromide **C** of the α-amino acid phenylalanine.

Alkylation of Ester Enolates

Ester enolates are generated by the reaction between an ester and LDA at –78 °C in THF, enolate formation usually being *"E"*-selective (Figure 13.16). The *"Z"*-enolate is obtained in analogous deprotonations of esters that carry an alkoxy group in the α-position relative to the C=O double bond (Figure 13.40). Product development control is the reason for the latter stereochemical outcome: the *"Z"*-enolate and the lithium form an energetically favored five-membered chelate ring.

Many ester enolates can be alkylated, and this is irrespective of whether they are *"E"*- or *"Z"*-configured. The example of Figure 13.40 shows the butylation of a *"Z"*-configured α-oxygenated ester enolate. The butylated ester **B** is both a benzyl ester and benzyl ether. The two benzylic C—O bonds in this compound can be removed subsequently by way of hydrogenolysis (see Figure 17.51). Overall, this reaction sequence represents a method that allows for the elongation of alkylating agents to α-hydroxycarboxylic acids **A.**

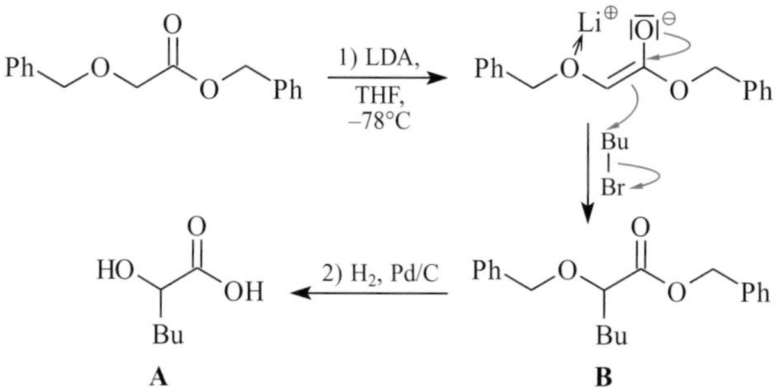

Ph / O / O / Ph — 1) LDA, THF, –78°C → $Li^{\oplus}$ / $\overline{O}^{\ominus}$ / Ph / O / O / Ph — Bu–Br → Ph / O / O / Ph / Bu / B — 2) H_2, Pd/C → HO / OH / Bu / A

Fig. 13.40. Alkylation of an ester enolate for the preparation of an α-hydroxycarboxylic acid (for the preparation of enantiomerically pure α-hydroxy carboxylic acid through alkylation of an enantiomerically pure ester enolate cf. Figure 13.41). The initially formed benzyl ester **B** contains two benzylic C—O bonds, which can be cleaved by means of hydrogenolysis.

Diastereoselective Alkylation of Chiral Ester and Amide Enolates: Generation of Enantiomerically Pure Carboxylic Acids with Chiral Centers in the α-Position

The alkylation of an *achiral* ester enolate to give an α-alkylated carboxylic ester can generate a new stereocenter. If so, this stereocenter is formed without stereocontrol. For such a substrate, the two half-spaces above and below the enolate plane are enantiotopic. Consequently, the reaction of an achiral alkylating reagent occurs from both faces with the same rate constant (cf. discussion of Section 3.4.1). Thus, one obtains the alkylated ester as a racemic mixture. In the alkylation of an achiral amide enolate, the outcome is entirely analogous: the resulting α-alkylated amide either is achiral or a racemate.

The situation changes when chiral ester enolates or chiral amide enolates are alkylated. There, the half-spaces on the two sides of the enolate planes of the substrates are diastereotopic, and alkylating reagents can react from one of the sides selectively (see discussion in Section 3.4.1). Stereogenic alkylations of such enolates therefore may take place diastereoselectively.

Side Note 13.4 presents the diastereoselective alkylation of a very special ester enolate in which one can easily understand what the stereocontrol observed is based upon. However, only *very* specific carboxylic acid derivatives are made accessible by those alkylations. Much more broadly applicable diastereoselective alkylations of chiral ester or amide enolates will be introduced in Figures 13.42 and 13.43. Figure 13.42 shows alkylations of a propionic acid ester—derived from an enantiomerically pure chiral alcohol—via the *"E"*- and *"Z"*-enolate. Figure 13.43 illustrates alkylations of two propionic acid amides, where in each case the N atom is part of an enantiomerically pure heterocycle, proceeding via the respective *Z*-configured amide enolates.

Side Note 13.4. An Enolate Alkylation with "Self-Reproduction of Chirality"

Compound **A** (Figure 13.41) was introduced in Section 9.2.2 as "a kind of *O,O* acetal." Here you will see how asymmetric synthesis may take advantage of the fact that the "acetal" carbon atom of this compound is a stereocenter that uniformly displays the absolute configuration given here. In the reaction with the base LDA, the compound **A** acts as an ester, i.e., it undergoes deprotonation to the ester enolate **B**. If the latter is alkylated, the bulky *tert*-butyl

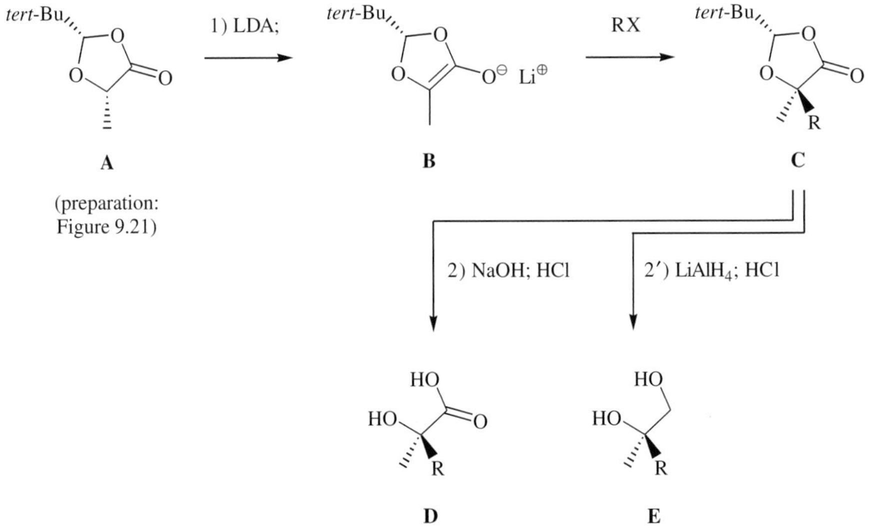

tert-Bu, O, O, O — A (preparation: Figure 9.21) — 1) LDA; — tert-Bu, O, O, $O^{\ominus}$ $Li^{\oplus}$ — B — RX — tert-Bu, O, O, O, R — C — 2) NaOH; HCl — 2′) $LiAlH_4$; HCl — HO, HO, O, R — D — HO, HO, R — E

Fig. 13.41. Alkylation of an enantiomerically pure lactone enolate (**B**) for the preparation of enantiomerically pure α-alkyl-α-hydroxycarboxylic acids and enantiomerically pure 1,2-diols, respectively. In the lactone **C** the carboxyl group may react with water or with a hydride donor. In any case, the leaving group released is a hemiacetal anion that decomposes to pivalaldehyde and the α-alkyl-α-hydroxycarboxylic acid **D** or the enantiomerically pure 1,2-diol **E**.

residue at the "acetal" carbon atom ensures that the alkylating agent exclusively approaches the enolate from the opposite side of the molecule. This is why the alkylation products **C** are obtained with high diastereoselectivity. They may then be hydrolyzed to yield the enantiomerically pure α-hydroxy carboxylic acid **D** (or be reduced to the enantiomerically pure 1,2-diols **E**).

The lactic acid, which initiates the reaction sequence *S*-lactic acid → "acetal" **A** → enolate **B** → "acetal" **C** → *R* hydroxycarboxylic acid **D**, has a stereocenter with a well-defined absolute configuration that is destroyed in the enolate intermediate **B**, but finally restored in the hydroxycarboxylic acid **C**. This is why the principle concerning the stereochemistry of the key step ("acetal" **A** → enolate **B** → "acetal" **C**) is referred to as the "self-reproduction of chirality."

Both enantiomers of camphor are commercially available. One of the camphor enantiomers can be converted into the enantiomerically pure, chiral alcohol contained in the propionic acid ester **A** in five steps. Ester **A** is employed in the **Helmchen synthesis** of Figure 13.42. The enantiomer of this ester can be obtained from the other camphor isomer. Each of these esters can be alkylated with high diastereoselectivity, as shown in Figure 13.42 for two alkylations of ester **A.** The highest selectivities are achieved if the ester is deprotonated at –78 °C with lithium cyclohexyl isopropyl amide. This reagent is an amide base with a somewhat higher steric demand than LDA. The deprotonation with this reagent is a stereogenic reaction just like the LDA deprotonation of the propionic acid esters of Figures 13.16 and 13.17, and the same stereoselectivities result: in pure THF propionic acid ester **A** yields the *"E"*-enolate with high diastereoselectivity (Figure 13.42, left). In THF/HMPA mixtures, on the other hand, the reaction of the same propionic acid ester **A** with the same base occurs with complete reversal of stereochemistry, i.e., yields the *"Z"*-enolate (Figure 13.42, right). Accordingly, THF/HMPA mixtures have the same effect on the stereoselectivity of ester enolate formation as we discussed for THF/DMPU mixtures in the context of Figure 13.17. Fortunately, in *this* case the HMPA (carcinogenic) can be replaced by DMPU (not carcinogenic). This option was not known at the time when the investigations described in Figure 13.42 were carried out.

The chiral alcohol group in Figure 13.42 was chosen to differentiate as much as possible between the half-spaces on both sides of the enolate plane. One half-space should be left entirely unhindered while the other should be blocked as completely as possible. The reaction of the alkylating reagent then occurs preferentially, and in the ideal case exclusively, from the unhindered half-space. The stereostructures of the two ester enolates of Figure 13.42 therefore model the enolate moieties of the (early!) transition states of these alkylations. The part of the transition state structure that contains the alkylating reagent is not shown.

It is assumed that the preferred conformation of the substructure C=C—O—C of the *"E"*-configured ester enolate in the preferred transition state of the alkylation is that depicted in the center of the left-hand column of Figure 13.42. In the projection shown, the alkylating reagent reacts with the enolate from the front side for the reasons just stated. The reaction occurs with a diastereoselectivity of 97:3. Chromatography allows for the complete separation of the main diastereoisomer from the minor diastereoisomer. Reduction of the main diastereoisomer (for the mechanism, see Section 17.4.3) affords the alcohol **B**, a derivative of *S*-α-benzyl propionic acid, with 100% *ee*. *Hydrolysis* of the benzylated esters without iso-

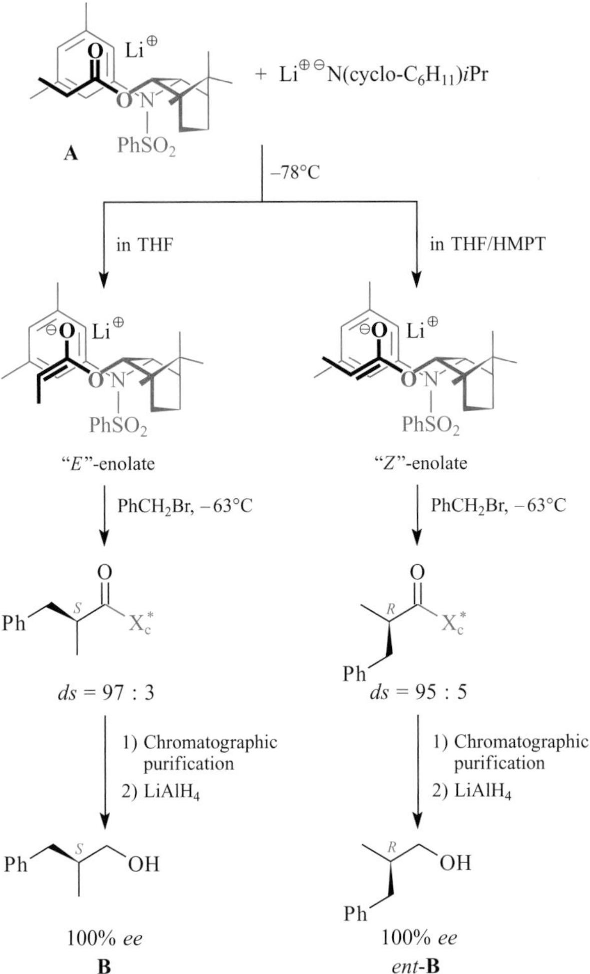

Fig. 13.42. Helmchen synthesis of enantiomerically pure α-alkylated carboxylic acids. The deprotonation of the propionic acid ester results in the "*E*"-enolate in the solvent THF and in the "*Z*"-enolate in the solvent mixture THF/HMPA. In these projections, both enolates react preferentially from the front. The "*E*"'-enolate results in a 97:3 mixture of *S*- and *R*-configured α-benzylpropionic acid esters (X_c^* marks the chiral alkoxide group), while the "*Z*"-enolate results in a 5:95 mixture. Chromatographic separation and reduction of the C(=O)—X_c^* groups afford alcohol **B** with 100% *ee* from the "*E*"-enolate and alcohol *ent*-**B** with 100% *ee* from the "*Z*"-enolate.

merization is impossible, however, so that optically active α-benzylpropionic *acids* cannot be obtained in this way. The center of the right-hand column of Figure 13.42 shows the assumed stereostructure of the substructure C=C—O—C of the "*Z*"-configured ester enolate in the preferred transition state of the alkylation. In the chosen projection, the alkylating reagent again reacts from the front side. With a diastereoselectivity of 95:5, the benzylated ester that was the minor product in the alkylation of the "*E*"-configured ester enolate is now formed as the main product. Again, chromatography allows one to separate the minor from the major diastereoisomer. The main product is then reduced without any isomerization to afford the *alcohol ent*-**B** with an *ee* value of 100%.

Why are the benzylated esters of Figure 13.42 not obtained with higher diastereoselectivities than 95 or 97%, respectively? One possible reason lies in the failure of both the "*E*"- and the "*Z*"-enolate to form with perfect stereocontrol. Small contaminations of these enolates by just 5 or 3% of the corresponding enolate with the opposite configuration would explain the

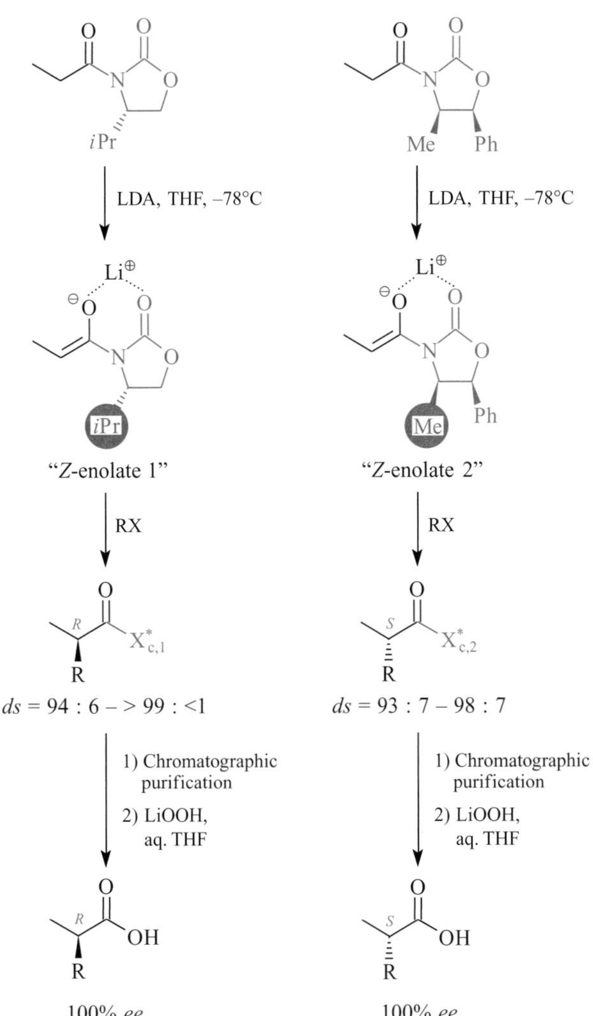

Fig. 13.43. Evans synthesis of enantiomerically pure α-alkylated carboxylic acids. The amides are derived from oxazolidinones and yield "Z"-enolates with high stereoselectivity. The alkylating agent reacts in both cases from the side that is opposite to the side of the substituent highlighted in red. Alkaline hydrolysis accelerated by hydrogen peroxide proceeds with retention of configuration and yields enantiomerically pure a-alkylated carboxylic acids; $X^*_{c,1}$ and $X^*_{c,2}$ are the chiral amide groups.

observed amounts of the minor diastereoisomers, even if every enolate were alkylated with 100% diastereoselectivity.

Only chiral propionic acid amides can be alkylated with still higher diastereoselectivity than chiral propionic acid esters. This is because according to Figure 13.18, the selectivity for the formation of a *"Z"*-configured amide enolate is higher than the selectivities that can be achieved in the conversion of esters to the *"E"*- and *"Z"*-enolates. The alkylation of the *"Z"*-configured lithium enolates of the two enantiomerically pure propionic acid amides in the **Evans synthesis** of Figure 13.43 proceeds with particularly high diastereoselectivity. These amide enolates contain an oxazolidinone ring, and the presence of this ring causes conformational rigidity of the enolates: lithium bridges between the enolate oxygen and the carbonyl O atom of the heterocycle to form a six-membered ring.

Both oxazolidinones in Figure 13.43 are selected such that the substituent marked by a red circle occupies one of the two half-spaces of the enolate. The oxazolidinone to the left in Figure 13.43 can be prepared from *S*-valine in two steps. The isopropyl group ensures that the most stable transition state for the alkylation of the "*Z*-enolate 1" involves reaction of the alkylating reagent from the front side (with respect to the selected projection). The alkylating agent thus reacts preferentially from the side opposite to the isopropyl group. Similar considerations apply to the most stable transition state structure of the alkylation of the oxazolidinone "*Z*-enolate 2" in Figure 13.43. This transition state results from the backside reaction of the alkylating agent (again with regard to the projection drawn). But again, this is a reaction from that side of the molecule that is opposite to the substituent marked by the red circle.

The alkylations of the oxazolidinone-containing amide enolate of Figure 13.43 occur with diastereoselectivities of 93:7 and > 99:1, respectively. The hydrogen peroxide-accelerated alkaline hydrolysis of these compounds occurs with complete retention of the previously established configuration at the α-stereocenter. To date, the Evans synthesis offers the most versatile access to enantiomerically pure α-alkylated carboxylic acids.

13.3 Hydroxyalkylation of Enolates with Carbonyl Compounds ("Aldol Addition"): Synthesis of β-Hydroxyketones and β-Hydroxyesters

An "aldol addition" involves the addition of the α-C atom of a carbonyl compound, a carboxylic acid, a carboxylic ester, or a carboxylic amide to the C=O double bond of an aldehyde or a ketone. The products of aldol additions are β-hydroxylcarbonyl compounds (**aldols**), β-hydroxycarboxylic acids, β-hydroxycarboxylic esters, or β-hydroxycarboxylic amides.

13.3.1 Driving Force of Aldol Additions and Survey of Reaction Products

The addition of an alkaline earth metal enolate **A** to a carbonyl compound is always an exergonic process irrespective of whether the enolate is derived from a ketone, an ester, or an amide and whether the carbonyl compound is an aldehyde or a ketone (Figure 13.44, top). One of the reasons for this exergonicity lies in the fact that the alkaline earth metal ion is part of a chelate in the alkoxide **B** of the aldol addition product. The driving forces for the additions of alkaline earth metal enolates of esters and amides to carbonyl compounds are further increased because the aldol adducts **B** are resonance-stabilized, whereas the enolates are not.

Table 13.6 shows the various aldol adducts that can be obtained if one reacts a quantitatively formed ester enolate or a quantitatively formed (kinetic) ketone enolate with three representative carbonyl compounds. **Crossed aldol adducts** are adducts that result from the addition of the enolate of *one* carbonyl compound to the C=O double bond of a *second* carbonyl compound (center column in Table 13.6).

Tab. 13.6 Representative Aldol Adducts Formed by Addition of Ketone or Ester Enolates to Selected Carbonyl Compounds

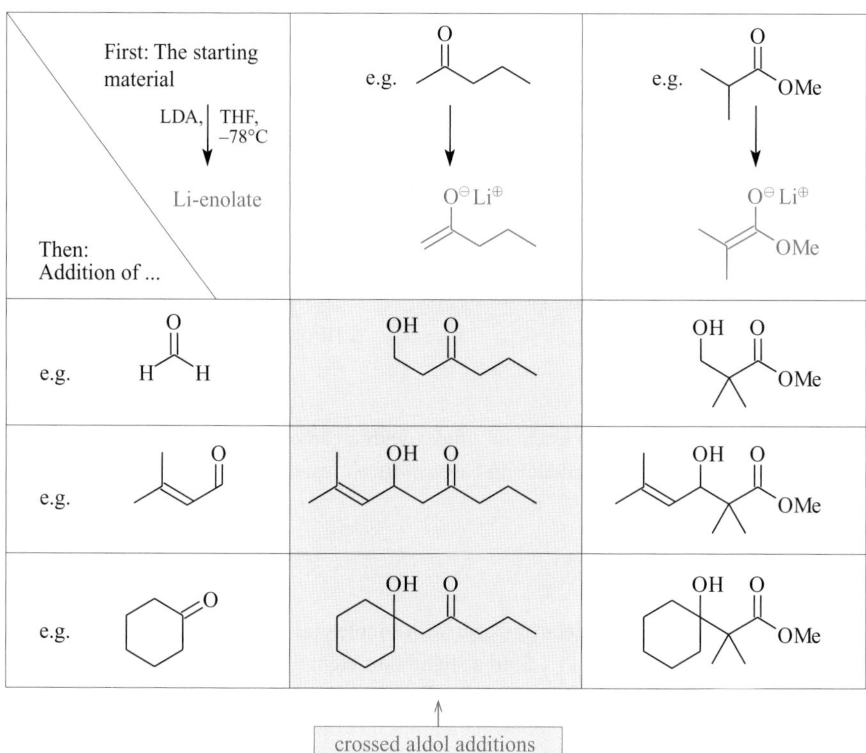

In principle it also is possible to obtain the β-hydroxycarbonyl compounds directly in neutral form rather than in form of their alkoxides (Figure 13.44, bottom). This is accomplished by the reaction of one carbonyl compound or of a mixture of two carbonyl compounds with a catalytic amount of MOH or MOR. Aldehyde enolates and ketone enolates are then formed in small amounts (see the Rule of Thumb at the beginning of Section 13.1.2). These enolates add to the C=O double bond of the starting substrate molecules or, if a mixture of carbonyl compounds is employed, they add to the C=O double bond of the more reactive of the carbonyl compounds. The alkoxides **B** of the aldol adducts are formed initially but are converted immediately and quantitatively into the aldols by way of protonation.

This base-catalyzed aldol addition is an equilibrium reaction, and all steps of this reaction are reversible. The free enthalpy of reaction $\Delta G_r°$ of such aldol reactions is close to zero. In fact, $\Delta G_r°$ is negative only if there are "many H atoms" among the substituents R^1, R^2, and R^3 of the two reacting components (structures in Figure 13.44, bottom). Otherwise, the formation of the aldol adduct is endergonic because of the destabilization due to the van-der-Waals repulsion between these substituents. A base-catalyzed aldol addition between two *ketones*, therefore, is never observed.

Esters and amides are much weaker C,H acids than aldehydes and ketones. Neither the ester nor the amide is deprotonated to any significant extent if a base such as MOH or MOR

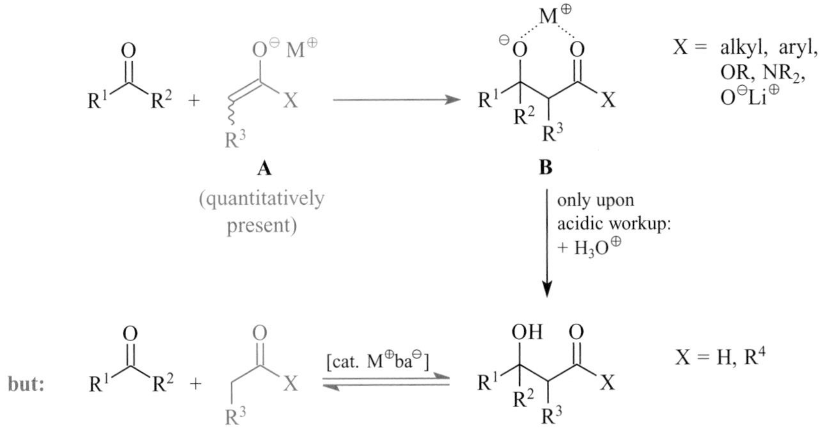

Fig. 13.44. Different driving forces of aldol additions depending on whether they involve quantitatively prepared enolates (top) or only small equilibrium concentrations of enolates (bottom).

is added to a mixture of these esters or amides with a carbonyl compound. Hence, neither esters nor amides afford aldol adducts in base-*catalyzed* reactions.

13.3.2 Stereocontrol

The preparation of aldol adducts may occur with simple diastereoselectivity. A definition of the term was given in Section 11.1.3. In a slightly different formulation, simple diastereoselectivity means that a single relative configuration is established at two neighboring C atoms that become stereocenters for the following reasons: (1) Both C atoms were sp^2-hybridized in the reactants; one was part of a nonhomotopic C=X double bond and the other was part of a nonhomotopic C=Y double bond. (2) The formation of a σ-bond between these C atoms causes them to be sp^3-hybridized in the reaction product.

The simple diastereoselectivity of aldol reactions was first studied in detail for the **Ivanov reaction** (Figure 13.45). The Ivanov reaction consists of the addition of a carboxylate enolate to an aldehyde. In the example of Figure 13.45, the diastereomer of the β-hydroxycarboxylic acid product that is referred to as the *anti*-diastereomer is formed in a threefold excess in comparison to the *syn*-diastereoisomer. Zimmerman and Traxler suggested a transition state model to explain this selectivity, and their transition state model now is referred to as the **Zimmerman–Traxler model** (Figure 13.46). This model has been applied ever since with good success to explain the simple diastereoselectivities of a great variety of aldol reactions.

The key idea of the Zimmerman–Traxler model is that aldol additions proceed via six-membered ring transition state structures. In these transition states, the metal (a magnesium

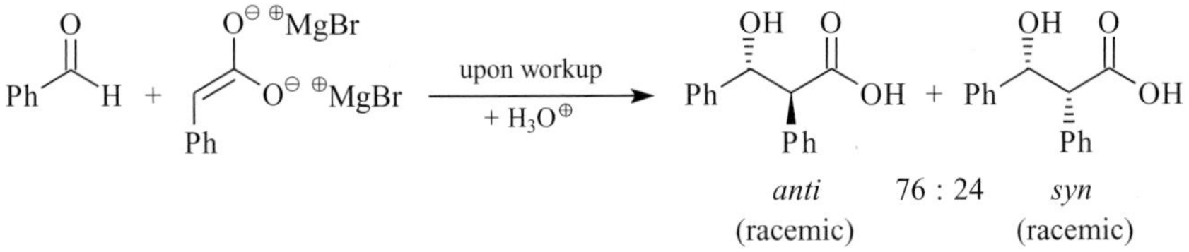

Fig. 13.45. The Ivanov reaction. For the generation of the carboxylate enolate see Figure 13.10.

cation in the case of the Ivanov reaction) coordinates both to the enolate oxygen and to the O atom of the carbonyl compound. By way of this coordination, the metal ion guides the approach of the electrophilic carbonyl carbon to the nucleophilic enolate carbon. The approach of the carbonyl and enolate carbons occurs in a transition state structure with a chair conformation. C—C bond formation is fastest in the transition state with the maximum number of pseudo-equatorially oriented and therefore sterically unhindered substituents.

The application of the Zimmerman–Traxler model to the specific case of the Ivanov reaction of Figure 13.45 is illustrated in Figure 13.46. The reaction proceeds preferentially through

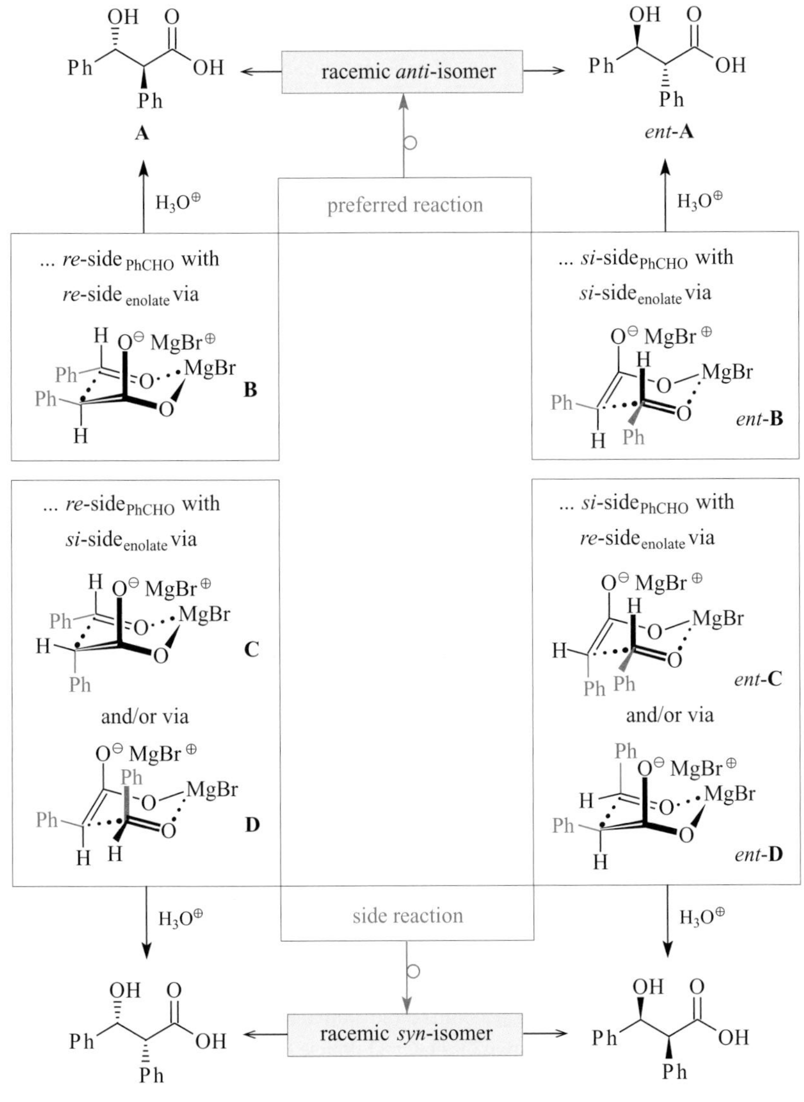

Fig. 13.46. Explanation of the *anti*-selectivity of the Ivanov reaction of Figure 13.45 by means of the Zimmerman–Traxler model. The stereodescriptors *Re* and *Si* are defined as follows. Suppose you are looking down on the plane of an alkene, in which an sp^2-hybridized C atom is connected to three different substituents. You are on the *Re* side of the double bond if the Cahn–Ingold–Prelog priorities of these substituents decrease going clockwise, and on the *Si* side otherwise.

transition state **B** and its mirror image *ent*-**B** and results in the formation of a racemic mixture of the enantiomers **A** and *ent*-**A** of the main diastereoisomer. Both phenyl groups are pseudo-equatorial in these transition states. All other transition state structures are less stable because they contain at least one phenyl group in a pseudo-axial orientation. For example, the phenyl group of the enolate is in a pseudoaxial position in transition state **C** and its mirror image *ent*-**C.** In transition state **D** and its mirror image *ent-D*, the phenyl groups of the benzaldehyde occupy pseudo-axial positions. The latter transition state structures are therefore just as disfavored as the pair **C** and *ent*-**C.** In fact, the *syn*-configured minor product of the Ivanov reaction, a racemic mixture of enantiomers **E** and *ent*-**E**, must be formed via the transition states **C** and *ent*-**C** or via **D** and *ent*-**D**, but it is not known which path is actually taken.

The *anti* adducts **A**/*ent*-**A** also could, in principle, result via the pair of transition states that contain *both* phenyl groups in pseudo-axial positions. This reaction path might contribute a small amount of the *anti* adduct, but this is rather improbable.

Lithium enolates of ketones and esters also add to aldehydes by way of Zimmerman–Traxler transition states. However, the Li—O bond is weaker and longer than the Mg—O bond. The lithium-containing transition state structures thus are less compact than those containing Mg. Therefore, in a Li-containing Zimmerman–Traxler transition state, a pseudo-axial and thus unfavorably positioned aldehyde substituent suffers only a weak *gauche* interaction with the skeleton of the six-membered ring. In fact, this destabilization generally is too small to render such a transition state structure inaccessible. Hence, the addition of lithium enolates to aldehydes only occurs with high diastereoselectivity if the aldehyde substituent does not assume a pseudo-axial position as the consequence of *another* destabilizing interaction.

Heathcock identified such a destabilizing interaction in the 1,3-diaxial interaction of the aldehyde substituent with a substituent at the C atom to which the enolate oxygen is attached. In spite of the relatively long Li—O distance, this 1,3-interaction can be sufficiently strong if the substituent of the aldehyde is extremely bulky. In that case, and only in that case, the aldehyde group is forced into the pseudo-equatorial orientation also in a lithium-containing Zimmerman–Traxler transition state. If, in addition, the lithium enolate of the aldehyde contains a homogeneously configured C=C(—O$^{\ominus}$) double bond, a highly diastereoselective aldol addition of a lithium enolate occurs. Two configurationally homogeneous lithium enolates with suitably bulky substituents are the *Z*-configured ketone enolate of Figure 13.47 with its $Me_2C(OSiMe_3)$ group and the *"E"*-configured ester enolate of Figure 13.48 with its (2,6-di-*tert*-butyl-4-methoxyphenyl)oxy group.

The ketone enolate **A** of Figure 13.47 is generated in a *Z*-selective fashion (as we saw in Figure 13.15). The bulky and branched enolate substituent destabilizes the Zimmerman–Traxler transition state **C** by way of the discussed 1,3-diaxial interaction, while the transition state structure **B** is not affected. Hence, the aldol addition of enolate **A** occurs almost exclusively via transition state **B**, and the *syn*-configured aldol adducts **D** (Figure 13.47) are formed with a near-perfect simple diastereoselectivity. The acidic workup converts the initially formed trimethysilyloxy-substituted aldol adducts into the *hydroxylated* aldol adducts.

It is for good reason that the bulky ketone substituent of enolate **A** contains a Me_3SiO group, which is carried on into the *syn*-configured aldol adduct: its acid-catalyzed hydrolysis affords an OH group in the α-position of the C=O double bond. Such an α-hydroxycarbonyl compound can be oxidatively cleaved with sodium periodate to afford a carboxylic acid (cf. Section 9.1.1). The mechanism of this oxidation is described later in connection with Fig-

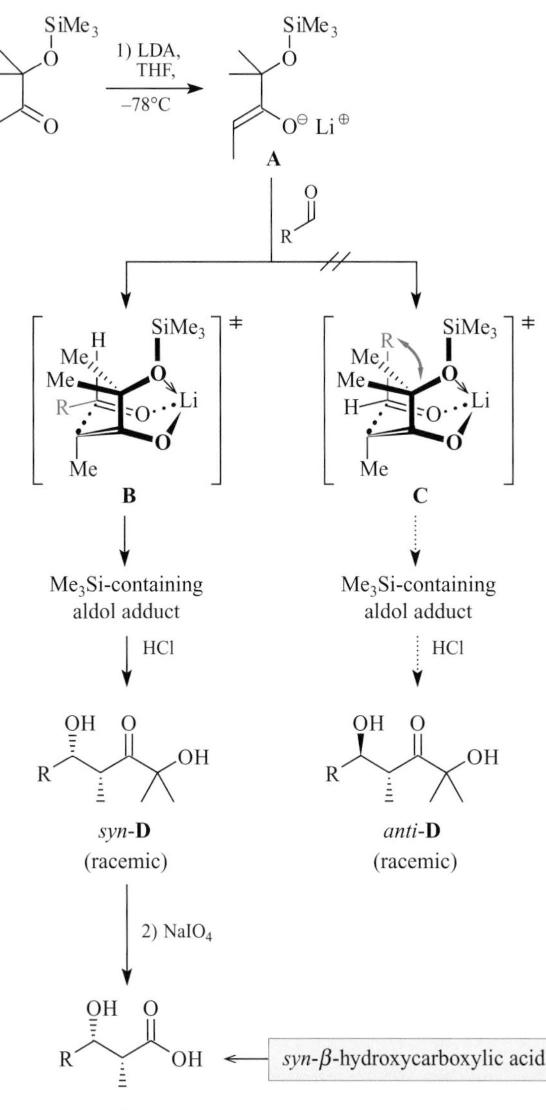

Fig. 13.47. *syn*-Selectivity of the aldol addition with a Heathcock lithium enolate including a mechanistic explanation. The Zimmerman–Traxler transition state **C** is destabilized by a 1,3-diaxial interaction, while the Zimmerman–Traxler transition state **B** does not suffer from such a disadvantage. The reaction thus occurs exclusively via transition state **B**.

ure 17.23. It involves the oxidation of the corresponding hydrate of the carbonyl compound. The hydrate of the α-hydroxycarbonyl compound is formed in an equilibrium reaction. This oxidation converts the *syn*-configured aldol adduct **D**—which contains a synthetically less useful ketone substituent—into a synthetically more valuable *syn*-configured β-hydroxycarboxylic acid, **E.**

The aldol addition of Figure 13.47 can also be carried out in such a fashion that the crude silyl ether-containing aldol adducts are treated directly with periodic acid without prior aqueous workup. In that case, the silyl ether and α-hydroxyketone cleavages both occur in one operation.

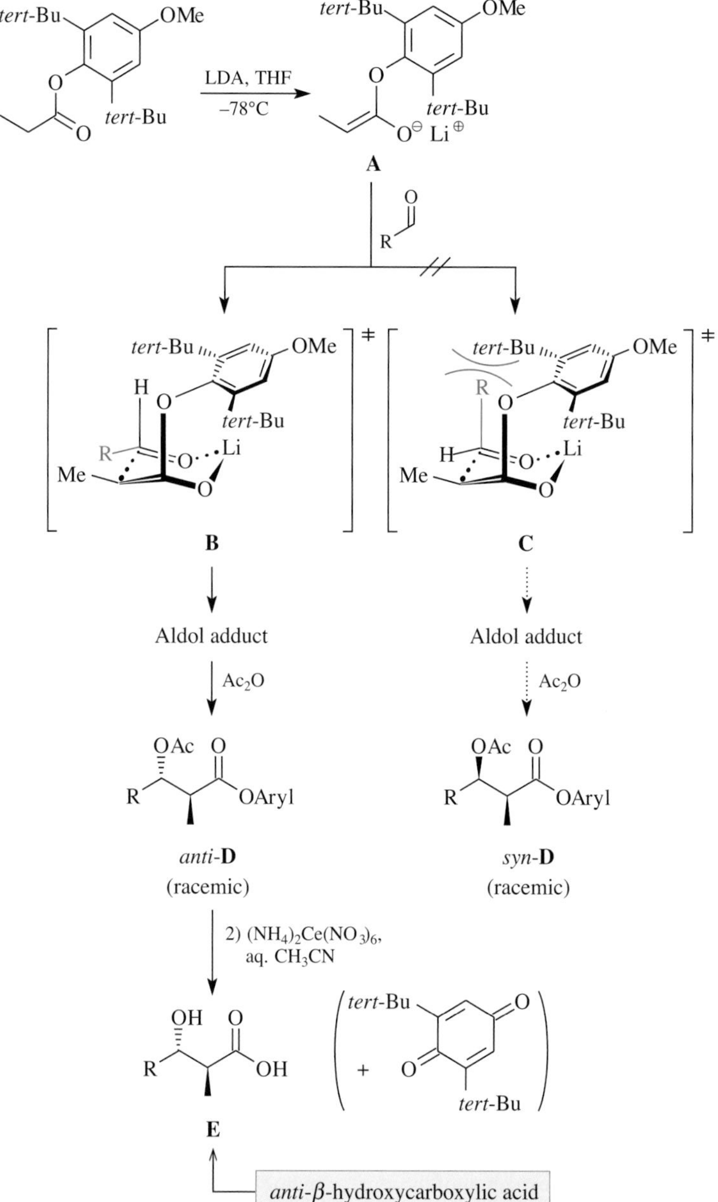

Fig. 13.48. *anti*-Selectivity of the aldol addition with a Heathcock lithium enolate including a mechanistic explanation. The Zimmerman–Traxler transition state **C** is destabilized by a 1,3-diaxial interaction, while the Zimmerman-Traxler transition state **B** does not suffer from such a disadvantage. The reaction thus occurs exclusively via transition state **B.**

Anti-configured *β*-hydroxycarboxylic acids are accessible via the reaction sequence depicted in Figure 13.48. The ester enolate **A** is generated by using LDA in THF with the usual *"E"*-selectivity (see Figure 13.16). The enolate contains a phenyl group with two *ortho*-attached *tert*-butyl groups. A phenyl substituent with such a substitution pattern must be twisted out of the plane of the enolate. This is true also in the Zimmerman–Traxler transition states **B** and **C.** One of the *tert*-butyl groups ends up directly on top of the chair structure. Being forced into this position above the ring, the *tert*-butyl group necessarily repels the non-hydrogen substituent of the aldehyde in transition state structure **C.** The associated destabilization of **C** does not occur in the diastereomeric transition state **B.** The aldol addition of Figure 13.48 thus proceeds exclusively via **B**, and perfect simple diastereoselectivity results.

The reaction mixture of Figure 13.48 is worked up such that it yields the β-*acetoxy*carboxylic acid esters **D** instead of the β-*hydroxy*carboxylic acid esters. These products are obtained with diastereoselectivities $ds > 98:2$; and this is independent of the nature of the aldehyde employed. The acetoxyesters **D** are prepared instead of the hydroxyesters because the latter would not survive the ester cleavage still to come. The problem is that this ester group cannot be taken off by way of hydrolysis because it is so bulky. However, it can be removed via oxidation with ceric ammonium nitrate, a quinone being the leaving group. The resulting β-hydroxycarboxylic acids **E** retain the high *anti*-stereochemistry established in **C.**

13.4 Condensation of Enolates with Carbonyl Compounds: Synthesis of Michael Acceptors

13.4.1 Aldol Condensations

An **aldol reaction** is a reaction between two carbonyl compounds in which one carbonyl compound plays the role of a nucleophile while the other carbonyl compound acts as an electrophile. The term "aldol reaction" covers two types of reactions, aldol additions (see Section 13.3) and aldol condensations. The aldol reactions that lead to β-hydroxycarbonyl compounds belong to the class of aldol *additions*. Aldol *condensations* start from the same substrates but result in α,β-unsaturated carbonyl compounds (Figure 13.49).

Aldol reactions often proceed as aldol condensations if the participating aldehyde or ketone enolates **C** are formed only in equilibrium reactions, i.e., incompletely (Figure 13.49). Under these reaction conditions an aldol addition occurs first: it leads to the formation of **D** proceeding by way of the mechanism shown in Figure 13.44 (bottom). Then an $E1_{cb}$ elimination takes place: in an equilibrium reaction, aldol **D** forms a small amount of enolate **E**, which eliminates NaOH or KOH.

If a stereogenic double bond is established by this $E1_{cb}$ elimination, one usually observes a *trans*- or an *E*-selectivity. This experimental finding could have two origins: (1) product development control (Section 4.1.3), if the stereoselectivity occurs under kinetic control, or (2) thermodynamic control. Thermodynamic control comes into play if the *cis,trans*- or *E,Z*-isomeric condensation products can be interconverted via a reversible 1,4-addition of NaOH or KOH. In the *trans*- or *E*-isomer of an α,β-unsaturated carbonyl compound the formyl or acyl group may lie *unimpeded* in the plane of the C=C double bond. This geometry allows one to take *full advantage* of the resonance stabilization {C=C—C=O $\leftrightarrow$ $^{\oplus}$C—C=C—O$^{\ominus}${.

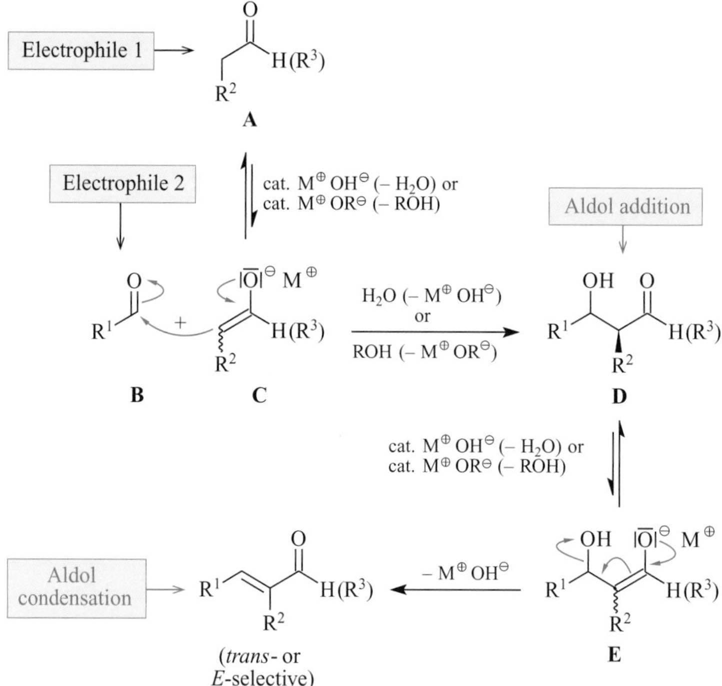

Fig. 13.49. Mechanisms of base-catalyzed aldol reactions: aldol addition (steps 1 and 2) and aldol condensation (up to and including step 4). $M^{\oplus}$ = $Na^{\oplus}$ or $K^{\oplus}$

On the other hand, in the *cis*- or *Z*-isomer of an α,β-unsaturated carbonyl compound the formyl or acyl group interferes to such an extent with the substituent at the other end of the C=C double bond that a planar geometry is no longer possible and the resonance stabilization consequently is reduced.

As with MOH- or MOH-*catalyzed* aldol additions (M = Na or K; Figure 13.44), MOH- or MOH-*catalyzed* aldol condensations (M = Na or K) can be carried out only with aldehyde or ketone enolates, not with ester or amide enolates. The reason for this is the same as discussed before, namely, that ester and amides are less acidic than carbonyl compounds and the amounts of enolate they form with the bases mentioned are much too small.

According to Figure 13.44, ketones often do not engage in base-catalyzed aldol *additions* because of a lack of driving force. Hence, ketones also are less suitable electrophiles than aldehydes in aldol condensations. However, for ketones, too, the elimination step is irreversible and they can therefore form α,β-unsaturated carbonyl compounds. It is not always possible to isolate the α,β-unsaturated carbonyl compounds thus formed. If the C-β atom is not sterically hindered, these products can act as electrophiles and add any residual ketone enolate; the α,β-unsaturated carbonyl compound acts as a Michael acceptor in this case (Section 13.6.1).

A broad spectrum of α,β-unsaturated carbonyl compounds becomes accessible in this way because a great variety of aldehydes is suited for aldol condensation. Table 13.7 exemplifies the broad scope by way of the reactions of an aldehyde enolate (center) and of the enolate of an unsymmetrical ketone (right). The right column of Table 13.7 also shows that the regioselectivity of the aldol condensation of the ketone is not easy to predict. Subtle substituent

Tab. 13.7 Representative α,β-Unsaturated Carbonyl Compounds Generated by Aldol Condensations of Carbonyl Compounds with Selected Aldehydes ($M^{\oplus}$ = $Na^{\oplus}$ or $K^{\oplus}$)

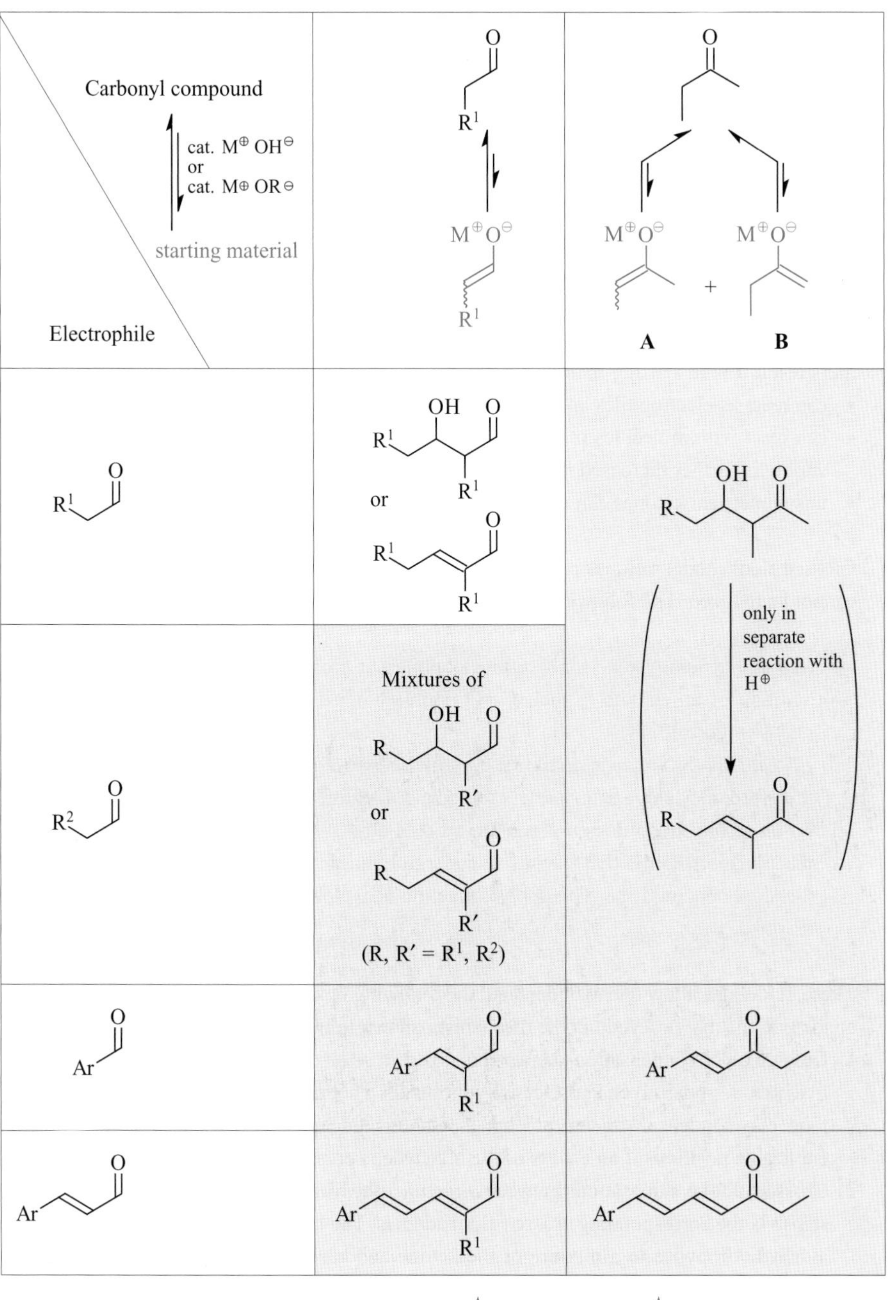

crossed aldol reactions

effects decide whether in a given case the thermodynamic enolate **A** or the kinetic enolate **B** (present in much smaller amounts) is responsible for the reaction. The thermodynamic enolate leads to the aldol addition, while the kinetic enolate leads to the aldol condensation.

Only one of the aldol condensations of Table 13.7 (top, center) concerns the reaction of a carbonyl compound *with itself.* In all other reactions of Table 13.7, the α,β-unsaturated carbonyl compounds are formed by two different carbonyl compounds. Such aldol condensations are referred to as **crossed aldol condensations** (cf. the discussion of crossed aldol additions in Section 13.3.1).

Of course, it is the goal of a crossed aldol condensation to produce a single α,β-unsaturated carbonyl compound. One has to keep in mind that crossed aldol condensations may result in up to four constitutionally isomeric condensation products (starting from two aldehydes or an aldehyde and a symmetric ketone) or even in eight constitutional isomers (starting with an aldehyde and an unsymetrical ketone). These maximum numbers of structural isomers result if both starting materials

- can react as electrophiles and as nucleophiles
- can react with molecules of their own kind as well as with other molecules. The product variety is further increased if
- unsymmetrical ketones can react via two regioisomeric enolates.

Crossed aldol condensations occur *with chemoselectivity* only if some of the foregoing options cannot be realized. The following possibilities exist.

Chemoselectivity of Crossed Aldol Condensations

1) *Ketones* generally react only as nucleophiles in crossed aldol additions because the addition of an enolate to their C=O double bond is thermodynamically disadvantageous (Figure 13.44).
2) *Benzaldehyde, cinnamic aldehyde, and their derivatives* do not contain any α-H atoms; therefore, they can participate in crossed aldol additions only as electrophiles.
3) *Formaldehyde* also does not contain an α-H atom. However, formaldehyde is such a reactive electrophile that it tends to undergo multiple aldol additions instead of a simple aldol condensation. This type of reaction is exploited in the pentaerythritol synthesis.

These guidelines allow one to understand the following observations concerning crossed aldol condensations that proceed via the mechanism shown in Figure 13.49.

1) Crossed aldol condensations between *benzaldehyde or cinnamic aldehyde or their derivatives ketones* pose no chemoselectivity problems. The least sterically hindered ketone, acetone, may condense with benzaldehyde, cinnamic aldehyde, and their derivatives with both enolizable positions if an excess of the aldehyde is employed.
2) Crossed aldol condensations between *aliphatic aldehydes and ketones* succeed only in two steps via the corresponding crossed aldol adducts. The latter can be obtained by adding the aldehyde dropwise to a mixture of the ketone and base. The aldol adducts subsequently must be dehydrated with acid catalysis.
3) Crossed aldol condensations between *aliphatic aldehydes* on the one hand and *benzaldehyde* or *cinnamic aldehyde* or their derivatives on the other also are possible. The reaction components can even be mixed together. The aldol adducts are formed without chemo-

selectivity, as a mixture of isomers, but their formation is reversible. The $E1_{cb}$ elimination to an α,β-unsaturated carbonyl compound is fast only if the newly created C=C double bond is conjugated to an aromatic system or to another C=C double bond already present in the substrate. This effect is due to product development control. All the starting materials thus react in this way via the most reactive aldol adduct.

4) Chemoselective crossed aldol condensations between *two different C,H-acidic aldehydes* are impossible. There is only a single exception, and that is the intramolecular aldol condensation of an unsymmetrical dialdehyde.

Side Note 13.5. Aldol Condensations of Ester Enolates

Ester enolates and aldehydes normally undergo aldol *additions*, as illustrated in the example of Figure 13.48, but not aldol *condensations*. Of course, these aldol *adducts* can subsequently be dehydratized to furnish aldol *condensates*. This Side Note, however, is intended to show that certain ester enolates may also undergo aldol *condensations* in a single step upon reaction with certain aldehydes.

Figure 13.50 outlines how esters in general (not shown) and especially lactones (shown) can be prepared for a one-step aldol condensation with an aldehyde: they are exposed to a mixed ("crossed") Claisen condensation with formic acid methyl ester (cf. Figure 13.59, first line). Like all Claisen condensations (Section 13.5.1), this also first leads to the formation of the enolate of the acylated ester. Unlike other Claisen condensations, this enolate is *isolated*. Resonance form **B** identifies it as a formylated ester- or lactone enolate and resonance form

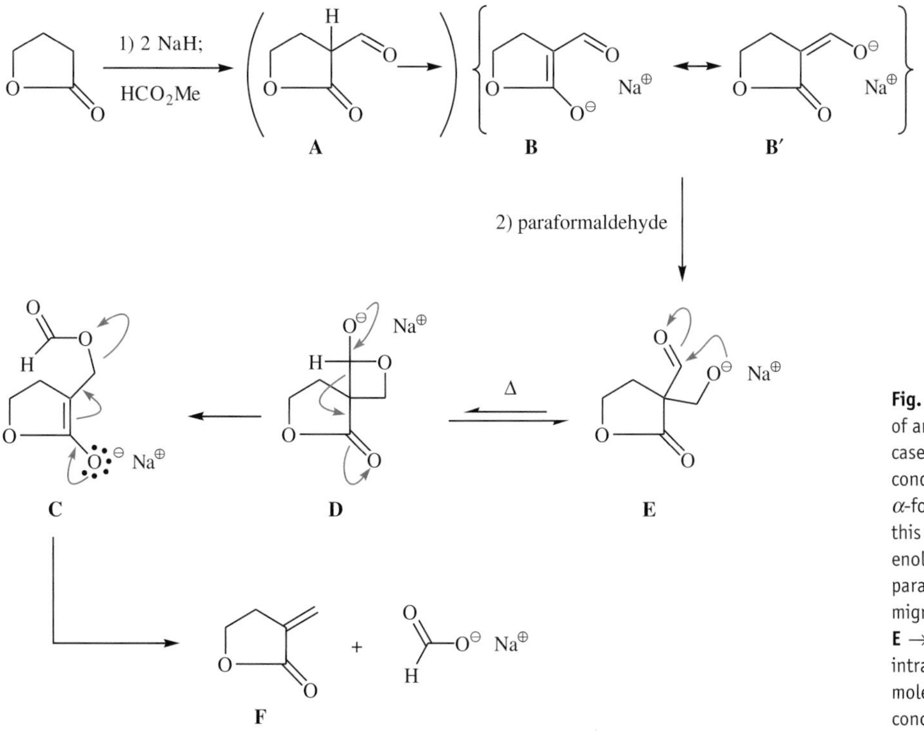

1) 2 NaH; HCO_2Me

A B B′

2) paraformaldehyde

Δ

C D E F

Fig. 13.50. α-Methylenation of an ester (in this special case: of a lactone) via aldol condensation of the resulting α-formylated ester enolate (in this special case: a lactone enolate) **B** ↔ **B′** with paraformaldehyde. Here, the migration of the formyl groups **E** → **C** is considered to proceed intramolecularly, but an intermolecular process is equally conceivable.

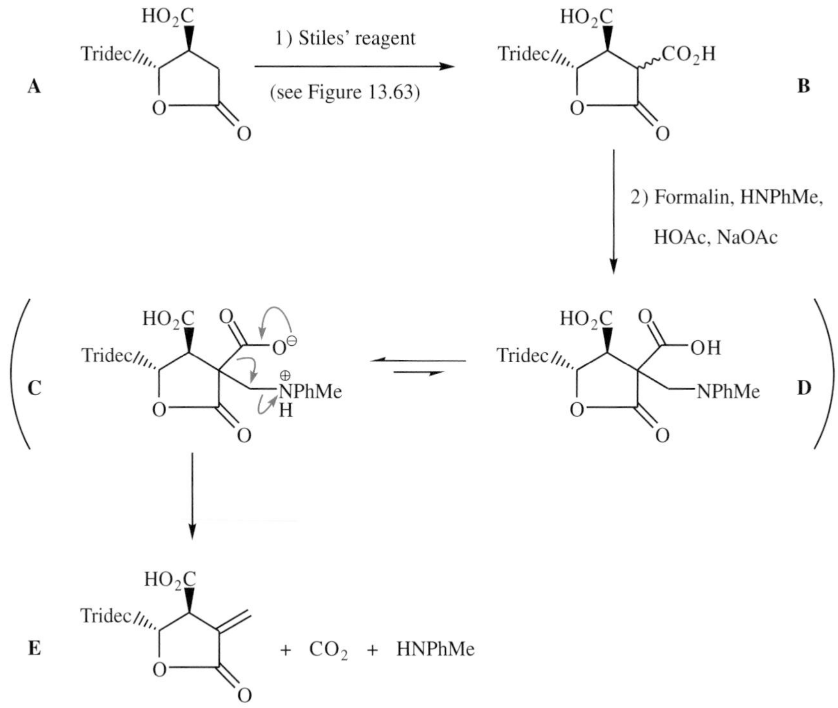

Fig. 13.51. α-Methylenation of a lactone via Mannich reaction (cf. Figure 12.14 and 12.15) of the α-carboxylated lactone enolate or lactone enol derived thereof. Another approach to α-methylenation of not only lactones, but normal carboxylic acid esters, is presented in Figure 13.50.

B′ as an ester- or lactone-substituted aldehyde enolate. Such enolates undergo condensations with all kinds of aldehydes, including paraformaldehyde. An *adduct* **E** is formed initially, acylating itself as soon as it is heated. The reaction could proceed intramolecularly via the tetrahedral intermediate **D** or intermolecularly as a retro-Claisen condensation. In both cases, the result is an acyloxy-substituted ester enolate. In the example given in Figure 13.50, this is the formyloxy-substituted lactone enolate **C**. As in the second step of an $E1_{cb}$ elimination, **C** eliminates the sodium salt of a carboxylic acid. The α,β-unsaturated ester (in Figure 13.50: the α,β-unsaturated lactone) remains as the aldol *condensation* product derived from the initial ester (here, a lactone) and the added aldehyde (here, paraformaldehyde).

Figure 13.51 shows a trick that allows the reaction of lactones with formaldehyde to yield the corresponding aldol condensation product in a single step. These lactones are first carboxylated with (methoxymagnesium)monomethyl carbonate. Figure 13.63 shows how this leads to the carboxylated lactone **B** shown here. The fact that only lactones can be activated in this manner is due to their elevated C,H acidity compared to normal esters (cf. Figure 13.7). In the second reaction of Figure 13.51, a change probably occurs from enolate to enol chemistry, since next a Mannich reaction (see Figures 12.14, 12.15) takes place to furnish the lactone **D**. This lactone is an amino acid and, like all amino acids, occurs as a zwitterion (Formula **C**). Upon heating, this zwitterion undergoes fragmentation—just like the pyridinium carboxylate in the last step of a decarboxylating Knoevenagel condensation (**F** → **G**, Figure 13.56) or like the sulfinate ion in the last step of the Julia–Kocienski olefination (**E** → **F**, Figure 11.23). Thus, the α-methylene lactone **E** is formed.

13.4.2 Knoevenagel Reaction

A Knoevenagel reaction is a condensation reaction between an active-methylene compound (or the comparably C,H-acidic nitromethane) and a carbonyl compound. The product of a Knoevenagel reaction is an alkene that contains two geminal acceptor groups (**B** in Figure 13.52) or one nitro group (**B** in Figure 13.53).

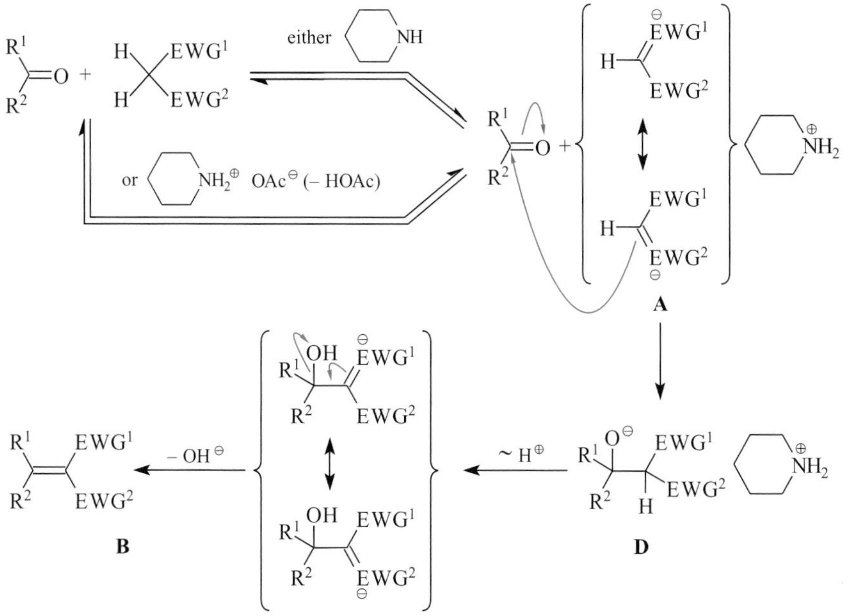

Fig. 13.52. Mechanism of the Knoevenagel reaction of active-methylene compounds; ~$H^{\oplus}$ indicates the migration of a proton.

Knoevenagel reactions are carried out in mildly basic media—in the presence of piperidine, for example—or in neutral solution—catalyzed by piperidinium acetate, for example. The basicity of piperidine or of acetate ions, respectively, suffices to generate a sufficiently high equilibrium concentration of the ammonium enolate of the active-methylene compound (**A** in Figure 13.52) or to generate a sufficiently high equilibrium concentration of the ammonium nitronate of the nitroalkane (**A** in Figure 13.53). The rather high acidity of the nitroalkanes (Table 13.1) alternatively allows the formation of nitroalkenes by way of reaction of nitroalkanes with aldehydes in the presence of basic aluminum oxide powder as base.

The enolate **A** or the nitronate **A**, respectively, initially adds to the C=O double bond of the aldehyde or the ketone. The primary product in both cases is an alkoxide, **D**, which contains a fairly strong C,H acid, namely, of an active-methylene compound or of a nitroalkane, respectively. Hence, intermediate **D** is protonated at the alkoxide oxygen and the C-β atom is deprotonated to about the same extent as in the case of the respective starting materials. An OH-substituted enolate **C** is formed (Figures 13.52 and 13.53), which then undergoes an $E1_{cb}$ elimination, leading to the condensation product **B.** The Knoevenagel condensation and the aldol condensation have in common that both reactions consist of a sequence of an enolate hydroxyalkylation and an $E1_{cb}$ elimination.

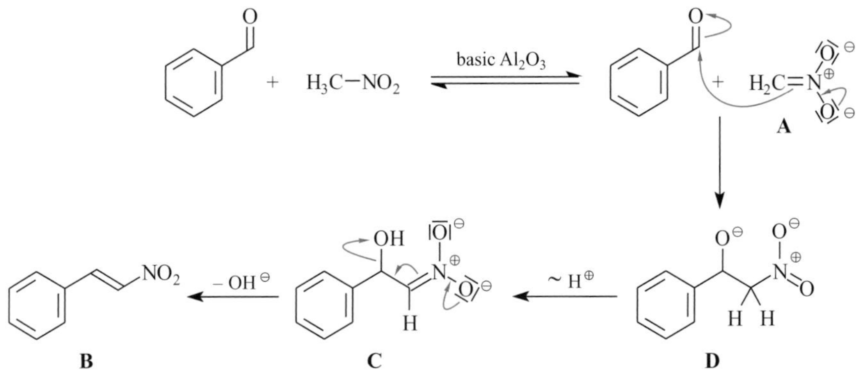

Fig. 13.53. Mechanism of a Knoevenagel reaction with nitromethane. Alkaline aluminum oxide powder is sufficiently basic to deprotonate nitromethane. The small amount of the anion generated from nitromethane suffices for the addition to aldehydes to proceed. The elimination of water via an $E1_{cb}$ mechanism follows quickly if a conjugated C=C double bond is formed, as in the present case.

The Knoevenagel reaction can be employed for the synthesis of a wide variety of condensation products—as shown in Figure 13.54—because the carbonyl component as well as the active-methylene component can be varied.

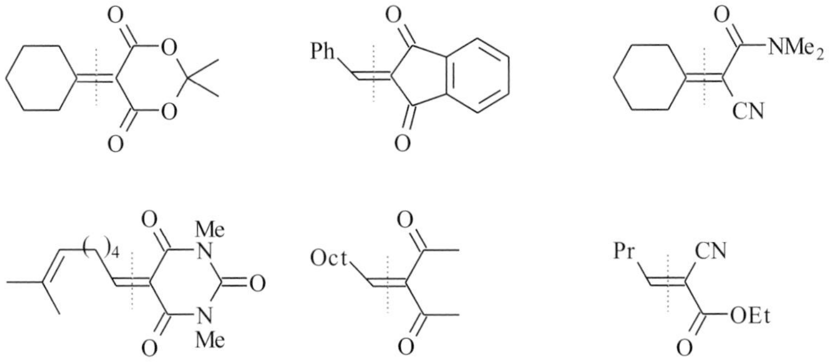

Fig. 13.54. Products of Knoevenagel condensations and their indication of their potential synthetic origin. The left molecule halves stem from the carbonyl compounds and the right fragments come from the active-methylene compounds.

Side Note 13.6. Knoevenagel Synthesis of Conjugated (Mono-) Carboxylic Acids

Malonic acid itself can react with aldehydes in the presence of piperidine by way of a Knoevenagel condensation. A decarboxylation occurs after the condensation, and this decarboxylation cannot be avoided. Figure 13.55 shows how the overall reaction can be employed for the synthesis of cinnamic or sorbic acid. This reaction sequence occurs under much milder conditions than the **Perkin synthesis of cinnamic acids**. (The Perkin synthesis consists of the condensation of aromatic aldehydes with acetic acid anhydride in the presence of sodium acetate.)

Mechanistic details about the decarboxylating Knoevenagel condensations between malonic acid and the unsaturated aldehydes of Figure 13.55 are given in Figure 13.56. There is no problem in assuming that the respective aldehyde itself is the electrophile in this Knoevenagel reaction. But it is also possible that the piperidinium salt derived thereof acts as the electrophile. This is why a question mark hovers above the first step of the mechanism in Figure 13.56. There is also a question as to the exact nature of the nucleophile. Certainly the

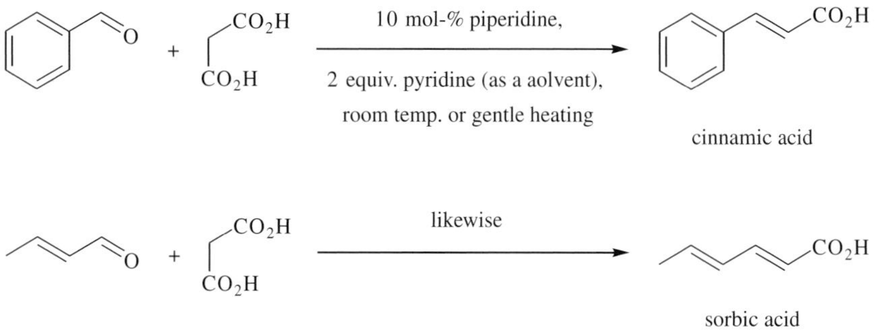

Fig. 13.55. A Knoevenagel reaction "with a twist": preparation of cinnamic acid and sorbic acid.

nucleophile has to be a species with an enolate carbon. The conceivable candidates include the malonic acid enolate **D** (malonic acid "monoanion") or the malonic acid "dianion". The malonic acid "trianion" cannot be generated by bases that are as weak as piperidine or pyridine. The nucleophilicity of the resulting enolates increases greatly in this order, while their concentrations in the corresponding deprotonation equilibria decrease drastically. The *combined* effect of nucleophilicity and abundance determines which nucleophile initiates the Knoevenagel condensation. It is therefore important to know the concentrations of the various species.

It can be assumed that the small amount of piperidine in the reaction mixture is completely protonated by malonic acid because piperidine is more basic than pyridine. Hence, only the less basic pyridine is available for the formation of the malonic acid enolate **D** from free malonic acid and for the formation of the malonic acid "dianion" from the malonic acid monocarboxylate **C.** The pK_a value of malonic acid *with regard to its C,H acidity* should be close to the pK_a value of malonic acid diethyl ester (pK_a = 13.3). The pK_a value of malonic acid monocarboxylate **C** *with regard to its C,H acidity* should be larger by at least a factor 10. Hence, the concentration of the malonic acid enolate **D** in the reaction mixture must be by many orders of magnitude higher than that of any malonic acid "dianion." Due to the advantages associated with this enormous concentration **D** could be the actual nucleophile in Knoevenagel condensations.

On this basis, Figure 13.56 presents a plausible mechanism for the transformations depicted in Figure 13.55. This mechanism is based on the assumption that the malonic acid enolate **D** initially reacts with benzaldehyde or crotonaldehyde just as the enolates of the active-methylene compounds. Thus, the malonic acid enolate and its unsaturated aldehydic reaction partners should first react to furnish the alkylidene malonic acids **B**. If pyridine were added to **B** in an equilibrium reaction (in terms of a Michael addition), the pyridinium-substituted malonic acid enolate **E** would be formed. This should undergo exergonic proton transfer to give the pyridinium carboxylate **F**. This zwitterion **F** could undergo fragmentation in the same way as described for the zwitterion **C** in Figure 13.51 to furnish the α,β-unsaturated carboxylic acids **G** as well as pyridine and carbon dioxide.

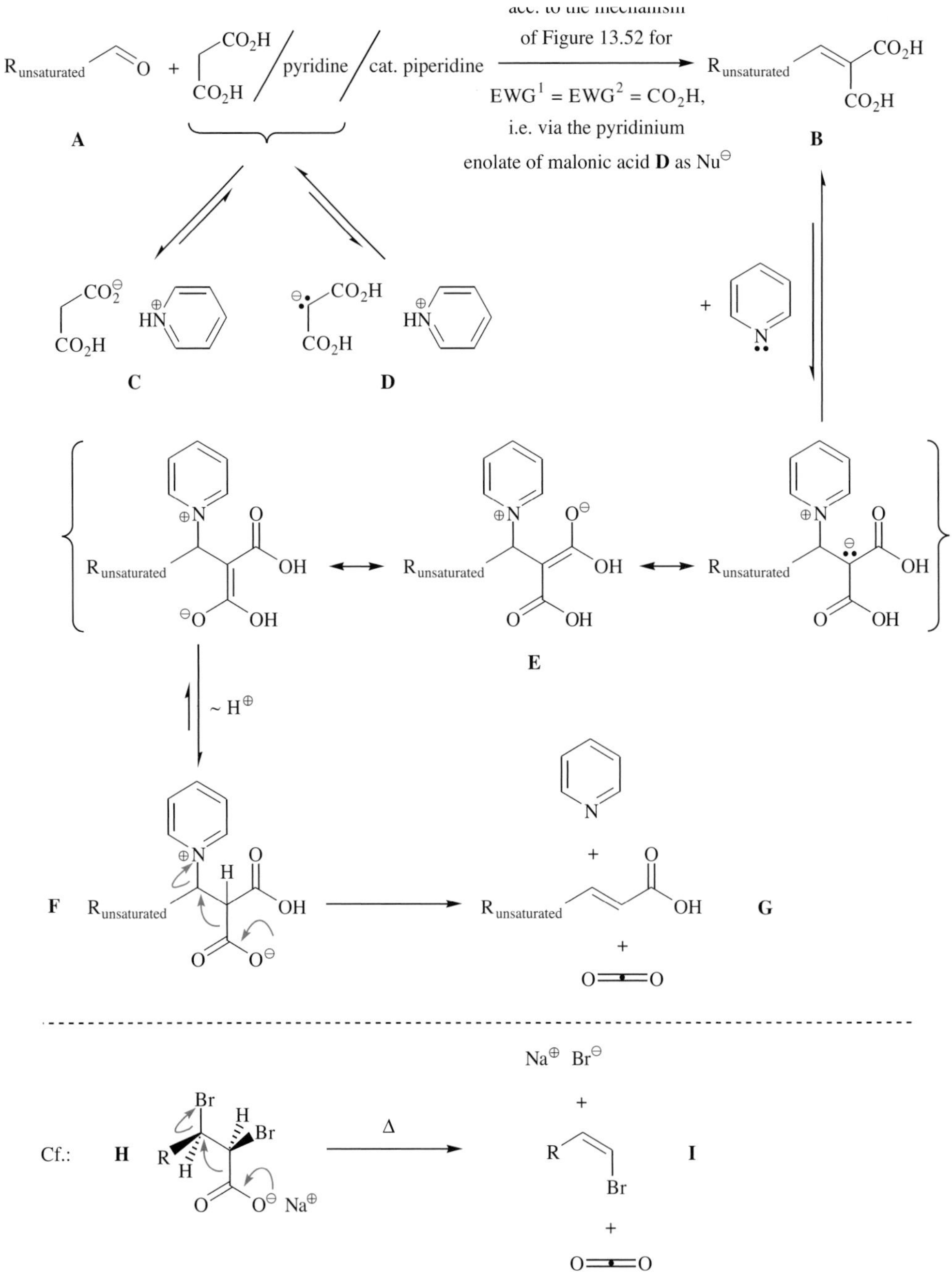

acc. to the mechanism
of Figure 13.52 for
$EWG^1 = EWG^2 = CO_2H$,
i.e. via the pyridinium
enolate of malonic acid **D** as $Nu^{\ominus}$

pyridine / cat. piperidine

Cf.:

Fig. 13.56. Mechanism of the Knoevenagel condensations in Figure 13.55. The C,H(!)-acidic reaction partner is malonic acid in the form of the malonic acid enolate **D** (malonic acid "monoanion"). The decarboxylation proceeds as a fragmentation of the pyridinium-substituted malonic acid carboxylate **F** to furnish the α,β-unsaturated ester (**G**) and pyridine. This fragmentation resembles the decomposition of the sodium salts **H** of α,β-dibrominated carboxylic acids to yield the α,β-unsaturated bromides **I** and sodium bromide.

13.5 Acylation of Enolates

13.5.1 Acylation of Ester Enolates

A **Claisen condensation** is the acylation of an ester enolate by the corresponding ester. By deprotonating an ester with MOR, only a small concentration of the ester enolate is generated and this enolate is in equilibrium with the ester (cf. Table 13.1). The mechanism of the Claisen condensation is illustrated in detail in Figure 13.57 for the example of the condensation of ethyl butyrate. Both the deprotonation of the ester to give enolate **A** and the subsequent acylation of the latter are reversible. This acylation occurs via a tetrahedral intermediate (**B** in Figure 13.57) just like the acylations of other nucleophiles (Chapter 6). The equilibrium between two molecules of ethyl butyrate and one molecule each of the condensation product **C** and ethanol does not lie completely on the side of the products. In fact, Claisen condensations go to completion only

- if a stoichiometric amount of alkoxide is present, or
- if a stoichiometric amount of alkoxide can be generated from the one equivalent of alcohol liberated in the course of the Claisen condensation and a stoichiometric amount of Na or NaH.

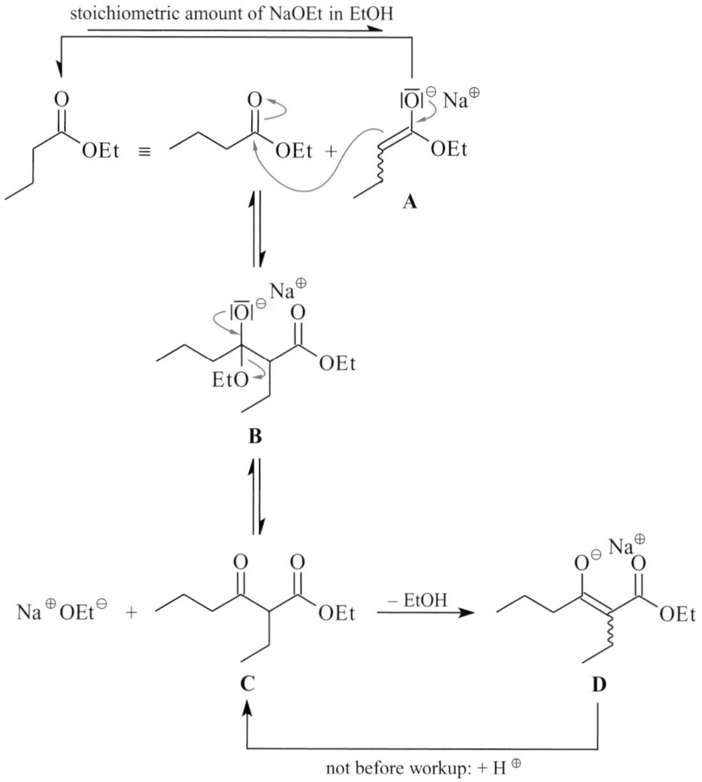

stoichiometric amount of NaOEt in EtOH

OEt ≡ OEt + OEt

A

B

$Na^{\oplus}OEt^{\ominus}$ + OEt — – EtOH → OEt

C D

not before workup: + $H^{\oplus}$

Fig. 13.57. Mechanism of a Claisen condensation. The deprotonation step $Na^{\oplus}OEt^{\ominus}$ + **C** → **D** + EtOH is irreversible, and it is for this reason that eventually all the starting material will be converted into the enolate **D**.

What is the effect of the stoichiometric amount of strong base that allows the Claisen condensation to proceed to completion? The β-ketoester **C**, which occurs in the equilibrium, is an active-methylene compound and rather C,H-acidic. Therefore, its reaction with the alkoxide to form the ester-substituted enolate **D** occurs *with considerable driving force.* This driving force is strong enough to render the deprotonation step **C** $\rightarrow$ **D** essentially irreversible. Consequently, the overall condensation also becomes irreversible. In this way, all the substrate is eventually converted into enolate **D.** The neutral β-ketoester can be isolated after addition of one equivalent of aqueous acid during workup.

Intramolecular Claisen condensations, called **Dieckmann condensations**, are ring-closing reactions that yield 2-cyclopentanone carboxylic esters (Figure 13.58) or 2-cyclohexanone carboxylic esters. The mechanism of the Dieckmann condensation is, of course, identical to the mechanism of the Claisen condensation (Figure 13.57). To ensure that the Dieckmann condensation goes to completion, the presence of a stoichiometric amount of base is required. As before, the neutral β-ketoester (**B** in Figure 13.58) is formed in a reversible reaction under basic conditions. However, the back-reaction of the β-ketoester **B** to the diester is avoided by deprotonation to the substituted enolate **A.** This enolate is the thermodynamic sink to which all the substrate eventually is converted. The β-ketoester **B** is regenerated in neutral form again during workup with aqueous acid.

Acylations of ester enolates with different esters are called **crossed Claisen condensations** and are carried out—just like normal Claisen condensations-in the presence of a stoichiometric amount of alkoxide, Na, or NaH. Crossed Claisen condensations can in principle lead to four products. In order that only a single product is formed in a crossed Claisen condensation, the esters employed need to be suitably differentiated: one of the esters must be prone to enolate formation, while the other must possess a high propensity to form a tetrahedral intermediate (see example in Figure 13.59).

The use of an ester without acidic α-H atoms ensures that this ester can act only as the electrophile in a crossed Claisen condensation. Moreover, this nonenolizable ester must be at least

Fig. 13.58. Mechanism of a Dieckmann condensation. The Dieckmann condensation is an intramolecular Claisen condensation.

Trace of MeOH and stoichiometric amount of Na or NaH or KH in THF

MeO, OMe, Na⊕, O⊖

– MeOH

$Na^{\oplus} OMe^{\ominus}$ +

A B C

not until workup: + $H^{\oplus}$

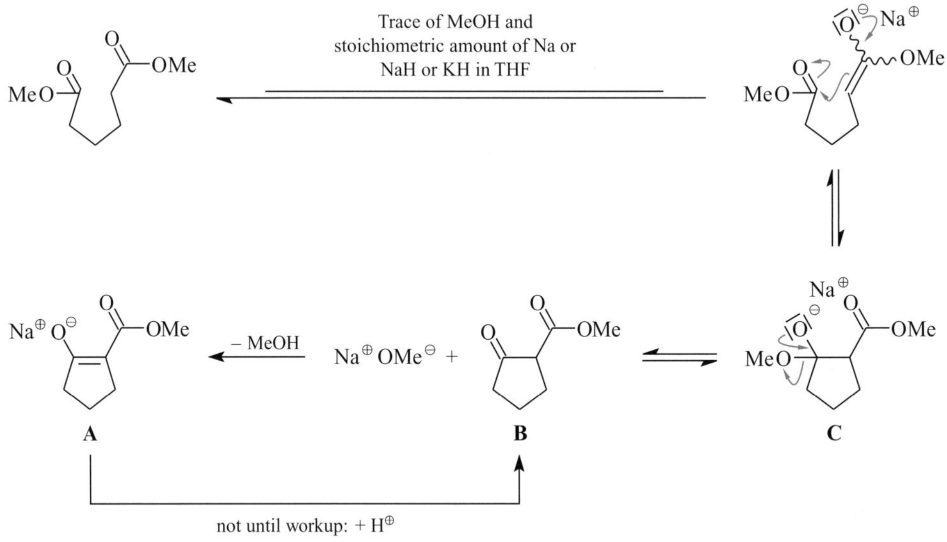

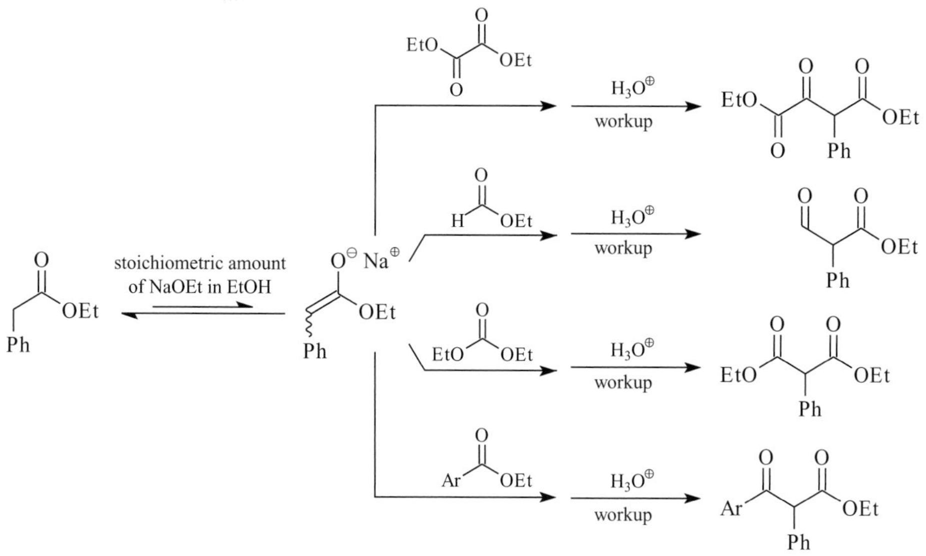

Fig. 13.59. Crossed Claisen condensation. Although the tautomers of the acylation products shown are not the major tautomer except for the third case from the top, they are presented because they show best the molecules from which these products were derived.

as electrophilic as the other ester. This is because the larger fraction of the latter is present in its nondeprotonated form; that is, it represents a possible electrophile, too, capable of forming a tetrahedral intermediate upon reacton with an enolate.

Accordingly, crossed Claisen condensations occur without any problems if the acylating agent is a better electrophile than the other, nondeprotonated ester. This is the case, for example, if the acylating agent is an oxalic ester (with an electronically activated carboxyl carbon) or a formic ester (the least sterically hindered carboxyl carbon).

Crossed Claisen condensations can be chemoselective even when the nonenolizable ester is *not* a better electrophile than the enolizable ester. This can be accomplished by a suitable choice of reaction conditions. The nonenolizable ester is mixed with the base and the enolizable ester is added slowly to that mixture. The enolate of the enolizable ester then reacts mostly with the nonenolizable ester for statistical reasons; it reacts much less with the nonenolized form of the enolizable ester, which is present only in rather small concentration. Carbonic acid esters and benzoic acid esters are nonenolizable esters of the kind just described.

Under different reaction conditions, esters other than the ones shown in Figure 13.59 can be employed for the acylation of ester enolates. In such a case, one *quantitatively* deprotonates *two* equivalents of an ester with LDA or a similar amide base. This is exemplified by the upper part of Figure 13.60, starting from two equivalents of acetic acid *tert*-butylester. Then 1.0 equivalent of the ester serving as the acylating agent is added. In the example given in Figure 13.60 (top). this may be any carboxylic acid methyl ester. The acylation product is a β-ketoester **B**, and thus a stronger C,H acid than the conjugate acid of the ester enolate employed. Therefore, the initially formed β-ketoester **B** reacts immediately in an acid/base reaction with the second equivalent of the ester enolate. The β-ketoester **B** protonates this ester enolate, consumes it completely and thereby is itself converted into the conjugate base, i.e., the enolate **C**. The β-ketoester **B** is reconstituted upon acidic workup.

In some acylations it may even be necessary to employ three equivalents of the ester enolate. The example at the bottom of Figure 13.60 is such a case: with its free OH group the acy-

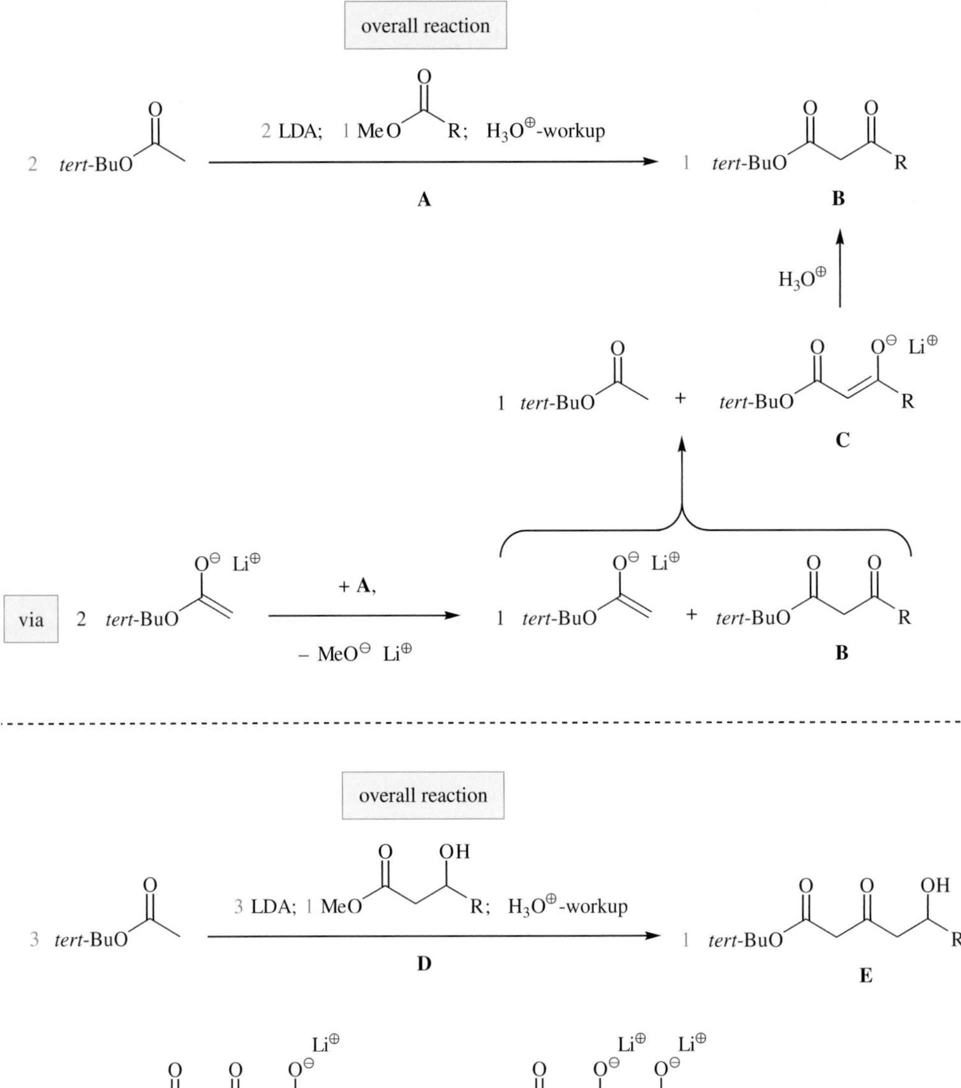

Fig. 13.60. Crossed ester condensation via acylation of a quantitatively prepared ester enolate. Three equivalents of ester enolate must be employed because the acylating ester contains a free OH group with an acidic H atom: one for the deprotonation of the OH group of the substrate, one for the substitution of the MeO group, and one for the transformation of the C,H-acidic substitution product into an enolate.

lating β-hydroxyester **D** protonates the first equivalent of the enolate of the acetic acid *tert*-butyl ester to furnish *tert*-butyl acetate. The second equivalent of the ester enolate undergoes a crossed Claisen condensation with the anion of the β-hydroxyester **D** that is present at this point. The resulting β-ketoester **F** protonates the third equivalent of the *tert*-butyl acetate ester enolate thereby forming its conjugate base, the enolate **G**. Only upon acidic workup does the latter afford the neutral Claisen condensation product, the δ-hydroxy-β-ketoester **E**.

A procedure that may be used to carboxylate lactone enolates is described in Figure 13.63.

13.5.2 Acylation of Ketone Enolates

Remember what we discussed in the context of Figure 13.44: *ketones* usually do not undergo aldol additions if they are deprotonated to only a small extent by an alkaline earth metal alkoxide or hydroxide. The driving force behind that reaction simply is too weak. In fact, only a very few ketones can react with themselves in the presence of alkaline earth metal alkoxides or alkaline earth metal hydroxides. And if they do, they engage in an aldol condensation. Cyclopentanone and acetophenone, for example, show this reactivity.

The relative inertness of ketone enolates toward ketones makes it possible to react non-quantitatively obtained ketone enolates with esters instead of with ketones. These esters—and reactive esters in particular—then act as acylating reagents.

In contrast to ketones, aldehydes easily undergo a base-catalyzed aldol reaction (Figure 13.44), and this reaction may even progress to an aldol condensation (Section 13.4.1). It is therefore not possible to acylate aldehyde enolates that are present only in equilibrium concentrations. Any such enolates would be lost completely to an aldol reaction.

Oxalic esters (for electronic reasons) and formic esters (because of their low steric hindrance) are reactive esters that can acylate ketone enolates formed with NaOR in equilibrium reactions. Formic esters acylate ketones to provide formyl ketones (for example, see Figure 13.61). It should be noted that under the reaction conditions the conjugate base of the active-methylene formyl ketone is formed. The neutral formyl ketone is regenerated upon acidic workup.

Most other carboxylic acid derivatives can acylate only ketone enolates that are formed quantitatively. In these reactions, the acylation product is a β-diketone, i.e., an active-methylene compound. As a consequence it is so acidic that it will be deprotonated quantitatively. This deprotonation will be effected by the ketone enolate. Therefore, a complete acylation of this type can be achieved only if two equivalents of the ketone enolate are reacted with one equivalent of the acylating agent. Of course, proceeding in that manner would mean an unacceptable waste in the case of a valuable ketone.

Fig. 13.61. Acylation of a ketone enolate with a formic ester to generate a formyl ketone. The ketone enolate intermediate (not shown) is formed in an equilibrium reaction.

O + H OEt (O) — stochiometric amount of KO*tert*-Bu in THF; $H_3O^{\oplus}$ → O O ⇌ O H O

Against Wastefulness: A Practical Hint Regarding the Acylation of Ketone Enolates

The following protocol requires no more than the stoichiometric amount of a ketone enolate to achieve a complete acylation. An ester is added dropwise to a 1:1 mixture of one equivalent each of the ketone enolate and LDA. The acidic proton of the β-diketone, which is formed, then is abstracted by the excess equivalent of LDA rather than by the ketone enolate.

The protocol described also can be used for the acylation of ketone enolates with carbonic acid derivatives (Figure 13.62). Especially good acylating agents are cyanocarbonic acid methyl ester (Mander's reagent, Figure 13.62, top) and dialkyl pyrocarbonates (bottom). Usually it is not possible to use dimethyl carbonate for the acylation of ketone enolates because dimethyl carbonate is a weaker electrophile than cyanocarbonic acid methyl ester or diethyl pyrocarbonates.

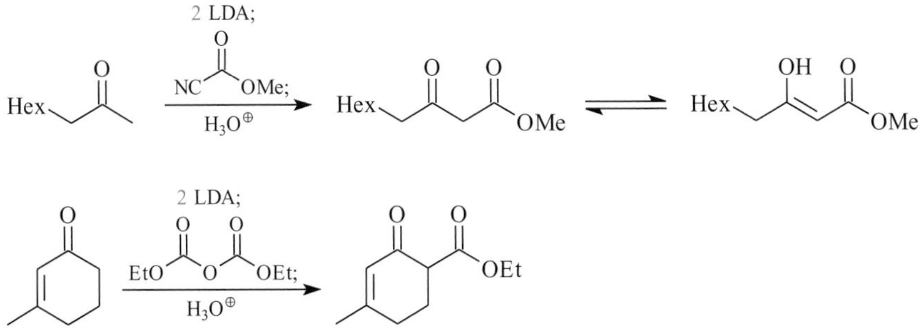

Fig. 13.62. Acylation of ketone enolates with carbonic acid derivatives. Especially good acylation reagents are cyanocarbonic acid methyl ester (top) and dialkyl pyrocarbonates (bottom).

A carbonic acid derivative which, surprisingly, also proves to be suitable for the acylation of ketone enolates, is Stiles' reagent, i.e., (methoxymagnesium) monomethyl carbonate. In Section 8.2, you saw how this reagent can be obtained. Ketone enolates are carboxylated by Stiles' reagent to furnish a β-keto carboxylic acid, as shown by the reaction equation below. As this keto acid is initially obtained as the (methoxymagnesium) carboxylate, such an acylation can easily proceed without the extra equivalents of enolate or base mentioned above.

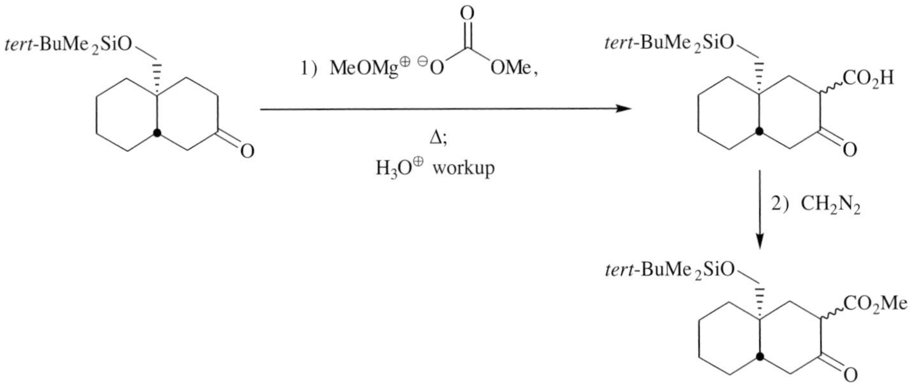

The reaction of the β-keto carboxylic acid resulting from the acidic workup of this carboxylation with diazomethane affords the corresponding β-ketoester (method: Figure 2.33).

Incidentally, lactones are also sufficiently C,H-acidic to be carboxylated by (methoxymagnesium) monomethyl carbonate. Figure 13.63 illustrates how this approach can be used to form an α-carboxylated lactone **B** and what the reaction mechanism might look like. An α-carboxylated lactone can react either with diazomethane to give an α-(methoxycarbonylated) lactone (not shown), or undergo a tandem reaction (consisting of Mannich reaction and fragmentation) to yield an α-methylene lactone (see Figure 13.51).

Weinreb amides are acylating agents that react according to the general mechanism outlined in Figure 6.4. Thus, the acylation product is not released from the tetrahedral intermediate as long as nucleophile is still present. Accordingly, the acylation of a ketone enolate by a Weinreb amide does not immediately result in the formation of the β-ketocarbonyl compound. Instead, the reaction proceeds just as an addition reaction and a tetrahedral intermediate is formed stoichiometrically (e. g., **C** in Figure 13.64). This tetrahedral intermediate is not an active-methylene compound but a donor-substituted ($O^{\ominus}$ substituent!) ketone. This intermediate therefore cannot act as a C,H acid with the ketone enolate or, in the present case, not even

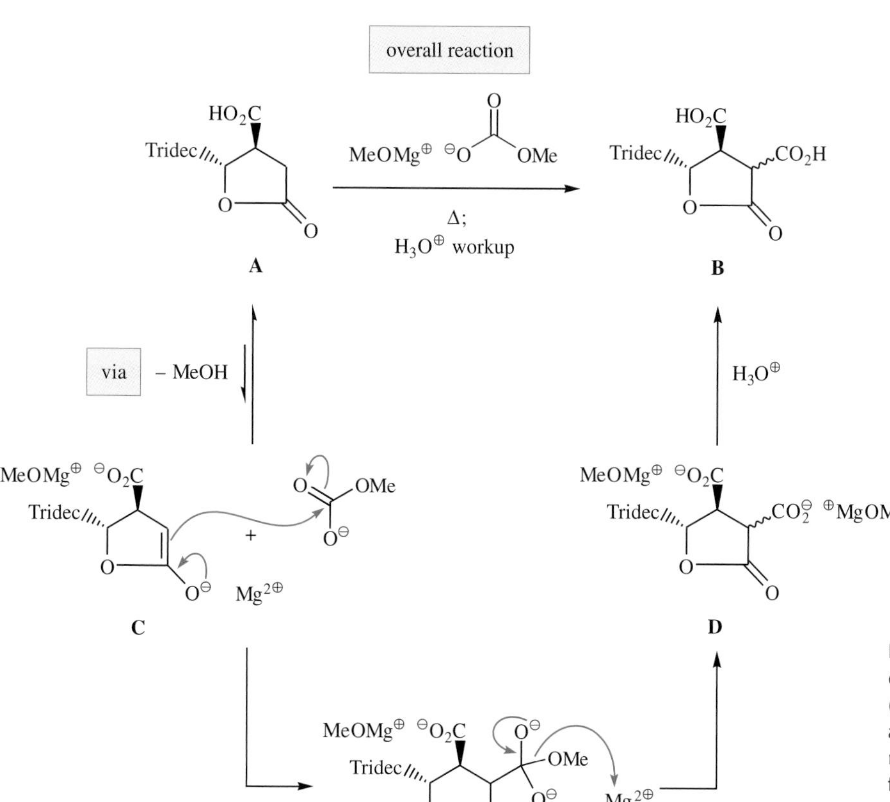

Fig. 13.63. α-Carboxylation of a lactone enolate with (methoxymagnesium) carbonic acid monomethyl ester ("Stiles' reagent"). In the C,C bond-forming step (**C** → **E**) this anion acts as an electrophile, since it is steered into close vicinity to the enolate carbon by a $Mg^{2\oplus}$ cation.

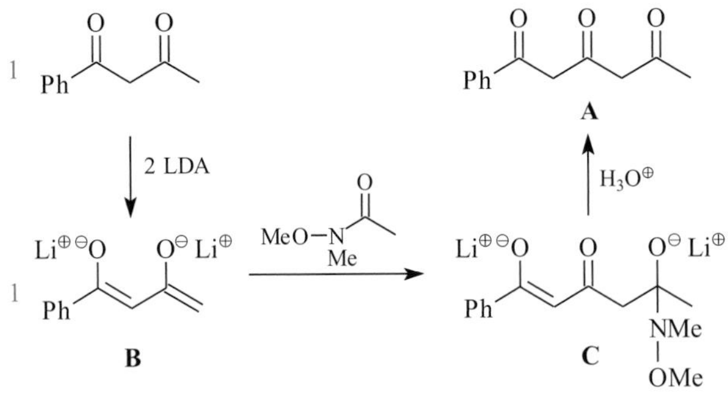

Fig. 13.64. Acylation of a bis(ketone enolate) with one equivalent of a Weinreb amide.

with the bis(ketone enolate) **B.** For reasons discussed in the context of Figure 6.42, the tetrahedral intermediate **C** is stable until it is protonated upon aqueous workup. Only then is the acylation product formed.

13.5.3 Acylation of the Enolates of Active-Methylene Compounds

When the enolate of an active-methylene compound undergoes acylation with a carboxylic acid chloride, an "active-methyne" compound is formed initially (Figure 13.65, 13.66). If the electron acceptors therein are solely acyl- or (alkoxycarbonyl) groups, the substructure mentioned suffers from steric hindrance and substantial electrostatic repulsion forces. "Active-methyne" compounds with such a substitution pattern will react to alleviate this destabilization.

This can occur through loss of an acyl- or (alkoxycarbonyl) group and formation of an active-methylene compound. Figure 13.65 shows three variants of how this principle can be exploited to synthesize β-ketoesters from acid chlorides. Naturally, the destabilization of "active-methyne" compounds is also removed if *two* of the three acyl- or (alkoxycarbonyl) groups are eliminated, leaving behind a compound with just *one* C=O double bond. Figure 13.66 presents two methods for the chain-elongating synthesis of ketones from carboxylic acid chlorides that are based on this principle.

The "active-methyne" compounds, which derive from the acylation of the enolates of active-methylene compounds with carboxylic acid chlorides, eliminate the extra acceptor(s) in an additional step or immediately *in situ*. The defunctionalizations involved include one or two decaboxylations depending on the nature of the reactants and subsequent processing steps (Figures 13.66 and 13.67)

Variant 1:

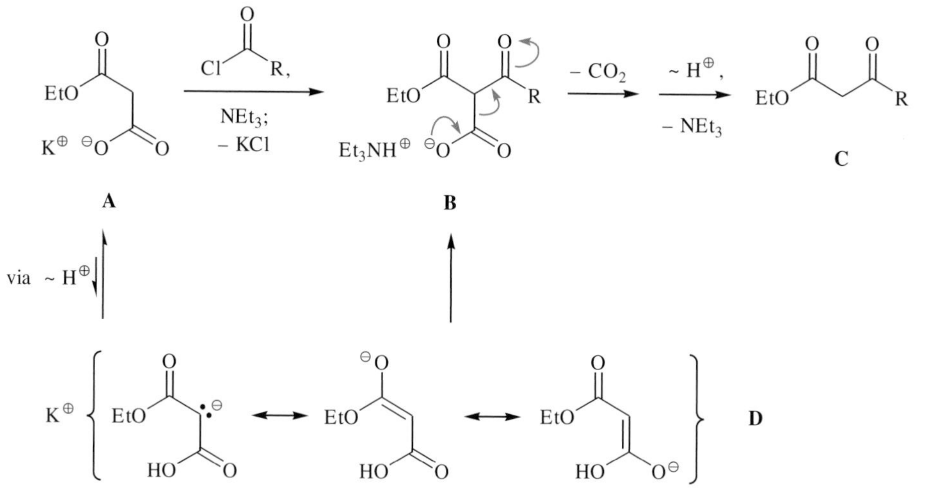

Variant 2:

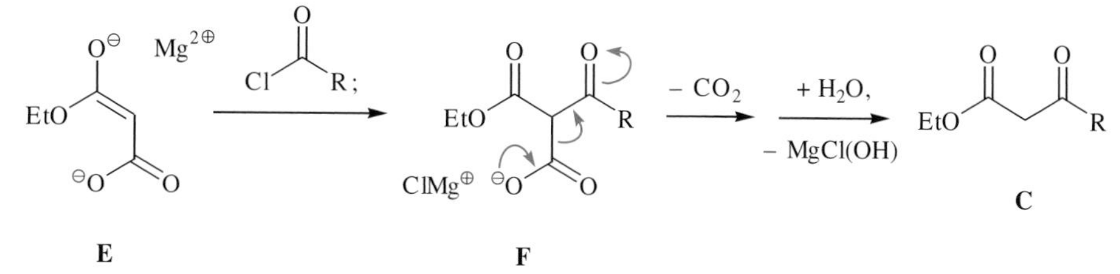

(preparation: Figure 13.10)

Variant 3:

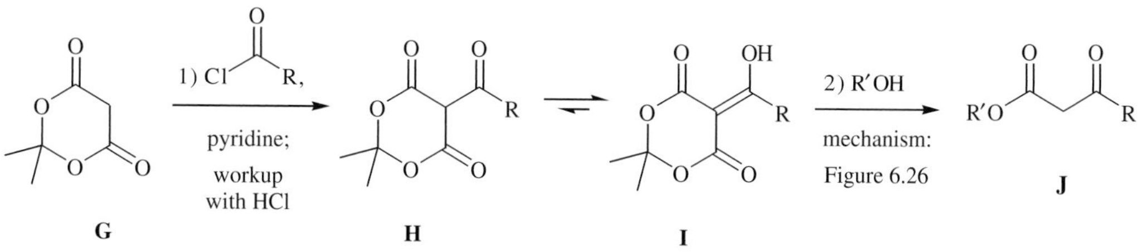

(preparation: Figure 9.21)

Fig. 13.65. Acylation of various malonic diester or malonic half-ester enolates with carboxylic acid chlorides. Spontaneous decarboxylation of the acylation products to furnish β-ketoesters (see variants 1 and 2) and transformation of the acylation products into β-ketoesters by way of alcoholysis/decarboxylation (see variant 3).

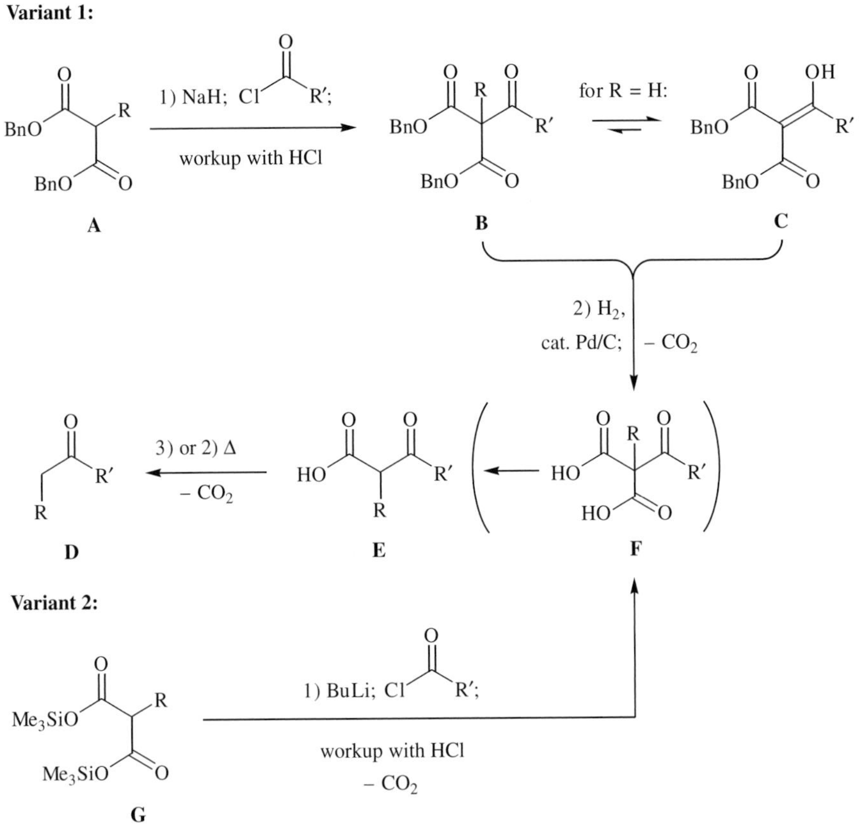

Variant 1:

1) NaH; Cl–C(O)–R′; workup with HCl

for R = H:

2) H_2, cat. Pd/C; – CO_2

3) or 2) Δ – CO_2

A B C D E F

Variant 2:

1) BuLi; Cl–C(O)–R′; workup with HCl – CO_2

G

Fig. 13.66. Acylation of two different malonic diester enolates with carboxylic acid chlorides. Transformation of the acylation products into ketones by heating after debenzylation (see variant 1) or just heating (variant 2).

13.6 Michael Additions of Enolates

13.6.1 Simple Michael Additions

A Michael addition consists of the addition of the enolate of an active-methylene compound, the anion of a nitroalkane, or a ketone enolate to an acceptor-substituted alkene. Such Michael additions can occur in the presence of catalytic amounts of hydroxide or alkoxide. The mechanism of the Michael addition is shown in Figure 13.67. The addition step of the reaction initially leads to the conjugate base of the reaction product. Protonation subsequently gives the product in its neutral and more stable form. The Michael addition is named after the American chemist Arthur Michael.

Acceptor-substituted alkenes that are employed as substrates in Michael additions include α,β-unsaturated ketones (for example, see Figure 13.68), α,β-unsaturated esters (Fig-

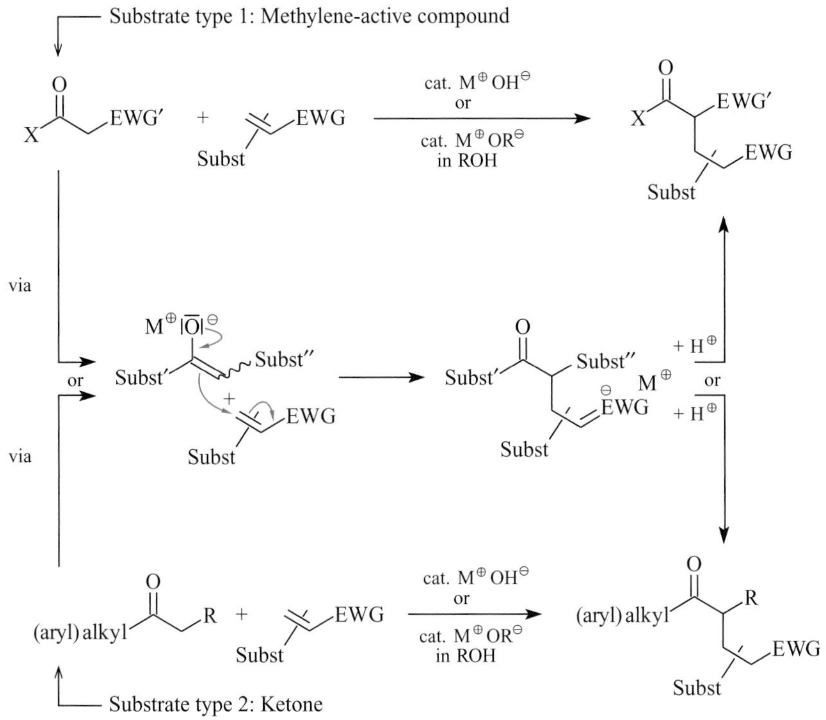

Fig. 13.67. Mechanism of the base-catalyzed Michael addition of active-methylene compounds (top) and of ketones (bottom), respectively. Subst refers to a substituent, and EWG stands for electron-withdrawing group.

ure 13.69), and α,β-unsaturated nitriles (Figure 13.70). The corresponding reaction products are bifunctional compounds with C=O and/or C≡N bonds in positions 1 and 5. Analogous reaction conditions allow Michael additions to vinyl sulfones or nitroalkenes. These reactions lead to sulfones and nitro compounds that carry a C=O and/or a C≡N bond at the C4 carbon.

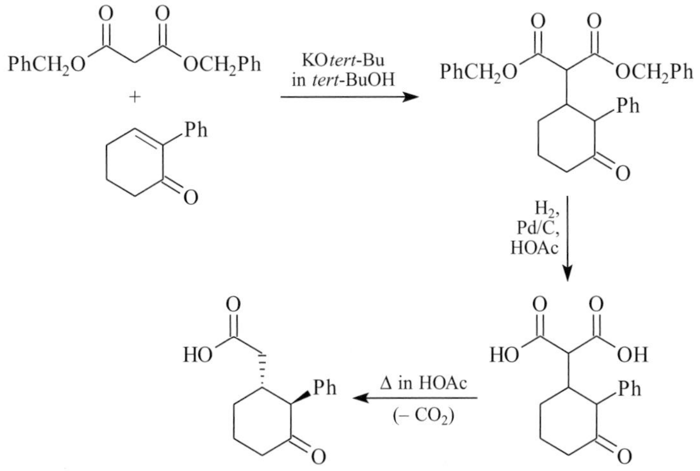

Fig. 13.68. Michael addition to an α,β-unsaturated ketone. A sequence of reactions is shown that effects the 1,4-addition of acetic acid to the unsaturated ketone. See Figure 17.51 regarding step 2 and Figure 13.37 for the mechanism of step 3. The stereochemistry of reaction steps 1 and 2 has not been discussed in the literature. The third step consists of a decarboxylation as well as an acid-catalyzed epimerization of the carbon in the position α to the carbonyl group. This epimerization allows for an equilibration between the *cis,trans*-isomeric cyclohexanones and causes the *trans*-configuration of the major product.

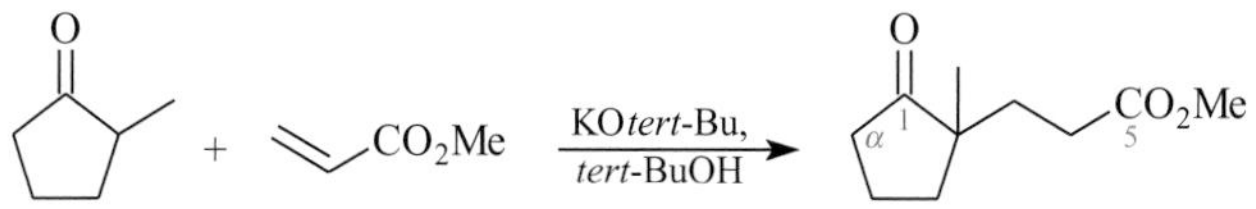

Fig. 13.69. Michael addition to an α,β-unsaturated ester.

Beyond the scope discussed so far, Michael additions also include additions of stoichiometrically generated enolates of ketones, SAMP or RAMP hydrazones, or esters to the C=C double bond of α,β-unsaturated ketones and α,β-unsaturated esters. These Michael additions convert one kind of enolate into another. The driving force stems from the C—C bond formation, not from differential stabilities of the enolates. It is important that the addition of the preformed enolate to the Michael acceptor is faster than the addition of the resulting enolate to another molecule of the Michael acceptor. If that reactivity order were not true, an anionic polymerization of the Michael acceptor would occur. In many Michael additions, however, the enolate *created* is more hindered sterically than the enolate *employed* as the starting material, and in these cases Michael additions are possible without polymerization.

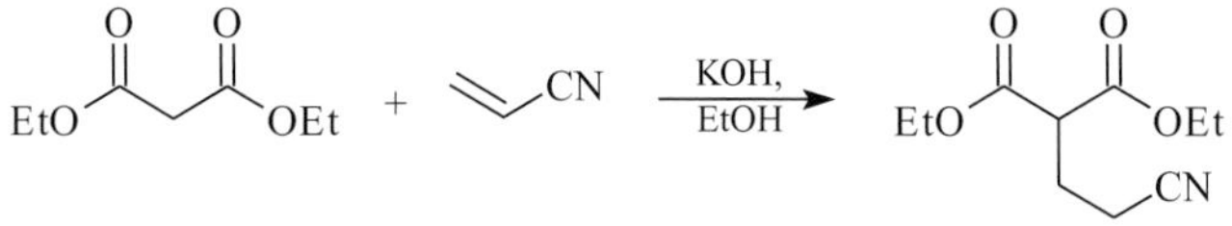

Fig. 13.70. Michael addition to an α,β-unsaturated nitrile.

13.6.2 Tandem Reactions Consisting of Michael Addition and Consecutive Reactions

If a Michael addition of an enolate forms a *ketone enolate* as the primary reaction product, this enolate will be almost completely protonated to give the respective ketone. The reaction medium is of course still basic, since it still contains $OH^{\ominus}$ or $RO^{\ominus}$ ions. The Michael adduct, a ketone, is therefore reversibly deprotonated to a small extent.

This deprotonation may reform the ketone enolate that was the intermediate en route to the Michael adduct. However, the regioisomeric ketone enolate also can be formed. Figures 13.71–13.74 show such enolate isomerizations **B** → **D**, which proceed via the intermediacy of a neutral Michael adduct **C.** This neutral adduct is a 1,5-diketone in Figure 13.71, a δ-ketoaldehyde in Figure 13.72, and a δ-ketoester in Figure 13.73.

The new enolate carbon is located in position 6 of intermediate **D.** In this numbering scheme, position 1 is the C=O double bond of the keto group (Figures 13.71, 13.72), the aldehyde group (Figure 13.73), or the ester group (Figure 13.74). Because of the distance between the enolate position and the C=O double bond, a bond might form between C1 and C6:

- Enolate **D** of Figure 13.71 can undergo an aldol reaction with the C=O double bond of the ketone. The bicyclic compound **A** is formed as the condensation product. It is often possible to combine the formation and the consecutive reaction of a Michael adduct in a one-pot reaction. The overall reaction then is an annulation of a cyclohexenone to an enolizable ketone. The reaction sequence of Figure 13.71 is the **Robinson annulation**, an extraordinarily important synthesis of six-membered rings.

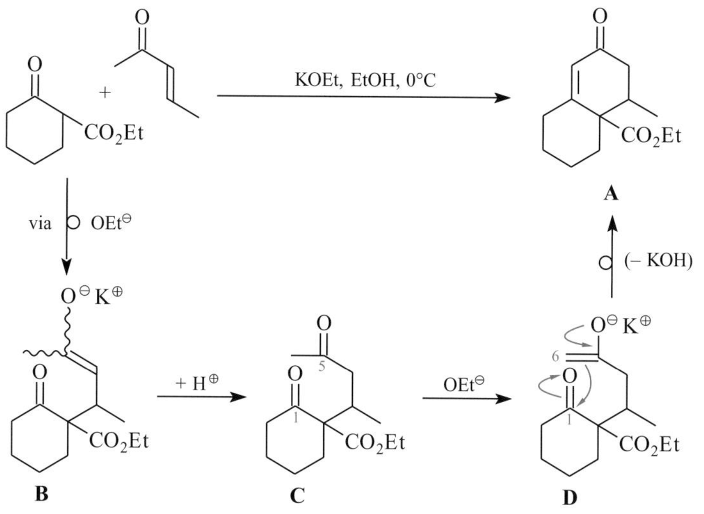

Fig. 13.71. Tandem reaction I, consisting of a Michael addition and an aldol condensation: Robinson annulation reaction for the synthesis of six-membered rings that are condensed to an existing ring.

- The enolate **D** in Figure 13.72 also undergoes an aldol condensation with a ketonic C=O double bond, furnishing the bicycle **A** as the condensation product. Here also, the formation and further reaction of the Michael adduct **C** can be combined in a one-pot reaction. As in the immediately preceding and following example, the final product is a cyclohexenone, but not one that would have been derived from an annulation. Despite the different number of C atoms that the respective reactants contribute to the resulting six-ring skeleton, a reaction like the one presented in Figure 13.72 is sometimes referred to as a **3+3 Robinson annulation** and thus distinguishes this one from the **2+4 Robinson annulations** shown in Figure 13.71.

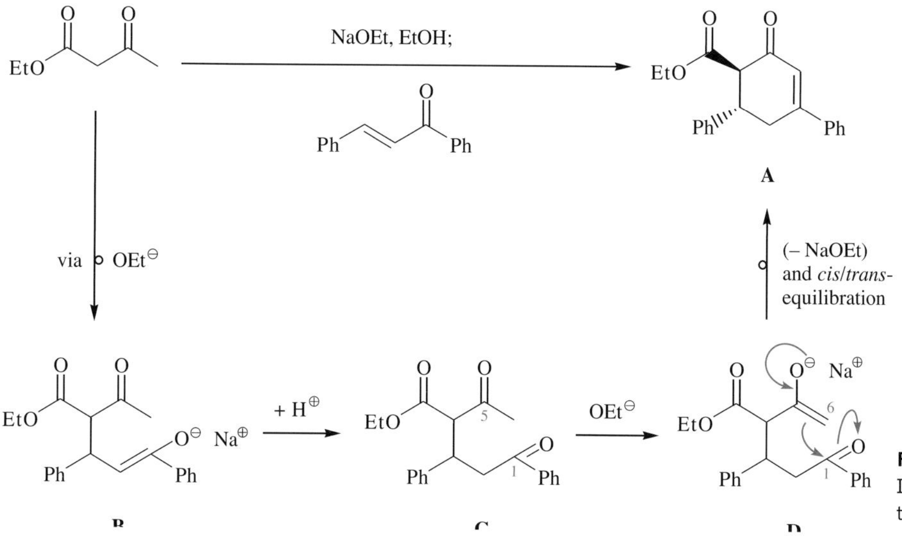

Fig. 13.72. Tandem reaction II, consisting of Michael addition and aldol condensation.

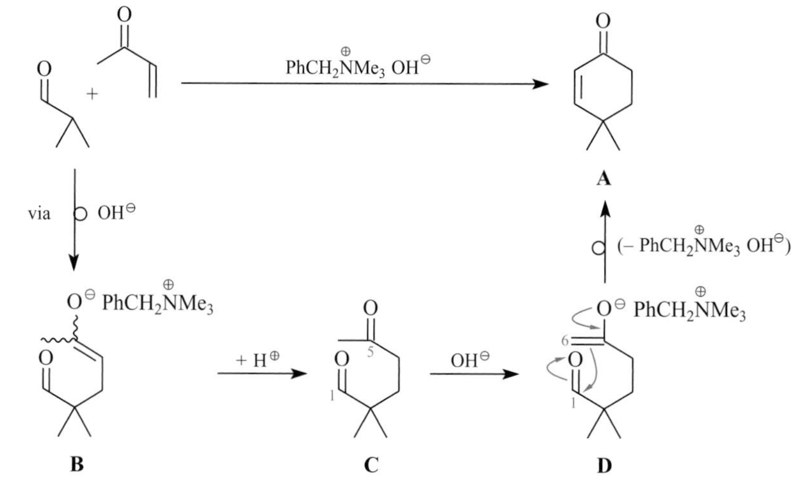

Fig. 13.73. Tandem reaction III, consisting of Michael addition and aldol condensation.

- Enolate **D** of Figure 13.73 undergoes an aldol condensation with the C=O double bond. The enone **A** is the condensation product. This reaction represents a six-membered ring synthesis even though it is not a six-ring annulation.
- Enolate **D** of Figure 13.74 is acylated by the ester following the usual mechanism. The bicyclic compound **A** is a product, which contains a new six-membered ring that has been annulated to an existing ring.

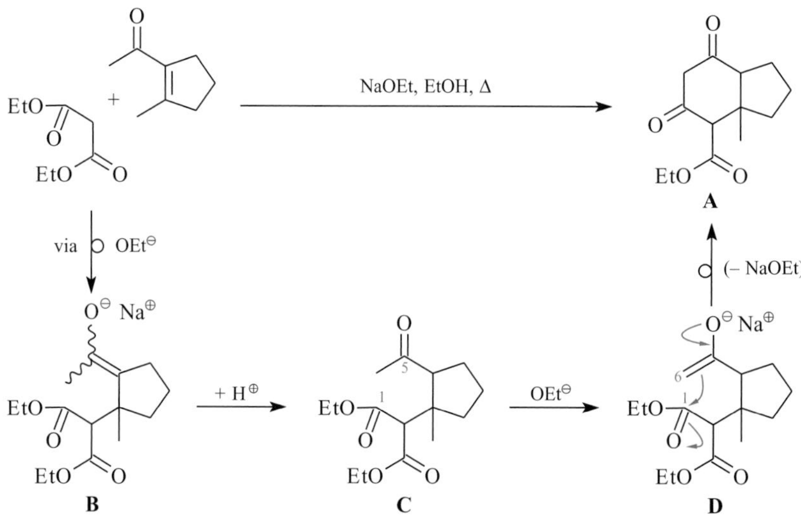

Fig. 13.74. Tandem reaction, consisting of a Michael addition and an enolate acylation (the major tautomer of the reaction product is not shown).

References

13.1

D. Seebach, "Structure and Reactivity of Lithium Enolates. From Pinacolone to Selective C-Alkylations of Peptides. Difficulties and Opportunities Afforded by Complex Structures," *Angew. Chem. Int. Ed. Engl.* **1988**, *27*, 1624–1654.

L. M. Jackman, J. Bortiatynski, "Structures of Lithium Enolates and Phenolates in Solution," in *Advances in Carbanion Chemistry* (V. Snieckus, Ed.), Vol. 1, 45, Jai Press Inc, Greenwich, **1992**.

G. Boche, "The Structure of Lithium Compounds of Sulfones, Sulfoximides, Sulfoxides, Thioethers and 1,3-Dithianes, Nitriles, Nitro Compounds and Hydrazones," *Angew. Chem. Int. Ed. Engl.* **1989**, *28*, 277–297.

R. Schwesinger, "Starke ungeladene Stickstoff-Basen," *Nachr. Chem. Techn. Lab.* **1990**, *38*, 1214–1226.

H. B. Mekelburger, C. S. Wilcox, "Formation of Enolates," in *Comprehensive Organic Synthesis* (B. M. Trost, I. Fleming, Eds.), Vol. 2, 99, Pergamon Press, Oxford, U. K., **1991**.

C. H. Heathcock, "Modern Enolate Chemistry: Regio- and Stereoselective Formation of Enolates and the Consequence of Enolate Configuration on Subsequent Reactions," in *Modern Synthetic Methods* (R. Scheffold, Ed.), Vol. 6, 1, Verlag Helvetica Chimica Acta, Basel, Switzerland, **1992**.

I. Kuwajima, E. Nakamura, "Reactive Enolates from Enol Silyl Ethers", *Acc. Chem. Res.* **1985**, *18*, 181.

D. Cahard, P. Duhamel, "Alkoxide-Mediated Preparation of Enolates from Silyl Enol Ethers and Enol Acetates – From Discovery to Synthetic Applications," *Eur. J. Org. Chem.* **2001**, 1023–1031.

13.2

A. C. Cope, H. L. Holmes, H. O. House, "The Alkylation of Esters and Nitriles," *Org. React.* **1957**, *9*, 107–331.

G. Frater, "Alkylation of Ester Enolates", in *Stereoselective Synthesis* (Houben-Weyl) 4th ed. **1996**, (G. Helmchen, R. W. Hoffmann, J. Mulzer, E. Schaumann, Eds.), **1996**, Vol. E21 (Workbench Edition), *2*, 723–790, Georg Thieme Verlag, Stuttgart.

H.-E. Högberg, "Alkylation of Amide Enolates", in *Stereoselective Synthesis* (Houben-Weyl) 4th ed. **1996**, (G. Helmchen, R. W. Hoffmann, J. Mulzer, E. Schaumann, Eds.), **1996**, Vol. E21 (Workbench Edition), *2*, 791–915, Georg Thieme Verlag, Stuttgart.

D. Caine, "Alkylations of Enols and Enolates," in *Comprehensive Organic Synthesis* (B. M. Trost, I. Fleming, Eds.), Vol. 3, 1, Pergamon Press, Oxford, **1991**.

T. Norin, "Alkylation of Ketone Enolates," in *Stereoselective Synthesis* (Houben-Weyl) 4th ed. **1996**, (G. Helmchen, R. W. Hoffmann, J. Mulzer, E. Schaumann, Eds.), **1996**, Vol. E 21 (Workbench Edition), *2*, 697–722, Georg Thieme Verlag, Stuttgart.

P. Fey, "Alkylation of Azaenolates from Imines," in *Stereoselective Synthesis* (Houben-Weyl) 4th ed. **1996**, (G. Helmchen, R. W. Hoffmann, J. Mulzer, E. Schaumann, Eds.), **1996**, Vol. E 21 (Workbench Edition), *2*, 973–993, Georg Thieme Verlag, Stuttgart.

P. Fey, "Alkylation of Azaenolates from Hydrazones," in *Stereoselective Synthesis* (Houben-Weyl) 4th ed. **1996**, (G. Helmchen, R. W. Hoffmann, J. Mulzer, E. Schaumann, Eds.), **1996**, Vol. E 21 (Workbench Edition), *2*, 994–1015, Georg Thieme Verlag, Stuttgart.

N. Petragnani, M. Yonashiro, "The Reactions of Dianions of Carboxylic Acids and Ester Enolates", *Synthesis* **1982**, 521.

S. Gil, M. Parra, "Dienediolates of Carboxylic Acids in Synthesis. Recent Advances," *Curr. Org. Chem.* **2002**, *6*, 283–302.

D. A. Evans, "Studies in Asymmetric Synthesis – The Development of Practical Chiral Enolate Synthons", *Aldrichimica Acta* **1982**, *15*, 23.

D. A. Evans, "Stereoselective Alkylation Reactions of Chiral Metal Enolates," in *Asymmetric Synthesis – Stereodifferentiating Reactions, Part B* (J. D. Morrison, Ed.), Vol. 3, 1, AP, New York, **1984**.

C. Spino, "Chiral Enolate Equivalents. A Review", *Org. Prep. Proced. Int.* **2003**, *35,* 1–140.

D. Enders, L. Wortmann, R. Peters, "Recovery of Carbonyl Compounds from *N,N*-Dialkylhydrazones," *Acc. Chem. Res.* **2000**, *33*, 157–169.

A. Job, C. F. Janeck, W. Bettray, R. Peters, D. Enders, "The SAMP/RAMP-Hydrazone Methodology in Asymmetric Synthesis," *Tetrahedron* **2002**, *58,* 2253–2329.

D. Seebach, A. R. Sting, M. Hoffmann, "Self-Regeneration of Stereocenters (SRS – Applications, Limitations, and Abandonment of a Synthetic Principle," *Angew. Chem. Int. Ed. Engl.* **1997**, *35*, 2708–2748.

J. E. McMurry, "Ester Cleavages via S_N2-Type Dealkylation," *Org. React.* **1976**, *24*, 187–224.

13.3

D. A. Evans, J. V. Nelson, T. R. Taber, "Stereoselective Aldol Condensations," *Top. Stereochem.* **1982**, *13*, 1.

C. H. Heathcock, "The Aldol Addition Reaction," in *Asymmetric Synthesis – Stereodifferentiating Reactions, Part B* (J. D. Morrison, Ed.), Vol. 3, 111, AP, New York, **1984**.

M. Braun, "Recent Developments in Stereoselective Aldol Reactions", in *Advances in Carbanion Chemistry* (V. Snieckus, Ed.), Vol. 1, 177, Jai Press Inc, Greenwich, CT, **1992**.

M. Braun, "Simple Diastereoselection and Transition State Models of Aldol Additions", in *Stereoselective Synthesis* (Houben-Weyl) 4th ed. **1996**, (G. Helmchen, R. W. Hoffmann, J. Mulzer, E. Schaumann, Eds.), **1996**, Vol. E 21 (Workbench Edition), *3*, 1603–1666, Georg Thieme Verlag, Stuttgart.

C. H. Heathcock, "The Aldol Reaction: Group I and Group II Enolates," in *Comprehensive Organic Synthesis* (B. M. Trost, I. Fleming, Eds.), Vol. 2, 181, Pergamon Press, Oxford, **1991**.

C. H. Heathcock, "Modern Enolate Chemistry: Regio- and Stereoselective Formation of Enolates and the Consequence of Enolate Configuration on Subsequent Reactions," in *Modern Synthetic Methods* (R. Scheffold, Ed.), Vol. 6, 1, Verlag Helvetica Chimica Acta, Basel, Switzerland, **1992**.

C. Palomo, M. Oiarbide, J. M. Garcia, "The Aldol Addition Reaction: An Old Transformation at Constant Rebirth", *Chem. Eur. J.* **2002**, *8,* 36–44.

13.4

A. T. Nielsen, W. J. Houlihan, "The Aldol Condensation", *Org. React.* **1968**, *15*, 1–438.

G. Jones, "The Knoevenagel Condensation", *Org. React.* **1967**, *15*, 204–599.

L. F. Tietze, U. Beifuss, "The Knoevenagel Reaction", in *Comprehensive Organic Synthesis* (B. M. Trost, I. Fleming, Eds.), Vol. 2, 341, Pergamon Press, Oxford, **1991**.

M. Braun, "Syntheses with Aliphatic Nitro Compounds", in *Organic Synthesis Highlights* (J. Mulzer, H.-J. Altenbach, M. Braun, K. Krohn, H.-U. Reißig, Eds.), VCH, Weinheim, New York, **1991**, 25–32.

13.5

B. R. Davis, P. J. Garratt, "Acylation of Esters, Ketones and Nitriles," in *Comprehensive Organic Synthesis* (B. M. Trost, I. Fleming, Eds.), Vol. 2, 795, Pergamon Press, Oxford, **1991**.

T. H. Black, "Recent Progress in the Control of Carbon Versus Oxygen Acylation of Enolate Anions", *Org. Prep. Proced. Int.* **1988**, 21, 179–217.

C. R. Hauser, B. E. Hudson, Jr., "The Acetoacetic Ester Condensation and Certain Related Reactions", *Org. React.* **1942**, *1,* 266–302.

W. S. Johnson, "The Formation of Cyclic Ketones by Intramolecular Acylation," *Org. React.* **1944**, *2,* 114–177.

J. P. Schaefer, J. J. Bloomfield, "The Dieckmann Condensation", *Org. React.* **1967**, *15*, 1–203.

C. R. Hauser, F. W. Swamer, J. T. Adams, "The Acylation of Ketones to Form β-Diketones or β-Keto Aldehydes", *Org. React.* **1954**, *8*, 59–196.

V. Kel'in, "Recent Advances in the Synthesis of 1,3-Diketones," *Curr. Org. Chem.* **2003**, *7,* 1691–1711.

S. Benetti, R. Romagnoli, C. De Risi, G. Spalluto, V. Zanirato, "Mastering β-Keto Esters," *Chem. Rev.* **1995**, *95*, 1065–1115.

13.6

Y. Yamamoto, S. G. Pyne, D. Schinzer, B. L. Feringa, J. F. G. A. Jansen, "Formation of C–C Bonds by Reactions Involving Olefinic Double Bonds – Addition to α,β-Unsaturated Carbonyl Compounds (Michael-Type Additions)," in *Methoden Org. Chem.* (Houben-Weyl) 4th ed. **1952**–, *Stereoselective Synthesis* (G. Helmchen, R. W. Hoffmann, J. Mulzer, E. Schaumann, Eds.), Vol. E21b, 2041, Georg Thieme Verlag, Stuttgart, **1995**.

P. Perlmutter, "Conjugate Addition Reactions in Organic Synthesis," Pergamon Press, Oxford, U. K., **1992**.

A. Bernardi, "Stereoselective Conjugate Addition of Enolates to α,β-Unsaturated Carbonyl Compounds", *Gazz. Chim. Ital.* **1995**, *125*, 539–547.

R. D. Little, M. R. Masjedizadeh, O. Wallquist (in part), J. I. McLoughlin (in part), "The Intramolecular Michael Reaction," *Org. React.* **1995**, *47*, 315–552.

M. J. Chapdelaine, M. Hulce, "Tandem Vicinal Difunctionalization: β-Addition to α,β-Unsaturated Carbonyl Substrates Followed by α-Functionalization," *Org. React.* **1990**, *38*, 225–653.

D. A. Oare, C. H. Heathcock, "Stereochemistry of the Base-Promoted Michael Addition Reaction", *Top. Stereochem.* **1989**, *19*, 227–407.

E. D. Bergmann, D. Ginsburg, R. Pappo, "The Michael Reaction", *Org. React.* **1959**, *10*, 179– 555.

R. D. Little, M. R. Masjedizadeh, "The Intramolecular Michael Reaction," *Org. React. (N. Y.)* **1995**, *47*, 315.

J. A. Bacigaluppo, M. I. Colombo, M. D. Preite, J. Zinczuk, E. A. Ruveda, "The Michael-Aldol Condensation Approach to the Construction of Key Intermediates in the Synthesis of Terpenoid Natural Products," *Pure Appl. Chem.* **1996**, *68*, 683.

J. H. Brewster, E. L. Eliel, "Carbon-Carbon Alkylations with Amines and Ammonium Salts", *Org. React.* **1953**, *7*, 99–197 [covers Michael Additions Starting from β-(Dialkylamino) ketones, -esters or Nitriles].

B. L. Feringa, J. F. G. A. Jansen, "Addition of Enolates and Azaenolates to α,β-Unsaturated Carbonyl Compounds", in *Stereoselective Synthesis* (Houben-Weyl) 4th ed. **1996**, (G. Helmchen, R. W. Hoffmann, J. Mulzer, E. Schaumann, Eds.), **1996**, Vol. E21 (Workbench Edition), *4*, 2104–2156, Georg Thieme Verlag, Stuttgart.

W. Nagata, M. Yoshioka, "Hydrocyanation of Conjugated Carbonyl Compounds," *Org. React.* **1977**, *25*, 255–476.

S. Arseniyadis, K. S. Kyler, D. S. Watt, "Addition and Substitution Reactions of Nitrile-Stabilized Carbanions", *Org. React.* **1984**, *31*, 1–364.

H. L. Bruson, "Cyanoethylation," *Org. React.* **1949**, *5*, 79–135.

F. F. Fleming, Q. Wang, "Unsaturated Nitriles: Conjugate Additions of Carbon Nucleophiles to a Recalcitrant Class of Acceptors," *Chem. Rev.* **2003**, *103,* 2035–2077.

M. Ihara, K. Fukumoto, "Syntheses of Polycyclic Natural Products Employing the Intramolecular Double Michael Reaction," *Angew. Chem. Int. Ed. Engl.* **1993**, *32*, 1010–1022.

E. V. Gorobets, M. S. Miftakhov, F. A. Valeev, "Tandem Transformations Initiated and Determined by the Michael Reaction," *Russ. Chem. Rev.* **2000**, *69*, 1001–1019.

A. Berkessel, "Formation of C-O Bonds by Conjugate Addition of *O*-Nucleophiles", in *Stereoselective Synthesis* (Houben-Weyl) 4th ed. **1996**, (G. Helmchen, R. W. Hoffmann, J. Mulzer, E. Schaumann, Eds.), **1996**, Vol. E21 (Workbench Edition), *8*, 4818–4856, Georg Thieme Verlag, Stuttgart.

R. Schwesinger, J. Willaredt, "Epoxidation of Nucleophilic C–C Double Bonds", in *Stereoselective Synthesis* (Houben-Weyl) 4th ed. **1996**, (G. Helmchen, R. W. Hoffmann, J. Mulzer, E. Schaumann, Eds.), **1996**, Vol. E21 (Workbench Edition), *8*, 4600–4648, Georg Thieme Verlag, Stuttgart.

M. Schäfer, K. Drauz, M. Schwarm, "Formation of C-N Bonds by Conjugate Addition of *N*-Nucleophiles", in *Stereoselective Synthesis* (Houben-Weyl) 4th ed. **1996**, (G. Helmchen, R. W. Hoffmann, J. Mulzer, E. Schaumann, Eds.), **1996**, Vol. E21 (Workbench Edition), *9*, 5588–5642, Georg Thieme Verlag, Stuttgart.

Further Reading

J. R. Johnson, "The Perkin Reaction and Related Reactions," *Org. React.* **1942**, *1,* 210–265.

H. E. Carter, "Azlactones", *Org. React.* **1946**, *3*, 198–239.

M. S. Newman, B. J. Magerlein, "The Darzens Glycidic Ester Condensation," *Org. React.* **1949**, *5*, 413–440.

W. S. Johnson, G. H. Daub, "The Stobbe Condensation," *Org. React.* **1951**, *6*, 1–73.

R. R. Phillips, "The Japp-Klingemann Reaction," *Org. React.* **1959**, *10*, 143–178.

S. J. Parmerter, "The Coupling of Diazonium Salts with Aliphatic Carbon Atoms," *Org. React.* **1959**, *10*, 1–142.

D. B. Collum, "Solution Structures of Lithium Dialkylamides and Related *N*-lithiated Species: Results from ^{6}Li-^{15}N Double Labeling Experiments", *Acc. Chem. Res.* **1993**, *26*, 227–234.

P. Brownbridge, "Silyl Enol Ethers in Synthesis", *Synthesis* **1983**, 85.

J. M. Poirier, "Synthesis and Reactions of Functionalized Silyl Enol Ethers," *Org. Prep. Proced. Int.* **1988**, *20*, 317–369.

H. E. Zimmerman, "Kinetic Protonation of Enols, Enolates, Analogues. The Stereochemistry of Ketonisation," *Acc. Chem. Res.* **1987**, *20*, 263.

S. Hünig, "Formation of C–H Bonds by Protonation of Carbanions and Polar Double Bonds", in *Stereoselective Synthesis* (Houben-Weyl) 4th ed. **1996**, (G. Helmchen, R. W. Hoffmann, J. Mulzer, E. Schaumann, Eds.), **1996**, Vol. E21 (Workbench Edition), *7*, 3851–3912, Georg Thieme Verlag, Stuttgart.

C. Fehr, "Enantioselective Protonation of Enolates and Enols", *Angew. Chem. Int. Ed. Engl.* **1996**, *35*, 2566–2587.

F. A. Davis, B.-C. Chen, "Formation of C-O Bonds by Oxygenation of Enolates," in *Stereoselective Synthesis* (Houben-Weyl) 4th ed. **1996**, (G. Helmchen, R. W. Hoffmann, J. Mulzer, E. Schaumann, Eds.), **1996**, Vol. E21 (Workbench Edition), *8*, 4497–4518, Georg Thieme Verlag, Stuttgart.

P. Fey, W. Hartwig, "Formation of C-C Bonds by Addition to Carbonyl Groups (C=O) – Azaenolates or Nitronates," in *Methoden Org. Chem.* (Houben-Weyl) 4th ed. **1952**–, *Stereoselective Synthesis* (G. Helmchen, R. W. Hoffmann, J. Mulzer, E. Schaumann, Eds.), Vol. E21b, 1749, Georg Thieme Verlag, Stuttgart, **1995**.

K. Krohn, "Stereoselective Reactions of Cyclic Enolates", in *Organic Synthesis Highlights* (J. Mulzer, H.-J. Altenbach, M. Braun, K. Krohn, H.-U. Reißig, Eds.), VCH, Weinheim, New York, etc., **1991**, 9–13.

K. F. Podraza, "Regiospecific Alkylation of Cyclohexenones. A Review", *Org. Prep. Proced. Int.* **1991**, *23*, 217–235.

S. K. Taylor, "Reactions of Epoxides with Ester, Ketone and Amide Enolates", *Tetrahedron* **2000**, *56*, 1149–1163.

T. M. Harris, C. M. Harris, "The α-Alkylation and α-Arylation of Dianions of β-Dicarbonyl Compounds", *Org. React.* **1969**, *17*, 155–212.

C. M. Thompson, D. L. C. Green, "Recent Aadvances in Dianion Chemistry," *Tetrahedron* **1991**, *47*, 4223–4285.

B. M. Kim, S. F. Williams, S. Masamune, "The Aldol Reaction: Group III Enolates", in *Comprehensive Organic Synthesis* (B. M. Trost, I. Fleming, Eds.), Vol. 2, 239, Pergamon Press, Oxford, **1991**.

M. Sawamura, Y. Ito, "Asymmetric Carbon-Carbon Bond Forming Reactions: Asymmetric Aldol Reactions," in *Catalytic Asymmetric Synthesis* (I. Ojima, Ed.), 367, VCH, New York, **1993**.

A. S. Franklin, I. Paterson, "Recent Developments in Asymmetric Aldol Methodology," *Contemporary Organic Synthesis* **1994**, *1*, 317.

C. J. Cowden, I. Paterson, "Asymmetric Aldol Reactions Using Boron Enolates," *Org. Prep. Proced. Int.* **1997**, *51*, 1–200.

E. M. Carreira, "Aldol Reaction: Methodology and Stereochemistry," in *Modern Carbonyl Chemistry*, Ed.: J. Otera, Wiley-VCH, Weinheim, **2000**, 227–248.

I. Paterson C. J. Cowden, D. J. Wallace, "Stereoselektive Aldol Reactions in the Synthesis of Polyketide Natural Products," in *Modern Carbonyl Chemistry*, Ed.: J. Otera, Wiley-VCH, Weinheim, **2000**, 249–297.

G. Casiraghi, F. Zanardi, G. Appendino, G. Rassu, "The Vinylogous Aldol Reaction: A Valuable, Yet Understated Carbon-Carbon Bond-Forming Maneuver", *Chem. Rev.* **2000**, *100*, 1929–1972.

M. Sawamura, Y. Ito, "Asymmetric Aldol Reactions-Discovery and Development", in *Catalytic Asymmetric Synthesis*, Ed.: I. Ojima, Wiley-VCH, New York, 2nd ed., **2000**, 493–512.

E. M. Carreira, "Recent Advances in Asymmetric Aldol Addition Reactions", in *Catalytic Asymmetric Synthesis*, Hrsg.: I. Ojima, Wiley-VCH, New York, 2nd ed., **2000**, 513–542.

T. D. Machajewski, C.-H. Wong, R. A. Lerner, "The Catalytic Asymmetric Aldol Reaction," *Angew. Chem. Int. Ed. Engl.* **2000**, *39*, 1352–1374.

A. Fürstner, "Recent Advancements in the Reformatsky Reaction," *Synthesis* **1989**, *8*, 571–590.

B. C. Chen, "Meldrum's Acid in Organic Synthesis," *Heterocycles* **1991**, *32*, 529–597.

C. F. Bernasconi, "Nucleophilic Addition to Olefins. Kinetics and Mechanism", *Tetrahedron* **1989**, *45*, 4017–4090.

F. A. Luzzio, "The Henry Reaction: Recent Examples," *Tetrahedron* **2001**, *57*, 915–945.

S. Arseniyadis, K. S. Kyler, D. S. Watt, "Addition and Substitution Reactions of Nitrile-Stabilized Carbanions", *Org. React.* **1984**, *31*, 1–364.

S. G. Pyne, "Addition of Metalated Allylic Phosphine Oxides, Phosphonates, Sulfones, Sulfoxides and Sulfoximines to α,β-Unsaturated Carbonyl Compounds", in *Stereoselective Synthesis* (Houben-Weyl) 4th ed. **1996**, (G. Helmchen, R. W. Hoffmann, J. Mulzer, E. Schaumann, Eds.), **1996**, Vol. E21 (Workbench Edition), *4*, 2068–2086, Georg Thieme Verlag, Stuttgart.

M. Kanai, M. Shibasaki, "Asymmetric Michael Reactions", in *Catalytic Asymmetric Synthesis*, Hrsg.: I. Ojima, Wiley-VCH, New York, 2nd ed., **2000**, 569–592.

K. Tomioka, "Asymmetric Michael-Type Addition Reaction," in *Modern Carbonyl Chemistry*, Ed.: J. Otera, Wiley-VCH, Weinheim, **2000**, 491–505.

N. Krause, A. Hoffmann-Roder, "Recent Advances in Catalytic Enantioselective Michael Additions," *Synthesis* **2001**, 171–196.

Rearrangements

14

The term "rearrangement" is used to describe two different types of organic chemical reactions. A rearrangement may involve the *one-step* migration of an H atom or of a larger molecular fragment. On the other hand, a rearrangement may be a *multistep reaction* that includes the migration of an H atom or of a larger molecular fragment as one of its steps. The Wagner–Meerwein rearrangement of a carbenium ion (Section 14.3.1) exemplifies a rearrangement of the first type. Carbenium ions are so short-lived that neither the starting material nor the primary rearrangement product can be isolated. The Claisen rearrangement of allyl alkenyl ethers also is a one-step rearrangement (Section 14.5). In contrast to the Wagner–Meerwein rearrangement, however, both the starting material and the product of the Claisen rearrangement are molecules that can be isolated. The ring expansion reaction shown in Figure 14.25 and the alkyne synthesis depicted in Figure 14.32 are examples of multistep rearrangement reactions.

14.1 Nomenclature of Sigmatropic Shifts

In many rearrangements, the migrating group connects to one of the direct neighbors of the atom to which it was originally attached. Rearrangements of this type are the so-called **[1,2]-rearrangements** or **[1,2]-shifts.** These rearrangements can be considered as **sigmatropic processes**, the numbers "1" and "2" characterizing the subclass to which they belong. The adjective "sigmatropic" emphasizes that a σ-bond formally migrates in these reactions. How far it migrates is described by specifying the positions of the atoms between that the bond is shifted. The atoms that are *initially* bonded are assigned positions 1 and 1'. The subsequent atoms in the direction of the σ-bond migration are labeled 2, 3, and so forth, on the side of center 1 and labeled 2', 3', and so forth, on the side of center 1'. After the rearrangement, the σ bond connects two atoms in positions n and m'. The rearrangement can now be characterized by the positional numbers n and m' in the following way: the numbers are written between brackets, separated by a comma, and the primed number is given without the prime. Hence, an $[n,m]$-rearrangement is the most general description of a sigmatropic process. A [1,2]-rearrangement is the special case with $n = 1$ and $m' = 2$ (Figure 14.1). **[3,3]-Rearrangements** occur when $n = m' = 3$ (Figure 14.2). Many other types of rearrangements are known, including [1,3]-, [1,4]-, [1,5]-, [1,7]-, [2,3]-, and [5,5]-rearrangements.

Bruckner R (author), Harmata M (editor) In: *Organic Mechanisms – Reactions, Stereochemistry and Synthesis*
Chapter DOI: 10.1007/978-3-642-03651-4_14, © Springer-Verlag Berlin Heidelberg 2010

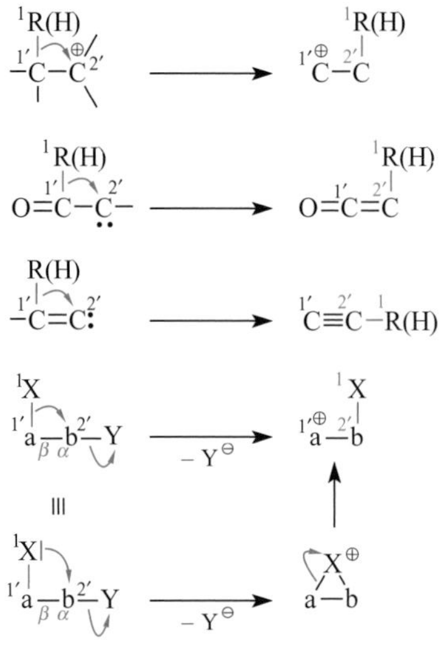

Fig. 14.1. The three reactions on top show [1,2]-rearrangements to a sextet carbon. The two reactions at the bottom show [1,2]-rearrangements to a neighboring atom that is coordinatively saturated but in the process of losing a leaving group.

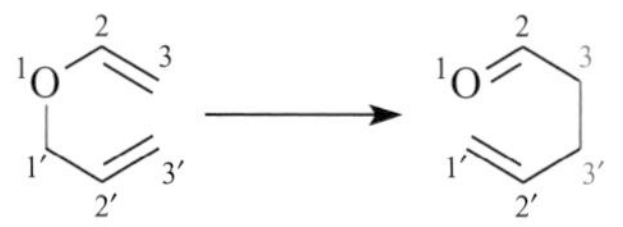

Fig. 14.2. A Claisen rearrangement as an example of a [3,3]-rearrangement.

14.2 Molecular Origins for the Occurrence of [1,2]-Rearrangements

Figure 14.1 shows the structures of the immediate precursors of one-step [1,2]-sigmatropic rearrangements. These formulas reveal two different reasons for the occurrence of rearrangements in organic chemistry. Rows 1–3 of Figure 14.1 reveal the first reason for [1,2]-rearrangements to take place, namely, the occurrence of a valence electron sextet at one of the C atoms of the substrate. This sextet may be located at the $C^{\oplus}$ of a carbenium ion or at the C: of a carbene. Carbenium ions are extremely reactive species. If there exists no good opportunity for an intermolecular reaction (i.e., no good possibility for stabilization), carbenium ions often undergo an intramolecular reaction. This intramolecular reaction in many cases is a [1,2]-rearrangement.

Suppose a valence electron sextet occurs at a carbon atom and the possibility exists for a [1,2]-rearrangement to occur. The thermodynamic driving force for the potential [1,2]-rearrangement will be *significant* if the rearrangement leads to a structure with octets on all atoms. It is for this reason that acyl carbenes rearrange quantitatively to give ketenes (row 2 in Figure 14.1) and that vinyl carbenes rearrange quantitatively to give acetylenes (row 3 in Figure 14.1). In contrast, another valence electron sextet species is formed if the [1,2]-rearrangement of a carbenium ion leads to another carbenium ion. Accordingly, the driving force of a [1,2]-rearrangement of a carbenium ion is much smaller than the driving force of a [1,2]-rearrangement of a carbene. The following rules of thumb summarize all cases for which nonetheless quantitative carbenium ion rearrangements are possible.

[1,2]-Rearrangements of carbenium ions occur quantitatively only

- if the new carbenium ion is substantially better stabilized electronically by its substituents than the old carbenium ion;
- if the new carbenium ion is substantially more stable than the old carbenium ion because of other effects such as reduced ring strain,
- or if the new carbenium ion is captured in a subsequent, irreversible reaction.

Notwithstanding these cases, many [1,2]-rearrangements of carbenium ions occur reversibly because of the small differences in the free enthalpies.

Consideration of rows 4 and 5 of Figure 14.1 suggests a second possible cause for the occurrence of [1,2]-rearrangements. In those cases the substrates contain a b—Y bond. The heterolysis of this bond would lead to a reasonably stable leaving group $Y^{\ominus}$ but would also produce a cation $b^{\oplus}$ with a sextet. Such a heterolysis would be possible only—even if assisting substituents were present—if the b—Y bond was a C—Y bond and the product of heterolysis was a carbenium ion. If the b—Y bond were an N—Y or an O—Y bond, the heterolyses would generate nitrenium ions $R^1R^2N^{\oplus}$ and oxenium ions $RO^{\oplus}$, respectively. Neither heterolysis has ever been observed. Nitrenium and oxenium ions have much higher heats of formation than carbenium ions because the central atoms nitrogen and oxygen are substantially more electronegative than carbon.

Heterolyses of N—Y and O—Y bonds are, however, not *entirely* unknown. This is because these heterolyses can occur concomitantly with a [1,2]-rearrangement. In a way, such a [1,2]-rearrangement avoids the formation of an unstable valence electron sextet. The b—Y bond thus undergoes heterolysis and releases $Y^{\ominus}$. However, the formation of an electron sextet at center b (=NR or O) is avoided because center b shares another electron pair by way of bonding to another center in the molecule. This new electron pair may be in either the β- or the γ-position relative to the position of the leaving group Y. If the b center binds to an electron pair in the β-position relative to the leaving group Y (row 4 in Figure 14.1), this electron pair is the bonding electron pair of the $\sigma_{a—X}$ bond in the substrate. On the other hand, if the b center binds to an electron pair in the γ-position relative to the leaving group Y (row 5 in Figure 14.1), then this electron pair is a nonbonding lone pair of the X group attached to the β-position. The engagement of this free electron pair leads to the formation of a positively charged three-membered ring. The substituent X exerts a so-called neighboring group effect if subsequently this b—X bond is again broken (Section 2.7). On the other hand, a [1,2]-rearrangement occurs if the a—X bond of the three-membered ring is broken. Such a neighboring-group-effect was discussed in connection with the deuteration experiment of Figure 2.26.

14.3 [1,2]-Rearrangements in Species with a Valence Electron Sextet

14.3.1 [1,2]-Rearrangements of Carbenium Ions

Wagner–Meerwein Rearrangements

Wagner–Meerwein rearrangements are [1,2]-rearrangements of H atoms or alkyl groups in carbenium ions that do not contain any heteroatoms attached to the valence-unsaturated center C1 or to the valence-saturated center C2. The actual rearrangement step consists of a reaction that cannot be carried out separately because both the starting material and the product are extremely short-lived carbenium ions that cannot be isolated. Wagner–Meerwein rearrangements therefore occur only as part of a reaction sequence in which a carbenium ion is generated in one or more steps, and the rearranged carbenium ion reacts further in one or several steps to give a valence-saturated compound. The sigmatropic shift of the Wagner–Meerwein rearrangement therefore can be embedded between a great variety of carbenium-ion-generating and carbenium-ion-annihilating reactions (Figures 14.3–14.11).

In Section 5.2.5, we discussed the Friedel–Crafts alkylation of benzene with 2-chloropentane. This reaction includes a Wagner–Meerwein reaction in conjunction with other elementary reactions. The Lewis acid catalyst $AlCl_3$ first converts the chloride into the 2-pentyl cation **A** (Figure 14.3). Cation **A** then rearranges into the isomeric 3-pentyl cation **B** (Figure 14.3), in part or perhaps to the extent that the equilibrium ratio is reached. The new carbenium ion **B** is not significantly more stable than the original one (**A**), but it also is not significantly less stable. In addition, both the cations **A** and **B** are relatively unhindered sterically, and each can engage in an Ar-S_E reaction with a comparable rate of reaction. Thus, aside from the alkyla-

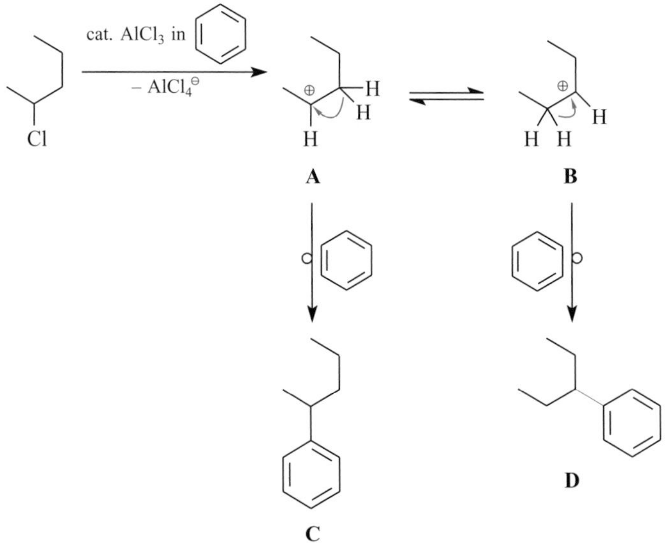

cat. $AlCl_3$ in [benzene]

$- AlCl_4^{\ominus}$

A **B** **C** **D**

Fig. 14.3. Mechanism of an Ar-SE reaction (details: Section 5.2.5), which includes a reversible Wagner–Meerwein rearrangement.

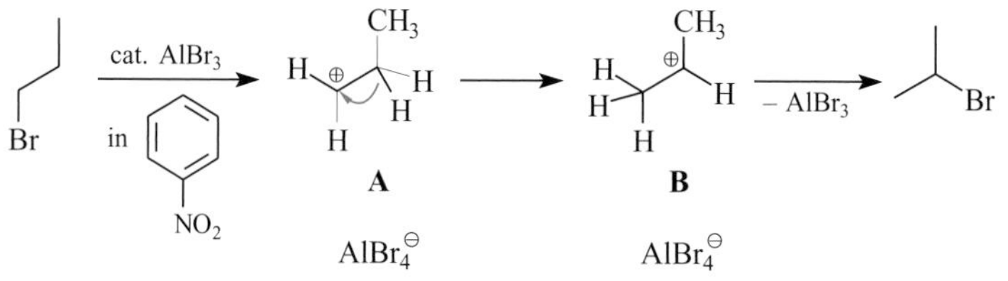

Fig. 14.4. Wagner–Meerwein rearrangement in the isomerization of an alkyl halide.

tion product **C** with its unaltered alkyl group, the isomer **D** with the isomerized alkyl group also is formed.

A Wagner–Meerwein rearrangement can be part of the isomerization of an alkyl halide (Figure 14.4). For example, 1-bromopropane isomerizes quantitatively to 2-bromopropane under Friedel–Crafts conditions. The [1,2]-shift **A** → **B** involved in this reaction again is an H atom shift. In contrast to the thermoneutral isomerization between carbenium ions **A** and **B** of Figure 14.3, in the present case an energy gain is associated with the formation of a secondary carbenium ion from a primary carbenium ion. Note, however, that the different stabilities of the carbenium ions are not responsible for the complete isomerization of 1-bromopropane into 2-bromopropane. The position of this isomerization equilibrium is determined by thermodynamic control at the level of the alkyl halides. 2-Bromopropane is more stable than 1-bromopropane and therefore formed exclusively.

There also are Wagner–Meerwein reactions in which alkyl groups migrate rather than H atoms (Figures 14.3 and 14.4). Of course, these reactions, too, are initiated by carbenium-ion-generating reactions, as exemplified in Figure 14.5 for the case of an E1 elimination from an alcohol (Section 4.5). The initially formed neopentyl cation—a primary carbenium ion—rearranges into a tertiary carbenium ion, thereby gaining considerable stabilization. An elimination of a β-H atom is possible only after the rearrangement has occurred. It terminates the overall reaction and provides an alkene (Saytzeff product).

The sulfuric acid-catalyzed transformation of pinanic acid into abietic acid shown in Figure 14.6 includes a Wagner–Meerwein shift of an alkyl group. The initially formed carbenium ion, the secondary carbenium ion **A**, which is a localized carbenium ion, is generated by protonation of one of the C=C double bonds. A [1,2]-sigmatropic shift of a methyl group occurs in **A**, and the much more stable, delocalized, and tetraalkyl-substituted allyl cation **B** is formed. Cation **B** is subsequently deprotonated and a 1,3-diene is obtained. Overall, Figure 14.6 shows the isomerization of a less stable diene into a more stable diene. The direction of this isomerization is determined by thermodynamic control. The product 1,3-diene is conjugated and therefore more stable than the unconjugated substrate diene.

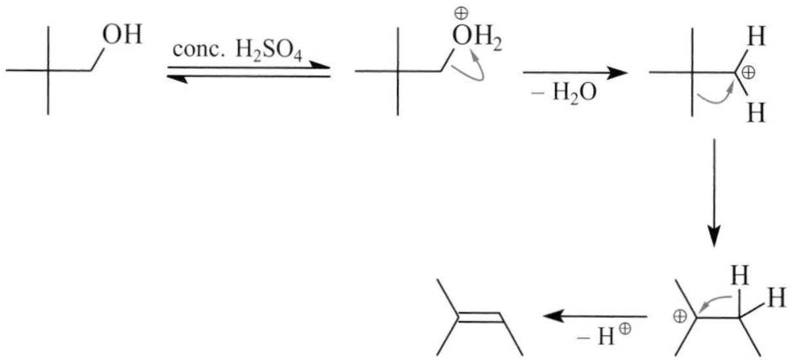

Fig. 14.5. Wagner–Meerwein rearrangement as part of an isomerizing E1 elimination.

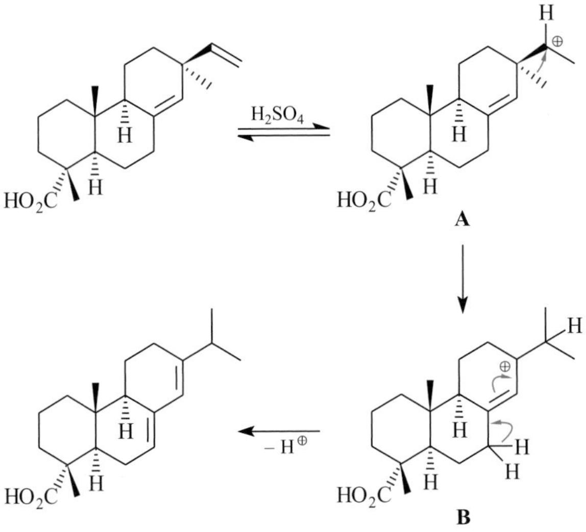

Fig. 14.6. Wagner–Meerwein rearrangement as part of an alkene isomerization.

In the carbenium ion **A** of Figure 14.6, there are three different alkyl groups in α-positions with respect to the carbenium ion center, and each one could in principle undergo the [1,2]-rearrangement. Yet, only the migration of the methyl group is observed. Presumably, this is the consequence of product development control. The migration of either one of the other two alkyl groups would have resulted in the formation of a seven-membered and therefore strained ring. Only the observed methyl shift **A** → **B** retains the energetically advantageous six-membered ring skeleton.

Even in rearrangements of carbenium ions that show no preference for the migration of a particular group that would be based on thermodynamic control or on product development control, one can observe chemoselectivity. This is because certain potentially migrating groups exhibit different intrinsic tendencies toward such a migration: in Wagner–Meerwein rearrangements, and in many other [1,2]-migrations, tertiary alkyl groups migrate faster than secondary, secondary alkyl groups migrate faster than primary, and primary alkyl groups in turn migrate faster than methyl groups. Thus cation **A** in Figure 14.7 rearranges

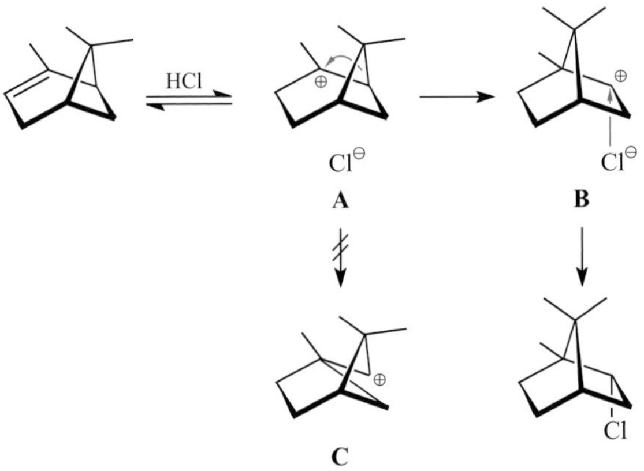

Fig. 14.7. Wagner–Meerwein rearrangement as part of an HCl addition to a C=C double bond:

into cation **B** by way of a C_{tert} migration rather than into the cation **C** via a C_{prim} migration. Both cations **B** and **C** are secondary carbenium ions, and both are bicyclo[2.2.1]heptyl cations; thus they can be expected to be comparable in stability. If there were no intrinsic migratory preference of the type $C_{tert} > C_{prim}$, one would have expected the formation of comparable amounts of **B** and **C.**

The [1,2]-alkyl migration **A** → **B** of Figure 14.7 converts a cation with a well-stabilized tertiary carbenium ion center into a cation with a less stable secondary carbenium ion center. This is possible only because of the driving force that is associated with the reduction of ring strain: a cyclobutyl derivative **A** is converted into a cyclopentyl derivative **B**.

Wagner–Meerwein Rearrangements in the Context of Tandem and Cascade Rearrangements

A carboxonium ion (an all-octet species) may become less stable than a carbenium ion (a sextet species) only when ring-strain effects dominate. In such cases carbenium ions can be generated from carboxonium ions by way of a Wagner–Meerwein rearrangement. Thus, the decrease of ring strain can provide a driving force strong enough to overcompensate for the conversion of a more stable into a less stable cationic center. In Figure 14.8, for example, the carboxonium ion **A** rearranges into the carbenium ion **B** because of the release of cyclobutane strain (of about 26 kcal/mol) in the formation of the cyclopentane (ring strain of about 5 kcal/mol). Cation **B** stabilizes itself by way of another [1,2]-rearrangement. The resulting cation **C** has comparably little ring strain but is an electronically favorable carboxonium ion. The pinacol and semipinacol rearrangements (see below) include [1,2]-shifts that are just like the second [1,2]-shift of Figure 14.8, the only difference being that the *β*-hydroxylated carbenium ion intermediates analogous to **B** are generated in a different manner.

Camphorsulfonic acid is generated by treatment of camphor with concentrated sulfuric acid in acetic anhydride (Figure 14.9). Protonation of the carbonyl oxygen leads to the formation of a small amount of the carboxonium ion **A.** **A** undergoes a Wagner–Meerwein rearrangement into cation **B.** However, this rearrangement occurs only to a small extent, since an all-octet species is converted into an intermediate with a valence electron sextet and there is no supporting release of ring strain. Hence, the rearrangement **A** → **B** is an endothermic pro-

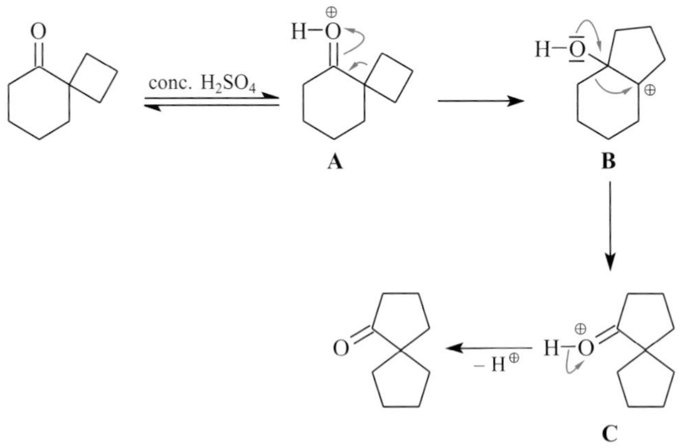

conc. H_2SO_4

A

B

C

$-H^{\oplus}$

Fig. 14.8. Tandem rearrangement comprising a Wagner–Meerwein rearrangement and a semipinacol rearrangement.

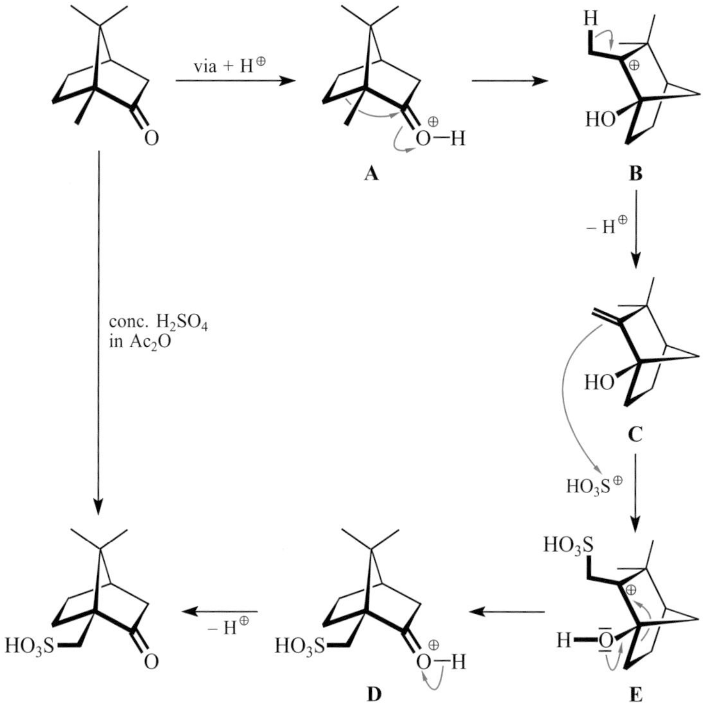

Fig. 14.9. Preparation of optically active camphorsulfonic acid via a path involving a Wagner–Meerwein rearrangement (**A** → **B**) and a semipinacol rearrangement (**E** → **D**).

cess. Nevertheless, the reaction ultimately goes *to completion* in this energy-consuming direction because the carbenium ion **B** engages in subsequent irreversible reactions.

Cation **B** is first deprotonated to give the hydroxycamphene derivative **C. C** reacts with an electrophile of unknown structure that is generated from sulfuric acid under these conditions. In the discussion of the sulfonylation of aromatic compounds (Figure 5.17), we mentioned protonated sulfuric acid $H_3SO_4^{\oplus}$ and its dehydrated derivative $HSO_3^{\oplus}$ as potential electrophiles, which might assume the same role here. In any case, the reaction results in the formation of carbenium ion **E.**

A carbenium ion with a *β*-hydroxy group, **E,** stabilizes itself by way of a carbenium ion → carboxonium ion rearrangement. Such a rearrangement occurs in the third step of the pinacol rearrangement (Figure 14.14) and also in many semipinacol rearrangements (Figures 14.19 and 14.21). The carbenium ion **E** therefore is converted into the carboxonium ion **D**. In the very last step, **D** is deprotonated and a ketone is formed. The final product of the rearrangement is camphorsulfonic acid.

While the sulfonation of C10 of camphor involves two [1,2]-rearrangements (Figure 14.9), the bromination of dibromocamphor involves four of these shifts, namely, **A** → **B, B** → **D, I** → **H**, and **H** → **G** (Figure 14.10). A comparison of the mechanisms of sulfonylation (Figure 14.9) and bromination (Figure 14.10) reveals that the cations marked **B** in the two cases react in different ways. **B** undergoes elimination in the sulfonylation (→ **C**, Figure 14.9) but not in the bromination (Figure 14.10). This difference is less puzzling than it might seem at first. In fact, it is likely that elimination also occurs intermittently during the course of the bromination (Figure 14.10) but the reaction simply is inconsequential.

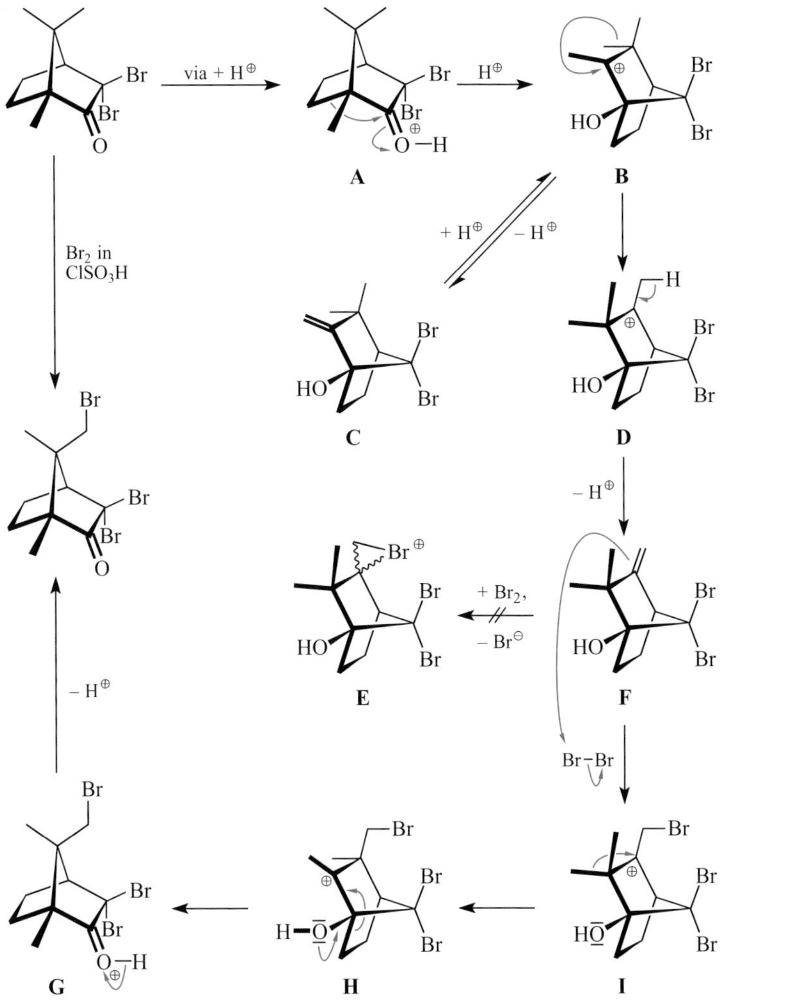

Fig. 14.10. Preparation of optically active tribromocamphor via a path involving three Wagner–Meerwein rearrangements (**A** → **B**, **B** → **D**, **I** → **E**) and a semipinacol rearrangement (**H** → **G**).

Molecular bromine, Br_2, is a weak electrophile and does not react with alkene **C** fast enough. It is for this reason and in contrast to the sulfonylation (Figure 14.9) that the carbenium ion **B** of Figure 14.10, which is in equilibrium with alkene **C**, has sufficient time to undergo another Wagner–Meerwein reaction that converts the β-hydroxycarbenium ion **B** into the carbenium ion **D.** Ion **D** is more stable than **B** because the hydroxy group in **D** is in the γ-position relative to the positive charge, while it is in the β-position in **B;** that is, the destabilization of the cationic center by the electron-withdrawing group is reduced in **D** compared to **B.** The carbenium ion **D** is now deprotonated to give the alkene **F**. In contrast to **C**, alkene **F** reacts with Br_2. The bromination results in the formation of the bromo-substituted carbenium ion **I.** The bromination mechanism of Section 3.5.1 might have suggested the formation of the bromonium ion **E**, but this is not formed. It is known that open-chain intermediates of type **I** may occur in brominations of C=C double bonds (see commentary in Section 3.5.1 regard-

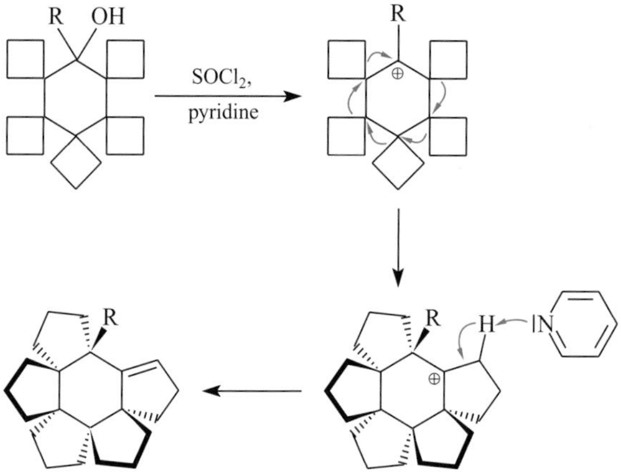

Fig. 14.11. An E1 elimination involving five Wagner–Meerwein rearrangements.

ing Figures 3.5 and 3.6). In the *present* case, the carbenium ion **I** is presumably formed because it is less strained than the polycyclic bromonium ion isomer **E**. In light of the reaction mechanism, one can now understand why Br_2 reacts with **F** faster than **C.** The hydroxyl group is one position farther removed from the cationic center in the carbenium ion **I** that is generated from alkene **F** in comparison to the carbenium ion that would be formed by bromination of **C.** Hence, **I** is more stable than the other carbenium ion so that the formation of **I** is favored by product development control.

The reactions shown in Figures 14.9 and 14.10 exemplify a **tandem rearrangement and a cascade rearrangement**, respectively. These terms describe sequences of two or more rearrangements taking place more or less directly one after the other. Cascade rearrangements may involve even more Wagner–Meerwein rearrangements than the one shown in Figure 14.10. The rearrangement shown in Figure 14.11, for example, involves five [1,2]-rearrangements, each one effecting the conversion of a spiroannulated cyclobutane into a fused cyclopentane.

Every polycyclic hydrocarbon having the molecular formula $C_{10}H_{16}$ can be isomerized to adamantane. The minimum number of [1,2]-rearrangements needed in such rearrangements is so high that it can be determined only with the use of a computer program. The rearrangements occur in the presence of catalytic amounts of $AlCl_3$ and *tert*-BuCl:

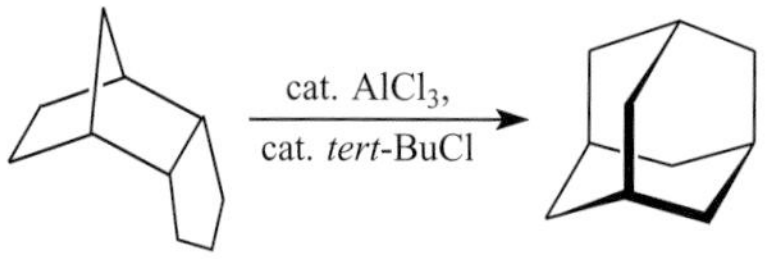

These isomerizations almost certainly involve [1,2]-shifts of H atoms as well as of alkyl groups. One cannot exclude that [1,3]-rearrangements may also play a role. The reaction product, adamantane, is formed under thermodynamic control under these conditions. It is the so-called **stabilomer** (the most stable isomer) of all the hydrocarbons having the formula $C_{10}H_{16}$.

This impressive cascade reaction begins with the formation of a small amount of the *tert*-butyl cation by reaction of $AlCl_3$ with *tert*-BuCl. The *tert*-butyl cation abstracts a hydride ion from the substrate $C_{10}H_{16}$. Thus, a carbenium ion with formula $C_{10}H_{15}^{\oplus}$ is formed. This and related carbenium ions are certainly substrates for Wagner–Meerwein rearrangements and also potentially substrates for [1,3]-rearrangements, thereby providing various isomeric cations *of the same formula*. Some of these cations can abstract a hydride ion from the neutral starting material $C_{10}H_{16}$. The saturated hydrocarbons *iso*-$C_{10}H_{16}$ obtained in this way are isomers of the original starting material $C_{10}H_{16}$. Hydride transfers and [1,2]- and [1,3]-shifts, respectively, are repeated until the reaction eventually arrives at adamantane by way of the adamantyl cation.

Side Note 14.1. Meerwein Rearrangement Cascades ("Tandem Wagner–Meerwein Rearrangements") in Biosynthesis

Mother Nature is an excellent organic chemist! This, of course, is known to anyone who shows even a little interest in natural products and is amazed at the amount and complexity of what is being synthesized by living organisms at temperatures that are essentially quite similar to our body temperature. The fascinating thing is that this kind of biosynthesis is just normal organic chemistry in the sense that everything that our understanding of organic chemistry is built upon, in particular reaction mechanisms, also forms the basis of all biosyntheses.

There are probably no other examples in support of the latter statement that are more impressive than the enzyme-catalyzed transformations of the acyclic triterpenes squalene oxide (Formula **A** in Figure 14.12) or squalene (Formula **A** in Figure 14.13) into the tetracycle lanosterol (Formula **G** in Figure 14.12) and the pentacycle hopene (Formula **I** in Figure 14.13), respectively. Both cyclizations constitute one-step reactions in the sense that they each require just a single enzyme to proceed. This is an oxidosqualene-lanosterol cyclase in the first case and a squalene-hopene cyclase in the second. Both cyclizations display a high degree of stereocontrol: lanosterol is formed as one of 64 conceivable stereoisomers and hopene as one of 512 possible stereoisomers. Essentially, this stereocontrol is based on the enzymatic control of the substrate conformation. Ultimately, the oxidosqualene-lanosterol cyclase forces its substrate, i.e., squalene oxide, to adopt a chair-boat-chair conformation (starting with Formulas **B** and **D** in Figure 14.12). So the asymmetric induction that derives from the stereocenter of squalene epoxide is irrelevant for the occurrence of stereocontrol in the formation of lanosterol. However, the existence of a "proper" versus a "false" absolute configuration determines whether enzymatic cyclization does or does not occur. In contrast, the squalene hopene cyclase only lets its substrate (squalene) attain a chair-chair-chair conformation (cf. Formulas **B** and **D** in Figure 14.13). A significant contribution to our very detailed knowledge of the enzymatic squalene oxide → lanosterol cyclization (Figure 14.12) and the enzymatic squalene → hopene cyclization (Figure 14.13) has been made by the many years of stereochemical studies and crystallographic analyses conducted by G. E. Schulz and E. J. Corey, respectively.

The enzymatic squalene oxide → lanosterol cyclization in Figure 14.12 is induced by protonation of the properly folded epoxide **B** (→ **D**), followed by three electrophilic additions, in each case of a *C* electrophile to the less alkylated carbon of the next neighboring C=C double bond (i.e., Markovnikov-selectively). There is still disagreement concerning the chronology of these additions. If the carbenium ion intermediate **C** has been obtained in this way, a Wagner–Meerwein rearrangement takes place (→ **E**), which is then followed by a fourth Markovnikov-selective ring closure reaction (→ **F**) and a series of four Wagner–Meerwein rearrangements.

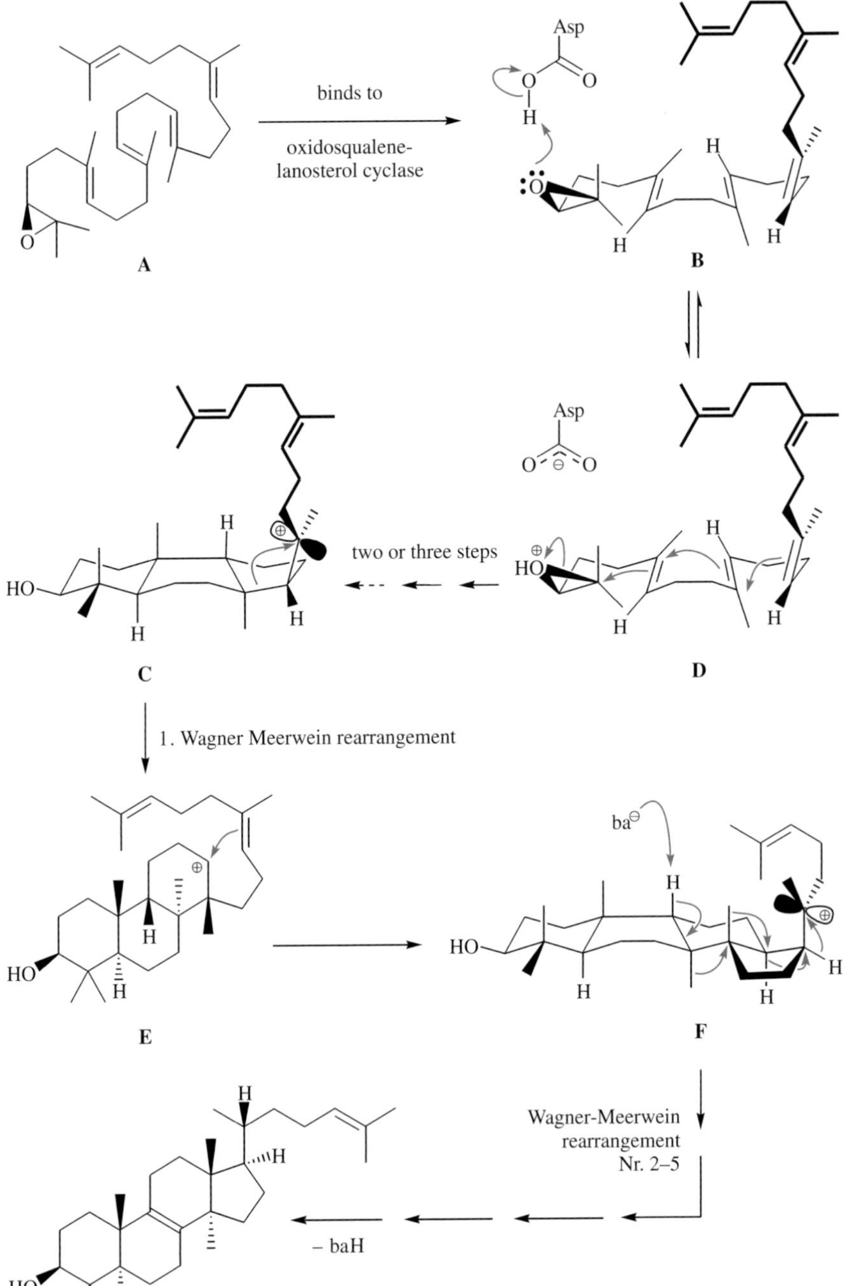

Fig. 14.12. Enzymatic transformation of acyclic squalene oxide (**A**) into tetracyclic lanosterol (**G**). The oxidosqualene-lanosterol cyclase controls the conformation of the substrate so effectively that only one out of 64 possible diastereomers is formed.

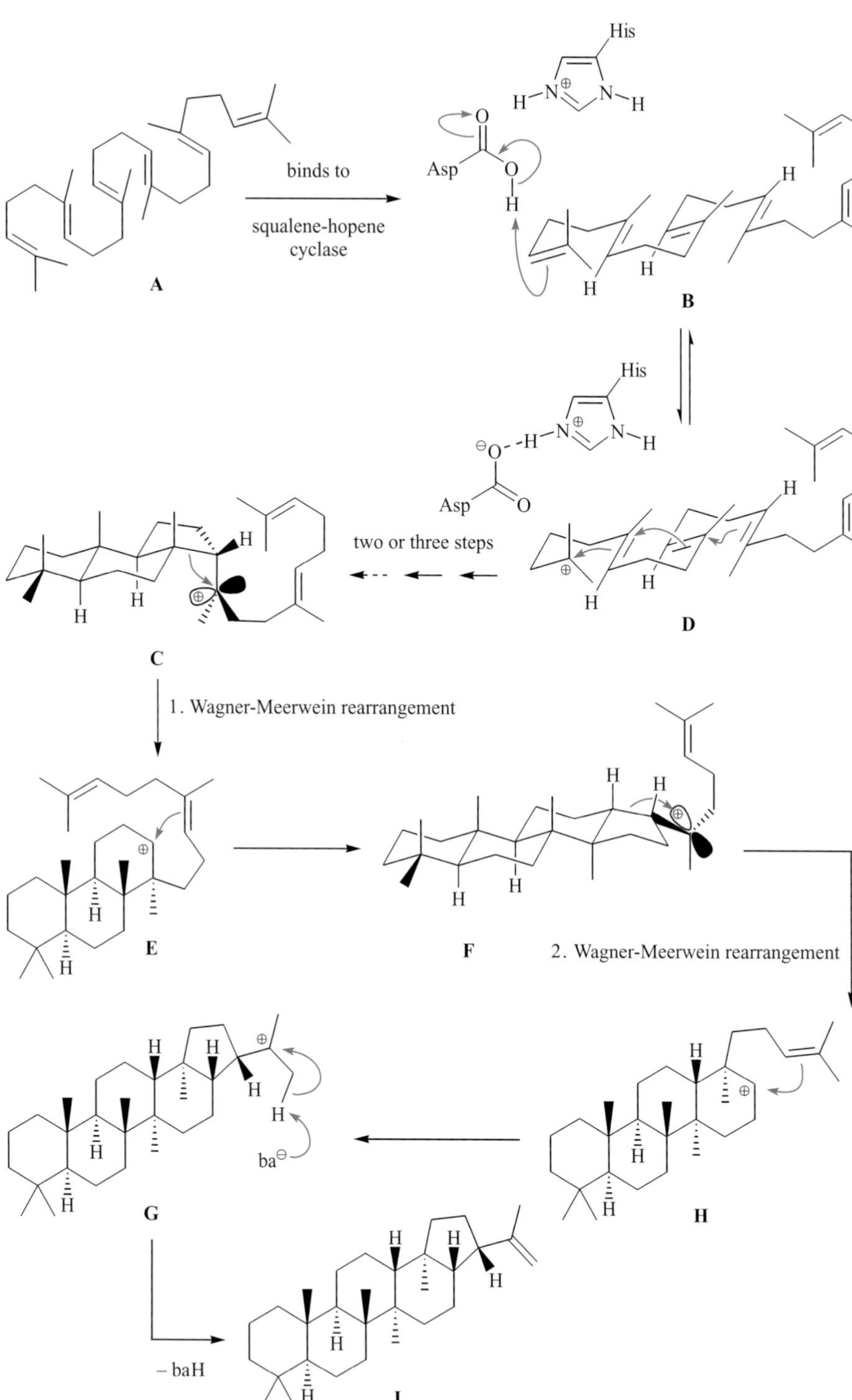

Fig. 14.13. Enzymatic transformation of acyclic squalene (**A**) into pentacyclic hopene (**I**). The squalene-hopene cyclase controls the conformation of the substrate so effectively that only one out of 512 possible stereoisomers is formed.

The carbenium ion formed last is deprotonated in the last step of this biosynthesis, thus furnishing the central C=C double bond of the final product lanosterol (**G**).

The enzymatic squalene → hopene cyclization in Figure 14.13 begins with the protonation of the properly folded substrate **B** to give the carbenium ion **D**. Three Markovnikov-selective ring closure reactions follow, and again we do not know whether these take place simultaneously or one after the other. In the carbenium ion **C**, a Wagner–Meerwein rearrangement (→ **E**) occurs, which is followed by the fourth Markovnikov-selective ring closure reaction (→ **F**), the second Wagner–Meerwein rearrangement (→ **H**), the fifth Markovnikov-selective ring closure (→ **G**), and finally the deprotonation to hopene (**I**).

Pinacol Rearrangement

Di-*tert*-glycols rearrange in the presence of acid to give α-tertiary ketones (Figure 14.14). The trivial name of the simplest glycol of this type is **pinacol**, and this type of reaction therefore is named pinacol rearrangement (in this specific case, the reaction is called a pinacol-pinacolone rearrangement). The rearrangement involves four steps. One of the hydroxyl groups is protonated in the first step. A molecule of water is eliminated in the second step, and a tertiary carbenium ion is formed. The carbenium ion rearranges in the third step into a more stable carboxonium ion via a [1,2]-rearrangement. In the last step, the carboxonium ion is deprotonated and the product ketone is obtained.

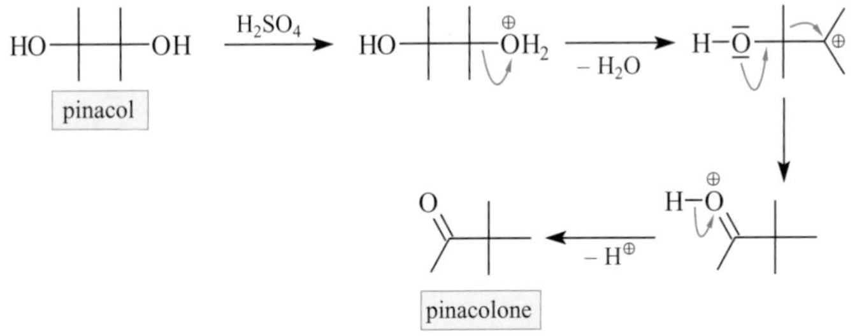

HO–pinacol–OH →(H_2SO_4) HO–…–$OH_2^{\oplus}$ →($-H_2O$) H–Ō–…$^{\oplus}$ → H–$\overset{\oplus}{O}$=… →($-H^{\oplus}$) O=… (pinacolone)

Fig. 14.14. Mechanism of the pinacol rearrangement of a symmetric glycol. The reaction involves the following steps: (1) protonation of one of the hydroxyl groups, (2) elimination of one water molecule, (3) [1,2]-rearrangement, and (4) deprotonation.

For *unsymmetrical* di-*tert*-glycols to rearrange under the same conditions to a single ketone, steps 2 and 3 of the overall reaction (Figure 14.14) must proceed chemoselectively. Only one of the two possible carbenium ions can be allowed to form and, once it is formed only one of the neighboring alkyl groups can be allowed to migrate. Product development control ensures the formation of the more stable carbenium ion in step 2. In the rearrangement depicted in Figure 14.15, the more stable cation is the benzhydryl cation **B** rather than the tertiary alkyl cation **D**.

The pinacol rearrangement **E** → **F** (Figure 14.16) can be carried out in an analogous fashion for the same reason. As with the reaction shown in Figure 14.15, this reaction proceeds exclusively via a benzhydryl cation. In a crossover experiment (see Section 2.4.3 regarding the "philosophy" of crossover experiments), one rearranged a mixture of di-*tert*-glycols **A** (Figure 14.15) and **E** (Figure 14.16) under acidic conditions. The reaction products were the α-tertiary cations **C** and **F**, already encountered in the respective reactions conducted sepa-

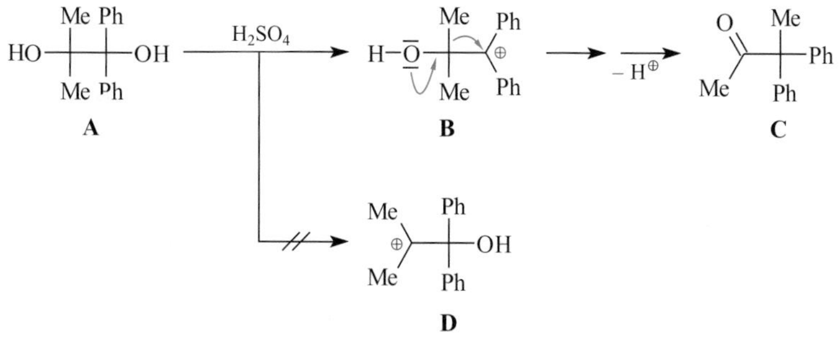

Fig. 14.15. Regioselectivity of the pinacol rearrangement of an unsymmetrical glycol. The more stable carbenium ion is formed under product development control. Thus, the benzhydryl cation **B** is formed here, while the tertiary alkyl cation **D** is not formed.

rately. The absence of the crossover products **G** and **H** proves the intramolecular nature of the pinacol rearrangement.

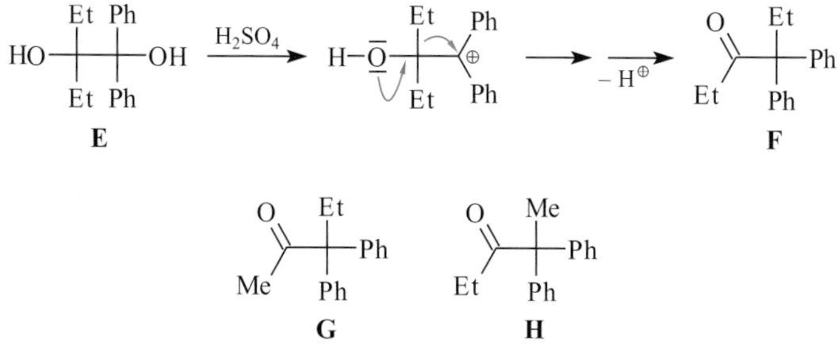

Fig. 14.16. Regioselective pinacol rearrangement of an asymmetric glycol. As in the case depicted in Figure 14.15, the reaction proceeds exclusively via the benzhydryl cation.

Semipinacol Rearrangements

Rearrangements that are not pinacol rearrangements but also involve a [1,2]-shift of an H atom or of an alkyl group from an *oxygenated* C atom to a neighboring C atom, that is, a carbenium ion → carboxonium ion rearrangement, are called semipinacol rearrangements. As shown later, however, there also are some semipinacol rearrangements that proceed without the intermediacy of carboxonium ions (Figures 14.20 and 14.22–14.24).

Lewis acids catalyze the ring opening of epoxides. If the carbenium ion that is generated is not trapped by a nucleophile, such an epoxide opening initiates a semipinacol rearrangement (Figures 14.17 and 14.18). Epoxides with different numbers of alkyl substituents on their ring C atoms are ideally suited for such rearrangements. In those cases, because of product

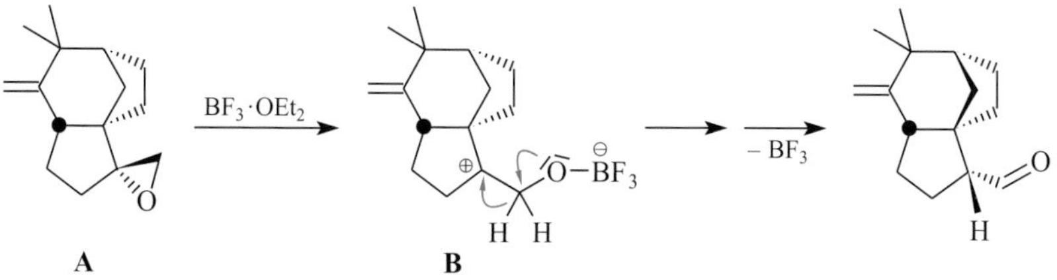

Fig. 14.17. "Accidental" diastereoselectivity in the semipinacol rearrangement of an epoxide. The more substituted carbenium ion is formed exclusively during ring-opening because of product development control. Only two H atoms are available for possible migrations, and no alkyl groups. In general, diastereoselectivity may or may not occur, depending on which one of the diastereotopic H atoms migrates in which one of the diastereotopic conformers. The present case exhibits diastereoselectivity.

development control, only one carbenium ion is formed in the ring-opening reaction. It is the more alkylated carbenium ion, hence the more stable one. There is only an H atom and no alkyl group available for a [1,2]-migration in the carbenium ion **B** in Figure 14.17. It is again an H atom that migrates in the carbenium ion **B** formed from the epoxide **A** in Figure 14.18. A carbenium ion with increased ring strain would be formed if an alkyl group were to migrate instead.

The [1,2]-rearrangements shown in Figures 14.17 and 14.18 are stereogenic and proceed stereoselectively (which is not true for many semipinacol rearrangements). In [1,2]-rearrangements, the migrating H atom always is connected to the same face of the carbenium ion from which it begins the migration. It follows that the [1,2]-shift in the carbenium ion **B** of Figure 14.18 *must* proceed stereoselectively. This is because the H atom can begin its migration only on one side of the carbenium ion. On the other hand, an H atom can in principle migrate on either side of the carbenium ion plane of carbenium ion **B** of Figure 14.17. However, only the migration on the top face (in the selected projection) does *in fact* occur.

The semipinacol rearrangements of Figures 14.17 and 14.18 are [1,2]-rearrangements in which the target atoms of the migrations are the higher alkylated C atoms of the 1,2-dioxygenated (here: epoxide) substrates. The opposite direction of migration—toward the less alkylated C atom of a 1,2-dioxygenated substrate—can be realized in *sec,tert*-glycols. These glycols can be toslyated with tosyl chloride at the less hindered secondary OH group. Glycol monosulfonates of the types shown in Figures 14.19 and 14.20 can be obtained in this way. With these glycol monosulfonates, semipinacol rearrangements with a reversed direction of the migration can be carried out.

Glycol monosulfonates of the type discussed eliminate a tosylate under solvolysis conditions, as illustrated in Figure 14.19. Solvolysis conditions are achieved in this case by carrying out the reaction in a solution of $LiClO_4$ in THF. Such a solution is more polar than water! The solvolysis shown first leads to a carbenium ion **A.** At this point, a rearrangement into a carboxonium ion (and subsequently into a ketone) could occur via the migration of a primary alkyl group or of an alkenyl group. The migration of the alkenyl group is observed exclusively. Alkenyl groups apparently possess a higher intrinsic propensity for migration than alkyl groups.

Figure 14.20 depicts a semipinacol rearrangement that is initiated in a different manner. This reaction proceeds *without* the occurrence of a sextet intermediate. We want to discuss this reaction here because this process provides synthetic access to molecules that are rather similar to molecules that are accessible via "normal" semipinacol rearrangements. The glycol monotosylate **A** is deprotonated by KO*tert*-Bu to give alkoxide **B** in an equilibrium reaction. Under these conditions other glycol monotosylates would undergo a ring closure delivering the epoxide. However, *this* compound cannot form an epoxide because the alkoxide O atom is incapable of a backside reaction on the C—OTs bond. Hence, the tosylated alkoxide **B** has the

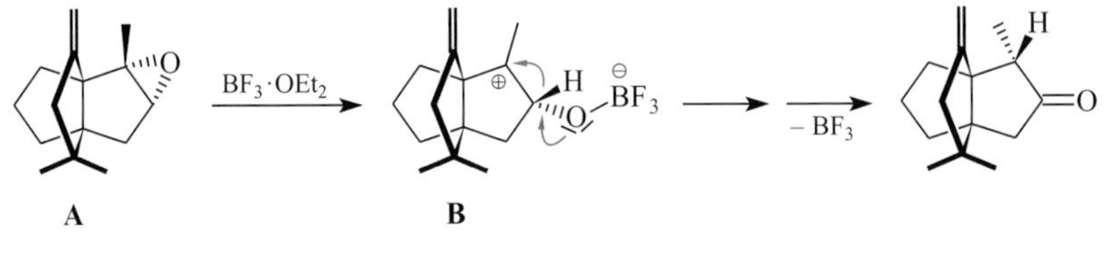

Fig. 14.18. Mechanism-based diastereoselectivity in the semipinacol rearrangement of an epoxide. This rearrangement is stereoselective, since there is only one H atom in the position next to the sextet center and the H atom undergoes the [1,2]-migration on the same face of the five-membered ring.

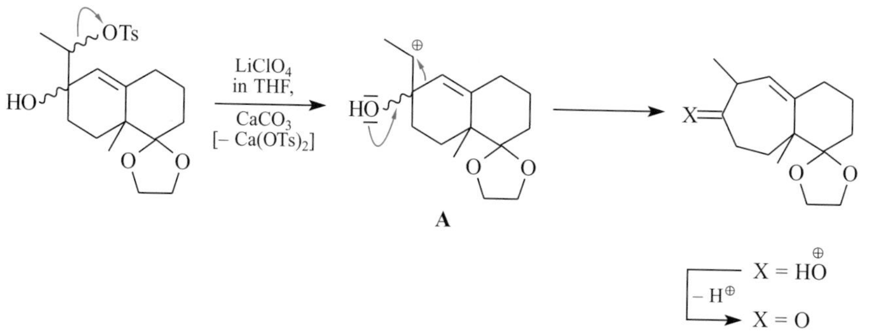

Fig. 14.19. First semipinacol rearrangement of a glycol monotosylate. The reaction involves three steps in neutral media: formation of a carbenium ion, rearrangement to a carboxonium ion, and deprotonation to the ketone.

opportunity for a [1,2]-alkyl shift to occur with *concomitant* elimination of the tosylate. This [1,2]-shift occurs stereoselectively in such a way that the C—OTs bond is broken by a backside reaction of the migrating alkyl group. Stereoelectronics control the migration. The bond antiperiplanar to the leaving group migrates.

Other leaving groups in other glycol derivatives facilitate other semipinacol rearrangements. Molecular nitrogen, for example, is the leaving group in

- the Tiffeneau–Demjanov rearrangement of diazotized amino alcohols—a reaction that is of general use for the ring expansion of cycloalkanones to their next higher homologs (for example, see Figure 14.21),
- analogous ring expansions of cyclobutanones to cyclopentanones (for example, see Figure 14.22), or
- the ring expansion of cyclic ketones to carboxylic esters of the homologated cycloalkanone (for example, see Figure 14.23); each of these ring expansions leads to one additional ring carbon, as well as
- a chain-elongating synthesis of β-ketoesters from aldehydes and diazomethyl acetate (Figure 14.24).

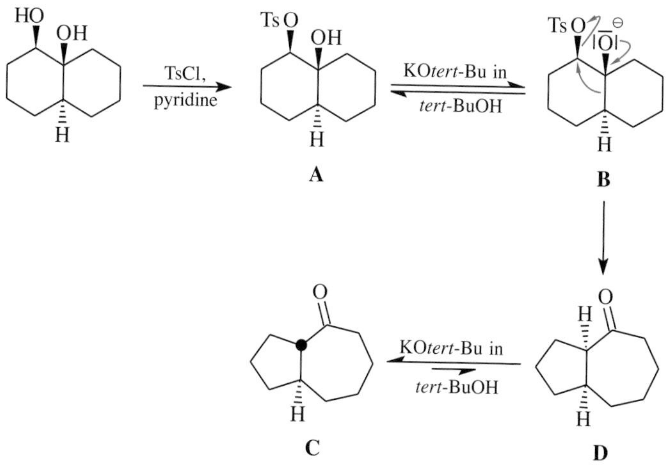

Fig. 14.20. Second semipinacol rearrangement of a glycol monotosylate. The reaction involves two steps in basic media, since the [1,2]-rearrangement and the dissociation of the tosylate occur at the same time. Under reaction conditions this rearrangement is followed by a base-catalyzed epimerization. This is why instead of the initially *cis*-annulated bicyclic ketone **D** its *trans*-isomer **C** is isolated.

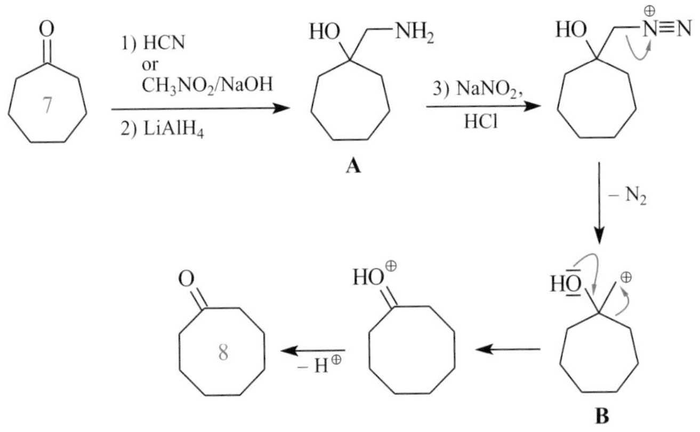

Fig. 14.21. Ring expansion of cyclic ketones via the Tiffeneau–Demjanov rearrangement. The first step consists of the additions of HCN or nitromethane, respectively, to form either the cyanohydrin or the *β*-nitroalcohol, respectively. The vicinal amino alcohol **A** is formed in the next step by reduction with $LiAlH_4$. The Tiffeneau–Demjanov rearrangement starts after diazotation with the dediazotation.

Figure 14.21 shows how the **Tiffeneau–Demjanov rearrangement** can be employed to insert an additional CH_2 group into the ring of cyclic ketones. Two steps are required to prepare the actual substrate of the rearrangement. A nitrogen-containing C_1 nucleophile is added to the substrate ketone. This nucleophile is either HCN or nitromethane, and the addition yields a cyanohydrin or a *β*-nitroalcohol, respectively. Both these compounds can be reduced with lithium aluminum hydride to the vicinal amino alcohol **A.** The Tiffeneau–Demjanov rearrangement of the amino alcohol is initiated by a diazotation of the amino group. The diazotation is achieved either with sodium nitrite in aqueous acid or with isoamyl nitrite in the absence of acid and water. The mechanism of these reactions corresponds to the usual preparation of aryldiazonium chlorides from anilines (Figure 17.40, top) or to a variation thereof.

Aliphatic diazonium salts are much less stable than aromatic diazonium salts (but even the latter tend to decompose when isolated!). The first reason for this difference is that aliphatic diazonium salts, in contrast to their aromatic counterparts, lack stabilization through resonance. Second, aliphatic diazonium salts release N_2 much more readily than their aromatic analogs since they thus react to give relatively stable alkyl cations. Hence, the decomposition of aliphatic diazonium ion is favored by product development control. Aromatic diazonium salts, on the other hand, form phenyl cations upon releasing N_2 and these are even less stable than alkenyl cations. The electron-deficient carbon atom in an alkenyl cation can be stabilized, at least to a certain degree, because of its linear coordination (which the phenyl cation cannot adopt). This can be rationalized with the MO diagrams of Figure 1.4. The orbital occupancy of a bent carbenium ion $=C^{\oplus}—R$ resembles that of the bent carbanion $=C^{\ominus}—R$ except that the n_{sp^2} orbital remains empty in the former. Nevertheless, even linear alkenyl cations are less stable than alkyl cations.

The molecular nitrogen in aliphatic diazonium ions is an excellent leaving group. In fact, nitrogen is eliminated from these salts so fast that an external nucleophile does not assist in the nucleophilic displacement of molecular nitrogen. Only an *internal* nucleophile, that is, a neighboring group, can provide such assistance in displacing nitrogen (example in Figure 2.28). Therefore, aliphatic diazonium ions *without* neighboring groups always form carbenium ions.

Unfortunately, carbenium ions often undergo a variety of consecutive reactions and yield undesired product mixtures. The situation changes significantly if the carbenium ion carries a

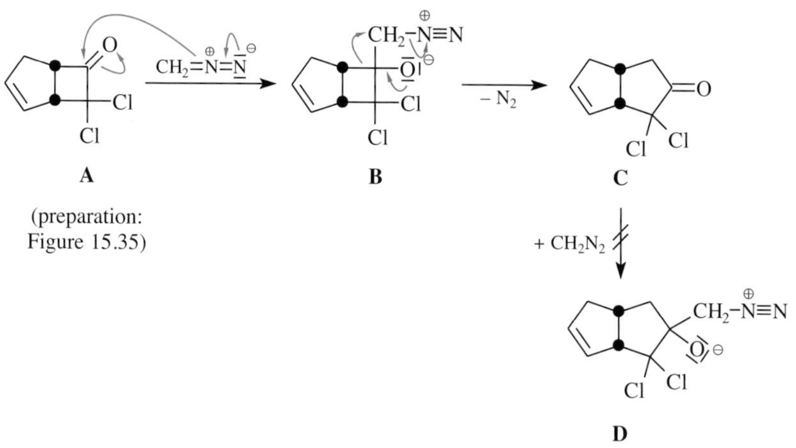

Fig. 14.22. Ring expansion of cyclobutanones. The cyclobutanones are accessible via [2+2]-cycloadditions of dichloroketene (for the mechanism, see Section 15.4).

hydroxyl group in the *β*-position or if the diazonium salt contains an $O^{\ominus}$, an $OBF_3^{\ominus}$, or an $OSnCl_2^{\ominus}$ substituent in the *β*-position with respect to the $N_2^{\oplus}$ group. The first of these structural requirements is fulfilled in the carbenium ion intermediate **B** of the Tiffeneau–Demjanov rearrangement of Figure 14.21; it contains a hydroxyl group in the *β*-position. The diazonium ion intermediate **B** of the ring-expansion of Figure 14.22 contains an $O^{\ominus}$ substituent in the *β*-position, that of Figure 14.23 an $OBF_3^{\ominus}$ substituent. The diazonium ion intermediate **A** of Figure 14.24 carries an $OSnCl_2^{\ominus}$ substituent in the *β*-position. The aforementioned O substituents in the *β*-positions of the diazonium ions and of the resulting carbenium ions allow for [1,2]-rearrangements that generate a carboxonium ion (Figure 14.21) or a ketone (Figures 14.22–14.24), respectively. Accordingly, a favorable all-octet species is formed in both cases, whether charged or neutral.

Let us take a closer look at the ring expansion of Figure 14.22. Cyclobutanones like **A** initially add diazomethane, which is a C nucleophile, and a tetrahedral intermediate **B** is formed. The cyclopentanone **C** is obtained by dediazotation and concomitant regioselective [1,2]-rearrangement. This rearrangement does not belong in the present section in the strictest sense, since the rearrangement **B** → **C** is a one-step process and therefore does not involve a sextet intermediate. Nevertheless, the reaction is described here because of its close similarity to the Tiffeneau–Demjanov rearrangement of Figure 14.21.

The tetrahedral intermediate **B** of Figure 14.22 does not build up because the ring expansion reaction that it undergoes is at least as fast as its formation. Once the *first* ring-expanded ketone is formed, there obviously still is plenty of unreacted substrate ketone **A** present. The question is now: which ketone reacts faster with diazomethane? The answer is that the cyclobutanone reacts faster with diazomethane because of product development control: the formation of the tetrahedral intermediate **B** results in a substantial reduction of the ring strain in the four-membered ring because of the rehybridization of the carbonyl carbon from sp^2 to sp^3. As is well known, carbon atoms with sp^2 hybridization prefer 120° bond angles, while sp^3-hybridized carbon atoms prefer tetrahedral bond angles (109°28'). Hardly any ring strain would be relieved in the addition of diazomethane to the newly formed cyclopentanone **C.** It thus follows that the ring expansion of the cyclobutanone occurs fast and proceeds to completion before the product cyclopentanone **C** in turn under goes its slower ring expansion via the tetrahedral intermediate **D.**

This line of argument also explains why most other cyclic ketones do not undergo chemoselective ring expansions with diazomethane. In the general case, both the *substrate* ketone and the *product* ketone would be suitable reaction partners for diazomethane, and multiple, i.e., consecutive ring expansions could not be avoided. Therefore, it is worth remembering that the Tiffeneau–Demjanov rearrangement of Figure 14.21 shows how to accomplish a ring expansion of *any* cycloalkanone by exactly one CH_2 group.

A process is shown in Figure 14.23 that allows for the ring expansion of any cycloalkanone by exactly one C atom, too, using ethyl diazoacetate. Ethyl diazoacetate is a relatively weak nucleophile because of its CO_2R substituent, and its addition to unstrained cycloalkanones is possible only in the presence of $BF_3 \cdot OEt_2$. In that case, the electrophile is the ketone–BF_3 complex **A.** The tetrahedral intermediate **B** formed from **A** and diazoacetic acid ethyl ester also is a diazonium salt. Thus it is subject to a semipinacol rearrangement. As in the cases described in Figures 14.20 and 14.22, this semipinacol rearrangement occurs *without* the intermediacy of a carbenium ion. A carbenium ion would be greatly destabilized because of the close proximity of a positive charge and the CO_2R substituent in the α-position.

The diazonium salt **B** of Figure 14.23 contains a tertiary and a primary alkyl group in suitable positions for migration. In contrast to the otherwise observed intrinsic migratory trends, only the primary alkyl group migrates in this case, and the β-ketoester **E** is formed by the rearrangement. Interestingly, the product does not undergo further ring expansion. This is because the Lewis acid BF_3 catalyzes the enolization of the keto group of **E**, and BF_3 subsequently complexes the resulting enol at the ester oxygen to yield the aggregate **C.** In contrast to the BF_3 complex **A** of the unreacted substrate, this species **C** is not a good electrophile. Hence, only the original ketone continues to react with the diazoacetic acid ethyl ester.

The cyclic β-ketoester **E** subsequently can be saponified in acidic medium. The acid obtained then decarboxylates via the mechanism of Figure 13.27 to provide the ester-free cycloalkanone **D.** Product **D** represents the product of a CH_2 insertion into the starting ketone.

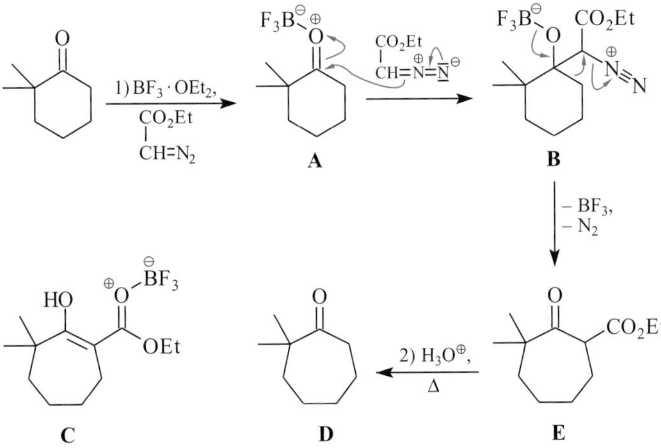

Fig. 14.23. Ring expansion of a cyclohexanone via addition of diazoacetic acid ethyl ester and subsequent [1,2]-rearrangement.

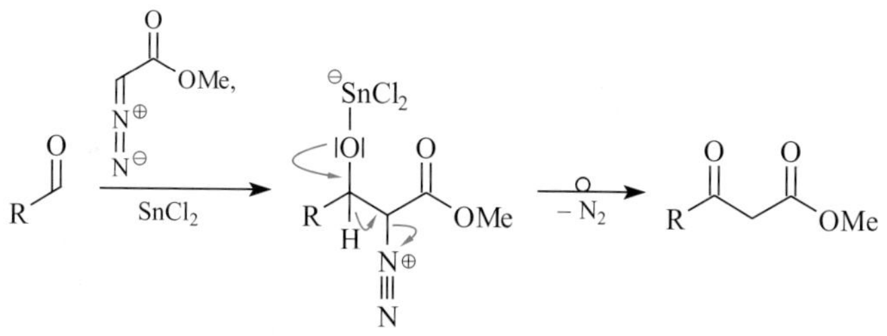

Fig. 14.24. C_2 extension of aldehydes to β-ketoesters via a semipinacol rearrangement.

It would be impossible to obtain this product with diazomethane. In Section 14.3.2 (Figure 14.25), a third method will be described for the insertion of a CH_2 group into a cycloalkanone. Again, a [1,2]-rearrangement will be part of that insertion reaction.

Finally, an interesting C_2 elongation of aldehydes to β-ketoesters is presented in Figure 14.24. This elongation reaction involves a semipinacol rearrangement that occurs in complete analogy to the one shown in Figure 14.23. The differences are merely that the Lewis acid $SnCl_2$ is employed instead of BF_3 in the first step, and an H atom undergoes the [1,2]-migration instead of an alkyl group.

14.3.2 [1,2]-Rearrangements in Carbenes or Carbenoids

A Ring Expansion of Cycloalkanones

Figure 14.25 shows how the ring expansion of cyclic ketones can be accomplished without the liberation of molecular nitrogen (in contrast to the ring expansions of Figures 14.21–14.23). A chemoselective **mono**insertion of CH_2 occurs because the product ketone is never exposed to the reaction condition to which the substrate ketone is subjected. This is a similarity between the present method and the processes described in Figures 14.21 and 14.23, and this feature is in contrast to the method depicted in Figure 14.22.

In the first step of the reaction shown in Figure 14.25, CH_2Br_2 is deprotonated by LDA and the organolithium compound Li-$CHBr_2$ is formed. This reagent adds to the C=O double bond

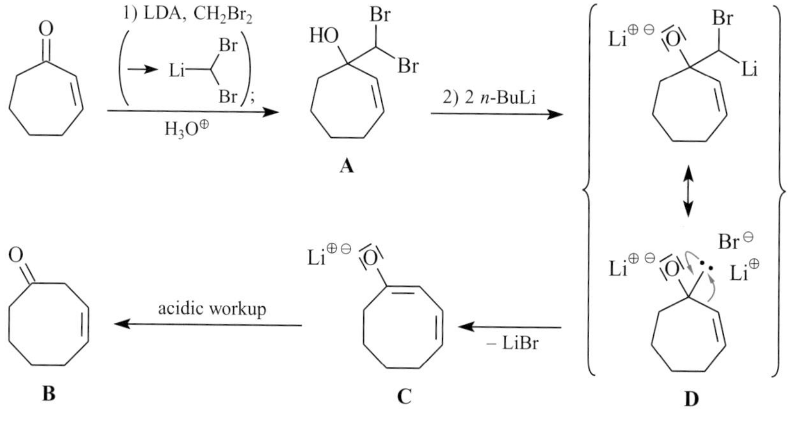

Fig. 14.25. Ring expansion of cycloheptenone via a carbenoid intermediate. The elimination of LiBr from the carbenoid occurs with or is followed by a [1,2]-alkenyl shift. The enolate **C** is formed and, upon aqueous workup, it is converted to the ring-expanded cycloalkenone **B.**

of the ketone substrate and forms an alkoxide. The usual acidic work up yields the corresponding alcohol **A.** The alcohol group of alcohol **A** is deprotonated with one equivalent of *n*-BuLi in the second step of the reaction. A bromine/lithium exchange (mechanism in Figure 16.16, top row) is accomplished in the resulting alkoxide with another equivalent of *n*-BuLi. The resulting organolithium compound **D** is a carbenoid. As discussed in Section 3.3.2, a carbenoid is a species whose reactivity resembles that of a carbene even though there is no free carbene involved. In the VB model, one can consider carbenoid **D** to be a resonance hybrid between an organolithium compound and a carbene associated with LiBr.

The elimination of LiBr from this carbenoid is accompanied or followed (the timing is not completely clear) by a [1,2]-rearrangement. The alkenyl group presumably migrates faster than the alkyl group, as in the case of the semipinacol rearrangement of Figure 14.19. The primary product most probably is the enolate **C**, and it is converted into the ring-expanded cyclooctanone **B** upon aqueous workup. The C=C double bond in **B** is not conjugated with the C=O double bond. This can be attributed to a kinetically controlled termination of the reaction. If thermodynamic control had occurred, some 20% of the unconjugated ketone would have isomerized to the conjugated ketone. Why not more? The formation of no *more* than 20% of the conjugated ketone is due to a medium-sized ring effect. Aligning the carbonyl and the double bond pi bonds (conjugation) would increase ring strain. On the other hand, 3-cyclohexenone contains a smaller ring and it would be converted completely to 2-cyclohexenone under equilibrium conditions.)

Wolff Rearrangement

Wolff rearrangements are rearrangements of α-diazoketones leading to carboxylic acid derivatives via ketene intermediates. Wolff rearrangements can be achieved with metal catalysis or photochemically. As Figure 14.26 shows, the α-diazoketone **D** initially loses a nitrogen molecule and forms a ketene **G.** Heteroatom-containing nucleophiles add to the latter in uncatalyzed reactions (see Figure 8.12 for the mechanism). These heteroatom-containing nucleophiles must be present during the ketene-forming reaction because only if one traps the ketenes immediately in this way can unselective consecutive reactions be avoided. Thus, after completion of the Wolff rearrangement, only the addition products of the transient ketenes are isolated. These addition products are carboxylic acid derivatives (cf. Section 8.2 and the keyword "Arndt–Eistert reaction"). In the presence of water, alcohols, or amines, the Wolff rearrangement yields carboxylic acids, carboxylic esters, or carboxylic amides, respectively.

Let us consider the mechanistic details of the Wolff rearrangement (Figure 14.26). If the rearrangement is carried out in the presence of catalytic amounts of silver(I) salts, the dediazotation of the α-diazoketone initially generates the ketocarbene **E** and/or the corresponding ketocarbenoid **F.** A [1,2]-shift of the alkyl group R^1 that stems from the acyl substituent R^1—C=O of the carbene or the carbenoid follows. This rearrangement converts each of the potential intermediates **E** and **F** into the ketene **G.** The same ketene **G** is obtained if the Wolff rearrangement is initiated photochemically. In this case, molecular nitrogen and a ketocarbene are formed initially. The ketocarbene (**E**) can then undergo the [1,2]-rearrangement discussed earlier. On the other hand, excited ketocarbenes **C** occasionally rearrange into an isomeric ketocarbene **B** via an *anti*-aromatic oxirene intermediate **A.** In that case the [1,2]-rearrange-

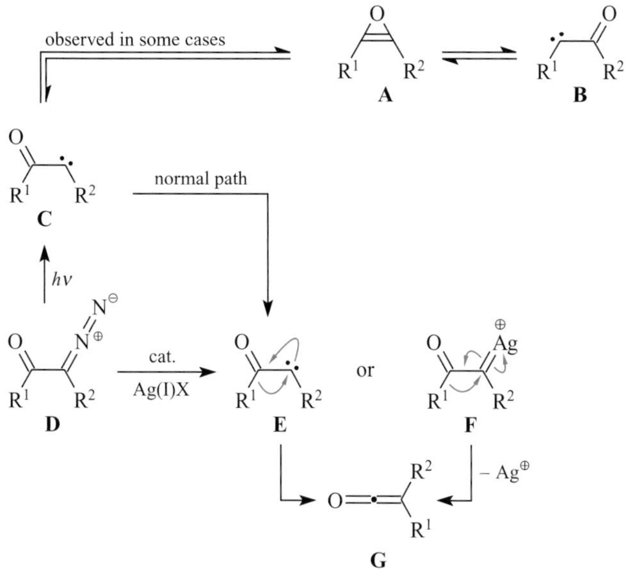

Fig. 14.26. Mechanisms of the photochemically initiated and Ag(**I**)-catalyzed Wolff rearrangements with formation of the ketocarbene **E** and/or the ketocarbenoid **F** by dediazotation of the diazoketene **D** in the presence of *catalytic amounts of Ag(I)*. **E** and **F** are converted into **G** via a [1,2]-shift of the alkyl group R^1. N_2 and a carbene **C** are formed in the *photochemically initiated* reaction. The carbene **E** continues to react to give **G.** The ketocarbene **C** may on occasion isomerize to **B** via an oxacyclopropene **A.** The [1,2-]-shift of **B** also leads to the ketene **G.**

ment would occur in **E** and in **B** as well or in **B** alone. The same ketene **G** is formed in any case.

The Wolff rearrangement is the third step of the Arndt–Eistert homologation of carboxylic acids. Figure 14.27 picks up an example that was discussed in connection with Figure 8.13, that is, the homologation of trifluoroacetic acid to trifluoropropionic acid. The first step of the Arndt–Eistert synthesis consists of the activation of the carboxylic acid via the acid chloride. The C_1 elongation to an α-diazoketone occurs in the second step.

If an alkyl group that migrates in a Wolff rearrangement contains a stereocenter at the C-α atom, the migration of the alkyl group proceeds with retention of configuration. An example of a reaction that allows one to recognize this stereochemical propensity is provided by the double Wolff rearrangement depicted in Figure 14.28. The bis(diazoketone) **A**, a *cis*-disubstituted cyclohexane, is the substrate of the rearrangement. The bisketene, which cannot be isolated, must have the same stereochemistry, because the dimethyl ester **B** formed from the bisketene by *in situ* addition of methanol still is a *cis*-disubstituted cyclohexane.

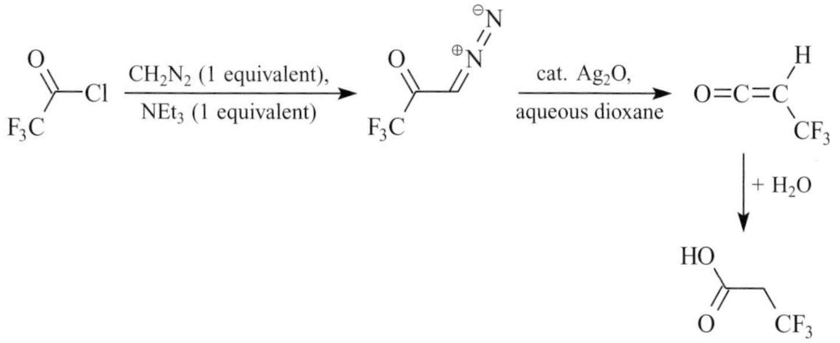

Fig. 14.27. Wolff rearrangement as the third step in the Arndt–Eistert homologation of carboxylic acids. This example shows the homologation of trifluoroacetic acid to trifluoropropionic acid. The conversion of the acid into the acid chloride is the first step (not shown).

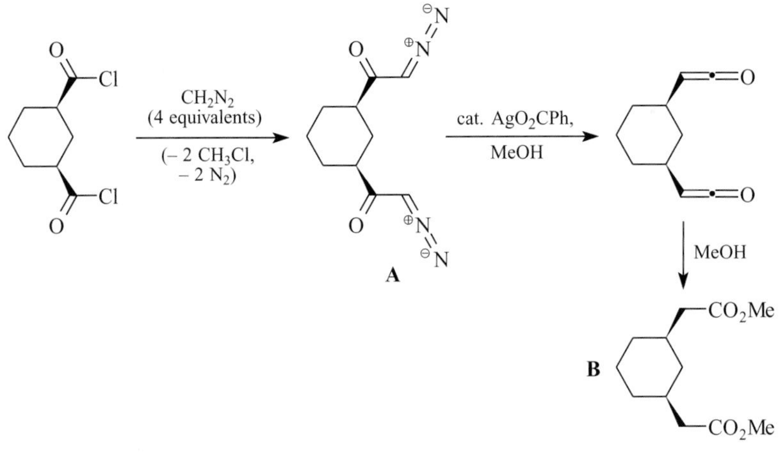

Fig. 14.28. Twofold Wolff rearrangement in the bishomologation of dicarboxylic acids according to Arndt and Eistert. Both alkyl group migrations occur with retention of configuration.

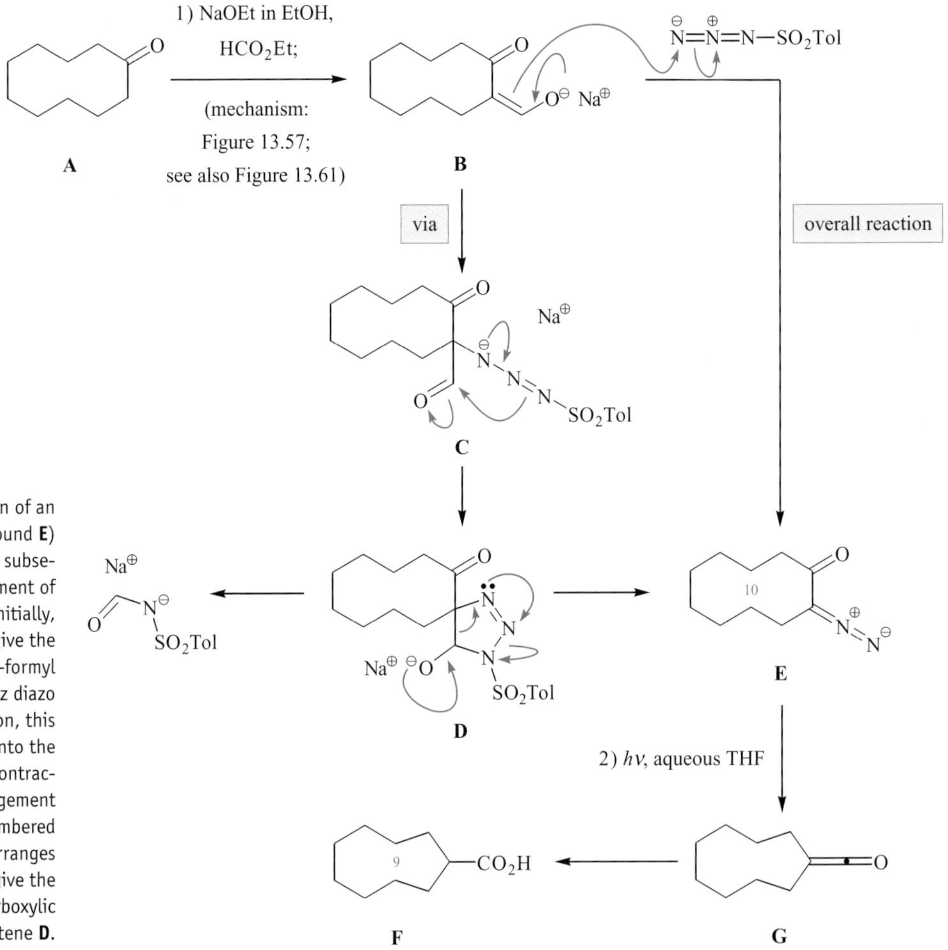

Fig. 14.29. Preparation of an α-diazoketone (compound **E**) from a ketone (**A**) and subsequent Wolff rearrangement of the α-diazoketone. Initially, **A** is transformed to give the enolate **B** of its α-formyl derivative. In a Regitz diazo group transfer reaction, this will then be converted into the α-diazoketone **E**. Ring contraction via Wolff rearrangement occurs and the 10-membered cyclic diazoketone **C** rearranges in aqueous media to give the nine-membered ring carboxylic acid **E** via the ketene **D**.

Cycloalkanones can be converted into the enolate of an α-formyl ketone by reaction with ethyl formate and one equivalent of sodium ethoxide (Figure 13.61). Such a reaction is shown in Figure 14.29 as transformation **A** → **B.** This reaction sets the stage for a diazo group transfer reaction. Depending on which enolate is used, diazo group transfer reactions may follow two different mechanisms. With a tertiary enolate carbon the mechanism given in Figure 14.29 applies; with a secondary enolate carbon the reaction proceeds via the mechanism shown in Figure 15.42. The sulfonamide anion, which is released in the course of these reactions, will be formylated via the mechanism of Figure 14.29, and not by the one depicted in Figure 15.42. At the end of the diazo group transfer in Figure 14.29, the cyclic diazoketone **E** has been formed. This compound can be converted into ketene **G** *via* a photochemical Wolff rearrangement. If this is done in an aqueous medium, the ketene hydrolyzes *in situ* to give the carboxylic acid **F**, which possesses a nine-membered ring, while its precursor **A** was a ten-membered ring. Hence, these kinds of Wolff rearrangements are ring-contraction reactions.

Aldehyde → Alkyne Elongation via Carbene and Carbenoid Rearrangements

Figures 14.30–14.32 show a reaction and a sequence of reactions, respectively, that allow for the conversion of aldehydes into alkynes that contain one more C atom. These transformations involve a [1,2]-rearrangement.

The one-step **Seyferth procedure** is shown in Figure 14.30. The reaction begins with the Horner–Wadsworth–Emmons olefination of the aldehyde to the alkene **A.** The mechanism of this olefination is likely to resemble the mechanism earlier presented in Figure 11.13 for a reaction that involved a different phosphonate anion. As can be seen, alkene **A** is an unsaturated diazo compound. As soon as the reaction mixture is allowed to warm to room temperature, compound **A** eliminates molecular nitrogen and the vinyl carbene **B** is generated. A [1,2]-rearrangement of **B** forms the alkyne. It is presumably the H atom rather than the alkyl group that migrates to the electron-deficient center.

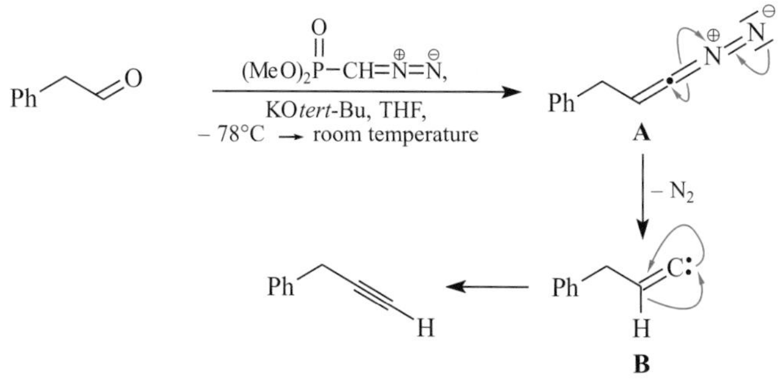

Ph⌒⌒O — $(MeO)_2P(=O)-CH=\overset{\oplus}{N}=\overset{\ominus}{N}$, KO*tert*-Bu, THF, – 78°C → room temperature → **A** — $-N_2$ → **B** → Ph⌒≡H

Fig. 14.30. Aldehyde → alkyne chain elongation via [1,2]-rearrangement of a vinyl carbene (Seyferth procedure). First, a Horner–Wadsworth–Emmons olefination of the aldehyde is carried out to prepare alkene **A.** Upon warming to room temperature, alkene **A** decomposes and gives the vinyl carbene **B.** From that, the alkyne is formed by way of a [1,2]-rearrangement.

Side Note 14.2. The Bestmann Modification of the Seyferth Reaction

Seyferth's diazo*methane* phosphonic acid dimethyl ester, which enables the aldehyde → alkyne elongation shown in Figure 14.30, is unstable. This is why Bestmann replaced it by a less sensitive synthetic equivalent, namely diazo *acetone* phosphonic acid dimethyl ester (Formula **C** in Figure 14.31), since with potassium methoxide—which is proportionately present in a solution/suspension of solid potassium carbonate in dry methanol—the diazophosphonic acid ester **C** may be converted *in situ* into Seyferth's diazophosphonic acid ester (Formula **D** in Figure 14.31) by a kind of retro-Claisen reaction. It proceeds via the tetrahedral intermediate **F** and the phosphonate anion **G**. The latter is just the Horner–Wadsworth–Emmons reagent of Seyferth's procedure.

The phosphonate anion **G** in Figure 14.31 causes the aldehyde → alkyne elongation **E** → **H**. In the context of this eight-step reaction sequence, this reaction has a key role in that it forms the C≡C triple bond of the *N*-Boc-protected amino acid ethynyl glycine (**I**). In this reaction sequence, the ethynyl glycine occurs as D-enantiomer and—in terms of configuration *and* constitution—is nonproteinogenic. Figure 14.31 contains all the cross-references you need to be able to follow the individual steps of this synthesis.

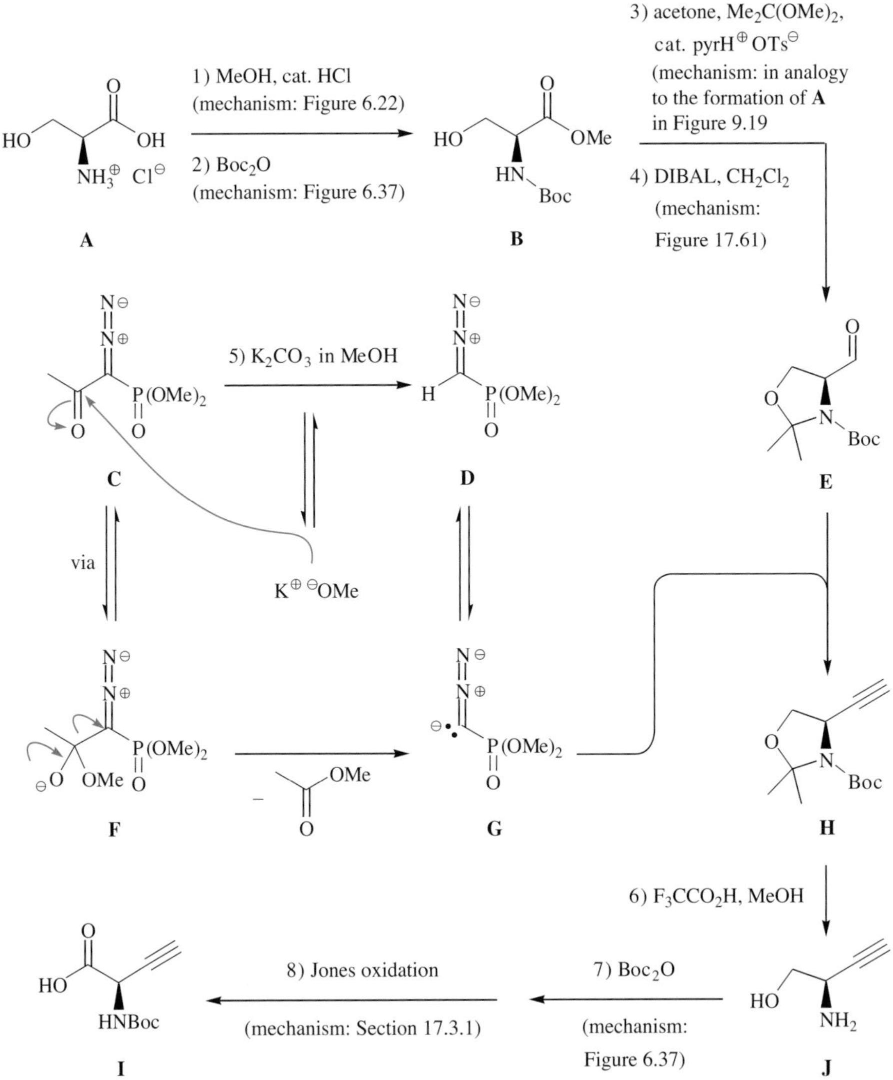

Fig. 14.31. Eight-step synthesis of D-configured N-Boc-ethynyl glycine (**I**) from L-serine (**A**). Here, the key step is the fifth step where a C_1 elongation aldehyde (**E**) → alkyne (**H**) takes place through [1,2]-rearrangement of a vinyl carbene (not shown), which is generated *in situ* from this aldehyde by way of the Bestmann modification of Seyferth's Horner–Wadsworth–Emmons reaction—i.e., starting from the diazoketophosphonate **C**. – Note: In the sixth step of the reaction sequence, the mixture of trifluoroacetic acid and methanol opens the ring, reversing the protection of step 3. However, the trifluoroacetic acid/methanol mixture also effects cleavage of the Boc group, which in this case is an undesired reaction.

The two-step **Corey–Fuchs procedure** offers an alternative aldehyde → alkyne elongation (Figure 14.32). In the first step, the dibrominated phosphonium ylide **A** is generated *in situ* by reaction of Ph_3P, CBr_4, and Zn. A Wittig reaction (for the mechanism, see Figure 11.2) between ylide **A** and an added aldehyde elongates the latter to give a 1,1-dibromoalkene (Formula **B** in Figure 14.32). In the second phase of the reaction, the 1,1-dibromoalkene is treated with two equivalents of *n*-BuLi. Then the *n*-BuLi first initiates a bromine/lithium exchange (for the mechanism, see the top row of Figure 16.16), which surprisingly—for a reason unknown—proceeds selectively at the sterically more hindered bromine atom, and generates the α-lithiated bromoalkene **D** instead of its isomer *iso*-**D.** Compound **D** is a carbenoid, as indicated by the resonance forms shown in Figure 14.32. It is unknown whether the carbenoid rearranges or whether the free carbene is formed prior to the rearrangement. It is known from analogous experiments, in which the carbene carbon was ^{13}C-labeled, that only an H atom migrates, not the alkyl group. The alkyne formed **(C)** is so acidic that it immediately reacts with a second equivalent of *n*-BuLi to give the corresponding lithium acetylide. Alkyne **C** is regenerated from the acetylide upon aqueous workup.

When the reaction of Figure 14.32 was carried out with less than two equivalents of *n*-BuLi, the alkyne **C** was deprotonated not only by *n*-BuLi but also by some of the carbenoid **D.** In this way, **D** was converted into a monobromoalkene, which could be isolated. This observation provided evidence that the reaction indeed proceeds via the carbenoid **D** and not by another path.

Fig. 14.32. Aldehyde → alkyne chain elongation via [1,2]-rearrangement of a vinyl carbenoid (Corey—Fuchs procedure). The aldehyde and phosphonium ylide **A** generated in situ undergo a Wittig olefination and form the 1,1-dibromoalkene (**B**). In the second stage, the dibromoalkene is reacted with two equivalents of n-BuLi and the vinyl carbenoid **D** is formed stereoselectively. The carbenoid undergoes H migration to form the alkyne **C**. The alkyne **C** reacts immediately with the second equivalent of n-BuLi to give the lithium acetylide and is reconstituted by reprotonation during aqueous workup.

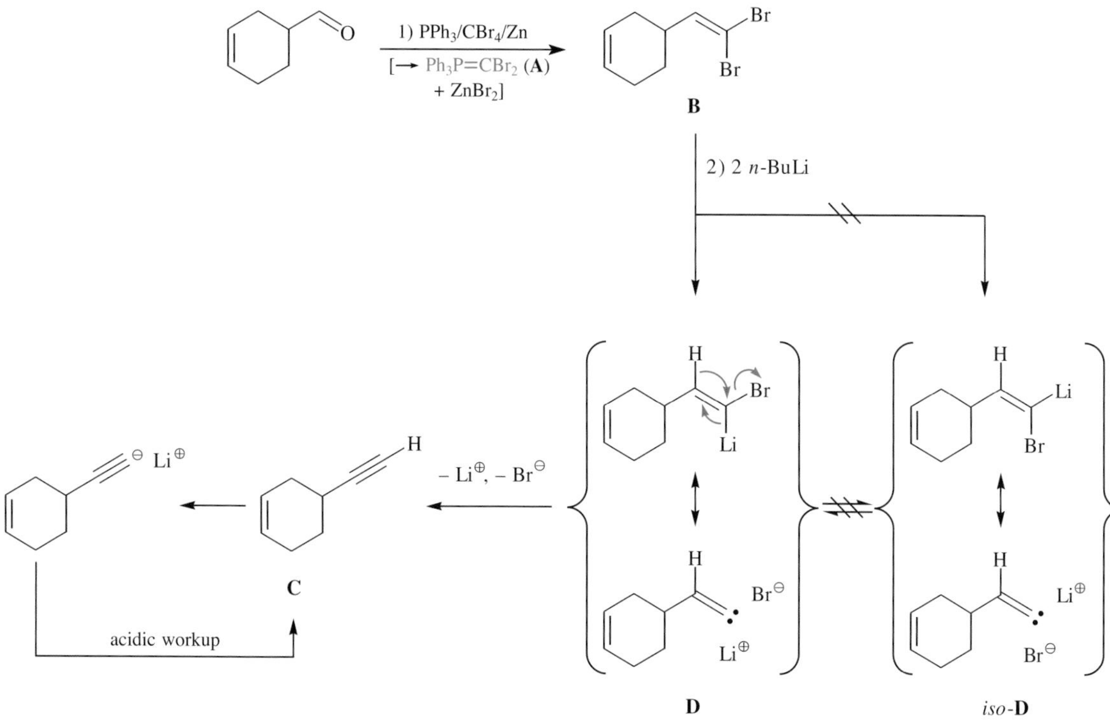

1) $PPh_3/CBr_4/Zn$ [→ $Ph_3P{=}CBr_2$ (**A**) + $ZnBr_2$]

B

2) 2 *n*-BuLi

$-Li^{\oplus}$, $-Br^{\ominus}$

acidic workup

C

D

iso-**D**

It is unclear why in the carbenoid **D** it is the H atom in *cis*-position relative to the bromine that undergoes migration, and not the cyclohexenyl residue, which is in *trans*-position to the bromine. In the **Fritsch–Buttenberg–Wiechell rearrangement** of the chlorine-containing vinyl carbenoid it is the aryl residue in *trans*-position to the chlorine atom that undergoes migration:

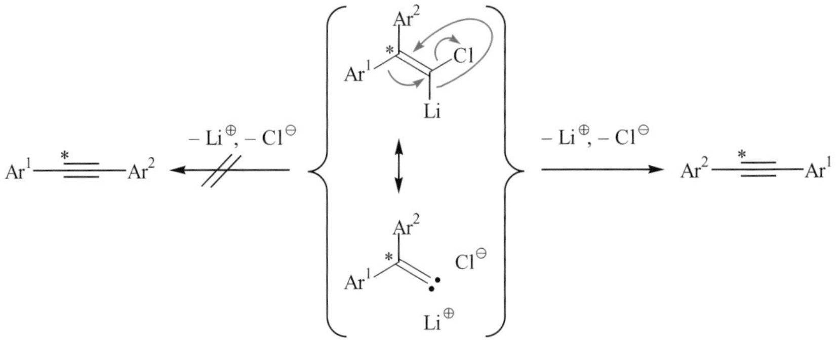

Therefore, a stereoelectronic reason cannot explain why hydrogen and not alkyl migration takes place in the bromine-containing vinyl carbenoid **D** of Figure 14.32.

14.4 [1,2]-Rearrangements without the Occurrence of a Sextet Intermediate

The reader has already encountered semipinacol rearrangements in which the elimination of the leaving group was not *followed* but instead was *accompanied* by the [1,2]-rearrangement (Figures 14.20 and 14.22–14.24). In this way, the temporary formation of an energetically unfavorable valence electron sextet could be avoided. These rearrangements are summarized in the first two rows of Table 14.1.

It was pointed out in the discussion of the lower part of Figure 14.1 that leaving groups cannot be dissociated from O or N atoms, respectively, if these dissociations resulted in the formation of oxenium or nitrenium ions, respectively. The same is true if a nitrene (R—N:) would have to be formed. These three sextet systems all are highly destabilized in comparison to carbenium ions and carbenes because of the high electronegativities of O and N. The O- and N-bound leaving groups therefore can be expelled only if *at the same time* either an α-H atom or an alkyl group undergoes a [1,2]-rearrangement to the O or N atom:

- The entries in rows 3–5 in Table 14.1 refer to one-step eliminations/rearrangements of this type in which oxenium ions are avoided.
- The Beckmann rearrangement (details: Figure 14.42) is a [1,2]-rearrangement in which the occurrence of a nitrenium ion is avoided via the one-step mode of elimination and rearrangement (second to last entry in Table 14.1).

Tab. 14.1 Survey of [1,2]-Rearrangements without Sextet Intermediates

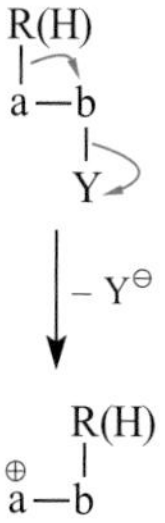

	a	b Y−	
R(H) / a−b / Y, $-Y^\ominus$ → R(H) / $\oplus$ a−b	$CR(O^\ominus)$	CRH−OTs	Semipinacol rearrangement
	$CR(O^\ominus)$	$CH_2-\overset{\oplus}{N}\equiv N$	Semipinacol rearrangement
	CR_2	$O-OH_2^\oplus$	Hydroperoxide-rearrangement
	CR(OH)	O−OC(=O)Ar	Baeyer–Villiger-oxidation
	$BR_{2-n}(OR)_n^\ominus$	O−OH	Borane-oxidation/ boronate-oxidation
	$C_{sp^2}R$	$N_{sp^2}-OH_2^\oplus$	Beckmann-rearrangement
	C(=O)	$\overset{\ominus}{N}-\overset{\oplus}{N}\equiv N$	Curtius-degradation

- Curtius rearrangements (details in Figures 14.43 and 14.44 occur as one-step reactions to avoid the intermediacy of nitrenes (last entry in Table 14.1).

14.4.1 Hydroperoxide Rearrangements

Tertiary hydroperoxides undergo a hydroperoxide rearrangement in acidic media, as exemplified in Figure 14.33 by the rearrangement of cumene hydroperoxide (for the preparation, see Figure 1.37). The cumene hydroperoxide rearrangement is employed for the synthesis of acetone and phenol on an industrial scale. The OH group of the hydroperoxide is protonated by concentrated sulfuric acid so that the carboxonium ion **A** can be generated by the elimination of water. The ion **A** immediately adds the water molecule again to form the protonated hemiacetal **C.** Tautomerization of **C** leads to **B**, and the latter decomposes to phenol and protonated acetone. The last reaction step is merely the deprotonation of the protonated acetone.

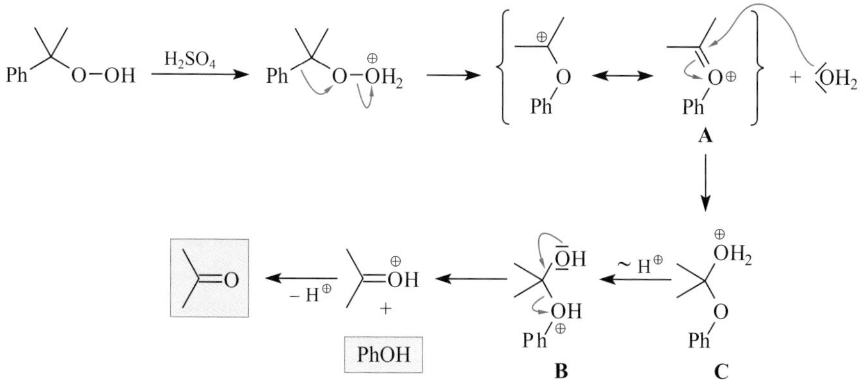

Fig. 14.33. Cumene hydroperoxide rearrangement.

14.4.2 Baeyer–Villiger Rearrangements

The Baeyer–Villiger rearrangement often is called Baeyer–Villiger oxidation (see the last subsection of Section 17.3.2). In the Baeyer–Villiger rearrangement, a carbonyl compound (ketones are almost always used) and an aromatic peracid react via insertion of the peroxo-O atom next to the C=O bond of the carbonyl compound. The Baeyer–Villiger rearrangement of *cyclic* ketones results in *lactones* (as in Figure 14.34).

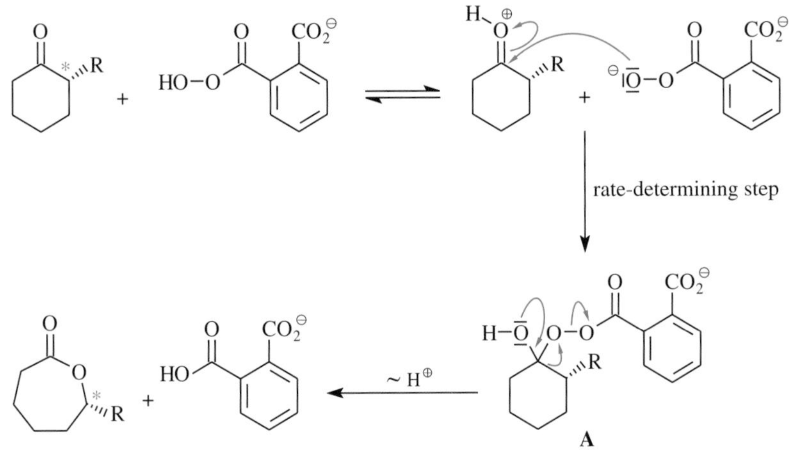

Fig. 14.34. Regioselective and stereoselective Baeyer–Villiger rearrangement of an unsymmetrical ketone with magnesium monoperoxophthalate hexahydrate (in the drawing, $Mg^{2\oplus}$ is omitted for clarity).

A Baeyer–Villiger rearrangement starts with the proton-catalyzed addition of the peracid to the C=O double bond of the ketone (Figure 14.34). This affords the α-hydroxyperoxoester **A.** The O—O bond of **A** is labile and breaks even *without* prior protonation of the leaving group. This is different from the fate of the O—O bond of a hydroperoxide, which does not break unless it is protonated. The different behavior is due to the fact that the Baeyer–Villiger rearrangement releases magnesium phthalate, and that this anion is rather stable. The cleavage of a hydroperoxide in the absence of an acid would result in the much more basic and therefore much less stable hydroxide ion.

The O—O bond cleavage of the α-hydroxyperoxoester intermediate of a Baeyer–Villiger rearrangement is accompanied by a [1,2]-rearrangement. One of the two substituents of the former carbonyl group migrates. In the example shown in Figure 14.34, either a primary or a secondary alkyl group in principle could migrate. As in the case of the Wagner–Meerwein rearrangements, the intrinsic propensity for migration in Baeyer–Villiger rearrangements follows the order $R_{tert} > R_{sec} > R_{prim}$. Hence, the secondary alkyl group migrates exclusively in intermediate **A.** It migrates with complete retention of configuration. This stereochemistry is quite common in [1,2]-rearrangements and was mentioned earlier in connection with the double Wolff rearrangement (Figure 14.28). If the substrate of the Baeyer–Villiger rearrangement of Figure 14.34 is an enantiomerically pure ketone (possible method for the preparation: in analogy to Figures 13.34 or 13.35), an enantiomerically pure lactone is formed.

Side Note 14.3. Chemoselectivity: Baeyer-Villiger Oxidation or Alkene Epoxidation?

If the α-substituent R of the cyclohexanone in Figure 14.34 were unsaturated, i.e., if it were, for instance, an allylic residue, this special cyclohexanone would also be oxidized by *m*-chloroperoxybenzoic acid (MCPBA) in a Baeyer–Villiger rearrangement. In Figure 14.35, this is illustrated by the formation of the corresponding seven-membered lactone **D**. But with the C=C double bond of their allylic substituent both the resulting lactone **D** and the cyclohexanone employed contain a structural element that is basically able to also undergo a reaction with MCPBA, namely the formation of an epoxide (cf. Figure 3.19). As Figure 14.35 reveals, nonconjugated unsaturated ketones are chemoselectively Baeyer–Villiger-oxidized (→ **D**) and not epoxidized (→ **C**) by MCPBA. The same applies to magnesium monoperoxyphthalate. In contrast, the same unsaturated ketones undergo chemoselective epoxidation (→ **C**) instead of a Baeyer–Villiger oxidation (→ **D**) by imido percarboxylic acids like imido perbenzoic acid (**A**) or imido peracetic acid (not shown). The imido peracids mentioned can be generated *in situ* by adding hydrogen peroxide to a solution of a 1:1 mixture of unsaturated ketone and benzonitrile in an inert solvent or, even more conveniently, by adding hydrogen peroxide to the solution of the unsaturated ketone in acetonitrile. The hydrogen peroxide then adds to the respective C≡N triple bond (for mechanistic details see Figure 7.10).

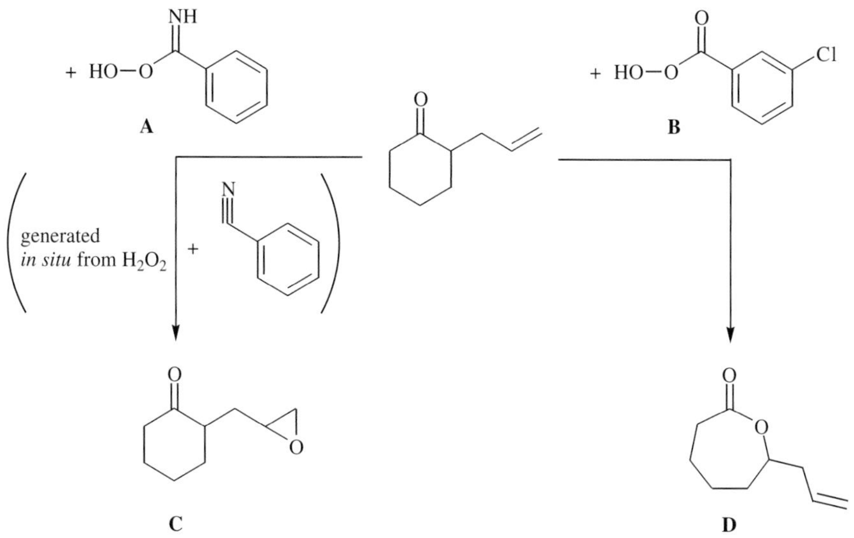

+ HO—O (NH) A
+ HO—O (O) Cl B
(generated *in situ* from H_2O_2 +)
C
D

Fig. 14.35. Chemoselective oxidations of an unsaturated ketone: the imido peracid **A** epoxidizes the C=C double bond, while the peracid **B** reacts with the C=O double bond causing a Baeyer–Villiger rearrangement.

The Baeyer–Villiger rearrangement of acetophenone, shown in Figure 14.36, also proceeds via the mechanism described in Figure 14.34. The aryl group migrates rather than the methyl group—this is true no matter whether the acetophenone is electron rich or electron poor. The product of this rearrangement is an aryl acetate. The hydrolysis of this aryl acetate occurs quickly (see discussion of Table 6.1). The combination of the method for preparing acetophenones (Section 5.2.7) with the Baeyer–Villiger rearrangement allows for the synthesis of phenols from aromatic compounds.

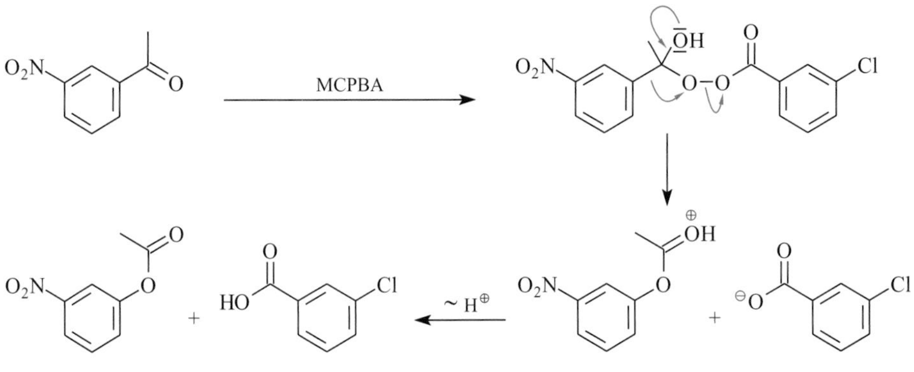

Fig. 14.36. Regioselective Baeyer–Villiger rearrangement of an unsymmetrical ketone with MCPBA (*meta*-chloroperoxybenzoic acid). The aryl group is [1,2]-shifted in all cases and irrespective of whether the acetophenone is electron-rich or electron-poor.

In Baeyer–Villiger rearrangements, electron-rich aryl groups migrate faster than H atoms, and H atoms in turn migrate faster than electron-poor aryl groups. Aldehydes, benzaldehydes, and *electron-poor* aromatic aldehydes thus react with peracids to afford *carboxylic acids* (Figure 14.37), while *electron-rich* aromatic aldehydes react with peracids to afford *aryl formates* (Figure 14.38). It must be emphasized that in contrast to Figures 14.34 and 14.36, the transition state of the "Baeyer–Villiger rearrangement" of Figure 14.37 perhaps might not even be that of a rearrangement at all. Instead, it is entirely possible that a β-elimination of benzoic acid from the α-hydroxyperoxoester occurs. This β-elimination might involve a cyclic transition state (see the *cis*-eliminations in Section 4.2).

Cyclobutanones are the only ketones that undergo Baeyer–Villiger rearrangements not only with peracids but even with alkaline H_2O_2 or alkaline *tert*-BuOOH (Figure 14.39). In this case, the driving forces of two crucial reaction steps are higher than normal because of the stepwise release of ring strain. The ring strain in the four-membered ring is reduced somewhat during the formation of the tetrahedral intermediate in step **A** → **B**, since the reacting C atom is rehybridized from sp^2 to sp^3. Accordingly, the preferred bond angle is reduced from 120° to

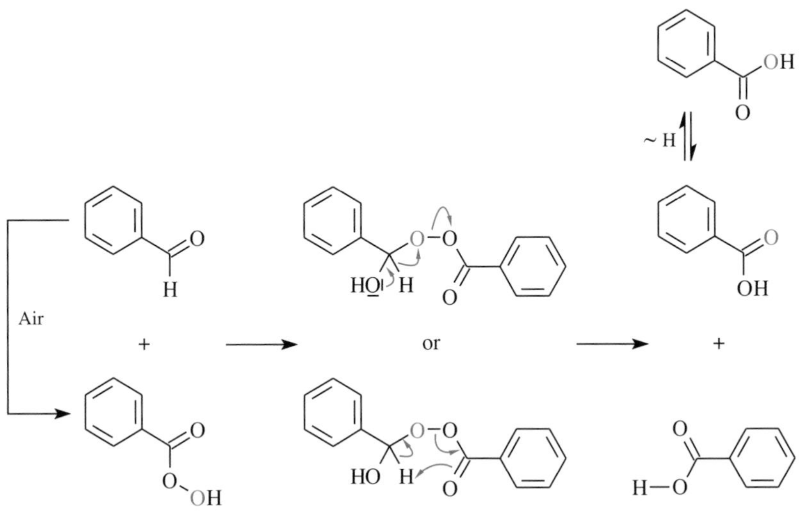

Fig. 14.37. Regioselective Baeyer–Villiger rearrangement of an electron-poor aromatic aldehyde. This reaction is part of the autoxidation of benzaldehyde to benzoic acid. Both alternative reaction mechanisms are shown: the [1,2]-rearrangement (top) and the β-elimination (bottom).

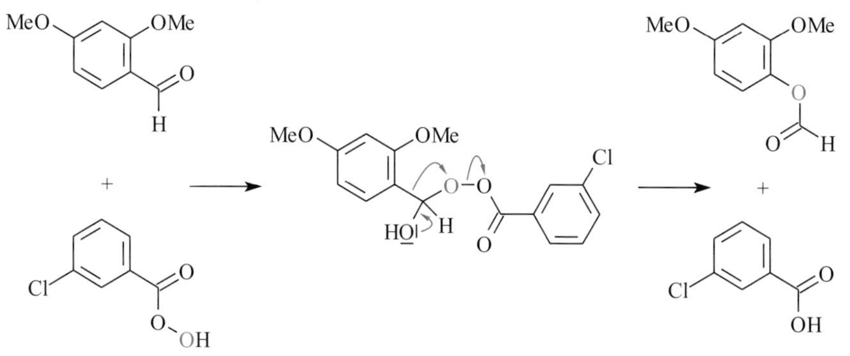

Fig. 14.38. Regioselective Baeyer–Villiger rearrangement of an electron-rich aromatic aldehyde.

109°28'. While still too large for **B** to be strain-free, some relief in comparison to **A** is provided. In the second step **B** → **C** of this particularly fast Baeyer–Villiger rearrangement, the four-membered tetrahedral intermediate is converted into the five-membered rearrangement product **C.** In the process, the ring strain is drastically reduced. The extra exothermicities of these two reaction steps are manifested in lowered activation barriers because of product development control.

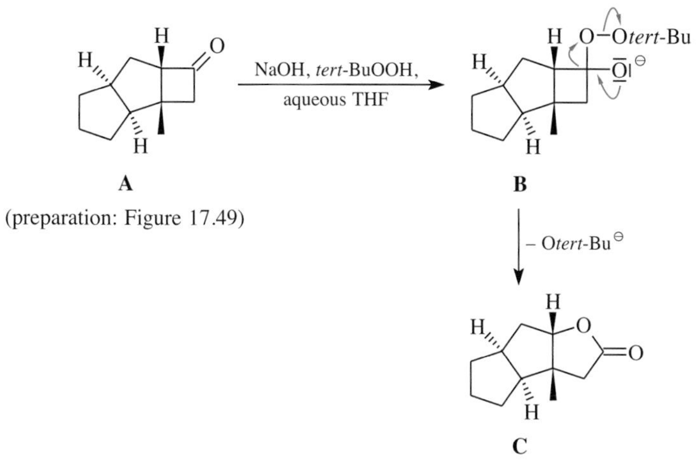

Fig. 14.39. Baeyer–Villiger rearrangement of a strained ketone with alkaline *tert*-BuOOH.

14.4.3 Oxidation of Organoborane Compounds

Rearrangements also are involved in the oxidations of trialkylboranes with H_2O_2/NaOH (Figure 14.40) and of arylboronic acid esters with H_2O_2/HOAc (Figure 14.41). These rearrangements are driven by O—O bond cleavages. The formation of energetically unacceptable oxenium ions is strictly avoided. The oxidation of trialkylboranes with H_2O_2/NaOH is the second step of the hydroboration/oxidation/hydrolysis sequence for the hydration of alkenes (see Section 3.3.3). The oxidation of arylboronic acid esters with H_2O_2 is the second step of the reac-

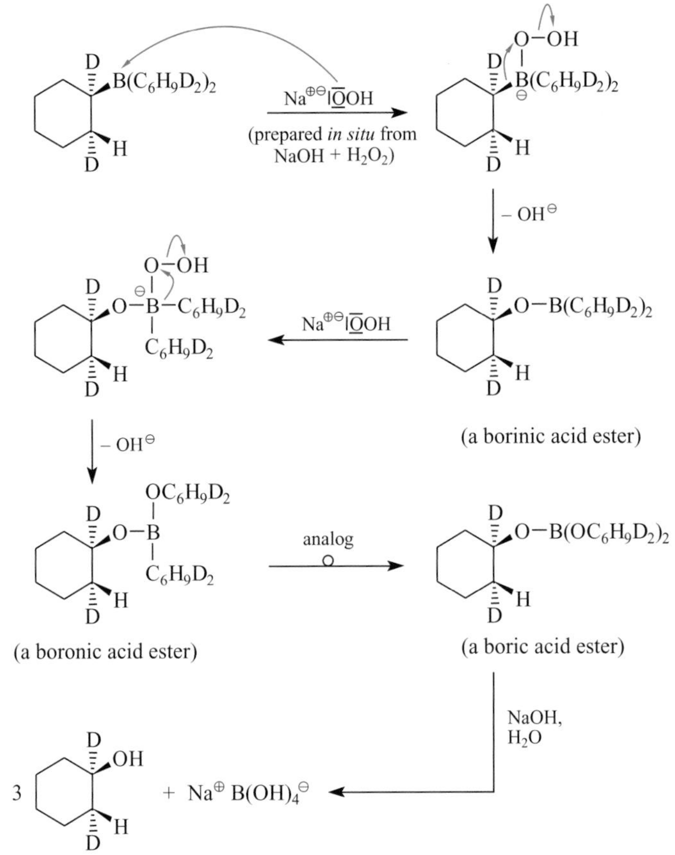

Na$^{\oplus\ominus}$|$\overline{O}$OH

(prepared *in situ* from NaOH + H_2O_2)

– OH$^{\ominus}$

Na$^{\oplus\ominus}$|$\overline{O}$OH

(a borinic acid ester)

– OH$^{\ominus}$

analog

(a boronic acid ester)

(a boric acid ester)

NaOH, H_2O

3 + Na$^{\oplus}$ B(OH)$_4^{\ominus}$

Fig. 14.40. H_2O_2/NaOH oxidation of a trialkylborane (see Figure 3.23 for the preparation of trialkylboranes **D** and **E** and for the mechanism of the hydrolysis of the resulting boric acid ester: Figure 3.23, bottom). Deuterium labeling studies show that the conversion of the C—B into the C—O bonds occurs with retention of configuration.

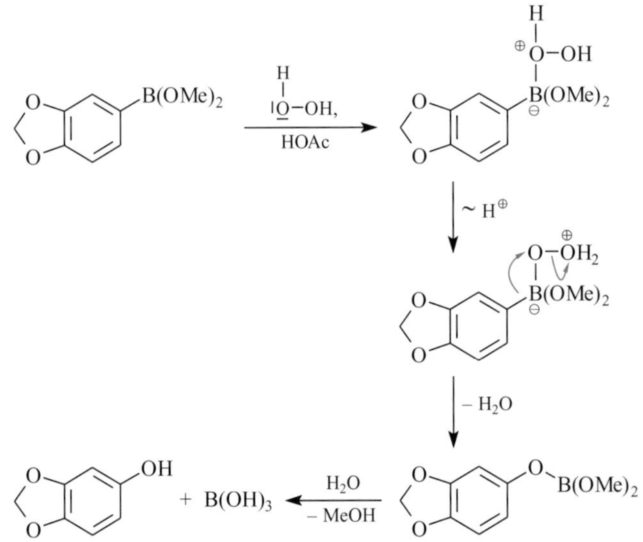

HOAc

~ H$^{\oplus}$

– H_2O

+ B(OH)$_3$

H_2O – MeOH

Fig. 14.41. H_2O_2/HOAc oxidation of an arylboronic acid ester (for the preparation of this compound, see Figure 5.45).

tion sequence Ar—Br → Ar—B(OMe)$_2$ → ArOH or of a similar reaction sequence *o*-DMG—Ar → *o*-MDG—Ar-B(OMe)$_2$ → *o*-DMG—ArOH (see Section 5.3.3; remember that DMG is the abbreviation for directed metalation group).

The mechanisms of these two oxidations are presented in detail in Figures 14.40 and 14.41. They differ so little from the mechanism of the Baeyer–Villiger rearrangement (Section 14.4.2) that they can be understood without further explanations. In the example depicted in Figure 14.40, all the [1,2]-rearrangements occur with complete retention of configuration at the migrating C atom just as we saw in other [1,2]-rearrangements [cf. Figures 14.28 (Wolff rearrangement) and 14.34 (Baeyer–Villiger rearrangement)]. The Curtius rearrangement shown later in Figure 14.44 also occurs with retention of configuration.

14.4.4 Beckmann Rearrangement

The OH group of ketoximes $R^1R^2C(=NOH)$ can become a leaving group. Tosylation is one way to convert this hydroxyl group into a leaving group. The oxime OH group also can become a leaving group if it is either protonated or coordinated by a Lewis acid in an equilibrium reaction. Oximes activated in this fashion may undergo N—O heterolysis. Since the formation of a nitrenium ion needs to be avoided (see discussion of Table 14.1), this heterolysis is accompanied by a simultaneous [1,2]-rearrangement of the group that is attached to the C=N bond in the *trans*-position with regard to the O atom of the leaving group. A nitrilium ion is formed initially (see **A** in Figure 14.42). It reacts with water to form an imidocarboxylic acid, which tautomerizes immediately to an amide. The overall reaction sequence is called the Beckmann rearrangement.

The Beckmann rearrangement of *cyclic* oximes results in *lactams.* This is exemplified in Figure 14.42 with the generation of ε-caprolactam, the monomer of nylon-6. The nitrilium ion

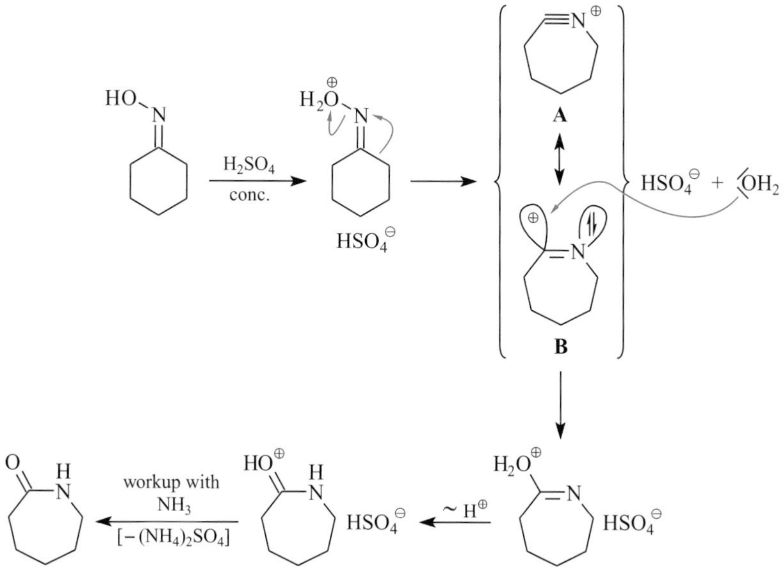

HO N; H_2SO_4 conc.; $H_2\overset{\oplus}{O}$ N; $HSO_4^{\ominus}$; N$^{\oplus}$ **A**; **B**; $HSO_4^{\ominus}$ + OH_2; $H_2O^{\oplus}$ N $HSO_4^{\ominus}$; ~ H$^{\oplus}$; $HO^{\oplus}$ H N $HSO_4^{\ominus}$; workup with NH_3 [–$(NH_4)_2SO_4$]; O H N

Fig. 14.42. Industrial synthesis of caprolactam via the Beckmann rearrangement of cyclohexanone oxime.

intermediate cannot adopt the preferred linear structure because it is embedded in a seven-membered ring. Therefore, in this case the intermediate might better be described as the resonance hybrid of the resonance forms **A** (C≡N$^{\oplus}$ triple bond) and **B** (C$^{\oplus}$=N double bond). The C,N multiple bond in this intermediate resembles the bond between the two C atoms in benzyne that do not carry H atoms.

14.4.5 Curtius Degradation

The Curtius degradation of acyl azides (Figure 14.43) expels molecular nitrogen and at the same time leads to the [1,2]-rearrangement of the substituent that is attached to the carboxyl carbon. It is the simultaneous occurrence of these two events that prevents the formation of an energetically unacceptably disfavored acylnitrene intermediate. The rearranged product is an isocyanate.

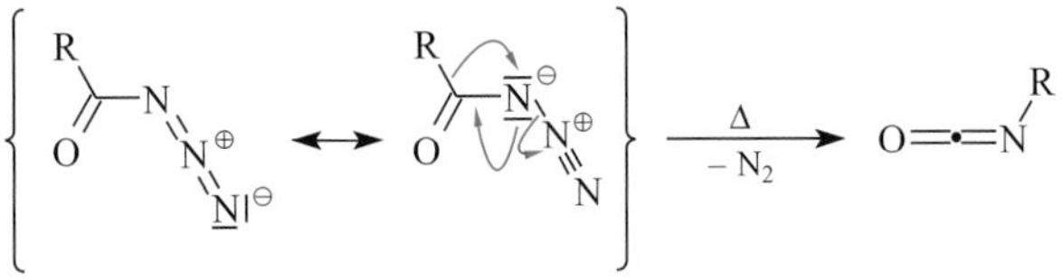

Fig. 14.43. Mechanism of the Curtius degradation.

The isocyanate can be isolated if the Curtius degradation is carried out in an inert solvent. The isocyanate also can be reacted with a heteroatom-nucleophile either subsequently or *in situ.* The heteroatom nucleophile adds to the C=N double bond of the isocyanate via the mechanism of Figure 8.12. In this way, the addition of water initially results in a carbamic acid. However, all carbamic acids are unstable and immediately decarboxylate to give amines (see Figure 8.5). Because of *this* consecutive reaction, the Curtius rearrangement represents a valuable amine synthesis. The amines formed contain one C atom less than the acyl azide substrates. It is due to this feature that one almost often refers to this reaction as Curtius *degradation*, not as Curtius *rearrangement.*

The reaction sequence of Figure 14.44 shows how a *carboxylic acid* can be subjected to a Curtius degradation *in a one-pot reaction.* This one-pot procedure is convenient because there is no need to isolate the potentially explosive acyl azide. The conversion of the carboxylic acid into the acyl azide occurs in the initial phase of the one-pot reaction of Figure 14.44 by means of a phosphorus(V) reagent. This reagent reacts in a manner analogous to the role of $POCl_3$ in the conversion of carboxylic acids into acid chlorides (see Figure 6.10). Accordingly, a mixed carboxylic acid/phosphoric acid anhydride **B** is formed *in situ.* It acylates the simultaneously formed azide ion to give the acyl azide **A**.

Compound **A** represents the immediate substrate of the Curtius degradation of Figure 14.44. Compound **A** contains a *trans*-configured cyclopropyl substituent. The *trans*-configuration of this substituent remains unchanged in the course of the [1,2]-rearrangement leading to the isocyanate **C.** Thus, it migrates with complete retention of the configuration at the migrating C atom. Since it is possible to synthesize α-chiral carboxylic acids with well-defined absolute configurations (for a possible preparation, see Figures 13.42 and 13.43), the

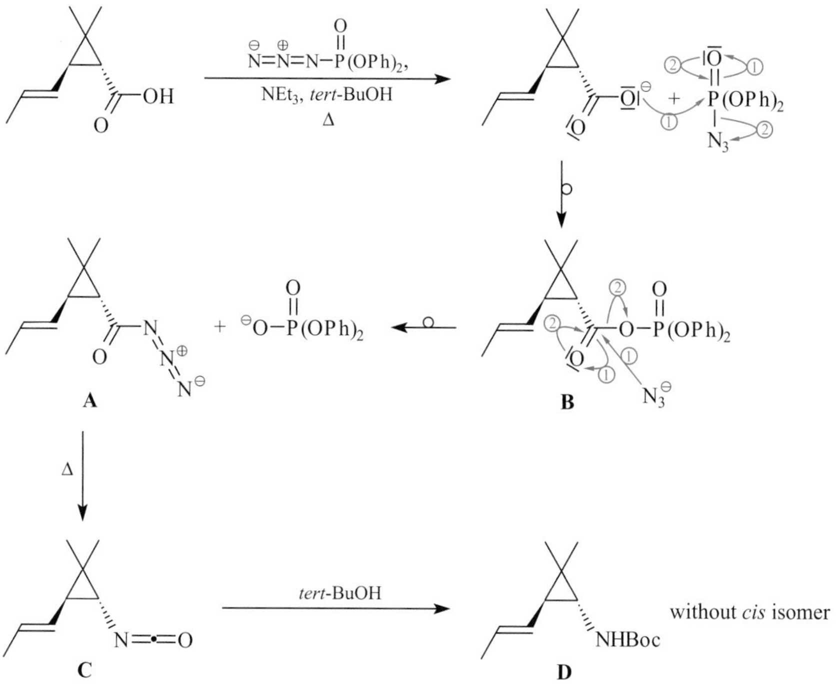

Fig. 14.44. A one-pot diastereoselective degradation of a carboxylic acid to a Boc-protected amine via a Curtius rearrangement; Boc refers to *tert*-butoxylcarbonyl. The mixed anhydride **B** is formed by a condensation of the phosphorus(**V**) reagent with the carboxyl group. The anhydride **B** acylates the concomitantly generated azide ion forming the acyl azide **A**. A Curtius degradation converts **A** to **C**, and the latter reacts subsequently with *tert*-butanol to the Boc-protected amine.

Curtius degradation represents an interesting method for their one-step conversion into enantiomerically pure amines of the structure $R^1R^2CH—NH_2$.

Figure 14.44 also shows how the Curtius degradation of an acyl azide can be combined with the addition of *tert*-butanol to the initially obtained isocyanate. This addition gives a carbamate. In the present case a *tert*-butoxycarbonyl-protected amine ("Boc-protected amine") is formed.

Side Note 14.4. A Curtius Route Leading to α-Aminophosphonic Acids

Let us take what in Figure 14.43 started with a Curtius degradation to an isocyanate and what in Figure 14.44 was continued with a Curtius degradation to a carbamate one step further to further demonstrate the Curtius degradation as an amine synthesis. In Figure 14.45, the Curtius degradation starting from the acyl azide **F** via the isocyanate (not shown) furnishes the carbamate **E**, which is then hydrolyzed to give the amine **D**. **D** is a special amine, namely an α-aminophosphonic acid. It is the phosphorus analog of an α-aminocarboxylic acid and also occurs as a zwitterion.

The first three steps show that the reaction sequence in Figure 14.45 actually provides a widely applicable approach to such α-aminophosphonic acids. The step leading to the acyl azide **F**, i.e., the nitrite oxidation of an acyl hydrazide (Formula **C** in Figure 14.45), is as commonly used for the preparation of an acyl azide synthesis as that shown in Figure 14.44.

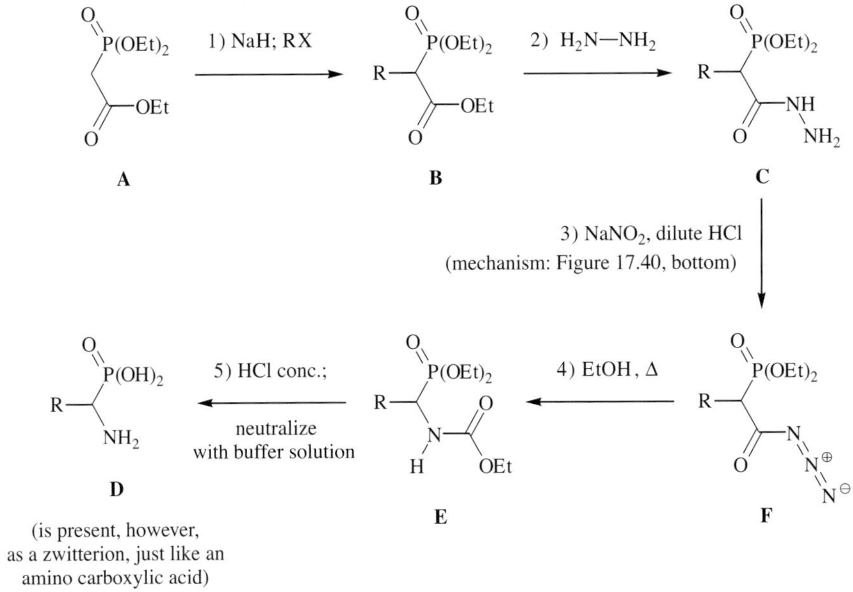

A B C D E F

1) NaH; RX

2) $H_2N{-}NH_2$

3) $NaNO_2$, dilute HCl (mechanism: Figure 17.40, bottom)

4) EtOH, Δ

5) HCl conc.; neutralize with buffer solution

(is present, however, as a zwitterion, just like an amino carboxylic acid)

Fig. 14.45. Transformation of an α-phosphonylcarboxylic acid ester (**B**) via the related carboxylic acid azide **F** and its Curtius degradation in ethanol to furnish an ethoxycarbonyl-protected α-aminophosphonic acid ester **E**. The N- and O-bound protective groups of the latter compounds are cleaved off under acidic conditions. In this manner α-aminophosphonic acids are synthesized. They are interesting analogs of the biologically important α-amino carboxylic acids.

14.5 Claisen Rearrangement

14.5.1 Classical Claisen Rearrangement

The classical Claisen rearrangement is the first and slow step of the isomerization of allyl aryl ethers to *ortho*-allylated phenols (Figure 14.46). A cyclohexadienone **A** is formed in the actual rearrangement step, which is a [3,3]-sigmatropic rearrangement. Three valence electron pairs are shifted simultaneously in this step. Cyclohexadienone **A**, a nonaromatic compound, cannot be isolated and tautomerizes immediately to the aromatic and consequently more stable phenol **B.**

Not only an aryl group—as in Figure 14.46—but also an alkenyl group can participate in the Claisen rearrangement of allyl ethers (Figure 14.47). Allyl enol ethers are the substrates in this case. Figure 14.47 shows how this kind of allyl alkenyl ether (**D**) can be prepared from an

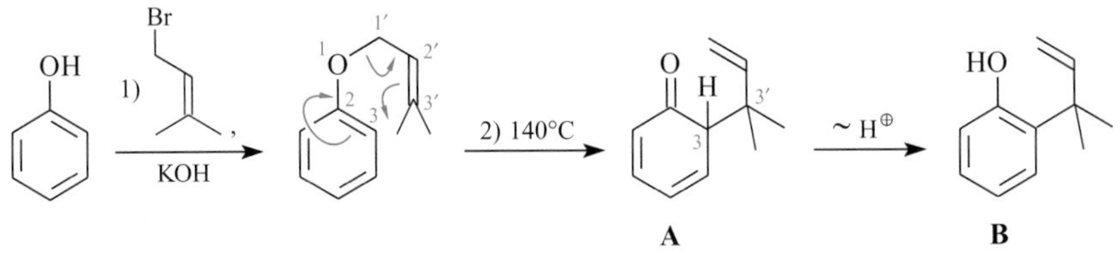

1) KOH

2) 140°C

~ $H^{\oplus}$

A B

Fig. 14.46. Preparation of an allyl aryl ether and subsequent Claisen rearrangement. (The rearrangement is named after the German chemist Ludwig Claisen. Basically, the name is pronounced the German way; but it is known that the family preferred the pronunciation /klæzn/.)

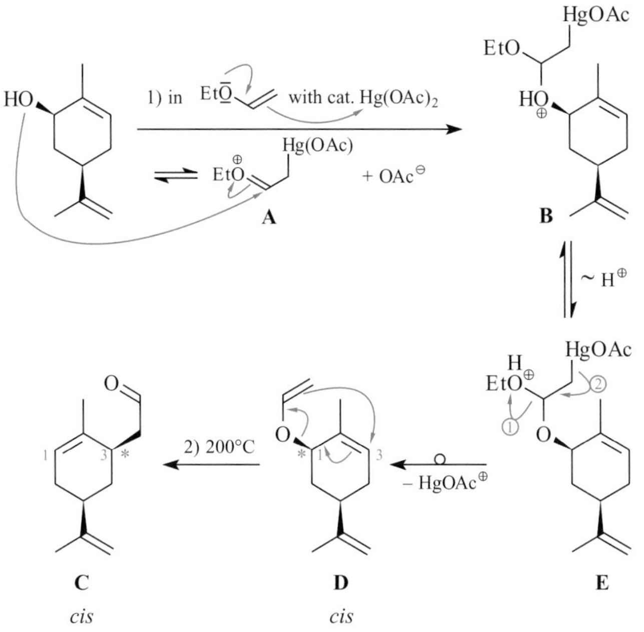

Fig. 14.47. Preparation of an allyl vinyl ether, **D**, from allyl alcohol and a large excess of ethyl vinyl ether. Subsequent Claisen rearrangement **D** → **C** of the allyl vinyl ether proceeding with chirality transfer.

allyl alcohol in a single operation. The allyl alcohol is simply treated with a large excess of ethyl vinyl ether in the presence of catalytic amounts of $Hg(OAc)_2$.

The preparation involves an oxymercuration (Section 3.5.3) of the C=C double bond of the ethyl vinyl ether. The $Hg(OAc)^{\oplus}$ ion is the electrophile as expected, but it forms an open-chain cation **A** as an intermediate rather than a cyclic mercurinium ion. The open-chain cation **A** is more stable than the mercurinium ion because it can be stabilized by way of oxocarbenium ion resonance. Next, cation **A** reacts with the allyl alcohol, and a protonated mixed acetal **B** is formed. Compound **B** eliminates EtOH and $Hg(OAc)^{\oplus}$ in an E1 process, and the desired enol ether **D** results. The enol ether **D** is in equilibrium with the substrate alcohol and ethyl vinyl ether. The equilibrium constant is about 1. However, the use of a large excess of the ethyl vinyl ether shifts the equilibrium to the side of the enol ether **D** so that the latter can be isolated in high yield.

Upon heating, the enol ether **D** is converted into the aldehyde **C** via a Claisen rearrangement as depicted in Figure 14.47. The product **C** and its precursor **D** both are *cis*-disubstituted cyclohexanes. The σ-bond that has migrated connects two C atoms in the product, while it connected a C and an O atom in the substrate. The σ-bond remains on the same side of the cyclohexane ring; hence this Claisen rearrangement occurs with *complete* transfer of the stereochemical information from the original, oxygenated stereocenter to the stereocenter that is newly formed. Such a stereocontrolled transformation of an old stereocenter into a new one is called a **chirality transfer.** The Claisen rearrangement **D** → **C** of Figure 14.47 represents the special case of a **1,3-chirality transfer**, since the new stereocenter is at position 3 with respect to the old stereocenter, which is considered to be at position 1.

14.5.2 Ireland-Claisen Rearrangements

As shown earlier (Figure 13.22), silyl ketene acetals can be prepared at –78 °C by the reaction of ester enolates with chlorosilanes. *O*-Allyl-*O*-silyl ketene acetals (**A** in Figure 14.48) are formed in this reaction if one employs *allyl* esters. Silyl ketene acetals of type **A** undergo [3,3]-rearrangements rapidly upon warming to room temperature. This variation of the Claisen rearrangement is referred to as the Ireland-Claisen rearrangement.

Ireland-Claisen rearrangements obviously occur under much milder conditions than the classical Claisen rearrangements of Figures 14.46 and 14.47. Among other things, this is due to product development control. The rearranged product of a Claisen–Ireland rearrangement is an α-allylated silyl ester, and its C=O bond is stabilized by ester resonance ($\approx$ 14 kcal/mol

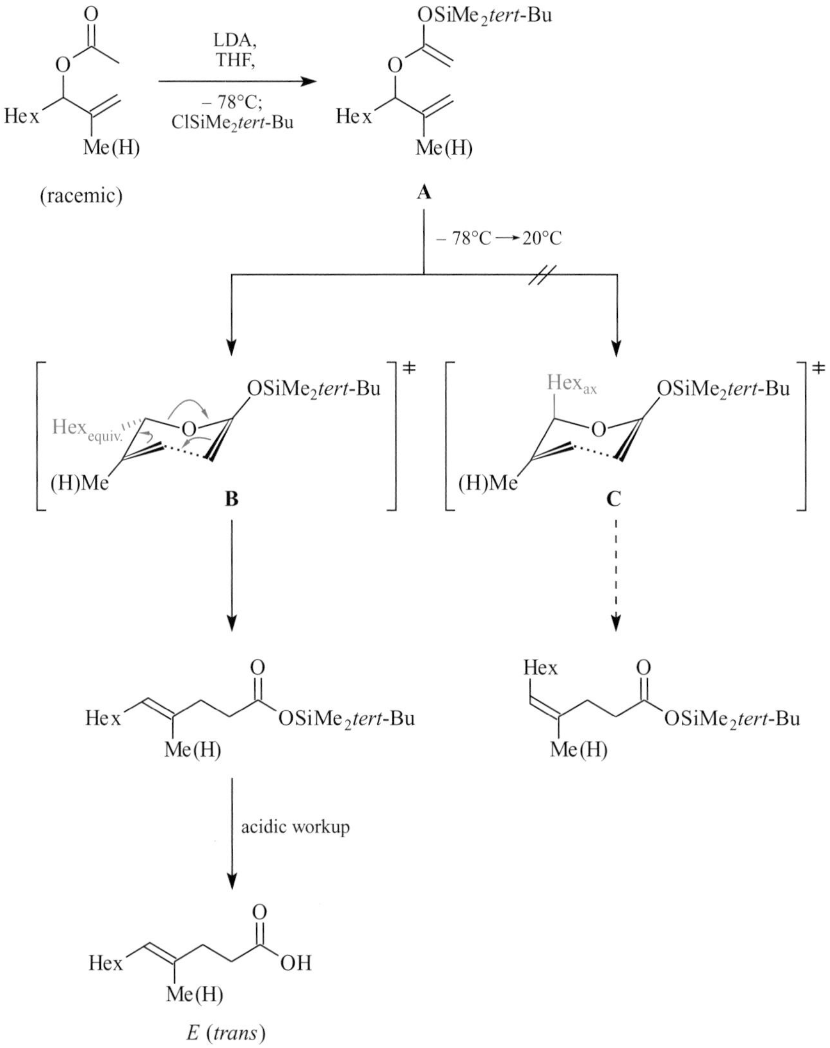

Fig. 14.48. Ireland-Claisen rearrangement of two *O*-allyl-*O*-silyl ketene acetals. *Trans*-selective synthesis of disubstituted and *E*-selective synthesis of trisubstituted alkenes.

according to Table 6.1). This resonance stabilization provides an additional driving force in comparison to the classical Claisen rearrangement: the primary products of classical Claisen rearrangements are ketones (Figure 14.46) or aldehydes (Figure 14.47), and the C=O double bonds of these species do not benefit from a comparably high resonance stabilization. The additional driving force corresponds, according to the Hammond postulate, to a lowered activation barrier, i.e., to an increased rearrangement rate.

The product of a Ireland-Claisen rearrangement is a silyl ester. However, silyl esters generally are so sensitive toward hydrolysis that one usually does not attempt to isolate them. Instead, the silyl esters are hydrolyzed completely during work-up. Thus, Ireland-Claisen rearrangements *de facto* afford carboxylic acids and, more specifically, they afford γ,δ-unsaturated carboxylic acids.

Ireland-Claisen rearrangements are extraordinarily interesting from a synthetic point of view for several reasons. First, the Ireland-Claisen rearrangement is an important C=C bond-forming reaction. Second, Ireland-Claisen rearrangements afford γ,δ-unsaturated carboxylic acids, which are valuable bifunctional compounds. Both of the functional groups of these acids can then be manipulated in a variety of ways.

Ireland-Claisen rearrangements frequently are used for the synthesis of alkenes. This works particularly well if the allyl ester is derived from a secondary allyl alcohol. In this case a stereogenic double bond is formed in the rearrangement. The examples in Figure 14.48 show that the alkene is mostly *trans*-configured if this C=C bond is 1,2-disubstituted and almost completely *E*-configured if it is trisubstituted.

This stereoselectivity of the Ireland-Claisen rearrangement is the result of kinetic control. The transition state structure of the Ireland-Claisen rearrangement is a six-membered ring, for which a chair conformation is preferred. In the case of the two Ireland-Claisen rearrangements shown in Figure 14.48, one can imagine two chair-type transition states **B** and **C.** These transition states differ only in the orientation of the substituent at the allylic center with respect to the chair: the substituent is pseudo-equatorial in **B** and pseudo-axial in **C.** Hence, the substituent is in a better position in **B** than in **C.** This is true even though in the rearrangement of the *methylated* substrate the allylic substituent and the sp^2-bound methyl group approach each other closer in the transition state **B** than in the transition state **C.**

Figure 14.49 shows how enantiomerically pure allyl alcohols (for a possible preparation, see Figure 3.39) can be converted first into allyl acetates **A** and then at –78 °C into *O*-allyl-*O*-silyl ketene acetals **C.** These will undergo Ireland–Claisen rearrangements as they are warmed slowly to room temperature. The ultimately formed unsaturated carboxylic acids then contain a stereogenic C=C bond as well as a stereocenter. Both stereoelements have defined configurations. The double bond is *trans*-configured, and the absolute configuration of the stereocenter depends only on the substitution pattern of the allyl alcohol precursor.

Structure **B** corresponds to the most stable transition state of the Ireland-Claisen rearrangement of Figure 14.49. In this transition state, the substituent at the allylic stereocenter is in a pseudo-equatorial orientation with respect to the chair-shaped skeleton. This is the same preferred geometry as in the case of the most stable transition state **B** of the Claisen rearrangement of Figure 14.48. The reason for this preference is as before: that is, an allylic substituent that is oriented in this way experiences the smallest possible interaction with the chair skeleton. The obvious similarity of the preferred transition state structures of the Ireland-Claisen rearrangements of Figures 14.49 and 14.48 causes the same *trans*-selectivity.

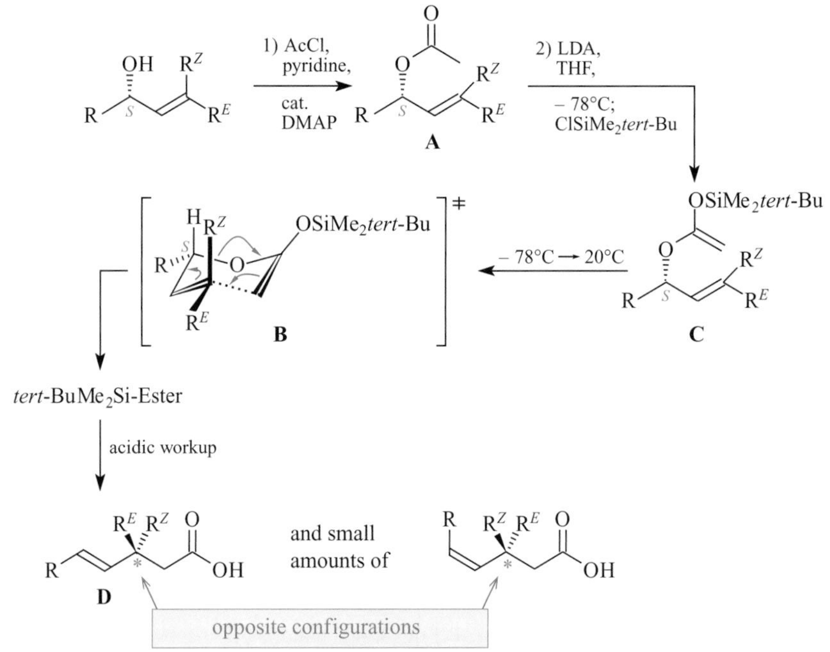

Fig. 14.49. *Trans*-selective Ireland-Claisen rearrangements with 1,3-chirality transfer. (DMAP refers to 4-**dim**ethyl**a**mino**p**yridine; see Figure 6.9 on DMAP-catalyzed ester formation.)

The pseudo-equatorial orientation of the allylic substituent in the preferred transition state **B** of the Ireland-Claisen rearrangement of Figure 14.49 also is responsible for a nearly perfect 1,3-chirality transfer. The absolute configuration of the new stereocenter depends both on the configuration of the stereocenter of the allyl alcohol (here *S*-configured) and also on the allyl alcohol's configuration about the C=C double bond (*cis* or *trans*). The main rearrangement product **D** is an *S*-enantiomer if a *trans*-configured *S*-allyl alcohol is rearranged. However, **D** will be an *R*-enantiomer if a *cis*-configured *S*-allyl alcohol is the starting material.

Ireland-Claisen rearrangements also allow for the realization of 1,4-chirality transfers. Two examples are shown in Figure 14.50, where the usual *trans*-selectivity also is observed (cf. Figures 14.48 and 14.49). In the rearrangements in Figure 14.50, the 1,4-chirality transfer is possible basically because propionic acid esters can be deprotonated in pure THF or in a THF/DMPU mixture, respectively, in a stereoselective fashion to provide *"E"*-configured (see Figure 13.16) or *"Z"*-configured (see Figure 13.17) ester enolates, respectively. This knowledge can be applied to the propionic acid allyl esters of Figure 14.50; that is, the propionic acid ester *syn*-**B** can be converted into an *"E"*-enolate and the ester anti-**B** into a *"Z"*-enolate. Each of these enolates can then be silylated with *tert*-$BuMe_2SiCl$ at the enolate oxygen.

The *O*-allyl-*O*-silyl ketene acetals *syn, E*-C and *anti, Z*-**C** of Figure 14.50 are thus formed isomerically pure. Each one undergoes a Ireland-Claisen rearrangement upon warming to room temperature. Again, the allylic substituent is oriented in a pseudo-equatorial fashion in the energetically most favorable transition state (**D** and **E**, respectively). It follows that this allylic substituent determines (a) the configuration of the newly formed C=C double bond (similar to the cases in Figures 14.48 and 14.49) and (b) the preferred configuration of the new

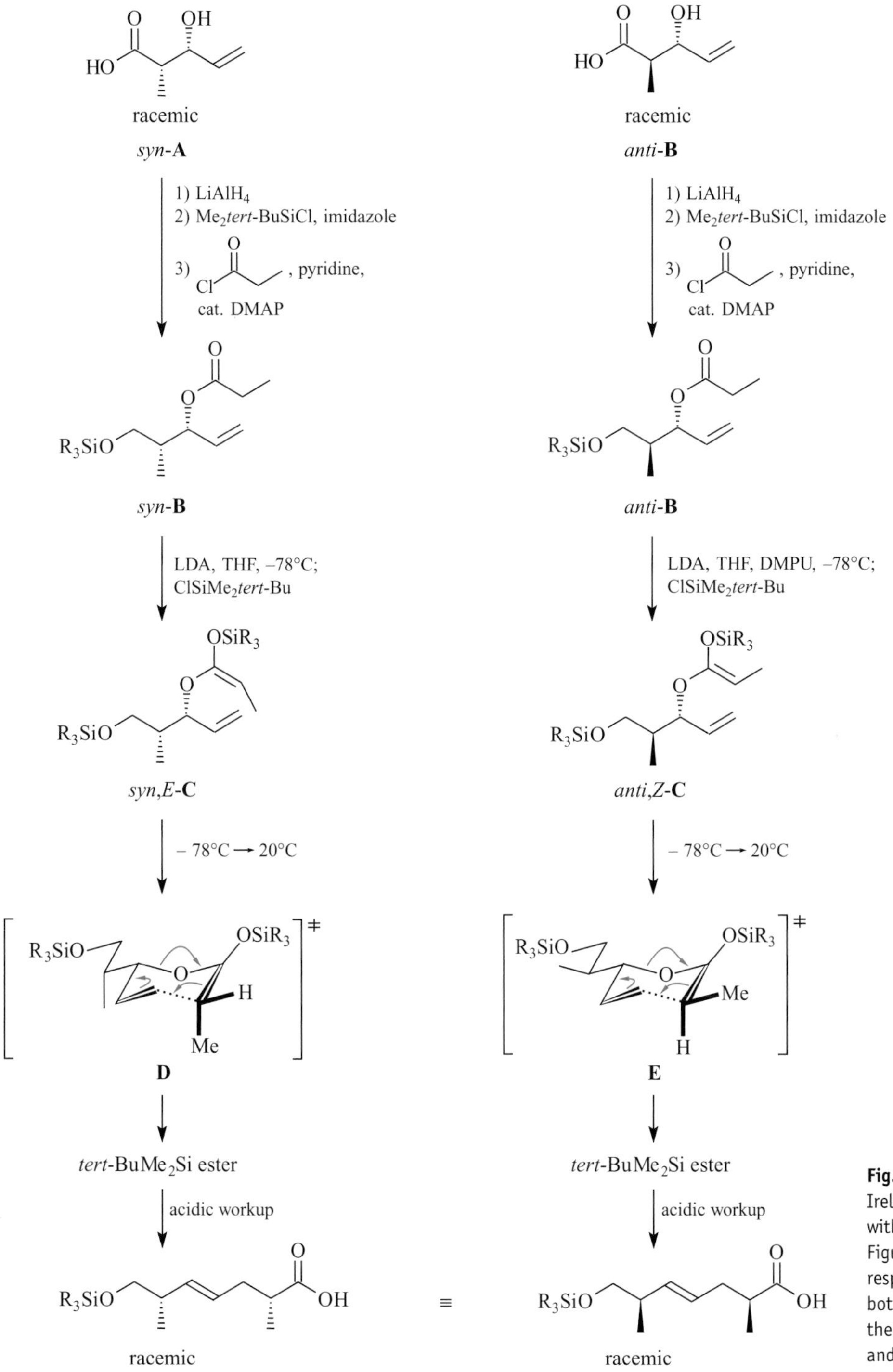

O OH

HO

racemic

syn-**A**

1) $LiAlH_4$
2) Me_2*tert*-BuSiCl, imidazole
3) Cl–C(O)–Et, pyridine,
cat. DMAP

R_3SiO

syn-**B**

LDA, THF, –78°C;
$ClSiMe_2$*tert*-Bu

$OSiR_3$

R_3SiO

syn,*E*-**C**

– 78°C → 20°C

R_3SiO $OSiR_3$ O H Me ‡

D

tert-$BuMe_2Si$ ester

acidic workup

R_3SiO O OH

racemic

≡

O OH

HO

racemic

anti-**B**

1) $LiAlH_4$
2) Me_2*tert*-BuSiCl, imidazole
3) Cl–C(O)–Et, pyridine,
cat. DMAP

R_3SiO

anti-**B**

LDA, THF, DMPU, –78°C;
$ClSiMe_2$*tert*-Bu

$OSiR_3$

R_3SiO

anti,*Z*-**C**

– 78°C → 20°C

R_3SiO $OSiR_3$ O Me H ‡

E

tert-$BuMe_2Si$ ester

acidic workup

R_3SiO O OH

racemic

Fig. 14.50. *Trans*-selective Ireland-Claisen rearrangements with 1,4-chirality transfer. (See Figures 13.47 and 13.48, respectively, with R = vinyl in both cases, for preparations of the starting materials *syn*-**A** and *anti*-**A**, respectively.)

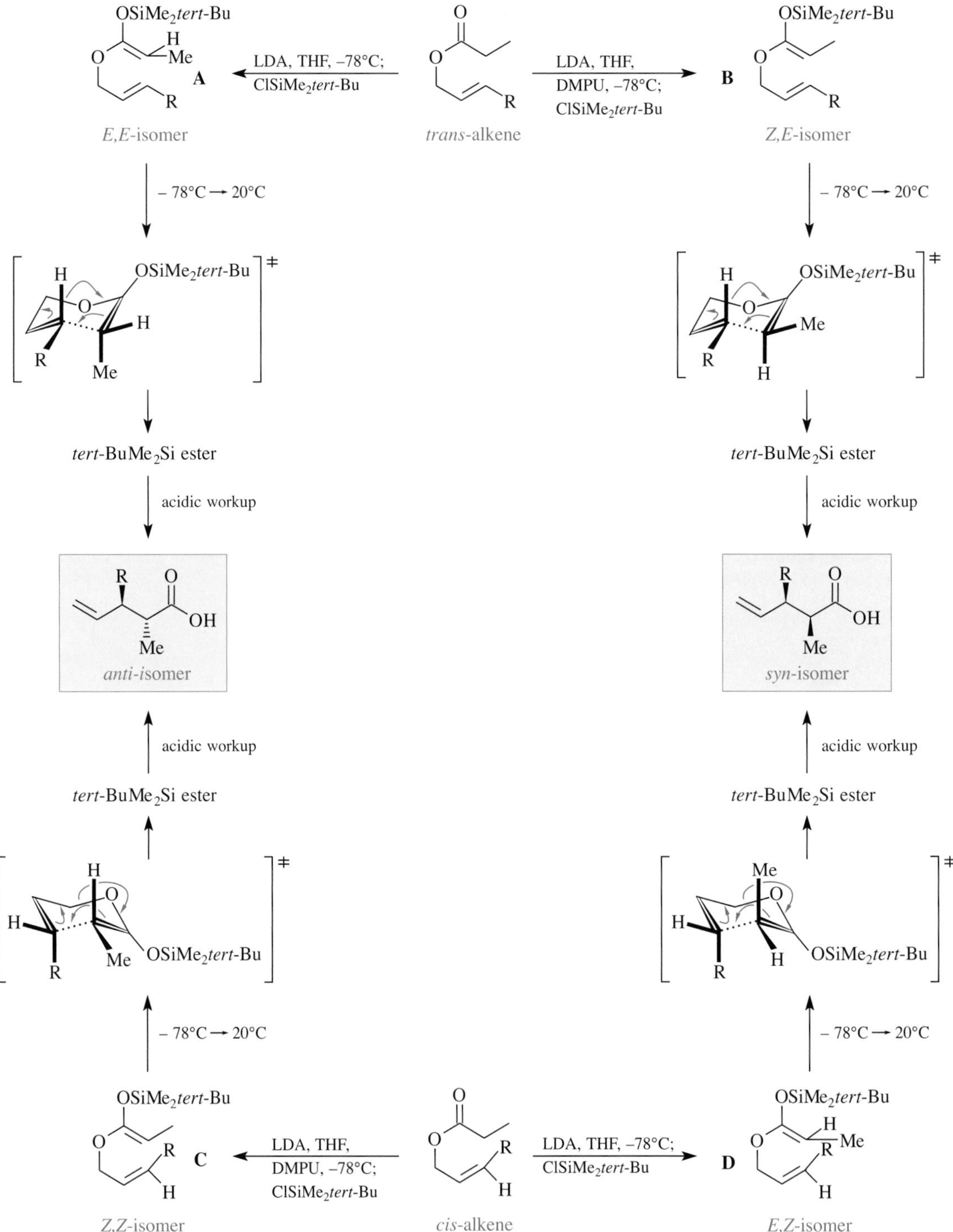

Fig. 14.51. Ireland-Claisen rearrangements with stereocontrol through simple diastereoselectivity.

stereocenter. These stereochemical relationships can be recognized if one "translates" the stereo-structures of the transition state structures **D** and **E** into the stereostructures of the respective rearrangement products by way of shifting three valence electron pairs simultaneously. Interestingly, *the same rearrangement product* is formed from the *stereoisomeric* silyl ketene acetals of Figure 14.50 via the *stereoisomeric* transition states **D** and **E**.

The rearrangement product of Figure 14.50 is synthetically useful. Heterogeneous catalytic hydrogenation allows for the conversion of this compound into a saturated compound with two methyl-substituted stereocenters with defined relative configurations. The racemic synthesis of a fragment of the vitamin E side chain (for the structure of this vitamin, see Figure 17.76) has been accomplished in this way. Ireland-Claisen rearrangements of this type also play a role in the stereoselective synthesis of other acyclic terpenes.

Figure 14.51 shows four Ireland-Claisen rearrangements that exhibit simple diastereoselectivity (see Section 11.1.3 for a definition of the term). The substrates are two *cis*, *trans*-isomeric propionic acid esters. The propionic acid esters in Figure 14.51 are derived from achiral allyl alcohols. This is different from the situation in Figure 14.50. However, these esters contain a stereogenic C=C double bond. Both the esters in Figure 14.51 can be converted into their *"E"*-enolates with LDA in pure THF (cf. Figure 13.16). Silylation affords the two *E*-configured *O*-allyl-*O*-silyl ketene acetals **A** and **D,** respectively. Alternatively, the two esters of Figure 14.51 can be converted into their *"Z"*-enolates with LDA in a mixture of THF and DMPU (cf. Figure 13.17). Treatment with *tert*-$BuMe_2SiCl$ then leads to the *Z*-isomers **B** and **C** of the *O*-allyl-*O*-silyl ketene acetals **A** and **D**, respectively.

Each of the four *O*-allyl-*O*-silyl ketene acetals **A**–**D** of Figure 14.51 undergoes a Ireland-Claisen rearrangement between –8 °C and room temperature. These reactions all are highly stereoselective. After aqueous workup, only one of the two possible diastereomeric carboxylic acids is formed in each case. These carboxylic acids contain two new stereocenters at the α- and β-C atoms. The two stereoisomers are *syn*- or *anti*-disubstituted. The *anti*-configured carboxylic acid is formed stereoselectively if one starts with the silyl ketene acetal **A** or its isomer **C.** In contrast, the Ireland-Claisen rearrangement of the silyl ketene acetal **B** or its isomer **D** gives the *syn*-configured carboxylic acid. These simple diastereoselectivities result from the fact that the transition states of the Ireland-Claisen rearrangements have a chair conformation.

It does not matter whether it is the *cis*- or the *trans*-isomer of the allyl alcohol that is more easily accessible. According to Figure 14.51, by selecting the appropriate solvent, enolate formation can be directed to convert both the *cis*- and the *trans*-allyl alcohols into rearranged products that contain either a *syn*- or an *anti*-arrangement of the vicinal alkyl groups.

References

L. M. Harwood (Ed.), "Polar Rearrangements," Oxford University Press, New York, **1992**.

I. Coldham, "One or More CH and/or CC Bond(s) Formed by Rearrangement", in *Comprehensive Organic Functional Group Transformations* (A. R. Katritzky, O. Meth-Cohn, C. W. Rees, Eds.), Vol. 1, 377, Elsevier Science, Oxford, U. K., **1995**.

H. McNab, "One or More C=C Bond(s) by Pericyclic Processes," in *Comprehensive Organic Functional Group Transformations* (A. R. Katritzky, O. Meth-Cohn, C. W. Rees, Eds.), Vol. 1, 771, Elsevier Science, Oxford, U. K., **1995**.

P. J. Murphy, "One or More =CH, =CC and/or C=C Bond(s) Formed by Rearrangement," in *Comprehensive Organic Functional Group Transformations* (A. R. Katritzky, O. Meth-Cohn, C. W. Rees, Eds.), Vol. 1, 793, Elsevier Science, Oxford, U. K., **1995**.

14.3

M. Saunders, J. Chandrasekhar, P. v. R. Schleyer, "Rearrangements of Carbocations", in *Rearrangements in Ground and Excited States* (P. D. Mayo, Ed.), Vol. 1, 1, AP, New York, **1980**.

W. Kirmse, "Umlagerungen von Carbokationen," *Chem. unserer Zeit*, **1982**, *16*, 197–206.

J. R. Hanson, "Wagner–Meerwein Rearrangements", in *Comprehensive Organic Synthesis* (B. M. Trost, I. Fleming, Eds.), Vol. 3, 705, Pergamon Press, Oxford, U. K., **1991**.

K. U. Wendt, G. E. Schulz, E. J. Corey, D. R. Liu, "Enzyme Mechanisms for Polycyclic Triterpene Formation," *Angew. Chem. Int. Ed. Engl.* **2000**, *29*, 2812–2833.

V. A. Chuiko, O. G. Vyglazov, "Skeletal Rearrangements of Monoterpenoids of the Carane Series", *Russ. Chem. Rev.* **2003**, *72,* 54–74.

B. Rickborn, "The Pinacol Rearrangement," in *Comprehensive Organic Synthesis* (B. M. Trost, I. Fleming, Eds.), Vol. 3, 721, Pergamon Press, Oxford, U. K., **1991**.

D. J. Coveney, "The Semipinacol and Other Rearrangements", in *Comprehensive Organic Synthesis* (B. M. Trost, I. Fleming, Eds.), Vol. 3, 777, Pergamon Press, Oxford, U. K., **1991**.

C. D. Gutsche, "The Reaction of Diazomethane and Its Derivatives with Aldehydes and Ketones," *Org. React.* **1954**, *8*, 364–429.

P. A. S. Smith, D. R. Baer, "The Demjanov and Tiffeneau-Demjanov Ring Expansions," *Org. React.* **1960**, *11*, 157–188.

T. Money, "Remote Functionalization of Camphor: Application to Natural Product Synthesis," in *Organic Synthesis: Theory and Applications* (T. Hudlicky, Ed.), **1996**, *3*, JAI, Greenwich, CT.

W. E. Bachmann and W. S. Struve, "The Arndt-Eistert Reaction", *Org. React.* **1942**, *1*, 38–62.

G. B. Gill, "The Wolff Rearrangement," in *Comprehensive Organic Synthesis* (B. M. Trost, I. Fleming, Eds.), Vol. 3, 887, Pergamon Press, Oxford, U. K., **1991**.

T. Ye, M. A. McKervey, "Organic Synthesis with α-Diazo Carbonyl Compounds," *Chem. Rev.* **1994**, *94*, 1091–1160.

M. P. Doyle, M. A. McKervey, T. Ye, "Reactions and Syntheses with α-Diazocarbonyl Compounds", Wiley, New York, **1997**.

Grandjean, D.; Pale, P., Selective debromination of 1,1-dibromoalkenes; a new access to di- or trisubstituted alkenes. *Tetrahedron Letters* (1993), 34(7), 1155–8.

14.4

C. H. Hassall, "The Baeyer-Villiger Oxidation of Aldehydes and Ketones", *Org. React.* **1957**, *9*, 73–106.

G. R. Krow, "The Baeyer-Villiger Reaction", in *Comprehensive Organic Synthesis* (B. M. Trost, I. Fleming, Eds.), Vol. 7, 671, Pergamon Press, Oxford, U. K., **1991**.

G. R. Krow, "The Bayer-Villiger Oxidation of Ketones and Aldehydes," *Org. React.* **1993**, *43*, 251–798.

L. G. Donaruma, W. Z. Heldt, "The Beckmann Rearrangement", *Org. React.* **1960**, *11*, 1–156.

D. Craig, "The Beckmann and Related Reactions", in *Comprehensive Organic Synthesis* (B. M. Trost, I. Fleming, Eds.), Vol. 7, 689, Pergamon Press, Oxford, U. K., **1991**.

R. E. Gawley, "The Beckmann Reactions: Rearrangements, Elimination-Additions, Fragmentations, and Rearrangement-Cyclizations," *Org. React.* **1988**, *35*, 1–420.

P. A. S. Smith, "The Curtius Reaction", *Org. React.* **1946**, *3,* 337–449.

14.5

D. S. Tarbell, "The Claisen Rearrangement," *Org. React.* **1944**, *2,* 1–48.

S. J. Rhoads, N. R. Raulins, "The Claisen and Cope Rearrangements," *Org. React.* **1975**, *22*, 1–252.

F. E. Ziegler, "The Thermal Aliphatic Claisen Rearrangement," *Chem. Rev.* **1988**, *88*, 1423.

P. Wipf, "Claisen Rearrangements", in *Comprehensive Organic Synthesis* (B. M. Trost, I. Fleming, Eds.), Vol. 5, 827, Pergamon Press, Oxford, U. K., **1991**.

H.-J. Altenbach, "Diastereoselective Claisen Rearrangements," in *Organic Synthesis Highlights* (J. Mulzer, H.-J. Altenbach, M. Braun, K. Krohn, H.-U. Reißig, Eds.), VCH, Weinheim, New York, **1991**, 111–115.

H.-J. Altenbach, "Ester Enolate Claisen Rearrangements", in *Organic Synthesis Highlights* (J. Mulzer, H.-J. Altenbach, M. Braun, K. Krohn, H.-U. Reißig, Eds.), VCH, Weinheim, New York, **1991**, 116–118.

S. Pereira, M. Srebnik, "The Ireland–Claisen Rearrangement," *Aldrichimica Acta* **1993**, *26*, 17–29.

B. Ganem, "The Mechanism of the Claisen Rearrangement: Déjâ vu All Over Again", *Angew. Chem. Int. Ed. Engl.* **1996**, *35*, 937–945.

H. Frauenrath, "Formation of C-C Bonds by [3,3] Sigmatropic Rearrangements," in *Stereoselective Synthesis* (Houben-Weyl) 4th ed., **1996**, (G. Helmchen, R. W. Hoffmann, J. Mulzer, E. Schaumann, Eds.), **1996**, Vol. E21 (Workbench Edition), *6*, 3301–3756, Georg Thieme Verlag, Stuttgart.

Y. Chai, S.-P. Hong, H. A. Lindsay, C. McFarland, M. C. McIntosh, "New Aspects of the Ireland and Related Claisen Rearrangements," *Tetrahedron* **2002**, *58,* 2905–2928.

Further Reading

E. L. Wallis, J. F. Lane, "The Hofmann Reaction", *Org. React.* **1946**, *3,* 267–306.

H. Wolff, "The Schmidt Reaction", *Org. React.* **1946**, *3,* 307–336.

A. H. Blatt, "The Fries Reaction", *Org. React.* **1942**, *1,* 342–369.

G. Magnusson, "Rearrangements of Epoxy Alcohols and Related Compounds", *Org. Prep. Proced. Int.* **1990**, *22*, 547.

A. Heins, H. Upadek, U. Zeidler, "Preparation of Aldehydes by Rearrangement with Conservation of Carbon Skeleton", in *Methoden Org. Chem.* (Houben-Weyl) 4th ed., **1952**–, *Aldehydes* (J. Falbe, Ed.), Vol. E3, 491, Georg Thieme Verlag, Stuttgart, **1983**.

W. Kirmse, "Alkenylidenes in Organic Synthesis", *Angew. Chem. Int. Ed. Engl.* **1997**, *36*, 1164– 1170.

A. Padwa, D. J. Austin, "Ligand Effects on the Chemoselectivity of Transition Metal Catalyzed Reactions of α-Diazo Carbonyl Compounds", *Angew. Chem. Int. Ed. Engl.* **1994**, *33*, 801–811.

G. Strukul, "Transition Metal Catalysis in the Baeyer-Villiger Oxidation of Ketones", *Angew. Chem. Int. Ed. Engl.* **1998**, *37*, 1198–1209.

F. P. J. T. Rutjes, L. B. Wolf, H. E. Schoemaker, "Applications of Aliphatic Unsaturated Non-Proteinogenic α-H-α-Amino Acids," *Perkin Trans. I* **2000**, *24*, 4197–4212.

J. Huang, R. Chen, "An Overview of Recent Advances on the Synthesis and Biological Activity of α-Aminophosphonic Acid Derivatives," *Heteroatom Chem.* **2000**, *11*, 480–492.

M. Braun, "α-Heteroatom Substituted 1-Alkenyllithium Reagents: Carbanions and Carbenoids for C–C Bond Formation", *Angew. Chem. Int. Ed. Engl.* **1998**, *37*, 430–451.

K. N. Houk, J. Gonzalez, Y. Li, "Pericyclic Reaction Transition States: Passions and Punctilios, 1935–1995", *Acc. Chem. Res.* **1995**, *28*, 81.

U. Nubbemeyer, "Recent Advances in Asymmetric [3,3]-Sigmatropic Rearrangements," *Synthesis* **2003**, 961–1008.

R. K. Hill, "Cope, Oxy-Cope and Anionic Oxy-Cope Rearrangements," in *Comprehensive Organic Synthesis* (B. M. Trost, I. Fleming, Eds.), Vol. 5, 785, Pergamon Press, Oxford, U. K., **1991**.

F. E. Ziegler, "Consecutive Rearrangements", in *Comprehensive Organic Synthesis* (B. M. Trost, I. Fleming, Eds.), Vol. 5, 875, Pergamon Press, Oxford, U. K., **1991**.

D. L. Hughes, "Progress in the Fischer Indole Reaction," *Org. Prep. Proced. Int.* **1993**, *25*, 607.

G. Boche, "Rearrangements of Carbanions", *Top. Curr. Chem.* **1988**, *146*, 1.

R. K. Hill, "Chirality Transfer via Sigmatropic Rearrangements," in *Asymmetric Synthesis. Stereodifferentiating Reactions – Part A* (J. D. Morrison, Ed.), Vol. 3, 502, AP, New York, **1984**.

C. J. Roxburgh, "Syntheses of Medium Sized Rings by Ring Expansion Reactions," *Tetrahedron* **1993**, *49*, 10749–10784.

15 Thermal Cycloadditions

15.1 Driving Force and Feasibility of One-Step [4+2]- and [2+2]-Cycloadditions

We dealt with [4+2]-cycloadditions very briefly in Section 3.3.1. As you saw there, a [4+2]-cycloaddition requires two different substrates: one of these is an alkene—or an alkyne—and the other is 1,3-butadiene or a derivative thereof. The reaction product, in this context also called the cycloadduct, is a six-membered ring with one or two double bonds. Some hetero analogs of alkenes, alkynes, and 1,3-butadiene also undergo analogous [4+2]-cycloadditions. In a [2+2]-cycloaddition an alkene or an alkyne reacts with ethene or an ethene derivative to form a four-membered ring. Again, hetero analogs may be substrates in these cycloadditions; allenes and some heterocumulenes also are suitable substrates.

Two new σ-bonds are formed in both the [4+2]- and the [2+2]-cycloadditions while two π-bonds are broken. It is for this reason that most cycloadditions exhibit a significant driving force, and this remains true even when the cycloadduct is strained. Having realized this, one is a bit surprised that only a few of the cycloadditions just mentioned occur quickly (see Figure 15.1, bottom). Others require rather drastic reaction conditions. The two simplest [4+2]-cycloadditions, the additions of ethene or acetylene to 1,3-butadiene (Figure 15.1, top), belong to the latter group of reactions. Some cycloadditions cannot be carried out in a one-step process at all. The [2+2]-cycloadditions between two alkenes or between an alkene and an alkyne (Figure 15.1, center) belong to this kind of cycloadditions.

[2+2]-Cycloadditions are less exothermic than [4+2]-cycloadditions, since the former result in a strained cycloadduct while the latter give unstrained rings. Thus, according to the Hammond postulate, the [2+2]-cycloadditions should occur more slowly than the [4+2]-cycloadditions. While these cycloadditions would be expected to be *slower*, thermochemistry does *not* explain why one-step cycloadditions between two alkenes or between an alkene and an alkyne—in the absence of light—cannot be achieved at all, *whereas other one-step*

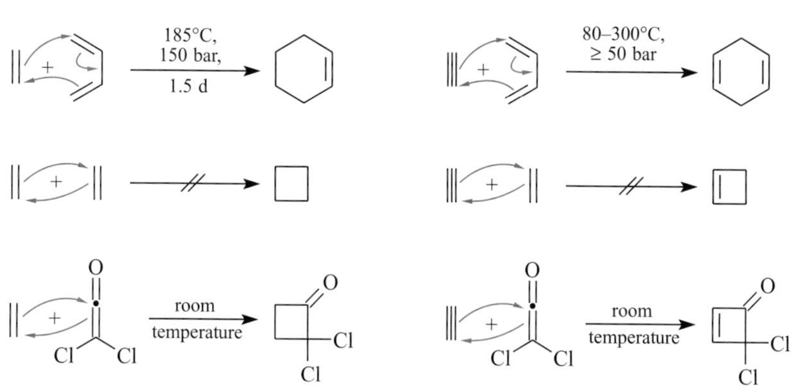

Fig. 15.1. One-step [4+2]- and [2+2]-cycloadditions and their feasibility in the absence of light. The [4+2]-cycloaddition between ethene or acetylene and 1,3-butadiene (top) requires drastic conditions. The one-step [2+2]-cycloaddition between two alkenes or between an alkene and an acetylene (center) cannot be achieved at all. The only [2+2]-cycloadditions that proceed at room temperature are those between ethene or an alkyne and dichloroketene.

Bruckner R (author), Harmata M (editor) In: *Organic Mechanisms – Reactions, Stereochemistry and Synthesis*
Chapter DOI: 10.1007/978-3-642-03651-4_15,

cycloadditions that lead to four-membered rings do occur remarkably fast at room temperature; the additions of dichloroketene to alkenes or acetylenes provide striking examples (Figure 15.1, bottom). The latter [2+2]-cycloadditions afford cyclobutanones and cyclobutenones as cycloadducts, which are even more strained than the cyclobutanes and cyclobutenes, which are inaccessible via one-step additions.

Evidently, the ring-size dependent exothermicities of one-step cycloadditions do not explain the differences in their reaction rates. In fact, these differences can be understood only by going beyond the simplistic "electron-pushing" formalism. To really understand these reactions, one needs to compare the transition states of these reactions in the context of molecular orbital (MO) theory. These comparisons—and the presentation of the requisite theoretical tools—are the subjects of Sections 15.2.2–15.2.4.

15.2 Transition State Structures of Selected One-Step [4+2]- and [2+2]-Cycloadditions

15.2.1 Stereostructure of the Transition States of One-Step [4+2]-Cycloadditions

The combination of powerful computers and modern methods of theoretical chemistry makes it possible to obtain detailed information about transition states and transition state structures. One can compute the Cartesian coordinates of all the atoms involved and all their bond lengths and angles. The energies of the transition states also can be computed. The theoretical estimate for the activation energy of a specific reaction can be determined by subtracting the energies of the starting materials from the energy of the transition state. Yet, we will not be concerned with such numerical data in this section.

The computed transition state of the [4+2]-cycloaddition between ethene and butadiene is shown in Figure 15.2 (top), along with the computed transition state of the [4+2]-cycloaddition between acetylene and butadiene. It is characteristic of the stereochemistry of these transition states that ethene or acetylene, respectively, approaches the *cis*-conformer of butadiene from a face (and not in-plane). Figure 15.2 also shows that the respective cycloadducts—cyclohexene or 1,4-cyclohexadiene—initially result in the twist-boat conformation.

The transition states of the two [4+2]-cycloadditions in Figure 15.2 are "early" because their geometries are more similar to the starting materials than to the cycloadducts (see tabular section of Figure 15.2). To begin with, consider the distances between atoms that are bonded in the starting materials and in the products via bonds that undergo a bond order *change* in the course of the [4+2]-cycloadditions. In going from the starting materials to the transition states, these distances are altered (increased or decreased) by *less* than half the overall difference between the starting materials and the products. Second, the newly formed σ-bonds in both transition states remain about 1.5 times longer than the respective bond lengths in the cycloadducts: the formation of these σ-bonds is only just starting in the transition states.

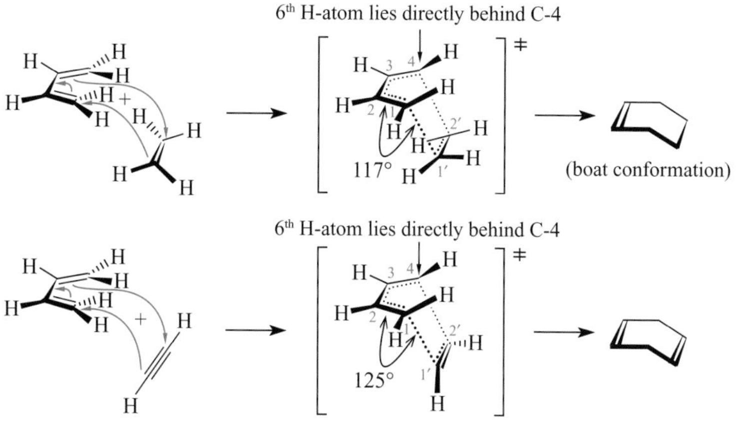

6th H-atom lies directly behind C-4

117°

(boat conformation)

6th H-atom lies directly behind C-4

125°

[4 + 2]-Addition of 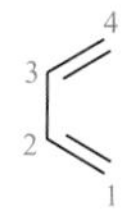 (butadiene, C-1 to C-4) and ...	C^1/C^2 or C^3/C^4	C^2/C^3	$C^{1'}/C^{2'}$	$C^1/C^{1'}$ and $C^4/C^{2'}$
	Extent of the bond length changes in the transition state (100% bond length change is realized when the cycloadduct has been reached):			Extent of the bond length elongations in the transition state (in percent of the respective bond length in the cycloadduct):
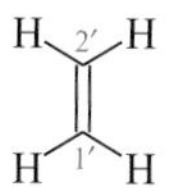 $H_2C^{2'}{=}C^{1'}H_2$ (ethene)	31%	43%	28%	152%
 $HC^{2'}{\equiv}C^{1'}H$ (acetylene)	28%	52%	29%	146%

Fig. 15.2. Perspective drawings of the transition state structures of the [4+2]-cycloadditions ethene + butadiene → cyclohexene and acetylene + butadiene → 1,4-cyclohexadiene.

Third, the hybridization change of the four C atoms that change their hybridization in the course of the [4+2]-cycloaddition has not progressed much: the bond angles at these C atoms have changed only very little in comparison to the bond angles in the starting materials.

15.2.2 Frontier Orbital Interactions in the Transition States of One-Step [4+2]-Cycloadditions

What Are the Factors Contributing to the Activation Energy of [4+2]-Cycloadditions?

The computation of the activation energy of the cycloaddition of ethene and butadiene requires that one sums up the cumulative effects of all the energy changes that are associated with the formation of the transition state of this reaction (Figure 15.2, top) from the starting materials. To begin with, there are the energy increases due to the changes of bond lengths and bond angles, which already were presented. A second contribution is due to the newly occurring inter- and intramolecular van der Waals repulsions. The energy lowering that is associated with incipient bond formations would be a third contribution to consider.

The last factor often is the one that determines the reaction rates of [4+2]-cycloadditions. This factor allows one to understand, for example, why the cycloadditions of ethene or acetylene with butadiene (cf. Figure 15.1) occur only under rather drastic conditions, while the analogous cycloadditions of tetracyanoethene or acetylenedicarboxylic acid esters are relatively rapid. As will be seen, a simple orbital interaction between the reagents at the sites where the new σ bonds are formed is responsible for this advantageous reduction of the activation energies of the latter two reactions.

One can, with surprisingly simple methods, establish whether such transition state stabilizations through orbital interactions occur, and one can even estimate their extent. These considerations are based on the knowledge that the transition states are "early" (Section 15.2.1), that is, that the transition states resemble the starting materials both structurally and energetically. It is for this reason that one can *model these transition states using the starting materials and discuss the MOs of the transition states by inspection of the MOs of the starting materials.* The stabilization of the transition states as the result of the incipient bond formations is thus due to additional orbital overlap, which does not occur in the separated starting materials. Note that these overlaps result from *intermolecular* orbital interactions.

These intermolecular orbital interactions, which are of the σ type, occur between the ends of the π-type MOs that are associated with the respective two reagents. We will see in the next subsection what the ends of these π-type MOs are like. In the subsequent subsection we will deal with the energy effects of the new orbital interactions. The energy effects associated with the special case of [4+2]-cycloadditions will be considered thereafter. Finally, these new orbital interactions will be discussed for [2+2]-cycloadditions.

The LCAO Model of π-MOs of Ethene, Acetylene, and Butadiene; Frontier Orbitals

There are a variety of methods for the computation of the MOs that interact in the transition states of [4+2]-cycloadditions. The **LCAO method** (**l**inear **c**ombination of **a**tomic **o**rbitals) is often employed, and the basic idea is as follows. The MOs of the π-systems of alkenes, conjugated polyenes, or conjugated polyenyl cations, radicals, or anions all are built by so-called linear combinations of $2p_z$ AOs. In a somewhat casual formulation, one might say that the MOs of these π-systems are constructed "with the help of the $2p_z$ AOs." These AOs are centered at the positions of the n C atoms that are part of the π-system. LCAO computations describe a conjugated π-electron system that extends over n sp^2-hybridized C atoms by way of n π-type MOs.

MOs that describe a π-system and have lower and higher energy, respectively, than a $2p_z$ AO, are called **bonding** and **antibonding MOs**, respectively. Figure 15.3 shows examples. In acyclic π-systems with an odd number of centers n, a **nonbonding MO** also occurs. This is illustrated in Figure 15.3 as well. The nonbonding MO has the same energy as the $2p_z$ AO. An **MO diagram** shows the n π-type MOs and their occupation.

The distribution of π-electrons over the π-MOs is regulated by the *Aufbau* principle and the Pauli rule. The highest energy occupied MO thereof is called the **HOMO** (**h**ighest **o**ccupied **m**olecular **o**rbital). The lowest unoccupied MO thereof is called the **LUMO** (**l**owest **u**noccu-

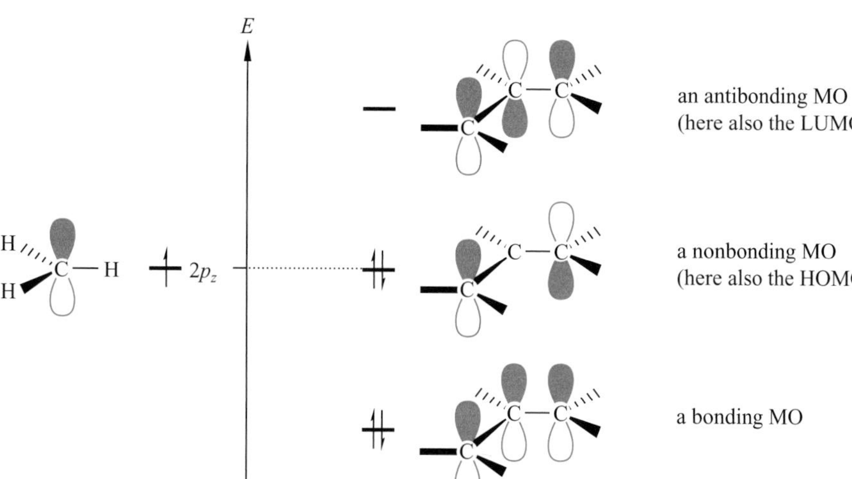

Fig. 15.3. Illustration of the term "MO diagram": left, the π-MO diagram of the methyl radical (the only π-MO is identical with the $2p_z$ AO of the trivalent C atom: cf. Section 1.2); right, the π-MO diagram of the allyl anion.

pied **m**olecular **o**rbital). HOMOs and LUMOs are the so-called **frontier orbitals** since they flank the borderline between occupied and unoccupied orbitals.

The application of the LCAO method to ethene yields two π-type MOs, since two sp^2-hybridized centers build up the π-electron system. Butadiene contains four π-type MOs because four sp^2-hybridized centers build up the conjugated π-electron system. The MOs of ethene and butadiene and their occupations are shown in Figure 15.4. In the π-MO diagrams

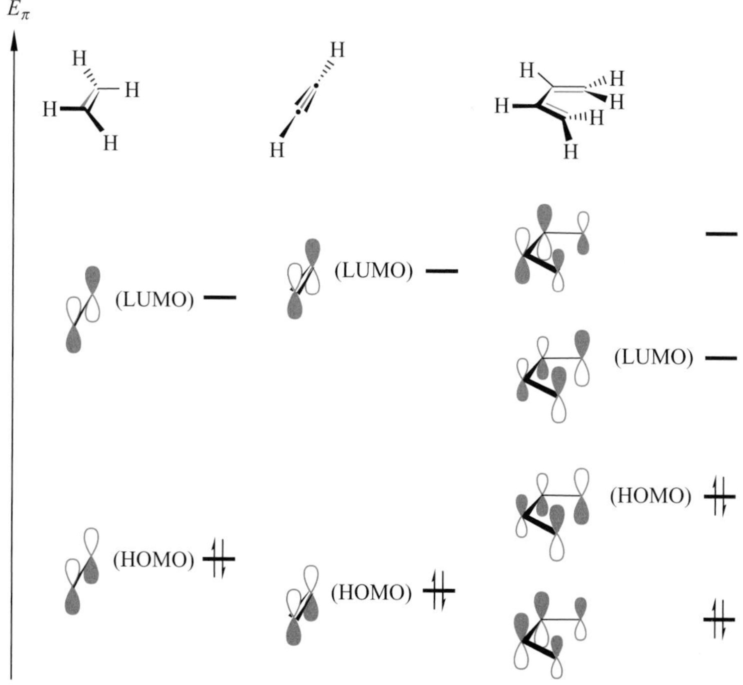

Fig. 15.4. The π-MOs of ethene, acetylene, and 1,3-butadiene and the respective energy-level diagrams. The sign of the $2p_z$ AOs is indicated by the open and shaded orbital lobes. The relative importance of each contributing AO is indicated by the size of the atomic orbital.

of ethene and butadiene, all the bonding MOs—one for ethene and two for butadiene—are completely occupied. The antibonding MO of ethene and the antibonding MOs of butadiene are unoccupied.

Frontier Orbital Interactions in Transition States of Organic Chemical Reactions and Associated Energy Effects

Generally, when two originally isolated molecules approach each other as closely as is the case in the transition state of a chemical reaction, interactions will occur between all the MOs of one molecule and all the MOs of the other molecule in those regions in which the molecules approach each other most closely. From now on, we shall refer to those moieties of the MOs of the substrates, which are directly involved in the mentioned interactions, as "orbital fragments" of the respective MOs. The MOs that used to be localized on the individual substrates now serve to create new MOs that are delocalized over both substrates. If each substrate has only one MO that contributes to the orbital interaction, the latter leads to *two* more delocalized MOs. Specifically, the result of this interaction is that one of the delocalized MOs lies lower in energy than the more stable MO of the isolated substrates while the other delocalized MO is higher in energy than the less stable MO of the isolated substrates. The stabilization of the one delocalized MO is proportional to the bonding overlap between the MOs of the substrates. The other delocalized MO becomes destabilized by about the same amount as a result of the antibonding overlap between the MOs of the substrates. Several factors determine whether the interactions between the substrate MOs—or, differently expressed, the concomitant formation of more delocalized MOs in the transition state—will lead to a stabilization or a destabilization of the transition state and what the extent of that (de)stabilization will be. These factors can be summarized as follows.

- The interaction between a fully occupied MO of the substrate and a fully occupied MO of the other starting material at a single center of each component leads neither to stabilization nor to destabilization (Figure 15.5a). The same is true if both interacting MOs are empty (Figure 15.5c).

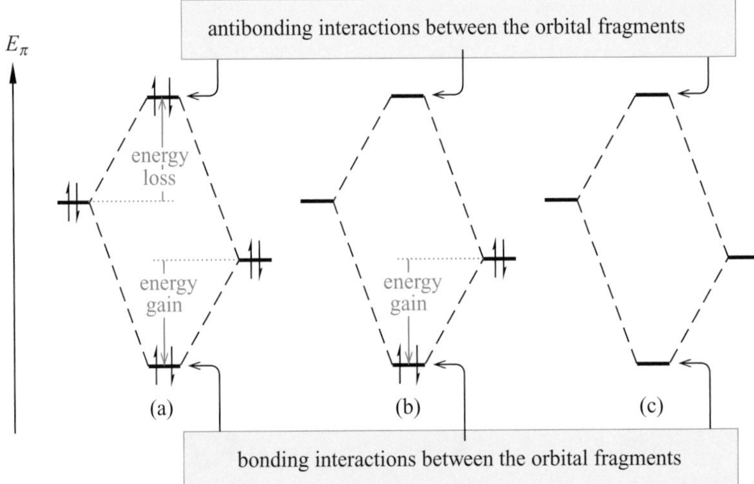

Fig. 15.5. σ-Type interaction between one end of a conjugated π-electron system and one end of another conjugated π-electron system; influence of the orbital occupancy of the π-MOs on the electronic energy.

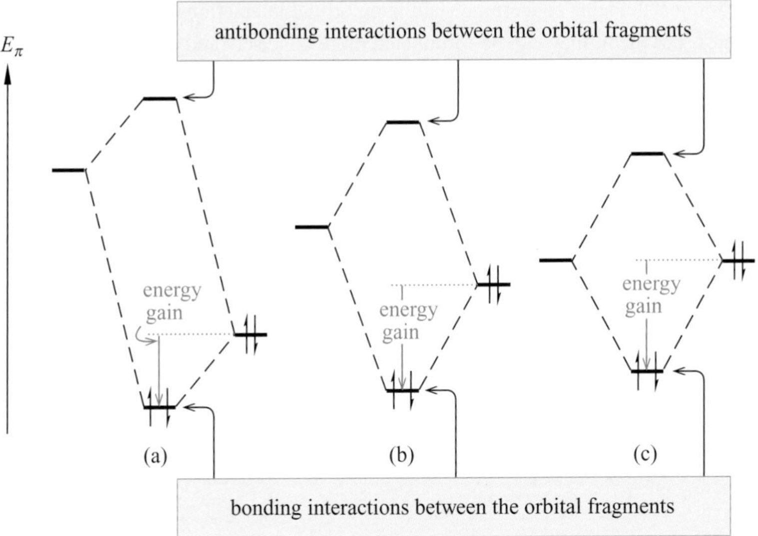

Fig. 15.6. σ-Type interaction between one end of a conjugated π-electron system and one end of another conjugated π-electron system; influence of the energy difference between the π-MOs on the energy gain.

- The interaction between a fully occupied MO of the substrate and an unoccupied MO of the other starting material at a single center of each component leads to stabilization if the interaction is bonding (Figure 15.5b) and to destabilization if the interaction is antibonding.
- The extent of the stabilization or destabilization, respectively, is inversely proportional to the energy difference between the localized MOs involved (Figure 15.6). The effect increases with decreasing energy difference.
- For a given energy difference between the interacting MOs, the magnitude of the stabilization or destabilization, respectively, is proportional to the amount of overlap between the orbital fragments (Figure 15.7 provides a plausible example). A large overlap results in a large stabilization or destabilization, respectively, and vice versa.

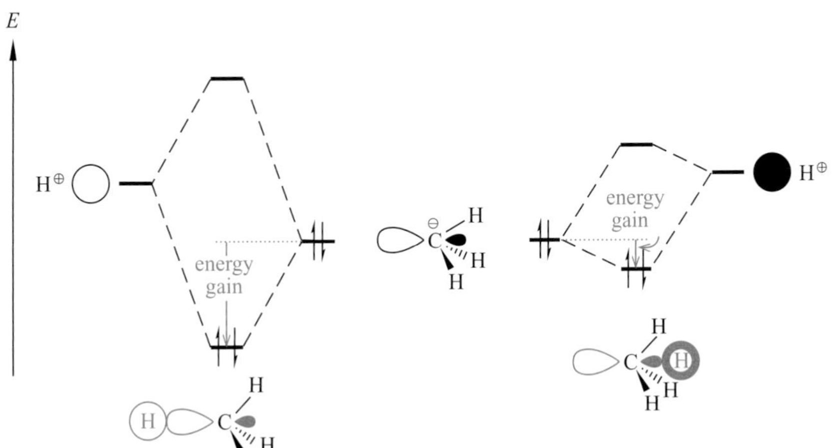

Fig. 15.7. σ-Type interactions between the unoccupied 1*s* AO of a proton and the doubly occupied sp^3 AO of a CH_3 anion; influence of the magnitude of the overlap on the stabilization of the transition states of two bond-forming reactions. Left, formation of tetrahedral methane; right, formation of a fictitious stereoisomer—an unsymmetrical trigonal bipyramid.

The foregoing statements concerning stabilizing orbital interactions can be extended to transition states. For them, they can be formulated more concisely and more generally using the term "frontier orbitals," as introduced in the discussion of Figure 15.3:

Essentially the only mechanism for the electronic stabilization of transition states is the bonding interaction between fully occupied and empty frontier orbitals.

Equation 15.1 offers a quantitative formulation of this statement. This equation makes a statement about the stabilization ΔE_{TS} of the transition state of a reaction between substrate I (reacting at its reactive center 1, where its frontier orbitals have the coefficients $C_{1,\mathrm{HOMO_I}}$ and $C_{1,\mathrm{LUMO_I}}$) and substrate II (reacting at its reactive center 1 as well, where its frontier orbitals have the coefficients $C_{1,\mathrm{HOMO_{II}}}$ and $C_{1,\mathrm{LUMO_{II}}}$) owing to the two frontier orbital interactions:

$$\Delta E_{\mathrm{TS}} \propto \frac{\left(C_{1,\mathrm{HOMO_I}} \cdot C_{1,\mathrm{LUMO_{II}}}\right)^2}{E_{\mathrm{HOMO_I}} - E_{\mathrm{LUMO_{II}}}} + \frac{\left(C_{1,\mathrm{LUMO_I}} \cdot C_{1,\mathrm{HOMO_{II}}}\right)^2}{E_{\mathrm{HOMO_{II}}} - E_{\mathrm{LUMO_I}}} \tag{15.1}$$

Here we are mostly interested in the transition states of one-step cycloadditions between two unsaturated molecules I and II. In this special case, the frontier orbitals will be π-type orbitals, and overlaps at *two* ends of each orbital fragment contribute to each frontier orbital interaction: overlap at the termini C-1_{I}/C-1_{II} and another overlap involving the termini C-ω_{I}/C-ω_{II}. (Substrate I reacts at its C atom 1 with C atom 1 of substrate II and at its C atom ω with C atom ω of substrate II). Substrate I possesses the frontier orbital coefficients $C_{1,\mathrm{HOMO_I}}$ and $C_{1,\mathrm{LUMO_I}}$ at its reactive center $\mathrm{C1_I}$ and the coefficients $C_{\omega,\mathrm{HOMO_I}}$ and $C_{\omega,\mathrm{LUMO_I}}$ at $\mathrm{C}\omega_I$. In analogy, the frontier orbital coefficients of substrate II are $C_{1,\mathrm{HOMO_{II}}}$ and $C_{1,\mathrm{LUMO_{II}}}$ at its reactive center $\mathrm{C1_{II}}$ and $C_{\omega,\mathrm{HOMO_{II}}}$ and $C_{\omega,\mathrm{LUMO_{II}}}$ at its reactive center $\mathrm{C}\omega_{II}$. The stabilization ΔE_{TS} of the transition state of such a one-step cycloaddition can be expressed in terms of the frontier orbitals as Equation 15.2.

$$\Delta E_{\mathrm{TS}} \propto \frac{\left[\left(C_{1,\mathrm{HOMO_I}} \cdot C_{1,\mathrm{LUMO_{II}}}\right) + \left(C_{\omega,\mathrm{HOMO_I}} \cdot C_{\omega,\mathrm{LUMO_{II}}}\right)\right]^2}{E_{\mathrm{HOMO_I}} - E_{\mathrm{LUMO_{II}}}} + \frac{\left[\left(C_{1,\mathrm{LUMO_I}} \cdot C_{1,\mathrm{HOMO_{II}}}\right) + \left(C_{\omega,\mathrm{LUMO_I}} \cdot C_{\omega,\mathrm{HOMO_{II}}}\right)\right]^2}{E_{\mathrm{HOMO_{II}}} - E_{\mathrm{LUMO_I}}} \tag{15.2}$$

Interestingly, according to Equation 15.2 the HOMO/LUMO interactions in cycloadditions are not necessarily stabilizing; they also can be nonbonding. Whether the interaction is bonding or nonbonding depends on the size and the sign of the fragment orbitals at the reacting centers. In contrast, according to Equation 15.1, the HOMO/LUMO interactions for reactions in which only one bond is formed are always bonding.

Frontier Orbital Interactions in Transition States of One-Step [4+2]-Cycloadditions

Figure 15.2 (Section 15.2.1) showed the stereostructures of the transition states of the [4+2]-cycloadditions between ethene or acetylene, respectively, and butadiene. The HOMOs and LUMOs of all substrates involved are shown in Figure 15.4. Figures 15.8 and 15.9 depict the corresponding HOMO/LUMO pairs in the transition states of the respective [4+2]-cycloadditions. Evaluation of Equation 15.2 reveals two new bonding HOMO/LUMO interactions of comparable size in both transition states. Therefore, the transition states of *both* cycloadditions benefit from a stabilization that is attenuated by a large energy difference between the frontier orbitals involved. That is why fairly drastic conditions are require for these specific processes.

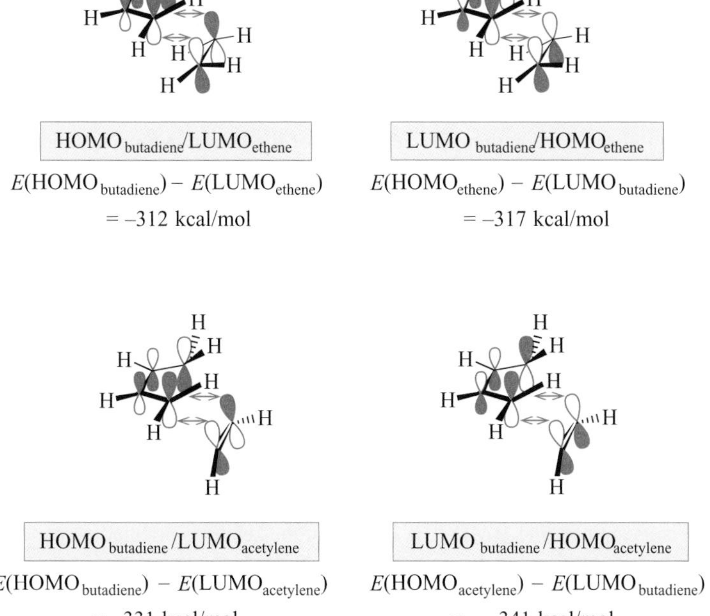

Fig. 15.8. Frontier orbital interactions in the transition state of the one-step [4+2]-cycloaddition of ethene and butadiene.

Fig. 15.9. Frontier orbital interactions in the transition state of the one-step [4+2]-cycloaddition of acetylene and butadiene.

15.2.3 Frontier Orbital Interactions in the Transition States of the Unknown One-Step Cycloadditions of Alkenes or Alkynes to Alkenes

The one-step cycloadditions ethene + ethene → cyclobutane and ethene + acetylene → cyclobutene are unknown (see Figure 15.1). One can understand why this is so by analyzing the frontier orbital interactions in the associated transition states (Figure 15.10). *Both*

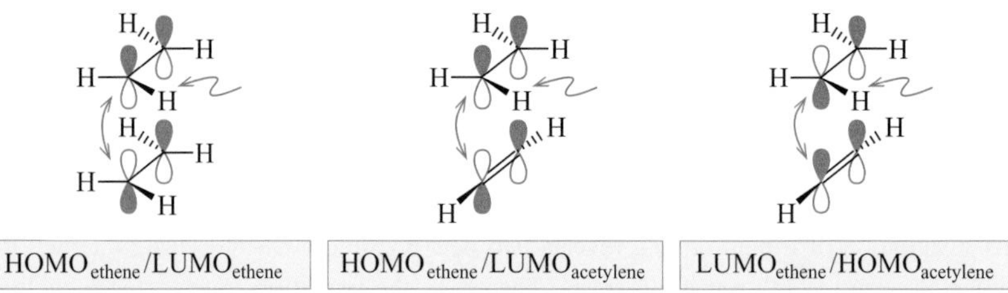

Fig. 15.10. Frontier orbital interactions in plausible transition states of the one-step [2+2]-cycloadditions of ethene and ethene (left) and of ethene and acetylene (center and right).

HOMO/LUMO interactions are nonbonding. This circumstance contributes to the fact that the respective transition states are energetically out of reach.

15.2.4 Frontier Orbital Interactions in the Transition State of One-Step [2+2]-Cycloadditions Involving Ketenes

The transition state of the [2+2]-cycloaddition of ketene and ethene is shown in Figure 15.11. In this transition state, the carbonyl C atom (C2) of the ketene approaches the ethene more closely than the methylene C atom (C1) does. The two C atoms and the O atom of the ketene fragment no longer are collinear. Yet, all five atoms of the ketene remain in one plane. The structural changes between ethene and the ethene moiety in the transition state are minor. Besides, in the transition state of the [2+2]-cycloaddition, the four atoms that will eventually form the cycloadduct are still far removed from their positions in the cycloadduct. All these structural features characterize this transition state as an early one. Therefore, much as in the case of the transition states of the one-step [4+2]-cycloadditions in Section 15.2.2, the bonding situation can be described by means of the MOs of the separated reagents.

The π-MO diagram of the substrate ethene is shown on the left of Figure 15.12; in the center and to the right is shown the MO diagram of the other reactant, the ketene. The HOMO of ethene (HOMO$_A$ in Figure 15.12) is its bonding π-MO and the LUMO (LUMO$_A$ in Figure 15.12) is its antibonding π-MO. The HOMO of the ketene is oriented perpendicular with regard to the plane of the methylene group (HOMO$_B$ in Figure 15.12). This MO extends over

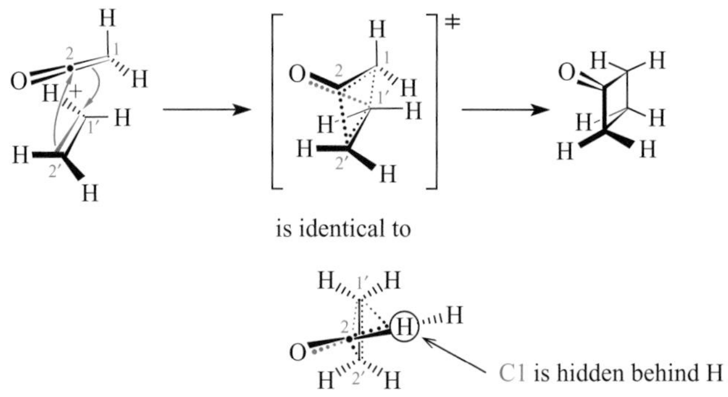

is identical to

C1 is hidden behind H

Fig. 15.11. Perspective drawing of the structure of the transition state of the [2+2]-cycloaddition ketene + ethene → cyclobutanone.

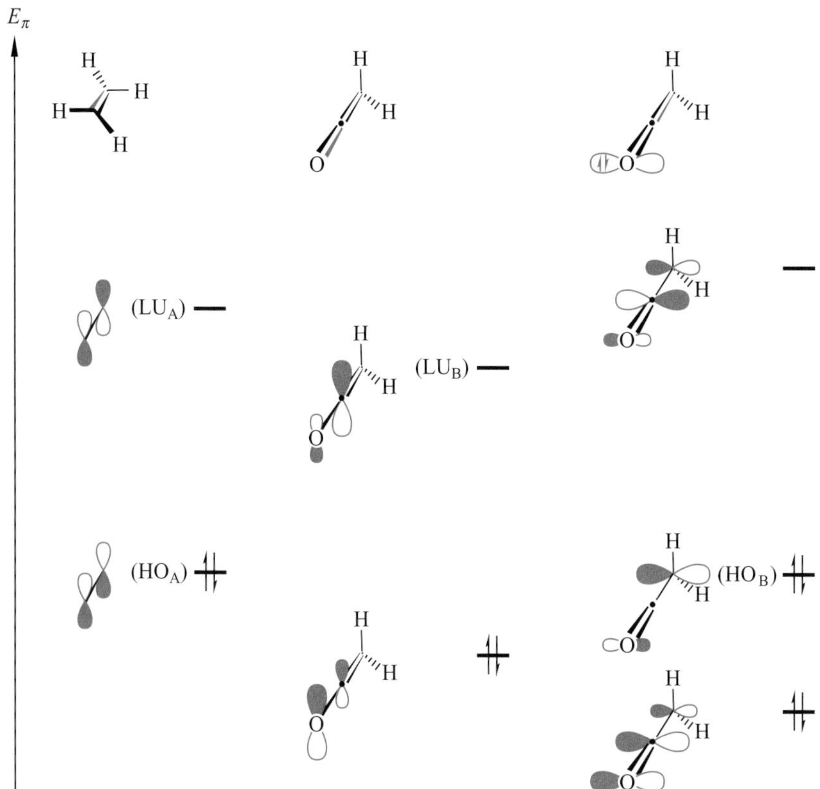

Fig. 15.12. π-Type MOs of ethene (left) and ketene (center and right); the subscripts A and B refer to ethene and ketene, respectively.

three centers; it has its largest coefficient at the methylene carbon, a small coefficient at the oxygen, and a near-zero coefficient at the central C atom. The LUMO of ketene is the antibonding π*-orbital of the C=O double bond (LUMO$_B$ in Figure 15.12). This MO lies in the plane of the methylene group: that is, it is perpendicular to the HOMO$_B$, and its largest coefficient by far is located at the carbonyl carbon.

As with the transition state of the [4+2]-addition of butadiene and ethene (Figure 15.8) *both* HOMO/LUMO interactions are stabilizing in the transition state of the [2+2]-addition of ketene to ethene (Figure 15.13). This explains why [2+2]-cycloadditions of ketenes to alkenes—and similarly to alkynes—can occur in one-step reactions while this is not so for the additions of alkenes to alkenes (Section 15.2.3).

In contrast to the [4+2]-additions of butadiene to ethene or acetylene (Figures 15.8 and 15.9), the two HOMO/LUMO interactions stabilize the transition state of the [2+2]- addition of ketenes to alkenes *to a very different extent.* Equation 15.2 reveals that the larger part of the stabilization is due to the LUMO$_{\text{ketene}}$/HOMO$_{\text{ethene}}$ interaction. This circumstance greatly affects the geometry of the transition state. If there were *only this one* frontier orbital interaction in the transition state, the carbonyl carbon of the ketene would occupy a position in the transition state that would be perpendicular above the midpoint of the ethene double bond. The Newman projection of the transition state (Figure 15.11) shows that this is almost the case but

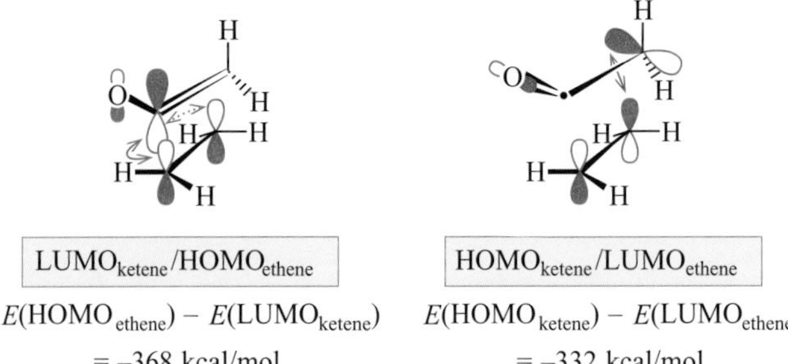

Fig. 15.13. Frontier orbital interactions in the transition state of the one-step [2+2]-cycloaddition of ketene and ethene.

not quite. The less stabilizing frontier orbital interaction—the one between the HOMO of the ketene and the LUMO of the ethene—is responsible for this small distortion. The *large* $2p_z$ lobe of the ketene's HOMO—located at the methylene carbon—overlaps best with the LUMO of ethene in a way that a banana bond of sorts results between this lobe and *one* of the $2p_z$ lobes of the ethene LUMO. It is for this reason that the carbonyl carbon of the ketene is slightly moved out of the π-orbital plane of the ethene and at the same time approaches one of the two C atoms of ethene (C2) more closely.

15.3 Diels–Alder Reactions

[4+2]-Cycloadditions are called Diels–Alder reactions in honor of Otto Diels and Kurt Alder, the chemists who carried out the first such reaction. The substrate that reacts with the diene in these cycloadditions is called the **dienophile.** As you saw in Figure 15.1, the simplest Diels–Alder reactions, i.e., the ones between ethene and butadiene and between acetylene and butadiene, respectively, occur only under drastic conditions. Well-designed Diels–Alder reactions, on the other hand, occur much more readily. In the vast majority of those cases, acceptor-substituted alkenes serve as dienophiles. In the present section, we will be concerned only with such Diels–Alder reactions (see Figures 15.16, 15.17, and 15.23 for exceptions).

Up to four stereocenters may be constructed in one Diels–Alder reaction, and the great number of possible substrates gives the reaction great scope. The enormous value of the Diels–Alder reaction for organic synthesis is thus easy to understand. It is not at all exaggerating to say that this reaction is the most important synthesis for six-membered rings and that, moreover, it is one of the most important stereoselective C,C bond-forming reactions in general.

15.3.1 Stereoselectivity of Diels–Alder Reactions

Essentially all Diels–Alder reactions are one-step processes. If the reactions are stereogenic, they often occur with predictable stereochemistry. For example, *cis,trans*-1,4-disubstituted 1,3-butadienes afford cyclohexenes with a *trans*-(!) arrangement of the substituents at C3 and C6 (Figure 15.14, top). In contrast, *trans,trans*-1,4-disubstituted 1,3-butadienes afford cyclohexenes in which the substituents at C3 and C6 are *cis* (!) with respect to each other (Figure 15.14, bottom). Accordingly, Diels–Alder reactions exhibit stereospecificity with regard to the diene.

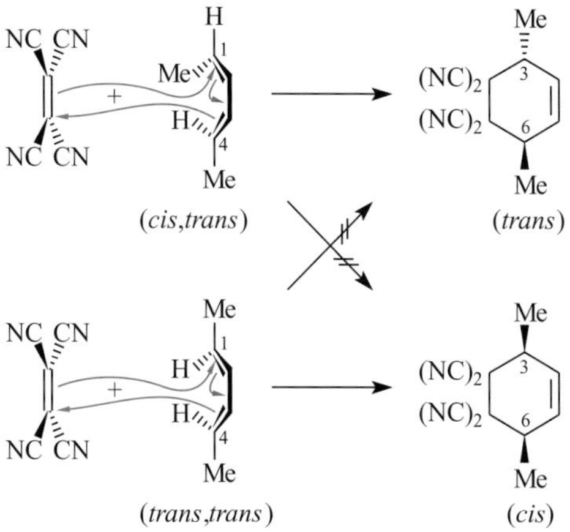

Fig. 15.14. Evidence for stereoselectivity and stereospecificity with regard to the butadiene moiety in a pair of Diels–Alder reactions. The *cis,trans*-1,4-disubstituted 1,3-butadiene forms cyclohexene with a *trans* arrangement of the methyl groups. The *trans,trans*-1,4-disubstituted 1,3-butadiene forms cyclohexene with *cis*-methyl groups.

cis,cis-1,4-Disubstituted 1,3-butadienes undergo Diels–Alder reactions only when they are part of a cyclic diene. Cyclopentadiene is an example of such a diene. In fact it is one of the most reactive dienes in general. In stark contrast, the transition states of Diels–Alder reactions of acyclic *cis,cis*-1,4-disubstituted 1,3-butadienes usually suffer from a prohibitively large repulsion between the substituents in the 1- and 4-positions, which arises when *these* substrates—as any 1,3-diene must—assume the *cis* conformation about the C2—C3 bond.

Stereoselectivity is also observed in Diels–Alder reactions of dienophiles, which contain a stereogenic C=C double bond (Figure 15.15). A *cis,trans*-pair of such dienophiles, moreover,

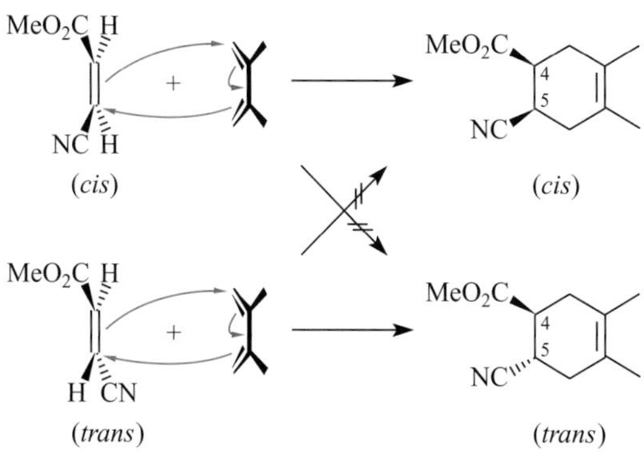

Fig. 15.15. Evidence for stereoselectivity and stereospecificity with regard to the dienophile in a pair of Diels–Alder reactions. The *cis*-configured dienophile affords a 4,5-*cis*-substituted cyclohexene, whereas the *trans* isomer results in a 4,5-*trans*-substituted cyclohexene.

react stereospecifically with 1,3-dienes (as long as the latter do not contain any stereogenic double bonds): the *cis*-configured dienophile affords a 4,5-*cis*-disubstituted cyclohexene (Figure 15.15, top), while its *trans*-isomer gives the 4,5-*trans*-disubstituted product (Figure 15.15, bottom).

Only very few [4+2]-cycloadditions are known that are not stereoselective with regard to the diene or the dienophile (see, e. g., Figure 15.17), or are stereoselective but not stereospecific (e. g., Figure 15.19). From these stereochemical outcomes, one can then safely conclude that the latter [4+2]-cycloadditions are *multistep* reactions.

Chloroprene undergoes three different [4+2]-cycloadditions with itself, proceeding as parallel reactions. One of these [4+2]-cycloadditions does not occur in a stereoselective fashion with respect to the dienophile. These cycloadditions are dimerizations that yield compounds **A–C** in Figure 15.16. Chloroprene plays two roles in these [4+2]-cycloadditions: it serves as diene and also as dienophile. In addition, small amounts of chloroprene dimerize (in a multistep process!) to give a [2+2]-cycloadduct **D** and to give a [4+4]-cycloadduct **E** (Figure 15.16).

The dimerization of chloroprene leading to the [4+2]-cycloadduct **C** (Figure 15.16) definitely is a multistep process. This has been demonstrated by analysis of the stereochemistry of a [4+2]-cycloaddition that led to the dideutero analogs of this cycloadduct (Figure 15.17). Instead of chloroprene, a monodeuterated chloroprene (*trans*-[D]-chloroprene) was dimerized. This monodeuterated chloroprene of course also underwent all five chloroprene dimerization reactions. The elucidation of the stereochemistry of the dideutero analog $[D]_2$-**C** (Figure 15.17) established how the [4+2]-cycloadduct **C** (Figure 15.16) is formed. Compound $[D]_2$-**C** was isolated as a mixture of four racemic diastereomers: ^{1,2}*trans,trans*-$[D]_2$-**C**, ^{1,2}*cis,trans*-$[D]_2$-**C**, ^{1,2}*trans,cis*-$[D]_2$-**C**, and ^{1,2}*cis, cis*-$[D]_2$-**C**.

Cl Cl Cl Cl Cl Cl

probably a onestep process

probably a onestep process

at least in part a multistep process (cf. Figure 15.17)

Chloroprene 50°C

A (21%) + **B** (15%) + **C** (22%) + **D** (23%) + **E** (19%)

Fig. 15.16. Thermal cycloadditions of chloroprene (2-chlorobutadiene) I: product distribution.

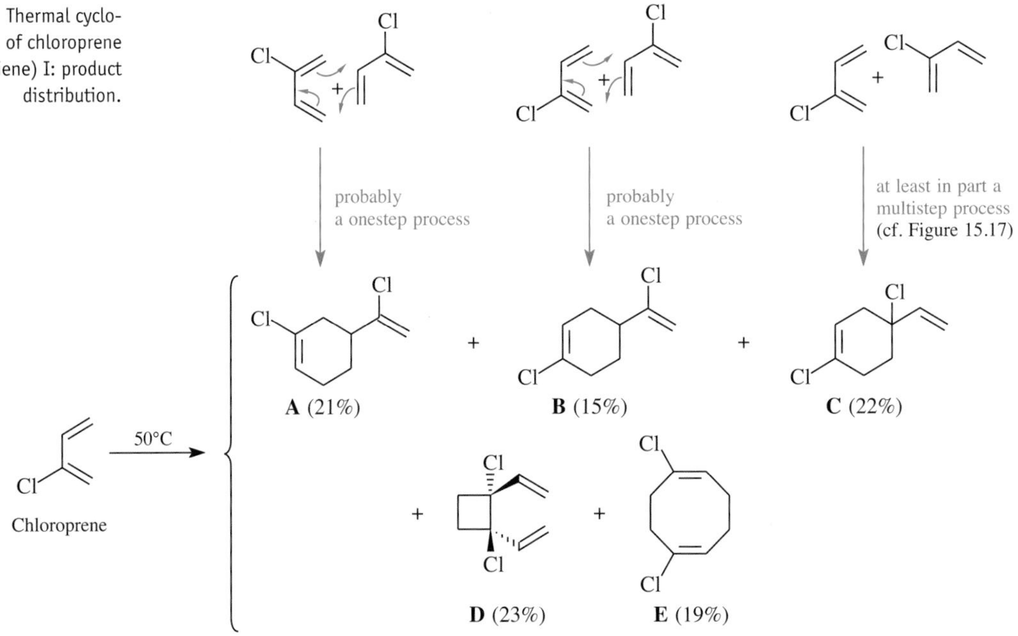

A *one-step* dimerization of the *trans*-[D]-chloroprene shown in Figure 15.17 would yield only the first two diastereomers, that is, ^{1,2}*trans, trans*-$[D]_2$-**C** and ^{1,2}*cis, trans*-$[D]_2$-**C**. In these isomers, the *trans*-arrangement between the D atom and the proximate Cl atom in the dienophile is conserved. However, the same two atoms are in a *cis* arrangement in the additionally formed [4+2]-cycloadducts ^{1,2}*trans, cis*-$[D]_2$-**C** and ^{1,2}*cis,cis*-$[D]_2$-**C**. These cycloadducts $[D]_2$-**C** can only have lost the original *trans*-arrangement of the atoms in question because they were formed via a multistep mechanism, assuming the diene maintaines it stereohemical integrity under the reaction conditions. Specifically, this mechanism must include an intermediate in which the *trans*-arrangement of the D and Cl atoms is partly lost. This is only conceivable if in this intermediate a rotation is possible about the C,C bond that connects the deuterated and the chlorinated C atoms since such a rotation is not possible in the starting material or product.

The most likely multistep mechanism of this type is shown in the lower part of Figure 15.17. It is a two-step mechanism where the diastereomeric diradicals **F** and **G** are the two intermediates that allow for rotation about the configuration-determining C—C bond. Each of the two radical centers is part of a well-stabilized allyl radical (cf. Section 1.2.1). Biradicals **F** and **G** cyclize *without* diastereocontrol to deliver the [4+2]-cycloadducts: biradical **F** forms a *mixture of* ^{1,2}*trans,cis*-$[D]_2$-C and ^{1,2}*trans,trans* $[D]_2$-**C**, since a rotation about the C2—C3 bond is possible but not necessary. For the same reason, biradical **G** forms a mixture of ^{1,2}*cis,cis*-$[D]_2$-**C** and ^{1,2}*cis,trans* $[D]_2$-**C**.

Side Note 15.1. A One-Step Dimerization of a 1,3-Diene via Diels–Alder Reaction

The fact that Section 15.3.1 presents two examples that are definitely two-step Diels–Alder reactions—namely the one in Figure 15.16 proceeding via a 1,6-biradical intermediate (mechansim:Figure 15.17) and the one in Figure 15.19 proceeding via a 1,6-dipole intermediate (mechanism: Figure 15.20)—is not meant to create a false impression: an extensive and targeted search was required in order to find any examples of two-step Diels–Alder reactions at all! In this section, we present almost all the hits that were obtained *cumulatively*. Keeping *this* emphasis in mind will give you the *proper* impression of the mechanism of the Diels–Alder reaction: almost certainly, it is a one-step mechanism—as long as we disregard hetero Diels–Alder reactions.

The reaction of Figure 15.18 also points to this conclusion. There you find a stereochemical study of the Diels–Alder reaction of 1,3-diphenylbutadiene "with itself." Obviously, 1,3-diphenylbutadiene/1,3-diphenylbutadiene is a diene/dienophile pair that is markedly similar to the diene/dienophile pair chloroprene/chloroprene in Figures 15.16/15.17, since in both reacting pairs it is a 1,3-butadiene that plays this double role. In this respect there actually is a "reasonable suspicion" that this corresponding dual function of a 1,3-butadiene will also be associated with a corresponding Diels–Alder mechanism, which in Figure 15.17 was a two-step mechanism proceeding via a (bis-)allyl radical intermediate. This suspicion of a two-step mechanism is being supported by the phenyl groups and their positioning in the substrate: because of these phenyl groups each of the four aromatic resonance forms, which can be drawn for the favorable diradical intermediates **D** or **E** of a two-step Diels–Alder process, would not only be a (bis-)allyl radical but also a bisbenzyl radical. None of the four resonance forms in Figure 15.17, which are used to describe the diradical intermediates **F** and **G** of the evidently two-step Diels–Alder reaction of chloroprene, have benefited from such an extra stabilization!

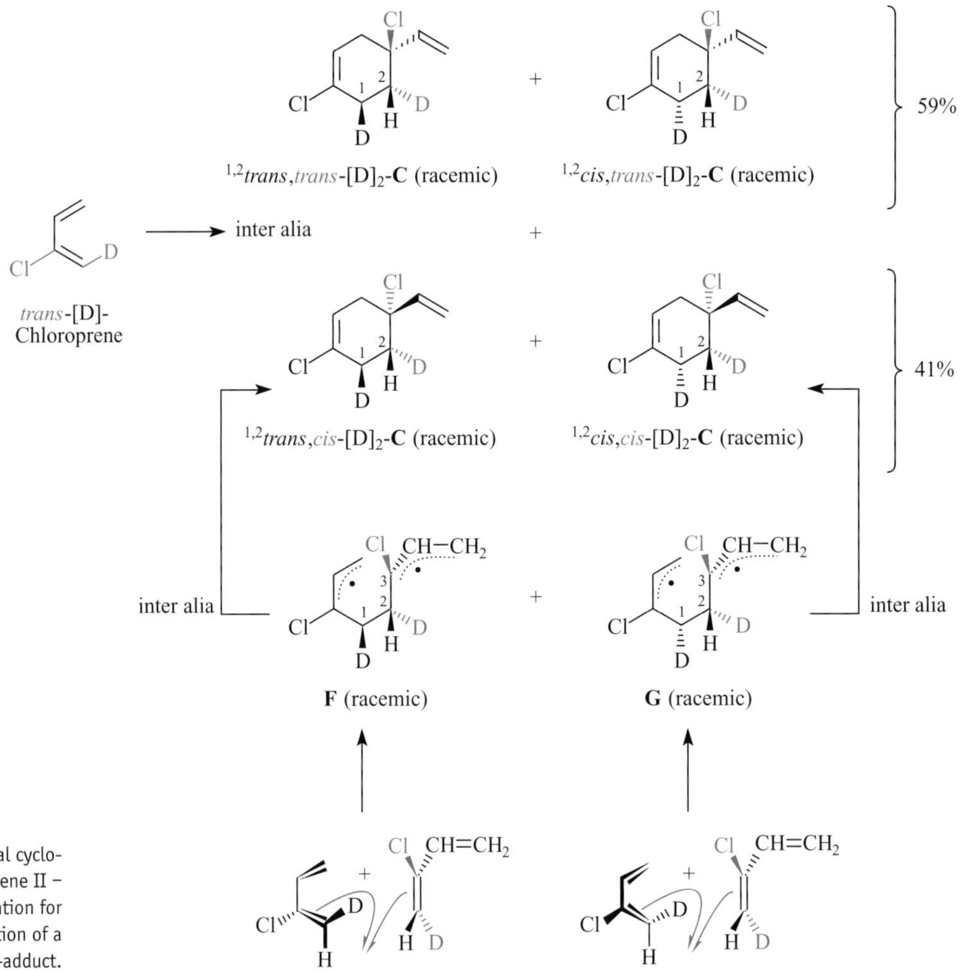

Fig. 15.17. Thermal cycloadditions of chloroprene II – evidence of and explanation for the two-step formation of a [4+2]-adduct.

But what does the stereochemical study of the Diels–Alder reaction of 1,3-diphenylbutadiene "with itself" indicate (Figure 15.18)? The question is whether the resulting cycloadduct can be assigned stereoformula **B** and/or **C**. In order to answer this question, deuterium-labeled *trans*-1,3-diphenylbutadiene was employed, i.e., sterically uniform 4-deutero-1,3-diphenylbutadiene (**A**). The question was whether the phenyl group at the only quarternary carbon atom of the cycloadduct contained deuterium or hydrogen as a *cis*-vicinal substituent. If this were deuterium the Diels–Alder adduct would have stereoformula **B**, in the latter case it would have to be stereoformula **C**. If the Diels–Alder adduct of stereoformula **B** had been produced, we would be able to retrace the *cis*-relation between deuterium and the vicinal phenyl substituent in the substrate **A**. In a Diels–Alder adduct of stereoformula **C, however,** this initial *cis*-relation between deuterium and the vicinal phenyl group would have been translated into a *trans*-relation.

Regarding the cycloaddition mechanism the conclusion is that the Diels–Alder adduct **C** must result from a multiple-step Diels–Alder reaction. The biradical **E** would be a plausible intermediate and **G** a plausible transition state. A Diels–Alder adduct **B** is the inevitable result of a one-step Diels–Alder reaction. Although a multiple-step Diels–Alder reaction via transition state **F** and the biradical intermediate **D** could also lead to cycloadduct **B**, this pathway should be taken with the same reaction rate as the alternative pathway via transition state **G** and the biradical **E** proceeding to cycloadduct **C**. In other words, if the Diels–Alder reaction in Figure 15.18 leads to the exclusive formation of the Diels–Alder adduct **B**, there is every reason to assume that it follows a one-step mechanism. But if a mixture of the Diels–Alder adducts **B** and **C** is formed this proves that a two-step mechanism is involved.

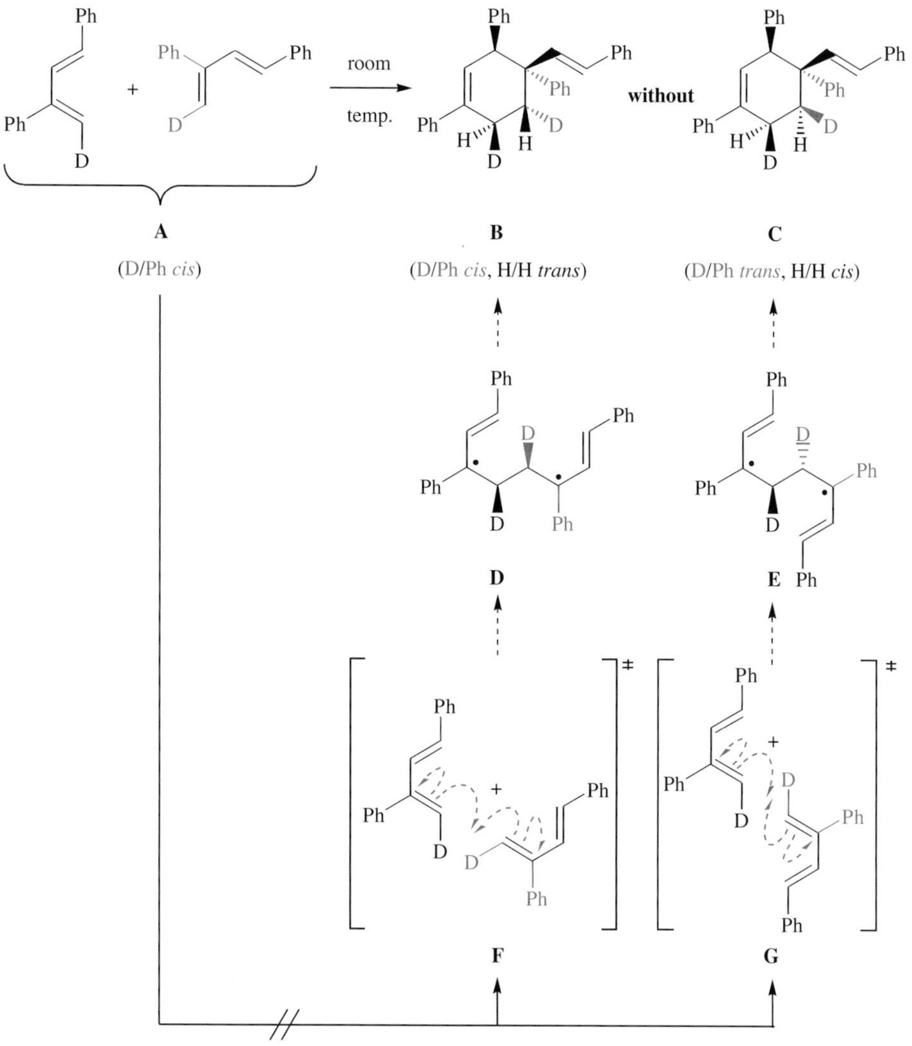

Fig. 15.18. Evidence for the one-step mechanism of the [4+2]-cycloaddition between ^{1,2}trans,^{3,4}cis-4-deutero-1,3-diphenyl-1,3-butadiene. This reaction leads to the exclusive formation of cycloadduct **B**, with a *cis*-relation between the deuterium atom (red) and the phenyl group (red) of the dienophile; in the (non-occurring) cycloadduct **C** these substituents would be in *trans*-position.

The problem is that according the above-mentioned criterion ("quarternary phenyl substituent *cis*-vicinal to D or H?") the stereostructures **B** and **C** cannot be distinguished with the standard workhorse of today's structure elucidation methods, i.e., by way of ^{1}H-NMR spectroscopy. **B** and **C** could be distinguished indirectly, though, by ^{1}H-NMR spectroscopy, since the two vicinal sp^3-bound H atoms in the cyclohexene substructures of the potential Diels–Alder adducts of Figure 15.18 are mutually *trans*-positioned in **B**, whereas they are *cis*-positioned in **C**. ^{1}H-NMR spectroscopy has demonstrated that these H atoms are exclusively *trans*-positioned. This means that a single Diels–Alder adduct has been formed, namely the stereoisomer **B**. As outlined above, this is a clear indication that the reaction presented is a single-step Diels–Alder reaction.

What do we learn from Figure 15.18? Even with Diels–Alder reactions of 1,3-dienes "with themselves" we may not assume that a two-step reaction course is likely!

1-(Dimethylamino)-1,3-butadiene and *trans*-dicyanoethenedicarboxylic acid diester react with each other in a stereoselective [4+2]-cycloaddition to give the cyclohexene *trans*,^{2,3}*trans*-**A** (Figure 15.19). 1-(Dimethylamino)-1,3-butadiene also undergoes a stereoselective [4+2]-cycloaddition with *cis*-dicyanoethenedicarboxylic acid diester (Figure 15.19). However, the latter reaction results in the same cyclohexene *trans*, ^{2,3}*trans*-**A** that is formed from the *trans*-configured ester. Thus, Figure 15.19 shows a pair of stereoselective [4+2]-cycloadditions that occur without stereospecificity but with **stereoconvergence** (see Section 3.2.2 for the introduction of this term). 1-(Dimethylamino)-1,3-butadiene and *trans*-dicyanoethenedicarboxylic acid diester thus form the [4+2]-cycloadduct with complete retention of the *trans*-relationship between the ester groups. The same would be true if a one-step addition mechanism were operative. 1-(Dimethylamino)-1,3-butadiene and *cis*-dicyanoethenedicarboxylic acid diester, on the other hand, form *trans*^{2,3}*trans*-**A** with complete inversion of the relative configuration of the two ester groups. This finding can be explained only by a multistep addition mechanism. A one-step mechanism could lead only to the cycloadducts *cis*^{2,3}*cis*-**A** and *cis*,^{2,3}*trans*-**A.**

Fig. 15.19. [4+2]-Cycloaddition between 1-(dimethylamino)-1,3-butadiene and the two isomeric dicyanoethenedicarboxylic acid diesters I: product distribution.

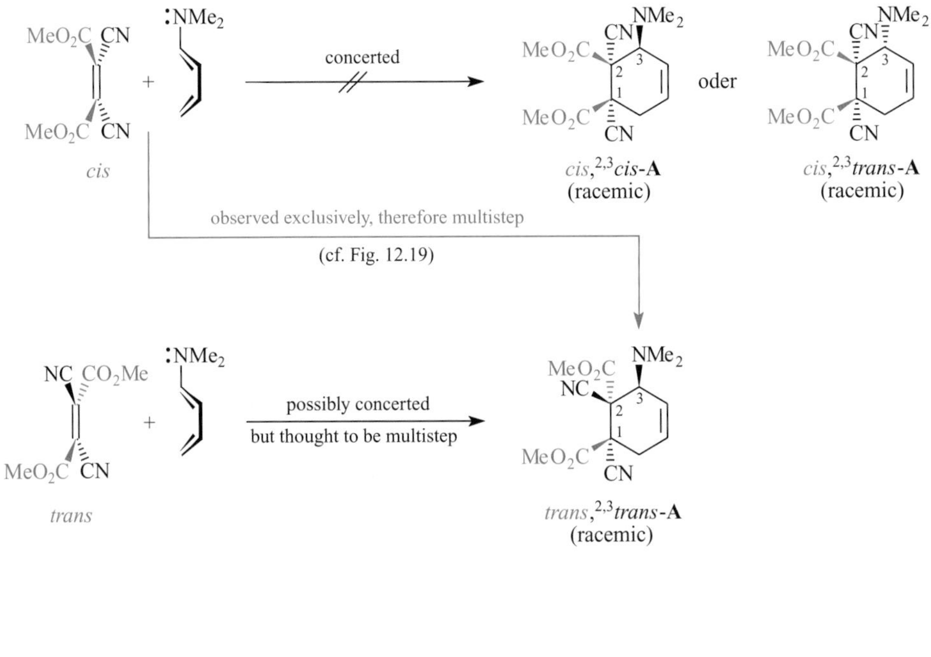

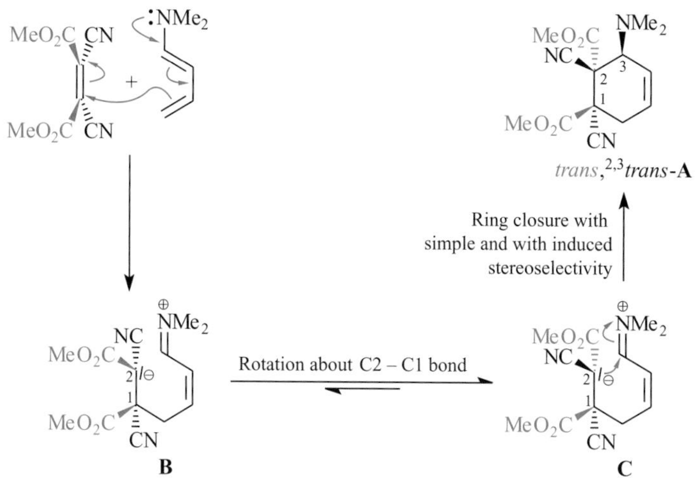

Fig. 15.20. [4+2]-Cycloaddition between 1-(dimethylamino)-1,3-butadiene and *cis*-dicyanoethene dicarboxylic acid diester II: explanation of the inversion of configuration in the dienophile moiety.

Figure 15.20 shows the multistep mechanism of the [4+2]-cycloaddition between 1-(dimethylamino)-1,3-butadiene and *cis*-dicyanoethenedicarboxylic acid diester. The reaction proceeds via an intermediate, which must be zwitterion conformer **B**. The anionic moiety of this zwitterion is well stabilized because it represents the conjugate base of a carbon-acidic compound (Section 13.1.2). The cationic moiety of zwitterion **B** also is well stabilized. It is an iminium ion (i.e., a species with valence electron octet) rather than a carbenium ion (which is a species with valence electron sextet). Moreover, the iminium ion is stabilized by conjugation to a C=C double bond.

The zwitterion intermediate of the [4+2]-cycloaddition depicted in Figure 15.20 is formed with stereostructure **B**. Therein the ester groups of the dienophile fragment maintain their original *cis* arrangement. However, this *cis*-arrangement is quickly lost owing to rotation about the C2—C1 bond. Zwitterion conformers with stereostructure **C**, i.e., with a *trans*-arrangement of the ester groups, are thus formed. Conformer **C** undergoes ring closure to the [4+2]-cycloadduct significantly faster than conformer **B.** In addition, the ring closure occurs with simple diastereoselectivity. Consequently, the zwitterion **C** only leads to the formation of the [4+2]-cycloadduct *trans*,^{2,3}*trans*-**A** and *trans*,^{2,3}*cis*-**A** does not form.

15.3.2 Substituent Effects on Reaction Rates of Diels–Alder Reactions

Cyclopentadiene is such a reactive 1,3-diene that it undergoes Diels–Alder reactions with all cyano-substituted ethenes. The rate constants of these cycloadditions (Table 15.1) show that each cyano substituent increases the reaction rate significantly and that geminal cyano groups accelerate more than vicinal cyano groups.

Diels–Alder reactions of the type shown in Table 15.1, that is, Diels–Alder reactions between electron-poor dienophiles and electron-rich dienes, are referred to as **Diels–Alder reactions with normal electron demand.** The overwhelming majority of known Diels–Alder reactions exhibit such a "normal electron demand." Typical dienophiles include acrolein, methyl vinyl ketone, acrylic acid esters, acrylonitrile, fumaric acid esters (*trans*-butenedioic

Tab. 15.1 Relative Rate Constants $k_{4+2,rel}$ of Analogous Diels–Alder Reactions of Polycyanoethenes

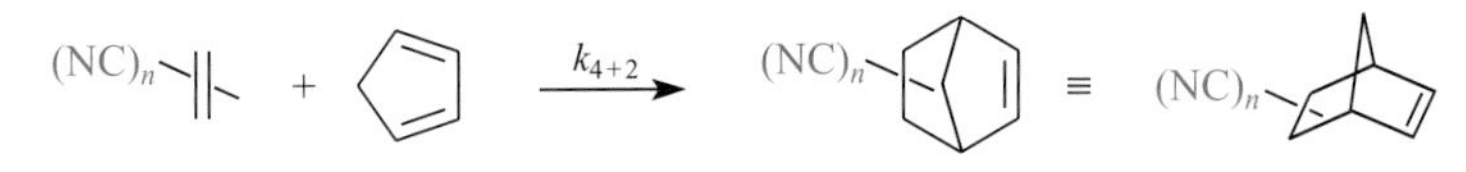

	NC-CH=CH$_2$	cis-NC-CH=CH-CN	trans-NC-CH=CH-CN	$(NC)_2C=CH_2$	$(NC)_2C=CH-CN$	$(NC)_2C=C(CN)_2$
$k_{4+2,rel}$	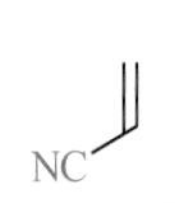 ≡ 1	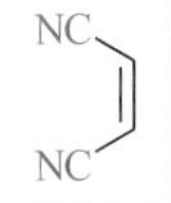 81	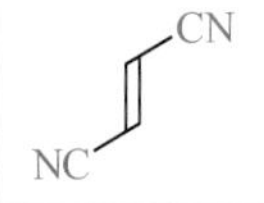 91	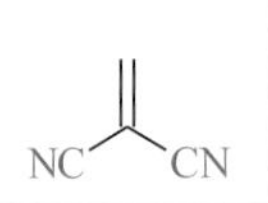 45 500	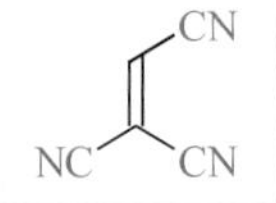 480 000	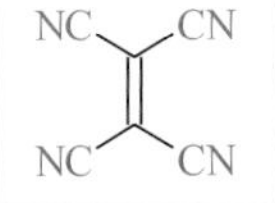 43 000 000

acid esters), maleic anhydride, and tetra-cyanoethene—all of which are acceptor-substituted alkenes. Typical dienes are cyclopentadiene and acyclic 1,3-butadienes with alkyl-, aryl-, alkoxy-, and/or trialkylsilyloxy substituents—all of which are dienes with a donor substituent.

The reaction rates for the cycloaddition of several of the mentioned dienophiles to electron-rich dienes are significantly increased upon addition of a catalytic amount of a Lewis acid. The $AlCl_3$ complex of methyl acrylate reacts 100,000 times faster with butadiene than pure methyl acrylate (Figure 15.21). Apparently, the C=C double bond in the Lewis acid complex of an acceptor-substituted dienophile is connected to a stronger acceptor substituent than in the Lewis-acid-free analog. A *better* acceptor increases the dienophilicity of a dienophile in a manner similar to the effect *several* acceptors have in the series of Table 15.1.

An increase in reactivity also can be observed in Diels–Alder reactions with normal electron demand if a given dienophile is reacted with a series of more and more electron-rich dienes. The reaction rates of the Diels–Alder reactions of Figure 15.22 show that the substituents MeO > Ph > alkyl are such reactivity-enhancing donors. The tabulated rate constants also show that a given donor substituent accelerates the Diels–Alder reaction more if located in position 1 of the diene than if located in position 2.

Diels–Alder reactions also may occur when the electronic situation of the substrates is completely reversed, that is, when electron-rich dienophiles react with electron-poor dienes. [4+2]-Cycloadditions of this type are called **Diels–Alder reactions with inverse electron demand.** 1,3-Dienes that contain heteroatoms such as O and N in the diene backbone are the

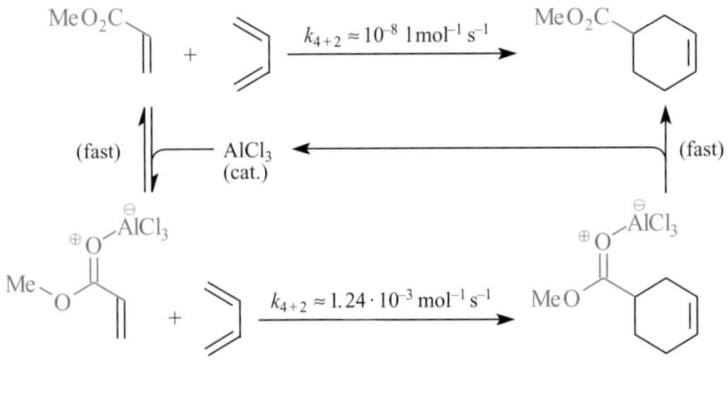

$k_{4+2} \approx 10^{-8}\ \mathrm{l\,mol^{-1}\,s^{-1}}$

(fast) $AlCl_3$ (cat.) (fast)

$k_{4+2} \approx 1.24 \cdot 10^{-3}\ \mathrm{mol^{-1}\,s^{-1}}$

Fig. 15.21. Diels–Alder reactions with normal electron demand; increase of the reactivity upon addition of a Lewis acid. The $AlCl_3$ complex of the acrylate reacts 100,000 times faster with butadiene than does the uncomplexed acrylate.

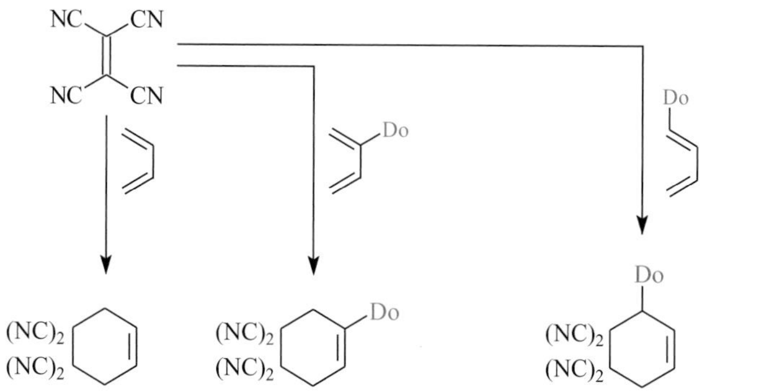

$k_{4+2,\,rel}$		Do = Me	Ph	OMe	Do = Me	Ph	OMe
$k_{4+2,\,rel}$	≡ 1	44	191	1 720	104	386	50 900

Fig. 15.22. Diels–Alder reactions with normal electron demand; reactivity increase by the use of donor-substituted 1,3-dienes (Do refers to a donor substituent).

dienes of choice for this kind of cycloaddition. The data in Figure 15.23 show the rate-enhancing effect of the presence of donor substituents in the dienophile.

Why do the Diels–Alder reactions with both normal and inverse electron demand occur under relatively mild conditions? And, in contrast, why can [4+2]-cycloadditions between ethene or acetylene, respectively, and butadiene be realized only under extremely harsh conditions (Figure 15.1)? Equation 15.2 described the amount of transition state stabilization of [4+2]-cycloadditions as the result of HOMO/LUMO interactions between the π-MOs of the diene and the dienophile. Equation 15.3 is derived from Equation 15.2 and presents a *simplified* estimate of the magnitude of the stabilization. This equation features a sum of two simple terms, and it highlights the essence better than Equation 15.2.

$$\Delta E_{\mathrm{TS}} \propto \frac{1}{E_{\mathrm{HOMO,diene}} - E_{\mathrm{LUMO,dienophile}}} + \frac{1}{E_{\mathrm{HOMO,dienophile}} - E_{\mathrm{LUMO,diene}}} \tag{15.3}$$

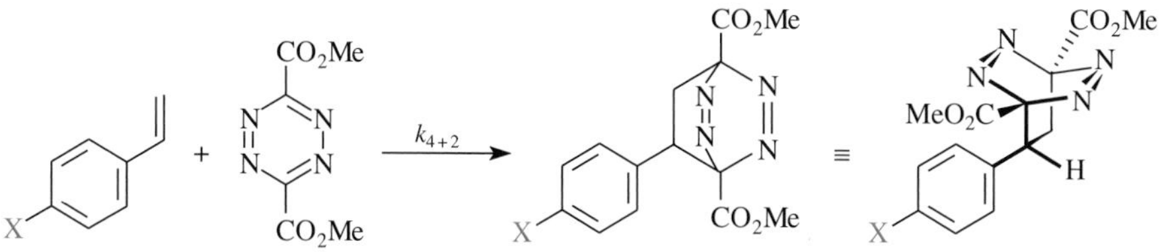

(products undergo subsequent reactions)

X	O_2N	H	MeO
$k_{4+2,\,rel}$	≡ 1	7.7	29

Fig. 15.23. Diels–Alder reactions with inverse electron demand; reactivity increase by the use of donor-substituted dienophiles (X refers to a substituent that may be a donor or an acceptor).

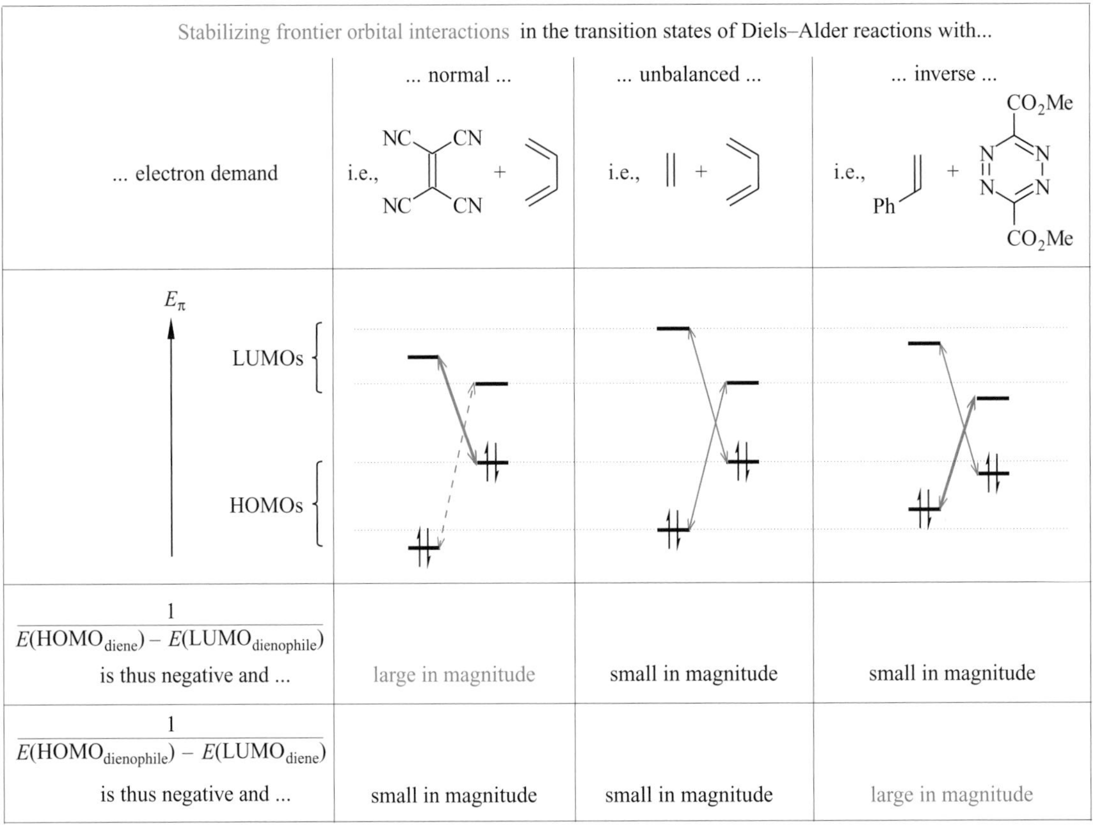

Fig. 15.24. Frontier orbital interactions in Diels–Alder reactions with varying electron demand.

The first term of Equation 15.3 is responsible for most of the transition state stabilization of a Diels–Alder reaction with normal electron demand. In this case, the first term is larger than the second term because the denominator is smaller. The denominator of the first term is smaller because the HOMO of an electron-rich diene is closer to the LUMO of an electron-poor dienophile than is the LUMO of the same electron-rich diene with respect to the HOMO of the same electron-poor dienophile (Figure 15.24, column 2). Acceptors lower the energy of all π-type MOs irrespective of whether these MOs are bonding or antibonding. This is all the more true the stronger the substituent effects and the more substituents are present.

The most important stabilizing interaction of the transition states of Diels–Alder reactions with inverse electron demand is due to the second term of Equation 15.3. In this case, the denominator of the second term is substantially smaller than that of the first term. This is because the HOMO of an electron-rich dienophile is closer to the LUMO of an electron-poor diene than is the HOMO of the same diene relative to the LUMO of the same dienophile (Figure 15.24, column 4). We saw the reason for this previously: acceptors lower the energies of all π-type MOs; donors increase these energies.

In summary, the transition states of Diels–Alder reactions with either normal or inverse electron demand are substantially stabilized because the HOMO of one reagent lies close to the LUMO of the other reagent. This stabilization of the transition states is responsible for the fast cycloadditions.

In the Diels–Alder reactions between ethene and butadiene and between acetylene and butadiene, respectively, the HOMOs are nearly isoenergetic and they are rather far away from the LUMOs (Figures 15.8 and 15.9). According to Equation 15.3, the transition states of these Diels–Alder reactions experience only a minor stabilization and, for *this* reason, these [4+2]-cycloadditions (Figure 15.24, column 3) are so much slower than the others (columns 2 and 4).

15.3.3 Regioselectivity of Diels–Alder Reactions

Diels–Alder reactions with symmetrically substituted dienophiles and/or with symmetrically substituted dienes afford cycloadducts that must be *constitutionally homogeneous*. In contrast, Diels–Alder reactions between an unsymmetrically substituted dienophile and an unsymmetrically substituted diene may afford two *constitutionally isomeric* cycloadducts. If only one of these isomers is actually formed, the Diels–Alder reaction is said to be **regioselective.**

1,3-Butadienes with alkyl substituents in the 2-position favor the formation of the so-called *para*-products (Figure 15.25, X = H) in their reactions with acceptor-substituted dienophiles. The so-called *meta*-product is formed in smaller amounts. This regioselectivity increases if the dienophile carries two geminal acceptors (Figure 15.25, X = CN). 2-Phenyl-1,3-butadiene exhibits a higher "*para*"-selectivity in its reactions with every unsymmetrical dienophile than any 2-alkyl-1,3-butadiene does. This is even more true for 2-methoxy-1,3-butadiene and 2-(trimethylsilyloxy)-1,3-butadiene. Equation 15.2, which describes the stabilization of the transition states of Diels–Alder reactions in terms of the frontier orbitals, also explains the "*para*"/"*meta*"-orientation. The numerators of both fractions assume different values depending on the orientation, while the denominators are independent of the orientation.

One can compute, for example, the stabilizations ΔE_{TS} for the transition states of the "*para*"- and "*meta*"-selective cycloadditions, respectively, of acrylonitrile and isoprene according to Equation 15.2 with the data provided in Figure 15.26 (HOMO/LUMO gaps, LCAO coefficients at the centers that interact with each other). The result for the Diels–Alder reaction of Figure 15.25 is shown in Equations 15.4 and 15.5:

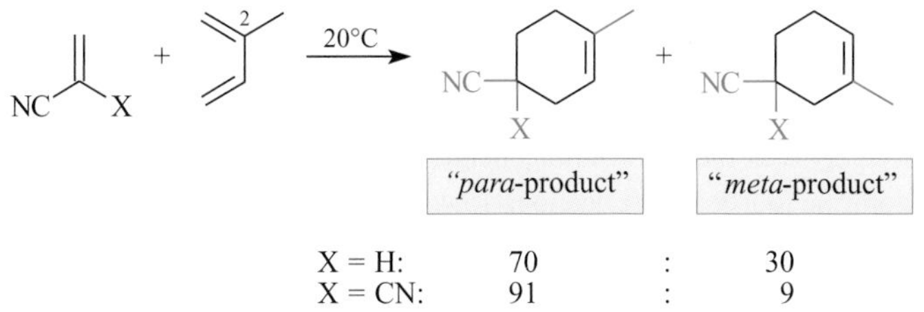

20°C

"*para*-product" "*meta*-product"

	"*para*-product"		"*meta*-product"
X = H:	70	:	30
X = CN:	91	:	9

Fig. 15.25. Regioselective Diels–Alder reactions with a 2-substituted 1,3-diene I: comparison of the effects exerted by one or two dienophile substituents.

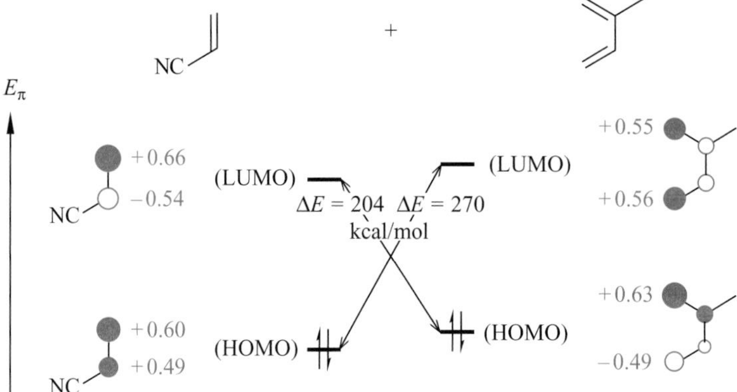

Fig. 15.26. Frontier orbital coefficients and energy difference of the HOMO–LUMO gaps in orientation-selective Diels–Alder reactions (cf. Figure 15.25, X = H).

$$\Delta E_{\mathrm{TS} \rightarrow \text{"}para\text{ product"}} \propto -0.0036 \text{ kcal/mol (63\% of which is due to the } \mathrm{HOMO}_{\mathrm{diene}} / \mathrm{LUMO}_{\mathrm{dienophile}} \text{ interaction)} \tag{15.4}$$

$$\Delta E_{\mathrm{TS} \rightarrow \text{"}meta\text{ product"}} \propto -0.0035 \text{ kcal/mol (61\% of which is due to the } \mathrm{HOMO}_{\mathrm{diene}} / \mathrm{LUMO}_{\mathrm{dienophile}} \text{ interaction)} \tag{15.5}$$

The transition state leading to the *"para"*-product is slightly more stabilized, and accordingly this product is favored. However, the 70:30 selectivity shows the preference to be small, as one would anticipate based on the minuscule energy difference computed by means of Equations 15.4 and 15.5.

The same kind of computation can be carried out for the substrate pair consisting of isoprene and 1,1-dicyanoethene. Based on the HOMO/LUMO gaps and with the orbital coefficients at the centers that interact with each other (see Figure 15.27), Equation 15.2 again gives a higher stabilization ΔE_{TS} for the "*para*"-transition state than for the "*meta*"-transition state (Equations 15.6 and 15.7).

$$\Delta E_{\mathrm{TS} \rightarrow \text{"}para\text{ product"}} \propto -0.0038 \text{ kcal/mol (67\% of which is due to the } \mathrm{HOMO}_{\mathrm{diene}} / \mathrm{LUMO}_{\mathrm{dienophile}} \text{ interaction)} \tag{15.6}$$

$$\Delta E_{\mathrm{TS} \rightarrow \text{"}meta\text{ product"}} \propto -0.0036 \text{ kcal/mol (66\% of which is due to the } \mathrm{HOMO}_{\mathrm{diene}} / \mathrm{LUMO}_{\mathrm{dienophile}} \text{ interaction)} \tag{15.7}$$

This difference between the stabilization energies is a bit larger than in the case of the addition of acrylonitrile to isoprene (Equations 15.4 and 15.5). This agrees with the data in Figure 15.27, which show a "*para/meta*"-selectivity of 91:9 for the addition of 1,1-dicyanoethene to isoprene—i.e., somewhat higher than the 70:30 ratio of the addition involving acrylonitrile.

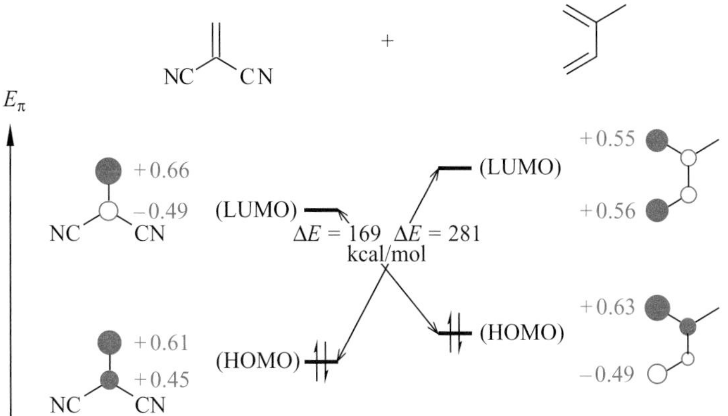

Fig. 15.27. Frontier orbital coefficients and energy difference of the HOMO–LUMO gaps in orientation-selective Diels–Alder reactions (cf. Figure 15.25, X = CN).

From Equation 15.2 one may derive the following general rules concerning the orientation selectivity of any one-step cycloaddition.

Rules for the Regioselectivity of Any One-Step Cycloaddition

The substrates preferentially bind each other with those atoms that exhibit the largest orbital coefficients (absolute values) in the closest energy pair of frontier orbitals. The regioselectivity generally increases, the larger the relative significance of one HOMO/LUMO interaction in comparison to the other and the greater in each of the two crucial frontier orbitals the difference in magnitude of the orbital coefficients (absolute values) at one terminus, compared to the other.

We can customize these general statements specifically for the case of the regioselectivity of Diels–Alder reactions with normal electron demand and make the following statement right away:

The LCAO coefficient in the HOMO of a 1,3-diene

- increases at C1 compared to C4, the better the donor in position 2,
- increases at C4 compared to C1, the better the donor in position 1, and
- is larger at C4 than at C1 when the same kind of donor is attached both to positions 1 and 2.

The consequences for the regioselectivity of Diels–Alder reactions can be summarized as follows.

Regioselectivity of Diels–Alder Reactions

Unsymmetrical dienophiles react with 1-donor-, 2-donor-, or 1,2-didonor-substituted 1,3-dienes, preferentially to the *"ortho"*-, *"para"*- or *"ortho,meta"*-cycloadduct, respectively. With regard to a given dienophile, the *"ortho"*-selectivity is larger than the *"para"*-selectivity. Comparison of the upper and the lower pairs of reactions in Figures 15.28 and 15.29 underscores the latter statement.

	"*para*-product"	:	"*meta*-product"
25°C, 41 d:	70	:	30
10–20°C, 1 mol% $AlCl_3$, 3 h:	95	:	5

Fig. 15.28. Regioselective Diels–Alder reactions with a 2-substituted diene II: selectivity increase by way of addition of a Lewis acid.

We go on to state that the orbital coefficient in the LUMO of a dienophile

- increases at C2 in comparison to C1, the more acceptors are bound to C1 [compare the orbital coefficients of acrylonitrile (Figure 15.26) compared to those of 1,1-dicyanoethene (Figure 15.27)] and
- increases at C2 in comparison to C1, the stronger the acceptor that is attached to C1.

What has just been stated regarding the LCAO coefficients of the dienophile LUMO combined with the "rules for the regioselectivity of any one-step cycloaddition" leads to the following consequences for the Diels–Alder reactions of isoprene:

1. Isoprene shows less "*para*"-selectivity in its addition to acrylonitrile than in its addition to 1,1-dicyanoethene (Figure 15.25).
2. Isoprene shows less "*para*"-selectivity in its additions to acrylate esters than to $AlCl_3$-complexed acrylic acid esters (Figure 15.28; the complex formation with $AlCl_3$ converts the CO_2Me group into a better acceptor than the uncomplexed CO_2Me group).

The increase in regioselectivity of Diels–Alder reactions upon addition of Lewis acid has a second cause aside from the one which was just mentioned. The reaction conditions described in Figure 15.28 indicate that $AlCl_3$ increases the rate of cycloaddition. The same effect also was seen in the cycloaddition depicted in Figure 15.21. In both instances, the effect is the consequence of the lowering of the LUMO level of the dienophile. According to Equation 15.2, this means that the magnitude of the denominator of the first term decreases and the first term therefore becomes larger than the second term. If, in addition, the numerators of these terms differ by a certain amount for the "*para-*" and "*meta*"-transition states (as determined by the combinations of the orbital coefficients), the effect is further enhanced. *This* also increases the "*para*"-selectivity.

Finally, the examples of the two Diels–Alder reactions in Figure 15.28 lead us to a general statement: in Diels–Alder reactions with normal electron demand, the addition of a Lewis acid such as $AlCl_3$ increases the reaction rate and the regioselectivity. This is a nice example of the failure of the reactivity-selectivity principle (Section 1.7.4), which is so often used in organic chemistry to explain reaction chemistry.

15.3.4 Simple Diastereoselectivity of Diels–Alder Reactions

We saw in Section 15.3.3 that 1,3-butadienes with a donor in the 1-position react with acceptor-substituted alkenes to form cycloadducts with high "*ortho*"-selectivity. The amount of

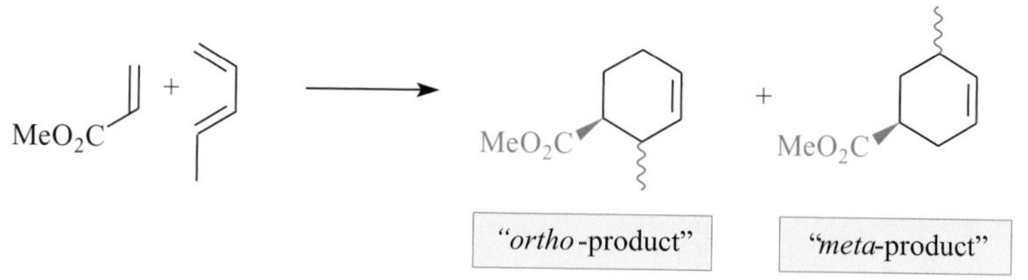

	"*ortho*-product"	:	"*meta*-product"
25°C, 70 d:	90 (*cis* : *trans* = 57 : 43)	:	10 (*cis* : *trans* = 73 : 27)
10–20°C, 10 mol% $AlCl_3$, 3 h:	98 (*cis* : *trans* = 95 : 5)	:	2 (mainly *cis*)

Fig. 15.29. Regioselectivity and simple diastereoselectivity of a Diels–Alder reaction with a 1-substituted diene; selectivity increase by way of addition of a Lewis acid.

"*meta*" products formed is usually less than 10% (example: Figure 15.29). This is particularly true for Diels–Alder reactions that are carried out in the presence of $AlCl_3$, which has the *same* effect of enhancing the regioselectivity as seen in Figure 15.28.

Here we are primarily concerned with the fact that this "*ortho*"-adduct may occur in the form of two diastereomers. The diastereomers are formed as a 57:43 *cis/trans*-mixture in the absence of $AlCl_3$, but a 95:5 *cis/trans*-mixture is obtained in the presence of $AlCl_3$. In the latter case, thus, one is dealing with a Diels–Alder reaction that exhibits a substantial "simple diastereoselectivity" (see Section 11.1.3 for a definition of the term). Here, the simple diastereoselectivity is due to kinetic rather than thermodynamic control, since the preferentially formed *cis*-disubstituted cyclohexene is less stable than its *trans*-isomer.

Simple diastereoselectivity may also occur in Diels–Alder reactions between electron-poor dienophiles and cyclopentadiene (Figure 15.30). Acrylic acid esters or *trans*-crotonic acid esters react with cyclopentadiene in the presence or absence of $AlCl_3$ with substantial selectivity to afford the so-called *endo*-adducts. When the bicyclic skeleton of the main product is viewed as a "roof," the prefix "*endo*" indicates that the ester group is below this roof, rather than outside (*exo*). However, methacrylic acid esters add to cyclopentadiene without any *exo,endo*-selectivity regardless whether the reaction is carried out with or without added $AlCl_3$ (Figure 15.30, bottom).

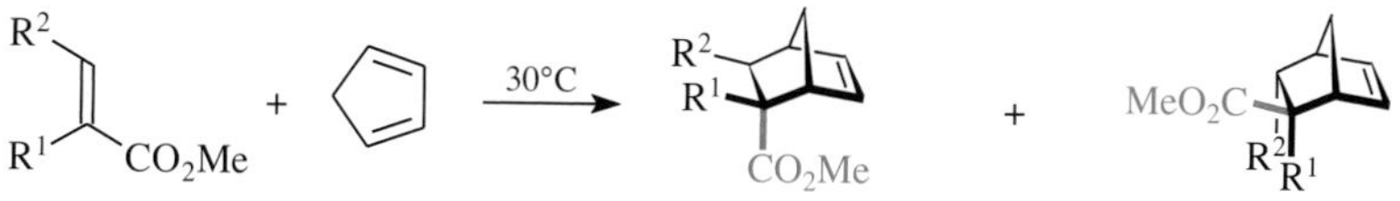

R^1	R^2	Conditions	*endo* product	:	*exo* product
H	H	7.5 h:	78	:	22
		1 equivalent $AlCl_3$, 30 min:	95	:	5
H	Me	7.5 h:	54	:	46
		1 equivalent $AlCl_3$, 30 min:	94	:	6
Me	H	7.5 h:	31	:	69
		1 equivalent $AlCl_3$, 30 min:	60	:	40

Fig. 15.30. Simple diastereoselectivity of the additions of various acrylic acid derivatives to cyclopentadiene.

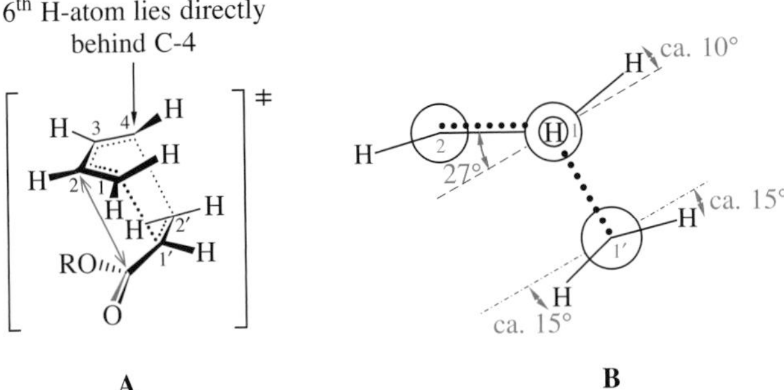

Fig. 15.31. Transition state structures of Diels–Alder additions of butadiene; **A**, side view of the addition of acrylic acid ester and **B**, Newman projection of the addition of ethene.

The high simple diastereoselectivities seen in Figures 15.29 and 15.30 are due to the same preferred orientation of the ester group in the transition states. The stereostructure of the cycloadduct shows unequivocally that the ester group points underneath the diene plane in each of the transition states of both cycloadditions and not away from that plane. Figure 15.31 exemplifies this situation for two transition states of simple Diels–Alder reactions of 1,3-butadiene: **A** shows a perspective drawing of the transition state of the acrylic acid ester addition, and **B** provides a side view of the addition of ethene, which will serve as an aid in the following discussion. Both structures were determined by computational chemistry.

The origin(s) for the preference of stereostructure **A** in the acrylic acid ester addition is not known with certainty. A steric effect may explain the observation. The bulky acceptor substituent of the dienophile might be less hindered—and this is quite counterintuitive—in the *endo*-orientation in the transition state shown in Figure 15.31 than in the alternative *exo*-position. One might use the structure **B** to suggest that the substituent of the dienophile in **A** does not try to avoid the C atoms C2 and C3 as much as it tries to stay away from the H atoms *cis*-H1 and *cis*-H4. The increase of *endo*-selectivity upon addition of a Lewis acid could then be explained by the premise that the complexing Lewis acid renders the ester group more bulky. This increased steric demand enhances its desire to avoid the steric hindrance in its *exo*-position.

Another consideration is something called a secondary orbital interaction, a bonding interaction due to orbital overlap that is not involved in new bond formation. There are several models for this, but one invokes an overlap between C2 of the diene and the ester carboxyl group as illustrated in structure A.

Side Note 15.2. Synthetic Applications of Diels–Alder Reactions

So many important things had to be considered about the mechanism of Diels–Alder reactions in this section that it is impossible to fully appreciate the preparative relevance of this reaction—despite the fact that it undoubtedly has to be included among the top ten of the "most important reactions of organic chemistry." So the two Diels–Alder reactions shown in Figure 15.32 must suffice to give you at least an inkling of what these reactions may effect in synthesis! The first step is an intermolecular Diels–Alder reaction providing a new ring. This is a cyclohexene ring with two *cis*-oriented substituents (Formula **C**). In a second step, an intramolecular Diels–Alder reaction follows that leads to the formation of two new rings, i.e., another cyclohexene, and a bridging eight-membered ring that is formed between these two

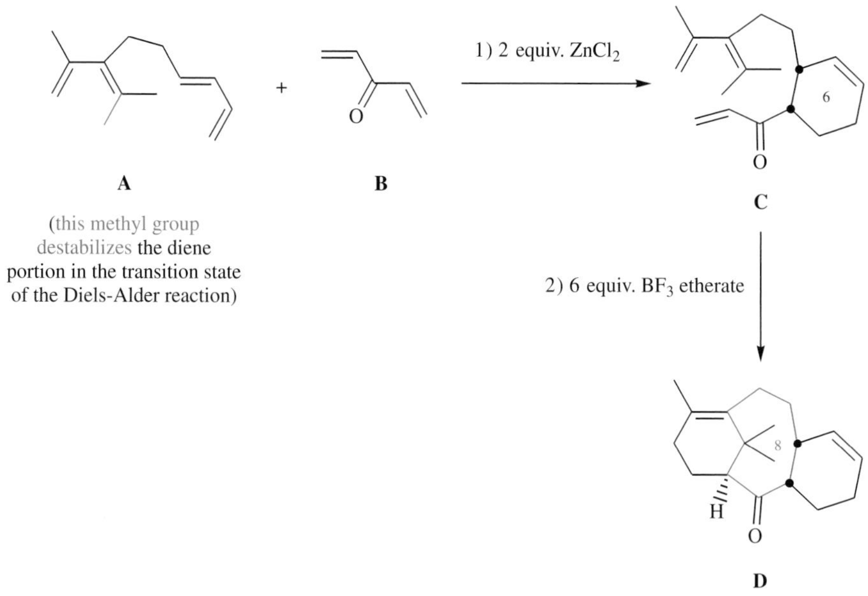

1) 2 equiv. $ZnCl_2$

2) 6 equiv. BF_3 etherate

(this methyl group destabilizes the diene portion in the transition state of the Diels-Alder reaction)

Fig. 15.32. Sequence of a Lewis-acid-catalyzed intermolecular and a Lewis-acid-catalyzed intramolecular Diels-Alder addition. The first Diels-Alder addition occurs with simple and the second with induced diastereoselectivity.

cyclohexenes (Formula **D**). In addition, high asymmetric induction in the second Diels Alder reaction was established by the two stereocenters present in the substrate **C** so that its product **D** exhibited just the relative configuration shown at its new stereocenter. The six-ring/eight-ring/six-ring skeleton of the bis(Diels–Alder adduct) **D,** which has been established in just two steps, can be found in the natural compound paclitaxel, which has been commercialized under the trademark Taxol® as one of the currently most potent antitumor agents.

15.4 [2+2]-Cycloadditions with Dichloroketene

Only a few ketenes can be isolated, and diphenylketene is one of those. The majority of the other ketenes dimerize quickly, as exemplified by the parent ketene $H_2C{=}C{=}O$. The fewer or smaller the substituents that are bound at the sp^2-hybridized carbon atom the quicker these dimerizations proceed:

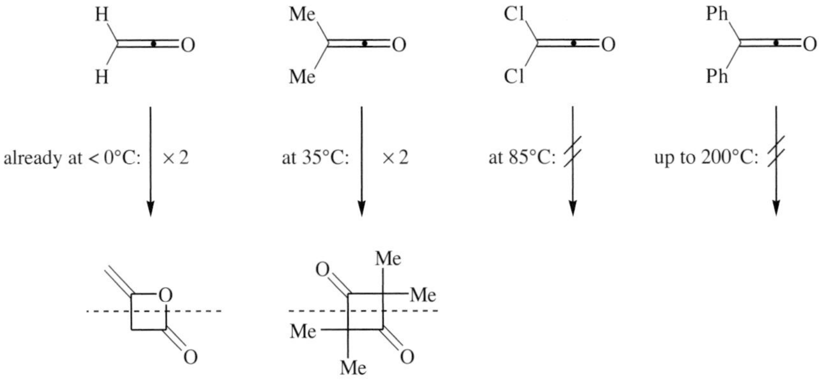

The dimerization of ketene leads to the formation of a lactone ("diketene"), the dimerization of dimethylketene furnishes a cyclobutanedione. In Figure 6.25, diketenes were introduced as reagents acylating alcohols to acetylacetates.

Cycloadditions with *reactive* ketenes therefore can be observed only when they are prepared *in situ* and in the presence of the alkene to which they shall be added. **Dichloroketene** generated *in situ* is the best reagent for intermolecular [2+2]-cycloadditions. Dichloroketene is poorer in electrons than the parent ketene and therefore more reactive toward the relatively electron-rich standard alkenes. The reason is that the dominating frontier orbital interaction between these reactants involves the LUMO of the ketene, not its HOMO (see Section 15.2.4).

Figure 15.33 is based on the first and most common method for the preparation of dichloroketene, i.e., the reductive β-elimination of chlorine from trichloroacetyl chloride with zinc (for the mechanism see Sections 4.7.1 and 17.4.1). Upon addition of the dichloroketene to the isomerically pure 2-butenes perfect stereoselectivity (and hence overall stereospecificity) occurs: *trans*-2-butene reacts to give *trans*-dichlorodimethylcyclobutanone and *cis*-2-butene to furnish its *cis*-isomer.

The [2+2]-cycloaddition in Figure 15.34 is also made possible through the aforementioned preparation of dichloroketene by zinc reduction of trichloroacetyl chloride. Regioselectivity

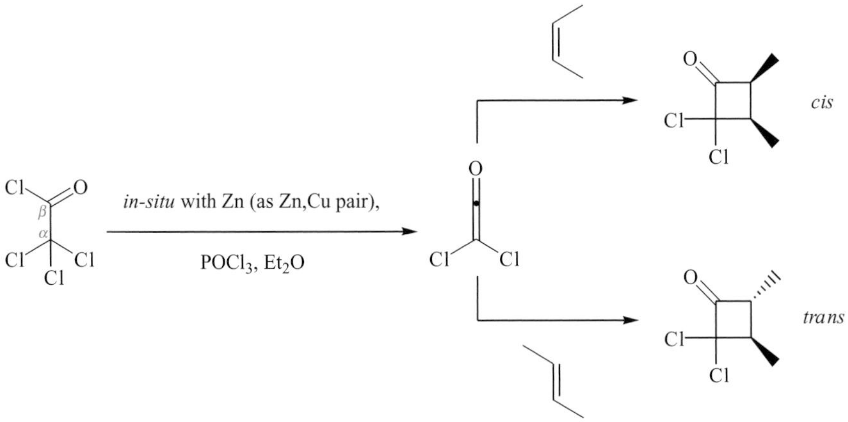

Fig. 15.33. Prototypical *in situ* generation of dichloroketene through reductive β-elimination of chlorine from trichloroacetyl chloride. Stereospecific [2+2]-cycloadditions of this dichloroketene with the stereoisomeric 2-butenes.

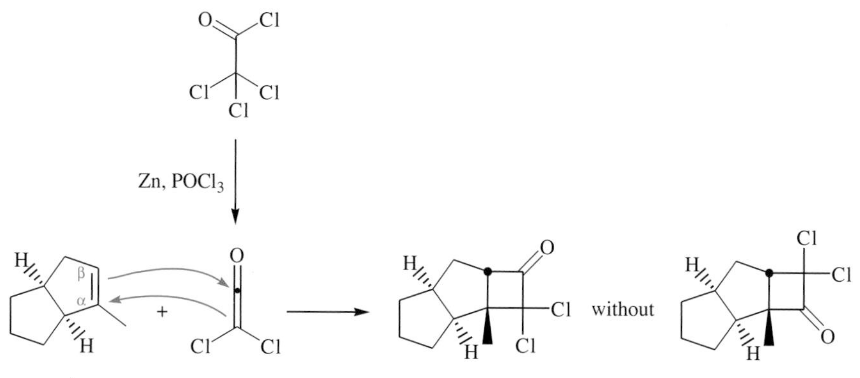

Fig. 15.34. *In situ* generation of dichloroketene according to the procedure described in Figure 15.33. Regioselective [2+2]-cycloaddition of this dichloroketene with an alkene whose C=C double bond contains more alkyl substituents at C-α than at C-β.

occurs with the addition of the dichloroketene to the trisubstituted alkene **A**. The carbonyl carbon of dichloroketene bonds to the C atom of the double bond that has the larger orbital coefficient in the HOMO of the alkene (C_β). Diastereoselectivity occurs with this [2+2]-addition as well. The bicyclic alkene **A** reacts with the ketene only on its more easily accessible convex side.

Figure 15.35 shows the second commonly employed method for the generation of dichloroketene, which involves an NEt_3-mediated β-elimination of HCl from dichloroacetyl chloride. Interestingly, dichloroketene does not participate as a dienophile in a Diels–Alder reaction with cyclopentadiene (Figure 15.35), preferring instead a [2+2]-cycloaddition that occurs with perfect regioselectivity. The preferred transition state is in line with the rule of thumb formulated in Section 15.3.3 for the orientation selectivity of one-step cycloadditions in general. It is the one in which the atoms with the largest orbital coefficient in the closest energy frontier orbital pair are connected together. Figure 15.13 demonstrates that the relevant frontier orbital pair includes the LUMO of dichloroketene and the HOMO of cyclopentadiene. The carbonyl carbon possesses the largest orbital coefficient in the LUMO of dichloroketene (Figure 15.12). The largest orbital coefficient in the HOMO of cyclopentadiene is located at Cl, not at C2, since cyclopentadiene is a 1,3-butadiene that bears an alkyl substituent both at C1 and C4 (see Section 15.3.3). Hence, the carbonyl carbon of the

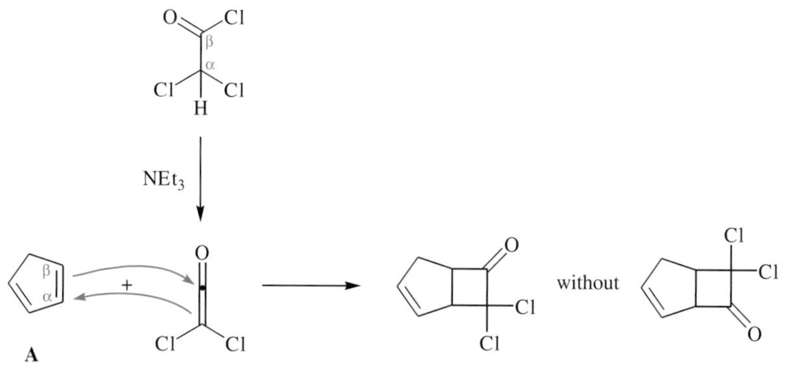

Fig. 15.35. Prototypical *in situ* generation of dichloroketene through NEt_3-mediated β-elimination of hydrogen chloride from dichloroacetyl chloride. Regioselective [2+2]-cycloaddition II of this dichloroketene with an alkene whose C=C double bond contains a conjugated substituent at C-α, but not at C-β.

dichloroketene binds to C1 of cyclopentadiene in the most stable transition state of the [2+2]-cycloaddition.

It is possible that a [4+2]-cycloaddition between the dichloroketene carbonyl group and the cyclopentadiene occurs. This would be followed by a Claisen rearrangement to give the product observed. The lesson is that you should learn what we think we know, but not be afraid to question it. That is science.

15.5 1,3-Dipolar Cycloadditions

15.5.1 1,3-Dipoles

A 1,3-dipole is a compound of the type a—Het—b that may undergo 1,3-dipolar cycloadditions with multiply bonded systems and can best be described with a zwitterionic all-octet Lewis structure. An unsaturated system that undergoes 1,3-dipolar cycloadditions with 1,3-dipoles is called **dipolarophile.** Alkenes, alkynes, and their diverse hetero derivatives may react as dipolarophiles. Since there is a considerable variety of 1,3-dipoles—Table 15.2 shows

Tab. 15.2 Important 1,3-Dipoles

1,3-dipoles of the propargyl/allenyl anion type	1,3-dipoles of the allyl anion type
$\{:N{\equiv}\overset{\oplus}{N}{-}\overset{\ominus}{C} \longleftrightarrow :\overset{\ominus}{N}{=}\overset{\oplus}{N}{=}C\}$ diazoalkane	$\{:O{-}\overset{\oplus}{O}{-}\overset{\ominus}{O}: \longleftrightarrow \overset{\ominus}{O}{-}\overset{\oplus}{O}{=}O:\}$ ozone
$\{-C{\equiv}\overset{\oplus}{N}{-}\overset{\ominus}{O}: \longleftrightarrow -\overset{\ominus}{C}{=}\overset{\oplus}{N}{=}O:\}$ nitrile oxide	$\{C{=}\overset{\oplus}{O}{-}\overset{\ominus}{O}: \longleftrightarrow \overset{\ominus}{C}{-}\overset{\oplus}{O}{=}O:\}$ carbonyl oxide
$\{N{\equiv}\overset{\oplus}{N}{-}\overset{\ominus}{N}- \longleftrightarrow :\overset{\ominus}{N}{=}\overset{\oplus}{N}{=}N-\}$ azide	
$\rightarrow :\overset{\ominus}{N}{=}\overset{\oplus}{N}{=}\overset{\ominus}{N}:$ azide ion	
c.f. $\{-C{\equiv}C{-}\overset{\ominus}{C}: \longleftrightarrow -\overset{\ominus}{C}{=}C{=}C\}$	c.f. $\{C{=}C{-}\overset{\ominus}{C} \longleftrightarrow \overset{\ominus}{C}{-}C{=}C\}$

a small selection—1,3-dipolar cycloadditions represent not only a general but also the most universal synthetic approach to five-membered heterocycles.

1,3-Dipoles are isoelectronic either to both propargyl and allenyl anions or to allyl anions. One may thus group these 1,3-dipoles into 1,3-dipoles of the propargyl/allenyl anion types (Table 15.2, left) and 1,3-dipoles of the allyl anion type (Table 15.2, right). The 1,3-dipoles of the propargyl/allenyl anion type contain a linearly coordinated central atom, as is the case for propargyl and allenyl groups in all sorts of compounds. The linearly coordinated central atom of these 1,3-dipoles is an N atom. This N atom carries a formal positive charge in both all-octet resonance forms. This central N atom is connected to C and/or to other heteroatoms, and these carry—one in each of the resonance forms—the formal negative charge. The azide ion is the only 1,3-dipole of the propargyl/allenyl anion type with a *net* negative charge.

1,3-Dipoles of the allyl anion type are bent just like allyl anions. The central atom can be an N, O, or S atom, and this atom carries a formal positive charge in both all-octet resonance forms. Centers 1 and 3 of the 1,3-dipole of the allyl anion type carry the formal negative charge, again one in each of the resonance forms. As with the dipoles of the propargyl/allenyl anion type, this negative charge may be located on C atoms and/or on heteroatoms.

15.5.2 Frontier Orbital Interactions in the Transition States of One-Step 1,3-Dipolar Cycloadditions; Sustmann Classification

Diazomethane adds to ethene to form Δ^1-pyrazoline (**A** in Figure 15.36). Its addition to acetylene first leads to the formation of the nonaromatic 3-*H*-pyrazole (**B** in Figure 15.36), which subsequently is converted into the aromatic 1-*H*-pyrazole (**C**) by way of a fast 1,5-hydrogen migration.

Do the transition states of the 1,3-dipolar cycloadditions with diazomethane benefit from a stabilizing frontier orbital interaction? Yes! Computations show that the $\text{HOMO}_{\text{diazomethane}}/\text{LUMO}_{\text{ethene}}$ interaction (orbital energy difference, –229 kcal/mol) stabilizes the transition state of the 1,3-dipolar cycloaddition to ethene (Figure 15.37) by about 11 kcal/mol. Moreover, computations also show that the $\text{HOMO}_{\text{ethene}}/\text{LUMO}_{\text{diazomethane}}$ interaction (orbital energy difference, –273 kcal/mol) contributes a further stabilization of 7 kcal/mol.

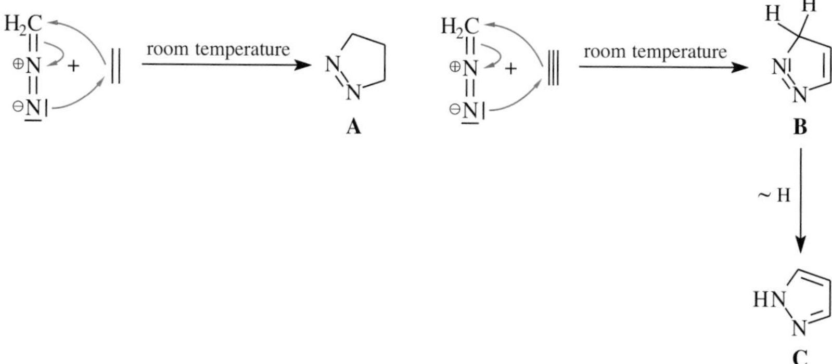

Fig. 15.36. The simplest 1,3-dipolar cycloadditions with diazomethane.

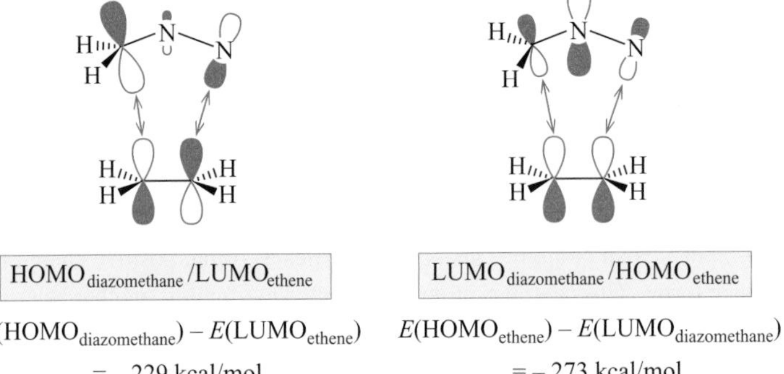

Fig. 15.37. Frontier orbital interactions in the transition state of the 1,3-dipolar cycloaddition of diazomethane to ethene.

Frontier orbital interactions are stabilizing the transition states of all the 1,3-dipolar cycloadditions. It is for this reason that one-step 1,3-dipolar cycloadditions are generally possible and, aside from some exotic exceptions, one does indeed observe one-step reactions.

As in the case of Diels–Alder reactions or of [2+2]-cycloadditions of ketenes, the rate of 1,3-dipolar cycloadditions is affected by donor and acceptor substituents in the substrates. Again, Equation 15.2 can be used to obtain a good approximation of their effects, since this equation applies to *any* one-step cycloaddition. We restated this equation once before as the approximation expressed in Equation 15.3. In that case, it was our aim to understand the rate of Diels–Alder reactions, and we are now faced with the task of making a statement concerning the rate of 1,3-dipolar cycloadditions. To this end it is advantageous to employ a different approximation to Equation 15.2, and that approximation is expressed in Equation 15.8.

$$\Delta E \propto \frac{1}{E_{\text{HOMO, dipole}} - E_{\text{LUMO, dipolarophile}}} + \frac{1}{E_{\text{HOMO, dipolarophile}} - E_{\text{LUMO, dipole}}} \tag{15.8}$$

This approximation again is a crude one, but it allows one to recognize the essentials more clearly.

- According to Equation 15.8, the transition state of a 1,3-dipolar cycloaddition is more stabilized and the reaction proceeds faster, the energetically closer the occupied and the unoccupied frontier orbitals are to each other.
- Especially fast 1,3-dipolar cycloadditions can be expected whenever the $\text{HOMO}_{\text{dipole}}/\text{LUMO}_{\text{dipolarophile}}$ interaction is particularly strong. In this case, the denominator of the first term in Equation 15.8 will be rather small, which makes the first term large. This scenario characterizes the so-called Sustmann type I additions (Figure 15.38, column 2).
- Especially fast 1,3-dipolar cycloadditions also can be expected whenever the denominator of the second term of Equation 15.8 is very small so that the magnitude of the second term is high. This scenario characterizes the so-called Sustmann type III additions. Therein, it is essentially the $\text{HOMO}_{\text{dipolarophile}}/\text{LUMO}_{\text{dipole}}$ interaction which stabilizes the transition state (Figure 15.38, column 4).
- Sustmann type II additions occur whenever both terms of Equation 15.8 contribute to a similar (and small) extent to the stabilization of the transition state of 1,3-cycloadditions (Figure 15.38, column 3). These reactions correspondingly represent a reactivity minimum.

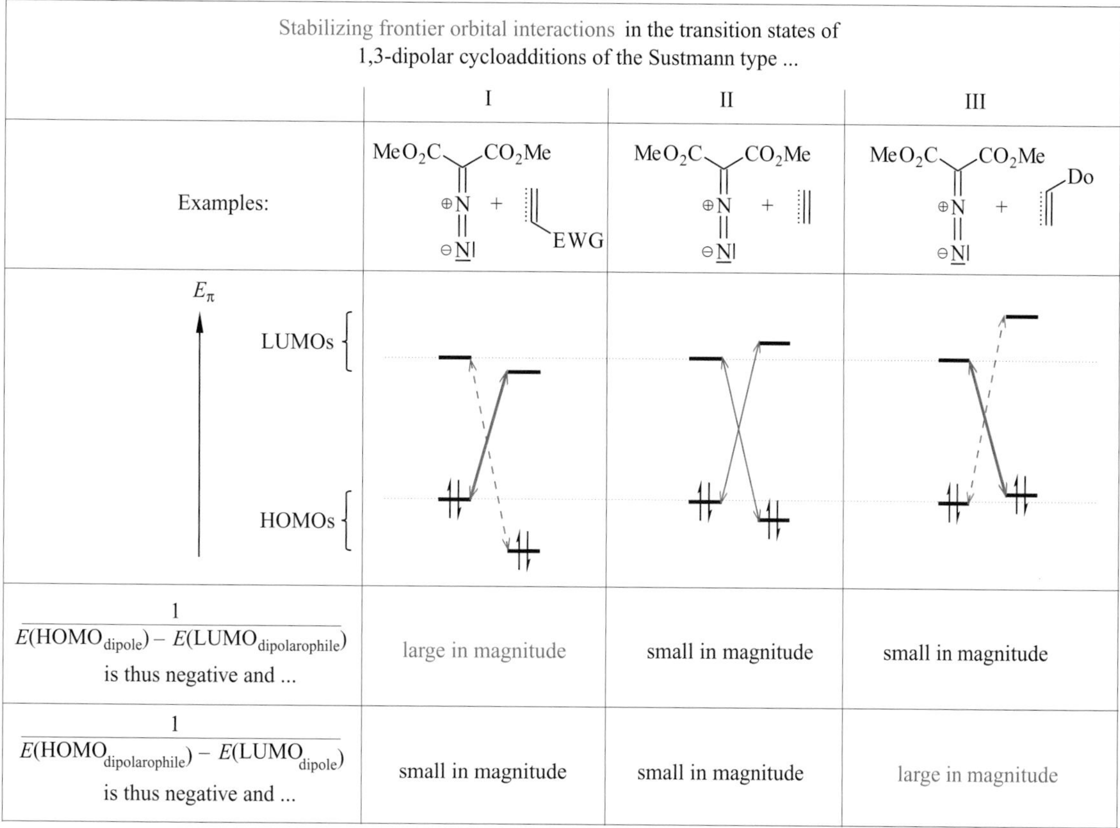

Fig. 15.38. Frontier orbital interactions of 1,3-dipolar cycloadditions with varying electron demands.

Some 1,3-dipoles possess HOMO and LUMO energies that allow for fast Sustmann type I additions with electron-poor dipolarophiles and for fast Sustmann type III additions with electron-rich dipolarophiles. In reactions with dipolarophiles with intermediate electron density, such dipoles merely are substrates of the much slower Sustmann type II additions. A plot of the rate constants of the 1,3-dipolar cycloadditions of such dipoles as a function of the HOMO energy of the dipolarophiles—or as a function of their LUMO energy that varies with the same trend—then has a U-shape. One such plot is shown in Figure 15.39 for 1,3-dipolar cycloadditions of diazomalonates.

15.5.3 1,3-Dipolar Cycloadditions of Diazoalkanes

The simplest 1,3-dipolar cycloadditions of **diazomethane** were presented in Figure 15.36. Diazomethane is generated from sulfonamides or alkyl carbamates of *N*-nitrosomethylamine. The preparation shown in Figure 15.40 is based on the commercially available *para*-toluenesulfonylmethylnitrosamide (Diazald®). In a basic medium, this amide forms a sulfonylated

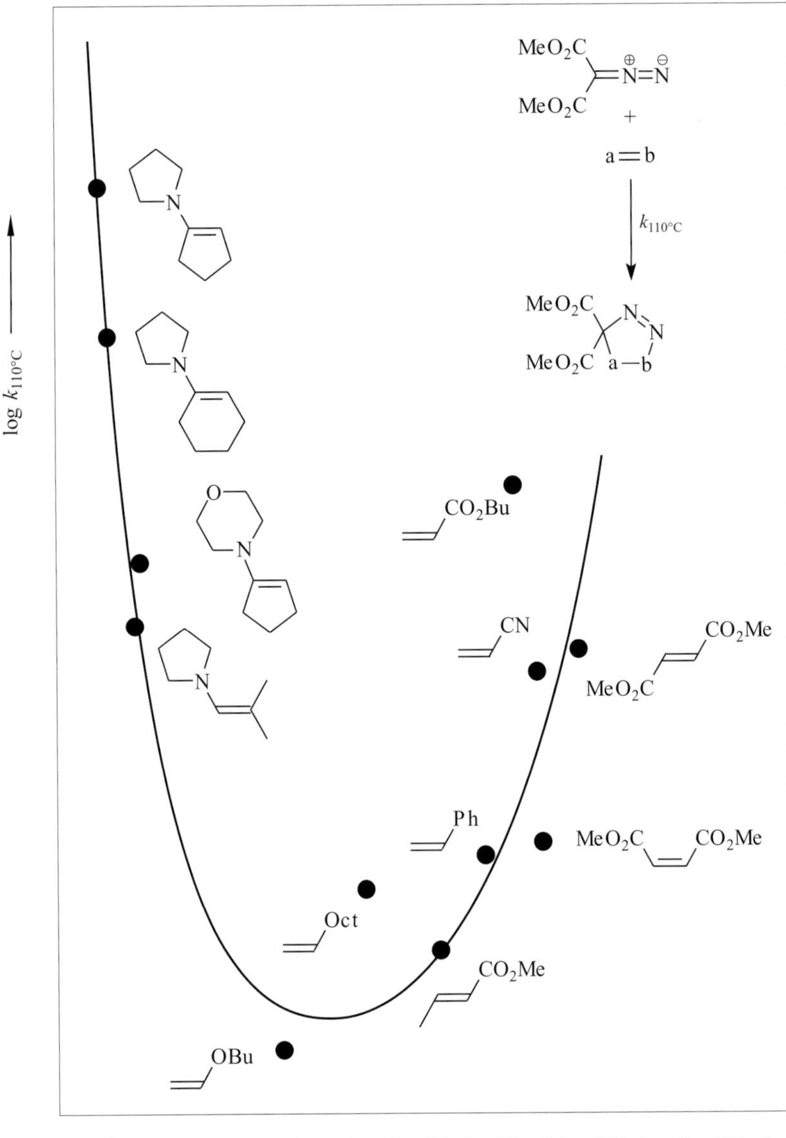

Fig. 15.39. Rate constants of 1,3-dipolar cycloadditions of diazomalonic ester as a function of the HOMO or LUMO energies, respectively, of the dipolarophile.

diazotate **A** by way of a [1,3]-shift. This then undergoes a base-mediated 1,3-elimination to form diazomethane.

Diazomethane is an electron-rich 1,3-dipole, and it therefore engages in Sustmann type I 1,3-dipolar cycloadditions. In other words, diazomethane reacts with acceptor-substituted alkenes or alkynes (e. g., acrylic acid esters and their derivatives) much faster than with ethene or acetylene (Figure 15.36). Diazomethane often reacts with unsymmetrical electron-deficient

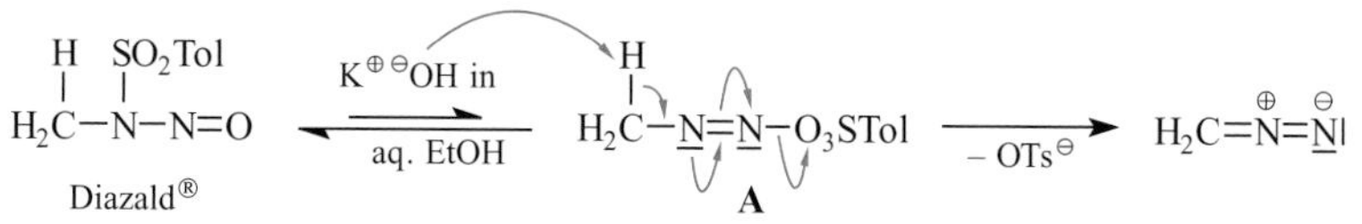

Fig. 15.40. Preparation of diazomethane by way of a 1,3-elimination.

dipolarophiles with regioselectivity, as exemplified in Figure 15.41 for the case of the 1,3-dipolar cycloaddition between diazomethane and the methyl ester of 2-methyl-2-butenoic acid.

This 1,3-dipolar cycloaddition also shows stereoselectivity. From the *trans*-configured ester the *trans*-configured cycloadduct is formed with a diastereoselectivity exceeding 99.997:0.003. This finding provides compelling evidence that this cycloaddition occurs in a single step. If the reaction were a two-step process, either dipole **A** or biradical **C** would occur as an intermediate. None of the intermediates would have to possess exactly the conformations shown in Figure 15.41, *but the conformation depicted certainly is correct in one detail:* the *trans*-configuration of the 2-methyl-2-butenoic acid moiety is initially conserved.

The reactive intermediates **A** and **C** could be so short-lived that they cyclize extremely fast. They certainly could cyclize so fast that the *trans*-configuration of the 2-methyl-2-butenoic acid moiety *by and large* is carried over into the cycloadduct. Intermediates **A** and **C** would therefore hardly have enough time to isomerize to conformers **B** and **D**, respectively, by way of a rotation about the C—C single bond between the two C atoms that used to be connected by a configurationally stable C=C double bond in the starting ester. However, it seems unbelievable that not even as little as 0.003% of either of the potential intermediates would have

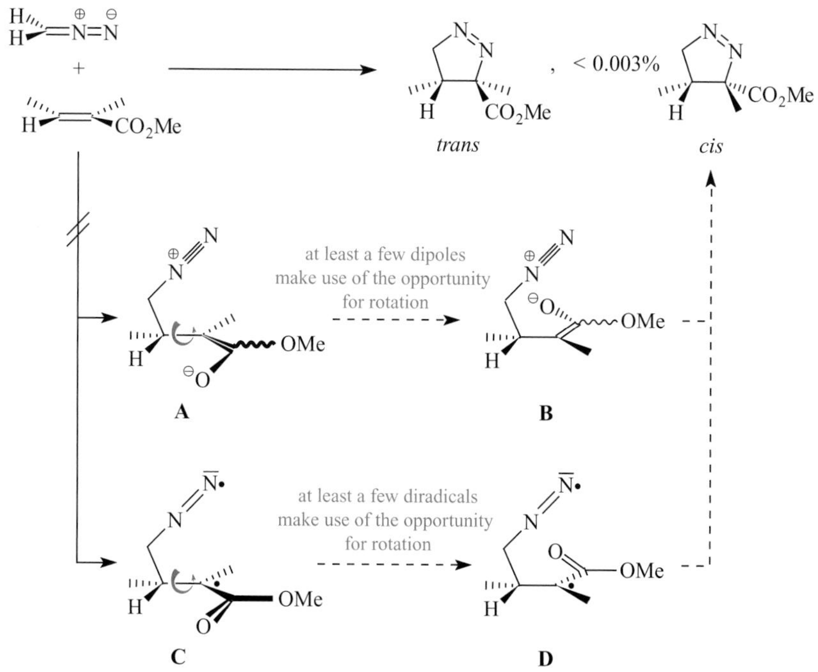

Fig. 15.41. Orientation-selective and stereoselective 1,3-dipolar cycloaddition of diazomethane. The *trans*-configured 2-methyl-2-butenoic acid is converted to the *trans*-configured cycloadduct with a diastereoselectivity of better than 99.997:0.003.

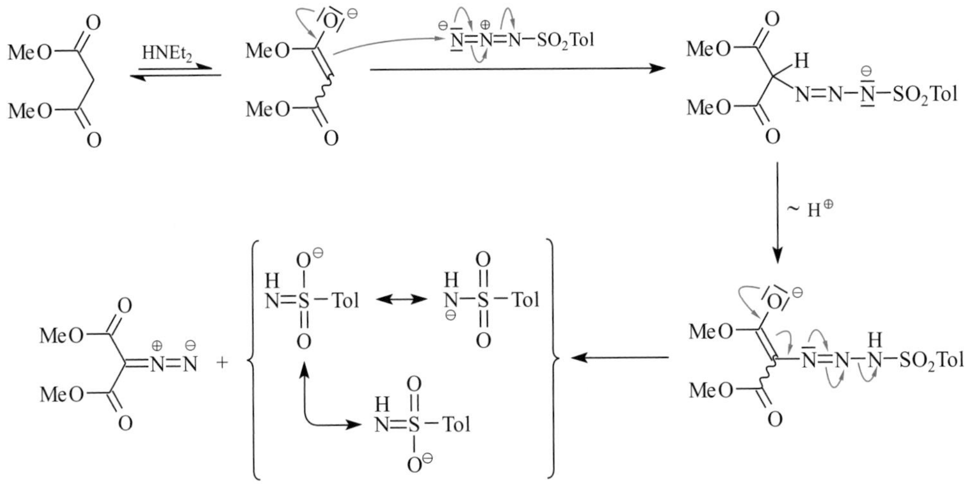

Fig. 15.42. Preparation of diazomalonic ester via diazo group transfer via the Regitz procedure.

an opportunity to rotate. The resulting isomerized intermediates **B** and **D** would then have cyclized to give the *cis*-cycloadduct about as fast as the original intermediates **A** and **C** would be assumed to give the *trans*-cycloadduct. Hence, for a multistep course of the 1,3-dipolar cycloaddition of Figure 15.41, one would expect more *cis*-cycloadduct to form than the amount (<<< 0.003%) that actually occurs.

Diazomalonic ester is another important 1,3-dipole for synthesis. We saw the kinetics of 1,3-dipolar cycloadditions of diazomalonic ester earlier, in the discussion of Figure 15.39. The preparation of this 1,3-dipole is accomplished most conveniently with the procedure shown in Figure 15.42.

15.5.4 1,3-Dipolar Cycloadditions of Nitrile Oxides

Nitrile oxides have the structure R—C≡N$^{\oplus}$—O$^{\ominus}$ or Ar—C≡N$^{\oplus}$—O$^{\ominus}$. Aliphatic nitrile oxides usually can be prepared only *in situ*, while the analogous aromatic compounds, which are resonance-stabilized, generally can be isolated. The most common preparation of nitrile oxides is the dehydration of aliphatic nitro compounds. Figure 15.43 shows in detail how this

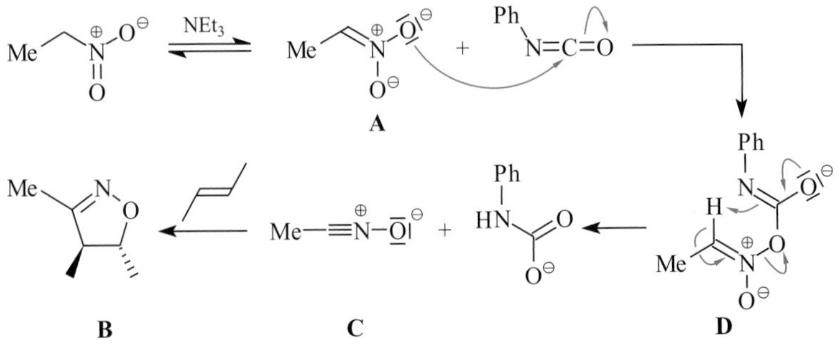

Fig. 15.43. Isoxazoline formation by way of a 1,3-dipolar addition of an *in situ* generated nitrile oxide to *trans*-butene.

dehydration can be effected by the reaction of nitroethane with a mixture of NEt_3 and Ph—N=C=O.

NEt_3 deprotonates a fraction of the nitro compound to give the nitronate **A.** One of its negatively charged O atoms adds to the C=O double bond of the isocyanate and the negatively charged adduct **D** is formed. This mode of reaction is reminiscent of the addition of carboxylic acids, alcohols, or water to isocyanates. Adduct **D** undergoes a β-elimination of phenyl carbamate. This elimination proceeds via a cyclic transition state that resembles the transition state of the Chugaev reaction (Figure 4.14). The nitrile oxide is formed and immediately adds to the *trans*-butene, which must be added to the reaction mixture before the nitrile oxide preparation is begun. The isoxazoline **B** is formed with complete *trans*-selectivity in accordance with a one-step 1,3-cycloaddition.

Benzonitrile oxide (**C** in Figure 15.44) is an isolable 1,3-dipole. It can be generated from benzaldoxime and an $NaOH/Cl_2$ solution. Under these reaction conditions the oxime/nitroso anion (**A** ↔ **B**) is initially formed and chlorine disproportionates into Cl—O$^\ominus$ and chloride. An S_N reaction of the negatively charged C atom of the anion **A** ↔ **B** at the Cl atom of Cl—O$^\ominus$ or of Cl—O—H affords the α-chlorinated nitroso compound **E**, which tautomerizes to the hydroxamic acid chloride **D.** From that species, the nitrile oxide **C** is generated via a base mediated 1,3-elimination. Isoxazoles are formed in the reactions of **C** with alkynes (Figure 15.44), while isoxazolines would be formed in its reactions with alkenes.

The sequence on the left in Figure 15.45 shows how the method described in Figure 15.43 can be used to convert unsaturated nitro compounds into nitrile oxides containing C=C double bonds. The sequence on the right in Figure 15.45 shows how unsaturated oximes can be converted into unsaturated nitrile oxides with the method described in Figure 15.44. The dipole and the dipolarophile contained in C=C-containing nitrile oxides can undergo an intramolecular 1,3-dipolar cycloaddition if these functional groups are located at a suitable distance—that is, if the possibility exists for the formation of a five- or six-membered ring. Two such cycloadditions are shown in Figure 15.45. Each is stereoselective and, when viewed in combination, the two reactions are stereospecific.

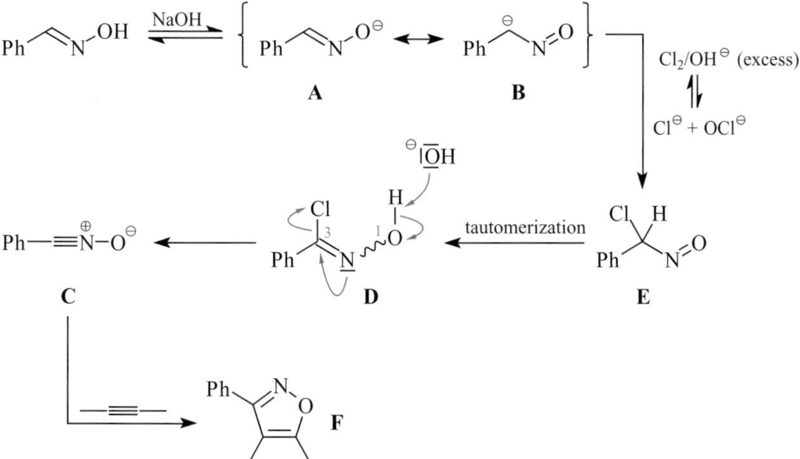

Fig. 15.44. Isoxazole formation by way of a 1,3-dipolar addition of an isolable nitrile oxide.

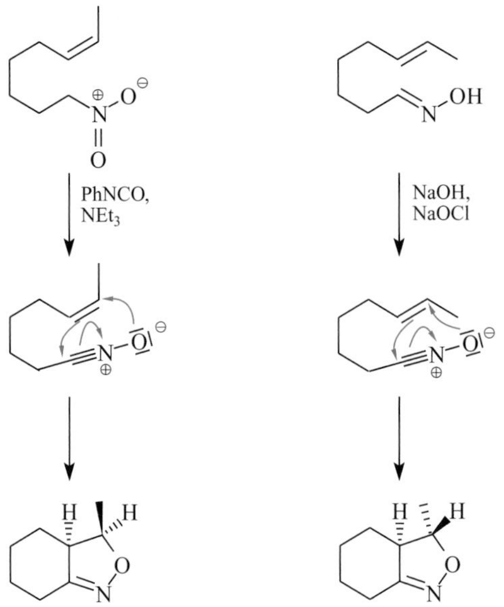

PhNCO, NEt_3

NaOH, NaOCl

Fig. 15.45. Intramolecular 1,3-dipolar additions of stereoisomeric nitrile oxides to form stereoisomeric isoxazolines.

Side Note 15.3. Isoxazolines as Synthetic Equivalents in Organic Chemistry

The N—O bond of isoxazolines can easily be cleaved via reduction. It is for this reason that isoxazolines are interesting synthetic intermediates. γ-Amino alcohols are formed by reduction with $LiAlH_4$ (for an example, see Figure 15.46, left). Hydrogenolysis of isoxazolines catalyzed by Raney nickel yields β-hydroxy imines, which undergo hydrolysis to β-hydroxycarbonyl compounds in the presence of boric acid (Figure 15.46, right).

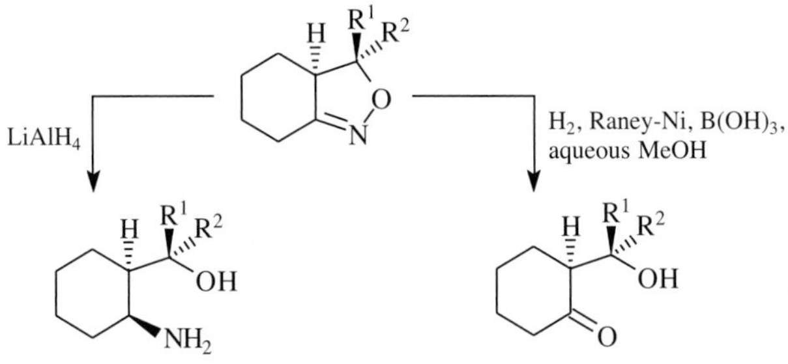

$LiAlH_4$

H_2, Raney-Ni, $B(OH)_3$, aqueous MeOH

Fig. 15.46. Isoxazolines (preparation: Figure 15.45) are synthetic equivalents for γ-amino alcohols and β-hydroxyketones.

Figures 15.45 and 15.46 illustrate impressively that the significance of 1,3-dipolar cycloadditions extends beyond the synthesis of five-membered heterocycles. In fact, these reactions can provide a valuable tool in the approach to entirely different synthetic targets. In the cases at hand, one can view the 1,3-dipolar cycloaddition of nitrile oxides to alkenes as a ring-closure reaction and more specifically, as a means of generating interestingly functionalized five- and six-membered rings in a stereochemically defined fashion.

15.5.5 1,3-Dipolar Cycloadditions and 1,3-Dipolar Cycloreversions as Steps in the Ozonolysis of Alkenes

The reaction of ozone with a C=C double bond begins with a 1,3-dipolar cycloaddition. It results in a 1,2,3-trioxolane, the so-called **primary ozonide:**

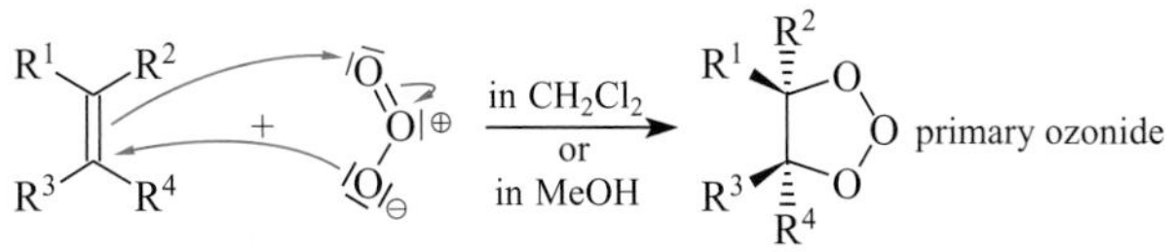

in CH_2Cl_2 or in MeOH

primary ozonide

The presence of two O—O bonds renders primary ozonides so unstable that they decompose immediately (Figures 15.47 and 15.48). The decomposition of the permethylated symmetric primary ozonide shown in Figure 15.47 yields acetone and a carbonyl oxide in a one-step reaction. The carbonyl oxide represents a 1,3-dipole of the allyl anion type (Table 15.2). When acetone is viewed as a dipolarophile, then the decomposition of the primary ozonide into acetone and a carbonyl oxide is recognized as the reversion of a 1,3-cycloaddition. Such a reaction is referred to as a **1,3-dipolar cycloreversion.**

The carbonyl oxide, a valence-unsaturated species, is not the final product of an ozonolysis. Rather, it will react further in one of two ways. Carrying out the ozonolysis in methanol leads to the capture of the carbonyl oxide by methanol to give a hydroperoxide, which is struc-

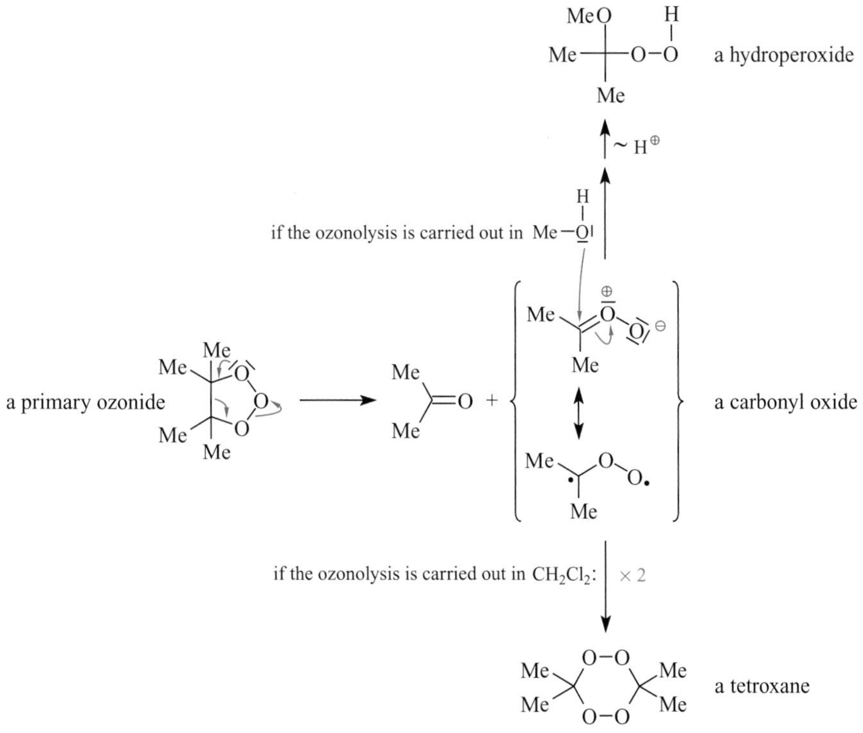

a hydroperoxide

if the ozonolysis is carried out in Me—O—H

a primary ozonide

a carbonyl oxide

if the ozonolysis is carried out in CH_2Cl_2: × 2

a tetroxane

Fig. 15.47. Solvent dependence of the decomposition of the symmetric primary ozonide formed in the ozonolysis of tetramethylethene (2,3-dimethyl-2-butene). The initially formed carbonyl oxide forms a hydroperoxide in methanol, while it dimerizes to give a 1,2,4,5-tetroxane in CH_2Cl_2.

turally identical to the "ether peroxide" of isopropyl methyl ether. However, if the same carbonyl oxide is formed in the absence of methanol (e. g., if the ozonolysis is carried out in dichloromethane) the carbonyl oxide undergoes a cycloaddition. If the carbonyl oxide is formed along with a ketone (Figure 15.47), as opposed to an aldehyde, it preferentially participates in a cycloaddition with another carbonyl oxide, i.e., a dimerization to a 1,2,4,5-tetroxane occurs. Such tetroxanes are extremely explosive compounds that one may not even attempt to isolate. Instead these compounds must be destroyed via reduction while they remain in solution. We will see later (Figures 17.27 and 17.28) how this is accomplished.

The chemistry of primary ozonides is more varied if they are less highly alkylated than the primary ozonide of Figure 15.47. This is particularly true if the primary ozonide is unsymmetrical, like the one shown in Figure 15.48. This is because its decay may involve two different 1,3-dipolar cycloreversions. Both of them result in one carbonyl oxide and one carbonyl compound. If the reaction is carried out in methanol, the two carbonyl oxides can react with the solvent (as in Figure 15.47) whereby each of them affords a hydroperoxide (an "ether peroxide" analog).

Obviously, this reaction channel is not an option for carbonyl oxides generated in dichloromethane (Figure 15.48). Instead, they can continue to react by way of a cycloaddition.

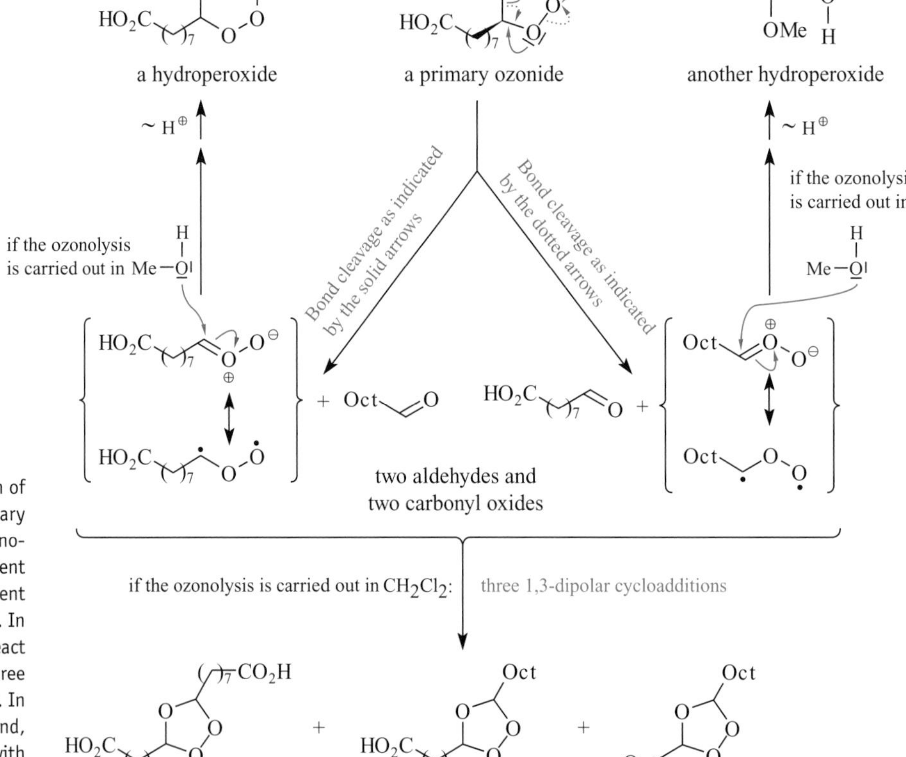

Fig. 15.48. Decomposition of the asymmetric primary ozonide formed in the ozonolysis of oleic acid. Two different aldehydes and two different carbonyl oxides are formed. In CH_2Cl_2 these molecules react with each other to form three secondary ozonides. In methanol, on the other hand, the carbonyl oxides react with the solvent to form hydroperoxides.

In contrast to the carbonyl oxide of Figure 15.47, they do not undergo a cycloaddition with each other. Instead, they undergo a 1,3-dipolar cycloaddition to the C=O double bond of the concomitantly formed aldehyde(s). The orientation selectivity is such that the trioxolane formed differs from the primary ozonide; the 1,2,4-trioxolane products are the so-called **secondary ozonides.**

Secondary ozonides are significantly more stable than primary ozonides, since the former contain only *one* O—O bond while the latter contain *two.* Some have been isolated and structurally characterized. However, they should be considered compounds that are explosive and dangerous. Therefore, secondary ozonides must be reduced to compounds without O—O bonds prior to reaction workup, too. Figures 17.27 and 17.28 show how this is accomplished.

15.5.6 A Tricky Reaction of Inorganic Azide

Aryldiazonium ions and sodium azide react to form aryl azides. We did not mention this reaction in the discussion of the S_N reactions of Ar—$N_2^{\oplus}$ (Section 5.4) because it belongs, at least in part, to the 1,3-dipolar cycloadditions.

Phenyl azide is formed from phenyldiazonium chloride and sodium azide by way of two competing reactions (Figure 15.49). The reaction path to the right begins with a 1,3-dipolar

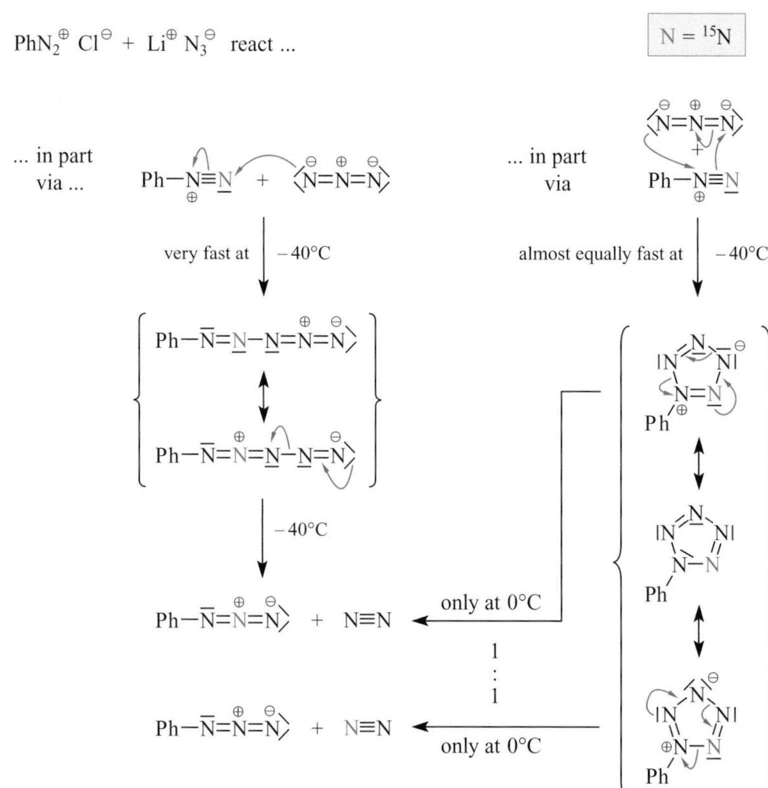

Fig. 15.49. The competing mechanisms of the "S_N reaction" Ar—$N_2^{\oplus}$ + $N_3^{\ominus}$ → Ar—N_3 + N_2. The reaction proceeds following two reaction channels: 1,3-dipolar cycloaddition via the intermediate phenylpentazole (right) and via the alternative phenylpentazene (left).

cycloaddition. At low temperature, this cycloaddition affords phenylpentazole, which decays above 0 °C via a 1,3-dipolar cycloreversion. This cycloreversion produces the 1,3-dipole phenyl azide as the *desired* product, and molecular nitrogen as a side product.

The rest of the phenyl azide produced in this reaction comes from an intermediate called 1-phenylpentazene, which is formed in competition with phenylpentazole (Figure 15.49, left). 1-Phenylpentazene is a compound that features a phenyl group at the end of a chain of five N atoms. The unavoidably ensuing reaction of this pentazene is an α-elimination of molecular nitrogen from the central N atom. It leads to the same organic product—phenyl azide—as the phenylpentazole route. That these two mechanisms are realized concomitantly was established by the isotopic labeling experiments reviewed in Figure 15.49.

References

W. Carruthers, "Cycloaddition Reactions in Organic Synthesis," Pergamon, Elmsford, NY, **1990**.

H. McNab, "One or More C=C Bond(s) by Pericyclic Processes," in *Comprehensive Organic Functional Group Transformations* (A. R. Katritzky, O. Meth-Cohn, C. W. Rees, Eds.), Vol. 1, 771, Elsevier Science, Oxford, U. K., **1995**.

15.1

F. Bernardi, M. Olivucci, M. A. Robb, "Predicting Forbidden and Allowed Cycloaddition Reactions: Potential Surface Topology and Its Rationalization," *Acc. Chem. Res.* **1990**, *23*, 405.

15.2

K. Fukui, "The Role of Frontier Orbitals in Chemical Reactions (Nobel Lecture)," *Angew. Chem. Int. Ed. Engl.* **1982**, *21*, 801–809.

R. Huisgen, "Cycloadditions – Definition, Classification, and Characterization," *Angew. Chem. Int. Ed. Engl.* **1968**, *7*, 321–328.

K. N. Houk, J. Gonzalez, Y. Li, "Pericyclic Reaction Transition States: Passions and Punctilios, 1935–1995," *Acc. Chem. Res.* **1995**, *28*, 81–90.

O. Wiest, K. N. Houk, "Density-Functional Theory Calculations of Pericyclic Reaction Transition Structures," *Top. Curr. Chem.* **1996**, *183*, 1–24.

K. N. Houk, Y. Li, J. D. Evanseck, "Transition Structures of Hydrocarbon Pericyclic Reactions," *Angew. Chem. Int. Ed. Engl.* **1992**, *31*, 682–708.

15.3

H. L. Holmes, "The Diels–Alder Reaction: Ethylenic and Acetylenic Dienophiles," *Org. React.* **1948**, *4*, 60–173.

M. C. Kloetzel, "The Diels–Alder Reaction with Maleic Anhydride," *Org. React.* **1948**, *4*, 1–59.

L. L. Butz, A. W. Rytina, "The Diels–Alder Reaction: Quinones and Other Cyclenones," *Org. React.* **1949**, *5*, 136–192.

J. Sauer, R. Sustmann, "Mechanistic Aspects of Diels–Alder Reactions: A Critical Survey," *Angew. Chem. Int. Ed. Engl.* **1980**, *19*, 779–807.

J. Jurczak, T. Bauer, C. Chapuis, "Intermolecular [4+2] Cycloadditions," in *Stereoselective Synthesis* (Houben-Weyl) 4th ed., **1996**, (G. Helmchen, R. W. Hoffmann, J. Mulzer, E. Schaumann, Eds.), **1996**, Vol. E21 (Workbench Edition), *5*, 2735–2871, Georg Thieme Verlag, Stuttgart.

F. Fringuelli, A. Taticchi, "Dienes in the Diels–Alder Reaction," Wiley, New York, **1990**.

V. D. Kiselev, A. I. Konovalov, "Factors that Determine the Reactivity of Reactants in the Normal and Catalyzed Diels–Alder Reaction," *Russ. Chem. Rev. (Engl. Transl.)* **1989**, *58*, 230.

W. Oppolzer, "Intermolecular Diels–Alder Reactions," in *Comprehensive Organic Synthesis* (B. M. Trost, I. Fleming, Eds.), Vol. 5, 315, Pergamon Press, Oxford, U. K., **1991**.

F. Fringuelli, A. Taticchi, E. Wenkert, "Diels–Alder Reactions of Cycloalkenones in Organic Synthesis. A Review;" *Org. Prep. Proced. Int.* **1990**, *22*, 131–165.

F. Fringuelli, L. Minuti, F. Pizzo, A. Taticchi, "Reactivity and Selectivity in Lewis Acid-Catalyzed Diels–Alder Reactions of 2-Cyclohexenones," *Acta Chem. Scand.* **1993**, *47*, 255–263.

E. Ciganek, "The Intramolecular Diels–Alder Reaction," *Org. React.* **1984**, *32*, 1–374.

W. R. Roush, "Stereochemical and Synthetic Studies of the Intramolecular Diels–Alder Reaction," in *Advances in Cycloaddition* (D. P. Curran, Ed.) **1990**, *2*, Jai Press, Greenwich, CT.

D. Craig, "Intramolecular [4+2] Cycloadditions," in *Stereoselective Synthesis* (Houben-Weyl) 4th ed., **1996**, (G. Helmchen, R. W. Hoffmann, J. Mulzer, E. Schaumann, Eds.), **1996**, Vol. E21 (Workbench Edition), *5*, 2872–2952, Georg Thieme Verlag, Stuttgart.

A. G. Fallis, "Harvesting Diels and Alder's Garden: Synthetic Investigations of Intramolecular [4 + 2] Cycloadditions," *Acc. Chem. Res.* **1999**, *32*, 464–474.

J. M. Coxon, R. D. J. Froese, B. Ganguly, A. P. Marchand, K. Morokuma, "On the Origins of Diastereofacial Selectivity in Diels–Alder Cycloadditions," *Synlett* **1999**, 1681–1703.

J. I. Garcia, J. A. Mayoral, L. Salvatella, "Do Secondary Orbital Interactions Really Exist?", *Acc. Chem. Res.* **2000**, *33*, 658–664.

15.4

W. E. Hanford, J. C Sauer, "Preparation of Ketenes and Ketene Dimers," *Org. React.* **1946**, *3*, 108–140.

S. Patai (Ed.), "The Chemistry of Ketenes, Allenes, and Related Compounds," Wiley, New York, **1980**.

W. T. Brady, "Synthetic Applications Involving Halogenated Ketenes," *Tetrahedron* **1981**, *37*, 2949.

P. W. Raynolds, "Ketene," in *Acetic Acid and Its Derivatives* (V. H. Agreda, J. R. Zoeller, Eds.), 161, Marcel Dekker, New York, **1993**.

E. Lee-Ruff, "New Synthetic Pathways from Cyclobutanones," in *Advances in Strain in Organic Chemistry* (B. Halton, Ed.) **1991**, *1*, Jai Press, Greenwich, CT.

J. A. Hyatt, P. W. Raynolds, "Ketene Cycloadditions," *Org. React.* **1994**, *45*, 159–646.

15.5

A. Padwa (Ed.), "1,3-Dipolar Cycloaddition Chemistry," Wiley, **1984**.

R. Huisgen, "Steric Course and Mechanism of 1,3-Dipolar Cycloadditions," in *Advances in Cycloaddition* (D. P. Curran, Ed.) **1988,** *1*, 11–31, Jai Press, Greenwich, CT.

R. Sustmann, "Rolf Huisgen's Contribution to Organic Chemistry, Emphasizing 1,3-Dipolar Cycloadditions," *Heterocycles* **1995**, *40*, 1–18.

A. Padwa, "Intermolecular 1,3-Dipolar Cycloadditions," in *Comprehensive Organic Synthesis* (B. M. Trost, I. Fleming, Eds.), Vol. 4, 1069, Pergamon Press, Oxford, U. K., **1991**.

M. Cinquini, F. Cozzi, "1,3-Dipolar Cycloadditions," in *Stereoselective Synthesis* (Houben-Weyl) 4th ed., **1996**, (G. Helmchen, R. W. Hoffmann, J. Mulzer, E. Schaumann, Eds.), **1996**, Vol. E21 (Workbench Edition), *5*, 2953–2987, Georg Thieme Verlag, Stuttgart.

A. Padwa, A. M. Schoffstall, "Intramolecular 1,3-Dipolar Cycloaddition Chemistry," in *Advances in Cycloaddition* (D. P. Curran, Ed.), Vol. 2, 1, JAI Press, Greenwich, CT, **1990**.

P. A. Wade, "Intramolecular 1,3-Dipolar Cycloadditions," in *Comprehensive Organic Synthesis* (B. M. Trost, I. Fleming, Eds.), Vol. 4, 1111, Pergamon Press, Oxford, U. K., **1991**.

T. H. Black, "The Preparation and Reactions of Diazomethane," *Aldrichimica Acta* **1983**, *16*, 3.

V. Dave, E. W. Warnhoff, "The Reactions of Diazoacetic Esters with Alkenes, Alkynes, Heterocyclic, and Aromatic Compounds," *Org. React.* **1970**, *18*, 217–402.

K. B. G. Torsell, "Nitrile Oxides, Nitrones and Nitronates in Organic Synthesis. Novel Strategies in Synthesis," VCH Verlagsgesellschaft, Weinheim, **1988**.

S. Kanemasa, O. Tsuge, "Recent Advances in Synthetic Applications of Nitrile Oxide Cycloaddition (1981-1989)," *Heterocycles* **1990**, *30*, 719–736.

D. P. Curran, "The Cycloaddition Approach to α-Hydroxy Carbonyls: An Emerging Alternative to the Aldol Strategy," in *Advances in Cycloaddition* (D. P. Curran, Ed.) **1988**, *1*, 129–189, Jai Press, Greenwich, CT.

R. P. Litvinovskaya, V. A. Khripach, "Regio- and Stereochemistry of 1,3-Dipolar Cycloaddition of Nitrile Oxides to Alkenes," *Russ. Chem. Rev.* **2001**, *70*, 464–485.

R. L. Kuczkowski, "Formation and Structure of Ozonides," *Acc. Chem. Res.* **1983**, *16*, 42.

R. L. Kuczkowski, "The Structure and Mechanism of Formation of Ozonides," *Chem. Soc. Rev.* **1992**, *21*, 79–83.

O. Horie, G. K. Moortgat, "Gas-Phase Ozonolysis of Alkenes. Recent Advances in Mechanistic Investigations," *Acc. Chem. Res.* **1998**, *31*, 387–396.

W. Sander, "Carbonyl Oxides: Zwitterions or Diradicals?", *Angew. Chem. Int. Ed. Engl.* **1990**, *29*, 344–354.

W. H. Bunnelle, "Preparation, Properties, and Reactions of Carbonyl Oxides," *Chem. Rev.* **1991**, *91*, 335–362.

K. Ishiguro, T. Nodima, Y. Sawaki, "Novel Aspects of Carbonyl Oxide Chemistry," *J. Phys. Org. Chem.* **1997**, *10*, 787–796.

Further Reading

S. Danishefsky, "Cycloaddition and Cyclocondensation Reactions of Highly Functionalized Dienes: Applications to Organic Synthesis," *Chemtracts: Org. Chem.* **1989**, *2*, 273–297.

T. Kametani, S. Hibino, "The Synthesis of Natural Heterocyclic Products by Hetero Diels–Alder Cycloaddition Reactions," *Adv. Heterocycl. Chem.* **1987**, *42*, 246.

J. A. Coxon, D. Q. McDonald, P. J. Steel, "Diastereofacial Selectivity in the Diels–Alder Reaction," in *Advances in Detailed Reaction Mechanisms* (J. M. Coxon, Ed.), Vol. 3, Jai Press, Greenwich, CT, **1994**.

J. D. Winkler, "Tandem Diels–Alder Cycloadditions in Organic Synthesis," *Chem. Rev.* **1996**, *96*, 167–176.

C. O. Kappe, S. S. Murphree, A. Padwa, "Synthetic Applications of Furan Diels–Alder Chemistry," *Tetrahedron* **1997**, *53* 14179–14231.

D. Craig, "Stereochemical Aspects of the Intramolecular Diels–Alder Reaction," *Chem. Soc. Rev.* **1987**, *16*, 187.

G. Helmchen, R. Karge, J. Weetman, "Asymmetric Diels–Alder Reactions with Chiral Enoates as Dienophiles," in *Modern Synthetic Methods* (R. Scheffold, Ed.), Vol. 4, 262, Springer, Berlin, **1986**.

L. F. Tietze, G. Kettschau, "Hetero Diels–Alder Reactions in Organic Chemistry," *Top. Curr. Chem.* **1997**, *189*, 1–120.

K. Neuschuetz, J. Velker, R. Neier, "Tandem Reactions Combining Diels–Alder Reactions with Sigmatropic Rearrangement Processes and Their Use in Synthesis," *Synthesis* **1998**, 227–255.

D. L. Boger, M. Patel, "Recent Applications of the Inverse Electron Demand Diels–Alder Reaction," in *Progress in Heterocyclic Chemistry* (H. Suschitzky, E. F. V. Scriven, Eds.) **1989**, *1*, 36–67, Pergamon Press, Oxford, U. K.

M. J. Tashner, "Asymmetric Diels–Alder Reactions," in *Organic Synthesis: Theory and Applications* (T. Hudlicky, Ed.) **1989**, *1*, Jai Press, Greenwich, CT.

H. B. Kagan, O. Riant, "Catalytic Asymmetric Diels–Alder Reactions," *Chem. Rev.* **1992**, *92*, 1007.

T. Oh, M. Reilly, "Reagent-Controlled Asymmetric Diels–Alder Reactions," *Org. Prep. Proced. Int.* **1994**, *26*, 129.

I. E. Marko, G. R. Evans, P. Seres, I. Chelle, Z. Janousek, "Catalytic, Enantioselective, Inverse Eectron-Demand Diels–Alder Reactions of 2-Pyrone Derivatives," *Pure Appl. Chem.* **1996**, *68*, 113.

H. Waldmann, "Asymmetric Hetero-Diels–Alder Reactions," *Synthesis* **1994**, 635–651.

H. Waldmann, "Asymmetric Aza-Diels–Alder Reactions," in *Organic Synthesis Highlights II* (H. Waldmann, Ed.), VCH, Weinheim, New York, **1995**, 37–47.

L. C. Dias, "Chiral Lewis Acid Catalysts in Diels Alder Cycloadditions: Mechanistic Aspects and Synthetic Applications of Recent Systems," *J. Braz. Chem. Soc.* **1997,** 8, 289–332.

D. Carmona, M. Pilar Lamata, L. A. Oro, "Recent Advances in Homogeneous Enantioselective Diels–Alder Reactions Catalyzed by Chiral Transition-Metal Complexes," *Coord. Chem. Rev.* **2000**, *200-2*, 717–772.

F. Fringuelli, O. Piermatti, F. Pizzo, L. Vaccaro, "Recent Advances in Lewis Acid Catalyzed Diels–Alder Reactions in Aqueous Media," *Eur. J. Org. Chem.* **2001**, *3*, 439–455.

B. R. Bear, S. M. Sparks, K. J. Shea, "The Type 2 Intramolecular Diels–Alder Reaction: Synthesis and Chemistry of Bridgehead Alkenes," *Angew. Chem. Int. Ed. Engl.* **2001**, *40*, 820–849.

E. J. Corey, "Catalytic Enantioselective Diels–Alder Reaction: Methods, Mechanistic Basis, Reaction Pathways, and Applications," *Angew. Chem. Int. Ed. Engl.* **2002**, *41*, 1650–1667.

K. C. Nicolaou, S. A. Snyder, T. Montagnon, G. E. Vassilikogiannakis, "The Diels–Alder Reaction in Total Synthesis," *Angew. Chem. Int. Ed. Engl.* **2002**, *41*, 1668–1698.

J. D. Roberts, C. M. Sharts, "Cyclobutane Derivatives from Thermal Cycloaddition Reactions", *Org. React.* **1962**, *12*, 1–56.

B. B. Snider, "Thermal [2+2] Cycloadditions," in *Stereoselective Synthesis* (Houben-Weyl) 4th ed., **1996**, (G. Helmchen, R. W. Hoffmann, J. Mulzer, E. Schaumann, Eds.), **1996**, Vol. E21 (Workbench Edition), *5*, 3060–3084, Georg Thieme Verlag, Stuttgart.

J. Mulzer, "Natural Product Synthesis via 1,3-Dipolar Cycloadditions," in *Organic Synthesis Highlights* (J. Mulzer, H.-J. Altenbach, M. Braun, K. Krohn, H.-U. Reißig, Eds.), VCH, Weinheim, New York, **1991**, 77–95.

K. V. Gothelf, K. A. Jürgensen, "Metal-Catalyzed Asymmetric 1,3-Dipolar Cycloaddition Reactions," *Acta Chem. Scand.* **1996**, *50*, 652.

P. N. Confalone, E. M. Huie, "The [3+2] Nitrone-Olefin Cycloaddition Reaction," *Org. React.* **1988**, *36*, 1–173.

Transition Metal-Mediated Alkenylations, Arylations, and Alkynylations

16

As a rule, halogens, OH groups, and all other O substituents that are attached to an sp^2-hybridized C atom of an alkene or an aromatic compound cannot be substituted by nucleophiles alone. One exception has already been discussed and that was the addition/elimination mechanism of the nucleophilic substitution. This mechanism occurs, among other substrates, with alkenes that carry a strong electron acceptor on the neighboring C atom (see Figures 10.47–10.49 for examples). The same mechanism is known to also operate for aromatic compounds—under the condition, again, that they carry an acceptor substituent at an appropriate position (Section 5.5). The benzyne mechanism of nucleophilic substitution, that is, an elimination/addition mechanism (Section 5.6), presents a second mode of nucleophilic substitution of a halogen at an sp^2-hybridized C atom.

Halogens or O-bound leaving groups, however, also can be detached from the sp^2-hybridized C atom of an alkene or an aromatic compound.

- if organometallic compounds act as nucleophiles, and
- if a transition metal is present in at least catalytic amounts. These SN reactions are designated as **C,C coupling reactions**.

The most important substrates for substitutions of this type are alkenyl and aryl triflates, bromides, or iodides (Sections 16.1–16.4). The most important organometallic compounds to be introduced into the substrates contain Cu, Mg, B, Zn or Sn. The metal-bound C atom can be sp^3-, sp^2-, or sp-hybridized in these compounds, and each of these species, in principle, is capable of reacting with unsaturated substrates. Organocopper compounds often (Section 16.1, 16.2), but not always substitute *without* the need for a catalyst (Section 16.4.5). Grignard compounds substitute in the presence of catalytic amounts of Ni complexes (Section 16.3), while organoboron (Section 16.4.2), organozinc (Section 16.4.3) and organotin (Section 16.4.4) compounds are typically reacted in the presence of Pd complexes (usually $Pd(PPh_3)_4$).

All these C,C coupling reactions can be carried out in an analogous fashion at *sp*-hybridized carbons, too, as long as this carbon binds to a Br or I atom as a leaving group. However, we will present this type of reaction only briefly (Figure 16.33).

In Section 16.5, a few *other* C,C coupling reactions of alkenes and of aromatic compounds, which contain an sp^2—OTf, an sp^2—Br, or an sp^2—Cl bond, will be discussed because these C,C couplings and the preceding ones are closely related mechanistically. These substrates, however, react with *metal-free* alkenes. Palladium complexes again serve as the catalysts.

Bruckner R (author), Harmata M (editor) In: *Organic Mechanisms – Reactions, Stereochemistry and Synthesis*
Chapter DOI: 10.1007/978-3-642-03651-4_16,

16.1 Alkenylation and Arylation of Gilman Cuprates

Me_2CuLi couples with a variety of alkenyl triflates (bromides, iodides) giving methyl derivatives (Table 16.1), and, in complete analogy, with aryl triflates (bromides, iodides). Occasionally, Me_2CuLi also couples with alkenyl or aryl phosphates.

Figure 16.1 depicts the presumed reaction mechanism for such C,C coupling reactions between Gilman cuprates and the C atom of suitable C_{sp^2}—X bonds. Some of the reaction steps (1–4) might well be comprised of more than one elementary reaction. It is very important to understand these four steps as a general reaction concept. In step 1, the heterosubstituted alkene or the aromatic compound enters the coordination sphere of copper as a two-electron π-donor. The reactive π-complex will have at least one vacancy at the metal. Step 2 consists of the **oxidative addition** of the triflate or the halide, respectively, to the metal. In this step the metal inserts into the C_{sp^2}—X bond and the metal's oxidation state is increased from + 1 to + 3. The reaction types of steps 1 and 2 reoccur in reverse order in steps 3 and 4. Step 3 is complementary to step 2 and represents a **reductive elimination**. The reaction product is a coordinatively unsaturated π-complex of—again—Cu(I). The C,C coupling product functions as a π-donor in this complex until the metal assumes a coordination number suitable for further reaction. The final step, 4, is complementary to step 1 and entails the dissociation of the π-bonded ligand. This step yields the final coupling product and an organocopper compound (in place of the cuprate employed originally). The devil is in the details, and many of these still have yet to be uncovered, for this and other reactions of organocuprates.

Many alkenyl triflates (-bromides, -iodides) may be prepared with *cis*- or *trans*- and *E*- or *Z*-configuration, respectively:

- This is illustrated in Figure 16.2 by the stereoselective synthesis of the *E*,*Z*-isomeric enol triflates **B**. If di-*n*-propylketone is deprotonated with LDA in THF at –78 °C, the *E*-enolate is selectively formed. The reason is the same as the one given for the selective formation of the "*E*"-configured ester enolate via the cyclic transition state **A** of the deprotonation shown in Figure 13.16, namely the avoidance of the *syn*-pentane strain in the cyclic transition state of the deprotonation. In contrast, the *Z*-enolate of di-*n*-propylketone is formed

Tab. 16.1 Product Spectrum of C,C Coupling Reactions with the Gilman Cuprate Me_2CuLi. Analogous products are obtained with the Gilman cuprates R_2CuLi, where R can be an alkyl (= Me), alkenyl or aryl substituent

Substrate	Me, OSO_2CF_3 (cyclohexenyl triflate)	Me, OSO_2CF_3 (cyclohexenyl triflate)	R–CH=CH–Br(I)	R–CH=CH–Br(I)	$ArOSO_2CF_3$ or Ar−Br(I)
Preparation according to	Fig. 13.23	Fig. 13.23	Fig. 16.15	Fig. 13.56 Fig. 16.15 Fig. 16.17	e. g. as in section. 5.2.1
Reaction with Me_2CuLi yields	Me, Me (cyclohexene)	Me, Me (cyclohexene)	R–CH=CH–Me	R–CH=CH–Me	Ar−Me

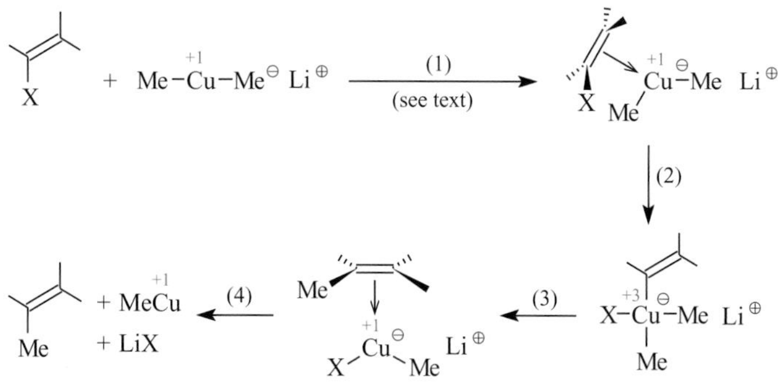

Fig. 16.1. Presumed elementary steps of a C,C coupling between a Gilman cuprate and an alkenyl or aryl triflate ($X = O_3S{-}CF_3$), bromide (X = Br), or iodide (X = I). The four elementary steps of the reaction, discussed in the text, are (1) complexation, (2) oxidative addition of the substrate to the metal, (3) reductive elimination, and (4) dissociation of the π-bound ligand.

selectively when LiHMDS instead of LDA is used for deprotonation. The difference is due to the much lower basicity of LiHMDS (see Section 13.1.2). Consequently, LiHMDS in THF might react as a solvent-separated ion pair which would make the lithium-free $(Me_3Si)_2N^{\ominus}$ anion available, while LDA in THF—just as a reminder—is a contact ion pair. The trifluoromethanesulfonation of the enolate intermediates in Figure 16.2 with

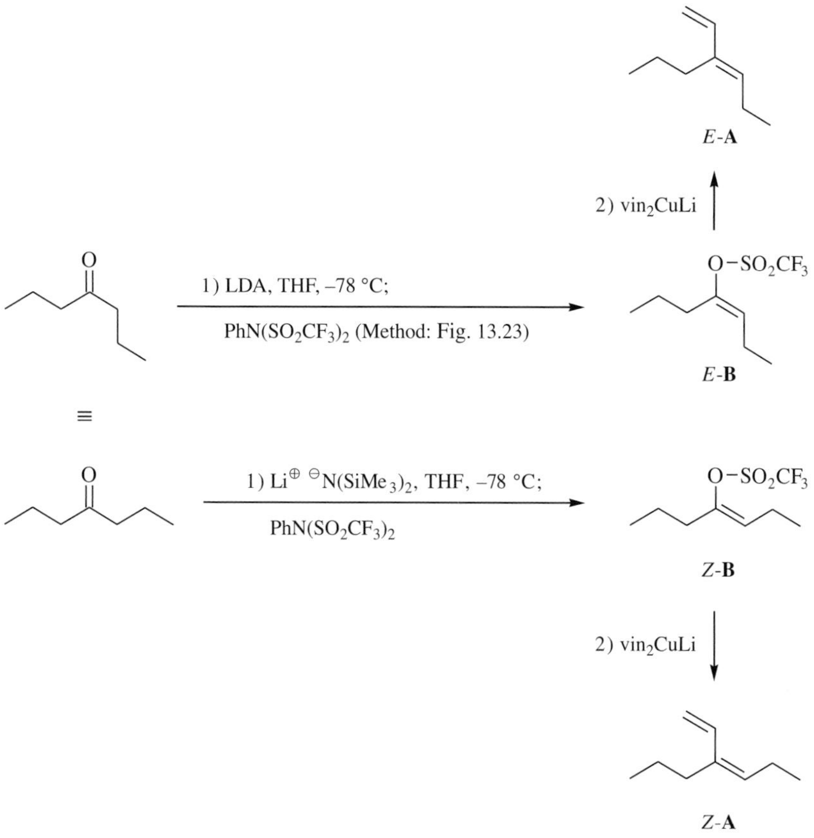

Fig. 16.2. Stereoselective synthesis of 1,3-dienes with a trisubstituted C=C double bond by way of C,C coupling reactions between the stereouniform alkenyl triflates **B** and an unsaturated Gilman cuprate. The C_{sp^2}–triflate bond is converted into the C_{sp^2}–C bond of the coupling product with full retention of configuration. According to the mechanism in Figure 16.1 all these couplings proceed via a Cu(III) intermediate with three unsaturated ligands: vinyl (twice) and (α-alkyl)alkenyl (once). Basically, reductive elimination of these could not only furnish the substituted 1,3-dienes shown, but also 1,3-butadiene itself. Cross-coupling is preferred, however. –The explanation of the *E*-selective formation of the enolate (and enol triflate) corresponds to the one given in Figure 13.16; the explanation of the *Z*-selective enolate (and enol triflate) formation is analogous to the example depicted in Figure 13.17.

$PhN(SO_2CF_3)_2$ at the enolate oxygen fixes the previously formed C=C double bond geometries, and the alkenyl triflates *E*- and *Z*-**B**, respectively, are obtained.

- Figure 13.25 presented the stereoselective synthesis of ester-substituted, *E*- or *Z*-configured alkenyl phosphates..
- Stereouniform haloalkenes can be prepared from terminal alkynes, for example—through hydroboration/halogenation (Figure 16.15) or carbocupration/halogenation (Figure 16.17). In this way, *cis*-alkenyl bromides (Figure 16.15), *cis*-alkenyl iodides (Figure 16.17) or *trans*-alkenyl iodides (Figure 16.15) are produced.

Each of the above-mentioned and stereochemically uniform alkenyl triflates (-phosphates, -bromides, -iodides) is capable of undergoing couplings with Gilman cuprates. This process always involves the transformation of the C_{sp^2}–X bond of the substrate into the C_{sp^2}–C bond of the product with full retention of configuration. An illustration is given in Figure 16.2 where the stereoselective syntheses of the 1,3-dienes **A** from vin_2CuLi and the diastereomorphic alkenyl triflates *E*- and *Z*-**B**, respectively, were used as an example. A pair of stereospecific substitution reactions of Me_2CuLi on stereoisomeric alkenyl phosphates was introduced in Figure 10.48; as mentioned earlier these are assumed to proceed in analogy to the C,C coupling reactions discussed in this section rather than according to the addition/elimination mechanism presented in Chapter 10. With Gilman cuprates, the C_{sp^2}–Br- and C_{sp^2}–I bonds of alkenyl halides are also stereoselectively transformed into C_{sp^2}–C bonds with full retention of configuration (Table 16.1, columns 4 and 5). Hence, the coupling reaction is stereospecific.

16.2 Arylation and Alkynylation of Neutral Organocopper Compounds I

In this section monometallic compounds RCu and CuX will be referred to as *neutral* organocopper compounds and *neutral* copper compounds, respectively, and conceptually distinguished from Gilman *cuprates* (R_2CuLi), cyanide-containing Gilman *cuprates* (better known as cyanocuprates), Normant cuprates and Knochel cuprates. Unlike the “Arylations, Alkenylations and Alkynylations of Neutral Organocopper Compounds II” in Section 16.4.5, the coupling reactions of neutral organocopper compounds discussed here need not be catalyzed by palladium(0) complexes.

Neutral arylcopper compounds are accessible from aryllithium compounds by way of transmetalation with one equivalent of CuI (example: Figure 16.4). Here, one equivalent of lithium iodide is released. This transmetalation also is a step in the formation of Gilman cuprates from two equivalents of the same organolithium compound and one equivalent of CuI (Figure 10.43, top left). In the latter case, the transmetalation is followed by the addition of a second equivalent of an organolithium compound: ArLi + ArCu → Ar_2CuLi. Neutral arylcopper compounds are also accessible from aryl iodides and recently from aryl bromides also, if these are reduced with elemental copper (example: Figure 16.5).

Neutral alkynylcopper compounds are not prepared by transmetalation of alkynyllithium compounds. Rather, they are obtained by partially deprotonating terminal alkynes with amines and capturing the ammonium acetylide formed at equilibrium with CuI (→ R–C≡C–Cu + $R_3NH^{\oplus}I^{\ominus}$; example: Figure 16.7). Copper(I) cyanide couples with aryl iodides and -bromides in a similar fashion as alkynylcopper compounds (which may well be conceived as their carba analogs).

The last C nucleophile among the copper(I) nucleophiles to be presented in this section is the copper(I) enolate of malonic acid diethyl ester. Unlike the "neutral" (organo-)copper compounds introduced above, this one is not employed in stoichiometric amounts, but is generated in small amounts by adding some CuI to sodium malonate with the copper(I) enolate being the reactive species. In the same way, a number of *O*-, *S*- and *N* nucleophiles that are present in stoichiometric amounts can be transformed into small amounts of the corresponding Cu(I) derivatives, which again are highly reactive, by adding different copper sources—ranging from copper powder to Cu(II)O.

All of the "neutral" (organo-)copper compounds just listed can be arylated by aryl iodides and/or -bromides; Ar–Cu and R–C≡C–Cu can also be alkynylated by alkynyl iodides and/or -bromides. Table 16.2 summarizes the whole spectrum of corresponding products. After having grasped the mechanism of these reactions by way of Figure 16.3, we will discuss their synthetic potential by means of the examples given in Figures 16.4–16.9.

If a suitable copper(I) compound (Ar–Cu, alkynyl–Cu, Cu–CN, Cu–O_{aryl}, Cu–S_{aryl}, Cu–NH_{aryl} or Cu malonate) is added to its reaction partner, i.e., the aryl or alkynyl iodide/bromide, or, in the presence of this reaction partner, is generated in an equilibrium reaction, the respective substitution product is always formed according to the mechanism depicted in Figure 16.3. *Conveniently, this mechanism involves the same steps that are already familiar from the arylation or alkenylation of Gilman cuprates according to Figure 16.1.* Again, step 1 consists of the formation of a π-complex of Cu(I), and step 2 includes an oxidative addition forming a Cu(III) intermediate. The copper here inserts into the C_{sp^2}–Hal- and C_{sp}–Hal bond of the arylating or alkynylating agent, respectively. Step 3 consists of a carbon-carbon bond formation via a reductive elimination, still leaving the resulting substitution product as a π-complex at the regenerated Cu(I). In step 4, the reaction product is then released from this π-complex.

Tab. 16.2 Range of Products of C,C and C,Het Couplings with Neutral Copper Compounds that do not Need a Catalyst (with some of these reactions requiring only catalytic instead of stoichiometric amounts of copper)

Substrate	Ar–Br(I)			R–≡–Br(I)
Preparation according to	Section 5.2.1			Section 16.2
Reaction with suitable CuR yields	Ar–Ar or Ar–Ar′	Ar–≡–R Ar–≡N	Ar–Het	R–≡–≡–R′

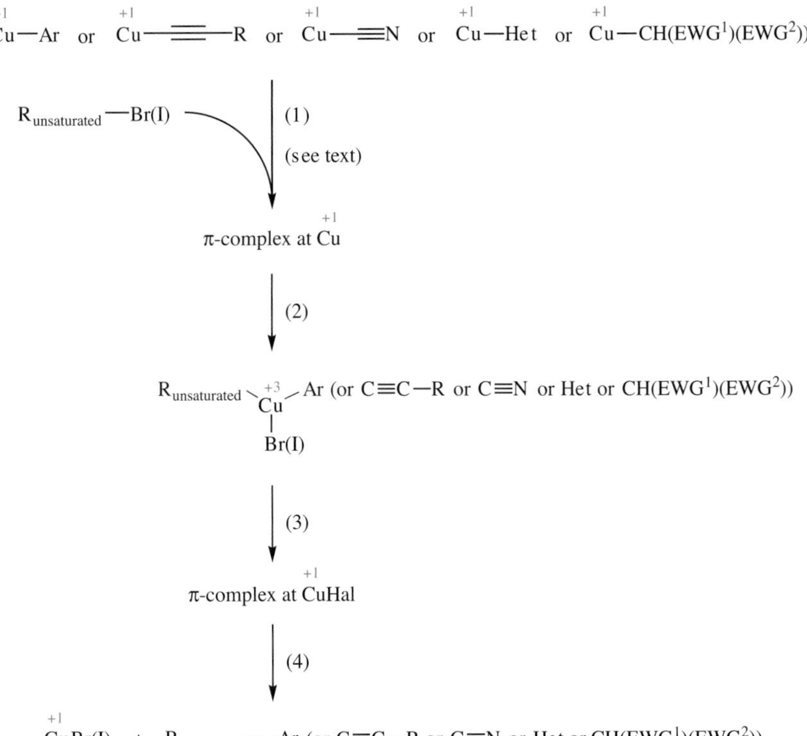

Fig. 16.3. Mechanism of the arylation or alkynylation of Cu(I) compounds shown in the top row, i.e. the introduction of $R_{unsaturated}$ into these Cu(I) compounds. It involves the same steps—(1) complexation, (2) oxidative addition of the aryl halide or alkynyl halide to the metal, (3) reductive elimination and (4) dissociation of the π-bound ligand—which in Figure 16.1 are shown for the arylation and alkenylation of Gilman cuprates.

At the same time a Cu(I) compound is formed. This is required in catalytic amounts only if it additionally acts to convert a less efficient nucleophile into a more efficient copper-containing nucleophile *in situ*.

You should feel challenged to explicitly identify the four steps described above in all the reaction examples following in this section—and you should be pleased to hear that there is no need to learn a new mechanism every single time!

Figure 16.4 presents the synthesis of an asymmetric biaryl. Here, the copper nucleophile is phenylcopper, which is accessible from iodobenzene by iodine/lithium exchange (mechanism: Figure 5.46) and transmetalation of the resulting phenyllithium with copper(I) iodide. The arylation is conducted with (2,6-dimethoxyphenyl)iodide that can be obtained from resorcinol dimethyl ether via *ortho*-lithiation and iodination. The mechanism is the same as that of Figure 16.3, which was discussed in detail (formation of a π-complex, insertion, C,C bond formation within the complex through reductive elimination and finally cleavage of the coupling product, i.e. a biaryl, from the π-complex).

The **Ullmann reaction** (Figure 16.5) represents another synthesis of biaryls that reliably leads to the formation of symmetric biaryls (upper and middle reaction examples), but, in particular cases, is also suitable for the synthesis of unsymmetrical biaryls (lower reaction example). To prepare a symmetric biaryl in the traditional way, an aryl iodide is heated with Cu powder; in order to prepare the parent compound of biaryls, i.e., biphenyl, one starts with iodobenzene. The metal reduces 50% of the substrate to phenylcopper *in situ*. The latter

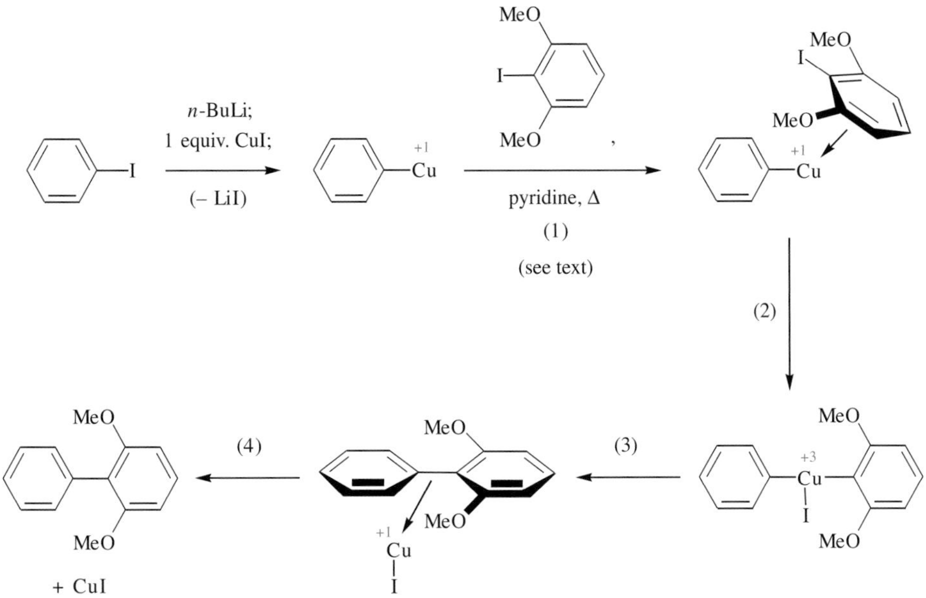

Fig. 16.4. Biaryl synthesis via arylcopper compounds I—preparation of the nucleophile in a separate reaction step.

reacts with the remaining iodobenzene via the mechanism in Figure 16.3, and biphenyl is formed.

A modern variant of the Ullmann coupling also uses the less expensive bromo- instead of iodobenzene as the starting material (Figure 16.5, middle). Here, copper(I)thiophene carboxylate (**A**) is used instead of elemental copper, and a new path is followed to deliver the known intermediate phenylcopper. First, bromobenzene forms a π-complex with **A**. This is followed by oxidative addition (→ **C**) and reductive elimination (→ phenylcopper + acyl hypohalogenite **B**).

In a few instances, it is also possible to generate unsymmetrical biaryls using Ullmann's arylcopper compound generated *in situ*. In these cases, one employs a mixture of an aryl iodide and another aryl halide (not an iodide!); *and* the other aryl halide *must* exhibit a higher propensity than the aryl iodide to couple to the arylcopper intermediate (example: Figure 16.5, bottom). This is referred to as a **crossed Ullmann coupling**.

Side Note 16.1. Ullmann Reactions for the Synthesis of Aromatic Compounds

The name Ullmann is not only associated with the biaryl synthesis (Figure 16.4, 16.5), but is also known from the synthesis of diaryl ethers (**Ullmann synthesis of diaryl ethers**). An example is given in the topmost reaction of Figure 16.6. Remember Side Note 5.6, where we asked the following question: "How can diphenyl ethers be prepared?" Now you are ready to give a correct answer, which is: "By way of Ullmann synthesis."

The term "Ullmann reaction" is often expanded to also include "copper-mediated nucleophilic substitution reactions (of almost any kind) to iodo- or bromoaromatic compounds." The discussion of Figure 16.3 provided all the necessary mechanistic information. Figure 16.6 shows that Ullmann reactions in the above sense allow a large number of functionalizations of

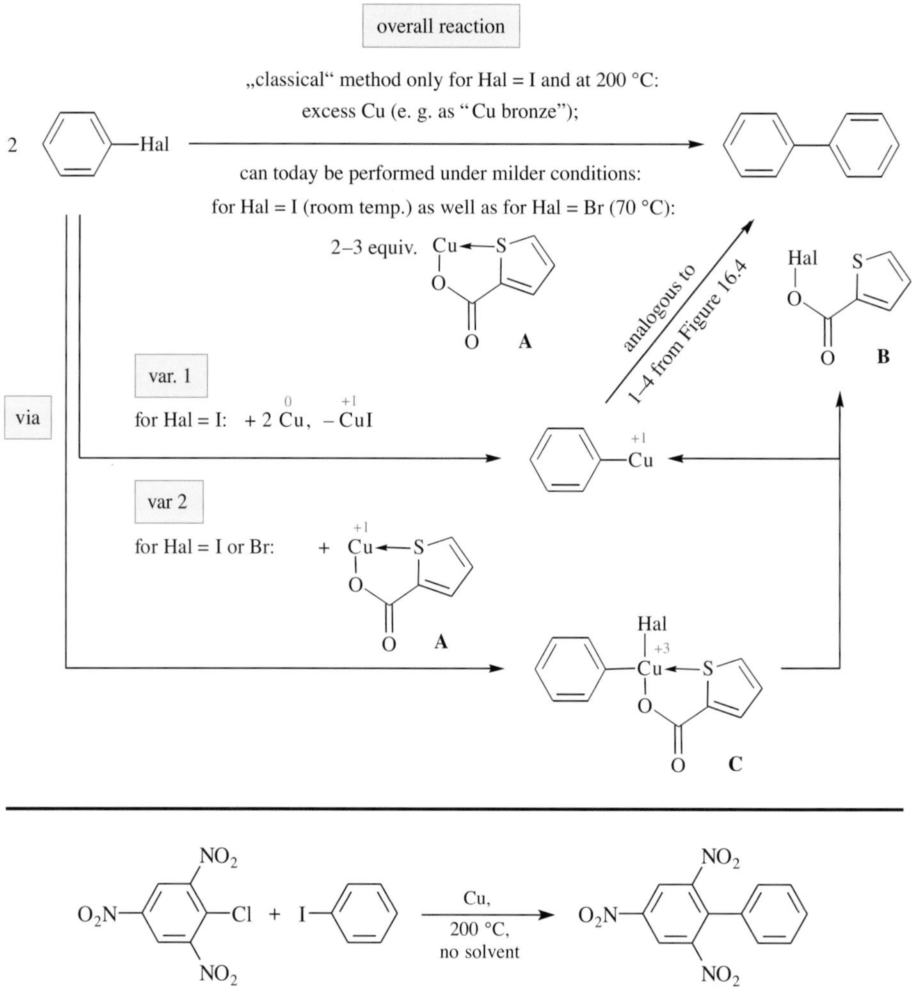

overall reaction

„classical" method only for Hal = I and at 200 °C:
excess Cu (e. g. as "Cu bronze");

2 ⟨Ph⟩–Hal → ⟨Ph⟩–⟨Ph⟩

can today be performed under milder conditions:
for Hal = I (room temp.) as well as for Hal = Br (70 °C):
2–3 equiv. A

analogous to 1–4 from Figure 16.4

B

via

var. 1
for Hal = I: + 2 $\overset{0}{Cu}$, – $\overset{+1}{Cu}I$

var 2
for Hal = I or Br: + A

C

O_2N–(2,6-dinitro)–Cl + I–Ph → (Cu, 200 °C, no solvent) → O_2N–(2,6-dinitro)biphenyl

Fig. 16.5. Biaryl synthesis via arylcopper compounds II—*in-situ* preparation of the nucleophile with the classical Ullmann procedure ("variant 1"), with a procedure that may also be applied to aryl bromides ("variant 2") and with a crossed classical Ullmann coupling (bottom-most reaction example).

aromatic compounds including, for example, their transformation into diaryl sulfides, diaryl amines, arylmalonic esters (**A**) and arylacetic acids (**B**). For this type of substitution reactions no method whatsoever was provided in Chapter 5 (preparation of a diaryl sulfide), or "more explosive" reagents were used (three equivalents of metallic lithium for the preparation of diphenylamine, see Figure 5.65). Without the Ullmann reaction, and based on the methods of the other chapters of this book alone, the formation of the *C*-functionalized aromatic compounds **A** and **B** would only be possible with a multistep sequence such as, for example, Ar–Br → Ar–CH_2–CH_2–OH → Ar–CH_2–CO_2H (**B**) → Ar–CH_2–CO_2Et → Ar–$CH(CO_2Et)_2$ (**A**).

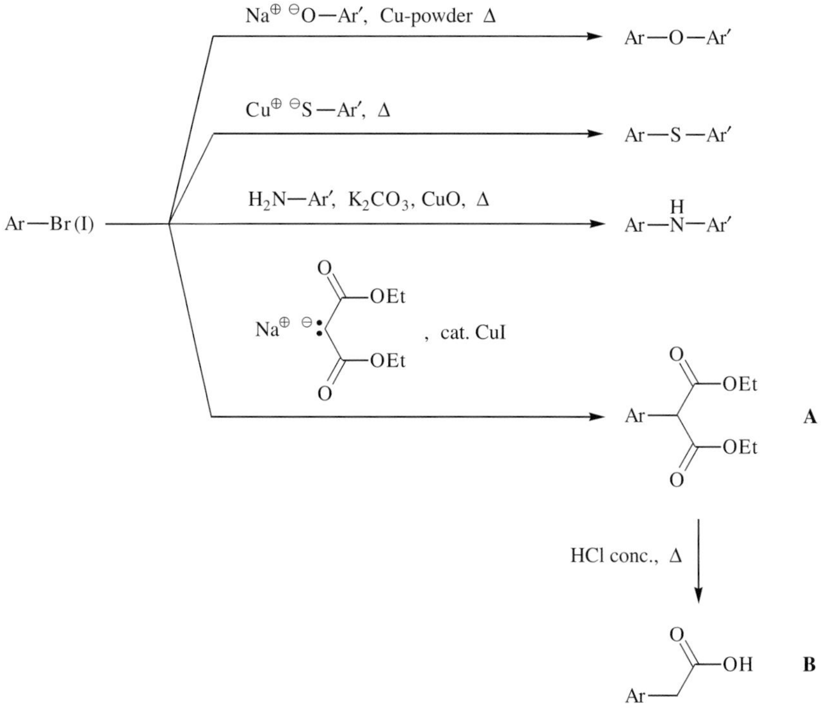

Fig. 16.6. Copper-mediated nucleophilic substititon reactions with aryl iodides or bromides for the preparation of diaryl ethers, diaryl sulfides, diaryl amines and arylmalonic esters (**A**). Often, these reactions are also referred to as Ullmann reactions.

Castro–Stephens couplings follow a totally analogous mechanism, but do not strictly fit into this section's title as far as its products are concerned. An example is given in Figure 16.7 where you can see how phenylacetylene can be arylated via its copper acetylide **A** according to the mechanism given in Figure 16.3. Castro–Stephens couplings are coupling reactions between aryl- or alkenyl halides and copper acetylides. They lead to the formation of conjugated aryl alkynes and enynes, respectively. Castro–Stephens couplings allow indirect arylations or alkenylations of terminal alkynes, since the latter must be employed in the form of their stoichiometrically generated Cu(I) derivatives. Both this and the fact that they are conducted in boiling pyridine distinguishes Castro–Stephens couplings from *direct* arylations and alkenylations of terminal alkynes that are also known and, according to Section 16.4.5, often proceed at room temperature. These are the Sonogashira–Hagihara coupling (example: Figure 16.30, 16.31) and the Cacchi coupling (example: Figure 16.32). However, these direct couplings often need to be catalyzed by palladium, while Castro–Stephens couplings proceed without a palladium catalyst.

As Figure 16.7 reveals, an aryl iodide reacts more rapidly with an alkynylcopper compound than an aryl bromide. The palladium-catalyzed C,C coupling reactions, which will be discussed later in the present chapter, also proceed more rapidly with an aryl iodide than with an aryl bromide (example: Suzuki coupling in Figure 16.22) or an aryl chloride (example: Stille reaction in Figure 16.27). There are still some details that are not fully understood; one is inclined to assume that in accordance with the Hammond postulate the weaker C–I bond (dissociation energy $DE = 51$ kcal/mol) breaks more rapidly with the initial oxidative addition

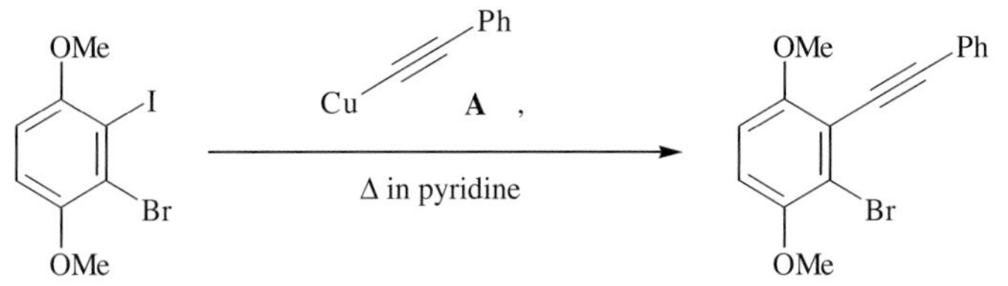

Fig. 16.7. Uncatalyzed Ar–Hal/Cu–C≡C-R coupling ("Castro–Stephens coupling"), which here exclusively involves the Ar–I- and not the Ar–Br bond, but basically can proceed with the latter also.

than the stronger C–Br bond (*DE* = 68 kcal/mol) or the C–Cl bond (*DE* = 81 kcal/mol). However, this explanation bears certain difficulties since aryl triflates are *much* better coupling partners than aryl chlorides despite their C–O bond that is much more difficult to break (*DE* = 85.5 kcal/mol) (example: Negishi coupling in Figure 16.25; Heck coupling in Figure 16.35). If a particular arylation agent Ar–X turns out to be "unreactive" in a palladium-catalyzed C,C coupling, it is often difficult to decide whether the respective Ar–X does not undergo oxidative addition at all or whether it does, but with the product Ar–$PdXL_{1-2}$ failing at a later step of the catalytic cycle. For example, the latter could be the rate-determining step of the catalytic cycle of reactive arylation agents: step 5—i.e., the transmetalation step—of the Stille reaction in Figure 16.27 or step 5—the carbopalladation—of the Heck coupling in Figure 16.35 (part II).

Side Note 16.2. Rosenmund–von-Braun Reaction

Figure 16.8 shows a nucleophilic aromatic substitution reaction conducted with copper cyanide. This **Rosenmund–von–Braun reaction** is not included among the numerous Ullmann reactions (Figure 16.4–16.6). Regarding the preparation of aromatic nitriles, this reaction presents an alternative to the Sandmeyer reaction of aryl diazonium salts with CuCN (cf. Figure 5.53).

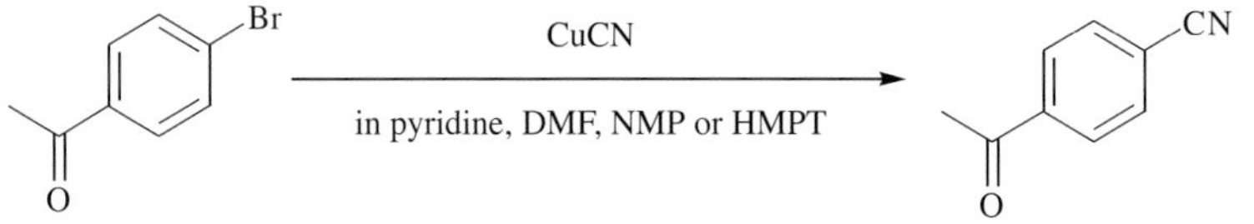

Fig. 16.8. Rosenmund–von–Braun synthesis of aromatic nitriles taking the preparation of *para*-acetylbenzonitrile as an example.

As a last example of an uncatalyzed C,C coupling of a "neutral" organocopper compound Figure 16.9 depicts the alkynylation of a copper acetylide with a bromoalkyne which is easily accessible via bromination of a terminal alkyne:

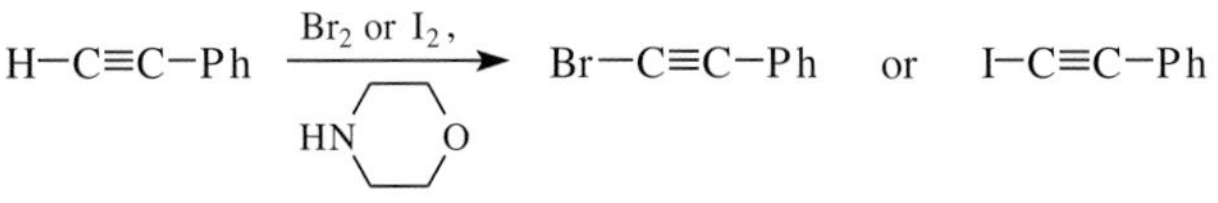

$$\text{H–C≡C–Ph} \xrightarrow[\text{morpholine}]{\text{Br}_2 \text{ or } \text{I}_2,} \text{Br–C≡C–Ph} \quad \text{or} \quad \text{I–C≡C–Ph}$$

Alternatively, copper acetylides can be alkynylated with iodoalkynes. Iodoalkynes can be prepared in the same way as bromoalkynes.

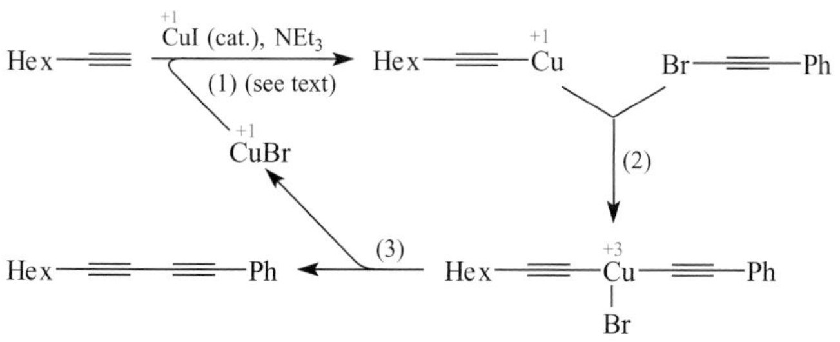

Fig. 16.9. Uncatalyzed, i.e. "true" Cadiot–Chodkiewicz coupling. The catalyzed variant is presented in Figure 16.33.

The alkynylation of copper acetylides is used in the **Cadiot–Chodkiewicz coupling** for the preparation of symmetric or used in the unsymmetrical conjugated diynes. In contrast to Figure 16.7, a copper acetylide is not employed explicitly here, but is generated *in situ* in step 1 of the diyne synthesis given in Figure 16.9 from a terminal alkyne, one equivalent of an amine and substoichiometric amounts of CuI. The following steps 2 and 3 display clear analogies to the mechanisms already discussed. Step 2 represents the oxidative addition of the haloalkyne to a Cu(I) species proceeding either via an intermediate π-complex or in a single step. In step 3, a reductive elimination occurs with concomitant formation of the new C–C bond, which directly or via an intermediate π-complex leads to the formation of the coupling product and CuBr.

16.3 Alkenylation and Arylation of Grignard Compounds (Kumada Coupling)

Alkenyl bromides and iodides as well as aryl triflates, bromides, and iodides can undergo substitution reactions with Grignard compounds in the presence of catalytic amounts of a nickel complex. Even though the catalytically active species is a Ni(0) complex, these reactions—also known as Kumada couplings—can be initiated by Ni(II) complexes, which are easier to handle. Starting with a different oxidation state is possible since the organometallic compound reduces the Ni(II) complex to Ni(0) *in situ*. The addition of about 1 mol% $NiCl_2$(dppe) to the reaction mixture suffices to catalyze the alkenylation (Figure 16.10) or arylation (Figure 16.11) of Grignard compounds.

These reactions provide representative examples and the polarity of the substrates is interchangeable:

- Primary alkyl Grignard compounds can be alkenylated (Figure 16.10) and arylated (in analogy to Figure 16.11).
- Aromatic Grignard compounds can be arylated (Figure 16.11) and alkenylated (in analogy to Figure 16.10).
- Alkenyl Grignard compounds, too, can be alkenylated as well as arylated.

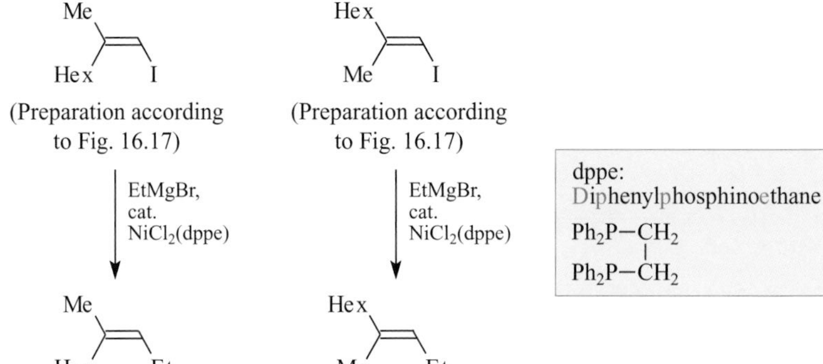

Fig. 16.10. Nickel-catalyzed alkenylation of Grignard compounds; occurrence of stereoselectivity and stereospecificity.

In all these reactions, any existing configuration of any stereogenic double bond—whether it be in the alkenyl bromide or iodide or in the alkenyl Grignard compound- is completely retained. The reactions in Figure 16.10 provide good examples.

A plausible reaction mechanism for such couplings is presented in Figure 16.12 for the specific example of the transformation of Figure 16.11. We do not specify the number *n* and the nature of the ligands L of the intermediate Ni complexes in Figure 16.12. Little is known about either one. Also, it is quite possible that more than one elementary reaction is involved in some of steps 1–5. In any case, these five steps are *certainly* involved in the overall reaction. In step 1, the aryl bromide enters the coordination sphere of the Ni(0) compound as a π-ligand. At least one metal coordination site must be vacated before an oxidative addition of the aryl bromide to the Ni atom may occur in step 2. The nickel inserts into the C_{sp^2}—Br bond, and its oxidation number is increased from +0 to +2. Step 3 is a transmetalation. The aryl Grignard compound is converted into an aryl nickel compound (a diaryl nickel compound in the specific example) by way of replacement of a bromide ion by a magnesium-bound aryl group. Step 4 is a reductive elimination of the coupling product. It yields a coordinatively unsaturated Ni(0) complex to which the product is attached as a π-ligand. Step 5 completes the reaction: the coupled aromatic compound leaves the complex, and the remaining Ni(0) complex is ready to enter the next passage through the catalytic cycle.

Fig. 16.11. Nickel-catalyzed arylation of Grignard compounds. The Grignard compound can be prepared *via* a substituent-directed *peri*-lithiation of a substituted naphthalene (see Section 5.3.1 for the analogous *ortho*-lithiation) and subsequent transmetalation with $MgBr_2$ (for the method cf. Table 10.1).

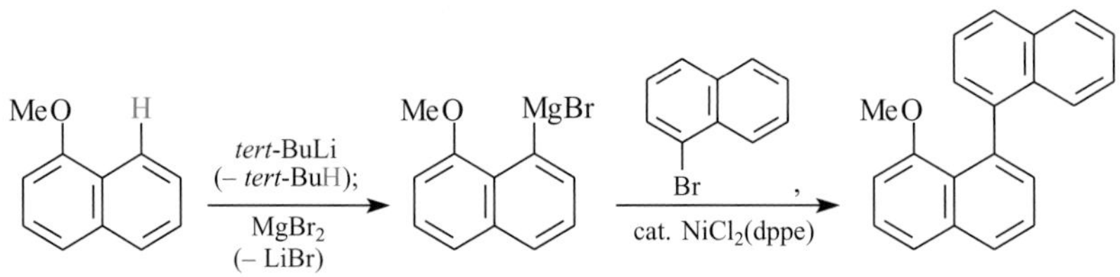

MeO H
tert-BuLi (– *tert*-BuH); $MgBr_2$ (– LiBr)
MeO MgBr
Br, cat. $NiCl_2$(dppe)
MeO

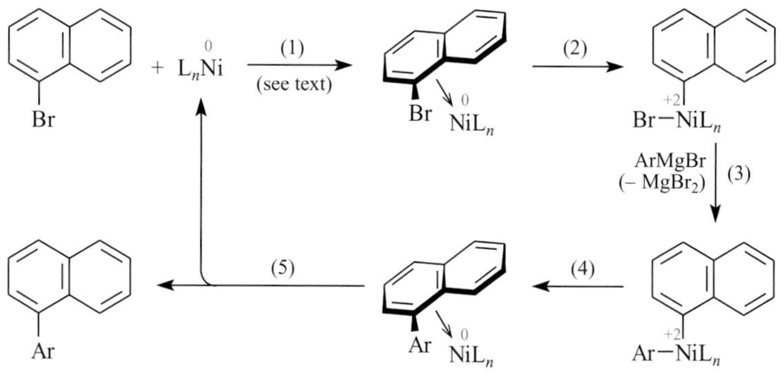

Fig. 16.12. Mechanism of a Ni-catalyzed C,C coupling between an aryl Grignard reagent and an aryl halide—using the example of the reaction in Figure 16.11. The elementary steps, discussed in the text, are (1) complexation, (2) oxidative addition, (3) transmetalation, (4) reductive elimination, and (5) dissociation. Note: The mechanistic analysis provided here includes less details than the analysis of both the Stille reaction (Figure 16.27) and the Heck reaction (Figure 16.35, part II) that have been explored in much more detail. Therefore, much less is known about the coupling in this figure. This is why the present scheme is less explicit with regard to the number and/or nature of ligands in the catalytically active metal species, the arrangement of the ligands at the metal center and the details of the transmetalation.

Side Note 16.3. Surprises with Kumada Couplings

C,C couplings of secondary alkyl Grignard reagents may occur in analogy to the reactions of the primary Grignard reagents (Figure 16.10), but they may also lead to unexpected reactions as demonstrated by Figures 16.13 and 16.14. Figure 16.13 shows Ni-catalyzed C,C couplings of *sec*-BuMgCl with chlorobenzene. Depending on the catalyst, either the expected *sec*-butylbenzene is obtained or the unexpected *n*-butylbenzene. How does such an isomerizing coupling occur?

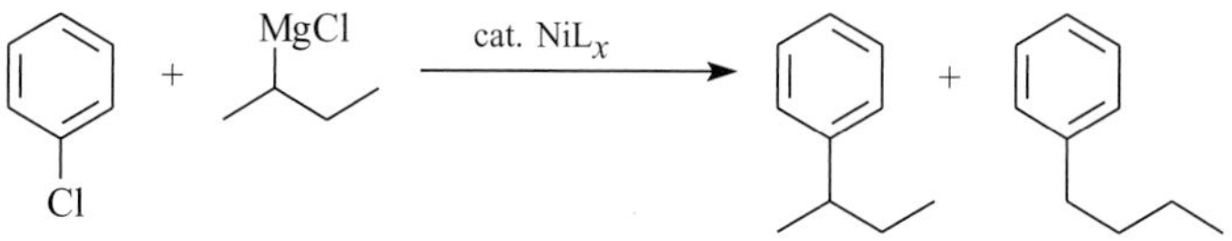

for $NiL_x = NiCl_2[Ph_2P{-}(CH_2)_3{-}PPh_2]$ 93 : 7

for $NiL_x = NiCl_2[Me_2P{-}CH_2{-}CH_2{-}PMe_2]$ 5 : 95

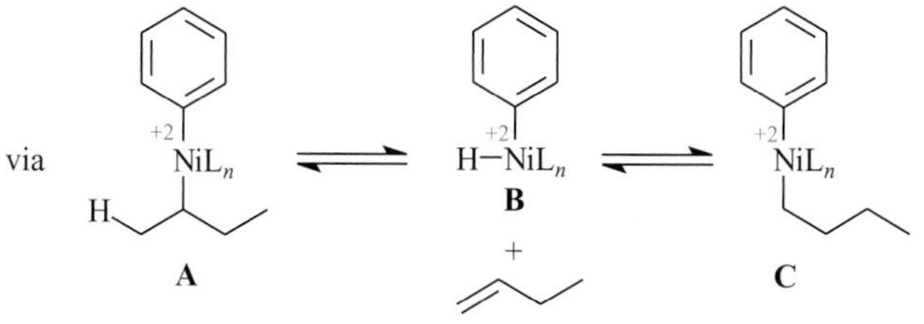

Fig. 16.13. Normal and isomerizing C,C couplings of a Grignard compound.

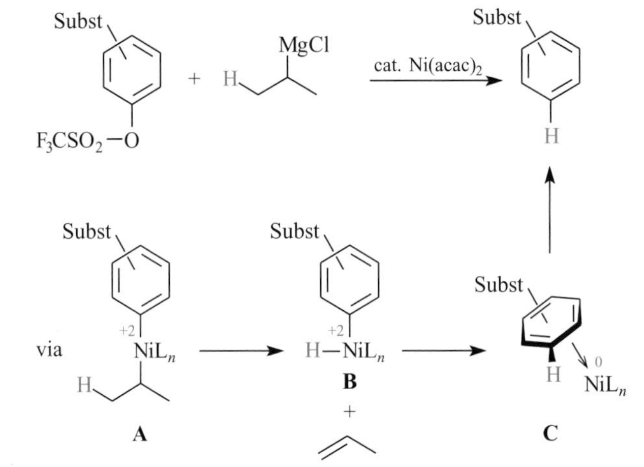

Fig. 16.14. Nickel-catalyzed reduction of an aryl triflate by a Grignard compound; acac refers to acetylacetonate, which is the enolate of pentane-2,4-dione.

Apparently, the next-to-last step of the respective catalytic cycle—the one that would correspond to step 4 in Figure 16.12—is slow due to steric hindrance. In Figure 16.13, this step would be the reductive elimination of *sec*-BuPh from the Ni complex **A**. For this elimination to occur as desired, a *secondary* C atom would have to be connected to the aromatic compound. As the result of steric hindrance and the associated slowing down of this reaction step a competing reaction becomes feasible. It consists of a β-elimination and leads to the hydrido-Ni complex **B** and 1-butene. This Ni complex **B** would again add to 1-butene in a reversal of its formation reaction. This addition corresponds to a hydronickelation of the C=C double bond (cf. the hydroboration of the C=C double bonds, Section 3.3.3). The hydronickelation of 1-butene (Figure 16.13) has two regioselectivity options. The original Ni complex **A** with its isobutyl residue can be reformed, in which case the overall reaction would not be affected because the expected coupling product with the branched side chain would still be obtained. However, the hydrido-Ni complex **B** and 1-butene may also combine with opposite regioselectivity to form the isomeric Ni complex **C** with its unbranched butyl group. The reductive elimination of an alkylbenzene from **C** then leads to the unexpected substitution product with the isomerized, that is, unbranched, side chain.

Figure 16.14 shows another case, namely, the reaction of isopropylmagnesium chloride with aryl triflates, in which a Ni complex does not effect the expected C,C coupling. Rather, it initiates a reduction and the deoxygenated aromatic compound is isolated (a valuable method for the deoxygenation of phenols!). This reaction mode again is a consequence of steric hindrance. The latter causes the interruption of the catalytic cycle of the Ni-catalyzed C,C coupling (Figure 16.12) prior to the reductive elimination of the coupling product (step 4 in Figure 16.12). This elimination would have to occur with the intermediate **A** of Figure 16.14, but it is slowed down because a *secondary* C atom would have to bind to the aromatic compound. This is why the Ni complex **A** undergoes β-elimination, which leads to a hydrido-Ni complex **B** and propene. In contrast to the hydrido–Ni complex **B** of Figure 16.13, the analogous complex **B** of Figure 16.14 undergoes a reductive elimination in which the aryl group picks up the H atom from the Ni. The elimination product is a π-complex of the *reduced* aromatic compound. Finally, this aromatic compound dissociates off the metal.

16.4 Palladium-Catalyzed Alkenylations and Arylations of Organometallic Compounds

The stereoselective synthesis of alkenes is basically a solved problem. Nowadays, all kinds of alkenes can be synthesized irrespective of whether their double bond is isolated or conjugated with another C=C double bond, a C≡C triple bond, or an aromatic ring. This state of affairs is largely due to the discovery and the development of a number of palladium-catalyzed alkenylation and arylation reactions of organometallic compounds.

16.4.1 A Prelude: Preparation of Haloalkenes and Alkenylboronic Acid Derivatives, Important Building Blocks for Palladium-Mediated C,C Couplings; Carbocupration of Alkynes

As detailed in Section 3.3.3, borane and its mono- and dialkyl derivatives add to C=C double bonds in a *cis*-selective fashion. These reagents also add to C≡C triple bonds *cis*-selectively, but the primary products formed, alkenylboranes, may react with the boranes once more. The second reaction is almost always faster than the first one. Consequently, alkenylboranes are not accessible in this way.

Fortunately, though, there is *one* borane, catecholborane (**A** in Figure 16.15), that adds to C≡C triple bonds but not to C=C double bonds (in the absence of transition metal catalysts). This reagent adds to alkynes with *cis*-selectivity, so that the reaction stops at the stage of *trans*-alkenylboronic acid esters (**B** in Figure 16.15).

These boronic esters are easily hydrolyzed to give *trans*-alkenylboronic acids with complete retention of their stereochemistry (**C** in Figure 16.15). Alkenylboronic esters and alkenylboronic acids are organometallic compounds that can be alkenylated and arylated in Pd-catalyzed reactions (Section 16.4.2). Aside from this, the *trans*-alkenylboronic acid esters as well as the *trans*-alkenylboronic acids are valuable precursors of haloalkenes (Figure 16.15).

Alkenylboronic acid esters **B** react with bromine to give the *trans*-addition product initially. This primary product is not isolated but is immediately reacted with a solution of NaOMe in MeOH. Addition of $MeO^{\ominus}$ ion to the B atom of the bromine adduct forms the borate complex **D**. This borate complex is converted into a *cis*-configured bromoalkene and a mixed boronic acid ester by way of a β-elimination.

trans-Alkenylboronic acids (**C** in Figure 16.15) exhibit the complementary diastereoselectivity in their reactions with elemental iodine: they form iodoalkenes with *trans*-selectivity. The conversion of the C—B bond into the C—I bond occurs with complete retention of configuration. This stereoselectivity is typically encountered in reactions regardless of the organometallic compound or electrophile, i.e., the C—M bond almost always is converted into a C—E bond with retention of configuration. The electrophile (here: I_2) may react at the location of highest electron density, that is, at the center of the C—M bond [here: C—$B(OH)_3^{\ominus}$ bond].

cis-Bromoalkenes—accessible, for instance, according to the procedure outlined in Figure 16.15 or according to the lower part of Figure 13.56—can be employed for the prepara-

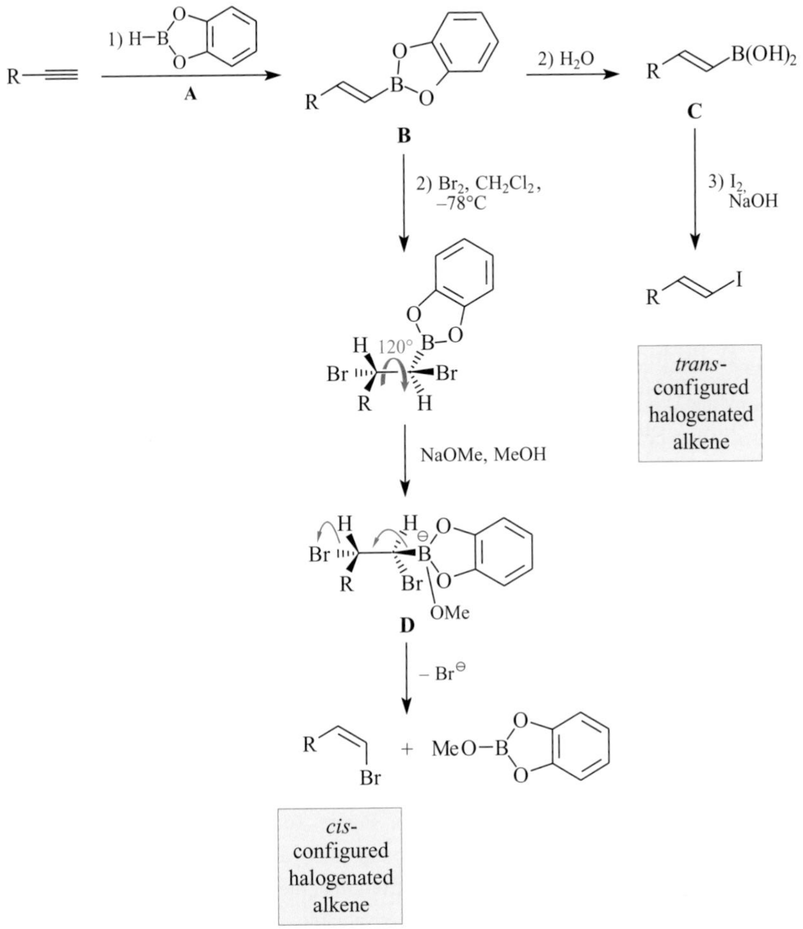

Fig. 16.15. Stereoselective preparations of *trans*-alkenylboronic acid esters (**B**) and *trans*-alkenylboronic acids (**C**) and their stereoselective conversion into *cis*-bromoalkenes and *trans*-iodoalkenes, respectively.

tion of alkenylboronic acid esters and alkenylboronic acids. To do so, one first carries out a bromine/lithium exchange by way of treating the bromide with *n*-BuLi (Figure 16.16). This exchange is analogous to the bromine/lithium exchange reaction of aryl bromides (Figure 5.41). In agreement with a foregoing statement, a key step consists of the reaction of $Li^{\oplus}$ at the center of the C—$Br^{\ominus}$—*n*-Bu bond of the transient complex **A**. Thus, complete retention of configuration is observed in the conversion of the C_{sp^2}—Br bond into the C_{sp^2}—Li bond of the lithioalkene **B**. The transmetalation of this lithioalkene with triisopropyl borate proceeds with complete retention of configuration. The reason for this retention is that the electrophile, $B(O\textit{i}Pr)_3$, reacts with the C—M bond—that is, the C_{sp^2}—Li bond—at the center. Hence, a *cis*-configured complex **C** with tetravalent and negatively charged boron is obtained. Complex **C** can be hydrolyzed under mildly acidic conditions to give a *cis*-configured alkenylboronic acid ester. This ester in turn can be hydrolyzed—now preferentially with base catalysis—to give the parent alkenylboronic acid with retention of the *cis* geometry.

The basicity of Gilman cuprates is so low that they do not undergo acid/base reactions with acetylene or higher terminal alkynes. Instead, they can add to their C≡C triple bond (Fig-

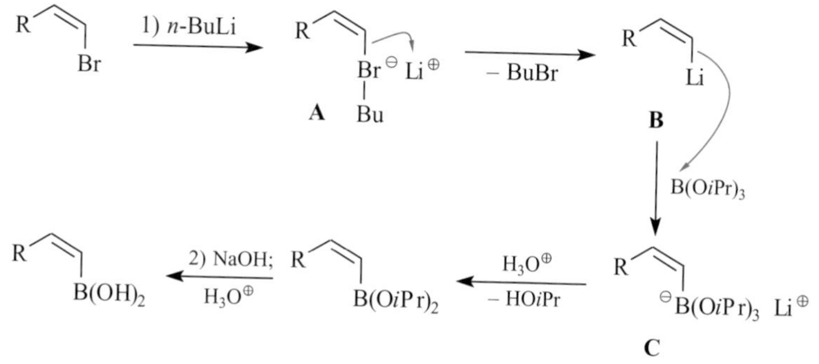

Fig. 16.16. Stereoselective preparations of *cis*-alkenylboronic acids and the corresponding diisopropyl ester starting with *cis*-bromoalkenes. The first step involves a Br/Li exchange to form the alkenyllithium compound **B**. This organolithium compound is subsequently transmetalated to give complex **C** by using $B(OiPr)_3$.

ure 16.17). During this addition reaction the initial cuprate transfers alkyl-, alkenyl- or aryl residues to the alkyne. A new Gilman cuprate **I** is produced via a series of intermediates (which will soon be discussed). Its structural formula shows that the addition must proceed *cis*-selectively and with the same regioselectivity as the hydroboration (example: formation of **A** in Figure 16.15) or hydrozirconation of terminal C≡C triple bonds (example: formation of **H** in Figure 16.26), i.e., the copper is bound to the unsubstituted terminus of the C≡C triple bond. Additions of organometallic compounds to the C≡C triple bond of alkynes or to the C=C double bond of alkenes are referred to as **carbometalations**. Depending on which metal is involved, a distinction is made between **carbocuprations** (Figure 16.17), **carbopalladations** (Section 16.5) and many other carbometalations (not dealt with in this book).

But let us now leave carbometalations in general and the carbocuprations of Figure 16.17 in particular to turn towards the actual topic of Section 16.4, namely the preparation of isomerically pure haloalkenes. Depending on the structure of the reacting alkyne, the carbocupration products **I** in Figure 16.17, which are Gilman cuprates, contain two sterically uniform, mono- or *di*substituted alkenyl groups. These cuprates **I** can be reacted with elemental iodine to give the iodoalkenes **J** without affecting the C=C double bond. The C_{sp^2}–Cu bonds of the respective cuprate are transformed into the C_{sp^2}–I bonds of the iodoalkenes **J** with complete retention of configuration. Mechanistically these iodolyses correspond to the conversion of the *trans*-alkenylboronic acid **C** of Figure 16.15 into a *trans*-iodoalkene.

We will now turn our attention to the mechanism of the carbocupration. This carbocupration begins with a "Gilman dimer" (a contact ion pair; see Figure 10.44) and not the "Gilman monomer" (a solvent-separated ion pair; see Figure 10.44). It is unclear, though, whether this "Gilman dimer" reacts as the eight-membered ring **A** or, after ring-opening—indicated as step 1 in Figure 16.17—as the open-chain form **B**. In step 2, **A** and/or **B** react(s) wtih the C≡C triple bond of the alkyne with their Cu(I), leading to the formation of a 1:1 adduct that can be described as the resonance hybrids **C** and **D**, respectively, of each a π-complex and a metallocyclopropene. This is followed by step 3, a *cis*-selective **cuprolithiation** of the (previous) C≡C triple bond. These reaction paths, which have so far been alternatives, converge in the resulting nine-membered ring intermediate **E**. **E** is a triorganocopper(III) compound that in step 4 undergoes a reductive elimination to a C,C coupling product and an organocopper(I) compound. However, the C,C coupling product is still an alkenyllithium compound. Both components are attached to each other as a π-complex (Formula **F** in Figure 16.17). In this respect, step 4 strongly resembles the reductive eliminations which in step 3 of Figures 16.1, 16.3 and 16.9 led to a C,C coupling product complexed by Cu(I) via a Cu(III) intermediate.

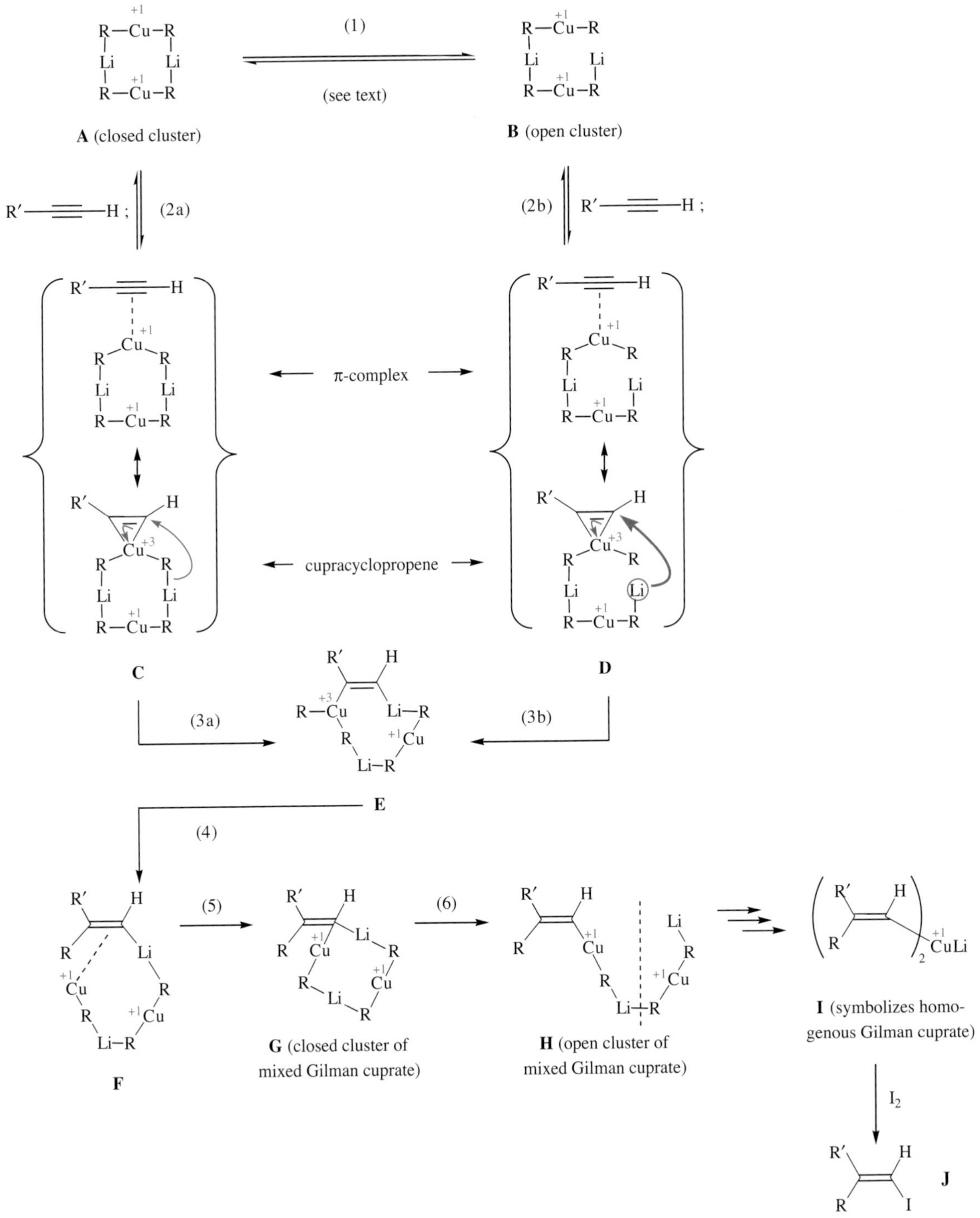

(1)

(see text)

A (closed cluster)

B (open cluster)

(2a) (2b)

π-complex

cupracyclopropene

C

D

(3a) (3b)

E

(4)

(5)

(6)

F

G (closed cluster of mixed Gilman cuprate)

H (open cluster of mixed Gilman cuprate)

I (symbolizes homogenous Gilman cuprate)

I_2

J

The C,C coupling product **F** complexed by Cu(I) in Figure 16.17 is *one* isomer of a **"mixed" Gilman cuprate** dimer. Unlike **simple Gilman cuprates** the mixed Gilman cuprates $(R_2CuLi)_{1\text{ or }2}$ contain different substituents R. Structurally the mixed Gilman cuprate dimer **F** deserves a center position between the open-chain form (not shown) and the eight-ring form **G** (which does not contain a pentavalent vinyl carbon atom, but a Cu- and Li-bridged carbon atom). Step 5 represents the *formation* of this eight-membered ring form **G**, which in step 6 is followed by ring-opening to form the acyclic "Gilman dimer" **H**. The latter is a mixed "Gilman dimer," too, since it contains the alkenyl residue (in single form) resulting from carbocupration *and* residues (in triplicate) that have not yet been transferred to the alkyne. In a series of ligand exchange reactions the mixed "Gilman dimer" **H** equilibrates with the simple "Gilman dimers" **I** and **A**, where **I** is the final carbocupration product, and **A** is further reacted until the carbocupration is completed.

16.4.2 Alkenylation and Arylation of Boron-Bound Groups (Suzuki Coupling)

Alkenylations and arylations of organoboron compounds can be realized with alkenyl and aryl triflates, bromides, and iodides. Suitable organoboron compounds include alkenyl or aryl boronic acids and esters as well as 9-BBN derivatives with a primary alkyl group. However, all these reactions succeed only in the presence of catalytic amounts of Pd complexes. $Pd(PPh_3)_4$ is most commonly used, that is, a Pd(0) complex. In solution, one or two PPh_3 ligands dissociate off to form the electron-deficient complexes $Pd(PPh_3)_3$ or $Pd(PPh_3)_2$. The latter initiates the respective reactions of all the alkenylations, arylations (and alkylations) of organoboron compounds known as **Suzuki couplings**.

All of the alkenylations and arylations discussed in this section follow the common mechanism exemplified in Figure 16.18. The steps shown in Figure 16.18 may involve more than one elementary reaction, as in the case of the mechanistic course of the Ni-catalyzed C,C coupling with Grignard compounds (Figure 16.12). It should be noted that the *basic* sequence of steps is very much the same in Figures 16.18 and 16.12.

In step 1 of Figure 16.18, one of the above-mentioned Pd(0) complexes, $Pd(PPh_3)_3$ or $Pd(PPh_3)_2$, forms a π-complex with the aryl triflate. Step 2 is an oxidative addition. The Pd inserts into the C_{sp^2}—O bond of the aryl triflate and the oxidation number of Pd increases from

Fig. 16.17. Mechanism of the carbocupration of acetylene (R' = H) or terminal alkynes (R' ≠ H) with a saturated Gilman cuprate. The unsaturated Gilman cuprate **I** is obtained via the cuprolithiation product **E** and the resulting carbolithiation product **F** in several steps—and stereoselectively. Iodolysis of **I** leads to the formation of the iodoalkenes **J** with complete retention of configuration. Note: The last step but one in this figure does not only afford **I**, but again the initial Gilman cuprate **A** ⇌ **B**, too. The latter reenters the reaction chain "at the top" so that in the end the entire saturated (and more reactive) initial cuprate is incorporated into the unsaturated (and less reactive) cuprate (**I**). – Caution: The organometallic compounds depicted here contain two-electron, *multi*-center bonds. Other than in "normal" cases, i.e., those with two-electron, *two*-center bonds, the lines cannot be automatically equated with the number of electron pairs. This is why only three electron shift arrows can be used to illustrate the reaction process. The fourth red arrow—in boldface—is not an electron shift arrow, but only indicates the site where the lithium atom binds next.

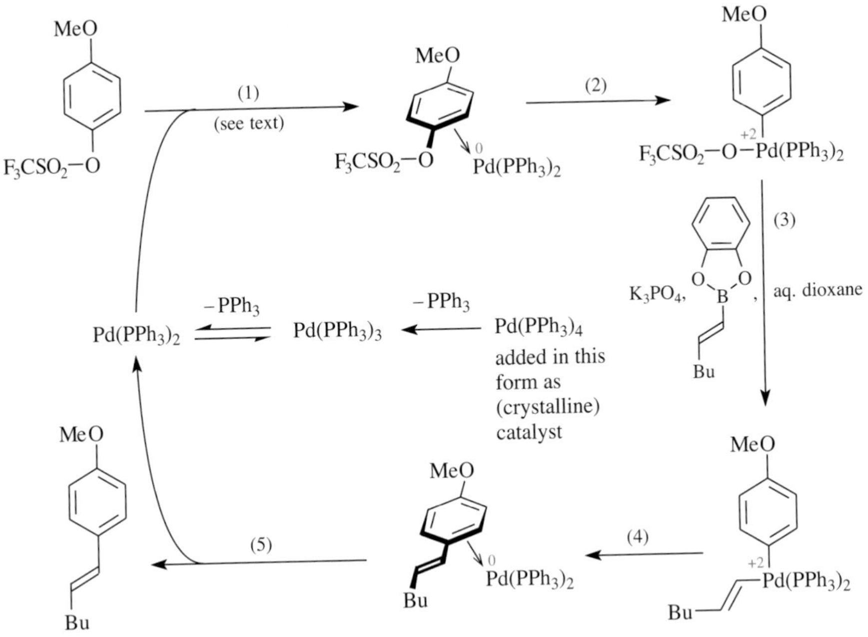

Fig. 16.18. Representative mechanism of the Pd-catalyzed C,C coupling of an organoboron compound. The elementary steps, discussed in the text, are (1) complexation, (2) oxidative addition, (3) transmetalation of the alkenylboron compound to afford an alkenylpalladium compound, (4) reductive elimination, and (5) dissociation of the coupled product from the metal. – Note: Regarding the arrangement of the ligands around the metal center of the individual intermediates and the details of the transmetalation the present mechanistic analysis is less complete than the mechanistic analysis of other Pd-catalyzed C,C couplings, namely the Stille coupling (Figure 16.27) or Heck reaction (Figure 16.35, part II), which have been investigated in great detail.

0 to +2. A transmetalation occurs in step 3: the alkenyl-B compound is converted into an alkenyl-Pd compound. This step corresponds to a ligand exchange reaction at Pd; concomitantly, the former leaving group of the electrophile—here the triflate—is replaced by the alkenyl group. The two σ-Pd-bonded organic moieties combine in step 4, which, since the oxidation number of Pd is reduced from +2 to 0, is a reductive elimination. At that stage, the couling product remains bound to Pd as a π-complex. It dissociates off the metal in step 5. This step reconstitutes the valence-unsaturated Pd(0) complex that can start another passage through the catalytic cycle.

Arylalkenes are accessible not only by way of an arylation of alkenylboron compounds (Figure 16.18) but also via an alkenylation of arylboron compounds. Figure 16.19 exemplifies

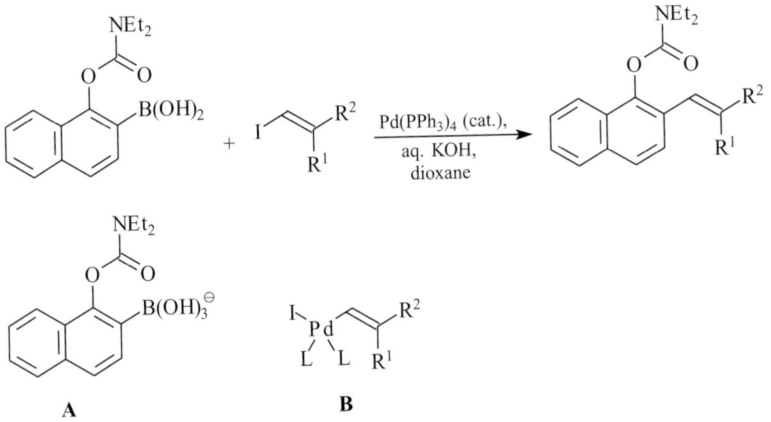

Fig. 16.19. Palladium-catalyzed, stereoselective alkenylation of an arylboronic acid (preparation according to Figure 5.39) with a variety of iodoalkenes. The boronic acid is converted into the boronate anion **A**. The ion **A** reacts with the Pd(II) intermediate **B** via transmetalation: subsequent reductive elimination leads to the coupling products.

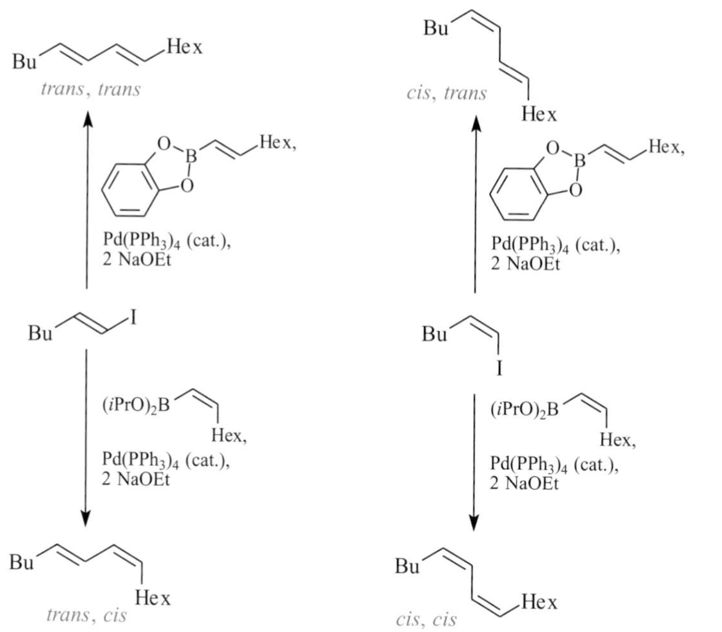
Bu
Hex
trans, trans
Hex,
Pd(PPh3)4 (cat.),
2 NaOEt
Bu
I
(iPrO)2B
Hex,
Pd(PPh3)4 (cat.),
2 NaOEt
Bu
Hex
trans, cis
Bu
cis, trans
Hex
Hex,
Pd(PPh3)4 (cat.),
2 NaOEt
Bu
I
(iPrO)2B
Hex,
Pd(PPh3)4 (cat.),
2 NaOEt
Bu
Hex
cis, cis

Fig. 16.20. Alkenylation of isomeric alkenylboronic acid esters with isomeric iodo-alkenes; stereoselective synthesis of isomeric 1,3-dienes.

this for a Pd-catalyzed reaction of an arylboronic acid with iodoalkenes with widely variable substitution patterns. The addition of KOH increases the reactivity of the arylboronic acid in this coupling and in similar ones. The base converts the boronic acid into a negatively charged boronate ion **A**. This ion **A** is transmetalated faster than the neutral boronic acid by the Pd(II) intermediate. The boronate ion is a superior nucleophile and replaces the iodide ion in the Pd(II) complex particularly fast.

Alkenylboronic acid esters can be prepared with *cis*-configuration (e. g., as in Figure 16.16) or with *trans*-configuration (e. g., as in Figure 16.15). Iodoalkenes are also easily synthetically accessible in both configurations (see Figures 16.17 and 16.15, respectively). Palladium-catalyzed C,C coupling reactions between two of these building blocks stereoselectively lead to each of the four geometrical isomers of the 1,3-diene depicted in Figure 16.20. The couplings shown in this figure are carried out in the presence of NaOEt. Thereby, the nucleophilicity of the organometallic compound can be increased just as in the case of the coupling of boronic acids by way of KOH addition (Figure 16.19). In both cases, one generates negatively charged and therefore more nucleophilic tetravalent boron complexes *in situ*. These complexes have the structure alkenyl–$B(OR)_3^{\ominus}$ in the reactions shown in Figure 16.20, while a complex **A** of composition aryl-$B(OH)_3^{\ominus}$ occurs in Figure 16.19.

Side Note 16.4. Stereoselective Synthesis of Vitamin A

As discussed, in the absence of catalysts catecholborane adds to C≡C triple bonds but not to C=C double bonds (see Figure 16.15). Accordingly, if enynes are reacted with catecholborane, the latter will add chemoselectively to their C≡C triple bond (Figure 16.21). As expected, the resulting boronic ester features the same regio- and stereochemistry seen in Fig-

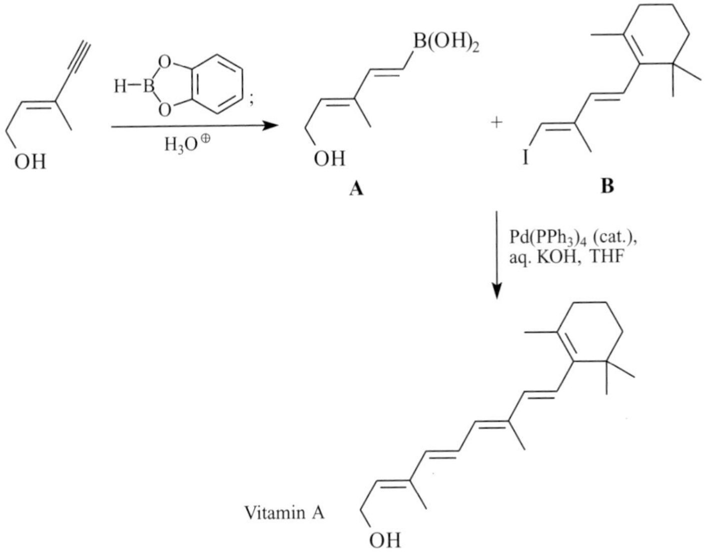

Fig. 16.21. Alkenylation of a dienylboronic acid with an iodinated triene; stereoselective synthesis of vitamin **A**. The enyne (top left) is added to catecholborane to prepare the *trans*-configured boronic ester in a chemoselective fashion. The latter affords *trans*-dienylboronic acid **A** upon acid-catalyzed hydrolysis.

ure 16.15. The acid-catalyzed hydrolysis of this boronic ester affords a *trans*-configured dienylboronic acid (**A** in Figure 16.21). This acid undergoes a reaction with the iodotriene **B** (see Figure 16.17 for a method of preparation) to provide a conjugated pentaene. The latter contains completely stereodefined C=C double bonds and it is none other than vitamin A.

In the case of aromatic compounds with two triflate groups and/or bromine or iodine substituents, it is sometimes possible to achieve chemoselective and sequential Pd-catalyzed substitutions of these leaving groups by organoboron compounds. In the ideal scenario, this reaction sequence is even possible in a one-pot reaction. One first adds *one* equivalent of *one* unsaturated boronic acid to the substrate/$Pd(PPh_3)_4$/base mixture. The leaving groups are cleaved from the aromatic compound in the order I > Br > triflate. Hence, the first boronic acid replaces the iodide in bromoiodobenzene (Figure 16.22). Subsequently, one adds a *second* unsaturated boronic acid *in excess* to replace the remaining bromine. Alkenylboronic acids are just as suitable for such tandem coupling reactions as the arylboronic acids shown in Figure 16.22. The fact that the reaction rate of an Ar–I- is higher than that of an Ar–Br bond is already familiar from Castro–Stephens couplings (Figure 16.7) and has been discussed in that context.

In Figures 16.18–16.22, we introduced the reaction principle of the Pd(0)-catalyzed alkenylation or arylation of *unsaturated* boronic esters and boronic acids, respectively. This reaction principle can be extended to another class of organoboron compounds, namely, to certain *trialkylboranes*. Specifically, the trialkylboranes resulting from the addition of 9-BBN to terminal alkenes can be combined with aryl–X or alkenyl–X compounds (X = OTf, Br, I) via Pd-catalyzed **Suzuki couplings**. 9-BBN adds to terminal alkenes in a regioselective fashion and, interestingly, this reaction proceeds uneventfully even if the alkene contains a series of other functional groups (see first two rows of Figures 10.39 and 10.49). In fact, a number

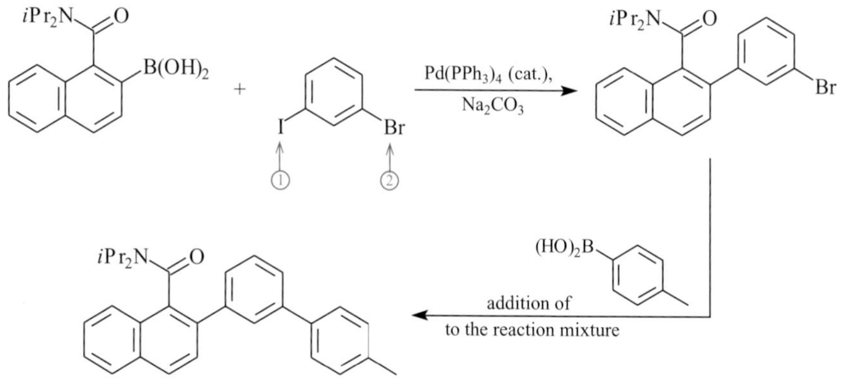

Fig. 16.22. Palladium-catalyzed C,C coupling with bromoiodobenzene: regioselective tandem coupling I.

of functional groups interfere neither with the formation of a 9-BBN adduct nor with the subsequent Pd(0)-catalyzed C,C coupling. Such arylating (Figures 16.23 and 16.24) or alkenylating coupling reactions of hydroborated alkenes are starting points for the synthesis of aromatic compounds and alkenes, respectively, that may contain a variety of functional groups in different parts of the molecules. The figures cited show compatibility with epoxides, ketones, and esters.

Finally, Figure 16.24 shows a particularly interesting one-pot reaction consisting of two consecutive Suzuki coupling reactions of an aromatic compound bearing two sp^2-bonded leaving groups. The Ar—Br moiety reacts faster than the Ar—OTf moiety. This is expected in the context of what was said regarding the regioselectivity observed in Figures 16.7 and 16.22.

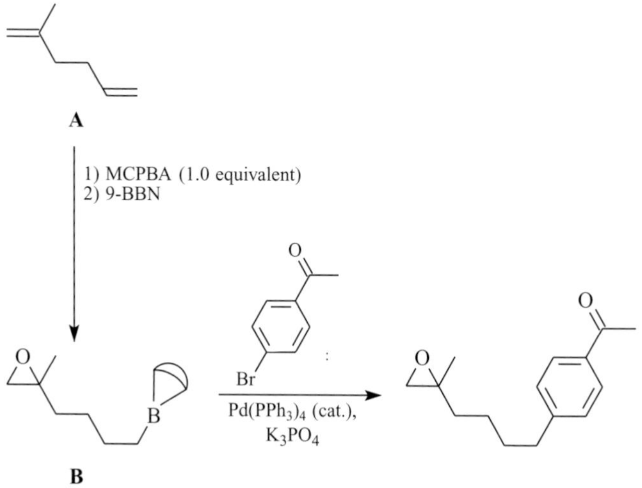

Fig. 16.23. Suzuki coupling for the synthesis of aromatic compounds with a functionalized side chain.

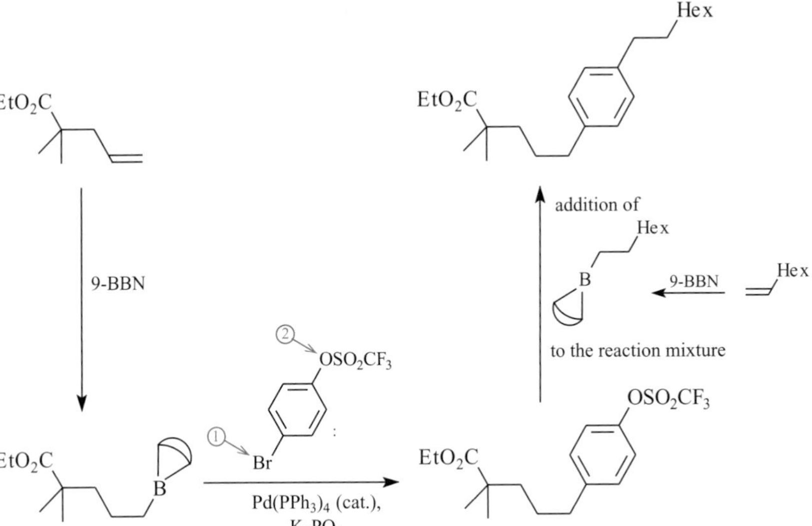

Fig. 16.24. Suzuki coupling for the synthesis of aromatic compounds with a functionalized side chain and regioselective tandem coupling II.

16.4.3 Alkenylation and Arylation of Organozinc Compounds (Negishi Couplings) and of Functionalized Organozinc Compounds

Alkynylzinc, arylzinc, alkenylzinc and alkylzinc iodides, -bromides and -chlorides can be arylated by aryl triflates (bromides, iodides) and alkenylated by alkenyl triflates (bromides, iodides) in the presence of catalytic amounts of $Pd(PPh_3)_4$. The *alkyl*zinc iodides are particularly interesting because they are accessible even if they contain other functional groups (see discussions in connection with Figures 10.43 and 17.45). For example, the organozinc compound shown in Figure 16.25 contains an ester group. The 9-BBN analog of this organozinc compound—potentially a substrate for a Suzuki coupling (Figure 16.24)—would *not* be accessible in a straightforward manner, that is, by the hydroboration of the C=C bond of ethyl acrylate. This is because 9-BBN and ethyl acrylate would react with each other in a 1,4-addition to yield a boron enolate of ethyl propionate. Palladium-catalyzed C,C couplings of unsaturated organozinc compounds are referred to as **Negishi couplings**.

The mechanism of the Pd(0)-catalyzed coupling of functionalized organozinc compounds is exemplified in Figure 16.25 by the arylation of the above-mentioned ester-functionalized alkylzinc iodide. The catalytic cycle consists of six reaction steps. Steps 1, 2, and 4–6 correspond exactly to the five steps of the C,C coupling reactions already discussed, that is, the Ni-catalyzed coupling of Grignard reagents (Figure 16.12) and the Pd-catalyzed coupling of organoboron compounds (Figure 16.18). Everything that was said in those discussions applies fully to the sequence of events Figure 16.25.

The only difference between the present mechanism and the mechanisms to which it was compared is that the couplings of the organozinc compounds in Figure 16.25—just like some Pd-catalyzed C,C coupling reactions between organotin compounds and arylating or alkeny-

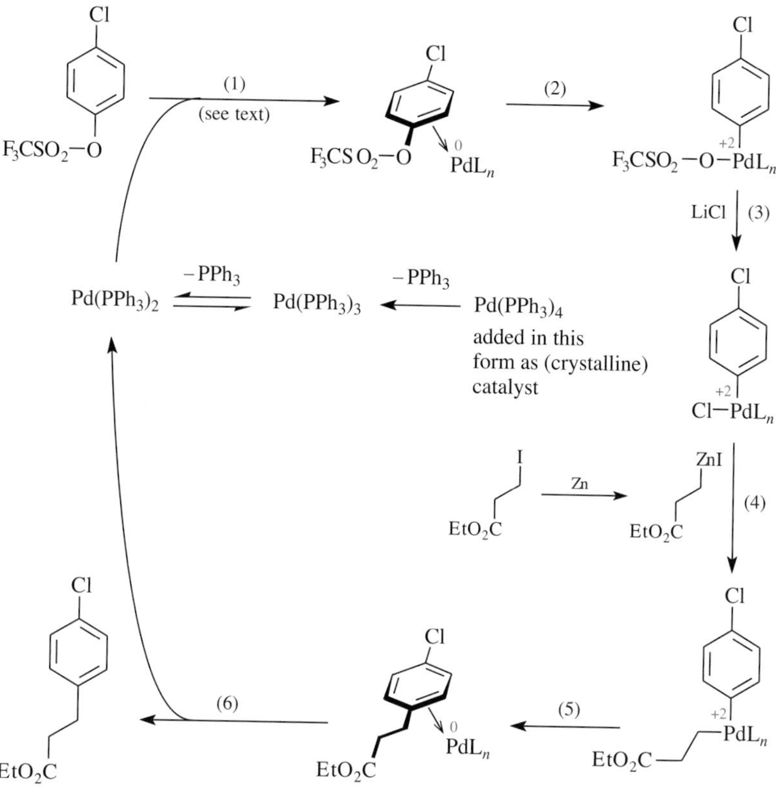

Fig. 16.25. Representative mechanism of the Pd(0)-catalyzed arylation and alkenylation of organozinc iodides. Steps 1, 2, and 4–6 correspond to steps that can also be found—in some cases with different numbers—in Figures 16.12 and 16.18. Step 3, which is new, represents a ligand exchange reaction of the aryl–Pd(II) complex. – Note: Concerning the arrangement of the ligands in the metal-containing intermediates and the details of the transmetalation the present mechanistic analysis is less complete than the mechanistic analysis of other Pd-catalyzed C,C couplings, namely the Stille coupling (Figure 16.27) or Heck reaction (Figure 16.35, part II), which have been investigated in great detail.

lating reagents—require the addition of LiCl. The role of LiCl may be to bind to the Pd catalyst to create an electron rich Pd species (not shown) that undergoes very rapid oxidative addition and/or to engage in ligand exchange to stabilize the product of oxidative addition.

Alkylzinc iodides are readily obtained by reducing alkyl iodides with zinc (example: Figure 10.43, 17.45). Arylzinc iodides and -bromides can basically be prepared in an analogous fashion. Alternatively—and quite often—advantage is taken of the fact that aryllithium compounds can be transmetalated with zinc chloride to access arylzinc chlorides, since the latter can also be employed to conduct Pd-catalyzed C,C coupling reactions. Likewise, alkenylzinc compounds are commonly—even preferentially—made accessible via the latter route, where a stereouniform bromoalkene undergoes a Br/Li exchange reaction (example: Figure 16.16) and transmetalation of the resulting lithioalkene with zinc chloride. Alkynylzinc compounds can generally be produced by transmetalation only. Two typical transmetalations of this kind are shown at the very top of Figure 16.26: the transmetalation of lithiated heptene diyne with zinc bromide to yield the alkynylzinc bromide **A** as well as the transmetalation of ethynyl magnesium bromide with zinc bromide to furnish ethynyl zinc bromide **B**.

Figure 16.26 provides an elegant synthesis of a fully conjugated polyalkene/polyalkyne like the target molecule **I** by way of Pd-catalyzed C,C couplings. The organometallic compounds involved include the already mentioned alkynylzinc bromides **A** and **B** and the alkenylzinc chloride **G**. The latter emerges from the hydrozirconation **F** → **H** of a terminal

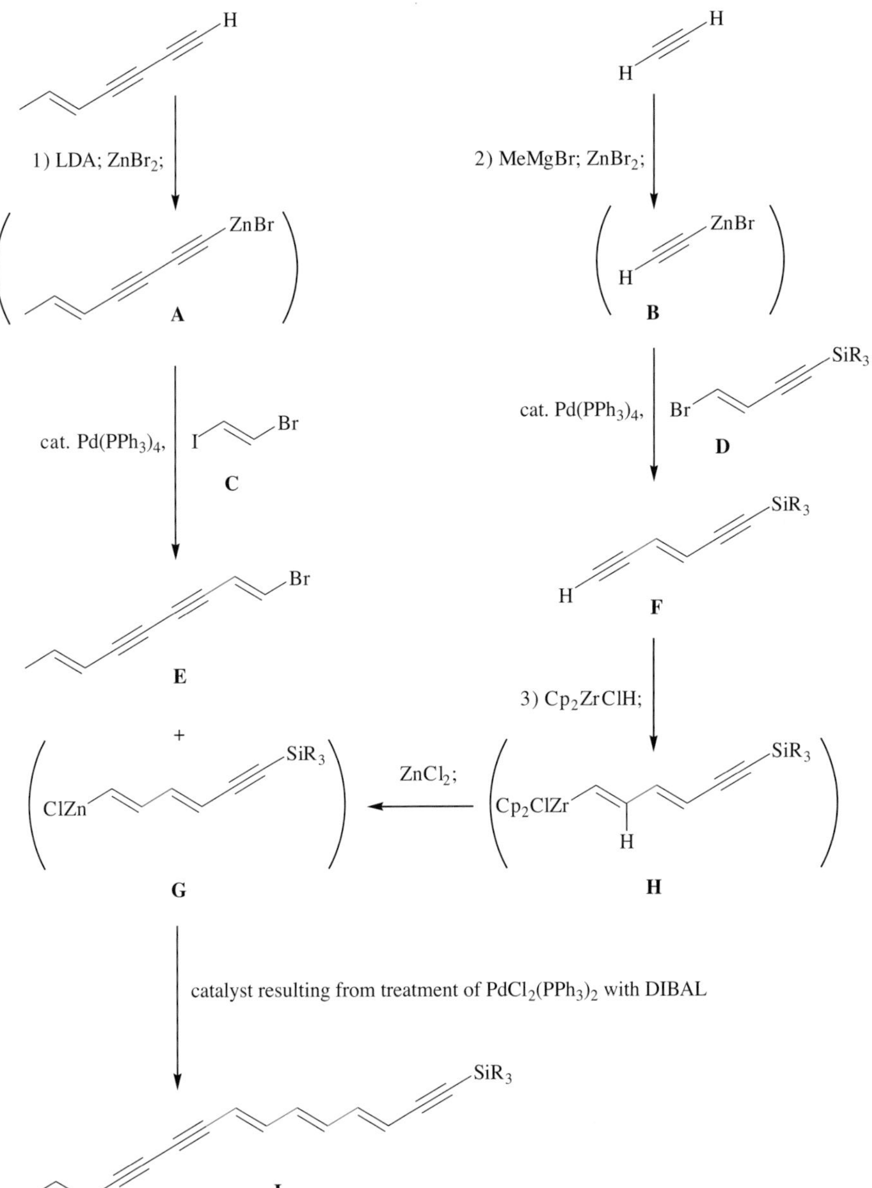

Fig. 16.26. Pd(0)-catalyzed alkenylations of the organozinc bromides **A** and **B** and the organozinc chloride **G** with the iodoalkene **C** and the bromoalkenes **D** and **E**, respectively, as an efficient method for the preparation of *trans*-configured C=C double bonds.

C≡C triple bond and the Zr/Zn exchange reaction **H** → **G**. The electrophilic reaction partners are the stereouniform haloalkenes **C** and **D**. Three C–ZnHal- and three C–Hal bonds are reacted with each other by pairs and always with complete retention of configuration (**A** + **C** → **E**, **B** + **D** → **F** and **G** + **E** → **I**). This results in a *trans*-configuration of all C=C double bonds of the central triene moiety of the target molecule.

16.4.4 Alkenylation and Arylation of Tin-bound Groups (Stille Reaction)

In this chapter, we have already seen that C,C couplings may involve organocopper, -magnesium, -boron and -zinc compounds. But the number of suitable organometallic compounds is not exhausted. A commonly used metal for coupling purposes is tin and organotin compounds are extensively used in **Stille coupling reactions**.

Unlike the previously discussed organocopper, -magnesium, -boron and -zinc compounds, organotin compounds can be handled and stored for long periods of time and undergo chromatography like any "normal" organic compound. And the problem of lacking functional group compatibility is totally alien to them. The only disadvantage of organotin compounds and the respective Stille reactions as compared to their most efficient competitors, i.e. organoboron compounds and their corresponding Suzuki couplings, is their toxicity and the fact that getting rid of tin byproducts in many reactions is a real pain in the…well, it can be quite difficult.

Figure 16.27 presents a rather special Stille reaction—a savvy combination of reactants and catalysts. It is the preparation of a dichlorotrifluorostyrene from an aromatic trifluorodichloro*iodine* compound and (tributylstannyl)ethylene. The choice of this of all aromatic compounds was the result of an extensive search for a pair of reactants that would allow elucidation of as many details of the coupling mechanism as possible. These many details nearly extended the bounds of Figure 16.27, but hopefully do not impede its understanding! The source of the catalysts in this figure given as $Pd(HetPh_3)_2$—Het = P and Het = As have been investigated—also departs from routine procedures: they are not generated by twofold cleavage of $HetPh_3$ from $Pd(HetPh_3)_4$ as in other reactions we have discussed.

The mechanism of the Stille reaction in Figure 16.27 can best be elucidated if you try to discover among its many reaction steps those that are known to you from the numerous mechanisms that have already been described in this chapter:

- Step 1 comprises the formation of a π-complex.
- Step 2 is an oxidative addition.
- Step 6 is a reductive elimination to form the π-complex of the coupling product.
- Step 7 includes the binding of the one equivalent of $HetPh_3$ that has been released to this π-complex (this step has so far been omitted from the discussion of the figures for the simple reason that the number and kind of the inert ligands at the palladium atom have largely been disregarded by the general term "Ln").
- Step 8 comprises a decomplexation yielding the free coupling product with concomitant regeneration of the catalyst.

In addition, steps 3–5 of Figure 16.27 so far offer the most precise model of how the transmetalation of the other stoichiometrically employed organometallic compounds proceeds in a Pd-catalyzed C,C coupling to give the organopalladium compound:

- Step 3 is a very fast *cis* → *trans*-isomerization of the organopalladium(II) compound that results from the oxidative addition. Naturally, the same step must follow all the oxidative additions of an arylating or alkenylating agent that have been discussed in Chapter 16 and that lead to the formation of a *cis*-configured Pd(II) complex with two $HetPh_3$ residues.

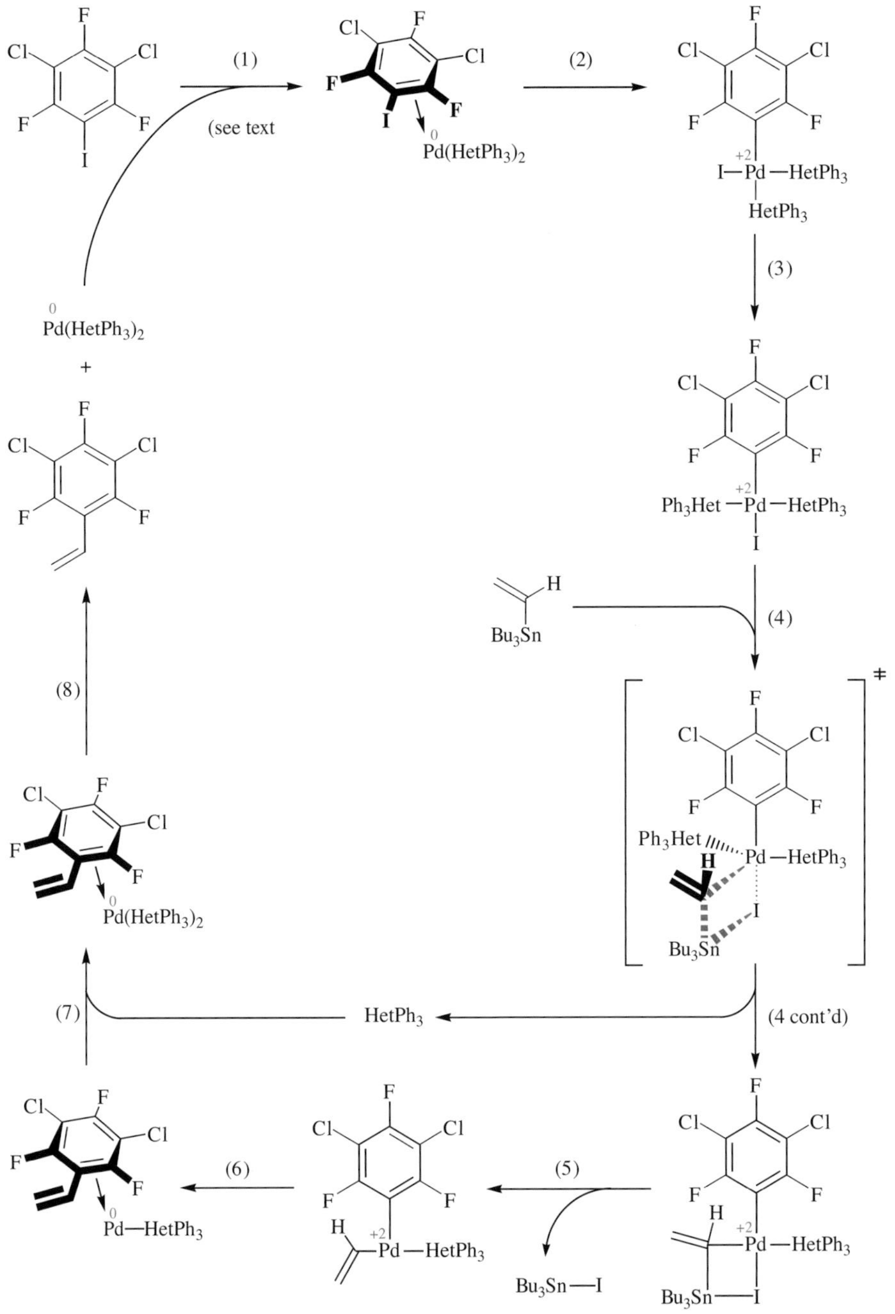

Fig. 16.27. Thoroughly explored mechanism of the Pd(0)-catalyzed arylation of an alkenyltributylstannane. Sequence of events: (1) formation of a π-complex from the arylating agent and $Pd(HetPh_3)_2$; (2) oxidative addition to give the *cis*-configured square planar Pd(II) complex; (3) isomerization to furnish the *trans*-configured Pd(II) complex; (4) displacement of a $HetPh_3$ ligand by the stannane; (5) completion of the vinyl residue transfer from Sn to Pd; (6) reductive elimination of the π-bound coupling product; (7) increase in the number of ligands at Pd(0); (8) dissociation of the coupling product/regeneration of $Pd(HetPh_3)_2$

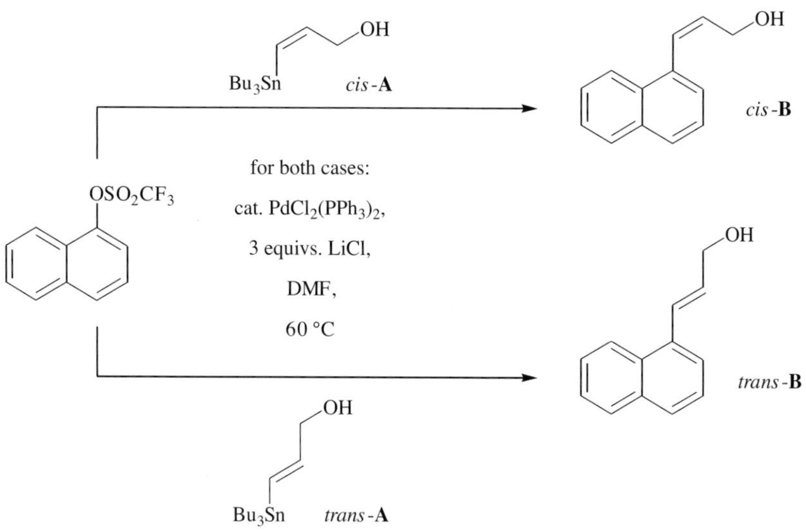

Fig. 16.28. Arylation of stereoisomeric alkenylstannanes: preparation of stereouniform cinnamylalcohols.

- Step 4 includes the beginning of the transmetalation of the vinyl-Sn bond into the vinyl-Pd bond. This transmetalation probably follows the two-step substitution mechanism presented here. The substitution product does not contain a pentavalent vinyl carbon atom, but a carbon atom bridging Pd with Sn.
- Step 5 completes the transmetalation of the vinyl-Sn bond into the vinyl-Pd bond and is assumed to be rate-determining, though hardly distinguishable from the alternative step 4.

$AsPh_3$ is a poorer σ-donor than PPh_3. Accordingly, Pd(II) with one (or two) $AsPh_3$ ligands is more electrophilic than Pd(II) with one (or two) PPh_3 ligands. This explains why Stille reactions proceed faster if palladium has access to $AsPh_3$ instead of PPh_3 as a ligand—irrespective of whether step 5 or 4 is the rate-determining step.

Figure 16.28 presents two stereoselective Stille reactions that—when considered as a pair—proceed stereospecifically and transfer a *cis*- or *trans*-configured allyl alcohol to the site of an aryl-bound triflate residue. $PdCl_2(PPh_3)_2$ is employed as a precatalyst that under the reaction conditions leads to the formation of the active catalyst $PdCl(PPh_3)_2^{\ominus}$. (This catalyst also occurs with the Sonogashira–Hagihara coupling (Figure 16.30); its structure corresponds to the active catalyst $Pd(OAc)(PPh_3)_2^{\ominus}$ of the "classical" Heck reactions, as can be seen in Figure 16.35.) Incidentally, the reaction rates are high enough only if several equivalents of lithium chloride are present. The effect of lithium chloride should be the same as previously mentioned (Figure 16.25).

Side Note 16.5. C,C Couplings via Acylpalladium Complexes

All the C,C couplings that have so far been introduced in Section 16.4 share the following features:

1. They are all homogeneously catalyzed by palladium(0) complexes.
2. A Pd(0) complex inserts into a $C_{sp^2 \text{ or } sp}$ bond of the leaving group of an arylating or alkenylating agent and thereby produces a Pd(II) complex.

3. The Pd(II) complex is modified in that at least the dissociation of a ligand occurs; often an initially present ligand L is exchanged even for a new ligand L' so that the complex displays the structure $R_{unsaturated}$–Pd(–X)L_n or $R_{unsaturated}$–Pd(–X)L_{n-1}L', where X is a halogen atom or a triflate group.
4. The organometallic reaction partner R'–M is transmetalated to give a Pd(II) complex, which from the point of view of palladium amounts to an exchange of ligand X for R'.
5. The arylation or alkenylation product $R_{unsaturated}$–R' is then released from this Pd(II) complex by reductive elimination.

With the C,C couplings that follow this pattern there are basically two modifications that result in a Pd-catalyzed ketone synthesis:

Palladium-catalyzed ketone synthesis A. In step 1, the Pd(0) complex is made to insert into the C_{sp^2}–Cl bond of an *acylating agent*. This leads to the formation of an acyl-Pd(II) com-

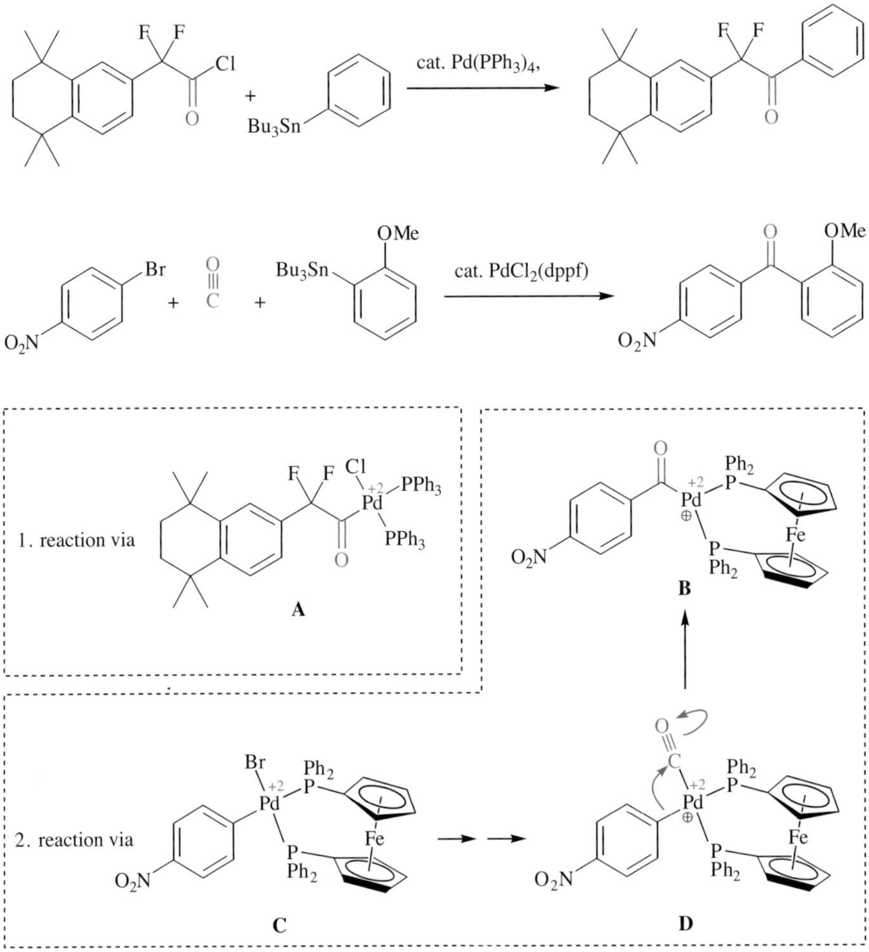

Fig. 16.29. Two-component variant (upper example) and three-component variant (lower example) of Stille reactions to be used for the preparation of unsaturated ketones. The phosphane ligand in the lower example is dppf [for bis(**di**phenyl**p**hosphino)**f**errocene], a diphosphane; the structure of this ligand can be derived from the Formulas **B–D**.

plex which, with regard to the above-mentioned reaction steps 2–4, basically behaves exactly like the aryl-Pd(II)- and alkenyl-Pd(II) complexes there. Accordingly, the structure of the product of the reductive elimination, i.e., of step 5, is RC (=O)–R'.

Palladium-catalyzed ketone synthesis B. The reaction mixture is saturated with carbon monoxide, which intervenes in step 2 by forming a palladium(II) carbonyl complex. Before the transmetalation (above referred to as step 3) takes place a rearrangement is interposed. The ligand $R_{unsaturated}$ undergoes a [1,2]-shift from Pd(II) to the carbon atom of carbon monoxide, leading to the formation of an acylpalladium(II) complex with the structure $R_{unsaturated}$–(C=O)–Pd(–X) L_{n-1}. With regard to the above-mentioned steps 3–4 it behaves like the acyl-Pd(II) complex of the ketone synthesis **A** and, after reductive elimination, i.e. in step 5, yields a ketone $R_{unsaturated}$–C (=O)–R'.

Figure 16.29 illustrates these procedures with tin compounds as the organometallic components. They are aromatic tin compounds, and they arylate the corresponding acylpalladium(II) intermediate. In the first ketone synthesis shown in Figure 16.29, the palladium catalyst inserts into the C–Cl bond of a fluorinated arylacetic acid chloride. This is how the acylpalladium(II) complex **A** shown in the lower part of Figure 16.29 is formed. This complex causes the transmetalation of the phenyl-Sn bond of the arylating agent into the phenyl-Pd bond of a new acylpalladium(II) complex. The phenylketone shown in the figure is released from the latter by reductive elimination. The second ketone synthesis in Figure 16.29 proceeds via the acylpalladium(II) complex **B**. The first step in its generation is the oxidative addition of *para*-nitrophenyl bromide to the Pd(0) catalyst (→ complex **C**), which is followed by the exchange of the bromide ion for carbon monoxide (→ palladium-carbonyl complex **D**) and the [1,2]-shift of the nitrophenyl residue. The acylpalladium(II) complex **B** formed reacts in the same way as the acylpalladium(II) complex **A** does.

16.4.5 Arylations, Alkenylations and Alkynylations of Neutral Organocopper Compounds II

Remember that as in Section 16.2, the term *neutral* organocopper compound will here also refer to monometallic compounds RCu and will be conceptually distinguished from Gilman *cuprates* (R_2CuLi), cyanide-containing Gilman *cuprates* (better known as cyanocuprates), Normant cuprates and Knochel cuprates. Unlike the "Arylations and Alkynylations of Neutral Organocopper Compounds I," which were discussed in Section 16.2, the coupling reactions of neutral organocopper compounds dealt with in this section need to be catalyzed by palladium(0) complexes. In return, the Pd-catalyzed reactions in this chapter also work with *catalytic* amounts of copper salts, whereas the reactions in Section 16.2 usually require (more than) stoichiometric amounts of these copper salts.

Terminal alkynes can be alkenylated by alkenyl triflates (bromides, iodides) in the presence of catalytic amounts of a palladium(0) complex (or a precursor thereof) and usually an additional substoichiometric amount of copper(I) iodide (CuI), and they can be arylated by aryl triflates (bromides, iodides). These reactions are called **Cacchi coupling** reactions if *triflate* reagents are employed, and **Sonogashira–Hagihara coupling** reactions if *halides* are used.

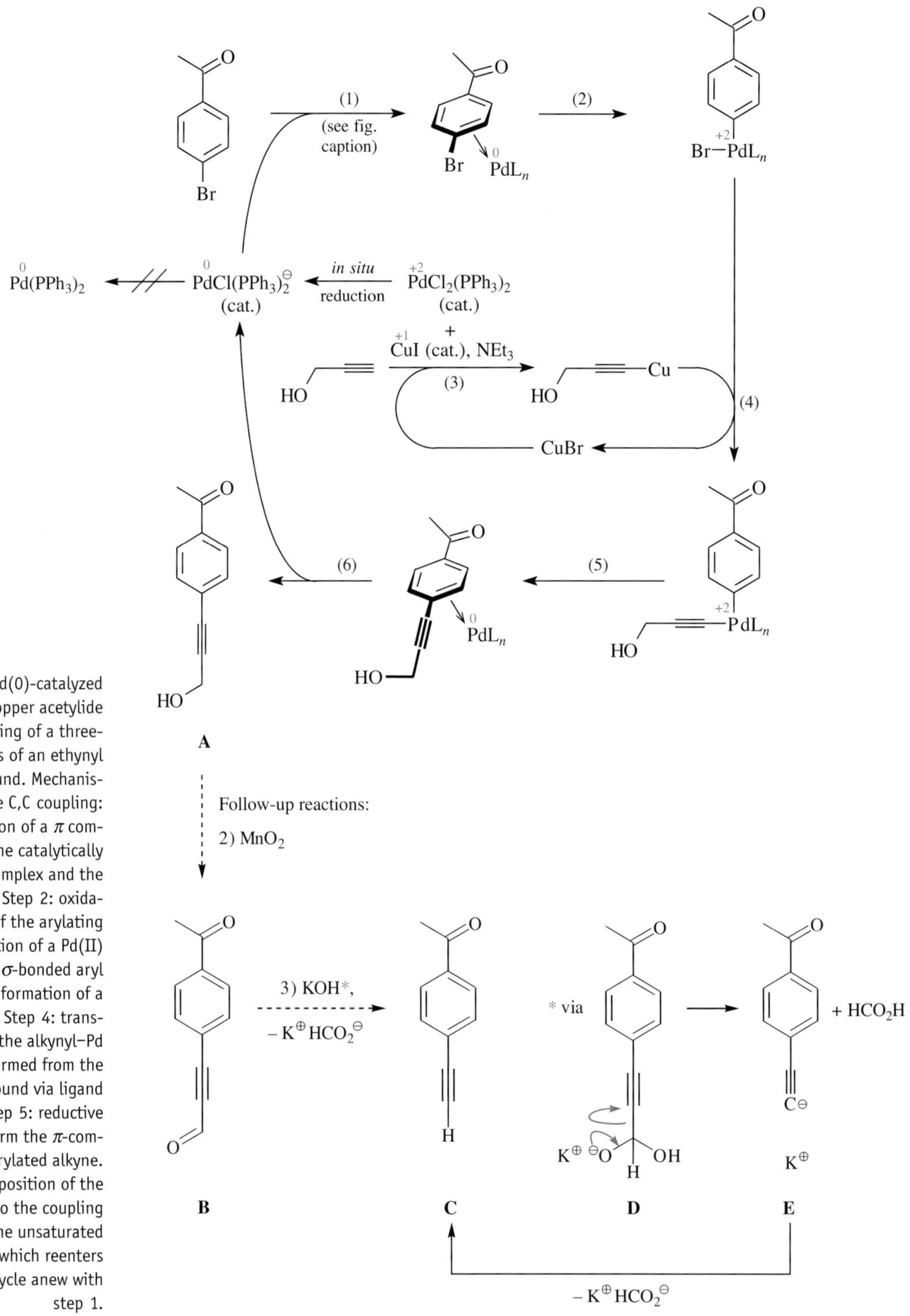

Fig. 16.30. Pd(0)-catalyzed arylation of a copper acetylide at the beginning of a three-step synthesis of an ethynyl aromatic compound. Mechanistic details of the C,C coupling: Step 1: formation of a π complex between the catalytically active Pd(0) complex and the arylating agent. Step 2: oxidative addition of the arylating agent and formation of a Pd(II) complex with a σ-bonded aryl moiety. Step 3: formation of a Cu-acetylide. Step 4: transmetalation; the alkynyl–Pd compound is formed from the alkynyl-Cu compound via ligand exchange. Step 5: reductive elimination to form the π-complex of the arylated alkyne. Step 6: decomposition of the complex into the coupling product and the unsaturated Pd(0) species, which reenters the catalytic cycle anew with step 1.

Cacchi- and Sonogashira–Hagihara couplings occur only if a primary, secondary, or tertiary amine is present, and it is best to have the amine present in large excess. Under these conditions the acetylene will at least form a small equilibrium concentration of the corresponding ammonium acetylide or copper complex thereof. The copper iodide serves to trap this species as a copper acetylide. The copper acetylide represents a substantially improved nucleophile in comparison to the free acetylene. Without the CuI addition, the acetylide content of the reaction mixture is so small that a reaction occurs only at higher temperatures.

Furthermore, Cacchi- and Sonogashira–Hagihara coupling reactions require catalysis by a complex of Pd(0); this (among other things) distinguishes them from the Castro–Stephens coupling (Figure 16.7). $PdCl_2(PPh_3)_2$, a Pd(II) complex, can be handled more conveniently and is usually employed, since under the reaction conditions this complex is reduced immediately to provide the catalytically active Pd(0) complex. The amine as well as the terminal alkyne or its respective acetylide ion can act as the reducing reagent. The amine would be oxidized to give an iminium ion while the acetylide would be converted into a 1,3-diyne, that is, the product of an oxidative dimerization (**Glaser coupling**).

The mechanism of the arylation of a copper acetylide by way of the Sonogashira–Hagihara coupling is illustrated in Figure 16.30 as a representative example. The catalytic cycle starts with the formation of the Cu-acetylide (step 3 in Figure 16.30) and comprises five other steps. These steps are basically the same as the steps in the Ni-catalyzed coupling of Grignard compounds (Figure 16.12), the Pd-catalyzed coupling of organoboron compounds (Figure 16.18) and the Pd-catalyzed couplings of organotin compounds (Figure 16.26). An equally pronounced similarity exists between this mechanism and the Pd-catalyzed coupling of organozinc compounds (Figure 16.25). The series of steps listed in the caption of Figure 16.30 thus suffices to fully describe the reaction mechanism.

No formula is given for Cacchi's palladium-catalyzed arylation of alkynes with aryl triflates as it proceeds mechanistically in analogy with the Sonogashira–Hagihara arylation.

Figure 16.31 shows Sonogashira–Hagihara *alkenylations* of an alkyne. The alkenylating agent is a *cis*-iodoalkene or its *trans*-isomer. These couplings are highly stereoselective reac-

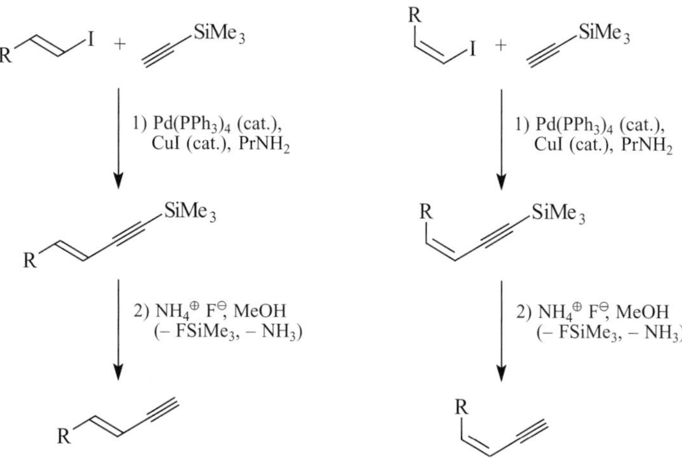

Fig. 16.31. Stereoselective and stereospecific Pd(0)-catalyzed alkenylations of copper acetylides with iodoalkenes at the beginning of a two-step synthesis of stereouniform ethynyl-substituted alkenes.

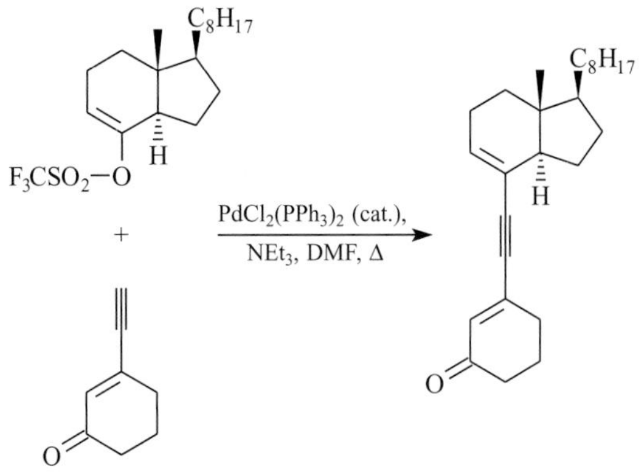

Fig. 16.32. Pd(0)-catalyzed alkenylation of a copper acetylide with an alkenyl triflate.

tions that proceed with retention of the respective double bond geometry and are stereospecific when considered as a pair.

Figure 16.32 represents the alkenylation of an acetylene according to the procedure developed by Cacchi. The coupling partners are an alkenyl triflate (prepared in analogy to Figure 13.23) and an alkyne. The coupling product of these compounds is a 1,3-enyne—in this case conjugated with a cyclohexenone. The C≡C triple bond of such an enyne can be hydrogenated in a *cis*-selective fashion by using Lindlar's Pd catalyst (cf. Figure 17.81). In this way 1,3-dienes are formed that contain at least one *cis*-configured C=C double bond.

In addition to the syntheses of the enynes just discussed (Figure 16.31 and 16.32), a diyne synthesis is also presented in Figure 16.33. Regarding reagents, reaction conditions and reaction mechanism this reaction corresponds to the Sonogashira–Hagihara coupling. Its synthetic performance is exactly the same as with the Cadiot–Chodkiewicz coupling shown in Figure 16.9. So is it really worthwhile to employ a palladium catalyst as in Figure 16.33, if we, according to Figure 16.9, can perfectly do without one? The answer is: "yes." The reason is that a side reaction, i.e., a halogen migration from one alkyne to the other, occurs and appears to be much more pronounced under the conditions of Figure 16.9 than under those of Figure 16.33. The result is that apart from the desired unsymmetrical diyne substantial amounts of the two symmetric diynes may frequently form with the Cadiot–Chodkiewicz coupling.

If you look back on the arylations and alkenylations that have so far been discussed in this section, you will find out that the acetylene itself was never used as the nucleophile, but always a higher alkyne. That is no coincidence, since in the presence of an amine, acetylene and CuI form the poorly soluble Cu_2C_2 most of which precipitates. If at all, the very small portion of this species that remains in solution would couple with the arylating and alkenylating agents on both C atoms. This would at best result in the formation of the respective bis-coupling product of acetylene.

Hence, monoarylation, monoalkenylation and monoalkynylation products of acetylene can only be obtained via monocoupling with alkynes that bear a substituent at one terminus that may be cleaved off after the coupling. In *this* section, we learned about three alkynes where this procedure is practical even if this fact has so far not been mentioned. Examples of alkynes with such "protecting groups" were presented in Figure 16.30 $(HO)H_2$ C–C≡C–H), in Fig-

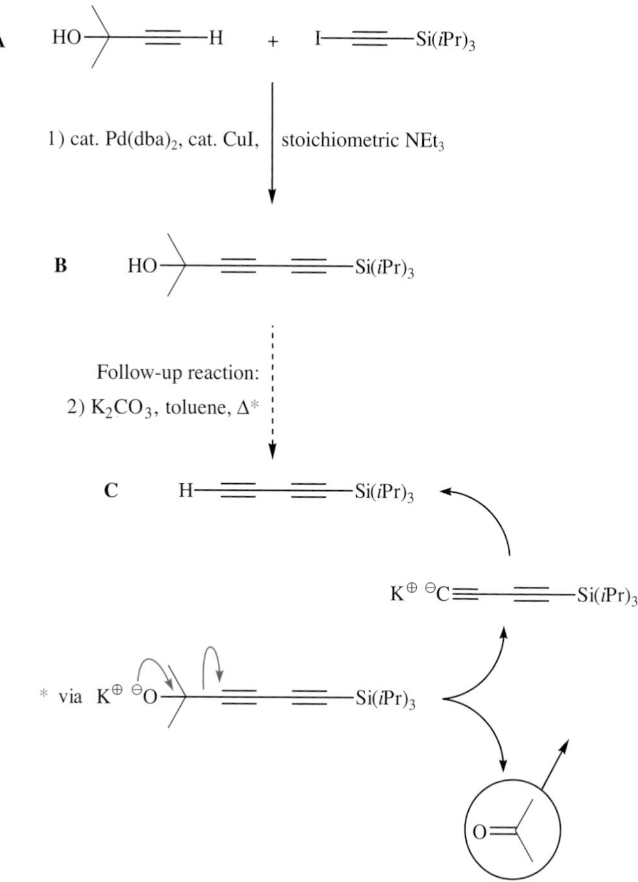

Fig. 16.33. Pd(0)-catalyzed alkynylation of a copper acetylide with a silylated ethynyl iodide in a two-step synthesis of a monosilylated 1,3-butadiyne. If the same copper acetylide is alkynylated with higher alkynyl iodides and subsequently heated with potassium carbonate in toluene, monoalkylated 1,3-butadiynes result. – The Pd-free alkynylation of a copper acetylide ("Cadiot–Chodkiewicz coupling") is shown in Figure 16.9.

ure 16.31 (Me_3Si–C=C–H) and in Figure 16.33 (HO)Me_2 C–C≡C–H). In each of the explicitly outlined follow-up reactions these "protecting groups" were removed as follows:

- The (HO)H_2C–C≡C group of the monocoupling product A in Figure 16.30 is oxidized so that the aldehyde (O=)HC–C≡C–Ar (B) is formed. Under basic conditions, this cleaves off formate and leaves the monocoupling product H–C≡C-Ar.
- The Me_3Si–C≡C groups of the monocoupling products in Figure 16.31 are desilylated with ammonium fluoride so that Me_3Si–F and the monocoupling products H–C≡C–CH=CH-R are formed.
- The (HO)Me_2C–C≡C group of the monocoupling product **B** in Figure 16.33 is split into two parts by KOH, namely acetone and the H–C≡C terminus of the monocoupling product H-C≡ C–C≡C-Si(iPr)$_3$.

16.5 Heck Reactions

Unlike the previous nucleophiles in Chapter 16, certain alkenes—even though they do not contain a metal—can be alkenylated with alkenyl bromides (-iodides, -triflates) and arylated with aryl bromides (-iodides, -triflates). If the price of the arylating agent matters one can switch to aryl chlorides (as long as they are reactive enough). In special cases the arylation can also be performed with aryldiazonium ions. Recently, certain aromatic carboxylic acids and their derivatives (anhydrides, active esters; Ar–C(=O)–Het) have also been used as arylating agents. With these, $Het^{\ominus}$ and C≡O become the leaving group, one after the other. All these reactions are referred to as **Heck reactions** (Figure 16.34).

For a **Heck reaction** to occur, catalytic amounts of (usually) palladium(II) acetate—though other palladium sources are also used—as well as other "additives" need to be added to the mixture of starting materials. Figure 16.34 contains three typical procedures: for the original

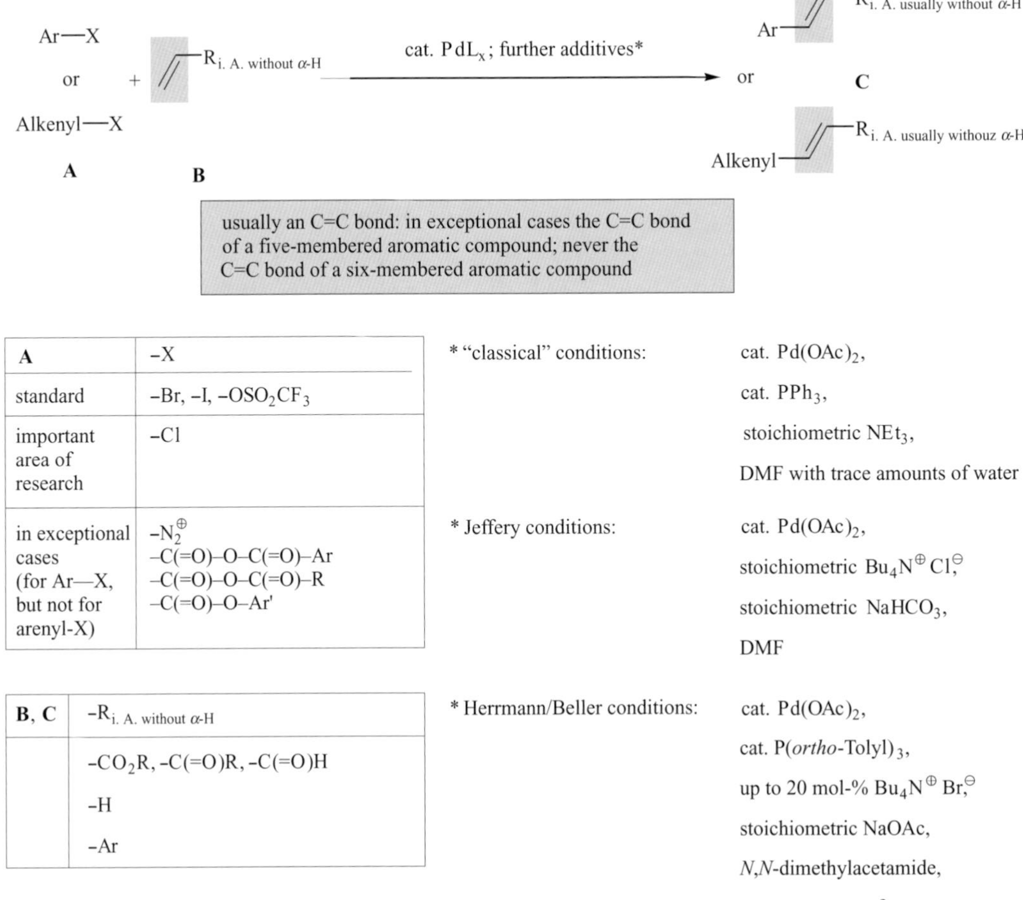

A	–X
standard	–Br, –I, $–OSO_2CF_3$
important area of research	–Cl
in exceptional cases (for Ar—X, but not for arenyl-X)	$-N_2^{\oplus}$ –C(=O)–O–C(=O)–Ar –C(=O)–O–C(=O)–R –C(=O)–O–Ar'

B, C	–R i. A. without α-H
	$-CO_2R$, –C(=O)R, –C(=O)H –H –Ar

* "classical" conditions: cat. $Pd(OAc)_2$, cat. PPh_3, stoichiometric NEt_3, DMF with trace amounts of water

* Jeffery conditions: cat. $Pd(OAc)_2$, stoichiometric $Bu_4N^{\oplus} Cl^{\ominus}$, stoichiometric $NaHCO_3$, DMF

* Herrmann/Beller conditions: cat. $Pd(OAc)_2$, cat. P(*ortho*-Tolyl)$_3$, up to 20 mol-% $Bu_4N^{\oplus} Br^{\ominus}$, stoichiometric NaOAc, *N*,*N*-dimethylacetamide, a trace amount of water

Fig. 16.34. Heck reactions: substrates, reaction conditions and products.

method by Heck, for the phosphane-free catalysis according to Jeffery, and finally for a catalyst system that is associated with the names of Herrmann and Beller. The latter has led to increases in efficiency—amongst other things—which have hardly been thought possible, especially with the initially slow "aromatic Heck compounds." Thanks to multiple research efforts the most reactive Heck catalysts look quite different today. The additives of a Heck reaction also include stoichiometric amounts of triethylamine, sodium hydrogen carbonate or sodium acetate. These additives neutralize the respective strong acid that is also released in stoichiometric amounts during the reaction. A triarylphosphine is frequently used as an additive in Heck reactions, but is not present, for example, under Jeffery conditions nor in those cases in which a Heck reaction can be catalyzed with finely divided elemental Pd (e. g., 10% Pd on carbon; or even a Pd cluster).

In light of its very broad range of substrates and products (Formula **A/B** and **C** in Figure 10.34), the Heck reaction can be considered a universal C,C coupling reaction. Alkenes *without* allylic H atoms—compounds like ethenes, acceptor-substituted ethenes, and styrenes—can be *alkenylated* by Heck reactions in a clearly predictable fashion to give 1,3-dienes, $\alpha,\beta,\gamma,\delta$-unsaturated carbonyl or carboxylic compounds (for an example of each cf. Figure 16.36) as well as aryl-substituted 1,3-dienes. Moreover, the same alkenes can be *arylated* with reliably predictable outcomes to give styrenes, α,β-unsaturated β-arylated carbonyl compounds, α,β-unsaturated β-arylated carboxylic compounds (example: Figure 16.35, part II) as well as stilbenes (example: Figure 16.37).

Alkenes *with* H atoms in the allyl position are also perfectly capable of a Heck reaction. Depending on their precise structure—among other things, e. g., on whether their C=C double bond is part of a ring system or not—the C=C double bond, however, need not necessarily occupy the same position in the coupling product as in the initial alkene, and in some cases cannot do so at all. We will not burden ourselves with this problem here and rather restrict our attention to the alkenes without allylic hydrogens as participants in Heck reactions.

Figure 16.35 presents the mechanistic details of a representative example of a Heck couplings performed under classical "Heck conditions". The arylating reagent is a benzene derivative that contains both a good leaving group (a triflate group) and a poor leaving group (chloride). The good leaving group is cleaved off, the poor one remains. The alkene that is being functionalized is acrylic acid methyl ester, and the coupling product is a *trans*-configured cinnamic acid ester.

Don't worry if you cannot immediately discover these compounds in the figure. The reaction process is so complicated and has, at the same time, been so fully elucidated that a two part figure is necessary to accommodate the more than twelve individual reactions involved in this process. Figure 16.35 (part I) shows the generation of the active Pd(0) catalyst, while Figure 16.35 (part II) starts with this active Pd(0) catalyst and shows the actual catalytic cycle of the coupling process. The actual final product, i.e., the *trans*-configured cinnamic acid ester that has briefly been mentioned above, can be found in Figure 16.35 (part II) at the end of the reaction arrow that is marked with the number (8).

Figure 16.35 (part I) shows how palladium(II) acetate, triphenylphosphane and water react to furnish the complex **F** as the active palladium(0) catalyst. For a long time, the presence of water had not been suspected as a solvent contaminant. It was unknown that it is essential for initiating the reaction process. In the mid-1990s, Amatore and Jutand demonstrated that the reduction Pd(II) $\rightarrow$ Pd(0), which is involved in the formation of **F**, is effected by a tri-

Fig. 16.35 (part I). Heck reaction of an aryl triflate with acrylic acid methyl ester under "classical," i.e., the Richard Heck conditions with a mechanistic analysis of the beginning of the reaction sequence, namely the formation of the catalytically active Pd(0) complex (= compound **F**).

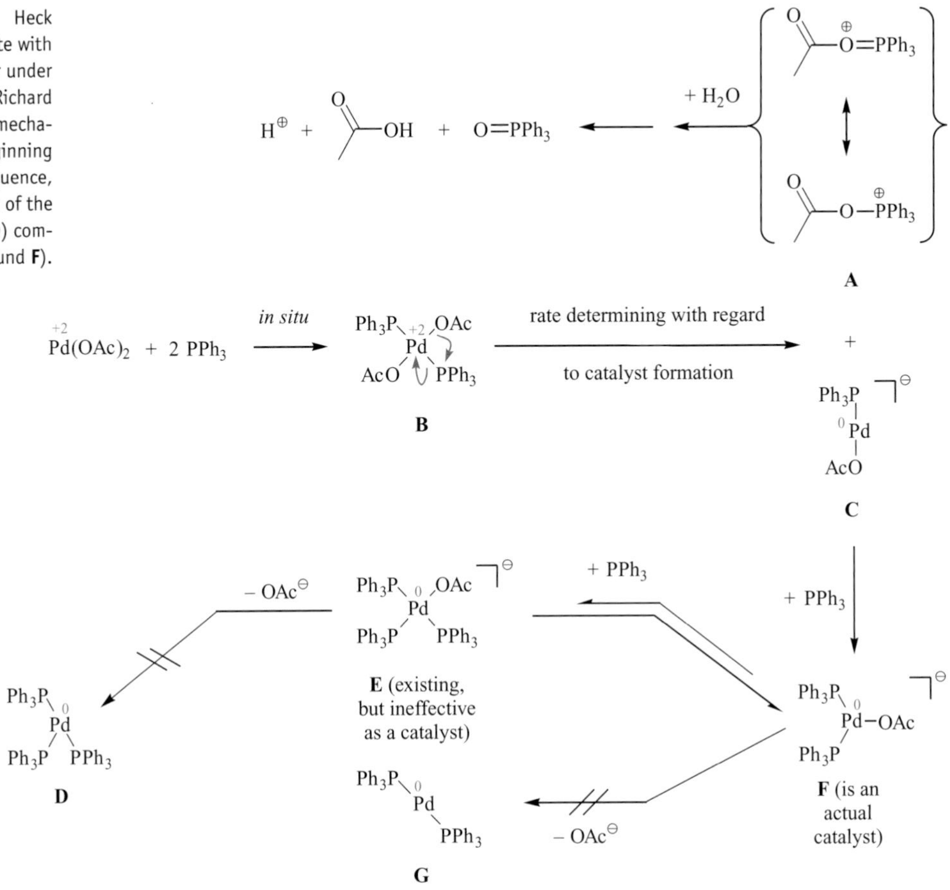

phenylphosphane ligand of the bis(triphenylphosphane)palladium(II) diacetate **B** consuming stoichiometric amounts of water. A triphenylphosphane transfers the initially free electron pair of its phosphorus atom to palladium and in its place picks up the electron pair by which one of the acetate ligands was bound to palladium. In this way, acetylated triphenylphosphane oxide (**A**) and (triphenylphosphane) palladium(0) acetate (**C**) are formed. The acetylated triphenylphosphane oxide is an acyloxonium ion and hence a good acylating agent, as you can see from the analogy with the acyliminium ions in Figure 6.9 and 6.28. It acetylates water to give acetic acid and releases triphenylphosphane oxide as the leaving group. The mono(triphenylphosphane) complex **C** irreversibly binds a second ligand of triphenylphosphane. The resulting bis(triphenylphosphane)palladium(0) acetate **F** is the active catalyst of the "classical" Heck reaction. Being an anion, it resembles the active catalyst—bis(triphenylphosphane)palladium(0) chloride—of the Sonogashira–Hagihara coupling in Figure 16.30.

The catalytically inactive tris(triphenylphosphane) complex **E** is formed in an equilibrium reaction from the bis(triphenylphosphane) complex F and a third ligand of triphenylphosphane oxide. Loss an acetate ion from complex **E** does not occur to give tris(triphenylphosphane)palladium **D**. Nor does the analogous acetate cleavage from **F** occur to give bis(tri-

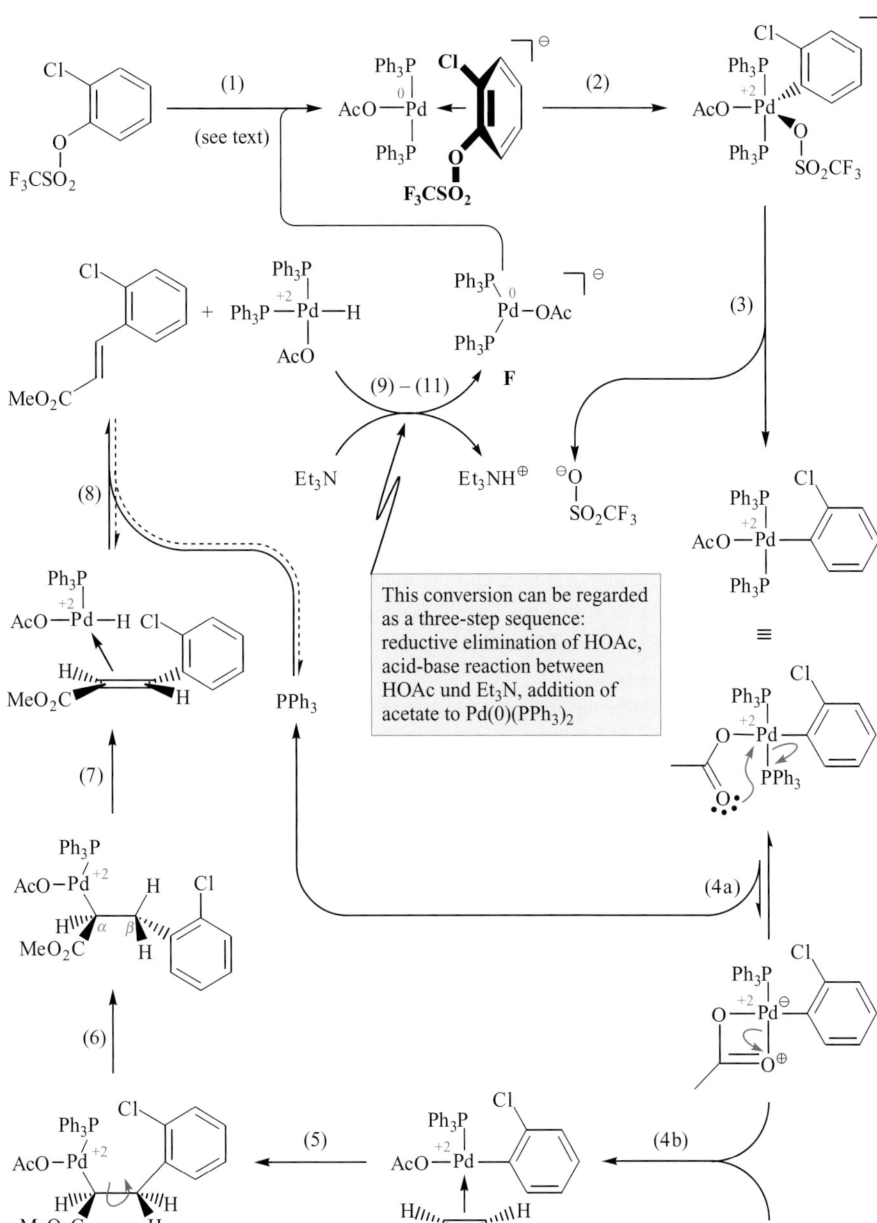

Fig. 16.35 (part II). Heck coupling of an aryl triflate with acrylic acid methyl ester under "classical," conditions: mechanistic analysis of the catalytic cycle, starting with the catalytically active Pd(0) complex **F** (for the formation of **F** see Figure 16.35, part I). The elementary steps 1 and 2 correspond to those in the Figures 16.12, 16.18, 16.25 and 16.27. The further course of the reaction is described in the text.

phenylphosphane)palladium **G**, although **D** and **G** are very well known as dissociation products of $Pd(PPh_3)_4$, and **G** can actually catalyze C,C couplings (Figure 16.18, 16.25).

Steps 1–4 of the catalytic cycle (Figure 16.35, part II) correspond to various steps of other catalytic cycles that have previously been discussed in this chapter. Step 1: *π-complex* formation by combination of the aryl triflate and the active Pd(0) species **F**. Step 2: oxidative addition of the aryl triflate to Pd with formation of a C_{sp^2}—Pd(II) bond. Admittedly, it is new that pentavalent palladium occurs here. Step 3, however, in the form of the cleavage of a triflate ion, leads back to a species with tetravalent palladium, just like what you are used to from the products of oxidative additions in Figures 16.18, 16.25, 16.27 and 16.30. Steps 4a and 4b: exchange of a PPh_3 ligand for acrylic acid methyl ester via a dissociation/association mechanism; the newly entered acrylic ester is bound as a π-complex.

Step 5 of the mechanism shown in Figure 16.35 (part II) is new. It consists of the *cis*-selective addition of the aryl-Pd complex to the C=C double bond of the acrylic acid methyl ester, i.e., a **carbopalladation** of this double bond. A related reaction, the *cis*-selective **carbocupration** of C≡C triple bonds, was mentioned in connection with Figure 16.17. The regioselectivity of the **carbopalladation** is such that the organic moiety is bonded to the methylene carbon and Pd to the methyne carbon of the reacting C=C double bond. The addition product is an alkyl-Pd(II) complex.

Prior to the next bond formation/bond cleavage event, a rotation about the newly formed C—C bond of this complex is required, which occurs in step 6. The rotation brings an β-H atom and the $Pd(PPh_3)(OAc)$ group into a *syn*-relationship. This conformation is essential for the formation of a hydrido-Pd(II) complex to occur by a β-elimination of this β-H atom and the $Pd(PPh_3)(OAc)$ group in step 7. This complex contains the arylated acrylic ester, i.e., the coupling product, as a π-bonded ligand. Mechanistically related β-eliminations were encountered earlier in Figures 16.13 and 16.14, where hydrido-Ni(II) complexes were formed from alkyl-Ni(II) complexes.

In step 8, the π-bonded coupling product dissociates from the hydrido–Pd(II) complex of Figure 16.35, and the hydrido-Pd(II) complex $Pd(PPh_3)_2(OAc)H$ is formed, from which the catalytically active Pd(0) complex **F** is regenerated. It is still unknown how this is effected. Conceivably, $Pd(PPh_3)_2(OAc)H$ could be deprotonated by triethylamine in a one-step process. However, a three-step sequence as shown in steps 9–11 of Figure 16.35 (part II) is more probable. In step 9, this sequence begins with a reductive elimination of acetic acid [→ bis(triphenylphosphane)palladium(0), Formula **G** in Figure 16.35 (part I)]. An acid/base reaction acetic acid + triethylamine → acetate ion + triethylammonium ion follows in step 10. In step 11 the resulting acetate ion finally binds to bis(triphenylphosphane)palladium(0). Thereby, the same valence-unsaturated Pd(0) complex **F** is formed that initiated the reaction sequence in step 1, so this complex is now available to begin another catalytic cycle.

In the Heck reaction, stereogenic C=C double bonds are always formed in a *trans*-selective fashion. A few examples are shown in Figures 16.35–16.37. This *trans*-selectivity is due to product development control in the β-elimination of the hydrido–Pd(II) complex (step 7 of Figure 16.35, part II). The more stable *trans*-alkene is formed faster than the *cis*-isomer. But there also is another reason. Step 7 can be reversible, and the hydrido–Pd(II) complex may react with the alkyl–Pd(II) complex (an analogous option occurred in the case of the hydrido–Ni(II) complex **B** in Figure 16.13). The coupling product and its C=C double bond in particular thus disappear again. Subsequently, step 7 will occur again in the forward direction, that is, as a β-elimination. Thus, the coupling product and its C=C double bonds are formed again—*but perhaps not with the same configuration as when formed for the first time.* A sufficiently high number of such readditions and renewed β-eliminations assure that the double bond geometry will be determined by thermodynamic control, and the *trans*-configuration is of course the thermodynamically preferred geometry. Therefore, the synthetic importance of the

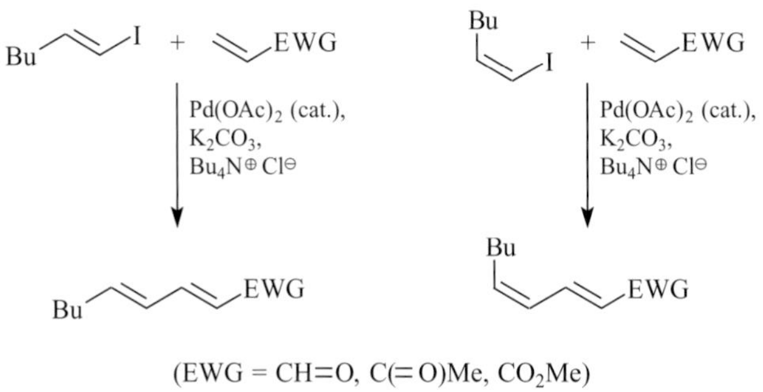

(EWG = CH=O, C(=O)Me, CO_2Me)

Fig. 16.36. Stereoselectivity and stereospecificity of Heck coupling reactions with isomeric iodoalkenes.

Heck reaction is not just to make possible the alkenylation or arylation of alkenes *at all*, but also to guarantee that these alkenylations and arylations occur in a *trans*-selective fashion.

Although the Heck reaction may be efficiently employed for synthesis, it has its limits that should not go unmentioned: the Heck reaction can*not*—at least not intermolecularly—couple alkenyl triflates (-bromides, -iodides) or aryl triflates (-bromides, -iodides) with metal-free *aromatic compounds* in the same way as it is possible with the same substrates and metal-free *alkenes*. The reason is step 4 of the mechanism in Figure 16.35 (part II). If an aromatic compound instead of an alkene was the coupling partner the aromaticity with this carbopalladation of a C=C double bond would have to be sacrificed in step 4. Typically, Heck reactions can only be run at a temperature of 100 °C even if they proceed *without* any such energetic effort. This is why this additional energetically demanding loss of aromaticity is not feasible.

A haloalkene that contains a stereogenic C=C double bond can usually be coupled with alkenes via the Heck reaction without isomerization. This is illustrated with the three reaction pairs in Figure 16.36. As can be seen, both the *cis*- and the *trans*-configured iodoalkenes react with acrolein or methyl vinyl ketone or acrylic acid methyl ester with complete retention of the C=C double bond configuration. These coupling reactions are thus *stereoselective* and—when considered as a pair—*stereospecific*.

Substrates with several sp^2-bound triflate groups, bromine or iodine atoms, respectively, may sometimes undergo several Heck reactions in a row. A threefold Heck reaction of 1,3,5-tribromobenzene with styrene is shown in Figure 16.37.

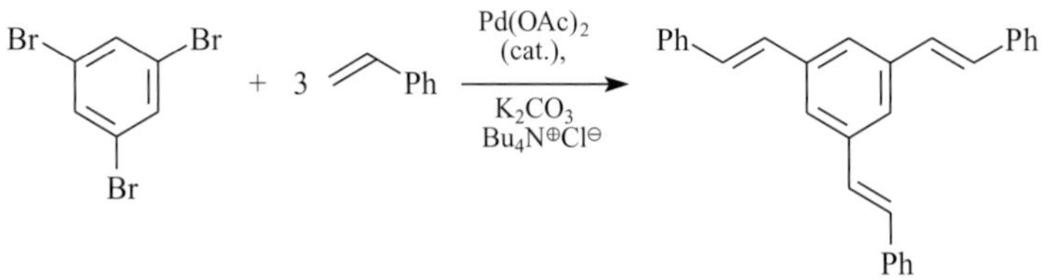

Fig. 16.37. Three "one-pot," *trans*-selective Heck coupling reactions.

References

K. Tamao, "Coupling Reactions Between sp^3 and sp^2 Carbon Centers," in *Comprehensive Organic Synthesis* (B. M. Trost, I. Fleming, Eds.), Vol. 3, 435, Pergamon Press, Oxford, U. K., **1991**.

D. W. Knight, "Coupling Reactions between sp^2 Carbon Centers," in *Comprehensive Organic Synthesis* (B. M. Trost, I. Fleming, Eds.), Vol. 3, 481, Pergamon Press, Oxford, U. K., **1991**.

R. F. Heck, "Palladium Reagents in Organic Synthesis," Academic Press, **1985**.

H.-J. Altenbach, "Regio- and Stereoselective Aryl Coupling," in *Organic Synthesis Highlights* (J. Mulzer, H.-J. Altenbach, M. Braun, K. Krohn, H.-U. Reißig, Eds.), VCH, Weinheim, New York, **1991**, 181–185.

K. Ritter, "Synthetic Transformations of Vinyl and Aryl Triflates," *Synthesis* **1993**, 735–762.

P. J. Stang, F. Diederich (Eds.), "Modern Acetylene Chemistry," VCH, Weinheim, Germany, **1995**.

J. Tsuji, "Palladium Reagents and Catalysis: Innovations in Organic Synthesis," Wiley, New York, **1995**.

G. Poli, G. Giambastiani, A. Heumann, "Palladium in Organic Synthesis: Fundamental Transformations and Domino Processes," *Tetrahedron* **2000**, *56*, 5959–5989.

S. P. Stanforth, "Catalytic Cross-Coupling Reactions in Biaryl Synthesis," *Tetrahedron* **1998**, *54*, 263–303.

G. Bringmann, R. Walter, R. Weirich, "Biaryls," in *Stereoselective Synthesis* (Houben-Weyl) 4th ed., **1996**, (G. Helmchen, R. W. Hoffmann, J. Mulzer, E. Schaumann, Eds.), **1996**, Vol. E21 (Workbench Edition), *1*, 567–588, Georg Thieme Verlag, Stuttgart.

F. Diederich, P. J. Stang (Eds.), "Metal-Catalyzed Cross-Coupling Reactions," Wiley-VCH, Weinheim, Germany, **1998**.

N. Miyaura (Ed.), "Cross-Coupling Reactions. A Practical Guide," *Top. Curr. Chem.* **2002**, *219,* 3–540.

16.2

P. Siemsen, R. C. Livingston, F. Diederich, "Acetylenic Coupling: A Powerful Tool in Molecular Construction," *Angew. Chem. Int. Ed. Engl.* **2000**, *39*, 2632–2657.

P. E. Fanta, "The Ullmann Synthesis of Biaryls," *Synthesis* **1974**, 9.

J. A. Lindley, "Copper Assisted Nucleophilic Substitution of Aryl Halogen," *Tetrahedron* **1984**, *40*, 1433.

J. S. Sawyer, "Recent Advances in Diaryl Ether Synthesis," *Tetrahedron* **2000**, *56*, 5045– 5065.

G. P. Ellis, T. M. Romsey-Alexander, "Cyanation of Aromatic Halides," *Chem. Rev.* **1987**, *87*, 779.

I. A. Rybakova, E. N. Prilezhaeva, V. P. Litvinov, "Methods of Replacing Halogen in Aromatic Compounds by RS-Functions," *Russ. Chem. Rev. (Engl. Transl.)* **1991**, *60*, 1331.

J. Hassan, M. Sevignon, C. Gozzi, E. Schulz, M. Lemaire, "Aryl-Aryl Bond Formation One Century after the Discovery of the Ullmann Reaction," *Chem. Rev.* **2002**, *102,* 1359–13469.

J.-P. Finet, A. Y. Fedorov, S. Combes, G. Boyer, "Recent Advances in Ullmann Reaction: Copper(II) Diacetate Catalyzed N-, O- and S-Arylation Involving Polycoordinate Heteroatomic Derivatives," *Curr. Org. Chem.* 2002, *6,* 597–626.

S. V. Ley, A. W. Thomas, "Modern Synthetic Methods for Copper-Mediated C(Aryl)-O, C(Aryl)-N, and C(Aryl)-S Bond Formation," *Angew. Chem. Int. Ed. Engl.* **2003**, *42,* 5400–5449.

K. Kunz, U. Scholz, D. Ganzer, "Renaissance of Ullmann and Goldberg Reactions – Progress in Copper Catalyzed C-N-, C-O- and C-S-Coupling," *Synlett* **2003**, 2428–2439.

16.3

H. Urabe, F. Sato, "Metal-Catalyzed Reactions," in *Handbook of Grignard Reagents* (G. S. Silverman, P. E. Rakita, Eds.), Marcel Dekker Inc., New York, **1996**, 577–632.

E.-i. Negishi, F. Liu, "Palladium- or Nickel-catalyzed Cross-coupling with Organometals Containing Zinc, Magnesium, Aluminium, and Zirconium," in *Metal-catalyzed Cross-coupling Reactions*, Eds.: F. Diederich, P. J. Stang, Wiley-VCH, Weinheim, **1998**, 1–42.

E. Alexander Hill, "Nucleophilic Displacements at Carbon by Grignard Reagents," in *Grignard Reagents – New Developments* (H. G. Richey, Jr., Ed.), John Wiley & Sons, Chichester, **2000**, 27–64.

16.4

D. S. Matteson, "Boronic Esters in Stereodirected Synthesis," *Tetrahedron* **1989**, *45*, 1859– 1885.

A. R. Martin, Y. Yang, "Palladium Catalyzed Cross-Coupling Reactions of Organoboronic Acids with Organic Electrophiles," *Acta Chem. Scand.* **1993**, *47*, 221–230.

A. Suzuki, "New Synthetic Transformations via Organoboron Compounds," *Pure Appl. Chem.* **1994**, *66*, 213–222.

N. Miyaura, A Suzuki, "Palladium-Catalyzed Cross-Coupling Reactions of Organoboron Compounds," *Chem. Rev.* **1995**, *95*, 2457–2483.

S. Suzuki, "Cross-Coupling Reactions of Organoboron Compounds with Organic Halides," in *Metal-catalyzed Cross-coupling Reactions* (F. Diederich, P. J. Stang, Eds.), Wiley-VCH, Weinheim, **1998**, 49–89.

N. Miyaura, "Synthesis of Biaryls via the Cross-Coupling Reaction of Arylboronic Acids," in *Advances in Metal-Organic Chemistry* (L. S. Liebeskind, Ed.), JAI Press, Stamford, **1998**, Vol. 6, 187–243.

S. R. Chemler, D. Trauner, S. J. Danishefsky, "The *B*-Alkyl Suzuki-Miayura Cross-Coupling Reaction: Development, Mechanistic Study, and Applications in Natural Product Synthesis," *Angew. Chem. Int. Ed. Engl.* **2001**, *40*, 4544–4568.

S. Kotha, K. Lahiri, D. Kashinath, "Recent Applications of the Suzuki-Miyaura Cross-Coupling Reaction in Organic Synthesis," *Tetrahedron* **2002**, *58*, 9633–9695.

M. Ishikura, "Applications of Heteroarylboron Compounds to Organic Synthesis," *Curr. Org. Chem.* **2002**, *6*, 507–521.

E. Tyrrell, P. Brookes, "The Synthesis and Applications of Heterocyclic Boronic Acids," *Synthesis* **2003**, 469–483.

E. Erdik, "Use of Activation Methods for Organozinc Reagents," *Tetrahedron* **1987**, *43*, 2203–2212.

E. Erdik, "Transition Metal Catalyzed Reactions of Organozinc Reagents," *Tetrahedron* **1992**, *48*, 9577–9648.

E. Erdik (Ed.), "Organozinc Reagents in Organic Synthesis," CRC, Boca Raton, FL, **1996**.

Y. Tamaru, "Unique Reactivity of Functionalized Organozincs," in *Advances in Detailed Reaction Mechanism. Synthetically Useful Reactions* (J. M. Coxon, Ed.), **1995**, *4*, JAI Press, Greenwich, CT.

P. Knochel, "Organozinc, Organocadmium and Organomercury Reagents," in *Comprehensive Organic Synthesis* (B. M. Trost, I. Fleming, Eds.), Vol. 1, 211, Pergamon Press, Oxford, U. K., **1991**.

P. Knochel, M. J. Rozema, C. E. Tucker, C. Retherford, M. Furlong, S. A. Rao, "The Chemistry of Polyfunctional Organozinc and Copper Reagents," *Pure Appl. Chem.* **1992**, *64*, 361–369.

P. Knochel, "Zinc and Cadmium: A Review of the Literature 1982–1994," in *Comprehensive Organometallic Chemistry II* (E. W. Abel, F. G. A. Stone, G. Wilkinson, Eds.), Vol. 11, 159, Pergamon, Oxford, U. K., **1995**.

P. Knochel, "Preparation and Application of Functionalized Organozinc Reagents," in *Active Metals* (A. Fürstner, Ed.), 191, VCH, Weinheim, Germany, **1996**.

P. Knochel, J. J. A. Perea, P. Jones, "Organozinc Mediated Reactions," *Tetrahedron*, **1998**, *54*, 8275–8319.

P. Knochel, "Carbon-Carbon Bond Formation Reactions Mediated by Organozinc Reagents," in *Metal-catalyzed Cross-coupling Reactions* (F. Diederich, P. J. Stang, Eds.), Wiley-VCH, Weinheim, **1998**, 387–416.

E.-i. Negishi, F. Liu, "Palladium- or Nickel-catalyzed Cross-coupling with Organometals Containing Zinc, Magnesium, Aluminium, and Zirconium," in *Metal-catalyzed Cross-coupling Reactions*, (F. Diederich, P. J. Stang, Eds.), Wiley-VCH, Weinheim, **1998**, 1–42.

P. Knochel, P. Jones, (Eds.) "Organozinc Reagents: A Practical Approach," Oxford University Press, Oxford, UK, **1999**.

A. Boudier, L. O. Bromm, M. Lotz, P. Knochel, "New Applications of Polyfunctional Organometallic Compounds in Organic Synthesis," *Angew. Chem. Int. Ed. Engl.* **2000**, *39*, 4414–4435.

J. K. Stille, "The Palladium-catalyzed Cross-Coupling Reactions of Organotin Reagents with Organic Electrophiles," *Angew. Chem. Int. Ed. Engl.* **1986**, *25*, 508–524.

V. Farina, G. P. Roth, "Recent Advances in the Stille Reaction," in *Advances in Metal-Organic Chemistry* (L. S. Liebeskind, Ed.), JAI Press, Greenwich, Connecticut, **1996**, Vol. 5, 1–53.

V. Farina, V. Krishnamurthy, W. J. Scott, "The Stille Reaction," *Org. React.* **1997**, *50,* 1–652.

A. G. Davies, "Organotin Chemistry," Wiley-VCH, Weinheim, Germany, **1997**.

T. N. Mitchell, "Organotin Reagents in Cross-Coupling," in *Metal-catalyzed Cross-coupling Reactions*, (F. Diederich, P. J. Stang, Eds.), Wiley-VCH, Weinheim, **1998**, 167–197.

R. Skoda-Foldes, L. Kollar, "Synthetic Applications of Palladium Catalyzed Carbonylation of Organic Halides," *Curr. Org. Chem.* **2002**, *6,* 1097–1119.

I. B. Campbell, "The Sonogashira Cu-Pd-Catalyzed Alkyne Coupling Reactions," in *Organocopper Reagents: A Practical Approach* (R. J. K. Taylor, Ed.), 217, Oxford University Press, Oxford, U. K., **1994**.

P. Siemsen, R. C. Livingston, F. Diederich, "Acetylenic Coupling: A Powerful Tool in Molecular Construction," *Angew. Chem. Int. Ed. Engl.* **2000**, *39*, 2632–2657.

16.5

R. F. Heck, "Vinyl Substitutions with Organopalladium Intermediates," in *Comprehensive Organic Synthesis* (B. M. Trost, I. Fleming, Eds.), Vol. 4, 833, Pergamon Press, Oxford, U. K., **1991**.

H.-U. Reißig, "Palladium-Catalyzed Arylation and Vinylation of Olefins," in *Organic Synthesis Highlights* (J. Mulzer, H.-J. Altenbach, M. Braun, K. Krohn, H.-U. Reißig, Eds.), VCH, Weinheim, New York, **1991**, 174–180.

L. E. Overman, "Application of Intramolecular Heck Reactions for Forming Congested Quaternary Carbon Centers in Complex Molecule Total Synthesis," *Pure Appl. Chem.* **1994**, *66*, 1423–1430.

A. de Meijere, F. E. Meyer, "Fine Feathers Make Fine Birds: The Heck Reaction in Modern Garb," *Angew. Chem. Int. Ed. Engl.* **1994**, *33*, 2379–2411.

W. Cabri, I. Candiani, "Recent Developments and New Perspectives in the Heck Reaction," *Acc. Chem. Res.* **1995**, *28*, 2–7.

S. E. Gibson, R. J. Middleton, "The Intramolecular Heck Reaction," *Contemp. Org. Synth.* **1996**, *3*, 447–472.

T. Jeffery, "Recent Improvements and Developments in Heck-Type Reactions and their Potential in Organic Synthesis," in *Advances in Metal-Organic Chemistry* (L. S. Liebeskind, Ed.), JAI Press, Greenwich, **1996**, Vol. 5, 153–260.

M. Shibasaki, C. D. J. Boden, A. Kojima, "The Asymmetric Heck Reaction," *Tetrahedron* **1997**, *53*, 7371–7393.

S. Bräse, A. de Meijere, "Palladium-Catalyzed Coupling of Organyl Halides to Alkenes – The Heck Reaction," in *Metal-catalyzed Cross-coupling Reactions* (F. Diederich, P. J. Stang, Eds.), Wiley-VCH, Weinheim, **1998**, 99–154.

J. T. Link, L. E. Overman, "Intramolecular Heck Reactions in Natural Product Chemistry," in *Metal-catalyzed Cross-coupling Reactions* (F. Diederich, P. J. Stang, Eds.), Wiley-VCH, Weinheim, **1998**, 231–266.

G. T. Crisp, "Variations on a Theme: Recent Developments on the Mechanism of the Heck Reaction and Their Implications for Synthesis," *Chem. Soc. Rev.* **1998**, *27*, 427–436.

Y. Donde, L. E. Overman, "Asymmetic Intramolecular Heck Reactions," in *Catalytic Asymmetric Synthesis* (I. Ojima, Ed.), Wiley-VCH, New York, 2nd ed., **2000**, 675–698.

C. Amatore, A. Jutand, "Anionic Pd(0) and Pd(II) Intermediates in Palladium-Catalyzed Heck and Cross-Coupling Reactions," *Acc. Chem. Res.* **2000**, *33*, 314–321.

I. P. Beletskaya, A. V. Cheprakov, "The Heck Reaction as a Sharpening Stone of Palladium Catalysis," *Chem. Rev.* **2000**, *100*, 3009–3066.

N. J. Whitcombe, K. K. (Mimi) Hii, S. E. Gibson, "Advances in the Heck Chemistry of Aryl Bromides and Chlorides," *Tetrahedron* **2001**, *57,* 7449–7476.

M. Toyota, M. Ihara, "Development of Palladium-Catalyzed Cycloalkenylation and its Application to Natural Product Synthesis," *Synlett* **2002**, 1211–1222.

B. Dounay, L. E. Overman, "The Asymmetric Intramolecular Heck Reaction in Natural Product Total Synthesis," *Chem. Rev.* **2003**, *103,* 2945–2963.

Further Reading

J. F. Hartwig, "Transition Metal Catalyzed Synthesis of Arylamines and Aryl Ethers from Aryl Halides and Triflates: Scope and Mechanism," *Angew. Chem. Int. Ed. Engl.* **1998**, *37*, 2046–2067.

A. Culkin, J. F. Hartwig, "Palladium-Catalyzed α-Arylation of Carbonyl Compounds and Nitriles," *Acc. Chem. Res.* **2003**, *36,* 234–245.

R. G. Jones, H. Gilman, "The Halogen-Metal Interconversion Reaction with Organolithium Compounds," *Org. React*. **1951**, *6*, 339–366.

V. Snieckus, "Combined Directed Ortho Metalation-Cross Coupling Strategies. Design for Natural Product Synthesis," *Pure Appl. Chem*. **1994**, *66*, 2155–2158.

G. Bringmann, R. Walter, R. Weirich, "The Directed Synthesis of Biaryl Compounds: Modern Concepts and Strategies," *Angew. Chem. Int. Ed. Engl.* **1990**, *29*, 977–991.

V. Fiandanese, "Sequential Cross-Coupling Reactions as a Versatile Synthetic Tool," *Pure Appl. Chem.* **1990**, *62*, 1987–1992.

R. Rossi, A. Carpita, F. Bellina, "Palladium- and/ or Copper-Mediated Cross-Coupling Reactions Between 1-Alkynes and Vinyl, Aryl, 1-Alkynyl, 1,2-Propadienyl, Propargyl and Allylic Halides or Related Compounds. A Review," *Org. Prep. Proced. Int*. **1995**, *27*, 127–160.

R. Rossi, F. Bellina, "Selective Transition Metal-Promoted Carbon-Carbon and Carbon-Heteroatom Bond Formation. A Review," *Org. Prep. Proced. Int*. **1997**, *29*, 139–176.

R. Franzen, "The Suzuki, the Heck, and the Stille Reaction; Three Versatile Methods for the Introduction of New C–C Bonds on Solid Support," *Can. J. Chem.* **2000**, *78*, 957–962.

S. Bräse, J. H. Kirchhoff, J. Kobberling, "Palladium-Catalyzed Reactions in Solid Phase Organic Synthesis," *Tetrahedron* **2003**, *59,* 885–939.

F. Littke, G. C. Fu, "Palladium-Catalyzed Coupling Reactions of Aryl Chlorides," *Angew. Chem. Int. Ed. Engl.* **2002**, *41,* 4176–4211.

C. Amatore, A. Jutand, "Role of DBA in the Reactivity of Palladium(0) Complexes Generated *in situ* from Mixtures of $Pd(DBA)_2$ and Phosphines," *Coord. Chem. Rev*. **1998**, *178–180*, 511–528.

K. Osakada, T. Yamamoto, "Mechanism and Relevance of Transmetalation to Metal Promoted Coupling Reactions," *Rev. Heteroatom Chem.* **1999**, *21*, 163–178.

K. Fagnou, M. Lautens, "Halide Effects in Transition Metal Catalysis," Angew. *Chem. Int. Ed. Engl.* **2002**, *41,* 26–47.

Oxidations and Reductions

17

17.1 Oxidation Numbers in Organic Chemical Compounds, and Organic Chemical Redox Reactions

Everybody learns early on how to determine the oxidation states of inorganic compounds by assigning oxidation numbers to the atoms of which they are composed. Let us review the examples of H_2O (Figure 17.1), H_2O_2 (Figure 17.3), and $NH_4^{\oplus}$ (Figure 17.5). After dealing with the inorganic molecules, we will analyze the oxidation numbers of comparable organic molecules, i.e., the oxidation numbers in CH_4 (Figure 17.2), C_2H_6 (Figure 17.4), and CH_3—NH_3^+ (Figure 17.6). It will be seen that the organic chemist's approach is entirely the same as the familiar approach taken in inorganic chemistry.

The principles for the assignment of oxidation numbers to atoms in covalently bound molecules are as follows:

Determination of the Oxidation Numbers in Molecules That Contain Covalent Bonds Only

1. In the case of a covalent bond between different atoms A and B, one assumes that all the bonding electrons are localized on the more electronegative atom B. Therefore, a diatomic molecule AB or a diatomic substructure AB of a molecule is considered
   - as $A^{\oplus}B^{\ominus}$, if it contains an A—B single bond,
   - as $A^{2\oplus}B^{2\ominus}$, if it contains an A=B double bond, and
   - as $A^{3\oplus}B^{3\ominus}$, if it contains an A≡B triple bond.
2. In the case of a covalent bond between *identical* atoms A, one assumes that 50% of the bonding electrons are localized on each atom A. In a diatomic molecule AA or in a diatomic substructure AA of a molecule, one therefore assigns
   - the two bonding electrons of an A—A single bond to the bonded atoms in the form $A^{\cdot}\,A^{\cdot}$,
   - the four bonding electrons of an A=A double bond to the bonded atoms in the form $A^{\cdot\cdot}\,A^{\cdot\cdot}$, and
   - the six bonding electrons of an A≡A triple bond to the bonded atoms in the form $A^{\cdot\cdot\cdot}\,A^{\cdot\cdot\cdot}$.
3. For each atom of the molecule under consideration, one determines the charge on the atom based on the foregoing electron assignments. This charge is the **oxidation number. It is not the same as formal charge, which based on valence electron count and a different formalism.** The sum of the oxidation numbers of all atoms must be zero for an uncharged molecule, and the sum of the oxidation numbers of all atoms equals the total charge of an ion.

Figure 17.1 shows how the oxidation numbers of the atoms in H_2O are obtained in this way. According to the foregoing rules, the bonding electrons of both O—H bonds are assigned to

Bruckner R (author), Harmata M (editor) In: *Organic Mechanisms – Reactions, Stereochemistry and Synthesis*
Chapter DOI: 10.1007/978-3-642-03651-4_17, © Springer-Verlag Berlin Heidelberg 2010

the more electronegative oxygen, not to the hydrogens. The O atom thus possesses oxidation number –2, and both H atoms have oxidation number +1. The formal charge on each atom is zero.

1. Step: Set oxidation state = + 1

H–O–H

2. Step: Set oxidation state = – 2

Fig. 17.1. Determination of the oxidation numbers in H_2O.

The same procedure is used in Figure 17.2 to determine the oxidation numbers in methane, the simplest organic molecule. Since carbon has a higher electronegativity than hydrogen, the two bonding electrons of each C—H bond are assigned to carbon. Hence, the oxidation number of carbon is –4 and that of all H atoms is +1.

1. Step: Set oxidation state = + 1

CH_4

2. Step: Set oxidation state = – 4

Fig. 17.2. Determination of the oxidation numbers in CH_4.

For the assignment of the oxidation numbers in hydrogen peroxide (Figure 17.3), the two electrons of each O—H bond count only for the O atoms, and the two bonding electrons of the O—O bond count 50% to each O atom. In this way, one finds the oxidation numbers in +1 for both H atoms and –1 for both O atoms.

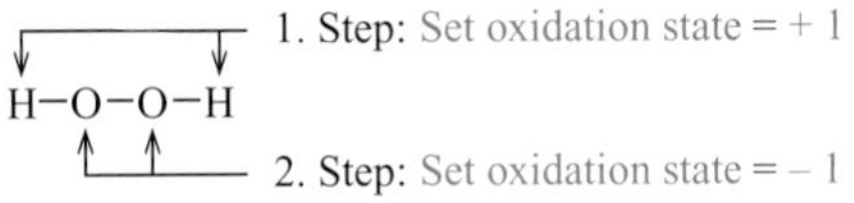

1. Step: Set oxidation state = + 1

H–O–O–H

2. Step: Set oxidation state = – 1

Fig. 17.3. Determination of the oxidation numbers in H_2O_2.

Applying the analogous approach to ethane (Figure 17.4) results in the oxidation number of +1 for each H atom and –3 for each C atom.

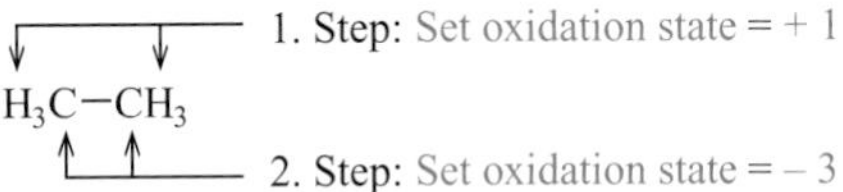

Fig. 17.4. Determination of the oxidation numbers in C_2H_6.

Figure 17.5 reminds us that the oxidation numbers of the atoms in ions and molecules are determined in the same way as in inorganic chemistry. In the ammonium ion, the four H atoms again possess the oxidation number +1, and the N atom has the oxidation number –3. But the same atom has a formal charge of +1.

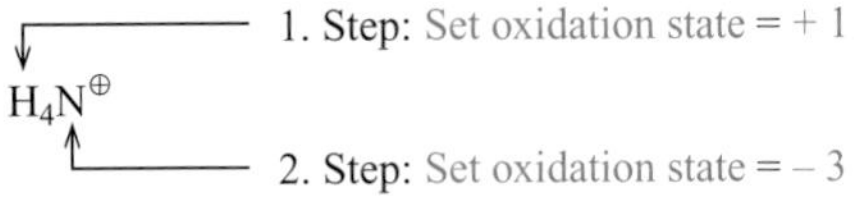

1. Step: Set oxidation state = + 1

$H_4N^{\oplus}$

2. Step: Set oxidation state = – 3

Fig. 17.5. Determination of the oxidation numbers in NH_4^+.

The simplest organic analog of the ammonium ion is the methylammonium ion (Figure 17.6). If one assigns the bonding electrons of the C—H bonds to carbon and those of the N—H and

C—N bonds to nitrogen, one obtains oxidation numbers of +1 for each of the H atoms, –3 for the N atom, and –2 for the C atom.

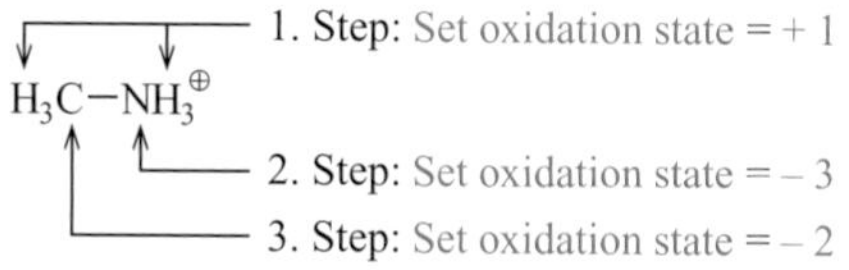

1. Step: Set oxidation state = + 1

$H_3C{-}NH_3^{\oplus}$

2. Step: Set oxidation state = – 3

3. Step: Set oxidation state = – 2

Fig. 17.6. Determination of the oxidation numbers in $CH_3{-}NH_3^+$.

The procedure exemplified for ethane (Figure 17.4) can be employed to assign oxidation numbers to the C and H atoms of all hydrocarbons. The oxidation number of every H atom is +1. However, the oxidation numbers in the C atoms depend on the structure, and they are summarized in Figure 17.7. The C atom of a methyl group always possesses the oxidation number –3 in any hydrocarbon. The C atom of a methylene group always possesses the oxidation number –2 in any hydrocarbon, the C atom of a methyne group always possesses –1, and every quaternary C atom possesses the oxidation number 0.

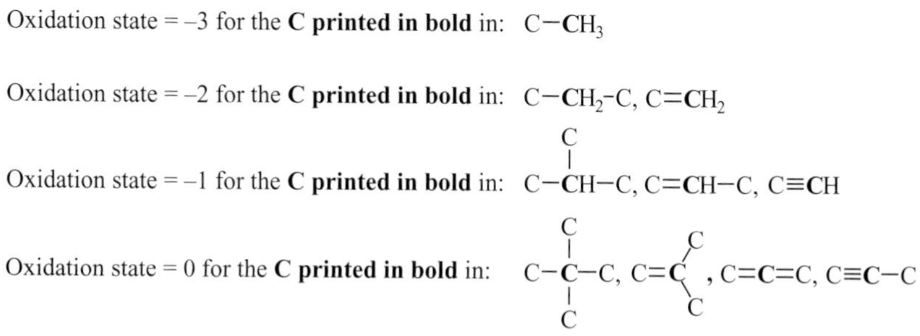

Oxidation state = –3 for the **C printed in bold** in: C−**C**H_3

Oxidation state = –2 for the **C printed in bold** in: C−**C**H_2-C, C=**C**H_2

Oxidation state = –1 for the **C printed in bold** in: C−**C**H(−C)−C, C=**C**H−C, C≡**C**H

Oxidation state = 0 for the **C printed in bold** in: C−**C**(C)(C)−C, C=**C**(C)C, C=**C**=C, C≡**C**−C

Fig. 17.7. Oxidation numbers of C atoms in selected hydrocarbon substructures.

With the data in Figure 17.7, oxidation numbers can be assigned to the C atoms of the two isomeric butenes of Figure 17.8. Hence, 1-butene possesses the oxidation number –3 at one C atom, the oxidation number –2 at two C atoms, and the oxidation number –1 at one C atom. On the other hand, 2-butene consists of two sets of two C atoms with oxidation numbers –3 and –1, respectively.

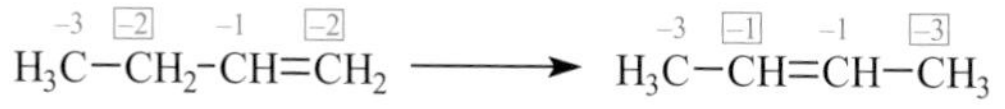

$$\overset{-3}{H_3C}-\overset{\boxed{-2}}{CH_2}-\overset{-1}{CH}=\overset{\boxed{-2}}{CH_2} \longrightarrow \overset{-3}{H_3C}-\overset{\boxed{-1}}{CH}=\overset{-1}{CH}-\overset{\boxed{-3}}{CH_3}$$

Fig. 17.8. Oxidation numbers of the C atoms of 1- and 2-butene.

This difference has one irritating consequence: the isomerization of 1-butene → 2-butene would change the oxidation numbers of two atoms. This isomerization thus would constitute a redox reaction or, more specifically, a redox disproportionation. That result, however, is not compatible with "good common sense."

What causes this problem? This is a formalism, so there is no problem. The assignment of oxidation numbers in organic chemistry should not be overly burdened by questions of whether the procedure really makes sense. The important feature of the butenes of Figure 17.8 lies with the fact that *the C atoms in the butenes on average possess the same oxidation number.* The average oxidation numbers are (–3 –2 –1 –2)/4 = –2 for 1-butene and (–3 –1 –1 – 3)/4 = –2 for 2-butene. The isomerization 1-butene → 2-butene leaves the average oxidation numbers of the atoms invariant, and the isomerization of butene rightly no longer needs to be viewed as a redox reaction. It is best to remember the following:

Two organic chemical compounds possess the *same* oxidation state if the average oxidation numbers of their C atoms are the same *and* if any heteroatoms that might be present possess their usual oxidation numbers (Li, +1; Mg, +2; B, +3; N and P, –3; O and S, –2; and –1 for halogen atoms).

The six C_3 skeletons shown in Figure 17.9 all have an average oxidation number of –1.33 of their C atoms (i.e, overall oxidation state of all carbon atoms is -4). Accordingly, all these compounds are representatives of the same oxidation state.

$H_3C{-}C{\equiv}CH$ (–3, 0, –1) $H_2C{=}C{=}CH_2$ (–2, 0, –2) cyclopropene H_2C (–2), $HC{=}CH$ (–1, –1)

$H_3C{-}C(Br){=}CH_2$ (–3, +1, –2) $H_3C{-}CH{-}CH_2$ epoxide (–3, 0, –1) $H_3C{-}C({=}O){-}CH_3$ (–3, +2, –3)

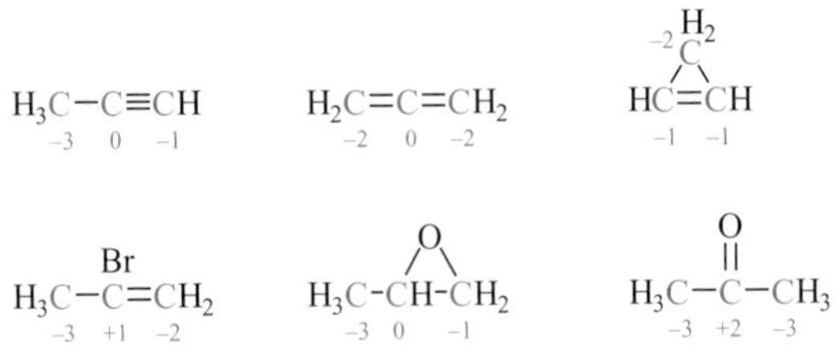

Fig. 17.9. A selection of compounds with the same average oxidation number of –1 ⅓ at every atom: oxidation number of the O atoms, –2; oxidation number of the Br atoms, –1.

Based on the preceding rule, the following can be stipulated:

The Terms "Oxidation" and "Reduction" in Organic Chemistry

Reactions that increase the average oxidation number of the C atoms of a substrate are **oxidations** (loss of electrons). The same is true if the oxidation number of one of the heteroatoms increases. Conversely, reactions that decrease the average oxidation number of the C atoms in the substrate or decrease the oxidation number of one of the heteroatoms are **reductions** (gain in electrons).

The columns of Table 17.1 contain characteristic substructures ordered by common average carbon oxidation numbers. Also, the average carbon oxidation number increases in going from left to right in this table. Accordingly, reactions are oxidations if they convert a substructure of one column into a substructure of a column that is further to the right. The opposite is true for reductions.

An analogous listing of N-containing substructures of organic compounds is given in Table 17.2.

Again, these substructures are organized in such a way that the oxidation number in the N atom or the average oxidation number of the N atoms, respectively, increases from left to right. From this it follows that here, too, oxidations are reactions that convert a substrate into a compound that occurs further to the right in the table, and vice versa for reductions.

Tab. 17.1 Organic Chemical Redox Reactions I: Change of the Average Oxidation Numbers of C Atoms*0

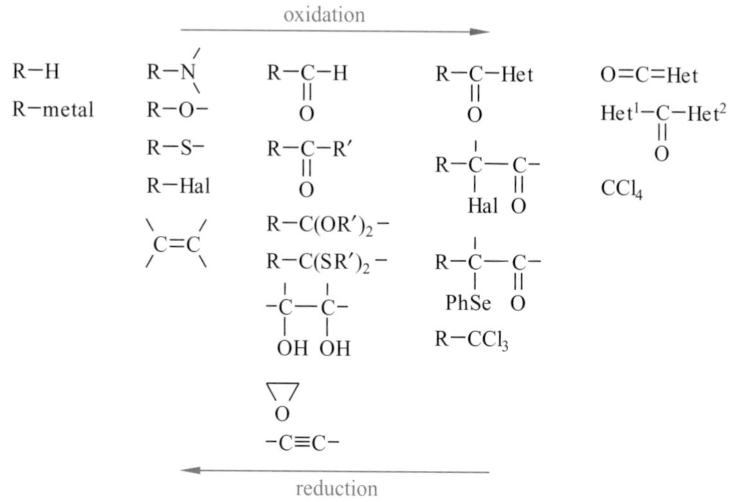

oxidation →

R−H	R−N<	R−C(=O)−H	R−C(=O)−Het	O=C=Het
R−metal	R−O−	R−C(=O)−R′	R−C(Hal)−C(=O)−	Het^1−C(=O)−Het^2
	R−S−	R−C(OR′)$_2$−	R−C(PhSe)−C(=O)−	CCl_4
	R−Hal	R−C(SR′)$_2$−	R−CCl_3	
	>C=C<	−C(OH)−C(OH)−		
		epoxide (O)		
		−C≡C−		

← reduction

Tab. 17.2 Organic Chemical Redox Reactions II: Change of the Average Oxidation Numbers of N Atoms

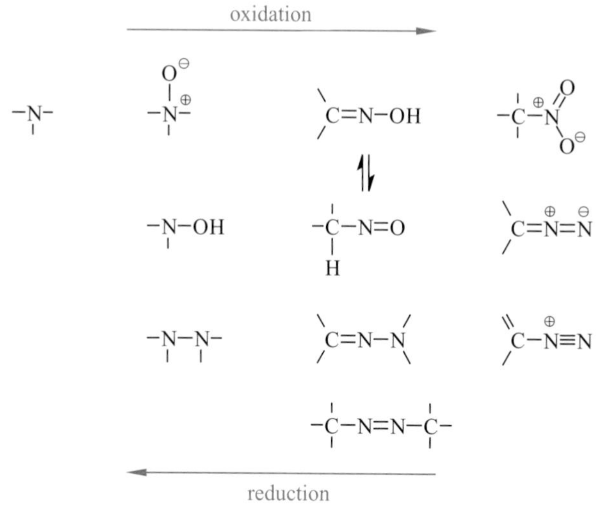

oxidation →

−N<	−N⊕(−O⊖)<	>C=N−OH ⇅ −C(H)−N=O	−C−N⊕(=O)−O⊖
	−N(−OH)−	>C=N−N<	>C=N⊕=N⊖
	−N−N−	−C−N=N−C−	>C−N⊕≡N

← reduction

17.2 Cross-References to Redox Reactions Already Discussed in Chapters 1–16

Many reactions that fit the definition of an organic chemical redox reaction (Section 17.1) have already been presented in Chapters 1–16. The presentation of these reactions in various other places—without alluding at all to their redox character—was done because they follow mechanisms that were discussed in detail in the respective chapters or because these reactions showed chemical analogies to reactions discussed there. Tables 17.3 and 17.4 provide cross-references to all oxidations and reductions discussed thus far.

Tab. 17.3 Compilation of Oxidation Reactions Presented Elsewhere in This Book

$R{-}Hg{-}O_2CCF_3 \xrightarrow[NaBH_4]{O_2,} R{-}OH$	Fig. 1.16
$R{-}H \xrightarrow{Hal_2} R{-}Hal$	Section 1.7
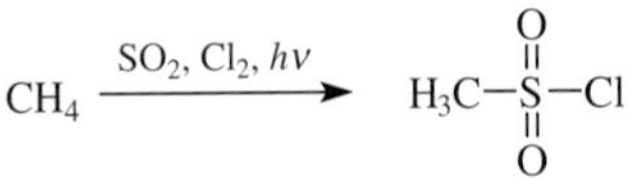 $CH_4 \xrightarrow{SO_2,\ Cl_2,\ h\nu} H_3C{-}SO_2{-}Cl$	Fig. 1.35
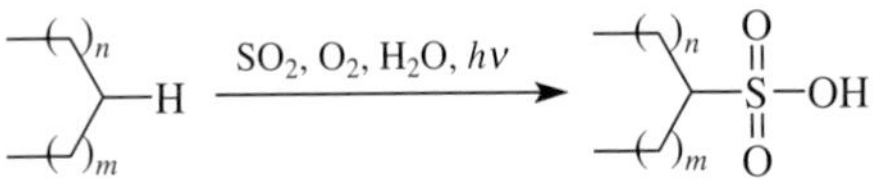 $(\ )_n(\ )_m CH{-}H \xrightarrow{SO_2,\ O_2,\ H_2O,\ h\nu} (\ )_n(\ )_m CH{-}SO_2{-}OH$	Fig. 1.36
$R{-}H \xrightarrow{O_2} R{-}O{-}O{-}H$	Fig. 1.37, 1.38
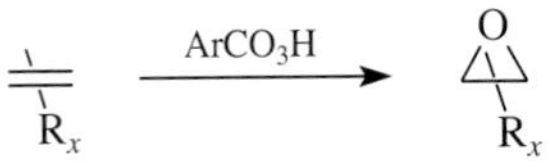 alkene (R_x) $\xrightarrow{ArCO_3H}$ epoxide (R_x)	Fig. 3.19
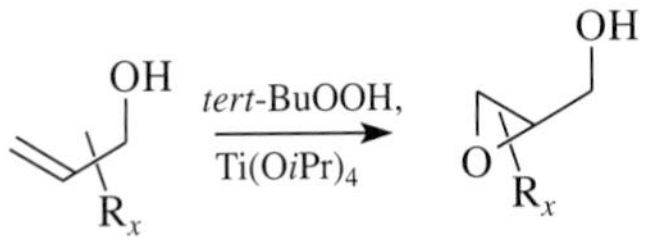 allylic alcohol (R_x, OH) $\xrightarrow[Ti(O\mathit{i}Pr)_4]{\mathit{tert}\text{-}BuOOH,}$ epoxy alcohol (R_x, OH)	Section 3.4.6
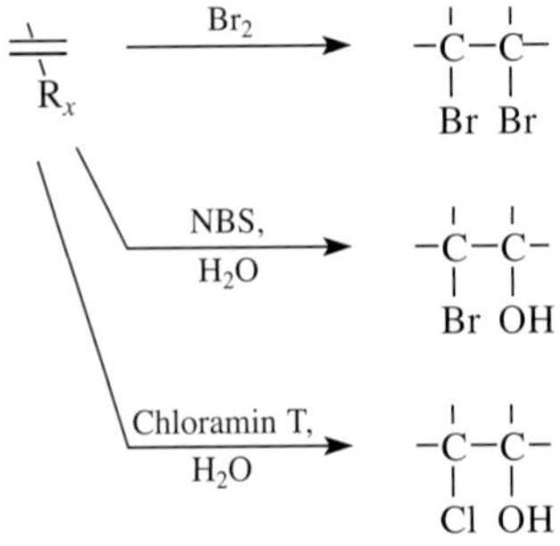 alkene (R_x) $\xrightarrow{Br_2}$ $-C(Br){-}C(Br)-$	Section 3.5.1
alkene (R_x) $\xrightarrow[H_2O]{NBS,}$ $-C(Br){-}C(OH)-$	Fig. 3.43–3.45
alkene (R_x) $\xrightarrow[H_2O]{\text{Chloramin T,}}$ $-C(Cl){-}C(OH)-$	Fig. 3.46

Tab. 17.3 (continued)

HO(O)C–CH₂CH₂–CH=CH–CH₃ $\xrightarrow{I_2}$ iodolactone	Fig. 3.47
cyclohexenyl ethanol $\xrightarrow{NBS}$ bromo-tetrahydrobenzofuran (Br)	Fig. 3.47
$R{-}Se{-}Ph \xrightarrow{H_2O_2} R{-}Se(=O){-}Ph$	Fig. 4.10
$Ar{-}H \xrightarrow[MHal_\chi]{Hal_2,} Ar{-}Hal$	Section 5.2.1
$Ar{-}H \xrightarrow{H_2SO_4} Ar{-}SO_3H$	Section 5.2.2
$Ar{-}H \xrightarrow[H_2SO_4]{HNO_3,} Ar{-}NO_2$	Section 5.2.3
$Ar{-}H \xrightarrow{N{\equiv}N^{\oplus}{-}Ar'} Ar{-}N{=}N{-}Ar'$	Fig. 5.24, 5.25
$Ar{-}Metall \xrightarrow{E^{\oplus}} Ar{-}E$	Section 5.3.1, 5.3.2
$Ar{-}B(OMe)_2 \xrightarrow[HOAc]{H_2O_2,} Ar{-}OH$	Fig. 5.48, 14.41
EtO(O)C–CH₂–C(O)–CH₃ $\xrightarrow{Br_2}$ EtO(O)C–CHBr–C(O)–CH₃	Fig. 12.4
PhC(O)CH₃ $\xrightarrow[cat.\ AlCl_3]{Br_2,}$ PhC(O)CH₂Br	Fig. 12.5
HO(O)C–CHR–C(O)OH $\xrightarrow[\varnothing]{Br_2,}$ R–CHBr–C(O)OH	Fig. 12.6
R–CH₂–C(O)OH $\xrightarrow[cat.\ PBr_3]{Br_2,}$ R–CHBr–C(O)OH	Fig. 12.7
R–CH₂–C(O)OH $\xrightarrow[stoich.\ PBr_3]{Br_2,}$ R–CHBr–C(O)Br	Fig. 12.8

Tab. 17.3 (continued)

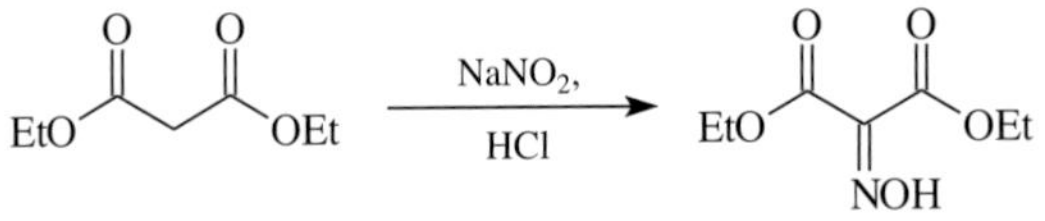

EtO–CO–CH$_2$–CO–OEt → EtO–CO–C(=NOH)–CO–OEt	$NaNO_2$, HCl	Fig. 12.9
cyclohexanone → 2-hydroxycyclohex-2-enone ⇌ cyclohexane-1,2-dione	SeO_2, aqueous dioxane	Fig. 12.10
Me_3SiO-dimethylcyclohexene → dimethylcyclohexenone	cat. $PdCl_2$, stoich. DDQ (Cl, Cl, CN, CN, O, O)	Fig. 12.24
2-methyl-1,3-dioxolane → 2-(bromomethyl)-1,3-dioxolane (Br)	pyridinium ($\overset{\oplus}{N}$H) $Br_3^{\ominus}$	Fig. 12.25
camphor → camphorsulfonic acid (SO_3H)	H_2SO_4, Ac_2O	Fig. 14.9
dibromocamphor (Br, Br) → tribromocamphor (Br, Br, Br)	Br_2, $ClSO_3H$	Fig. 14.10
sodium enolate ($O^{\ominus}$ $Na^{\oplus}$) of formyl cyclododecanone → diazo ketone ($N^{\oplus}$, $N^{\ominus}$)	$\overset{\ominus}{N}=\overset{\oplus}{N}=N-SO_2Tol$	Fig. 14.29
Ph–C(CH$_3$)$_2$–O–OH → Ph–OH + acetone	H_2SO_4	Fig. 14.33
$R^1-C(=O)-R^2$ → $R^1-C(=O)-OR^2$	$ArCO_3H$	Section 14.4.2
cyclobutanone–R_x → γ-lactone–R_x	H_2O_2 or *tert*-BuOOH	Section 14.4.2
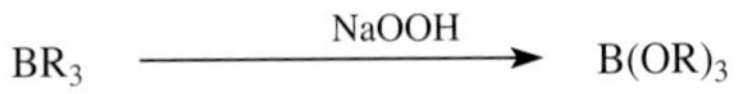 BR_3 → $B(OR)_3$	NaOOH	Fig. 14.40

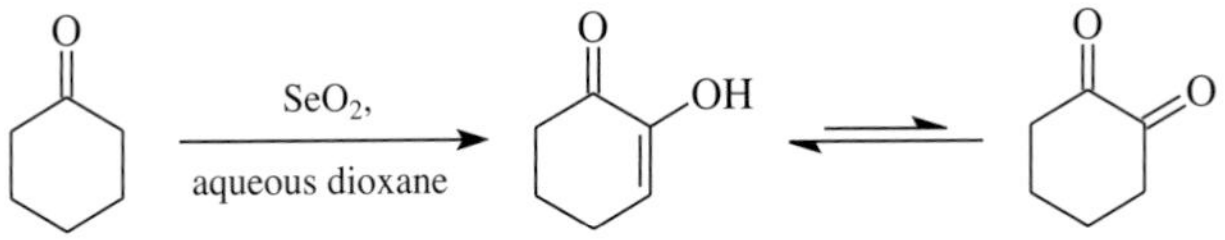

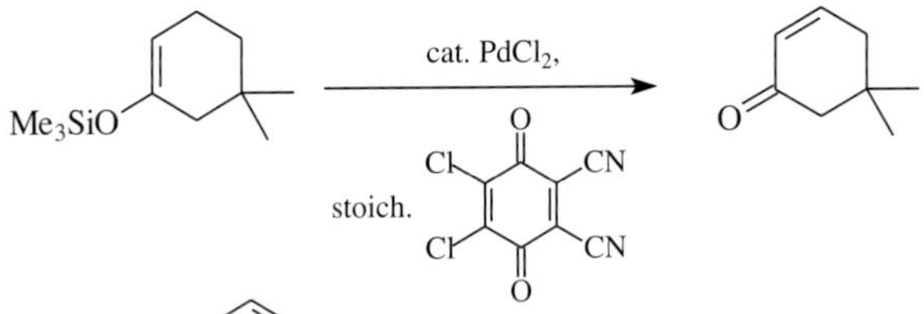

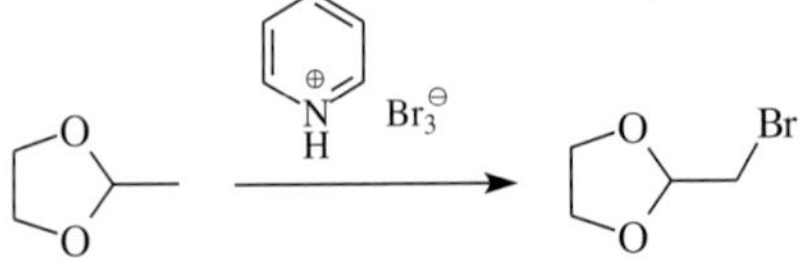

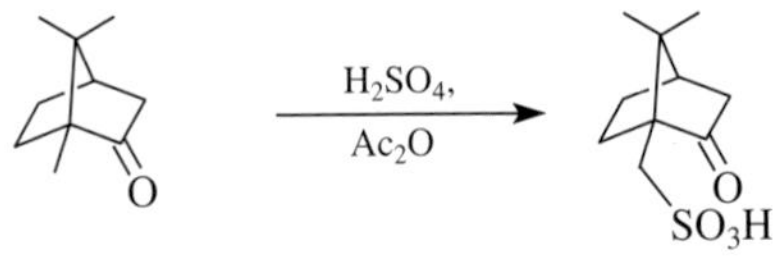

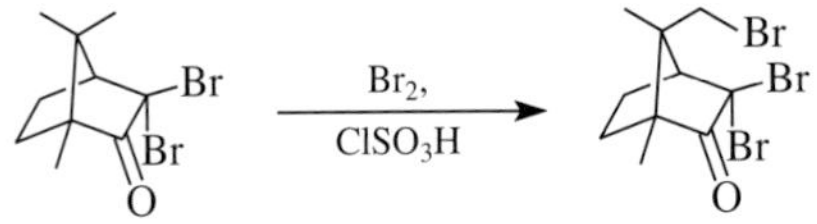

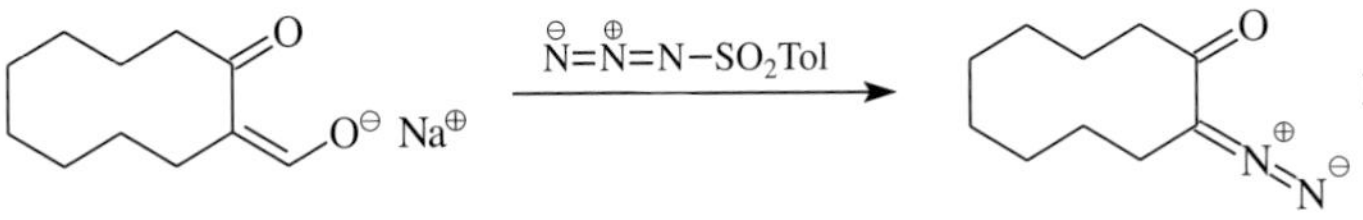

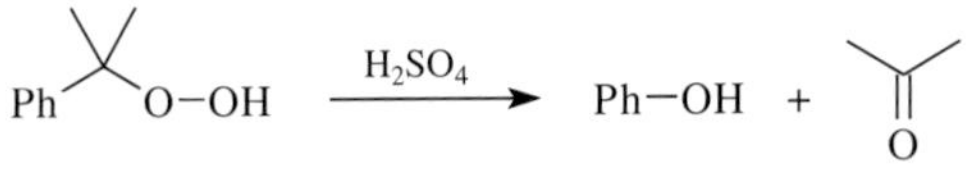

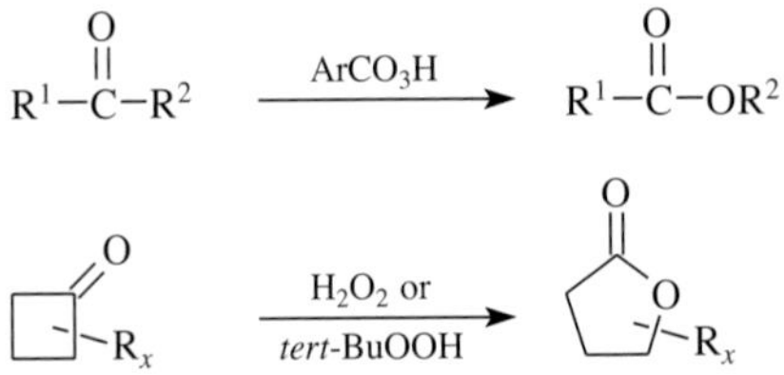

Tab. 17.3 (continued)

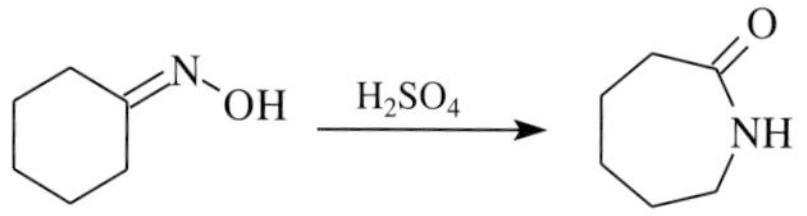 Cyclohexanone oxime $\xrightarrow{H_2SO_4}$ caprolactam (NH, O)	Fig. 14.42
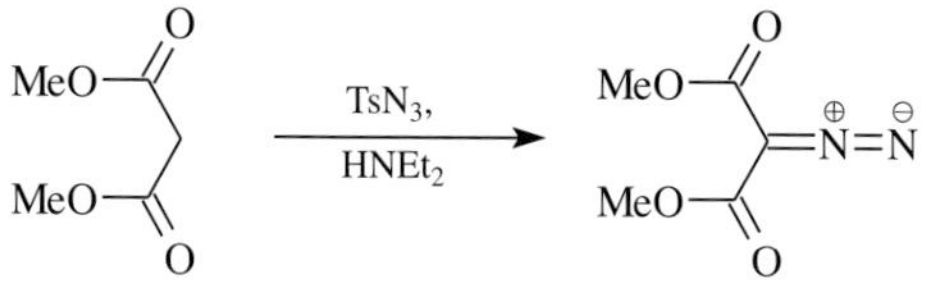 Dimethyl malonate (MeO, O) $\xrightarrow{TsN_3,\ HNEt_2}$ dimethyl diazomalonate ($=\overset{\oplus}{N}=\overset{\ominus}{N}$)	Fig. 15.42
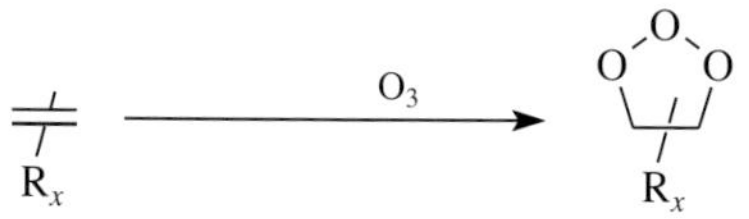 Alkene (R_x) $\xrightarrow{O_3}$ ozonide (O–O–O, R_x)	Section 15.5.5
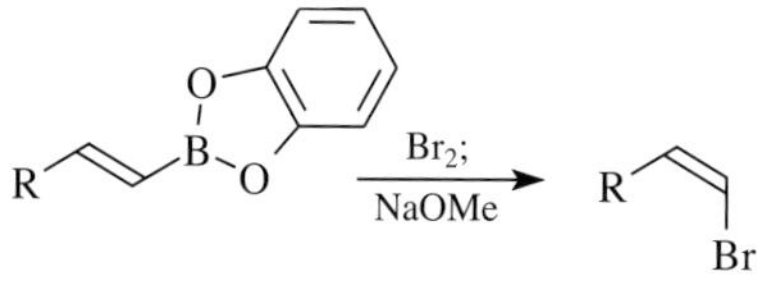 R–CH=CH–B(catecholate) $\xrightarrow{Br_2;\ NaOMe}$ R–CH=CH–Br	Fig. 16.15
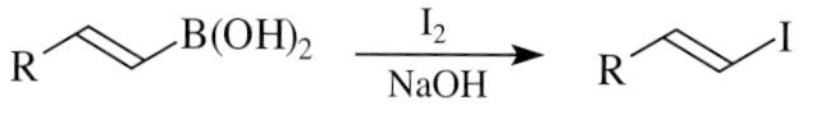 R–CH=CH–B(OH)$_2$ $\xrightarrow{I_2,\ NaOH}$ R–CH=CH–I	Fig. 16.15
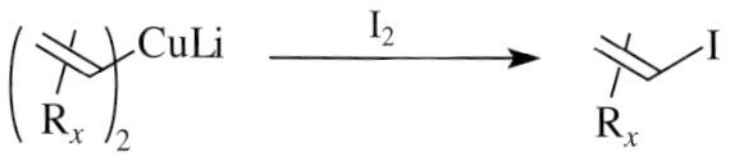 (alkenyl(R_x))$_2$CuLi $\xrightarrow{I_2}$ alkenyl–I (R_x)	Fig. 16.17
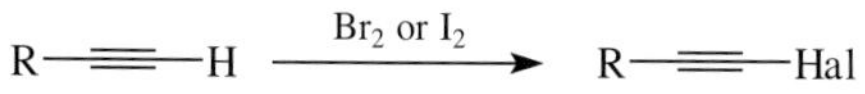 R–≡–H $\xrightarrow{Br_2\ or\ I_2}$ R–≡–Hal	Section 16.2

Tab. 17.4 Compilation of Reduction Reactions Presented Elsewhere in This Book

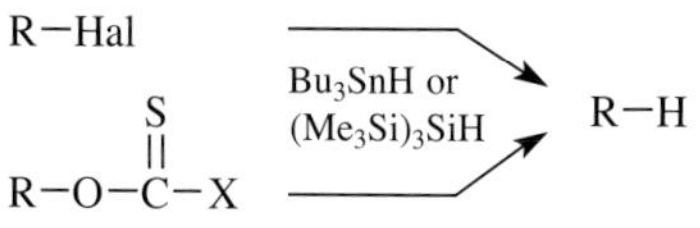 R–Hal or R–O–C(=S)–X $\xrightarrow{Bu_3SnH\ or\ (Me_3Si)_3SiH}$ R–H	Section 1.10.1
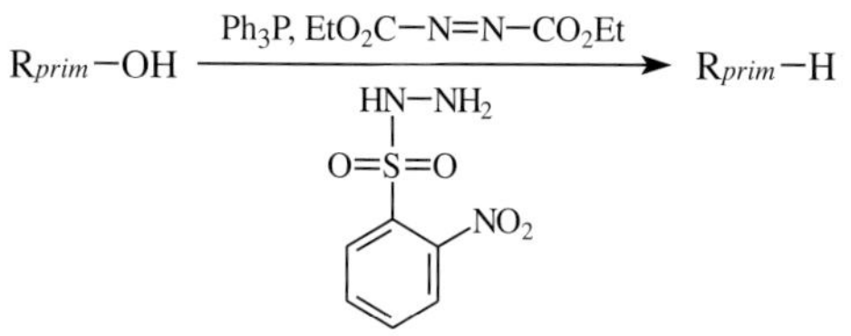 R_{prim}–OH $\xrightarrow{Ph_3P,\ EtO_2C–N=N–CO_2Et;\ HN–NH_2(O=S=O)(2\text{-}NO_2C_6H_4)}$ R_{prim}–H	Fig. 1.48 and Fig. 2.38
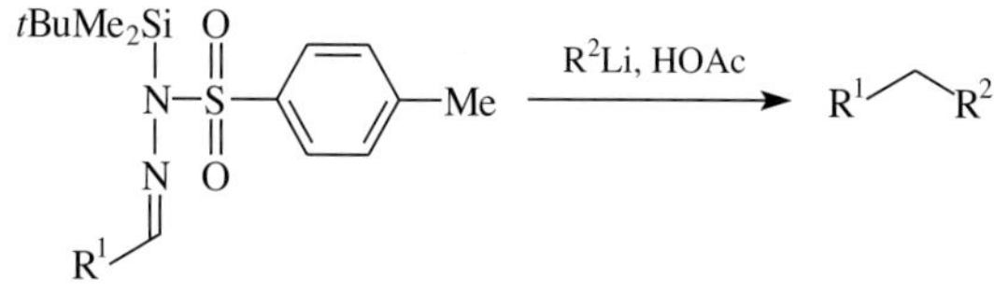 tBuMe$_2$Si–N(–S(=O)$_2$–C$_6$H$_4$–Me)–N=CH–R^1 $\xrightarrow{R^2Li,\ HOAc}$ R^1–CH$_2$–R^2	Fig. 1.49

Tab. 17.4 (continued)

Reaction	Reagent	Product	Reference
alkene (R_x)	$H-BL_n$	$-C(H)-C(BL_n)-$	Fig. 3.21
alkene (R_x)	H_2, Pd/C	$-C(H)-C(H)-$	Fig. 3.28
1-pyrazoline (N=N)	Δ	cyclopropane	Fig. 4.1
1,3-dibromopropane (Br, Br)	Mg	cyclopropane	Fig. 4.1
1,4-dibromobut-2-ene (R_x), Br	Zn	1,3-butadiene (R_x)	Fig. 4.1
1,2-dibromoethane (Br, Br)	Mg	ethylene	Section 4.7.1
$Cl_3C-C(=O)-Cl$	Zn	$Cl_2C=C=O$ (see also Figures 15.33, 15.34)	Section 4.7.1
Br–C–C–OR (R_x)	Mg or Li	alkene (R_x)	Section 4.7.1
R^1–CH($PhSO_2$)–CH(OAc)–R^2	$NaHg_x$	R^1–CH=CH–R^2	Fig. 4.40 and Fig. 17.85
R_x–aryl–I	*i*Pr−MgBr	R_x–aryl–MgHal	Fig. 5.46
R_x–aryl–I(Br)	1 *n*Bu−Li or 2 *tert*-Bu−Li	R_x–aryl–Li	Fig. 5.46
$Ar-\overset{\oplus}{N}\equiv N\ X^{\ominus}$	H_3PO_2/Cu_2O	Ar−H	Fig. 5.55
$R-C(=O)-Het$	„$H^{\ominus}$"	$R-C(=O)-H$	Section 6.5.2
$R^1-C(=O)-R^2$	„$H^{\ominus}$"	$R^1-C(OH)(H)-R^2$	Section 10.2–10.4, Fig. 11.11

Tab. 17.4 (continued)

Reaction	Reference
O → (iPrMgBr) → OH	Fig. 10.29
$R_{funct}-I$ → (Zn) → $R_{funct}-ZnI$	Fig. 10.37
R → ($HBEt_2$; $ZnEt_2$) → $(R)_2Zn$	Fig. 10.38, 10.49
O → (Li, liquid NH_3, THF) → $Li^{\oplus}$ $^{\ominus}O$, H	Fig. 13.20
O → ($Li^{\oplus}$ $^{\ominus}HB(sec\text{-}Bu)_3$) → $Li^{\oplus}$ $^{\ominus}O$	Fig. 13.20
OH, Br, Br, n → (2 n-BuLi) → O, n+1 or O, n+1	Fig. 14.25
R, Br, Br → (2 n-BuLi) → R≡	Fig. 14.32
R_x, O, N → ($LiAlH_4$) → R_x, OH, NH_2	Fig. 15.46
R_x, O, N → (H_2, Raney nickel, $B(OH)_3$) → R_x, OH, O	Fig. 15.46
$Ph-\overset{\oplus}{N}\equiv N\ X^{\ominus}$ → (NaN_3) → $Ph-N=\overset{\oplus}{N}=\overset{\ominus}{N}$	Fig. 15.49
Ph–I → (Cu bronze) → Ph–Ph	Fig. 16.5
OTf, R_x → (iPrMgCl, $Ni(acac)_2$) → H, R_x	Fig. 16.14

Tab. 17.4 (continued)

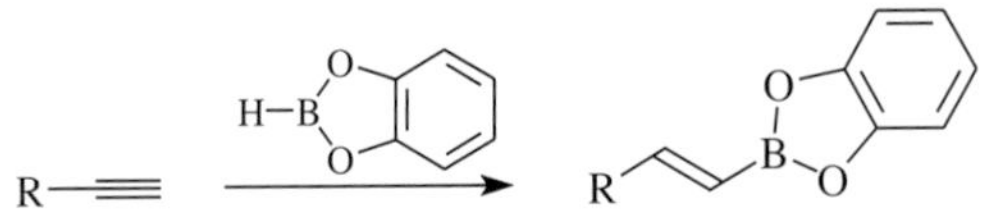

R—≡ —(H–B(catecholborane))→ R–CH=CH–B(catecholborane)	Fig. 16.15
R–CH=CH–Br —(n-BuLi)→ R–CH=CH–Li	Fig. 16.16

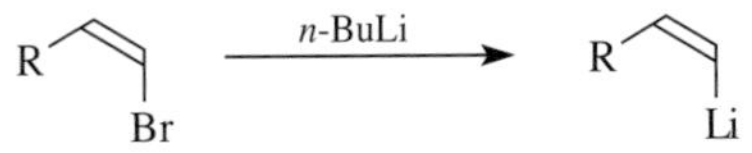

The plethora of entries in Tables 17.3 and 17.4 emphasizes that organic chemical redox reactions are not limited to one mechanism and are not even based on a small number of mechanistic principles. Hence, one should not expect any mechanistic homogeneity among the reactions to be discussed in Chapter 17. Sections 17.3 (oxidations) and 17.4 (reductions) are thus not organized on the basis of mechanistic considerations. Instead, the ordering principle reflects preparative aspects: Which classes of compounds can be oxidized or reduced into which other classes of compounds, and how can these transformations be accomplished?

17.3 Oxidations

17.3.1 Oxidations in the Series Alcohol → Aldehyde → Carboxylic Acid

Survey

For the oxidation of primary or secondary alcohols on a laboratory scale, one usually employs one of the reagents listed in Table 17.5. The reactivity of these six reagents toward alcohols and aldehydes can be summarized as follows:

1) *All* reagents listed in Table 17.5 can be used for the oxidation of secondary alcohols to ketones.
2) Primary alcohols can be oxidized to the respective carboxylic acids. These oxidations can be achieved with aqueous potassium dichromate ($K_2Cr_2O_7$) or the **Jones reagent** as chromium(VI)-containing oxidizing agents, and with aqueous ruthenium tetroxide (RuO_4) as a chromium-free oxidizing agent.
3) Alternatively, it is possible to oxidize a primary alcohol no further than to give the aldehyde. This can be accomplished with the **Collins reagent**, PCC, PDC—all of them being Cr(VI)-containing oxidizing agents—as well as with **activated dimethyl sulfoxide**. Several more methods have been developed, as the transformation primary alcohol → aldehyde is of great preparative importance and because all four oxidation methods that have

Tab. 17.5 Standard Reagents for Oxidations in the Series Alcohol → Aldehyde → Carboxylic Acid and Alcohol → Ketone

Oxidizing agent	Can be used in selective oxidations: $R^1CH(OH)R^2$ → $R^1C(=O)R^2$	RCH_2OH → $RCHO$	RCH_2OH → $RCOOH$	$RCHO$ → $RCOOH$	Mechanistic details and/or example of the respective oxidation
$\overset{+6}{K_2Cr_2O_7}$, dilute H_2SO_4	✓	no	✓	✓	Fig. 17.10, 17.11
$\overset{+8}{RuO_4}$ (in situ from cat. $\overset{+3}{RuCl_3}$ + stoich. $NaIO_4$)	✓	no	✓	✓	Fig. 17.12
$\overset{+6}{CrO_3}$, dilute H_2SO_4, acetone [1]	✓	sometimes[10]	✓	✓	Fig. 17.10, 17.11
Pyridine $\cdot \frac{1}{2}$ $\overset{+6}{CrO_3}$ [2]	✓	✓	no	no	Fig. 17.10
Pyridinium ($N^{\oplus}$–H) $Cl-\overset{+6}{Cr}O_3^{\ominus}$ (≡ PCC[3])	✓	✓	no	no	Fig. 17.10
(Pyridinium $N^{\oplus}$–H$)_2$ $\overset{+6}{Cr_2O_7}{}^{2\ominus}$ (≡ PDC[4])	✓	✓	no	no	Fig. 17.10
$H_3C-\overset{\pm 0}{S}(=O)-CH_3$, $Cl-C(=O)-C(=O)-Cl$; NEt_3 [5]	✓	✓	no	no	Fig. 17.13
Benzo-fused iodolactone with $\overset{+3}{I}(OAc)_3$ (AcO, AcO, OAc) [6]	✓	✓	no	no	Fig. 17.14
cat. $Pr_4N^{\oplus}$ $\overset{+7}{Ru}O_4^{\ominus}$ (TPAP[7]), stoich. NMO	✓	✓	no	no	Fig. 17.15
cat. 2,2,6,6-tetramethylpiperidine $N-\overset{-1}{O}^{\bullet}$ (TEMPO[8]), stoich. NaOCl	✓	✓	no	no	Fig. 17.16
$Na\overset{+3}{Cl}O_2$ + another reducing or oxidizing agent[9]	no	no	no	✓	Fig. 17.17
$\overset{+4}{MnO_2}$, NaCN, MeOH	usually no[11]	usually no[11]	no	✓ [12]	Fig. 17.18

[1]Jones reagent, [2]Collins reagent, [3]PCC, pyridinium chlorochromate, [4]PDC, pyridinium dichromate, [5]Swern oxidation, [6]Dess–Martin periodinane, [7]TPAP, tetrapropylammonium perrruthenate, [8]TEMPO, tetramethylpiperidine-N-oxyl, [9]The second redox-active agent destroys the reduction product of sodium chlorite, i.e., the initially formed sodium hypochlorite. [10]For R = aryl or alkenyl (in the latter case, *cis* → *trans* and *E* ⇌ *Z*-isomerations are possible). [11]MnO_2 can be used to oxidize allyl alcohols to give α,β-unsaturated carbonyl compounds and to oxidize benzyl alcohols to form aromatic aldehydes or ketones. [12]The carboxylic methyl ester is formed instead of the free carboxylic acid.

so far been introduced may very well fail in particular cases. The most three relevant of these have been summed up in Table 17.5 in the last but one box and will be dealt with in more detail in Figures 17.14–17.16 below.

4) The oxidation of primary alcohols *with $K_2Cr_2O_7$ in aqueous solution* to nothing but the aldehyde, (i.e., without further oxidation to the carboxylic acid) is possible only if a volatile aldehyde results and is distilled off as it is formed. This is the only way to prevent the further oxidation of the aldehyde in the (aqueous) reaction mixture. Selective oxidations of primary alcohols to aldehydes with the Jones reagent succeed only for allylic and benzylic alcohols. Otherwise, the Jones reagent directly converts alcohols into carboxylic acids (see above).
5) Aside from $K_2Cr_2O_7$ in aqueous H_2SO_4 and the Jones reagent (the first and second box in Table 17.5), there also exist three chromium-free methods (second entry and lowest box in Table 17.5) for the oxidation of aldehydes to carboxylic acids. This method is of interest because Cr(VI) compounds are recognized to be carcinogens. For example, many aldehydes can be oxidized "chromium-free" to carboxylic acids with inexpensive sodium chlorite ($NaClO_2$). Conjugated aldehydes can be oxidized "chromium-free" with manganese dioxide/sodium cyanide in methanol to give the methyl ester of the respective carboxylic acid.

In the following passages and related Figures 17.10–17.18, you will learn more about the course of the oxidations and the reasons for the chemoselectivities listed in Table 17.5.

Cr(VI) Oxidation of Alcohol and Aldehydes

The oxidation of alcohols to carbonyl compounds with Cr(VI) occurs via chromium(VI) acid monoesters (**A** in Figure 17.10). These esters yield chromium(IV) acid by way of a β-elimination via a cyclic transition state (Figure 17.10). Alternatively, one could also imagine that an acyclic transition state might be involved. The chromium(IV) acid could then disproportionate giving Cr(III) and Cr(VI) (see Figure 17.10, center); that is, the inorganic Cr(III) and Cr(VI) products would be obtained without the participation of an organic molecule. On the other hand, the chromium(IV) acid itself also is capable of oxidizing the alcohol while being reduced to Cr(III), presumably via a radical mechanism (Figure 17.10, bottom). The hydroxy-substituted radical **B** is formed from the alcohol by a one- or multistep transfer of an H atom from the α-position onto the Cr(IV) species. Subsequently, this radical is oxidized further to the carbonyl compound. Cr(VI) might act as the oxidizing reagent, and it would be reduced to a chromium(V) acid. This species is the third oxidizing agent capable of reacting with the alcohol in Cr(VI) oxidations. Presumably, the alcohol and the Cr(V) acid form a chromium(V) acid monoester **C**. This ester may undergo a β-elimination, much like the Cr(VI) analog **A**, and thereby form the carbonyl compound and a Cr(III) compound.

These oxidations are complicated, but it is probably best to remember that elimination step. It is common in many alcohol oxidation processes.

Oxidations of alcohols with *water-free* Cr(VI) reagents, such as the ones in rows 3–5 in Table 17.5, always result in the formation of carbonyl compounds. In particular, a carbonyl compound is obtained even if it is an aldehyde and therefore in principle could be oxidized to give the carboxylic acid. On the other hand, aldehydes *are* oxidized further (rows 1 and 3 in Table 17.5) if one uses *water-containing* Cr(VI) reagents (e. g., oxidations with $K_2Cr_2O_7$ in sulfuric acid) or the Jones reagent (unless the latter is used under really mild conditions). Fig-

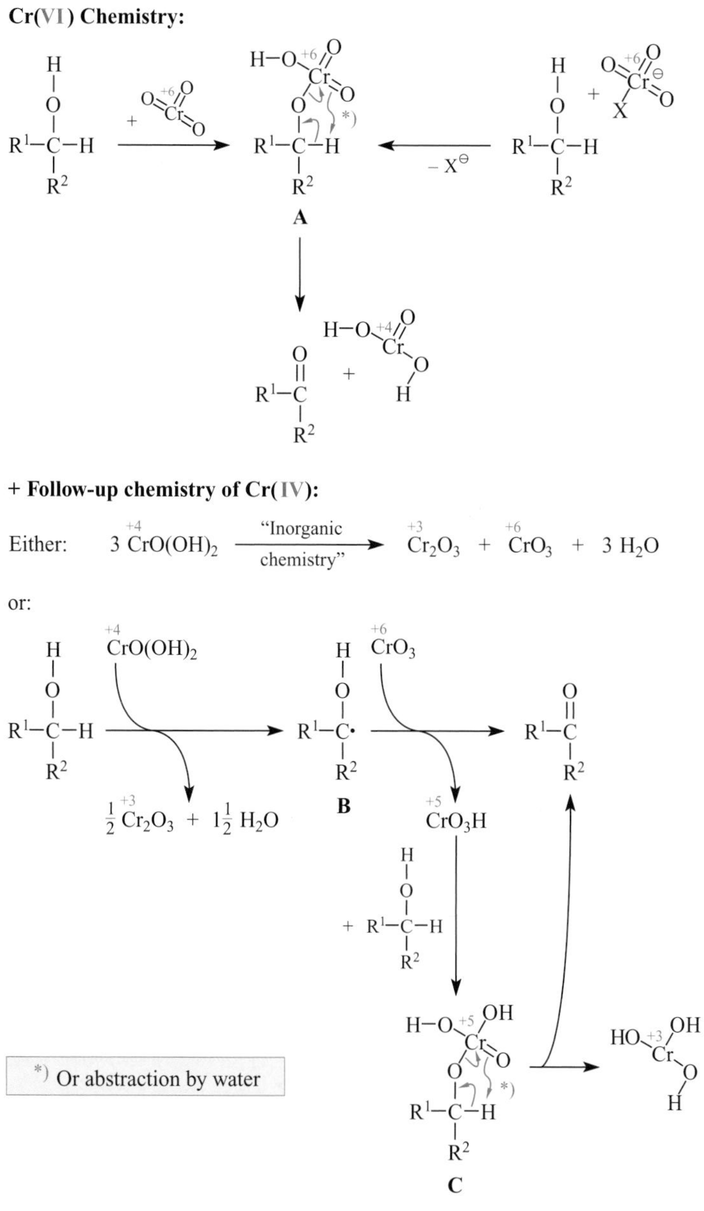

Cr(VI) Chemistry:

+ Follow-up chemistry of Cr(IV):

Either: $3\ \overset{+4}{CrO}(OH)_2$ —"Inorganic chemistry"→ $\overset{+3}{Cr_2}O_3 + \overset{+6}{Cr}O_3 + 3\ H_2O$

or:

*) Or abstraction by water

Fig. 17.10. Mechanism of the Cr(VI) oxidation of alcohols to carbonyl compounds. The oxidation proceeds via the chromium(VI) acid ester **A** ("chromic acid ester") and yields chromium(IV) acid. The chromium(IV) acid may either disproportionate in an "inorganic" reaction or oxidize the alcohol to the hydroxy-substituted radical **B.** This radical is subsequently oxidized to the carbonyl compound by Cr(VI), which is reduced to Cr(V) acid in the process. This Cr(V) acid also is able to oxidize the alcohol to the carbonyl compound while it is undergoing reduction to a Cr(III) compound.

ure 17.11 provides an explanation of this effect of water. Cr(VI) compounds cannot react with aldehydes at all. Only the hydrates of these aldehydes react, and these hydrates are formed in an equilibrium reaction from the aldehydes in the presence of water (see Figure 9.1). The hydrate of an aldehyde behaves toward the Cr(VI) compound just like any ordinary alcohol. Hence, the hydrate of the aldehyde is oxidized by the mechanism discussed for alcohols (Figure 17.10) after the hydrate has been converted into a chromium(VI) acid monoester **A** (Figure 17.11). This ester undergoes the same reactions described in Figure 17.10.

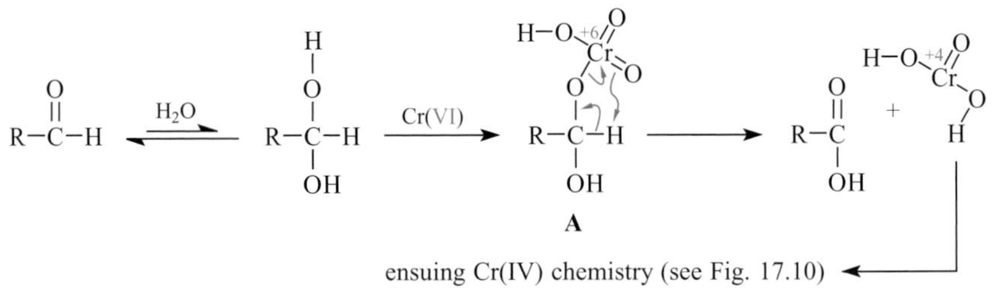

Fig. 17.11. Mechanism of the Cr(VI) oxidation of an aldehyde. As in the case described in Figure 17.10, the oxidation can proceed via three different paths.

Side Note 17.1. Chromium-free Direct Oxidation Alcohol → Carboxylic Acid

Ruthenium tetroxide oxidizes an alcohol to furnish an aldehyde. However, the oxidation does not stop at this stage. As soon as the corresponding aldehyde hydrate has been formed from the aldehyde at equilibrium—which is ensured by the high water content in the reaction mixture—the oxidation continues to form the carboxylic acid (example: Figure 17.12). Mecha-

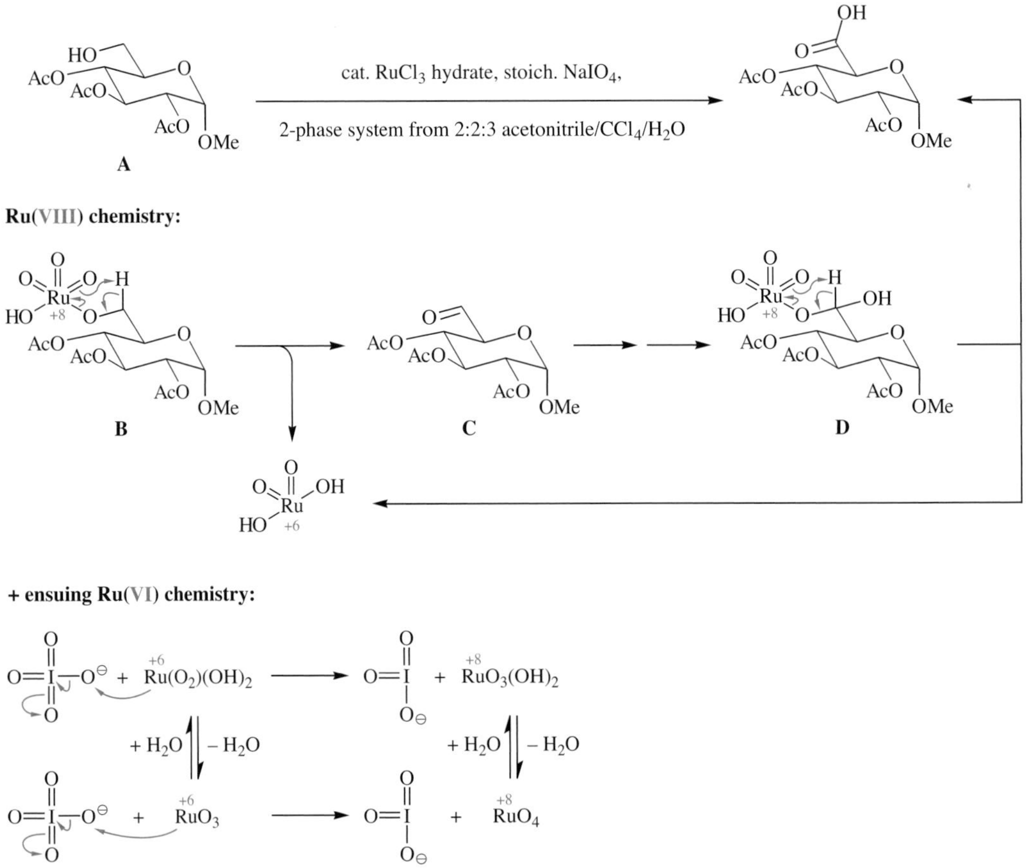

Fig. 17.12. Mechanism of the Ru(VIII) oxidation of the alcohol (**A**). Since the reaction medium contains water a carboxylic acid is formed instead of the aldehyde **C**.

nistically, these reactions proceed in complete analogy to the oxidations with chromium trioxide (CrO_3) shown in Figures 17.10 and 17.11. This means that the oxidative step from the alcohol to the aldehyde **C** in Figure 17.12 proceeds via the Ru(VIII) analog **B** of the Cr(VI) ester **A** in Figure 17.10 and that the step involving the oxidation from the aldehyde to carboxylic acid proceeds via the Ru(VIII) analog **D** of the Cr(VI) ester **A** of Figure 17.11. The reduction product of both steps is Ru $(=O)_2(OH)_2$. The latter, however, is reoxidized by sodium periodate, which has been added as a co-oxidant, by transferring an oxo ligand *in situ* to form ruthenium tetroxide and Ru(VIII) acid, respectively. This reoxidation is so fast that the oxidation alcohol → carboxylic acid can be conducted with catalytic amounts of ruthenium tetroxide. Usually, one does not even use the very expensive ruthenium tetroxide to start the reaction, but the slightly less expensive ruthenium trichloride (Figure 17.12). Sodium periodate initially oxidizes this compound, too, to give ruthenium tetroxide.

Oxidations of Alcohols with Activated Dimethyl Sulfoxide

In this subsection we want to consider oxidations that employ dimethyl sulfoxide (DMSO) as the oxidizing reagent. These oxidations, which almost always are carried out in the presence of first oxalyl chloride and then NEt_3, are referred to as **Swern-oxidations**. The mechanism of this reaction is known in detail (Figure 17.13). In the prelude, the O atom of DMSO acts as

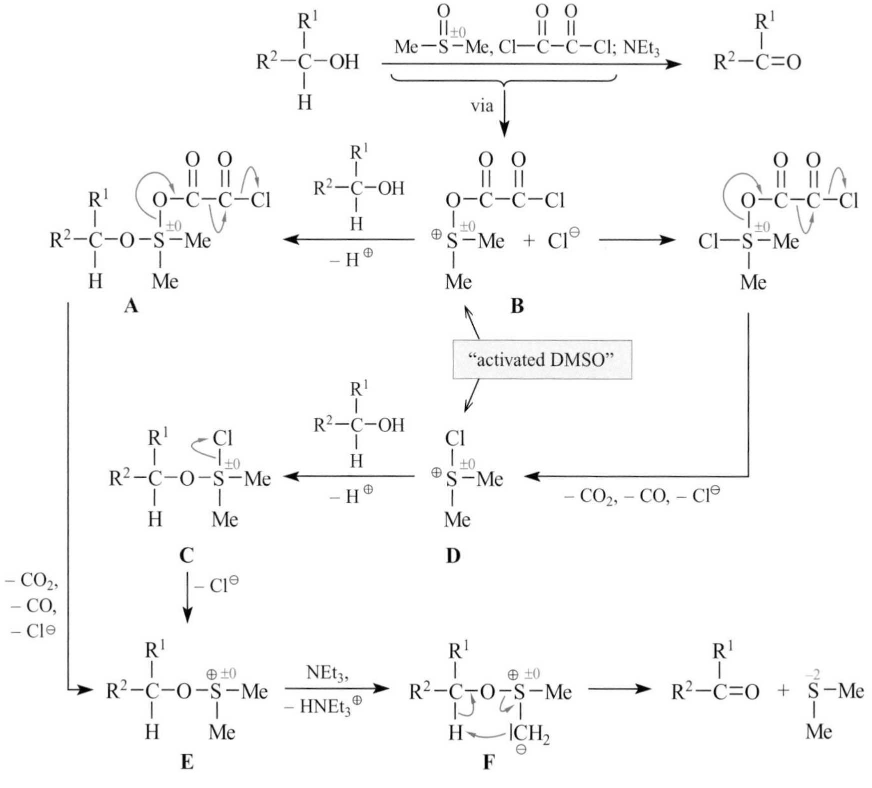

Fig. 17.13. Mechanism of the Swern oxidation of alcohols. The actual reagent is an "activated DMSO" (compound **B** or **D**), which reacts with an alcohol with formation of **A** or **C**, respectively. Dissociation leads to the sulfonium salt **E**, which is then converted into the sulfonium ylide **F** after NEt_3 addition and raising the temperature from −60 to −45 °C. *β*-Elimination via a cyclic transition state generates the carbonyl compound and dimethyl sulfide from **F**.

a nucleophile and, following the mechanism of Figure 6.2, undergoes an S_N reaction at one of the carboxyl carbons of the oxalyl chloride to form the sulfonium ion **B**. This is *one* form of the so-called activated DMSO. Addition of a chloride ion yields a sulfurane intermediate. Loss of the —O—C(=O)—C(=O)—Cl group, which fragments, results in the formation of the sulfonium ion **D**. This ion **D** represents *another* "activated DMSO."

Irrespective of whether the initially formed sulfonium ion **B** or the subsequently formed sulfonium ion **D** reacts with the alcohol, the alcohol is taken up by such a sulfonium ion with formation of sulfuranes **A** (first case) or **C** (second case). Any of these sulfuranes would yield the sulfonium salt **E** after dissociation. Once this sulfonium salt has formed, five equivalents of NEt_3 are added to the reaction mixture, which then is allowed to warm up from –60 to –45°C. Under these conditions, the sulfonium salt **E** is deprotonated to give the sulfonium ylide **F**. This ylide undergoes a β-elimination via a cyclic transition state to form the desired carbonyl compound and dimethyl sulfide as a side-product.

Special Oxidation Methods for R—CH_2OH → R—CH(=O)

It might seem that the methods presented thus far for the oxidation of alcohols to aldehydes offer more than enough options. In practice, however, situations may well arise in which *none* of these methods works. It is because of such failures that there has been and still is a constant need for the development of alternatives. Three of such alternative procedures for R—CH_2OH → R—CH(=O) transformations will be presented (Figures 17.14–17.16).

The oxidation of alcohol **A** in Figure 17.14 to give aldehyde **E** fails under Swern conditions because NEt_3 catalyzes an ensuing E2 elimination leading to the conjugated aldehyde **D**. In contrast, the oxidation of the same aldehyde **A** with the **Dess–Martin reagent (B** in Figure 17.14; preparation, Figure 17.39) occurs without base. **B** is a mixed anhydride of an aryliodo(III) acid and two different carboxylic acids. One of the three acetoxy groups of

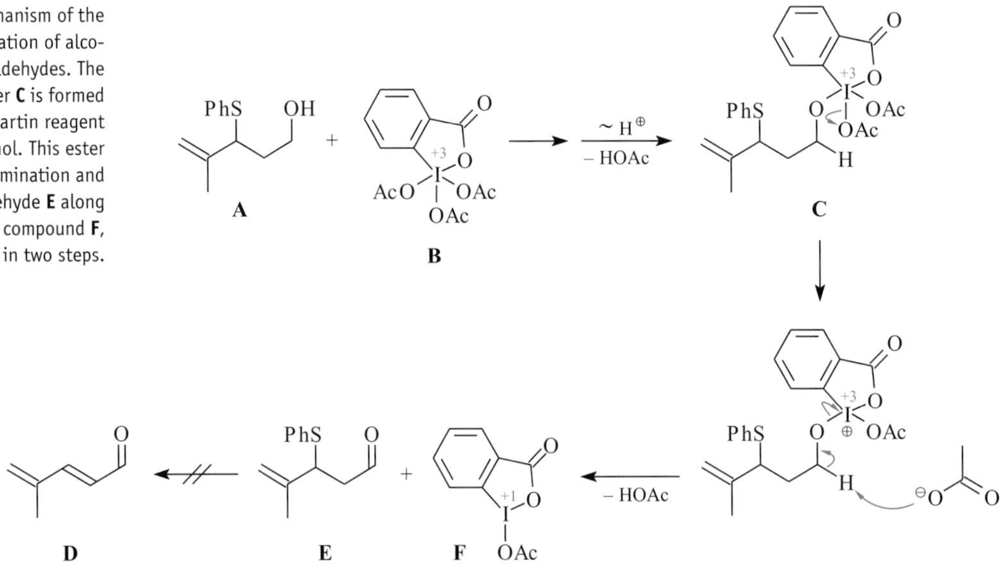

Fig. 17.14. Mechanism of the Dess–Martin oxidation of alcohols to aldehydes. The aryliodo(III) ester **C** is formed from the Dess–Martin reagent **B** and the alcohol. This ester undergoes a β-elimination and forms the aldehyde **E** along with the iodo(I) compound **F**, presumably in two steps.

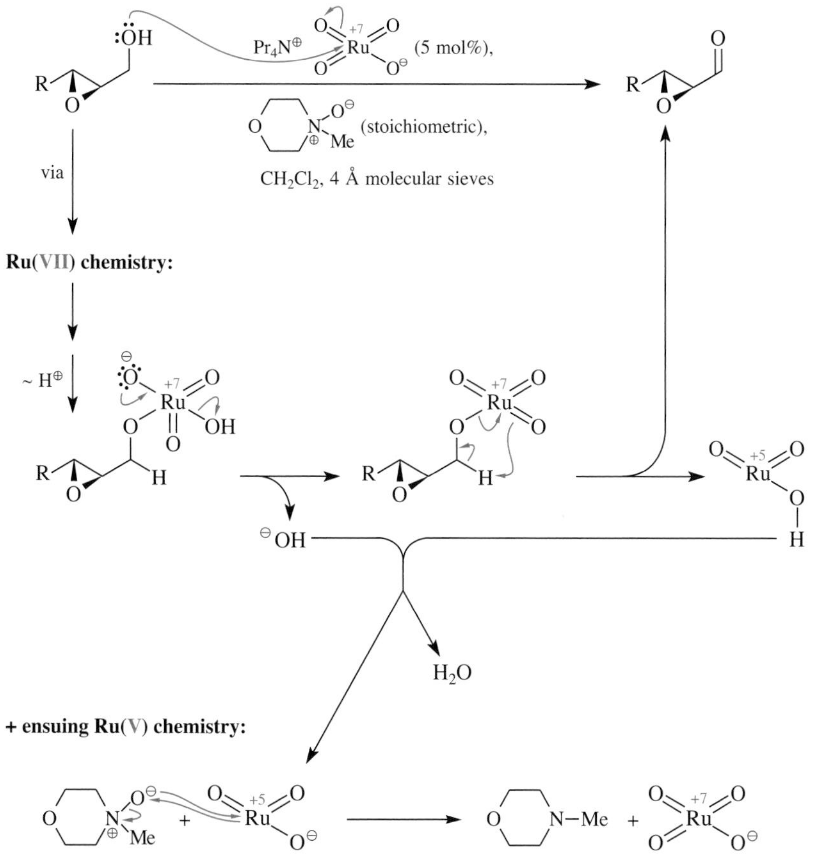

Fig. 17.15. Mechanism of the TPAP oxidation of an alcohol to an aldehyde (TPAP stands for tetrapropylammonium perruthenate). The effective oxidant is a Ru(VII) oxide, other than in Figure 17.12 where a Ru(VIII) oxide is employed. Here, the stoichiometrically used oxidizing agent is *N*-methylmorpholin-*N*-oxide ("NMO"), whereas in Figure 17.12 $NaIO_4$ is used.

reagent **B** is replaced by the substrate alcohol via a dissociation/association mechanism. The aryliodo(III) acid ester **C** is formed in this way. **C** undergoes a *β*-elimination, presumably in two steps. The first step consists of the cleavage of an acetate ion from the iodine. The second step is an E2 elimination. The products of elimination are the desired aldehyde **E** and the iodo(I) derivative **F**. See a pattern? Functionalize the –OH group of the alcohol, then perform an elimination.

The enantiomerically pure epoxy alcohols **A** shown in Figure 17.15 can be oxidized with a combination of two oxidizing reagents. This combination consists of a stoichiometric amount of *N*-methylmorpholine-*N*-oxide (see Figure 17.36) and a catalytic amount of tetrapropylammonium perruthenate (TPAP). This oxidation works much better than either the standard oxidizing agent PDC or the Swern reagent. The water formed is removed by added molecular sieves to prevent the formation of an aldehyde hydrate and, via the latter, a progression of the oxidation toward the carboxylic acid (see the mechanism in Figure 17.11 or 17.12).

The effective oxidant in the TPAP oxidation of alcohols is the perruthenate ion, a Ru(VII) compound. This compound is employed only in catalytic amounts but is continuously replenished (see below). The mechanism of the alcohol → aldehyde oxidation with TPAP presum-

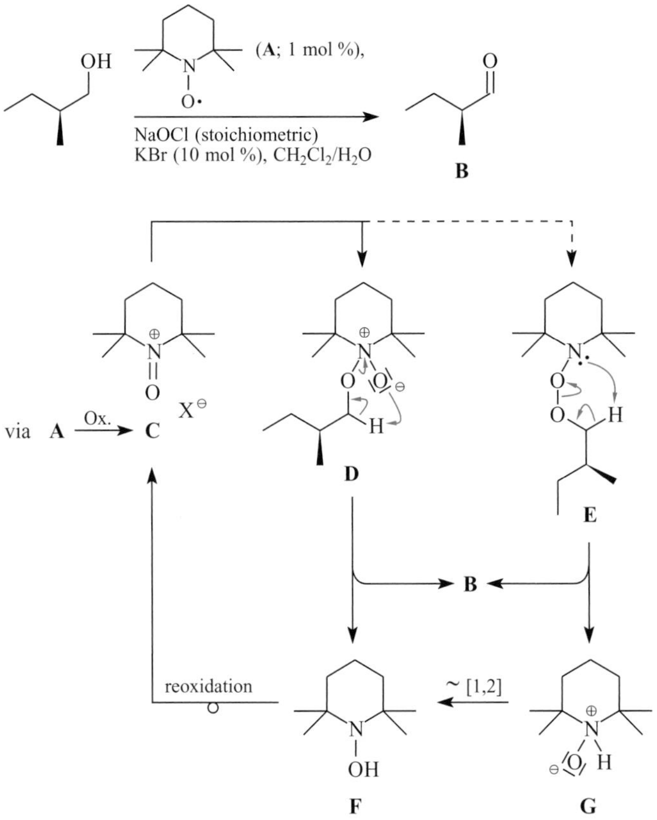

Fig. 17.16. Mechanism of the TEMPO oxidation of alcohols to aldehydes (TEMPO stands for tetramethylpiperidine nitroxyl).

ably corresponds to the nonradical pathway of the same oxidation with Cr(VI) (Figure 17.10, top) and to the course of the Ru(VIII) oxidation of Figure 17.12. Accordingly, the key step of the TPAP oxidation is a β-elimination of the ruthenium(VII) acid ester **B**. The metal is reduced in the process to ruthenium(V) acid.

Figure 17.16 illustrates a third alternative procedure for the alcohol → aldehyde oxidation with the example of a racemization-free oxidation of an enantiomerically pure alcohol. Two oxidizing agents are employed, a stoichiometric amount of NaOCl and a catalytic amount of the nitroxyl **A**. These two components form the actual oxidizing reagent, the nitrosonium ion **C** (X = Cl). A small amount of KBr is added as a third component to the mixture. KBr serves to increase the solubility of the nitrosonium ion in the organic phase, where the bromide salt **C** (X = Br) is more soluble than the chloride salt **C** (X = Cl). In the organic phase, the nitrosonium ion combines with the alcohol, which acts as a nucleophile and adds to the $^{\oplus}$N=O double bond. This reaction of the alcohol is more likely to occur at the N atom (→ **D**) than at the O atom (→ **E**) of the $^{\oplus}$N=0 bond. However, each of these intermediates would produce the same desired aldehyde **B** via a β-elimination. The accompanying product could be the hydroxylamine **F** or its tautomer, the *N*-oxide **G**. Certainly, **G** would immediately isomerize to give hydroxylamine **F**. Subsequently, **F** is reoxidized by NaOCl to restore, via the TEMPO radical **A**, the nitrosonium salt **C**. This completes the catalytic cycle.

Special Oxidation Methods R–CH=O → R–CO_2H or R–CO_2Me

In your basic chemistry courses you have learned that primary alcohols are very easily overoxidized under aqueous conditions, i.e., react beyond the aldehyde to form a carboxylic acid. This leaves the impression that it might be a cakewalk to oxidize an aldehyde to give carboxylic acids. But experienced chemists know that this will not necessarily occur and they are glad that in some more difficult cases they can resort to one of the procedures shown in Figures 17.17 and 17.18.

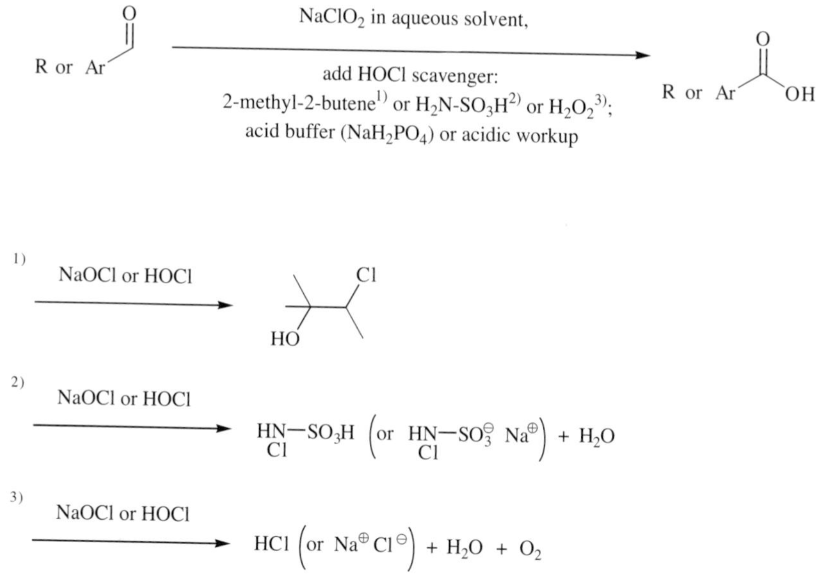

R or Ar–CH=O → ($NaClO_2$ in aqueous solvent, add HOCl scavenger: 2-methyl-2-butene[1)] or H_2N-SO_3H[2)] or H_2O_2[3)]; acid buffer (NaH_2PO_4) or acidic workup) → R or Ar–COOH

1) NaOCl or HOCl → HO–C(CH$_3$)$_2$–CHCl–CH$_3$

2) NaOCl or HOCl → HN(Cl)–SO_3H (or HN(Cl)–$SO_3^{\ominus}$ $Na^{\oplus}$) + H_2O

3) NaOCl or HOCl → HCl (or $Na^{\oplus}$ $Cl^{\ominus}$) + H_2O + O_2

Fig. 17.17. Sodium chlorite oxidation of aliphatic or aromatic aldehydes to form a carboxylic acid. The extra additive destroys the reduction product of the oxidant, i.e., sodium hypochlorite or hypochloric acid.

Saturated aldehydes and aromatic aldehydes undergo oxidation by sodium chlorite to form a carboxylic acid (Figure 17.17). Apart from the carboxylic acid, hypochloric acid (HOCl) and sodium hypochlorite (NaOCl), respectively, are formed, too. The mechanism of the oxidation is unclear. One could imagine that the nucleophilic addition of the chlorite ion to the carbonyl group and a protonation yield the chlorous acid half-ester of an aldehyde hydrate. in analogy to the corresponding chromium(VI)- and ruthenium(VIII) acid half-esters of aldehyde hydrates, which were already encountered in this section (as intermediates of the full oxidation of primary alcohols to carboxylic acid). This aldehyde hydrate could decompose into carboxylic acid and hypochloric acid or sodium hypochlorite. Hypochloric acid and sodium hypochlorite are stronger oxidants than sodium chlorite and need to be decomposed *in situ* by suitable additives before they are able to react with the substrate or the reaction product. Figure 17.17 shows three well-established reagents as well as the respective HOCl- or NaOCl follow-up products.

Conjugated—α,β-unsaturated or aromatic—aldehydes are oxidized by manganese dioxide in the presence of one equivalent of sodium cyanide to give an acyl cyanide (Formula **C** in Figure 17.18) via an intermediate cyanohydrin (**B** in Figure 17.18), which is also either an allylic or a benzylic alcohol. Alcohols of this kind can be oxidized by a relatively weak oxi-

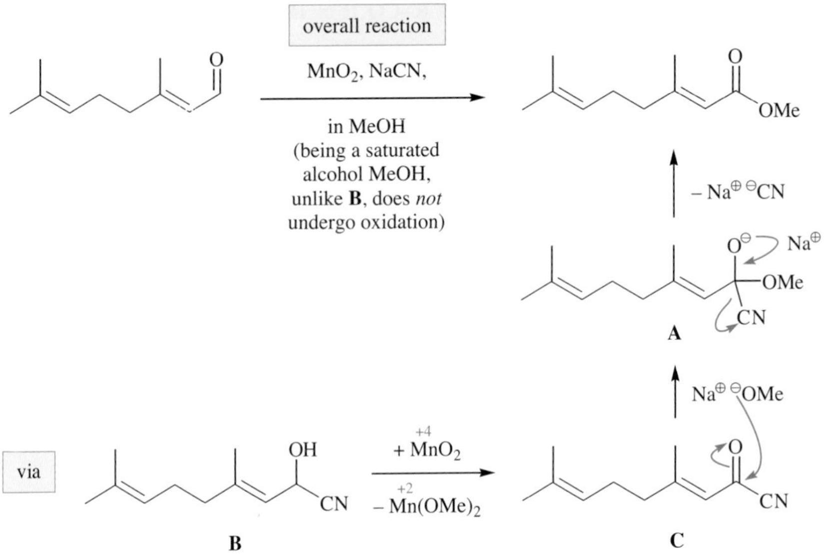

overall reaction

MnO_2, NaCN, in MeOH (being a saturated alcohol MeOH, unlike **B**, does *not* undergo oxidation)

$- Na^{\oplus}\,{}^{\ominus}CN$

$Na^{\oplus}\,{}^{\ominus}OMe$

via

$+ \overset{+4}{MnO_2}$ / $- \overset{+2}{Mn}(OMe)_2$

A **B** **C**

Fig. 17.18. Mechanism of the MnO_2 oxidation of conjugated aldehydes proceeding via cyanohydrin (**B**), which is generated *in situ*. Subsequently, the actual oxidation product (**C**) undergoes methanolysis (according to the $B_{AC}2$ mechanism, Figure 6.24) to the finally isolated methyl ester.

dant like manganese dioxide, while the same oxidant does not react with a "normal" alcohol like methanol, which in the same reaction is employed as a solvent (Figure 17.18). (A "normal" alcohol here refers to an alcohol whose OH group is not in an allylic or benzylic position.) If this oxidation proceeds via an addition product formed at equilibrium of the alcohol to MnO_2—i.e., via a manganese(IV) acid half-ester—and the decomposition of this adduct to the carbonyl- and a Mn(II) compound as the rate-determining step, substrate selectivity could be accounted for by "product development control": only allylic or benzylic alcohols lead to the formation of carbonyl compounds whose C=O double bond is resonance-stabilized by a conjugated substituent. MnO_2 oxidation of allylic or benzylic cyanohydrins yields acyl cyanides (**C** in Figure 17.18). These behave as acylating agents towards the solvent methanol acylating it via a tetrahedral intermediate **A** to the isolated unsaturated carboxylic acid ester.

17.3.2 Oxidative Cleavages

The *cis-vic*-Dihydroxylation of Alkenes: No Oxidative Cleavage, but an Important Prelude

Alkenes can be dihydroxylated *cis*-selectively by reaction with a stoichiometric amount of *N*-methylmorpholine-*N*-oxide (NMO), a catalytic amount of a suitable Os(VIII) reagent, and some water. From a preparative point of view, this reaction is closely related to oxidative cleavages, and for this reason it is introduced now (Figure 17.19).

Osmium tetroxide (OsO_4) is the reactive Os(VIII) species in the *cis-vic*-dihydroxylation of alkenes. This compound is a solid but is not easy to handle because it has a rather high vapor pressure. Therefore, it is often preferable to prepare the compound *in situ*. Such a preparation involves the oxidation of the potassium salt $K_2OsO_4 \cdot 2H_2O$ ($K_2OsO_2(OH)_4$) of osmium(VI) acid with NMO as the oxidizing agent. This salt also is a solid but has a much lower vapor pressure than OsO_4. The oxidant most commonly used in this case is ***N*-methyl-morpholine-*N*-oxide** (NMO).

Extensive investigations have demonstrated that OsO_4 reacts with alkenes in a 1,3-dipolar cycloaddition while forming the five-membered ring **A** of Figure 17.19 as an intermediate. **A** is a cyclic diester of an osmium(VI) acid. As such it will be hydrolyzed by the pre-

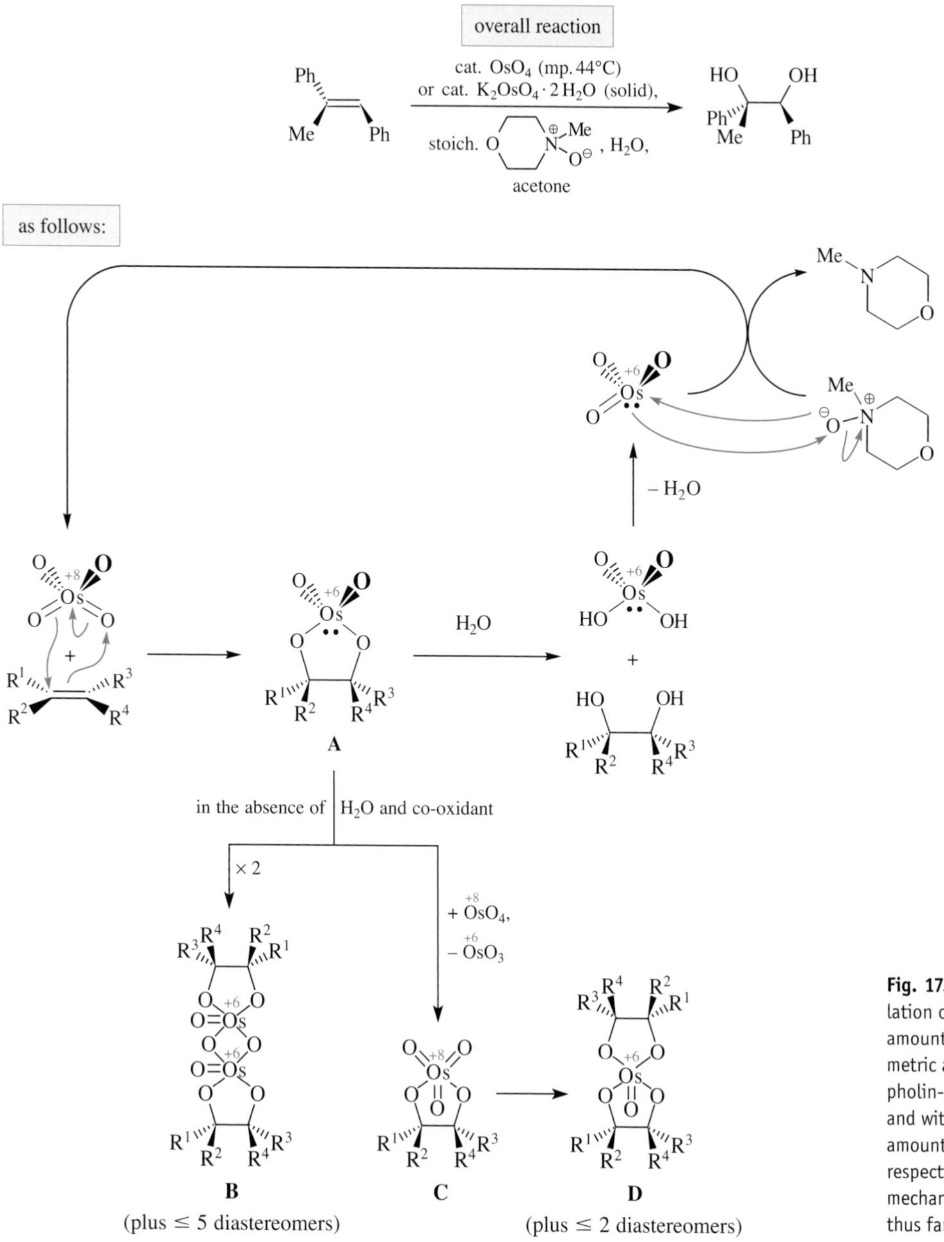

overall reaction

cat. OsO_4 (mp. 44°C) or cat. $K_2OsO_4 \cdot 2\,H_2O$ (solid), stoich. NMO, H_2O, acetone

as follows:

$-H_2O$

H_2O

A

in the absence of H_2O and co-oxidant

× 2

$+\overset{+8}{OsO_4}$, $-\overset{+6}{OsO_3}$

B (plus ≤ 5 diastereomers)

C

D (plus ≤ 2 diastereomers)

Fig. 17.19. *cis-vic*-Dihydroxylation of alkenes with catalytic amounts of Os(VIII)/stoichiometric amounts of **m**ethylmor-pholin-*N*-**o**xide [NMO]; top) and with stoichiometric amounts of Os(VIII) (bottom), respectively, and pertinent mechanistic insights obtained thus far.

sent H_2O, releasing the *cis*-diol and the osmium(VI) acid. *cis*-*vic*-Dihydroxylations with the osmium(VI) acid as a final product would be far too expensive. Fortunately, there is no need for the stoichiometric use of OsO_4 or $K_2OsO_4 \cdot 2\ H_2O$. If NMO is added to the reaction mixture in stoichiometric amounts, it will act as a co-oxidant and ensure that Os(VIII) will be continuously regenerated from the Os(VI) that is formed during the alkene oxidation.

In connection with mechanistic studies, how OsO_4 reacts with alkenes in the absence of both water and a co-oxidant was also explored. The outcome is shown in Figure 17.19 at the bottom. The osmium(VI) derivatives **B** and **D** are formed. The occurrence of **B** and **D** led to the conclusion that the osmium(VI) acid ester **A** is an intermediate in the *cis*-*vic*-dihydroxylation of alkenes when water is both present *and* absent. In the *absence* of water, of course, the Os(VI) ester **B** cannot hydrolyze to the *cis*-*vic*-diol, and dimerization to **B** presents a possible alternative reaction path. On the other hand, the Os(VI) ester **A** can be oxidized to the Os(VIII) analog **C** by some of the remaining OsO_4. Then **C** adds to the C=C double bond of another alkene, and the tetraester **D** is formed. Similar to the addition of OsO_4, this addition to a C=C double bond presumably takes place as a 1,3-dipolar cycloaddition. It is assumed that *in the presence of* H_2O—which represents the usual procedure—the osmium(VI) diester **A** undergoes hydrolysis to give the *cis*-*vic*-diol. In a strict sense, though, the possibility cannot be excluded that the further reaction proceeds via the intermediate **C** and perhaps even via **D**. Hydrolysis of either of these species would yield the same diol as the hydrolysis of the osmium(VI) diester **A**.

If a tertiary amine like pyridine is added in the *cis*-*vic*-dihydroxylation of an alkene, a 1:1-adduct **A** of OsO_4 and pyridine is formed:

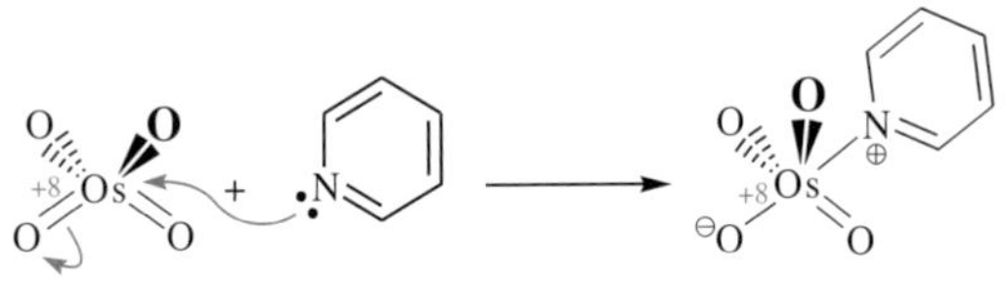

A

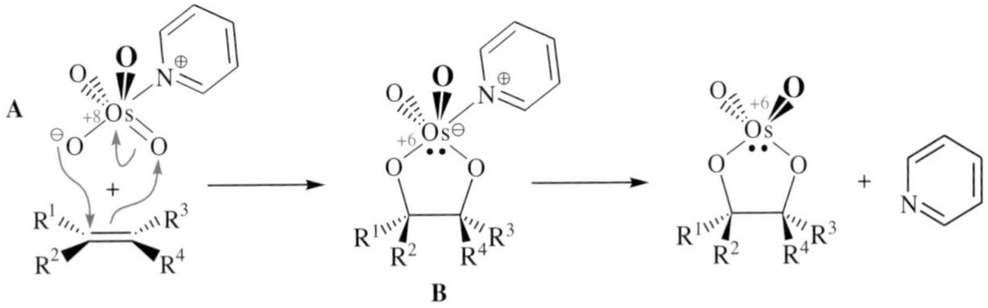

A

B

A reacts with the alkene in a 1,3-dipolar cycloaddition to form a pyridine-associated cyclic osmium(VI)acid diester **B**. The latter decomposes to pyridine and the same osmium(VI) acid diester that is directly generated in a pyridine-free reaction as an intermediate (**A** in Figure 17.19). However, the formation of the osmium(VI) acid diester via these three steps is much faster than via the one-step reaction shown in Figure 17.19. In other words, tertiary amines like pyridine are potent catalysts of *cis*-*vic*-dihydroxylations of alkenes with OsO_4.

Thus, in *cis-vic*-dihydroxylations of alkenes with OsO_4 tertiary amines, like pyridine, have "ligand acceleration" effects (this term was introduced in Section 3.4.6, using the Sharpless epoxidation as an example).

Side Note 17.2. Alternatives to the Osmium-mediated *cis-vic*-Dihydroxylation

For reasons of toxicity, cost and/or reactivity, other reagents than OsO_4 are also employed for the *cis-vic*-dihydroxylation of alkenes. **Potassium permanganate** as a stoichiometric oxidant is inexpensive and can also effect dihydroxylations in a *cis*-selective and often quite rapid fashion (example: Figure 17.20, top). With this reagent, careful control of the reaction conditions is called for. In contrast to OsO_4, potassium permanganate is capable of effecting a subsequent oxidation of the diols to give two carbonyl and/or carboxyl compounds. Hence, potassium permanganate by itself may occasionally have the same effect as the $KMnO_4/NaIO_4$ mixture in the Lemieux–von Rudloff oxidation of alkenes (see Figure 17.26, top right).

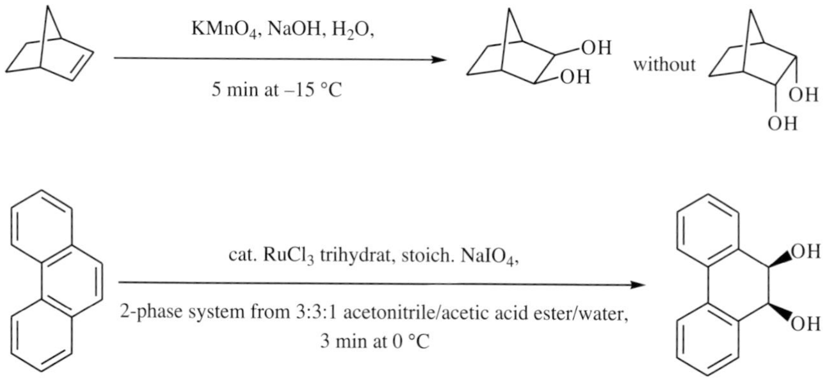

KMnO$_4$, NaOH, H$_2$O, 5 min at –15 °C; OH, OH; without; OH, OH

cat. RuCl$_3$ trihydrat, stoich. NaIO$_4$, 2-phase system from 3:3:1 acetonitrile/acetic acid ester/water, 3 min at 0 °C; OH, OH

Fig. 17.20. *cis-vic*-Dihydroxylation of alkenes with stoichiometric amounts of Mn(VII) (top) and catalytic amounts of Ru(VIII)/stoichiometric amounts of $NaIO_4$ (bottom), respectively.

Ruthenium tetroxide is also very expensive even if only catalytic amounts are needed. As with OsO_4, catalytic amounts are sufficient if stoichiometric amounts of the co-oxidant sodium periodate are employed (this co-oxidant makes other RuO_4 oxidations catalytic as well: Figures 17.12, 17.26, 17.29). Figure 17.20 (bottom) presents a dihydroxylation with RuO_4, where even an aromatic C=C double bond reacts. OsO_4 is not capable of doing this reaction. $KMnO_4$ would also react with the aromatic compound in Figure 17.20, but the dihydroxylated aromatic compound would immediately undergo further reaction in terms of a Lemieux–von-Rudloff oxidation (Figure 17.26, bottom right): the central ring would be cleaved to yield *ortho,ortho′*-biphenyl dicarboxylic acid.

The Nobel Prize in Chemistry 2001 was awarded to three researchers for their pioneering work in the field of asymmetric catalysis. One of them, K. Barry Sharpless, was honored for the epoxidations named after him (Section 3.4.6). The second reason for the award was his development of the **asymmetric dihydroxylation** (AD; Figure 17.21). The Sharpless reactions that were honored with the Nobel Prize have three things in common: first, they are oxidations, second, they are catalytic asymmetric syntheses, and third, they owe their high enantiocontrol to the additive control of stereoselectivity. In the introductory passages to

Section 3.4.6—which, by the way, referred to Sharpless epoxidations—you learned that *catalytic* asymmetric syntheses are among the most elegant asymmetric syntheses and that they can rely on a substrate reacting (exclusively) with an enantiomerically pure reagent that is formed *in situ* from an achiral precursor molecule and a *catalytic* amount of an enantiomerically pure additive. It was emphasized that an *exclusive* reaction of this precursor molecule/additive complex with the substrate takes place if it is much more reactive than the achiral precursor molecule without the enantiomerically pure additive. If under these circumstances **additive control of stereoselectivity** occurs, this principle even allows recovery of the enantiomerically pure additive after the conversion, which is more convenient than if it could only be released from the product via a chemical reaction.

Using Sharpless' asymmetric dihydroxylation, the *cis-vic*-dihydroxylation of C=C double bonds succeeds in a highly enantioselective fashion for most of the substitution patterns in Figure 17.21 (part I) with a large variety of double bond substituents R^1 to R^3. As little as 0.2 mole percent of the Os(VI) species $K_2OsO_2(OH)_4$ suffices for the precursor of the reactive Os(VIII) species. if an amount of $K_3Fe(CN)_6$ is added as a co-oxidant in excess. Once more, the solvent is strongly hydrous (but different from the racemic dihydroxylation in Figure 17.19). The enantioselectivity of the asymmetric dihydroxylation is the result of the ligand acceleration effect, which is due to the addition of tertiary amines. In order to dihydroxylate alkenes from their α-side (relating to a projection of the alkene that is difficult and not worthwhile to remember), 1 mole percent $(DHQ)_2$-PHAL is added as an enantiomerically pure additive. If alkenes are to be dihydroxylated from their β-side, success is guaranteed with 1 mole percent $(DHQD)_2$-PHAL as the enantiomerically pure additive. The two additives used are alkaloid derivatives, and both of them are enantiomerically pure hexaamines (for a brief explanation of the abbreviations that are generally used see the figure caption). The bottom part of Figure 17.21 (part I) provides the structures of these alkaloid derivatives. Note that the additives $(DHQ)_2$-PHAL and $(DHQD)_2$-PHAL are not enantiomorphic, but diastereomeric. This is not due to design, but reflects their origin from two different (i.e., diastereomeric) natural products, namely quinine and quinidine. Each of these additives requires just one out of its six nitrogen atoms to initiate ligand-accelerated *cis-vic*-dihydroxylations, which will soon become clear. Three of the "superfluous" nitrogen atoms derive from the cited natural compounds, two other stem from the heterocycle attached by laboratory synthesis.

Figure 17.21 (part I) also shows the most stable conformer each of $(DHQ)_2$-PHAL and $(DHQD)_2$-PHAL. Its conformation first reflects the propensity of the oxygen atoms at the central heterocycle to conjugate with its π-electron system leading to an sp^2-hybridization of these oxygen atoms. Second, the bulky *sec*-alkyl substituents are in search for the least hindered position. The result is that they try to avoid *peri*-interaction with the labeled H atoms of the central heterocycle (for the term *peri*-interaction see the discussion of Figure 5.8), but do not tolerate a U-conformation in the non-hydrogen substituent–C_{sec}–O–C=N substructure.

At the end of this conformational analysis it is essential to realize that the preferential conformer of both $(DHQ)_2$-PHAL and $(DHQD)_2$-PHAL contains a twofold rotation axis coinciding with the longitudinal axis of the central heterocycle. The existence of this rotation axis implies, for example, that the two dihydroquinuclidine subunits of each compound can be mapped onto each other through 180° rotation. This means that in a given compound one dihydroquinuclidine subunit has exactly the same properties as the other.

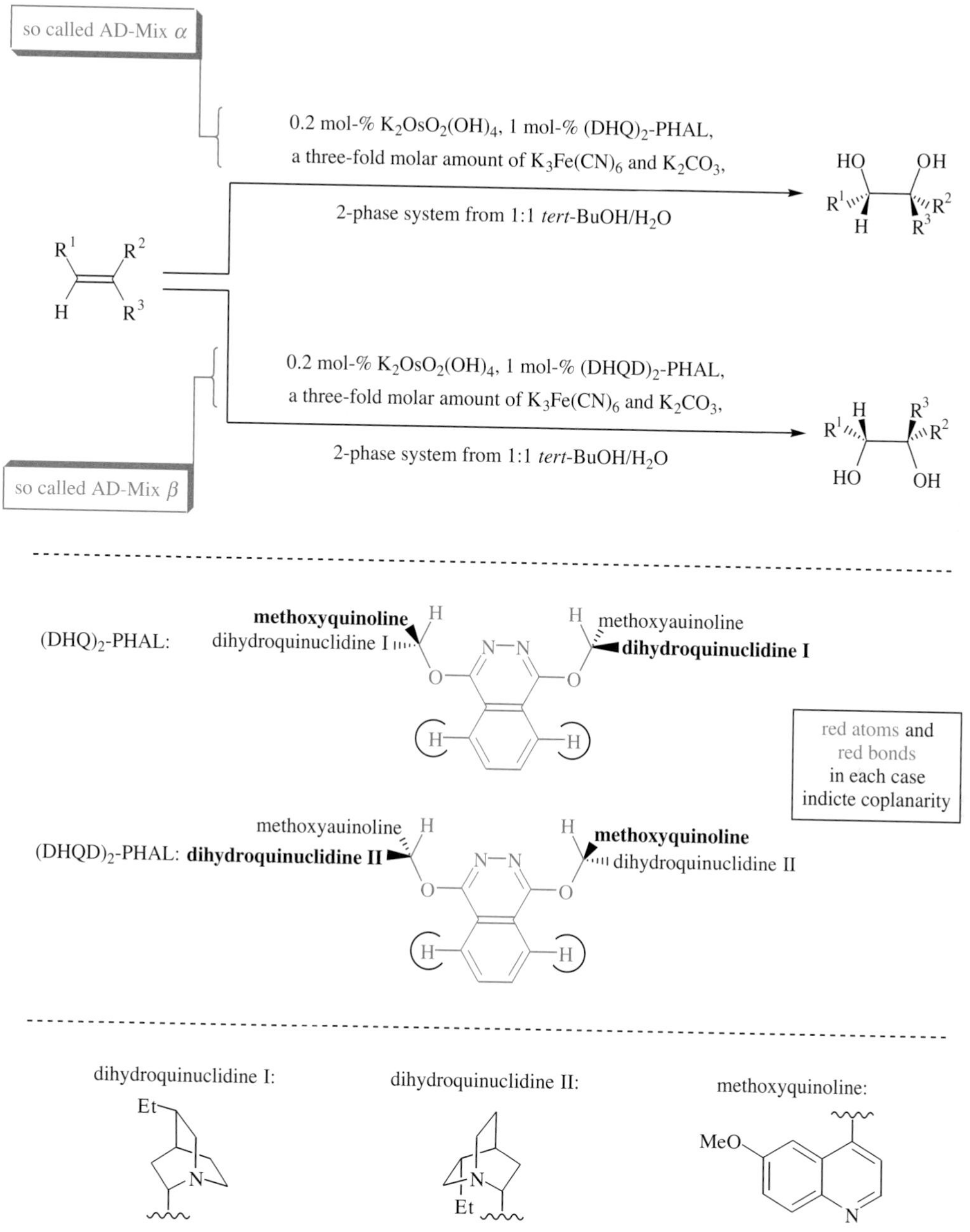

Fig. 17.21 Part I. Asymmetric *cis-vic*-**d**ihydroxylation ("AD") of alkenes with catalytic amounts of Os(VIII), stoichiometric amounts of $K_3Fe(CN)_6$ and an enantiomerically pure hexaamine. The latter is derived from the hydrogenation products of enantiomerically pure alkaloids, i.e., from **dihy**droquinine ("DHQ") or from **dih**ydro**q**uini**d**ine (DHQD), as specified in between the dashed horizontal lines: as phthalazine $(DHQ)_2$-PHAL and $(DHQD)_2$-PHAL, respectively. The structural formulas presented here become true structural formulas if the text is replaced by the fragments "dihydroquinuclidine I," "dihydroquinuclidine II" and "methoxyquinoline," which are explained at the bottom of Figure 17.21.

The more enantioselective dihydroxylation in Figure 17.21 (part I)

R1 R2 H R3 —AD-Mix β→ R1 H R3 R2 HO OH ... via ...

methoxyquinoline MeO R3 R2 O O Os +8 O H R1 O H H O N N–N O O H dihydroquinuclidine II H Et

methoxyquinoline MeO R3 R2 O O Os +6 O R1 H H H O N N–N O O H dihydroquinuclidine II H Et

red molekule moieties = passive ("by-standers")

Fig. 17.21 part II. Mechanism of the asymmetric *cis-vic*-dihydroxylation of alkenes with catalytic amounts of Os(VIII), stoichiometric amounts of $K_3Fe(CN)_6$ and one of the enantiomerically pure hexaamines $(DHQ)_2$-PHAL and $(DHQD)_2$-PHAL, respectively, of part I of Figure 17.21.

Ligand acceleration of asymmetric dihydroxylations as described above occurs if the nitrogen atom of *one* of these dihydroquinuclidine subunits binds to OsO_4. Rotational symmetry in $(DHQ)_2$-PHAL and $(DHQD)_2$-PHAL then implies that the same asymmetric induction takes place, irrespective of whether OsO_4 is bound by the left or by the right dihydroquinuclidine subunit —in the projection of Figure 17.21 (part I). It still has to be explained why it is the nitrogen atom of dihydroquinuclidine that binds to OsO_4 and not one of the aromatic nitrogen atoms. The reason is the hybridization of the atomic orbital accommodating the free electron pair that establishes the bond with osmium. The free electron pair of the nitrogen atom of dihydroquinuclidine resides in an sp^3 AO and the free electron pair of the nitrogen atom of an aromatic compound in an sp^2 AO. An sp^3 AO and an sp^2 AO consist of 25% and 33% *s*-character, respectively. Hence, an electron pair in an sp^3 AO is more remote from the nucleus, and thus less strongly bound and more easily accessible to OsO_4 than an electron pair in an sp^2 AO.

Figure 17.21 (part II) shows the 1:1 complex from $(DHQD)_2$-PHAL and OsO_4 together with the stereostructure, which is derived from the previous discussion, in the transition state of the asymmetric dihydroxylation. Here, the alkene nestles between the amine-complexed OsO_4 on the one side and the methoxyquinoline residue on the other. The enantioselectivity of the dihydroxylation is the result of the alkene's preference to nestle in this niche with the orientation shown here. This orientation is characterized by the fact that no repulsion may occur between the alkene and the "bottom" of this niche, i.e., the central heterocycle of $(DHQD)_2$-PHAL. This is the case if and only if the sp^2-bound hydrogen atom (as the smallest double bond substituent in the substrate) points in the direction of the central heterocycle.

Once this and the principle of stereocontrol of Sharpless' asymmetric dihydroxylations is understood it is clear that both *cis*- and *trans*-disubstituted alkenes allow for two orientations in the transition state. See for yourself by explicitly writing down the corresponding dihydroxylations according to the general format laid out in Figure 17.21 (part II): based on *cis*-

alkenes the coexistence of these orientations leads to low *ee* values; *trans*-disubstituted alkenes, however, produce a totally different picture!

Side Note 17.3. *cis-vic*-Aminohydroxylation of C=C Double Bonds

Starting from catalytic amounts of osmium tetroxide, the aza analog of the *cis-vic-* dihydroxylation of alkenes, namely the *cis-vic*-aminohydroxylation (Figure 17.22), can be realized. This reaction relies first on the generation of catalytic amounts of osmium trioxide in the reaction mixture and second on its oxidation to osmium(VIII) occurring in such a way that $O_3Os{=}N$–EWG instead of OsO_4 ("$O_3Os{=}O$") is formed (EWG stands for electron-withdrawing group). Oxidations of OsO_3 with this kind of chemoselectivity can be conducted with deprotonated *N*-chlorosulfonic acid amides, *N*-halogen carboxylic acid amides or *N*-chlorocarbamidic acid esters. In the reaction mixture, these oxidants must be present in stoichiometric amounts. They are either employed as such—for example $Na^{\oplus}Cl{-}N^{\ominus}{-}SO_2Tol$ ("chloroamine T")—or generated *in situ* via deprotonation of the corresponding neutral compound.

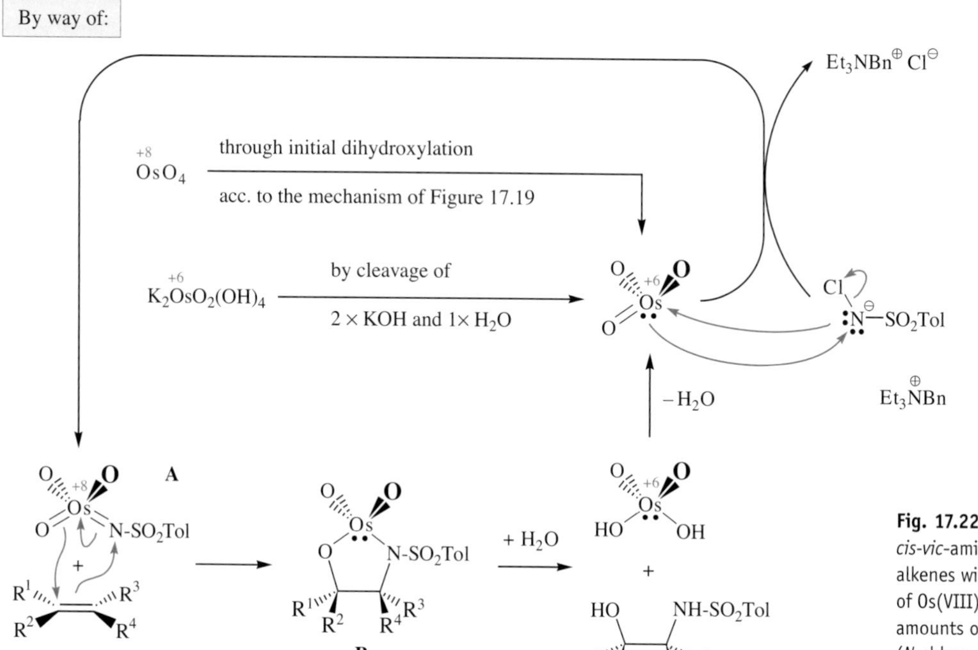

Fig. 17.22. Mechanism of the *cis-vic*-aminohydroxylation of alkenes with catalytic amounts of Os(VIII)/stoichiometric amounts of chloramine T (*N*-chloro-*para*-toluene sulfonamide sodium salt).

(The mechanism of the *cis-vic*-aminohydroxylation (Figure 17.22) resembles that of the *cis-vic*-dihydroxylation (Figure 17.19). If OsO_3 undergoes oxidation with chloramine T to $O_3Os{=}N{-}SO_2Tol$ (**A**), a 1,3-dipole is obtained that reacts with its Os=N- and an Os=O bond in a 1,3-dipolar cycloaddition **A** → **B**. The resulting osmium(VI) amido ester **B** hydrolyzes to form the *cis-vic*-amino alcohol with an acceptor substituent at the N atom and to OsO_3, which maintains the catalytic cycle until the reaction is completed.

With *cis-vic*-aminohydroxylations of unsymmetrical alkenes, however, it may be a problem that two regioisomers occur—a complication that does not occur with *cis-vic*-dihydroxylations. The addition of $(DHQ)_2$-PHAL or $(DHQD)_2$-PHAL (Figure 17.21, part I) in a *cis-vic*-aminohydroxylation will also cause asymmetric catalysis. The related reactions are known as **asymmetric aminohydroxylations**.

Oxidative Cleavage of Glycols

$NaIO_4$ or H_5IO_6 can be used to cleave vicinal glycols into two carbonyl compounds or into a single compound with two carbonyl groups (Figure 17.23). The formation of a diester of an iodo(VII) acid as an intermediate is key to the success of this reaction. Such a diester formation is even possible starting from *trans*-configured 1,2-cyclohexanediols. The diester intermediate decomposes in a one-step reaction in which three valence electron pairs are shifted simultaneously. One of these shifted electron pairs ends up as a lone pair on iodine, and the iodine(VII) initially present is thereby reduced to iodine(V).

Glycols are essential structural elements of sugars. Figure 17.24 shows the oxidative cleavages of two sugar derivatives **A** and **B**, both monoglycols, with $Pb(OAc)_4$ (mechanism below). Glycol **A** is formed in acidic environment from D-mannitol and acetone. It is formed chemoselectively and regioselectively as the least sterically hindered (see discussion of Figure 9.18) D-mannitol bisacetonide. Glycol **B** is obtained in an analogous fashion in the presence of acid as the most stable monoacetonide of the dithioacetal of L-arabinose. This dithioacetal can be prepared starting from L-arabinose with the procedure described in Section 9.2.3 for the preparation of the dithioacetal of D-glucose. The oxidative cleavages of glycols **A** and **B** (Figure 17.24) result in acetonides of the *R*- and *S*-configured glycerol aldehydes. Taking into account their origin, these compounds belong to the "chiral pool," that is, the pool of naturally occurring chiral compounds.

The glycol cleavages shown in Figure 17.24 are carried out with $Pb(OAc)_4$ in an anhydrous solvent. If one wanted to use $NaIO_4$ as the oxidizing agent, the same cleavages would have to

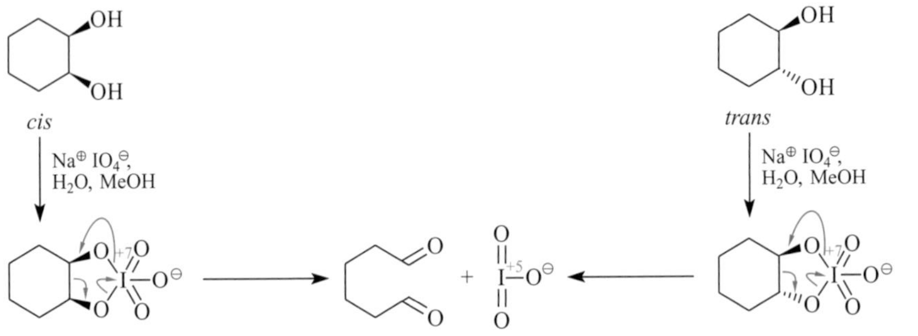

OH OH *cis* $Na^{\oplus}\ IO_4^{\ominus}$, H_2O, MeOH

OH OH *trans* $Na^{\oplus}\ IO_4^{\ominus}$, H_2O, MeOH

O +7 O I O⊖ O O → O O + O I +5 O⊖ O ← O +7 O I O⊖ O O

Fig. 17.23. Mechanism of the glycol cleavage with $NaIO_4$ or H_5IO_6, respectively. A diester of iodo(VII) acid (periodic acid) is formed initially. The ester decomposes in a one-step reaction in which three valence electron pairs are shifted simultaneously.

be carried out in water-containing solvents for reasons of solubility. In such water-containing media, however, the desired aldehydes—α-oxygenated aldehydes—tend to form aldehyde hydrates (Figure 9.1), from which the aldehydes can be recovered only by way of an azeotropic (i.e., water-removing) distillation. With H_5IO_6, though, glycol cleavages can be achieved in the absence of water. Hence, α-oxygenated aldehydes such as the ones in Figure 17.24 could be obtained *without hydrate formation* in principle by way of the H_5IO_6-mediated glycol cleavage. Yet, this procedure would be prone to fail because of the acid sensitivity of the acetal groups in the aldehydes as well as in their precursors.

Glycol cleavages with $Pb(OAc)_4$ preferentially proceed via five-membered lead(IV) acid diesters. Such esters are shown in Figure 17.24 as **C** and **D**. Each one of these esters decomposes in a one-step reaction to furnish $Pb(OAc)_2$ and two equivalents of the carbonyl compound. The cleavage is caused by the concerted shift of three valence electron pairs. One of these becomes a lone pair at Pb. The oxidation number of lead is thereby reduced from +4 to +2.

Neither $NaIO_4$ nor H_5IO_6 can convert a conformationally fixed *trans*-glycol (such as compound **A** in Figure 17.25) into a cyclic iodo(VII) acid ester (of the type shown in Figure 17.23). Such glycols therefore cannot be cleaved oxidatively by these reagents (Figure 17.25, top left). $Pb(OAc)_4$, on the other hand, does cleave the same diol oxidatively even though a cyclic lead(IV) acid diester of the type shown in Figure 17.24 also cannot be formed. The reason is that $Pb(OAc)_4$ can react with glycols by way of a second cleavage mechanism that involves the formation of a lead(IV) acid monoester (**B** in Figure 17.25). The cleavage products are formed by a fragmentation reaction in which four electron pairs are shifted simultaneously.

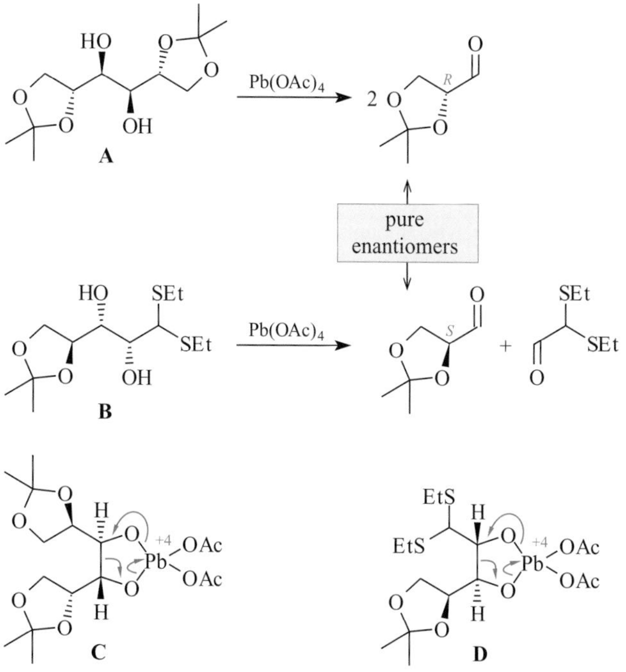

Fig. 17.24. Standard mechanism of the glycol cleavage with $Pb(OAc)_4$. The reaction proceeds preferentially via a cyclic diester of Pb(IV) acid, which decomposes in a one-step reaction to $Pb(OAc)_2$ and two equivalents of the carbonyl compound.

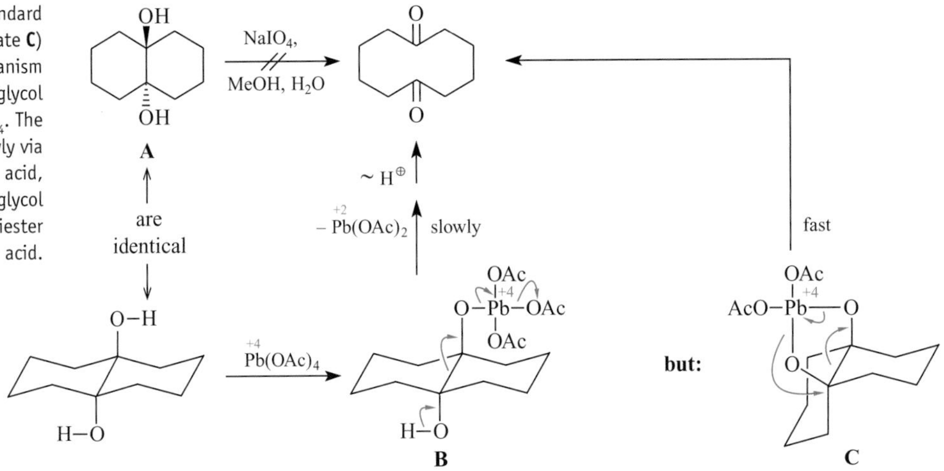

Fig. 17.25. The standard mechanism (transition state **C**) and the alternative mechanism (transition state **B**) of glycol cleavages with $Pb(OAc)_4$. The *trans*-glycol **A** reacts slowly via the monoester **B** of Pb(IV) acid, while the isomeric *cis*-glycol reacts fast via the cyclic diester **C** of lead(IV) acid.

The second mechanism described for the $Pb(OAc)_4$ cleavage of glycols is only "second rate." Accordingly, $Pb(OAc)_4$ cleaves the *cis* isomer of *trans*-glycol **A** (Figure 17.25) much faster because the *cis*-glycol can react with $Pb(OAc)_4$ via the mechanism shown in Figure 17.24 involving a cyclic intermediate (**C** in Figure 17.25).

Oxidative Cleavage of Alkenes

The glycol formation from alkenes (Figure 17.19) and the glycol cleavage with $NaIO_4$ (Figure 17.23) can be combined in a one-pot reaction. As mentioned in the discussion of Figure 17.19, such a one-pot oxidation/cleavage of alkenes can be accomplished with $KMnO_4$. Under these conditions, however, controlling the oxidation state of the cleavage products might pose problems. Aldehydes *or* carboxylic acids may be formed, and ketones might perhaps even be subject to further oxidation (see Figure 17.33).

Well-defined cleavage products will *definitely* be obtained if one employs the pairs of oxidizing reagents shown in Figure 17.26 instead of $KMnO_4$ alone. The stoichiometric oxidizing reagent is still $NaIO_4$, and catalytic amounts of an appropriate co-oxidant are added. The basic approach to achieving the various cleavages in Figure 17.26 always is the same: the alkenes first are converted into *cis*-configured diols, which are then subjected to glycol cleavage. The further oxidation of an aldehyde is most likely a subsequent reaction.

In the **Lemieux–Johnson oxidation** of alkenes (Figure 17.26, left), the dihydroxylation of the C=C double bond is achieved by using catalytic OsO_4. $NaIO_4$, employed in stoichiometric amounts, plays two roles: it effects the reoxidation of Os(VI) to Os(VIII), and it cleaves the glycols to the aldehydes and/or ketones (mechanism in Figure 17.23).

In the **Lemieux–von Rudloff oxidation** of alkenes (Figure 17.26, right), stoichiometric amounts of the oxidizing reagent $NaIO_4$ are used. In the classical procedure, a catalytic amount of $KMnO_4$ is employed as the co-oxidizing reagent; more modern variants employ RuO_2 or $RuCl_3$. The *cis* hydroxylation is effected by the MnO_4^- ion in the first case and by RuO_4, formed *in situ*, in the other two cases. Both reactions formally resemble the corresponding OsO_4 reaction (Figure 17.19) and are mechanistically rather similar. The periodate

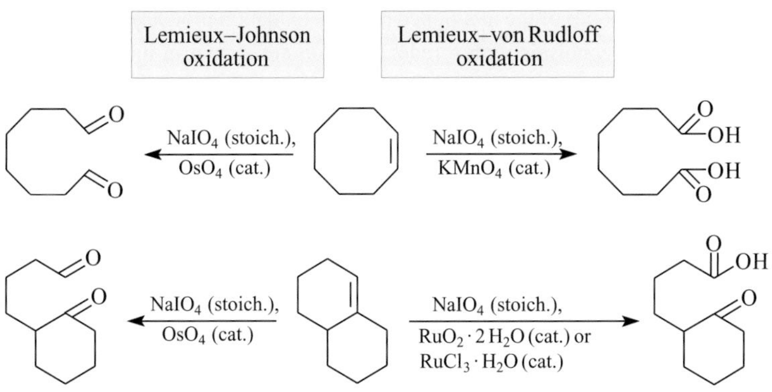

Fig. 17.26. Oxidative cleavages of alkenes with symmetrically (top) and asymmetrically (bottom) substituted C=C double bonds.

subsequently cleaves the formed glycol to aldehydes and/or ketones (for the mechanism, see Figure 17.23). The MnO_4^- anion or RuO_4 further oxidizes any initially formed aldehydes to carboxylic acids via the aldehyde hydrate (see the mechanism shown in Figure 17.11).

Alkene C=C double bonds also can be cleaved by **ozone**. The mechanism of this reaction was discussed in connection with Figures 15.47 and 15.48, and it has been explained that

- mixtures of hydroperoxides and carbonyl compounds are formed in MeOH, and
- mixtures of tetroxane(s) and ketone(s) or mixtures of secondary ozonides are formed in CH_2Cl_2.

These findings are summarized in the upper part of Figure 17.27.

The ozonolyses need to be *terminated* with a redox reaction (or by the *β*-elimination of an acylated hydroperoxide as shown, e. g., in Figure 17.28, right). Each of these reactions breaks

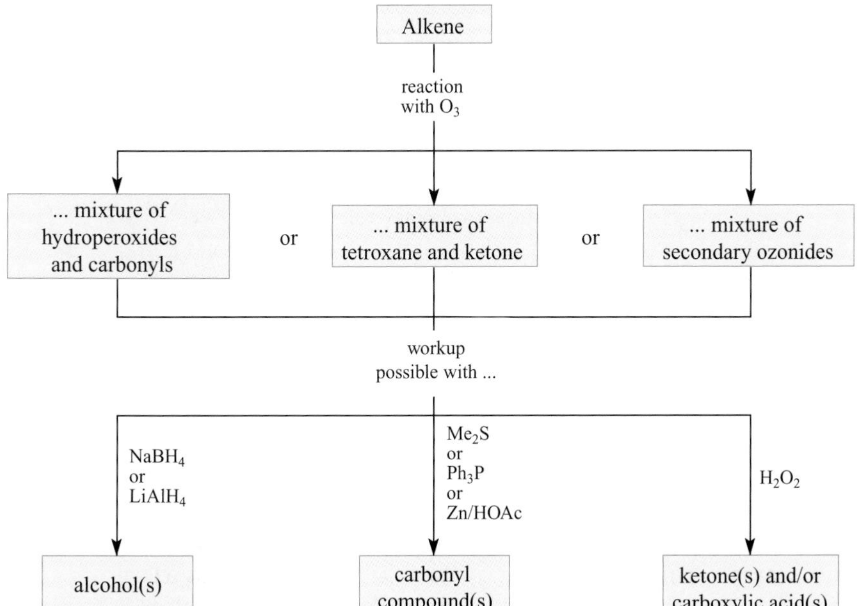

Fig. 17.27. Intermediates (center row) and final products (bottom row) of the ozonolysis of alkenes.

the weak O—O single bonds of the mentioned primary oxidation products. The usual methods for the workup of an ozonolysis with a redox reaction are shown in the lower part of Figure 17.27:

- The use of $NaBH_4$ or $LiAlH_4$ results in the reduction of the O—O and C=O bonds in the primary products to alcohol groups.
- The use of Me_2S or Ph_3P or Zn/HOAc leaves the molecules containing a C=O bond unaffected, while the molecules containing an O—O bond are reduced to carbonyl compounds.
- The use of H_2O_2 leaves ketones intact and converts all initially formed aldehydes and their derivatives into carboxylic acids.

The ozonolysis of cyclohexene to 1,6-dioxygenated compounds is shown in Figure 17.28. Other cycloalkenes similarly afford other 1,γ-dioxygenated cleavage products. With the three methods for workup (Figure 17.27), this ozonolysis provides access to 1,6-hexanediol, 1,6-hexanedial, or to 1,6-hexanedicarboxylic acid. Each of these compounds contains two functional groups of the same kind.

Cyclohexene also can be cleaved into a 1,6-dioxygenated C_6 chain with *different* termini, and this option is shown at the far right in Figure 17.28. Cyclohexene is subjected to ozonolysis in methanol to afford the hydroperoxide **A**. The peroxoacetate **B** is obtained by treatment of **A** with acetic anhydride/NEt_3 (mechanism analogous to Figure 6.2). NEt_3 might cause the elimination of HOAc from **B** in an E2-type fashion. One could also imagine that the peroxoacetate **B** undergoes a *cis*-elimination via the cyclic transition state shown (this transition

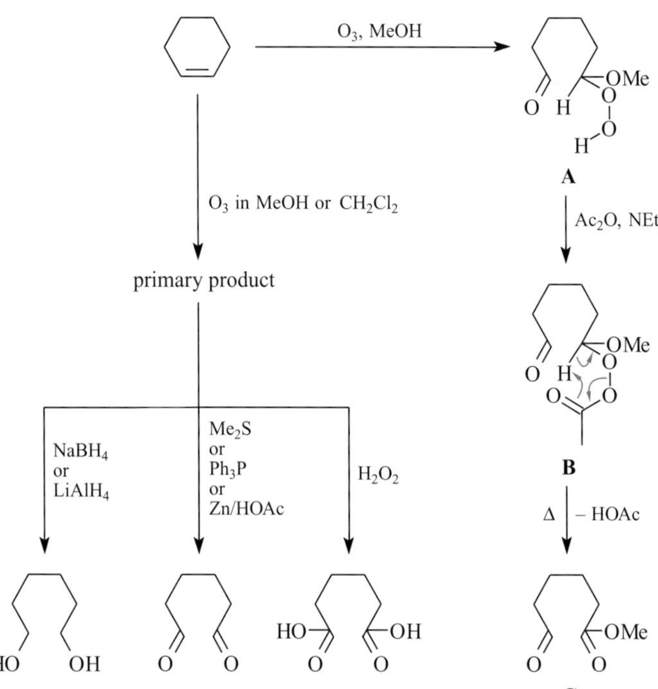

Fig. 17.28. Conversion of cyclohexene into symmetric cleavage products (left side) and into one asymmetric product (right side).

state resembles the one of the Baeyer–Villiger rearrangement shown in Figure 14.37). In any case, the aldehyde ester **C** is formed as the elimination product.

Oxidative Cleavage of Aromatic Compounds

The reagents that effect the oxidative cleavage of *alkene* C=C double bonds (Figures 17.26–17.28) in principle also are suitable for the cleavage of *aromatic* C=C double bonds (Figures 17.29–17.31). The mechanism is unchanged.

Ru(VIII) cleaves monoalkylbenzene to alkanecarboxylic acid (Figure 17.29). The Ru(VIII) is generated *in situ* by using a stoichiometric amount of NaOCl or $NaIO_4$ and a catalytic

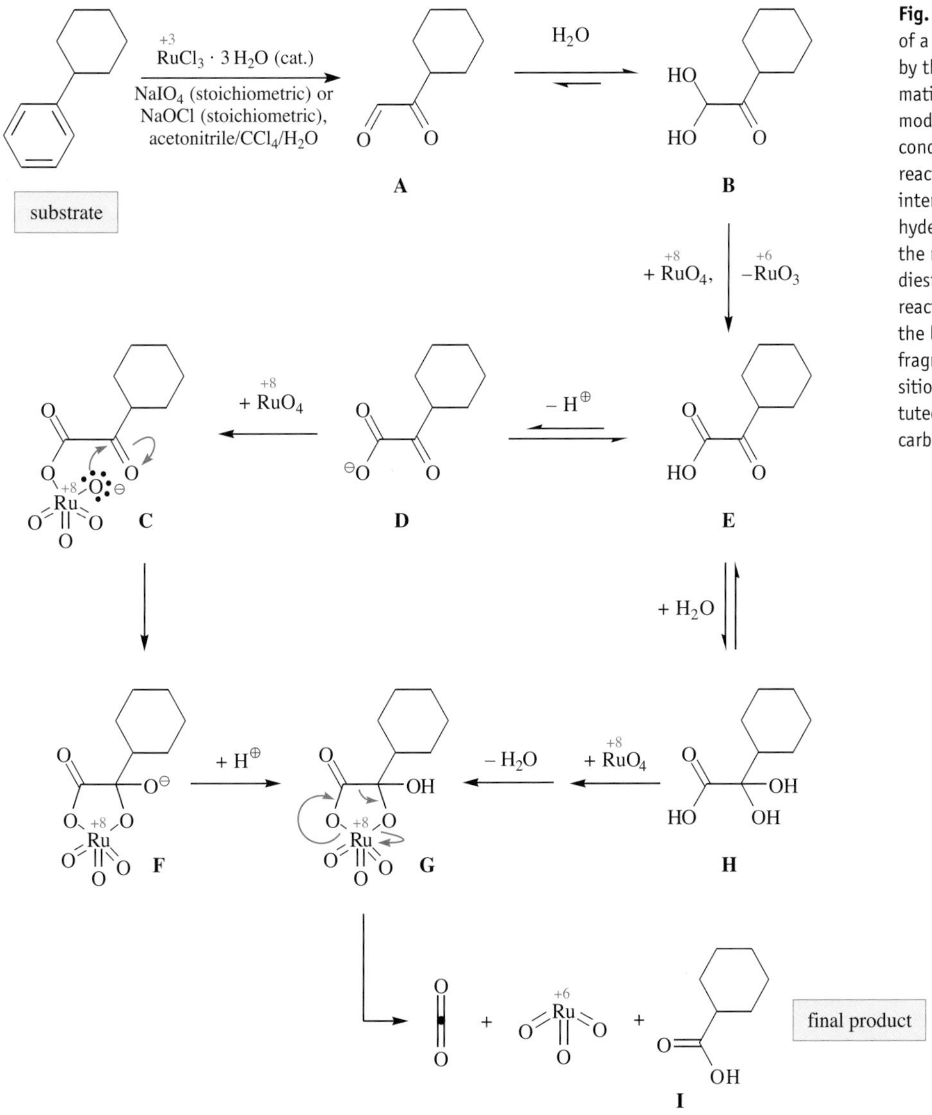

Fig. 17.29. The RuO_4 cleavage of a phenyl ring is exemplified by the case of an alkylated aromatic compound using the modified Lemieux-von Rudloff conditions (Figure 17.26). The reaction involves three key intermediates: the α-ketoaldehyde **A**, the α-keto acid **E**, and the ruthenium(VIII) acid diester **G**. It is unclear how the reaction of **E** → **G** proceeds. In the last step, **G** undergoes a fragmentation via a cyclic transition state and forms a substituted acetic acid **I**, RuO_3 and carbon dioxide.

amount of RuO_2. The Ru(VIII) is continuously regenerated. The α-ketoaldehyde **A** is formed as the first key intermediate. Its two C=O double bonds stem from the oxidative cleavage of a C=C bond that the substrate contained in their place. In an aqueous environment, this α-ketoaldehyde **A** should to a large extent further react to yield the aldehyde hydrate **B** (see Section 9.1.1). As shown in Figure 17.12, aldehyde hydrates are oxidized by RuO_4 to give a carboxylic acid. If the α-keto acid **E** of Figure 17.29 has been formed in this way there are two options available for its further reaction. It is not known which of them is used; only one thing is certain, namely that in either case the same cyclic ruthenium(VIII) acid diester **G** is ultimately obtained. The first variant would start with the deprotonation of the α-keto acid **E** to give the carboxylate **D** and with its nucleophilic addition to a Ru=O "double bond" of RuO_4. This would lead to the mixed anhydride **C**. After an addition to the C=O double bond in terms of a ring closure reaction and a protonation the ruthenium(VIII) acid *di*ester **G** would be formed. In a second variant the formation of **G** from the α-keto acid **E** would begin with an energetically favorable (cf. Table 9.1) addition of water to the ketone hydrate **H**. The latter could undergo condensation with RuO_4 to form the ruthenium(VIII) acid *di*ester **G**.

The Ru(VIII) acid diester **G** is the second key intermediate of the oxidative cleavage shown in Figure 17.29. **G** fragments by way of a concerted shift of three valence electron pairs in analogy to the reactions of cyclic iodo(VII) acid diesters (Figure 17.23) and of cyclic lead(IV) acid diesters (Figure 17.24). One of the participating valence electron pairs becomes a nonbonding electron pair at Ru, and the oxidation number of Ru therefore is lowered from +8 to +6. The organic products of this cleavage are an alkane carboxylic acid **I** remaining in the reaction mixture and carbon dioxide escaping from it.

Only *one* ring of condensed aromatic systems is cleaved with either of the two mixtures of oxidizing agents shown in Figure 17.29—stoichiometric $NaIO_4$/catalytic $RuCl_3$ or stoichiometric NaOCl/catalytic $RuCl_3$. In the laboratory, naphthalene, for example, can be oxidized to obtain phthalic acid anhydride by using such a catalytic oxidation (Figure 17.30). The industrial variant of this oxidation uses air as the stoichiometric oxidant and a V_2O_5 catalyst. It is unknown whether the mechanisms of these two oxidations are analogous.

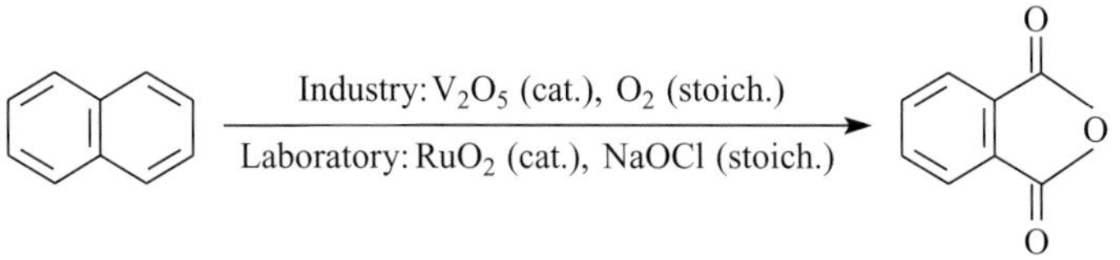

Industry: V_2O_5 (cat.), O_2 (stoich.)
Laboratory: RuO_2 (cat.), NaOCl (stoich.)

Fig. 17.30. Transition metal-mediated cleavage of naphthalene.

Ozone, too, can be used to oxidize a benzene ring into a carboxylic acid (Figure 17.31). In this case, the primary product of the ozonolysis is worked up under oxidizing conditions (see Figure 17.27). The fact that phenyl rings undergo ozonolysis is synthetically useful because it allows one to introduce a comparatively inert phenyl ring into a synthetic intermediate instead of a more reactive carboxylic acid. The carboxyl group can be generated subsequently, as illustrated by the example shown in Figure 17.31. In this context, the phenyl ring plays the role of a **masked** or **latent** carboxyl group.

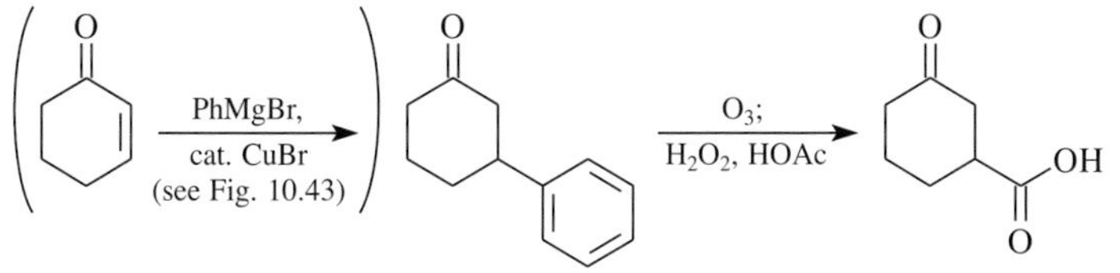

Fig. 17.31. Ozonolysis of a phenyl ring.

Oxidative Cleavage of Ketones

The C_α—C(=O) bond of ketones can be cleaved with peracids, and esters are formed via a Baeyer–Villiger rearrangement and the insertion of an O atom (for mechanism, see Figures 14.34–14.38). Cyclic ketones produce lactones in this way (Figure 14.34). Cyclobutanones react in a rather similar fashion with H_2O_2 or *tert*-BuOOH and form γ-butyrolactones (Figure 14.39). The Baeyer–Villiger rearrangement of unsymmetrical ketones breaks the C_α—C(=O) bond that leads to the higher substituted C atom in the α-position (Section 14.4.2). If this atom is a stereocenter, one observes complete retention of configuration (Figure 14.34). The top reaction in Figure 17.32 illustrates the corresponding synthetic potential of the Baeyer–Villiger oxidation with the example of the regioselective and stereoselective cleavage of menthone.

The bottom reaction of Figure 17.32 shows an oxidative cleavage of the same menthone with the opposite regiochemistry. In a preparatory step, menthone is converted into the silyl enol ether **B** via its kinetic enolate (cf. Figure 13.11). The C=C double bond of this compound is subsequently cleaved with ozone. This cleavage (cf. Figures 15.47 and 15.48) proceeds in MeOH as if the Me_3SiO substituent was not even present, and the ozonolysis results in an α-methoxyhydroperoxide that is a part of a trimethylsilyl ester. $NaBH_4$ is added to reduce the hydroperoxide to an alcohol. Of course, $NaBH_4$ does not react with the trimethylsilyl ester. The *reduced* cleavage product is thus the hydroxy-substituted silyl ester **C**. Upon acidic workup, ester **C** hydrolyzes, and the hydroxycarboxylic acid that is formed cyclizes rapidly to provide lactone **D** (cf. Figure 6.27). **D** is a constitutional isomer of lactone **A**, which was obtained earlier via the Baeyer–Villiger oxidation of menthone (cf. Figure 17.32, top).

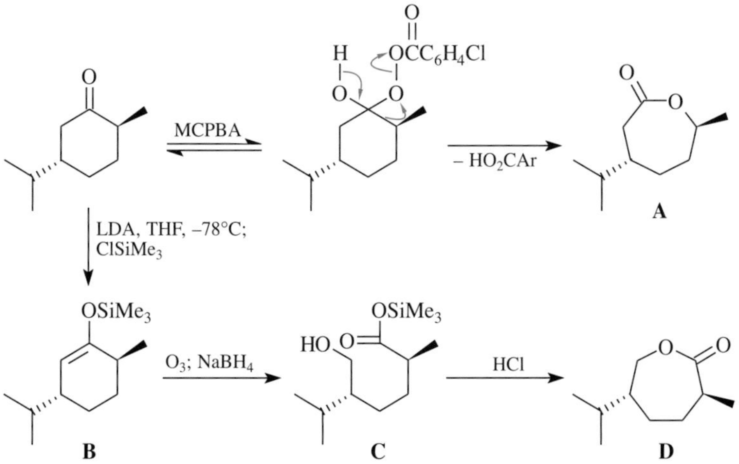

Fig. 17.32. Oxidative cleavage of an asymmetric ketone with complementary regioselectivities. Lactone **A** is obtained by Baeyer–Villiger oxidation of menthone [2-methyl-5-(1-methylethyl)cyclohexanone]. Alternatively, one may first convert menthone into the silylenol ether **B** and cleave its C=C double bond with ozone to obtain a silyl ester containing an α-methoxyhydroperoxide group as a second functional group (which resembles the unstable structural element of the so-called ether peroxides; cf. Figure 1.38). The latter is reduced with $NaBH_4$ to the hydroxylated silyl ester **C**. The hydroxycarboxylic acid is obtained by acid-catalyzed hydrolysis. It cyclizes spontaneously to give lactone **D**.

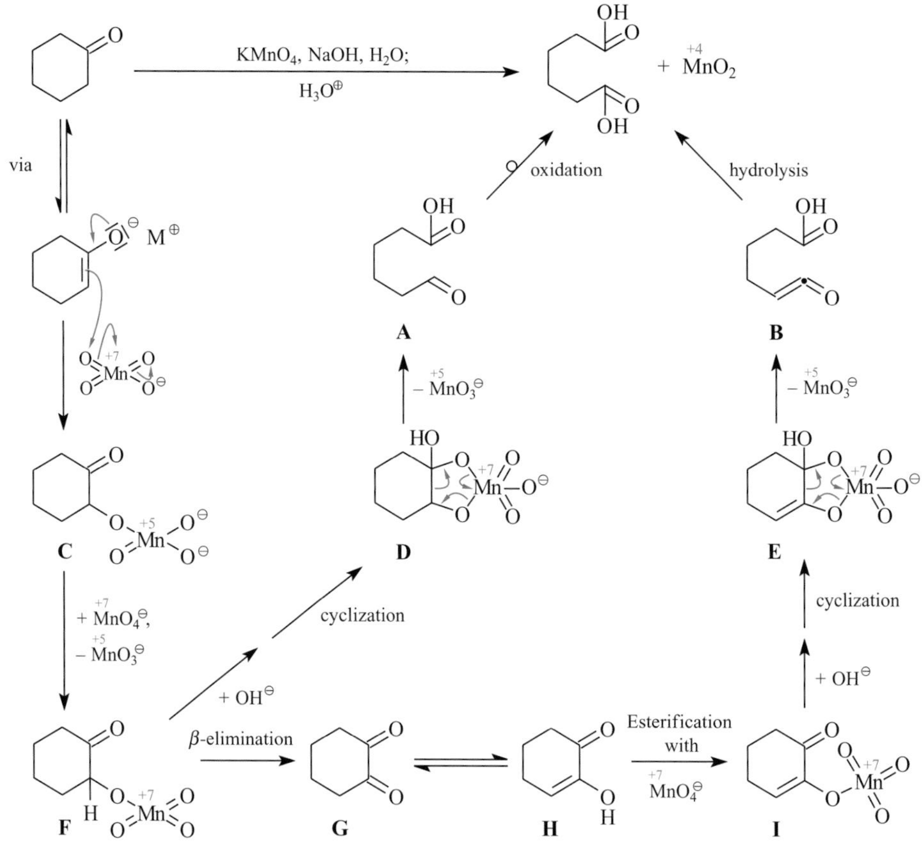

Fig. 17.33. Mechanism of the $KMnO_4$ cleavage of cyclohexanone to adipic acid.

Cyclic ketones also can be cleaved by oxidation with $KMnO_4$ (Figure 17.33). Under these conditions, 1,ω-ketocarboxylic acids or 1,ω-dicarboxylic acids are formed. The mechanism of this reaction is not known in detail. It is known only that $KMnO_4$ serves as a three-electron oxidizing reagent and that it is reduced to MnO_2.

A plausible mechanism is shown in Figure 17.33. It is assumed that the enolate, not the ketone itself, reacts with the permanganate. The MnO_4^- anion presumably reacts with the enolate carbon electrophilically with one of its doubly bonded O atoms. The manganese(V) acid *mono*ester **C** is formed and is oxidized by a permanganate ion to give the analogous manganese(VII) acid *mono*ester **F**. The subsequent chemistry of **F** is difficult to predict.

- **Variant 1.** The manganese(VII) acid monoester **F** is hydrated at the carbonyl group or at the metal. Subsequently, the ring closes and leads to the manganese(VII) acid diester **D**. Its heterocyclic ring fragments by a concerted shift of three valence electron pairs, as in the cases of the esters of the iodo(VII) acid, lead(VI) acid, and ruthenium(VIII) acid (Figures 17.23, 17.24, and 17.29, respectively). The aldehyde acid **A** is the organic product of this fragmentation reaction, and it is oxidized subsequently to the dicarboxylic acid.
- **Variant 2.** The manganese(VII) acid monoester **F** undergoes a β-elimination of a manganese(V) acid, and diketone **G** is formed. Compound **G** is in equilibrium with enol **H**.

Enol **H** and a permanganate ion form the manganese(VII) enol ester **I** by way of a condensation reaction. Ester **I** reacts further to form the manganese(VII) acid diester **E**. Then **E** might undergo fragmentation to yield the ketene carboxylic acid **B**. The sequence **I** → **E** → **B** is step for step analogous to the sequence **F** → **D** → **A** of variant 1. Finally, the ketene carboxylic acid **B** would have to be hydrated to the dicarboxylic acid (for the mechanism, see Figure 8.13).

17.3.3 Oxidations at Heteroatoms

The oxidations shown in Figures 17.34–17.37 and the primary reaction in Figure 17.38 all involve the same pattern. In all cases, the heteroatom reacts as the nucleophile in an S_N2-type substitution reaction with an O atom of the O—O bond of H_2O_2 or a peracid.

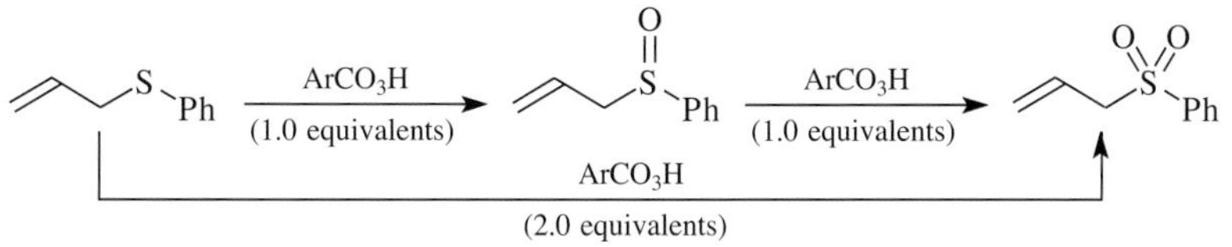

Fig. 17.34. Sulfide → sulfoxide → sulfone oxidation(s).

Sulfoxides are most commonly prepared via the oxidation of sulfides with peracids, *tert*-BuOOH, or H_2O_2 (Figure 17.34). Sulfoxides are obtained in a chemoselective fashion if one equivalent of the respective oxidizing reagent is employed. Sulfones also can be obtained chemoselectively from the same sulfides if two equivalents of the oxidizing reagent are employed (Figure 17.34). The two oxidations as well as the oxidation of a sulfoxide to a sulfone with one equivalent of the same oxidizing agent can also be carried out in the presence of C=C double bonds.

Fig. 17.35. A one-pot combination of a selenide → selenoxide oxidation and a selenoxide "pyrolysis" (see Figures 4.10–4.12 for the mechanism).

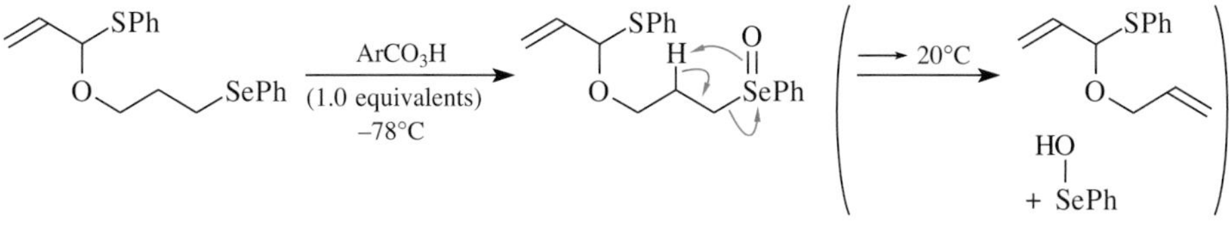

The oxidation of selenides to selenium oxides (Figure 17.35) is faster than the oxidation of sulfides to sulfoxides. The former reaction also succeeds in the presence of C=C double bonds, and even a sulfide group that might be present in the substrate will not react. Tertiary amines yield amine *N*-oxides in a similar fashion (Figure 17.36).

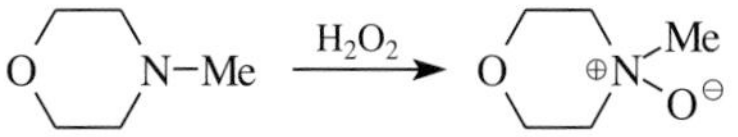

Fig. 17.36. A *tert*-amine → amine *N*-oxide oxidation as preparation of the oxidizing reagent NMO (see Figures 17.15 and 17.19 for synthetic applications).

Hydrazone **A** in Figure 17.37 undergoes oxidation just like an amine, and amine *N*-oxide **B** is formed. Immediately, **B** undergoes a *β*-elimination via a cyclic transition state, and nitrile **C**

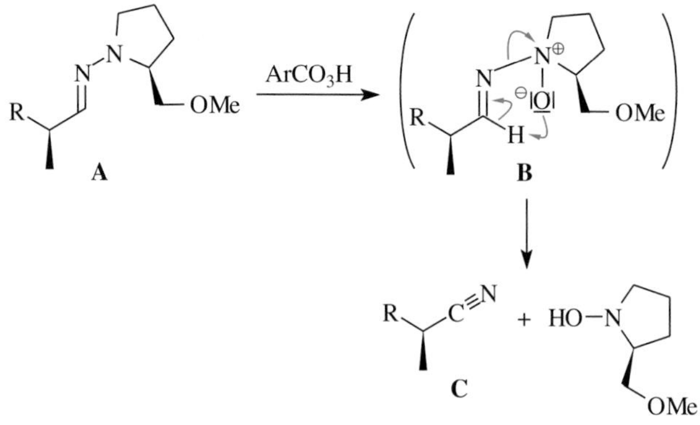

Fig. 17.37. A hydrazone → hydrazone *N*-oxide oxidation as part of a one-pot reaction for the conversion of SAMP hydrazones into enantiomerically pure nitriles.

and a hydroxylamine are formed. Since the hydrazone precursor is accessible as a pure enantiomer (cf. Figure 13.34), the nitrile also can be generated as a pure enantiomer.

Side Note 17.4. Preparation of TEMPO

Peroxo compounds initially oxidize secondary amines to the corresponding hydroxyl amines (e. g., **A** in Figure 17.38) but the oxidation generally does not stop at this stage. The hydroxylamine loses the H atom of the hydroxyl group and a nitroxyl radical, a rather resonance-stabilized radical, is formed. The nitroxyl radical shown in Figure 17.38 is so stable that it can be stored. It is commercially available as the oxidizing agent TEMPO (tetramethylpiperidine nitroxyl free radical).

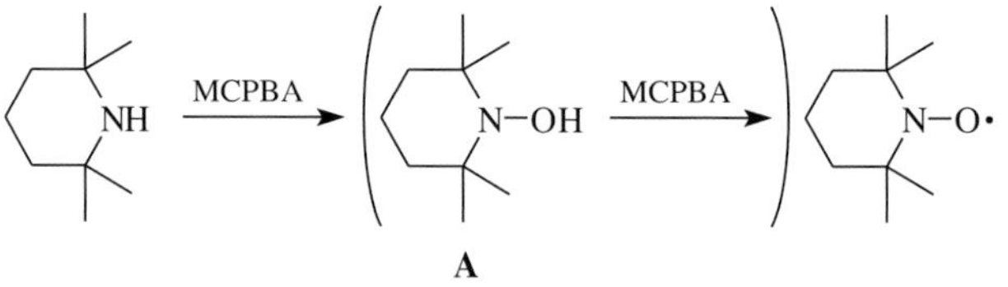

Fig. 17.38. A *sec*-amine → nitroxyl radical oxidation exemplified by the preparation of the oxidizing agent TEMPO (see Figure 17.16 for a synthetics application).

The oxidation of *ortho*-iodobenzoic acid to give the **Dess–Martin reagent** is shown in Figure 17.39. This oxidation is a synthetically important oxidation of an iodine atom. The mechanism of the oxidation is not yet understood.

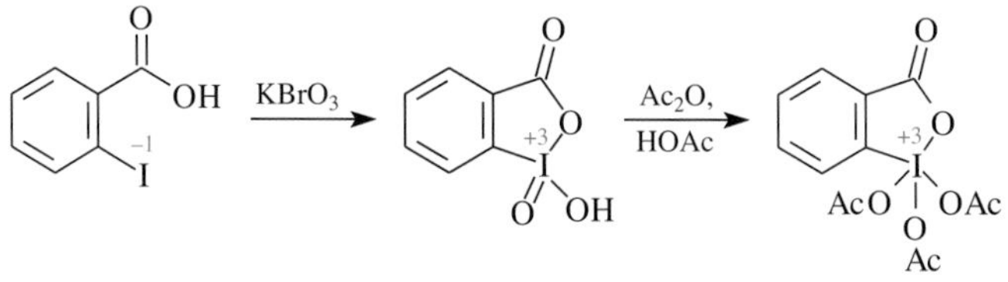

Fig. 17.39. An iodide → periodinane oxidation exemplified by the preparation of the Dess–Martin reagent (see Figure 17.14 for a synthetics application).

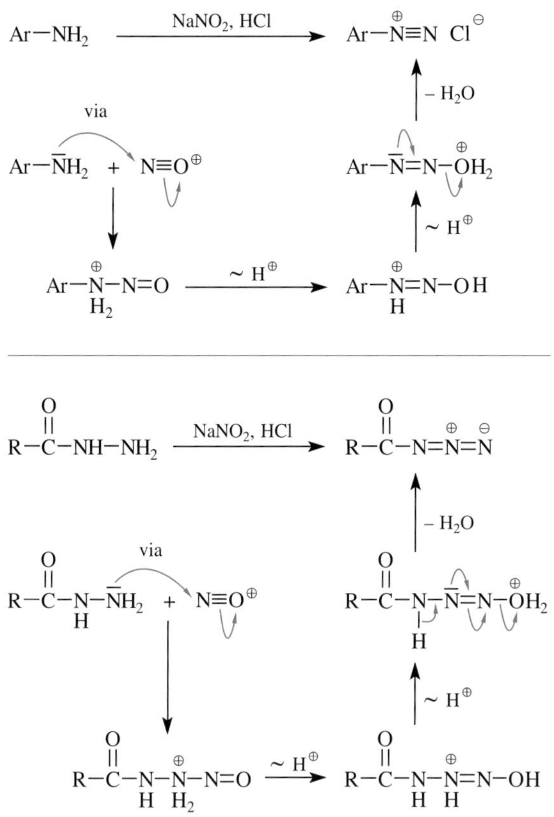

$Ar{-}NH_2 \xrightarrow{NaNO_2,\ HCl} Ar{-}\overset{\oplus}{N}{\equiv}N\ Cl^{\ominus}$

via

$Ar{-}\overline{N}H_2 + N{\equiv}O^{\oplus} \longrightarrow Ar{-}\overset{\oplus}{N}H_2{-}N{=}O \xrightarrow{\sim H^{\oplus}} Ar{-}\overset{\oplus}{N}H{=}N{-}OH \xrightarrow{\sim H^{\oplus}} Ar{-}\overline{N}{=}N{-}\overset{\oplus}{O}H_2 \xrightarrow{-H_2O} Ar{-}\overset{\oplus}{N}{\equiv}N$

$R{-}C({=}O){-}NH{-}NH_2 \xrightarrow{NaNO_2,\ HCl} R{-}C({=}O){-}N{=}\overset{\oplus}{N}{=}\overset{\ominus}{N}$

via

$R{-}C({=}O){-}NH{-}\overline{N}H_2 + N{\equiv}O^{\oplus} \longrightarrow R{-}C({=}O){-}NH{-}\overset{\oplus}{N}H_2{-}N{=}O \xrightarrow{\sim H^{\oplus}} R{-}C({=}O){-}NH{-}\overset{\oplus}{N}H{=}N{-}OH \xrightarrow{\sim H^{\oplus}} R{-}C({=}O){-}NH{-}\overline{N}{=}N{-}\overset{\oplus}{O}H_2 \xrightarrow{-H_2O} R{-}C({=}O){-}N{=}\overset{\oplus}{N}{=}\overset{\ominus}{N}$

Fig. 17.40. Comproportionation of two nitrogen compounds to give a diazonium ion (top) or an acyl azide (bottom), respectively.

As our last examples of oxidation reactions of heteroatoms, we consider the reactions depicted in Figure 17.40. These reactions are oxidations and, in contrast to the reactions in Figures 17.34–17.38, these oxidations also are condensation reactions, since the oxidizing reagent remains in the product. The mechanistic details outlined in Figure 17.40 are so familiar by now that no further explanation is needed.

17.4 Reductions

In inorganic chemistry, the term **reduction** indicates a process in which a substrate gains electrons. Of course, the same is true in organic chemistry as well. Additional orientation is provided by the classes of compounds compiled in Tables 17.1 and 17.2. Because of their particular order, reductions can be described as "transformations that convert any given compound into a compound in a column further to the left." As the tables reveal, reductions of organic chemical compounds often are associated with a net-uptake of hydrogen. In accordance with these statements, there are the following reducing reagents in organic chemistry:

- electron donors (metals, which dissolve in suitable solvents in the presence or absence of a proton donor);
- elemental hydrogen (in catalytic hydrogenations or hydrogenolyses, respectively);
- H-atom transfer reagents [Bu_3SnH, $(Me_3Si)_3SiH$; see Section 1.10]; and
- reagents that transfer nucleophilic hydrogen.

The kinds of reagents that belong to the last group of reducing reagents were mostly already discussed in Section 10.1:

- covalent neutral metal hydrides such as BH_3, DIBAL, or Et_3SiH (in the special case of the reduction of oxocarbenium ions or benzyl cations; not yet discussed reductant);
- soluble ionic complex metal hydrides derived from tetravalent boron or from tetravalent aluminum; and
- organometallic compounds that contain a *β*-H-atom that can be transferred onto an organic substrate.

17.4.1 Reductions R_{sp^3}—X → R_{sp^3}—H or R_{sp^3}—X → R_{sp^3}—M

Primary and secondary alkyl bromides, iodides, and sulfonates can be reduced to the corresponding alkanes with $LiBHEt_3$ (**superhydride**) or with lithium aluminum hydride ($LiAlH_4$, other names: lithium tetrahydridoaluminate or lithium alanate). If such a reaction occurs at a stereocenter, the reaction proceeds with substantial or often even complete stereoselectivity via backside attack by the hydride transfer reagent. The reduction of alkyl chlorides to alkanes is much easier with superhydride than with $LiAlH_4$. The same is true for sterically hindered halides and sulfonates:

$$\text{Bu–C(Me)}_2\text{–CH}_2\text{–OTs} \xrightarrow{\text{LiBHEt}_3} \text{Bu–C(Me)}_2\text{–CH}_3$$

Several options can be considered when it comes to the mechanisms of these reductions and of other reductions with complex metal hydrides. It is convenient to imagine that a hydrogen atom with hydride character is detached from the reducing agent in the transition state. However, $LiAlH_4$ appears to be capable of effecting a single electron transfer onto organic substrates.

The same reducing agents also react with epoxides in S_N2-type reactions converting them into alcohols (Figure 17.41). The sterically less hindered C—O bond reacts in unsymmetrical epoxides regioselectively.

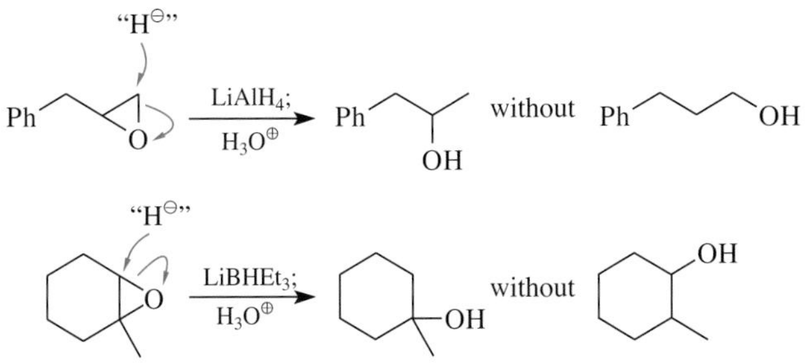

Fig. 17.41. Reduction of epoxides with hydride-transferring agents. The sterically less hindered C—O bond reacts regioselectively, i.e., the C_{prim}–O- instead of the C_{sec}–O bond in the first and the C_{sec}–O- instead of the C_{tert}–O bond in the second example.

Side Note 17.5. Radical Epoxide Reduction: Borane-free Two-step Synthesis of anti-Markownikow Alcohols from Alkenes

Unsymmetrical epoxides can be reduced to alcohols with the reverse regioselectivity compared to hydrdride reduction by means of bis(cyclopentadienyl)titanium(III) chloride (Figure 17.42). The upper half of the figure shows the reaction of the same epoxide **A**, a substrate in Figure 17.41. Ring-opening is achieved with a stoichiometric amount of the above-mentioned Ti(III) reagent and regioselectively leads to the most stable radical possible, i.e., the *secondary* radical **C**. The radical center of **C** abstracts a hydrogen atom from the 1,4-cyclohexadiene that has been added in stoichiometric amounts. This reaction has a considerable driving force since a delocalized radical (**D**) is generated from a localized radical (**C**). Upon workup, the resulting Ti(IV) alkoxide **E** is protonated to form the phenylpropanol isomer **B**, an isomer of the alcohol formed in Figure 17.41.

The bottom half of Figure 17.42 shows a further development of the epoxide reduction by way of radical intermediates. It illustrates nicely that every now and then interesting variants of familiar reactions may be discovered. Until the formation of the valence-saturated Ti(IV) alkoxide **I**, this reduction proceeds exactly like the first one—except for the use of only catalytic amounts of the expensive bis(cyclopentadienyl)titanium(III) chloride and the additional use of stoichiometric amounts of collidinium hydrochloride and manganese powder, which are much cheaper. Collidinium hydrochloride decomposes the Ti(IV) alkoxide **I** by way of protonating the organic moiety to the final alcoholic product **G** and converting the inorganic moiety into bis(cyclopentadienyl)titanium(IV) dichloride. The resulting collidine is the first stoichiometrically obtained side product of this reaction. Bis(cyclopentadienyl)titanium(IV) dichloride is reduced by manganese again to yield bis(cyclopentadienyl)titaniumium(III) chloride, which is ready to reductively cleave the next epoxide ring. At the same time, manganese(II) chloride is formed as the second stoichiometrically obtained side product of this reaction. The regioselectivity of the second epoxide reduction in Figure 17.42 resembles that of the first: the less α-branched of the two conceivable alcohols is formed since the most stable possible, i.e., the most *substituted*, radical is generated in the ring-opening step (which in this case is the *tertiary* radical **H**).

In a synthetic context this means that if an alkene is first epoxidized and then reduced according to the methods described, the same *anti*-Markovnikov hydration occurs as if the initial alkene had been hydroborated/oxidized (cf. Section 3.3.3).

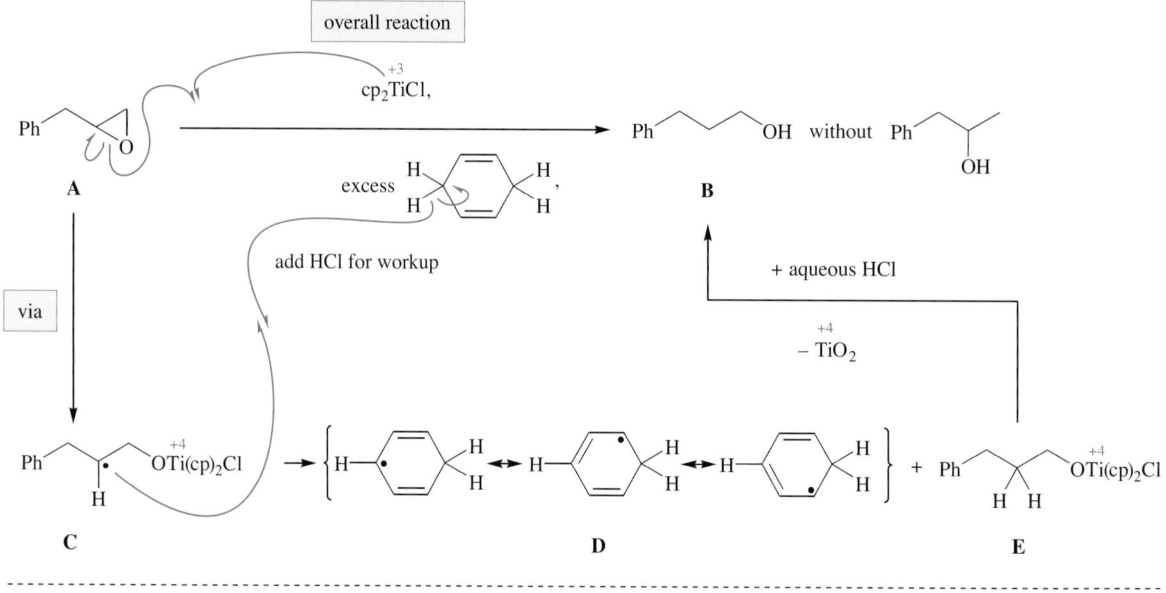

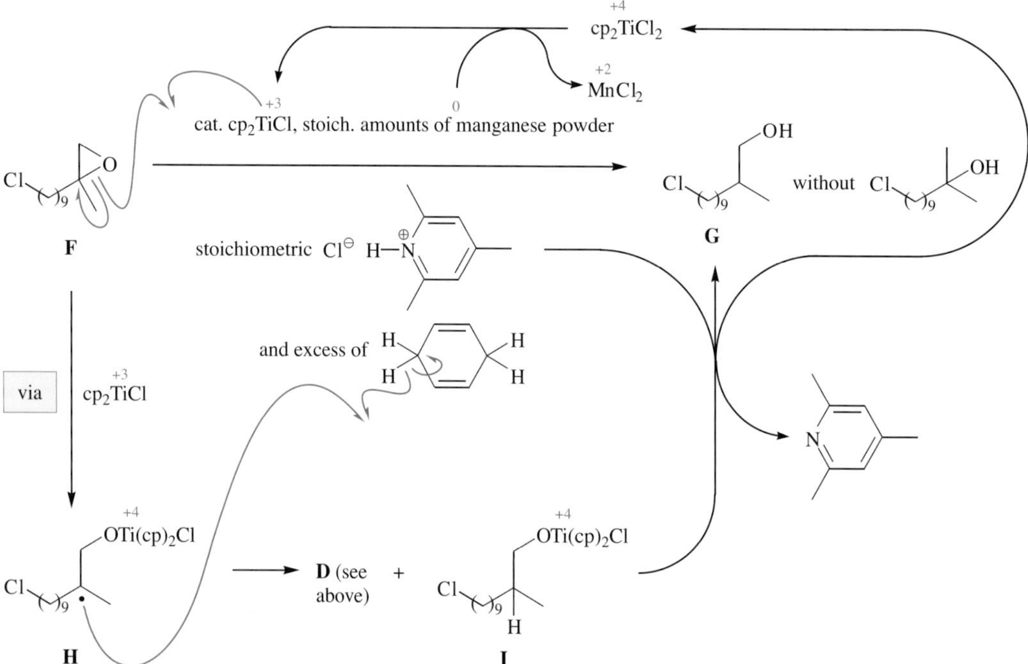

Fig. 17.42. Reduction of epoxides with electron-transferring agents and 1,4-cyclohexadiene as the H atom donor. The more sterically hindered C–O bond cleaves in a regioselective fashion, namely the C_{sec}–O- instead of the C_{prim}–O bond in the upper example—stoichiometric amounts of bis(cyclopentadienyl)titanium(III) chloride (Cp_2TiCl) act as electron carrier—or the C_{tert}–O- instead of the C_{prim}–O bond in the bottom example, respectively (electron carrier: catalytic amounts of Cp_2TiCl/stoichiometric amounts of manganese powder).

Epoxy *alcohols* (**A** in Figure 17.43) are accessible in enantiomerically pure form via the Sharpless oxidation (Figure 3.35). The epoxide substructure of these compounds can be reduced to obtain an alcohol in a regiocontrolled fashion to afford an enantiomerically pure 1,3-diol, **B**, or an enantiomerically pure 1,2-diol, **C** (Figure 17.43). The 1,3-diols **B** and the 1,2-diols **C** are best accessible via reduction of **A** with Red-Al [$NaAlH_2(OCH_2CH_2\ OCH_3)_2$] or DIBAL, respectively.

Red-Al® first generates one equivalent of hydrogen gas from epoxy alcohols **A** (Figure 17.43). An O—Al bond forms in the resulting trialkoxyaluminate **D**. The epoxy fragment in **D** then is reduced via an intramolecular reaction. The transfer of a hydride ion from aluminum leads selectively to the formation of a 1,3-diol, since the approach path that would lead to the 1,2-diol cannot be collinear to the C—O bond that would have to be broken (stereoelectronics!, cf. Section 2.4.3).

The treatment of epoxy alcohols **A** with DIBAL also first liberates one equivalent of H_2 (Figure 17.43, right). An O—Al bond is formed, which in the resulting intermediate **E** is part of a chelating ring. This intermediate can be reduced only in an *intermolecular* fashion, since it does not contain any hydridic hydrogen. The binding of the epoxy oxygen to the Al atom in the chelate should remain intact as much as possible during this reaction, and it is for this reason that the 1,2-diol is formed preferentially.

Tertiary iodides and bromides usually are reduced with Bu_3SnH or $(Me_3Si)_3SiH$ via a radical chain reaction, as discussed in Section 1.10. Primary and secondary alkyl iodides, bromides, and xanthates can be reduced with the same reagents under the same conditions via the same radical mechanism.

Primary, secondary, and tertiary alkyl halides also can be reduced with dissolving metals. The primary reduction product is an organometallic compound. Whether the latter is formed quantitatively or whether it is converted into the corresponding hydrocarbon by protonation depends on the solvent. The organometallic compound is stable in aprotic solvents (hexane, ether, THF), while it is protonated in protic solvents (HOAc, alcohols).

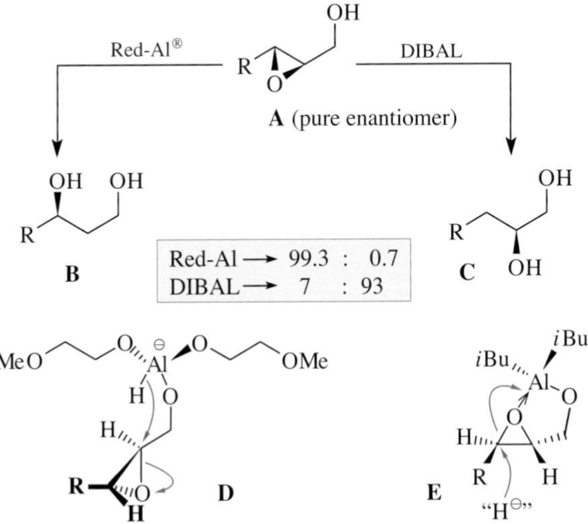

Fig. 17.43. Regioselective reduction of enantiomerically pure epoxy alcohols to enantiomerically pure diols. 1,3-Diols are formed with Red-Al and 1,2-diols are formed with DIBAL.

The reduction of alkyl halides to organometallic compounds in aprotic solvents involves a heterogeneous reaction on the metal surface. This metal surface must be pure metal if it is to react efficiently in the desired way. If the surface of the metal has reacted with oxygen (Li, Mg, Zn) or even with nitrogen (Li), one must first remove the metal oxide or lithium nitride layer. This can be accomplished in the following ways:

- mechanically (stirring of Mg shavings overnight; pressing of Li through a fine nozzle into an inert solvent),
- chemically (etching of Mg shavings with I_2 or 1,2-dibromoethylene; etching of Zn powder with Me_3SiCl),
- or by an *in situ* preparation of the respective metal via reduction of a solution of one of its salts under an inert atmosphere [$MgCl_2$ + lithium naphthalenide → "Rieke-Mg"].

The mechanism of the dissolving metal reduction of alkyl halides presumably is the same for reductions with Li, Mg, or Zn. Alkenyl and aryl halides generally can also be reduced in this fashion via the corresponding alkenyl or aryl organometallic compounds. The mechanism shown in Figure 17.44 for the reduction of MeI by Mg to give a Grignard compound is representative; the individual steps of this reaction are described in the figure caption.

Alkyllithium compounds can be prepared in a heterogeneous reaction between the organic halide and lithium metal (Figure 17.45, center). This is similar to the preparations of Grignard compounds (Figure 17.45, top) and alkylzinc iodide (Figure 17.45, bottom). However, a heterogeneous reaction is not the only way to make alkyllithium compounds. They also can be prepared by homogeneous reactions of alkyl phenyl sulfides or alkyl chlorides (Figure 17.46). This method for the preparation of alkyllithium compounds (and other organolithium compounds as well, cf. Figure 17.47) is known as **reductive lithiation**. Independent of the substrate, the reducing agent is the soluble lithium salt of a radical anion derived from naphthalene, 1-(dimethylamino)naphthalene, or 4,4'-di-*tert*-butylbiphenyl.

The first step in the reductive lithiation of alkyl phenyl sulfides (**Screttas–Cohen process**) consists of preparing the reducing reagent in stoichiometric amounts. Lithium naphthalenide, for example, is made from lithium and naphthalene. The sulfide is added dropwise to this reducing reagent. The mechanism of reduction corresponds step by step to the one outlined in Figure 17.44, except that the dissolved radical anion is the source of the electrons, as opposed

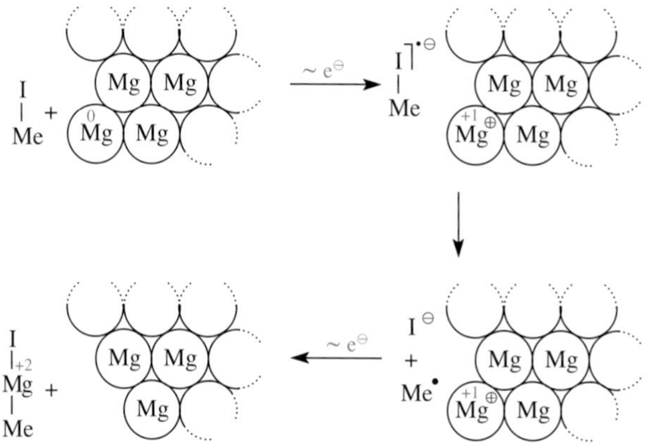

Fig. 17.44. Mechanism of the formation of a Grignard compound (~e⁻ indicates electron migration). The reaction is initiated by an electron transfer from the metal to the substrate. The extra electron occupies the σ^*(C—I) orbital, whereby the C—I bond is weakened and breaks. This cleavage leads to the formation of a methyl radical and an iodide ion on the metal surface. In the third step of the reaction, the valence electron septet of the methyl radical is converted into an octet by formation of a covalent bond between the methyl radical and a metal radical. The Grignard reagent is thus formed.

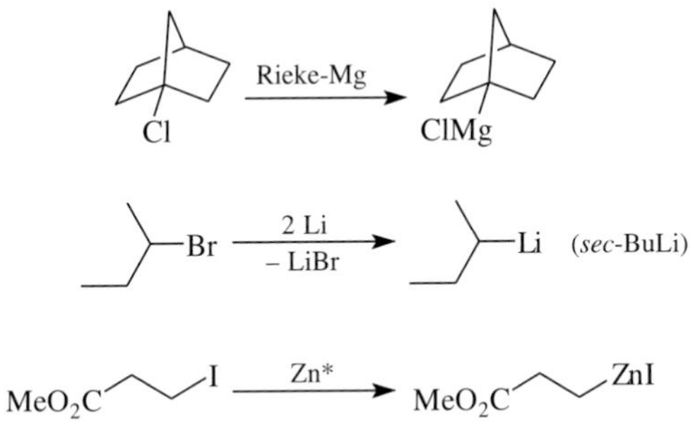

Fig. 17.45. Heterogeneous reduction of alkyl halides to Grignard compounds, organolithium compounds, and—possibly functionalized—organozinc compounds; Zn* refers to surface-activated metallic zinc.

to a metal surface. Furthermore, a C_{sp^3}—S bond breaks instead of a C_{sp^3}—I bond. In the reductive lithiation of alkyl chlorides (**Screttas–Yus process**), the radical anion is generated *in situ* and only in catalytic amounts, starting with a stoichiometric amount of Li powder and a few mole percent of di-*tert*-butylbiphenyl. The lithium di-*tert*-butylbiphenylide reduces the alkyl chloride. This reaction and the reductive lithiation of alkyl phenyl sulfides follow the same mechanism.

Lithium di-*tert*-butylbiphenylide in homogeneous solution is a *very* strong reducing reagent. This reagent allows for easy metalations in cases that are more difficult with metallic lithium and would be impossible to accomplish with magnesium. The reduction of chloride **A** in Figure 17.46 provides a good example. **A** is a rather weak electron acceptor because it is an alkoxide, that is, an anion.

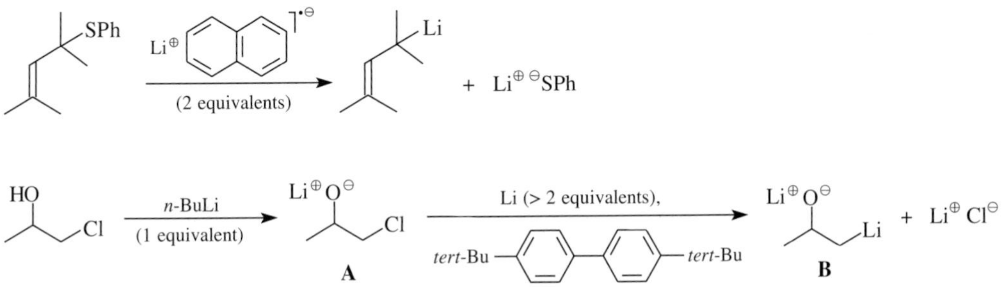

Fig. 17.46. Reductive lithiation of an alkyl aryl sulfide (Screttas–Cohen process) and an alkyl chloride (Screttas–Yus process).

The reducing power of lithium di-*tert*-butylbiphenylide is so high that it is even capable of the reductive lithiation of a C_{sp^2}—Cl bond (Figure 17.47; C_{sp^2}—Cl bonds are stronger than C_{sp^3}—Cl bonds; see Section 1.2). In the example given, a lithium (dialkylamino)carbonyl compound **A** is formed that *is not accessible in any other way*. If one generates the organolithium com-

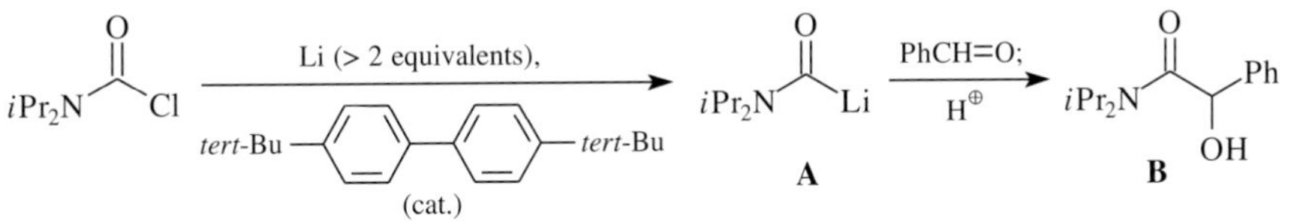

Fig. 17.47. Reductive lithiation of a carbamoyl chloride to the (dialkylamino)carbonyl lithium compound **A** and its immediately following reaction with a carbonyl compound (Barbier reaction) leading to alcohol **B**.

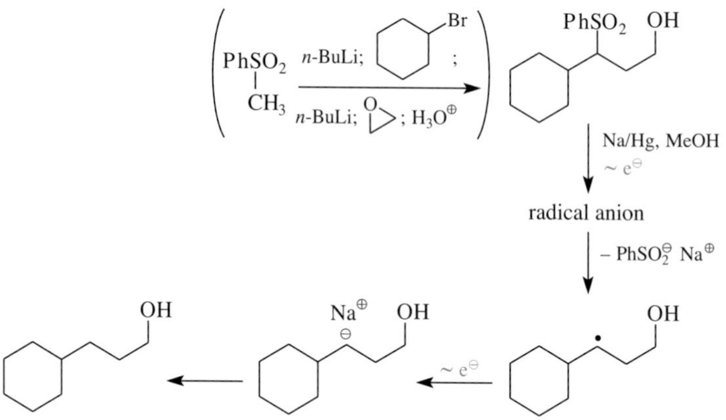

Fig. 17.48. The reduction of an alkyl aryl sulfone to an alkane. The preparation of the alkyl aryl sulfone (top) is followed by a reduction with Na amalgam (right and bottom).

pound **A** in this way and in *the presence of a carbonyl compound*, the former adds immediately to the C=O double bond of the latter. For example, **A** reacts *in situ* with benzaldehyde to give alcohol **B**. The generation of organometallic compounds from halides in the presence of a carbonyl compound followed by an *in situ* reaction of these species with each other is referred to as the **Barbier reaction**.

In some cases, a metal reduces a C_{sp^3}-heteroatom bond faster to a C_{sp^3}-M bond than it reduces the hydrogen of an OH group to H_2. In such cases, protic solvents can be used. Instead of the organometallic compound, one then obtains the product resulting from its protonation. This kind of reduction is used, for example,

- in the defunctionalization of alkyl phenyl sulfones with sodium amalgam in MeOH. (Figure 17.48);
- in the dehalogenation of 2,2-dichlorocyclobutanones (Figure 17.49, top) with Zn/HOAc;
- in the dehalogenation of dichlorocyclopropane (Section 3.3.1) with Na/*tert*-BuOH;
- and in the deoxygenation of acyloin to ketones (Figure 17.49, bottom) with Zn/HCl.

The C_{sp^3}—O bonds of benzyl alcohols, benzyl ethers, benzyl esters, and benzyl carbamates also can be reduced to a C—H bond (Figures 17.50 and 17.51). Lithium or sodium in liquid ammonia are good reducing agents for this purpose. One usually adds an alcohol, such as *tert*-BuOH, as a weak proton source. Lithium and sodium dissolve in liquid ammonia. The resulting solutions contain metal cations and solvated electrons. Hence, it is assumed that the elec-

Fig. 17.49. Reductions of α-heterosubstituted ketones to α-unsubstituted ketones (see Figures 15.34 and 17.59 for the preparation of compounds **A** and **B**, respectively). Here, a ketyl is formed as a radical anion intermediate (for more details about ketyls see Section 17.4.2). The ketyl obtained from **A** releases a chloride ion, the ketyl resulting from **B** releases a hydroxide ion. In each case, an enol radical is formed thereby which picks up an electron. This leads to the formation of a zinc enolate from which the final product is generated by protonation.

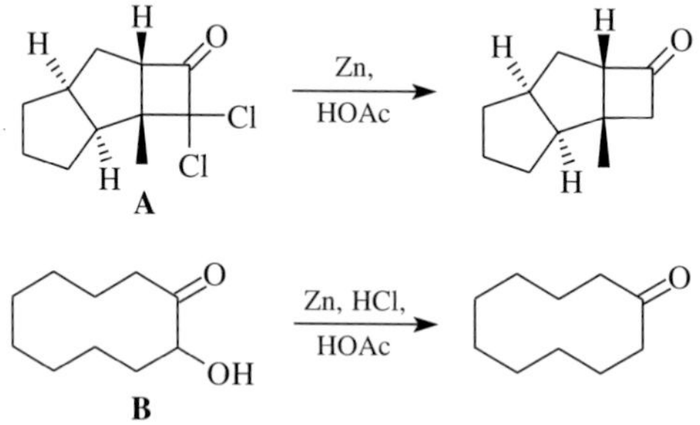

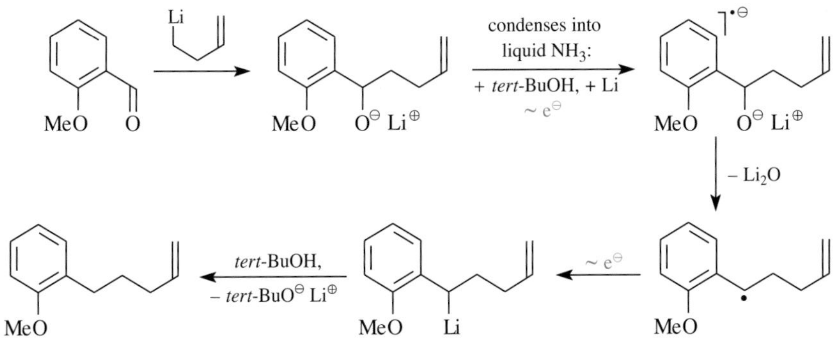

Fig. 17.50. One-pot synthesis of an alkyl-substituted aromatic compound that involves a dissolving lithium reduction of a benzyl alkoxide.

trons are not transferred from the solid metal onto the substrate but that instead the electron transfer occurs in homogeneous solution.

Figure 17.50 shows how a benzyl alkoxide can be reduced in this way. Here, the reduction is part of the synthesis of an alkylated aromatic system. The target molecule contains a primary alkyl group. It is not possible to introduce this alkyl group by way of a Friedel–Crafts alkylation into the *ortho*-position of anisole in a regioselective fashion and without rearrangements (see Section 5.2.5).

The reduction of a benzylic C—O bond to a C—H bond often is employed to remove benzyl-containing protecting groups in benzyl ethers, benzyl carbonates, or *O*-benzyl carbamates. One can use the procedure shown in Figure 17.50 and react the respective compound with Li or Na in an NH_3/*tert*-BuOH mixture. On the other hand, benzylic C—O bonds of the substrates mentioned can be cleaved via Pd-catalyzed hydrogenolyses. Figure 17.51 presents an example of the removal of a benzyloxycarbonyl group from an amino group in this way. The

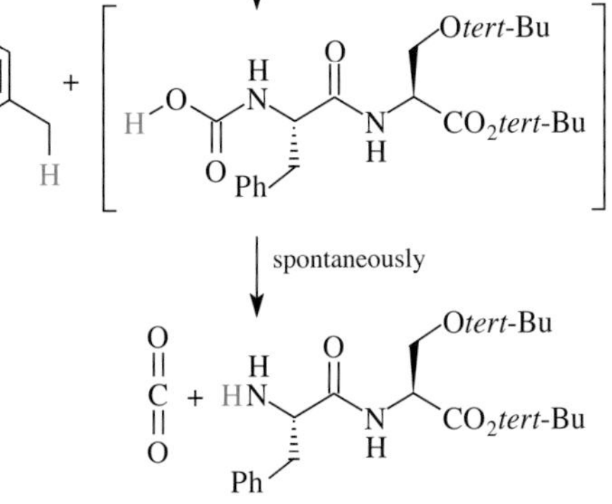

(Preparation: Fig. 6.31)

Fig. 17.51. The *O*-benzylcarbamate → toluene reduction for the removal of a protecting group.

cleavage product is a carbamic acid. Such an acid decarboxylates spontaneously via a mechanism shown in Figure 8 to give the corresponding amine.

17.4.2 One-Electron Reductions of Carbonyl Compounds and Esters; Reductive Coupling

Dissolving metals initially convert aldehydes, ketones, and esters into radical anions. Subsequently, proton donors may react with the latter, which leads to neutral radicals. This mode of reaction is used, for example, in the drying of THF or ether with potassium (or sodium) in the presence of the indicator benzophenone. Potassium and benzophenone react to give the deep-blue potassium ketyl radical anion **A** (Figure 17.52). Water then protonates ketyl **A** to the hydroxylated radical **B** as long as traces of water remain. Further potassium reduces **B** via another electron transfer to the hydroxy-substituted organopotassium compound **C**. **C** immediately tautomerizes to the potassium alkoxide **D**. Once all the water has been consumed, no newly formed ketyl **A** can be protonated so that its blue color indicates that drying is complete.

In the drying of THF or ether (Figure 17.52), the sequence ketone → ketyl → hydroxylated radical → hydroxylated organometallic compound → alkoxide is of course not intended to convert *all* the ketone into "product." The reaction depicted in Figure 17.53 features the same sequence of steps as Figure 17.52. In the reaction of Figure 17.53, however, the reaction is intended to run to *completion* until all of the ketone has been consumed. The reason for this is that it is the purpose of this reaction to reduce the ketone to the alcohol. The substrate in Figure 17.53 is a conformationally fixed cyclohexanone and the reducing agent is sodium.

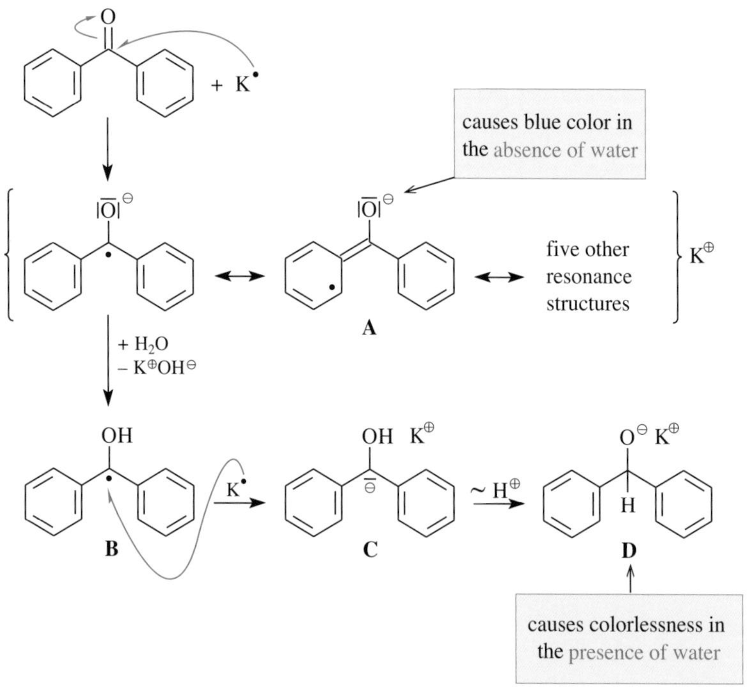

Fig. 17.52. Chemistry of the drying of THF or ether with potassium and benzophenone featuring the ketone → ketyl reduction and the trapping reaction of the ketyl with residual water.

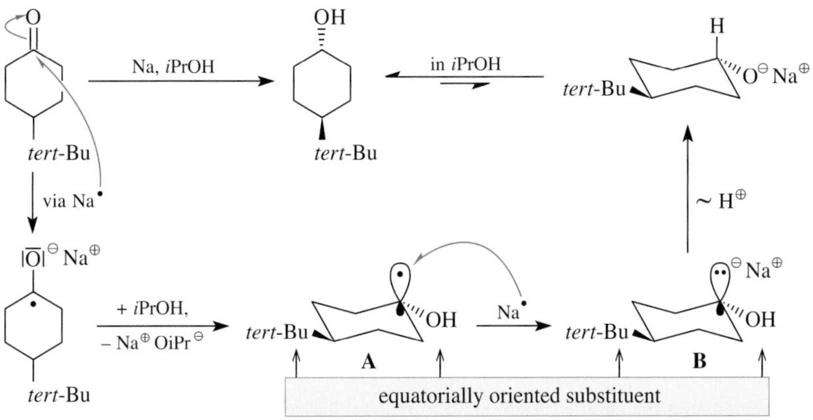

Fig. 17.53. Diastereoselective reduction of a cyclohexanone with dissolving sodium.

Playing a dual role, this solvent also acts as a proton source. Since the supply of this proton source—unlike water with the drying of THF (Figure 17.52)—is unlimited in this case, *all* of the ketone is converted into alkoxide.

Interestingly, the reduction shown in Figure 17.53 is highly diastereoselective. Only the *trans*-configured cyclohexanol is formed, that is, the equatorial alcohol. Such a level of diastereoselectivity cannot be achieved with hydride transfer reagents (cf. Figure 10.11).

The diastereoselectivity of the reduction depicted in Figure 17.53 is determined when the hydroxylated radical **A** is reduced to the hydroxylated organosodium compound **B**. For steric reasons, the OH group assumes a pseudoequatorial position in the trivalent and moderately pyramidalized C atom of the radical center of the cyclohexyl ring of intermediate **A**. Consequently, the unpaired electron at that C atom occupies a pseudoaxially oriented AO. This preferred geometry is fixed with the second electron transfer. It gives rise to the organosodium compound **B**. **B** isomerizes immediately to afford the equatorial sodium alkoxide.

In reactions of carbonyl compounds other than the ones shown in Figures 17.52 and 17.53 with nonprecious metals, ketyls are initially formed as well. However, the Mg ketyl of acetone (Figure 17.54) is sterically much less hindered than the K ketyl of benzophenone (Figure 17.52). It also does not benefit from resonance stabilization by phenyl groups. The former therefore dimerizes while the latter does not. The magnesium salt of glycol is thus formed. It yields a glycol with the trivial name pinacol upon workup with weak acid. Hence, this kind of reductive coupling of carbonyl compounds is called **pinacol coupling**.

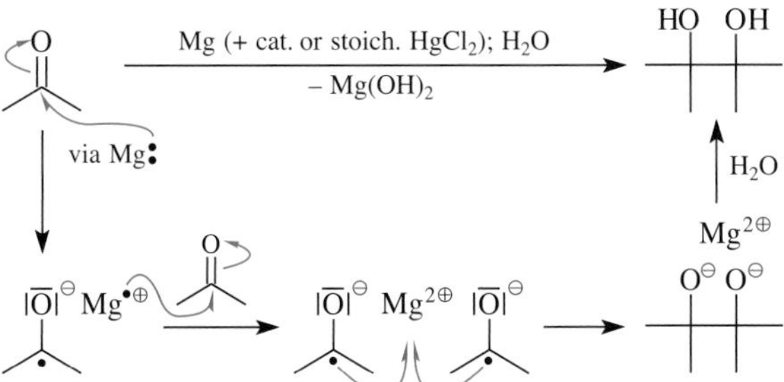

Fig. 17.54. Reductive dimerization of acetone (pinacol coupling).

Carbonyl compounds also can undergo a reductive coupling with so-called **low-valent titanium**, which is a commonly used but vague collective term for titanium compounds in which Ti possesses an oxidation number between 0 and +2. Using a dialdehyde as an example, Figure 17.55 demonstrates that such a reductive coupling can also proceed intramolecularly, i.e., as a ring closure reaction. Depending on the temperature, different products are obtained:

- At *low* temperature a pinacol coupling occurs (similar to the reaction with magnesium in Figure 17.54). If under the reaction conditions of Figure 17.55 low-valent titanium is formed from titanium(III) chloride and a copper/zinc couple, the pinacol coupling leads to the formation of a dititanium(III) derivative **B** of a glycol. These coupling reactions have only one disadvantage, namely that these Ti(III) glycolates **B** are generally formed without simple diastereoselectivity, that is, they are formed as mixtures of diastereoisomers. The coupling shown in Figure 17.55, for instance, gives the *cis*- and *trans*-dihydroxylated 14-membered ring **B** in a 30:70 ratio. And of course, the corresponding glycol **A** is obtained after aqueous workup as a *cis*:*trans* mixture in a ratio of 30:70 as well.
- At *higher* temperatures, the same carbonyl compounds and the same "low-valent titanium" lead to the formation of alkenes instead of glycolates/glycols (dicarbonyl compounds

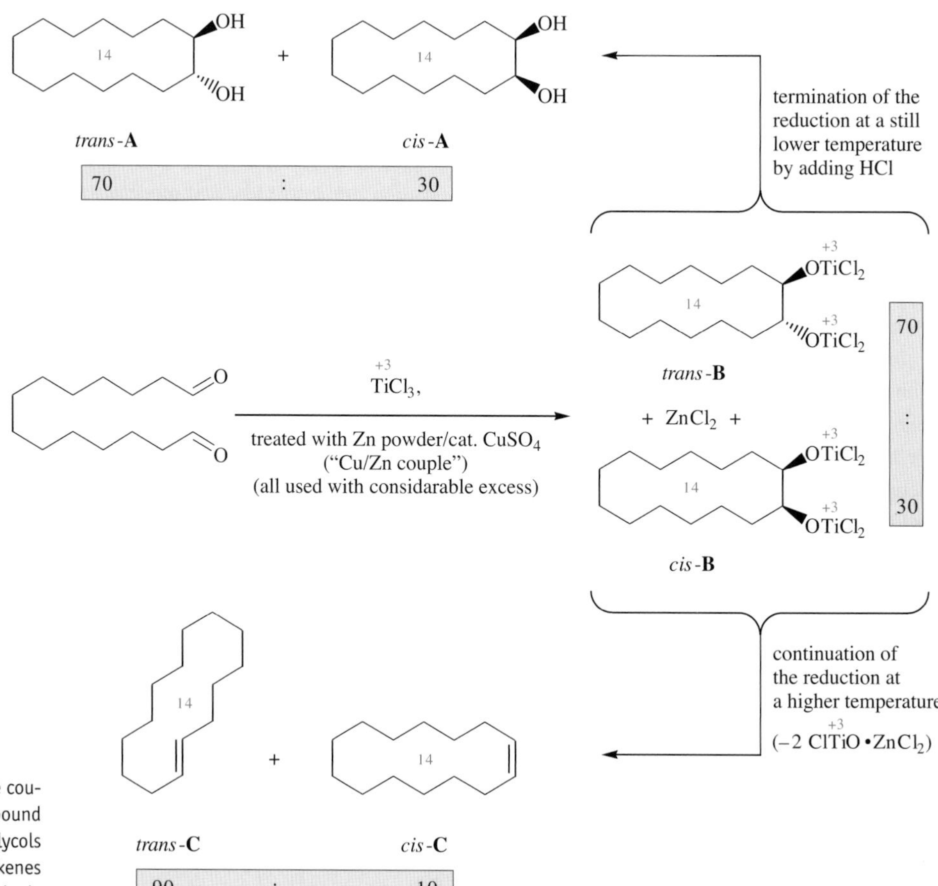

Fig. 17.55. Reductive coupling of a dicarbonyl compound to afford diastereomeric glycols or diastereomeric alkenes (McMurry reaction).

cyclize under these conditions giving cycloalkenes). This type of reductive coupling is called a **McMurry reaction**. As with the Ti-mediated glycol formation, this reaction generally is not diastereoselective. This is illustrated by the example in Figure 17.55. Here, the reaction starts with the above-mentioned C_{14}-dialdehyde, which affords the 14-membered cycloalkene **C**. **C** is formed in a 10:90 *cis:trans* ratio.

The diastereoselectivities for a Ti-mediated "low-temperature" coupling (such as the pinacol coupling in Figure 17.55, top: *ds* = 30:70) versus a Ti-mediated "high-temperature" coupling (like the McMurry reaction in Figure 17.55, bottom: *ds* = 10:90) are rather different. Upon heating the glycolates of the "low-temperature" coupling are thus not stereospecifically converted into the alkenes of the "high-temperature" coupling—a clear indication of a multiple-step conversion mechanism. We can only propose that diastereomeric glycolates react via a common intermediate to form the alkene mixture. However, this would imply that the same alkene mixture would be produced from two diastereomeric titanium glycolates. This clearly is not the case as the reaction pair in Figure 17.56 demonstrates. The alkene **C** is formed as a

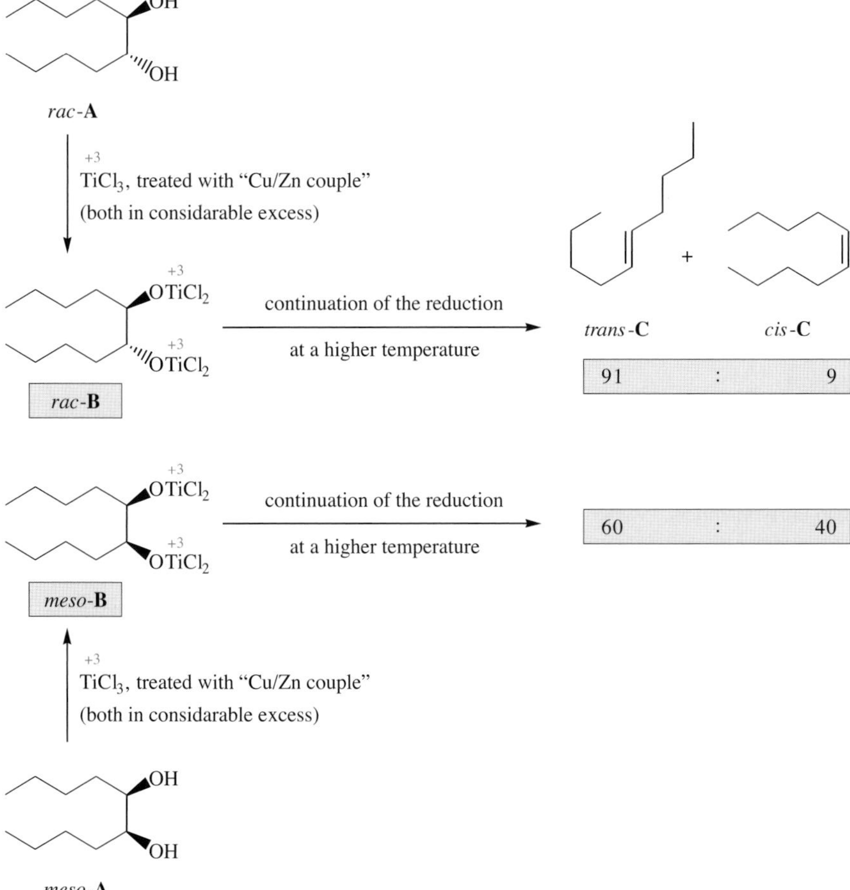

Fig. 17.56. Mechanistically relevant observations on the McMurry reaction. At higher temperatures, the diastereomeric dititanium glycolates *rac*-**B** and *meso*-**B** are reduced to *cis,trans*-mixtures of the alkene **C**. The stereoselectivity of the alkene formation is low, but different for each of the dititanium glycolates. This proves that the reduction of *rac*-**B** and *meso*-**B** to the alkene does not proceed via the same intermediate. The mechanistic implications of these findings are shown in Figure 17.58.

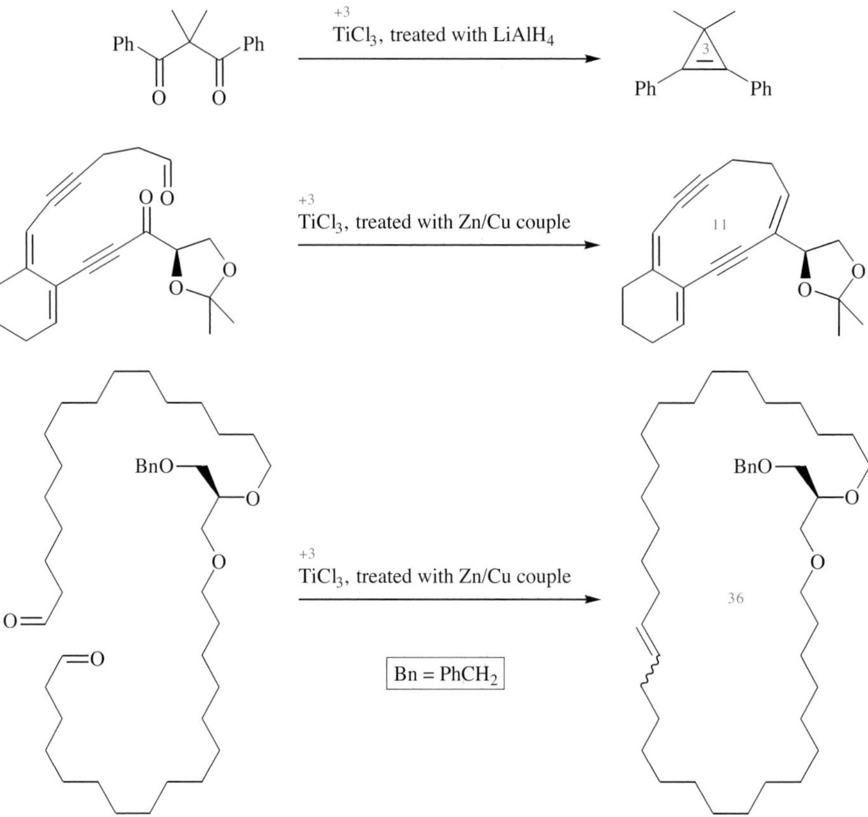

Fig. 17.57. Ring closure through McMurry reaction—suitable for small rings (top example) and normal rings (not shown) as well as for medium-sized rings (center) and large rings (bottom).

9:91 *cis:trans*-mixture from titanium(III) glycolate *rac*-**B** and a 40:60 *cis:trans*-mixture from *meso*-**B**.

Figure 17.57 reveals why the McMurry reaction has become famous. It leads to the formation of all kinds of cycloalkenes; from small rings, "normal" rings, medium rings and large rings, often in excellent yields. The variety of accessible ring sizes ranges from three to more than thirty members.

For many years the elucidation of the mechanism of the McMurry reaction has been complicated by the fact that the most commonly used low-valent titanium was derived from hour-long "pre-reducing DME-complexed $TiCl_3$ with a zinc/copper couple" (DME is 1,2-dimethoxyethane). Apparently, however, the zinc/copper couple reacts with $TiCl_3$ only if the carbonyl compound is present, too. This fact has become part of the current mechanistic view of the McMurry reaction (Figure 17.58). At the same time, it became the starting point of the following variants of the McMurry reaction:

1) With the **instant McMurry reaction** time can be saved. This is important, particularly since the "pre-reduction" doesn't take place as once believed:

iPrO OiPr iPrO OiPr + Zn/Cu-couple $\xrightarrow[\text{(as DME-complex*)}]{\text{addition of } \overset{+3}{\text{Ti}}\text{Cl}_3}$ iPrO OiPr iPrO OiPr

* DME = 1,2-Dimethoxyethane (= Ethylene Glycol Dimethyl Ether)

2) With a cheaper variant—which we will simply call **"low-budget" McMurry reaction**—insensitive and inexpensive titanium powder is used as a titanium source. This avoids the use of the moisture-sensitive and expensive DME-complexed $TiCl_3$ whose oxidation level is much too high anyway. Titanium powder needs to be slightly etched superficially with trimethylchlorosilane (which is also inexpensive):

2 [retinal] $\xrightarrow{\overset{0}{\text{Ti}}\text{-Powder, } Me_3SiCl}$ [β-carotene]

3) With the **catalytic McMurry reaction** a few mole percent of DME-free $TiCl_3$ are sufficient to run the reaction to completion. The trick is that stoichiometrically added zinc continuously provides "low-valent titanium" and the oxygen that is released is bound by a superstoichiometric amount of chlorosilane as a disilyl ether ("disiloxane"). This procedure has been utilized for many indole syntheses based on the McMurry reaction (Side Note 17.6).

Side Note 17.6. Fürstner Indole Synthesis – Catalytic McMurry Reaction

The intramolecular McMurry coupling between the C=O group of an amide and the C=O double bond of a ketone leads to the formation of a heterocycle with an enamine-like substructure. The most important heterocycle of this type is the indole skeleton:

The cyclization of a ketoamide illustrates one of the most versatile syntheses of 2,3-disubstituted indoles. The mechanism underlying these reactions corresponds exactly to that of the ketone/ketone coupling in Figure 17.58.

As we have seen McMurry reactions and Ti-mediated pinacol couplings can be performed with various "low-valent titanium species." This precludes the possibility that the respective mechanisms resemble each other in every detail. The best established mechanism is given as an example in Figure 17.58, namely the reaction of acetophenone with titanium(III) chloride and copper/zinc couple. Both in the intramolecular (Figure 17.55) and intermolecular reaction in Figure 17.58, the McMurry coupling takes place only at higher temperatures, while the pinacol coupling occurs at a lower temperature. As in the intramolecular case (Figure 17.55), there is at best a partial correlation between the diastereomeric ratio in the McMurry product of the intermolecular reaction (Figure 17.58)—**72**:28 ***E***:*Z* in THF, but **92**:8 ***Z***:*E* in DME—and the diastereomeric ratio in the pinacol.

Figure 17.58 shows two reduction steps that are both initiated by activated zinc. The initial reductive step takes place at lower temperatures and involves the conversion of Ti(III)-complexed acetophenone **A** into the Ti(II)-complexed acetophenone **B**. If you wonder why titanium(III) chloride can be reduced while bearing a negative charge in complex **A**, but not while it is free and uncharged, you should take a look at the second resonance form of **B** and think of product development control. This resonance form reveals that the reduced titanium/acetophenone complex also represents a well-stabilized—i.e., phenyl- and oxygen-substituted—radical. The same resonance form is able to explain the further path that leads to the titanium(III) glycolate **C** and is labeled "variant 1" in Figure 17.58, namely, a radical dimerization. The alternative pathway according to "variant 2" leading to the same titanium(III) glycolate **C** begins with the ring closure reaction of the Ti(II)-complexed acetophenone **B** to form the three-membered ring **D**. The latter may also be conceived as a $TiCl_2^{\oplus}$-bridged acetophenone ketyl that forms the Lewis acid/Lewis base complex **F** with one equivalent of acetophenone. An intramolecular nucleophilic addition of the C–Ti- to the C=O double bond then leads to titanium(IV) glycolate **E**. The latter combines with the $TiCl_2$ moiety of a suitable intermediate—**B** or **D**—to furnish the titanium(III) glycolate **C**. Upon workup the corresponding glycol is formed in a 8:92- *meso:rac*-mixture if THF is used as a solvent or in a 4:96-*meso:rac*-mixture if DME is employed.

The second reductive step in Figure 17.58 only occurs if the mixture is heated when the copper/zinc pair is still present and titanium(II) glycolate **G** is formed from the titanium(III) glycolate **C**. A C–O bond in **G** breaks homolytically, forming $ZnCl_2$-complexed Ti(III)ClO and the radical **H**. However, the two diastereomers of the titanium(III) glycolate **C** do not yield quite the same radical intermediate **H**—otherwise it would be a *common* intermediate, which

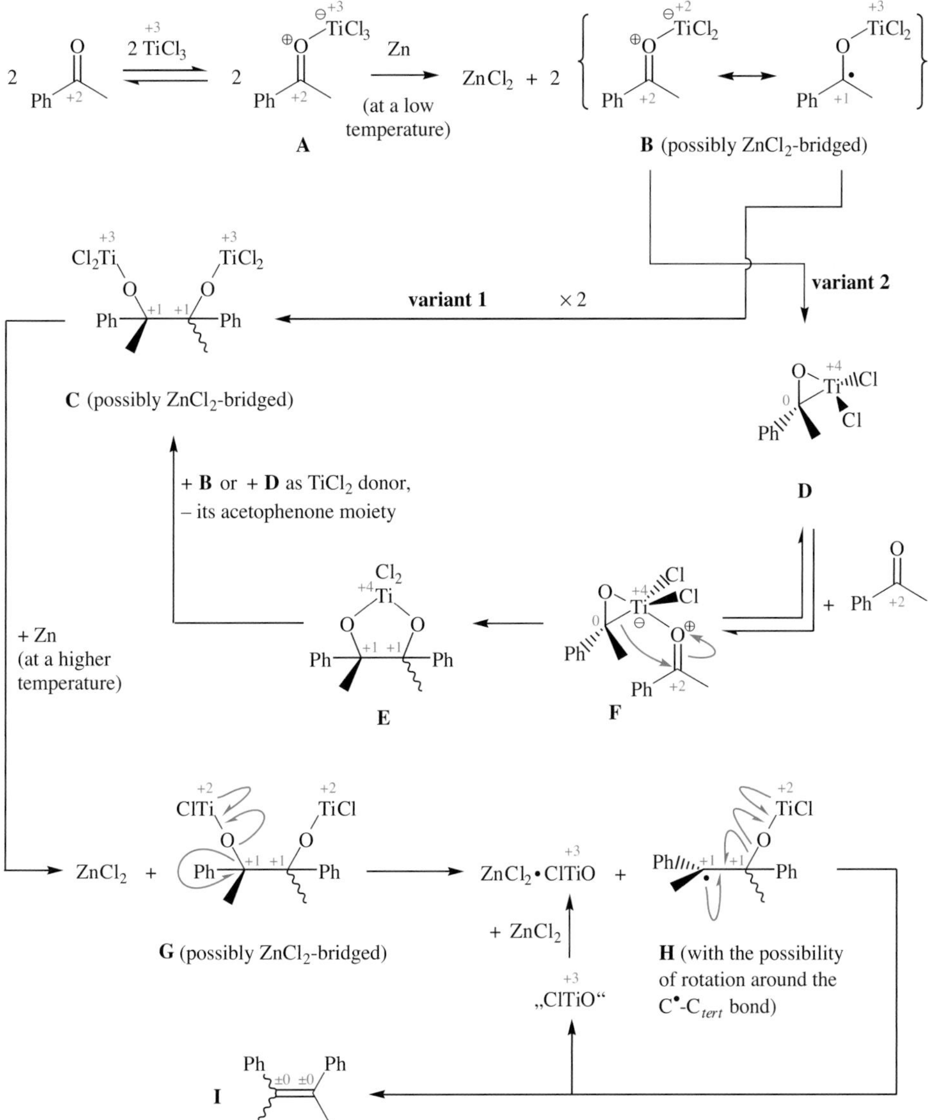

Fig. 17.58. The mechanism of the McMurry reaction. A mixture of diastereomers of the dititanium(III) glycolate **C** is either directly generated via the initially formed Ti(III) ketyl **B** (variant 1) or in multiple steps (variant 2). At sufficiently high temperatures, **C** is then reduced to a mixture of diastereomers of the corresponding dititanium(III) glycolate **G**. The latter decomposes via homolytic cleavage of *one* of its C—O bonds to furnish the radical intermediate **H**. If the only C—O bond left in this radical also breaks homolytically—which partly occurs without (!) prior rotation around the • C–C(OTiCl) bond—the alkene is formed as an *E,Z*-mixture. Its composition may (somewhat) depend on the configuration of the dititanium(II) glycolate precursor **G** (cf. Figure 17.56).

can generally be precluded based on the fact that different glycolates lead to different diastereomeric mixtures of alkenes. Upon further reaction the latter violate the Curtin–Hammett principle (see Figure 10.17). They can do so because they need more time to lose their initial conformation—despite the basically free rotation of their C–C(O) bond—than their C–O bond needs to homolytically decompose in the very last step of the reaction. It accounts for the fact that a McMurry product like **I** does not usually represent a stereouniform alkene, that its double bond configuration can depend on the stereostructure of its titanium(III) glycolate precursor and that $ZnCl_2$-complexed Ti(III)ClO is formed as an inorganic side product.

Carboxylic esters also can be reduced with dissolving sodium (Figure 17.59). Very different products are obtained depending on whether the reduction is carried out in ethanol or xylene.

The reaction of esters with sodium *in ethanol* is referred to as the **Bouveault–Blanc reaction**. Prior to the discovery of complex metal hydrides, this reaction was the only method for the reduction of esters to alcohols. The *di*esters shown in Figure 17.59 produce a *diol* in this way.

The mechanism of the Bouveault–Blanc reduction is shown in rows 1 and 2 of Figure 17.59. It starts with the sequence ester → radical anion **C** → hydroxylated radical

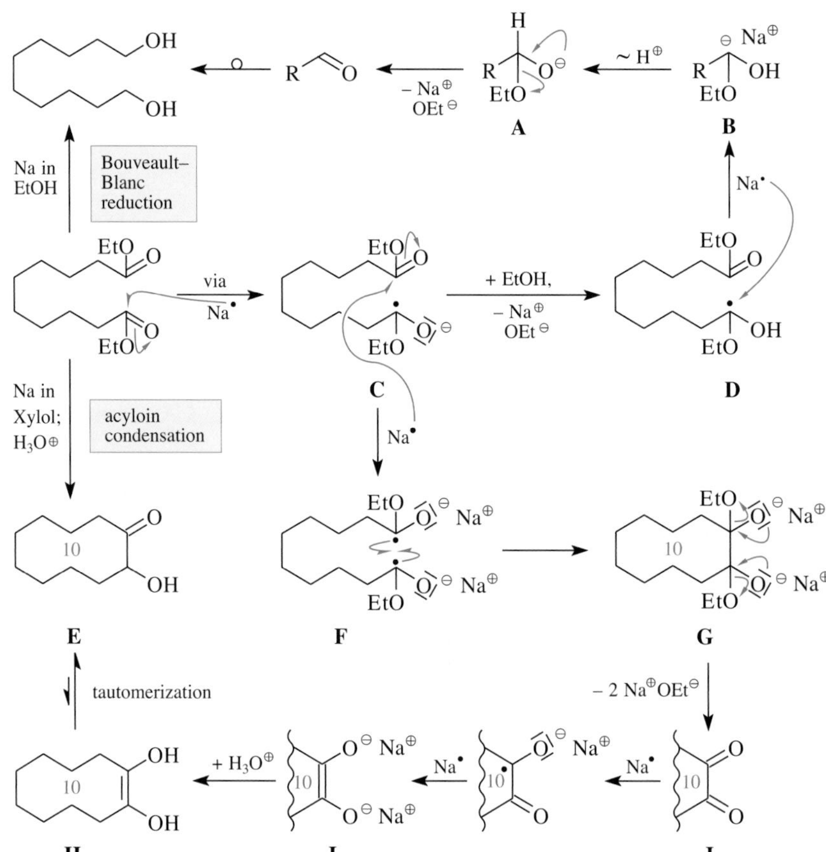

Fig. 17.59. Reduction of a carboxylic ester with dissolving sodium. Branching of the reduction paths in the presence (Bouveault–Blanc reduction) and absence (acyloin condensation) of protons.

D → hydroxylated organosodium compound **B** → hemiacetal anion **A**. This sequence is completely analogous to the sequence ketone → ketyl → hydroxylated radical **A** → hydroxylated organosodium compound **B** → sodium alkoxide that occurs in the reduction of a ketone with Na in *i*PrOH (Figure 17.53).

Intermediate **A** of the Bouveault–Blanc reduction of Figure 17.59 is not a simple alkoxide but rather the anion of a hemiacetal. Accordingly, it decomposes into an alkoxide anion and an aldehyde. In the further course of the Bouveault–Blanc reduction, this aldehyde is reduced by Na/EtOH just as the ketone of Figure 17.53 is reduced by Na/*i*PrOH.

The **acyloin condensation** consists of the reduction of esters—and the reduction of diesters in particular—with sodium in xylene. The reaction mechanism of this condensation is shown in rows 2–4 of Figure 17.59. Only the first of these intermediates, radical anion **C**, occurs as an intermediate in the Bouveault–Blanc reduction as well. In xylene, of course, the radical anion **C** cannot be protonated. As a consequence, it persists until the second ester also has taken up an electron while forming the bis(radical anion) **F**. The two radical centers of **F** combine in the next step to give the sodium "glycolate" **G**. Compound **G**, the dianion of a bis(hemiacetal), is converted into the 1,2-diketone **J** by elimination of two equivalents of sodium alkoxide. This diketone is converted by two successive electron transfer reactions into the enediolate **I**, which is stable in xylene until it is converted into the enediol **H** during acidic aqueous workup. This enediol subsequently tautomerizes to furnish the α-hydroxyketone—or acyloin—**E**. The acyloin condensation is of special value because this method allows for the synthesis of medium and large rings in good yields without the need for high dilution techniques.

17.4.3 Reductions of Carboxylic Acid Derivatives to Alcohols or Amines

Complex or soluble neutral metal hydrides are usually employed for the reduction of carboxylic acid derivatives to alcohols or amines. The standard reagents for the most important transformations are shown in Table 17.6. For completeness, various reagents also are listed for the reduction of carboxylic acid derivatives to aldehydes. The latter mode of reduction was discussed in Section 6.5.2.

The mechanism of the $LiAlH_4$ reduction of carboxylic esters to alcohols (Figure 17.60) can be readily understood based on the discussions of Section 6.5.1. A tetrahedral intermediate **A**

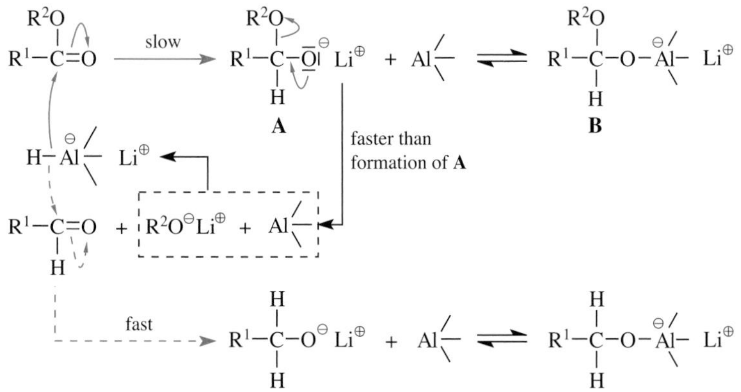

Fig. 17.60. Mechanism of the $LiAlH_4$ reduction of carboxylic esters to alcohols via aldehydes.

is formed by addition of a hydride ion to the ester. Intermediate **A** decomposes into an aldehyde and a lithium alkoxide. The aldehyde is a better electrophile than the remaining ester, and therefore the aldehyde reacts faster by the reductant than the ester. Consequently, $LiAlH_4$ effectively reduces esters to alcohols without there being a way of scavenging the aldehyde in this process.

A similar mechanism is operative in the reduction of carboxylic esters with DIBAL (Figure 17.61). The tetrahedral intermediate **A** is formed by addition of an Al—H bond of the reducing agent to the ester C=O bond. This tetrahedral intermediate **A** does *not necessarily* decompose immediately to an aldehyde and ROAl(*i*Bu)$_2$. In *nonpolar* media, **A** definitely decomposes quite slowly. In fact, at very low temperatures **A** remains unchanged until it is protonated to a hemiacetal during aqueous workup. The latter eliminates water to give the aldehyde.

In *polar* solvents, however, the tetrahedral intermediate **A** of Figure 17.61 decomposes faster than it is generated, with formation of an aldehyde and ROAl(*i*Bu)$_2$. This solvent effect is explained in Figure 17.61, using THF as an example. The tetrahedral intermediate **A** contains a trivalent Al and forms a Lewis acid–Lewis base complex with the solvent. The intermediate **A** is thus converted into the aluminate complex **B**, which contains tetravalent Al. The Al atom in **A** is bonded to only one O atom, which used to be the carbonyl oxygen of the ester. In complex **B**, the Al atom still binds to the same O atom, but it also binds weakly to a THF molecule. The bond between Al and the O atom that originated from the ester is stronger in **A** than in **B** because of the additional contact with THF in the latter. The Al—O bond thus breaks relatively easily in **B** and gives rise to a situation in which the aldehyde is formed faster from **B** than the tetrahedral intermediate **A** is formed from DIBAL and the ester. The aldehyde and the still unconsumed ester therefore compete for the remaining DIBAL, and the aldehyde wins because of its higher electrophilicity. Therefore, DIBAL reductions of carboxylic esters in polar solvents go "all the way" to the alcohol.

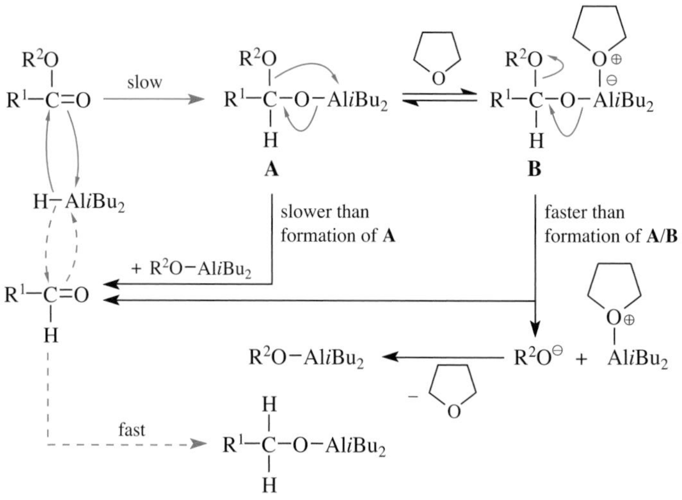

Fig. 17.61. Mechanism of the DIBAL reduction of carboxylic esters to aldehydes and further to alcohols. In nonpolar solvents, the reaction stops with the formation of the tetrahedral intermediate **A**. During aqueous workup, **A** is converted into the aldehyde via the hemiacetal. In polar solvents, however, the tetrahedral intermediate **A** quickly decomposes forming the aldehyde via complex **B**. In the latter situation the aldehyde successfully competes with unreacted ester for the remaining DIBAL. The aldehyde is reduced preferentially, since the aldehyde is the better electrophile, and it is converted into the alcohol.

The particular stability that characterizes the tetrahedral intermediate **A** of the addition of DIBAL to carboxylic acid ester in methylene chloride (Figure 17.61) is the basis of a two-step reduction of carboxylic acid esters to ethers (Figure 17.62) since such tetrahedral intermediates can be acylated with acetic anhydride at the hemiacetal oxygen atom. In this way α-acetoxyethers are formed (Formula **A** in Figure 17.62). These acetoxyethers can be reduced with triethylsilane if boron trifluoride etherate is present as a Lewis-acid catalyst. In this process, the acetoxy residue is replaced by Si-bound hydrogen and an ether is generated. This AcO → H exchange takes place as an S_N1 reaction. Initially the substrate (**A**) is complexed at the carboxyl oxygen by BF_3 (→ **B**), which is followed by heterolysis to yield the oxocarbenium ion **C** and the BF_3-bound acetate ion. The latter may may reversibly engage in the formation of small amounts of complex D. The resulting –ate complex **D** is intercepted by the oxocarbenium ion **C** as the latter abstracts a hydride ion from the –ate complex (an –ate complex was first encountered in connection with the discussion of Figure 5.41). This completes the S_N1 reaction.

A related (tandem) $S_N1$reduction will be introduced in Figure 17.72. There a benzyl(triethylsilyl) ether undergoes reduction by triethylsilane as a reducing agent and trifluoroacetic acid as a Brønsted acid.

Side Note 17.7. An Ester → Ether Reduction

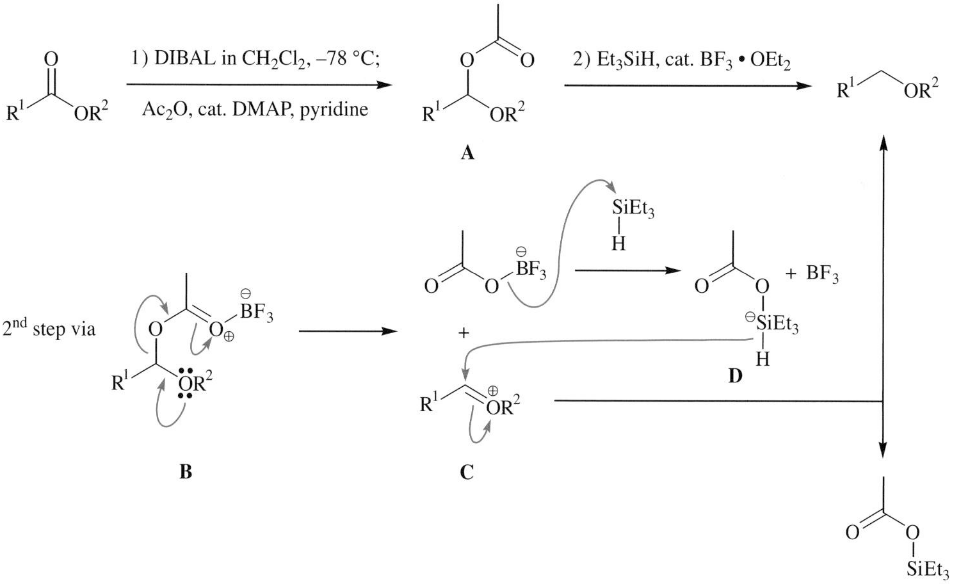

Fig. 17.62. Two-step reduction of carboxylic acid esters to ethers. In the first step, a reduction via hydride transfer occurs and leads to a tetrahedral intermediate that is stable enough to be acylated to the (alpha))-acetoxyether **A.** The second step involves a so-called "ionic reduction" (details: see text) of the oxocarbenium ion **C**, which derives from the acetoxyether by cleavage of an acetate residue.

1) DIBAL in CH_2Cl_2, –78 °C; Ac_2O, cat. DMAP, pyridine

2) Et_3SiH, cat. $BF_3 \cdot OEt_2$

2nd step via

There is *one* type of ester → alcohol reduction for which one employs DIBAL (in a polar solvent) rather than $LiAlH_4$ (in ether of THF). This reduction is the reduction of α,β-unsaturated esters to allyl alcohols (example in Figure 17.63). The reaction of this kind of substrate with $LiAlH_4$ sometimes results in a partial reduction of the C=C double bond to a C—C single bond in addition to the desired transformation —C(=O)OR → —CH_2OH.

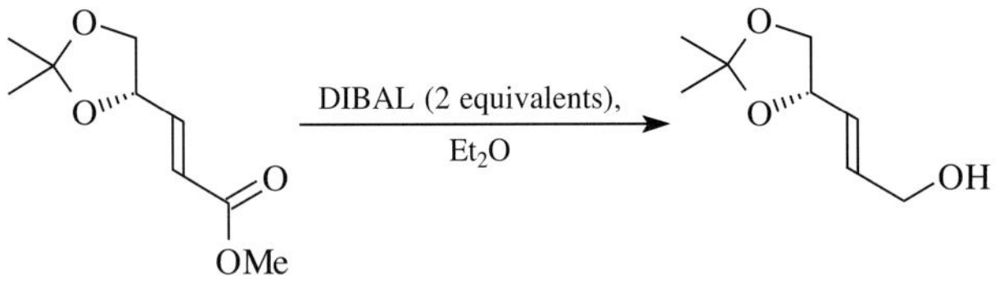

Fig. 17.63. DIBAL reduction of an α,β-unsaturated ester to an allylic alcohol. (See Figure 11.6 for a preparation of the substrate.)

The two reducing agents considered so far in this section—$LiAlH_4$ and DIBAL—also are the reagents of choice for the reduction of nitriles (Figure 17.64). The mechanistic details of these reactions can be gathered from the figure, and the result can be summarized as follows.

Products of the Reduction of Nitriles with $LiAlH_4$ and DIBAL

- The reduction of nitriles with $LiAlH_4$ leads to iminoalanate **B** via iminoalanate **A**. The hydrolytic workup affords an amine.
- The reduction of nitriles with DIBAL can be stopped at this stage of the iminoalane **C**. The hydrolysis of this iminoalane gives the aldehyde. The iminoalane **C** also can be reacted—more slowly—with another equivalent of DIBAL to the aminoalane **D**. The latter yields the amine upon addition of water.

$LiAlH_4$ or DIBAL reduces carboxylic amides at *low* temperatures to such an extent that aldehydes are obtained after aqueous workup (Figure 17.65). This works most reliably, according to Figure 6.42, if the amides are Weinreb amides. At *higher* temperatures, the treatment of carboxylic amides with either $LiAlH_4$ or DIBAL results in amines. Accordingly, *N,N*-disubstituted amides give tertiary amines, mono-substituted amides give secondary amines, and unsubstituted amides give primary amines (Figure 17.65).

It is noteworthy that all of the latter reductions yield amines rather than alcohols. This is due to the favored decomposition path of the initially formed tetrahedral intermediates.

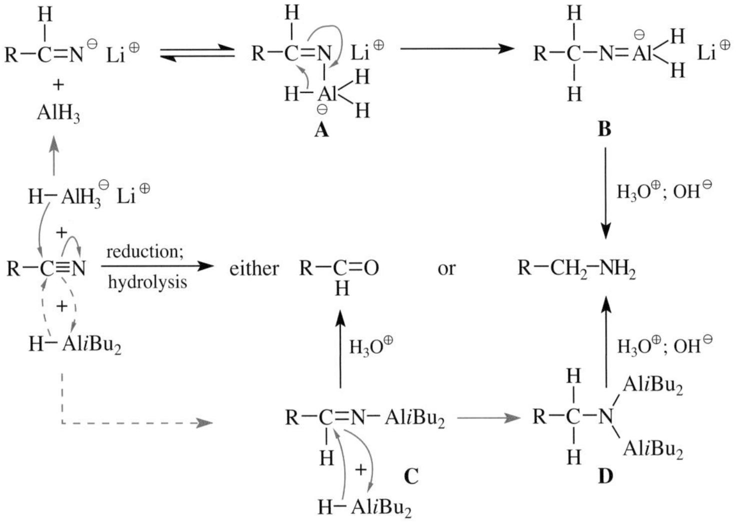

Fig. 17.64. Mechanism of the $LiAlH_4$ reduction (top) and the DIBAL reduction (bottom) of nitriles.

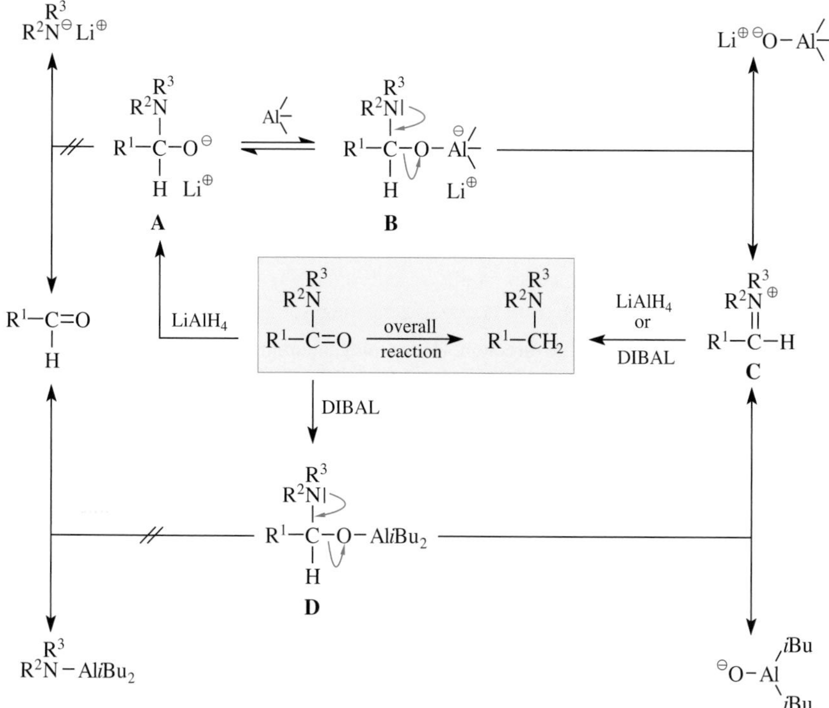

Fig. 17.65. Chemoselectivity in the reduction of amides.

Specifically, the tetrahedral intermediates **B** and **D** are formed in amide reductions with $LiAlH_4$ and DIBAL, respectively. Their decomposition in principle could affect the C—O bond (→ → → amine) or the C—N bond (→ → → alcohol). There are two factors that provide an advantage for the C—O bond cleavage:

1) The strong O—Al bond of the respective tetrahedral intermediate is replaced by a still stronger O—Al bond in the leaving group ($R_2Al—O^\ominus$ or $R_3Al^\ominus—O^\ominus$).
2) The amino group, a donor substituent, facilitates loss of the leaving group ($R_2Al—O^\ominus$ or $R_3Al^\ominus—O^\ominus$), because an iminium ion is formed.

The competing C—N bond cleavage of the tetrahedral intermediates B and D of the reductions in Figure 17.65 does not occur for the reasons just given and because C—N bond cleavage has further disadvantages:

1) The amide anion is a much poorer leaving group than the aluminum alkoxide ($R_2Al—O^\ominus$ or $R_3Al^\ominus—O^\ominus$). This is true even if the amide group is complexed by $i$$Bu_2Al$ since the anion center in the latter is an oxygen instead of a nitrogen atom. Due to the difference in electronegativity (O > N) a negatively charged oxygen is preferred over a similarly charged nitrogen. Finally, the N—Al bond is not as strong as the O—Al bond.
2) The putative cation that would be generated by C—N bond cleavage in **D** would contain an O atom with a very weak pi electron donating ability (+M effect). This is because the $i$$Bu_2Al$ group, which binds to this O atom, acts as a Lewis acid. Consequently, the cation in question would be an acceptor-substituted carbenium ion (with a C—O single bond) rather than an oxocarbenium ion (with a C=O double bond).

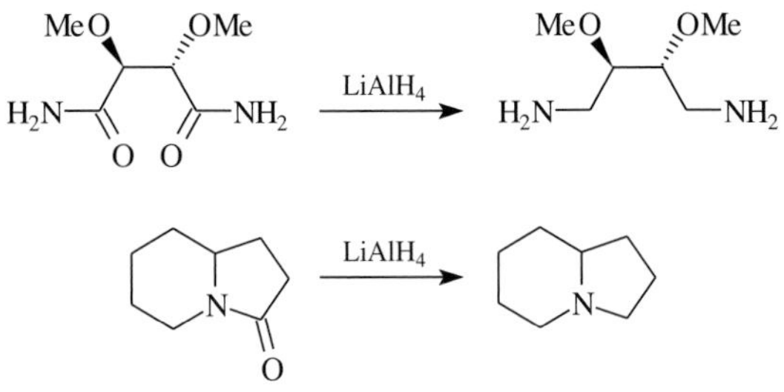

Fig. 17.66. Amide → amine reduction (top) and lactam → cycloamine reduction (bottom) with $LiAlH_4$.

All four factors push in the same direction. Hence, *only* the iminium ion **C** is formed from the tetrahedral intermediates **B** and **D** of Figure 17.65. This iminium ion then consumes a second hydride ion from the reducing agent and yields the Al derivative of the product amine.

The two specific examples shown in Figure 17.66 illustrate the general concept of the amide → amine reduction depicted in Figure 17.65. The reduction of a diamide to a diamine is shown on top, and the reduction of a lactam to a bicyclic amine is shown on bottom.

17.4.4 Reductions of Carboxylic Acid Derivatives to Aldehydes

These reductions were discussed in Section 6.5.2.

17.4.5 Reductions of Carbonyl Compounds to Alcohols

The reader is already familiar with the reductions of carbonyl compounds to alcohols, since they were described in Sections 10.2–10.4. (To complete the overall picture: in α,β-unsaturated carbonyl compounds the C=C- instead of the C=O double bond can also be reduced according to Figure 13.20, through the Birch and L-Selectride® reduction of such carbonyl compounds.)

17.4.6 Reductions of Carbonyl Compounds to Hydrocarbons

The Wolff–Kishner reduction is an old and still often used method for the reduction of a ketone to the corresponding alkane. The **Huang–Minlon modification** of this reduction is commonly employed. It entails the treatment of the ketone with hydrazine hydrate and KOH in diethylene glycol, first at low temperature and then at reflux (200 °C).

Cram's room temperature (!) **variant** is less frequently employed. This reaction relies on a much stronger base, that is, potassium-*tert*-butoxide, whose basicity is yet increased by using dimethyl sulfoxide (DMSO) as a solvent. Being a dipolar aprotic solvent DMSO does not provide any notable stabilization to the *tert*-butoxide ion and thus considerably increases its basicity (as well as its nucleophilicity; see Section 2.2, pp. 58–59).

The mechanism of Wolff–Kishner reductions is exemplified in Figure 17.67 with the second step **A** → **E** of the **Haworth naphthalene synthesis** (steps 1 and 3 of this synthesis: Figure 5.34). This reduction includes the following steps: (1) formation of the hydrazone **B** (for the mechanism, see Table 9.2), (2) base-catalyzed tautomerization of hydrazone **B** to the azo compound **C**, (3) and (4) $E1_{cb}$ elimination with liberation of N_2 and formation of the benzylpotassium **G**; (5) protonation of **G** in the benzylic position, leading to the formation of the carboxylate **F**. The carboxylic acid **E** is formed by protonation of this carboxylate during acidic workup of the reaction mixture.

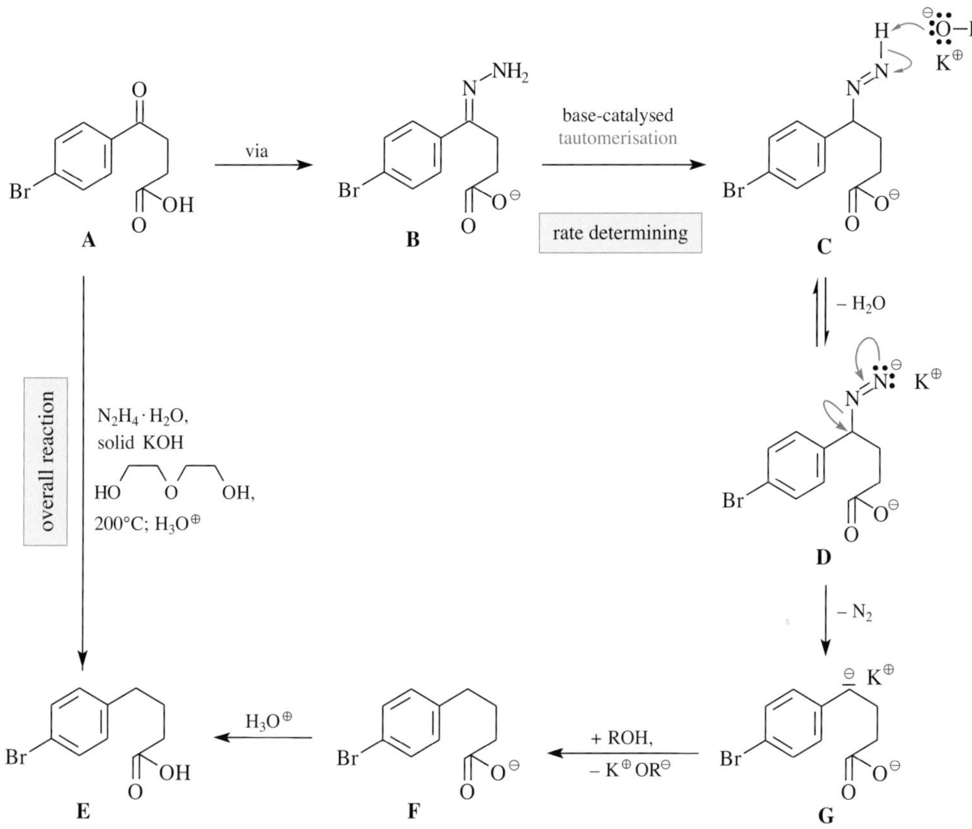

Fig. 17.67. Wolff–Kishner reduction of a ketone. The example shows the second step of the five-step Haworth synthesis of naphthalene.

Starting from isolated hydrazones, reduction to the corresponding hydrocarbons by treatment with base in an *aprotic* solvent takes place at temperatures significantly below the 200 °C of the Huang–Minlon modification of the Wolff–Kishner reduction. However, hydrazones cannot be prepared in a one-step reaction between a ketone and hydrazine, since usually azines ($R^1R^2C{=}N{-}N{=}CR^1R^2$) are formed instead. However, semicarbazones are hydrazone derivatives that are easily accessible by the reaction of a ketone with semicarbazide (for the mechanism, see Table 9.2). **Semicarbazones** can be converted into alkanes with KO*t*Bu in toluene at temperatures as low as 100 °C. This method provides an alternative to the Wolff–Kishner reduction when much lower than usual reduction temperatures are desirable.

The mechanism of such a semicarbazone reduction is exemplified in Figure 17.68 by the deoxygenation of the α,β-unsaturated ketone **A**. Since the substrate contains a C=C double

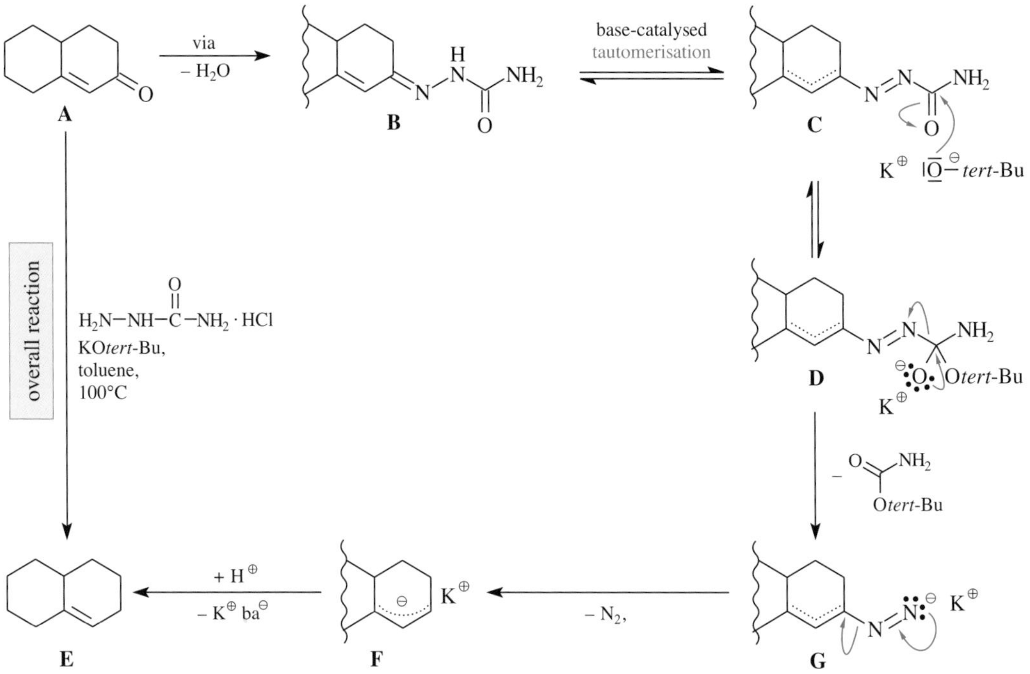

Fig. 17.68. Alternative I to the Wolff–Kishner reduction: reductive decomposition of a semicarbazone.

bond—which is, however, not a prerequisite for the feasibility of such a reaction—one obtains a product with a C=C double bond. Hence, an *alkene* **E** is formed, not an alkane. The first two steps **A** → **B** → **C** of the semicarbazone reduction of Figure 17.68 correspond to the introductory steps of the Wolff–Kishner reduction (Figure 17.67). They comprise a hydrazone formation and a tautomerization to an azo compound **C**. The position of the C=C double bond in **C** is not known, which is indicated by the dotted lines in the formula (as well as the follow-up steps **D** and **G**). The *tert*-butoxide ion adds to the C=O double bond of the carbonic acid moiety of the azo compound **C** and forms the tetrahedral intermediate **D** (the position of the C=C double bond remains unknown). The fragmentation of **D** yields the diazene anion **G** and carbamic acid *tert*-butylester. With this diazene anion **G** we have caught up with the mechanism of the original Wolff–Kishner reduction (cf. Formula **D** in Figure 17.67). It decomposes to N_2 and the allyl potassium compound **F**. The protonation of **F**—presumably under kinetic control—occurs in a regioselective fashion and results in the alkene **E**.

Side Note 17.8. Reductive Cyanation

Diazene anions have a key role in both the Wolff–Kishner reduction and its alternative, the semicarbazone reduction (Formula **D** in Figure 17.67 and Formula **G** in Figure 17.68, respectively): *they* decompose into elemental nitrogen and an organometallic compound. The italics immediately explain what happens in the second step of the "reductive cyanation of a ketone" shown in Figure 17.69.

The first step involves the formation of the (mesitylenesulfonyl)hydrazone **A** (mechanism: Table 9.2), and the second comprises its reaction with potassium cyanide in acetonitrile. Being

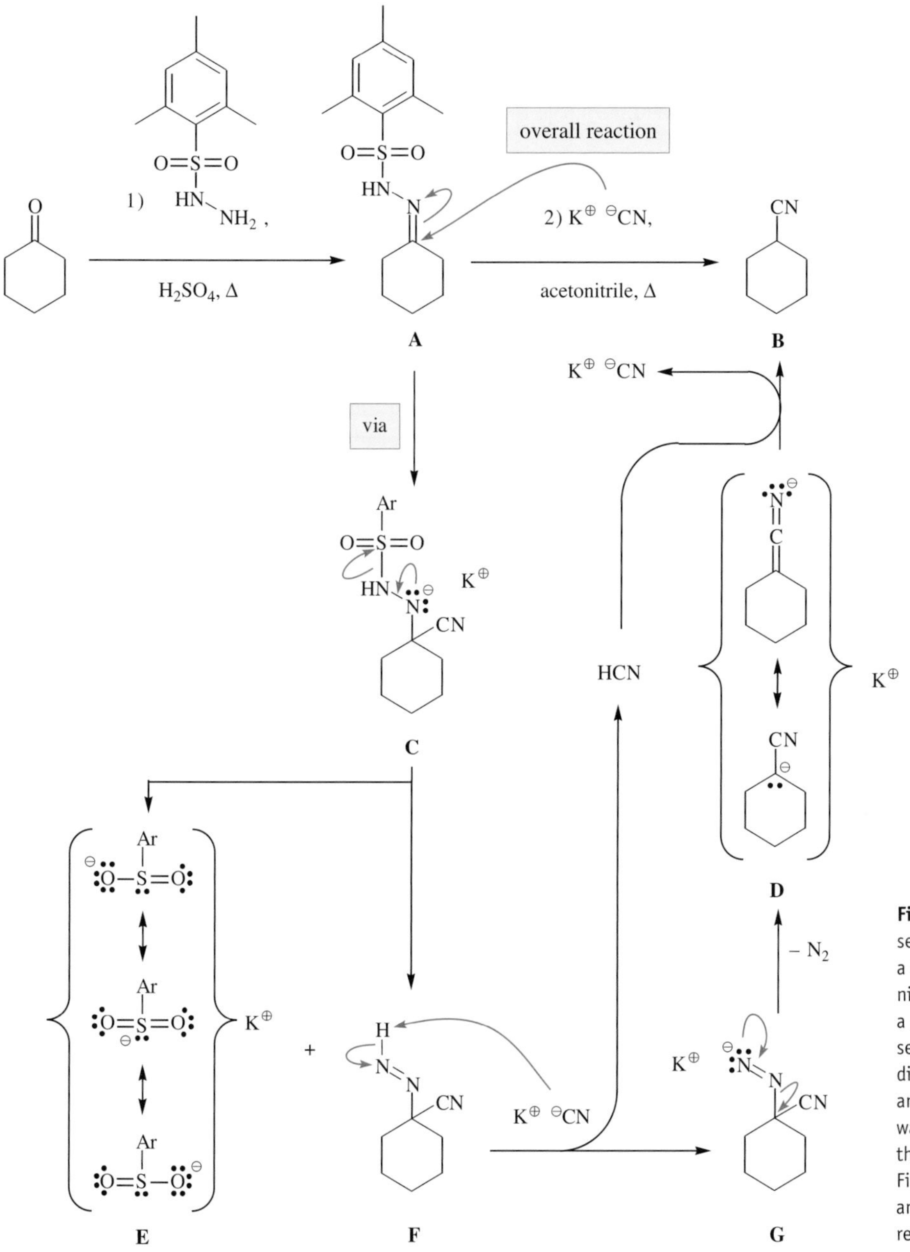

Fig. 17.69. Two-step sequence for the conversion of a ketone into the homologous nitrile ("reductive cyanation of a carbonyl compound"). In the second step of the reaction the diazene anion **G** is generated and decomposes in a similar way as the diazene anion **D** in the Wolff–Kishner reduction of Figure 17.67 and the diazene anion **G** of the semicarbazone reduction in Figure 17.68.

a dipolar aprotic solvent, acetonitrile solvates the $K^{\oplus}$ ions very well, but has no noticeable effect on the cyanide ions. This enables the cyanide ions (cf. Section 2.2, pp. 58–59) to undergo nucleophilic addition to the C=N double bond of the hydrazone **A**. The resulting anion **C** eliminates a sulfinate ion—as in the second step of a β-elimination according to the $E1_{cb}$ mechanism. The cyano-substituted diazene **F** remains. Since potassium cyanide is more basic in acetonitrile than in a protic solvent, it is able to deprotonate the diazene **F** to the diazene anion **G**. Like its analogs, this diazene anion tends to decompose with liberation of nitrogen. As a result, the "cyano-substituted carbanion"—or aza enolate **D**—is formed, which undergoes protonation to the nitrile **B**.

Tosylhydrazones can be reduced to the corresponding alkanes under milder conditions compared to the reduction of carbonyl compounds by the Wolff–Kishner method. This is illustrated in Figure 17.70 by the reduction of the aldhydrazone **A** to the alkane **C**. The reduction is carried out with $NaBH_4$ in MeOH. The effective reducing agent, formed *in situ*, is $NaBH(OMe)_3$. This reductant delivers a hydride ion for addition to the C=N double bond of the tosyl-hydrazone **A**. Thereby the hydrazide anion **B** is formed. Much as in the second step of an $E1_{cb}$ elimination, the anion **B** eliminates a *para*-tolylsulfinate and a diazene **D** is generated. The conversion of the diazene **D** into the hydrocarbon product **C** requires the same structural changes as the conversion of the diazene intermediate of the alkane synthesis according to Myers (cf. Formula **C** in Figure 1.48) into its final product (R_{prim}–H) and follows the radical chain mechanism described there. Due to the lack of a base the diazene intermediate **D** in Figure 17.70 is not deprotonated to a diazene anion.

Conjugated tosylhydrazones also can be reduced to hydrocarbons with the method depicted in Figure 17.70, that is, with $NaBH(OMe)_3$. The C=C double bond is retained but it is shifted. Figure 17.71 exemplifies this situation for tosylhydrazone **A**. The sequence of initial steps **A** → **B** → **D** resembles the one shown in Figure 17.70. However, the diazo compound **D** undergoes a different reaction, namely, **a retro-ene reaction**. A retro-ene reaction is a one-step fragmentation reaction of an unsaturated compound of type **A** or **B**, which affords two unsaturated compounds according to the following pattern:

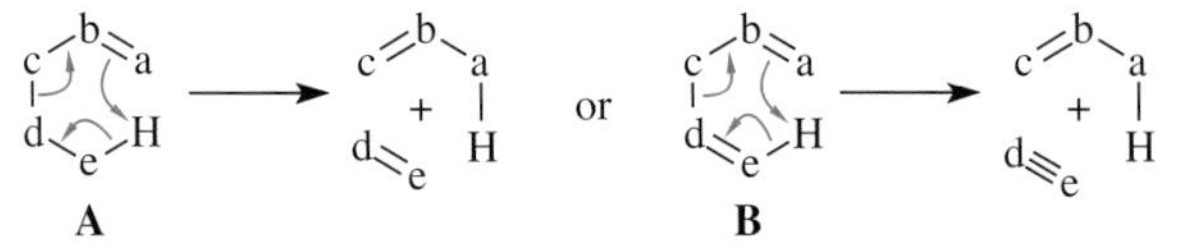

The retro-ene reaction of Figure 17.71 proceeds from **D** and directly leads to the reaction product **C** and the release of elemental nitrogen.

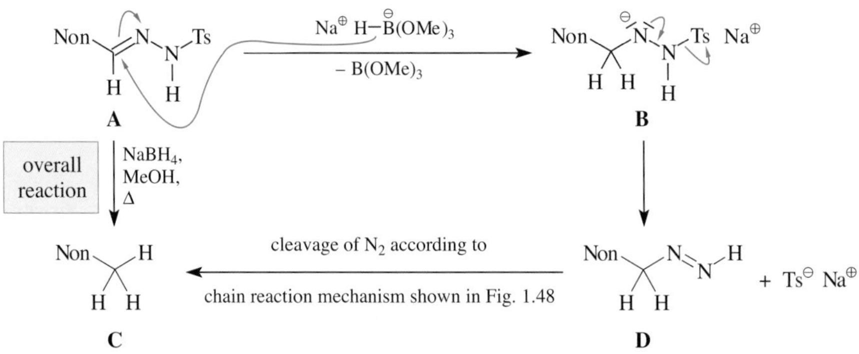

Fig. 17.70. Alternative II to the Wolff–Kishner reduction: reduction of a tosylhydrazone. Due to the lack of a sufficiently strong base, the diazene intermediate **D** does *not* react further via a diazene *anion*, which was encountered in the reactions of the Figures 17.67–17.69, but undergoes a *radical* reaction. "Non" refers to a nonyl group.

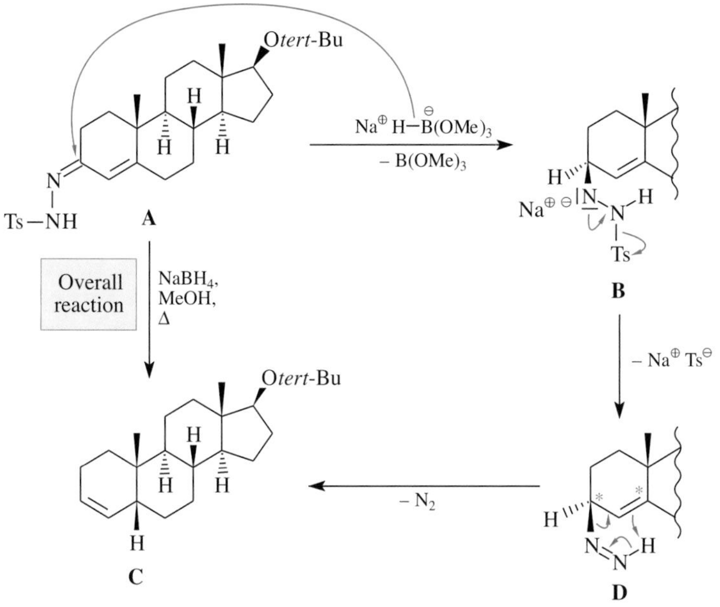

Fig. 17.71. Reduction of a conjugated tosylhydrazone. Since the diazene intermediate **D** cannot decompose via a diazene *anion* (cf. Figure 17.67–17.69) for lack of base and the *radical* decomposition mechanism of Figure 17.70 is not employed either, the *third* of a total of three mechanisms of the diazene → hydrocarbon transformation is presented here.

Via the so-called **ionic hydrogenation** or **ionic hydrogenolysis** aromatic aldehydes and aromatic ketones can be reduced to hydrocarbons, i.e., alkyl aromatic compounds, in a completely different manner as well (Figure 17.72). This kind of reduction is possible only if it can proceed via resonance-stabilized cationic intermediates. This resonance stabilization is readily achieved in a benzylic position, and it is therefore advantageous to employ aromatic carbonyl compounds in this kind of reduction. The ion **A** is the first cationic intermediate in the reduction of *meta*-nitroacetophenone and is generated via partial reversible protonation with CF_3CO_2H. To a small extent the trifluoroacetate ion released adds to the silicon atom of the first equivalent of the reducing agent, forming an –ate complex. The ion **A** abstracts a hydride ion from this –ate complex, and a benzyl alcohol and (triethylsilyl)trifluoroacetate are thus obtained. Immediately, a transfer of the silyl group between these compounds occurs. The resulting silyl ether then undergoes reduction via an S_N1 mechanism. To this end, it is protonated and eventually converted into the benzyl cation **B** in an equilibrium reaction. The latter abstracts another hydride ion from the second equivalent of the triethylsilanate complex. This leads to the final product.

A reduction with triethylsilane, which is a pure S_N1 reaction, is familiar from Figure 17.62, namely as the second step of a two-step reduction of a carboxylic acid ester to a simple ether via an α-acetoxyether.

The reduction of a ketone to an alkene is feasible not only for unsaturated ketones (Figures 17.68 and 17.71) but for saturated ketones as well. To this end, the latter can be converted into enol phosphonoamidates or enol dialkylphosphonates via suitable lithium enolates. One substrate of each type is shown in Figure 17.73 (**A** and **B**). Lithium dissolved in $EtNH_2$/*tert*-BuOH mixtures is a suitable reducing agent for both these compounds. Their C_{sp^2}—O bond is cleaved by a sequence of three elementary steps with which you are famil-

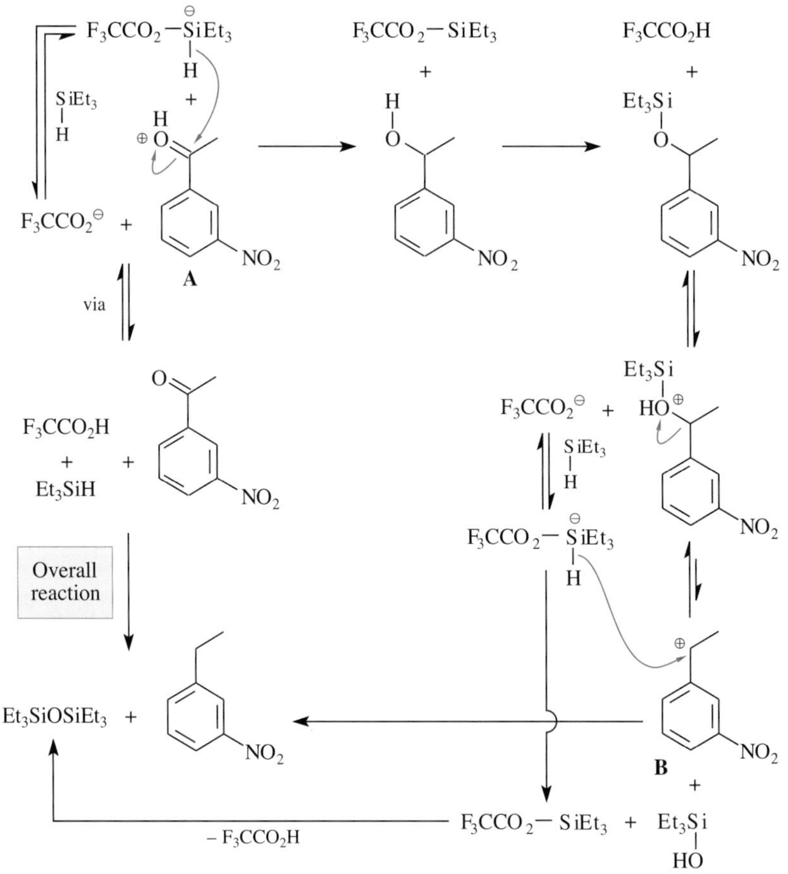

Fig. 17.72. Ionic hydrogenation/hydrogenolysis of an aromatic ketone (*meta*-nitroacetophenone). CF_3COOH causes a reversible protonation of the ketone to the ion **A.** The reducing agent triethylsilane then transfers a hydride ion onto **A** to form a benzylic alcohol. This alcohol presumably is silylated, protonated, and converted into the benzyl cation **B.** A second hydride transfer yields the final product.

iar from the formation of methyl-magnesium iodide (Figure 17.44): (1) electron transfer, (2) dissociation of the radical anion obtained to a vinyl radical and a negatively charged phosphoric acid derivative, and (3) electron transfer onto the vinyl radical and formation of an alkenyllithium compound. In the final and irreversible fourth reaction step, the alkenyllithium compound is protonated by *tert*-BuOH to furnish the alkene product. The C=C double bond remains in the same position as in the precursors **A** and **B**, respectively (Figure 17.73). This means that two different alkenes are formed, since the double bonds were in different positions in compounds **A** and **B**.

17.4.7 Hydrogenation of Alkenes

It was mentioned in Section 3.3.4 that alkenes react with H_2 on the surface of elemental Pd or elemental Pt to form alkanes. Similar hydrogenations occasionally also can be accomplished using Raney nickel as a catalyst. Raney nickel is prepared from a 1:1 Ni/Al alloy and aqueous KOH.

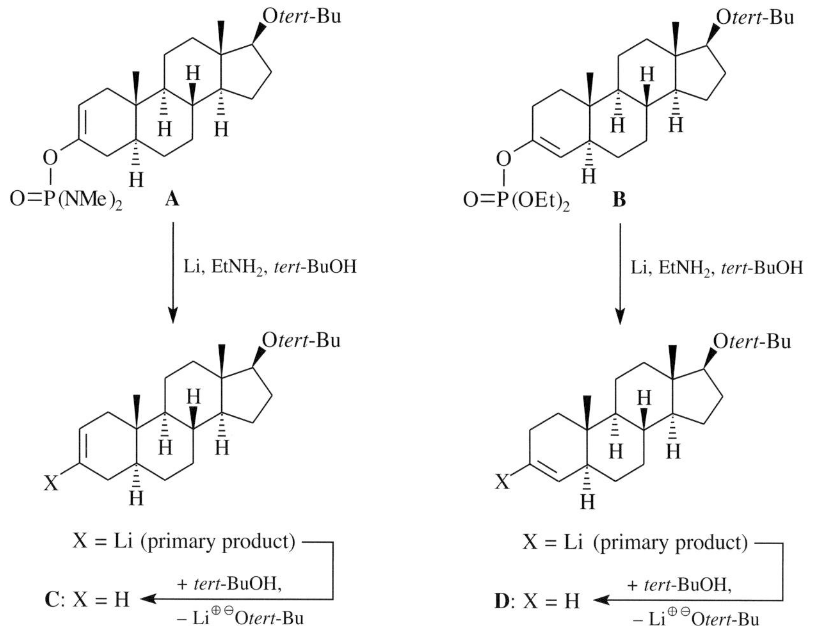

Li, $EtNH_2$, *tert*-BuOH

X = Li (primary product)

C: X = H

+ *tert*-BuOH, – $Li^{\oplus\ominus}$O*tert*-Bu

D: X = H

Fig. 17.73. Ketone → alkane reduction via enol phosphonoamidates (for one way to prepare **A**, see Figure 13.24) and enol dialkylphosphates (one way to prepare **B** is to use a combination of the methods depicted in Figures 13.20 and 13.25). The cleavage of the C_{sp^2}—O bond of the substrates occurs in analogy to the electron transfers in the formation of methylmagnesium iodide (Figure 17.44). The alkenyllithium intermediates are protonated in the terminating step to afford the target alkenes.

The examples of Figure 3.28 illustrate standard scenarios of heterogeneous catalytic hydrogenations of alkenes. Such reductions usually are highly stereoselective *cis* additions. Exceptions can be observed occasionally. For example, the hydrogenations shown in Figure 17.74 produce up to 30% of the *trans*-hydrogenated product. The mechanism outlined in Figure 17.75 shows that hydrogenations involve *several* steps. This fact explains why heterogeneous catalytic hydrogenations usually are *cis*-selective (part I) and why *trans* product may occur occasionally (part II).

Part I of Figure 17.75 shows the beginning and the *cis*-selective path of the hydrogenation. First, an H_2 molecule dissolved in the liquid phase is bonded covalently to the surface of (or absorbed into) the metal catalyst. The alkene **B** also is bonded to that surface. This bonding is accomplished by reversible π-complex formation. Occasionally, an alkene will thus be bonded to the catalyst surface in proximity to a Pd—H bond as shown in **C**. What follows is a kind of *cis*-selective hydropalladation of the alkene. A Pd atom binds to one end of a C=C bond and an H atom that was bonded to a proximate Pd adds to the other end of the C=C bond. The hydropalladation product is described by stereostructure **E**. Compound **E** can react further

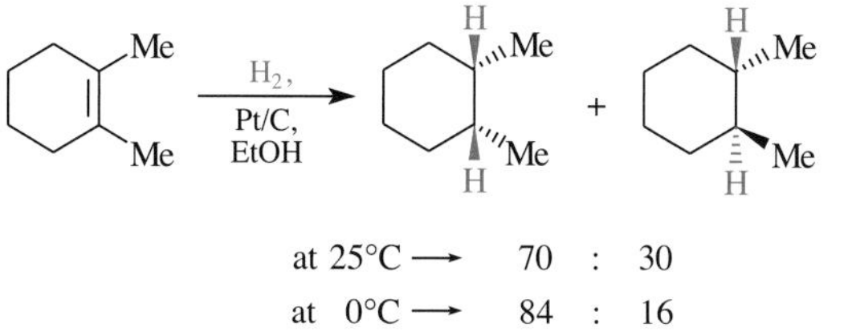

H_2, Pt/C, EtOH

at 25°C → 70 : 30

at 0°C → 84 : 16

Fig. 17.74. Examples of heterogeneous catalytic hydrogenations of C=C double bonds that occur with incomplete *cis*-selectivity.

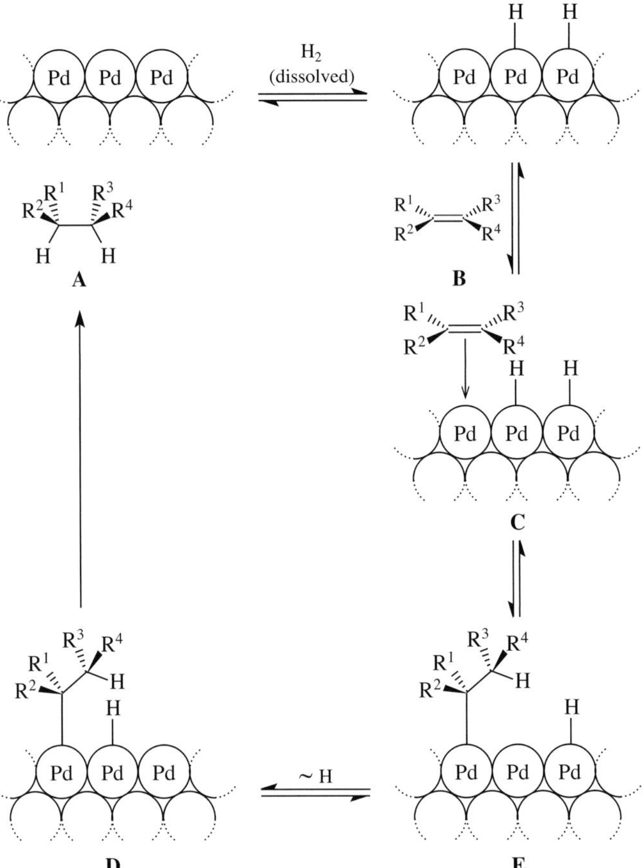

Fig. 17.75. Part I Mechanism of the *cis*-selective heterogeneous Pd-catalyzed hydrogenation of C=C double bonds.

only if an H atom migrates to a Pd atom that is right next to the Pd atom that was involved in the hydropalladation. This migration (of an H atom that is already bonded elsewhere on the surface) occurs by way of surface diffusion. The intermediate **D** is then formed. It releases alkane **A**, the product of a stereoselective *cis*-hydrogenation.

Part II of Figure 17.75 shows the side reactions that occur when the Pd-catalyzed hydrogenation is not completely *cis*-selective. The start is the formation of the π-complex **F** from the hydropalladation product **E**. In a way, this reaction is the reverse of the reaction type that formed **E** from the π-complex **C** (cf. part I of Figure 17.75). In an equilibrium reaction, the isomerized π-complex **F** subsequently releases the alkene *iso*-**B**, which is a double bond isomer of the substrate alkene **B**—and represents a type of compound that could well be the side product of an alkene hydrogenation, too.

Under the reaction conditions described the new alkene *iso*-**B** is usually hydrogenated immediately. In principle, it may add hydrogen from either of its diastereotopic faces. However, even if the addition were 100% *cis*-selective (according to the mechanism outlined in Figure 17.75, part I), a mixture of alkane **A** and *iso*-**A** could be formed, as shown by the structures depicted in the bottom half of Figure 17.75 (part II). With respect to the alkene plane of

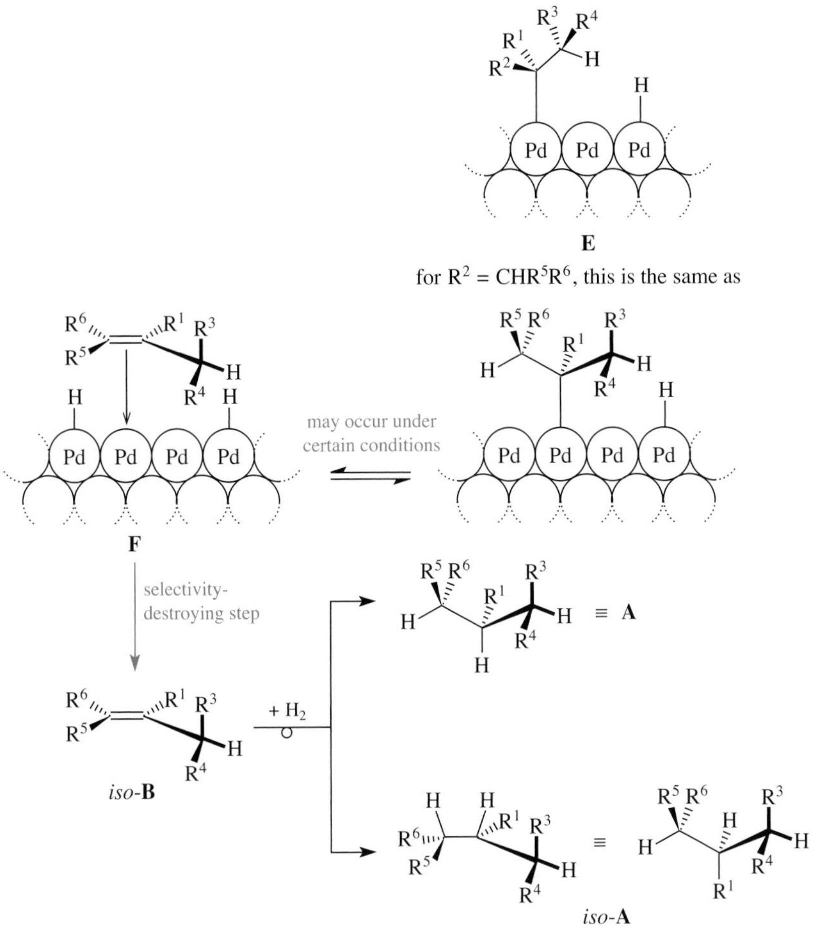

E

for R^2 = CHR^5R^6, this is the same as

F

may occur under certain conditions

selectivity-destroying step

iso-B

$+ H_2$

≡ A

iso-A

Fig. 17.75. Part II Mechanism of the non-stereoselective heterogeneous Pd-catalyzed hydrogenation of C=C double bonds.

the original substrate **B**, the newly added hydrogen atoms in *iso*-**A** are oriented in the way that would have resulted from a *trans*-hydrogenation of **B**.

The alkene isomerization **B** → *iso*-**B** also may have another stereochemical consequence (part II of Figure 17.75): the destruction of the configurational homogeneity of a stereocenter $C^*HR^5R^6$ in the allylic position of the substrate. In that case, the hydrogenation results in a mixture containing the stereoisomers **A** and *iso*-**A**.

Catalytic hydrogenations of C=C double bonds can be carried out not only in a heterogeneous fashion—on metal surfaces—but also in **homogeneous phase** using *soluble* metal complexes as catalysts. This possibility is of special significance if the hydrogenation is carried out with an enantiomerically pure chiral catalyst. It allows for the *enantioselective* addition of hydrogen to certain functionalized alkenes. Horner, Pracejus, Kagan and Knowles discovered highly efficient catalysts for such **asymmetric hydrogenations** as early as the end of the 1950s—far ahead of their time. Knowles' catalyst soon gained particular importance when it was used in the first asymmetric synthesis that was conducted on industrial scale: in the Monsanto synthesis of the nonproteinogenic amino acid L-DOPA (3,4-dihydroxy-*S*-phenylalanine). It allowed for the construction of its stereocenter with uniform configuration. In the early 1980s, Noyori initiated a significant upturn when he introduced the enantiomerically pure ligand BINAP, a phosphine derivative of 2,2'-binaphthyl, into asymmetric catalysis. The racemic synthesis of this ligand from BINOL was described in Figures 5.61 and 5.44. BINAP-containing noble metal complexes catalyze more than a dozen fundamentally important stereogenic reactions. These include—among others—the following asymmetric hydrogenations:

- enantioselective hydrogenations of the C=C double bonds of certain allyl alcohols (example in Figure 17.76),
- enantioselective hydrogenations of the C=C double bonds of α-(acylamino)acrylic acids (example in Figure 17.77), and
- enantioselective hydrogenations of the C=O double bonds of β-ketoesters (not shown).

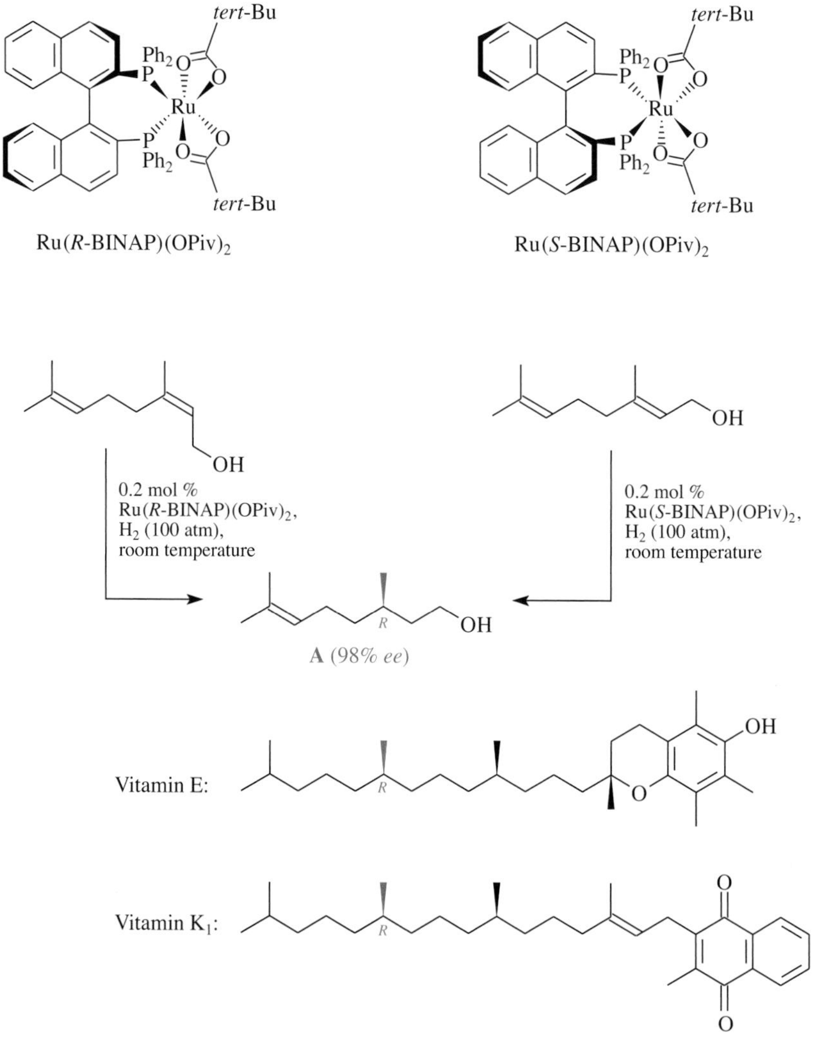

Fig. 17.76. Enantioselective homogeneous catalytic hydrogenations of nerol (left) and geraniol (right) to form the same *R*-configured alcohol **A**. [OPiv refers to the pivaloyloxy group *tert*-Bu—C(=O)—O—]. The aldehyde, which can be obtained by oxidation of the alcohol **A**, can be elongated by way of the Still–Gennari- (Figure 11.14, right) or the Ando variant (Section 11.3.1) of the Horner–Wadsworth–Emmons reaction to furnish a *Z*-configured α,β-unsaturated carboxylic acid ester, or by means of a Wittig reaction to provide an *E*-configured α,β-unsaturated carboxylic acid ester (Figure 11.7). The reduction of these esters with diisobutylaluminum hydride furnishes the allylic alcohols (Figure 17.63). Like nerol or geraniol, they can also undergo enantioselective hydrogenation, thus establishing the second stereocenter of important compounds such as the vitamins E and K$_1$. – Mechanism of the hydrogenation of nerol: Figure 17.79.

In 2001, Knowles' and Noyori's pioneering achievements in this field were honored with the award of the Nobel Prize in Chemistry.

The asymmetric hydrogenation of geraniol or nerol is the most elegant method for the stereoselective construction of the side chains of vitamins E and K_1 (Figure 17.76). In this manner, the stereocenter (highlighted in red) of these vitamins can be established with the correct configuration by the incorporation of the precursor molecule **A**. One can also achieve comparably high stereocontrol at the neighboring stereocenter by means of asymmetric hydrogenations that are analogous to the ones shown, namely, which start with trisubstituted allylic alcohols. Such allylic alcohols can be constructed from **A** by way of chain-elongating syntheses (for details see the caption of Figure 17.76).

The enantioselective hydrogenation of α-(acylamino)acrylic acids (Figure 17.77) or α-(acylamino)acrylic acid esters, usually catalyzed by rhodium complexes, is mainly used for

$Rh(R\text{-}BINAP)(MeOH)_2^{\oplus}\ BF_4^{\ominus}$ $\quad$ $Rh(S\text{-}BINAP)(MeOH)_2^{\oplus}\ BF_4^{\ominus}$

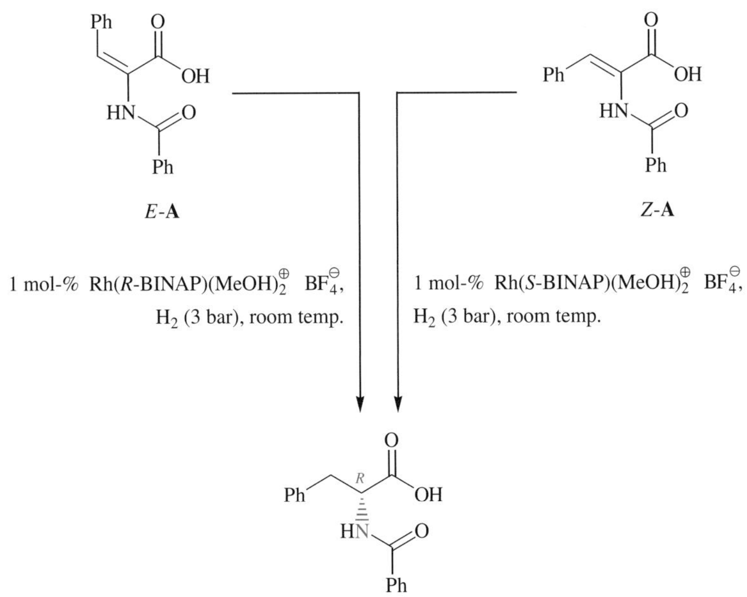

Fig. 17.77. Enantioselective homogeneous catalytic hydrogenations of two stereoisomeric α-(acylamino)acrylic acids to one and the same *R*-amino acid. – The mechanism of the hydrogenation of these acids is exemplified by their *Z*-isomer: Figure 17.78.

the preparation of *R*- or (non-proteinogenic) *S*-configured amino acids. Such enantiomerically pure amino acids are needed for the synthesis of antibiotic peptides and peptidomimetics.

The asymmetric hydrogenation of nitrogen-free acrylic acids or acrylic acid esters proceeds in a similar enantioselective fashion. For these substrates Rh- and Ru-catalysts are used with the same frequency.

You will find that the absolute configuration of the newly formed stereocenter will change in said asymmetric hydrogenations upon alteration of one (no matter which) of the following variables (1) to (3):

(1) choice of a rhodium or ruthenium complex,
(2) use of *R*- or *S*-BINAP as the stereo-inducing ligand and
(3) use of an *E*- or Z-configured alkene as the substrate.

This has the following interesting implications:

- If variable (1) is maintained, but variables (2) and (3) are altered, the same absolute configuration is induced at the newly formed stereocenter. This is evidenced by the reaction pairs in Figure 17.76 and Figure 17.77.
- If only variable (1) is changed, the amidocarboxylic acid *Z*-**A** in Figure 17.77 is hydrogenated in the presence of Rh(*S*-BINAP)$(MeOH)_2^{\oplus}BF_4^{\ominus}$ to furnish N-benzoyl-***R***-phenylalanine (as shown). In the presence of Ru(*S*-BINAP)$(OPiv)_2$, however, *N*-benzoyl-***S***-phenylalanine (not shown) is formed. This surprising effect can be attributed to the fact that Rh- and Ru-catalyzed asymmetric hydrogenation of C=C double bonds follow fundamentally different mechanisms as will soon be seen.

Homogeneous catalytic hydrogenations—both racemic and asymmetric—proceed via catalytic cycles that involve many steps, as is illustrated below taking the Rh catalysis in Figure 17.78 and the Ru catalysis in Figure 17.79 as examples. Most probably, the intermediates given there are the appropriate ones. Not all the species discussed in these catalytic cycles have been proven completely, but it is very likely that these cycles are basically correct.

For enantioselectivity to occur with homogeneous hydrogenations, the unsaturated substrate must bind to the catalytic center in such a way that a complex with well-defined stereostructure is formed. Accordingly, a highly enantioselective hydrogenation is assured—at least in most cases—if the substrate forms *two* bonds to the metal. The substrate is π-bonded to the metal via the C=C double bond that is to be hydrogenated. It is also σ-bonded to the metal via a heteroatom that is close enough to this C=C double bond.

Achieving successful enantioselective hydrogenations of C=C double bonds requires that one more prerequisite be met. It must be possible to convert the substrate/metal complex into a complex that contains a hydrido ligand as well. Hydrometalation of the π-bonded C=C double bond is possible only in such a hydrido complex. The first C—H bond together with a C-metal bond is formed in this fashion. A reductive elimination of this C-metal bond forms the second C—H bond. This completes what is an overall "hydrogenation" of the substrate. Figures 17.78 and 17.79 depict the *specific* intermediates of the hydrogenations in Figures 17.77 and 17.76, respectively.

The initially formed rhodium/substrate (**B**) complex of the hydrogenation of *Z*-α-(acylamino)acrylic acid in Figure 17.77 possesses structure **C** (Figure 17.78). The reaction starts with the reversible dissociation of two MeOH ligands from the central atom Rh(I). The two now

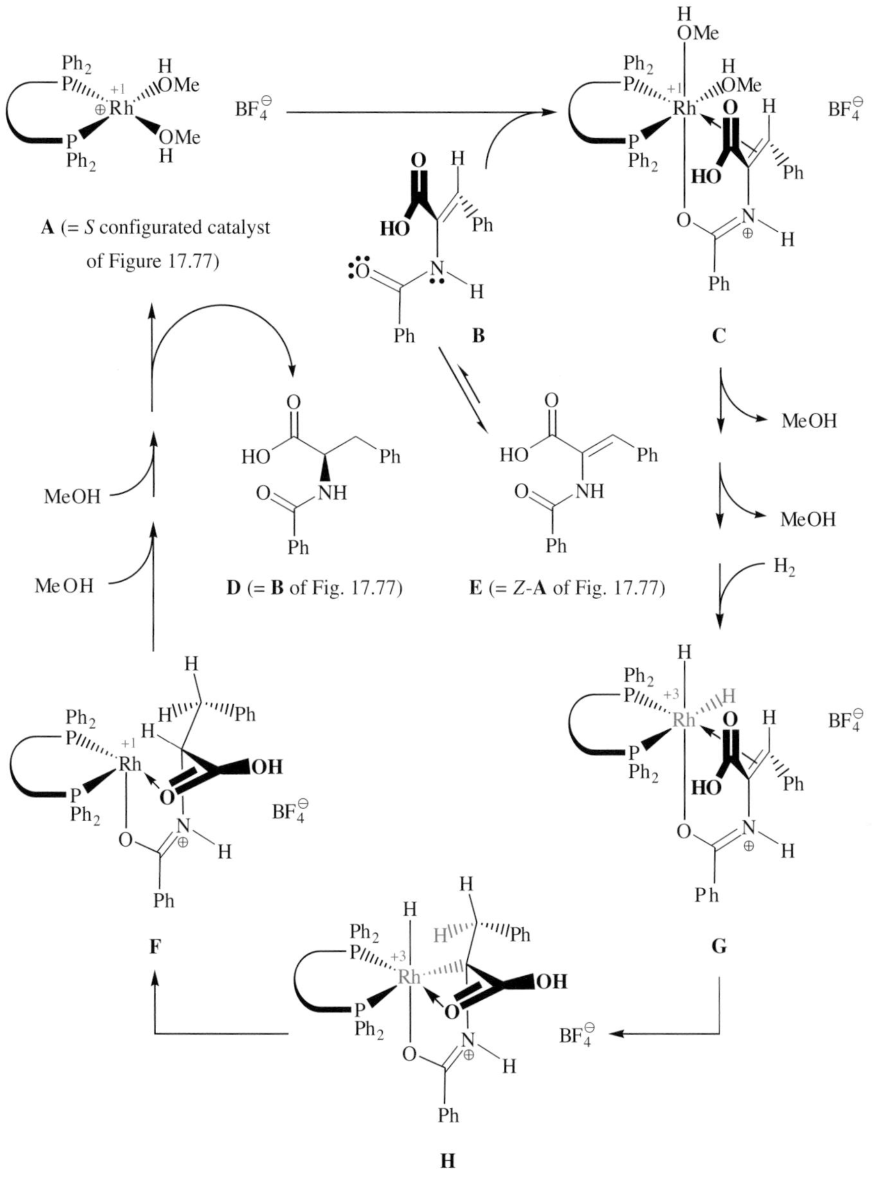

Fig. 17.78. Key intermediates in the enantioselective hydrogenations of the Z-configurated stereoisomeric α-(acylamino)acrylic acids in Fig. 17.77 (right). The BINAP ligand is shown schematically as U-shaped, with two PPh_2 substituents. Rh-phosphine complexes undergo hydrogenation via a Rh(I)/Rh(III) cycle, in which the double bond is hydrometalated in a dihydrodo metal complex. MeOH occupies the ligand sphere of the metal Rh(I) (**F** →→→ **A**), while it oxidatively adds to Ru(II) in Figure 17.79.

vacant coordination sites allow for the oxidative addition of a hydrogen molecule. The dihydrido Rh(III) complex **G** is formed. The hydrometalation of the C=C double bond, that is, the *cis*-addition of L_nRh—H to that double bond, occurs *in this complex*. This reaction affords the monohydrido Rh(III) complex **H**. The next step consists of the reductive elimination of the Rh(III)-bonded H atom jointly with the Rh(III)-bonded alkyl group. The C—Rh bond in **H** thereby is converted into a C—H bond with retention of configuration. Two MeOH molecules add to the new Rh(I) complex, and the hydrogenated product—the *N*-benzoylphenylalanine (**D**)—dissociates.

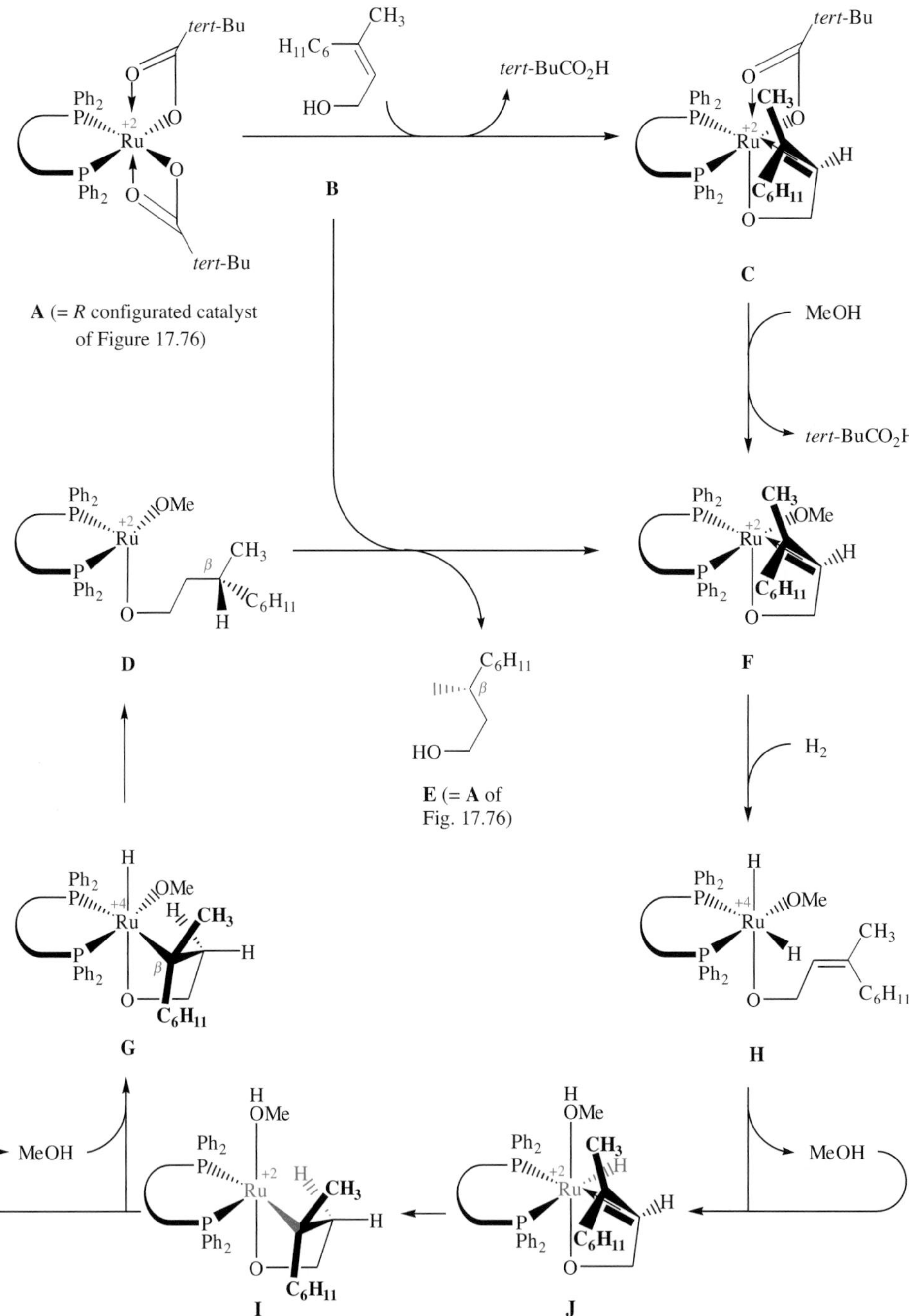

Fig. 17.79. Key intermediates in the enantioselective hydrogenations of Figure 17.76 (left). The BINAP ligand is shown schematically as U-shaped, with two PPh_2 substituents. Ru-phosphine complexes undergo hydrogenation via Ru(II)/Ru(IV) intermediates, and the double bond of the substrate is hydrometalated in a monohydrido metal complex. Here, MeOH oxidatively adds to Ru(II) (**I** → **G**), while it completes the ligand sphere of the metal Rh(I) in Figure 17.78.

The structure **C** in Figure 17.79 describes the initially formed metal/substrate (**B**) complex of the hydrogenation of nerol in Figure 17.76. A molecule of hydrogen can add oxidatively only if the bidentate pivaloyloxy group is displaced by a monodentate methoxy residue and the C=C double bond of the allylic alcohol liberates another coordination site at Ru(II) (→ **F**). The dihydrido complex **H** of Ru(IV) formed by this addition undergoes a reductive elimination with methanol. The resulting monohydrido Ru(II) complex is not shown in the figure because it immediately binds the released methanol as a σ-donor ligand, and the monohydrido Ru(II) complex **J** is obtained. It is *this complex* in which the hydrometalation of the substrate occurs via a *cis*-addition of L_nRu—H to the C=C double bond, forming the first C—H bond. Complex **I**, which still contains Ru(II), is produced. The σ-donor ligand MeOH of this complex **I** dissociates in the next step, leaving an alkyl Ru(II) complex with two vacant metal coordination sites. In the figure this complex is not shown since the methanol that dissociates from **I** immediately adds to it oxidatively. The resulting presence of an alkyl and a hydrido ligand in the Ru(IV) complex **G** initiates the last step of the catalytic cycle, namely the reductive elimination of the hydrogenation product [→ Ru(II) complex **D**]. The C—Ru bond, which has formed *cis* to the first C—H bond, is converted into the second C—H bond with retention of configuration. Therefore, this H_2 addition to the C=C double bond again occurs with *cis*-selectivity.

17.4.8 Reductions of Aromatic Compounds and Alkynes

Benzene and its derivatives can be hydrogenated on the surface of precious metals only under pressure and even then only at elevated temperatures (Figure 17.80), at least in most cases. Obviously, these hydrogenations require more drastic conditions than the hydrogenations of alkenes—as one would expect, based on the Hammond postulate. The first step in the hydrogenation of an aromatic compound is the conversion into a cyclohexadiene. This step is associated with the loss of the benzene resonance energy of 36 kcal/mol. It is for this reason, and in contrast to the hydrogenation of alkenes, that the hydrogenation of the first aromatic C=C double bond is endothermic. Once the benzene has been converted into the cyclohexadiene, the consecutive hydrogenations via cyclohexene on to cyclohexane are exothermic reactions. In agreement with Hammond's postulate, the latter two hydrogenations are faster than the initial hydrogenations of the aromatic system. Consequently, benzene derivatives are hydrogenated "all the way" to cyclohexanes.

The two examples on top of Figure 17.80 show that these kinds of hydrogenations have a certain significance for the stereoselective synthesis of *cis*-1,3,5-trisubstituted cyclohexanes. The third example in Figure 17.80 shows that aromatic compounds can be hydrogenated with metallic Rh as the catalyst in a *chemoselective* fashion even though C—O bonds that are rather sensitive to hydrogenolysis are present in the molecule (cf. Figure 17.51).

Alkynes are hydrogenated "all the way" to alkanes if the usual heterogeneous catalysts (Pd, Pt, Raney Ni) are used. If a suitable deactivated catalyst is used, however, it is possible to stop these reactions after monohydrogenation. The so-called **Lindlar catalyst** is a commonly used deactivated catalyst of this type (Figure 17.81). To prevent an overhydrogenation, it is still necessary to monitor the rate of hydrogen consumption and to interrupt the reaction after one equivalent of hydrogen gas has been absorbed even when the deactivated catalyst is used. The

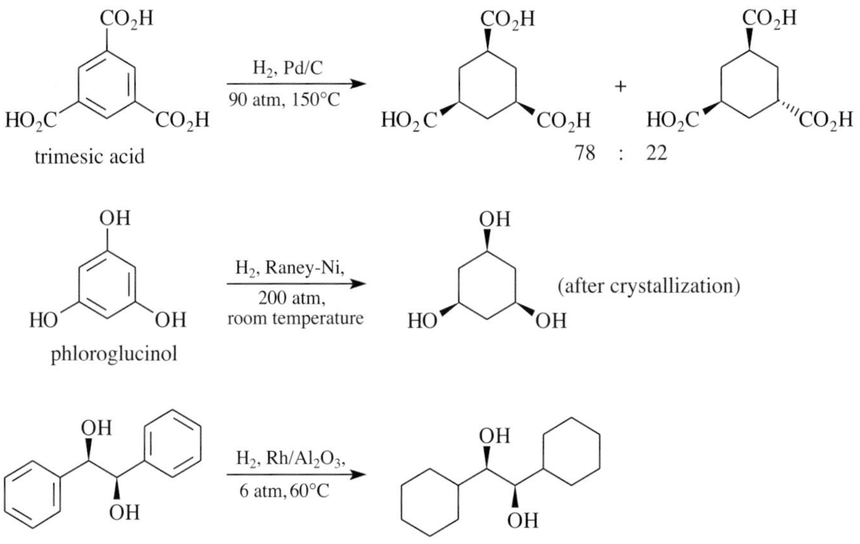

Fig. 17.80. Heterogeneous catalytic hydrogenations of aromatic compounds.

Lindlar hydrogenation depicted in Figure 17.81 shows that C=C double bonds already present in the substrate also are not hydrogenated under these conditions.

The mechanism of the Lindlar hydrogenation is analogous to the mechanism of the alkene hydrogenation (Figure 17.75). It is for this reason that hydrogen additions to C≡C triple bonds usually also exhibit high *cis*-selectivity. Lindlar hydrogenations present a good synthetic route to *cis*-alkenes. Only occasionally is this *cis*-selectivity not perfect, for the reasons discussed in the case of the hydrogenations of C=C double bonds (Figure 17.75, part II) with heterogeneous catalysts.

The **Birch reduction** is another method for the conversion of benzene and its derivatives into nonaromatic six-membered rings (Figure 17.82). Solvated electrons in liquid NH_3 or in liquid $EtNH_2$ are the effective reducing agent. The solvated electrons are generated by dissolving Li or Na in these media. Birch reductions of aromatic compounds can be carried out in the presence of several equivalents of alcohol without the formation of hydrogen gas. This is important because some of the later steps of the Birch reduction require an alcohol as a proton donor.

One of the solvated electrons is transferred into an antibonding π^*-orbital of the aromatic compound, and a radical anion of type **C** is formed (Figure 17.82). The alcohol protonates this radical anion in the rate-determining step with high regioselectivity. In the case under scrutiny, and starting from other *donor-substituted* benzenes as well, the protonation occurs in the *ortho* position relative to the donor substituent. On the other hand, the protonation of the radical anion intermediate of the Birch reduction of *acceptor-substituted* benzenes occurs in the *para*-position relative to the acceptor substituent.

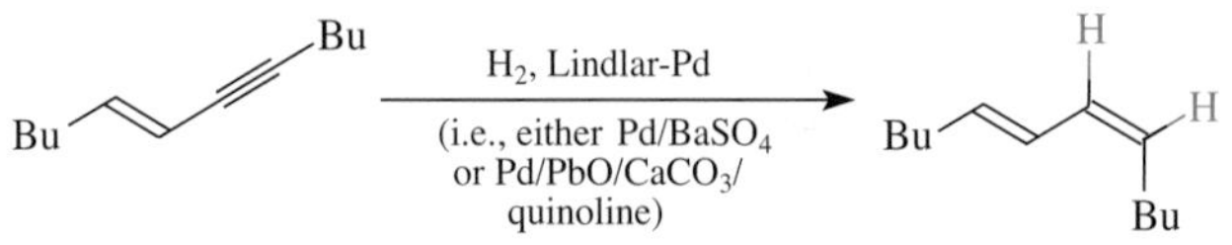

Fig. 17.81. Lindlar hydrogenation of a C≡C triple bond. (The starting material can be prepared in analogy to Figure 16.25.)

Protonation of the radical anion **C** of Figure 17.82 results in the radical **D**. This radical regains its valence electron octet by capturing another solvated electron to form the carbanion **E**. Carbanions of this type are protonated regioselectively by the second equivalent of alcohol. The regioselectivity is independent of the substituents: protonation forms a 1,4-dihydroaromatic compound rather than a 1,2-dihydroaromatic compound. Hence, the 1,4-dihydrobenzene **A** is the reduction product in the example of Figure 17.82. Reasons for the selectivity in the protonation of the pentadienylic anion have ranged from "least motion" considerations to more easily understood charge density arguments. Simply put, there is a higher charge density in the central carbon of a pentadienylic anion, favoring protonation at that site.

1,4-dihydroaromatic compounds obtained in this way can be used for the preparation of six-membered rings that would be difficult to synthesize otherwise. For example, the dihydrobenzene **A** of Figure 17.82 can be converted into the 4-substituted 2-cyclohexenone **B** via an acid-catalyzed hydrolysis and an acid-catalyzed migration of the remaining C=C double bond.

The Birch reduction of aromatic compounds in NH_3/ROH or $EtNH_2$/ROH mixtures includes the following steps (Figure 17.82): electron transfer to the substrate, protonation of the resulting radical anion by the alcohol, electron transfer to the resulting radical, and protonation of the formed organometallic compound. The same four elementary reactions also can be used for the reduction of disubstituted alkynes to give alkenes (Figure 17.83). Protonation of the radical anion intermediate **A** or of the alkenylsodium compound **B** is accomplished in this case by the weakly acidic solvent NH_3.

The rate-determining step in the Na/NH_3 reduction of alkynes is the protonation of the radical anion **A**. The next step, the reaction of the alkenyl radical **C** to the alkenyl-sodium inter-

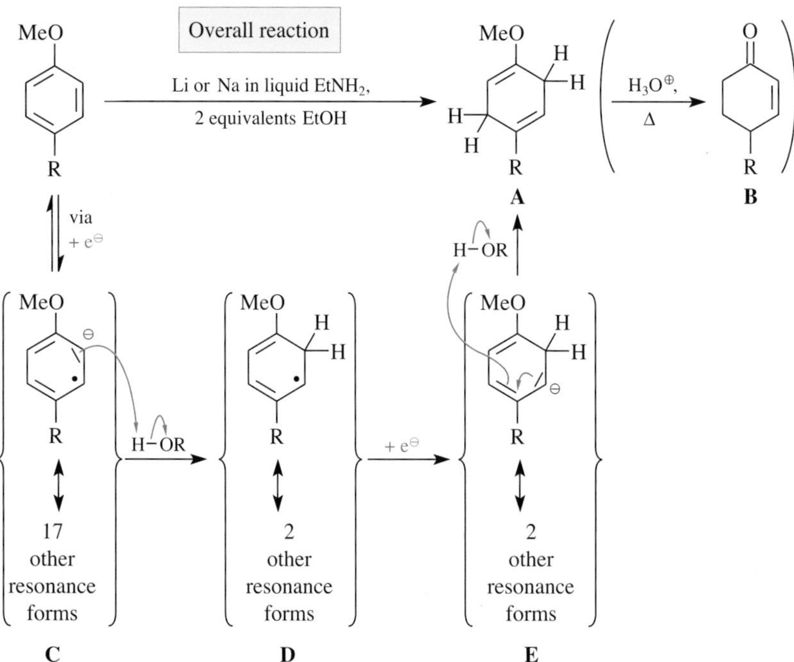

Fig. 17.82. Birch reduction of benzenes give 1,4-cyclohexadienes. The radical anion **C** is formed by capture of a solvated electron in an antibonding π^*-orbital of an aromatic compound. The alcohol protonates this radical anion to the radical **D**, which captures another electron from the solution to form the carbanion **E**. The carbanion is protonated by a second equivalent of the alcohol, and the 1,4-dihydroaromatic compound results.

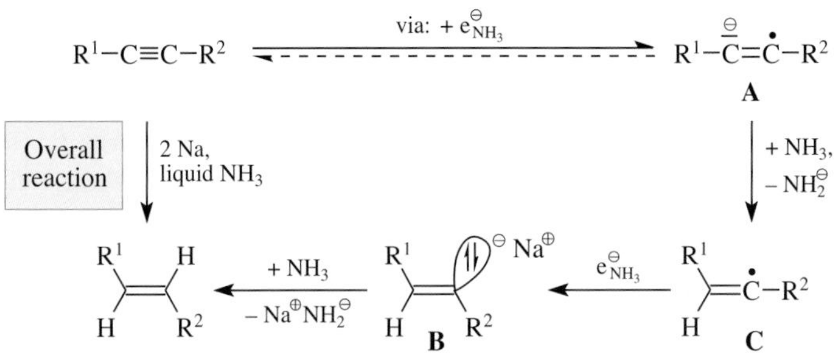

Fig. 17.83. *trans*-Selective reduction of C≡C triple bonds.

mediate **B**, determines the stereochemistry. The formation of **B** occurs such that the substituents of the C=C double bond are in *trans*-positions. This *trans*-selectivity can be explained by product development control in the formation of **B** or perhaps also by the preferred geometry of radical **C** provided it is nonlinear at the radical carbon. The alkenylsodium compound **B** is protonated with retention of configuration, since alkenylsodium compounds are configurationally stable (cf. Section 1.1.1). The Na/NH_3 reduction of alkynes therefore represents a synthesis of *trans*-alkenes.

It is difficult to determine directly whether the electron transfer to an alkyne is a reversible or an irreversible reaction (Figure 17.83, top). The current thinking on the matter is that it is a reversible process. However, compounds containing two C≡C triple bonds that are not too far apart can be reduced chemoselectively with Na in liquid ammonia such that only one of the C≡C triple bonds is reduced. Figure 17.84 depicts the example of the dialkyne **A**. Such a

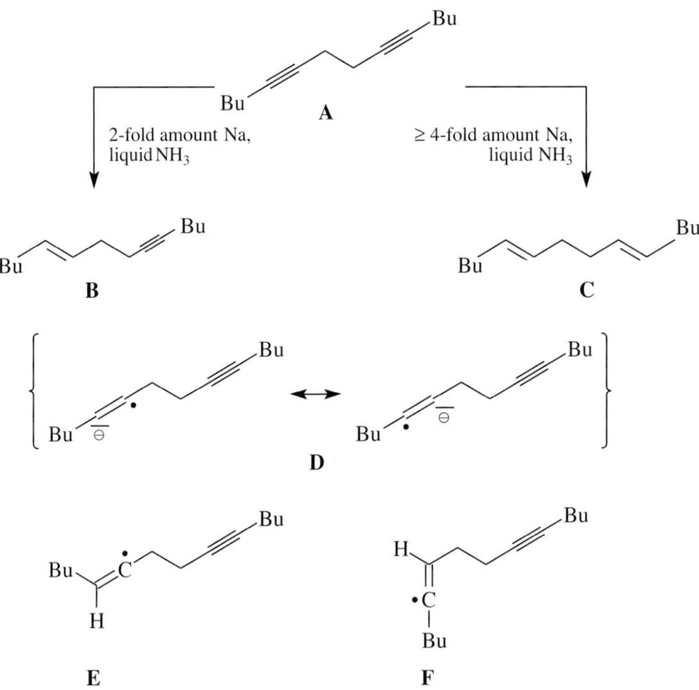

Fig. 17.84. Chemoselective and *trans*-selective reduction of exactly one C≡C triple bond of a diyne.

reduction to the monoalkyne can be accomplished if one employs only the amount of sodium required for monoreduction. If one employs twice the amount of sodium, one obtains the dialkene **C**, as expected. The first reduction **A** → **B** consequently must be faster than the subsequent reduction **B** → **C**. This can be understood immediately if one assumes that the electron transfer to the C≡C triple bond is *irreversible and fast*. If that is true, then the radical anion **D** would be formed quantitatively upon consumption of exactly 50% of the reducing reagent. After protonation in the rate-determining step, the radical anion would yield a mixture of the isomeric alkenyl radicals **E** and **F**. These radicals would quickly consume the remaining 50% of the reducing reagent. Thus, their reaction to form the monoalkyne **B** would be faster than the reduction of the monoalkyne to the dialkene **C**.

17.4.9 The Reductive Step of the Julia–Lythgoe Olefination

The Julia–Lythgoe olefination has already been addressed twice as an important C=C double bond-forming two- or three-step synthesis of *trans*-alkenes (*trans*-**B** in Figure 17.85). The step

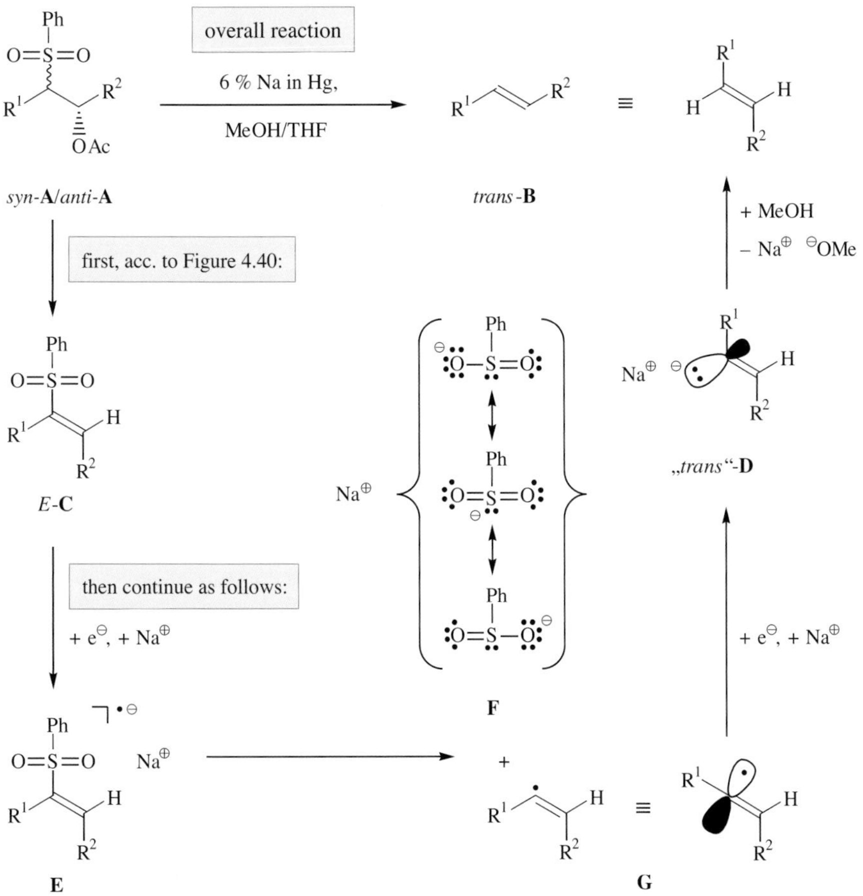

Fig. 17.85. Mechanistic analysis of the second part of the reaction process where the treatment of the acetoxy sulfones *syn*- and *anti*-**A** with sodium amalgam completes the Julia–Lythgoe olefination. Series of a first electron transfer (→ alkenyl phenylsulfone radical anion **E**), homolysis (→ alkenyl radical **G** + sodium benzene sulfinate), second electron transfer (→ alkenyl anion "*trans*"-**D**) and *in-situ* protonation.

that forms the double bond is the $E1_{cb}$ elimination of the acetylated sulfones *syn*- and *anti*-**A** to afford the *E*-configured alkenyl sulfones (*E*-**C** in Figure 17.85). This forms the basis for our third and last reference to the Julia–Lythgoe olefination at this point. All we have to do now is explain how the actual reduction of these *in situ* generated sulfones *E*-**C** proceeds by treatment with sodium amalgam in MeOH/THF to give the final alkene product. Figure 17.85 picks up the thread with cleavage of the alkenyl sulfone *E*-**C** via its radical anion **E** into the benzene sulfinate anion **F** and the alkenyl radical **G**. The latter accepts an electron to achieve the valence electron octet. As in the fully analogous step **C** → **B** of the **Birch reduction** in Figure 17.83, a "*trans*"-configured alkenyl sodium **D** is formed, which is protonated by MeOH with retention of configuration to yield the alkene *trans*-**B**.

References

S. D. Burke, R. L. Danheiser, (Eds.), "Handbook of Reagents for Organic Synthesis: Oxidizing and Reducing Agents," Wiley, New York, **1999**.

17.1

G. Calzaferri, "Oxidation Numbers," *J. Chem. Ed.* **1999,** *76,* 362–363.

17.3

A. H. Haines, "Methods for the Oxidation of Organic Compounds: Alkanes, Alkenes, Alkynes, Arenes," AP, New York, **1985**.

A. H. Haines, "Methods for the Oxidation of Organic Compounds: Alcohols, Alcohol Derivatives, Alkyl Halides, Nitroalkanes, Alkyl Azides, Carbonyl Compounds, Hydroxyarenes, and Aminoarenes," Academic Press, **1988**.

M. Hudlicky, "Oxidations in Organic Chemistry," American Chemical Society, Washington, DC, **1990**.

H. Bornowski, D. Döpp, R. Jira, U. Langer, H. Offermans, K. Praefcke, G. Prescher, G. Simchen, D. Schumann, "Preparation of Aldehydes by Oxidation," in *Methoden Org. Chem.* (Houben-Weyl) 4th ed., **1952**–, *Aldehydes* (J. Falbe, Ed.), Bd. E3, 231, Georg Thieme Verlag, Stuttgart, **1983**.

A. J. Mancuso, D. Swern, "Activated Dimethyl Sulfoxide: Useful Reagents for Synthesis," *Synthesis* **1981**, 165.

T. T. Tidwell, "Oxidation of alcohols by activated dimethyl sulfoxide and related reactions: an update," *Synthesis* **1990**, 857–870.

T. T. Tidwell, "Oxidation of Alcohols to Carbonyl Compounds via Alkoxysulfonium Ylides: The Moffatt, Swern, and Related Oxidations," *Org. React.* **1990**, *39*, 297–572.

F. A. Luzzio, "The Oxidation of Alcohols by Modified Oxochromium(VI)-Amine Reagents," *Org. React.* **1998**, *53*, 1–221.

E. J. de Nooy, A. C. Besemer, H. van Bekkum, "On the Use of Stable Organic Nitroxyl Radicals for the Oxidation of Primary and Secondary Alcohols," *Synthesis*, **1996**, 1153– 1174.

R. M. Moriarty, O. Prakash, "Oxidation of Carbonyl Compounds with Organohypervalent Iodine Reagents," *Org. React.* **1999**, *54*, 273–418.

J. B. Arterburn, „Selective Oxidation of Secondary Alcohols," *Tetrahedron* **2001**, *57,* 9765–9788.

D. V. Deubel, G. Frenking, "[3+2] Versus [2+2] Addition of Metal Oxides Across C=C Bonds. Reconciliation of Experiment and Theory," *Acc. Chem. Res.* **2003**, *36,* 645–651.

R. A. Johnson, K. B. Sharpless, "Catalytic Asymmetric Dihydroxylation-Discovery and Development," in *Catalytic Asymmetric Synthesis* (I. Ojima, Ed.), Wiley-VCH, New York, 2nd ed., **2000**, 357–389.

C. Bolm, J. P. Hildebrand, K. Muniz, "Recent Advances in Asymmetric Dihydroxylation and Aminohydroxylation," in *Catalytic Asymmetric Synthesis*, (I. Ojima, Ed.), Wiley-VCH, New York, 2nd ed., **2000**, 399–428.

D. Nilov, O. Reiser, "The Sharpless Asymmetric Aminohydroxylation – Scope and Limitation," *Adv. Synth. Catal.* **2002**, *344,* 1169–1173.

J. A. Bodkin, M. D. McLeod, "The Sharpless Asymmetric Aminohydroxylation," *J. Chem. Soc. Perkin Trans. I* **2002**, 2733–2746.

R. L. Kuczkowski, "The structure and mechanism of formation of ozonides," *Chem. Soc. Rev.* **1992**, *21*, 79–83.

E. L. Jackson, "Periodic Acid Oxidation," *Org. React.* **1944**, *2,* 341–375.

C. H. Hassall, "The Baeyer–Villiger Oxidation of Aldehydes and Ketones," *Org. React.* **1957**, *9*, 73–106.

G. R. Krow, "The Baeyer–Villiger Reaction," in *Comprehensive Organic Synthesis* (B. M. Trost, I. Fleming, Eds.), Vol. 7, 671, Pergamon Press, Oxford, U. K., **1991**.

G. R. Krow, "The Bayer–Villiger Oxidation of Ketones and Aldehydes," *Org. React.* **1993**, *43*, 251–798.

17.4

M. Hudlicky, "Reductions in Organic Chemistry," The Royal Society of Chemistry, Cambridge, U. K., **1996**.

A. F. Abdel-Magid (Ed.), "Reductions in Organic Synthesis: Recent Advances and Practical Applications," ACS Symposium Series, The Royal Society of Chemistry, Cambridge, U. K., **1996**.

A. Hajos, "Reduction with Inorganic Reducing Agents – Metal Hydrides and Complex Hydrides," in *Methoden Org. Chem.* (Houben-Weyl) 4th ed., **1952**–, *Reduction Part II* (H. Kropf, Ed.), Bd. 4/1d, 1, Georg Thieme Verlag, Stuttgart, **1981**.

W. G. Brown, "Reductions by Lithium Aluminum Hydride," *Org. React.* **1951**, *6*, 469–509.

J. Malek, "Reductions by Metal Alkoxyaluminum Hydrides," *Org. React.* **1985**, *34*, 1–317.

J. Malek, "Reduction by Metal Alkoxyaluminum Hydrides. Part II. Carboxylic Acids and Derivatives, Nitrogen Compounds, and Sulfur Compounds," *Org. React.* **1988**, *36*, 249–590.

J. Seyden-Penne, "Reductions by the Alumino- and Borohydrides in Organic Synthesis," VCH, New York, **1991**.

A. J. Downs, C. R. Pulham, "The Hydrides of Aluminum, Gallium, Indium, and Thallium – A Reevaluation," *Chem. Soc. Rev.* **1994**, *23*, 175.

N. M. Yoon, "Selective Reduction of Organic Compounds with Aluminum and Boron Hydrides," *Pure Appl. Chem.* **1996**, *68*, 843.

J. Seyden-Penne,"Reductions by the Alumino- and Borohydrides in Organic Synthesis," 2nd ed., Wiley, New York, **1997**.

G. W. Gribble, "Sodium Borohydride in Carboxylic Acid Media: A Phenomenal Reduction System," *Chem. Soc. Rev*. **1998**, *27*, 395–404.

L. K. Keefer, G. Lunn, "Nickel-Aluminum Alloy as a Reducing Agent," *Chem. Rev.* **1989**, *89*, 459–502.

T. Imamoto, "Reduction of Saturated Alkyl Halides to Alkanes," in *Comprehensive Organic Synthesis* (B. M. Trost, I. Fleming, Eds.), Vol. 8, 793, Pergamon Press, Oxford, U. K., **1991**.

S. W. McCombie, "Reduction of Saturated Alcohols and Amines to Alkanes," in *Comprehensive Organic Synthesis* (B. M. Trost, I. Fleming, Eds.), Vol. 8, 811, Pergamon Press, Oxford, U. K., **1991**.

A. G. Sutherland, "One or More CH Bond(s) Formed by Substitution: Reduction of C-Halogen and C-Chalcogen Bonds," in *Comprehensive Organic Functional Group Transformations* (A. R. Katritzky, O. Meth-Cohn, C. W. Rees, Eds.), Vol. 1, 1, Elsevier Science, Oxford, U. K., **1995**.

W. H. Hartung, R. Simonoff, "Hydrogenolysis of Benzyl Groups Attached to Oxygen, Nitrogen, or Sulfur," *Org. React.* **1953**, *7*, 263–326.

A. Gansäuer, S. Narayan, "Titanocene-Catalyzed Electron Transfer-Mediated Opening of Epoxides," *Adv. Synth. Catal.* **2002**, *344,* 465–475.

C. Blomberg, "The Barbier Reaction and Related One-Step Processes," Springer-Verlag, Heidelberg, **1994**.

C. G. Screttas, B. R. Steele, "Organometallic Carboxamidation. A Review," *Org. Prep. Proced. Int.* **1990**, *22*, 269–314.

T. Cohen, M. Bhupathy, "Organoalkali Compounds by Radical Anion Induced Reductive Metalation of Phenyl Thioethers," *Acc. Chem. Res.* **1989**, *22*, 152–161.

M. Yus, "Arene-Catalyzed Lithiation Reactions," *Chem. Soc. Rev.* **1996**, *25*, 155–162.

M. Yus, F. Foubelo, "Reductive Opening of Saturated Oxa-, Aza- and Thia-Cycles by Means of an Arene-Promoted Lithiation: Synthetic Applications," *Rev. Heteroatom Chem.* **1997**, *17,* 73–108.

L. Eberson, "Problems and Prospects of the Concerted Dissociative Electron-Transfer Mechanism," *Acta Chem. Scand.* **1999**, *53*, 751–764.

D. J. Ramon, M. Yus, "New Methodologies Based on Arene-Catalyzed Lithiation Reactions and Their Application to Synthetic Organic Chemistry," *Eur. J. Org. Chem.* **2000**, 225–237.

C. Najera, M. Yus, "Functionalized Organolithium Compounds: New Synthetic Adventures," *Curr. Org. Chem.* **2003**, *7,* 867–926.

Y. H. Lai, "Grignard Reagents from Chemically Activated Magnesium," *Synthesis* **1981**, 585.

C. Walling, "The Nature of Radicals Involved in Grignard Reagent Formation," *Acc. Chem. Res.* **1991**, *24*, 255.

H. M. Walborsky, "Mechanism of Grignard Reagent Formation. The Surface Nature of the Reaction," *Acc. Chem. Res.* **1990**, *23*, 286–293.

H. M. Walborsky, "Wie entsteht eine Grignard-Verbindung?", *Chem. unserer Zeit*, **1991**, *25*, 108–116.

J. F. Garst, "Grignard Reagent Formation and Freely Diffusing Radical Intermediates," *Acc. Chem. Res.* **1991**, *24*, 95–97.

R. D. Rieke, M. S. Sell, "Magnesium Activation," in *Handbook of Grignard Reagents* (G. S. Silverman, P. E. Rakita, Eds.), Marcel Dekker Inc., New York, **1996**, 53–78.

C. Humdouchi, H. M. Walborsky, "Mechanism of Grignard Reagent Formation," in *Handbook of Grignard Reagents* (G. S. Silverman, P. E. Rakita, Eds.), Marcel Dekker Inc., New York, **1996**, 145–218.

J. F. Garst, F. Unváry, "Mechanisms of Grignard Reagent Formation," in *Grignard Reagents – New Developments*, Ed. by H. G. Richey, Jr., John Wiley & Sons, Chichester, U. K., **2000**, 185–275

R. D. Rieke, "The Preparation of Highly Reactive Metals and the Development of Novel Organometallic Reagents," *Aldrichimica Acta* **2000**, *33*, 52–60.

R. D. Rieke, "Preparation of Organometallic Compounds from Highly Reactive Metal Powders," *Science* **1989**, *246*, 1260–1264.

A. Gansäuer, H. Bluhm, "Reagent-Controlled Transition-Metal-Catalyzed Radical Reactions," *Chem. Rev.* **2000**, *100*, 2771–2788.

J. S. Thayer, "Not for Synthesis Only: The Reactions of Organic Halides with Metal Surfaces," *Adv. Org. Chem.* **1995**, *38*, 59–78.

D. Caine, "Reduction and Related Reactions of α,β-Unsaturated Compounds with Metals in Liquid Ammonia," *Org. React.* **1976**, *23*, 1–258.

J. W. Huffman, "Reduction of C=X to CHXH by Dissolving Metals and Related Methods," in *Comprehensive Organic Synthesis* (B. M. Trost, I. Fleming, Eds.), Vol. 8, 107, Pergamon Press, Oxford, U. K., **1991**.

A. M. El-Khawaga, H. M. R. Hoffmann, "Formation of C-H Bonds by the Reduction of C=C Double Bonds and of Carbonyl Groups with Metals ('Dissolving Metal Reduction')," in *Stereoselective Synthesis* (Houben-Weyl) 4th ed., **1996**, (G. Helmchen, R. W. Hoffmann, J. Mulzer, E. Schaumann, Eds.), **1996**, Vol. E21 (Workbench Edition), *7*, 3967–3987, Georg Thieme Verlag, Stuttgart.

S. M. McElvain, "The Acyloins", *Org. React.* **1948**, *4,* 256–268.

J. J. Bloomfield, D. C. Owsley, J. M. Neike, "The Acyloin Condensation," *Org. React.* **1976**, *23*, 259–403.

R. Brettle, "Acyloin Coupling Reactions," in *Comprehensive Organic Synthesis* (B. M. Trost, I. Fleming, Eds.), Vol. 3, 613, Pergamon Press, Oxford, U. K., **1991**.

G. M. Robertson, "Pinacol Coupling Reactions," in *Comprehensive Organic Synthesis* (B. M. Trost, I. Fleming, Eds.), Bd. 3, 563, Pergamon Press, Oxford, U. K., **1991**.

G. C. Fu, "Pinacol Coupling," in *Modern Carbonyl Chemistry* (J. Otera, Ed.), Wiley-VCH, Weinheim, **2000**, 69–91.

O. Hammerich, M. F. Nielsen, "The Competition Between the Dimerization of Radical Anions and Their Reactions with Electrophiles," *Acta Chem. Scand.* **1998,** *52*, 831–857.

D. Lenoir, "The application of low-valent titanium reagents in organic synthesis," *Synthesis* **1989**, *12*, 883–897.

J. E. McMurry, "Carbonyl-Coupling Reactions Using Low-Valent Titanium," *Chem. Rev.* **1989**, *89*, 1513–1524.

T. Lectka, "The McMurry Recation," in *Active Metals* (A. Fürstner, Ed.), 85, VCH, Weinheim, Germany, **1996**.

A. Fürstner, B. Bogdanovic, "New Developments in the Chemistry of Low-Valent Titanium," *Angew. Chem. Int. Ed. Engl.* **1996**, *35*, 2442–2469.

A. G. M. Barrett, "Reduction of Carboxylic Acid Derivatives to Alcohols, Ethers and Amines," in *Comprehensive Organic Synthesis* (B. M. Trost, I. Fleming, Eds.), Vol. 8, 235, Pergamon Press, Oxford, U. K., **1991**.

E. Mosettig, R. Mozingo, "The Rosenmund Reduction of Acid Chlorides to Aldehydes," *Org. React.* **1948**, *4,* 362–377.

E. Mosettig, "The Synthesis of Aldehydes from Carboxylic Acids," *Org. React.* **1954**, *8*, 218–257.

J. S. Cha, "Recent Developments in the Synthesis of Aldehydes by Reduction of Carboxylic Acids and their Derivatives with Metal Hydrides," *Org. Prep. Proced. Int.* **1989**, *21*, 451– 477.

A. P. Davis, "Reduction of Carboxylic Acids to Aldehydes by Other Methods," in *Comprehensive Organic Synthesis* (B. M. Trost, I. Fleming, Eds.), Vol. 8, 283, Pergamon Press, Oxford, U. K., **1991**.

N. Greeves, "Reduction of C=O to CHOH by Metal Hydrides," in *Comprehensive Organic Synthesis* (B. M. Trost, I. Fleming, Eds.), Vol. 8, 1, Pergamon Press, Oxford, U. K., **1991**.

H. Brunner, "Formation of C-H Bonds by Reduction of Carbonyl Groups (C=O) – Hydrogenation," in *Methoden Org. Chem.* (Houben-Weyl) 4th ed., **1952**–, *Stereoselective Synthesis* (G. Helmchen, R. W. Hoffmann, J. Mulzer, E. Schaumann, Eds.), Vol. E21d, 3945, Georg Thieme Verlag, Stuttgart, **1995**.

A. P. Davis, M. M. Midland, L. A. Morell, "Formation of C-H Bonds by Reduction of Carbonyl Groups (C=O), Reduction of Carbonyl Groups with Metal Hydrides," in *Methoden Org. Chem.* (Houben-Weyl) 4th ed., **1952**–, *Stereoselective Synthesis* (G. Helmchen, R. W. Hoffmann, J. Mulzer, E. Schaumann, Eds.), Vol. E21d, 3988, Georg Thieme Verlag, Stuttgart, **1995**.

M. M. Midland, L. A. Morell, K. Krohn, "Formation of C-H Bonds by Reduction of Carbonyl Groups (C=O) – Reduction with C-H Hydride Donors," in *Methoden Org. Chem.* (Houben-Weyl) 4th ed., **1952**–, *Stereoselective Synthesis* (G. Helmchen, R. W. Hoffmann, J. Mulzer, E. Schaumann, Eds.), Vol. E21d, 4082, Georg Thieme Verlag, Stuttgart, **1995**.

D. Todd, "The Wolff-Kishner Reduction," *Org. React.* **1948**, *4,* 378–422.

R. O. Hutchins, "Reduction of C=X to CH_2 by Wolff-Kishner and Other Hydrazone Methods," in *Comprehensive Organic Synthesis* (B. M. Trost, I. Fleming, Eds.), Vol. 8, 327, Pergamon Press, Oxford, U. K., **1991**.

E. L. Martin, "The Clemmensen Reduction," *Org. React.* **1942**, *1,* 155–209.

H. Meerwein, K. Wunderlich, K. F. Zenner, "Ionic Hydrogenations and Dehydrogenations," *Angew. Chem. Int. Ed. Engl.* **1962**, *1*, 613–617.

E. Vedejs, "Clemmensen Reduction of Ketones in Anhydrous Organic Solvents," *Org. React.* **1975**, *22*, 401–422.

G. R. Pettit, E. E. van Tamelen, "Desulfurization with Raney Nickel," *Org. React.* **1962**, *12*, 356–529.

P. N. Rylander, "Hydrogenation Methods," in *Best Synthetic Methods*, Academic Press, **1985**.

H. Brunner, "Hydrogenation with Molecular Hydrogen," in *Stereoselective Synthesis* (Houben-Weyl) 4th ed., **1996**, (G. Helmchen, R. W. Hoffmann, J. Mulzer, E. Schaumann, Eds.), **1996**, Vol. E21 (Workbench Edition), *7*, 3945–3966, Georg Thieme Verlag, Stuttgart.

U. Kazmaier, J. M. Brown, A. Pfaltz, P. K. Matzinger, H. G. W. Leuenberger, "Formation of C–BH Bonds by Reduction of Olefinic Double Bonds – Hydrogenation," in *Methoden Org. Chem.* (Houben-Weyl) 4th ed., **1952**–, *Stereoselective Synthesis* (G. Helmchen, R. W. Hoffmann, J. Mulzer, E. Schaumann, Eds.), Vol. E21d, 4239, Georg Thieme Verlag, Stuttgart, **1995**.

V. A. Semikolenov, "Modern approaches to the preparation of 'palladium on charcoal' catalysts," *Russ. Chem. Rev.* **1992**, *61*, 168–174.

N. M. Dmitrievna, K. O. Valentinovich, "Heterogeneous Catalysts of Hydrogenation," *Russ. Chem. Rev.* **1998**, *67*, 656–687.

A. J. Birch, D. H. Williamson, "Homogeneous Hydrogenation Catalysts in Organic Solvents," *Org. React.* **1976**, *24*, 1–186.

H. Takaya, "Homogeneous Catalytic Hydrogenation of C=C and Alkynes," in *Comprehensive Organic Synthesis* (B. M. Trost, I. Fleming, Eds.), Vol. 8, 443, Pergamon Press, Oxford, U. K., **1991**.

R. Noyori, H. Takaya, "BINAP: An Efficient Chiral Element for Asymmetric Catalysis," *Acc. Chem. Res.* **1990**, *23*, 345–350.

R. Noyori, "Binaphthyls as Chiral Elements for Asymmetric Synthesis," in *Stereocontrolled Organic Synthesis* (B. M. Trost, Ed.), Blackwell Scientific Publications, Oxford, U. K., **1994**, 1–15.

T. Ohkuma, M. Kitamura, R. Noyori, "Asymmetric Hydrogenation," in *Catalytic Asymmetric Synthesis* (I. Ojima, Ed.), Wiley-VCH, New York, 2nd ed., **2000**, 1–110.

V. Ratovelomanana-Vidal, J.-P. Genet, "Synthetic Applications of the Ruthenium-Catalyzed Hydrogenation via Dynamic Kinetic Resolution," *Can. J. Chem.* **2000**, *78*, 846–851.

K. Rossen, "Ru- and Rh-Catalyzed Asymmetric Hydrogenations: Recent Surprises from an Old Reaction," *Angew. Chem. Int. Ed. Engl.* **2001**, *40*, 4611–4613.

J. M. Hook, L. N. Mander, "Recent Developments in the Birch Reduction of Aromatic Compounds: Applications to the Synthesis of Natural Products", *Nat. Prod. Rep.* **1986**, *3*, 35.

P. W. Rabideau, "The Metal-Ammonia Reduction of Aromatic Compounds", *Tetrahedron* **1989**, *45*, 1579–1603.

L. N. Mander, "Partial Reduction of Aromatic Rings by Dissolving Metals and by Other Methods," in *Comprehensive Organic Synthesis* (B. M. Trost, I. Fleming, Eds.), Vol. 8, 489, Pergamon Press, Oxford, U. K., **1991**.

P. W. Rabideau, Z. Marcinow, "The Birch Reduction of Aromatic Compounds," *Org. React.* **1992**, *42*, 1–334.

A. J. Birch, "The Birch Reduction in Organic Synthesis," *Pure Appl. Chem.* **1996**, *68*, 553–556.

H.-J. Deiseroth, "Alkalimetall-Amalgame," *Chem. unserer Zeit*, **1991**, *25*, 83–86.

Further Reading

A. B. Jones, "Oxidation Adjacent to X=X Bonds by Hydroxylation Methods," in *Comprehensive Organic Synthesis* (B. M. Trost, I. Fleming, Eds.), Vol. 7, 151, Pergamon Press, Oxford, U. K., **1991**.

J. Cason, "Synthesis of Benzoquinones by Oxidation," *Org. React.* **1948**, *4,* 305–361.

P. T. Gallagher, "The Synthesis of Quinones," *Contemp. Org. Synth.* **1996**, *3*, 433–446.

V. D. Filimonov, M. S. Yusubov, Ki-WhanChi, "Oxidative Methods in the Synthesis of Vicinal Di- and Poly-Carbonyl Compounds," *Russ. Chem. Rev.* **1998**, *67*, 803–826.

S. Akai, Y. Kita, "Recent Progress in the Synthesis of *p*-Quinones and *p*-Dihydroquinones Through Oxidation of Phenol Derivatives," *Org. Prep. Proceed. Int.* **1998**, *30*, 603–629.

J. Tsuji, "Synthetic Applications of the Palladium-Catalysed Oxidation of Olefins to Ketones," *Synthesis* **1984**, 369.

J. Tsuji, "Addition Reactions with Formation of Carbon-Oxygen Bonds – The Wacker Oxidation and Related Reactions," in *Comprehensive Organic Synthesis* (B. M. Trost, I. Fleming, Eds.), Vol. 7, 469, Pergamon Press, Oxford, U. K., **1991**.

C. Limberg, "On the Trail of CrO_2Cl_2 in its Reactions with Organic Compounds," *Chemistry – Eur. J.* **2000**, *6*, 2083–2089.

A. Fatiadi, "The Classical Permanganate Ion/Still a Novel Oxidant in Organic Chemistry," *Synthesis* **1987**, 85.

H. Kropf, E. Müller, A. Weickmann, "Ozone as an Oxidation Agent," in *Methoden Org. Chem.* (Houben-Weyl) 4th ed., **1952**–, *Oxidation Part I* (H. Kropf, Ed.), Vol. 4/1a, 3, Georg Thieme Verlag, Stuttgart, **1981**.

H. Heaney, "Oxidation reactions using magnesium monoperphthalate and urea hydrogen peroxide," *Aldrichimica Acta* **1993**, *26*, 35–45.

W. P. Griffith, S. V. Ley, "TPAP: Tetra-*n*-Propylammonium Perruthenate, a Mild and Convenient Oxidant for Alcohols," *Aldrichimica Acta* **1990**, *23*, 13–19.

S. V. Ley, J. Norman, W. P. Griffith, S. P. Maraden, "Tetrapropylammonium Perruthenate, $Pr_4N^+RuO_4^-$, TPAP: A Catalytic Oxidant for Organic Synthesis," *Synthesis* **1994**, 639–666.

P. Langer, "Tetra-*n*-propyl Ammonium Perruthenate (TPAP) – An Efficient and Selective Reagent for Oxidation Reactions in Solution and on the Solid Phase," *J. Prakt. Chem.* **2000**, *342*, 728–730.

N. Rabjohn, "Selenium Dioxide Oxidation," *Org. React.* **1949**, *5*, 331–386.

N. Rabjohn, "Selenium Dioxide Oxidation," *Org. React.* **1976**, *24*, 261–415.

W.-D. Woggon, "Formation of C-O Bonds by Allylic Oxidation with Selenium(IV) Oxide," in *Stereoselective Synthesis* (Houben-Weyl) 4th ed., **1996**, (G. Helmchen, R. W. Hoffmann, J. Mulzer, E. Schaumann, Eds.), **1996**, Vol. E21 (Workbench Edition), *8*, 4947–4956, Georg Thieme Verlag, Stuttgart.

H. Waldmann, "Hypervalent Iodine Reagents," in *Organic Synthesis Highlights II* (H. Waldmann, Ed.), VCH, Weinheim, New York, **1995**, 223–230.

A. Varvoglis (Ed.), "Hypervalent Iodine in Organic Synthesis," Academic, San Diego, CA, **1996**.

C. Djerassi, "The Oppenauer Oxidation," *Org. React.* **1951**, *6*, 207–272.

M. Nishizawa, R. Noyori, "Reduction of C=X to CHXH by Chirally Modified Hydride Reagents," in *Comprehensive Organic Synthesis* (B. M. Trost, I. Fleming, Eds.), Vol. 8, 159, Pergamon Press, Oxford, U. K., **1991**.

R. M. Kellogg, "Reduction of C=X to CHXH by Hydride Delivery from Carbon," in *Comprehensive Organic Synthesis* (B. M. Trost, I. Fleming, Eds.), Vol. 8, 79, Pergamon Press, Oxford, U. K., **1991**.

A. L. Wilds, "Reduction with Aluminum Alkoxides (The Meerwein-Ponndorf-Verley Reduction)," *Org. React.* **1944**, *2*, 178–223.

J. Martens, "Formation of C-H Bonds by Reduction of Imino Groups (C=N)," in *Stereoselective Synthesis* (Houben-Weyl) 4th ed., 1996, (G. Helmchen, R. W. Hoffmann, J. Mulzer, E. Schaumann, Eds.), **1996**, Vol. E21 (Workbench Edition), *7*, 4199–4238, Georg Thieme Verlag, Stuttgart.

R. O. Hutchins, "Reduction of C=N to CHNH by Metal Hydrides," in *Comprehensive Organic Synthesis* (B. M. Trost, I. Fleming, Eds.), Vol. 8, 25, Pergamon Press, Oxford, U. K., **1991**.

R. H. Shapiro, "Alkenes from Tosylhydrazones," *Org. React.* **1976**, *23*, 405–507.

A. R. Chamberlin, S. H. Bloom, "Lithioalkenes from Arenesulfonylhydrazones," *Org. React.* **1990**, *39*, 1–83.

U. Kazmaier, "General (Nondirected) Hydrogenations," in *Stereoselective Synthesis* (Houben-Weyl) 4th ed., **1996**, (G. Helmchen, R. W. Hoffmann, J. Mulzer, E. Schaumann, Eds.), **1996**, Vol. E21 (Workbench Edition), *7*, 4239–4316, Georg Thieme Verlag, Stuttgart.

J. M. Brown, "Directed Hydrogenations," in *Stereoselective Synthesis* (Houben-Weyl) 4th ed., **1996**, (G. Helmchen, R. W. Hoffmann, J. Mulzer, E. Schaumann, Eds.), **1996**, Vol. E21 (Workbench Edition), *7*, 4317–4333, Georg Thieme Verlag, Stuttgart.

H. Adkins, "Catalytic Hydrogenation of Esters to Alcohols," *Org. React.* **1954**, *8*, 1–27.

H. Kumobayashi, "Industrial Application of Asymmetric Reactions Catalyzed by BINAP-Metal Complexes," *Rec. Trav. Chim. Pays-Bas* **1996**, *115*, 201–210.

K. Inoguchi, S. Sakuraba, K. Achiwa, "Design Concepts for Developing Highly Efficient Chiral Bisphosphine Ligands in Rhodium-Catalyzed Asymmetric Hydrogenations," *Synlett* **1992**, 169–178.

C. Rosini, L. Franzini, A. Raffaelli, P. Salvadori, "Synthesis and Applications of Binaphthylic C_2-Symmetry Derivatives as Chiral Auxiliaries in Enantioselective Reactions," *Synthesis* **1992**, 503–517.

P. Kocovsky, S. Vyskocil, M. Smrcina, "Non-Symmetrically Substituted 1,1'-Binaphthyls in Enantioselective Catalysis," *Chem. Rev.* **2003**, *103,* 3213–3245.

W. Tang, X. Zhang, "New Chiral Phosphorus Ligands for Enantioselective Hydrogenation," *Chem. Rev.* **2003**, *103,* 3029–3069.

D. J. Ager, S. A. Laneman, "Reductions of 1,3-Dicarbonyl Systems with Ruthenium-Biarylbisphosphine Catalysts," *Tetrahedron Asymmetry* **1997**, *8*, 3327–3355.

S. Otsuka, K. Tani, "Catalytic Asymmetric Hydrogen Migration of Allylamines," *Synthesis* **1991**, 665–680.

W. S. Knowles, "Asymmetric Hydrogenations (Nobel Lecture)," *Angew. Chem. Int. Ed. Engl.* **2002**, *41,* 1998–2007.

W. S. Knowles, "Asymmetric Hydrogenations (Nobel Lecture 2001)," *Adv. Synth. Catal.* **2003**, *345,* 3–13.

R. Noyori, "Asymmetric Catalysis: Science and Opportunities (Nobel Lecture 2001)," *Adv. Synth. Catal.* **2003**, *345,* 15–32.

E. Block, "Olefin Synthesis via Deoxygenation of Vicinal Diols," *Org. React. (N. Y.)* **1984**, *30*, 457.

M. M. Midland, "Asymmetric Reduction with Organoborane Reagents," *Chem. Rev.* **1989**, *89*, 1553.

H. C. Brown, P. V. Ramachandran, "Asymmetric Reduction with Chiral Organoboranes Based on α-Pinene," *Acc. Chem. Res.* **1992**, *25*, 16–24.

V. Ponec, "Selective De-Oxygenation of Organic Compounds," *Rec. Trav. Chim. Pays-Bas* **1996**, *115*, 451–455.

V. K. Singh, "Practical and Useful Methods for the Enantioselective Reduction of Unsymmetrical Ketones," *Synthesis* **1992**, 607–617.

M. Wills, J. R. Studley, "The Asymmetric Reduction of Ketones," *Chem. Ind.* **1994**, 552–555.

S. Itsuno, "Enantioselective Reduction of Ketones," *Org. React.*, **1998**, *52*, 395–576.

R. Noyori, T. Ohkuma, "Asymmetric Catalysis by Architectural and Functional Molecular Engineering: Practical Chemo- and Stereoselective Hydrogenation of Ketones," *Angew. Chem. Int. Ed. Engl.* **2001**, *40*, 40–73.

Subject Index

B

E

I

M

N

O

S

T

U

V